THE PERIODIC TABLE

Subshells being completed: ns (groups I A–II A), $(n-1)d$ (transition metals), np (groups III A–VIII A)

Principal quantum number n	I A	II A	III B	IV B	V B	VI B	VII B	VIII B	VIII B	VIII B	I B	II B	III A	IV A	V A	VI A	VII A	VIII A
1	1 H	2 He																2 He
2	3 Li	4 Be											5 B	6 C	7 N	8 O	9 F	10 Ne
3	11 Na	12 Mg											13 Al	14 Si	15 P	16 S	17 Cl	18 Ar
4	19 K	20 Ca	21 Sc	22 Ti	23 V	24 Cr	25 Mn	26 Fe	27 Co	28 Ni	29 Cu	30 Zn	31 Ga	32 Ge	33 As	34 Se	35 Br	36 Kr
5	37 Rb	38 Sr	39 Y	40 Zr	41 Nb	42 Mo	43 Tc	44 Ru	45 Rh	46 Pd	47 Ag	48 Cd	49 In	50 Sn	51 Sb	52 Te	53 I	54 Xe
6	55 Cs	56 Ba	57 La	72 Hf	73 Ta	74 W	75 Re	76 Os	77 Ir	78 Pt	79 Au	80 Hg	81 Tl	82 Pb	83 Bi	84 Po	85 At	86 Rn
7	87 Fr	88 Ra	89 Ac	104 Rf	105 Ha	106												

Light metals: I A, II A

Transition metals: III B – II B

Nonmetals: III A – VIII A (right of the stepped line)

Posttransition metals: 13 Al, 31 Ga, 32 Ge, 49 In, 50 Sn, 51 Sb, 81 Tl, 82 Pb, 83 Bi, 84 Po

$(n-2)f$

Lanthanide series	58 Ce	59 Pr	60 Nd	61 Pm	62 Sm	63 Eu	64 Gd	65 Tb	66 Dy	67 Ho	68 Er	69 Tm	70 Yb	71 Lu
Actinide series	90 Th	91 Pa	92 U	93 Np	94 Pu	95 Am	96 Cm	97 Bk	98 Cf	99 Es	100 Fm	101 Md	102 No	103 Lr

INORGANIC CHEMISTRY
THIRD EDITION

Principles of structure and reactivity

James E. Huheey
UNIVERSITY OF MARYLAND

HARPER INTERNATIONAL SI EDITION
Cambridge, Philadelphia, San Francisco,
London, Mexico City, São Paulo, Sydney

Figures from the following journals are copyright © to the American Chemical Society: *Accounts of Chemical Research, Chemical and Engineering News, Chemical Reviews, Inorganic Chemistry, Journal of Chemical Education, Journal of Physical Chemistry,* and *Journal of the American Chemical Society.* Grateful acknowledgment is also made to The American Association for the Advancement of Science, The Chemical Society, The International Union of Crystallography, *Nature,* The Nobel Foundation of the Royal Academy of Science, Sweden, and *Zeitschrift für Naturforschung* for the use of materials that are copyright © to them. Individual acknowledgments are given on the page where the material appears.

Permission for the publication herein of Sadtler Standard Spectra ® has been granted, and all rights are reserved, by Sadtler Research Laboratories, Inc.

Sponsoring Editor: Malvina Wasserman
Project Editor: Cynthia L. Indriso
Designer: Madalyn Hart
Production Assistant: Debi Forrest-Bochner
Compositor: Syntax International Pte. Ltd.
Printer and Binder: The Murray Printing Company
Art Studio: Danmark & Michaels, Inc.
SI Conversion by Henry Heikkinen, University of Maryland

Inorganic Chemistry: Principles of Structure and Reactivity, Third Edition

Library of Congress Cataloging in Publication Data

Huheey, James E.
Inorganic chemistry.

Bibliography: p.
Includes index.
1. Chemistry, Inorganic. I. Title.
QD151.2.H84 1983 546 83-253
ISBN 0-06-042987-9
Harper International SI Edition
ISBN 0-06-350352-2

To those closest to me—Mom, Dad, Cathy, Terry, and Martha

Contents

Preface

In writing this book I have tried to portray inorganic chemistry as an exciting field of research rather than as a closed body of knowledge. No doubt my various enthusiasms and prejudices have colored my presentation. Enthusiasms and prejudices aside, however, I hope that my readers will appreciate the diversity of different research areas that fall within the rubric "inorganic chemistry." As the editor of the journal *Inorganic Chemistry* stated when it was founded about twenty years ago: "The limits of inorganic chemistry are difficult to define. Subject matter may range from the limits of physical and organic chemistry to the edges of theoretical physics." This statement is as true today as then, but one might add "to the limits of biology."

I believe that a textbook suitable for the traditional course in introductory inorganic chemistry cannot be encyclopedic. The discussion in the present book is therefore essentially topical in nature. Every effort has been made to present the many facets of inorganic chemistry rather than an exhaustive treatment of the subject. Although the emphasis is on structural and bonding principles, it is believed that sufficient "descriptive" chemistry has been included to give the student a good appreciation of the range of properties and reactions encountered in inorganic chemistry. In this edition increased emphasis has been placed on the redox chemistry of metals through the use of Latimer diagrams, a useful tool that has been somewhat neglected of late.

Students using this book come from exceedingly diverse backgrounds: Many will have had extensive experience in physical and organic chemistry, perhaps even a previous inorganic course. For most, however, this will be the first contact with inorganic chemistry, and some may have had only limited contact with bonding theory in other courses. For this reason, Chapters 2 and 3 present the fundamentals of atomic and molecular structure from the inorganic chemist's perspective. The well-prepared reader may use them as a brief review as well as mortar to chink between previous blocks of knowledge, and then move quickly on to Chapters 4, 5, and 6 where new material on solids, molecular structure, and chemical forces is introduced.

The middle chapters of the book, Chapters 7, 8, 9, and 10 present the "heart of inorganic chemistry," acid–base theory, nonaqueous solvents, and coordination chemistry. In line with the philosophy of a topical approach and flexible course content, the last eight chapters of the book are essentially independent of each other and one or more may readily be omitted depending on the inclination of the instructor and the time available.

A complete chapter has been devoted to the burgeoning field of bioinorganic chemistry. Added emphasis has been given to organometallic catalysis, solid electrolytes, batteries, hydrometallurgy, and recent advances in acid–base and coordination chemistry.

I have included an increased number of stereoviews in this edition. Those unfamiliar with stereoviews, or who have had difficulty resolving them, should turn to Appendix H where the "tricks" of stereopsis are discussed.

It is said that everyone feels that he "knows everything" in his field for almost fifteen minutes after his Ph.D. final exam, and that his estimate of his own knowledge decreases steadily thereafter. Even so, there are two times when a professor feels keenly aware of the fact that he does *not* know everything: When he steps before his first class of eager students, and when he sits down to write a book. Especially depreciating to one's ego is the thought that, creative and original as one might try to be, 99 percent of one's ideas, lectures, notes, and research work come directly or indirectly from stimulating associates. I am indebted to such a very large number of friends and colleagues who have helped in this endeavor that I hope that none will feel slighted by the physical impossibility of listing every name. All may rest assured that I certainly appreciate their help in sending papers prior to publication, in providing artwork for direct reproduction, in giving advice in their areas of expertise, in critically reading portions of the manuscript, and perhaps most appreciated of all, by providing support and encouragement. Many of you may check "All of the above!" I also appreciate the large number of people who took the time to complete the feedback loop: Student → Professor → Author to improve the text.

Despite the impossibility of giving credit to everyone who has been of help, I should like to single out certain people who have been especially instrumental, directly or indirectly, in the writing of this book. It has been my very good fortune to have had contact with exceptional teachers and researchers when I was an undergraduate (Thomas B. Cameron and Hans H. Jaffé, University of Cincinnati) and a graduate student (John C. Bailar, Jr., Theodore L. Brown, and Russell S. Drago, University of Illinois); and to have had stimulating and helpful colleagues where I have taught (William D. Hobey and Robert C. Plumb, Worcester Polytechnic Institute; Jon M. Bellama, Alfred C. Boyd, Samuel O. Grim, James V. McArdle, Gerald Ray Miller, Carl L. Rollinson, Nancy S. Rowan, and John A. Tossell, University of Maryland). I have benefited by having had a variety of students, undergraduate, graduate, and thesis advisees, who never let me relax with a false feeling that I "knew it all." Finally, it has been my distinct privilege to have had the meaning of research and education exemplified to me by my graduate thesis advisor, Therald Moeller, and to have had a most patient and understanding friend, Hobart M. Smith, who gave me the joys of a second profession while infecting me with the "*mihi* itch:" Professors Moeller and Smith, through their teaching, research, and writing, planted the seeds that grew into this volume.

Several reviewers provided careful and helpful critiques for the various editions of this book. For their help with the third edition, I am especially grateful to Thomas B. Brill, Jeremy K. Burdett, Mary G. Enig, William E. Hatfield, Jay A. Labinger, Joel F. Liebman, Mary C. McKenna, Joseph A. Stanko, Galen D. Stucky, and Kenneth S. Suslick, who read parts or all of the present manuscript, as well as earlier editions of the book, and helped greatly in the difficult decisions of what to add and what to delete. Four librarians, George W. Black, Jr., of Southern Illinois University at Carbondale, and Sylvia D. Evans,

Elizabeth W. McElroy, and Elizabeth K. Tomlinson, of the University of Maryland, helped greatly with retrieval and use of the literature.

There is a group of people of whom most textbook readers are only dimly aware, if at all. Even textbook *authors* learn to recognize their work rather tardily. I refer to those who perform the usually unappreciated miracle of accepting an author's partly formed idea and eventually producing a finished book; who are tolerant of his idiosyncrasies, yet intolerant of his peccadillos, and, as the old saying goes, have the wisdom to know the difference. I have enjoyed working with and learning from Malvina Wasserman, who cajoled and cudgeled, as appropriate, to get this edition to press; Ann Brody, Lois Lombardo, and Cindy Indriso, three superb project editors who taught me how one turns a tattered typescript into a well-produced text, and Rita Naughton and Madalyn Hart, designers, whose work on art and layout made this book much clearer and more useful to the reader. The professional expertise of all of these is reflected many times in the following pages.

I should like to give special thanks to Gerald Ray Miller who read the entire manuscript and proofs at the very beginning, and who has been a ready source of consultation through all editions. Germán de la Fuente, Cindy Hawthorne, Joanne Kla, Jeff Robinson, and Timothy Watt helped to correct proof at a time when the schedule was tight. Caroline L. Evans made substantial contributions to the contents of this book and will always receive my appreciation. Finally, the phrase "best friend and severest critic" is so hackneyed through casual and unthinking use, paralleled only by its rarity in the reality, that I hesitate to proffer it. The concept of two men wrangling over manuscripts, impassioned to the point of literally (check Webster's) calling each other's ideas "poppycock" may seem incompatible with a friendship soon to enter its second quarter-century. If you think so, you must choose to ignore my many trips to Southern Illinois University to work with Ron Brandon, to visit with him and his family, to return home with both my emotional and intellectual "batteries" recharged.

My family has contributed much to this book, both tangible and intangible, visible and (except to me) invisible. My parents have tolerated and provided much over the years, including love, support, and watching their dining room become an impromptu office, often the same week as holiday dinners. My sister, Cathy Donaldson, and her husband, Terry, themselves university teachers, have both answered and posed questions ranging from biology to chemical engineering. More important, they "have been there" when I needed their unique help. Martha G. Price is a nutritional biochemist who has better fish to fry (in *polyunsaturated lipids*—sorry about that!) than inorganic chemistry, but who found time to draw the new catalytic and enzymic cycles, as well as to help me comprehend the profundities of biochemistry. To all of these, go my deepest gratitude and thanks. If you have never sat on this side of the typewriter, you may not know what I mean, but I fervently hope *they* do.

James E. Huheey

Introduction to SI Version

The SI version of this text incorporates all major features of The International System of Units, and should thus serve the needs of the growing population of chemistry students internationally who work with this modernized metric system.

Several SI-related style points are worth noting here, to avoid confusion. The use of the electronvolt (eV), even though not part of SI, is permitted with SI, since its value expressed in joules must be obtained by experiment, and is therefore not known exactly ($1 \text{ eV} = 1.60219 \times 10^{-19}$ J approximately). The reader will thus find some references to ionization energies in electronvolt terms, in cases where source data were reported in these units. The spectroscopist's unit of wave number (reciprocal of wavelength) is also retained in its cm^{-1} form, due to its continued popularity in literature.

One difficulty in creating an SI version of a text containing useful and abundant references to source literature (such as this) is that the original literature is often expressed in non-SI terms. In instances where an exact order-of-magnitude conversion of original units in tables or graphs to SI values could be made (angstroms to picometres, for example), the source material was brought to SI standards by simple decimal point moves. However, in the few cases where such changes would require re-drawing of graph coordinates (kcal mol^{-1} to kJ mol^{-1}, for example) the original versions were retained.

During these years of world-wide transition to SI units, we all need to cultivate at least a modest degree of 'non-SI unit literacy' to read the technical literature with understanding. Appendix C is offered as an attempt to make the SI user of this book a bit more comfortable with the array of units still on view in the relevant literature.

As a confirmed SI advocate within a nation which still inches its way toward metric use, I am delighted to be associated with an SI version of a text which has had a prominent place on my desk over its past two editions. Jim Huheey's encouragement and long-standing friendship are also greatly appreciated.

Henry Heikkinen

To the student

As I have explained in the preface, my intention has been to provide you with an appreciation of inorganic chemistry as a growing body of knowledge rather than as a static science. Most people using this book will have advanced undergraduate or graduate standing and I have therefore attempted to write more in the style of a professional review rather than as a lower level text, especially in the later chapters. For example, I have followed the practice of inserting pertinent references directly in the body of the text rather than as a compendium of "Suggested Readings" at the end of each chapter. My reason for this has been to make it as easy as possible for the reader to follow up a particularly interesting point and to get the student involved in reading the original literature. Also, it has been my experience that when the "Literature Cited" section is not encountered until the end of the chapter it often becomes a list of "Literature Slighted." A common criticism of footnote citations is that they are distracting to the student. I don't believe that this is really true but if it is you may as well get used to it. For the rest of your professional life you will encounter this style when reading research journals!

The reader's attention is called to those sections of the text set in smaller type. These sections consist of material not necessary to the main discussion but which discuss interesting side issues or carry a point a little further in its ramifications.

The problems at the ends of the chapters are an integral part of the book. Some of the problems present the usual review of material presented in the chapter. In others, the student will provide substantiation for discussions in the text. A third type, admittedly more difficult, presents the reader with the opportunity to explore topics beyond the scope of the present book. Many of the problems have references to discussions in the text or the original literature to provide help in solving the problems. It is suggested that these references, set in square brackets, should not be consulted until an attempt has been made to solve the problem without help.

There may be some of you who feel that I have introduced too much mathematics and quantum mechanics into inorganic chemistry. To you, I can only plead the argument of Lord Kelvin:

> I often say that when you can measure what you are speaking about, and express it in numbers, you know something about it; but when you cannot express it in numbers, your knowledge is of a meagre and unsatisfactory kind; it may be the beginning of knowledge, but you have scarcely, in your thoughts, advanced to the stage of *Science,* whatever the matter may be.

On the other hand, there may be those who feel that my explanations are too much on the qualitative, "hand-waving" side, that I should state the appropriate rigorous mathematical equation and let it go at that. To this I can only reply that application of the appropriate equation does not necessarily imply understanding of the physical nature of the system. As one of the most successful synthesizers of theory and practice has said (R. Hoffmann, *Acc. Chem. Res.,* **1971,** *4,* 1):

> In principle one could go ahead and calculate each molecule . . . however . . . even if the results were in excellent agreement with experiment, the resultant predictability would not necessarily imply understanding. True understanding implies a knowledge of the various physical factors, the mix of different physical mechanisms, that go into making an observable.

Finally, there may be some of you who wonder, "Why did he write so *much*?!" I apologize for the increased drain on your pocketbook and the added grams to your bookbag, but I *cannot* apologize for the *diversity* of inorganic chemistry. Nay, I revel in it! As mentioned in the preface, I do not expect your professor to cover every chapter—I certainly do not when I teach the course. And each of us has his own enthusiams and prejudices as I said before. I know one inorganic chemist who studies minerals, a traditional subject of inorganic chemistry. However he does all of his work not with blowpipe and charcoal block, but by trying to make a computer tell him more about the electronic structure of various minerals. Another wades out in the tide, up to his knees in the surf, not a traditional abode of the inorganic chemist, to collect sipunculid worms for his studies on hemerythrin, molecules of which contain only about four parts per thousand of iron—the rest is "organic" or "biological." A third works for a large chemical company trying to find better catalysts for the petrochemical industry, but another might scoff, "That's *organic* chemistry!" There are many (including me) who would like to know why *cis*-diamminedichloroplatinum(II) is active against certain cancers, but a colleague may exclaim, "Platinum, atomic number 78, was discovered by de Ulloa in 1748, over a quarter of a century before the historic discovery of oxygen by Priestley; there are two common oxidation states, +2 and +4 forming square planar and octahedral complexes, respectively; but there are 107 other elements in the periodic chart, all equally interesting!" as he strokes his very *non*Mendeleevian beard. Yet another is fascinated by the fact that she finds some of the reactions of cobalt and rhodium seemingly "nonperiodic." Who's right? They *all* are, of course! As you read this book, I hope some item will fire your imagination as it has these friends of mine.

Lastly, I hope you will enjoy reading this book as much as I have enjoyed writing it. If you do, drop me a postcard.

James E. Huheey

1

Introductory survey of inorganic chemistry

WHAT IS INORGANIC CHEMISTRY?

It is customary for chemistry books to begin with questions of this type—questions are usually unanswerable in simple "twenty-five-words-or-less" definitions. If a definition of inorganic chemistry *must* be given, the following tautological one might be ventured: *Inorganic chemistry is any phase of chemistry of interest to an inorganic chemist.* While adding little to one's comprehension of inorganic chemistry,[1] this definition does imply that the subject matter is diverse and overlaps that of other chemical disciplines, and furthermore there is a species of chemist known as the inorganic chemist that may often be more readily recognized and defined than the subject "inorganic chemistry" itself. In this introductory chapter an attempt will be made to illustrate inorganic chemistry in terms of the concepts and compounds that the inorganic chemist studies. Some of the concepts, such as that of periodicity, are over 100 years old, yet are, curiously, still quite "current" topics in inorganic chemistry. Other areas, such as bioinorganic chemistry, were born just a few short years ago. Most of the compounds discussed here have either been synthesized or adequately studied only in the past decade. In an effort to be brief, only a limited number of molecules can be presented here and it is a certainty that if another inorganic chemist were asked to choose a dozen molecules, many (if not all) would be different. Such is the diversity and catholicity of viewpoints in inorganic chemistry. But if pressed, every inorganic chemist would find at least one molecule of the selection given here which would be of interest in relation to his own particular sphere of research interests.

[1] Indeed, the definition has drawn some flak. It was made partially tongue-in-cheek and partially as an admission that inorganic chemistry is not readily definable. Another light-hearted definition, based on the supposed uniqueness of carbon, is "the chemistry of the *other* 107 elements." One of the better serious definitions has recently been offered: "Inorganic chemistry is the experimental investigation and theoretical interpretation of the properties and reactions of all the elements and of all their compounds, except the hydrocarbons and most of their derivatives." [T. Moeller, "Inorganic Chemistry, A Modern Introduction," Wiley, New York, **1982**, p. 2.]

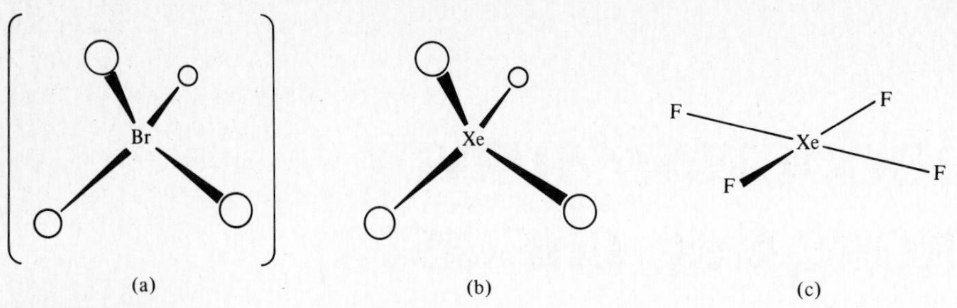

Fig. 1.1 Structures of unexpectedly synthesized species: (a) the perbromate ion, BrO_4^-; (b) xenon tetroxide, XeO_4; (c) xenon tetrafluoride, XeF_4.

One purpose which the undergraduate inorganic chemistry course serves is to "fine-tune" one's understanding of basic chemistry principles learned in earlier courses. Unfortunately, given the nature of the learning process, it is impossible to learn everything immediately. Actually, this is a blessing in disguise because it means that one spends one's entire life learning, and I can think of nothing more boring than an existence without new revelations and new surprises. But this does present problems to the student who needs to adjust his ideas and, rarely, to overhaul completely what he has learned previously.

Take the periodic chart, for example. Formulated by Mendeleev essentially in its present form in 1869, it packs a tremendous amount of information into a small amount of space. It is the most useful tool available to a chemist, especially an inorganic chemist, to systematize his knowledge. It is therefore introduced early in high school and freshman chemistry courses. However, it is impossible in such courses to teach *everything* about the 108 elements that make up the periodic table. In attempts to simplify and generalize, implicit half-truths and an occasional misintentioned falsehood are allowed to creep in. The learning process involves continually refining ideas, correcting partial misconceptions, and improving the discernment of small differences. Which is more electronegative, silicon or lead? Which has the higher electron affinity, oxygen or sulfur? If you chose silicon, you will get a hot debate from many inorganic chemists; if you chose oxygen, there's no doubt about it—you're wrong! Experimental measurements clearly show that the electron affinity of sulfur is greater. Why this seeming contradiction of straightforward principles presented in introductory courses? The physical universe does not deign to be as simple as some people would like. That does not make it less logical or less interesting; it means we have to work harder to unravel its nature and obtain a greater reward for so doing.

The first chemical species to be discussed here is one of the simplest, the perbromate ion, BrO_4^- (Fig. 1.1a). Although both the perchlorate and periodate ions have been known for quite a long time, the perbromate ion eluded synthesis until rather recently.[2] This apparent reluctance to be synthesized is not unique to BrO_4^-. Certain other molecules containing elements in the first long row of the periodic chart in high oxidation states are either unknown or unstable, although both heavier and lighter analogues are known.

[2] Because this is a survey chapter, direct references to the literature will not be given here—each of these topics is discussed and cited elsewhere in the text. They are listed in the index.

For example, PCl_5 and $SbCl_5$ are well known, $AsCl_5$ is not. Many attempts to prepare it from $AsCl_3$ and Cl_2 failed and only very recently has it been synthesized. In addition, SeO_3 is much less stable than either SO_3 or TeO_3; the Se—X bond energies are less than would be expected from interpolation of S—X and Te—X values. Therefore, most chemists were surprised when perbromate *was* synthesized since its apparent instability was paralleled by similar compounds. It is of particular interest to note that although chemists were not sure whether perbromate could or could not exist or, if the latter, why it was unstable, there was no uncertainty about the most stable structure for it. If the ion existed, it was virtually certain to be tetrahedral.

A second molecule of interest is xenon tetroxide, XeO_4 (Fig. 1.1b). In some ways it is similar to the perbromate ion. First, it is isoelectronic with a perfectly stable, well-known species, the periodate ion, IO_4^-. Although the first stable, isolable noble gas compound was not discovered until 1962, all that have been well characterized have been isoelectronic with previously known interhalogen or halogen–oxygen compounds. This is not to imply that the chemistry of the noble gases is identical with that of the halogens (it certainly is not!) but to point out that no unique problem in bonding was presented by the noble gas compounds. The same difficulties had already existed but had been conveniently "overlooked" in compounds known long before 1962. For example, XeF_4 (Fig. 1.1c) is isoelectronic and isostructural with the tetrafluoroiodate anion, IF_4^-. The second connection between the noble gas compounds and the perbromate ion is that the former are convenient oxidizing agents for the synthesis of difficultly prepared materials. The only by-product in the use of xenon fluorides as oxidizers and fluorinators is gaseous xenon, considerably simplifying the purification of the product. Perbromate ion is relatively easy to synthesize with XeF_2! Krypton difluoride may be used to synthesize AuF_5, another truly unexpected compound. Thus a discovery in one area of chemistry greatly aids progress in another.

The discovery of the compounds of the noble gases cannot be overemphasized in its effect upon chemical thinking. I realize that most of the users of this book had not yet learned to read, much less become interested in chemistry, in 1962 when Neil Bartlett reported on some of these compounds. It is therefore probably impossible for them to appreciate the mystique that surrounded the so-called "inert gases." The supposed nobility of these elements was well enshrined, supported by chemical experimentation, theory, and folklore. Looking backward, we can see that the amount of chemical *folklore* far outshadowed good experimentation and good theory. The successful synthesis was stimulated by some shrewd thinking and careful prediction by Bartlett. Likewise Sir Ronald Nyholm predicted that Cs^+Au^- might exist, and ten years later several workers showed how to make it, the first compound containing a simple metal *anion*.

One of the most interesting and unexpected discoveries in this general area is that of polymeric sulfur nitride, $(SN)_x$. This material has been known for a long time and can be prepared rather simply from easily accessible materials:

$$S_4N_4 \xrightarrow{Ag} S_2N_2 \longrightarrow (SN)_x \tag{1.1}$$

We might expect the resulting polymer, composed of nonmetallic elements, to resemble polyethylene, $(CH_2)_x$, and other plastics. In some respects it does, but in other ways it has some unique properties. For example, it has a metallic appearance and it conducts electricity about as well as mercury metal! Furthermore, when chilled almost to zero kelvins, it becomes a superconductor. It is worth noting that these properties appear only in very pure material that must be prepared from extremely pure starting materials, which requires a very careful preparative technique.

Substances with even lower resistivities can be made from unlikely materials. Even though graphite is a conductor, its conductivity is not high—not in the range of metal conductivities. However, when treated with nitryl salts, $NO_2^+X^-$ (where $X^- = BF_4^-$, PF_6^-, AsF_6^-, and SbF_6^-) graphite undergoes the following reaction:

$$nC + NO_2^+X^- \longrightarrow nC^+X^- + NO_2 \tag{1.2}$$

The nitryl ion, NO_2^+, is reduced to nitrogen dioxide, which is visibly released as a red-brown gas. The anions become intercalated between the layers of carbon atoms. The resultant compound has a conductivity that is even *higher* than mercury metal, not from the anions, but from the cationic charges, or "holes," in bands of the graphite.

One area of active work in recent years has been the improvement of bonding theories to accommodate an ever-growing variety of compounds. A stimulating interplay exists between synthesis and theory that leads to results such as (1) a new compound is synthesized; (2) it does not fit adequately into any known schemes of bonding; (3) theory is extended to treat the new compound; (4) the amplified theory suggests many new ideas for synthesis and characterization.

An example of the extension of bonding schemes is $[Re_2Cl_8]^{-2}$. Single, double, and triple bonds have long been used to explain the properties of organic and inorganic molecules. However, certain features of $[Re_2Cl_8]^{-2}$, such as its diamagnetism, an exceptionally short bond, and an eclipsed conformation, have forced acceptance of the idea of a quadruple bond in this species (Fig. 1.2). This ion is but one example from a wide range of chemistry dealing with metal–metal bonds.

Not all new discoveries are made with recently synthesized compounds or expensive equipment. Consider chromium(II) acetate. It has been known since 1844, and its synthesis has been assigned for years in preparative inorganic laboratory courses as a test of skill and patience. The synthesis is straightforward: Reduce Cr^{+3} with a zinc amalgam and precipitate the beautiful brick-red crystals with excess acetate ion. However, unless the

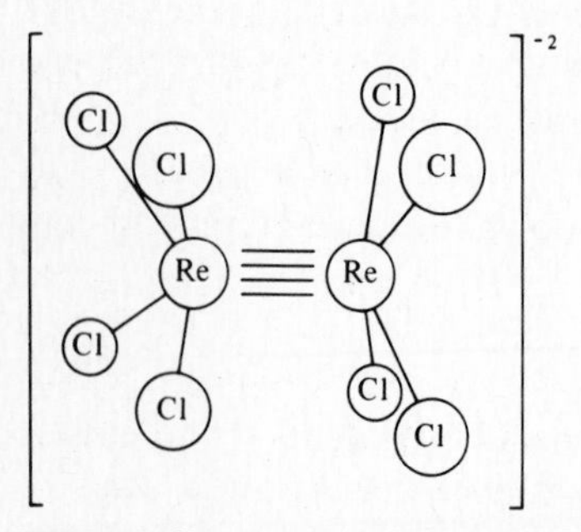

Fig. 1.2 Structure of the octachlorodirhenate(III) anion.

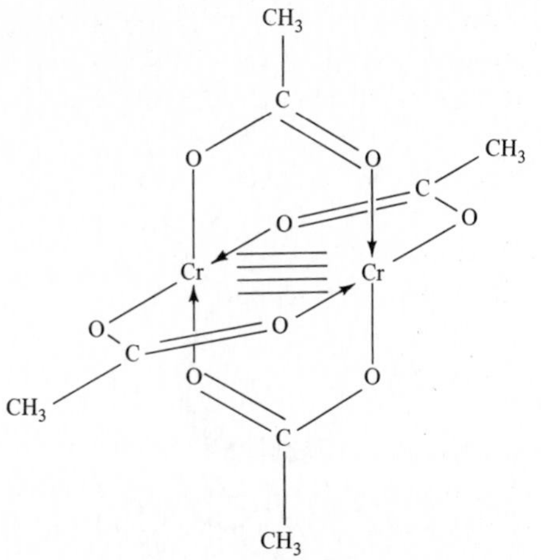

Fig. 1.3 Structure of chromium(II) acetate.

precipitate is carefully washed with solvents such as ethanol and ethyl ether to remove *all* traces of moisture, the product spontaneously oxidizes to the Cr(III) compound. There is something fascinating, if frustrating, in watching your beautiful brick-red crystals turn to a gray-green powder! What does this have to do with the other compounds discussed in this chapter? It has recently been determined that $Cr_2(OOCCH_3)_4$ has a quadruple bond (Fig. 1.3) and that it was the first member of a series of quadruply bonded dichromium compounds, some of which contain "super-short" bonds.

Two metal complexes of the "sandwich" type are shown in Fig. 1.4. The ferrocene molecule, $Fe(C_5H_5)_2$ (Fig. 1.4a), was discovered in 1951 and was the first example of the now well-known metallocene or sandwich compounds. A series of rather interesting sandwiches has been prepared in which a carborane cage (Fig. 1.4b) replaces the cyclopentadienyl (C_5H_5) ring. An example is shown in Fig. 1.4c. This ion, discovered by Hawthorne, is very similar to ferrocene-type molecules in that the orbitals that arise from the $B_9C_2H_{11}^{-2}$ ion are very similar to those in the cyclopentadienyl radical. Figure 1.5c represents a structure having an interesting combination of coordination chemistry and nonmetal chemistry since borohydride molecules and ions such as $B_9C_2H_{11}^{-2}$ are of considerable interest in their own right. In these compounds the boron atoms often have a coordination number of 5 or 6 (i.e., they are "bonded" to five or six neighboring atoms) yet possess only three valence electrons, enough to form only three "normal" bonds. Here we have a problem which is seemingly completely opposed to that presented by noble gas compounds. In the latter there seem to be too many electrons, since noble gas atoms already have filled octets. In the boron hydrides there are too few electrons, only three per boron, to form four, five, or six bonds. Yet, surprisingly, the same type of bonding scheme may be used to explain both types of bonding.

While we are considering the subject of unusual bonding and unusual structures, Fig. 1.5 should be discussed. The structure of this molecule is perfectly "normal" for a polynuclear carbonyl compound except for the fact that the carbon at the center of the base of the square pyramid of iron atoms is pentacoordinate and presumably pentacovalent.

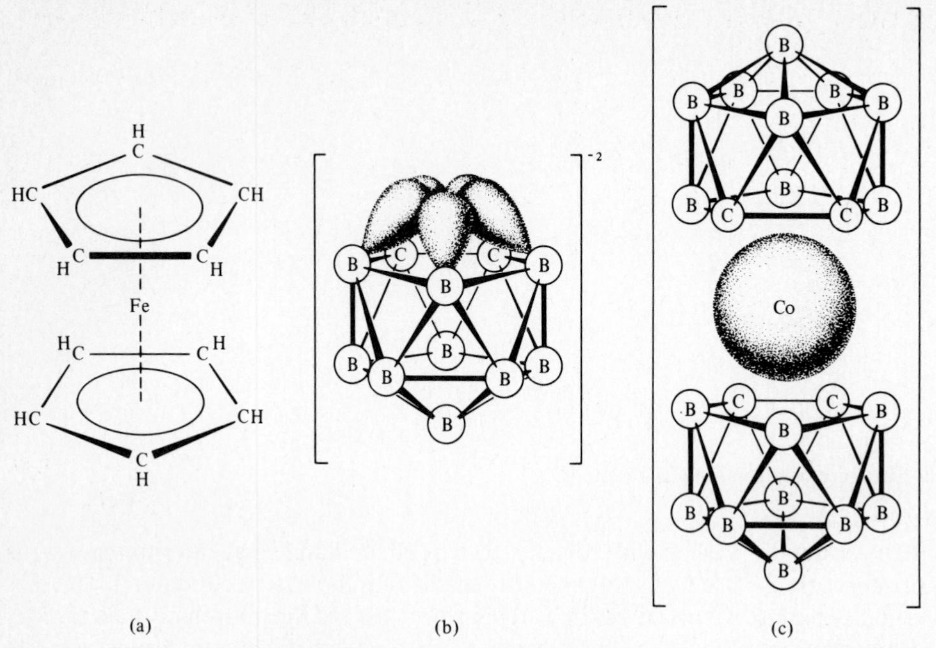

Fig. 1.4 (a) Structure of the sandwich compound ferrocene, $Fe(C_5H_5)_2$; (b) the carborane anion $Bg_9C_2H_{11}^{-2}$, illustrating the presumed orientation of orbitals used in metallocarboranes; (c) the structure of the metallocarborane anion $Co(Bg_9C_2H_{11})_2^-$. Hydrogen atoms have been omitted for clarity. [Structures (b) and (c) from R. G. Adler and M. F. Hawthorne, *J. Am. Chem. Soc.*, **1970**, *92*, 6174. Reproduced with permission.]

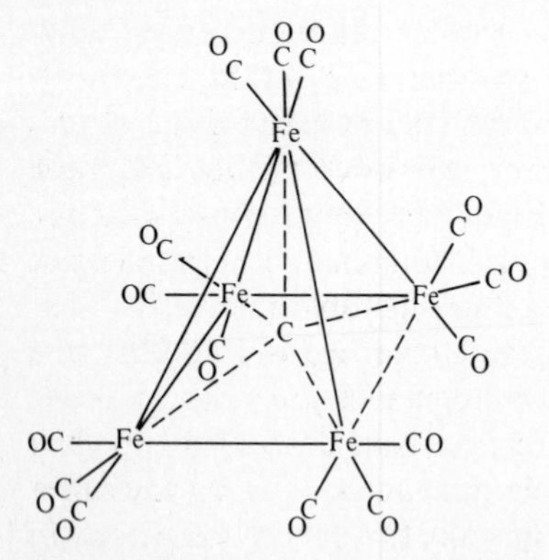

Fig. 1.5 Structure of the iron carbonyl carbide $Fe_5(CO)_{15}C$.

Although classical organic chemists seemingly delighted in deriding pentacovalent carbon structures which neophyte chemists presented on examinations, modern organic chemistry will often include both pentacovalent (e.g., CH_5^+ ion) and tricovalent (e.g., carbocation) carbon atoms. Thus the pentacovalent nature of the carbon in this complex is not unique and even hexacovalent carbon atoms are known. One, shown in Eq. 1.3, may be oxidized to the "butterfly" carbonyl carbide which serves as a model for organometallic catalysis:[3]

$$Fe_6C(CO)_{16}^{-2} \xrightarrow{[O]} Fe_4C^{\delta+} \xrightarrow{CO} Fe_4(CO)_{12}C{-}CO \xrightarrow{MeOH} Fe_4(CO)_{12}C{-}C({=}O)OCH_3 \qquad (1.4)$$

Most metal clusters do *not* contain unusual carbon atoms like those shown above. Nevertheless, they form a fascinating series of geometric diversity—polyhedra of all shapes and sizes to tax the bonding theory of the inorganic chemist. Two species, $[Rh_{13}(CO)_{24}H_{5-n}]^{-n}$, (where $n = 2, 3$; see p. 599), are effectively minute chunks of the crystal lattice of metallic rhodium with the surface bonded to carbon monoxide molecules.

Are rhodium compounds of interest then simply because they form laboratory curiosities with peculiar structures? By no means! A plant capable of producing more than 100 000 metric tons per year of acetic acid from methanol and carbon monoxide has recently gone on-stream in Texas using an organorhodium catalyst (see p. 659). The use of catalysts of this sort becomes increasingly important in a world facing shortages and high prices of petroleum, until recently the cheapest source of both feedstocks *and* energy.

An inorganic synthesis of great industrial importance, one that requires a catalyst, is the Haber synthesis of ammonia from nitrogen and hydrogen. Even with the best man-made catalyst, the process requires very high temperatures and pressures. Yet we know that better catalysts are possible since we can watch a clover plant fixing atmospheric nitrogen at room temperature and pressure! Much current research is aimed at elucidating processes that operate at the interface of biochemistry and inorganic chemistry (see p. 892).

The final compound discussed, hemoglobin, might seem to stretch the limits of inorganic chemistry. Yet we all know that hemoglobin (Fig. 1.6) contains iron (every molecule of hemoglobin contains four iron atoms), and the chemistry of iron is certainly

[3] As indicated by the empirical formulas listed below each molecule, there are several carbon monoxide molecules attached to the iron atoms on the periphery of the complex. They have been omitted for clarity.

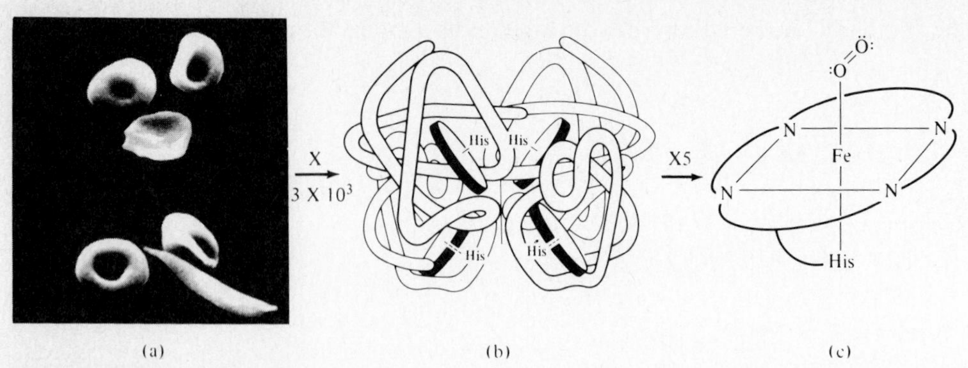

Fig. 1.6 Relative scale of (a) red blood cells, the *biological* unit of oxygen transport; (b) the hemoglobin molecule, the *biochemical* unit of oxygen transport; and (c) the heme group, the *inorganic* unit of oxygen transport. The relative sizes are given by the factors over the arrows. [Scanning electron micrograph courtesy of M. Barnhart, Wayne State University of Medicine.]

a subject of interest to the inorganic chemist. Furthermore, hemoglobin is an excellent example of a *coordination compound* formed between a *metal* and *ligands*, in this case a huge protein molecule in which four nitrogen atoms in a porphyrin ring and one from the amino acid histidine bind each iron atom. The beauty of the hemoglobin molecule is that each iron atom can bind *one more ligand*, the oxygen molecule, a process necessary for sustaining life in higher animals. Again, we know the results of this process, but it takes considerable insight to understand completely. For example, we all know that upon oxygenation blood becomes bright red. But why?

More fundamental, and more important, is the fact that the hemoglobin molecule must have an avid appetite for oxygen in the lungs, where it is plentiful, yet turn around and release it equally readily in the tissues, where it is needed to metabolize food for the production of energy. If you were presented the task of designing such a chameleon-like molecule, where would you begin? How does this molecule "know" from its environment whether it is expected to accept or release oxygen molecules?

The title of this book implies that two important aspects of inorganic chemistry (in fact, of chemistry in general) are *structure* and *reactivity*. Someone has said that all of physical science consists of this duality. In physics we could speak of *statics* and *dynamics*, and by analogy in everyday life, the *status quo* and *change*. What properties of atoms and molecules can we expect to be important in determining the structure and reactivity of molecules such as those described in this chapter? Obviously the structure of individual atoms will determine their valence capabilities, and these will then determine the structure of molecules and crystals. The static structures will, in turn, be an important factor in the possible reactivity to form new compounds.

Just as we need to be able to handle the structural determinants of chemistry, both atomic and molecular, in order to discuss reactivity, we also need to have an appreciation of the role played by energy. And don't forget that every "static" molecule results from some reaction in the first place. So the second factor of fundamental concern is energy and how it interacts with the structural constraints to allow or forbid reactions, to stabilize or destabilize molecules, and to make possible the science we call chemistry.

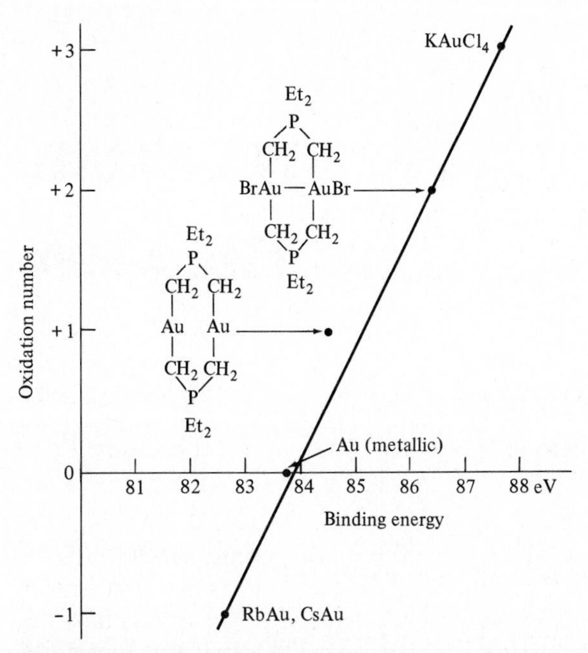

Fig. 1.7 ESCA binding energies for gold atoms in various oxidation states. Note that the value for CsAu corresponds to that expected for a −1 oxidation state. [From J. Knecht et al., *Chem. Commun.*, **1978**, 905. Reproduced with permission.]

This brief overview of some inorganic systems could not possibly cover the entire area of inorganic chemistry. Much of the work is not readily portrayed by structures. Reaction rates, thermodynamics, etc., and the employment of tools such as IR, UV, NQR, and NMR spectroscopy are methods used by inorganic chemists to determine other important properties as well as to elucidate structures. As a very simple example, note the graph pictured in Fig. 1.7. While the methodology is involved and the theoretical interpretation subject to various factors, it may readily be seen that these data provide strong and straightforward support for the formation of cesium auride, Cs^+Au^-, mentioned earlier.

Finally, it is impossible in such a brief discussion to explain the *reasons* for the interesting behavior of the atoms in these molecules and ions. To do so would require more than a single chapter. In fact, to explain *all* of the interesting features of inorganic chemistry would require many, many books. However, the following chapters in this book attempt to provide the reader with sufficient basic knowledge of the structure and reactivity of inorganic systems to provide an understanding of the molecules and ions presented here, as well as many others.

PROBLEMS

1.1. Carefully read Chapter 1, and wherever an idea is presented that puzzles or intrigues, jot it down for future reference at the end of the semester.

1.2. On the basis of your limited exposure, define inorganic chemistry. Save your definition for future reference.

2

The structure of the atom

At the turn of the last century chemistry found itself in the position of being able to correlate a considerable body of data by means of the gas laws, the ideas of valence, and similar empirical relationships. Inorganic chemistry had progressed for several decades under the framework of the periodic table. Yet there was no knowledge of the actual structure of the atoms involved or of the basis of the periodic law. It was not until 1911 that Rutherford (as a result of experiments performed in his laboratory by Geiger and his young assistant Marsden) proposed the nuclear atom with almost the entire mass concentrated in a relatively small volume. At that time physics was in a ferment. The classical (i.e., Newtonian) mechanics which had served so well in describing the motions of planets and billiard balls was becoming increasingly beleaguered by seemingly unsolvable problems at the microscopic level. In solving these problems in physics, men such as Planck, Einstein, and Bohr opened the way to a complete (in principle at least) understanding of the structure of the atom and of chemical bonding. Although their work was fundamental to the development of theories of atomic and molecular structure, it is now mainly of historical interest. However, the early discovery of the interaction of energy and matter has been of lasting importance. The development of modern inorganic chemistry is based on the ever-increasing use of spectroscopic methods.

SPECTROSCOPY

When excited, hydrogen atoms emit light of definite frequencies (Table 2.1). The lines of the spectrum were named after the men who discovered them: Lyman (ultraviolet); Balmer (visible); and Paschen, Brackett, and Pfund (infrared). In 1885 Balmer related the frequencies of the spectral lines of the series which bears his name in terms of a single, variable integer.[1]

[1] The frequency of light may be expressed in many ways—convenient for the practicing spectroscopist but, unfortunately, often confusing to the student. The frequency, ν, of light is related to its wavelength, λ, inversely through the speed of light ($c = 3.00 \times 10^8$ m s^{-1}):

$$\nu = c/\lambda$$

Table 2.1 Hydrogen spectrum

Series	Wavelength (nm)	Frequency (cm^{-1})	n_2	n_1
Lyman	93.8	106 600	6	1
	95.0	105 300	5	1
	97.3	102 800	4	1
	102.6	97 480	3	1
	121.6	82 260	2	1
Balmer	397.0	25 190	7	2
	410.2	24 380	6	2
	434.0	23 040	5	2
	486.1	20 570	4	2
	656.3	15 240	3	2
Paschen	954.6	10 470	8	3
	1 005.0	9 950	7	3
	1 093.8	9 142	6	3
	1 281.8	7 801	5	3
	1 875.1	5 333	4	3
Brackett	2 630	3 800	6	4
	4 050	2 470	5	4
Pfund	7 400	1 350	6	5

Rydberg then showed that Balmer's formula was but a special case of a more general equation applicable to all the lines of the hydrogen spectrum:

$$\bar{\nu} = 109\,677\left(\frac{1}{n_1^2} - \frac{1}{n_2^2}\right) \qquad \text{where } n_2 > n_1$$

Example

If $n_2 = 3$ and $n_1 = 2$, then:

$$\bar{\nu} = 109\,677(\tfrac{1}{4} - \tfrac{1}{9}) = 15\,240 \text{ cm}^{-1}$$

Or if $n_2 = 4$ and $n_1 = 2$, then:

$$\bar{\nu} = 109\,677(\tfrac{1}{4} - \tfrac{1}{16}) = 20\,570 \text{ cm}^{-1}$$

It will be seen that the Lyman series consists of frequencies generated by $n_1 = 1$ and $n_2 = n_1 + 1, n_1 + 2, \ldots$. Likewise, for the Balmer series, $n_1 = 2$; Paschen, $n_1 = 3$; Brackett,

The unit of frequency is the hertz (Hz). The wavelength is often expressed in centimetres, or nanometres (10^{-9} m). The non-SI Angstrom unit (1 Å = 10^{-10} m) is sometimes found in literature as a wavelength unit also. Since frequencies are derived from the speed of light, which is a very large number, they tend to become cumbersome, and often the reciprocal of wavelength expressed as wave numbers ($\bar{\nu}$, cm^{-1}) is used instead. Unfortunately, there is no universal agreement on the use of units; visible and ultraviolet spectra are usually reported in units of wavelength, but infrared spectra are often reported in wave numbers.

$n_1 = 4$; and Pfund, $n_1 = 5$. The frequencies calculated in the above example are representative of the Balmer series and are observed in the Fraunhofer lines in the visible spectrum of the sun. They arise as atoms in the cool outer atmosphere of the sun absorb photons corresponding to these electronic transitions. The discovery of helium came about by the chance observation of a similar emission line in the sun's chromosphere.[2]

THE WAVE EQUATION

Based on the spectroscopic observations of his predecessors, in 1913 Niels Bohr attempted to rationalize the behavior of atoms in terms of a simple, dynamic model. Although his model was successful in reproducing the spectrum of the hydrogen atom in the absence of a magnetic field, other cases proved troublesome. Thus, if one examines the spectrum of the hydrogen atom in a magnetic field, the Zeeman effect (a more complicated spectrum) is observed. Sommerfeld modified the original Bohr model to overcome this defect. When larger, polyelectronic atoms were considered, additional modifications were necessary, and the model became cumbersome, or even impossible, to use.

Meanwhile, Louis de Broglie had suggested a dual wave/particle nature to the electron, which was confirmed experimentally by the diffraction of electrons by a crystal. Furthermore, Heisenberg's uncertainty principle stated that even if an electron were a particle, we could never describe it in so precise a manner as implied by the Bohr model. The stage was set for a completely different perspective on the atom.[3]

In 1926 Schrödinger proposed the wave equation which now bears his name. Its purpose was to describe the behavior of a subatomic particle in the same way that macroscopic particles are described by classical mechanics. The wave equation in three dimensions is:

$$\frac{\partial^2 \Psi}{\partial x^2} + \frac{\partial^2 \Psi}{\partial y^2} + \frac{\partial^2 \Psi}{\partial z^2} + \frac{8\pi^2 m}{h^2}(E - V)\Psi = 0 \quad \textbf{(2.1)}$$

where

Ψ = wave function

x, y, z = coordinates in space

m = mass

h = Planck's constant

E = total energy

V = potential energy

The solution to the wave equation is the wave function Ψ. For real waves Ψ corresponds to the amplitude of the wave and thus has no physical reality in the present applica-

---

[2] See M. E. Weeks and H. M. Leicester, "Discovery of the Elements," 7th ed., Chemical Education Publ. Co., Easton, Pa., **1968**.

[3] The pre-Schrödinger work has been given rather short shrift to make more room for inorganic chemistry. Good discussions of it may be found in most physical chemistry textbooks.

tion. However, just as the *intensity* of a light wave, for example, is given by the *amplitude squared*, so is the *probability of finding a particle* proportional to the *square of its wave function.*

Particle in a box

In general, solution of the wave equation for an atom may be quite difficult or impossible. However, it is possible to illustrate such a solution by treating a similar problem, that of a particle in a one-dimensional box. A particle confined in such a box is in some ways analogous to an electron constrained in a three-dimensional atom. An even closer analogy may be made to a linear, conjugated molecule in which an electron is free to move the entire length of the molecule. The particle in a box thus provides not only a simple application of the methods of quantum mechanics but also a recurring and valuable model of the electronic behavior of molecules.

Consider a particle in a box as shown in Fig. 2.1. In regions I and III (outside the box) the potential energy is taken to be infinite (the particle cannot escape the box) and the potential energy inside the box is taken to be zero. Under these conditions, Newtonian mechanics predicts that the particle has an equal probability of being in any part of the box and the kinetic energy of the particle is allowed to have any value. In contrast, as we shall see, wave mechanics leads to quite different results.

We shall use the wave equation in one dimension, inserting appropriate values of the potential energy, V:

$$\text{I, III; } V = \infty \qquad \frac{d^2\Psi}{dx^2} + \frac{8\pi^2 m}{h^2}(E - \infty)\Psi = 0 \tag{2.2}$$

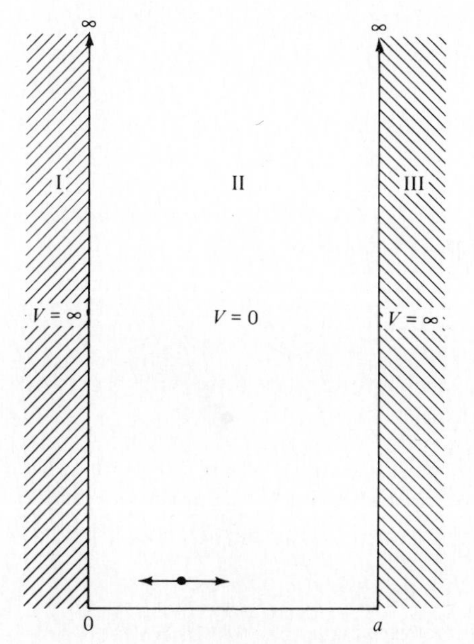

Fig. 2.1 Particle in a box, one-dimensional case. The particle is free to move in region II (from 0 to a) but not in regions I and III.

$$\text{II; } V = 0 \qquad \frac{d^2\Psi}{dx^2} + \frac{8\pi^2 m}{h^2}(E - 0)\Psi = 0 \tag{2.3}$$

We shall impose the following limitations on the wave function, Ψ:

1. *Ψ must be single-valued.* This arises from the fact that the probability of finding the particle will be proportional to Ψ^2; if it were not single-valued, there would be two or more distinct probabilities of finding the particle at the same point in space—obviously impossible.
2. *Ψ must be finite.* The value of Ψ^2 must be finite since it represents a probability.
3. *Ψ must be continuous.* As all real waves are continuous, we make this a condition for our system.

In regions I and III the only value of Ψ that will satisfy Eq. 2.2 is $\Psi = 0$ since for any finite value of Ψ the left-hand side of Eq. 2.2 will equal infinity, not zero.

For region II we have the problem of finding a solution which satisfies Eq. 2.3 and at the same time will give a value of $\Psi = 0$ at $x = 0$ and $x = a$.

One now assumes a trial solution and tries to prove mathematically that it satisfies the three conditions imposed above. For example, a trial solution might be:

$$\Psi = A \cos \beta x + B \sin \beta x \tag{2.4}$$

When we do this we find that we can let $A = 0$, and we obtain the following solution:[4]

$$\Psi = B \sin\left(\frac{n\pi x}{a}\right) \tag{2.5}$$

where n is an integer. This solution combined with the original equations for the energy of the particle (Eqs. 2.2 and 2.3) provides the allowed energies of the particle in the box:

$$\beta^2 = \frac{n^2\pi^2}{a^2} = \frac{8\pi^2 mE}{h^2} \tag{2.6}$$

$$E = \frac{n^2h^2}{8ma^2} \tag{2.7}$$

For the more general case of a three-dimensional box the energy is given by:

$$E = \frac{h^2}{8m}\left(\frac{n_x^2}{a^2} + \frac{n_y^2}{b^2} + \frac{n_z^2}{c^2}\right) \tag{2.8}$$

The results of the particle-in-a-box calculation are of interest mainly with regard to the following two points:

1. Unlike the classical prediction, the probability of finding the particle is not constant but is a function of x. Furthermore, the probability of finding the particle in a particular portion of the box depends upon the energy of the particle (see Fig. 2.2).

[4] To conserve space, the requisite mathematical manipulations, which appeared in the first edition of this text, have been omitted from this edition. The discussion of the particle in a box by F. L. Pilar, "Elementary Quantum Mechanics," McGraw-Hill, New York, **1968**, is especially recommended.

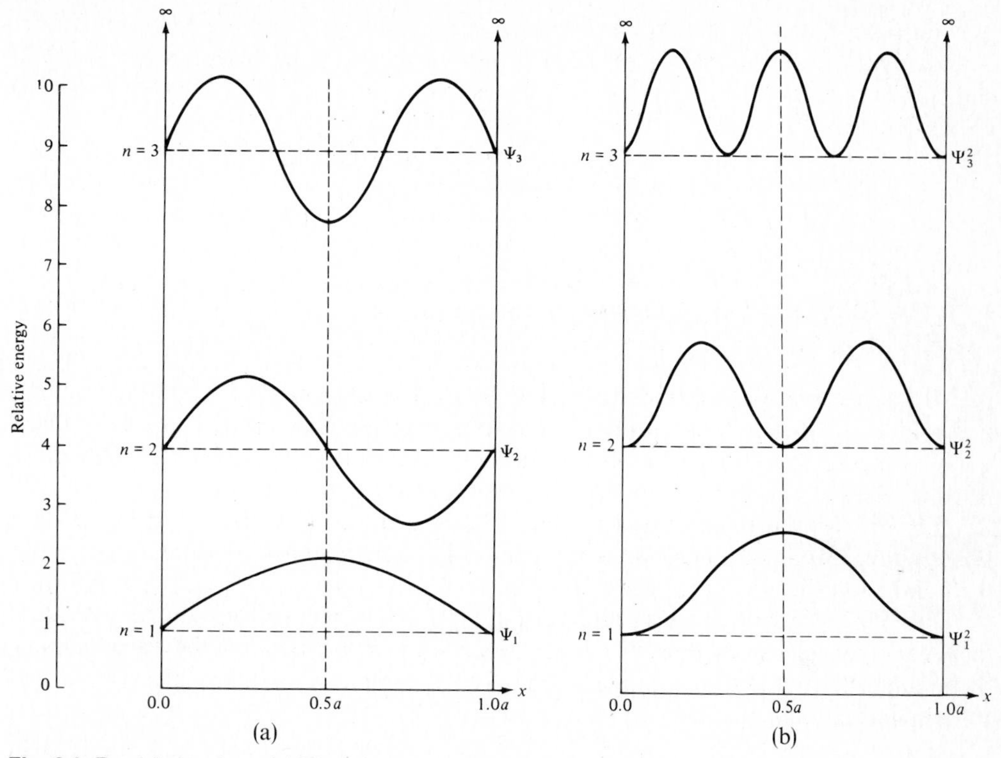

Fig. 2.2 Particle in a box: (a) Wave functions. Note that the energy is actually independent of x and given by the dashed line for each value of n. (b) Probability functions of a particle in a box. [Adapted from F. L. Pilar, "Elementary Quantum Chemistry," McGraw-Hill, New York, **1968**. Reproduced with permission.]

2. Another difference from classical predictions is the fact that only certain energies, related to the quantum number n (for a one-dimensional box), are allowed. Zero energy ($n = 0$) is not allowed (otherwise $\Psi = 0$ and the solution is trivial—the probability of finding the particle is zero, $\Psi^2 = 0$, and therefore it doesn't exist!), and the energy increases as n^2.

A further item of interest is the fact (see Eqs. 2.7 and 2.8) that a quantum number is required for each degree of freedom ("dimension") allowed the particle. For a three-dimensional atom we should thus expect three quantum numbers related to the spatial position of the electron.

THE HYDROGEN ATOM

Although solution of the Schrödinger equation for the hydrogen atom is not overly difficult, we shall not go into the problem here since we are more interested in the results than in the method. It may be pointed out, however, that the method is not unlike that used previously for a particle in a box. In the case of the hydrogen atom, the "box" is a sphere with sloping rather than vertical potential "walls." As with the particle-in-a-box solution, constraints or boundary conditions are imposed which make possible the solution of the wave equation. These conditions are:

1. The wave function must be single-valued.
2. The wave function and its first derivative must be continuous.
3. The wave function must go to zero at infinity. This is a necessary condition so that the atom be finite.
4. The probability of finding the electron summed over all space must be one, that is, the wave function must be normalized.

It is found that the solution for the hydrogen atom contains three quantum numbers n, l, and m_l (as expected for a three-dimensional system). The allowed values for these quantum numbers are discussed below. Each solution found for a different set of n, l, m_l is called an *eigenfunction* and represents an orbital in the hydrogen atom.

In order to plot the complete wave functions, one would in general require a four-dimensional graph with coordinates for each of the three spatial dimensions (x, y, z; or r, θ, ϕ) and a fourth value of the wave function.

In order to circumvent this problem and also to make it easier to visualize the actual distribution of electrons within the atom, it is common to break down the wave function, Ψ, into three parts, each of which is a function of but a single variable. It is most convenient to use polar coordinates, so one obtains:

$$\Psi(r, \theta, \phi) = R(r) \cdot \Theta(\theta) \cdot \Phi(\phi) \tag{2.9}$$

where $R(r)$ gives the dependence of Ψ upon distance from the nucleus and Θ and Φ give the angular dependence.

The radial wave function, R

The radial functions for the first three orbitals[5] in the hydrogen atom are

$$n = 1, l = 0, m_l = 0 \qquad R = 2\left(\frac{Z}{a_0}\right)^{3/2} e^{-Zr/a_0} \qquad 1s \text{ orbital}$$

$$n = 2, l = 0, m_l = 0 \qquad R = \left(\frac{1}{2\sqrt{2}}\right)\left(\frac{Z}{a_0}\right)^{3/2}\left(2 - \frac{Zr}{a_0}\right)e^{-Zr/2a_0} \qquad 2s \text{ orbital}$$

$$n = 2, l = 1, m_l = 0 \qquad R = \left(\frac{1}{2\sqrt{6}}\right)\left(\frac{Z}{a_0}\right)^{3/2} \frac{Zr}{a_0} e^{-Zr/2a_0} \qquad 2p \text{ orbital}$$

[5] The general formulas for radial eigenfunctions in terms of the quantum numbers n and l are given by L. Pauling, "The Nature of the Chemical Bond," Cornell University Press, Ithaca, N.Y., **1960**, and by D. J. Royer, "Bonding Theory," McGraw-Hill, New York, **1968**. The complete wave functions including both the radial and angular parts through the nth level are given by Pauling ($n = 6$), Royer ($n = 3$), and F. A. Cotton and G. Wilkinson, "Advanced Inorganic Chemistry," 2nd ed., Wiley, New York, **1966** ($n = 3$).

where Z is the nuclear charge, e is the base of natural logarithms, and a_0 is the radius of the first Bohr orbit. According to the Bohr theory, this was an immutable radius, but in wave mechanics it is simply the "most probable" radius for the electron to be at. Its value, 52.9 pm, is determined by $a_0 = h^2/4\pi^2 me^2$, where h is Planck's constant and m and e are the mass and charge of the electron, respectively. In hydrogen, $Z = 1$, but similar orbitals may be constructed where $Z > 1$ for other elements. For many-electron atoms, exact solutions of the wave equation are impossible to obtain, and these "hydrogen-like" orbitals are often used as a first approximation.[6]

Although the radial functions may appear formidable, the important aspects may be made apparent by grouping the constants. For a given atom, Z will be constant and may be combined with the other constants, resulting in considerable simplification:

$$n = 1,\ l = 0,\ m_l = 0 \qquad R = K_{1s}e^{-Zr/a_0} \qquad 1s \text{ orbital}$$

$$n = 2,\ l = 0,\ m_l = 0 \qquad R = K_{2s}\left(2 - \frac{Zr}{a_0}\right)e^{-Zr/2a_0} \qquad 2s \text{ orbital}$$

$$n = 2,\ l = 1,\ m_l = 0 \qquad R = K_{2p}re^{-Zr/2a_0} \qquad 2p \text{ orbital}$$

The most apparent feature of the radial wave functions is that they all represent an exponential "decay" (mathematically similar in form to the radioactive decay of radioisotopes), and that for $n = 2$ the decay is slower than for $n = 1$. This may be generalized for all radial functions: They decay as e^{-Zr/na_0}. For this reason, the radius of the various orbitals (actually, the most *probable* radius) increases with increasing n. A second feature is the presence of a *node* in the $2s$ radial function. At $r = 2a_0/Z$, $R = 0$ and the value of the radial function changes from positive to negative. Again, this may be generalized: s orbitals have $n - 1$ nodes, p orbitals have $n - 2$ nodes, etc. The radial functions for the hydrogen $1s$, $2s$, and $2p$ orbitals are shown in Fig. 2.3.

Since we are principally interested in the *probability* of finding electrons at various points in space, we shall be more concerned with the *square* of the radial functions than with the functions themselves. A useful way of looking at the problem is to consider the atom to be composed of "layers" much like an onion and to examine the probability of finding the electron in the "layer" which extends from r to $r + dr$, as shown in Fig. 2.4. The volume of the thin shell may be considered to be dV. Now the volume of the sphere is:

$$V = \frac{4\pi r^3}{3} \tag{2.10}$$

$$dV = 4\pi r^2\, dr \tag{2.11}$$

$$R^2\, dV = 4\pi r^2 R^2\, dr \tag{2.12}$$

[6] The use of hydrogen-like orbitals for multielectron atoms neglects electron-electron repulsion, which may often cause serious problems (see pp. 27–29).

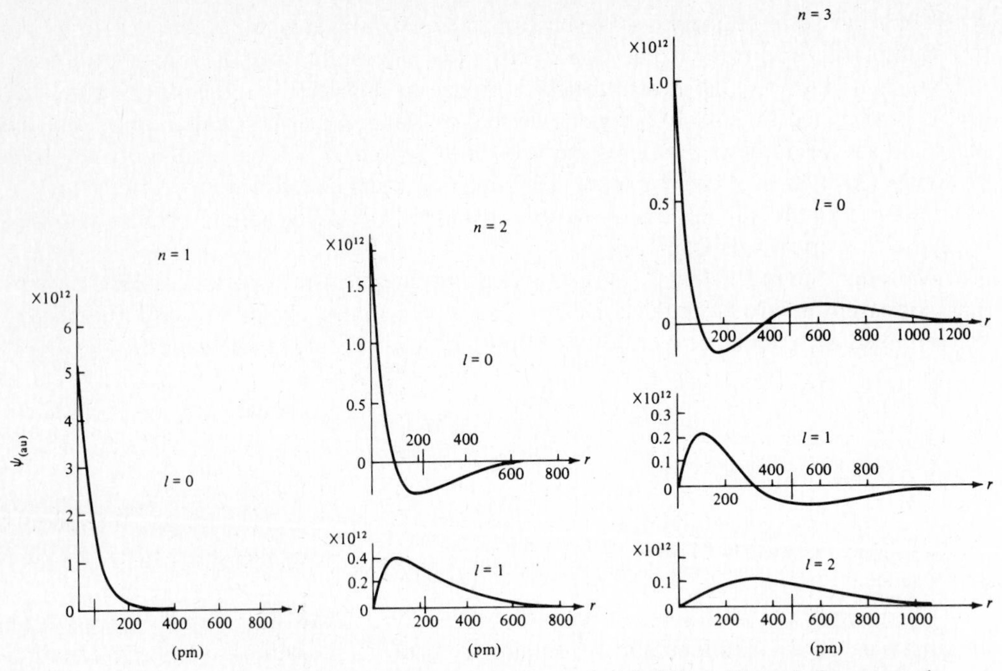

Fig. 2.3 Radial part of the hydrogen eigenfunctions for $n = 1, 2, 3$. [From "Atomic Spectra and Atomic Structure," by Gerhard Herzberg, Dover Publications, Inc., New York, **1944**. Reprinted through permission of the publisher.]

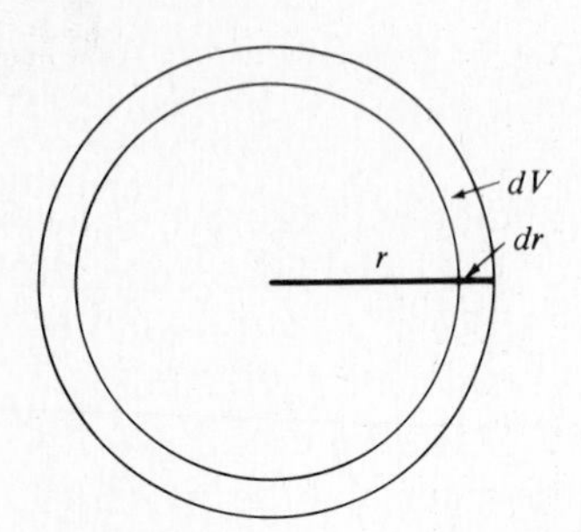

Fig. 2.4 Volume of shell of thickness *dr*.

Consider the radial portion of the wave function for the 1*s* orbital as plotted in Fig. 2.3. When it is squared and multiplied by $4\pi r^2$, we obtain the *probability function* shown in Fig. 2.5. The essential features of this function may be obtained qualitatively as follows:

1. At $r = 0$, $4\pi r^2 R^2 = 0$; hence the value at the nucleus must be zero.
2. At large values of r, R approaches zero rapidly and hence $4\pi r^2 R^2$ must approach zero.
3. In between, r and R both have finite values, so there is a maximum in the plot of probability ($4\pi r^2 R^2$) as a function of r. This maximum occurs at $r = a_0$, the value of the Bohr radius.

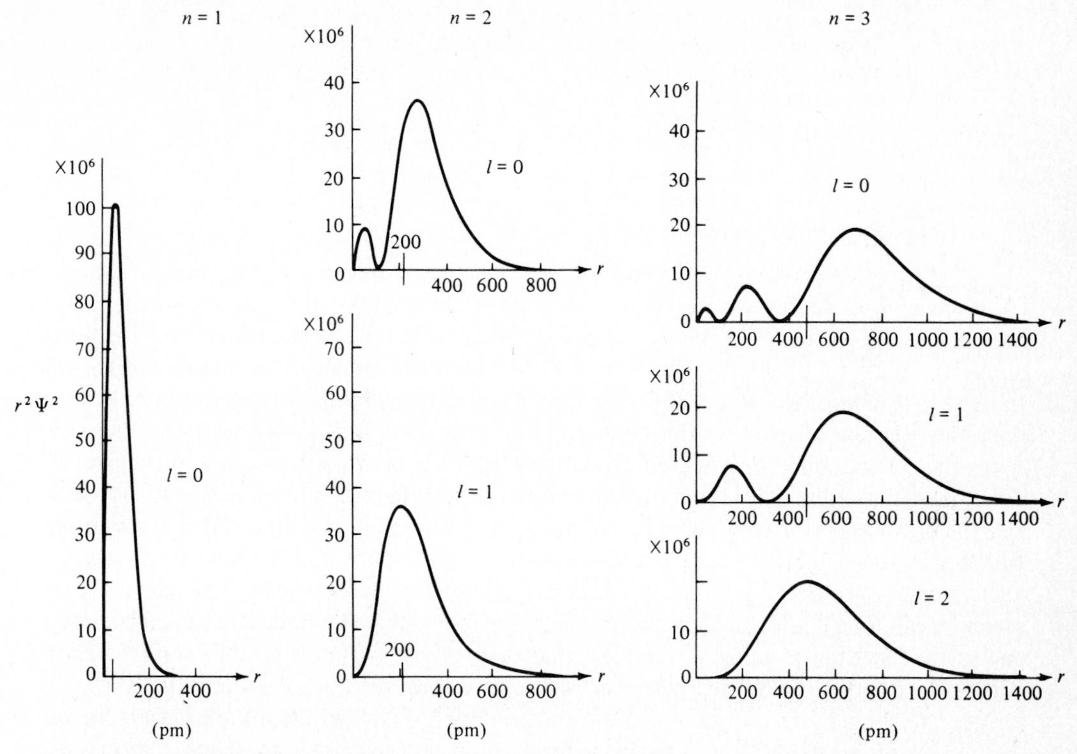

Fig. 2.5 Radial probability functions for $n = 1, 2, 3$ for the hydrogen atom. The function gives the probability of finding the electron in a spherical shell of thickness dr at a distance r from the nucleus. [From "Atomic Spectra and Atomic Structure," by Gerhard Herzberg, Dover Publications, Inc., New York, **1944**. Reprinted through permission of the publisher.]

Similar probability functions (including the factor $4\pi r^2$) for the 2*s*, 2*p*, 3*s*, 3*p*, and 3*d* orbitals are also shown in Fig. 2.5. Note that although the radial function for the 2*s* orbital is both positive ($r < 2a_0/Z$) and negative ($r > 2a_0/Z$), the probability function is everywhere *positive* (as of course it must be to have any physical meaning) as a result of the squaring operation.

The presence of a node in the wave function indicates a point in space at which the probability of finding the electron has gone to zero. This raises the interesting question, "How does the electron get from one side of the node to the other if it can never be found exactly *at* the node?" This is not a valid question as posed, since it presupposes our macroscopically prejudiced view that the electron is a particle. If we consider the electron to be a standing wave, no problem arises because it simultaneously exists on both sides of a node. Consider a vibrating string on an instrument such as a guitar. If the string is fretted at the twelfth fret the note will go up one octave because the wavelength has been shortened by one-half. Although it is experimentally difficult (a finger is not an infinitesimally small point!), it is possible to sound the same note on either half of the octave-fretted string. This vibration can be continuous through the node at the fret. In fact, on the open string overtones occur at the higher harmonics such that nodes occur at various points along the string. Nodes are quite common to wave behavior, and conceptual problems arise only when we try to think of the electron as a "hard" particle with a definite position.

Dirac[7] has provided a relativistic treatment that differs from the "standard" theory in a number of ways. (1) There are no nodes, either radial or angular in a Dirac atom; however, there are regions in which the electron density becomes very small.[8] (2) Electron "spin" is shown to be a natural property of the electron. (3) The electrons of given value of l (see p. 25) are not *exactly* degenerate but differ slightly in energy. (4) The usual treatment of spin–orbit interactions is shown to be artificial. Nevertheless, almost all of the current discussions in quantum mechanics are in nonrelativistic terms because of mathematical simplicity and because the errors incurred are small; however, a strongly dissenting opinion on this practice has been presented by Powell.[9]

The presence of one or more nodes causes small maxima in electron density between the nucleus and the largest maximum. It is often stated that these nodes and maxima have no chemical effect, but this is slightly misleading. There are two ways in which these nodes and maxima could affect bonding. (1) In a covalent bond, as we shall see later, we shall be very concerned with total overlap of the atomic orbitals which combine to form the bond. Regions of space where there are low electron densities, and especially where the sign of the wave function is changing, are not conducive to good overlap. Conceivably, if the node in a radial wave function fell at the appropriate place it could seriously weaken a covalent bond. Prior to the synthesis[10] of the perbromate ion such a radial node in the $4d$ orbitals was suggested[11] as a possible reason for its apparent nonexistence. The subsequent synthesis of BrO_4^- obviates the necessity for such an explanation, and in every case in which careful calculations have been made it is found that the nodes lie too close to the nucleus to affect the bonding. (2) The fact that electrons in s orbitals spend a small fraction of their time very close to the nucleus is extremely important in determining the energy of the orbitals. For a given value of n the ionization energies of s electrons are always higher than those of corresponding p electrons (and the trend continues to d and f electrons) because the s orbitals are more *penetrating*; i.e., they have considerable electron density in the region of the nucleus. This is the fundamental reason for the ordering of the energy levels: $1s$, $2s$, $2p$, $3s$, $3p$, etc. (see pp. 27 and 37).

Angular wave functions

The angular part of the wave function determines the shape of the electron cloud and varies depending upon the type of orbital involved (s, p, d, or f) and its orientation in space. However, for a given type of orbital, such as s or p_z, the angular wave function is independent of the principal quantum number or energy level. Some typical angular functions are:

$l = 0, m_l = 0 \qquad \Theta\Phi = (1/4\pi)^{1/2}$ $\qquad s$ orbital

$l = 1, m_l = 0 \qquad \Theta\Phi = (3/4\pi)^{1/2} \cos\theta$ $\qquad p_z$ orbital

$l = 2, m_l = 0 \qquad \Theta\Phi = (5/16\pi)^{1/2}(3\cos^2\theta - 1)$ $\qquad d_{z^2}$ orbital

[7] P. A. Dirac, *Proc. R. Soc. London, Ser. A*, **1928**, *117*, 610; **1928**, *118*, 351.
[8] For contour diagrams of relativistic orbitals, see A. Szabo, *J. Chem. Educ.*, **1969**, *46*, 678.
[9] R. E. Powell, *J. Chem. Educ.*, **1968**, *45*, 558.
[10] E. H. Appelman, *J. Am. Chem. Soc.*, **1968**, *90*, 1900.
[11] D. S. Urch, *J. Inorg. Nucl. Chem.*, **1963**, *25*, 771.

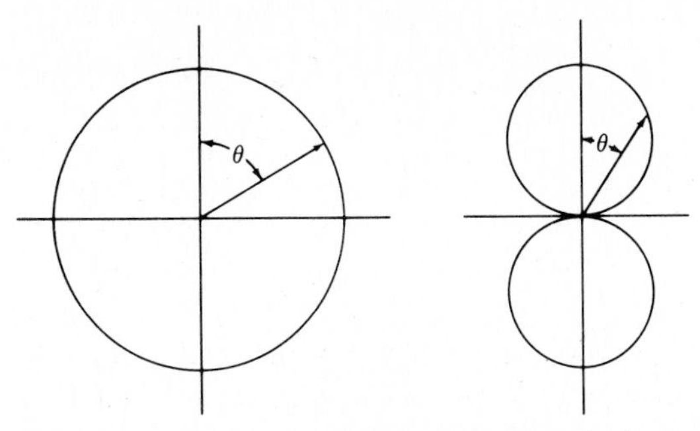

Fig. 2.6 Angular part of the wave function for hydrogen-like *s* orbitals (left) and *p* orbitals (right). Only two dimensions of the three-dimensional function have been shown.

The angular functions for the *s* and p_z orbital are illustrated in Fig. 2.6. For an *s* orbital, $\Theta\Phi$ is independent of angle and is of constant value. Hence this graph is circular or, more properly, in three dimensions—spherical. For the p_z orbital we obtain two tangent spheres. The p_x and p_y orbitals are identical in shape but are oriented along the *x* and *y* axes, respectively. We shall defer extensive treatment of the *d* orbitals (Chapter 9) and *f* orbitals (Chapter 16) until bond formation in coordination compounds is discussed, simply noting here that the basic angular function for *d* orbitals is four-lobed and that for *f* orbitals is six-lobed (see Fig. 2.11).

We are most interested in the probability of finding an electron, and so we shall wish to examine the function $\Theta^2\Phi^2$ since it corresponds to the angular part of Ψ^2. When the angular functions are squared, different orbitals change in different ways. For an *s* orbital squaring causes no change in shape since the function is everywhere the same; thus another sphere is obtained. For both *p* and *d* orbitals, however, the plot tends to become more elongated (see Fig. 2.7).

The meaning of Figs. 2.6 and 2.7 is easily misinterpreted. Neither one has any direct physical meaning. Both are graphs of mathematical functions, just as Figs. 2.3 and 2.5 are. Both may be used to obtain information about the probable distribution of electrons, but

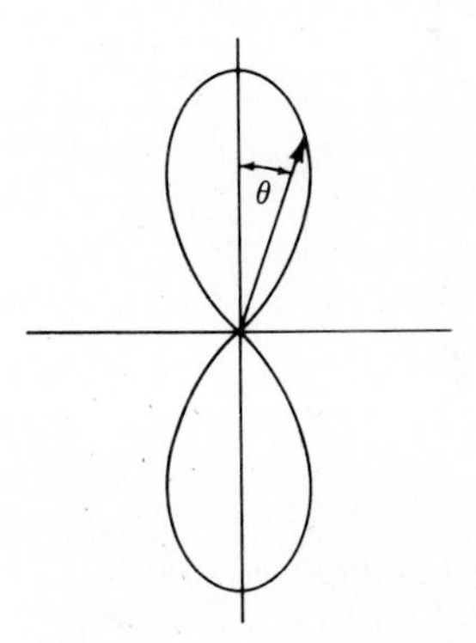

Fig. 2.7 Angular probability function for hydrogen-like *p* orbitals. Only two dimensions of the three-dimensional function have been shown.

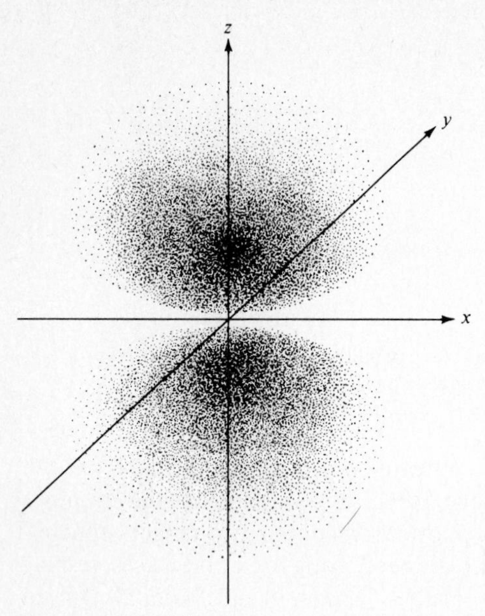

Fig. 2.8 Pictorial representation of electron density in a hydrogen-like $2p$ orbital.

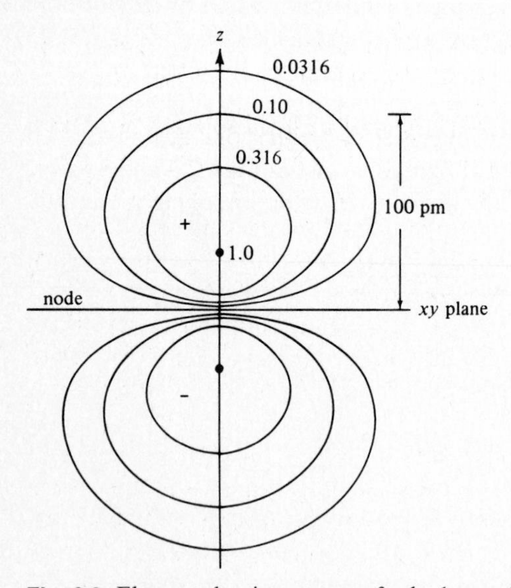

Fig. 2.9 Electron density contours for hydrogen-like $2p_z$ orbital of carbon. Contour values are relative to the electron density maximum. The xy plane is a nodal surface. The signs (+ and −) refer to the original wave function, Ψ. [From E. A. Ogryzlo and G. B. Porter, *J. Chem. Educ.*, **1963**, *40*, 258. Reproduced with permission.]

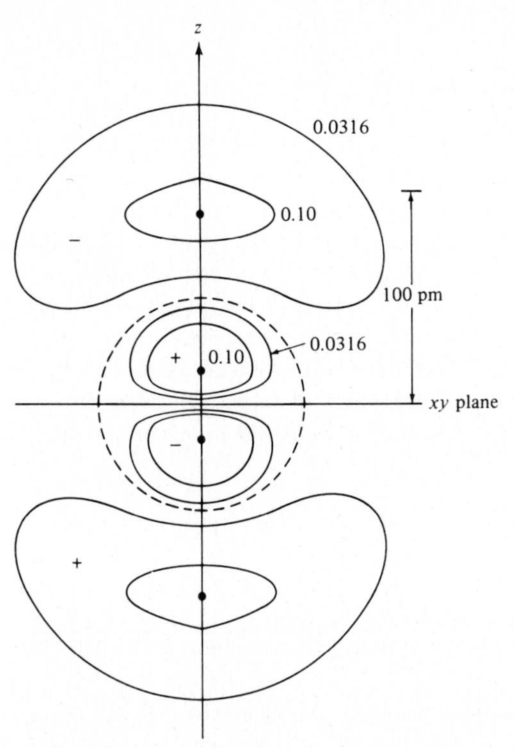

Fig. 2.10 Electron density contours for hydrogen-like $3p_z$ orbital of chlorine. Contour values are relative to the electron density maximum. The xy plane and a sphere of radius 52 pm (dashed line) are nodal surfaces. The signs (+ and −) refer to the original wave function, Ψ. [From E. A. Ogryzlo and G. B. Porter, *J. Chem. Educ.*, **1963**, *40*, 258. Reproduced with permission.]

neither may in any way be regarded as a "picture" of an orbital. It is an unfortunate fact that fuzzy drawings of Figs. 2.6 or 2.7 are often presented as "orbitals." Now one can define an orbital in any way one wishes, corresponding to Ψ, Ψ^2, R, R^2, $\Theta\Phi$, or $\Theta^2\Phi^2$, but it should be realized that Figs. 2.3, 2.5, 2.6, and 2.7 are mathematical functions and drawing them fuzzily does *not* represent an atom. Chemists tend to think in terms of electron clouds, and hence Ψ^2 probably gives the best intuitive "picture" of an orbital. Methods of showing the total probability of finding an electron including *both* radial and angular probabilities are shown in Figs. 2.8–2.10. Although electron density may be shown either by shading (Fig. 2.8) or by contours of equal electron density (Figs. 2.9 and 2.10), only the latter method is quantitatively accurate.[12]

Since $\Theta^2\Phi^2$ is termed an angular probability function, the question may properly be asked what its true meaning is, if not a "picture" of electron distribution. Like any other graph, it simply plots the value of a function ($\Theta^2\Phi^2$) versus the variable (θ or θ, ϕ). If one chooses an angle θ, the probability that the electron will be found in that direction (summed over all distances) is proportional to the magnitude of the vector connecting the origin with the functional plot at that angle.

[12] For further discussion of this point, see B. Perlmutter-Hayman, *J. Chem. Educ.*, **1969**, *46*, 428.

Symmetry of orbitals

In Fig. 2.11 are shown sketches of the angular part of the wave functions for *s*, *p*, and *d* orbitals. The signs in the lobes represent the sign of the wave function in those directions. For example, in the p_z orbital, for $\theta = 90°$, $\cos\theta = 0$, and for $90° < \theta < 270°$, $\cos\theta$ is negative. The signs of the wave functions are very important when considering the overlap of two bonding orbitals. It is customary to speak of the symmetry of orbitals as *gerade* or *ungerade*. These German words meaning even and uneven refer to the operation shown in the sketches—inversion about the center. If on moving from any point A to the equivalent point B on the opposite side of the center the sign of the wave function does not change, the orbital is said to be *gerade*. The *s* orbital is a trivial case in which the sign of the angular wave function is everywhere the same. The *d* orbitals (only two of which are shown here) are also *gerade*. The *p* orbitals, however, are unsymmetrical with respect to inversion and the sign changes on going from A to B; hence the symmetry is *ungerade*. Likewise, *f* orbitals are *ungerade*. These terms are often abbreviated *g* and *u*.

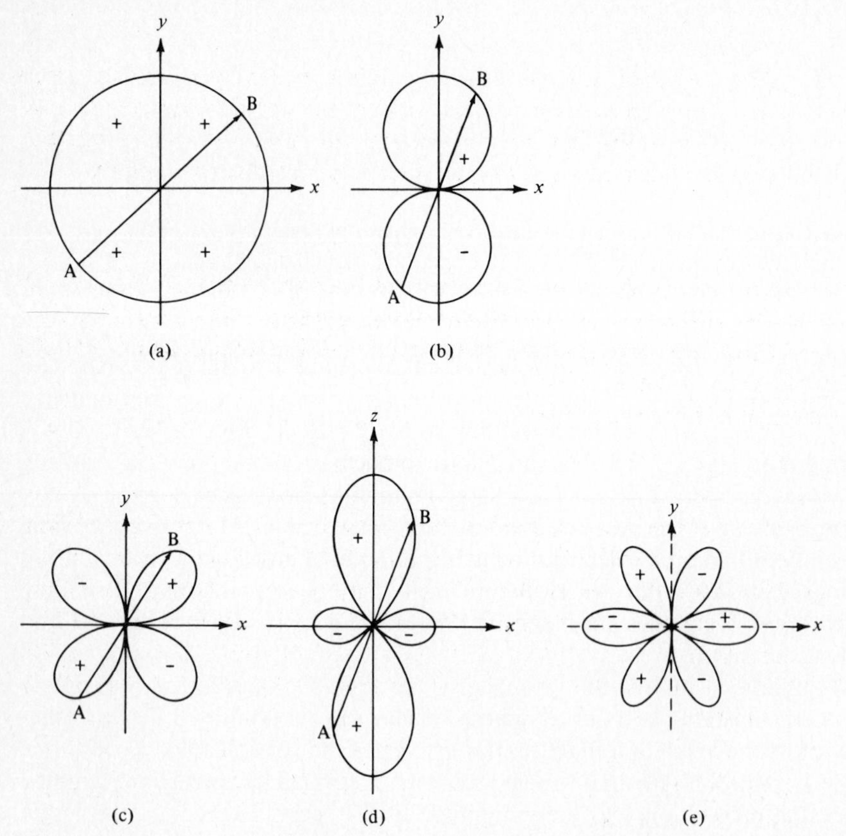

Fig. 2.11 Angular wave functions of *s*, *p*, *d*, and *f* orbitals illustrating *gerade* and *ungerade* symmetry: (a) *s* orbital, *gerade*; (b) *p* orbital, *ungerade*; (c) d_{xy} orbital and (d) d_{z^2} orbital, both *gerade*; (e) f_{z^3} orbital, *ungerade*.

In addition to symmetry with respect to inversion about the center, orbitals have other symmetry properties with respect to rotation about the various coordinate axes. An introduction to symmetry operations and symmetry is given in Appendix B.

Attention should be called to a rather confusing practice which chemists commonly use. In Figs. 2.9 and 2.10 it will be noted that small plus and minus signs appear. Although the figure refers to the *probability* of finding the electron and thus must be everywhere positive, the signs + and − refer to the sign of the original wave function, Ψ, in these regions of space. In Fig. 2.10, for example, in addition to the inversion resulting from the *ungerade p* orbital, there is a second node (actually a spherical nodal surface) at a distance of 6 a_0/Z resulting from the radial wave function. Although this practice may seem confusing, it is useful and hence has been accepted. The Ψ^2 plot is useful in attempting to visualize the physical "picture" of the atom, but the sign of Ψ is important with respect to bonding.[13] The energy levels of the hydrogen atom are found to be determined solely by the principal quantum number,[14] and their relationship is the same as found for a Bohr atom:

$$E_n = -\frac{2\pi^2 me^4}{n^2 h^2} \tag{2.13}$$

where m is the mass of the electron, e is the electronic charge, n is the principal quantum number, and h is Planck's constant. Quantization of energy and angular momentum were introduced as assumptions by Bohr, but they follow naturally from the wave treatment. The quantum number n may have any positive, integral value from one to infinity:

$$n = 1, 2, 3, 4, \ldots, \infty$$

The lowest (most negative) energy corresponds to the minimum value of n ($n = 1$) and the energies increase (become less negative) with increasing n until the *continuum* is reached ($n = \infty$). Here the electron is no longer bound to the atom and thus is no longer quantized, but may have any amount of kinetic energy.

The allowed values of l range from zero to $n - 1$:

$$l = 0, 1, 2, 3, \ldots, n - 1$$

The quantum number l is a measure of the orbital angular momentum of the electron and determines the "shape" of the orbital. The types of orbitals are designated by the letters *s, p, d, f, g, . . .*, corresponding to the values of $l = 0, 1, 2, 3, 4, \ldots$. The first four letters originate in spectroscopic notation (see p. 32) and the remainder follow alphabetically. In the previous section we have seen the various angular wave functions and the resulting distribution of electrons. The nature of the angular wave function is determined by the value of the quantum number l.

---

[13] See M. Orchin and H. H. Jaffé, "The Importance of Antibonding Orbitals," Houghton Mifflin, Boston, **1967**, pp. 5, 6, 9, for a good discussion of this point.

[14] This statement is true only for the nonrelativistic treatment. If relativity effects are included, a small dependence on l is also found. This should not be confused with the much *larger* differences in orbital energies that result from electron-electron interactions in *polyelectronic atoms*.

For orbitals with $l > 0$, there are a number $(2l + 1)$ of equivalent ways in which the orbitals may be oriented in space. In the absence of an electric or magnetic field these orientations are *degenerate*; i.e., they are identical in energy. Consider, for example, the p orbital. It is possible to have a p orbital in which the maximum electron density lies on the z axis and the xy plane is a nodal plane. Equivalent orientations have the maximum electron density along the x or y axis. Application of a magnetic field splits the degeneracy of the set of three p orbitals. The *magnetic quantum number*, m_l, is related to the component of angular momentum along a chosen axis—for example, the z axis—and determines the orientation of the orbital in space. Values of m_l range from $-l$ to $+l$:

$$m_l = -l, -l+1, \ldots, -1, 0, +1, +2, \ldots, +l$$

Thus for $l = 1$, $m_l = -1, 0, +1$, and there are three p orbitals possible, p_x, p_y, and p_z. Similarly, for $l = 2$ (d orbitals), $m_l = -2, -1, 0, +1, +2$, and for $l = 3$ (f orbitals), $m_l = -3, -2, -1, 0, +1, +2, +3$.[15]

It is an interesting fact that just as the single s orbital is spherically symmetric, the summation of a set of three p orbitals, five d orbitals, or seven f orbitals is also spherical (Unsöld's theorem). Thus, although it might appear as though an atom such as neon with a filled set of s and p orbitals would have a "lumpy" electron cloud, the total probability distribution is perfectly spherical (see Problem 2.16).

From the above rules we may obtain the allowed values of n, l, and m_l. We have seen previously (p. 16) that a set of particular values for these three quantum numbers determines an eigenfunction or orbital for the hydrogen atom. The possible orbitals are therefore

$n = 1$	$l = 0$	$m_l = 0$	1s orbital
$n = 2$	$l = 0$	$m_l = 0$	2s orbital
$n = 2$	$l = 1$	$m_l = -1, 0, +1$	$2p_{(x,y,z)}$ orbitals
$n = 3$	$l = 0$	$m_l = 0$	3s orbital
$n = 3$	$l = 1$	$m_l = -1, 0, +1$	$3p_{(x,y,z)}$ orbitals
$n = 3$	$l = 2$	$m_l = -2, -1, 0, +1, +2$	$3d_{(z^2, x^2-y^2, xy, xz, yz)}$ orbitals[16]
$n = 4$	$l = 0$	$m_l = 0$	4s orbital

We can now summarize the relation between the quantum numbers n, l, and m_l and the physical pictures of electron distribution in orbitals by a few simple rules. It should be emphasized that these rules are no substitute for a thorough understanding of the previous discussion, but merely serve as handy guides to recall some of the relations.

1. Within a given atom, the lower the value of n, the more stable (lower in energy) will be the orbital.

[15] Although the p_z and d_{z^2} orbitals correspond to $m = 0$, there is no similar one-to-one correspondence for the other orbitals and other values of m. See J. C. Davis, Jr., "Advanced Physical Chemistry," Ronald, New York, **1965**, pp. 170–171, or B. N. Figgis, "Introduction to Ligand Fields," Wiley, New York, **1966**, pp. 9–15.

[16] These orbitals are sketched and discussed further in Chapter 9.

2. There are n types of orbitals in the nth energy level (e.g., the third energy level has s, p, and d orbitals).

3. There are $2l + 1$ orbitals of each type (e.g., one s, three p, five d, and seven f). This is also equal to the number of values which m_l may assume for a given l value.

4. There are $n - l - 1$ nodes in the radial distribution functions of all orbitals (e.g., the 3s orbital has two nodes, the 4d orbitals each have one).

5. There are l nodal surfaces in the angular distribution function of all orbitals (e.g., s orbitals have none, d orbitals have two).

THE POLYELECTRONIC ATOM

With the exception of Unsöld's theorem, above, *everything discussed thus far has dealt only with the hydrogen atom, the only atom for which the Schrödinger wave equation has been solved exactly.* This treatment can be extended readily to one-electron ions isoelectronic with hydrogen, such as He^+, Li^{+2}, and Be^{+3}, by using the appropriate value of the nuclear charge, Z. The next simplest atom, helium, consists of a nucleus and two electrons. We thus have three interactions: the attraction of electron 1 for the nucleus, the attraction of electron 2 for the nucleus, and the repulsion between electrons 1 and 2. This is an example of the classic three-body problem in physics and cannot be solved exactly. We can, however, approximate a solution to a high degree of accuracy using successive approximations. For simple atoms such as helium this is not too difficult, but for heavier atoms the number of interactions which must be considered rises at an alarming rate and the calculations become extremely laborious. A number of methods of approximation have been used, but we shall not explore them here beyond describing in conceptual terms one of the more accurate methods. It is referred to as the Hartree-Fock method, after the men who developed it, or as the *self-consistent field* (SCF) method. It consists of (1) assuming a reasonable wave function for each of the electrons in an atom except one, (2) calculating the effect which the field of the nucleus and the remainder of the electrons exert on the chosen electron, and (3) calculating a wave function for the last electron, including the effects of the field of the other electrons. A different electron is then chosen, and using the field resulting from the other electrons (including the contribution from the improved wave function of the formerly chosen electron), an improved wave function for the second electron is calculated. This process is continued until the wave functions for all of the electrons have been improved, and the cycle is then started over to improve further the wave function of the first electron in terms of the field resulting from the improved wave functions of the other electrons. The cycle is repeated as many times as necessary until a negligible change takes place in improving the wave functions. At this point it may be said that the wave functions are self-consistent and are a reasonably accurate description of the atom.

Such calculations indicate that orbitals in atoms other than hydrogen do not differ in any radical way from the hydrogen orbitals previously discussed. The principal difference lies in the consequence of the increased nuclear charge—all the orbitals are somewhat contracted. It is common to call such orbitals which have been adjusted by an appropriate nuclear charge *hydrogen-like orbitals.* Within a given major energy level it is found that the energy of these orbitals increases in the order $s < p < d < f$. For the higher energy levels

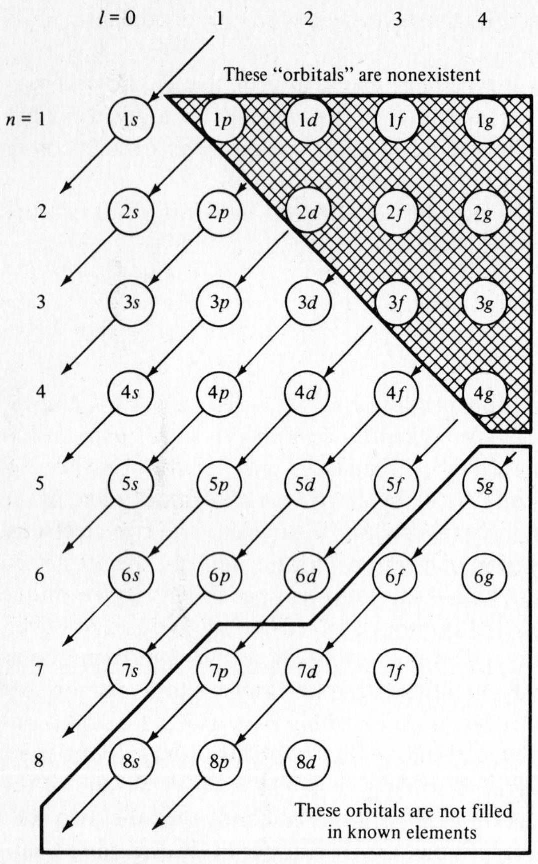

Fig. 2.12 Mnemonic for determining the order of filling of orbitals (approximate). [Adapted from T. Moeller, "Inorganic Chemistry," Wiley, New York, **1952**. Reproduced with permission.]

these differences are sufficiently pronounced that a staggering of orbitals may result, such as $6s < 5d \simeq 4f < 6p$, etc. The energy of a given orbital depends on the nuclear charge (atomic number) and different types of orbitals are affected to different degrees. Thus there is no single ordering of energies of orbitals which will be universally correct for all elements.[17] Nevertheless, the order $1s < 2s < 2p < 3s < 3p < 4s < 3d < 4p < 5s < 4d < 5p < 6s < 5d \simeq 4f < 6p < 7s < 6d < 5f$ is found to be extremely useful. This complete order is correct for *no* single element; yet, paradoxically, with respect to *placement of the outermost or valence electron,* it is remarkably accurate for all elements. For example, the valence electron in potassium must choose between the $3d$ and $4s$ orbitals, and as predicted by this series it is found in the $4s$ orbital. The above ordering should be assumed to be only a rough guide to the filling of energy levels (see "*aufbau* principle," p. 29). In many cases the orbitals are very similar in energy and small changes in atomic structure can invert two levels and change the order of filling. Nevertheless, the above series is a useful

[17] For a complete discussion, see R. N. Keller, *J. Chem. Educ.*, **1962**, *39*, 289; D. DeVault, ibid., **1944**, *21*, 526, 575; or K. B. Harvey and G. B. Porter, "Introduction to Physical Inorganic Chemistry," Addison-Wesley, Reading, Mass, **1963**, pp. 79–83.

guide to the building up of electronic structure if it is realized that exceptions may occur. A useful mnemonic diagram has been suggested by Moeller[18] (Fig. 2.12). To recall the order of filling, merely follow the arrows and the numbers from one orbital to the next.

Electron spin and the Pauli principle

As expected from our experience with a particle in a box, three quantum numbers are necessary to describe the spatial distribution of electrons in atoms. To describe an electron in an atom completely, a fourth quantum number, m_s, called the spin quantum number must be specified.[19] This is because every electron has associated with it a magnetic moment which is quantized in one of two possible orientations: parallel with or opposed to an applied magnetic field. The magnitude of the magnetic moment is given by the expression:

$$\mu = 2.00\sqrt{s(s+1)} \qquad \textbf{(2.14)}$$

where the moment (μ) is expressed in Bohr magnetons $(eh/4\pi m)$[20] and $s = |m_s|$. The allowed values of the spin quantum number are $\pm\frac{1}{2}$. For an atom with two electrons the spins may be either parallel ($s = 1$) or opposed and thus cancel ($s = 0$). In the latter situation the electrons are referred to as *paired.* Atoms having only paired electrons ($s = 0$) are repelled slightly when placed in a magnetic field and are termed *diamagnetic.* Atoms having one or more unpaired electrons ($s \neq 0$) are strongly attracted by a magnetic field and are termed *paramagnetic.*

Electrons having the same spin strongly repel each other and tend to occupy different regions of space. This is a result of a fundamental law of nature known as the *Pauli exclusion principle.* It states that total wave functions (including spin) must change their signs on exchange of any pair of electrons in the system. Briefly, this means that if two electrons have the same spin they must have different spatial wave functions (i.e., different orbitals) and if they occupy the same orbital they must have paired spins. The Pauli principle and the so-called Pauli repulsive forces[21] have far-reaching consequences in chemistry, but for our present discussion it may be stated as follows: *In a given atom no two electrons may have all four quantum numbers identical.* This means that in a given orbital specified by n, l, and m_l, a maximum of two electrons may exist ($m_s = +\frac{1}{2}$ and $m_s = -\frac{1}{2}$).

The *aufbau* principle

The *electron configuration,* or distribution of electrons among orbitals, may be determined by application of the Pauli principle and the ordering of energy levels suggested above. The method of determining the appropriate electron configuration of minimum energy

[18] T. Moeller, "Inorganic Chemistry," Wiley, New York, **1952**, p. 97.

[19] In the Dirac relativistic treatment, mentioned previously, electron "spin" follows naturally as the *fourth* quantum number necessary to define electron behavior.

[20] In SI the Bohr magneton has a value of 9.27×10^{-24} with units of $A\ m^2$ or $J\ T^{-1}$.

[21] The Pauli "force" corresponds to no classical interaction but results from the nature of quantum mechanics. Although it is common in chemistry to speak of "repulsions" and "stabilizing energies" resulting from the Pauli principle, these do not arise directly from the energetics of spin–spin interactions but from the *electrostatic energies* resulting from the spatial distribution of electrons as a result of the requirements of the Pauli exclusion principle. See W. J. Kauzman, "Quantum Chemistry," Academic, New York, **1957**, pp. 319–320; F. A. Matsen, *J. Am. Chem. Soc.*, **1970**, *92*, 3525.

(the *ground state*) makes use of the *aufbau* principle, or "building up" of atoms one step at a time. Protons are added to the nucleus and electrons are added to orbitals to build up the desired atom. It should be emphasized that this is only a formalism for arriving at the desired electron configuration, but an exceedingly useful one.

The quantum numbers n, l, and m_l in various permutations describe the possible orbitals of an atom. These may be arranged according to their energies. The ground state for the hydrogen atom will be the one with the electron in the lowest orbital, the 1s. The spin of the electron may be of either orientation with neither preferred. We would thus expect a random distribution of spins; indeed, if a stream of hydrogen atoms were introduced into a magnetic field, half would be deflected in one direction, the other half in the opposite direction. Thus the four quantum numbers (n, l, m_l, m_s) for a hydrogen atom are $(1, 0, 0, \pm\frac{1}{2})$. For the helium atom we can start with a hydrogen atom and add a proton to the nucleus and a second electron. The first three quantum numbers of this second electron will be identical to those from a hydrogen atom (i.e., the electron will also seek the lowest possible energy, the 1s orbital), but the spin must be opposed to that of the first electron. So the quantum numbers for the two electrons in a helium atom are $(1, 0, 0, +\frac{1}{2})$ and $(1, 0, 0, -\frac{1}{2})$. The 1s orbital is now filled, and the addition of a third electron to form a lithium atom requires that the 2s orbital, the next lowest in energy, be used. The electron configuration of the first five elements together with the quantum numbers of the last electron are:[22]

$$\begin{array}{ll}
{}_1\mathrm{H} = 1s^1 & 1, 0, 0, \pm\frac{1}{2} \\
{}_2\mathrm{He} = 1s^2 & 1, 0, 0, \pm\frac{1}{2} \\
{}_3\mathrm{Li} = 1s^2 2s^1 & 2, 0, 0, \pm\frac{1}{2} \\
{}_4\mathrm{Be} = 1s^2 2s^2 & 2, 0, 0, \pm\frac{1}{2} \\
{}_5\mathrm{B} = 1s^2 2s^2 2p^1 & 2, 1, 1, \pm\frac{1}{2}
\end{array}$$

This procedure may be continued, one electron at a time, until the entire list of elements has been covered. A complete list of electron configurations of the elements is given in Table 2.2. It will be seen that there are only a few differences between these configurations obtained experimentally and a similar table which might be constructed on the basis of the *aufbau* principle. In every case in which an exception occurs the energy levels involved are exceedingly close together and factors not accounted for in the above discussion invert the energy levels. For example, the $(n-1)d$ and ns levels tend to lie very close together when these levels are filling, with the latter slightly lower in energy. If some special stability arises, such as a filled or half-filled subshell (see pp. 41 and 379), the most stable arrangement may not be $(n-1)d^x ns^2$. In Cr and Cu atoms the extra stability associated with half-filled and filled subshells is apparently sufficient to make the ground-state configuration of the isolated atoms $3d^5 4s^1$ and $3d^{10} 4s^1$ instead of $3d^4 4s^2$ and $3d^9 4s^2$, respectively. Too much importance should not be placed on this type of deviation, however. Its effect on the chemistry of these

---

[22] The m_s values for the unpaired electron in H, Li, and B are, or course, undefined and may be either $\pm\frac{1}{2}$. It is merely necessary that the values for He and Be be opposite of those for H and Li. Likewise, the last electron in boron may enter the p_x, p_y, or p_z orbital, all equal in energy, and so the m_l value given above is arbitrary.

Table 2.2 Electron configurations of the elements

Z	Element	Electron configuration	Z	Element	Electron configuration
1	H	$1s$	53	I	$[Kr]4d^{10}5s^25p^5$
2	He	$1s^2$	54	Xe	$[Kr]4d^{10}5s^25p^6$
3	Li	$[He]2s$	55	Cs	$[Xe]6s$
4	Be	$[He]2s^2$	56	Ba	$[Xe]6s^2$
5	B	$[He]2s^22p$	57	La	$[Xe]5d6s^2$
6	C	$[He]2s^22p^2$	58	Ce	$[Xe]4f5d6s^2$
7	N	$[He]2s^22p^3$	59	Pr	$[Xe]4f^36s^2$
8	O	$[He]2s^22p^4$	60	Nd	$[Xe]4f^46s^2$
9	F	$[He]2s^22p^5$	61	Pm	$[Xe]4f^56s^2$
10	Ne	$[He]2s^22p^6$	62	Sm	$[Xe]4f^66s^2$
11	Na	$[Ne]3s$	63	Eu	$[Xe]4f^76s^2$
12	Mg	$[Ne]3s^2$	64	Gd	$[Xe]4f^75d6s^2$
13	Al	$[Ne]3s^23p$	65	Tb	$[Xe]4f^96s^2$
14	Si	$[Ne]3s^23p^2$	66	Dy	$[Xe]4f^{10}6s^2$
15	P	$[Ne]3s^23p^3$	67	Ho	$[Xe]4f^{11}6s^2$
16	S	$[Ne]3s^23p^4$	68	Er	$[Xe]4f^{12}6s^2$
17	Cl	$[Ne]3s^23p^5$	69	Tm	$[Xe]4f^{13}6s^2$
18	Ar	$[Ne]3s^23p^6$	70	Yb	$[Xe]4f^{14}6s^2$
19	K	$[Ar]4s$	71	Lu	$[Xe]4f^{14}5d6s^2$
20	Ca	$[Ar]4s^2$	72	Hf	$[Xe]4f^{14}5d^26s^2$
21	Sc	$[Ar]3d4s^2$	73	Ta	$[Xe]4f^{14}5d^36s^2$
22	Ti	$[Ar]3d^24s^2$	74	W	$[Xe]4f^{14}5d^46s^2$
23	V	$[Ar]3d^34s^2$	75	Re	$[Xe]4f^{14}5d^56s^2$
24	Cr	$[Ar]3d^54s$	76	Os	$[Xe]4f^{14}5d^66s^2$
25	Mn	$[Ar]3d^54s^2$	77	Ir	$[Xe]4f^{14}5d^76s^2$
26	Fe	$[Ar]3d^64s^2$	78	Pt	$[Xe]4f^{14}5d^96s$
27	Co	$[Ar]3d^74s^2$	79	Au	$[Xe]4f^{14}5d^{10}6s$
28	Ni	$[Ar]3d^84s^2$	80	Hg	$[Xe]4f^{14}5d^{10}6s^2$
29	Cu	$[Ar]3d^{10}4s$	81	Tl	$[Xe]4f^{14}5d^{10}6s^26p$
30	Zn	$[Ar]3d^{10}4s^2$	82	Pb	$[Xe]4f^{14}5d^{10}6s^26p^2$
31	Ga	$[Ar]3d^{10}4s^24p$	83	Bi	$[Xe]4f^{14}5d^{10}6s^26p^3$
32	Ge	$[Ar]3d^{10}4s^24p^2$	84	Po	$[Xe]4f^{14}5d^{10}6s^26p^4$
33	As	$[Ar]3d^{10}4s^24p^3$	85	At	$[Xe]4f^{14}5d^{10}6s^26p^5$
34	Se	$[Ar]3d^{10}4s^24p^4$	86	Rn	$[Xe]4f^{14}5d^{10}6s^26p^6$
35	Br	$[Ar]3d^{10}4s^24p^5$	87	Fr	$[Rn]7s$
36	Kr	$[Ar]3d^{10}4s^24p^6$	88	Ra	$[Rn]7s^2$
37	Rb	$[Kr]5s$	89	Ac	$[Rn]6d7s^2$
38	Sr	$[Kr]5s^2$	90	Th	$[Rn]6d^27s^2$
39	Y	$[Kr]4d5s^2$	91	Pa	$[Rn]5f^26d7s^2$
40	Zr	$[Kr]4d^25s^2$	92	U	$[Rn]5f^36d7s^2$
41	Nb	$[Kr]4d^45s$	93	Np	$[Rn]5f^46d7s^2$
42	Mo	$[Kr]4d^55s$	94	Pu	$[Rn]5f^67s^2$
43	Tc	$[Kr]4d^55s^2$	95	Am	$[Rn]5f^77s^2$
44	Ru	$[Kr]4d^75s$	96	Cm	$[Rn]5f^76d7s^2$
45	Rh	$[Kr]4d^85s$	97	Bk	$[Rn]5f^97s^2$
46	Pd	$[Kr]4d^{10}$	98	Cf	$[Rn]5f^{10}7s^2$
47	Ag	$[Kr]4d^{10}5s$	99	Es	$[Rn]5f^{11}7s^2$
48	Cd	$[Kr]4d^{10}5s^2$	100	Fm	$[Rn]5f^{12}7s^2$
49	In	$[Kr]4d^{10}5s^25p$	101	Md	$[Rn]5f^{13}7s^2$
50	Sn	$[Kr]4d^{10}5s^25p^2$	102	No	$[Rn]5f^{14}7s^2$
51	Sb	$[Kr]4d^{10}5s^25p^3$	103	Lr	$[Rn]5f^{14}6d7s^2$
52	Te	$[Kr]4d^{10}5s^25p^4$			

SOURCE: C. E. Moore, "Ionization Potentials and Ionization Limits Derived from the Analyses of Optical Spectra," NSRDS-NBS 34, National Bureau of Standards, Washington, D.C., **1970**, except for the data on the actinides, which are from G. T. Seaborg, *Ann. Rev. Nucl. Sci.*, **1968**, *18*, 53 and references therein.

two elements is minimal. It is true that copper has a reasonably stable +1 oxidation state (corresponding to $3d^{10}4s^0$), but the +2 state is even *more* stable in most chemical environments. For chromium the most stable ion in aqueous solution is Cr^{+3}, with the Cr^{+2} ion and the Cr^{VI} oxidation state (as in CrO_4^{-2}) reasonably stable; the Cr^I oxidation state is practically unknown. For both Cu^{+2} and Cr^{+3} (as well as many other transition metal ions) ligand field effects in their complexes (see Chapter 9) are sufficient to permit the easy loss of *d* electrons.

In the case of the lanthanide elements (elements 58–71) and those immediately following, the 5*d* and 4*f* levels are exceedingly close. In the lanthanum atom it appears that the 57th electron enters the 5*d* level rather than the 4*f*. Thereafter the 4*f* level starts to fill, and some lanthanides appear not to have any 5*d* electrons. Here again, too much attention to details of the electron configuration is not rewarding from a chemist's point of view—indeed it may be quite misleading. The difference in energy between a $5d^{n+1}4f^m$ configuration and a $5d^n4f^{m+1}$ configuration is very small. For mnemonic purposes all lanthanide elements behave as though they had an electron configuration: $6s^25d^14f^n$; i.e., the most stable oxidation state is always that corresponding to loss of three electrons (the 6*s* and 5*d*). There are some other "abnormalities" in the electron configurations of various elements, but they are of minor importance from a chemical point of view.

Although the *aufbau* principle and the ordering of orbitals given previously may be used reliably to determine electron configurations, it must again be emphasized that the device is a formalism and may lead to serious error if overextended. For example, in the atoms of the elements potassium, calcium, and scandium the 4*s* level is lower in energy than the 3*d* level. This is not true for heavier elements or for charged ions. The energies of the various orbitals are sensitive to changes in nuclear charge and to the occupancy of other orbitals by electrons (see "Shielding," p. 36), and this prevents the designation of an absolute ordering of orbital energies. It happens that the ordering suggested by Fig. 2.12 is reasonably accurate when dealing with orbitals corresponding to the valence shell of an atom; that is, the energies $3d > 4s$ and $5p > 4d$ are correct for elements potassium and yttrium, for example, but not necessarily elsewhere.

Atomic states, term symbols, and Hund's rule

It is convenient to be able to specify the energy, angular momentum, and spin multiplicity of an atom by a symbolic representation. For example, for the hydrogen atom we may define *S*, *P*, *D*, and *F* states, depending upon whether the single electron occupies an *s*, *p*, *d*, or *f* orbital.[23] The ground state of hydrogen, $1s^1$, is a 1*S* state; a hydrogen atom excited to a $2p^1$ configuration is in a 2*P* state; etc. For polyelectronic atoms, an atom in a *P* state has the same total angular momentum (for all electrons) as a hydrogen atom in a *P* state. Corresponding to states *S*, *P*, *D*, *F*, . . . are quantum numbers $L = 0, 1, 2, 3, 4, \ldots$, which parallel the *l* values for *s*, *p*, *d*, *f*, . . . orbitals.[24] Likewise, there is quantum number *S* (not

[23] Since $_2He = 1s^2$, chemists sometimes "abbreviate" as follows: $_5B = [He]2s^22p^1$. This saves considerable space when writing electron configurations of the heavier elements.

[24] This is the reverse of the historical process. *S*, *P*, *D*, and *F* states were observed spectroscopically and named after *sharp*, *principal*, *diffuse*, and *fundamental* characteristics of the spectra. Later the symbols *s*, *p*, *d*, and *f* were applied to orbitals.

to be confused with the S state just mentioned) that is the summation of all the electronic spins. For a closed shell or subshell, obviously $S = 0$, since all electrons are paired. Somewhat less obviously, under these conditions $L = 0$, since all of the orbital momenta cancel. This greatly simplifies working with states and term symbols.

The chemist frequently uses a concept known as *multiplicity*, originally derived from the number of lines shown in a spectrum. It is related to the number of unpaired electrons and, in general, is given by the expression $2S + 1$. Thus, if $S = 0$, the multiplicity is one and the state is called a singlet; if $S = \frac{1}{2}$ the multiplicity is two and the state is a doublet; $S = 1$ is a triplet; etc. Hund's *rule of maximum multiplicity* states that the ground state of an atom will be the one having the greatest multiplicity (i.e., the greatest value of S). Consider a carbon atom ($= 1s^2\ 2s^2\ 2p^2$). We may ignore the closed $1s^2$ and $2s^2$. The two $2p$ electrons may be paired in the same orbital ($S = 0$) or have parallel spins in different orbitals ($S = 1$). Hund's rules predict that the latter will be the ground state, i.e., a triplet state. It happens that in this state $L = 1$, so we may say that the ground state of carbon is ^3P (pronounced "triplet-P"). The "^3P" is said to be the *term symbol*.[25]

It is convenient for many purposes to draw "box diagrams" of electron configurations in which boxes represent individual orbitals, and electrons and their spins are indicated by arrows:

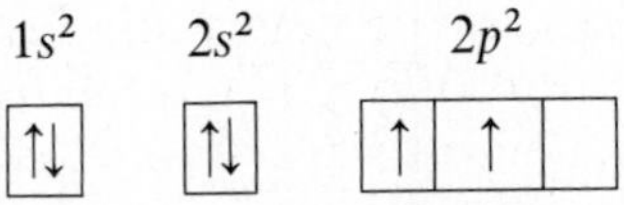

$1s^2$	$2s^2$	$2p^2$		
↑↓	↑↓	↑	↑	

Such devices can be very useful for bookkeeping, providing pigeonholes in which to place electrons. However, the reader is warned that they can be misleading if improperly used, especially with respect to term symbols.

Periodicity of the elements

For chemists working with several elements the periodic chart of the elements is so indispensable that one is apt to forget that, far from being divinely inspired, it resulted from the hard work of countless chemists. True, there is a quantum-mechanical basis for the periodicity of the elements, as we shall see shortly. But the inspiration of such scientists as Mendeleev and the perspiration of a host of nineteenth-century chemists provided the chemist with the benefits of the periodic table about half a century before the existence of the electron was proved! The confidence that Mendeleev had in his chart and his predictions based on it[26] should make fascinating reading for any chemist.

The common long form of the periodic chart (Fig. 2.13) may be considered a graphic portrayal of the rules of atomic structure given previously. The arrangement of the atoms follows naturally from the *aufbau* principle. The various groups of the chart may be

---

[25] The methods for ascertaining the various possible values of L, as well as several other aspects of atomic states and term symbols, are given in Appendix B.

[26] See M. E. Weeks and H. M. Leicester, "Discovery of the Elements," 7th ed., Chemical Education Publ. Co., Easton, Pa., **1968**.

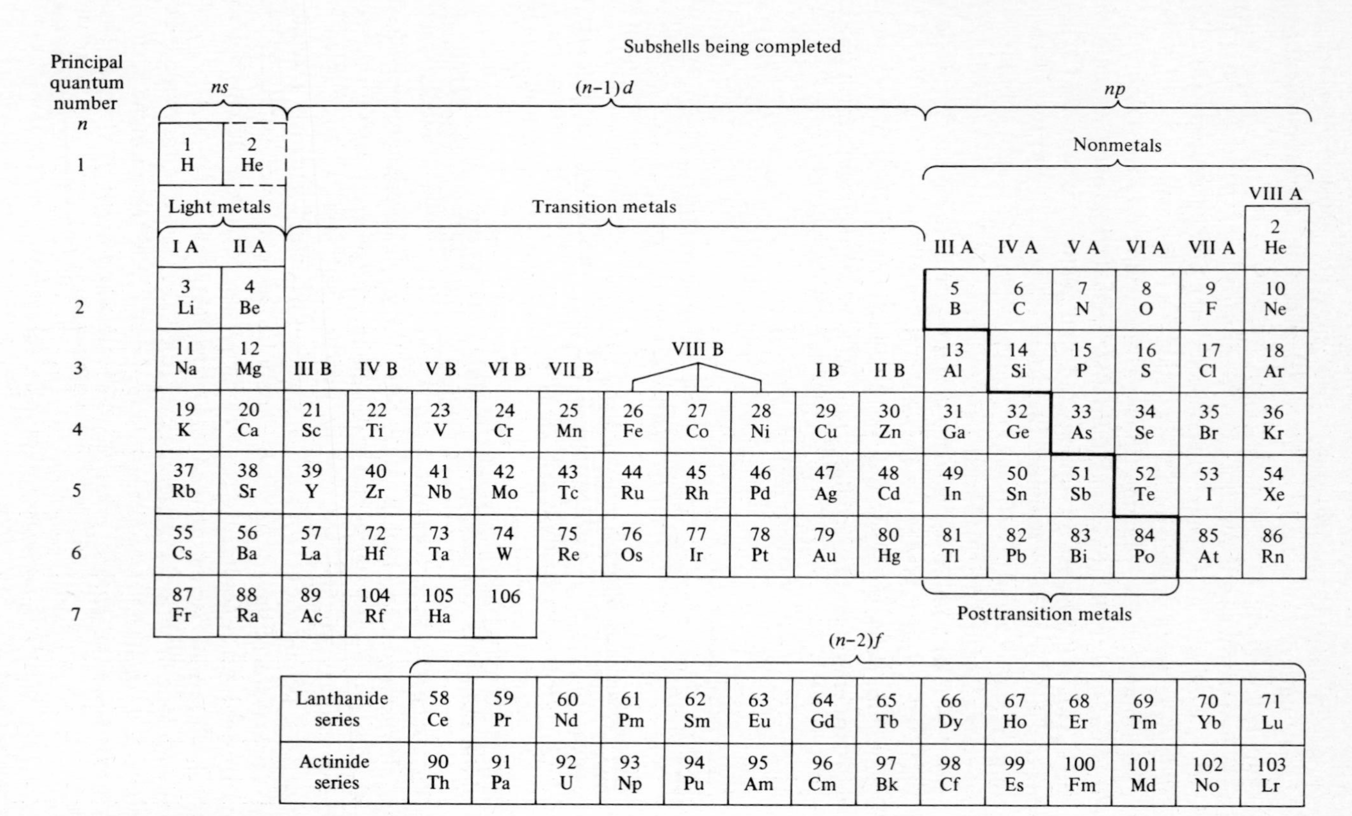

Fig. 2.13 Periodic chart of the elements.

classified as follows:

1. *Vertical Groups IA and IIA, the alkali and alkaline earth metals.* The elements are characterized by an ns^1 or ns^2 electron configuration, respectively.

2. *The "B" block of elements, the transition metals.* Characteristically, atoms of these elements in their ground states have partially filled *d* orbitals.[27] For example, the first transition series proceeds from Sc($4s^2 3d^1$) through Zn($4s^2 3d^{10}$). Each of these ten elements stands at the head of a family of congeners (e.g., the chromium family, VIB).

3. *The lanthanide and actinide elements.* Technically these two series belong to Group IIIB, but because of their remarkable electronic and chemical properties they are usually set apart. Each series starts with a predicted Ce($4f^1$) and Th($5f^1$)[28] and ends with Lu($4f^{14}$) and Lr($5f^{14}$).

4. *The nonmetals and posttransition metals, Groups IIIA to VIIIA.* This block of elements contains six vertical families corresponding to the maximum occupancy of six electrons in a set of *p* orbitals. The classification here is imprecise, principally because the distinction between metal and nonmetal is somewhat arbitrary, though usually associated with a "stair-step" dividing line running from boron to astatine. All these elements (except He) share the feature of filling *p* orbitals, the noble gases representing a completely filled set of *p* orbitals.

Semantics

Some chemists would define transition metals strictly as those elements whose ground-state atoms have partially filled *d* orbitals. This excludes zinc from the first transition series. One must admit that Zn^{+2} does have several properties that distinguish it from "typical" transition metal behavior—it is never paramagnetic, it is never colored, it forms rather weak complexes (for its small size), etc. However, if zinc (along with its congeners, cadmium and mercury) is excluded from the transition series, then the noble gases must be excluded from the nonmetals. Some might favor that, but the point is made elsewhere[29] that the separation of the noble gases from the, in some ways, closely related halogens impeded the development of noble gas chemistry. Finally, to be internally logical, lutetium would have to be removed from the lanthanides and lawrencium from the actinides. I know of no one who has suggested that the latter would be a useful action.

Likewise, the designation of groups as "A" or "B" is purely arbitrary. In fact, many inorganic chemistry books use a different system from that shown here, although the present system is that found in most wall charts in current use. A definite advantage of the present system is the grouping of all transition metals together as "B" groups.[29a]

It is possible to trace the *aufbau* principle simply by following the periodic chart. Consider the elements from Cs($Z = 55$) to Rn($Z = 86$). In the elements Cs and Ba the electrons enter (and fill) the 6*s* level. The next electron enters the 5*d* level and La($Z = 57$) may be considered a transition element. In the elements Ce through Lu the electrons are

---

[27] There are problems with any simple definition of "transition metal." See the discussion under "Semantics."

[28] Here again, thorium "should" have a $5f^1 6d^1 7s^2$ configuration. Table 2.2 indicates that experimentally it is found to be otherwise. Again, as far as the chemist is concerned, this is merely troublesome "noise" in the regularity of the periodic chart—the characteristic oxidation state is Th(IV) in any event.

[29] See Chapter 15.

[29a] The usage as given here has historical precedence. See W. C. Fernelius and W. H. Powell, *J. Chem. Educ.*, **1982**, *59*, 504.

added to $4f$ levels and these elements are *lanthanide* or *inner transition elements*. After the $4f$ level is filled with Lu, the next electrons continue to fill the $5d$ orbitals (the transition elements Hf to Hg), and finally, the $6p$ level is filled in the elements Tl to Rn, in accord with Table 2.2. The periodic chart may thus be used to derive the electron configuration of an element as readily as use of the rules given above. It should be quite apparent, however, that the chart can give us back only the chemical knowledge that we have used in composing it; it is not a source of knowledge in itself. It is useful in portraying and correlating the information which has been obtained with regard to electron configurations and other atomic properties.

Shielding

The energy of an electron in an atom is a function of Z^2/n^2. Since the nuclear charge (= atomic number) increases more rapidly than the principal quantum number, one might be led to expect that the energy necessary to remove an electron from an atom would continually increase with increasing atomic number. This is not so, as can be shown by comparing hydrogen ($Z = 1$) with lithium ($Z = 3$). The ionization energies are 1312 kJ mol^{-1} (H) and 520 kJ mol^{-1} (Li). The ionization energy of lithium is lower for two reasons: (1) The average distance of a $2s$ electron is greater than that of a $1s$ electron (see Fig. 2.5); (2) the $2s^1$ electron in lithium is repelled by the inner core $1s^2$ electrons, so that the former is more easily removed than if the core were not there. Another way of treating this inner core repulsion is to view it as "shielding" or "screening" of the nucleus by the inner electrons, so that the valence electron actually "sees" only part of the total charge. Thus, the ionization energy for lithium corresponds to an *effective nuclear charge* of between one and two units. The radial probability functions for hydrogen-like orbitals have been discussed previously (Fig. 2.5). The bulk of the electron density of the $1s$ orbital lies between the nucleus and the bulk of the $2s$ density. The laws of electrostatics state that when a test charge is outside of a "cage" of charge such as that represented by the $1s$ electrons, the potential is exactly the same as though the latter were located at the center (nucleus). In this case the valence electron in the $2s$ orbital would experience a potential equivalent to a net nuclear charge of one ($Z^* = 1.0$). A charge which penetrates the cage will be unshielded and would experience a potential equivalent to the full nuclear charge, $Z^* = 3.0$. This is not meant to imply that the energy of the $2s$ electron varies as it penetrates the $1s$ orbital, but that the energy is determined by an effective nuclear charge, Z^*, which is somewhat less than the actual nuclear charge, Z:

$$Z^* = Z - S \tag{2.15}$$

where S is the shielding or screening constant.

As a result of the presence of one or more maxima near the nucleus, s orbitals are very penetrating and are somewhat less shielded by inner-shell electrons than are orbitals with larger values of l. In turn, they tend to shield somewhat better than other orbitals. Orbitals with high l values, such as d and f orbitals, are much less penetrating and are far poorer at shielding.

In a similar manner the radial distributions of $3s$, $3p$, and $3d$ orbitals may be compared (Fig. 2.5). Although the d orbitals are "smaller" in the sense that the most probable radius decreases in the order $3s > 3p > 3d$, the presence of one node and an intranodal maximum in the $3p$ orbital and the presence of two nodes and two intranodal maxima in the $3s$ orbital cause them to be affected more by the nucleus. Hence the energies of these orbitals lie $3d > 3p > 3s$ as we have seen in filling the various energy levels previously.

In order to estimate the extent of shielding, a set of empirical rules has been proposed by Slater.[30] It should be realized that these rules are simplified generalizations based upon the *average* behavior of the various electrons. Although the electronic energies estimated by Slater's rules are often not very accurate, they permit simple estimates to be made and will be found useful in understanding related topics such as atomic size and electronegativity.

To calculate the shielding constant for an electron in an np or ns orbital:

1. Write out the electronic configuration of the element in the following order and groupings: $(1s)$ $(2s, 2p)$ $(3s, 3p)$ $(3d)$ $(4s, 4p)$ $(4d)$ $(4f)$ $(5s, 5p)$, etc.
2. Electrons in any group to the right of the (ns, np) group contribute nothing to the shielding constant.
3. All of the other electrons in the (ns, np) group shield the valence electron to an extent of 0.35 each.[31]
4. All electrons in the $n - 1$ shell shield to an extent of 0.85 each.
5. All electrons $n - 2$ or lower shield completely; i.e., their contribution is 1.00 each.

When the electron being shielded is in an nd or nf group, rules 2 and 3 are the same but rules 4 and 5 become:

6. All electrons in groups lying to the left of the nd or nf group contribute 1.00.

Examples

1. Consider the valence electron in the atom ${}_7N = 1s^22s^22p^3$. Grouping of the orbitals gives $(1s)^2(2s, 2p)^5$. $S = (2 \times 0.85) + (4 \times 0.35) = 3.10$. $Z^* = Z - S = 7.0 - 3.1 = 3.9$.
2. Consider the valence $(4s)$ electron in the atom ${}_{30}Zn$. The grouped electron configuration is $(1s)^2(2s, 2p)^8(3s, 3p)^8(3d)^{10}(4s)^2$. $S = (10 \times 1.00) + (18 \times 0.85) + (1 \times 0.35) = 25.65$. $Z^* = 4.35$.
3. Consider a $3d$ electron in Zn. The grouping is as in example 2, but the shielding is $S = (18 \times 1.00) + (9 \times 0.35) = 21.15$. $Z^* = 8.85$.

It can be seen that the rules are an attempt to generalize and to quantify those aspects of the radial distributions discussed previously. For example, d and f electrons are screened more effectively ($S = 1.00$) than s and p electrons ($S = 0.85$) by the electrons lying immediately below them. On the other hand, Slater's rules assume that all electrons, s, p, d, or f, shield electrons lying above them equally well (in computing shielding the nature of the *shielding* electron is ignored). This is not quite true, as we have seen above and will lead to some error. For example, in the Ga atom ($= \ldots 3s^23p^63d^{10}4s^24p^1$) the rules imply that the $4p$ electron is shielded as effectively by the $3d$ electrons as by the $3s$ and $3p$ electrons, contrary to Fig. 2.5.

---

[30] J. S. Slater, *Phys. Rev.*, **1930**, *36*, 57.

[31] Except for the $1s$ orbital for which a value of 0.30 seems to work better.

Slater formulated these rules in proposing a set of orbitals for use in quantum-mechanical calculations. Slater orbitals are basically hydrogen-like but differ in two important respects:

1. They contain no nodes. This simplifies them considerably but of course makes them less accurate.

2. They make use of Z^* in place of Z, and for heavier atoms, n is replaced by n^*, where for $n = 4$, $n^* = 3.7$; $n = 5$, $n^* = 4.0$; $n = 6$, $n^* = 4.2$. The difference between n and n^* is referred to as the quantum defect.

To remove the difficulties and inaccuracies in the simplified Slater treatment of shielding, Clementi and Raimondi[32] have obtained effective nuclear charges from self-consistent field wave functions for atoms from hydrogen to krypton and have generalized these into a set of rules for calculating the shielding of any electron. The shielding which an electron in the nth energy level and lth orbital (S_{nl}) experiences is given by:

$$S_{1s} = 0.3(\mathrm{N}_{1s} - 1) + 0.0072(\mathrm{N}_{2s} + \mathrm{N}_{2p}) + 0.0158(\mathrm{N}_{3s,p,d} + \mathrm{N}_{4s,p}) \quad \mathbf{(2.16)}$$

$$S_{2s} = 1.7208 + 0.3601(\mathrm{N}_{2s} - 1 + \mathrm{N}_{2p}) + 0.2062(\mathrm{N}_{3s,p,d} + \mathrm{N}_{4s,p}) \quad \mathbf{(2.17)}$$

$$S_{2p} = 2.5787 + 0.3326(\mathrm{N}_{2p} - 1) - 0.0773\mathrm{N}_{3s} - 0.0161(\mathrm{N}_{3p} + \mathrm{N}_{4s}) - 0.0048\mathrm{N}_{3d} + 0.0085\mathrm{N}_{4p} \quad \mathbf{(2.18)}$$

$$S_{3s} = 8.4927 + 0.2501(\mathrm{N}_{3s} - 1 + \mathrm{N}_{3p}) + 0.0778\mathrm{N}_{4s} + 0.3382\mathrm{N}_{3d} + 0.1978\mathrm{N}_{4p} \quad \mathbf{(2.19)}$$

$$S_{3p} = 9.3345 + 0.3803(\mathrm{N}_{3p} - 1) + 0.0526\mathrm{N}_{4s} + 0.3289\mathrm{N}_{3d} + 0.1558\mathrm{N}_{4p} \quad \mathbf{(2.20)}$$

$$S_{4s} = 15.505 + 0.0971(\mathrm{N}_{4s} - 1) + 0.8433\mathrm{N}_{3d} + 0.0687\mathrm{N}_{4p} \quad \mathbf{(2.21)}$$

$$S_{3d} = 13.5894 + 0.2693(\mathrm{N}_{3d} - 1) - 0.1065\mathrm{N}_{4p} \quad \mathbf{(2.22)}$$

$$S_{4p} = 24.7782 + 0.2905(\mathrm{N}_{4p} - 1) \quad \mathbf{(2.23)}$$

where N_{nl} represents the number of electrons in the nl orbital. For the examples given above, the effective nuclear charges obtained are $Z^*_{\mathrm{N}} = 3.756$, $Z^*_{\mathrm{Zn},4s} = 5.965$, and $Z^*_{\mathrm{Zn},3d} = 13.987$. The shielding rules of Clementi and Raimondi explicitly account for penetration of outer orbital electrons. They are thus more realistic than Slater's rules, at the expense, however, of more complex computation with a larger number of parameters. If accuracy greater than that afforded by Slater's rules is necessary, it would appear that direct application of the effective nuclear charges from the SCF wave functions is not only simple but also accurate. Such values are listed in Table 2.3. With the accurate values of Table 2.3 available, the chief justification of "rules," whether Slater's or those of Clementi and Raimondi, is the insight they provide into the phenomenon of shielding.

The sizes of atoms

Atomic size is at best a rather nebulous quantity since an atom can have no well-defined boundary similar to that of a billiard ball. In order to answer the question, "How big is an

---

[32] E. Clementi and D. L. Raimondi, *J. Chem. Phys.*, **1963**, *38*, 2868.

Table 2.3 Effective nuclear charges for elements 1 to 36

Element	1*s*	2*s*	2*p*	3*s*	3*p*	4*s*	3*d*	4*p*
H	1.000							
He	1.688							
Li	2.691	1.279						
Be	3.685	1.912						
B	4.680	2.576	2.421					
C	5.673	3.217	3.136					
N	6.665	3.847	3.834					
O	7.658	4.492	4.453					
F	8.650	5.128	5.100					
Ne	9.642	5.758	5.758					
Na	10.626	6.571	6.802	2.507				
Mg	11.619	7.392	7.826	3.308				
Al	12.591	8.214	8.963	4.117	4.066			
Si	13.575	9.020	9.945	4.903	4.285			
P	14.558	9.825	10.961	5.642	4.886			
S	15.541	10.629	11.977	6.367	5.482			
Cl	16.524	11.430	12.993	7.068	6.116			
Ar	17.508	12.230	14.008	7.757	6.764			
K	18.490	13.006	15.027	8.680	7.726	3.495		
Ca	19.473	13.776	16.041	9.602	8.658	4.398		
Sc	20.457	14.574	17.055	10.340	9.406	4.632	7.120	
Ti	21.441	15.377	18.065	11.033	10.104	4.817	8.141	
V	22.426	16.181	19.073	11.709	10.785	4.981	8.983	
Cr	23.414	16.984	20.075	12.368	11.466	5.133	9.757	
Mn	24.396	17.794	21.084	13.018	12.109	5.283	10.528	
Fe	25.381	18.599	22.089	13.676	12.778	5.434	11.180	
Co	26.367	19.405	23.092	14.322	13.435	5.576	11.855	
Ni	27.353	20.213	24.095	14.961	14.085	5.711	12.530	
Cu	28.339	21.020	25.097	15.594	14.731	5.858	13.201	
Zn	29.325	21.828	26.098	16.219	15.369	5.965	13.878	
Ga	30.309	22.599	27.091	16.996	16.204	7.067	15.093	6.222
Ge	31.294	23.365	28.082	17.760	17.014	8.044	16.251	6.780
As	32.278	24.127	29.074	18.596	17.850	8.944	17.378	7.449
Se	33.262	24.888	30.065	19.403	18.705	9.758	18.477	8.287
Br	34.247	25.643	31.056	20.218	19.571	10.553	19.559	9.028
Kr	35.232	26.398	32.047	21.033	20.434	11.316	20.626	9.769

atom?" one must first pose the questions, "How are we going to measure the atom?" and "How hard are we going to push?" If we measure the size of a xenon atom resting in the relatively relaxed situation obtaining in solid xenon, we might expect to get a different value than if the measurement is made through violent collisions. A sodium ion should be compressed more if it is tightly bound in a crystal lattice (e.g., NaF) than if it is loosely solvated by molecules of low polarity. The question of how hard we are going to push is particularly important because measuring atoms is analogous to measuring an overripe grapefruit with a pair of calipers—the value we get depends on how hard we squeeze. For this reason it is impossible to set up a single set of values called "atomic radii" applicable under all conditions. It is necessary to define the conditions under which the atom (or ion) exists and also our method of measurement. These will be discussed in Chapter 6. Nevertheless, it will be useful now to discuss *trends* in atomic sizes without becoming too specific at the present time about the actual sizes involved.

As we have seen from the radial distribution functions, the most probable radius tends to increase with increasing n. Counteracting this tendency is the effect of increasing effective nuclear charge, which tends to contract the orbitals. From these opposing forces we obtain the following results:

1. Atoms in a given family tend to increase in size from one period (= horizontal row of the periodic chart) to the next. Because of shielding, Z^* increases very slowly from one period to the next. For example, using Slater's rules we obtain the following values for Z^*:

$$H = 1.0 \quad Li = 1.3 \quad Na = 2.2 \quad K = 2.2 \quad Rb = 2.2 \quad Cs = 2.2$$

The result of the opposing tendencies of n and Z^* is that atomic size increases as one progresses down Group I. This is a general property of the periodic chart with but few minor exceptions which will be discussed later.

2. Within a given series, the principal quantum number does not change. (Even in the "long" series in which the filling may be in the order ns, $(n - 1)d$, np, the outermost electrons are always in the nth level.) The effective nuclear charge increases steadily, however, since electrons added to the valence shell shield each other very ineffectively. For the second series:

$$Li = 1.3 \quad Be = 1.95 \quad B = 2.60 \quad C = 3.25 \quad N = 3.90 \quad O = 4.55 \quad F = 5.20 \quad Ne = 5.85$$

As a result there is a steady contraction from left to right. The net effect of the top-to-bottom and the left-to-right trends is a discontinuous variation in atomic size. There is a steady contraction with increasing atomic number until there is an increase in the principal quantum number. This causes an abrupt increase in size followed by a further decrease.

Ionization energy

The energy necessary to remove an electron from an isolated atom in the gas phase is the *ionization energy* (often called ionization potential) for that atom. It is the energy difference between the highest occupied energy level and that corresponding to $n = \infty$, i.e., complete removal. It is possible to remove more than one electron, and the succeeding ionization energies are the second, third, fourth, etc.[33] Ionization energies are always endothermic and thus are always assigned a positive value in accord with common thermodynamic convention (see Table 2.4A). The various ionization energies of an atom are related to each other by a polynomial equation which will be discussed in detail later in this chapter.

For the nontransition elements (alkali and alkaline earth metals and the nonmetals) there are fairly simple trends with respect to ionization energy and position in the periodic

[33] There is often confusion regarding whether the second ionization potential or second ionization energy relates to the removal of the second valence electron or both the first and second valence electrons. If *ionization potential* is meant, then the answer is that both the first and second will be removed—a *potential* necessary to remove the second will assuredly remove the first also. However, in most cases one is interested in the total *energy*, and in this case to remove both electrons requires $IE_1 + IE_2$.

chart. Within a given family, increasing n tends to cause reduced ionization energy because of the combined effects of size and shielding. The transition and posttransition elements show some anomalies in this regard which will be discussed in Chapters 16 and 17. Within a given series there is a general tendency for the ionization energy to increase with increase in atomic number. This is a result of the tendency for Z^* to increase progressing from left to right in the periodic chart. There are two other factors which prevent this increase from being monotonic. One is the change in type of orbital which occurs as one goes from Group IIA (s orbital) to Group IIIA (p orbital). The second is the exchange energy between electrons of like spin. This stabilizes a system of parallel electron spins because electrons having the same spin tend to avoid each other as a result of the Pauli exclusion principle. The electrostatic repulsions between electrons are thus reduced. We have seen previously that this tends to maximize the number of unpaired electrons (Hund's principle of maximum multiplicity) and also accounts for the "anomalous" behavior of Cu and Cr. It also tends to make it more difficult to remove the electron from the nitrogen atom than would otherwise be the case. As a result of this stabilization, the ionization energy of nitrogen is greater than that of oxygen (see Fig. 2.14).

The ionization energies of a few groups are known (Table 2.4B). Although not generally as useful as atomic values, they can be used in Born-Haber calculations (see Chapter 3) involving polyatomic cations, such as NO^+ and O_2^+. They also provide a rough estimate of the electron-donating or -withdrawing tendencies of groups.

Ionization

The electrons which are lost on ionization are those that lie at higher energies and therefore require the least energy to remove. One might expect, therefore, that electrons would be lost on ionization in the reverse order in which they were filled (see "The *aufbau* Principle"). There is a tendency for this to be true. However, there are some very important exceptions, notably in the transition elements, which are responsible for the characteristic chemistry of these elements. In general, transition elements react as follows:

$$\underset{\text{Atom}}{1s^22s^22p^63s^23p^63d^n4s^2} \longrightarrow \underset{\text{Dipositive cation}}{1s^22s^22p^63s^23p^63d^n}$$

This is true not only for the first transition series but also for the heavier metals: The ns^2 electrons are lost before the $(n-1)d$ or $(n-2)f$ electrons. This gives a common $+2$ oxidation state to transition metals, although in many cases there is a more stable higher or lower oxidation state.

This phenomenon is puzzling because it appears to contradict simple energetics: If the $4s$ level is lower and fills first, then its electrons should be more stable and be ionized last, shouldn't they? One might ask if there is a possible reversal of energy levels within the transition series. If the relative energies of the $3d$ and $4s$ levels are examined, it is found that they lie very close together and that the energy of the $3d$ level decreases with increasing atomic number. This is often advanced as the explanation for the electronic configuration of Cu. If the $3d$ level has dropped below the $4s$ at atomic number 29, then the ground state must be $3d^{10}4s^1$. Nevertheless, this can have no effect on the phenomena we are investigating since we are inquiring as to the difference in configuration between ground state of the neutral atom and the ionic states of the same element. Since *all* of the transition metals in

Table 2.4A Ionization energies (MJ mol^{-1})

Z	Element	I	II	III	IV	V	VI	VII	VIII	IX	X
1	H	1.3120									
2	He	2.3723	5.2504								
3	Li	0.5203	7.2981	11.8149							
4	Be	0.8995	1.7571	14.8487	21.0065						
5	B	0.8006	2.4270	3.6598	25.0257	32.8266					
6	C	1.0864	2.3526	4.6205	6.2226	37.8304	47.2769				
7	N	1.4023	2.8561	4.5781	7.4751	9.4449	53.2664	64.3598			
8	O	1.3140	3.3882	5.3004	7.4693	10.9895	13.3264	71.3345	84.0777		
9	F	1.6810	3.3742	6.0504	8.4077	11.0227	15.1640	17.8677	92.0378	106.4340	
10	Ne	2.0807	3.9523	6.122	9.370	12.178	15.238	19.999	23.069	115.3791	131.4314
11	Na	0.4958	4.5624	6.912	9.544	13.353	16.610	20.115	25.490	28.934	141.3626
12	Mg	0.7377	1.4507	7.7328	10.540	13.628	17.995	21.704	25.656	31.643	25.462
13	Al	0.5776	1.8167	2.7448	11.578	14.831	18.378	23.295	27.459	31.861	38.457
14	Si	0.7865	1.5771	3.2316	4.3555	16.091	19.785	23.786	29.252	33.877	38.733
15	P	1.0118	1.9032	2.912	4.957	6.2739	21.269	25.397	29.854	35.867	40.959
16	S	0.9996	2.251	3.361	4.564	7.013	8.4956	27.106	31.670	36.578	43.138
17	Cl	1.2511	2.297	3.822	5.158	6.54	9.362	11.0182	33.605	38.598	43.962
18	Ar	1.5205	2.6658	3.931	5.771	7.238	8.7810	11.9952	13.8417	40.760	46.187
19	K	0.4189	3.0514	4.411	5.877	7.976	9.649	11.343	14.942	16.964	48.576
20	Ca	0.5898	1.1454	4.9120	6.474	8.144	10.496	12.32	14.207	18.192	20.3849
21	Sc	0.631	1.235	2.389	7.089	8.844	10.72	13.32	15.31	17.370	21.741
22	Ti	0.658	1.310	2.6525	4.1746	9.573	11.517	13.59	16.26	18.64	20.833
23	V	0.650	1.414	2.8280	4.5066	6.299	12.362	14.489	16.760	19.86	22.24
24	Cr	0.6528	1.496	2.987	4.74	6.69	8.738	15.54	17.82	20.19	23.58
25	Mn	0.7174	1.5091	2.2484	4.94	6.99	9.2	11.508	18.956	21.40	23.96
26	Fe	0.7594	1.561	2.9574	5.29	7.24	9.6	12.1	14.575	22.678	25.29
27	Co	0.758	1.646	3.232	4.95	7.67	9.84	12.4	15.1	17.959	26.6
28	Ni	0.7367	1.7530	3.393	5.30	7.28	10.4	12.8	15.6	18.6	21.66
29	Cu	0.7455	1.9579	3.554	5.33	7.71	9.94	13.4	16.0	19.2	22.4
30	Zn	0.9064	1.7333	3.8327	5.73	7.97	10.4	12.9	16.8	19.6	23.0
31	Ga	0.5788	1.979	2.963	6.2						
32	Ge	0.7622	1.5372	3.302	4.410	9.02					
33	As	0.944	1.7978	2.7355	4.837	6.043	12.31				
34	Se	0.9409	2.045	2.9737	4.1435	6.59	7.883	14.99			
35	Br	1.1399	2.10	3.5	4.56	5.76	8.55	9.938	18.60		
36	Kr	1.3507	2.3503	3.565	5.07	6.24	7.57	10.71	12.2	22.28	
37	Rb	0.4030	2.633	3.9	5.08	6.85	8.14	9.57	13.1	14.5	26.74
38	Sr	0.5495	1.0643	4.21	5.5	6.91	8.76	10.2	11.80	15.6	17.1
39	Y	0.616	1.181	1.980	5.96	7.43	8.97	11.2	12.4	14.11	18.4
40	Zr	0.660	1.267	2.218	3.313	7.86					
41	Nb	0.664	1.382	2.416	3.69	4.877	9.900	12.1			
42	Mo	0.6850	1.558	2.621	4.477	5.91	6.6	12.23	14.8		
43	Tc	0.702	1.472	2.850							
44	Ru	0.711	1.617	2.747							
45	Rh	0.720	1.744	2.997							
46	Pd	0.805	1.875	3.177							
47	Ag	0.7310	2.074	3.361							
48	Cd	0.8677	1.6314	3.616							
49	In	0.5583	1.8206	2.705	5.2						
50	Sn	0.7086	1.4118	2.9431	3.9303	6.974					
51	Sb	0.8316	1.595	2.44	4.26	5.4	10.4				
52	Te	0.8693	1.79	2.698	3.610	5.669	6.82	13.2			
53	I	1.0084	1.8459	3.2							
54	Xe	1.1704	2.046	3.10							
55	Cs	0.3757	2.23								
56	Ba	0.5029	0.96526								

To obtain values in electron volts, multiply table values by 10.364.

XI	XII	XIII	XIV	XV	XVI	XVII	XVIII	XIX	XX	XXI
159.0745										
169.9914	189.3671									
42.654	201.2707	222.3143								
45.934	50.511	235.2046	257.9208							
46.272	54.072	59.036	271.7990	296.1928						
48.705	54.482	62.874	68.230	311.0590	337.1359					
51.067	57.118	63.362	72.340	78.096	352.9913	380.7572				
52.002	59.652	66.199	72.918	82.472	88.6	397.6024	427.0635			
54.431	60.699	68.894	75.948	83.150	93.4	99.77	444.8982	476.0613		
57.048	63.333	70.053	78.792	86.368	94.0	104.9	111.6	494.8873	527.7598	
24.1055	66.180	72.893	80.064	89.347						
25.591	28.1257	75.967	83.107	90.733						
24.608	29.742	32.4455	86.412	93.980						
26.13	28.75	34.3	37.080	97.5138						
27.60	30.34	33.15	39.0	42.00	109.63					
28.02	31.92	34.83	37.84	44.1	47.23	122.16				
29.4	32.4	36.6	39.7	42.8	49.4	52.76	135.37			
30.99	34.0	37.1	41.5	44.8	48.1	55.1	58.59	149.3		
25.7	35.58	38.7	42.0	46.7	50.2	53.7	61.1	64.7	163.8	
26.4	29.99	40.50	43.8	47.3	52.3	55.9	59.7	67.3	71.2	179.1
31.27										
19.9	36.09									

Table 2.4A *(Continued)*

Z	Element	I	II	III	IV	V	VI	VII	VIII	IX	X
57	La	0.5381	1.067	1.8503	4.820						
58	Ce	0.528	1.047	1.949	3.543						
59	Pr	0.523	1.018	2.086	3.761	5.552					
60	Nd	0.530	1.034	2.13	3.900	5.790					
61	Pm	0.536	1.052	2.15	3.97	5.953					
62	Sm	0.543	1.068	2.26	4.00	6.046					
63	Eu	0.547	1.085	2.40	4.11	6.101					
64	Gd	0.592	1.17	1.99	4.24	6.249					
65	Tb	0.564	1.112	2.11	3.84	6.413					
66	Dy	0.572	1.126	2.20	4.00	5.990					
67	Ho	0.581	1.139	2.20	4.10	6.169					
68	Er	0.589	1.151	2.19	4.11	6.282					
69	Tm	0.5967	1.163	2.284	4.12	6.313					
70	Yb	0.6034	1.175	2.415	4.22	6.328					
71	Lu	0.5235	1.34	2.022	4.36	6.445					
72	Hf	0.654	1.44	2.25	3.21	6.596					
73	Ta	0.761									
74	W	0.770									
75	Re	0.760									
76	Os	0.84									
77	Ir	0.88									
78	Pt	0.87	1.7911								
79	Au	0.8901	1.98								
80	Hg	1.0070	1.8097	3.30							
81	Tl	0.5893	1.9710	2.878							
82	Pb	0.7155	1.4504	2.0815	4.083	6.64					
83	Bi	0.7033	1.610	2.466	4.37	5.40	8.62				
84	Po	0.812									
85	At										
86	Rn	1.0370									
87	Fr										
88	Ra	0.5094	0.97906								
89	Ac	0.49	1.17								
90	Th	0.59	1.11	1.93	2.78						
91	Pa	0.57									
92	U	0.59									
93	Np	0.60									
94	Pu	0.585									
95	Am	0.578									
96	Cm	0.581									
97	Bk	0.601									
98	Cf	0.608									
99	Es	0.619									
100	Fm	0.627									
101	Md	0.635									
102	No	0.642									

SOURCE: C. E. Moore, "Ionization Potentials and Ionization Limits Derived from the Analyses of Optical Spectra," NSRDS-NBS 34, National Bureau of Standards, Washington, D. C., **1970** and personal communication. Data for the lanthanides and actinides from W. C. Martin et al., *J. Phys. Chem. Ref. Data*, **1974**, *3*, 771 and J. Sugar, *J. Opt. Soc. Am.*, **1975**, *65*, 1366.

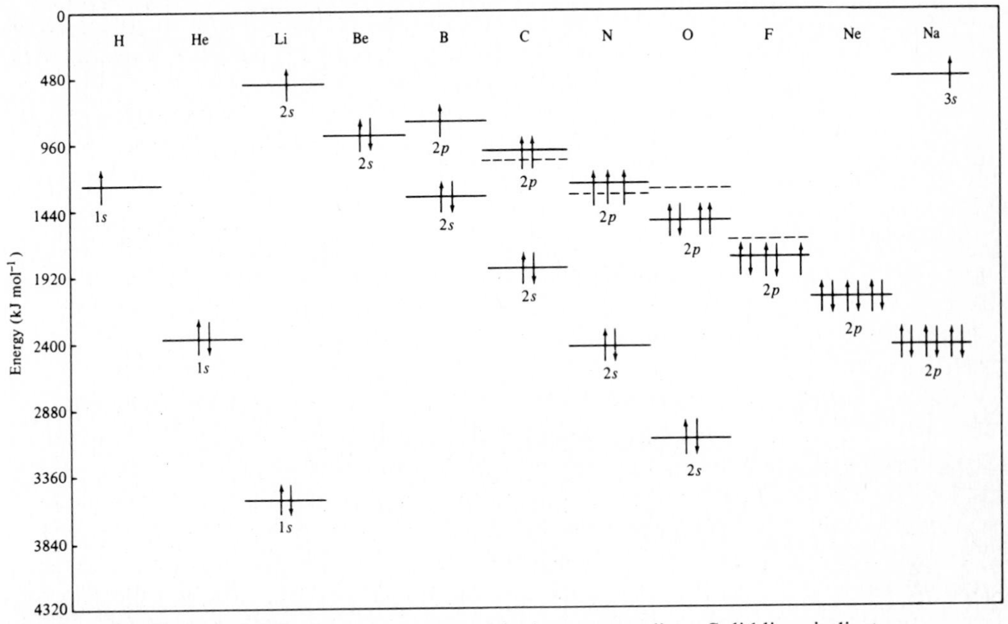

Fig. 2.14 Relative orbital energies of the elements hydrogen to sodium. Solid lines indicate *one-electron orbital energies*. Dashed lines represent experimental ionization energies which differ as a result of electron–electron interactions.

Table 2.4B Ionization energies

Molecule	I (MJ mol^{-1})	I (eV)
CH_3	0.949	9.84
C_2H_5	0.81	8.4
CH_3O	1.19	12.3
CN	1.40	14.5
CO	1.352	14.01
CF_3	0.9745	10.10
N_2	1.503	15.58
NH_2	1.10	11.4
NO	0.894	9.27
NO_2	0.944	9.78
O_2	1.164	12.06
OH	1.271	13.17
F_2	1.51	15.7

SOURCE: J. L. Franklin et al., "Ionization Potentials, Appearance Potentials, and Heats of Formation of Gaseous Positive Ions," NSRDS-NBS 26, National Bureau of Standards, Washington, D.C., **1969**.

the first series (with exception of Cr and Cu) have a $3d^n4s^2$ ground state for the neutral atom and a stable $3d^n4s^0$ state for the dipositive ion, the source of our problem must be sought in the difference between atom and ion, not in trends along the series.

Before the question can be answered adequately, it is necessary to define what is meant by "orbital energy." For most purposes, that given by Koopmans' theorem is adequate: If the *n*th orbital is the highest orbital used to describe the ground-state wave function of an atom X, then the energy of orbital *n* is approximated by the ionization energy of the atom. These energies can be calculated by the self-consistent-field method described previously (see p. 27). The calculations always indicate that the 4*s* level lies *above* the 3*d* level.[34] Why, then, does the 4*s* level fill first? Simply because the total energy of the atom is the quantity that is important, not the single energy of the electron entering the 4*s* level. When all the energies are summed and *all of the electron–electron repulsions* summed, it is found that the [Ar] $4s^1$ configuration is lower in energy than the [Ar] $3d^1$ configuration.[35]

Unfortunately, this does not answer our original question completely, for we cannot simply say that 4*s* electrons ionize before the 3*d* electrons because the 4*s* orbital lies higher in energy—we have seen that we must inspect the total energy. One could postulate (incorrectly) that because of electron–electron repulsions the correct configuration for Ti^{+2} was [Ar] $4s^2$ instead of [Ar] $3d^2$, which runs counter to experimental fact. In order to account for the apparent change in stability, depending on whether the 4*s* or the 3*d* orbital is occupied, we must compare the two systems involved, Ti versus Ti^{+2}, or more generally, M versus M^{+2}. A clue may be sought from the trend along the 3*d* series mentioned above. Although it is not responsible for the effect (as discussed previously), it does give an indication. It appears that as the atomic number goes up, and hence as Z^* increases, the energy levels approach more closely to those in a hydrogen atom, namely, all levels having the same principal quantum number being degenerate and lying below those of the next quantum number. Now the effective nuclear charge in the ion increases markedly because of the net ionic charge and the reduced shielding. It is not unreasonable to suppose, then, that the formation of a dipositive ion accomplishes more than the gradual changes of the entire transition series were able to—that is, lowering the 3*d* level so far below the 4*s* that the repulsion energies are overcome and the total energy is minimized if the 3*d* level, rather than the 4*s*, is occupied. This tendency towards hydrogen-like orbitals is dramatic with increasing effective nuclear charge. For example, core electrons are scarcely differentiated energetically according to type of orbital—they closely approach the hydrogenic degeneracy.[36]

Electron affinity

Electron affinity is conventionally defined as the energy *released* when an electron is added to the valence shell of an atom. Unfortunately, this is in contradiction to the universal thermodynamic convention that enthalpies of exothermic reactions shall be assigned *negative* signs. Since it seems impossible to overthrow the electron affinity convention at

[34] F. L. Pilar, *J. Chem. Educ.*, **1978**, *55*, 2.

[35] For a lucid discussion of this somewhat confusing problem, consult the reference in footnote 34.

[36] For further discussion of the relative energies and the reasons, see R. N. Keller, *J. Chem. Educ.*, **1962**, *39*, 289, and M. Karplus and R. N. Porter, "Atoms and Molecules," Benjamin, New York, **1970**, pp. 269–271.

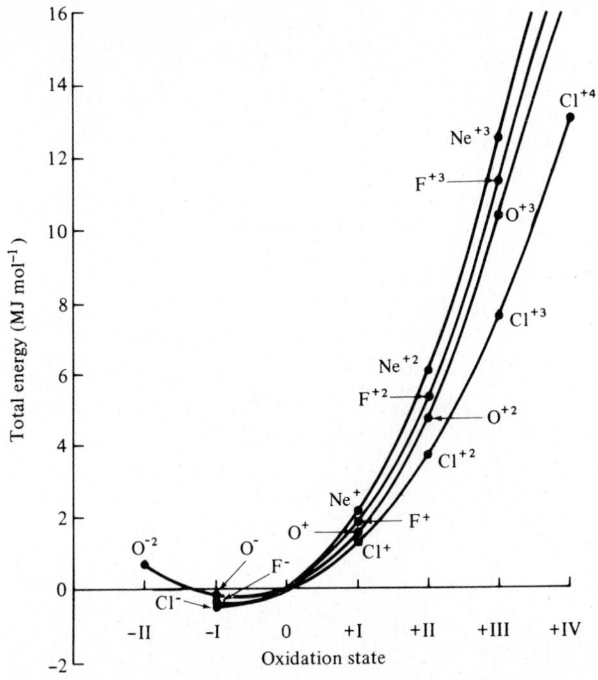

Fig. 2.15 Ionization energy–electron affinity curves for oxygen, fluorine, neon, and chlorine.

this late date without undue confusion, one can adopt one of two viewpoints to minimize confusion. One is to let the electron affinities of the most active nonmetals be *positive*, even though in thermodynamic calculations the enthalpies are *negative*:

$$F + e^- \rightarrow F^- \qquad EA = +337 \text{ kJ mol}^{-1} \qquad \Delta H = -337 \text{ kJ mol}^{-1}$$

A slightly different approach is to consider the electron affinities of the atoms to be the same as the ionization energies of the anion. Now the positive electron affinity corresponds to an endothermic reaction:

$$F^- \rightarrow F + e^- \qquad IE = +337 \text{ kJ mol}^{-1} \qquad \Delta H = +337 \text{ kJ mol}^{-1}$$

This second approach has the added benefit of calling attention to the very close relationship between electron affinity and ionization potential. In fact, when the ionization energies and electron affinities of atoms are plotted, a smooth curve results and the function may be described rather accurately by the quadratic formula:[37]

$$E = \alpha q + \beta q^2 \qquad \textbf{(2.24)}$$

where E is the total energy of the ion (Σ_{IE} or Σ_{EA}) and q is the ionic charge. See Fig. 2.15.

[37] Actually a more accurate expression is a polynomial of the type $E = \alpha q + \beta q^2 + \gamma q^3 + \delta q^4$. The constants γ and δ, however, are small and Eq. 2.24 is a good approximation. See R. P. Iczkowski and J. L. Margrave, *J. Am. Chem. Soc.*, **1961**, *83*, 3547.

Table 2.5 Electron affinities (kJ mol^{-1})[a]

Element	Value	Element	Value
1. H	72.7651	35. Br	324.6
2. He	0	36. Kr	0
3. Li	59.8	37. Rb	46.89
4. Be	0	38. Sr	0
5. B	27	39. Y	0.0
6. C	122.3	40. Zr	50
7. N $\rightarrow N^{-1}$	−7	41. Nb	100
$N^{-1} \rightarrow N^{-2}$	−800[b]	42. Mo	100
$N^{-2} \rightarrow N^{-3}$	−1290[b]	43. Tc	70
8. O $\rightarrow O^{-1}$	141.0	44. Ru	110
$O^{-1} \rightarrow O^{-2}$	−780[c]	45. Rh	120
9. F	327.9	46. Pd	60
10. Ne	0	47. Ag	125.7
11. Na	52.7	48. Cd	0?
12. Mg	0	49. In	29
13. Al	44	50. Sn	121
14. Si	133.6	51. Sb	101
15. P	71.7	52. Te	190.14
16. S $\rightarrow S^{-1}$	200.40	53. I	295.3
$S^{-1} \rightarrow S^{-2}$	−590	54. Xe	0
17. Cl	348.8	55. Cs	45.49
18. Ar	0	56. Ba	0
19. K	48.36	57. La	50
20. Ca	0	58–71. Ln	50
21. Sc	0	72. Hf	0
22. Ti	20	73. Ta	60
23. V	50	74. W	60
24. Cr	64	75. Re	15
25. Mn	0	76. Os	110
26. Fe	24	77. Ir	160
27. Co	70	78. Pt	205.3
28. Ni	111	79. Au	222.73
29. Cu	118.3	80. Hg	0
30. Zn	0	81. Tl	30
31. Ga	29	82. Pb	110
32. Ge	120	83. Bi	110
33. As	77	84. Po	180
34. Se $\rightarrow Se^{-1}$	194.9	85. At	270
$Se^{-1} \rightarrow Se^{-2}$	−420	86. Rn	0

[a] Unless otherwise noted, all values are from H. Hotop and W. C. Lineberger, *J. Phys. Chem. Ref. Data*, **1975**, *4*, 539.

[b] E. C. Baughan, *Trans. Faraday Soc.*, **1961**, *57*, 1863.

[c] A. P. Ginsberg and J. M. Miller, *J. Inorg. Nucl. Chem.*, **1958**, *7*, 351.

Table 2.6 Electron affinities of molecules[a]

Molecule	Experimental (kJ mol^{-1})	Molecule	Experimental (kJ mol^{-1})
CH_3	8	O-*neo*-C_5H_{11}	193
C_2	300	OC_6H_5	227
C_3	170	O_2H	300
C≡CH	360	F_2	297
C_5H_5	172	SiH_3	140
C_6H_5	212	PH_2	120
$C_6H_5CH_2$	87	SH	220
CN	369	SO_2	100
N_3	300–340[b]	SCN	206
NH_2	139	SF_5	353
NO_2	230	SF_6	144
NO_3	380	Cl_2	232
O_2	48	Br_2	250
O_3	190	TeF_6	322
OH	177	I_2	240
OCH_3	153	WF_6	357
O-*t*-C_4H_9	180	UF_6	500

[a] B. K. Janousek and J. I. Brauman, "Gas Phase Ion Chemistry," Vol. 2, M. T. Bowers, Ed., Academic, New York, **1979**, Chapter 10.
[b] R. S. Berry, *Chem. Rev.*, **1969**, *69*, 533.

It may readily be seen that whereas the acceptance of electrons by active nonmetals is initially *exothermic*, the atoms become "saturated" relatively quickly, the energy reaches a minimum, and further addition of electrons is *endothermic*. In fact, for dinegative ions such as O^{-2} and S^{-2}, the total electron affinity is *negative*; that is, their enthalpy of formation is *positive*. Such ions cannot exist except through stabilization by their environment, either in a crystal lattice or by solvation in solution.

As might be supposed, electron affinity trends in the periodic chart parallel those of ionization energies (Table 2.5). Elements with large ionization energies tend to have large electron affinities as well. There are a few notable exceptions, however. Fluorine has a lower electron affinity than chlorine, and this apparent anomaly is even more pronounced for N/P and O/S. It is a result of the smaller size of the first-row elements and consequent greater electron-electron repulsion in them. Although they initially have greater tendencies to accept electrons (note the slopes of the lines as they pass through the origin, or neutral atom, in Fig. 2.15), they quickly become "saturated" as the electron-electron repulsion rapidly dominates (the flat, bottom portion of the curve).[38] Fewer data are available for neutral molecules (Table 2.6). Free radicals made up of electronegative atoms, such as CN, NO_2, NO_3, SF_5, etc., have the expected high electron affinities, and we shall see later that they are among the most electronegative of groups.

[38] For further discussion of electron affinities, together with useful charts and graphs, see E. C. M. Chen and W. E. Wentworth, *J. Chem. Educ.*, **1975**, *52*, 486.

PROBLEMS

2.1. What is the frequency of light emitted when an electron drops from $n = 3$ to $n = 1$? From $n = 6$ to $n = 2$?

2.2. Write out the electron configurations for the following elements. Determine the number of unpaired electrons in the ground state.

$_{6}C$ $_{8}O$ $_{15}P$ $_{20}Ca$ $_{23}V$ $_{64}Gd$ $_{75}Re$

2.3. Determine the possible states for the elements listed in Problem 2.2, give the term symbols, and choose the ground state.

2.4. Write out the electron configurations for the following ions. Determine the number of unpaired electrons in the ground state.

$_{22}Ti^{+3}$ $_{25}Mn^{+2}$ $_{29}Cu^{+2}$ $_{78}Pt^{+2}$ $_{64}Gd^{+3}$ $_{65}Tb^{+4}$

2.5. Determine the appropriate term symbols for the ground states of the ions in Problem 2.4.

2.6. Clearly distinguish the following aspects of the structure of an atom and sketch the appropriate function for 1*s*, 2*s*, 2*p*, 3*s*, and 3*p* orbitals:
a. radial wave function
b. radial probability function
c. angular wave function
d. angular probability function
e. contour map of electron density

2.7. Using Slater's rules, calculate Z^* for the following electrons:
a. the valence (most easily ionizable) electron in Ca
b. the valence electron in Mn
c. a 3*d* electron in Mn
d. the valence electron in Br

Compare the values of Z^* thus obtained with those of Clementi and Raimondi.

2.8. Which has the higher first ionization energy:

Li or Cs? Li or F? Cs or F? F or I?

2.9. Which has the higher electron affinity:

C or F? F or I? Te or I?

2.10. Which has the higher electron affinity:

F or Cl? Cl or Br? O or S? S or Se?

2.11. In the hydrogen atom the 3*s* and 3*p* orbitals have identical energies, but in the chlorine atom the 3*s* orbital lies at a considerably lower energy than the 3*p*. Explain.

2.12. Slater proposed his shielding rules "tentatively" in 1930, yet they have never been modified. Look at the Clementi-Raimondi results (Eqs. 2.12–2.19), and suggest some modifications of Slater's rules that would improve them.

2.13. Plot the *total ionization energies* of Al^{+n} as a function of n from $n = 1$ to $n = 8$. Make two plots, one using linear and the other using log–log graph paper. Explain the source of the discontinuity in your curve.

2.14. The Pauli exclusion principle forbids certain combinations of m_l and m_s in determining the term symbols for the states of the carbon atom. Consider an excited carbon atom in which the electronic configuration is $1s^2 2s^2 2p^1 3p^1$. What states now are possible?

2.15. **a.** Calculate the third ionization energy of lithium. [*Hint:* This requires no approximations or assumptions.]

b. Calculate the first and second ionization energies of lithium using Slater's rules.

c. Calculate the first and second ionization energies of lithium using the rules of Clementi and Raimondi.

2.16. Prove that the statement made on p. 26 concerning the shape of the sum of a set of three p orbitals is true. [*Hint:* Work first in two dimensions and add the appropriate values from Fig. 2.6 or 2.7 for a p_x and p_y orbital.]

2.17. The statement is made on p. 49 that the second electron affinity for Group VIA elements, such as O and S, is always endothermic. Which element might be an exception to this rule? Why?

2.18. Plot the third ionization energy for the lanthanide elements, $Z = 57–72$. Explain any trends or features of your plot.

2.19. The symbol S has been used in this chapter to represent three completely distinct quantities or phenomena. Name the three and clearly differentiate among them by discussing each.

2.20. Assign the correct values for the quantum numbers n, l, and m_l for the following orbitals: $2p_z$, $3d_{z^2}$, $4s$.

2.21. As a rough approximation, conjugated polyenes may be considered to be linear, one-dimensional boxes. Consider the three conjugated polyenes butadiene (I), vitamin A (II), and carotene (III):

$$CH_2{=}CH{-}CH{=}CH_2$$

(I)

H_3C CH_3

$CH{=}CHC{=}CHCH{=}CHC{=}CHCH_2OH$

CH_3 CH_3

CH_3

(II)

H_3C CH_3 H_3C CH_3

$CH{=}CHC{=}CHCH{=}CHC{=}CHCH{=}CHCH{=}CCH{=}CHCH{=}CCH{=}CH$

CH_3 CH_3 CH_3 CH_3

CH_3 H_3C

(III)

a. The colors of these polyenes are: butadiene, colorless; vitamin A, orange-yellow; carotene, ruby red. Explain the colors of these compounds *qualitatively* in terms of the length of the conjugated chain. [*Note:* We are here dealing with light *absorbed*, not emitted.]

b. Vitamin A absorbs most strongly at 332 nm. Using this value to obtain the energy of the electronic transition estimate the length of the vitamin A "box." Note that the π electrons in II will fill the first five energy levels. Don't be disturbed if this crude approximation gives a poor estimate when compared with an average bond distance of 139 pm for conjugated single–double bonds. What are some sources of error in this approximation?

c. Why do vitamin A and carotene give broad absorptions rather than the sharp lines found in atomic spectra?

2.22. Show that the equation $E = \dfrac{h^2}{8ma^2}$ is dimensionally correct.

2.23. Which of the halogens, X_2, would you expect to be most likely to form a cation, X^+? It is now felt that X^+ ions are not known in chemical systems, but that X_2^+, X_3^+, and X_5^+ ions do exist. Why should the latter be more stable than X^+?

2.24. Table 2.4A lists the first through twenty-first ionization energies of various atoms. Many workers have argued that the electron affinity (Table 2.5) should be considered the *zero*th ionization energy. Discuss.

2.25. Consider the facts that:

a. The electron affinity of SF_5 is higher than that of S, F, or any other single atom.

b. Although the electron affinity of SF_5 is among the highest known, that of SF_6 is quite modest. Explain.

3

Bonding models in inorganic chemistry

Structure and bonding lie at the heart of modern inorganic chemistry. It is not too much to say that the renaissance of inorganic chemistry following World War II was concurrent with the development of a myriad of spectroscopic methods of structure determination. Methods of rationalizing and predicting structures soon followed. In this chapter we shall encounter the simplest methods of explaining and predicting the bonding in a variety of compounds. Subsequent chapters will elaborate upon these basic themes.

THE IONIC BOND

Although there is no sharp boundary between ionic bonding and covalent bonding, it is convenient to consider each of these as a separate entity before attempting to discuss molecules and lattices, in which *both* are important. Furthermore, since the purely ionic bond may be treated by a simple electrostatic model, it is advantageous to discuss it first. The simplicity of the electrostatic model has caused chemists to think of many solids as systems of ions. We shall see that this view needs some modification, and there are, of course, many solids, ranging from diamond to metals, which require alternative theories of bonding.

Properties of ionic substances

Several properties distinguish ionic compounds from covalent compounds. These may be related rather simply to the crystal structure of ionic compounds, namely, a lattice composed of positive and negative ions in such a way that the attractive forces between oppositely charged ions are maximized and the repulsive forces between ions of the same charge are minimized. Before discussing some of the possible geometries, a few simple properties of ionic compounds may be mentioned:

1. Ionic compounds tend to have very low electrical conductivities as solids[1] but conduct electricity quite well when molten. This conductivity is attributed to the presence of *ions*, atoms charged either positively or negatively, which are free to move under the influence of an electric field. In the solid, the ions are bound tightly in the lattice and are not free to migrate and carry electrical current. It should be noted that we have no absolute *proof* of the existence of ions in solid sodium chloride, for example, though our best evidence will be discussed later in this chapter (pp. 71–72). The fact that ions are found when sodium chloride is melted or dissolved in water does not *prove* that they existed in the solid crystal. However, their existence in the solid is usually assumed, since the properties of these materials may readily be interpreted in terms of electrostatic attractions.

2. Ionic compounds tend to have high melting points. Ionic bonds usually are quite *strong* and they are *omnidirectional*. The second point is quite important, since ignoring it could lead one to conclude that ionic bonding was much stronger than covalent bonding—which is *not* the case. We shall see that substances containing strong, multidirectional covalent bonds, such as diamond, also have very high melting points. The high melting point of sodium chloride, for example, results from the strong electrostatic attractions between the sodium cations and the chloride anions, and from the lattice structure, in which each sodium ion attracts six chloride ions, each of which in turn attracts six sodium ions, etc., throughout the crystal. The relation between bonding, structure, and the physical properties of substances will be discussed at greater length in Chapter 6.

3. Ionic compounds usually are very hard but brittle substances. The hardness of ionic substances follows naturally from the argument presented above, except in this case we are relating the multivalent attractions between the ions with *mechanical* separation rather than separation through thermal energy. The tendency toward brittleness results from the nature of ionic bonding. If one can apply sufficient force to displace the ions slightly (e.g., the length of one-half of the unit cell in NaCl), the formerly attractive forces become repulsive as anion–anion and cation–cation contacts occur; hence the crystal flies apart. This accounts for the well-known cleavage properties of many minerals.

4. Ionic compounds are often soluble in polar solvents with high permittivities (dielectric constants). The energy of interaction of two charged particles is given by

$$E = \frac{q^+ q^-}{4\pi r \varepsilon} \tag{3.1}$$

where q^+ and q^- are the charges, r is the distance of separation, and ε is the permittivity of the medium. The permittivity of a vacuum, ε_0, is 8.85×10^{-12} C^2 m^{-1} J^{-1}. For common polar solvents, however, the permittivity values are considerably higher. For example, the permittivity is 7.25×10^{-10} C^2 m^{-1} J^{-1} for water, 2.9×10^{-10} C^2 m^{-1} J^{-1} for acetonitrile, and 2.2×10^{-10} C^2 m^{-1} J^{-1} for ammonia, giving relative permittivities of 82 ε_0 (H_2O), 33 ε_0 (CH_3CN), and 25 ε_0 (NH_3). Since the permittivity of ammonia is 25 times that of a vacuum, the attraction between ions dissolved in ammonia, for example, is only

[1] For a discussion of some very interesting exceptions, ionic compounds with high conductivities as solids, see p. 191.

4% as great as in the absence of solvent. For solvents with higher permittivities the effect is even more pronounced.

Another way of looking at this phenomenon is to consider the interaction between the dipole moments of the polar solvent and the ions. Such solvation will provide considerable energy to offset the otherwise unfavorable energetics of breaking up the crystal lattice (see Chapter 6).

Occurrence of ionic bonding

Simple ionic compounds form only between very active metallic elements and very active nonmetals.[2] Two important requisites are that the ionization energy to form the cation and the electron affinity to form the anion must be energetically favorable. This does not mean that these two reactions must be exothermic (an impossibility—see Problem 3.8), but means, rather, that they must not cost too much energy. Thus the requirements for ionic bonding are: (1) The atoms of one element must be able to lose one or two (rarely three) electrons without undue energy input and (2) the atoms of the other element must be able to accept one or two electrons (almost never three) without undue energy input. This restricts ionic bonding to compounds between the most active metals—Groups IA, IIA, part of IIIA, and some lower oxidation states of the transition metals (forming cations)—and the most active nonmetals—Groups VIIA, VIA, and nitrogen (forming anions).[3] All ionization energies are endothermic, but for the metals named above they are not prohibitively so. Electron affinities are exothermic only for the halogens but are not excessively endothermic for the chalcogens and nitrogen.

Structures of crystal lattices

Before discussing the energetics of lattice formation, it will be instructive to examine some of the most common arrangements of ions in crystals. Although only a few of the many possible arrangements are discussed, they indicate some of the possibilities available for the formation of lattices. We shall return to the subject of structure after some basic principles have been developed.

The first four structures described below contain equal numbers of cations and anions, i.e., the 1:1 and 2:2 salts. Most simple ionic compounds with such formulations crystallize in one of these four structures. They differ principally in the coordination number, i.e., the number of counterions grouped about a given ion, in these examples four, six, and eight.

The sodium chloride structure. Sodium chloride crystallizes in a face-centered cubic structure (Fig. 3.1a). To visualize the face-centered arrangement, consider *only* the sodium

[2] It is true that ionic compounds such as $[NH_4]^+ [B(C_6H_5)_4]^-$ are known in which there are no extremely active metals or nonmetals. Nevertheless, the above statement is for all practical purposes correct, and we can consider compounds such as ammonium tetraphenylborate to result from the particular *covalent* bonding properties of nitrogen and boron.

[3] Since the transition between ionic bonding and covalent bonding is not a sharp one, it is impossible to define precisely the conditions under which it will occur. However, the generalization is helpful and does not rule out the possibility of unusual ionic bonds, as, for example, between two metals: $Cs^+ Au^-$. See p. 531.

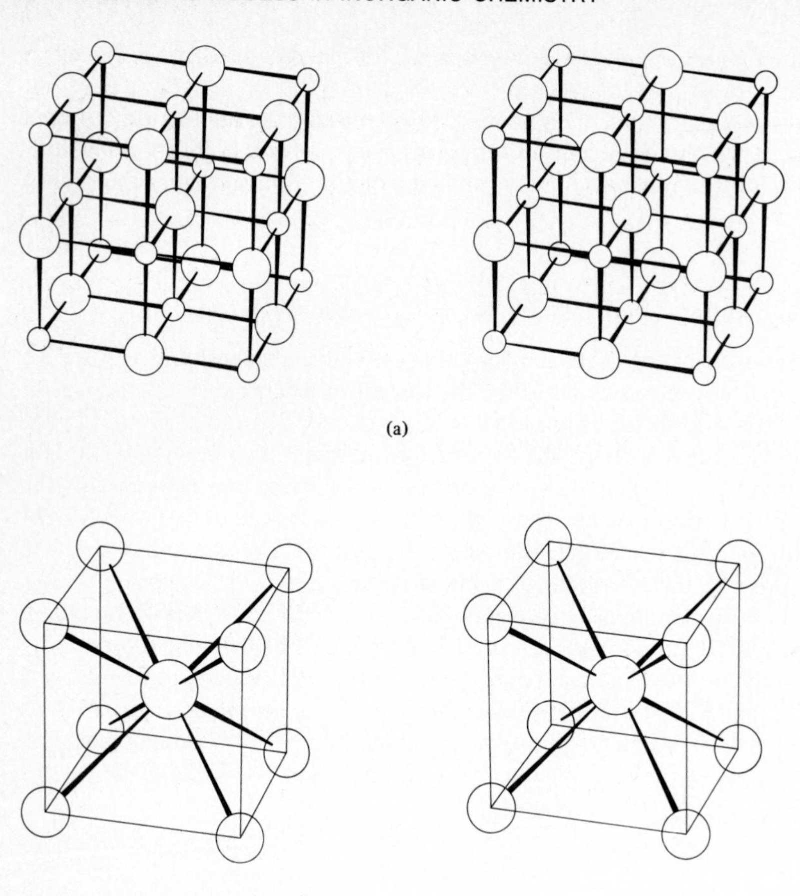

Fig. 3.1 Crystal structures of two 1:1 ionic compounds. (a) Unit cell of sodium chloride; (b) unit cell of cesium chloride. [From M. F. C. Ladd, "Structure and Bonding in Solid State Chemistry," Wiley, New York, **1979**. Reproduced with permission.]

ions *or* the chloride ions (this will require extensions of the sketch of the lattice). Eight sodium ions form the corners of a cube and six more are centered on the faces of the cube. The chloride ions are similarly arranged, so that the sodium chloride lattice consists of two interpenetrating face-centered cubic lattices. The coordination number (C.N.) of both ions in the sodium chloride lattice is 6; i.e., there are six chloride ions about each sodium ion and six sodium ions about each chloride ion.

The cesium chloride structure. Cesium chloride crystallizes in the cubic arrangement shown in Fig. 3.1b. The cesium (or chloride) ions occupy the eight corners of the cube and the counterion occupies the center of the cube.[4] Again, we must consider a lattice

[4] The structure of CsCl is sometimes referred to as "body-centered cubic," but this is incorrect. True body-centered cubic lattices have the *same* species on the *corners* and the *center* of the unit cell, as in the alkali metals, for example.

composed either of the cesium ions or of the chloride ions, both of which have simple cubic symmetry. The coordination number of both ions in the cesium chloride structure is 8; i.e., there are eight anions about each cation and eight cations about each anion.

The zinc blende and wurtzite structures. Zinc sulfide crystallizes in two distinct lattices: the wurtzite (Fig. 3.2a) and the zinc blende (Fig. 3.2b). We shall not elaborate upon them now (see p. 82), but simply note that in both the coordination number is 4 for both cations and anions.

All the following structures have twice as many anions as cations (1:2 structures); thus the coordination number of the cation *must* be twice that of the anion: 8:4, 6:3, 4:2, etc. The inverse structures are also known where the cations outnumber the anions by two to one.

The fluorite structure. Calcium fluoride crystallizes in the fluorite structure (Fig. 3.3). The coordination numbers are 8 for the cation (eight fluoride ions form a cube about each

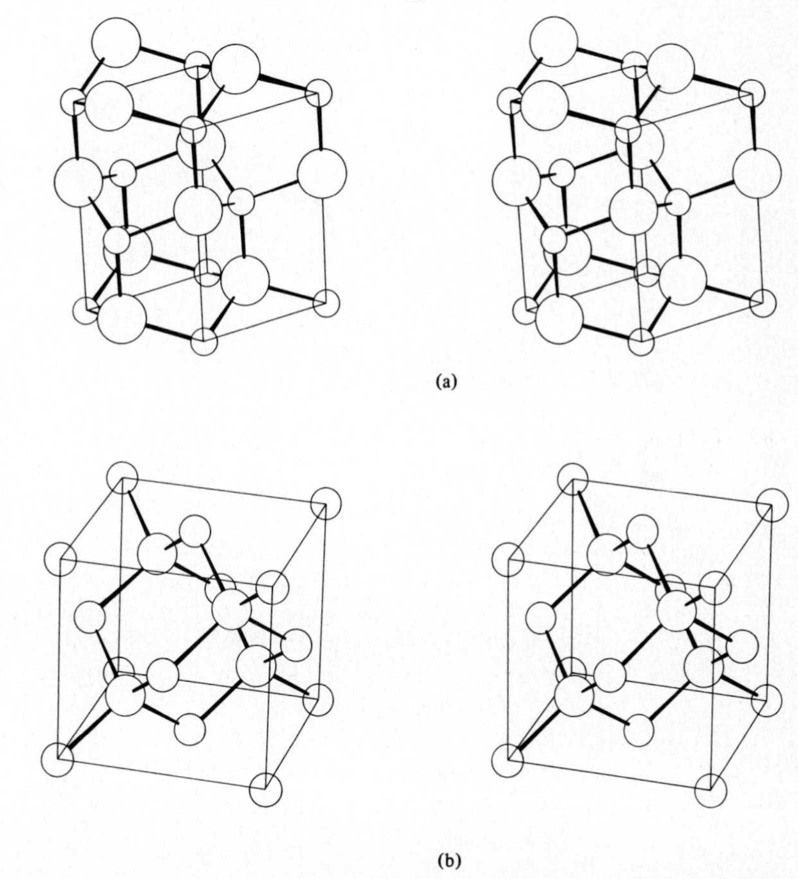

Fig. 3.2 Unit cells of two zinc sulfide (2:2) structures; circles in order of decreasing size are S and Zn. (a) Wurtzite; (b) zinc blende. [From M. F. C. Ladd, "Structure and Bonding in Solid State Chemistry," Wiley, New York, **1979**. Reproduced with permission.]

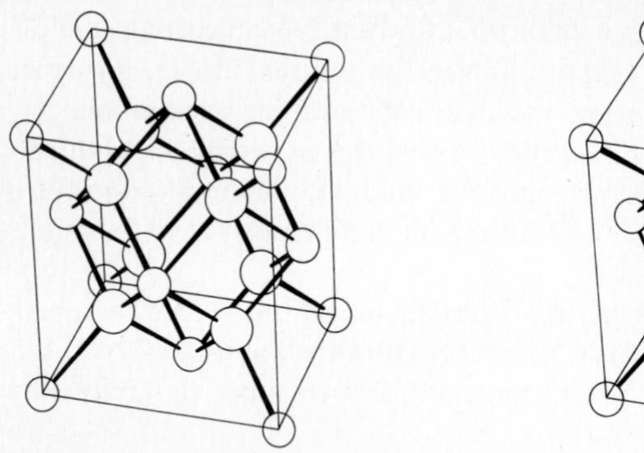

Fig. 3.3 Unit cell of the fluorite structure; circles in order of decreasing size are F and Ca. [From M. F. C. Ladd, "Structure and Bonding in Solid State Chemistry," Wiley, New York, **1979**. Reproduced with permission.]

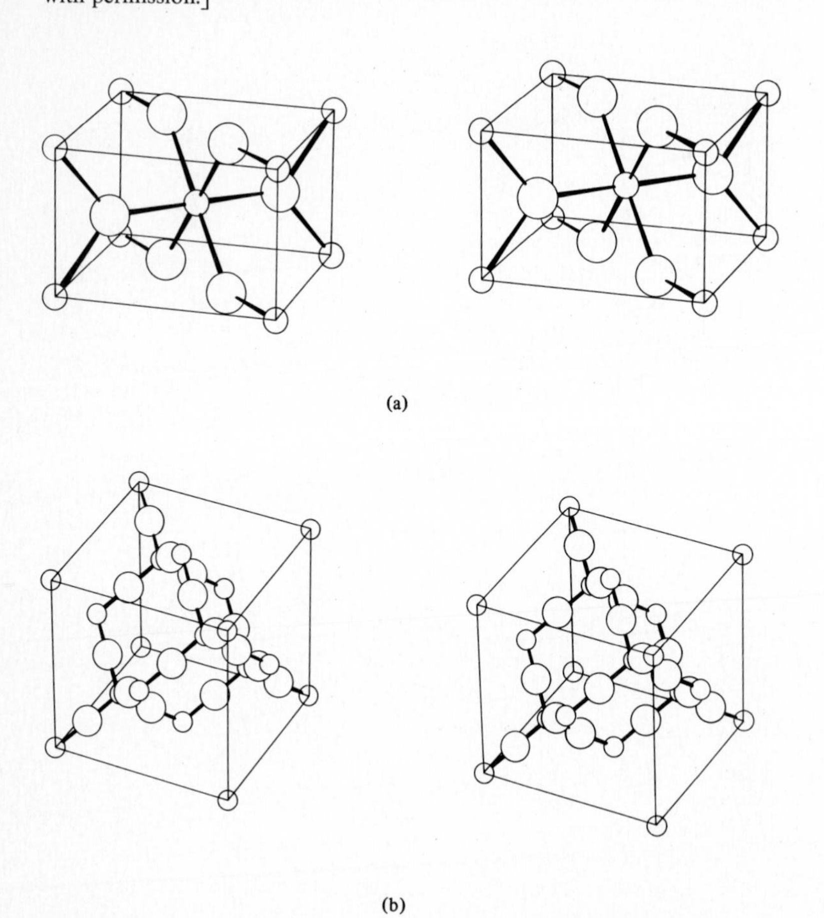

(a)

(b)

Fig. 3.4 Crystal structures of two more 1:2 compounds; oxygen is the larger circle in both. (a) Unit cell of the rutiles, TiO_2; (b) unit cell of β-cristobalite, SiO_2. [From M. F. C. Ladd, "Structure and Bonding in Solid State Chemistry," Wiley, New York, **1979**. Reproduced with permission.]

calcium ion) and 4 for the anion (four Ca^{+2} ions tetrahedrally arranged about each F^- ion). If the positions and numbers of the cations and anions are reversed, one obtains the *antifluorite structure* which is adopted by such compounds as Li_2O and Na_2O.

The rutile structure. Titanium dioxide crystallizes in two crystal forms: anatase and rutile (Fig. 3.4a). The coordination numbers are 6 for the cation (six oxide anions arranged approximately octahedrally about the titanium) and 3 for the anion (three Ti^{+4} ions trigonally about the anion).

The β-cristobalite structure. Silicon dioxide crystallizes in several forms (some of which are stabilized by foreign atoms). One is β-cristobalite (Fig. 3.4b), which is related to zinc blende (Fig. 3.2b), having a silicon atom where every zinc *and* sulfur atom is in zinc blende, and with oxygen atoms between the silicon atoms. Another form, tridymite, is similarly related to the wurtzite structure. The coordination numbers in β-cristobalite and tridymite are 4 for silicon and 2 for oxygen.

LATTICE ENERGY

The energy of the crystal lattice of an ionic compound is the energy released when ions come together from infinite separation to form a crystal:

$$M^+_{(g)} + X^-_{(g)} \longrightarrow MX_{(s)} \tag{3.2}$$

It may be treated adequately by a simple electrostatic model. Although we shall include nonelectrostatic energies, such as the repulsions of closed shells, and more sophisticated treatments include such factors as dispersion forces and zero-point energy, simple electrostatics accounts for about 90% of the bonding energies. The theoretical treatment of the ionic lattice energy was initiated by Born and Landé, and a simple equation for predicting lattice energies bears their names. The derivation follows.

Consider the energy of an ion pair, M^+, X^-, separated by a distance r. The electrostatic energy of attraction is obtained from Coulomb's Law:[5]

$$E = \frac{Z^+Z^-}{4\pi\varepsilon_0 r} \tag{3.3}$$

Since one of the charges is negative, the energy is negative (with respect to the energy at infinite separation) and becomes increasingly so as the interionic distance decreases. Figure 3.5 shows the coulombic energy of an ion pair (dotted line). Because it is common to express Z^+ and Z^- as multiples of the electronic charge, $e = 1.6 \times 10^{-19}$ coulomb, we may write:

$$E = \frac{Z^+Z^-e^2}{4\pi\varepsilon_0 r} \tag{3.4}$$

Now in the crystal lattice there will be more interactions than the simple one in an ion pair. In the sodium chloride lattice, for example, there are attractions to the six nearest

[5] Note that these are *ionic charges* and not nuclear charges, for which Z is also used.

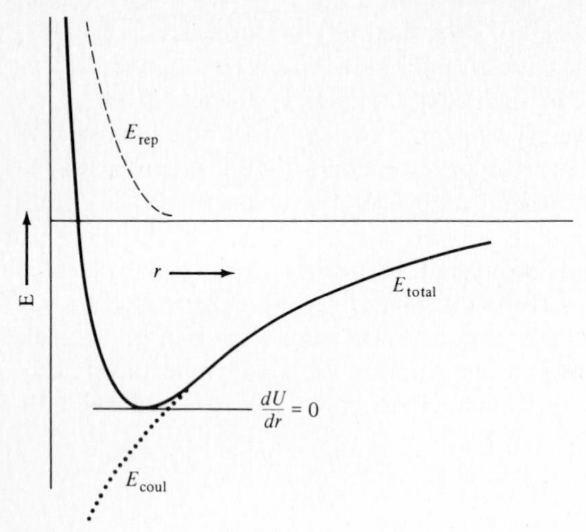

Fig. 3.5 Energy curves for an ion pair.

neighbors of opposite charge, repulsions by the twelve next nearest neighbors of like charge, etc. The summation of all of these geometrical interactions is known as the *Madelung constant*, *A*. The energy of a pair of ions in the crystal is then:

$$E_c = \frac{AZ^+Z^-e^2}{4\pi\varepsilon_0 r} \tag{3.5}$$

The evaluation of the Madelung constant for a particular lattice is straightforward. Consider the sodium ion (⊗) at the center of the cube in Fig. 3.6. Its nearest neighbors are the six face-centered chloride ions (●), each at a characteristic distance determined by the size of the ions involved. The next nearest neighbors are the twelve sodium ions (⊙) centered on the edges of that unit cell (cf. Fig. 3.1a inverted). The distance of these repelling ions can be related to the first distance by simple geometry, as can the distance of eight chloride ions in the next shell (those at the corners of the cube). If this process is followed until every ion in the crystal is included, the Madelung constant, *A*, may be obtained from the summation of all interactions. The first three terms for the interactions described above are:

$$A = 6 - \frac{12}{\sqrt{2}} + \frac{8}{\sqrt{3}} \cdots \tag{3.6}$$

Fortunately, the Madelung constant may be obtained mathematically from a converging series, and there are computer programs that converge rapidly. However, we need not delve into these procedures, but may simply employ the values obtained by other workers (Table 3.1). The value of the Madelung constant is determined only by the geometry of the lattice and is independent of ionic radius and charge. Unfortunately, previous workers have often incorporated ionic charge into the value which they used for the Madelung constant. The practice appears to have arisen from a desire to consider

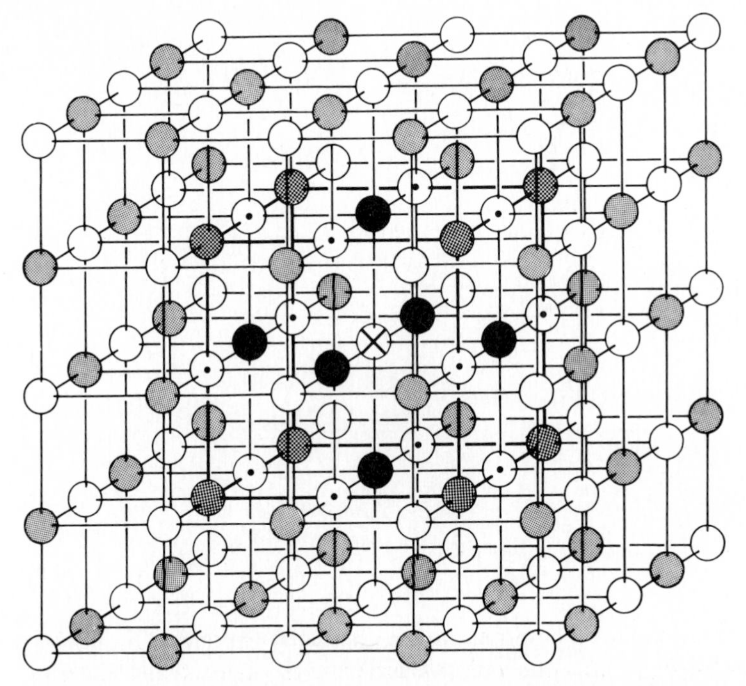

Fig. 3.6 An extended lattice of sodium chloride. Starting with the sodium ion marked ⊗, there are six nearest neighbors (●), twelve next nearest neighbors (⊙), eight next, next nearest neighbors (darkly shaded), and so on.

the energy of the "molecule" (e.g., MX_2) as:

$$E = \frac{\mathrm{A} Z_{\pm}^2 e^2}{4\pi\varepsilon_0 r} \tag{3.7}$$

where $\mathrm{A} = 2A$ and $Z_{\pm}^2$ is the highest common factor of Z^+ and Z^- (1 for NaCl, CaF_2, and Al_2O_3; 2 for MgO, TiO_2, and ReO_3; etc.). We could ignore this confusing practice and use the geometric Madelung constant, A, only, except that values reported in the literature are almost invariably given in terms of Eq. 3.7. Values for both A and A are given in Table 3.1, and the reader may readily confirm that use of either Eq. 3.5 or 3.7 yields identical results.[6]

Returning to Eq. 3.5 we see that unless there is a repulsion energy to balance the attractive coulombic energy no stable lattice can result. The attractive energy becomes infinite at infinitesimally small distances. Ions are, of course, not point charges but consist of electron clouds which repel each other at very close distances. This repulsion is shown by the dashed line in Fig. 3.5. It is negligible at large distances but increases very rapidly as the ions approach *each other closely*.

[6] For further discussion of the problem of defining Madelung constants, see D. Quane, *J. Chem. Educ.*, **170**, *47*, 396.

Table 3.1 Madelung constants of some common crystal lattices

Structure	Coordination number	Geometrical factor, A	Conventional[a] factor, A
Sodium chloride	6:6	1.74756	1.74756
Cesium chloride	8:8	1.76267	1.76267
Zinc blende	4:4	1.63806	1.63806
Wurtzite	4:4	1.64132	1.64132
Fluorite	8:4	2.51939	5.03878
Rutile	6:3	2.408[b]	4.816[b]
Corundum	6:4	4.1719[b]	25.0312[b]

[a] Use $Z_{\pm}$ = highest common factor.
[b] Exact values depend upon details of structure.

Born suggested that this repulsive energy could be expressed by

$$E_R = \frac{B}{r^n} \tag{3.8}$$

where B is a constant. Experimentally, information on the Born exponent, n, may be obtained from compressibility data, since the latter measure the resistance which the ions exhibit when forced to approach each other more closely. The total energy for a mole of the crystal lattice containing an Avogadro's number, N_A, of units is

$$U = E_C + E_R = \frac{AN_AZ^+Z^-e^2}{4\pi\varepsilon_0 r} + \frac{NB}{r^n} \tag{3.9}$$

The total lattice energy is shown by the solid line in Fig. 3.5. The minimum in the curve, corresponding to the equilibrium situation, may be found readily:

$$\frac{dU}{dr} = 0 = -\frac{AN_AZ^+Z^-e^2}{4\pi\varepsilon_0 r^2} - \frac{nNB}{r^{n+1}} \tag{3.10}$$

Physically this corresponds to equating the *force* of electrostatic attraction with the repulsive forces between the ions. It is now possible to evaluate the constant B and remove it from Eq. 3.8. Since we have fixed the energy at the minimum, we shall use U_0 and r_0 to represent this energy and the equilibrium distance. From Eq. 3.10:

$$B = \frac{-AZ^+Z^-e^2r^{n-1}}{4\pi\varepsilon_0 n} \tag{3.11}$$

$$U_0 = \frac{AZ^+Z^-Ne^2}{4\pi\varepsilon_0 r_0} - \frac{AN_AZ^+Z^-e^2}{4\pi\varepsilon_0 r_0 n} \tag{3.12}$$

$$U_0 = \frac{AN_AZ^+Z^-e^2}{4\pi\varepsilon_0 r_0}\left(1 - \frac{1}{n}\right) \tag{3.13}$$

Table 3.2 Values of the Born exponent, n

Ion configuration	n
He	5
Ne	7
Ar, Cu^+	9
Kr, Ag^+	10
Xe, Au^+	12

This is the Born-Landé equation for the lattice energy of an ionic compound. As we shall see, it is quite successful in predicting accurate values, although it omits certain energy factors to be discussed below. It requires only a knowledge of the crystal structure (in order to choose the correct value for A) and the interionic distance, r_0, both of which are readily available from X-ray diffraction studies.

The Born exponent depends upon the type of ion involved, with larger ions having relatively higher electron densities and hence larger values of n. For most calculations the generalized values suggested by Pauling (see Table 3.2) are sufficiently accurate for ions with the electron configurations shown.

The use of Eq. 3.13 to predict the lattice energy of an ionic compound may be illustrated as follows. For sodium chloride the various factors are:

$A = 1.74756$ (Table 3.1)
$N_A = 6.022 \times 10^{23}$ ion pairs mol^{-1}, Avogadro's number
$Z^+ = +1$, the charge of the Na^+ ion
$Z^- = -1$, the charge of the Cl^- ion
$e = 1.60210 \times 10^{-19}$ C, the charge on the electron (Appendix C)
$\pi = 3.14159$
$\varepsilon_0 = 8.854185 \times 10^{-12}\ \text{C}^2\ \text{J}^{-1}\ \text{m}^{-1}$ (Appendix C)
$r_0 = 2.81 \times 10^{-10}$ m, the sum of radii of Na^+ and Cl^- (Table 3.6)
$n = 8$, the average of the values for Na^+ and Cl^- (Table 3.2)

Performing the arithmetic, we obtain $U_0 = -755\ \text{kJ mol}^{-1}$, which may be compared with the best experimental value (Table 3.3) of $-770\ \text{kJ mol}^{-1}$. Eq. 3.13 accounts for 98% of the total energy, giving us confidence in using it in situations for which we have no experimental values.

For more precise work several other functions have been suggested to replace the one given above for the repulsion energy. In addition, there are three other energy terms which affect the result by a dozen or so kJ mol^{-1}: van der Waals or London forces (see Chapter 6), zero-point energy, and correction for heat capacity. Zero-point energy arises because even at zero kelvins the ions will vibrate in the lattice—they cannot be motionless (see p. 15 for the particle-in-a-box analogy). Lastly, we are usually interested in applying the results to calculations at temperatures higher than zero kelvins, in which case we must add a quantity:

$$\Delta E = \int_0^T (C_{v(\mathrm{MX})} - C_{v(\mathrm{M}^+)} - C_{v(\mathrm{X}^-)})\,dT \tag{3.14}$$

where the C_v terms are the heat capacities of the species involved.[7]

The best calculated values, taking into account these factors, increase the accuracy somewhat: $U_0 = -777$, overestimating the experimental value by slightly less than 1%. Unless one is interested in extreme accuracy, Eq. 3.13 is quite adequate.

The Born-Haber cycle

Hess's law states that the enthalpy of a reaction is the same whether the reaction takes place in one or several steps; it is a necessary consequence of the first law of thermodynamics concerning the conservation of energy. If this were not true, one could "manufacture" energy by an appropriate cyclic process. Born and Haber[8] applied Hess's law to the enthalpy of formation of an ionic solid. For the formation of an ionic crystal from the elements, the Born-Haber cycle may most simply be depicted as

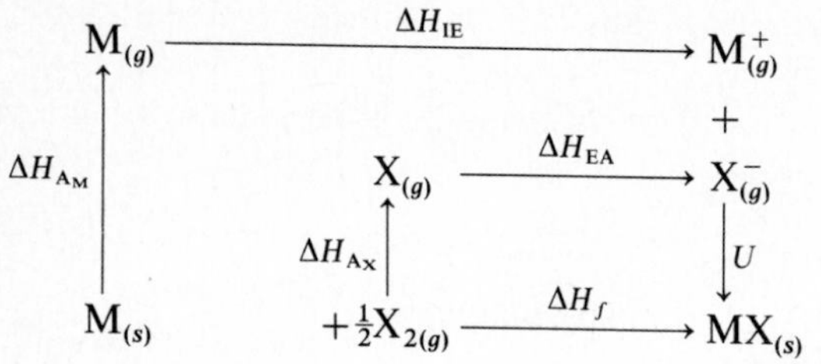

$$\begin{array}{ccccc}
\mathrm{M}_{(g)} & & \xrightarrow{\qquad\qquad\Delta H_{\mathrm{IE}}\qquad\qquad} & & \mathrm{M}^+_{(g)} \\
\uparrow & & & & + \\
\Delta H_{\mathrm{A_M}} & & \mathrm{X}_{(g)} & \xrightarrow{\Delta H_{\mathrm{EA}}} & \mathrm{X}^-_{(g)} \\
 & & \uparrow \Delta H_{\mathrm{A_X}} & & \downarrow U \\
\mathrm{M}_{(s)} & & +\tfrac{1}{2}\mathrm{X}_{2(g)} & \xrightarrow{\Delta H_f} & \mathrm{MX}_{(s)}
\end{array}$$

It is necessary that

$$\Delta H_f = \Delta H_{\mathrm{A_M}} + \Delta H_{\mathrm{A_X}} + \Delta H_{\mathrm{IE}} + \Delta H_{\mathrm{EA}} + U \tag{3.15}$$

The terms $\Delta H_{\mathrm{A_M}}$ and $\Delta H_{\mathrm{A_X}}$ are the enthalpies of atomization of the metal and the nonmetal, respectively. For gaseous diatomic nonmetals, ΔH_{A} is the enthalpy of dissociation (bond energy plus RT) of the diatomic molecule. For metals which vaporize to form monatomic gases, ΔH_{A} is identical to the enthalpy of sublimation. If sublimation occurs to a diatomic molecule, M_2, then the dissociation enthalpy of the reaction must also be included:

$$\mathrm{M}_2 \longrightarrow 2\mathrm{M} \tag{3.16}$$

---

[7] It is commonly assumed that M^+ and X^- will behave as ideal monatomic gases with heat capacities (at constant volume) of $\frac{3}{2}R$.

[8] M. Born, *Verhandl. Deut. Physik. Ges.*, **1919**, *21*, 13; F. Haber, ibid., **1919**, *21*, 750.

Values for the ionization energy, IE, and the electron affinity may be obtained from Tables 2.4A and 2.5. Bond dissociation energies for many molecules are given in Appendix E. A useful source of many data of use to the inorganic chemist has been written by Ball and Norbury.[9]

Uses of Born-Haber-type calculations

The enthalpy of formation of an ionic compound can be calculated with an accuracy of a few percent by means of the Born equation (Eq. 3.13) and the Born-Haber cycle. Consider NaCl, for example. We have seen that by using the predicted internuclear distance of 281 pm (or the experimental value of 281.4 pm), the Madelung constant of 1.748, the Born exponent n, and various constants, a value of -755.2 kJ mol^{-1} could be calculated for the lattice energy. The heat capacity correction is 2.1 kJ mol^{-1}, which yields $U_0^{298} = -757.3$ kJ mol^{-1}. The Born-Haber summation is then

$$\begin{aligned}
U_0^{298} &= -757.3 \text{ kJ mol}^{-1} \\
\Delta H_{\text{IE}} &= +495.4 \text{ kJ mol}^{-1} \\
\Delta H_{\text{EA}} &= -348.5 \text{ kJ mol}^{-1} \\
\Delta H_{\text{A}_{\text{Cl}}} &= +120.9 \text{ kJ mol}^{-1} \\
\Delta H_{\text{A}_{\text{Na}}} &= +108.4 \text{ kJ mol}^{-1} \\
\hline
\sum &= -381.1 \text{ kJ mol}^{-1}
\end{aligned}$$

This can be compared with an experimental value for the enthalpy of formation, $\Delta H_f^{298} = -410.9$ kJ mol^{-1}.

Separation of the energy terms in the Born-Haber cycle gives us some insight into their relative importance in chemical bonding. For example, the ΔH_{A} terms are always positive, but are usually of relatively small size compared with the other terms and do not vary greatly from compound to compound.[10] The ionization energies are always greatly endothermic. Electron affinities for the halogens are exothermic, but for the chalcogens they are endothermic as a result of forcing the second electron into the negatively charged X^- ion. In either case, the summation of ionization energy and electron affinity is *always* endothermic, and it is only the overwhelming exothermicity of the attraction of the ions for each other that makes ionic compounds stable with respect to dissociation into the elements. At room temperature this energy appears as the lattice energy. It should not be supposed, however, that at temperatures above the boiling point of the compound (1413 °C for NaCl, for example) no reaction would occur between an active metal and nonmetal. Even in the gas phase there will be electrostatic stabilization of the ions through

[9] M. C. Ball and A. H. Norbury, "Physical Data for Inorganic Chemists," Longman, London, **1974**.

[10] This statement is strictly true only for the halogens. The dissociation energies of O_2 and N_2 are considerably larger.

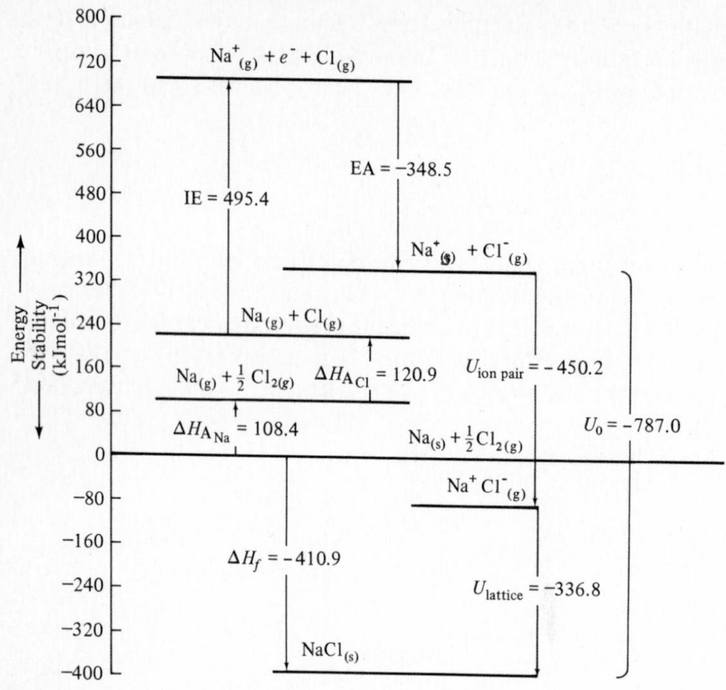

Fig. 3.7 Born-Haber diagram showing relative magnitudes of various terms for sodium chloride. [Adapted from G. P. Haight, *J. Chem. Educ.*, **1968**, *45*, 420. Reproduced with permission.]

the formation of ion pairs, M^+X^-. The latter should be added to the Born-Haber cycle, and to clarify the nature of the energy relationships, it is best to draw it in more explicit form as in Fig. 3.7. In such a diagram the individual enthalpies can be portrayed and related to the original enthalpy of the starting materials.[11]

Most of the enthalpies associated with steps in the cycle can be estimated, to a greater or lesser accuracy, by experimental methods. The lattice energy, however, is always obtained theoretically rather than from experimental measurement. It might be supposed that the "enthalpy of ionization" of a lattice could be measured in the same way as the ΔH_A of the metal and nonmetal, i.e., by heating the crystal and determining how much energy is necessary to dissociate it into ions. Unfortunately, this is experimentally impossible. When a crystal sublimes (ΔH_S), the result is not isolated gaseous ions but ion pairs and other clusters. For this reason it is necessary to use Eq. 3.13 or some more accurate version of it. We can then use the Born-Haber cycle to check the accuracy of our predictions if we can obtain accurate data on every other step in the cycle. Values computed from the Born-Haber cycle are compared with those predicted by Eq. 3.13 and modifications are shown in Table 3.3.

---

[11] For a discussion of this point as well as several others concerning Born-Haber-type cycles, see G. P. Haight, Jr., *J. Chem. Educ.*, **1968**, *45*, 420.

Table 3.3 Experimental and calculated lattice energies (kJ mol^{-1})

Salt	Experimental (Born-Haber cycle)	Simple model (Eq. 3.13)	"Best values"[a]	Kapustinskii approximation (Eq. 3.17)
LiF	1034	1008	1033	952.7
LiCl	840.1	811.3	845.2	803.7
LiBr	781.2	766.1	797.9	792.9
LiI	718.4	708.4	739.7	713.0
NaF	914.2	902.0	915.0	884.9
NaCl	770.3	755.2	777.8	752.9
NaBr	728.4	718.8	739.3	713.4
NaI	680.7	663.2	692.0	673.6
KF	812.1	797.5	813.4	788.7
KCl	701.2	687.4	708.8	680.7
KBr	671.1	659.8	679.5	674.9
KI	632.2	623.0	640.2	613.8
RbF	780.3	761.1	777.8	760.2
RbCl	682.4	661.5	686.2	661.9
RbBr	654.0	636.4	659.0	626.3
RbI	616.7	602.5	622.2	589.9
CsF	743.9	723.0	747.7	713.0
CsCl	629.7	622.6	652.3	625.1
CsBr	612.5	599.6	632.2	602.1
CsI	584.5	568.2	601.2	563.6

[a] Calculated using a modified Born equation with corrections for polarization effects, repulsion between nearest and next nearest neighbors, and zero-point energy (D. Cubicciotti, *J. Chem. Phys.*, **1959**, *31*, 1646; ibid., **1961**, *34*, 2189).

Kapustinskii[12] has noted that the Madelung constant, the internuclear distance, and the empirical formula of an ionic compound are all interrelated.[13] He has suggested that in the absence of knowledge of crystal structure (and hence of the appropriate Madelung constant) a reasonable estimation of the lattice energy can be obtained from the equation:

$$U = \frac{120\,200 v Z^+ Z^-}{r_0}\left(1 - \frac{34.5}{r_0}\right) \tag{3.17}$$

where v is the number of ions per "molecule" of the compound and r_0 is estimated as the sum of the ionic radii (Table 3.4), $r_+ + r_-$ (pm). For the sodium chloride example given previously, $v = 2$ and $r_0 = 281$ pm, yielding a lattice energy of 753 kJ mol^{-1}, or about 98% of the experimental value, comparing favorably with that obtained from Eq. 3.13.

[12] A. F. Kapustinskii, *Z. Physik. Chem. (Leipzig)*, **1933**, *B22*, 257; *Zhur. Fiz. Khim.*, **1943**, *5*, 59; *Quart. Rev. Chem. Soc.*, **1956**, *10*, 283.

[13] This follows from the fact that given a certain number of ions of certain sizes, the number of ways of packing them efficiently is severely limited. Simple cases of this are discussed in the sections entitled "Efficiency of Packing and Crystal Lattices" and "Radius Ratio." For a more thorough discussion of Kapustinskii's work, see T. C. Waddington, *Adv. Inorg. Chem. Radiochem.*, **1959**, *1*, 157.

Once we have convinced ourselves that we are justified in using theoretical values for U, we can use the cycle to help obtain information on any other step in the cycle which is experimentally difficult to measure. For many years electron affinities were obtained almost exclusively by this method since accurate estimates were difficult to obtain by direct experiment.

Finally, it is possible to predict the heat of formation of a new and previously unknown compound. Reasonably good estimates of enthalpies of atomization, ionization energies, and electron affinities are now available for most elements. It is then necessary to make some good guesses as to the most probable lattice structure, including internuclear distances and geometry. The internuclear distance can be estimated with the aid of tables of *ionic radii.* Sometimes it is also possible to predict the geometry (in order to know the correct Madelung constant) from a knowledge of these radii (see next section). In such a case it is possible to predict the lattice energy and the enthalpy of formation (the latter almost as accurately as it could be measured if the compound were available). Examples of calculations on hypothetical compounds are given below, and a final example utilizing several methods associated with ionic compounds is given on p. 88.

Consideration of the terms in a Born-Haber cycle helps rationalize the existence of certain compounds and the nonexistence of others. For example, consider the hypothetical sodium dichloride, Na^{+2}, $2Cl^-$. Because of the $+2$ charge on the sodium ion, we might expect the lattice energy to be considerably larger than that of NaCl, adding to the stability of the compound. But if all the terms are evaluated, it is found that the increased energy necessary to ionize sodium to Na^{+2} is more than that which is returned by the increased lattice energy. We can make a very rough calculation assuming that the internuclear distance in $NaCl_2$ is the same as[14] in NaCl and that it would crystallize in the fluorite structure with a Madelung constant of $A = 2.52$. The lattice energy is then $U_0 = -2155$ kJ mol^{-1}. The summation of Born-Haber terms is:

$$
\begin{aligned}
U_0 &= -2155 \\
\Delta H_{A_{Na}} &= \ \ +109 \\
IE_1 &= \ \ +494 \\
IE_2 &= +4561 \\
2EA &= \ \ -699 \\
\Delta H_{A_{Cl}} &= \ \ +247 \\
\hline
\Delta H_f &= +2557 \text{ kJ mol}^{-1}
\end{aligned}
$$

Although the estimation of U_0 by our crude approximation may be off by 10–20%, it cannot be in error by over 100%, or 2500 kJ mol^{-1}. Hence we can see why $NaCl_2$ does not exist: *The extra stabilization of the lattice is insufficient to compensate for the very large second ionization energy.*

A slightly different problem arises when we consider the *lower* oxidation states of metals. We know that CaF_2 is stable. Why not CaF as well? Assuming that CaF would crystallize in the same geometry as KF and that the internuclear distance would be about the same, we can calculate a lattice energy for CaF, $U_0 = -795$ kJ mol^{-1}. The terms in

[14] Later we shall see that this overestimates the distance, but for the present approximation it should be adequate.

the Born-Haber cycle are:

$$\begin{aligned} U_0 &= -795 \\ \Delta H_{A_{Ca}} &= +201 \\ \text{IE} &= +590 \\ \text{EA} &= -335 \\ \Delta H_{A_F} &= +79 \\ \hline \Delta H_f &= -260 \text{ kJ mol}^{-1} \end{aligned}$$

An enthalpy of formation of -260 kJ mol^{-1}, though not large, is perfectly acceptable since it is about the same as that of LiI, for example. Why then does CaF not exist? Because if one *were* able to prepare it, it would spontaneously disproportionate into CaF_2 and Ca exothermically:[15]

$$\begin{array}{llll} 2\text{CaF} \longrightarrow & \text{CaF}_2 & + \text{Ca} & \\ 2\,\Delta H_f = -520 & \Delta H_f = -1243 & \Delta H_f = 0 & \Delta H_r = -723 \text{ kJ mol}^{-1} \end{array} \quad \textbf{(3.18)}$$

An examination of the ionic compounds of the main group elements would show that all of the ions present have electronic configurations that are isoelectronic with noble gases; hence the supposed "stability of noble gas configurations." But what type of stability? It is true that the halogens are from 295 to 350 kJ mol^{-1} lower in energy as halide ions than as free atoms. But the formation of the O^{-2}, S^{-2}, N^{-3}, Li^+, Na^+, Mg^{+2}, and Ca^{+2} ions is *endothermic* by 400 to 2300 kJ mol^{-1}. Even though these ions possess noble gas configurations, they represent *higher* energy states than the free atoms. The "stability" of noble gas configurations is meaningless unless one considers the stabilization of the ionic lattice. For the main group elements the noble gas configuration is that which maximizes the gain from high charges (and large lattice energies) while holding the cost (in terms of ionization potential–electron affinity energies) as low as possible. This is shown graphically in Fig. 3.8. Although the second ionization energy for a metal is always larger than the first, and the third larger than the second, the increase is moderate except when a noble gas configuration is broken. Then the ionization energy increases markedly because the electron is being removed from the $n-1$ shell. Below this limit the lattice energy increases faster with oxidation state than does the ionization energy, so that the most stable oxidation state is the one that maximizes the charge without breaking the noble gas configuration. This is why aluminum always exists as Al^{+3} when in ionic crystals despite the fact that it costs 5140 kJ mol^{-1} to remove three electrons from the atom!

For transition metals all electrons lost on ionization are either ns or $(n-1)d$ electrons, which, as we have seen, are very similar in energy. Hence there are no abrupt increases in ionization energy and often several oxidation states can be stable.

In summary, the Born-Haber cycle provides interesting insights into the energy factors operating in ionic compounds. Furthermore, it is an excellent example of the

[15] The direction of chemical reaction will be determined by the *free energy*, ΔG, not the enthalpy, ΔH. However, in the present reaction the *entropy* term, ΔS, is apt to be comparatively small and since $\Delta G = \Delta H - T\,\Delta S$, the free energy will be dominated by the enthalpy at moderate temperatures.

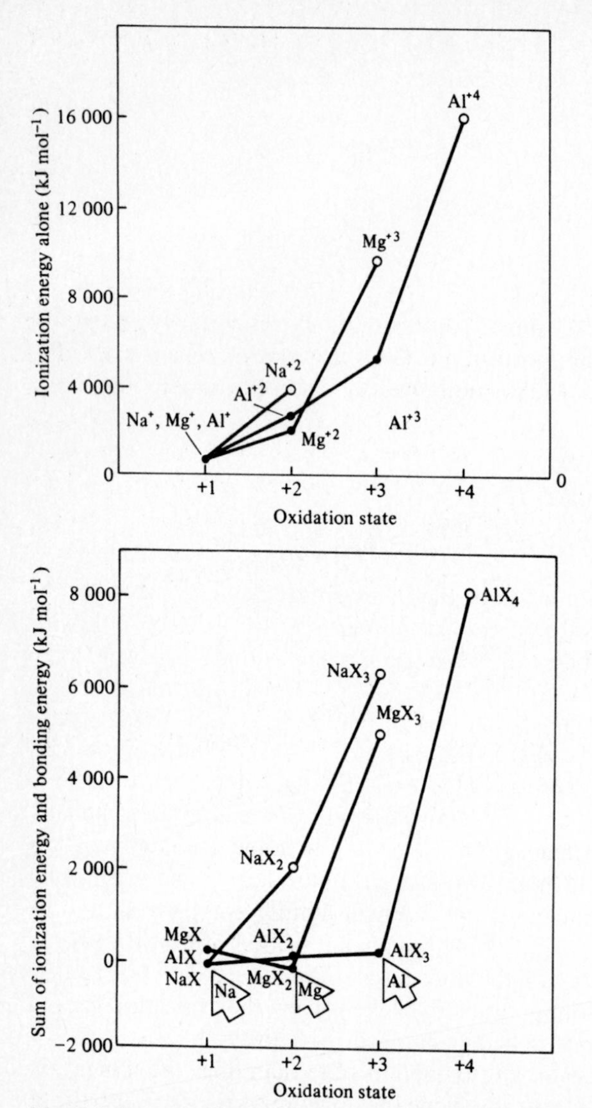

Fig. 3.8 Energy of free cations and ionic compounds as a function of oxidation state. *Top*: Line represents the ionization energy necessary to form the +1, +2, +3, and +4 cations of sodium, magnesium, and aluminum. Note that although the ionization energy increases most sharply when a noble gas configuration is "broken," *isolated ions are always less stable in higher oxidation states. Bottom*: Line represents the sum of ionization energy and ionic bonding energy for hypothetical molecules MX, MX_2, MX_3, and MX_4 in which the interionic distance, r_0, has been arbitrarily set at 200 pm. Note that the most stable compounds (identified by arrows) are NaX, MgX_2, and AlX_3. (All of these molecules will be stabilized additionally by the electron affinity of X.)

application of thermodynamic methods to inorganic chemistry and serves as a model for other, similar calculations.

SIZE EFFECTS

Ionic radii

From the success of our simple electrostatic model in accounting for the energy of an ionic compound we may feel justified in assuming the existence of ions and attempt to determine their size. We would be fortunate indeed if there were some way we could take a "picture" of an ionic lattice and find something resembling the drawing in Fig. 3.9. This is not possible, though we can come close. For many years the nemesis of the inorganic chemist was the impossibility of measuring more than the *total distance* between the centers of ions, not the relative *apportionment* between one ion and the other. In other words, although it was possible to measure the distance r_0 with a high degree of accuracy, experimentally we knew only that $r_0 = r_+ + r_-$ and could not tell where each cation ends and the anion begins. In fact, since there is some covalent bonding and overlap of electron clouds in the crystal, it was even questioned whether it was proper to ask that question. Nevertheless, it is convenient to think of crystals in mental pictures similar to Fig. 3.9, and so there have been many ingenious attempts to calculate the answer to this question. These methods are extremely interesting for their inventiveness and insight into atomic properties, but because they have been superseded by direct experimental measurements, they are relegated to Appendix I. Although they provide keen and interesting insights into the properties of ionic lattices, they are no longer necessary to the understanding of ionic radii.

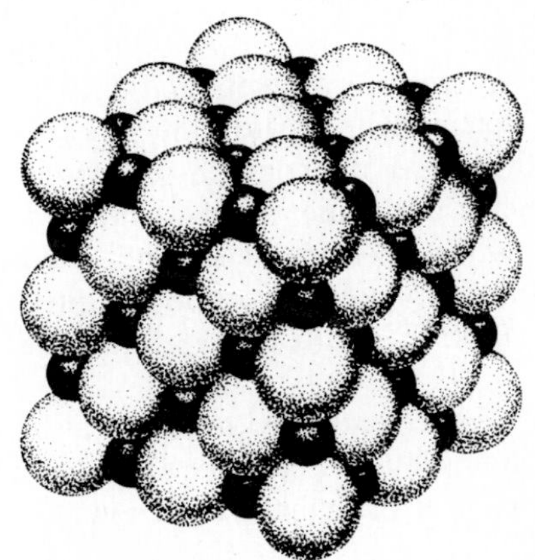

Fig. 3.9 Sodium chloride lattice showing packing of hard spheres representing discrete ions. This structure was suggested by W. Barlow (1898) prior to the development of methods of determining crystal structures. [Reprinted from Linus Pauling, "The Nature of the Chemical Bond." Copyright **1939** and **1940** by Cornell University. Third edition © **1960** by Cornell University. Used by permission of Cornell University Press.]

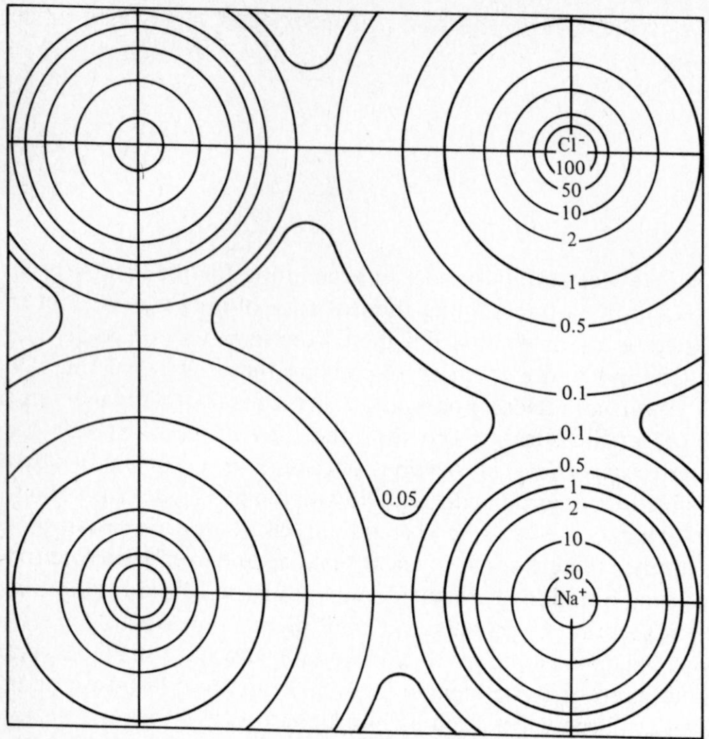

Fig. 3.10 Electron density contours in sodium chloride. Numbers indicate the electron density (electrons/10^{-30} m^3) along each contour line. The "boundary" of each ion is defined as the minimum in electron density between the ions. The internuclear distance is 281 pm. [Modified from G. Schoknecht, *Z. Naturforsch.*, **1957**, 12*A*, 983. Reproduced with permission.]

When an X-ray crystallographer determines the structure of a compound such as NaCl (Fig. 3.1a), usually only the *spacing* of ions is determined, because the repeated spacings of the atoms diffract the X rays as the grooves on a phonograph record diffract visible light. However, if very careful measurements are made, accurate maps of electron density can be constructed since, after all, it is the electrons of the individual atoms that scatter the X rays. The result is Fig. 3.10. One may now apportion the interatomic distance in NaCl, 281 pm, using the minimum in electron density as the operational definition of "where one ion stops and the other starts."

Although not many simple ionic compounds have been studied with the requisite accuracy to provide data on ionic radii, there are enough to provide a basis for a complete set of ionic radii. Such a set has been provided in the crystal radii of Shannon and Prewitt.[16] Values of these radii are given in Table 3.4.

[16] R. D. Shannon and C. T. Prewitt, *Acta Crystallogr.*, **1969**, *B25*, 925; R. D. Shannon, ibid., **1976**, *A32*, 751. Most inorganic chemistry books, including the first edition of the present one, have given some set of "traditional" ionic radii. The Shannon and Prewitt *crystal radii* given in Table 3.4 are 14 pm larger for cations and 14 pm smaller for anions than the "best" set of traditional ionic radii.

Table 3.4 Effective ionic radii of the elements[a]

Ion	Coordination number	pm
Ac^{+3}	6	126
Ag^{+1}	2	81
	4	114
	4 SQ	116
	5	123
	6	129
	7	136
	8	142
Ag^{+2}	4 SQ	93
	6	108
Ag^{+3}	4 SQ	81
	6	89
Al^{+3}	4	53
	5	62
	6	67.5
Am^{+2}	7	135
	8	140
	9	145
Am^{+3}	6	111.5
	8	123
Am^{+4}	6	99
	8	109
As^{+3}	6	72
As^{+5}	4	47.5
	6	60
At^{+7}	6	76
Au^{+1}	6	151
Au^{+3}	4 SQ	82
	6	99
Au^{+5}	6	71
B^{+3}	3	15
	4	25
	6	41
Ba^{+2}	6	149
	7	152
	8	156
	9	161
	10	166
	11	171
	12	175
Be^{+2}	3	30
	4	41
	6	59
Bi^{+3}	5	110
	6	117
	8	131
Bi^{+5}	6	90
Bk^{+3}	6	110
Bk^{+4}	6	97
	8	107
Br^{-1}	6	182
Br^{+3}	4 SQ	73
Br^{+7}	4	39
	6	53
C^{+4}	3	6
	4	29
	6	30
Ca^{+2}	6	114
	7	120
	8	126
	9	132
	10	137
	12	148
Cd^{+2}	4	92
	5	101
	6	109
	7	117
	8	124
	12	145
Ce^{+3}	6	115
	7	121
	8	128.3
	9	133.6
	10	139
	12	148
Ce^{+4}	6	101
	8	111
	10	121
	12	128
Cf^{+3}	6	109
Cf^{+4}	6	96.1
	8	106
Cl^{-1}	6	167
Cl^{+5}	3 PY	26
Cl^{+7}	4	22
	6	41
Cm^{+3}	6	111
Cm^{+4}	6	99
	8	109
Co^{+2}	4 HS[b]	72
	5	81
	6 LS[c]	79
	HS	88.5
	8	104
Co^{+3}	6 LS	68.5
	HS	75
Co^{+4}	4	54
	6 HS	67
Cr^{+2}	6 LS	87
	HS	94
Cr^{+3}	6	75.5
Cr^{+4}	4	55
	6	69
Cr^{+5}	4	48.5
	8	71
Cr^{+6}	4	40
	6	58
Cs^{+1}	6	181
	8	188
	9	192
	10	195
	11	199
	12	202
Cu^{+1}	2	60
	4	74
	6	91
Cu^{+2}	4	71
	4 SQ	71
	5	79
	6	87
Cu^{+3}	6 LS	68
D^{+1}	2	4
Dy^{+2}	6	121
	7	127
	8	133
Dy^{+3}	6	105.2
	7	111
	8	116.7
	9	122.3
Er^{+3}	6	103
	7	108.5
	8	114.4
	9	120.2
Eu^{+2}	6	131
	7	134
	8	139
	9	144
	10	149
Eu^{+3}	6	108.7
	7	115
	8	120.6
	9	126
F^{-1}	2	114.5
	3	116
	4	117
	6	119
F^{+7}	6	22
Fe^{+2}	4 HS	77
	4 SQ HS	78
	6 LS	75
	HS	92
	8 HS	106
Fe^{+3}	4 HS	63
	5	72
	6 LS	69
	HS	78.5

Table 3.4 (*Continued*)

Ion	Coordination number	pm
Br^{+5}	3 PY	45
Fe^{+4}	6	72.5
Fe^{+6}	4	39
Fr^{+1}	6	194
Ga^{+3}	4	61
	5	69
	6	76
Gd^{+3}	6	107.8
	7	114
	8	119.3
	9	124.7
Ge^{+2}	6	87
Ge^{+4}	4	53
	6	67
H^{+1}	1	−24
	2	−4
Hf^{+4}	4	72
	6	85
	7	90
	8	97
Hg^{+1}	3	111
	6	133
Hg^{+2}	2	83
	4	110
	6	116
	8	128
Ho^{+3}	6	104.1
	8	115.5
	9	121.2
	10	126
I^{-1}	6	206
I^{+5}	3 PY	58
	6	109
I^{+7}	4	56
	6	67
In^{+3}	4	76
	6	94
	8	106
Ir^{+3}	6	82
Ir^{+4}	6	76.5
Ir^{+5}	6	71
K^{+1}	4	151
	6	152
	7	160
	8	165
	9	169
	10	173
	12	178
La^{+3}	6	117.2
	7	124
	8	130
	9	135.6
	6	63
Li^{+1}	4	73
	6	90
	8	106
Lu^{+3}	6	100.1
	8	111.7
	9	117.2
Mg^{+2}	4	71
	5	80
	6	86
	8	103
Mn^{+2}	4 HS	80
	5 HS	89
	6 LS	81
	HS	97
	7 HS	104
	8	110
Mn^{+3}	5	72
	6 LS	72
	HS	78.5
Mn^{+4}	4	53
	6	67
Mn^{+5}	4	47
Mn^{+6}	4	39.5
Mn^{+7}	4	39
	6	60
Mo^{+3}	6	83
Mo^{+4}	6	79
Mo^{+5}	4	60
	6	75
Mo^{+6}	4	55
	5	64
	6	73
	7	87
N^{-3}	4	132
N^{+3}	6	30
N^{+5}	3	4.4
	6	27
Na^{+1}	4	113
	5	114
	6	116
	7	126
	8	132
	9	138
	12	153
Nb^{+3}	6	86
Nb^{+4}	6	82
	8	93
Nb^{+5}	4	62
	6	78
	7	83
	8	88
	8 HS	92
Nd^{+3}	6	112.3
	8	124.9
	9	130.3
	12	141
Ni^{+2}	4	69
	4 SQ	63
	5	77
	6	83
Ni^{+3}	6 LS	70
	HS	74
Ni^{+4}	6 LS	62
No^{+2}	6	124
Np^{+2}	6	124
Np^{+3}	6	115
Np^{+4}	6	101
	8	112
Np^{+5}	6	89
Np^{+6}	6	86
Np^{+7}	6	85
O^{-2}	2	121
	3	122
	4	124
	6	126
	8	128
OH^{-1}	2	118
	3	120
	4	121
	6	123
Os^{+4}	6	77
Os^{+5}	6	71.5
Os^{+6}	5	63
	6	68.5
Os^{+7}	6	66.5
Os^{+8}	4	53
P^{+3}	6	58
P^{+5}	4	31
	5	43
	6	52
Pa^{+3}	6	118
Pa^{+4}	6	104
	8	115
Pa^{+5}	6	92
	8	105
	9	109
Pb^{+2}	4 PY	112
	6	133
	7	137
	8	143
	9	149
	10	154
	11	159

Table 3.4 *(Continued)*

Ion	Coordination number	pm
	10	141
	12	150
	5	87
	6	91.5
	8	108
Pd^{+1}	2	73
Pd^{+2}	4 SQ	78
	6	100
Pd^{+3}	6	90
Pd^{+4}	6	75.5
Pm^{+3}	6	111
	8	123.3
	9	128.4
Po^{+4}	6	108
	8	122
Po^{+6}	6	81
Pr^{+3}	6	113
	8	126.6
	9	131.9
Pr^{+4}	6	99
	8	110
Pt^{+2}	4 SQ	74
	6	94
Pt^{+4}	6	76.5
Pt^{+5}	6	71
Pu^{+3}	6	114
Pu^{+4}	6	100
	8	110
Pu^{+5}	6	88
Pu^{+6}	6	85
Ra^{+2}	8	162
	12	184
Rb^{+1}	6	166
	7	170
	8	175
	9	177
	10	180
	11	183
	12	186
	14	197
Re^{+4}	6	77
Re^{+5}	6	72
Re^{+6}	6	69
Re^{+7}	4	52
	6	67
Rh^{+3}	6	80.5
Rh^{+4}	6	74
Rh^{+5}	6	69
Ru^{+3}	6	82
Ru^{+4}	6	76
Ru^{+5}	6	70.5
Nd^{+2}	8	143
	9	149
S^{+6}	4	26
	6	43
Sb^{+3}	4 PY	90
	5	94
	6	90
Sb^{+5}	6	74
Sc^{+3}	6	88.5
	8	101
Se^{-2}	6	184
Se^{+4}	6	64
Se^{+6}	4	42
	6	56
Si^{+4}	4	40
	6	54
Sm^{+2}	7	136
	8	141
	9	146
Sm^{+3}	6	109.8
	7	116
	8	121.9
	9	127.2
	12	138
Sn^{+4}	4	69
	5	76
	6	83
	7	89
	8	95
Sr^{+2}	6	132
	7	135
	8	140
	9	145
	10	150
	12	158
Ta^{+3}	6	86
Ta^{+4}	6	82
Ta^{+5}	6	78
	7	83
	8	88
Tb^{+3}	6	106.3
	7	112
	8	118
	9	123.5
Tb^{+4}	6	90
	8	102
Tc^{+4}	6	78.5
Tc^{+5}	6	74
Tc^{+7}	4	51
	6	70
Te^{-2}	6	207
	12	163
Pb^{+4}	4	79
	6	70
Th^{+4}	6	108
	8	119
	9	123
	10	127
	11	132
	12	135
Ti^{+2}	6	100
Ti^{+3}	6	81
Ti^{+4}	4	56
	5	65
	6	74.5
	8	88
Tl^{+1}	6	164
	8	173
	12	184
Tl^{+3}	4	89
	6	102.5
	8	112
Tm^{+2}	6	117
	7	123
Tm^{+3}	6	102
	8	113.4
	9	119.2
U^{+3}	6	116.5
U^{+4}	6	103
	7	109
	8	114
	9	119
	12	131
U^{+5}	6	90
	7	98
U^{+6}	2	59
	4	66
	6	87
	7	95
	8	100
V^{+2}	6	93
V^{+3}	6	78
V^{+4}	5	67
	6	72
	8	86
V^{+5}	4	49.5
	5	60
	6	68
W^{+4}	6	80
W^{+5}	6	76
W^{+6}	4	56
	5	65

Table 3.4 (*Continued*)

Ion	Coordination number	pm	Ion	Coordination number	pm	Ion	Coordination number	pm
Ru^{+7}	4	52	Te^{+4}	3	66		6	74
Ru^{+8}	4	50		4	80	Xe^{+8}	4	54
S^{-2}	6	170		6	111		6	62
S^{+4}	6	51	Te^{+6}	4	57	Y^{+3}	6	104
	7	110		7	106.5	Zr^{+4}	4	73
	8	115.9		8	112.5		5	80
	9	121.5		9	118.2		6	86
Yb^{+2}	6	116	Zn^{+2}	4	74		7	92
	7	122		5	82		8	98
	8	128		6	88		9	103
Yb^{+3}	6	100.8		8	104			

[a] Values of crystal radii from R. D. Shannon, *Acta Crystallogr.*, **1976**, *A32*, 751.
[b] High spin.
[c] Low spin.

Factors affecting the radii of ions

A comparison of the values given in Table 3.4 allows one to make some conclusions regarding the various factors that affect ionic size. We have already seen that progressing to the right in a periodic series should cause a decrease in size. If the ionic charge remains constant, as in the +3 lanthanide cations, the decrease is smooth and moderate. Progressing across the main group metals, however, the ionic charge is increasing as well, which causes a precipitous drop in cationic radii: Na^+ (116 pm), Mg^{+2} (86 pm), Al^{+3} (67.5 pm). In the same way, for a given metal, increasing oxidation state causes a shrinkage in size, not only because the ion becomes smaller as it loses electron density, but also because the increasing cationic charge pulls the anions in closer. This change can be illustrated by comparing the bond lengths in the complex anions $FeCl_4^{-2}$ and $FeCl_4^-$. The Fe(III)—Cl bond length is 11 pm shorter than the Fe(II)—Cl bond length.[17]

For transition metals the multiplicity of the spin state affects the way in which the anions can approach the cation; this alters the effective radius. Although this is an important factor in determining cationic radii, it is beyond the scope of the present chapter and will be deferred to Chapter 9 (pp. 387–389).

For both cations and anions *the crystal radius increases with the increase in coordination number*. As the coordination number increases, the repulsions among the coordinating counterions become greater and cause them to "back off" a bit. Alternatively, one can view a *lower* coordination number as allowing the counterions to compress the central ion and reduce its crystal radius.

As we shall see over and over again, the simple picture of billiard-ball-like ions of invariant radius is easy to describe but generally unrealistic. The fluorides and oxides

[17] J. W. Lauher and J. A. Ibers, *Inorg. Chem.*, **1975**, *14*, 348.

come closest to this picture, and so the values in Table 3.4 work best with them. Larger, softer anions in general will present more problems. Little work has been done in this area, but Shannon[17a] has presented a table, analogous to Table 3.4, for sulfides.

Radii of polyatomic ions

The sizes of polyatomic ions such as NH_4^+ and SO_4^{-2} are of interest for the understanding of the properties of ionic compounds such as $(NH_4)_2SO_4$, but the experimental difficulties attending their determination exceed those of simple ions. In addition, the problem of constancy of size from one compound to the next—always a problem even in simple ions—often becomes much worse. For example one set of data indicates that the radius of the ammonium ion is consistently 175 pm, but a different set indicates that it is the same size as Rb^+, 166 ppm.[18] This is not a serious discrepancy, but it is a disturbing one since its source is not obvious.

Yatsimirskii[19] has provided an ingenious method for estimating the radii of polyatomic ions. A Born-Haber calculation utilizing the enthalpy of formation and related data can provide an estimate of the lattice energy. It is then possible to find what value of the radius of the ion in question is consistent with this lattice energy. These values are thus termed *thermochemical radii.* The most recent set of such values is given in Table 3.5. In many cases the fact that the ions (such as CO_3^{-2}, CNS^-, CH_3COO^-) are markedly nonspherical limits the use of these radii. Obviously they can be reinserted into further thermochemical calculations and thus provide such data as the anticipated lattice energy of a new (sometimes hypothetical) compound.

Occasionally some unexpected trends are observed. The radii of the halide ions follow the expected trend: Cl^- (167 pm), Br^- (182 pm), I^- (206 pm). However, the oxyanions show a reverse trend: ClO_3^- (157 pm), BrO_3^- (140 pm), IO_3^- (108 pm). The absolute values for I^- versus IO_3^-, for example, are not strictly comparable because they were obtained by different methods, but the *trend* in the halate ions is probably significant and worth considering (see Chapter 17).

In the case of tetrahedral ions the symmetry is sufficiently high that the ions may be considered pseudospherical and the comparisons are more meaningful. For example:

$$BeF_4^{-2} > BF_4^-;\ CrO_4^{-2} \sim MnO_4^- > ClO_4^-$$

Efficiency of packing and crystal lattices

If we consider atoms and ions to be hard spheres, we find that there are certain geometric arrangements for packing them which are more efficient than others. This can be confirmed readily in two dimensions with a handful of coins. For example, if a set of coins of the same

[17a] R. D. Shannon, "Structure and Bonding in Crystals," Vol. II, M. O'Keeffe and A. Navrotsky, Academic, New York, **1981**, Chapter 16.

[18] R. D. Shannon, *Acta Crystallogr.*, **1976**, *A32*, 751.

[19] K. B. Yatsimirskii, *Izvest. Akad. Nauk SSSR, Otdel, Khim. Nauk*, **1947**, 453; **1948**, 398.

Table 3.5 Thermochemical radii of polyatomic ions[a]

Ion	pm	Ion	pm
Cations		*Anions*	
NH_4^+	151	$MnCl_6^{-2}$	308
Me_4N^+	215	MnF_6^{-6}	242
PH_4^+	171	MnO_4^-	215
		N_3^-	181
Anions		NCO^-	189
		$H_2CH_2CO_2^-$	176
$AlCl_4^-$	281	NO_2^-	178
BCl_4^-	296	NO_3^-	165
BF_4^-	218	O_2^-	144
BH_4^-	179	O_2^{-2}	159
BrO_3^-	140	OH^-	119
CH_3COO^-	148	$PbCl_6^{-2}$	334
ClO_3^-	157	$PdCl_6^{-2}$	305
ClO_4^-	226	$PtBr_6^{-2}$	328
CN^-	177	$PtCl_4^{-2}$	279
CNS^-	199	$PtCl_6^{-2}$	299
CO_3^{-2}	164	PtF_6^{-2}	282
$CoCl_4^-$	305	PtI_6^{-2}	328
CoF_6^{-2}	230	$SbCl_6^-$	337
CrF_6^{-2}	238	SeO_3^{-2}	225
CrO_4^{-2}	242	SeO_4^{-2}	235
$CuCl_4^{-2}$	307	SiF_6^{-2}	245
$FeCl_4^-$	344	$SnBr_6^{-2}$	349
$GaCl_4^-$	275	$SnCl_6^{-2}$	335
$GeCl_6^-$	314	SnI_6^{-2}	382
GeF_6^-	252	SO_4^{-2}	244
HCl_2^-	187	$TiBr_6^{-2}$	338
HCO_2^-	155	$TiCl_6^{-2}$	317
HCO_3^{-2}	142	TiF_6^{-2}	275
HF_2^-	158	VO_3^-	168
HS^-	193	VO_4^{-3}	246
HSe^-	191	$ZnBr_4^{-2}$	285
IO_3^-	108	$ZnCl_4^{-2}$	272
$IO_2F_2^-$	163	ZnI_4^{-2}	309
$IrCl_6^{-2}$	221		

[a] Data from H. D. B. Jenkins and K. P. Thakur, *J. Chem. Educ.*, **1979**, *56*, 576, adjusted to be compatible with Shannon-Prewitt crystal radii. Used with permission.

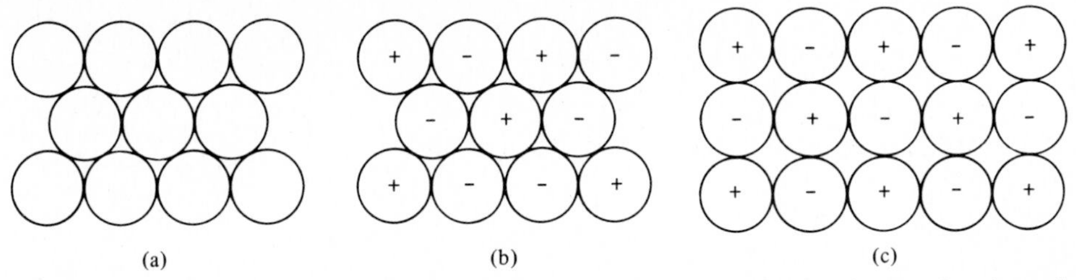

Fig. 3.11 Two-dimensional lattices. (a) Stable, 6-coordinate, closest-packed lattice of uncharged atoms; (b) unstable, 6-coordinate lattice of charged ions; (c) stable, 4-coordinate lattice of charged ions.

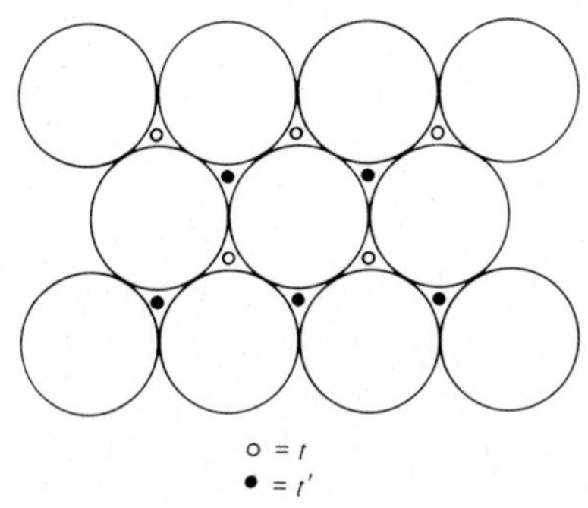

Fig. 3.12 Sites created by layer 1 and available to accept atoms in layer 2.

size (dimes, for example) is arranged, it will be found that six of them fit perfectly around another (i.e., touching each other and the central dime), giving a coordination number of 6. However, only five quarters or four silver dollars will fit around a dime,[20] illustrating the importance of size in determining the optimum coordination number. The effect of charge can also be illustrated. If all of the atoms are the same, the most efficient two-dimensional lattice is the closest-packed, 6-coordinate arrangement. If they are of the same size but opposite charge, the 6-coordinate structure is not stable since it will have too many repulsions of like-charge ions. This can also be readily shown with coins (using heads and tails to represent charge), and it can be seen that the most stable arrangement is a square lattice of alternating charge (Fig. 3.11).

The same principles hold for three-dimensional lattices. Consider first a lattice composed only of uncharged atoms as in a metal or a crystal of noble gas atoms. The first layer will consist of a two-dimensional, closest-packed layer (Fig. 3.11a). The second layer will be of the same type but centered over the "depressions" that exist where three atoms in the first layer come in contact (Fig. 3.12).[21] A layer containing n atoms will have $2n$ such sites capable of accepting atoms (marked t and t' in Fig. 3.12), but once an atom

[20] Similar observations can be made with various sized coins in other monetary systems, of course. For perspective, these diameters apply to U.S. coins: Dime = 17 mm; Quarter = 24 mm; Silver Dollar = 38 mm.

[21] See A. F. Wells, "Structural Inorganic Chemistry," 4th ed., Clarendon Press, Oxford, 1975.

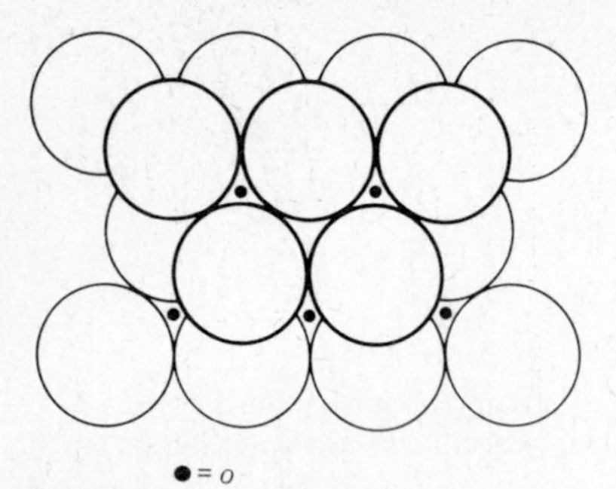

Fig. 3.13 Covering of all t sites by atoms in the second layer, making the t' sites (relabeled o) unavailable for occupancy by close-packed atoms.

has been placed in either of the two equivalent sets (t and t') the remainder of that layer must continue to utilize that type of site (Fig. 3.13), and the remaining n sites (labeled o) are not utilized by the packing atoms.

The third layer again has a choice of n sites out of a possible $2n$ available (t and t' types again). One alternative places the atoms of the third layer over those of the first; the other places the atoms of the third layer over the o sites of the first layer (Fig. 3.14). In the first type the layers alternate ABABAB and the lattice is known as the hexagonal closest-packed (*hcp*) system. Alternatively, the cubic closest-packed (*ccp*) system has three different layers, ABCABC. Both lattices provide a coordination number of 12 and are equally efficient at packing atoms into a volume.

It is easy to see the unit cell and the origin of the term *hexagonal closest packed.* In Fig. 3.14a the unit cell can be constructed by drawing a hexagon through the nuclei of the six outer atoms in layer A and a parallel hexagon in the next A layer above, and by then connecting the corresponding vertices of the hexagons with perpendicular lines to form a hexagonal prism (Fig. 3.15a).

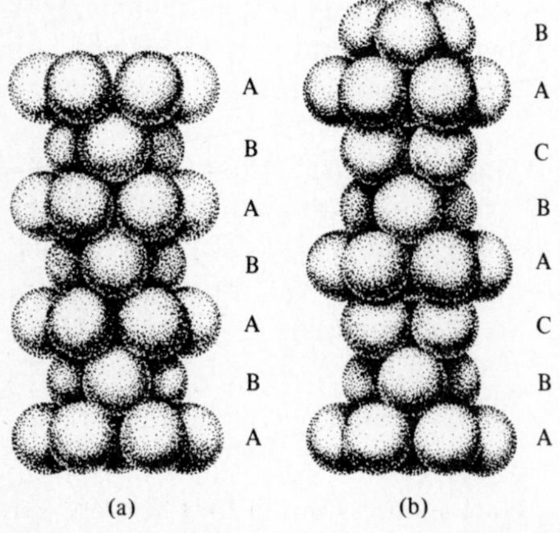

Fig. 3.14 Arrangement of layers in hexagonal closest-packed (a) and cubic closest-packed (b) structures. These are "side views" compared to the "top views" shown in the preceding figures.

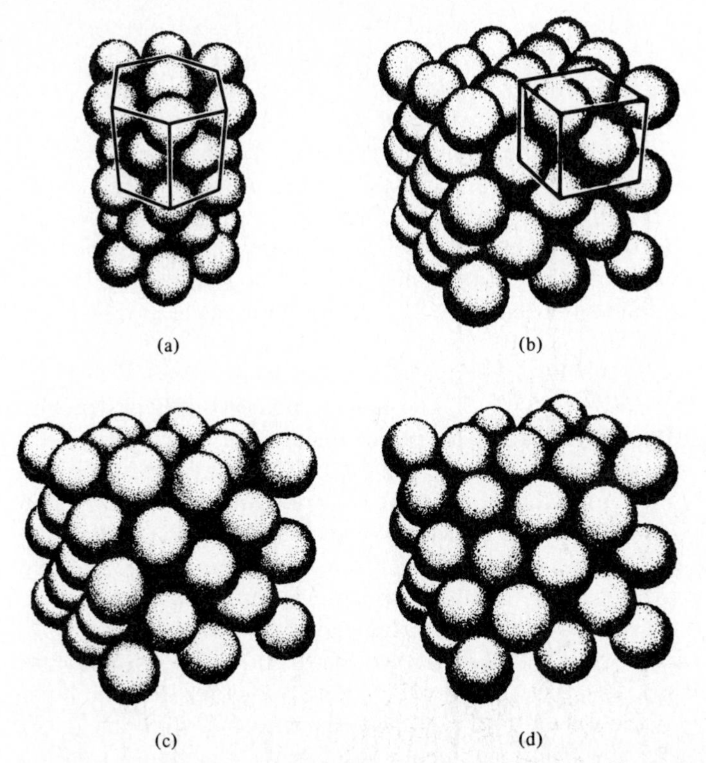

Fig. 3.15 Unit cells in closest-packed systems. (a) Unit cell of hexagonal closest-packed system. (b) Face-centered array of atoms. Unit cell outlined upper right. (c) One atom removed. This atom belongs to a "C" layer of the closest-packed system. (d) Six more atoms removed. They belong to a "B" layer. The exposed layer is the "A" layer. Note that the single atom that composed the C "layer" does not lie above any atom in the A layer (as it would if this were *hcp*). Note that "above and below," corresponding to a vertical line in Fig. 3.14b, here corresponds to a line running from lower right, rear to upper left, front (atom C), i.e., a diagonal of the cube. [From A. F. Wells, "Structural Inorganic Chemistry," 4th ed., Oxford University Press, Oxford, **1975**. Reproduced with permission.]

One could follow a similar practice and construct a similar hexagonal "sandwich" with two layers (B, C) of "filler," but a cubic cell of higher symmetry can be constructed; the second system is thus characterized as *cubic closest packed*. The relation between the cubic unit cell (which is identical to the face-centered cubic cell we have already seen) is not easy to visualize unless one is quite familiar with this system. The easiest way is to take a face-centered cubic array (Fig. 3.15b), and by removing an atom (Fig. 3.15c), then a few more (Fig. 3.15d), reveal the closest-packed layers corresponding to A, B, and C in Fig. 3.14b.

The noble gases and almost all metals crystallize in either the *hcp* or the *ccp* structure as would be expected for neutral atoms. The alkali metals, barium, and a few transition metals crystallize in the *body-centered cubic* system, though the reasons for this choice are unknown.

If all the packing atoms are no longer neutral (e.g., half are cations and half are anions), the closest-packed structures are no longer the most stable, as can be seen from the similar two-dimensional case (see above). However, these structures may still be useful when considered as limiting cases for certain ionic crystals. Consider lithium iodide, in which the iodide anions are so much larger than the lithium cations that they may be assumed to touch or nearly touch (see Appendix I). They can be considered to provide the framework for the crystal. The much smaller lithium ions can then fit into the small interstices between the anions. If they expand the lattice slightly to remove the anion–anion contact, the anionic repulsion will be reduced and the crystal stabilized, but the simple model based on a closest-packed system of anions may still be taken as the limiting case and a useful approximation.

Where the lithium ions fit best will be determined by their size relative to the iodide ions. Note from above that there are two types of interstices in a closest-packed structure. These represent tetrahedral (*t*) and octahedral (*o*) holes because the coordination of a small ion fitted into them is either tetrahedral or octahedral (see Figs. 3.12 and 3.13). The octahedral holes are considerably larger than the tetrahedral holes and can accommodate larger cations without severe distortion of the structure. In lithium iodide the lithium ions fit into the octahedral holes in a cubic closest-packed lattice of iodide ions. The resulting structure is the same as found in sodium chloride and is face-centered (note that face-centered cubic and cubic closest-packed describe the same lattice).

Consider a closest-packed lattice of sulfide ions. Zinc ions tend to occupy tetrahedral holes in such a framework since they are quite small (74 pm) compared with the larger sulfide ions (170 pm). If the sulfide ions form a *ccp* array, the resulting structure is zinc blende; if they form an *hcp* array, the resulting structure is wurtzite.[22] See Fig. 3.16.

Although in the present discussion size is the only parameter considered in determining the choice of octahedral versus tetrahedral sites, the presence of covalent bonding (d^2sp^3 versus sp^3 hybridization) and/or ligand field stabilization (see Chapter 9) can affect the stability of ions in particular sites. Size will usually be the determining factor when these additional factors are of small importance—for example, when considering alkali and alkaline earth ions. The concept of closest-packing of anions is also very useful in considering polar covalent macromolecules such as the silicates and iso- and heteropoly-anions.[23]

If the cations and anions are of approximately the same size, the limiting case of the framework being determined by the larger ion is inappropriate, and we simply determine the most efficient lattice for oppositely charged ions of equal size. This turns out to be the CsCl lattice, which maximizes cation–anion interaction (C.N. = 8) and is the most stable structure when the sizes of the cation and anion are comparable.

---

[22] For a more comprehensive discussion of the broad usefulness of classifying structures in terms of closest-packed structures, see S-M. Ho and B. E. Douglas, *J. Chem. Educ.*, **1968**, *45*, 474.

[23] A. F. Wells, "Structural Inorganic Chemistry," 4th ed., Clarendon Press, Oxford, **1975**; W. E. Addison, "Structural Principles in Inorganic Compounds," Wiley, New York, **1961**.

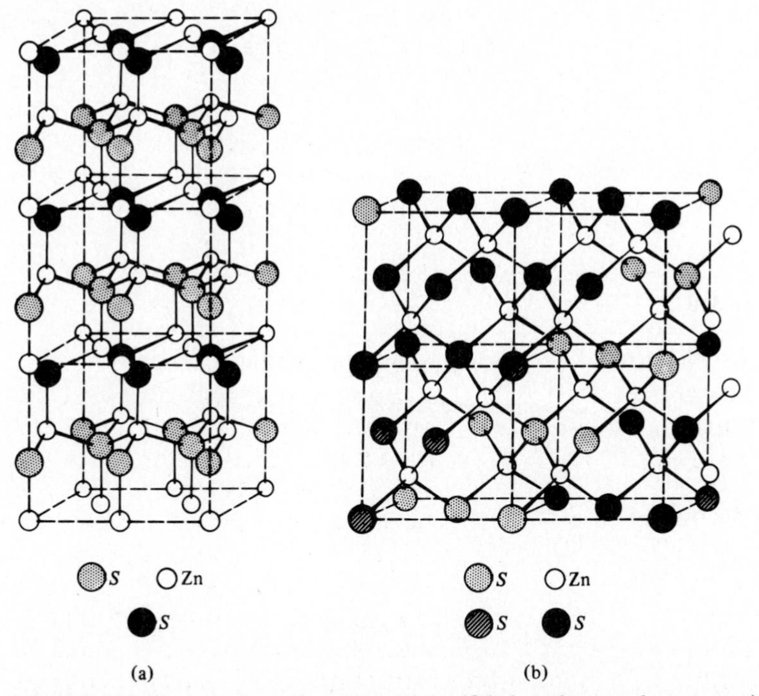

Fig. 3.16 (a) The structure of wurtzite. The sulfide ions form an *hcp* array with A (gray) and B (black) alternating layers. (Cf. Fig. 3.14a.) (b) The structure of zinc blende. The sulfide ions form a *ccp* array with A (white), B (black), and C (gray) layers. (Cf. Figs. 3.14b and 3.15c, d.) Note that in both structures the zinc atoms (*small* white circles) occupy tetrahedral holes.

Radius ratio

It is not difficult to calculate the size of the octahedral hole in a lattice of closest-packed anions. Figure 3.17 illustrates the geometric arrangement resulting from six anions in contact with each other and with a cation in the octahedral hole. Simple geometry allows us to fix the diagonal of the square as $2r_- + 2r_+$. The angle formed by the diagonal in the

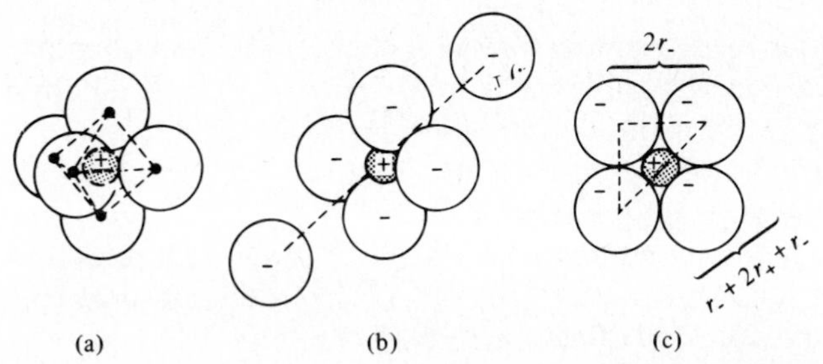

Fig. 3.17 (a) Small cation (dashed line) in octahedral hole formed by six anions. (b) Dissection of octahedron to illustrate geometric relationships shown in (c).

Table 3.6 Radius ratio and coordination number

Coordination number	Geometry	Limiting radius ratio[a]	Possible lattice structures
4	Tetrahedral		Wurtzite, zinc blende
		0.414; 2.42	
6	Octahedral		NaCl, rutile
		0.732; 1.37	
8	Cubic		CsCl, fluorite
		1.000	
12	Dodecahedral		[b]

[a] The second ratio is merely the reciprocal of the first. It is often convenient to have both values.

[b] Coordination number 12 is not found in simple ionic crystals. It occurs in complex metal oxides and in closest-packed lattices of atoms.

corner must be 45°, so we can say:

$$\frac{2r_-}{2r_- + 2r_+} = \cos 45^\circ = 0.707 \tag{3.19}$$

$$r_- = 0.707r_- + 0.707r_+ \tag{3.20}$$

$$0.293r_- = 0.707r_+ \tag{3.21}$$

$$\frac{r_+}{r_-} = \frac{0.293}{0.707} = 0.414 \tag{3.22}$$

This will be the limiting ratio since a cation will be stable in an octahedral hole only if it is at least large enough to keep the anions from touching, i.e., $r_+/r_- > 0.414$. Smaller cations will preferentially fit into tetrahedral holes in the lattice. By a similar geometric calculation it is possible to determine that the lower limit for tetrahedral coordination is $r_+/r_- = 0.225$. For radius ratios ranging from 0.225 to 0.414, tetrahedral sites will be preferred. Above 0.414, octahedral coordination is favored. By similar calculations it is possible to find the ratio when one cation can accommodate eight anions (0.732) or twelve anions (1.000). A partial list of limiting radius ratio values is given in Table 3.6.

The use of radius ratios to rationalize structures and to predict coordination numbers may be illustrated as follows.[24] Consider beryllium sulfide, in which $r_{Be^{+2}}/r_{S^{-2}} = 59$ pm/170 pm $= 0.35$. We should thus expect a coordination number of 4 as the Be^{+2} ion fits most readily into the *tetrahedral* holes of the closest-packed lattice, and indeed this is found experimentally: BeS adopts a wurtzite structure.

In the same way we can predict that sodium ions will prefer *octahedral* holes in a closest-packed lattice of chloride ions ($r_{Na^+}/r_{Cl^-} = 116$ pm/167 pm $= 0.69$), forming the well-known sodium chloride lattice with a coordination number of 6 (Fig. 3.1a).

[24] Since crystal radii vary slightly with coordination number, values from Table 3.4 were taken for C.N. = 6 as an "average" value.

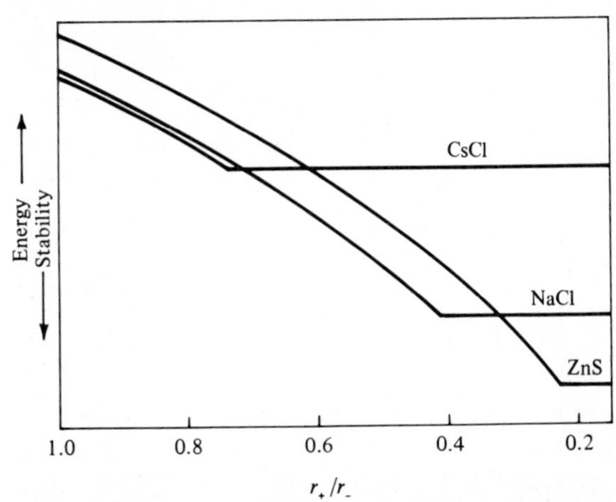

Fig. 3.18 The total energy of a cubic lattice of rigid anions and cations as a function of r_+ with r_- fixed, for different coordination configurations. When the anions come into mutual contact as a result of decreasing r_+ their repulsion determines the lattice constant and the cohesive energy becomes constant when expressed in terms of r_-. Thus near the values of r_+/r_- at which anion–anion contact takes place the radius ratio model predicts phase transitions to structures of successively lower coordination numbers. Note that the "breaks" in the curves correspond to the values listed in Table 3.6. [From "Treatise on Solid State Chemistry, Vol. I," N. B. Hannay, ed., Plenum, New York. Copyright © **1973**, Bell Telephone Laboratories, Inc. Reproduced with permission.]

With larger cations, such as cesium, the radius ratio (r_{Cs^+}/r_{Cl^-} = 181 pm/167 pm = 1.08) increases beyond the acceptable limit for a coordination number of 6; the coordination number of the cations (and anions) increases to 8, and the cesium chloride lattice (Fig. 3.1b) results. As we have seen, although this is an efficient structure for cations and anions of about the same size, it cannot be directly related to a closest-packed structure of anions.

Table 3.6 indicates that a coordination number of 12 should be possible when the radius ratio is 1.00. Geometrically it is possible to fit 12 atoms about a central atom (see the discussion of closest packing in metals, p. 76), but it is impossible to obtain mutual 12-coordination of cations and anions because of the limitations of geometry. Twelve-coordination does occur in complex crystal structures of mixed metal oxides in which one metal acts as one of the closest-packing atoms and others fit into octahedral holes, but a complete discussion of such structures is more appropriate in a book devoted to the structures of solids.[25]

The change in coordination number as a result of the ratio of ionic radii is shown graphically in Fig. 3.18. In general, as the cation decreases in size the lattice is stabilized (lattice energy becomes more negative) until anion–anion contact occurs. Further shrinkage of the lattice is impossible without a reduction in coordination number; therefore, zinc sulfide adopts the wurtzite or zinc blende structure, gaining additional energy over what would be possible in a structure with a higher coordination number. Note that although there is a significant difference between coordination numbers 4 and 6, there is little difference between 6 and 8. We might therefore expect small factors to tip the balance for or against 6- and 8-coordination and the radius ratio rules to be obeyed inexactly.

---

[25] See A. F. Wells, "Structural Inorganic Chemistry," 4th ed., Clarendon Press, Oxford, **1975**, pp. 480–589.

In a 1:1 or 2:2 salt the appropriate radius ratio is obviously the ratio of the smaller ion (usually the cation) to the larger to determine how many of the latter will fit around the smaller ion. In compounds containing different numbers of cations and anions (e.g., SrF_2, TiO_2, Li_2O, Rb_2S) it may not be immediately obvious how to apply the ratio. In such cases it is usually best to perform two calculations. For example, consider SrF_2:

$$\frac{r_{Sr^{+2}}}{r_{F^-}} = \frac{132}{119} = 1.11 \qquad \text{maximum C.N. of } Sr^{+2} = 8$$

$$\frac{r_{F^-}}{r_{Sr^{+2}}} = \frac{119}{132} = 0.90 \qquad \text{maximum C.N. of } F^- = 8$$

Now there must be twice as many fluoride ions as strontium ions, so the coordination number of the strontium ion must be twice as large as that of fluoride. Coordination numbers of 8 (Sr^{+2}) and 4 (F^-) are compatible with the maximum allowable coordination numbers and with the stoichiometry of the crystal. Strontium fluoride crystallizes in the fluorite lattice (Fig. 3.3).

A second example is SnO_2:

$$\frac{r_{Sn^{+4}}}{r_{O^{-2}}} = \frac{83}{126} = 0.66 \qquad \text{maximum C.N. of } Sn^{+4} = 6$$

$$\frac{r_{O^{-2}}}{r_{Sn^{+4}}} = \frac{126}{83} = 1.52 \qquad \text{maximum C.N. of } O^{-2} = 6$$

Considering the stoichiometry of the salt, the only feasible arrangement is with $\text{C.N.}_{O^{-2}} = 3$, $\text{C.N.}_{Sn^{+4}} = 6$; tin dioxide assumes the TiO_2 or rutile structure of Fig. 3.4. Note that the radius ratio would allow three more tin(IV) ions in the coordination sphere of the oxide ion, but the stoichiometry forbids it.

One final example is K_2O:

$$\frac{r_{K^+}}{r_{O^{-2}}} = \frac{152}{126} = 1.21 \qquad \text{maximum C.N. of } K^+ = 8$$

$$\frac{r_{O^{-2}}}{r_{K^+}} = \frac{126}{152} = 0.83 \qquad \text{maximum C.N. of } O^{-2} = 8$$

Considering the stoichiometry of the salt, the structure must be antifluorite (Fig. 3.3, reversed) with $\text{C.N.}_{O^{-2}} = 8$, $\text{C.N.}_{K^+} = 4$.

The radius ratio quite often predicts the correct coordination numbers of ions in crystal lattices. It must be used with caution, however, when covalent bonding becomes important. The reader may have been puzzled as to why beryllium sulfide was chosen to illustrate the radius ratio rule for coordination number 4 (p. 84) instead of zinc sulfide, which was used repeatedly earlier in this chapter to illustrate 4-coordinate structures such as wurtzite and zinc blende. The reason is simple. If ZnS had been used, it would have caused more confusion than enlightenment: It violates the radius ratio rule! Proceeding as above, we have $r_+/r_- = 88 \text{ pm}/170 \text{ pm} = 0.52$, indicating a coordination number of 6, yet both forms of ZnS, wurtzite and zinc blende, have a C.N. of 4, both cations and anions. If one argues that 0.52 does not differ greatly from 0.41, the point is well taken, but there exist more vexing cases. The radius ratio for mercury(II) sulfide, HgS, is 0.68, yet it crystallizes in the zinc blende structure. In both of these examples the sp^3-hybridized *covalent* bonding seems to be the dominant factor. Both ZnS and especially HgS are better regarded as infinite covalent lattices (see p. 195) than as ionic lattices.

It should be kept clearly in mind that the radius ratio rules apply strictly only to the packing of hard spheres of known size. As this is seldom the case, it is surprising that the rules work as well as they do. Anions are not "hard" like billiard balls, but polarizable under the influence of cations. To whatever extent such polarization or covalency occurs, errors are apt to result from application of the radius ratio rules. Covalent bonds are directed in space unlike electrostatic attractions, and so certain orientations are preferred.

There are, however, other exceptions that are difficult to attribute to directional covalent bonds. The heavier lithium halides only marginally obey the rule, and perhaps a case could be made for C.N. = 4 for LiI (Fig. 3.19). Much more serious, however, is the

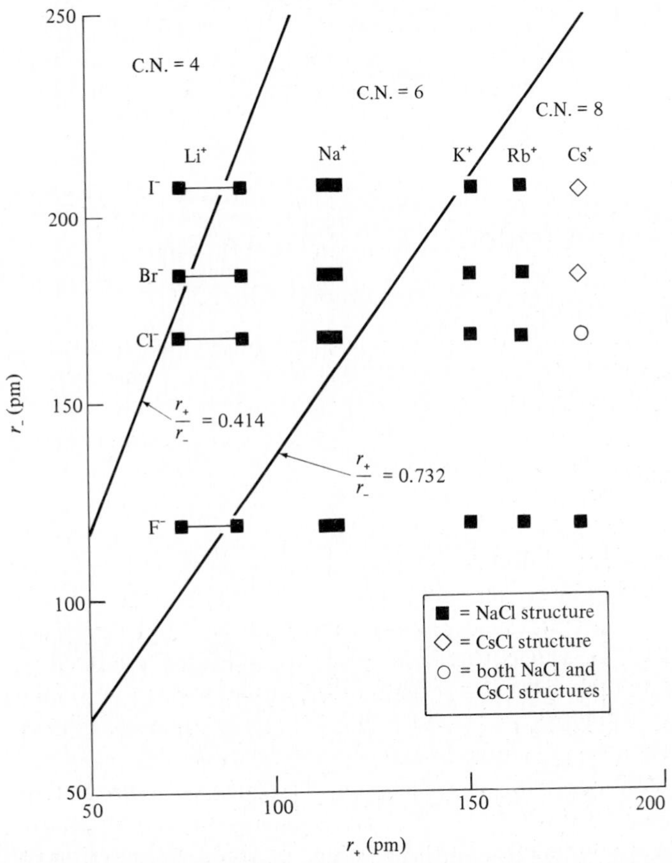

Fig. 3.19 Actual crystal structures of the alkali halides (as shown by the symbols) contrasted with the predictions of the radius ratio rule: The figure is divided into three regions by the lines $r_+/r_- = 0.414$ and $r_+/r_- = 0.732$, predicting coordination number 4 (wurtzite or zinc blende, upper left), coordination number 6 (rock salt, NaCl, middle), and coordination number 8 (CsCl, lower right). The crystal radius of lithium, and to a lesser extent that of sodium, changes with coordination number, so both the radii with C.N. = 4 (left) and C.N. = 6 (right) have been plotted.

problem of coordination number 6 versus 8. The relative lack of 8-coordinate structures—CsCl, CsBr, and CsI being the only known alkali metal examples—is commonly found, if hard to explain. There are no 8-coordinate oxides, MO, even though the larger divalent metal ions, such as Sr^{+2}, Ba^{+2}, and Pb^{+2}, are large enough that the radius ratio rule would predict the CsCl lattice. There is no simple explanation for these observations. Some solace can be obtained from the observation that the Madelung constant for C.N. = 8 (1.763) is hardly larger than the value for C.N. = 6 (1.748). This small difference means that the predicted lattice energies from the electrostatic model are less than 1% greater for the higher coordination number (see Fig. 3.18, where the lines for NaCl and CsCl almost coincide for $r_+/r_- > 0.732$). Thus, small energies coming from other sources could tip the balance.

What, then, are we to make of this problem? From a pragmatic point of view we can accept the radius ratio rule as a useful, if imperfect, tool in our arsenal for predicting and understanding the behavior of ionic compounds. From a theoretical point of view it rationalizes the choice of lattice for various ionic or partially ionic compounds. Its failings call our attention to forces in solids other than purely electrostatic ones acting on billiard-ball-like ions. We shall encounter modifications and improvements of the model in Chapter 4.

THE PREDICTIVE POWER OF THERMOCHEMICAL CALCULATIONS ON IONIC COMPOUNDS

The following example will illustrate the way in which the previously discussed parameters, such as ionic radii and ionization energies, can be used advantageously to explore the possible existence of an unknown compound. Suppose one were interested in dioxygenyl tetrafluoroborate, $[O_2]^+[BF_4]^-$. At first thought it might seem an unlikely candidate for existence since oxygen tends to gain electrons rather than lose them. However, the ionization energy of molecular oxygen is not excessively high (1164 kJ mol^{-1}; cf. Hg, 1009 kJ mol^{-1}), so some trial calculations might be made as follows.

The first values necessary are some estimates of the ionic radii of O_2^+ and BF_4^-. For the latter we may use the value obtained thermochemically by Yatsimirskii, 218 pm. An educated guess has to be made for O_2^+, since if we are attempting to make it for the first time (as was assumed above), we will not have any experimental data available for this species. However, we note that the CN^- ion, a diatomic ion which should be similar in size, has a thermochemical radius of 177 ppm. Furthermore, an estimate based on covalent and van der Waals radii (see Chapter 6) gives a similar value. Since O_2^+ has lost one electron and is positively charged, it will probably be somewhat smaller than this. We can thus take 177 pm as a conservative estimate, since if the cation is smaller than this, the compound will be more stable than our prediction and even more likely to exist. Adding the radii we obtain an estimate of 395 pm for the interionic distance.

Next the lattice energy can be calculated. One method would be to assume that we know nothing about the probable structure and use the Kapustinskii equation (Eq. 3.17) and $r_0 = 395$ pm. The resulting lattice energy is calculated to be -539 kJ mol^{-1}.

Alternatively, we might examine the radius ratio of $O_2^+BF_4^-$ and get a crude estimate of $177/218 = 0.8$. The accuracy of our values does not permit us to choose between coordination number 6 and 8, but since the value of the Madelung constant does not differ appreciably between the sodium chloride and cesium chloride structures, a value of 1.75 may be taken which will suffice for our present rough calculations.[26] We may then use the Born-Landé equation (Eq. 3.13), which provides an estimate of -590 kJ mol^{-1} for the attractive energy, which will be decreased by about 10% (if $n = 10$) to 20% (if $n = 5$). The two calculations thus agree that the lattice energy will probably be in the range -480 to -525 kJ mol^{-1}. This is a quite stable lattice and might be sufficient to stabilize the compound.

Next we might investigate the possible ways of producing the desired compound. Since the oxidation of oxygen is expected to be difficult to accomplish we might choose vigorous oxidizing conditions, such as the use of elemental fluorine:

$$O_2 + \tfrac{1}{2}F_2 + BF_3 \longrightarrow [O_2]^+[BF_4]^- \qquad \textbf{(3.23)}$$

It is possible to evaluate each term in a Born-Haber cycle based on Eq. 3.23.

The usual terms we have encountered in previous Born-Haber cycles may be evaluated readily:

Ionization energy of O_2 = 1164 kJ mol^{-1}

Dissociation of $\tfrac{1}{2}F_2$ = 79 kJ mol^{-1}

Electron affinity of F = -326 kJ mol^{-1}

One additional term occurs in this Born-Haber cycle, the formation of the tetrafluoroborate ion in the gas phase:

$$BF_{3(g)} + F^-_{(g)} \longrightarrow BF^-_{4(g)} \qquad \textbf{(3.24)}$$

Fortunately, the enthalpy of this reaction has been calculated to be -385 kJ mol^{-1}. Adding in a value of -500 ± 20 kJ mol^{-1} for the lattice energy provides an estimate of the heat of reaction (Eq. 3.23) of $+30$ kJ mol^{-1}. This is somewhat discouraging, since if Eq. 3.23 is endothermic, it would not be expected to proceed to form dioxygenyl tetrafluoroborate. Recall, however, that our estimates were on the conservative side and that 30 kJ mol^{-1} is well within the rather wide error limits on a calculation of this kind. We would therefore expect that dioxygenyl tetrafluoroborate is either energetically unfavorable or may form with a relatively low stability. It certainly is worth an attempt at synthesis.

In fact, dioxygenyl tetrafluoroborate *has* been synthesized by a reaction similar to Eq. 3.23, although in two steps: the formation of intermediate oxygen fluorides and then combination with boron trifluoride.[27] It is a white crystalline solid that slowly decomposes

[26] An exact value for the Madelung constant cannot be assigned in any event since there is a good possibility that the structure is distorted.

[27] I. J. Solomon et al., *Inorg. Chem.*, **1964**, *3*, 457; J. N. Keith et al., ibid., **1968**, *7*, 230; C. T. Goetschel, *J. Am. Chem. Soc.*, **1969**, *91*, 4702.

at room temperature. Energy calculations of this type are exceedingly useful in guiding research on the synthesis of new compounds. Usually it is not necessary to start with the complete absence of knowledge assumed in the present example. Often one or more factors can be evaluated from similar compounds. It was the observation of the formation of dioxygenyl hexafluoroplatinate(V) and similar calculations that led Bartlett to perform his first experiment in an attempt to synthesize compounds of xenon. This successful synthesis overturned prior chemical dogma (see Chapter 15).

Now that we have seen that dioxygenyl compounds can be prepared, we might be interested in preparing the exotic and intriguing compound dioxygenyl superoxide, $O_2^+O_2^-$. Using methods similar to those discussed above, we can set up a Born-Haber cycle and evaluate the following terms:[28]

$$O_2 \longrightarrow O_2^+ + e^- \qquad \Delta H = 1164 \text{ kJ mol}^{-1}$$
$$O_2 + e^- \longrightarrow O_2^- \qquad \Delta H = -176 \text{ kJ mol}^{-1}$$
$$\text{Lattice energy} \qquad \Delta H = \sim -500 \text{ kJ mol}^{-1}$$
$$\overline{\Delta H_f = \sim +488 \text{ kJ mol}^{-1}}$$

The calculations support our intuitive feelings about this compound. If it were somehow possible to make an ionic compound $O_2^+O_2^-$, it would decompose with the release of a large quantity of energy:

$$O_2^+O_2^- \longrightarrow 2O_2 \qquad \Delta H = \sim -488 \text{ kJ mol}^{-1}$$

Dioxygenyl superoxide is not a likely candidate for successful synthesis.

THE COVALENT BOND: A PRELIMINARY APPROACH

This section and those following will be devoted to a preliminary analysis of covalent bonding. Most of the ideas presented here may be found elsewhere and with greater rigor. However, since they form the basis for subsequent chapters, a brief presentation is in order here. Covalent bonding will be discussed in greater depth in Chapters 4, 5, and 9.

The Lewis structure

This method of thinking about bonding, learned in high school and too often forgotten in graduate school or before, is a most useful *first step* in thinking about molecules. Before delving into whether valence bond theory or molecular orbital theory is likely to be more helpful, whether valence shell electron pair repulsion theory should be used, or even whether a compound is likely to be composed of molecules, simple ions, or molecular ions, a Lewis structure should be sketched. The following is a brief review of the rules

[28] The ionization energy and electron affinity data are given in Chapter 2. The lattice energy was approximated as above.

for Lewis structures:

1. *Normally two electrons pair up to form each bond.* This is a consequence of the Pauli exclusion principle—two electrons must have paired spins if they are both to occupy the same region of space between the nuclei and attract both. The definition of a bond as a shared *pair* of electrons, however, is overly restrictive, and we shall see that the early emphasis on electron pairing in bond formation is unnecessary and even misleading.

2. *For most molecules there will be a maximum of eight electrons in the valence shell (=Lewis octet structure).* This is absolutely necessary for atoms of the elements lithium through fluorine since they have only four orbitals (an *s* and three *p* orbitals) in the valence shell. It is quite common, as well, for atoms of other elements that often utilize only their *s* and *p* orbitals. Under these conditions the sum of shared pairs (bonds) and unshared pairs (lone pairs) must equal the number of orbitals—four. This is the *maximum*, and for elements having fewer than four valence electrons, the octet will usually not be filled. The following compounds illustrate these possibilities:

```
   ..                                 
  :Cl:        H                        
..  ..  ..    ..       ..                                      
:Cl:C:Cl:   :P:H     :O:H    Cl:B:Cl    H:Be:H    Li:CH3
..  ..  ..    ..       ..         ..
  :Cl:        H        H          Cl
   ..
```

3. *For elements with available d orbitals, the valence shell can be expanded beyond an octet.* Since *d* orbitals first appear in the third energy level, these elements are all in period 3 or beyond and are nonmetals in their higher valence compounds and transition metals in complexes. In the nonmetals, where the number of valence electrons is usually the limiting factor, we have maximum covalencies of 5, 6, 7, and 8 in Groups VA, VIA, VIIA, and VIIIA. Factors determining coordination number in complexes are of several kinds, and discussion of them will be deferred. Examples of molecules or ions containing more than eight electrons in the valence shell are:

```
                                              ┌          NH3           ┐+3
    F          F          F                   │           |            │
    |  F       |  F    F  |  F                │           |  NH3       │
    | /        | /      \ | /                 │           | /          │
 F—P       F—S—F        I—F                   │ H3N——————Co——————NH3   │
    | \      / |        / | \                 │      H3N/ |            │
    |  F    F  |       F  |  F                │           |            │
    F          F          F                   │          NH3           │
                                              └                        ┘
```

4. *It has been assumed implicitly in all of these rules that the molecule will seek the lowest overall energy.* This means that, in general, the maximum number of bonds will form, that the strongest possible bonds will form, and that the arrangement of the atoms in the molecule will be such as to minimize adverse repulsion energies.

Bonding theory

The two "contenders for the throne" of bonding theory are *valence bond theory* (VBT) and *molecular orbital theory* (MOT). The allusion is an apt one since it seems that much of the history of these two theories consisted of contention between their respective proponents as to which was "best." This is indeed unfortunate inasmuch as they are but opposite sides of the same coin. Sometimes overzealous proponents of one theory give the impression that the other is "wrong." Granted that any theory can be used unwisely, it remains nonetheless true that neither theory can be "true" to the *exclusion* of the other since both reduce to the same common and fundamental quantum-mechanical ideas. As used in inorganic chemistry, however, they do have distinctive aspects. Given a specific question one theory may prove distinctly superior in insight, ease of calculation, or simplicity and clarity of results, but a different question may reverse the picture completely. Surely the inorganic chemist who does not become thoroughly familiar with *both* theories is like the carpenter who refuses to carry a saw because he already has a hammer! He is severely limiting his skills by limiting his tools.

VALENCE BOND THEORY

The valence bond (VB) theory grew directly out of the ideas of electron pairing by Lewis and others. In 1927 W. Heitler and F. London proposed a quantum-mechanical treatment of the hydrogen molecule. Their method has come to be known as the valence bond approach and was developed extensively by men such as Linus Pauling and J. C. Slater. The following discussion is adapted from the works of Pauling and Coulson.[29]

Suppose we have two isolated hydrogen atoms. We may describe them by the wave functions Ψ_A and Ψ_B, each having the form given on page 16 for a 1*s* orbital. If the atoms are sufficiently isolated so that they do not interact, the wave function for the system of two atoms is:

$$\Psi = \Psi_{A(1)}\Psi_{B(2)} \tag{3.25}$$

where A and B designate the atoms and the numbers 1 and 2 designate electrons number 1 and 2. Now we know that when the two atoms are brought together to form a molecule they will affect each other and that the individual wave functions Ψ_A and Ψ_B will change, but we may assume that Eq. 3.25 is a good starting place as a trial function for the hydrogen molecule and then try to improve it. When we solve for energy as a function of distance, we find that the energy curve for Eq. 3.25 does indeed have a minimum (curve *a*, Fig. 3.20) of about -24 kJ mol^{-1} at a distance of about 90 pm. The actual observed bond distance is 74 pm, which is not too different from our first approximation, but the experimental bond energy of H_2 is -458 kJ mol^{-1}, almost 20 times greater than our first approximation.

---

[29] L. Pauling, "The Nature of the Chemical Bond," 3rd ed., Cornell University Press, Ithaca, N.Y., **1960**; C. A. Coulson, "Valence," 2nd ed., Oxford University Press, London, **1961**; R. McWeeny, "Coulson's Valence," Oxford University Press, **1979**.

If we examine Eq. 3.25, we must decide that we have been overly restrictive in using it to describe a hydrogen molecule. First, we are not justified in labeling electrons since all electrons are indistinguishable from each other. Moreover, even if we could, we would not be sure that electron 1 will always be on atom A and electron 2 on atom B. We must alter Eq. 3.25 in such a way that the artificial restrictions are removed. We can do this by adding a second term in which the electrons have changed positions:

$$\Psi = \Psi_{A(1)}\Psi_{B(2)} + \Psi_{A(2)}\Psi_{B(1)} \tag{3.26}$$

This improvement was suggested by Heitler and London. If we solve for the energy associated with Eq. 3.26, we obtain curve *b* in Fig. 3.20. The energy has improved greatly (-303 kJ mol^{-1}) and also the distance has improved slightly. Since the improvement is a result of our "allowing" the electrons to exchange places, the increase in bonding energy is often termed the *exchange energy*. One should not be too literal in ascribing this large part of the bonding energy to "exchange," however, since the lack of exchange in Eq. 3.25 was merely a result of our inaccuracies in approximating a correct molecular wave function. If a physical picture is desired to account for the exchange energy, it is probably best to ascribe the lowering of energy of the molecule to the fact that the electrons now have a larger volume in which to move. Recall that the energy of a particle in a box is inversely related to the size of the box; i.e., as the box increases in size, the energy of the particle is lowered. By providing two nuclei at a short distance from each other, we have "enlarged the box" in which the electrons are confined.

A further improvement can be made if we recall that electrons shield each other (p. 36) and that the effective atomic number Z^* will be somewhat less than Z. If we adjust

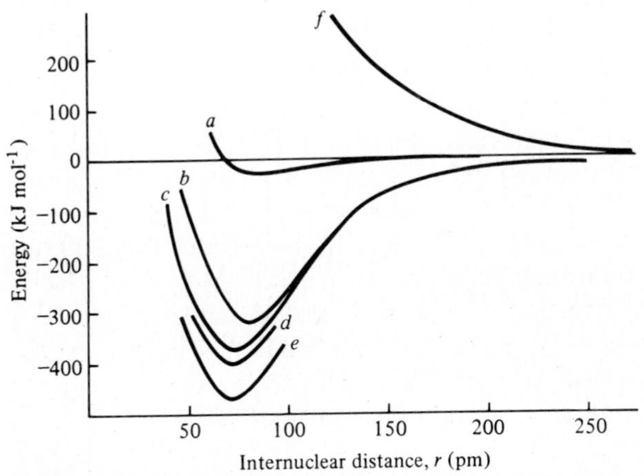

Fig. 3.20 Theoretical energy curves (*a–d, f*) for the hydrogen molecule, H_2, compared with the experimental curve (*e*). Curves (*a–d*) show successive improvements in the wave function as discussed in the text. Curve (*f*) is the repulsive interaction of two electrons of like spin. [Modified from F.A. Cotton and G. Wilkinson, "Advanced Inorganic Chemistry," 2nd ed., Wiley, New York, **1966**. Reproduced with permission.]

our wave functions, Ψ_A and Ψ_B, to account for the shielding from the second electron, we obtain energy curve *c*—a further improvement.

Lastly, we must again correct our molecular wave function for an overrestriction which we have placed upon it. Although we have allowed the electrons to exchange in Eq. 3.26, we have demanded that they must exchange simultaneously, i.e., that only one electron can be associated with a given nucleus at a given time. Obviously this is too restrictive. Although we might suppose that the electrons would tend to avoid each other because of mutual repulsion and thus tend to stay one on each atom, we cannot go so far as to say that they will *always* be in such an arrangement. It is common to call the arrangement given by Eq. 3.26 the "covalent structure" and to consider the influence of "ionic structures" on the overall wave function:

$$\underset{\text{Covalent}}{\mathrm{H{-}H}} \longleftrightarrow \underbrace{\mathrm{H^+H^-} \longleftrightarrow \mathrm{H^-H^+}}_{\text{Ionic}}$$

We then write

$$\Psi = \Psi_{A(1)}\Psi_{B(2)} + \Psi_{A(2)}\Psi_{B(1)} + \lambda\Psi_{A(1)}\Psi_{A(2)} + \lambda\Psi_{B(1)}\Psi_{B(2)} \tag{3.27}$$

where the first two terms represent the covalent structure and the second two terms represent ionic structures in which both electrons are on atom A or B. Since the electrons tend to repel each other somewhat, there is a smaller probability of finding them both on the same atom than on different atoms, so the second two terms are weighted somewhat less ($\lambda < 1$). Equation 3.27 can be expressed more succinctly as

$$\Psi = \Psi_{cov} + \lambda'\Psi_{H^+H^-} + \lambda'\Psi_{H^-H^+} \tag{3.28}$$

When we investigate the energetics of the wave function in Eq. 3.27, we find further improvement in energy and distance (curve *d*, Fig. 3.20).

This is the first example we have had of the phenomenon of *resonance*, which we shall discuss at some length later on. It should be pointed out now, however, that the hydrogen molecule has one structure which is described by *one* wave function, Ψ. However, it may be necessary because of our approximations, to write Ψ as a combination of two or more wave functions, each of which only partially describes the hydrogen molecule. Table 3.7 lists values for the energy and equilibrium distance for the various stages of our approximation, together with the experimental values.

Now, if one wishes, additional "corrections" can be included in our wave function, to make it more nearly descriptive of the actual situation obtaining in the hydrogen molecule.[30,31] However, the present simplified treatment has included the three important contributions to bonding: delocalization of electrons over two or more nuclei, mutual screening, and partial ionic character.

---

[30] A 100-term function (see ref. 31) has reproduced the experimental value to within 0.01 kJ mol^{-1}. See C. A. Coulson, "Valence," 2nd ed., Oxford University Press, London, **1961**, pp. 124–125. For a more rigorous discussion, see F. L. Pilar, "Elementary Quantum Chemistry," McGraw-Hill, New York, **1968**, pp. 234–249.

[31] W. Kolos and L. Wolniewicz, *J. Chem. Phys.*, **1968**, *49*, 404. See G. Herzberg, *J. Mol. Spectr.*, **1970**, *33*, 147, for the experimental value.

Table 3.7 Energies and equilibrium distances for VB wave functions

Type of wave function	Energy (kJ mol^{-1})	Distance (pm)
Uncorrected, $\Psi = \Psi_A \Psi_B$	24	90
"Heitler-London"	303	86.9
Addition of shielding	365	74.3
Addition of ionic contributions	388	74.9
Observed values	458.0	74.1

SOURCE: C. A. Coulson, "Valence," 2nd ed., Oxford University Press, London, **1961**, p. 125. Used with permission.

The reader may question the apparently *ad hoc* corrections for shielding and ionic contributions and wonder if it is not possible to "overcorrect" for a given physical factor. The answer is given by the *variation principle* which states that any trial wave function can never yield a lower energy (i.e., give a greater bonding energy) than the true energy of the system. All of our adjustments allow us to approach the true wave function for the molecule; we can never surpass it. As each adjustment is made (i.e., for shielding, etc.) the parameters in the trial function can be adjusted to yield the best energy. In fact, it is not necessary even to make the adjustments in terms of exchange, shielding, or ionicity—introduction of sufficient parameters can improve the wave function to any desired degree of accuracy. Again, this should remind us of the artificiality of such things as "exchange energy."

There is an implicit assumption contained in all of the above: *The two bonding electrons are of opposite spin.* If two electrons are of parallel spin, no bonding occurs, but repulsion instead (curve *f*, Fig. 3.20). This is a result of the Pauli exclusion principle. Because of the necessity for pairing in each bond formed, the valence bond theory is often referred to as the electron pair theory, and it forms a logical quantum-mechanical extension of Lewis's theory of electron pair formation.

Molecular orbital theory

A second approach to bonding in molecules is known as the molecular orbital (MO) theory. The assumption here is that if two nuclei are positioned at an equilibrium distance, and electrons are added, they will go into molecular orbitals that are in many ways analogous to the atomic orbitals discussed in Chapter 2. In the atom there are *s*, *p*, *d*, *f*, . . . orbitals determined by various sets of quantum numbers and in the molecule we have σ, π, δ, . . . orbitals determined by quantum numbers. We should expect to find the Pauli exclusion principle and Hund's principle of maximum multiplicity obeyed in these molecular orbitals as well as in the atomic orbitals.

When we attempt to solve the Schrödinger equation to obtain the various molecular orbitals, we run into the same problem found earlier for atoms heavier than hydrogen. We are unable to solve the Schrödinger equation exactly and therefore must make some approximations concerning the form of the wave functions for the molecular orbitals.

Of the various methods of approximating the correct molecular orbitals, we shall discuss only one, the linear combination of atomic orbitals (LCAO) method. We assume that we can approximate the correct molecular orbitals by combining the atomic orbitals of the atoms that form the molecule. The rationale is that most of the time the electrons will be nearer and hence "controlled" by one or the other of the two nuclei, and when this is so, the molecular orbital should be very nearly the same as the atomic orbital for that atom. We therefore combine the atomic orbitals, Ψ_A and Ψ_B, to obtain two molecular orbitals:[32]

$$\Psi_b = \Psi_A + \Psi_B \tag{3.29}$$

$$\Psi_a = \Psi_A - \Psi_B \tag{3.30}$$

The one-electron molecular orbitals thus formed consist of a *bonding* molecular orbital (Ψ_b) and an *antibonding* molecular orbital (Ψ_a).[33] If we allow a single electron to occupy the bonding molecular orbital (as in H_2^+, for example), the approximate wave function for the molecule is

$$\Psi = \Psi_{b(1)} = \Psi_{A(1)} + \Psi_{B(1)} \tag{3.31}$$

For a two-electron system such as H_2, the total wave function is the product of the wave functions for each electron:

$$\Psi = \Psi_{b(1)}\Psi_{b(2)} = [\Psi_{A(1)} + \Psi_{B(1)}][\Psi_{A(2)} + \Psi_{B(2)}] \tag{3.32}$$

$$\Psi = \Psi_{A(1)}\Psi_{A(2)} + \Psi_{B(1)}\Psi_{B(2)} + \Psi_{A(1)}\Psi_{B(2)} + \Psi_{A(2)}\Psi_{B(1)} \tag{3.33}$$

The results for the MO treatment are similar to those obtained by VB theory. Equation 3.33 is the same (when rearranged) as Eq. 3.27 except that the ionic terms ($\Psi_{A(1)}\Psi_{A(2)}$ and $\Psi_{B(1)}\Psi_{B(2)}$) are weighted as heavily as the covalent ones ($\Psi_{A(1)}\Psi_{B(2)}$ and $\Psi_{A(2)}\Psi_{B(1)}$). This is not surprising, since we did not take into account the repulsion of electrons in obtaining Eq. 3.32. This is a general result: Simple molecular orbitals obtained in this way from the linear combination of atomic orbitals (LCAO–MO theory) tend to exaggerate the ionicity of molecules, and the chief problem in adjusting this simple method to make the results more realistic consists of taking into account *electron correlation.* As in the case of VB theory it is possible to optimize the wave function by the addition of correcting terms. Some typical results for the hydrogen molecule are listed in Table 3.8.

The two orbitals Ψ_b and Ψ_a differ from each other as follows. In the bonding molecular orbital the wave functions for the component atoms reinforce each other in the region between the nuclei (Fig. 3.21a, b), but in the antibonding molecular orbital they cancel, forming a node between the nuclei (Fig. 3.21d). We are, of course, interested in learning of the *electron distribution* in the hydrogen molecule, and will therefore be interested in the *square* of the wave functions:

$$\Psi_b^2 = \Psi_A^2 + 2\Psi_A\Psi_B + \Psi_B^2 \tag{3.34}$$

$$\Psi_a^2 = \Psi_A^2 - 2\Psi_A\Psi_B + \Psi_B^2 \tag{3.35}$$

---

[32] The combination $\Psi_B - \Psi_A$ does not represent a third MO, but is another form of Ψ_a.

[33] Note that the subscripts a, b and A, B bear no relation to one another; the former refer to antibonding and bonding whereas the latter merely identify the atomic orbitals on the atoms forming the molecule.

Table 3.8 Energies and equilibrium distances for MO wave functions

Type of wave function	Energy ($kJ\ mol^{-1}$)	Distance (pm)
Uncorrected, $\Psi = \Psi_A + \Psi_B$	260	85
Addition of shielding	337	73
Addition of electron–electron repulsions	397	71
Observed values	458.0	74.1

SOURCE: C. A. Coulson, "Valence," 2nd ed., Oxford University Press, London, **1961**, p. 125. Used with permission.

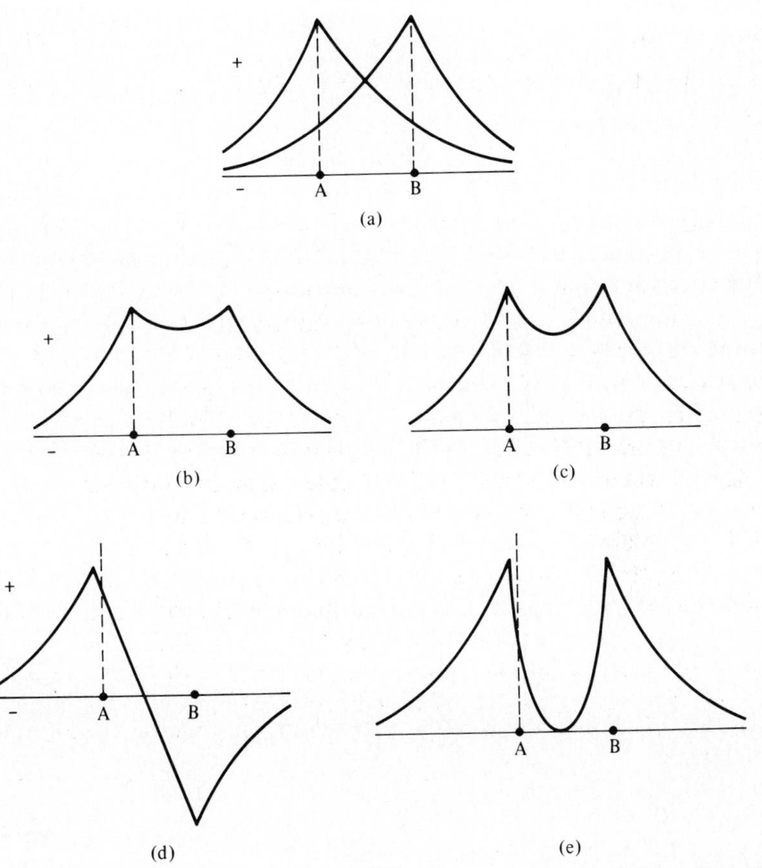

Fig. 3.21 (a) Ψ_A and Ψ_B for individual hydrogen atoms (cf. Fig. 2.3). (b) $\Psi_b = \Psi_A + \Psi_B$. (c) Probability function for the bonding orbital, Ψ_b^2. (d) $\Psi_a = \Psi_A - \Psi_B$. (e) Probability function for the antibonding orbital, Ψ_a^2. Note that the bonding orbital increases the electron density between the nuclei (c) but that the antibonding orbital decreases the electron density between the nuclei (e). [Adapted from H. H. Jaffé, in "Comprehensive Biochemistry," M. Florkin and E. H. Stotz, eds., Elsevier, Amsterdam, **1961**. Reproduced with permission.]

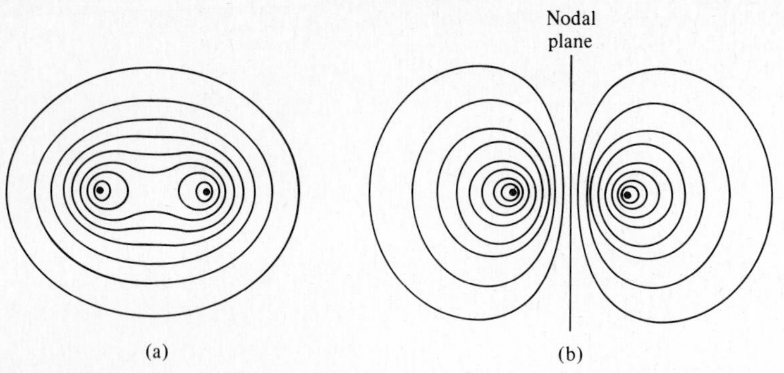

Fig. 3.22 Electron density contours for the H_2^+ ion: bonding (a) and antibonding (b) orbitals. [From D. R. Bates et al., *Phil. Trans. Roy. Soc.*, **1953**, *246*, 215. Reproduced with permission.]

The difference between the two probability functions lies in the cross term $2\Psi_A\Psi_B$. The integral $\int\Psi_A\Psi_B\,d\tau$ is known as the *overlap integral, S*, and is very important in bonding theory. In the bonding orbital the overlap is positive and the electron density between the nuclei is *increased*, whereas in the antibonding orbital the electron density between the nuclei is *decreased*. (See Fig. 3.21c, e.) In the former case the nuclei are shielded from each other and the attraction of both nuclei for the electrons is enhanced. This results in a *lowering* of the energy of the molecule and is therefore a *bonding* situation. In the second case the nuclei are partially bared toward each other and the electrons tend to be in those regions of space in which mutual attraction by both nuclei is severely reduced. This is a repulsive, or *antibonding*, situation. An electron density map for the hydrogen molecule ion, H_2^+, is shown in Fig. 3.22 illustrating the differences in electron densities between the bonding and antibonding conditions.[34]

We have postponed normalization of the molecular orbitals until now. Since $\int\Psi^2\,d\tau = 1$ for the probability of finding an electron somewhere in space, Eq. 3.34 becomes

$$\int N_b^2\Psi_b^2\,d\tau = N_b^2\left[\int\Psi_A^2\,d\tau + \int\Psi_B^2\,d\tau + 2\int\Psi_A\Psi_B\,d\tau\right] = 1 \tag{3.36}$$

where N_b is the *normalizing constant*. If we let S be the overlap integral, $\int\Psi_A\Psi_B\,d\tau$, we have

$$\int\Psi_b^2\,d\tau = \left[\int\Psi_A^2\,d\tau + \int\Psi_B^2\,d\tau + 2S\right] \tag{3.37}$$

Now since the atomic wave functions Ψ_A and Ψ_B were previously normalized, $\int\Psi_A^2\,d\tau$ and $\int\Psi_B^2\,d\tau$ each equal one. Hence

$$N_b^2 = \frac{1}{2 + 2S} \tag{3.38}$$

$$N_b = \sqrt{\frac{1}{2 + 2S}} \tag{3.39}$$

[34] The electron density map is easier to obtain for H_2^+ than for H_2 because of the lack of necessity for correction of electron interactions. The differences are not great.

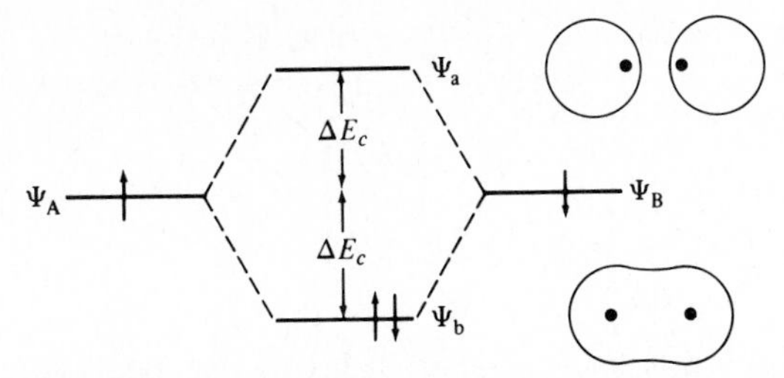

Fig. 3.23 Energy levels for the H_2 molecule with neglect of overlap. The quantity of ΔE_c represents the difference in energy between the energy levels of the separated atoms and the bonding molecular orbital. It is equal to 458 kJ mol^{-1}.

and

$$N_a = \sqrt{\frac{1}{2 - 2S}} \tag{3.40}$$

For most simple calculations the value of the overlap integral, S, is numerically rather small and may thus be neglected without incurring too great an error. This simplifies the algebra considerably and is sufficiently accurate for most purposes. With complete neglect of overlap, our molecular wave functions become

$$\Psi_b = \frac{1}{\sqrt{2}}(\Psi_A + \Psi_B) \tag{3.41}$$

$$\Psi_a = \frac{1}{\sqrt{2}}(\Psi_A - \Psi_B) \tag{3.42}$$

The idea of "complete neglect of overlap" is one which may readily lead to confusion. It is a common simplification often encountered in molecular orbital calculations. Yet in the next section great stress will be placed on the importance of overlap, and the generalization will be made that the strength of a bond is roughly proportional to overlap, and that if overlap is zero or negative, a bond will not form. This seeming paradox may be resolved as follows: The bond energy, ΔE_c, is proportional to the extent that the atomic orbitals overlap. Hence, in our qualitative attempts to rationalize covalent bonds we shall need to pay exceedingly close attention to the symmetry of orbitals and their overlap. "Neglect of overlap" simply means that the numerical value of the overlap integral, S, is neglected in the *normalization* procedure. If the overlap had been included in the calculation the normalization coefficients would have been: $N_a = 1.11$ and $N_b = 0.56$ instead of $N_a = N_b = 0.71$. Other molecules have less overlap than H_2, and so the effect is less.

The relative energies (assuming neglect of overlap) of the two molecular orbitals are shown in Fig. 3.23. The bonding orbital is stabilized relative to the energy of the isolated atoms by the quantity ΔE_c. The antibonding orbital is destabilized by an equivalent quantity.[35]

[35] If overlap is included in the calculation, the antibonding orbital is somewhat more destabilized than the bonding orbital is stabilized. This is an extremely important phenomenon which we shall encounter over and over, from the nonexistence of He_2 to various atomic repulsions.

The quantity ΔE_c is termed the *exchange energy* and corresponds to the exchange energy we observed previously in VB theory (p. 93). Here, too, one should be cautious about trying to interpret it literally as resulting from "exchange" or a "resonance."

SYMMETRY AND OVERLAP[36]

As we have seen from Eqs. 3.34 and 3.35 the only difference between the electron distribution in the bonding and antibonding molecular orbitals and the atomic orbitals is in those regions of space for which both Ψ_A and Ψ_B have appreciable values, so that their product ($S = \int \Psi_A \Psi_B \, d\tau$) has an appreciable nonzero value. Furthermore, for bonding, $S > 0$, and for antibonding, $S < 0$. The condition $S = 0$ is termed *nonbonding* and corresponds to no interaction between the orbitals. We may make the generalization that the strength of a bond will be roughly proportional to the extent of the overlap of the atomic orbitals. This is known as the *overlap criterion of bond strength* and indicates that bonds will form in such a way as to maximize overlap.

In *s* orbitals the sign of the wave function is everywhere the same (with the exception of small, intranodal regions for $n > 1$), and so there is no problem with matching the sign of the wave functions to achieve positive overlap. With *p* and *d* orbitals, however, there are several possible ways of arranging the orbitals, some resulting in positive overlap, some in negative overlap, and some in which the overlap is exactly zero (Fig. 3.24). Bonding can take place only when the overlap is positive.

It may occur to the reader that it is always possible to bring the orbitals together in such a way that the overlap is positive. For example, in Fig. 3.24g, h if negative overlap is obtained, one need only invert one of the atoms to achieve positive overlap. This is true for diatomic molecules or even for polyatomic linear molecules. However, when we come to cyclic compounds, we no longer have the freedom arbitrarily to invert atoms to obtain proper overlap matches. One example will suffice to illustrate this.

There is a large class of compounds, the phosphonitrilic halides (X = F, Cl, Br), containing the phosphazene ring system, $(PNX_2)_n$ (see Chapter 14). The trimer, $P_3N_3X_6$, is illustrated in Fig. 3.25. Note the resemblance to benzene in the alternating single and double bonds. Like benzene, the phosphonitrilic trimer is aromatic, i.e., the π electrons are delocalized over a conjugated system with resonance stabilization.[37] Unlike benzene, however, there is a node in the lowest lying molecular orbital which tends to disrupt the aromaticity and decrease the resonance energy of the system. This is illustrated in Fig. 3.26, which represents the phosphonitrilic ring split open and arranged linearly for clarity. We start on nitrogen atom number one (N_1) and assume an arbitrary assignment of the positive and negative lobes of the *p* orbital. The phosphorus atom π bonds through its *d* orbitals, and so for the P_1 atom we draw a *d* orbital with appropriate symmetry such that the overlap between N_1 and P_1 is greater than zero. We continue with N_2, P_2, N_3, and P_3, each time matching the orbital symmetries to achieve positive overlap. However, when we come to the overlap between

[36] A minimum of symmetry is presented here to allow discussion of overlap. The reader is encouraged to turn to Appendix B for a more lengthy discussion of the subject.

[37] A number of new terms are used here such as π electrons, aromaticity, delocalization, etc. If the reader is not familiar with these terms from previous work (e.g., organic chemistry) it will be easiest simply to note the mismatch in orbital symmetry and return to this discussion after completing the chapter.

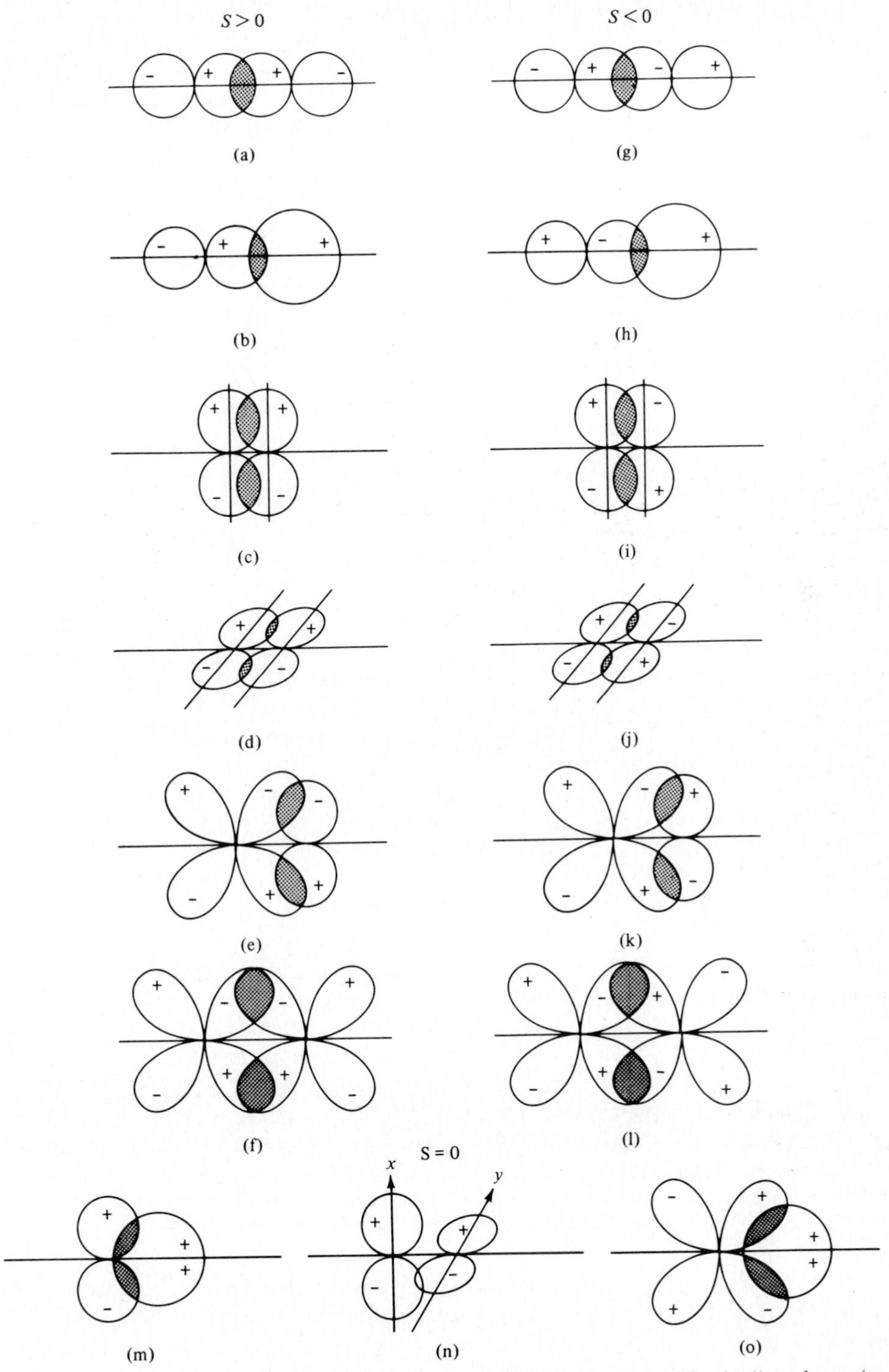

Fig. 3.24 Arrangements of atomic orbitals resulting in positive (a–f), negative (g–l), and zero (m–o) overlap.

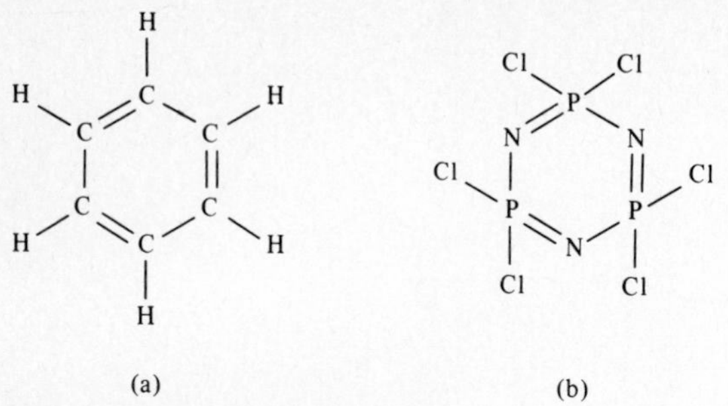

(a) (b)

Fig. 3.25 Comparison of bonding in the ring systems of (a) benzene and (b) phosphonitrilic chloride trimer.

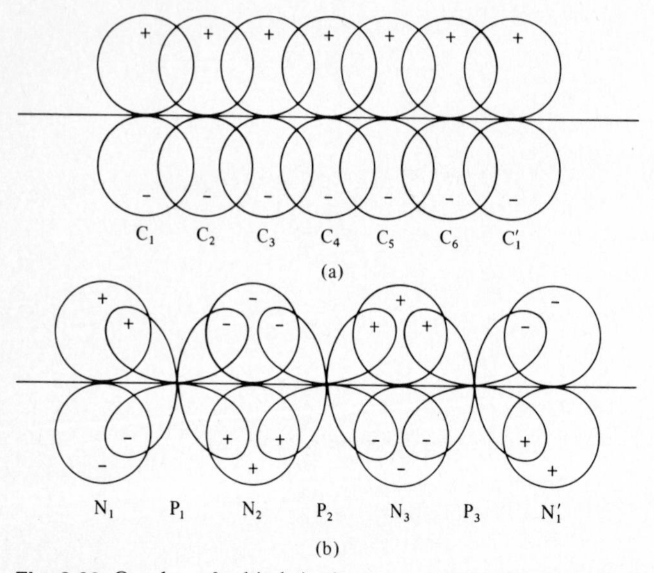

Fig. 3.26 Overlap of orbitals in the *p–p* π system in benzene (a) and *p–d* π system in the phosphonitrilic ring (b). Note the mismatch of orbital symmetry in the latter.

P_3 and N_1 (to close the ring) we find that we would like to have the N_1 orbital lie as shown on the right, but we have previously assigned it the arrangement shown on the left. It is impossible to draw the six orbitals in such a way as to avoid a mismatch or node in the system.

Symmetry of molecular orbitals

Some of the possible combinations of atomic orbitals are shown in Fig. 3.27. Those orbitals which are cylindrically symmetrical about the internuclear axis are called σ orbitals, analogous to an *s* orbital, the atomic orbital of highest symmetry. If the internuclear axis lies in a nodal plane, a π bond results. In δ bonds the internuclear axis lies in two mutually perpendicular nodal planes. All antibonding orbitals possess an additional

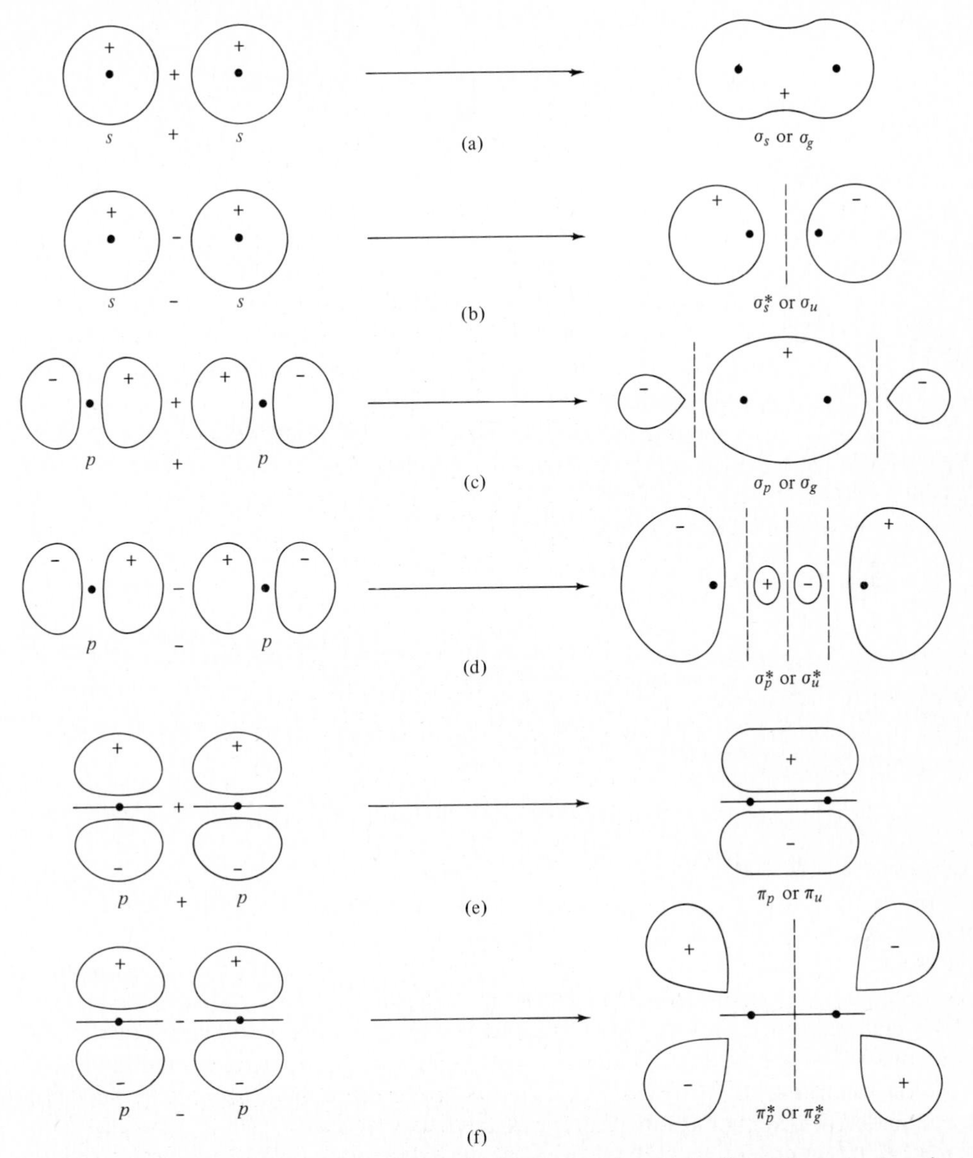

Fig. 3.27 Symmetry of molecular orbitals formed from atomic orbitals illustrating σ (a–d) and π (e, f) orbitals, and bonding (a, c, e) and antibonding (b, d, f) orbitals. The orbitals are depicted by electron density sketches with the sign of Ψ superimposed.

nodal plane perpendicular to the internuclear axis and lying between the nuclei. In addition to the presence or absence of one or more nodal planes, we shall also be interested in the symmetry with respect to inversion about the center of the molecule (see p. 24). Of interest in this regard are π_{p-p} orbitals, which are *ungerade*, and π_{p-p}^* orbitals, which are *gerade*.

Molecular orbitals in diatomic molecules

Molecules containing two atoms of the same element are the simplest molecules to discuss. We have already seen the results for the hydrogen molecule (p. 99; Fig. 3.23) and for the linear combination of *s* and *p* orbitals (Fig. 3.24). We shall now investigate the general case for molecular orbitals formed from two atoms having atomic orbitals 1*s*, 2*s*, 2*p*, 3*s*, etc.

There are two criteria which must be met for the formation of molecular orbitals which are more stable,[38] i.e., lower in energy than the contributing atomic orbitals. One is that the overlap between the atomic orbitals must be positive. Furthermore, in order that there be effective interaction between orbitals on different atoms, the energies of the two atomic orbitals must be approximately the same. This topic will be discussed at greater length later (pp. 137, 144), but for now it may be concluded that molecular orbitals will form from corresponding orbitals on the two atoms (i.e., $1s + 1s$; $2s + 2s$; etc.). We shall see later that this is an oversimplification that can occasionally lead us astray. However, it is the simplest way to approach the topic and we can easily modify it when necessary. When we combine the atomic orbitals in this way, the energy levels shown in Fig. 3.28 are obtained. The appropriate combinations are:[39]

$$\sigma_{1s} = 1s_A + 1s_B \qquad \pi_{2py} = 2p_{yA} + 2p_{yB}$$

$$\sigma^*_{1s} = 1s_A - 1s_B \qquad \pi_{2px} = 2p_{xA} + 2p_{xB}$$

$$\sigma_{2s} = 2s_A + 2s_B \qquad \pi^*_{2py} = 2p_{yA} - 2p_{yB}$$

$$\sigma^*_{2s} = 2s_A - 2s_B \qquad \pi^*_{2px} = 2p_{xA} - 2p_{xB}$$

$$\sigma_{2p} = 2p_{zA} + 2p_{zB}$$

$$\sigma^*_{2p} = 2p_{zA} - 2p_{zB}$$

The σ_{1s} and σ^*_{1s} orbitals correspond to the molecular orbitals seen previously for the hydrogen molecule. The atomic 2*s* orbitals form a similar set of σ and σ^* orbitals. The atomic *p* orbitals can form σ bonds from direct ("head on") overlap of the p_z orbitals and two π bonds from parallel overlap of the p_y and p_x orbitals. Since the overlap is greater in the former case, we should expect the exchange energy to be greater also (p. 100), and σ bonds are generally stronger than π bonds. Hence the σ_{2p} orbital is stabilized (lowered in energy) more than the π_{2p} orbitals, and conversely the corresponding antibonding orbitals are raised accordingly.[40] By analogy with atomic electron configurations, we can write molecular electron configurations. For H_2 we have:

$$H_2 = \sigma^2_{1s}$$

Using Fig. 3.28 as a guide, we can proceed to build up various diatomic molecules in much the same way as the *aufbau* principle was used to build up atoms.

---

[38] This statement applies to *bonding orbitals*, of course. *Antibonding* orbitals are always *less* stable than the corresponding atomic orbitals.

[39] Symbols A and B represent atoms, *x*, *y*, and *z* represent orientation of the *p* orbitals, and the asterisk (*) identifies antibonding orbitals.

[40] This discussion is somewhat oversimplified as we shall see later.

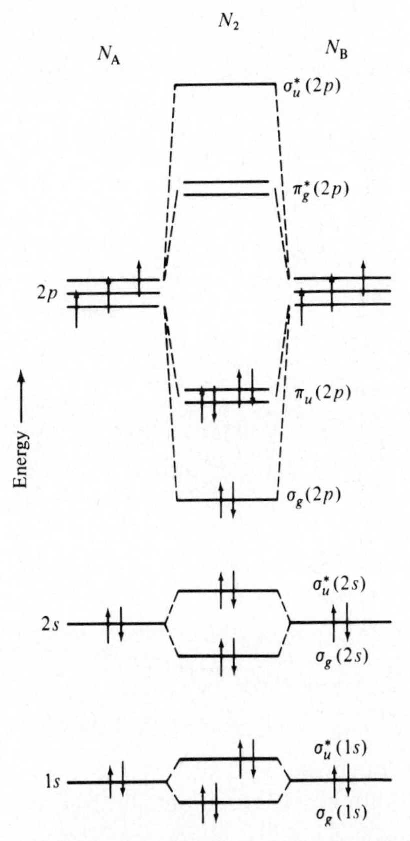

Fig. 3.28 Simplified molecular orbital energy levels assuming no mixing of s and p orbitals. The three $2p$ orbitals are degenerate, i.e., they all have the same energy and might better be shown as —— —— ——, but the latter is more cumbersome. The molecule shown is N_2. [Adapted from M. Orchin and H. H. Jaffé, "The Importance of Antibonding Orbitals," Houghton Mifflin, Boston, **1967**. Reproduced with permission.]

1. *Molecules containing one to four electrons.* We have already seen the H_2 molecule in which there are two electrons in the σ_{1s} orbital. Two bonding electrons constitute a chemical bond. The molecular orbital theory does not restrict itself to even numbers of bonding electrons, and so the bond order is given as one-half the difference between the number of bonding electrons and the number of antibonding electrons:

$$\text{Bond order} = \tfrac{1}{2}(N_b - N_a) \qquad \textbf{(3.43)}$$

The molecule He_2 is unknown since the number of antibonding electrons (2) is equal to the number of bonding electrons (2) and the net bond order is zero. With no bond energy to overcome[41] the dispersive tendencies of entropy, two helium atoms in a "molecule" will not remain together but fly apart. If it existed, molecular helium would have the electron

[41] If overlap is not neglected in the calculation, the antibonding orbital is *more* destabilizing than the bonding orbital is stabilizing and so He_2 actually has a repulsive energy forcing it apart.

configuration:

$$He_2 = \sigma_{1s}^2 \sigma_{1s}^{*2}$$

If helium is ionized, it is possible to form diatomic helium molecule–ions, He_2^+. Such a molecule will contain three electrons, two bonding and one antibonding, for a net bond order of one-half. Such a species, although held together with only about one-half the bonding energy of the hydrogen molecule, should be expected to exist. In fact it does, and it has been observed spectroscopically in highly energetic situations sufficient to ionize the helium. That it is not found under more familiar chemical situations as, for example, in salts, $He_2^+X^-$, is not a result of any unusual weakness in the He—He bond, but because contact with just about any substance will supply the missing fourth electron with resultant conversion into helium atoms.

Isoelectronic in a formal sense, but quite different in the energies involved is the Xe_2^+ ion believed to exist in certain very acidic solvents (see Chapter 15). The energetics of the situation are not completely understood, but presumably the much lower ionization energy of xenon can more readily be compensated by the solvation energy of the polar solvent, thus stabilizing the Xe_2^+ cation.

2. *Lithium and beryllium.* Two lithium atoms contain six electrons. Four will fill the σ_{1s} and σ_{1s}^* orbitals with no bonding. The last two electrons will enter the σ_{2s} orbital, giving a net bond order of one in the Li_2 molecule. The electron configuration will be:

$$Li_2 = KK\sigma_{2s}^2$$

where K stands for the K (1*s*) shell.[42]

Eight electrons from two beryllium atoms fill the four lowest energy levels, σ_{1s}, σ_{1s}^*, σ_{2s}, σ_{2s}^*, yielding a net bond order of zero, as in He_2, with an electron configuration of:

$$Be_2 = KK\sigma_{2s}^2 + \sigma_{2s}^{*2}$$

Like the latter molecule, Be_2 is not expected to exist. The experimental facts are that lithium is diatomic in the gas phase, but that beryllium is monatomic.

3. *Boron and carbon.* The molecules B_2 and C_2 are not commonly observed chemical species. Nevertheless, their properties are important in determining the correctness of our energy level diagram, and they will be discussed later.

4. *Nitrogen, oxygen, fluorine, and neon.* The nitrogen molecule contains fourteen electrons. Four of these lie in the σ_{1s} and σ_{1s}^* orbitals, which cancel and may thus be ignored. The next four occupy the σ_{2s} and σ_{2s}^* orbitals and contribute nothing to the net bonding. The remaining six electrons form a σ bond and two π bonds. The electronic configuration for the N_2 molecule may therefore be written:

$$N_2 = KK\sigma_{2s}^2\sigma_{2s}^{*2}\sigma_{2p}^2\pi_{2p}^4$$

---

[42] The inner shells of core electrons are often abbreviated since no net bonding takes places in them. The symbols used, K, L, M, etc., refer to the older system of designating the principal energy levels, $n = 1$(K), $n = 2$(L), etc. Thus Na_2 = KK LL σ_{3s}^2.

The bond order is three (one σ and two π bonds) agreeing with the experimentally observed large dissociation energy of 942 kJ mol^{-1}.

The oxygen molecule was one of the first applications of the molecular orbital theory in which it proved more successful than the simple valence bond theory. The oxygen molecule contains two more electrons than the N_2 molecule. The electron configuration is therefore:

$$O_2 = KK\sigma_{2s}^2\sigma_{2s}^{*2}\sigma_{2p}^2\pi_{2p}^4\pi_{2p}^{*2}$$

However, examination of the energy level diagram in Fig. 3.28 indicates that the π_{2p}^* level is doubly degenerate from the two equivalent π orbitals π_{2py}^* and π_{2px}^*. Hund's rule of maximum multiplicity predicts that the two electrons entering the π^* level will occupy two *different* orbitals, so the electronic configuration can be written more explicitly as:

$$O_2 = KK\sigma_{2s}^2\sigma_{2s}^{*2}\sigma_{2p}^2\pi_{2p}^4\pi_{2px}^{*1}\pi_{2py}^{*1}$$

This has no effect on the bond order, which is still two $[\frac{1}{2}(6-2)]$, as anticipated by valence bond theory. The difference lies in the *paramagnetism* of molecular oxygen resulting from the two unpaired electrons. (In this regard O_2 is analogous to atomic carbon in which the last two electrons remained unpaired by entering different, degenerate orbitals.) The simple valence bond theory predicts that all electrons in oxygen will be paired; in fact, the formation of two bonds *demands* that the maximum number of electrons be paired. This is the first case of several we shall encounter in which the stress placed on *paired bonding electrons* is exaggerated by the valence bond theory. The molecular orbital theory does not require such pairing as it merely counts the number of bonding versus antibonding electrons. The experimentally measured paramagnetism of O_2 confirms the accuracy of the MO treatment.

For the fluorine molecule, there will be a total of 18 electrons distributed:

$$F_2 = KK\sigma_{2s}^2\sigma_{2s}^{*2}\sigma_{2p}^2\pi_{2p}^4\pi_{2p}^{*4}$$

The net bond order is one, corresponding to the σ bond, and agreeing with the valence bond picture.

The addition of two more electrons to form the Ne_2 molecule will result in filling the last antibonding orbital, the σ_{2p}^* orbital. This will reduce the bond order to zero and Ne_2, like He_2, will not exist.

The results of this brief discussion of molecular orbitals in diatomic molecules are summarized in Table 3.9 along with appropriate experimental data.

Ionization of diatomic molecules, energetics, and bond length

Further support for the advantages of the MO viewpoint comes from investigation of the bond lengths in some diatomic molecules and ions. For example, consider the oxygen molecule. As we have seen previously, it contains a double bond resulting from two σ-bonding electrons, four π-bonding electrons, and two π-antibonding electrons. The bond

Table 3.9 Selected diatomic molecules

Molecular orbital predictions				Experimental data		
Molecule	Electrons	Net bonds	Unpaired electrons	Bond energy ($kJ\ mol^{-1}$)	Dia- or paramagnetic	Bond length (pm)
H_2	2	1σ	0	432.00	D	74.2
He_2	4	0	0	0	—	—
Li_2	6	1σ	0	105	D	267.2
Be_2	8	0	0	—	—	—
N_2	14	$1\sigma, 2\pi$	0	941.69	D	109.8
O_2	16	$1\sigma, 1\pi$	2	493.59	P	120.7
F_2	18	1σ	0	155	D	141.8
Ne_2	20	0	0	0	—	—

length is 121 pm. Addition of two electrons to the oxygen molecule results in the well-known peroxide ion, O_2^{-2}:

$$O_2 + 2e^- \longrightarrow O_2^{-2} \tag{3.44}$$

According to Fig. 3.28 these two electrons will enter the π^* orbitals, decreasing the bond order to one. Since the compressive forces (bond energy) are reduced and the repulsive forces (nonbonding electron repulsions) remain the same, the bond length is increased to 149 pm. If only *one* electron is added to an oxygen molecule, the superoxide ion, O_2^-, results. Since there is one less antibonding electron than in O_2^{-2}, the bond order is $1\frac{1}{2}$ and the bond length is 126 pm.

Furthermore, ionization of O_2 to a cation:

$$O_2 \longrightarrow O_2^+ + e^- \tag{3.45}$$

causes a *decrease* in bond length to 112 pm. The electron ionized is a π^* antibonding electron and the bond order in O_2^+ is $2\frac{1}{2}$.

The nitric oxide molecule, NO, has a bond length of 115 pm and a bond order of $2\frac{1}{2}$. Ionization to the nitrosyl ion, NO^+, removes an antibonding π^* electron and results in a bond order of three (isoelectronic with N_2) and a shortening of the bond length to 106 pm. In contrast, addition of an electron (to a π^* orbital) causes a decrease in bond order and an increase in bond length.

The fact that the formation of the nitrosyl ion results from the removal of an *antibonding electron* makes the ionization energy (IE) for the reaction

$$NO \longrightarrow NO^+ + e^- \qquad IE = 894\ kJ\ mol^{-1} \tag{3.46}$$

lower than would otherwise be the case for the unbound atoms of nitrogen (IE = 1402 kJ mol^{-1}) and oxygen (IE = 1314 kJ mol^{-1}). The nitrosyl ion is thus stabilized and exists in several compounds, such as $NO^+HSO_4^-$ and $NO^+BF_4^-$.

A comparison of the ionization energies of molecular oxygen and nitrogen illustrates the same point. The ionization energy of molecular nitrogen is 1503 kJ mol^{-1}, greater than that of atomic nitrogen, in agreement with Fig. 3.28 that a *bonding* (and therefore more stable) electron is removed. In contrast, the ionization energy of molecular oxygen is 1164 kJ mol^{-1}, *less* than that of atomic oxygen. In this case the ionized electron is removed from an antibonding orbital, requiring less energy.

A closer look at diatomic energy levels

Thus far we have employed the simple energy level diagram of Fig. 3.28 with considerable success in accounting for magnetic properties, bond energies, and ionization energies of various diatomic molecules. Nevertheless, some molecules indicate that the order of energy levels shown in Fig. 3.28 is not quite right. According to Fig. 3.28, the B_2 molecule would be predicted to have a single σ bond and be diamagnetic. Experimentally the B_2 molecule is found to have two unpaired electrons. The C_2 molecule would be predicted to have an electron configuration $KK\sigma_{2s}^2\sigma_{2s}^{*2}\sigma_{2p}^2\pi_{2p}^1\pi_{2p}^1$ and be paramagnetic. The experimental evidence indicates that the ground state of C_2 is diamagnetic.

The problem here is that in constructing Fig. 3.28 mixing was allowed only between orbitals on atoms A and B which were identical in energy. Actually, mixing will take place between *all* orbitals of proper symmetry, inhibited only by the fact that if the mismatch between orbitals is large, mixing will be reduced. We are therefore justified in dismissing mixing between the 1*s* and 2*s* orbitals. The energy difference between the 2*s* and 2*p* orbitals is less and varies with the effective atomic number of the element. With a larger effective atomic number, as in fluorine, the difference between *s* and *p* orbitals is more pronounced, and the mixing may be sufficiently small to be neglected.[43] The difference in energy between the 2*s* and 2*p* levels dramatically increases from about 200 kJ mol^{-1} in the lithium atom to about 2500 kJ mol^{-1} in the fluorine atom. In the case of the elements to the left of the series, the lower effective nuclear charge allows the 2*s* and 2*p* orbitals to come sufficiently close to mix. This phenomenon is the equivalent of hybridization in the valence bond theory discussed in the next section.

We shall not investigate the details of *s*–*p* mixing, but merely perform a thought experiment that indicates the nature of the results. We have seen above that Fig. 3.28 is approximately true if the energy gap between the *s* and *p* orbitals is large, as in F_2. Consider narrowing the gap between the 2*s* and 2*p* orbitals. As this gap narrows, mixing (hybridization) of the *s* and *p* orbitals becomes energetically feasible. However, more important in terms of the final energy-level diagram, molecular orbitals of the same symmetry will interact with one another, with the lower-energy orbital becoming more stabilized and the higher-energy orbital being destabilized. This is a result of the "no-crossing rule" (Fig. 3.29). In the present circumstance, as the $\sigma_g(2s)$ and the $\sigma_g(2p)$ orbitals approach (because of the lessening difference in energy between 2*s* and 2*p*) they interact. Because of the mixing of the

[43] There is no experimental evidence confirming or denying this possibility.

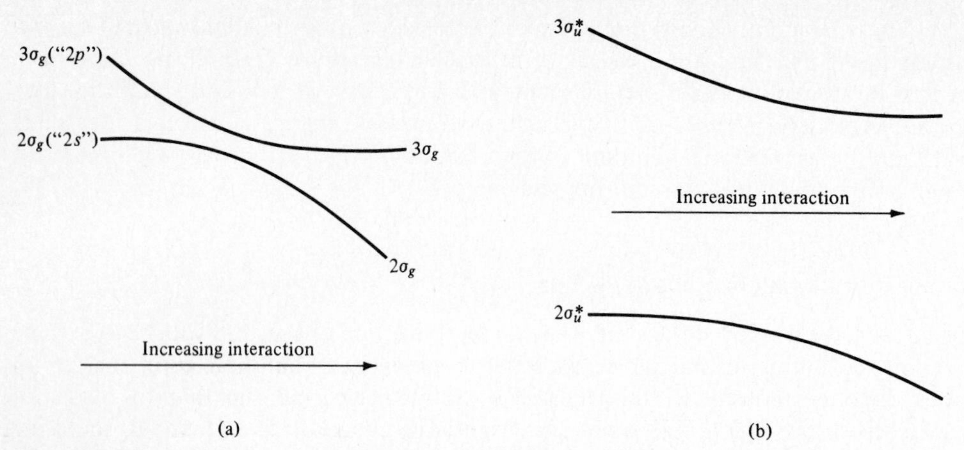

Fig. 3.29 Effects of the no-crossing rule. (a) Strong interaction between orbitals of the same symmetry. (b) Weak interaction between orbitals of the same symmetry.

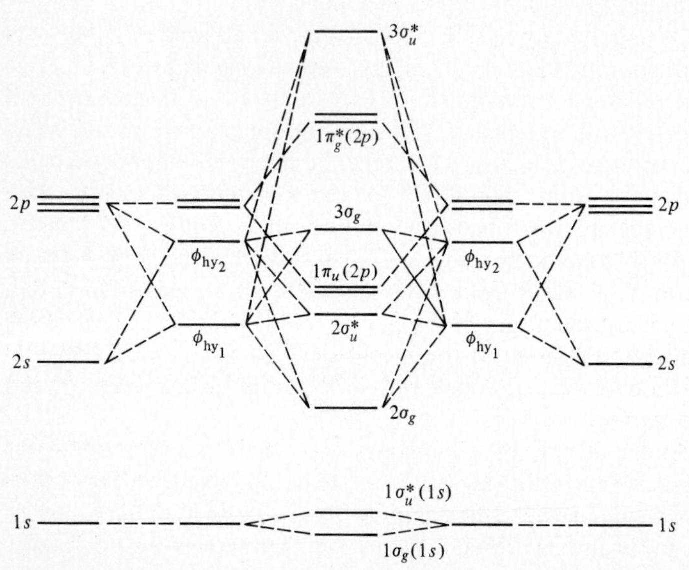

Fig. 3.30 Correct molecular orbital energy levels. Some mixing (hybridization) has occurred between the 2*s* and 2*p* orbitals. Note that it is somewhat more difficult to "keep books" and determine the bond order here than in Fig. 3.28: $2\sigma_g$ and $1\pi_u$ (2*p*) are clearly bonding (they lie below the atomic orbitals contributing to them); $2\sigma_u^*$ and $3\sigma_g$ are essentially nonbonding since they lie between the atomic orbitals contributing to them and roughly symmetrically spaced about the "center of gravity." The maximum net bond order is therefore one σ bond plus two π bonds. [Adapted from M. Orchin and H. H. Jaffé, "The Importance of Antibonding Orbitals," Houghton Mifflin, Boston, **1967**. Reproduced with permission.]

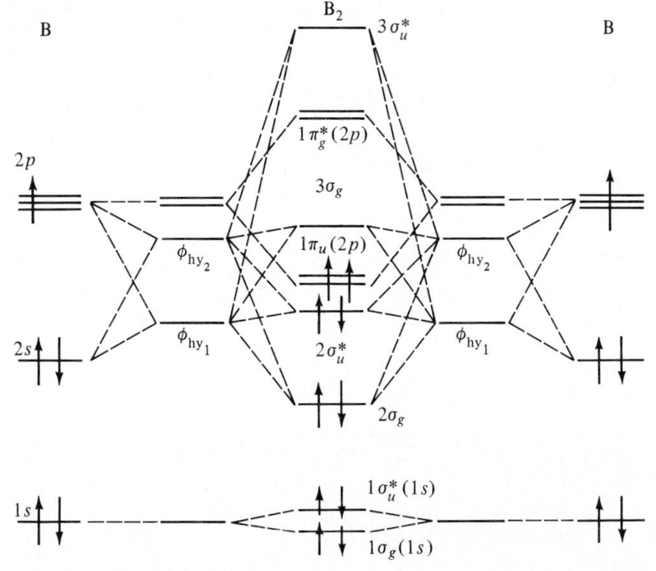

Fig. 3.31 Energy levels in the B_2 molecule. Note unpaired π electrons. [Adapted from M. Orchin and H. H. Jaffé, "The Importance of Antibonding Orbitals," Houghton Mifflin, Boston, **1967**. Reproduced with permission.]

s and *p* orbitals it is no longer appropriate to use the 2*s* and 2*p* labels, so we merely label the orbitals according to their symmetry and in order from the most stable. Of the two σ_g orbitals, the stabilized one becomes the $2\sigma_g$, and the fact that it is lowered stabilizes the molecule considerably. The destabilization of $3\sigma_g$ does not effect the lighter molecules, such as B_2 and C_2, since it is not populated until N_2. There is some interaction between the $\sigma_u^*(2s)$ and the $\sigma_u^*(2p)$ to form the $2\sigma_u^*$ and $3\sigma_u^*$, but because these orbitals do not approach each other closely, the interaction is negligible. The correct energy-level diagram, including both *s–p* mixing and orbital interaction, is shown in Fig. 3.30.

For most molecules discussed previously, such as Li_2, N_2, O_2, and F_2, the differences between electron configurations based on Figs. 3.28 and 3.30 are only qualitative and not susceptible to experimental verification. In the case of B_2 (Fig. 3.31) and C_2, the arrangement based on mixing of orbitals accounts for the experimentally observed magnetic properties. There is also some evidence indicating that in N_2^+ the odd electron is in a σ orbital indicating that the alternative set of energy levels is more realistic.

Electron density in molecules Li_2 through F_2

The approximate shapes of molecular orbitals have been given previously (Fig. 3.27). These give a general idea of the electron distribution in diatomic molecules. Wahl[44] has computed electron density contours for the molecular orbitals of diatomic molecules for

[44] A. C. Wahl, *Science*, **1966**, *151*, 961.

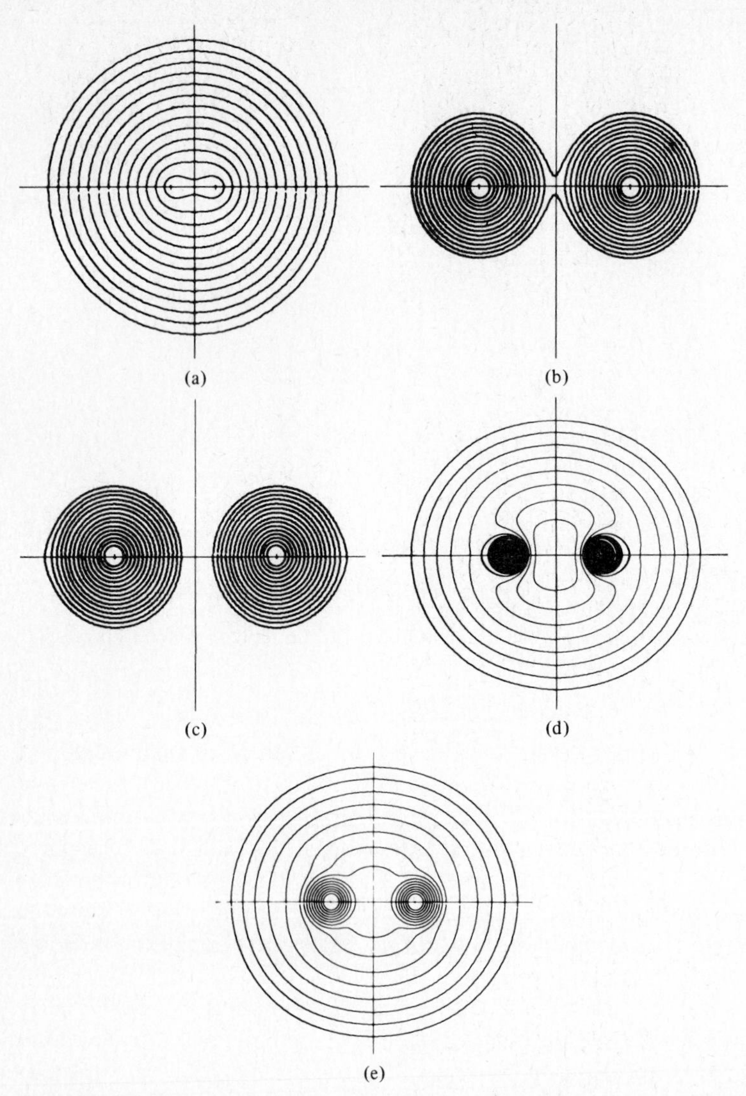

Fig. 3.32 Electron density contours for (a) H_2; (b) Li_2 σ_{1s} core; (c) Li_2 σ_{1s}^* core; (d) Li_2 σ_{2s}; (e) Li_2, total electron density. [From A. C. Wahl, *Science*, **1966**, *151*, 961. Copyright © 1966 by the American Association for the Advancement of Science. Reproduced with permission.]

H_2 to Ne_2. Some examples are shown in Figs. 3.32 and 3.33. Note particularly that: (1) Bonding orbitals cause an increase in electron density between the nuclei; (2) antibonding orbitals have nodes and reduced electron density between nuclei; and (3) inner shells (1*s* in Li, for example) are so contracted from the higher effective nuclear charge that they are nearly spherical with almost no overlap and thus contribute little to the overall bonding. We are thus justified in ignoring these core electrons in determining the molecular electron configuration (p. 106).

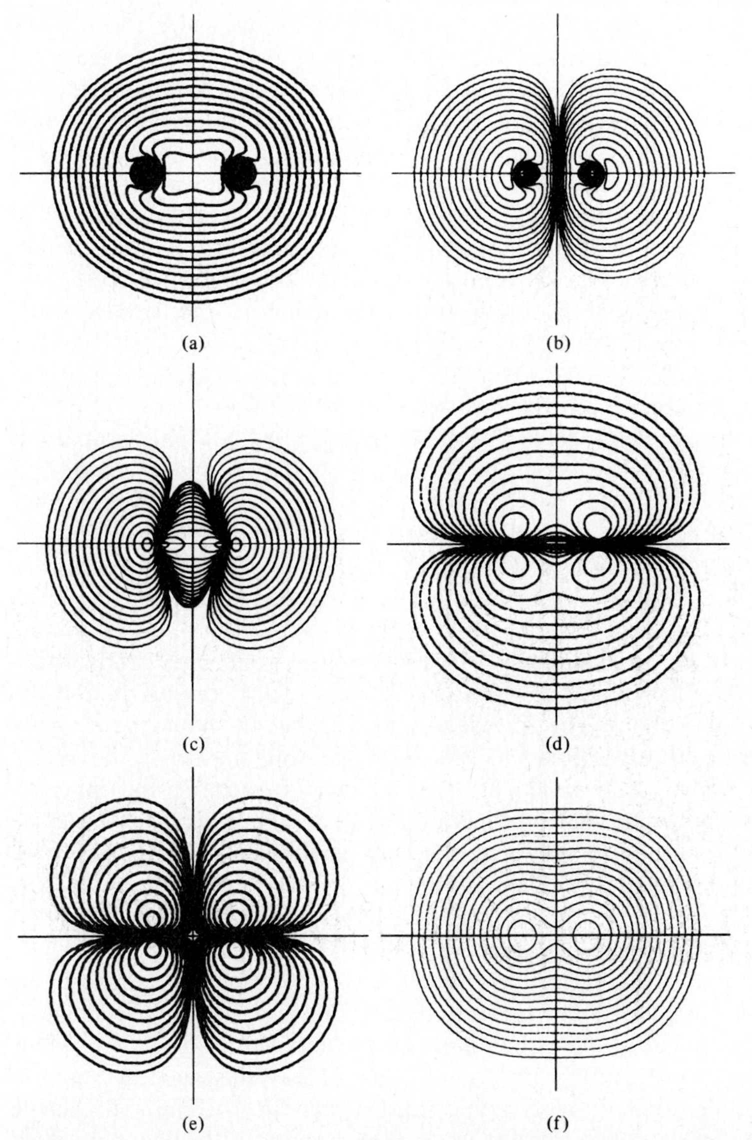

Fig. 3.33 Electron density contours for various orbitals in the O_2 molecule. (a) σ_{2s}; (b) σ_{2s}^*; (c) σ_{2p}; (d) π_{2p}; (e) π_{2p}^*; (f) total electron density. [From A. C. Wahl, *Science*, **1966**, *151*, 961. Copyright © 1966 by the American Association for the Advancement of Science. Reproduced with permission.]

HYBRIDIZATION

Until now, our discussion of the formation of bonds has involved only one type of orbital on an atom at a time, e.g., an $s + s \rightarrow \sigma + \sigma^*$, $p + p \rightarrow \pi + \pi^*$. We have seen the results of "mixing" of orbitals in molecular orbital theory, and hybridization has been mentioned but left unexplained. The mixing of orbitals enters automatically in molecular orbital theory, and theoreticians who tend to work in the molecular orbital context sometimes question the necessity of invoking hybridization. It is an integral part of the valence-bond approach. But even in a molecular orbital context it often aids in visualizing what is actually going on in the bonding process. In this section the topic of hybridization will be discussed in some detail, beginning with the basic concepts and proceeding to some more involved arguments.

Consider the methane molecule, CH_4. The ground state of a carbon atom is ^3P corresponding to the electron configuration of $1s^22s^22p_x^12p_y^1$. Carbon in this state would be divalent because only two unpaired electrons are available for bonding in the p_x and p_y orbitals.[45] Although divalent carbon is well known in methylene and carbene intermediates in organic chemistry, stable carbon compounds are tetravalent. In order for four bonds to form, the carbon atom must be raised to its *valence state*. This requires the *promotion* of one of the electrons from the 2*s* orbital to the formerly empty 2*p* orbital. This excited ^5S state has an electron configuration $1s^22s^12p_x^12p_y^12p_z^1$. This promotion costs 406 kJ mol^{-1}. Since the valence state, V_4, is defined as the state of an atom in a molecule, but without the addition of bonded atoms, it is necessary to supply a further quantity of energy to randomize the spins of the ^5S state, i.e., to supply enough energy to overcome the normal tendency toward parallel spins.[46] Despite all of the energy necessary to reach the valence state, the formation of two additional bonds makes CH_4 895 kJ mol^{-1} more stable than $CH_2 + 2H$.

It might be supposed that upon addition of four hydrogen atoms four bonds would form—three from overlap of the three 2*p* orbitals of carbon with hydrogen 1*s* and the fourth from the carbon 2*s* orbital. The three equivalent 2*p* bonds should be strongly directional since maximum overlap requires that the bonds lie along the axes of the three *p* orbitals involved. The fourth bond would be nondirectional because of the spherical symmetry of the carbon 2*s* orbital, but presumably this bond would tend to lie on the side of the carbon atom opposite to the three mutually perpendicular "*p*" bonds. It is, of course, a well-known fact that a molecule of methane answering this description has never been experimentally observed. All four bonds in methane are identical and all bond angles are $109\frac{1}{2}°$; we call this process *hybridization*.

Hybridization consists of a mixing or linear combination of the "pure" *s* and *p* orbitals of an atom in such a way as to form new hybrid orbitals. Thus we say that the single 2*s* orbital plus the three 2*p* orbitals of the carbon atom have combined to form a set of four spatially and energetically equivalent sp^3 hybrid orbitals. This is illustrated in Fig. 3.34 for the conceptually simpler case of the *sp* hybrid formed from an *s* orbital and a single *p* orbital. Combination of the *s* and *p* orbital causes a reinforcement in the region in which the sign of the wave function is the same, cancellation where the signs are opposite.

[45] The choice of p_x and p_y here instead of, for example, p_y and p_z is, of course, completely arbitrary.

[46] The existence of this extra valence state excitation energy may be clearer if the reverse process is considered. If (in a thought experiment) four hydrogen atoms are removed from methane but the carbon is not allowed to change in any way, the resulting spins will be perfectly randomized. Energy would then be released if the spins were allowed to become parallel. See C. A. Coulson, "Valence," 2nd ed., Oxford University Press, London, **1961**, pp. 206–208. It should be noted that unlike ^3P, ^1S, etc., V_4 is not an observable spectroscopic state but is calculated by adding to the spectroscopic states energies for parallel spins.

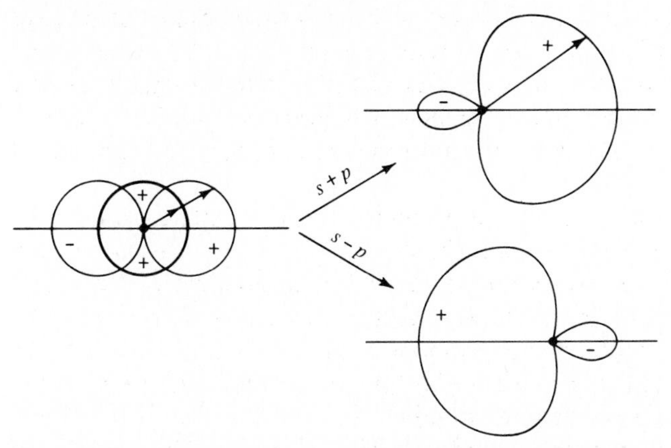

Fig. 3.34 Formation of *sp* hybrid orbitals by the addition and subtraction of angular wave functions.

If we let Ψ_s and Ψ_p represent the wave functions of an *s* and a *p* orbital, then we combine them to make two equivalent orbitals as follows:

$$\Psi_{di_1} = \frac{1}{\sqrt{2}}(\Psi_s + \Psi_p) \tag{3.47}$$

$$\Psi_{di_2} = \frac{1}{\sqrt{2}}(\Psi_s - \Psi_p) \tag{3.48}$$

where $1/\sqrt{2}$ is the normalizing coefficient and Ψ_{di_1} and Ψ_{di_2} are the new *digonal* (*di*) or *sp* orbitals. If this process appears reminiscent of the formation of molecular orbitals, it is because in both cases linear combinations of atomic orbitals are being made. It should be kept in mind, however, that in the present case we are combining two or more orbitals on the *same atom* to form a new set of hybrid atomic orbitals.

Mathematically, the formation of sp^3 or tetrahedral orbitals for methane is more complicated but not basically different. The results are four equivalent hybrid orbitals, each containing one part *s* to three parts *p* in each wave function, directed to the corners of a tetrahedron. As in the case of *sp* hybrids, the hybridization of *s* and *p* has resulted in one lobe of the hybrid orbital being much larger than the other (see Fig. 3.35). Hybrid orbitals may be pictured in many ways: by several contour surfaces (Fig. 3.35); a single, "outside" contour surface (Fig. 3.36a); cloud pictures (Fig. 3.36b); or by simpler, diagrammatic sketches which ignore the small, nonbonding lobe of the orbital and picture the larger, bonding lobe (Fig. 3.36c). The latter, though badly distorted, are commonly used in drawing molecules containing several hybrid orbitals.

It is possible to form a third type of *s–p* hybrid containing one *s* orbital and two *p* orbitals. This is called an sp^2 or *trigonal* (*tr*) hybrid. It consists of three identical orbitals, each of which does not differ appreciably in shape from Fig. 3.35 and is directed toward the corner of an equilateral triangle. The angles between the axes of the orbitals in a trigonal hybrid are thus all 120°.

Although promotion and hybridization are connected in the formation of methane from carbon and hydrogen, care should be taken to distinguish between them. Promotion

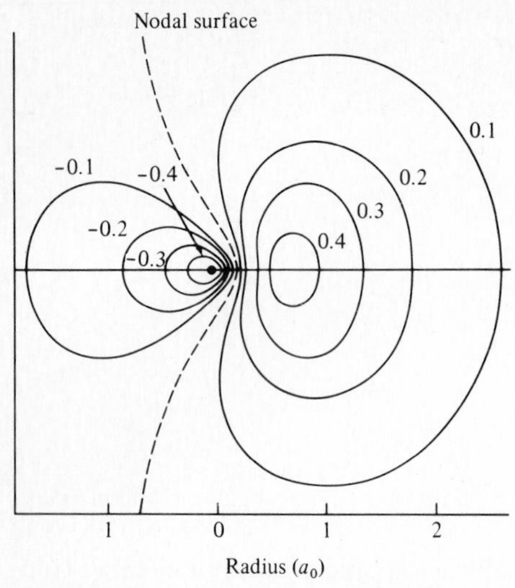

Fig. 3.35 Electron density contours for an sp^3 hybrid orbital. Note that the nodal surface does not pass through the nucleus. [Adapted from C. A. Coulson, "Valence," 2nd ed., Clarendon Press, Oxford, **1961**. Reproduced with permission.]

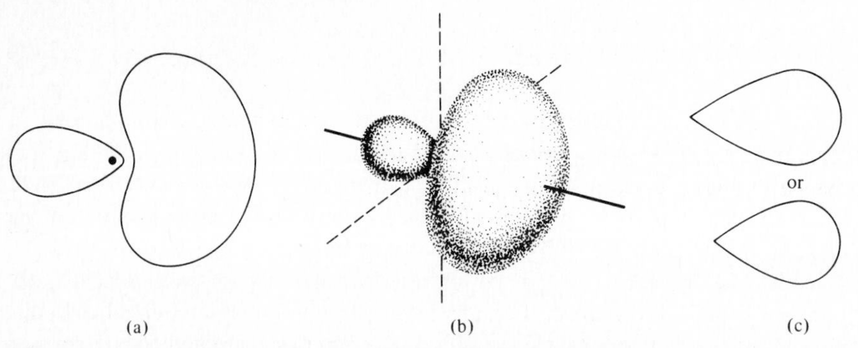

Fig. 3.36 Other ways of representing hybrid orbitals. (a) Orbital shape shown by single contour. (b) Cloud representation. (c) Simplified representation. The small back lobes have been omitted and the shape streamlined to make it easier to draw molecules containing several hybrid orbitals.

involves the addition of energy to raise an electron to a higher energy level in order that the two additional bonds may form. It is conceivable that after promotion the carbon atom could have formed three bonds with the three *p* orbitals and the fourth with the *s* orbital. That carbon forms tetrahedral bonds instead is a consequence of the greater stability of the latter, not a necessary result of promotion. Thus, although promotion and hybridization often occur together, either could occur without the other.

A second point to be made with regard to hybrids is the source of the driving force resulting in hybridization. Statements are often made to the effect that "methane is tetrahedral because the carbon is hybridized sp^3." This is very loose usage and gets the cart before the horse. The methane molecule is tetrahedral because the energy of the molecule is lowest in that configuration, principally because of increased bond energies and decreased repulsion energies. For this molecule to be tetrahedral, VB theory demands that sp^3 hybridization take place. Thus it is incorrect to attribute the shape of a molecule to hybridization—the latter *prohibits* certain configurations and *allows* others but does not indicate a *preferred* one. For example, consider the following possibilities for the methane molecule:

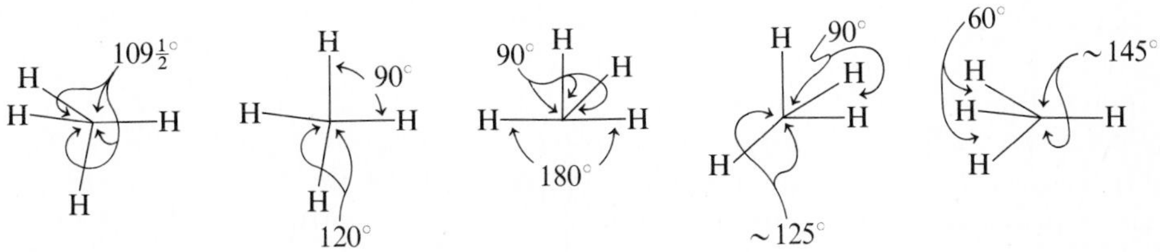

(I) sp^3 (II) $sp^2 + p$ (III) $sp + p^2$ (IV) $s + p^3$ (V)

The first three geometries involve the tetrahedral, trigonal, and digonal hybrids discussed above and the fourth involves the use of pure s and p orbitals as discussed on p. 114. The last structure contains three equivalent bonds at mutual angles of 60° and a fourth bond at an angle of approximately 145° to the others. It is impossible to construct s–p hybrid orbitals with angles less than 90°, and so structure (V) is ruled out. In this sense it may be said that hybridization does not "allow" structure (V), but it may *not* be said that it "chooses" one of the others. Carbon hybridizes sp, sp^2, and sp^3 in various compounds, and the choice of sp^3 in methane is a result of the fact that the tetrahedral structure is the most stable possible.

Although we shall not make explicit use of them, the reader may be interested in the form of the s–p hybrids we have seen:[47]

$$\Psi_{tr_1} = \frac{1}{\sqrt{3}}\Psi_s + \sqrt{\tfrac{2}{3}}\Psi_{p_x} \tag{3.49}$$

$$\Psi_{tr_2} = \frac{1}{\sqrt{3}}\Psi_s - \frac{1}{\sqrt{6}}\Psi_{p_x} + \frac{1}{\sqrt{2}}\Psi_{p_y} \tag{3.50}$$

$$\Psi_{tr_3} = \frac{1}{\sqrt{3}}\Psi_s - \frac{1}{\sqrt{6}}\Psi_{p_x} - \frac{1}{\sqrt{2}}\Psi_{p_y} \tag{3.51}$$

$$\Psi_{te_1} = \tfrac{1}{2}\Psi_s + \tfrac{1}{2}\Psi_{p_x} + \tfrac{1}{2}\Psi_{p_y} + \tfrac{1}{2}\Psi_{p_z} \tag{3.52}$$

$$\Psi_{te_2} = \tfrac{1}{2}\Psi_s - \tfrac{1}{2}\Psi_{p_x} - \tfrac{1}{2}\Psi_{p_y} + \tfrac{1}{2}\Psi_{p_z} \tag{3.53}$$

$$\Psi_{te_3} = \tfrac{1}{2}\Psi_s + \tfrac{1}{2}\Psi_{p_x} - \tfrac{1}{2}\Psi_{p_y} - \tfrac{1}{2}\Psi_{p_z} \tag{3.54}$$

$$\Psi_{te_4} = \tfrac{1}{2}\Psi_s - \tfrac{1}{2}\Psi_{p_x} + \tfrac{1}{2}\Psi_{p_y} - \tfrac{1}{2}\Psi_{p_z} \tag{3.55}$$

[47] The forms for common hybrids are given in Eqs. 3.49 to 3.55. The percent s and p character is proportional to the *square* of the coefficients since the wave function must be squared to have physical meaning. Taken from C. Y. Hsu and M. Orchin, *J. Chem. Educ.*, **1973**, *50*, 114.

It is not necessary to limit hybridization to *s* and *p* orbitals. The only criterion is that the radial parts of the wave functions of the orbitals being hybridized be similar; otherwise the radial wave function of the hybrid would be unsuited for bonding since the electron density would be spread too thinly. In practice this means that hybrids are formed among orbitals lying in the same principal energy level or, occasionally, in adjacent energy levels.

Some hybrid orbitals containing *s*, *p*, and *d* orbitals are listed in Table 3.10. The structural aspects of various hybrid orbitals will be discussed in Chapter 5, but the bond angles between orbitals of a given hybridization are also listed in Table 3.10 for reference.

Most sets of hybrid orbitals are equivalent and symmetric, i.e., four sp^3 orbitals directed to the corners of a regular tetrahedron, six d^2sp^3 orbitals to the corners of an octahedron, etc. In the case of sp^3d hybrids the resulting orbitals are not equivalent. In the trigonal bipyramidal arrangement three orbitals directed trigonally form one set of equivalent orbitals (these may be considered sp^2 hybrids) and two orbitals directed linearly (and perpendicular to the plane of the first three) form a second set of two (these may be considered *dp* hybrids). The former set is known as the *equatorial* orbitals and the latter as the *axial* orbitals. Because of the nature of the different orbitals involved, bonds formed from the two are intrinsically different and will have different properties even when bonded to identical atoms. For example, in molecules like PF_5 bond lengths differ for axial and equatorial bonds (see pp. 224–227).

Even in the case of *s*–*p* orbitals it is not necessary that all the orbitals be equivalent. Consider the water molecule, in which the H—O—H angle is $104\frac{1}{2}°$, which does not correspond to any of the hybrids described above, but lies between the $109\frac{1}{2}°$ angle for sp^3 and 90° for pure *p* orbitals. Presumably the two bonding orbitals in water are approximately tetrahedral orbitals but contain a little more *p* character, which correlates with the tendency of the bond angle to diminish toward the 90° of pure *p* orbitals. The driving forces for this effect will be discussed in Chapter 5.

The relationship between *p* or *s* character and bond angle will also be discussed in Chapter 5. For now we need only consider the possibility of *s*–*p* hybridization other than *sp*, sp^2, and sp^3. If we take the ratio of the *s* contribution to the total orbital complement in these hybrids, we obtain 50%, 33%, and 25% *s* character, respectively, for

Table 3.10 Bond angles of hybrid orbitals

Hybrid	Geometry	Bond angle(s)
sp(*di*)	Linear (digonal)	180°
$sp^2(tr)$	Trigonal	120°
$sp^3(te)$	Tetrahedral	$109\frac{1}{2}°$
dsp^3	Trigonal bipyramidal	90°, 120°
	or	
	Square pyramidal[a]	$>90°$, $<90°$
d^2sp^3	Octahedral	90°

[a] Not common.

Table 3.11 Effect of hybridization on overlap and bond properties

Molecule	Hybridization	C—H bond energy ($kJ\ mol^{-1}$)	C—H bond length (pm)
H—C≡C—H	sp	~506	106
$H_2C{=}CH_2$	sp^2	~444	107
CH_4	sp^3	410	109
CH radical	~p	~335	112

these hybrids. A pure *s* orbital would be 100% *s*, and a *p* orbital would have 0% *s* character. Since hybrid orbitals are constructed as linear combinations of *s* and *p* orbital wave functions,

$$\phi = a\Psi_s + b\Psi_p \tag{3.56}$$

there is no constraint that *a* and *b* must have values such that the *s* character is exactly 25%, 33%, or 50%. A value of 20% *s* character is quite acceptable, for example, and indeed this happens to be the value in water. When the hybridization is defined as above, the % *p* character is always the complement of % *s*, in the case of water, 80%.

Hybridization and overlap

Both pure *s* and pure *p* orbitals provide relatively inefficient overlap compared with that of hybrid orbitals. The relative overlap of hybrid orbitals decreases in the order $sp > sp^2 > sp^3 \gg p$. The differences in bonding resulting from hybridization effects on overlap can be seen in Table 3.11. The C—H bond in acetylene is *shorter* and *stronger* than in hydrocarbons having less *s* character in the bonding orbital. The hybridization in the hydrocarbons listed in Table 3.11 is dictated by the stoichiometry and stereochemistry. In molecules where variable hybridization is possible, various possible hybridizations, overlaps, and bond strengths are possible. Other things being equal,[48] we should expect molecules to maximize bond energies through the use of appropriate hybridizations.

Molecular orbital equivalent of hybridization

Hybridization does not enter explicitly into MO theory, but mixing of orbitals does occur as in VB theory. Since several examples of the application of MO methods to molecules will be discussed in later chapters, only a very simple example will be given here—that of the linear triatomic molecule BeH_2. The molecular orbitals for this molecule are constructed from the 1*s* orbitals on the hydrogen atoms (labeled H and H′) and the 2*s* and one of the 2*p* orbitals of the beryllium. The remaining two 2*p* orbitals of the beryllium

[48] As someone has sagely remarked, "other things" are *seldom* equal, but we can expect to see the effects of the idea expressed here on the molecular properties.

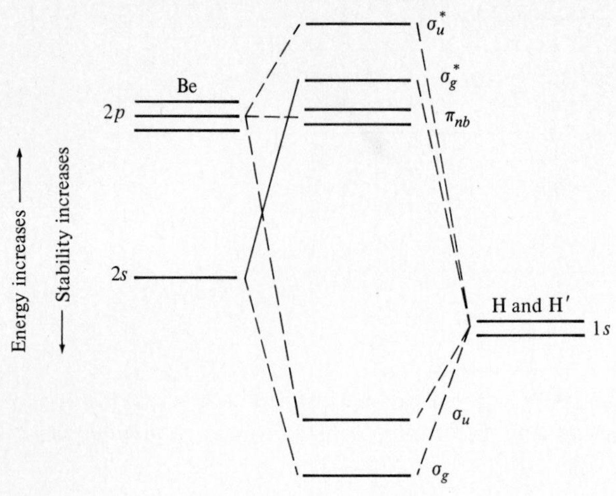

Fig. 3.37 Molecular orbital energy levels in the BeH_2 molecule.

cannot enter into bonding because they have zero overlap with the hydrogen orbitals. Since four atomic orbitals enter into the bonding, we anticipate the formation of four molecular orbitals. The bonding molecular orbitals are formed by linear combination of the atomic orbitals to give maximum overlap:

$$\Psi_g = a\Psi_{2s} + b(\Psi_{\mathrm{H}} + \Psi_{\mathrm{H'}}) = \sigma_g \tag{3.57}$$

$$\Psi_u = c\Psi_{2p} + d(\Psi_{\mathrm{H}} - \Psi_{\mathrm{H'}}) = \sigma_u \tag{3.58}$$

$$\Psi_g^* = b\Psi_{2s} - a(\Psi_{\mathrm{H}} + \Psi_{\mathrm{H'}}) = \sigma_g^* \tag{3.59}$$

$$\Psi_u^* = d\Psi_{2p} - c(\Psi_{\mathrm{H}} - \Psi_{\mathrm{H'}}) = \sigma_u^* \tag{3.60}$$

The subscripts g and u refer to the symmetry of the molecular orbitals (see p. 103) but may be considered useful labels. The parameters a, b, c, and d are weighting coefficients and are necessary because of differences in electronegativity between Be and H. They will be discussed in the section on heteropolar bonds. The energies of these molecular orbitals are shown in Fig. 3.37 and electron density boundary surfaces are sketched in Fig. 3.38. Both of the bonding molecular orbitals are *delocalized* over all three atoms. This is a general result of the MO treatment of polyatomic molecules. It is possible to convert these two delocalized molecular orbitals into localized molecular orbitals that resemble those obtained from VB treatment.[49] In any event, the mixing of the atomic orbitals occurs naturally in MO treatment. Prior hybridization of the atomic orbitals is not necessary but may be a convenience. Whether one goes through the formal step of hybridization or not, one should not lose sight of the fact that terms such as "25% *s* character" are just as appropriate in MO treatments as in VB treatments.

[49] W. A. Bernett, *J. Chem. Educ.*, **1969**, *46*, 746; I. Cohen and J. Del Bene, *J. Chem. Educ.*, **1969**, *46*, 487; D. K. Hoffman et al., *J. Chem. Educ.*, **1977**, *54*, 590.

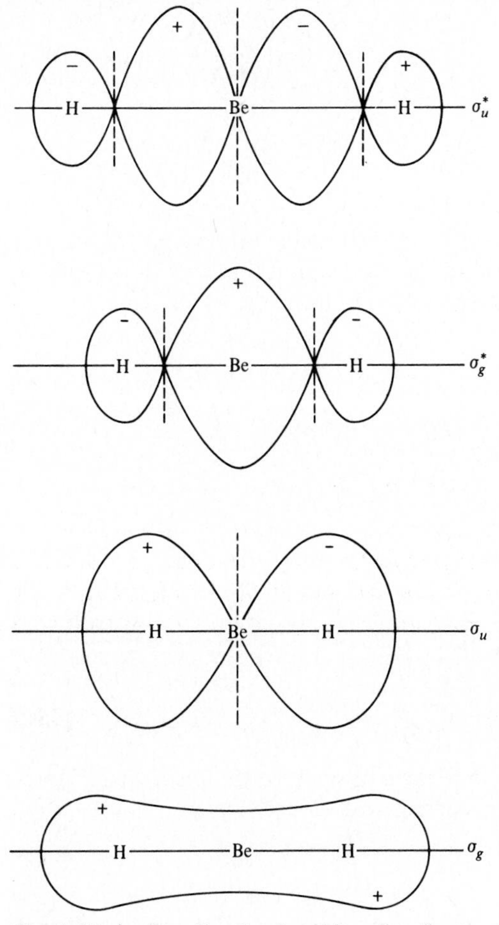

Fig. 3.38 Antibonding (top) and bonding (bottom) molecular orbitals in the BeH_2 molecule.

DELOCALIZATION

Resonance

When using valence bond theory it is often found that more than one acceptable structure can be drawn for a molecule or, more precisely, more than one wave function can be written. We have already seen in the case of the hydrogen molecule that we could formulate it either as H—H or as H^+H^-. Both are acceptable structures, but the second or ionic form would be considerably higher in energy than the "covalent" structure (because of the high ionization energy and low electron affinity of hydrogen). However, we may write the wave function for the hydrogen molecule as a linear combination of the ionic and covalent functions:

$$\Psi = (1 - \lambda)\Psi_{cov} + \lambda\Psi_{ion} \tag{3.61}$$

where λ determines the contribution of the two wave functions. When this is done, it is found that the new wave function is *lower* in energy than either of the *contributing structures*. This is a case of covalent–ionic resonance which will be discussed at greater length in the section on *electronegativity*.

Another type of resonance arises in the case of the carbonate ion. A simple Lewis structure suggests that the ion should have three σ bonds and one π bond. However, when it comes to the placement of the π bond, it becomes obvious that there is no unique way to draw the π bond. There is no *a priori* reason for choosing one oxygen atom over the other two to receive the π bond. We also find experimentally that it is impossible to distinguish one oxygen atom as being in any way different from the other two.

We can draw three equivalent contributing structures for the carbonate ion:

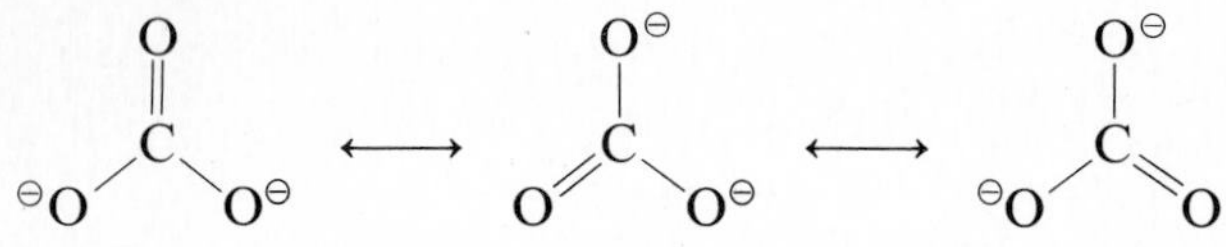

(I) ⟷ (II) ⟷ (III)

Each of these structures may be described by a wave function, Ψ_{I}, Ψ_{II}, or Ψ_{III}. The actual structure of the carbonate ion is none of the above, but a *resonance hybrid* formed by a linear combination of the three *canonical structures*:

$$\Psi = a\Psi_{\mathrm{I}} + b\Psi_{\mathrm{II}} + c\Psi_{\mathrm{III}} \tag{3.62}$$

There is no simple Lewis structure that can be drawn to picture the resonance hybrid, but the following gives a qualitative idea of the correct structure:

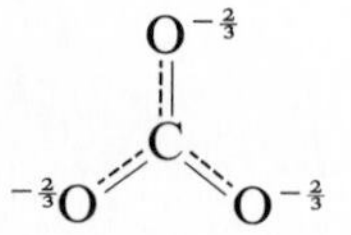

(IV)

It is found that the energy of (IV) is lower than that of (I), (II), or (III). It is common to speak of the difference in energy between (I) and (IV) as the *resonance energy* of the carbonate ion. One should realize however, that the resonance energy arises only because of the fact that our wave functions Ψ_{I}, Ψ_{II}, and Ψ_{III} are rather poor descriptions of the actual structure of the ion. In a sense, then, the resonance energy is simply a measure of our ignorance of the true wave function. More accurately, the resonance energy and the entire phenomenon of resonance are merely a result of the overly restrictive approach we have adopted in valance bond theory in insisting that a "bond" be a *localized pair* of electrons between two nuclei. When we encounter a molecule or ion in which one or more pairs of electrons are *delocalized* we must then remedy the situation by invoking resonance. We should not conclude, however, that valence bond theory is wrong—it merely gets cumbersome sometimes when we have many delocalized electrons to consider. In contrast, in molecules in which the electrons are localized, the valence bond theory often proves to be especially useful.

In the carbonate ion, the energies of the three contributing structures are identical, and so all three contribute equally ($a = b = c$) and the hybrid is exactly intermediate between the three. In many cases, however, the energies of the contributing structures differ (the hydrogen molecule was an example), and in these cases we find that the contribution of a canonical structure is *inversely* proportional to its energy; i.e., high energy, unstable structures contribute very little, and for resonance to be appreciable, the energies of the contributing structures must be comparable. Using the energy of the contributing structures as a basis, we can draw up a set of general rules for determining the possibility of contribution of a canonical structure.

1. The proposed canonical structure should have a maximum number of bonds, consistent, of course, with the other rules. In the carbon dioxide molecule, for example, the structure

```
 ..          
:O:⁻C:⁺:O:
 ..      ..
```

plays no appreciable role because of its much higher energy resulting from loss of the π-bonding stabilization. In general, application of this rule is simply a matter of drawing Lewis structures and using good chemical sense in proposing contributing structures.

2. The proposed canonical structures must be consistent with the location of the atoms in the actual molecule (resonance hybrid). The most obvious consequence of this rule is the elimination of tautomers as possible resonance structures. Thus the following structures for phosphorous acid represent an equilibrium between two distinct chemical species, *not* resonance:

```
    H                    
 .. .. ..             .. .. ..
H:O:P:O:H   ⇌   H:O:P:O:H
 .. .. ..             .. .. ..
   :O:                 :O:
    ..                  ..
                        H
```

A less obvious result of this criterion is that when contributing structures differ in bond angle, resonance will be reduced. Consider, for example, the following hypothetical resonance for nitrous oxide:[50]

```
 ..       ..
:N::N::O:   ⟷   :N::N:
 -   +             :O:
                    ..
   (I)             (II)
```

Aside from the fact that (II) is a strained structure and therefore less stable than (I), it will not contribute to the resonance of N_2O since the bond angle is 180° in (I) and 60° in (II).

[50] For the moment the charges should be ignored; they will be discussed in the next section.

For any intermediate hybrid, the contribution of either (I) or (II) would be unfavorable because of the high energy cost when (I) is bent or when (II) is opened up.

A few words should be said about the difference between resonance and molecular vibrations. Although vibrations take place, they are oscillations about an equilibrium position determined by the structure of the resonance hybrid, and *they should not be confused with the resonance among the contributing forms*. The molecule does not "resonate" or "vibrate" from one canonical structure to another. In this sense the term "resonance" is unfortunate since it has caused unnecessary confusion by invoking a picture of "vibration." The term arises from a mathematical analogy between the molecule and the classical phenomenon of resonance between coupled pendulums, or other mechanical systems.

3. Distribution of charges in a contributing structure must be reasonable. This rule may be considered in two parts. First, canonical forms in which *adjacent-like charges* appear will probably be unstable as a result of the electrostatic repulsion. A structure such as $A^-—B^+—C^+—D^-$ is therefore unlikely to play a major role in hybrid formation.

In the case of charges in which adjacent charges are *not* of the same sign, one must use some chemical discretion in estimating the contribution of a particular structure. This is best accomplished by examining the respective electronegativities of the atoms involved. A structure in which a positive charge resides on an electropositive element and a negative charge resides on an electronegative element may be quite stable, but the reverse will represent an unstable structure. For example, in the following two cases

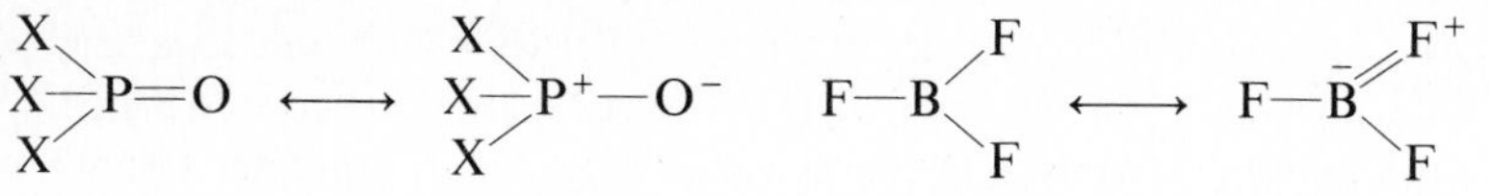

$$X_3P{=}O \longleftrightarrow X_3P^+{—}O^- \qquad F{—}B(F)_2 \longleftrightarrow F{—}\bar{B}(F)({=}F^+)$$

(I) (II) (I) (II)

the former contributes very much to the actual structure of phosphoryl compounds, but (II) contributes much less to BF_3 and, indeed, the actual contribution in compounds of this sort is still a matter of some dispute.

Furthermore, placement of adjacent charges of *opposite sign* will be more favorable than when these charges are separated. When adjacent, charges of opposite sign contribute electrostatic energy toward stabilizing a molecule (similar to that found in ionic compounds), but this is reduced when the charges are far apart.

4. Contributing forms must have the same number of unpaired electrons. For molecules of the type discussed previously, structures having unpaired electrons should not be considered since they usually involve loss of a bond

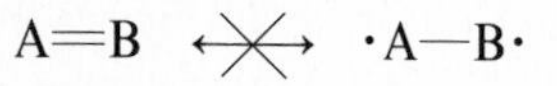

$$A{=}B \;\not\longleftrightarrow\; \cdot A{—}B\cdot$$

(I) (II)

and higher energy for structure (II). We shall see when considering coordination compounds, however, that complexes of the type ML_n (where M = metal, L = ligand) can exist with varying numbers of unpaired electrons but comparable energies. Nevertheless, resonance between such structures is still forbidden since spin of electrons is quantized and a molecule either has its electrons paired or unpaired (an intermediate or "hybrid" situation is impossible).

These rules may be applied to nitrous oxide, N_2O. Two structures which are important are:

$$:\underset{-}{\ddot{N}}::\underset{+}{N}::\ddot{O}: \longleftrightarrow :N:::\underset{+}{N}:\underset{\overset{..}{-}}{\ddot{O}}:$$

(I) (II)

Both of these structure have four bonds and the charges are reasonably placed. A third structure

$$:\underset{\overset{..}{-2}}{\ddot{N}}:\underset{+}{N}:::\underset{+}{O}:$$

(III)

is unfavorable because it places a positive charge on the electronegative oxygen atom and also has adjacent positive charges.

Other possibilities are:

$$:\ddot{N}:\ddot{N}::\ddot{O}: \qquad :\underset{-}{\ddot{N}}::\ddot{N}:\underset{\overset{..}{+}}{\ddot{O}}:$$

(IV) (V)

and the cyclic structure discussed under Rule 2. This last has been shown above to be unfavorable. Likewise (IV) and (V) should be bent and are energetically unfavorable when forced to be linear to resonate with (I) and (II). In addition, both have only three bonds instead of four and are therefore less stable. Furthermore, (V) has widely separated charges, and they are exactly opposite to those expected from electronegativity considerations.

It is almost impossible to overemphasize the fact that the resonance hybrid is the only structure which is actually observed and that the canonical forms are merely constructs which enable us to describe accurately the experimentally observed molecule. The analogy is often made that the resonance hybrid is like a mule, which is a genetic hybrid between a horse and a donkey. The mule is a mule and does not "resonate" back and forth between being a horse and a donkey. It is as though one were trying to describe a mule to someone who had never seen one before and had available only photographs of a jackass and a mare. One could then explain that their offspring, intermediate between them, was a mule.[51] There is perhaps a better analogy, though one that will be unfamiliar to those not versed in ancient mythology: Consider a falcon (a real animal) described as a hybrid of Rē (the falcon-headed Egyptian sun god) and a harpy (a creature with a woman's head and the body of a raptor), although neither of the latter has an independent existence.

---

[51] This analogy, like any other, can be pushed too far. The contributing structures should not be considered as "parents" of the hybrid.

Formal charge

In the discussion of nitrous oxide, charges were introduced without explanation. These charges are *formal charges*, which may be defined as the charge that an atom in a molecule would have if all of the atoms had the same electronegativity. It may correspond rather closely to a real *ionic charge* as in the ammonium ion, NH_4^+. In the ammonium ion the formal charge on each of the hydrogen atoms is zero and the nitrogen atom carries a single positive, formal charge corresponding to the ionic charge. On the other hand, some molecules, such as N_2O, exhibit formal charges on otherwise neutral atoms.

To obtain the formal charge on an atom, it is assumed that all electrons are shared equally and that each atom "owns" one-half of the electrons it shares with neighboring atoms. The formal charge is then:

$$Q_F = N_A - N_M = N_A - N_{LP} - \tfrac{1}{2}N_{BP} \tag{3.63}$$

where N_A is the number of electrons in the valence shell in the free atom and N_M is the number of electrons "belonging to the atom in the molecule" or N_{LP} and N_{BP} are the numbers of electrons in *unshared* pairs and *bonding* pairs, respectively.

Applied to the Lewis structure of the *ammonium ion*,

$$\begin{array}{c} \mathrm{H} \\ \mathrm{H{:}\ddot{\underset{\cdot\cdot}{N}}{}^{+}{:}H} \\ \mathrm{H} \end{array} \quad \text{or} \quad \begin{array}{c} \mathrm{H} \\ | \\ \mathrm{H{-}N^{+}{-}H} \\ | \\ \mathrm{H} \end{array}$$

we obtain the following formal charges:

$$Q_N = 5 - 4 = 5 - 0 - \tfrac{1}{2}(8) = +1 \tag{3.64}$$

$$Q_H = 1 - 1 = 1 - 0 - \tfrac{1}{2}(2) = 0 \tag{3.65}$$

This is in contrast to the *ammonia molecule*,

$$\begin{array}{c} \mathrm{H} \\ \mathrm{H{:}\ddot{\underset{\cdot\cdot}{N}}{:}H} \end{array} \quad \text{or} \quad \begin{array}{c} \mathrm{H} \\ | \\ \mathrm{H{-}\underset{\cdot\cdot}{N}{-}H} \end{array}$$

for which the formal charges are

$$Q_N = 5 - 5 = 5 - 2 - \tfrac{1}{2}(6) = 0 \tag{3.66}$$

$$Q_H = 1 - 1 = 1 - 0 - \tfrac{1}{2}(2) = 0 \tag{3.67}$$

To return now to nitrous oxide, N_2O, specifically structure (I), we have a Lewis structure:

$$:\ddot{N}_t::N_c::\ddot{O}: \quad \text{or} \quad N_t = N_c = O$$

where t = *terminal* (or left) and c = *central*, merely as identifying labels. Hence:

$$Q_{N_t} = 5 - 6 = 5 - 4 - \tfrac{1}{2}(4) = -1 \quad \textbf{(3.68)}$$

$$Q_{N_c} = 5 - 4 = 5 - 0 - \tfrac{1}{2}(8) = +1 \quad \textbf{(3.69)}$$

$$Q_{O} = 6 - 6 = 6 - 4 - \tfrac{1}{2}(4) = 0 \quad \textbf{(3.70)}$$

Formal charges are used less frequently today than formerly since better methods of treating electronegativity have developed, but they are often useful in a qualitative way as in the resonance example above.

Although formal charges do not represent real, ionic charges, they do represent a tendency for buildup of positive and negative charges. For example, consider carbon monoxide. The only reasonable Lewis structure that maximizes the bonding is the normal triple-bonded one:

$$:\overset{-}{C}:::\overset{+}{O}:$$

Note, however, that this places a formal positive charge on the oxygen and a formal negative charge on the carbon. If the electronegativities of carbon and oxygen were the same, carbon monoxide would have a sizable dipole moment (see pp. 160–162) in the direction

$$\underset{\longleftarrow\!\!\!+}{:C:::O:}$$

but since the electronegativity difference draws electron density back from the carbon atom to the oxygen atom, the effect is canceled and carbon monoxide has one of the lowest dipole moments known, 0.4×10^{-30} C m.

Molecular orbital equivalent of resonance

Resonance is a necessity in valence bond theory whenever *delocalized* electrons are present, i.e., electrons distributed over three or more nuclei. This is not necessary in molecular orbital theory since bonding electrons in molecular orbitals are not necessarily localized between two nuclei in the first place. In other words, valence bond theory overemphasizes the localized electron pair bond and uses resonance to compensate. In this section we shall look at the alternative MO method of treating delocalized electrons.

Consider the nitrite ion, NO_2^-. The σ system consists of the two N—O single bonds and the four lone pairs on the oxygens in sp^2 hybrid orbitals.[52] This leaves parallel p orbitals on the nitrogen and the oxygen atoms (Fig. 3.39). These orbitals will interact to form bonding and antibonding orbital combinations:

$$\Psi_b = p_{O_1} + p_{O_2} + p_N \qquad (\pi) \quad \textbf{(3.71)}$$

$$\Psi_a = p_{O_1} + p_{O_2} - p_N \qquad (\pi^*) \quad \textbf{(3.72)}$$

[52] Hybridization is, of course, completely unnecessary when using molecular orbital theory, but is a convenience here since we are primarily concerned with the π system. The VB and MO treatments of the σ system are not significantly different in their results.

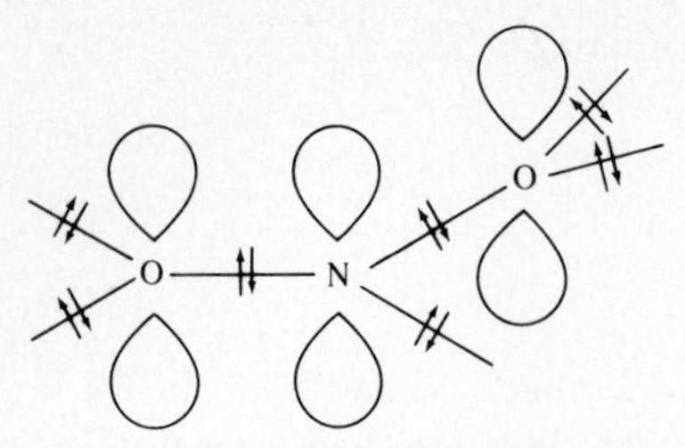

Fig. 3.39 Sigma bonds and lone pairs in the nitrite ion, NO_2^-.

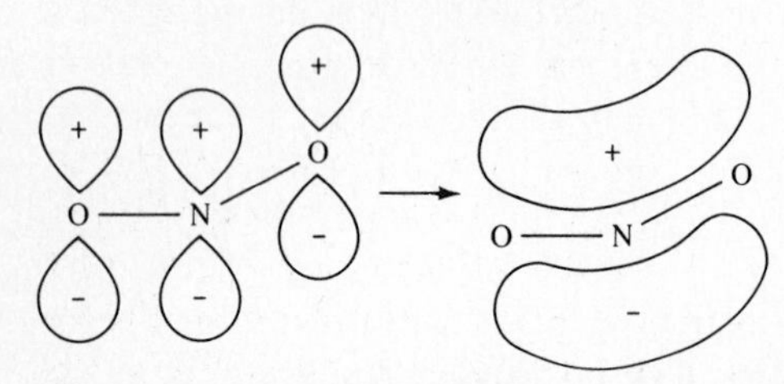

Fig. 3.40 Atomic orbitals (left) and resulting *bonding* molecular orbital (right) for π bond in the nitrite ion.

As in the cases we have seen before, the bonding orbital represents a concentration of electron density between the atoms (Fig. 3.40) and the antibonding orbital has nodes between the atoms (Fig. 3.41). There is a third combination possible:

$$\Psi_n = p_{O_1} - p_{O_2} \tag{3.73}$$

This represents a nonbonding situation since the positive overlap between p_{O_1} and p_N is cancelled by the negative overlap between p_{O_2} and p_N (Fig. 3.42).

The result is similar to that obtained by the valence bond picture with resonance. There is a bonding pair of π electrons spread over the nitrogen and two oxygen atoms. The second pair of electrons is nonbonding and effectively localized on the oxygen atoms (the node cuts through the nitrogen atom), but as in the VB picture, they, too, are "smeared" over both oxygen atoms rather than occupying discrete atomic orbitals. The MO method has the disadvantage of being somewhat less straightforward, in that the appropriate molecular orbitals must be constructed rather than the contributing resonance structures, but has its advantages in seeming less artificial.

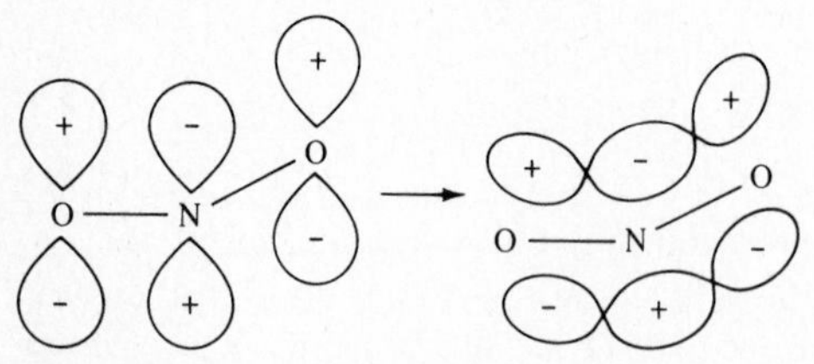

Fig. 3.41 Atomic orbitals (left) and resulting *antibonding* molecular orbital (right) for π bond in the nitrite ion.

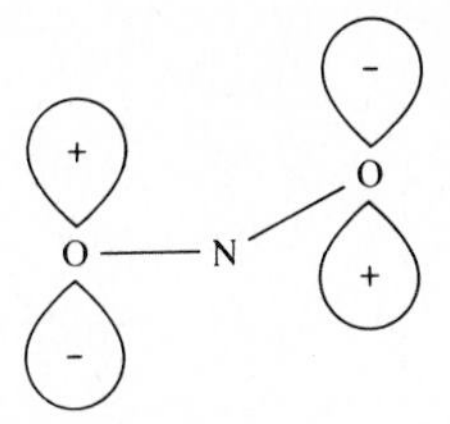

Fig. 3.42 *Nonbonding* orbitals in the nitrite ion.

COVALENT CHARACTER IN PREDOMINANTLY IONIC BONDS

As stated previously, it is probable that every heteronuclear bond the chemist has to deal with contains a mixture of covalent and ionic character. Ordinarily we speak glibly of an ionic compound or a covalent compound as long as the compound in question is predominantly one or the other. In many cases, however, it is convenient to be able to say something about intermediate situations. In general, there are two ways of treating ionic–covalent bonding. The method that has proved most successful is to consider the bond to be covalent and then consider the effect of increasing charge displacement from one atom toward another. This method will be discussed later in this chapter. Another method is to consider the bond to be ionic and then allow for a certain amount of covalency to occur. The second method was championed by Kasimir Fajans[53] in his quanticule theory. The latter theory has found no place in the repertoire of the theoretical chemist largely because it has not proved amenable to the quantitative calculations which other theories have developed. Nevertheless, the qualitative ideas embodied in "Fajans' rules" offer simple if inexact approaches to the problem of partial covalent character in ionic compounds.

Fajans considered the effect which a small, highly charged cation would have on an anion. If the anion were large and "soft" enough, the cation should be capable of polarizing it, and the extreme of this situation would be the cation actually penetrating the anionic electron cloud giving a covalent (shared electron) bond (Fig. 3.43). Fajans suggested the following rules to estimate the extent to which a cation could polarize an anion and thus

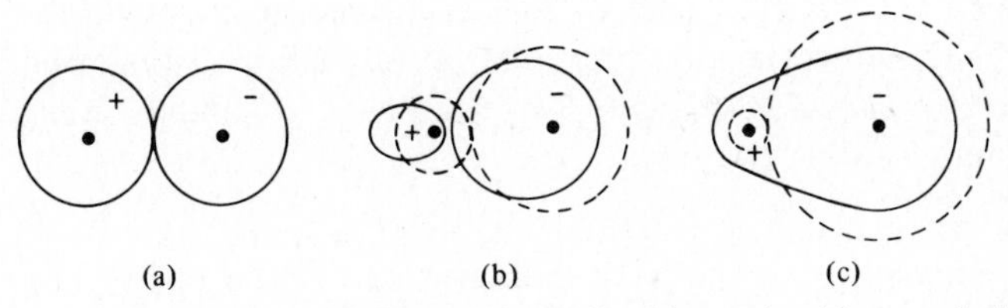

Fig. 3.43 Polarization effects: (a) idealized ion pair with no polarization; (b) mutually polarized ion pair; (c) polarization sufficient to form covalent bond. Dashed lines represent hypothetical unpolarized ions.

[53] K. Fajans, *Naturwissenschaften*, **1923**, *11*, 165. For a more recent discussion of the same subject, see K. Fajans, *Struct. Bonding*, **1967**, *2*, 88.

induce covalent character. Polarization will be increased by:

1. *High charge and small size of the cation.* Small, highly charged cations will exert a greater effect in polarizing anions than large and/or singly charged cations. This is often expressed by the *ionic potential*[54] of the cation: $\phi = Z^+/r$. For some simple ions, ionic potentials are as follows (r in nm):

$Li^+ = 17$	$Be^{+2} = 64$	$B^{+3} = 150$
$Na^+ = 10$	$Mg^{+2} = 31$	$Al^{+3} = 60$
$K^+ = 8$	$Ca^{+2} = 20$	$Ga^{+3} = 48$

Obviously there is no compelling reason for choosing Z/r instead of Z/r^2 or several other functions that could be suggested, and the values above are meant merely to be suggestive.[55] Nevertheless, polarization does follow some charge-to-size relationship, and those cations with large ionic potentials are those which have a tendency to combine with polarizable anions to yield partially covalent compounds. The ionic potentials above also rationalize an interesting empirical observation indicated by the dashed arrows: The first element in any given family of the periodic chart tends to resemble the second element in the family to the right. Thus lithium and magnesium have much in common (the best known examples are the organometallic compounds of these elements) and the chemistry of beryllium and aluminum is surprisingly similar despite the difference in preferred oxidation state. This relationship extends across the periodic chart; e.g., phosphorus and carbon resemble each other in their electronegativities (see pp. 823–824).

A word should be said here concerning unusually high ionic charges often found in charts of ionic radii. Ionic radii are often listed for Si^{+4}, P^{+5}, and even Cl^{+7}. Although at one time it was popular, especially among geochemists (see below), to discuss silicates, phosphates, and chlorates as though they contained these highly charged ions, no one today believes that such highly charged ions have any physical reality. The only possible meaning such radii can have is to indicate that if an ion such as P^{+5} or Cl^{+7} could exist, its high charge combined with small size would cause it immediately to polarize some adjacent anion and form a covalent bond.

2. *High charge and large size of the anion.* The polarizability of the anion will be related to its "softness," i.e., to the deformability of its electron cloud. Both increasing charge and increasing size will cause this cloud to be less under the influence of the nuclear charge of the anion and more easily influenced by the charge on the cation. Thus large anions such as I^-, Se^{-2}, and Te^{-2} and highly charged ones such as As^{-3} and P^{-3} are especially prone to polarization and covalent character.

A question naturally occurs: What about the polarization of a large cation by a small anion? Although this occurs, the results are not apt to be so spectacular as in the reverse

[54] G. H. Cartledge, *J. Am. Chem. Soc.*, **1928**, *50*, 2855, 2863; ibid., **1930**, *52*, 3076.

[55] It is true that the value of the ionic potential of Li^+ is closer to Ca^{+2} than Mg^{+2}. We could probably choose some other function to fit the experimental data in this section better, but why bother?

situation. Even though large, a cation is not likely to be particularly "soft" because the cationic charge will tend to hold on to the electrons. Likewise, a small anion can tend to polarize a cation, i.e., repel the outside electrons and thus make it possible to "see" the nuclear charge better, but this is not going to lead to covalent bond formation. No convincing examples of reverse polarization have been suggested.

3. *Electron configuration of the cation.* The simple form of the ionic potential considers only the net ionic charge of the ion with respect to its size. Actually an anion or polarizable molecule will feel a potential resulting from the total positive charge minus whatever shielding the electrons provide. To use the ionic charge is to assume implicitly that the shielding of the remaining electrons is perfect, i.e., 100% effective. The most serious problems with this assumption occur with the transition metal ions since they have one or more d electrons which shield the nucleus poorly. Thus for two ions of the same size and charge, one with an $(n-1)d^x\, ns^0$ electronic configuration (typical of the transition elements) will be more polarizing than a cation with a noble gas configuration $(n-1)s^2$ $(n-1)p^6ns^0$ (alkali and alkaline earth metals, for example). As an example, Hg^{+2} has an ionic radius (C.N. = 6) of 116 pm, yet it is considerably more polarizing and its compounds are considerably more covalent than those of Ca^{+2} with almost identical size (114 pm) and a charge.

Results of polarization

One of the most common examples of covalency resulting from polarization can be seen in the melting and boiling points of compounds of various metals.[56] Comparing the melting points of compounds having the same anion, but cations of different size, we have $BeCl_2$ = 405 °C, $CaCl_2$ = 772 °C; for cations of different charge, we have NaBr = 755 °C, $MgBr_2$ = 700 °C, $AlBr_3$ = 97.5 °C; for a constant cation, but anions of different sizes, we have LiF = 870 °C, LiCl = 613 °C, LiBr = 547 °C, LiI = 446 °C; and for ions having the same size and charge, the effect of electron configuration can be seen from $CaCl_2$ = 772 °C, $HgCl_2$ = 276 °C. Care must be taken not to interpret melting points and boiling points too literally as indicators of the degree of covalent bonding; there are many effects operative in addition to covalency and these will be discussed at some length in Chapter 6.

A second area in which polarization effects show up is the solubility of salts in polar solvents such as water. For example, consider the silver halides, in which we have a polarizing cation and increasingly polarizable anions. Silver fluoride, which is quite ionic, is soluble in water, but the less ionic silver chloride is soluble only with the inducement of complexing ammonia. Silver bromide is only slightly soluble and silver iodide is insoluble even with the addition of ammonia. Increasing covalency from fluoride to iodide is expected and decreased solubility in water is observed.

[56] One learns in General Chemistry courses that ionic compounds have high melting points and covalent compounds have low melting points. Although this oversimplification can be misleading, it may be applied to the present discussion. A more thorough discussion of the factors involved in melting and boiling points will be found in Chapter 6.

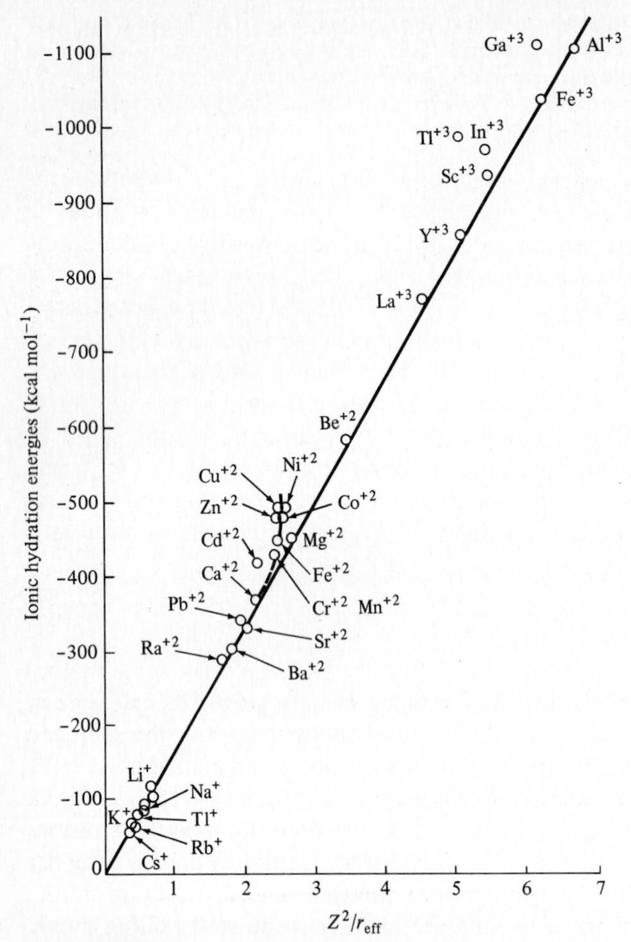

Fig. 3.44 Hydration energies as a function of size and charge of cations. [From C. S. G. Philips and R. J. P. Williams, "Inorganic Chemistry," Clarendon Press, Oxford, **1965**. Reproduced with permission.]

Silver halide	K_{sp}
Silver fluoride	Soluble
Silver chloride	2×10^{-10}
Silver bromide	5×10^{-13}
Silver iodide	8×10^{-17}

As in the case of melting points, solubility is a complex process, and there are many factors involved in addition to covalency.

Closely related to solubility are the hydration enthalpies of ions. It has been found[57] that it is possible to correlate the hydration enthalpies of cations with their "effective

[57] W. M. Latimer et al., *J. Chem. Phys.*, **1939**, *7*, 108.

ionic radii" by the expression (see Fig. 3.44):

$$\Delta H = -69\,500(Z^2/r_{\text{eff}})\ \text{kJ mol}^{-1} \qquad (r_{\text{eff}} \text{ in pm}) \tag{3.74}$$

In this case the reason for the correlation is fairly obvious. The parameter r_{eff} is equal to the ionic radius plus a constant, 85 pm, the radius of the oxygen atom in water. Therefore, r_{eff} is effectively the interatomic distance in the hydrate, and the Born-Landé equation (Eq. 3.13) can be applied.

The basis for other correlations between size, charge, and chemical properties is not so clearcut. Chemical reactions can often be rationalized in terms of the polarizing power of a particular cation. In the alkaline earth carbonates, for example, there is a tendency toward decomposition with the evolution of carbon dioxide:

$$MCO_3 \longrightarrow MO + CO_2 \tag{3.75}$$

The ease with which this reaction proceeds (as indicated by the temperature necessary to induce it) decreases with increasing cation size: $BeCO_3$, 100 °C; $MgCO_3$, 400 °C; $CaCO_3$, 900 °C; $SrCO_3$, 1290 °C; $BaCO_3$, 1360 °C. The effect of *d* electrons is also clear: Both $CdCO_3$ and $PbCO_3$ decompose at approximately 350 °C despite the fact that Cd^{+2} and Pb^{+2} are approximately the same size as Ca^{+2}. The decomposition of these carbonates occurs as the cation polarizes the carbonate ion, splitting it into an O^{-2} ion and CO_2.

Sterns[58] has extended the qualitative argument on decomposition by showing that the enthalpies of decomposition of carbonates, sulfates, nitrates, and phosphates are linearly related to a charge/size function, in this case $r^{1/2}/Z^*$ (see Fig. 3.45). Although the exact theoretical basis of this correlation is not clear, it provides another interesting example of the general principle that size and charge are the important factors that govern the polarizing power of ions and, consequently, many of their chemical properties.

From the preceding, it might be supposed that covalent character in predominantly ionic compounds always destabilizes the compound. This is not so. Instability results from polarization of the anion causing it to split into a more stable compound (in the above cases the oxides) with the release of gaseous acidic anhydrides. As will be seen in Chapter 14, many very stable, very hard minerals have covalent-ionic bonding.

Covalent bonding in "ionic" solids

Although the details of silicate structures will be reserved for Chapter 14, they will be discussed here briefly because they illustrate well the presence of covalent character in compounds with high ionicity as well. Historically, there has been a tendency to treat solid compounds in terms of the ionic model whenever possible because of its simplicity. For compounds such as the alkali halides it is quite adequate, although, as Waddington[59] has said, the history of the development of the theory of lattice energies is largely an account of the development of the ideas about nonelectrostatic forces. For many years it was common to treat minerals as purely ionic systems. For example, olivine, an important

[58] K. H. Sterns, *J. Chem. Educ.*, **1969**, *46*, 645.

[59] T. C. Waddington, *Adv. Inorg. Chem. Radiochem.*, **1959**, *1*, 159.

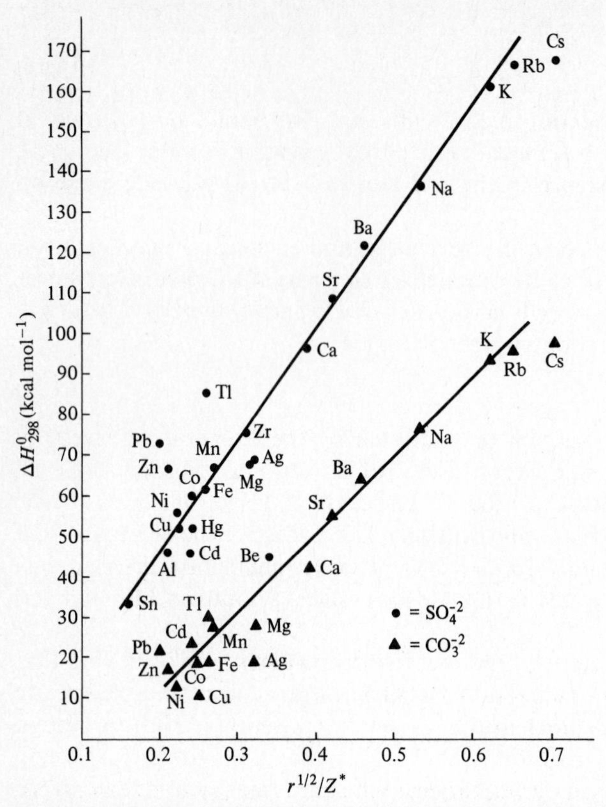

Fig. 3.45 Enthalpy of decomposition of sulfates and carbonates as a function of size and charge of the metal ion. [From K. H. Sterns, *J. Chem. Educ.*, **1969**, *46*, 645. Reproduced with permission.]

constituent of basalt rocks, has the approximate chemical composition $9Mg_2SiO_4 \cdot Fe_2SiO_4$. It may be pictured as a hexagonal closest packed (*hcp*) array of oxide ions with Si^{+4} ions in tetrahedral holes and Mg^{+2} and Fe^{+2} ions in octahedral holes. Most chemists today, however, would probably describe the system as composed of discrete SiO_4^{-4} anions and metallic cations. One should not lose sight of the fact that all of the bonds in silicates are highly polar. Pauling has given rules by which one can use an electrostatic approach to rationalize and predict structures.[60]

As discussed previously (pp. 71–77), it is very difficult to determine the exact sizes of atoms and the charge distributions in solids directly. However, it is possible to make some estimates based on the properties of ionic and covalent bonds. Both ionic and covalent bonds can be very strong (this is discussed at greater length on pp. 263–264), and so the force necessary to stretch a bond can be assumed to be independent of covalency. On

[60] L. Pauling, "The Nature of the Chemical Bond," 3rd ed., Cornell University Press, Ithaca, N.Y. **1960**, pp. 547 ff. See also, A. F. Wells, "Structural Inorganic Chemistry," 4th ed., Clarendon Press, Oxford, **1975**, pp. 274–285.

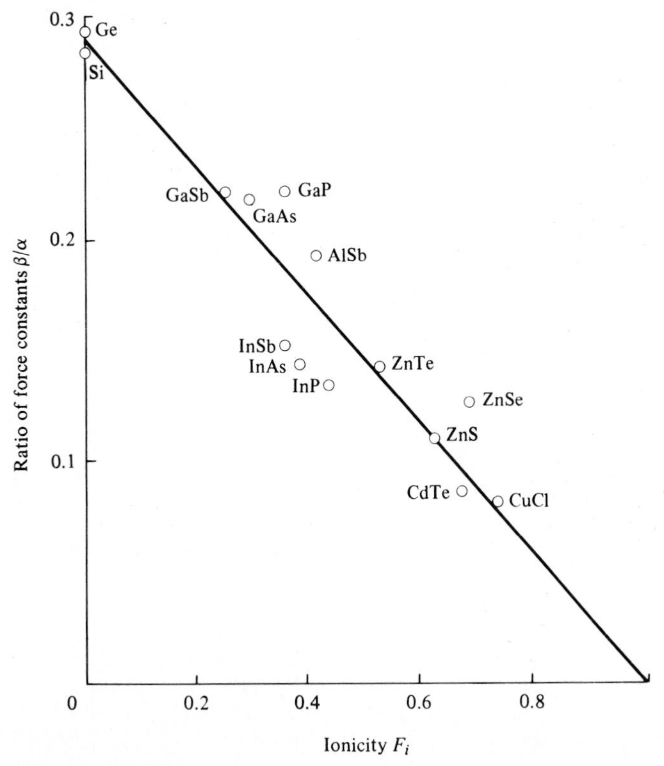

Fig. 3.46 Interatomic forces in $A^{+n}B^{n-8}$ compounds can be separated into bond-stretching forces (α) and bond-bending forces (β). The bond-stretching forces are central forces and change very little with ionicity for fixed bond length (e.g., in an isoelectronic sequence such as Ge–GaAs–ZnSe). The bond-bending forces are directional (noncentral) and are a direct measure of the extent of covalency $F_c = 1 - F_i$ of the AB bond. The ratio β/α is dimensionless, and is proportional to F_c, as shown here for many compounds. [From "Treatise on Solid State Chemistry," Vol. I, N. B. Hannay, ed., Plenum, New York. Copyright © **1973**, Bell Laboratories, Inc. Reproduced with permission.]

the other hand, bending forces *do* relate to the ionicity. Since covalent bonds are directional, they resist deformation more than nondirectional ionic bonds. For example, when a covalent bond is deformed in a bending mode, the overlap will decrease rapidly and it will cost a lot of energy; hence the bond will resist such deformation. If the bond were purely ionic, such deformation would have little effect since the electrostatic attractions are omnidirectional. If one plots the ratio of *bending* force constants,[61] β, to the stretching force constants, α, versus the expected ionicity,[62] a very good correlation is obtained (Fig. 3.46).[63]

[61] A force constant with respect to a molecular deformation is simply the proportionality constant in the Hooke's Law treatment of the bond as a spring.

[62] The ionicity here is that obtained by Phillips (*Rev. Mod. Phys.*, **1970**, *42*, 313).

[63] R. M. Martin, *Phys. Rev.*, **1970**, *B1*, 4001. The details of the method will not be given here. They may be obtained from the original paper or from J. C. Phillips, in "Treatise on Solid State Chemistry," Vol. I. The Chemical Structure of Solids, N. B. Hannay, ed., Plenum, New York, **1973**, pp. 34–39.

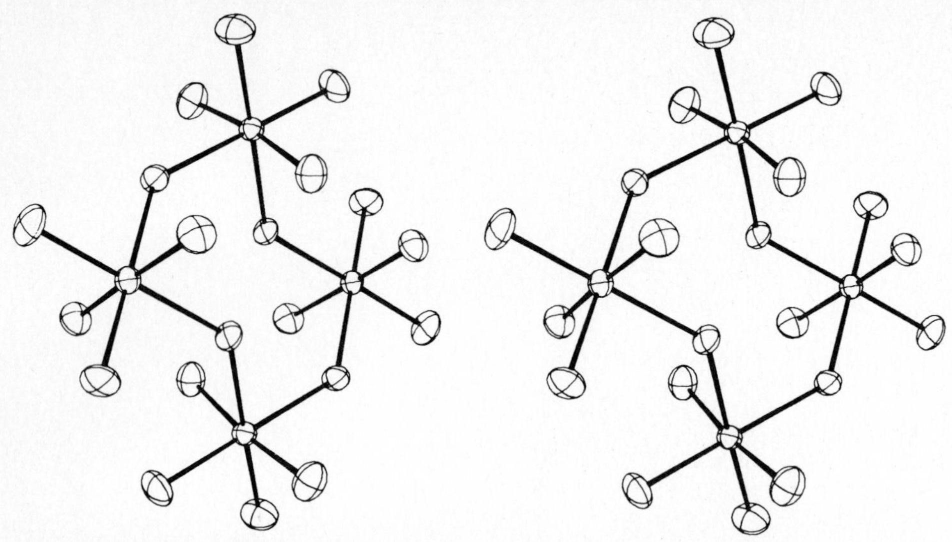

Fig. 3.47 Stereoview of tetrameric unit of Rh_4F_{20}. The rhodium atoms are at the centers of the octahedra of fluorine atoms. Note bridging fluorine atoms. [From B. K. Morrell et al., *Inorg. Chem.*, **1973**, *12*, 2640. Reproduced with permission.]

Since coordination compounds are usually considered to be covalently bonded to a first approximation (see Chapter 9), extended complex structures in the solid can readily be related to them. Consider, for example, rhodium pentafluoride. Obviously it could be considered as an ionic structure, Rh^{+5} $5F^-$, and indeed the crystal structure[64] consists, in part, of *hcp*-arranged fluoride ions with rhodium in octahedral holes. However, closer inspection of the structure reveals that it consists of tetrameric units, Rh_4F_{20}, that are distinct from one another (Fig. 3.47). The environment about each rhodium atom, an octahedron of six fluorine atoms, is what we should expect for a complex such as $[RhF_6]^-$. Bridging halide ions are well known in coordination compounds. Furthermore, according to Fajans' rules, we should be suspicious of an "ionic" structure containing a cation with a +5 charge.

Sanderson[65] would go so far as to say that even crystals such as alkali halides should be considered as infinite coordination polymers with each cation surrounded by an octahedral coordination sphere of six halogen atoms, which in turn "bridge" to five more alkali metal atoms. Although this point of view is probably of considerable use when discussing transition metal compounds, most chemists would not go so far as to extend it to all "ionic" lattices.

This brings us to a class of compounds too often overlooked in the discussion of simple "ionic" compounds: the transition metal halides. In general, these compounds (except fluorides) crystallize in structures that are hard to reconcile with the structures of simple ionic compounds seen previously (Figs. 3.1–3.3). For example, consider the

[64] B. K. Morrell et al., *Inorg. Chem.*, **1973**, *12*, 2640.

[65] R. T. Sanderson, *J. Chem. Educ.*, **1967**, *44*, 516. See also R. T. Sanderson, "Chemical Bonds and Bond Energy," 2nd ed., Academic Press, New York, **1976**, pp. 105–110.

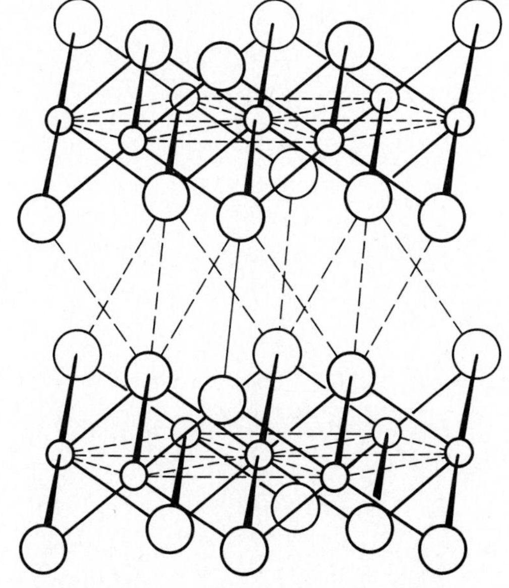

Fig. 3.48 Portions of two layers of the CdI_2 structure. The smaller circles represent the cadmium atoms. [From A. F. Wells, "Structural Inorganic Chemistry," 4th ed., Oxford University Press, Oxford, **1975**. Reproduced with permission.]

cadmium iodide structure (Fig. 3.48). It is true that the cadmium atoms occupy octahedral holes in a hexagonal closest packed structure of iodine atoms, but in a definite *layered structure* that can only be described adequately in terms of covalent bonding and infinite layer molecules.

CHARGE DISTRIBUTION IN MOLECULES

As has been noted previously, the distribution of electronic charge within a molecule is an important factor in determining its properties. Most of chemistry, whether inorganic, organic, or biochemistry, is dominated by the interaction of polar molecules and ions with each other. Such topics as dipole moments and physical properties related to them, acid–base reactions, coordination chemistry, and nonmetal chemistry are intimately associated with charge separation. As we have seen for other types of bonding problems, valence bond theory and molecular orbital theory provide different and complementary approaches. Likewise, we can use rough-and-ready, "back-of-the-envelope" means of attacking these problems, or if necessary and warranted, we may employ sophisticated computer calculations.

Molecular orbitals in polar molecules

Different types of atoms have different capacities to attract electrons. The ionization potential of fluorine is considerably greater than that of lithium. Likewise, the electron affinity of fluorine is strongly exothermic but that of lithium is much less so, and some

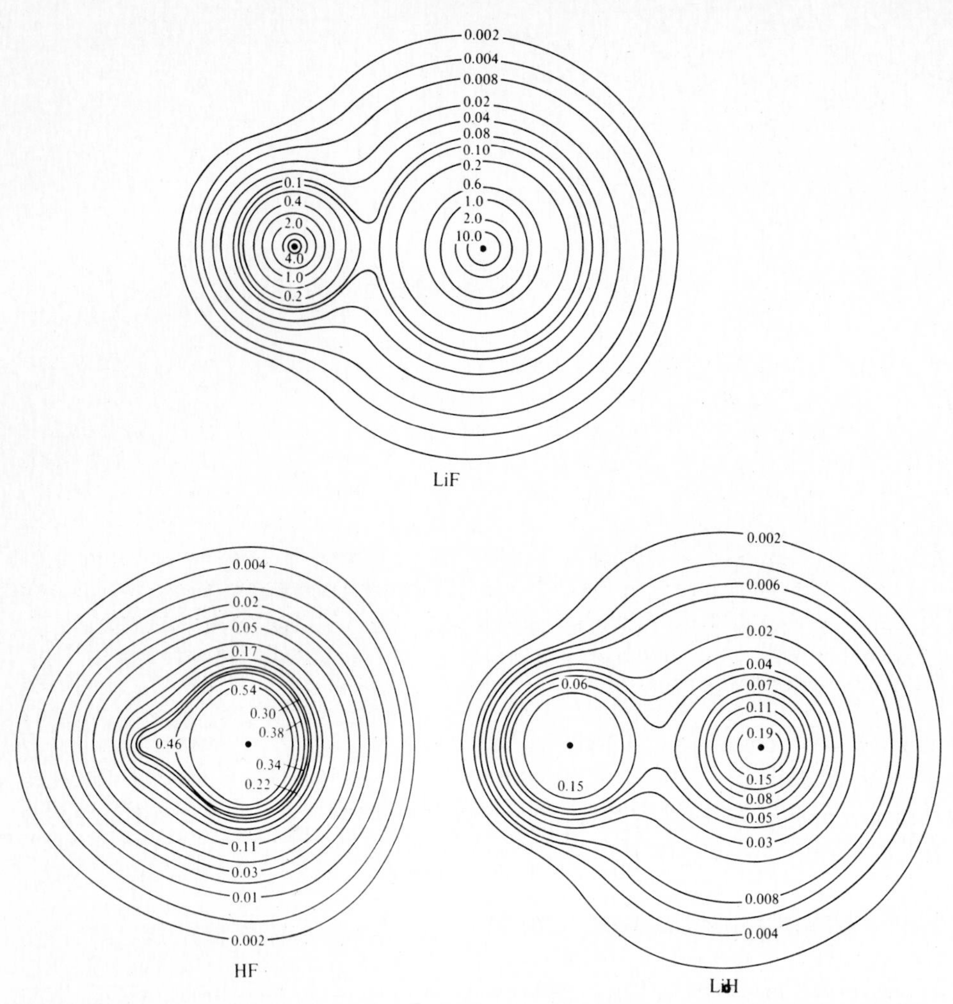

Fig. 3.49 Electron density contours for LiF, HF, and LiH molecules. All molecules drawn to the same scale. The inner contours of F in HF and Li in LiH have been omitted for clarity. [From R. F. W. Bader et al., *J. Chem. Phys.*, **1967**, *47*, 3381; **1968**, *49*, 1653. Reproduced with permission.]

metals have endothermic electron affinities. A bond between lithium and fluorine is predominantly ionic, consisting (to a first approximation) of transfer of an electron from the lithium atom to the fluorine atom. Hydrogen is intermediate in these properties between lithium and fluorine. When it bonds with lithium the hydrogen atom accepts electron density, but when it bonds with fluorine it loses electron density. All of these bonds, LiH, HF, and LiF, are more or less polar in nature in contrast to the bonds discussed previously (p. 111). Charge density distributions for these molecules are shown in Fig. 3.49, which may be compared with the nonpolar, homonuclear bonds in Figs. 3.32 and 3.33. Cross-sectional density profiles of several homonuclear and heteronuclear molecules are shown in Fig. 3.50. Although LiF gives an appearance of being (again, to a first approximation)

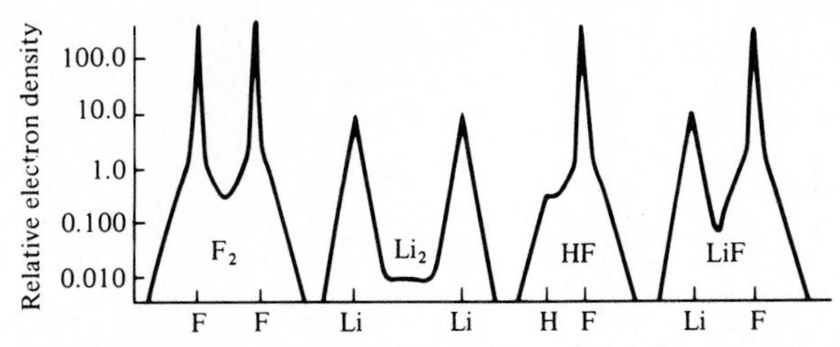

Fig. 3.50 Total charge density profiles of simple molecules as a function of internuclear distance. [From B. J. Ransil and J. J. Sinai, *J. Chem. Phys.*, **1967**, *46*, 4050. Reproduced with permission.]

an ion pair, in HF the hydrogen proton is deeply embedded in the electron cloud of the fluoride ion as predicted by Fajans' rules (pp. 129, 131).

The treatment of heteronuclear bonds revolves around the concept of *electronegativity*. This is simultaneously one of the most important and one of the most difficult problems in chemistry. In the previous discussion of molecular orbitals it was assumed that the atomic orbitals of the bonding atoms were at the same energy. In general, this will be true only for homonuclear bonds. Heteronuclear bonds will be formed between atoms with orbitals at different energies. When this occurs, the bonding electrons will be more stable when in the presence of the nucleus of the atom having the greater attraction (greater electronegativity), i.e., the atom having the lower atomic energy levels. They will thus spend more time nearer that nucleus. The electron cloud will be distorted toward that nucleus (see Fig. 3.49) and the bonding MO will resemble that AO more than the AO on the less electronegative atom.

Consider the carbon monoxide molecule, CO, isoelectronic with the N_2 molecule. Oxygen is more electronegative than carbon, so the bonding electrons are more stable if they can spend a larger proportion of their time in the region of the oxygen nucleus. The electron density on the oxygen atom is greater than that on the carbon atom (Fig. 3.51),

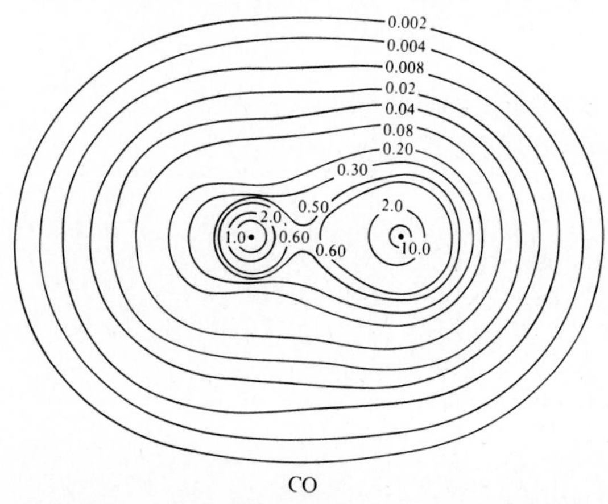

Fig. 3.51 Total charge density contours for the carbon monoxide molecule. (The carbon atom is on the left.) Compare to the isoelectronic nitrogen molecule, Fig. 3.52. [From R. F. W. Bader and A. D. Bandrauk, *J. Chem. Phys.*, **1968**, *49*, 1653. Reproduced with permission.]

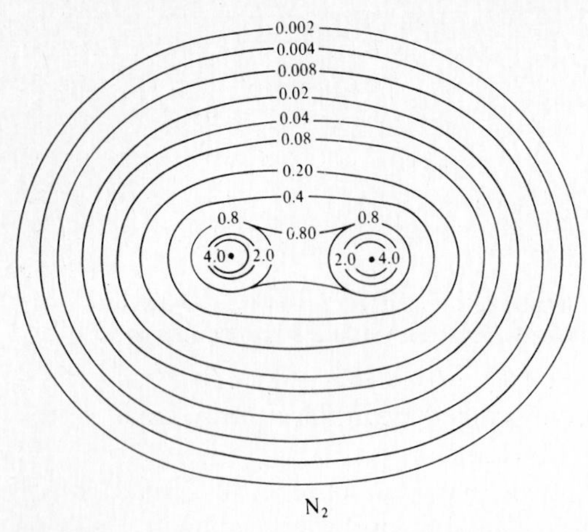

Fig. 3.52 Total charge density contours for the nitrogen molecule. [From R. F. W. Bader and A. D. Bandrauk, *J. Chem. Phys.*, **1968**, *49*, 1653. Reproduced with permission.]

in contrast to the symmetrical distribution in the N_2 molecule (Fig. 3.52). For homonuclear diatomic molecules we have seen that the molecular orbitals are

$$\Psi_b = \Psi_A + \Psi_B \tag{3.76}$$

$$\Psi_a = \Psi_A - \Psi_B \tag{3.77}$$

Both orbitals contribute equally. Now if one atomic orbital is lower in energy than the other, it will contribute *more* to the bonding orbital:

$$\Psi_b = a\Psi_A + b\Psi_B \tag{3.78}$$

where $b > a$ if atom B is more electronegative than atom A. Conversely, the more stable orbital contributes *less* to the antibonding orbital:

$$\Psi_a = b\Psi_A - a\Psi_B \tag{3.79}$$

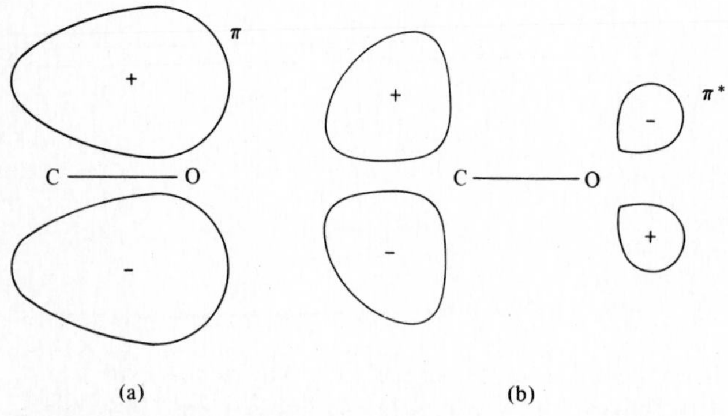

Fig. 3.53 Diagrammatic sketches of the molecular orbitals in carbon monoxide: (a) one π bonding orbital; (b) one π antibonding orbital.

In carbon monoxide the bonding molecular orbitals will resemble the atomic orbitals of oxygen more than they resemble those of carbon. The antibonding orbitals resemble the *least* electronegative element more, in this case the carbon (see Fig. 3.53). This results from what might be termed the conservation of orbitals. The number of molecular orbitals obtained is equal to the total number of atomic orbitals combined, and each orbital must be used to the same extent. Thus, if the carbon atomic orbital contributes *less* to the *bonding* molecular orbital, it must contribute *more* to the *antibonding* molecular orbital. The energy level diagram for CO is shown in Fig. 3.54.

A second feature of heteronuclear molecular orbitals which has been mentioned previously is the diminished covalent bond energy of bonds formed from atomic orbitals of different energies. Specifically the exchange energy is reduced if the energies of the atomic orbitals do not match. This may be shown qualitatively by comparing Fig. 3.55

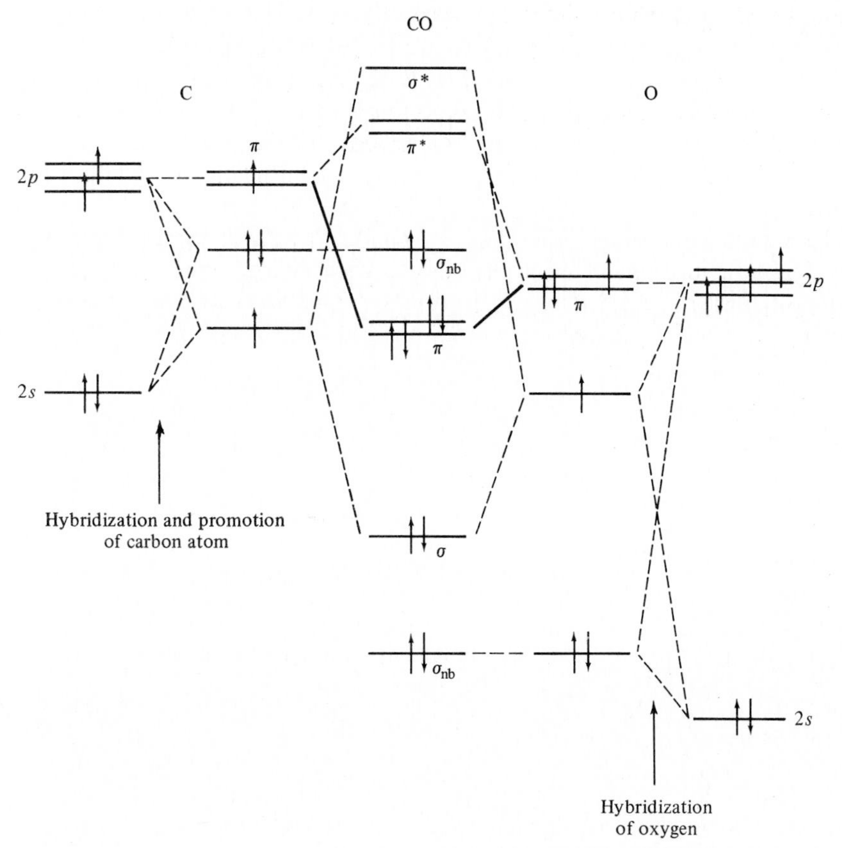

Fig. 3.54 Energy level diagram for the molecular orbitals of carbon monoxide. Note that upon bond formation electrons occupy orbitals that are more oxygen-like than carbon-like. Note carefully the bond order: The lowest and the fifth MOs are essentially nonbonding since they do not combine with orbitals on the opposite atom and do not change energy. The bond order, as in the N_2 molecule (Fig. 3.28) is three. [From H. H. Jaffé and M. Orchin, *Tetrahedron*, **1960**, *10*, 212. Reproduced with permission of Pergamon Press.]

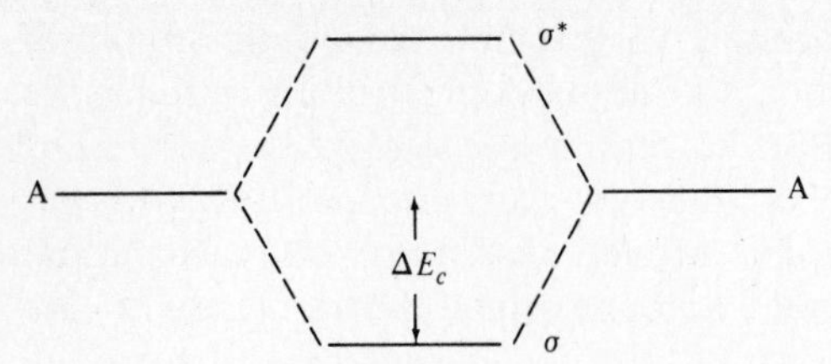

Fig. 3.55 Homonuclear diatomic molecule, A_2. The covalent energy is maximized.

with Fig. 3.56. This can be seen even more readily in Fig. 3.57, in which the electronegativity difference between atoms A and B is so great as to preclude covalent bonding. In this case the bonding MO does not differ significantly from the atomic orbital of B, and so transfer of the two bonding electrons to the bonding MO is indistinguishable from the simple picture of an ionic bond; the electron on A has been transferred completely to B.

This extreme situation in which the energy level of B is so much lower than that of A that the latter cannot contribute to the bonding may be visualized as follows: If the energy of atomic orbital B is very much lower than A, the electron will spend essentially all of its time in the vicinity of nucleus B. Although this may be a very stable situation, it hardly qualifies as a *covalent* bond or sharing of electrons. In this case, the exchange of electrons has been drastically reduced, and the exchange energy is negligible. All chemical bonds lie somewhere on the spectrum defined by Figs. 3.55–3.57.

There has been some confusion in the literature concerning the strength of bonds in situations such as shown in Figs. 3.56 and 3.57. Since good energy match is necessary for a large exchange energy and good *covalent* bonding, some workers have concluded that Figs. 3.56 and 3.57 represent increasingly weak bonding. This is not true, for the loss

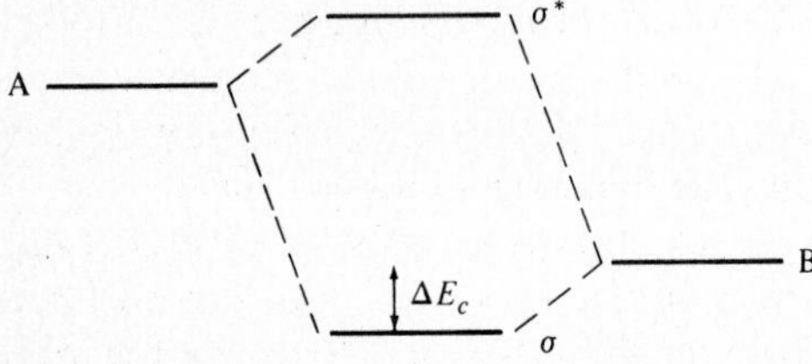

Fig. 3.56 Heteronuclear molecule, $A^{\delta+}B^{\delta-}$, with relatively small electronegativity different between A and B. Covalency reduced with respect to A_2 but still more important than the ionic contribution.

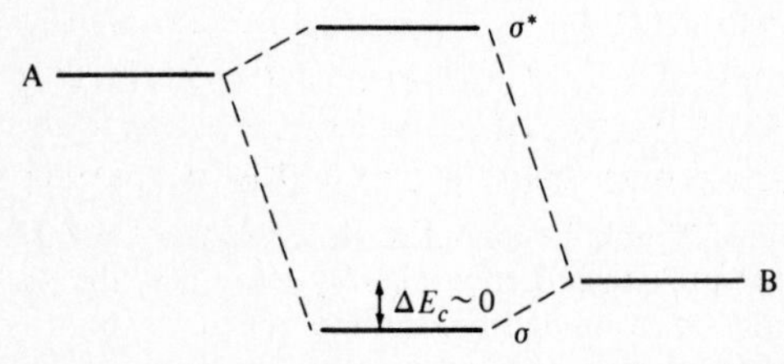

Fig. 3.57 Heteronuclear molecule, A^+B^-, with large electronegativity difference. Covalency is insignificant; the bond is essentially ionic.

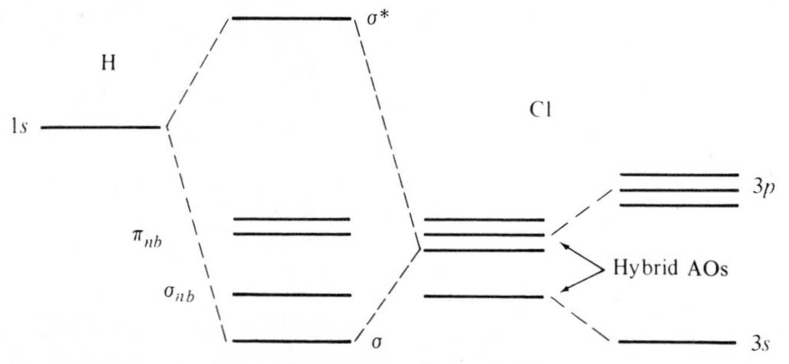

Fig. 3.58 Energy level diagram for the hydrogen chloride molecule, HCl. The mixing of the *s* and *p* orbitals has been emphasized for clarity.

of covalent bonding may be compensated by an increase in ionic bonding,[66] which, as we have seen previously, can be quite strong. In fact, the total of ionic and covalent bonding may make a very strong bond in the intermediate situation shown in Fig. 3.56. (Note that the *ionic* contribution to the bonding does not appear in these figures.) In fact, the strengthening of a polar bond over a corresponding purely covalent one is an important phenomenon (see below).

A second example of a heteronuclear diatomic molecule is hydrogen chloride. In this molecule the attraction of the chlorine nucleus for electrons is greater than that of the hydrogen nucleus. The energy of the 3*s* and 3*p* orbitals on the chlorine atom is less than that of the 1*s* orbital on hydrogen as a result of the imperfect shielding of the much larger nuclear charge of chlorine. The molecular orbitals for the hydrogen chloride molecule are illustrated in Fig. 3.58. There is one σ bond holding the atoms together. The remaining six electrons from chlorine occupy *nonbonding* orbitals which are almost unchanged atomic orbitals of chlorine. These nonbonding molecular orbitals correspond to the lone pair electrons of valence bond theory. They represent the two *p* orbitals on the chlorine atom which lie perpendicular to the bond axis. They are therefore orthogonal to and have a net overlap of zero with the hydrogen 1*s* orbital (Fig. 3.24m). As such, they cannot mix with the hydrogen orbital to form bonding and antibonding molecular orbitals. The third nonbonding MO is the second of the hybridized atomic orbitals resulting from some *s*/*p* mixing. If the mixing were complete (50% *s* character, which is *far* from the case in HCl), it would be the second *di* orbital directed *away* from the bond. Since little mixing of the *s* orbital of the chlorine into the bonding MO occurs, the third lone pair is essentially a largely *s* orbital, a distorted sphere of electron density with the major portion behind the chlorine atom. Favoring hybridization mixing of *s* and *p* are an increase in overlap and a decrease in repulsions; opposing hybridization is the increased promotion energy (see pp. 227–230). The same problem arises in molecular orbital theory: The chlorine can use both *s* and *p* orbitals in bonding to the hydrogen atom. Molecular orbital theory does not, in general, talk in terms of hybridization, because the requisite mixing of atomic orbitals can be accomplished in one step in the construction of the molecular orbitals. For example, the bonding molecular orbital in hydrogen chloride may be considered

[66] R. Ferreira, *Chem. Phys. Letters*, **1968**, *2*, 233.

to be formed

$$\Psi_b = a\Psi_{3s} + b\Psi_{3p} + c\Psi_{1s} \tag{3.80}$$

where Ψ_{3s}, Ψ_{3p}, Ψ_{1s} are the atomic orbitals on the chlorine ($3s$ and $3p$) and hydrogen ($1s$) atoms. Now a and b can be varied relative to each other in such a way that any amount of p character can be involved in the molecular orbital. For example, if $a = 0$, the chlorine atom uses a pure p orbital, but if $a^2 = \frac{1}{3}b^2$, the p character will be 75% (an "sp^3 hybrid" in VB terminology).[67] And, of course, the relative weighting of a and b versus c indicates relative contribution of chlorine versus hydrogen wave functions to the bonding molecular orbital.

Electronegativity and molecular orbital theory

As we have seen, the "distortion" that takes place as one progresses from Figs. 3.55 and 3.56 results from differences in electronegativity between the bonding atoms. In general, MO language does not speak of the "electronegativity difference" but of the magnitude of the *coulomb integrals* of the constituent atoms. These may be taken as the *valence state ionization energies* (VSIE), i.e., the energy necessary to remove an electron from an atom *under the conditions that exist in the molecule.* The VSIE will therefore not be identical to the usual ionization energy of an atom though related to it (see p. 152). A discussion of the various methods of estimating coulomb integrals would be beyond the scope of this text, but it may be mentioned that the VB methods of evaluating electronegativity are related to this problem. Inasmuch as more work has probably been done in the valence bond context, this aspect of electronegativity will be discussed at greater length with the reminder that the results can be transferred qualitatively to MO theory.

Pauling's electronegativity and valence bond theory

Pauling first defined electronegativity and suggested methods for its estimation. Pauling's definition[68] has not been improved upon: *The power of an atom in a molecule to attract electrons to itself*. It is evident from this definition that electronegativity is not a property of the isolated atom (although it may be related to such properties) but rather a property of an atom in a molecule, in the environment of and under the influence of surrounding atoms.

Pauling based his electronegativity scale on thermochemical data. It had been observed that bonds between dissimilar atoms were almost always stronger than might have been expected from the strength of bonds of the same elements when bonded in homonuclear bonds. For example, the bond energy of chlorine monofluoride, ClF, is about 255 kJ mol^{-1}, greater than either Cl_2 (242 kJ mol^{-1}) or F_2 (153 kJ mol^{-1}).[69] Pauling suggested that molecules formed from atoms of different electronegativity would be stabilized by *ionic*

[67] The percent s and p character is proportional to the *square* of the coefficients. In the case of Cl and H, the difference in energy between the $3s$ and $3p$ orbitals is so great as to preclude very much mixing of s and p character, but it is always possible; the closer the energy levels are, the more likely it is to occur.

[68] L. Pauling, "The Nature of the Chemical Bond," 3rd ed., Cornell University Press, Ithaca, N.Y., **1960**, p. 88.

[69] See Appendix E.

resonance energy resulting from resonance of the sort:

$$\Psi_{AB} = a\Psi_{A-B} + b\Psi_{A^+B^-} + c\Psi_{A^-B^+} \quad (3.81)$$

For molecules in which atoms A and B are identical, $b = c \ll a$ (see p. 94 for the H_2 molecule), and the contribution of the ionic structures is small. If B is more electronegative than A, then the energy of the contributing structure A^+B^- approaches more nearly that of the purely covalent structure A—B and resonance is enhanced. On the other hand, the energy of B^+A^- is so prohibitively high that this structure may be dismissed from further consideration. For a predominantly covalent, but polar, bond, $a > b \gg c$. The greater the contribution of the ionic structure (i.e., the closer it comes to being equivalent in energy to the covalent structure) the greater the resonance between the contributing structures and the greater the stabilizing resonance energy. Pauling suggested that electronegativity could be estimated from calculations involving this *ionic resonance* energy. The interested reader is referred to Pauling's discussions of the subject for the details of the methods he used,[70] but an outline follows.

Pauling assumed that if the ClF bond were completely covalent, its bond energy would be simply the average of the Cl_2 and F_2 bond energies:

$$\frac{242 + 153}{2} = 198 \text{ kJ mol}^{-1} \quad (3.82)$$

The ionic resonance energy is the difference between the experimental bond energy of ClF, 255 kJ mol^{-1}, and the calculated value, 198 kJ mol^{-1}, or 57 kJ mol^{-1}. Pauling defined the difference in electronegativity between chlorine and fluorine as the square root of the ionic resonance energy:[71]

$$\sqrt{\frac{57}{96.5}} = 0.77 \quad (3.83)$$

This may be compared with the tabular value for the difference in electronegativities of fluorine and chlorine, $3.98 - 3.16 = 0.82$, which is based on many experimental data, not just the single calculation illustrated here. Once again, the details of the calculation are not particularly important since Pauling's method of obtaining electronegativity data is probably mainly of historical interest. There are now improved methods of estimating electronegativities.[72] The concept of covalent–ionic resonance is still quite useful, however. As pointed out above, the fact that a bond with partial ionic character can be stronger than either a purely covalent or purely ionic bond has often been overlooked as alternative methods of treating electronegativity have developed. Energies associated with electronegativity differences can be useful in accounting for the total bonding energy of molecules.[73] Pauling-type values are listed in Table 3.12.

[70] L. Pauling, "The Nature of the Chemical Bond," 3rd ed., Cornell University Press, Ithaca, N.Y., **1960**, Chapter 3.

[71] The conversion factor 96.5 kJ mol^{-1} eV^{-1} (see Appendix C) is included because Pauling set up his scale based on electron volts.

[72] Not all chemists would agree with this statement.

[73] See pp. 319–322.

Table 3.12 Electronegativities of the elements

				Mulliken-Jaffé[d]			
					a		*b*
Element	Pauling[a]	Sanderson[b]	Allred-Rochow[c]	Orbital or hybrid[e]	Pauling scale	Volts	Volts/electron
1. H	2.20	2.31	2.20	*s*	2.21	7.17	12.85
2. He			5.50[f]	*s*	4.86[f]	15.08[f]	—
3. Li	0.98	0.86	0.97	*s*	0.84	3.10	4.57
4. Be	1.57	1.61	1.47	*sp*	1.40	4.78	7.59
5. B	2.04	1.88	2.01	sp^3	1.81	5.99	8.90
				sp^2	1.93	6.33	9.91
6. C	2.55	2.47	2.50	*p*	1.75	5.80	10.93
				sp^3	2.48	7.98	13.27
				sp^2	2.75	8.79	13.67
				sp	3.29	10.39	14.08
7. N	3.04	2.93	3.07	*p*	2.28	7.39	13.10
				23% *s*	3.56	11.21	14.64
				sp^3	3.68	11.54	14.78
				sp^2	4.13	12.87	15.46
				sp	5.07	15.68	16.46
8. O	3.44	3.46	3.50	*p*	3.04	9.65	15.27
				20% *s*	4.63	14.39	17.65
				sp^3	4.93	15.25	18.28
				sp^2	5.54	17.07	19.16
9. F	3.98	3.92	4.10	*p*	3.90	12.18	17.36
10. Ne		4.50[f]	4.84[f]	*p*	4.26[f]	13.29[f]	—
11. Na	0.93	0.85	1.01	*s*	0.74	2.80	4.67
12. Mg	1.31	1.42	1.23	*sp*	1.17	4.09	6.02
13. Al	1.61	1.54	1.47	sp^2	1.64	5.47	6.72
14. Si	1.90	1.74	1.74	sp^3	2.25	7.30	9.04
15. P	2.19	2.16	2.06	*p*	1.84	6.08	9.31
				sp^3	2.79	8.90	11.33
16. S	2.58	2.66	2.44	*p*	2.28	7.39	10.01
				sp^3	3.21	10.14	10.73
17. Cl	3.16	3.28	2.83	*p*	2.95	9.38	11.30
18. Ar		3.31[f]	3.20[f]	*p*	3.11[f]	9.87[f]	—
19. K	0.82	0.74	0.91	*s*	0.77	2.90	2.88
20. Ca	1.00	1.06	1.04	*sp*	0.99	3.30	4.74
21. Sc	1.36	1.09	1.20				
22. Ti(II)	1.54	1.13	1.32				
23. V(II)	1.63	1.24	1.45				
24. Cr(II)	1.66	1.35	1.56				
25. Mn(II)	1.55	1.44	1.60				
26. Fe(II)	1.83	1.47	1.64				
Fe(III)	1.96						
27. Co(II)	1.88	1.47	1.70				
28. Ni(II)	1.91	1.47	1.75				
29. Cu(I)	1.90	1.74	1.75	*s*	*1.36*	*4.31*	*6.82*
Cu(II)	2.00						
30. Zn(II)	1.65	1.86	1.66	*sp*	*1.49*	*4.71*	*6.43*
31. Ga(III)	1.81	2.10	1.82	sp^2	1.82	6.02	7.48
32. Ge(IV)	2.01	2.31	2.02	sp^3	2.50	8.07	6.82

Table 3.12 *(Continued)*

					Mulliken-Jaffé[d]		
					a		*b*
Element	Pauling[a]	Sanderson[b]	Allred-Rochow[c]	Orbital or hybrid[e]	Pauling scale	Volts	Volts/electron
33. As(III)	2.18	2.53	2.20	*sp*	1.59	5.34	8.03
				sp^3	2.58	8.30	8.99
34. Se	2.55	2.76	2.48	*p*	2.18	7.10	9.16
				sp^3	3.07	9.76	11.05
35. Br	2.96	2.96	2.74	*p*	2.62	8.40	9.40
36. Kr	3.00[f]	2.91[f]	2.94[f]	*p*	2.77[f]	8.86[f]	—
37. Rb	0.82	0.70	0.89	*s*	0.50	2.09	4.18
38. Sr	0.95	0.96	0.99	*sp*	0.85	3.14	4.41
39. Y	1.22	0.98	1.11				
40. Zr(II)	1.33	1.00	*1.22*				
41. Nb	1.60	1.12	*1.23*				
42. Mo(II)	2.16						
Mo(III)	2.19						
Mo(IV)	2.24	1.24	*1.30*				
Mo(V)	2.27						
Mo(VI)	2.35						
43. Tc	1.90	1.33	*1.36*				
44. Ru	2.20	1.40	*1.42*				
45. Rh	2.28	1.47	*1.45*				
46. Pd	2.20	1.57	*1.35*				
47. Ag	1.93	1.72	1.42				
48. Cd	1.69	1.73	1.46				
49. In	1.78	1.88	1.49	sp^2	1.57	5.28	6.79
50. Sn(II)	1.80	1.58		30% *s*	2.67	8.55	5.06
Sn(IV)	1.96	2.02	1.72	sp^3	2.44	7.90	5.01
51. Sb	2.05	2.19	1.82	*p*	1.46	4.96	7.57
				sp^3	2.64	8.48	9.37
52. Te	2.10	2.34	2.01	*p*	2.08	6.81	8.46
				sp^3	3.04	9.66	10.91
53. I	2.66	2.50	2.21	*p*	2.52	8.10	9.15
54. Xe	2.60[f]	2.34[f]	2.40[f]	*p*	2.40[f]	7.76[f]	—
55. Cs	0.79	0.69	0.86				
56. Ba	0.89	0.93	0.97				
57. La	1.10	0.92	1.08				
58. Ce	1.12	0.92	*1.08*				
59. Pr	1.13	0.92	*1.07*				
60. Nd	1.14	0.93	*1.07*				
61. Pm		0.94	*1.07*				
62. Sm	1.17	0.94	*1.07*				
63. Eu		0.94	*1.01*				
64. Gd	1.20	0.94	*1.11*				
65. Tb		0.94	*1.10*				
66. Dy	1.22	0.94	*1.10*				
67. Ho	1.23	0.96	*1.10*				
68. Er	1.24	0.96	*1.11*				
69. Tm	1.25	0.96	*1.11*				
70. Yb		0.96	*1.06*				
71. Lu	1.27	0.96	*1.14*				

Table 3.12 *(Continued)*

				Mulliken-Jaffé[d]			
					a		*b*
Element	Pauling[a]	Sanderson[b]	Allred-Rochow[c]	Orbital or hybrid[e]	Pauling scale	Volts	Volts/electron
72. Hf	1.30	0.98	*1.23*				
73. Ta	1.50	1.04	*1.33*				
74. W	2.36	1.13	*1.40*				
75. Re	1.90	1.19	*1.46*				
76. Os	2.20	1.26	*1.52*				
77. Ir	2.20	1.33	*1.55*				
78. Pt	2.28	1.36	*1.44*				
79. Au	2.54	1.72	*1.42*				
80. Hg	2.00	1.92	*1.44*				
81. Tl(I)	1.62	1.36 }	*1.44*				
Tl(III)	2.04	1.96 }					
82. Pb(II)	1.87	1.61					
Pb(IV)	2.33	2.01	*1.55*				
83. Bi	2.02	2.06	*1.67*				
84. Po	2.00		*1.76*				
85. At	2.20		*1.90*				
86. Rn			*2.06*[f]	*p*	2.12[f]	6.92[f]	—
87. Fr	0.70		*0.86*				
88. Ra	0.90		*0.97*				
89. Ac	1.10		*1.00*				
90. Th	1.30		*1.11*				
91. Pa	1.50		*1.14*				
92. U	1.70		*1.22*				
93. Np	1.30		*1.22*				
94. Pu	1.30		*1.22*				
95. Am	1.30		(*1.2*)				
96. Cm	1.30		(*1.2*)				
97. Bk	1.30		(*1.2*)				
98. Cf	1.30		(*1.2*)				
99. Es	1.30		(*1.2*)				
100. Fm	1.30		(*1.2*)				
101. Md	1.30		(*1.2*)				
102. No	1.30		(*1.2*)				

[a] Values to two decimal places are by A. L. Allred, *J. Inorg. Nucl. Chem.*, **1961**, *17*, 215, using Pauling's thermochemical method and recent data. Values to one decimal place are by L. Pauling, "The Nature of the Chemical Bond," 3d ed., Cornell University Press, Ithaca, N.Y., **1960**, p. 93.

[b] Calculated by R. T. Sanderson, "Inorganic Chemistry," Van Nostrand-Reinhold, New York, **1967**, pp. 72–76.

[c] Calculated by A. L. Allred and E. G. Rochow, *J. Inorg. Nucl. Chem.*, **1958**, *5*, 264, except for italicized values for heavier transition metals by E. J. Little and M. M. Jones, *J. Chem. Educ.*, **1960**, *37*, 231.

[d] Calculated from ionization energy–electron affinity data. All values from data of Jaffé and co-workers (J. Hinze and H. H. Jaffé, *J. Amer. Chem. Soc.*, **1962**, *84*, 540; *J. Phys. Chem.*, **1963**, *67*, 1501; J. Hinze et al., *J. Amer. Chem. Soc.*, **1963**, *85*, 148, except italicized values given by H. O. Pritchard and H. A. Skinner, *Chem. Rev.*, **1955**, *55*, 745. Values in the two right-hand columns are on the electron volt scale of Mulliken. The values of *a* have also been converted to the Pauling scale for comparison purposes.

[e] This column lists the hybridization of the bonding orbital. The numerical values represent hybridization in particular situations: 23% s = N in NH_3; 20% s = O in H_2O; 30% s = Sn in SnX_2 molecules (estimated).

[f] Values for the noble gases from L. C. Allen and J. E. Huheey, *J. Inorg. Nucl. Chem.*, **1980**, *42*, 1523.

Other methods of estimating electronegativity

Many other methods have been suggested for determining the electronegativity values of the elements. Only two general methods will be discussed here. The first was proposed by Mulliken[74] shortly after Pauling proposed his method. Mulliken suggested that the attraction of an atom for electrons should be an average of the ionization energy and electron affinity of the atom. This method was not used extensively until rather recently. It will be discussed further on pp. 152–155.

The other general method is to consider electronegativity to be some function of size and charge. These methods differ among themselves only in the choice of function (energy, force, etc.) and the method of estimating the effective charge. Allred and Rochow[75] defined electronegativity as the electrostatic *force* exerted by the nucleus on the valence electrons. They used effective nuclear charges obtained by Slater's rules[76] and obtained the formula:

$$\chi_{AR} = (3590\, Z^*/r^2) + 0.744 \qquad \textbf{(3.84)}$$

where r is the covalent radius (pm). The Allred-Rochow scale has been widely accepted as an alternative to Pauling's thermochemical method for determining electronegativities. Allred-Rochow values are listed in Table 3.12.

Another definition that is based on size and charge, but in a unique way, is the definition of Sanderson[77] based on relative electron density. This method has never been accepted widely, although Sanderson has applied it successfully to a wide variety of problems,[78] and it was the first to illustrate the interesting electronegativity properties of the posttransition elements (see Chapter 17).

Variation of electronegativity

Although electronegativity is often treated as though it were an invariant property of an atom, it actually depends upon the valence state of the atom in the molecule. In fact, this is probably the most important aspect of modern electronegativity theory to be appreciated: *The ability and capacity of an atom* (or more specifically an orbital on an atom) *to attract electrons will be rigorously defined by the environment of that atom.* Two factors will determine the attraction of atoms for electrons: the charge on the atom and the hybridization of the atom. An atom which has achieved a positive charge (either an integral charge as an

[74] R. S. Mulliken, *J. Chem. Phys.*, **1934**, *2*, 782; **1935**, *3*, 573; W. Moffitt, *Proc. Roy. Soc.* (*London*), **1950**, *A202*, 548.

[75] A. L. Allred and E. G. Rochow, *J. Inorg. Nucl. Chem.*, **1958**, *5*, 264.

[76] Allred and Rochow counted *all* of the electrons in the atom in the shielding; their Z^* values are thus 0.35 higher than those obtained by the usual application of Slater's rules. Differences of this sort are not important if one is consistent in application.

[77] R. T. Sanderson, *J. Chem. Educ.*, **1952**, *29*, 539; **1954**, *31*, 2, 238.

[78] R. T. Sanderson, "Inorganic Chemistry," Van Nostrand-Reinhold, New York, **1967**; *J. Inorg. Nucl. Chem.*, **1966**, *28*, 1553; **1968**, *30*, 375; R. T. Sanderson, "Chemical Bonds and Bond Energy," 2nd ed. Academic Press, New York, **1976**.

ion or a partial charge as an atom in a molecule) will tend to attract electrons to it more readily than will a neutral atom. In turn, a negatively charged atom (either an anion or an atom with a partial negative charge in a molecule) will attract electrons less than a neutral atom.

Hybridization affects electronegativity because of the lower energy and hence greater electron-attracting power of *s* orbitals. We might expect the electronegativity of an atom to vary slightly with hybridization, with those orbitals having greater *s* character being more electronegative. Some results of the variation in electronegativity have been given by Bent.[79] One factor affecting the acidity of hydrogen is the difference in electronegativity between the hydrogen atom and the atom to which it is bonded.[80] Methane, CH_4, with sp^3 hybridization and 25% *s* character is rather unreactive. The electronegativity of tetrahedral carbon is nearly the same as that of hydrogen. In ethylene, the carbon atom is hybridized approximately sp^2 and the hydrogen atom is somewhat more reactive, reflecting the increased electronegativity of carbon with 33% *s* character. Finally, acetylene has hydrogen atoms which are definitely acidic; salts such as $Ca^{+2}C{\equiv}C^{-2}$ form rather easily. In this case the digonally hybridized carbon atom (50% *s* character) has about the same electronegativity as a chlorine atom.

The basicity of amines is a function of the hybridization of the nitrogen atom.[80] The more electronegative the nitrogen atom, the less readily it will share its lone pair electrons and act as a base. The series of nitrogen bases, aliphatic amines, pyridine, and nitriles, exhibits this property:

$$\sim 25\%\ s \qquad R\ddot{N}H_2 + H_2O \rightleftharpoons RNH_3^+ + OH^- \qquad pK_b = 3\text{–}4 \tag{3.85}$$

$$33\%\ s \qquad C_5H_5N\colon + H_2O \rightleftharpoons C_5H_5NH^+ + OH^- \qquad pK_b = 8.8 \tag{3.86}$$

$$50\%\ s \qquad R{-}C{\equiv}N\colon + H_2O \rightleftharpoons \text{No reaction} \tag{3.87}$$

The electronegativity of the nitrogen atom increases as the *s* character of the hybridization increases, and hence its basicity decreases.

Another interesting case has been given by Streitwieser and co-workers.[81] It has been found that strained ring systems of the type shown in Fig. 3.59 are much more reactive at position No. 1 than at No. 2 in reactions involving loss of positive hydrogen. The strain in the four-membered ring results in the use of more *p* character in these bonds by the C_{10} atom (the shaded orbitals in Fig. 3.59). The corresponding increased *s* character in the bond to C_1 causes a greater electronegativity, an induced positive charge, and a greater acidity of the hydrogen atom. In the related pyridine derivative with a nitrogen atom in place of C_1, the same phenomenon results in reduced electron density on the nitrogen atom and reduced basicity compared to the unconstrained analogues.[82]

---

[79] H. A. Bent, *Chem. Rev.*, **1961**, *61*, 275.

[80] Acidity, basicity, and other properties depend on other properties in addition to electronegativity. Nevertheless, variation in electronegativity is important in determining the observed properties.

[81] A. Streitwieser, Jr., et al., *J. Am. Chem. Soc.*, **1968**, *90*, 1357.

[82] J. H. Markgraf and R. J. Katt, *J. Org. Chem.*, **1972**, *37*, 717.

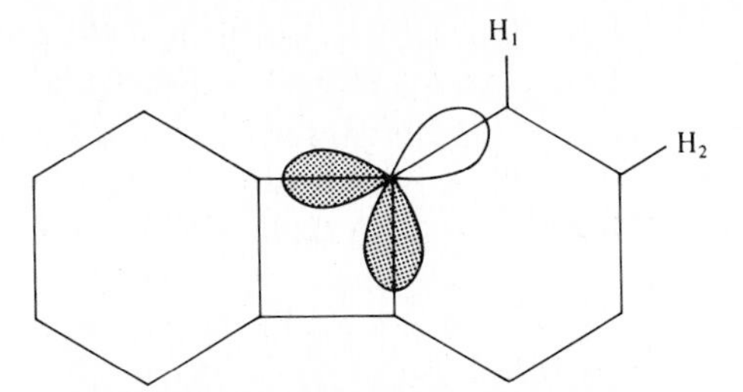

Fig. 3.59 Biphenylene. Shaded orbitals have increased *p* character; hence unshaded orbital has increased *s* character, increased electronegativity. [From A. Streitwieser, Jr., et al., *J. Am. Chem. Soc.*, **1968**, *90*, 1357. Reproduced with permission.]

The electronegativity of an atom can vary in response to the partial charge induced by substituent atoms or groups. For example, methyl iodide hydrolyzes as expected for alkyl halides, but trifluoromethyl iodide gives unusual products:

$$CH_3I + OH^- \longrightarrow CH_3OH + I^- \tag{3.88}$$

$$CF_3I + OH^- \longrightarrow CF_3H + IO^- \tag{3.89}$$

Although the products differ considerably in these two reactions, presumably the mechanisms are not drastically different. The negative hydroxide ion attacks the most positive atom in the organic iodide. In methyl iodide this is the carbon atom ($\chi_I > \chi_C$) and the iodide ion is displaced. In the trifluoromethyl iodide the fluorine atoms induce a positive charge on the carbon which increases its electronegativity until it is greater than that of iodine and thus induces a positive charge on the iodine. The latter is thus attacked by the hydroxide ion with the formation of hypoiodous acid, which then loses an H^+ in the alkaline medium to form IO^-.

It may seem paradoxical that the carbon atom can induce a greater positive charge on the iodine than that which the carbon itself bears but a simple calculation based on electronegativity equalization (see pp. 158–160) indicates that the charges are $\delta_I = +0.21$, $\delta_C = +0.15$, and $\delta_F = -0.12$. Although it is exceedingly unlikely that the real charges have these exact values, they are probably *qualitatively* accurate. This is an example of the importance of the ability of an atom to donate or accept charge. Iodine is the most polarizable atom in this molecule. The large, soft, polarizable nature of the iodine atom allows it to accept the larger charge (see p. 155).

A similar reaction of more interest to inorganic chemists is the reaction between carbonylate anions and alkyl iodides:

$$CH_3I + Na^+[Mn(CO)_5]^- \longrightarrow NaI + CH_3Mn(CO)_5 \tag{3.90}$$

$$2CF_3I + Na^+[Mn(CO)_5]^- \longrightarrow NaI + C_2F_6 + Mn(CO)_5I \tag{3.91}$$

In this reaction also, the polarity of the C—I bond depends upon the substituents on the carbon atom.

It is an interesting paradox that most of the examples of variable electronegativity come from organic chemistry, although it is probable that electronegativity variation is much more important in inorganic chemistry. For example, there must be a large difference in electronegativity between d^2sp^3 Cr(III) in $[Cr(NH_3)_6]^{+3}$ and sp^3 Cr(VI) in CrO_4^{-2}. Although the phenomenon is not so well documented as yet in inorganic chemistry, some examples (in addition to the above) include differences in basicity such as NH_3 versus NF_3, the oxidation state of oxyacids, the tendency of metal ions to hydrolyze,[83] and the effect of ring strain on acidity and basicity.[84]

Mulliken-Jaffé electronegativities

The experimental evidence for the importance of hybridization and partial charge on electronegativity is ample, and many workers have attempted to improve electronegativity theory to include these factors. Mulliken was able to include hybridization in his method based on ionization energy and electron affinity. Calculation of electronegativity for various hybridizations involves the computation of *valence state ionization energies* and *valence state electron affinities* by adjusting for the promotion energy from the ground state. The valence state ionization energy and electron affinity are not the experimentally observed values but those calculated for the atom in its *valence state* as it exists in a molecule. Two short examples will clarify the nature of these quantities.

Divalent beryllium bonds through two equivalent *sp*, or *digonal*, hybrids. The appropriate ionization energy therefore is not that of ground state beryllium, $1s^22s^2$, but an average of those necessary to remove electrons from the promoted, valence state:

$$1s^22s^12p^1 \longrightarrow 1s^22s^12p^0\ (\mathrm{IE}_p) \text{ and } 1s^22s^12p^1 \longrightarrow 1s^22s^02p^1\ (\mathrm{IE}_s)$$

It is thus possible to calculate the hypothetical energy necessary to remove an electron from an *sp* hybrid orbital. This VSIE (and the corresponding valence state electron affinity) can be used to calculate the electronegativity of an *sp* (*di*) orbital.

A fluorine atom may well be assumed to use a pure *p* orbital with hybridization neglected, but still the valence state ionization energy does not correspond to the experimentally observed quantity. We may consider that the use of ionization energies and electron affinities relates to the occurrence of covalent–ionic resonance as shown previously in Eq. 3.81. Now one of the requisites for resonance to occur is that all contributing forms have the same number of unpaired electrons (p. 124), so the wave function and energy for any contributions from F^+ must be for zero spin (all simple molecules containing fluorine are diamagnetic). The ground state ionization energy corresponds to the process $F(^2P) \rightarrow F^+(^3P)$, but the VSIE is for ionization to a singlet state for F^+, a suitably weighted average of ^1S and ^1D. We need not concern ourselves with the mechanics of calculating the necessary promotion energies for either beryllium or fluorine, but should remember that it is not possible to calculate accurate electronegativities simply from ground state ionization energies and electron affinities.

Mulliken electronegativities can be estimated by the equation:

$$\chi_M = 0.168\ (\mathrm{IE_v} + \mathrm{EA_v} - 1.23) \qquad \textbf{(3.92)}$$

---

[83] See p. 294.

[84] See p. 150.

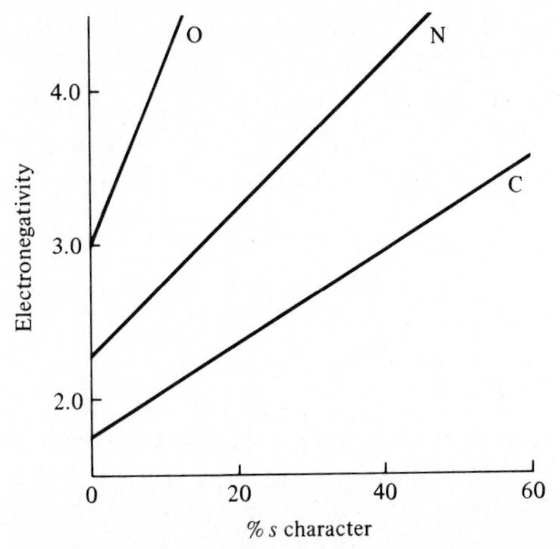

Fig. 3.60 Electronegativities of carbon, nitrogen, and oxygen as a function of *s* character.

where the valence state ionization energy (IE_v) and electron affinity (EA_v) are in electron volts.[85]

Since *s* electrons are held more tightly than *p* electrons, the electronegativity of an atom increases with *s* character (Fig. 3.60). For the atoms C and N discussed above, the electronegativities on Pauling's scale become:

	sp^3	sp^2	sp
C	2.48	2.75	3.29
N	3.68	3.94	4.67

The commonly accepted value of 2.5 for carbon appears to be based on tetrahedral (sp^3) hybridization. The usual value for nitrogen is about 3.0, corresponding to a hybridization between sp^3 and pure *p*. In their higher *s*-character hybrids (*sp*) nitrogen becomes more electronegative than fluorine (Pauling value = 3.98) and carbon approaches oxygen (Pauling value = 3.44).

Jaffé and co-workers[86] have provided data which allow the calculation of electronegativity as a function not only of hybridization but also of charge.[87] It was noted earlier (pp. 47–49) that the ionization energy–electron affinity curve for an element is approximately quadratic:

$$E = \alpha q + \beta q^2 \tag{3.93}$$

[85] Equation 3.92 gives Mulliken electronegativities in Pauling units. It is also common to use Mulliken values merely as the average of IE_V and EA_V. In this case, there are three options to conform to SI usage: (1) Convert the energies to kJ mol^{-1}; (2) Argue that the derivation of Mulliken–Jaffé electronegativities (Eqs. 3.93–3.95) leads logically to units of volts, an acceptable SI unit; (3) Argue that although the Mulliken scale, like the Pauling scale, is derived from energies, it is unitless. Methods 2 and 3 give the same result and since inorganic chemists have not yet confronted this question, they will be used in lieu of a consensus.

[86] J. Hinze and H. H. Jaffé, *J. Am. Chem. Soc.*, **1962**, *84*, 540; *J. Phys. Chem.*, **1963**, *67*, 1501; J. Hinze, M. A. Whitehead, and H. H. Jaffé, *J. Am. Chem. Soc.*, **1963**, *85*, 148.

[87] J. E. Huheey, *J. Phys. Chem.*, **1965**, *69*, 3284.

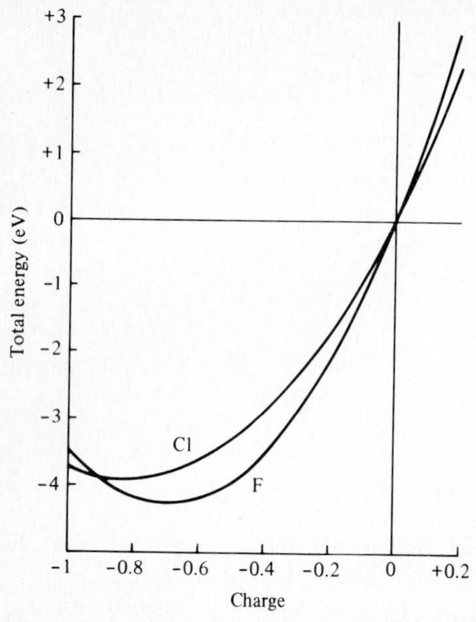

Fig. 3.61 Ionization energy–electron affinity curves for fluorine and chlorine. The electronegativities are given by the slopes of these curves. This figure is an enlarged portion of Fig. 2.15.

where E is the total energy of the atom and q is the ionic charge (see Fig. 3.61). The slope of this curve, dE/dq, has been suggested as a measure of electronegativity.[88] For the neutral atom this corresponds exactly to the Mulliken definition of electronegativity but has the added advantage of expressing the electronegativity (χ) as a function of charge:

$$\chi = \frac{dE}{dq} = \alpha + 2\beta q \tag{3.94}$$

which may be expressed in terms of partial charge (δ) instead of ionic charge (q) and the constants may be changed for convenience ($a = \alpha$; $b = 2\beta$):

$$\chi = \frac{dE}{d\delta} = a + b\delta \tag{3.95}$$

Electronegativity is thus seen to be a linear function of the partial charge on an atom (Fig. 3.62).

The importance of Eq. 3.95 lies in illustrating the large effect that charge can have on the electronegativity of an atom. For example, an iodine atom with a partial charge of about $+0.4$ is almost as electronegative as a neutral fluorine atom.

[88] R. P. Iczkowski and J. L. Margrave, *J. Am. Chem. Soc.*, **1961**, *83*, 3547.

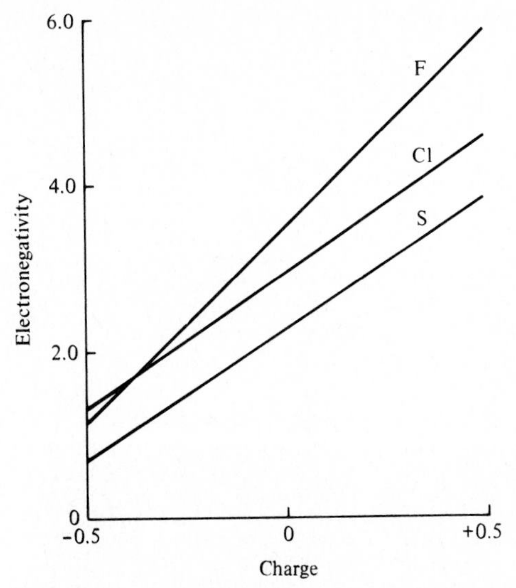

Fig. 3.62 Electronegativities of sulfur, chlorine, and fluorine as a function of charge, $\chi = a + b\delta$.

The significance of the parameters a and b is clear. The *inherent* or *neutral atom electronegativity* of an atom is given by a. This is the electronegativity in a particular valence state as estimated by the Mulliken method and corresponds to similar estimates of electronegativity by Pauling, Allred and Rochow, and others. The parameter b may be termed the *charge coefficient* and measures the rate of change of electronegativity with charge. Large, soft, polarizable atoms have low values of b, and small, hard, nonpolarizable atoms tend to have higher values. An atom with a large charge coefficient will change electronegativity much more rapidly than one with a lower value of b.

The charge coefficient, b, is an inverse measure of the *charge capacity* of an atom. A small atom (large b) has only a limited ability to donate or absorb electron density before its electronegativity changes too much for further electron transfer to take place. One of the most important examples of charge capacity is the very electronegative, but very small, fluorine atom. Although initially very electronegative (note the very steep slope at the origin of Fig. 3.61), it rapidly becomes "saturated" as it accepts electron density (note how quickly the slope flattens out between -0.4 and -0.6), and beyond ~ -0.7 it would be necessary to "push" to get more electron density on a fluorine atom. This is closely related to the "anomalous electron affinity" of fluorine (see pp. 49, 848).

The charge capacity effect in electronegativity is responsible for the well-known inductive effect of alkyl groups (see below). It also causes otherwise "unexpected" effects in basicity,[89] dipole moments, and related phenomena.[90]

[89] J. E. Huheey, *J. Org. Chem.*, **1971**, *36*, 204.

[90] P. Politzer, in "Homoatomic Rings, Chains and Macromolecules of Main-Group Elements," A. L. Rheingold, ed., Elsevier, New York, **1977**, Chapter 4; *Int. J. Quantum Chem.*, in press.

Table 3.13 Estimates of the electronegativities of some common groups

	Experimental methods		Calculations			
Group	"Best"[a]	Other[b]	χ_C[c]	χ_G[d]	a'[e]	b'[e]
CH_3	2.3	2.34, 2.63	2.28	2.30	7.45	4.64
CH_2CH_3			2.29	2.32	7.52	3.78
CF_3	3.35		3.55	3.32	10.50	5.32
CCl_3	3.0	2.76, 3.03	2.83	3.19	10.12	4.33
CBr_3		2.65	2.59	3.10	9.87	3.96
CI_3			2.51	2.96	9.43	3.77
NH_2		1.7, 2.99, 3.40	2.50	2.78	8.92	5.92
NF_2			3.62	3.78	11.89	6.77
$N(CH_3)_2$			2.37	2.61	8.50	3.79
OH		2.3, 3.51, 3.6, 3.89	2.86	3.42	10.80	8.86
CN	3.3	3.11, 3.17, 3.22, 3.27	4.17	3.76	11.83	6.47
COOH	2.85	2.84, 2.88, 2.97, 3.50		3.36	10.62	5.43
NO_2	3.4	3.2, 3.45, 3.92		4.32	13.51	6.93
$C(O)OCH_3$				3.50	10.05	4.68
C_6H_5	3.0	3.01, 3.13, 3.18	2.50	2.58	8.31	3.60

[a] P. R. Wells, *Progr. Phys. Org. Chem.*, **1968**, *6*, 111.
[b] Various experimental methods, mostly infrared spectroscopy, to give a range of values; for details see J. E. Huheey, *J. Phys. Chem.*, **1965**, *69*, 3284; **1966**, *70*, 2086.
[c] A. F. Clifford, *J. Phys. Chem.*, **1959**, *63*, 1227.
[d] Pauling scale value calculated by electronegativity equalization, see note *e*.
[e] a' and b' values correspond approximately to a and b values of Mulliken-Jaffé system. From J. C. Watts, Doctoral Dissertation, University of Maryland, College Park, 1971. See also J. E. Huheey, *J. Org. Chem.*, **1966**, *31*, 2365.

Group electronegativity

It is often convenient to have an estimate of the inductive ability of a substituent group. As we have seen previously, we cannot use a single value of carbon (~ 2.5) to represent the electronegativity of both CH_3 and CF_3. The electronegativities of these groups will be the electronegativity of carbon *as it is adjusted by the presence of three hydrogen or three fluorine atoms.* Estimation of group electronegativities has been approached from a variety of ways. Organic chemists have developed sets of substituent constants from kinetic data,[91] and these have proven useful in certain inorganic systems as well. Other values have been obtained from physical measurements and directly from atomic electronegativities.[92] Some comparative values are listed in Table 3.13.

[91] For a review of various sets of substituent constants, see C. G. Swain and E. C. Lupton, Jr., *J. Am. Chem. Soc.*, **1968**, *90*, 4328.

[92] For methods of calculating group electronegativities from Mulliken-Jaffé electronegativity values of the constituent atoms, see J. E. Huheey, *J. Phys. Chem.*, **1965**, *69*, 3284; **1966**, *70*, 2086; *J. Org. Chem.*, **1966**, *31*, 2365. Brief reviews of previous work by other methods are also included in these papers and in P. E. Wells, *Progr. Phys. Org. Chem.*, **1968**, 6, 111.

One feature of group electronegativity is readily apparent from the table: Groups have much lower b values than the corresponding elements. This means that groups can donate or accept charge far better than would be indicated by their inherent electronegativities (a values) alone. For example, the methyl group is slightly (though not significantly) *more* electronegative than hydrogen. Yet the methyl (and other alkyl) group is generally considered a better donor than is a hydrogen atom. It is the greater charge capacity, which results from ability to spread the charge around

$$\begin{array}{ccc} & H^{\delta+} & \\ & | & \\ H^{\delta+}- & C^{\delta+} & \longrightarrow \\ & | & \\ & H^{\delta+} & \end{array}$$

that allows the methyl group to donate more electron density than the smaller hydrogen atom.

Choice of electronegativity system

With more than one valid system available, the choice of the "best" one is not always easy. We can arbitrarily divide the various methods into two groups. One consists only of the Mulliken-Jaffé method or modifications thereof. It may be termed a "theoretical" or "absolute" scale since it is based only on the fundamental energies of isolated atoms. All of the other scales are "empirical" and "relative" since they utilize experimentally obtained data such as enthalpies of formation, covalent radii, etc. Both systems have advantages. In general, the Mulliken-Jaffé system is more satisfying since it is, in a sense, more fundamental and basic. In addition, the values are as accurate as the measurements of energies. Its one disadvantage stems from its specificity: It is necessary to know the proper hybridization to choose. Often the proper hybridization can be inferred from bond angles. In other cases there is no unambiguous method of choosing the proper hybridization (e.g., in transition metals both sp^3 and sd^3 give tetrahedral hybrids, but differ greatly in electronegativity). The empirical methods have an advantage here resulting directly from their indirect derivation. In other words, variables (hybridization, etc.) are often "built-in" as long as the atom under consideration is in a fairly typical environment. But if the atom is in an unusual hybridization, the empirical value based on some more typical, "average" environment is apt to be inaccurate. The choice of a proper value of electronegativity then reduces to this: If it is possible to specify the hybridization accurately, the Mulliken-Jaffé values are more accurate and more powerful in most cases. Otherwise, one of the empirical systems must be used. Each has advantages and disadvantages, adherents and detractors, and they do not really differ greatly among themselves. If the situation is sufficiently nonspecific to make it necessary to use an empirical system, it probably will not make a great deal of difference which is chosen. However, one must be consistent and avoid picking the value for one element from Pauling, another from Allred-Rochow, and a third from Sanderson and comparing the three. By judicious mixing of systems like this, one could probably "prove" anything!

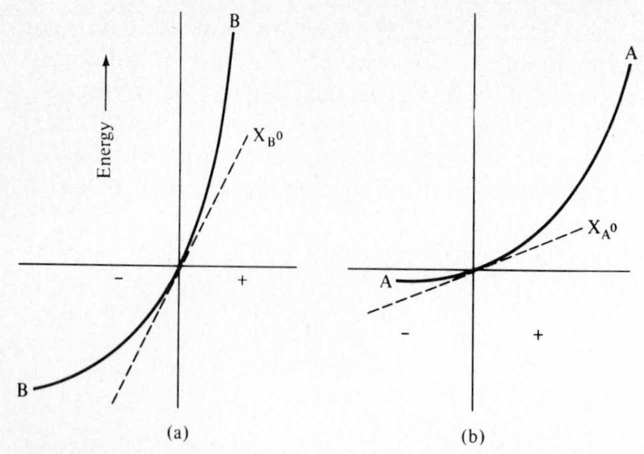

Fig. 3.63 Relation between ionization energy–electron affinity curve (solid line) and inherent electronegativity (dashed line) for a more electronegative element (B) and a less electronegative element (A).

Methods of estimating charges: Electronegativity equalization

For many reasons, chemists would like to be able to estimate the charges on the constituent atoms in a molecule. There have been many attempts to do this, but none has proved to be completely successful as yet. The best way would be to solve the wave equation for a molecule without the use of simplifying assumptions, and then to calculate the electron distribution. Such *ab initio* calculations are possible for small molecules[93] but become increasingly difficult as the number of atoms increases. Even when the calculations are possible, there is not complete agreement among chemists as to the best way of apportioning the charge density among the atoms in the molecule.[94]

Some workers have suggested methods of estimation based on various physical measurements of molecular properties. These measurements are discussed in the next section, and we shall see that generally one or more assumptions must be made; therefore, uncertainties are involved. With this failure of both experiment and theory to yield the desired results, several workers have suggested semiempirical methods based on electronegativity. Only one method will be discussed here. Sanderson[95] has suggested that when a bond forms between two atoms electron density will shift from one atom to the other until the electronegativities have become equalized. Initially the more electronegative element will have a greater attraction for electrons (Fig. 3.63), but as the electron density shifts toward that atom it will become negative and tend to attract electrons less. Conversely, the atom which is losing electrons becomes somewhat positive and attracts electrons better than it did when neutral. This process will continue until the two atoms attract the

[93] Figures 3.32, 3.33, and 3.49–3.52 were obtained from such calculations.

[94] P. Politzer and R. R. Harris, *J. Am. Chem. Soc.*, **1970**, *92*, 6451; P. Politzer and P. H. Reggio, ibid., **1972**, *94*, 8308; R. S. Evans and J. E. Huheey, *Chem. Phys. Lett.*, **1973**, *19*, 114.

[95] R. T. Sanderson, *J. Chem. Educ.*, **1954**, *31*, 2; see also footnote 78.

electrons equally, at which point the electronegativities will have been equalized and charge transfer will cease (Fig. 3.64):

$$\chi_A = a_A + b_A\delta_A = \chi_B = a_B - b_B\delta_A \tag{3.96}$$

$$\delta_A = \frac{a_B - a_A}{b_A + b_B} \tag{3.97}$$

The partial charges in the HCl molecule may be estimated with Eq. 3.97 by using the appropriate a and b values from Table 3.12: $a_H = 7.17$, $b_H = 12.85$, $a_{Cl(p)} = 9.38$, and $b_{Cl(p)} = 11.30$.

$$\delta_H = \frac{9.38 - 7.17}{11.30 + 12.85} = +0.09 \tag{3.98}$$

It will be noted that the charge estimated by this method is somewhat lower than a similar estimate using Eq. 3.99. This is a general result: Eq. 3.97 consistently produces values lower than Eq. 3.99. If the total ionization energy (including the electron affinity) were the only energy involved in the charge distribution, Eq. 3.97 would be rigorously correct. In a molecule, however, other energy terms are important. The exchange energy associated with overlap of orbitals tends to be reduced if the charge transfer is too great. The Madelung energy (so named because of resemblance to that found in ionic crystals) resulting from the electrostatic attraction of A^+ for B^- (*within the molecule*) tends to increase ionicity. Although these energies tend to cancel each other in effect because they work in opposite directions, Eq. 3.97 can only be considered a useful, qualitative approximation.

Although there is no universal agreement on the "real" charges in molecules (see p. 158 concerning charges in *ab initio* calculations), various attempts have been made to

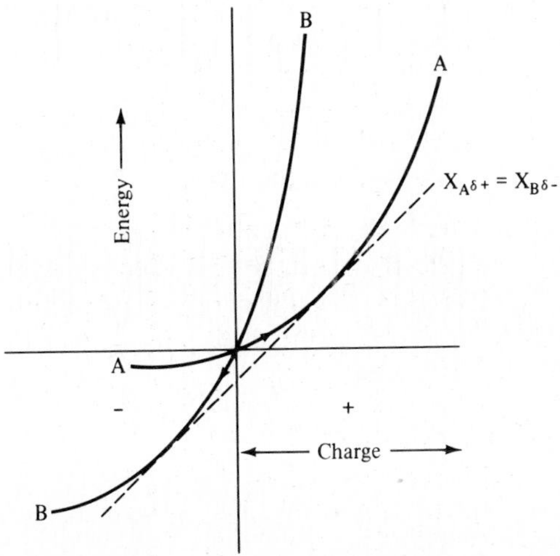

Fig. 3.64 Superposition of ionization energy–electron affinity curves for a more electronegative (B) and less electronegative (A) element. The common tangent (= equalized electronegativity) is given by the dashed line.

improve upon simple electronegativity equalization calculations. One method is to estimate the exchange and Madelung energies by simple bonding models, and then to use them to adjust the values obtained by the electronegativity equalization method. This modification has been found to correlate well with some *ab initio* calculations for some simple molecules.[94,96] Recently Parr and co-workers have advanced electronegativity equalization in a quantum-mechanical context.[97]

EXPERIMENTAL MEASUREMENT OF CHARGE DISTRIBUTION IN MOLECULES

Qualitatively it is simple to state that when a bond forms between two atoms, the bonding electrons will tend to move toward the atom of greater electronegativity, and *that* atom will tend to become more negative while the other atom will be more positive. Thus the *qualitative* relation between *ionicity* or *ionic character* and electronegativity is clear. Attempts to obtain a simple *quantitative* relationship have been much less successful. The primary reason for this is that a correlation can be no better than the data upon which it is based. Unfortunately, there is only *one* completely unambiguous way in which the electron density of an atom can be measured, namely careful X-ray analysis (p. 72). The technique is difficult enough to apply to highly symmetrical crystals such as NaCl. It is even more difficult to apply to molecules of lower symmetry, and it has only been in the last few years that X-ray analysis has been developed to the point that it could be usefully applied to molecules. The number of data currently available is severely limited (see p. 175). Although there is a host of chemical and physical properties which depend upon electron density, such as dipole moments, NMR chemical shifts and coupling constants, NQR values, and various reaction data, all depend on factors in addition to electronegativity. This makes it impossible to correlate theoretical predictions with experimental data and determine the accuracy of the predictions. Nevertheless, they do provide useful information and will be discussed briefly.

Dipole moments

One of the first correlations of a molecular property with electronegativity was between the dipole moment of a molecule and the difference in electronegativity of the constituent atoms.[98] On the basis of the dipole moments of the hydrogen halides, various expressions have been suggested, one of the simplest of which is[99]

$$P = 16(\Delta\chi) + 3.5(\Delta\chi)^2 \tag{3.99}$$

[96] R. S. Evans and J. E. Huheey, *J. Inorg. Nucl. Chem.*, **1970**, *32*, 777.

[97] R. G. Parr et al., *J. Chem. Phys.*, **1978**, *68*, 380; R. A. Donnelly and R. G. Parr, ibid., **1978**, *69*, 4431; P. Politzer and H. Weinstein, ibid., **1979**, *71*, 4218.

[98] J. G. Malone, *J. Chem. Phys.*, **1933**, *1*, 197; L. Pauling, "The Nature of the Chemical Bond," 3rd ed., Cornell University Press, Ithaca, N. Y., **1960**, pp. 78–79.

[99] N. B. Hannay and C. P. Smyth, *J. Amer. Chem. Soc.*, **1946**, *68*, 171.

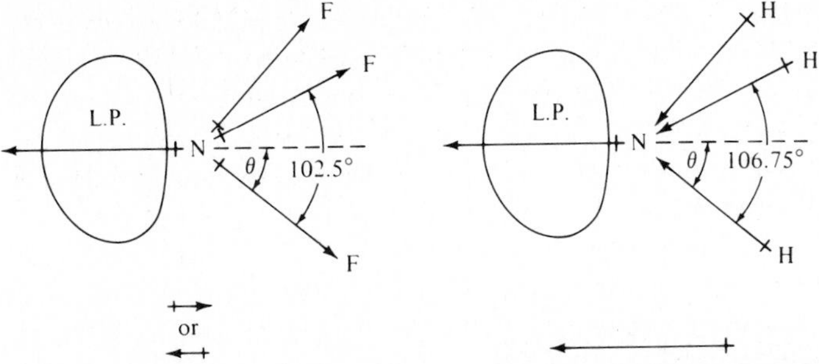

Fig. 3.65 Bond moments and total molecular dipole moments in the ammonia and nitrogen trifluoride molecules: $\mu_{\text{total}} = \mu_{\text{lone pair}} + 3 \cos \theta \mu_{\text{bond}}$, 5.0×10^{-30} C m (NH_3), 0.7×10^{-30} C m (NF_3).

where $\Delta\chi$ is the difference in electronegativity between two bonding atoms and P is the percent ionic character, assumed to be zero for a "purely covalent" bond and 100% for a "purely ionic" bond.

The relation between ionic character and dipole moment in a molecule such as HCl appears to be straightforward if the bond length (127 pm) and dipole moment (3.44×10^{-30} C m) are known:

$$\underset{+\!\!\longrightarrow}{\text{H—Cl}} \quad \mu = qr = 3.44 \times 10^{-30} \text{ C m} \tag{3.100}$$

$$q = \frac{3.44 \times 10^{-30} \text{ C m}}{127 \times 10^{-12} \text{ m}} \times \frac{1 \text{ electron}}{1.6 \times 10^{-19} \text{ C}} = 0.17 \text{ electron} \tag{3.101}$$

$$\delta_{\text{H}} = +0.17 \qquad \delta_{\text{Cl}} = -0.17 \tag{3.102}$$

The simplicity of this argument has made it a popular one in textbooks over the past quarter of a century. Unfortunately the situation is considerably more complicated. The *ionic bond moment* shown is only one of several contributions to the total *molecular dipole*. Two other important contributions are moments which arise from size differences between atoms and the hybridization of lone pairs. Estimates of these range from 3×10^{-30} C m toward the hydrogen to 6×10^{-30} C m toward the chlorine, irrespective of any additional moment resulting from a polar bond. Under these conditions assigning the total molecular dipole to the ionic character of a bond is extremely dubious.

A single example of the effect of hybridization and lone pair dipoles will be given.[100] Both NH_3 and NF_3 have approximately tetrahedral geometries (Fig. 3.65) with bond angles of 106.75° and 102.5°, respectively. Using accepted values for the electronegativities of N (3.0), H (2.2), and F (4.0), Eq. 3.99 predicts ionic characters of about 15% (NH) and 19% (NF). Combining the supposed ionic character in the N—H bond with the bond length (101 pm) yields a predicted dipole of about 5.7×10^{-30} C m, reasonably close to

[100] The interested reader is referred to R. McWeeny, "Coulson's Valence," Oxford University Press, London, **1979**, pp. 153–161, and F. A. Cotton and G. Wilkinson, "Advanced Inorganic Chemistry," 3rd ed., Wiley, New York, **1972**, pp. 120–122.

the experimental value of 5.0×10^{-30} C m. In contrast, the same type of calculation using a charge of −0.19 on the fluorine and the longer N—F bond distance (137 pm) predicts a dipole moment of about 10.3×10^{-30} C m for NF_3. The experimental value for the dipole of nitrogen trifluoride is actually 0.7×10^{-30} C m. The observed discrepancy arises from the neglect of the lone pair moment. A more accurate description would probably be to say that the lone pair moment and the bond moments are each about $2\text{–}3 \times 10^{-30}$ C m, and that in one case they reinforce each other to yield 5×10^{-30} C m, and in the other they nearly cancel each other to yield a net moment of 0.7×10^{-30} C m. It is the neglect of possible hybridization and lone pair moments, together with other effects, that renders the above treatment of HCl of little value.

Despite the uncertainties surrounding the relationship between dipole moments and the charge distribution in molecules, molecular dipole moments can be handled in a self-consistent way by the use of bond moments (Table 3.14). The treatment assumes that the properties of an individual bond are unaffected by the nature of the surrounding bonds (an obvious oversimplification), and thus one can predict the total molecular moment from the vector sum of the component bond moments. For example, if we assume methyl fluoride to be tetrahedral, we would predict a dipole moment of 5.9×10^{-30} (H—C) + 4.64×10^{-30} (C—F) = 5.9×10^{-30} C m (see Problem 3.49). The experimental value is 5.97×10^{-30} C m. Generally, we cannot expect such close agreement.

Nuclear quadrupole resonance

Another method which provides useful information on the ionic character of bonds is the nuclear quadrupole coupling constant. The reader is referred to extended discussions of the topic for the appropriate theory and experimental design,[101] but for the present discussion we may assume that the nuclear quadrupole coupling constant of a given atom is a measure of the asymmetry of the electric field at the nucleus. A purely ionic bond, were one to exist, would provide a completely spherical electric field about such a nucleus, like the chlorine nucleus in a chloride ion, for example. The chlorine atom in such an environment would exhibit no quadrupole coupling. To the extent that a bond is very ionic the electron distribution will approach a completely spherical situation and the coupling will be low. The more the atom bonded to the chlorine atom pulls electrons away from it, the more asymmetric will be the electric field gradient at the chlorine nucleus. This is shown nicely in Table 3.15. The coupling constant for lithium chloride (−6.14) has the smallest absolute magnitude, the alkyl chlorides are somewhat intermediate, and the chlorine molecule itself (in which the chlorine atom must have zero partial charge) is higher. Highest of all is chlorine fluoride in which the fluorine atom has induced a *positive* charge on the chlorine atom and perturbed the electric field drastically.

If ionicity were the single parameter affecting the magnitude of the coupling constant, it would be a simple matter to convert the data in Table 3.15 into a quantitative measure

[101] M. Kubo and D. Nakamura, *Adv. Inorg. Chem. Radiochem.*, **1966**, *8*, 257; E. A. C. Lucken, "Nuclear Quadrupole Coupling Constants," Academic Press, New York. **1969**; J. A. S. Smith. *J. Chem. Educ.*, **1971**, *48*, 39; L. Ramakrishan et al., *Coord. Chem. Rev.*, **1977**, *22*, 123.

Table 3.14 Selected bond moments[a]

Bond	Bond moment ($\times 10^{30}$ C m)	Bond	Bond moment ($\times 10^{30}$ C m)
Covalent single bonds		*Double and triple bonds*	
H—C	1.3	C=N	3.0
H—N	4.44	C=O	7.7
N—O	5.04	C=S	6.7
H—P	1.2	P=O	9.0
H—S	2.3	P=S	10.3
C—N	0.73	P=Se	10.7
C—O	2.5	As=O	14.0
C—S	3.0	Sb=S	15.0
C—Se	2.0	S=O	10.0
C—F	4.64	Se=O	10.3
C—Cl	4.90	Te=O	7.7
C—Br	4.74	C≡N	11.8
C—I	4.17	N≡C	10.0
Si—H	3.3		
Si—C	2.0[b]		
Si—N	5.17		
Ge—Br	7.0		
Sn—Cl	10.0	*Coordinate covalent bonds*	
Pb—I	11.0		
Hg—Br	11.8	N → B	8.51
N—F	0.57	N → O	14.3
P—Cl	2.70	P → B	14.7
As—F	6.77	O → B	12.0
S—Cl	2.0	S → B	12.7

[a] With δ^+ on left-hand atom. Values from G. J. Moody and J. D. R. Thomas, "Dipole Moments in Inorganic Chemistry," Edward Arnold, London, **1971**, and from V. I. Minkin, O. A. Osipov, and Y. A. Zhdanov, "Dipole Moments in Organic Chemistry," Plenum, New York and London, **1970**.

[b] A. P. Altschuller and L. Rosenblum, *J. Am. Chem. Soc.*, **1955**, *77*, 272.

of ionicity. Unfortunately (for such attempts at least) there is a second parameter which affects the asymmetry of the electric field gradient: hybridization. An *s* orbital is spherically symmetrical and can contribute nothing to the field gradient whether fully occupied or not. On the other hand, *p* orbitals, as a result of their nonspherical nature, create a large gradient at the nucleus if only partially filled (i.e., if the bond is not completely ionic). Hence in order to make a quantitative estimate of the ionicity of a particular Cl—X bond, it is necessary to know the hybridization of the orbital which the chlorine atom is using, i.e., how much *s* character is involved, since *s* character will decrease the coupling just as will ionicity and one must be known in order to determine the other. The equation relating

Table 3.15 Nuclear quadrupole coupling constants for ^{35}Cl in various compounds

Compound	Coupling constant (MHz)[a]
Free Cl^- ion	0.00 (theoretical value)
LiCl	−6.14
$(CH_3)_3CCl$	−62.13
$(CH_3)_2CHCl$	−64.13
CH_3CH_2Cl	−65.78
CH_3Cl	−68.06
CH_2Cl_2	−71.98
$CHCl_3$	−76.70
CCl_4	−81.2
OCl	−82.5
BrCl	−103.6
Cl_2	−108.95[b]
ClF	−146.0

[a] The negative sign of the coupling constant is determined by the nature of the ^{35}Cl nucleus and may be ignored for purposes of the present discussion.

[b] Solid state at 20 K. All other values are for the gas phase. There are small differences (~5%) between solid-phase and gas-phase values for a given molecule.

these two parameters to the coupling constant (eQq) is:[102]

$$eQq_{X-Cl} = (1 - S)(1 - i)(eQq_{Cl_2}) \qquad \textbf{(3.103)}$$

where S and i are the s character and ionic character, respectively. Since chlorine is monovalent, no estimate of the s character can be made from bond angles and little can be said except that the s character probably does not vary *too* much and the data in Table 3.15 *probably* show the ionicity trends *qualitatively*.

Mössbauer spectroscopy

Mössbauer spectroscopy[103] also provides us with certain information about the electronic environment at the nucleus of an atom. In Mössbauer spectroscopy fluorescence resulting from the absorption and emission of a γ ray is observed. As is the normal case for absorption

[102] There are several simplifying assumptions contained in Eq. 3.103 such as that there is no d orbital participation and no π bonding. It is also assumed that chlorine is the negative end of the polar bond. The components of the coupling constant are the electronic charge (e), the nuclear quadrupole moment (Q), and the electric field gradient (q), the second derivative of potential with respect to distance ($\partial^2 V/\partial x^2$).

[103] E. Fluck, *Adv. Inorg. Chem. Radiochem.*, **1964**, *6*, 433; G. M. Bancroft and R. H. Platt, ibid., **1972**, *15*, 59; R. H. Herber, *Progr. Inorg. Chem.*, **1967**, *8*, 1; V. I. Goldanskii and R. H. Herber, Eds., "Chemical Applications of Mössbauer Spectroscopy," Academic Press, New York, **1968**.

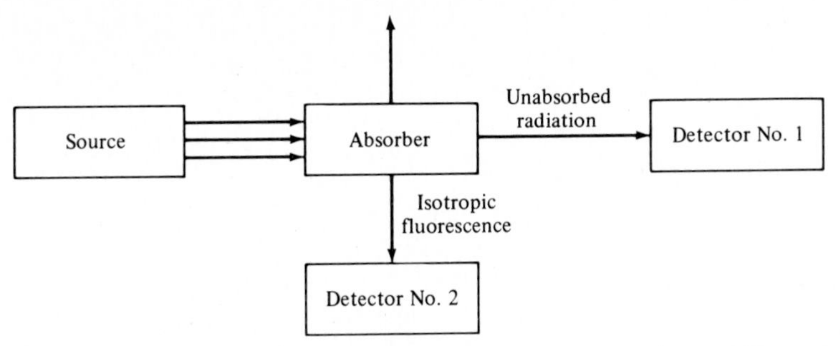

Fig. 3.66 Diagrammatic representation of Mössbauer experiment. When resonance occurs photons are detected at No. 2 but fewer at No. 1.

and emission of photons, the frequency of the photon will be related to the difference in energy of the two states before and after emission (or absorption):

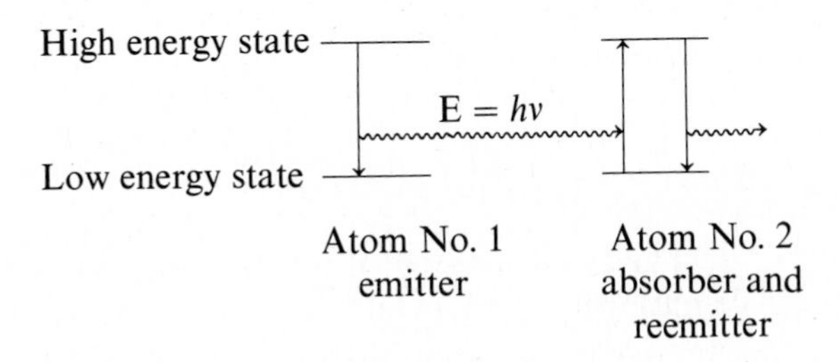

Since γ rays are at the extreme high-energy end of the spectrum, a significant fraction of the energy of transition would be dissipated in recoil (and hence unavailable as $h\nu$ and thus too low in energy to excite nucleus No. 2) if the atom were free to move.[104] Mössbauer spectroscopy therefore *must* be performed on a solid sample, often at low temperatures, with crystalline forces holding the emitting atom firmly in place to prevent recoil. This restricts the information obtainable from Mössbauer work. Although it might be desirable to obtain data from compounds in the gaseous or liquid state, they are experimentally inaccessible.

Two methods of detecting Mössbauer resonance are available (Fig. 3.66). If the detector is placed at position No. 1, a decrease in intensity will be observed as the radiation is absorbed at resonant frequencies. Alternatively, one can measure the isotropic radiation as the absorber is excited by the resonant γ-ray photons and reradiates them.

Now the nuclear energy levels responsible for the transitions giving rise to the γ-ray emissions and absorptions will be perturbed to a slight extent by the presence of electron density at the nucleus. Hence the Mössbauer effect serves as a useful probe of the electronic environment about the nucleus of an atom in a molecule. If atom No. 1 is in an identical

[104] Since both momentum (mv) and kinetic energy ($\frac{1}{2}mv^2$) must be conserved, the maximum energy for the projectile (photon) is obtained for minimum recoil of the emitter. As an analogy, consider the elimination of the "kick" of a rifle by butting the stock against a brick wall—the velocity of the bullet will be somewhat higher than for an unsupported rifle.

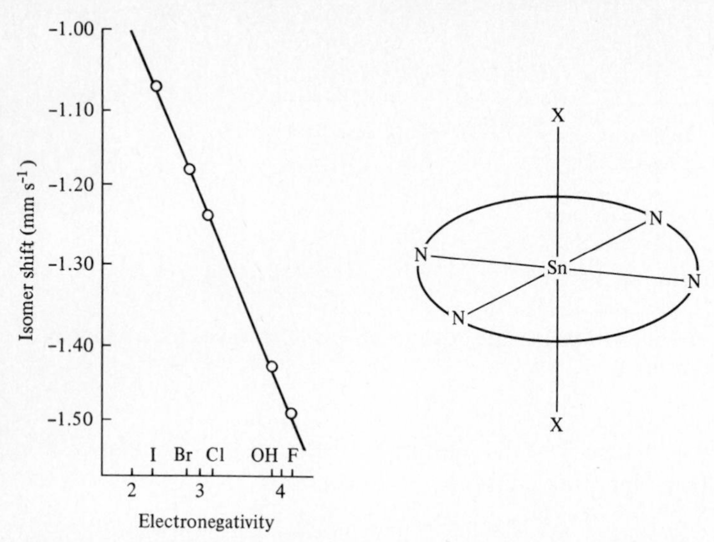

Fig. 3.67 Mössbauer isomer shifts in tin phthalocyanines as a function of the electronegativity of the axial substituents, X. [Taken, in part, from M. O'Rourke and C. Curran, *J. Am. Chem. Soc.*, **1970**, *92*, 1501. Reproduced with permission.]

environment to that of atom No. 2, absorption and subsequent fluorescence will occur; otherwise it will not (unless the two environments accidentally perturb the nucleus in exactly the same way). To obtain quantitative measures of the difference between two nuclear environments it is possible to cause absorption and fluorescence to occur in cases in which it would otherwise not by means of the Doppler effect. The latter increases or decreases the frequency of a wave passing between two objects that are moving toward or away from each other. If either the emitter or absorber is moved *toward the other* while the measurement is taking place, the photons will have higher energy than that corresponding to the transition in the emitter and *can activate fluorescence of a higher energy in the absorber* (and conversely for moving the two apart). This relative motion, measured in millimetres per second, is called the *isomer shift* and is a measure of the difference in the chemical environment of the two nuclei.

Once again two parameters are involved, and hence simple answers cannot be given. One parameter is the *s* character since only electron density which can actually be present at the nucleus can perturb it and only *s* orbitals have a finite electron density at the nucleus. The second parameter is the charge that has been built up on the atom in question which concentrates or disperses electron density. As long as one is aware of the possible effects of hybridization and makes an effort to minimize them, it is possible to correlate some Mössbauer data with electronegativity effects. For example, in the series of tin complexes formed from tin phthalocyanine (a complex organic molecule having four coplanar nitrogen atoms attached to the metal; see p. 854) and two additional ligands, X^-, there is a very nice correlation[105] between the isomer shift and the electronegativity of X (Fig. 3.67). Further, by using Mulliken-Jaffé electronegativities to calculate partial charge (see p. 159),

[105] M. O'Rourke and C. Curran, *J. Am. Chem. Soc.*, **1970**, *92*, 1501.

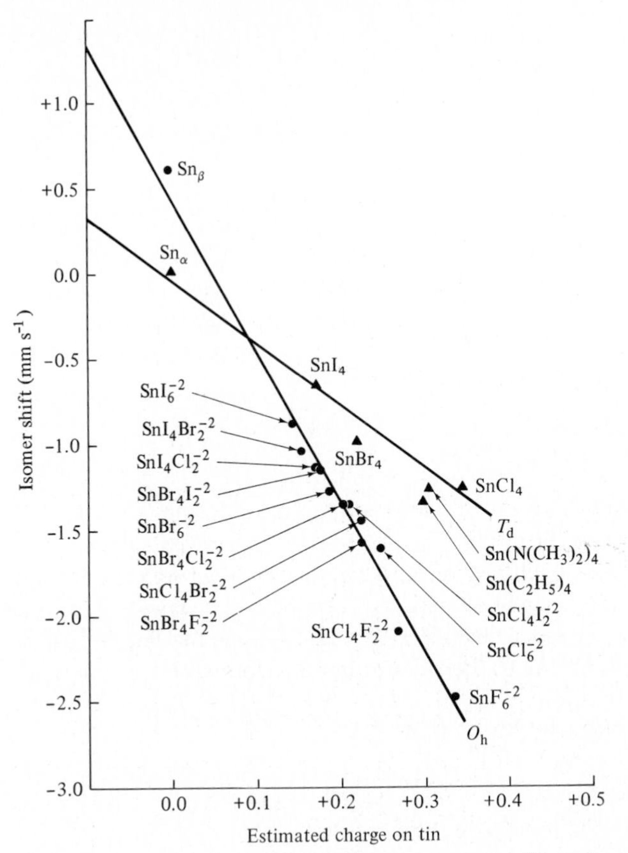

Fig. 3.68 Mössbauer isomer shift in SnX_4 and SnX_6^{-2} species as a function of the charge of the tin atom. [From J. E. Huheey and J. C. Watts, *Inorg. Chem.*, **1971**, *10*, 1553. Reproduced with permission.]

it is possible to correlate the isomer shifts of two series of tin compounds, SnX_4 and SnX_6^{-2}, with the partial charge induced on the tin atom by the halogen substituents (Fig. 3.68).[106] Nevertheless, it is not possible to use the Mössbauer effect to calculate the absolute charges on atoms; it can be used merely to correlate relative charges.

Nuclear magnetic resonance

Nuclear magnetic resonance once appeared to offer promising potential for the determination of relative charges on atoms, but this anticipation has not been realized. Although the chemical shift of an atom (such as a proton, for example) does measure the extent to which the nucleus is "deshielded" by the pulling of electrons away by electronegative atoms attached to it, it is now felt that little information can be gained on the ionicity of bonds from this experimental parameter. The reason for this is that the chemical shift, as measured experimentally, is determined by the sum of all of the magnetic fields about the nucleus,

[106] J. E. Huheey and J. C. Watts, *Inorg. Chem.*, **1971**, *10*, 1553.

not only those of the bonding electrons in the immediate vicinity but the anisotropic effects of magnetic fields arising from atoms that may be several bonds away.[107] Hence, although the chemical shift provides valuable information on the kinds of protons (or other nuclei) present in a molecule, it does not provide us with any quantitative information concerning the ionicity of bonds.

A second parameter derivable from NMR spectroscopy is the spin-spin coupling constant. It is a measure of the interaction of two nuclei through the electrons between them.[108] As such it is sensitive to the electron density at the nucleus and might be a useful probe of ionicity effects. However, as in the case of the Mössbauer effect, in addition to the effects of electronegativity the *s* character plays an important role (being the only orbital with electron density at the nucleus), and there is no uniform agreement as to the relative importance of hybridization and electronegativity effects.[109]

Infrared spectroscopy

Infrared spectroscopy is usually applied to the "fingerprinting" of molecules through the use of characteristic group frequencies. However, in order for a vibrational mode to be active in the infrared, it must involve a changing dipole moment. Diatomic molecules such as Cl_2 will therefore be inactive as will symmetric stretches of molecules such as O=C=O. More important for the present discussion is the fact that the characteristic frequencies of certain groups shift according to their molecular environment. For example, the carbonyl group is a resonance hybrid of

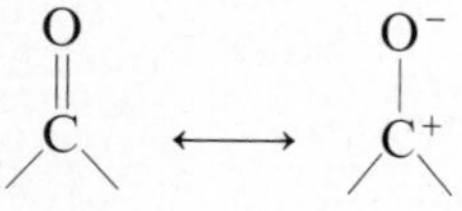

$$\underset{\text{(I)}}{>C{=}O} \longleftrightarrow \underset{\text{(II)}}{>C^{+}{-}O^{-}}$$

Electron-donating substituents stabilize canonical form II and cause more single-bond character; electronegative substituents favor canonical form I and double-bond character. Since the carbonyl stretching frequency will increase with increasing double-bond character, electronegative substituents will increase the stretching frequency and electropositive substituents will lower it (Table 3.16). This characteristic has been used to estimate group electronegativities,[110] and it also provides support for variable electronegativity as a function of hybridization. As with biphenylene, smaller rings cause more *p* character to be diverted into intracyclic orbitals, resulting in more electronegative *s* character in the

[107] For further discussion, see A. Carrington and A. D. McLachlan, "Introduction to Magnetic Resonance," Harper & Row, New York, **1967**, Chapter 5.

[108] See pp. 438–439 for further discussions and references to this phenomenon.

[109] For discussion of this subject, see M. Karplus and D. M. Grant, *Proc. Natl. Acad. Sci.* (*U.S.A.*), **1959**, *45*, 1269; N. Muller and D. E. Pritchard, *J. Chem. Phys.*, **1959**, *31*, 768, 1471; C. Juan and H. S. Gutowsky, *J. Chem. Phys.*, **1962**, *37*, 2198; D. M. Grant and W. M. Litchman, *J. Am. Chem. Soc.*, **1965**, *87*, 3994; J. E. Huheey, *J. Chem. Phys.*, **1966**, *45*, 405.

[110] R. E. Kagarise, *J. Am. Chem. Soc.*, **1955**, *77*, 1377.

Table 3.16 Effect of electronegative substituents on the frequency of the infrared absorption of ketones

Acetone, $CH_3C(O)CH_3$	1738 cm^{-1}
1,1,1-trifluoroacetone, $CF_3C(O)CH_3$	1780 cm^{-1}
Acetyl chloride, $CH_3C(O)Cl$	1822 cm^{-1}
Carbonyl chloride, ClC(O)Cl	1827 cm^{-1}
Acetyl fluoride, $CH_3C(O)F$	1872 cm^{-1}
Trifluoroacetyl fluoride, $CF_3C(O)F$	1901 cm^{-1}
Carbonyl fluoride, FC(O)F	1928 cm^{-1}

Table 3.17 Effect of ring size on the frequency of the infrared absorption of ketones

Cyclobutanone, $n = 4$	1820 cm^{-1}
Cyclopentanone, $n = 5$	1780 cm^{-1}
Cyclohexanone, $n = 6$	1745 cm^{-1}
Acyclic ketone	1737 cm^{-1}

orbitals directed toward the oxygen atom and thus favoring canonical form I. Increasing carbonyl stretching frequencies are thus observed in the smaller ring compounds (Table 3.17).[111] This interpretation is further strengthened by the observation that the basicity of the oxygen atom in the small ring compounds is reduced.[112] We shall see (Chapter 17) that other groups, perhaps of more interest to the average inorganic chemist, behave in a similar manner.

X-ray photoelectron spectroscopy

A physical method of growing importance in determining charges on atoms in molecules is X-ray Photoelectron Spectroscopy (XPS), sometimes called ESCA (Electron Spectroscopy for Chemical Analysis).[113] Although the principles involved in this type of spectroscopy have been known for some time and there was some early experimental work, it has been only in the last few years that the method has been extensively applied. The method involves the ionization of the inner, core electrons from an atom by X-radiation. The energy of the ionizing photons is known from their frequency ($E = h\nu$) and the kinetic energy of the

[111] There are also mechanical effects that operate in small ring compounds to increase the stretching frequency. See, e.g., J. I. Brauman and V. W. Laurie, *Tetrahedron*, **1968**, *24*, 2595.

[112] F. Kaplan, personal communication; C. A. L. Filgueiras and J. E. Huheey, *J. Org. Chem.*, **1976**, *41*, 49.

[113] J. M. Hollander and W. L. Jolly, *Acc. Chem. Res.*, **1970**, *3*, 193; K. Siegbahn et al., "*ESCA*; Atomic, Molecular and Solid State Structure by Means of Electron Spectroscopy," Almqvist & Wiksell, Uppsala, Sweden, **1967**,

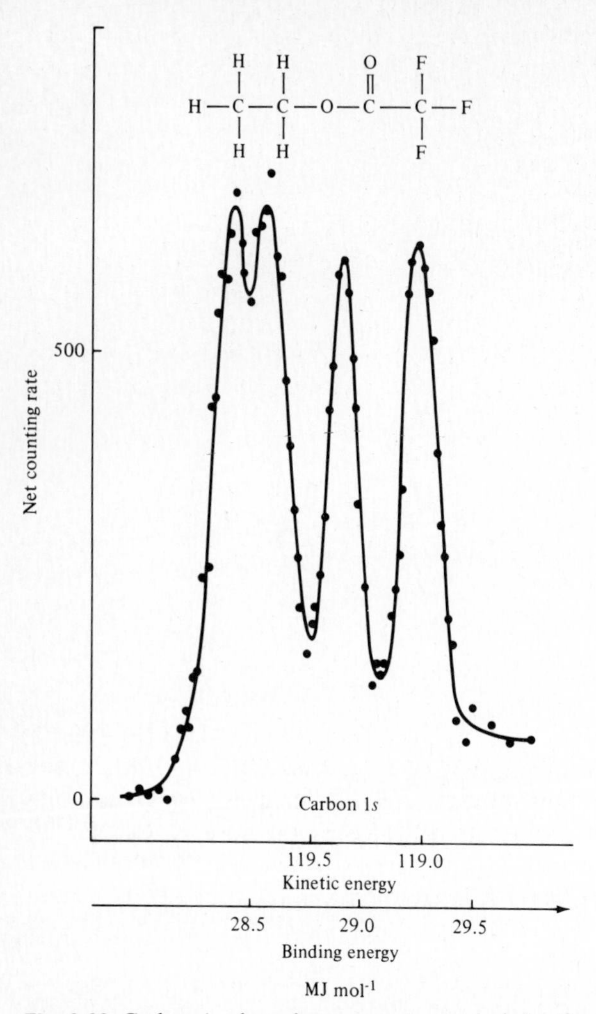

Fig. 3.69 Carbon 1*s* photoelectron spectrum of ethyl trifluoroacetate illustrating relationship between binding energy and the chemical environment of the carbon. [From J. M. Hollander and W. L. Jolly, *Acc. Chem. Res.*, **1970**, *3*, 193. Reproduced with permission.]

photoionized electrons may be measured. The difference between these two quantities[114] is the quantity of energy (the binding energy) that must be provided to overcome the attraction of the nucleus for the electron:

$$E_b = E_{h\nu} - E_k \tag{3.104}$$

It is normally assumed that the core electrons have little or no effect on the bonding properties of an atom and are therefore of no chemical interest. Although they may have no important effect on the bonding, the converse is not true. It appears that the chemical

[114] A small correction must be made for the energy necessary to get the ionized electron out of the solid sample from which it originates. See p. 174.

environment of an atom is reflected rather accurately by the binding energy. Consider, for example, Fig. 3.69, which illustrates the photoelectron spectrum obtained for the carbon 1*s* electrons in ethyl trifluoroacetate. Each of the four carbon atoms is seen as a distinct peak in the spectrum, though, to be sure, the splitting of the methyl and methylene carbon atoms is small. The relative binding energies may be correlated with the environment about each carbon atom. The one in the most electropositive environment (CH_3) has the lowest energy, and that in the most electronegative environment (CF_3) has the highest binding energy. This is to be expected since electronegative substituents should increase the partial positive charge on the carbon atom and thus increase the effective nuclear charge and the ionization energy.

Figure 3.70 illustrates the nitrogen 1*s* photoelectron spectrum of a cobalt(III) complex containing three different kinds of nitrogen. Again, each species has a distinct peak and the

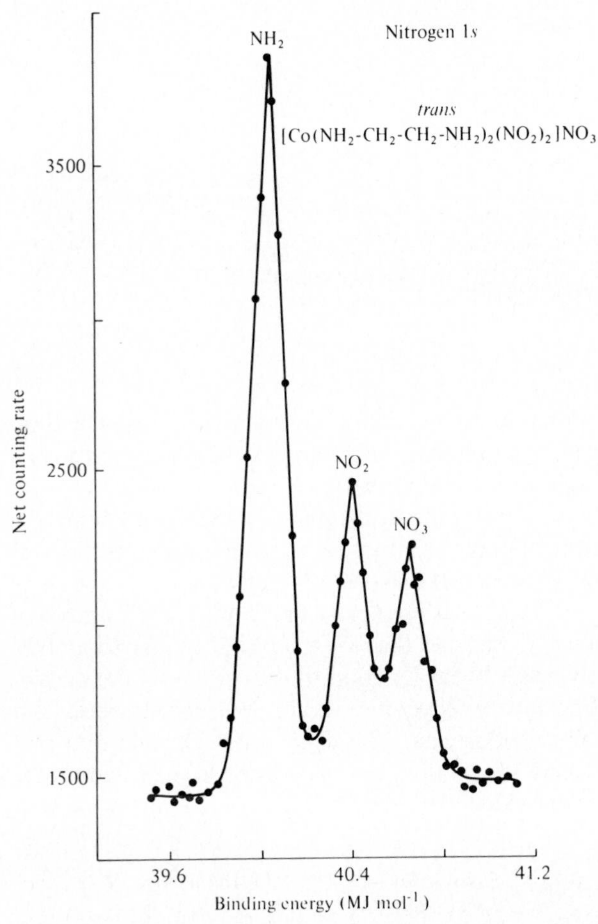

Fig. 3.70 Nitrogen 1*s* photoelectron spectrum of *trans*-dinitrobis(ethylenediamine)cobalt(III)nitrate. Note correlation of the binding energy and the oxidation state of the nitrogen. [From J. M. Hollander and W. L. Jolly, *Acc. Chem. Res.*, **1970**, *3*, 193. Reproduced with permission.]

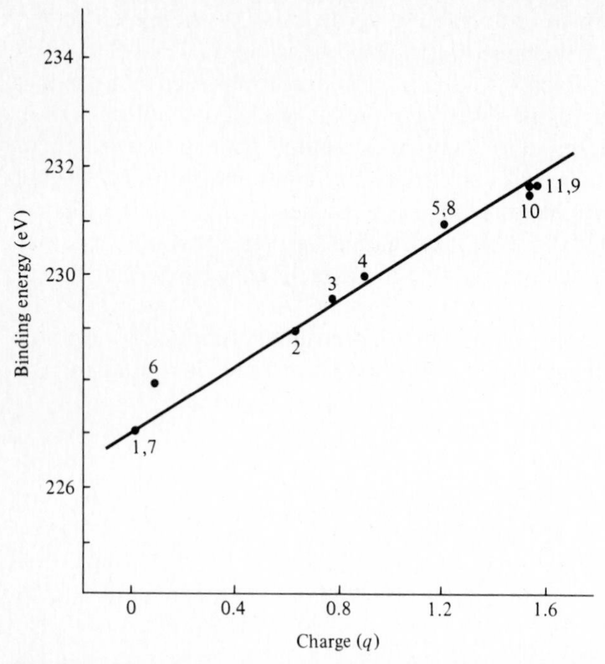

Fig. 3.71 Plot of Mo ($3d_{5/2}$) binding energy (eV) vs. calculated charge. Samples are: (1) Mo; (2) $MoCl_3$; (3) $MoCl_4$; (4) $MoCl_5$; (5) MoO_2; (6) MoS_2; (7) $MoSe_2$; (8) $MoO_2(acac)_2$; (9) MoO_3; (10) $Na_2MoO_4 \cdot 2H_2O$; (11) $(NH_4)_6Mo_7O_{24} \cdot 4H_2O$. [From S. O. Grim and L. J. Matienzo, *Inorg. Chem.*, **1975**, *14*, 1014. Reproduced with permission.]

areas of the peaks are roughly proportional to the number of nitrogen atoms of each kind. Again, this is to be expected since the number of electrons ionized should be roughly proportional to the number of atoms present to be bombarded by X rays.

This technique promises to have wide application in chemistry, but the present discussion will be limited to its use as a measure of partial charges on atoms in molecules. Several studies[115] have shown correlations of this type. In one, the binding energy of carbon 1*s* electrons in the halomethanes correlated nicely with the electronegativity and number of halogen atoms present.[116] Another study[117] showed that for a series of 11 simple molybdenum compounds, the binding energy was a linear function of the charge on the molybdenum atom as calculated from Pauling electronegativities (Fig. 3.71). Furthermore, data can be used to argue beyond simple charge densities. This same study showed that in a series of sustituted molybdenum carbonyls, the binding energies varied as expected on the basis of strong π donation to carbon monoxide ligands (see Chapter 9).

[115] S. Hagström, C. Nordling, and K. Siegbahn, *Z. Physik*, **1964**, *178*, 439; J. M. Hollander et al., *J. Chem. Phys.*, **1968**, *49*, 3315; D. N. Hendrickson et al., *Inorg. Chem.*, **1969**, *8*, 2642; **1970**, *9*, 612.

[116] T. D. Thomas, *J. Am. Chem. Soc.*, **1970**, *92*, 4184.

[117] S. O. Grim and L. J. Matienzo, *Inorg. Chem.*, **1975**, *14*, 1014.

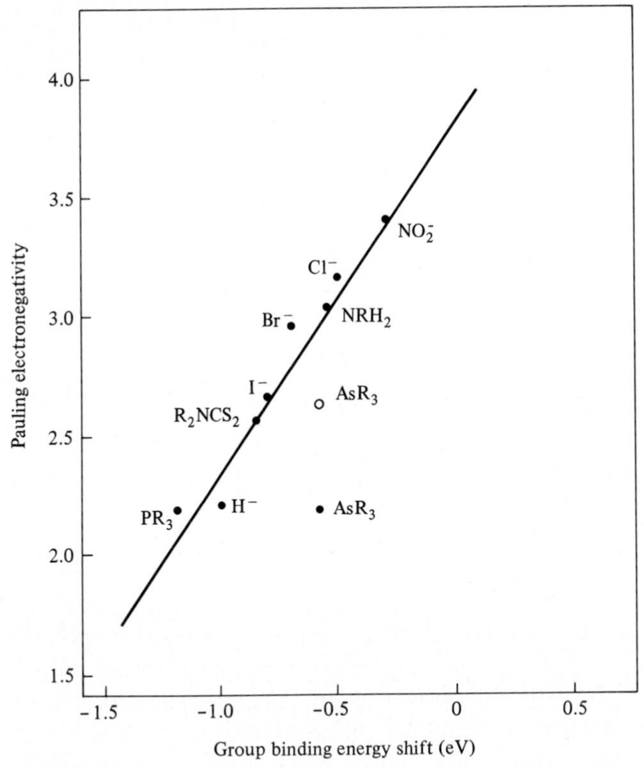

Fig. 3.72 Linear relationship between ligand group shifts (ESCA) and the Pauling electronegativities of the ligands (●). Only AsR_3 deviates significantly from the line. If Sanderson electronegativity value is used for As (○), AsR_3 fits much better. [From R. D. Feltham and P. Brant, *J. Am. Chem. Soc.*, **1982**, *104*, 641. Reproduced with permission.]

A similar study of over 100 complexes of thirteen transition metals provided similar evidence on σ and π bonding in these complexes (see Chapter 9), and it also showed the effect of ligand electronegativity on the binding energy of the metal.[118] The shift in binding energy between the "bare" metal atom and the complexed metal atom was found to be a linear function of the electronegativity of the ligand (see Fig. 3.72).

Experimental ESCA values can be used to augment theoretical studies. For example, if one is using a semiempirical approach in calculating atomic charges, the ESCA data can be used to parameterize ("calibrate") the computational method. For example, Jolly and co-workers[119] have used ESCA measurements in conjunction with the electronegativity equalization method (see pp. 158–160) to calculate charges on atoms in a large series of molecules.

[118] R. D. Feltham and P. Brant, *J. Am. Chem. Soc.*, **1982**, *104*, 641.

[119] W. L. Jolly and W. B. Perry, *J. Am. Chem. Soc.*, **1973**, *95*, 5442; *Inorg. Chem.*, **1974**, *13*, 2686.

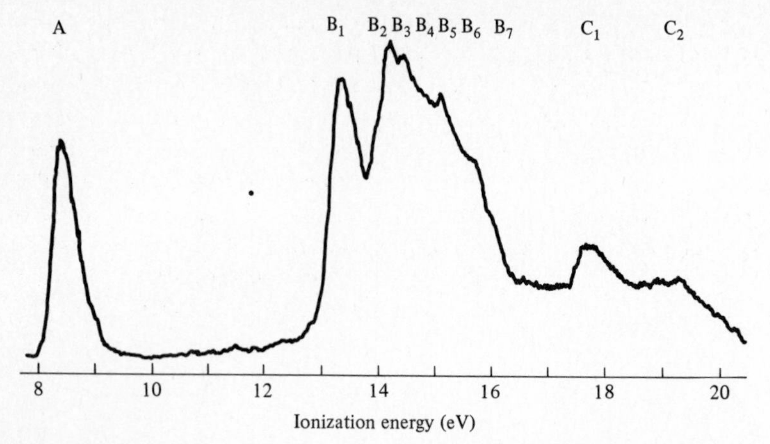

Fig. 3.73 He(I) ultraviolet photoelectron spectrum (UV PES) of $Cr(CO)_6$. Peaks indicate relative energies of molecular orbitals in the complex. [From B. R. Higginson et al., *J. Chem. Soc., Faraday Trans. 2*, **1973**, *69*, 1659.]

Unlike most of the methods discussed previously in this section, X-ray photoelectron spectroscopy appears to relate directly to partial charge without the complication of a second, indeterminable parameter such as hybridization. However, it should be pointed out that this technique is still relatively new and that there is some disagreement concerning its use as a measure of charge.[120] In general, however, most workers accept some form of Eq. 3.105 to describe the binding energy in a charged atom compared to the neutral one:

$$\Delta E = kq_i + \sum_{i \neq j} q_j(e^2/r_{ij}) + l \tag{3.105}$$

where q_i is the charge on the atom in question, the second term is the summation of all Madelung interactions with surrounding charges q_j at distances r_{ij}, and k and l are empirical constants to fit the data.[121] Recent studies indicate that for larger molecules and ions the effect of the Madelung potential may be neglected.[118,122]

It was predicted that X-ray photoelectron spectroscopy would play an increasingly important role in the experimental determination of the distribution of charge density in molecules.[123] It seems now that it is the best single method for this purpose.

If the bombarding photons are of a somewhat longer wavelength and lower energy (vacuum ultraviolet), their energy corresponds to that necessary to ionize the electrons

[120] M. T. Barber and D. T. Clark, *Chem. Commun.*, **1970**, 22, 23, 24; A. Van der Avoird, *Chem. Commun.*, **1970**, 727.

[121] See K. Siegbahn et al., "ESCA Applied to Free Molecules," North-Holland Publishing Co., Amsterdam, **1969**; G. D. Stucky et al., *J. Am. Chem. Soc.*, **1972**, *94*, 8009; T. X. Carroll et al., ibid., **1974**, *96*, 1989; W. L. Jolly and W. B. Perry, *Inorg. Chem.*, **1974**, *13*, 2686, and references therein.

[122] P. Grant and R. D. Feltham, *Inorg. Chem.*, **1980**, *19*, 2673.

[123] J. E. Huheey, "Inorganic Chemistry: Principles of Structure and Reactivity," Harper & Row, New York, **1972**.

in the valence shell rather than the lower-lying core electrons. This type of photoelectron spectroscopy is thus useful in determining the energy of the molecular orbitals involved in bonding. An example of the experimental data is the He(I) ultraviolet photoelectron spectrum shown in Fig. 3.73. The peaks represent the absorption of photons as electrons are ionized from the various occupied molecular orbitals of the molecule, in this case $Cr(CO)_6$. As in the case of ESCA, the energy necessary to remove the electron is measured and, according to Koopmans' theorem, gives the energy of the molecular orbital from which the electron was ejected. It is unprofitable to go into too much detail prior to discussion of the molecular orbitals of octahedral complexes (see Chapter 9), but we may observe that peak A corresponds to a molecular orbital at a higher level than B, which is, in turn, higher than C.[124]

X-ray diffraction

The use of X-ray diffraction is normally a structure-determining method and as such will be discussed in Chapter 5. However, the diffraction of the X rays is caused by the electrons in the sample and therefore very careful measurements can provide information on electron density distribution in the compound under study. However, since most of the electrons in atoms are core electrons rather than the valence electrons, they dominate the scattering and make analysis of valence electron density difficult.

We have repeatedly seen methods that promise to give information on electron density but have other variables involved such that simple interpretation of electron density is impossible. Careful X-ray studies can compensate for their difficulty by yielding unambiguous electron-density data; it was this method that provided the data in Table 3.4. Accurate X-ray diffraction studies should make it possible to obtain much information on electron densities, but unfortunately not much work of this type has yet been done. However, from the results obtained thus far, Coppens[125] has made the following generalizations:

1. Net atomic charges in molecular crystals are fractions of electron units in a variety of compounds, such as the mineral kernite, a nickel complex, and the zwitterion glycylglycine.
2. Oxygen atoms are negatively charged by 0.2–0.5 electron, and carbon atoms are positively charged in carboxyl and carbonyl groups, but they are negative in an aliphatic environment. Boron atoms are positive in kernite (+0.4) and borazine (+0.8).
3. Atomic charges correlate with atomic electronegativity rather than with the number of bonds—there are no significant differences between trigonal and tetrahedral boron atoms or between bridging and hydroxylic oxygen atoms in kernite.

In addition, it has been possible to use X-ray-determined charge densities and net atomic charges to calculate molecular dipole moments that are in good agreement with

[124] A. H. Cowley, *Progr. Inorg. Chem.*, **1979**, *26*, 45.

[125] P. Coppens, in "Electronic Structure of Polymers and Molecular Crystals," J-M. André and J. Ladik, eds., Plenum, New York and London, **1974**, p. 249.

both theoretical and experimental values.[126] In addition, careful measurement of electron densities in transition metal complexes confirms predictions of bonding theory in coordination chemistry (see Chapter 9).

CONCLUSION

Ionic crystals may be viewed quite simply in terms of an electrostatic model of lattices of hard-sphere ions of opposing charges. Although conceptually simple, this model is not completely adequate, and we have seen that modifications must be made in it. First, the bonding is not completely ionic with compounds ranging from the alkali halides, for which complete ionicity is a very good approximation, to compounds for which the assumption of the presence of ions is rather poor. Secondly, the assumption of a perfect, infinite mathematical lattice with no defects is an oversimplification. As with all models, the use of the ionic model does not necessarily imply that it is "true"—merely that it is convenient and useful—and, if proper caution is taken and adjustments are made, it proves to be a fruitful approach.

Covalent bonding is considerably more complicated than ionic bonding. Although the latter can be treated to a first approximation by classical electrostatics, treatment of the covalent bond requires a quantum-mechanical approach at the outset. Because of the impossibility of describing molecules with an exact Schrödinger equation, chemists must resort to approximations. These approximations range from very good in the case of simple, diatomic molecules such as H_2, to only gross approximations in the case of complex molecules. The two main approaches to approximate solutions, the valence bond and the molecular orbital method, in many ways complement each other rather than compete. The student should be conversant with both methods and be ready to apply whichever one appears to be most profitable in a particular case.

The problem of polarity of heteronuclear bonds is one for which there are no easy answers. Many approaches have been proposed because of the importance of treating molecular polarity. None is completely successful, and here too it is often necessary to apply approximations depending upon the situation.

Interest in probing the secrets of the covalent bond has led to the development of many techniques for measuring molecular properties. None is perfect; each has its advantages and disadvantages. Often two or more methods complement each other. It should be stressed that each of the techniques discussed here is a powerful tool for the study of the properties of molecules other than ionicity and charge distribution. Each may be used to distinguish atoms in one chemical environment from those in another, and this and related information is often of more direct usefulness than the limited information on charge distribution they provide. Examples of such uses are discussed in Chapter 5 and elsewhere in this book.[127]

---

[126] P. Coppens et al., *Acta Crystallogr., Sect. A*, **1979**, *35*, 63.

[127] The interested reader should also refer to R. S. Drago, "Physical Methods in Chemistry," Saunders, Philadelphia, **1977**.

PROBLEMS

3.1. Both CsCl and CaF_2 exhibit a coordination number of 8 for the cations. What is the structural relationship between these two lattices?

3.2. The contents of the unit cell of any compound must contain an integral number of formula units. (Why?) Note that unit cell boundaries "slice" atoms into fragments: An atom on a face will be split in *half* between *two* cells; one on an *edge* will be split into *quarters* among *four* cells, etc. Identify the number of Na^+ and Cl^- ions in the unit cell of sodium chloride illustrated in Fig. 3.1a and state how many formula units of NaCl the unit cell contains. Give a complete analysis.

3.3. The measured density of sodium chloride is 2.167 g cm^{-3}. From your answer to Problem 3.2 and your knowledge of the relationships among density, volume, Avogadro's number, and formula weight, calculate the volume of the unit cell and thence the length of the edge of the cell. Calculate the length $r_+ + r_-$. Check your answer, $r_+ + r_-$, against values from Table 3.4.

3.4. Show your understanding of the Born-Haber cycle by calculating the heat of formation of potassium fluoride analogous to the one in the text for sodium chloride.

3.5. Using any necessary data from appropriate sources (all present in this text), predict the enthalpy of formation of KCl by means of a Born-Haber cycle. You can check your lattice energy against Table 3.3.

3.6. Using any necessary data from appropriate sources (all present in this text), predict the enthalpy of formation of CaS by means of a Born-Haber cycle.

3.7. Show your understanding of the meaning of the Madelung constant by calculating A for the isolated $F^-Be^{+2}F^-$ fragment considered as a purely ionic species.

3.8. The ionic bond is often described as "the metal wants to lose an electron and the nonmetal wants to accept an electron, so the two react with each other." Criticize this statement quantitatively using appropriate thermodynamic quantities.

3.9. Why is the thermit reaction:

$$2Al + M_2O_3 = 2M + Al_2O_3 \qquad (M = Fe, Cr, etc.) \qquad \textbf{(3.106)}$$

so violently exothermic? (The ingredients start at room temperature and the desired product, iron, etc., is *molten*, up to 2700 °C, at the end of the reaction.)

3.10. Berkelium is currently available in microgram quantities—sufficient to determine structural parameters but not enough for thermochemical measurements.

a. Using the tabulated ionic radii and using the radius ratio rule, estimate the lattice energy of berkelium dioxide, BkO_2.

b. Assume that the radius ratio rule is violated (it is!). How much difference does this make in your answer?

3.11. The crystal structure of LaF_3 is different from those discussed. Assume it is unknown. Using the equation of Kapustinskii, estimate the lattice energy.

3.12. All the alkali halides crystallize in the sodium chloride structure with the exception of cesium chloride, cesium bromide, and cesium iodide which have the CsCl structure.

Using the tabulated values of ionic radii, calculate the radius ratio for each alkali halide. Which compounds violate the radius ratio rules?

3.13. To ionize Mg to Mg^{+2} costs *three* times as much energy as to form Mg^{+}. The formation of O^{-2} is *endothermic* rather than exothermic as for O^{-}. Nevertheless, magnesium oxide is always formulated as $Mg^{+2}O^{-2}$ rather than as $Mg^{+}O^{-}$.

a. What theoretical reason can be given for the $Mg^{+2}O^{-2}$ formulation?

b. What simple experiment could be performed to prove that magnesium oxide was not $Mg^{+}O^{-}$?

3.14. Some experimental values of the Born exponent are: LiF, 5.9; LiCl, 8.0; LiBr, 8.7; NaCl, 9.1; NaBr, 9.5. What is the percent error incurred in the calculation of lattice energies by Eq. 3.13 when Pauling's generalization (He = 5, Ne = 7, etc.) is used instead of the experimental value of n?

3.15. Using Fig. 3.6, generate the first five terms of the series for the Madelung constant for NaCl. How close is the summation of these terms to the limiting value given in Table 3.1?

3.16. The enthalpy of formation of sodium fluoride is -571 kJ mol^{-1}. Estimate the electron affinity of fluorine. Compare your value with those given in Table 2.5.

3.17. Footnote b to Table 7.4 states "Calculated from bond energies of HX and the electron affinities of X." The enthalpies in question are those of reactions of the type:

$$X^- + H^+ \longrightarrow HX$$

Perform these calculations and compare your answers with those given in Table 7.4.

3.18. All known copper(I) halides crystallize in a zinc blende structure. Copper(II) fluoride crystallizes in a distorted rutile structure (for the purposes of this problem assume there is no distortion). Calculate the enthalpies of formation of CuF and CuF_2. Discuss. (All of the necessary data should be readily available either in this text or in the reference book cited, but if you have difficulty finding a quantity, see how much of an argument you can make without it.)

3.19. Thallium has two stable oxidation states, +1 and +3. Use the Kapustinskii equation to predict the lattice energies of TlF and TlF_3. Predict the enthalpies of formation of these compounds. Discuss.

3.20. Plot the radii of Ln^{+3} ions from Table 3.4. Discuss.

3.21. All of the alkaline earth oxides, MO, except one crystallize in the rock salt (NaCl) structure. What is the exception and what is the likely structure for it? [A. F. Wells, "Structural Inorganic Chemistry," 4th ed., Oxford University Press, Oxford, **1975**, p. 445.]

3.22. Calculate the enthalpy of the reaction $CuI_2 \rightarrow CuI + \frac{1}{2}I_2$. Carefully list any assumptions.

3.23. It is not difficult to show mathematically that with the hard sphere model, anion–anion contact occurs at $r^+/r^- = 0.414$ for C.N. = 6. Yet Wells ("Structural Inorganic Chemistry," 4th ed., Oxford University Press, Oxford, **1975**, p. 263) states that even with the hard sphere model, we should not expect the change to take place until $r_+/r_- \simeq 0.35$. Rationalize this apparent contradiction. [Hint: Cf. Fig. 3.18.]

3.24. There exists the possibility that a certain circularity may develop in the radius ratio arguments on p. 86. By assuming a coordination number of 6 were the calculations biased? Discuss.

3.25. Draw Lewis structures for CS_2, PF_3, SnH_4, and $HONH_2$.

3.26. Draw Lewis structures for H_2CO_3, HNO_3, NO, $Be(CH_3)_2$.

3.27. Draw Lewis structures for BF_3, SF_6, XeF_2, PF_5, IF_7.

3.28. Show that there is no mismatch of the sign of the wave function in the π system of $(PNCl_2)_4$ in contrast to $(PNCl_2)_3$.

3.29. Write the MO electron configuration for the NO^- ion.
a. What is the bond order?
b. Will the bond length be shorter or longer than in NO?
c. How many unpaired electrons will be present?
d. Will the unpaired electrons be concentrated more on the N or the O? Explain.

3.30. Consider the hypothetical dioxygenyl superoxide, $O_2^+O_2^-$, discussed in this chapter. If this compound did exist, what would be the electronic structures of the ions? Discuss bond orders, bond lengths, and unpaired electrons.

3.31. The resonance of BF_3 (p. 124) is still a matter of some dispute because one chemist will point to the double bond in structure II (favorably); another will point to F^+ (unfavorably). Suggest a molecule for which charges completely rule out resonance.

3.32. Write resonance structures, including formal charges, for O_3, SO_3, NO_2.

3.33. The assumption was made that the carbon–carbon σ bond in $CH_2{=}CH_2$ is the same as that in CH_3CH_3. In reality, it is probably somewhat stronger. Discuss.

3.34. The NNO molecule was discussed on p. 125. Consider the isomeric NON molecule. Would you expect it to be more stable or less stable than NNO? Why? Why does CO_2 have the OCO arrangement rather than COO?

3.35. The cyanate ion, OCN^-, forms a stable series of salts, but many fulminates, CNO^-, are explosive (L. *fulmino*, to flash). Explain. [For a lead, see p. 124; for a slightly different approach, see L. Pauling, *J. Chem. Educ.*, **1975**, *52*, 577.]

3.36. Which of the following will exhibit the greater polarizing power?
a. K^+ or Ag^+
b. K^+ or Li^+
c. Li^+ or Be^{+2}
d. Cu^{+2} or Ca^{+2}
e. Ti^{+2} or Ti^{+4}

3.37. As one progresses across a transition series (e.g., Sc to Zn) the polarizing power of M^{+2} ions increases perceptibly. In contrast, in the lanthanides, the change in polarizing power of M^{+3} changes much more slowly. Suggest two reasons for this difference.

3.38. Calculate the electronegativity of hydrogen from the ionization potential and the electron affinity.

3.39. In later chapters you will find examples of the stabilization of covalent bonds through ionic resonance energy. For now, show its importance by predicting whether the molecules NX_3(X = hydrogen or halogen) are stable, i.e., whether the reaction

$$N_2 + 3X_2 \longrightarrow 2NX_3$$

is exothermic. Assume that neither ammonia nor any of the nitrogen halides has yet been synthesized, so you are permitted to look up bond energies for N≡N, N—N, and X—X (Appendix E), but you must predict the bond energy of N—X.

3.40. Which do you expect to be more acidic:

$CH_3-P(=O)(CH_3)-OH$ or (cyclobutyl)$-P(=O)-OH$?

Explain. [A. G. Cook and G. W. Mason, *J. Org. Chem.*, **1972**, *37*, 3342.]

3.41. This dipole moment of methysilane, CH_3SiH_3, is 2.4×10^{-30} C m. Since carbon is more electronegative than silicon, one might suppose that the negative end of the molecular dipole is the methyl group. In fact, the dipole moment lies the other way, CH_3SiH_3 ($+\!\!\longrightarrow$). Consult the table of bond moments (Table 3.14) and rationalize these results. [J. M. Bellama, R. S. Evans, and J. E. Huheey, *J. Am. Chem. Soc.*, **1973**, *95*, 7242.]

3.42. Predict the dipole moments of the following molecules:

H—F H—C≡N Cl—C≡N PH_3 ($\angle$H—P—H $\simeq 90°$)

3.43. In Table 3.12 the electronegativities of the noble gases are, as a group, the highest known, being higher even than those of the halogens. Yet we all know that the noble gases do *not* accept electrons from elements of low electronegativity:

$$Na + A \longrightarrow Na^+A^-$$

Discuss the meaning of the electronegativities of the noble gases.

3.44. In discussing ionic resonance, $AB \leftrightarrow A^+ + B^-$, where:

$$\Psi = a\Psi_{cov} + b\Psi_{ionic}$$

Pauling assumed that Ψ_{ionic} made a negligible contribution if $\chi_a = \chi_b$. The bond energy of Cl_2 is 240 kJ mol^{-1} and the bond length is 199 pm. In the Cl_2 molecule, $\chi_a \equiv \chi_b$. Show by means of a Born-Haber–type calculation that the canonical structure, Cl^+Cl^-, cannot contribute appreciably to the stability of the molecule. [You may check your answer with L. Pauling, "The Nature of the Chemical Bond," 3rd ed., Cornell University Press, Ithaca, N.Y., **1960**, p. 73.]

3.45. The energy necessary to break a bond is not always constant from molecule to molecule. For example:

$$NCl_3 \longrightarrow NCl_2 + Cl \qquad \Delta H = \sim 375 \text{ kJ mol}^{-1}$$

$$ONCl \longrightarrow NO + Cl \qquad \Delta H = 158 \text{ kJ mol}^{-1}$$

Suggest a reason for the difference of ~200 kJ mol^{-1} between these two enthalpies.

3.46. From what you know of the relationship between ionization energies, electron affinities, and electronegativities, would you expect the addition of some *d* character to a hybrid to raise or lower the electronegativity, e.g., will sulfur be more electronegative when hybridized sp^3 or sp^3d^2?

3.47. Predict the dipole moments of NF_3, F_2NNF_2, *cis*-FN=NF, *trans*-FN=NF, and CFClBrH. [For methods of treating molecules such as CFClBrH, see *J. Chem. Educ.*, **1976**, *53*, 23.]

3.48. On p. 162 in the calculation of the predicted moment of CH_3F, the implicit assumption was made that the vector sum of three H—C bonds at angles of 109.5° could be set equal to the bond moment of *one* H—C bond lying on the 3-fold axis of the methyl group. Show that this is rigorously true. [*Hint:* Does methane have a dipole moment?]

3.49. Using your results from Problem 3.49, show that methyl fluoride and trifluoromethane should have identical dipole moments if bond moments were exactly additive. In fact, their moments are 5.97×10^{-30} C m and 5.47×10^{-30} C m, respectively. Discuss.

3.50. The dipole moment of H—C≡C—Cl is in the direction ⟵+. Explain, *carefully*.

3.51. Prussian blue is formed by the addition of ferric salts to ferrocyanides:

$$K^+ + Fe^{+3} + [Fe(CN)_6]^{-4} \longrightarrow [KFe(CN)_6Fe]_x$$

Turnbull's blue is formed by addition of ferrous salts to ferricyanides:

$$K^+ + Fe^{+2} + [Fe(CN)_6]^{-3} \longrightarrow [KFe(CN)_6Fe]_x$$

The ^{57}Fe Mössbauer spectrum of Prussian blue shows but a single absorption, as does the spectrum of Turnbull's blue. Furthermore, the two spectra are identical. Interpret. [See pp. 521–523.]

3.52. Simple metal carbonyls such as $Ni(CO)_4$ differ from organic carbonyls in that they contain triple bonds: M—C≡O. Pure carbon monoxide absorbs in the infrared at 2143 cm^{-1}. Dicobalt octacarbonyl, $Co_2(CO)_8$, shows two absorption bands in the infrared, at ~ 2000 cm^{-1} and 1800 cm^{-1}. Suggest an interpretation. [See p. 592.]

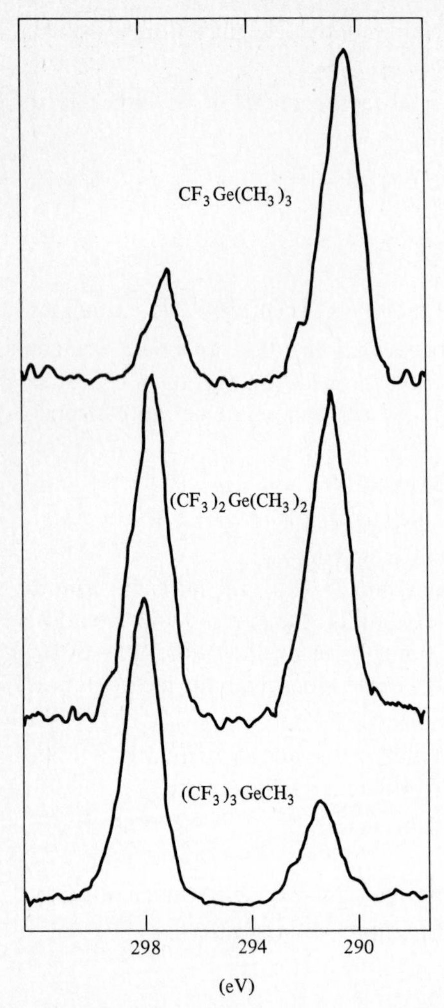

Fig. 3.74 ESCA carbon 1*s* spectra of trimethyltrifluoromethylgermane, dimethyl*bis*(trifluoromethyl)-germane, and methyl*tris*(trifluoromethyl)germane. [From J. E. Drake et al., *Inorg. Chem.*, **1982**, *21*, 1784. Reproduced with permission.]

3.53. The ESCA carbon 1*s* spectra of mixed methyltrifluoromethyl germanes are shown in Fig. 3.74. From the binding energies and other features, say as much as you can about these spectra and the corresponding molecules.

3.54. In Eq. 3.13, the lower the value of the Born exponent, n, the less is the loss of energy due to repulsion, $\frac{1}{n}$: Yet "soft" ions (low n) can come closer owing to less stringent repulsions and can thus progress further down the negative energy curve of Fig. 3.5 before being brought to a halt—thereby *increasing* the lattice energy. Resolve this apparent paradox.

3.55. In the legend to Fig. 3.54 it says "the lowest and the fifth molecular orbitals are essentially nonbonding. . . ." Describe these nonbonding orbitals more explicitly, perhaps in VB terms.

3.56. Look at Figs. 3.33 and 3.34 carefully. Identify:

a. the nodal planes responsible for the symmetry of the MOs, i.e., sigma, pi, etc.

b. the nodal planes responsible for bonding versus nonbonding orbitals

c. any changes in electron density that you can ascribe to bonding versus antibonding situations

4

The solid state

In the previous chapter we have seen how simple bonding models—the electrostatic one for ionic compounds, various theories of covalent bonding, partial ionic and covalent character, etc.—can be applied to the chemical and physical properties of compounds of interest to the inorganic chemist. Of course, there are other important factors such as dipole moments and van der Waals forces that influence these properties, and we shall encounter them later. In this chapter we shall examine examples of the solids held together by ionic or covalent bonds or mixtures of the two. Crystals held together by predominantly ionic forces (e.g., magnesium oxide, which has the NaCl structure; Fig. 3.1) and those held together by purely covalent forces (e.g., diamond; Fig. 4.1) are surprisingly similar in their physical properties. Both types of crystals are mechanically strong and hard, are insulators, and have very high melting points (e.g., MgO = 2852 °C; diamond = 3550 °C). Neither type is soluble in most solvents. The conspicuous difference between the two types of crystals is that there are a few solvents of high permittivity (water is the most notable,

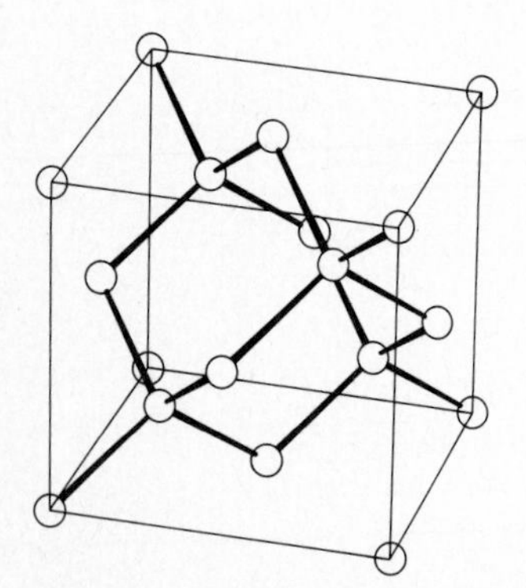

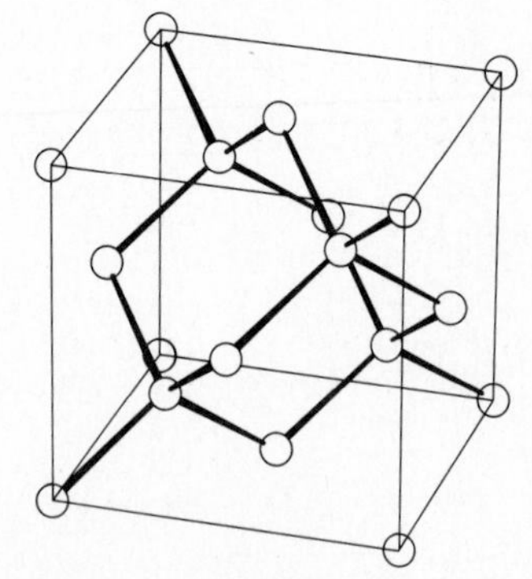

Fig. 4.1 Unit cell of the structure of diamond (carbon). Note the tetrahedral (sp^3) configuration about each atom. Cf. to Fig. 3.2b. [From M. F. C. Ladd, "Structure and Bonding in Inorganic Chemistry," Ellis Horwood, Chicester, **1979**. Reproduced with permission.]

but see Chapter 8) that will dissolve *some* ionic compounds (see p. 280). The second difference is that these solutions as well as the molten ionic compound conduct electricity, but that is not a property of the *solid* itself.

PREDICTION OF THE STRUCTURES OF COMPLEX IONIC COMPOUNDS

Given the difficulties and exceptions that we have seen with the radius ratio rule, we might despair that any predictive power was available to the inorganic chemist studying crystal structures. This is not the case. The chief difficulty is that the radius ratio rule is based purely on *geometric* considerations, not chemical ones. If we include *chemical* factors, such as partial covalency, our predictive power is considerably enhanced. There are several approaches to this problem, but only two methods will be mentioned here. The simplest is the purely empirical approach: One takes a list of compounds of known structures and the radii of the ions present in them. The radii of two of the ions present in a given compound are plotted against each other (this is graphically related to the arithmetic radius ratio approach). It is often found that compounds with similar structures are grouped together. In Fig. 4.2, for example, compounds of the type A_2BO_4 (where A is a metal and B is a higher-valent metal or nonmetal) are plotted as a function of the radii of A and B. To a first approximation, we can consider the oxide ions to form a closest-packed array and the sizes of A and B will determine how they can fit in (sometimes grossly distorting the closest-packed structure). For example, many minerals of interest to inorganic chemists (see Chapter 9), called *spinels* after the parent compound, $MgAl_2O_4$, cluster

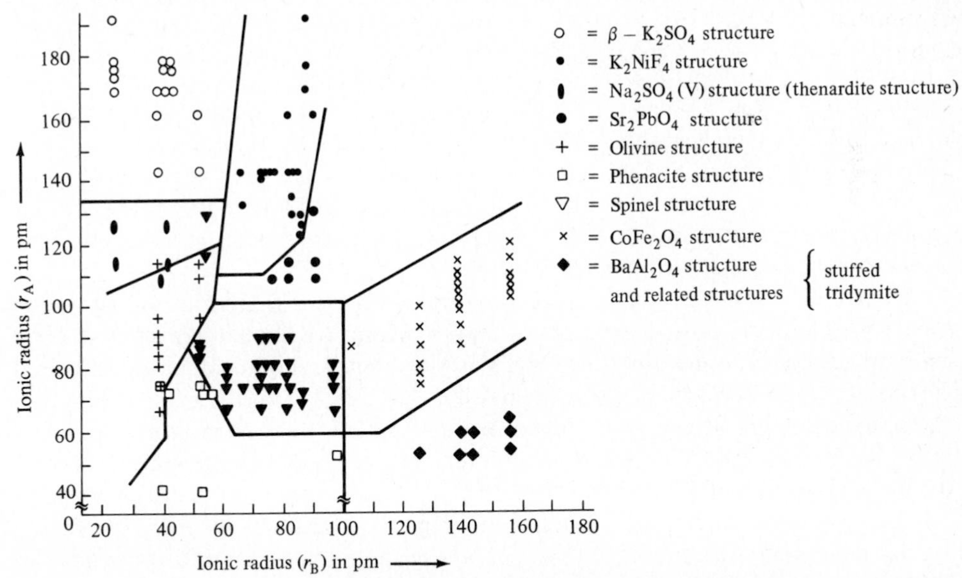

Fig. 4.2 Structure field map for A_2BO_4 compounds as a function of cation size. Note that only the more common structures are plotted. Each point on this plot represents at least one compound having the indicated structure and size of cations $A(r_A)$ and $B(r_B)$. [From O. Muller and R. Roy, "The Major Ternary Structural Families," Springer-Verlag, New York, **1974**. Reproduced with permission.]

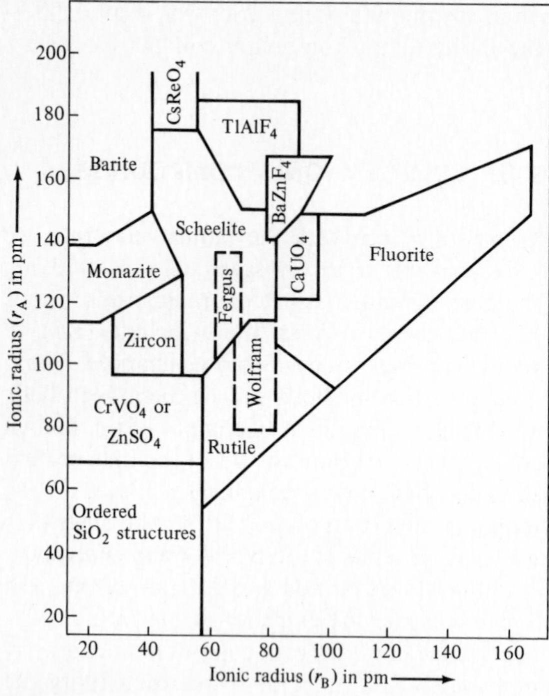

Fig. 4.3 Composite structure field map for ABX_4 structures. X = F or O. [From O. Muller and R. Roy, "The Major Ternary Structural Families," Springer-Verlag, New York, **1974**. Reproduced with permission.]

around the point $r_A = r_B = 60$ pm. If the univalent cation (A) has a radius of about 80 pm while the other (B) is smaller, about 40 pm, the structure is most likely to be an olivine. A graph such as the one in Fig. 4.2 is called a *structure field map* and is remarkably accurate. That exceptions *do* occur, usually on the borders, should not be surprising. Serious errors are relatively rare; however, three spinels—sodium molybdate, sodium tungstate, and silver molybdate—fall well outside their field.[1]

Once we have established the fields shown in Fig. 4.2, we can use it as follows: If we discover a new mineral with $r_A = 90$ pm and $r_B = 30$ pm, we should expect it to have the same structure as the mineral olivine, $(Mg, Fe)_2SiO_4$, but we should not be too surprised if it turned out to be isomorphous with thenardite, Na_2SO_4.

A second structure field map is shown in Fig. 4.3. This is a much more ambitious and generalized undertaking since the oxidation states for species A range from I to IV and for B from II to VI, with X = O or F. By combining such a large number of compounds and somewhat oversimplifying the resultant diagram, we lose some accuracy in predictability, but we gain in the knowledge that such a large and diverse set of compounds can be understood in terms of such simple parameters as relative sizes.

[1] O. Muller and R. Roy, "The Major Ternary Structural Families," Springer-Verlag, New York, **1974**, pp. 76–78.

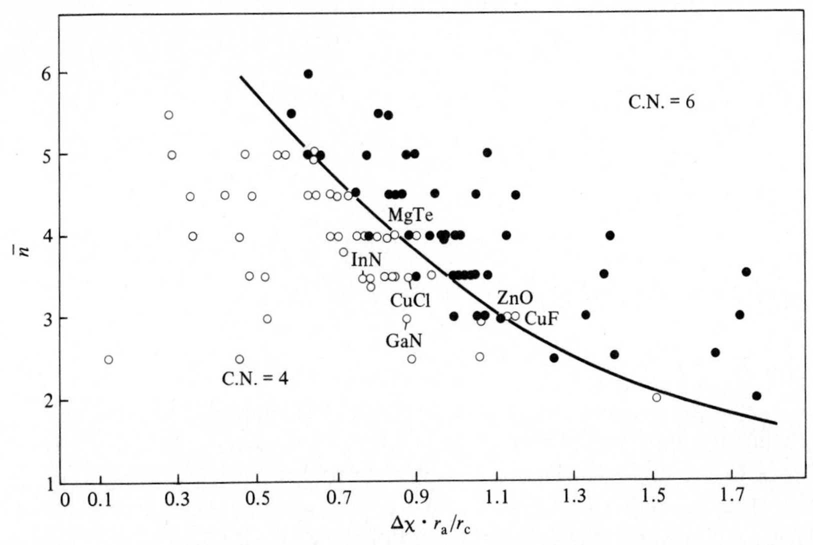

Fig. 4.4 Choice of coordination number of compounds A^nB^{n-8} as a result of the interaction of relative sizes of ions (r_+/r_-), differences in electronegativity ($\Delta\chi$), and principal quantum number, n. [From "Treatise on Solid State Chemistry," Vol. I, N. B. Hannay, ed., Plenum, New York. Copyright © 1973, Bell Telephone Laboratories, Inc. Reproduced with permission.]

If we wish to improve the predictability of our methods, as well as to increase our understanding of the forces at work, we can formulate various semiempirical models in which we take other factors into account. For example, we have previously seen that covalent character might be expected to cause a switch from coordination number 6 to coordination number 4. Therefore, we might attempt to bring in covalent corrections to improve a mechanical or radius ratio approach. Two factors increase covalency: (1) small differences in electronegativity; (2) other things being equal, atoms of the lighter elements form stronger covalent bonds than larger atoms (see p. 319). To incorporate these two variables, Pearson[2] has plotted the principal quantum number, n, (a rough indicator of size) versus a function of electronegativity difference ($\Delta\chi$) and radius ratio, r_+/r_-, and has shown that compounds with coordination number 4 segregate quite well from those with coordination number 6 (Fig. 4.4). Increased precision, however, has been purchased at the expense of simplicity. Other workers have developed the ideas presented here to increased levels of understanding.[3]

IMPERFECTIONS IN CRYSTALS

To this point the discussion of crystals has implicitly assumed that the crystals were perfect. Obviously, a perfect crystal will maximize the cation–anion interactions and minimize the cation–cation and anion–anion repulsions, and this is the source of the very

[2] W. B. Pearson, *J. Phys. Chem. Sol.*, **1962**, *23*, 103.

[3] For a very recent discussion of the prediction of the structures of solids, starting with material presented here and going far beyond, see J. K. Burdett, *Adv. Chem. Phys.*, **1982**, *49*, 47.

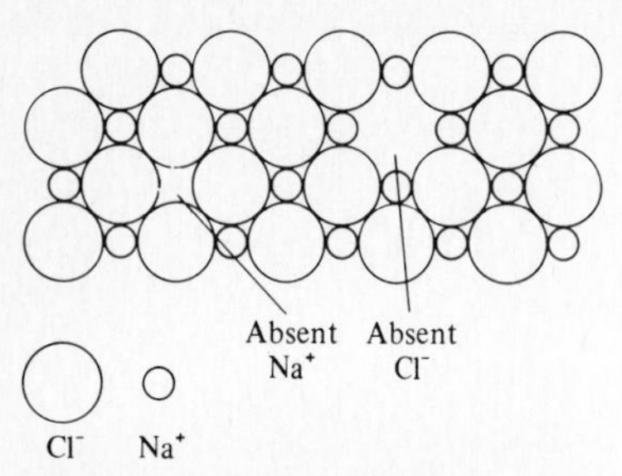

Fig. 4.5 Two Schottky defects balancing each other for no net charge.

strong driving force that causes gaseous sodium chloride, for example, to condense to the solid phase. In undergoing this condensation, however, it suffers a loss of entropy from the random gas to the highly ordered solid. This enthalpy–entropy antagonism is largely resolved in favor of the enthalpy because of the tremendous crystal energies involved, but the entropy factor will always result in equilibrium defects at all temperatures above zero kelvins.

The simplest type of defect is called the *Schottky* or *Schottky-Wagner* defect. Consider a lattice of neutral metal atoms (to be discussed on p. 195) with a single atom removed from the lattice. This vacancy is a Schottky defect. In an ionic crystal, electrical neutrality requires that the missing charge be balanced in some way. The simplest way is for the missing cation, for example, to be balanced by another Schottky defect, a missing anion, elsewhere (Fig. 4.5).

Alternatively, the missing ion can be balanced by the presence of an impurity ion of higher charge. For example, if a crystal of silver chloride is "doped" with a small amount of cadmium chloride, the Cd^{++} ion fits easily into the silver chloride lattice (cf. ionic radii,

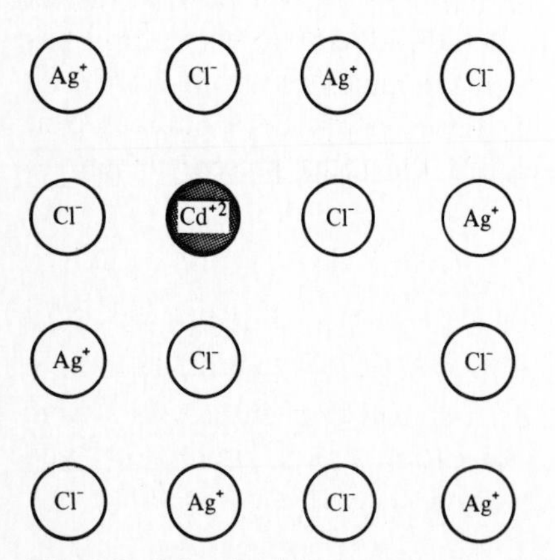

Fig. 4.6 Schottky defect (cation vacancy) induced and balanced by presence of higher valence cation. No net charge. [From N. B. Hannay, "Solid-state Chemistry," Prentice-Hall, Englewood Cliffs, N.J., **1967**. Reproduced with permission.]

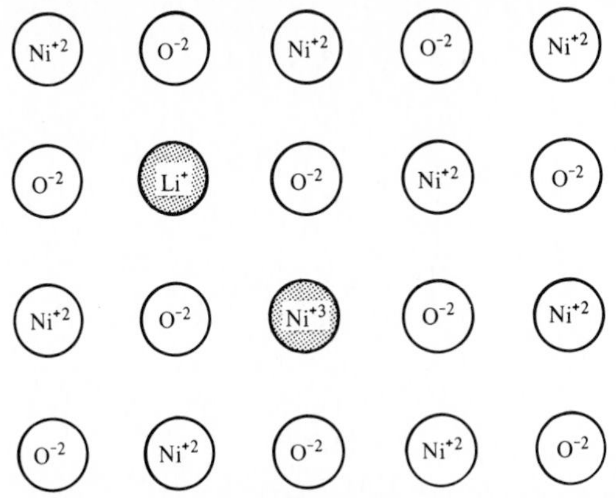

Fig. 4.7 Controlled valency ($Ni^{+2} \rightarrow Ni^{+3}$) by addition of Li^+ ions to NiO. [From N. B. Hannay, "Solid-state Chemistry," Prentice-Hall, Englewood Cliffs, N.J., **1967**. Reproduced with permission.]

Table 3.4). The dipositive charge necessitates the presence of a vacancy to balance the change in charge (Fig. 4.6). Closely related is the concept of "controlled valency," in which a differently charged, stable cation is introduced into a compound of a transition metal. Since the latter has a variable oxidation state, balance is achieved by gain or loss of electrons by the transition metal. For example, consider Fig. 4.7. Stoichiometric nickel(II) oxide, like aqueous solutions containing the Ni^{+2} ion, is pale green. Doping it with a little Li_2O induces a few Ni^{+2} ions to lose electrons and become Ni^{+3} ions, thus preserving the electrical neutrality of the crystal. The properties of the NiO change drastically: The color changes to gray-black, and the former insulator (to be expected of an ionic crystal, see p. 54) is now a semiconductor.[4]

A rather similar effect can occur with the formation of nonstoichiometric compounds. For example, copper(I) sulfide may not have the exact ratio of 2:1 expected from the formula, Cu_2S. Some of the Cu^+ ions may be absent if they are compensated by an equivalent number of Cu^{+2} ions. Since both Cu^+ and Cu^{+2} ions are stable, it is possible to obtain stochiometries ranging from the "ideal" to $Cu_{1.77}S$.

If the "vacancy" is not a true vacancy but contains a trapped electron at that site, the imperfection is called an *F* center. For example, if a small amount of sodium metal is doped into a sodium chloride crystal, the crystal energy causes the sodium to ionize to Na^+e^- and the electron occupies a site that would otherwise be filled by a chloride ion (Fig. 4.8). The resulting trapped electron can absorb light in the visible region and the compound is colored (*F* = Ger. *Farbe*, color). The material may be considered a nonstoichiometric compound, $Na_{1+\delta}Cl$,[5] or as a dilute solution of "sodium electride."[6]

[4] For a discussion of both of these effects as well as related topics, see N. B. Hannay, "Solid-State Chemistry," Prentice-Hall, Englewood Cliffs, N.J., **1967**, pp. 47–51, and A. F. Wells, "Structural Inorganic Chemistry," 4th ed., Clarendon Press, Oxford, **1975**, p. 445. Semiconductors are discussed on pp. 199–203 of this chapter.

[5] Where δ is small with respect to 1.

[6] See pp. 331–333 for solutions of the "compound" sodium electride in liquid ammonia.

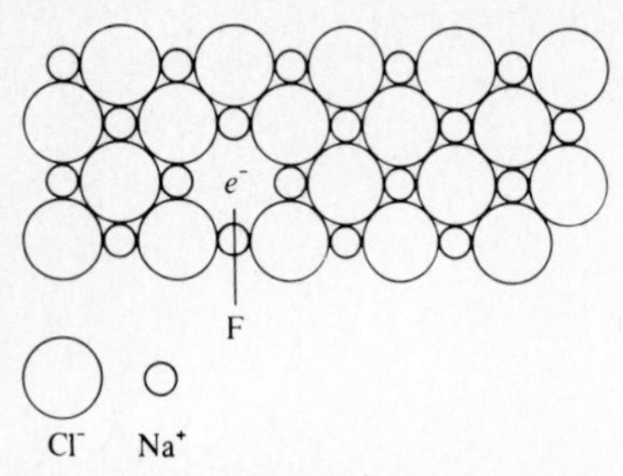

Fig. 4.8 An *F* center: an electron occupying an anionic site.

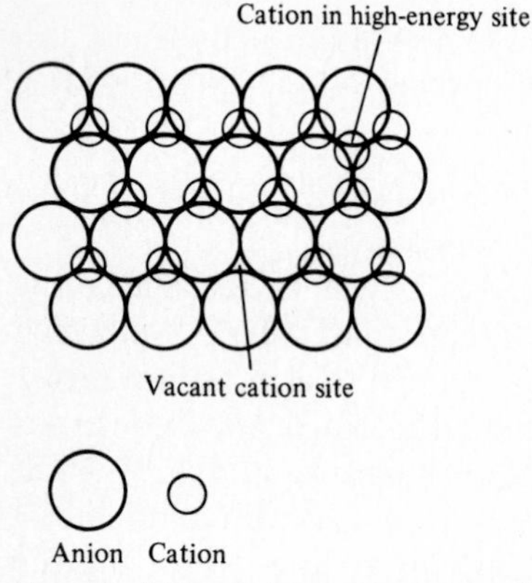

Fig. 4.9 A Frenkel defect: a cation displaced from its "normal" site.

If electrical neutrality is achieved by not completely removing the ion (in a Schottky-type defect) but by simply moving it to a nearby interstitial site, the result is termed a Frenkel defect (Fig. 4.9). The vacancy and corresponding interstitial ion may be caused by a cation or an anion, but since the cation is generally smaller than the anion, it will usually be easier to fit a cation into an interstitial hole other than the one in which it belongs. For the same reason, although it is theoretically possible to have both interstitial cations *and* anions at the same time, at least one will ordinarily be energetically unfavorable because of size.[7]

Finally, in addition to the point defects discussed above there are linear, planar, and volumetric effects.[8] For example, an edge dislocation is shown in Fig. 4.10. It is fairly obvious that such a defect cannot occur in an ionic crystal because ions of the same charges would come in contact. It can, however, occur in metals, and in fact it is quite important to the properties of metals. Since the results are normally of more interest to the metallurgist than the inorganic chemist, such defects will not be discussed further here.

[7] M. E. Fine, in "Treatise on Solid State Chemistry. Vol. I. The Chemical Structure of Solids," N. B. Hannay, ed., Plenum, New York, **1973**, pp. 287–290.

[8] See ref. 7, pp. 291–329.

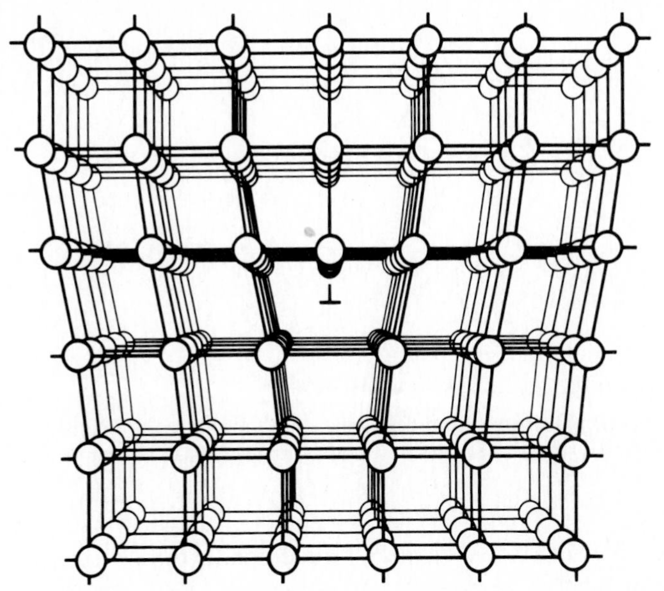

Fig. 4.10 An edge dislocation. [From N. B. Hannay, "Solid-state Chemistry," Prentice-Hall, Englewood Cliffs, N.J., **1967**. Reproduced with permission.]

CONDUCTIVITY IN IONIC SOLIDS

Normally, ionic solids have very low conductivities (see p. 54). Unlike metals or semiconductors (see p. 195) they cannot conduct by *electronic conduction* and must conduct, if at all, by *ionic conduction*. The conductivities that *do* obtain usually relate to the defects discussed in the previous section. The migration of ions may be classified into three types:[9]

1. *Vacancy mechanism.* If there is a vacancy in a lattice, it may be possible for an adjacent ion of the type that is missing (normally a cation) to migrate into it, the difficulty of migration being related to the sizes of the migrating ion and the ions that surround it and tend to impede it.
2. *Interstitial mechanism.* As we have seen with regard to Frenkel defects, if an ion is small enough (again, usually a cation), it can occupy an interstitial site (such as a tetrahedral hole in an octahedral lattice). It may then move to other interstitial sites.
3. *Interstitialcy mechanism.* This mechanism is a combination of the two above. It is a concerted mechanism, with one ion moving into an interstitial site and another ion moving into the vacancy thus created. These three mechanisms are shown in Fig. 4.11.

In purely ionic compounds the conductivity from these mechanisms is *intrinsic* and relates only to the entropy-driven Boltzmann distribution; the conductivity will thus increase with increase in temperature. Because the number of defects is quite limited, the

[9] G. C. Farrington and J. L. Briant, *Science*, **1979**, *204*, 1371.

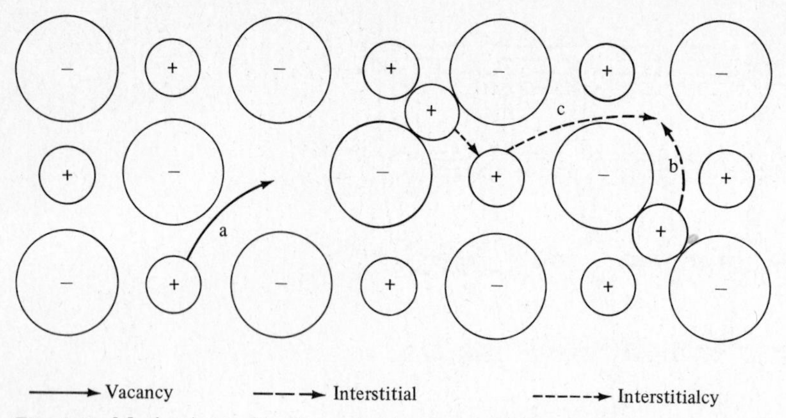

Fig. 4.11 Mechanisms of ionic conduction in crystals from defect structures: (a) Vacancy (Schottky defect) mechanism; (b) Interstitial (Frenkel defect) mechanism; (c) Interstitialcy (concerted Schottky-Frenkel) mechanism.

conductivities are low—of the order of $10^{-4}\ \Omega^{-1}\ m^{-1}$. In addition, *extrinsic* vacancies will be induced by ions of different charge (see p. 188).

There exist, however, a few ionic compounds that *as solids* have conductivities several orders of magnitude higher. One of the first to be studied and the one with the highest room-temperature conductivity ($27\ \Omega^{-1}\ m^{-1}$) is rubidium silver iodide, $RbAg_4I_5$.[9,10] The conductivity may be compared with that of a 35% aqueous solution of sulfuric acid, $80\ \Omega^{-1}\ m^{-1}$. The structure consists of a complex (*not* a simple closest-packed) arrangement of iodide ions with Rb^+ ions in octahedral holes and Ag^+ ions in tetrahedral holes.[11] Of the 56 tetrahedral sites available to the Ag^+ ions, only 16 are occupied, leaving many vacancies. The relatively small size of the silver ion (114 pm) compared with the rubidium (166 pm) and iodide (206 pm) ions gives the silver ion more mobility in the relatively rigid lattice of the latter ions. Furthermore, the vacant sites are arranged in "channels," down which the Ag^+ can readily move (Fig. 4.12).

Another solid electrolyte that may lead to important practical applications is sodium beta alumina.[9] Its unusual name comes from a misidentification and an uncertain composition. It was first thought to be "β-alumina," a polymorph of the common γ-alumina. Its actual composition is close to the ideal $Na_2Al_{22}O_{34}$, but there is always an excess of sodium, as, for example, $Na_{2.58}Al_{21.8}O_{34}$. The structure is closely related to spinel, with 50 of the 58 atoms in the unit cell arranged in exactly the same position as in the spinel structure.[12] In fact, sodium beta-alumina may be thought of as infinite sandwiches composed of slices of spinel structure with a filling of sodium ions. It is the presence of the sodium *between* the spinel-like layers that provides the high conductivity of sodium beta alumina. The Al—O—Al linkages between layers act like pillars in a parking garage (Fig. 4.13) and keep the layers far enough apart that the sodium ions can move readily, yielding conductivities as high as $3.0\ \Omega^{-1}\ m^{-1}$. There is a related structure called "sodium beta" alu-

[10] B. B. Owens and G. R. Argue, *Science*, **1967**, *157*, 308; S. Geller, *Acc. Chem. Res.*, **1978**, *11*, 87.
[11] S. Geller, *Acc. Chem. Res.*, **1967**, *157*, 310.
[12] A. F. Wells, "Structural Inorganic Chemistry," 4th ed., Clarendon Press, Oxford, **1975**, pp. 494–495.

Fig. 4.12 Structure of $RbAg_4I_5$ crystal. Iodide ions are represented by large spheres, rubidium ions by small white spheres. Tetrahedral sites suitable for silver ions marked with short sleeves on horizontal arms. (The easiest to see is perhaps the one formed by the triangle of iodide ions front left with the fourth iodide behind and to the right.) Conduction is by movement of Ag^+ ions from one tetrahedral site to the next, down channels in the crystal. One channel may be seen curving downward from upper center to lower left. [From S. Geller, *Science*, **1967**, *157*, 310. Reproduced with permission.]

mina with the layers held farther apart and with even higher conductivities of up to $18\ \Omega^{-1}\ m^{-1}$.[9]

There are many potential uses for solid electrolytes, but perhaps the most attractive is in batteries. Recall that a battery consists of two very reactive substances (the more so, the better), one a reducing agent and one an oxidizing agent (see Chapter 8 for a discussion of inorganic electrochemistry). To prevent them from reacting directly, these reactants must be separated by a substance that is unreactive towards both and is an *electrolytic*

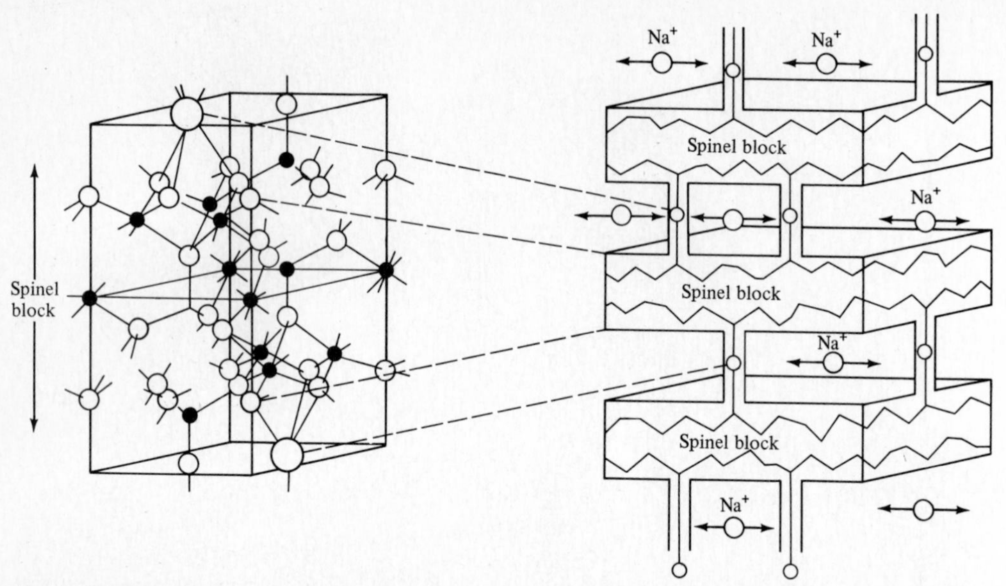

Fig. 4.13 Relation of the spinel structure (left) to the structure of sodium beta alumina (right). The sodium ions are free to move in the open spaces between spinel blocks, held apart by Al-O-Al pillars in the "parking garage" structure. [In part from A. F. Wells, "Structural Inorganic Chemistry," 4th ed., Oxford University Press, Oxford, **1975**. Reproduced with permission.]

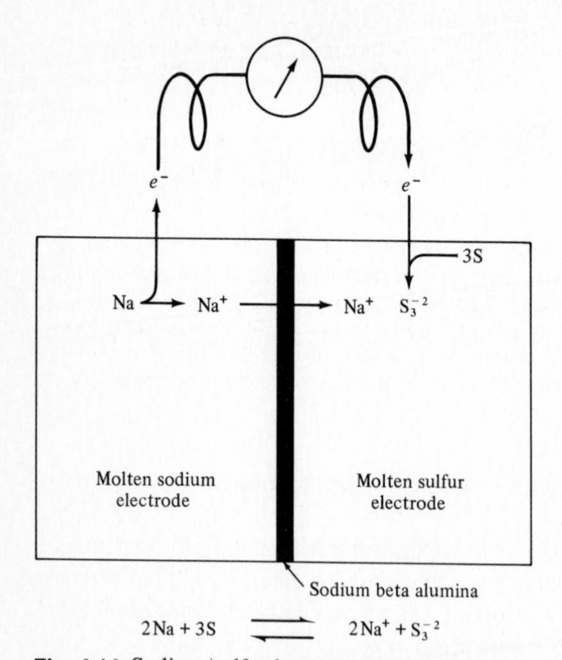

Fig. 4.14 Sodium/sulfur battery with a sodium beta alumina solid electrolyte.

conductor but an *electronic insulator*. Generally (as in the lead storage battery, the "dry" cell, and the nickel alkaline battery), solutions of electrolytes in water serve the last purpose, but in most common batteries this reduces the mass efficiency of the battery at the expense of reactants. The attractiveness of solid electrolytes is that they might provide more efficient batteries.

Consider the battery in Fig. 4.14. The sodium beta alumina barrier allows sodium ions formed at the anode to flow across to the sulfur compartment, where, together with the reduction products of the sulfur, it forms a solution of sodium trisulfide in the sulfur. The latter is held at 300 °C to keep it molten. The sodium beta alumina also acts like an electronic insulator to prevent short circuits, and it is inert toward both sodium and sulfur. The reaction is reversible. At the present state of development, when compared with lead storage cells, batteries of this sort develop twice the power on a volume basis or four times the power on a mass basis.

SOLIDS HELD TOGETHER BY COVALENT BONDING

Because some of the properties of solids that contain no ionic bonds may be conveniently compared with those of ionic solids, it is useful to include them here despite the fact that this chapter deals primarily with ionic compounds.

Types of solids

We may classify solids broadly into three types based on their electrical conductivity. Metals conduct electricity very well. In contrast, insulators do not. Insulators may consist of discrete small molecules, such as phosphorus triiodide, in which the energy necessary to ionize an electron from one molecule and transfer it to a second is too great to be effected under ordinary potentials.[13] We have seen that most ionic solids are nonconductors. Finally, solids that contain infinite covalent bonding such as diamond and quartz are usually good insulators (but see Problem 4.5).

The third type of solid comprises the group known as semiconductors. These are either elements on the borderline between metals and nonmetals, such as silicon and germanium, compounds between these elements, such as gallium arsenide, or various nonstoichiometric or defect structures. In electrical properties they fall between conductors and nonconductors (insulators).

Band theory

In order to understand the bonding and properties of an infinite array of atoms of a metallic element in a crystal, we should first examine what happens when a small number of metal atoms interact. For simplicity we shall examine the lithium atom, since it has but a single valence electron, $2s^1$, but the principles may be extended to transition and posttransition

[13] Given sufficient energy, of course, any insulator can be made to break down.

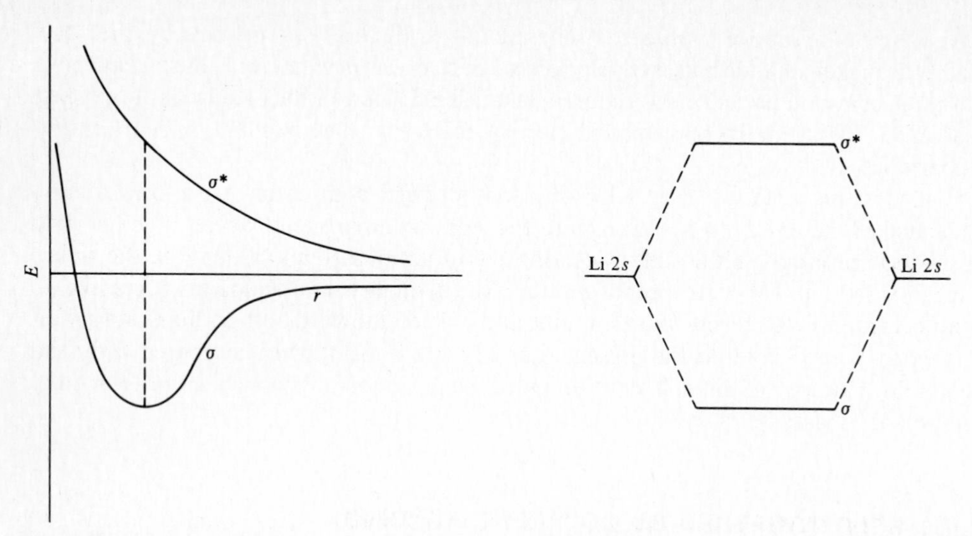

Fig. 4.15 Interaction of the 2*s* orbitals of two lithium atoms to form σ and σ^* molecular orbitals. (Cf. Figs. 3.20 and 3.23.)

metals as well. When two wave functions interact, one of the resultant wave functions is raised in energy and one is lowered. This is discussed for the hydrogen molecule on pp. 95–99. Similarly, interaction between two 2*s* orbitals of two lithium atoms would provide the bonding σ energy level and the antibonding σ^* energy level shown in Fig. 4.15. Interaction of *n* lithium atoms will result in *n* energy levels, some bonding and some antibonding (Fig. 4.16). A mole of lithium metal will provide an Avogadro's number of closely spaced

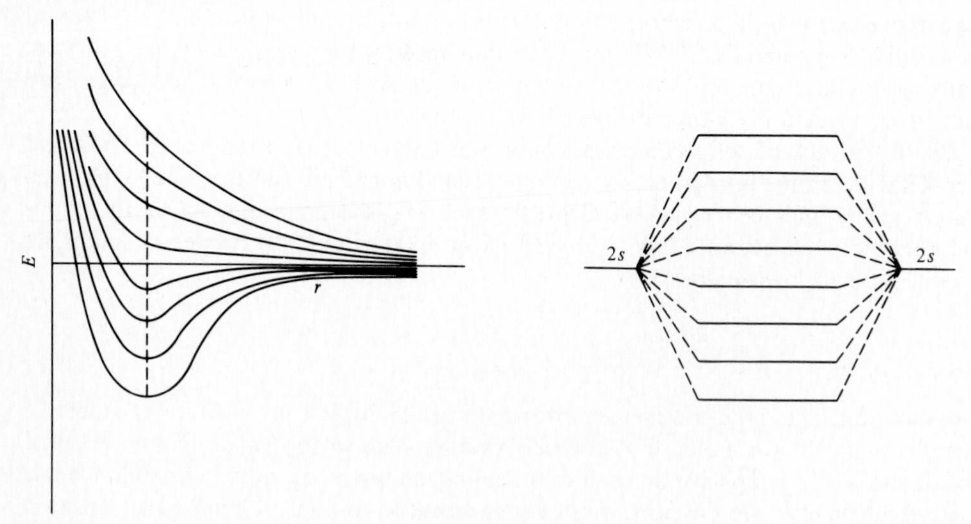

Fig. 4.16 Interaction of eight 2*s* orbitals of eight lithium atoms. The spacing of the energy levels depends upon the geometry of the cluster.

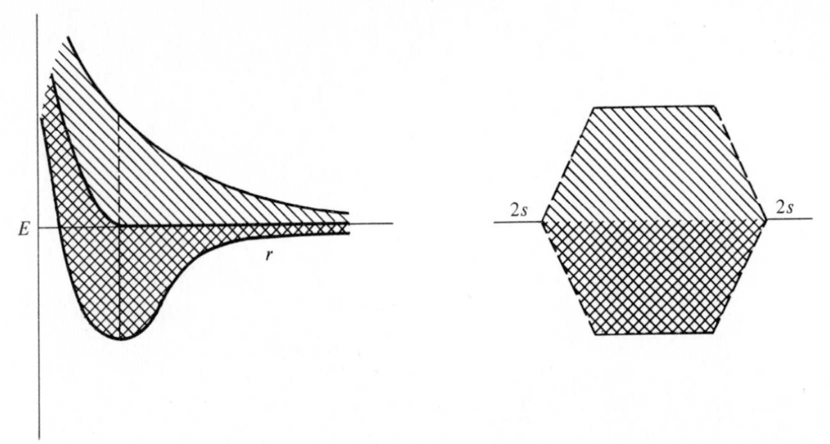

Fig. 4.17 Bonding of a mole of lithium atoms to form a half-filled "band." Heavy shading indicates the filled portion of the band. Real situation is more complicated because the $2p$ orbitals can interact as well.

energy levels (the aggregate is termed a band), the more stable of which are bonding and the less stable, antibonding.[14] Since each lithium atom has one electron and the number of energy levels is equal to the number of lithium atoms, half of the energy levels will be filled whether there are two, a dozen, or N lithium atoms. Thus, in the metal the band will be half-filled (Fig. 4.17), with the most stable half of the energy levels doubly occupied and the least stable, "upper" half empty. The preceding statement is absolutely true only for zero kelvins. At all real temperatures the Boltzmann[15] distribution together with the closely spaced energy levels in the band will ensure a large number of half-filled energy levels, and so the sharp cutoff shown in Fig. 4.17 should actually be somewhat "fuzzy."

Each energy state has associated with it a wave momentum either to the left or to the right. If there is no potential on the system, the number of states with electrons moving left is exactly equal to the number with electrons moving right, so that there is no net flow of current (Fig. 4.18a). However, if an electrostatic potential is applied to the metal, the potential energy of the states with the electron moving toward the positive charge is lower than those with it moving toward the negative charge; thus, the occupancy of the states is no longer 50:50 (Fig. 4.18b). The occupancy of states will change until the energies of the highest "left" and "right" states are equal. Thus, there is a net transfer of electrons into states moving toward the positive charge, the metal is a conductor. If the band is completely filled (Fig. 4.19), there is no possibility of transfer of electrons and, despite the presence of a potential, an equal number of electrons flow either way; therefore, the net current is zero and the material is an insulator.

[14] The levels near the center of the band are essentially nonbonding.

[15] The Boltzmann distribution states the population of higher energy states will be related to the value of the expression $e^{-E/kT}$ where e is the base of natural logarithms, E is the energy of the higher state, k is Boltzmann's constant, and T is the absolute temperature.

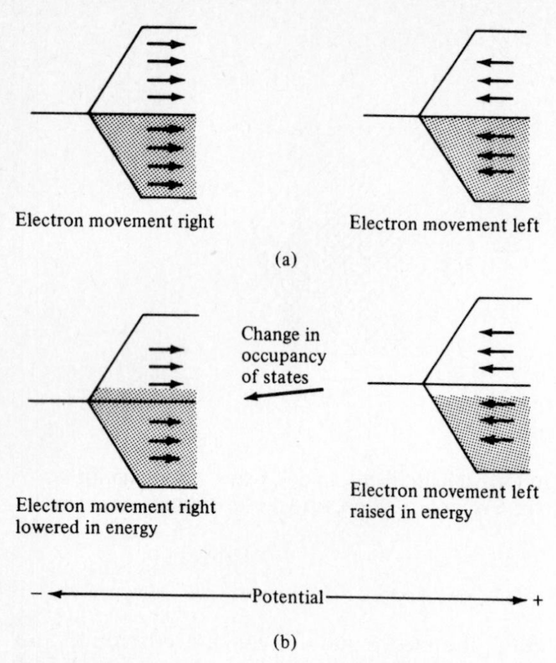

Fig. 4.18 Effect of an electric field on the energy levels in a metal: (a) No field, no net flow of electrons; (b) field applied, net flow of electrons to the right.

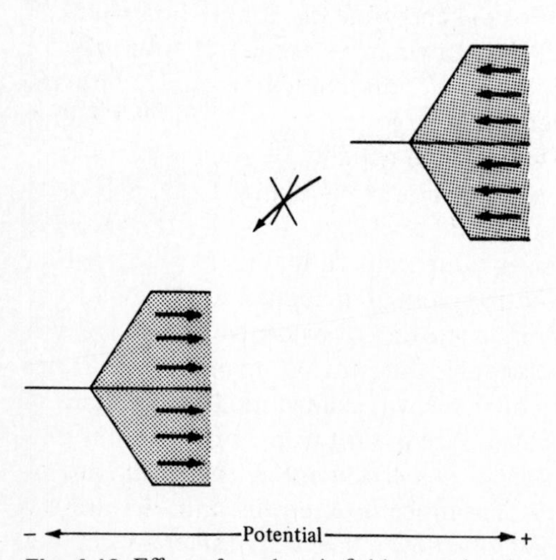

Fig. 4.19 Effect of an electric field on an insulator. Even with applied potential, flow is equal in both directions.

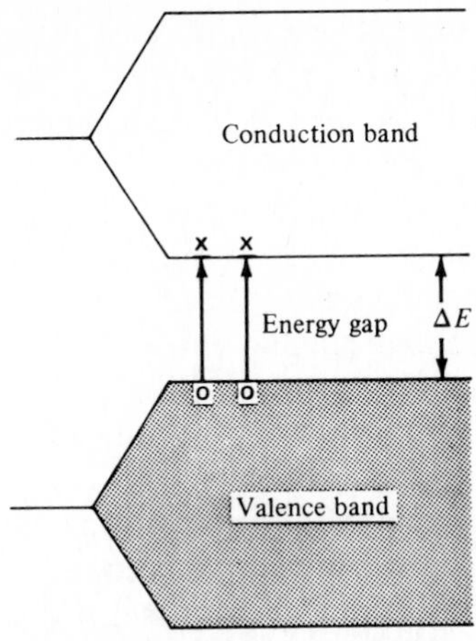

Fig. 4.20 Thermal excitation of electrons in an intrinsic semiconductor. The x's represent electrons and the o's holes.

Intrinsic and photoexcited semiconductors

All insulators will have a filled valence band plus a number of completely empty bands at higher energies, which arise from the higher-energy atomic orbitals. For example, the silicon atom will have core electrons in essentially atomic orbitals $1s^2$, $2s^2$, and $2p^6$, and a valence band composed of the $3s$ and $3p$ orbitals. Then there will be empty orbitals arising out of combinations of $3d$, $4s$, $4p$, and higher atomic orbitals. If the temperature is sufficiently high, some electrons will be excited thermally from the valence band to the lowest-lying empty band, termed the conduction band (Fig. 4.20). The number excited will be determined by the Boltzmann distribution as a function of temperature and band gap, ΔE. Before discussing the magnitude of the energy gap, let us note typical band-gap and resistivity values for insulators (diamond, C), semiconductors (Si, Ge), and an "almost metal," gray tin.[16]

	C	Si	Ge	Sn (gray)
Band gap, kJ mol^{-1}	580	105	58	7
Resistivity, Ω cm	10^6	6×10^4	50	1

For every electron excited to the antibonding conduction band, there will remain behind a hole, or vacancy, in the valence band. The electrons in both the valence band and the conduction band will be free to move under a potential by the process shown in Fig. 4.18b, but since the number of electrons (conduction band) and holes (valence band) is limited, only a limited shift in occupancy from "left-bound" states to "right-bound" states can occur and the conductivity is not high as in a metal. This phenomenon, known as *intrinsic semiconduction*, is the basis of *thermistors* (temperature-sensitive resistors).

[16] L. H. Van Vlack, "Elements of Materials Science," 2nd ed., Addison-Wesley, Reading, Mass., **1967**, p. 113.

An alternative picture of the conductivity of the electrons and holes in intrinsic semiconductors is to consider the electrons in the conduction band as migrating, as expected, toward the positive potential, and to consider the holes as discrete, positive charges migrating in the opposite direction. Although electrons are responsible for conduction in both cases, the hole formalism represents a convenient physical picture (Fig. 4.21).

If, instead of thermal excitation, a photon of light excites an electron from the valence band to the conduction band, the same situation of electron and hole carriers obtains, and one observes the phenomenon of *photoconductivity*, useful in photocells and similar devices.

Instead of silicon or germanium with four valence electrons (to yield a filled band of $4 + 4 = 8$ electrons on band formation), we can form a compound from gallium (three valence electrons) and arsenic (five valence electrons) to yield gallium arsenide with a filled valence band. In general, however, the ΔE for the band gap will differ from those of elemental semiconductors. The band gap will increase as the tendency for electrons to become increasingly localized on atoms increases and thus is a function of substituent electronegativity (Fig. 4.22). Note that conductivity is a continuous property ranging from metallic conductance (Sn) through elemental semiconductors (Ge, Si), compound semiconductors (GaAs, CdS) to insulators, both elemental (diamond, C) and compounds (NaCl).

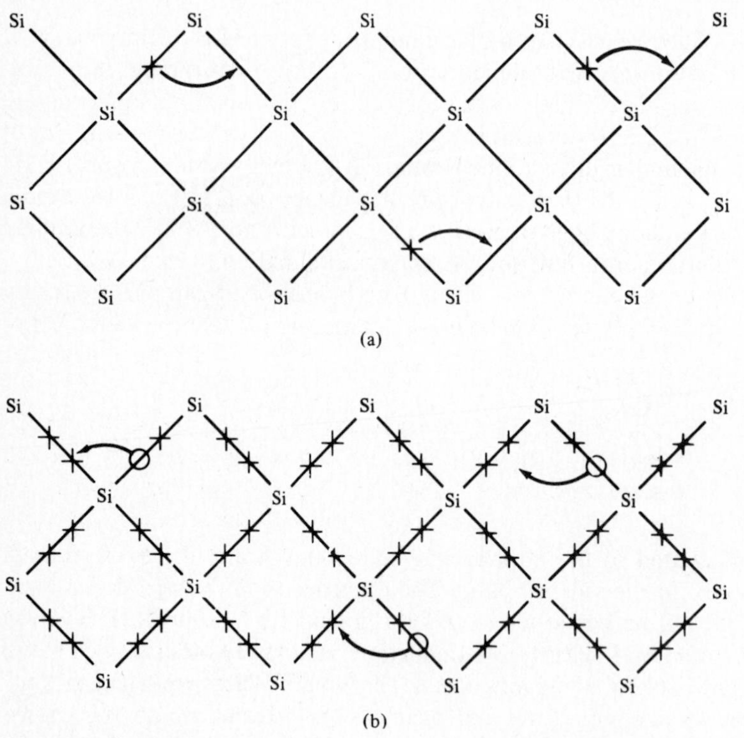

Fig. 4.21 Conduction by (a) electrons in the conduction band and (b) holes in the valence band of an intrinsic semiconductor.

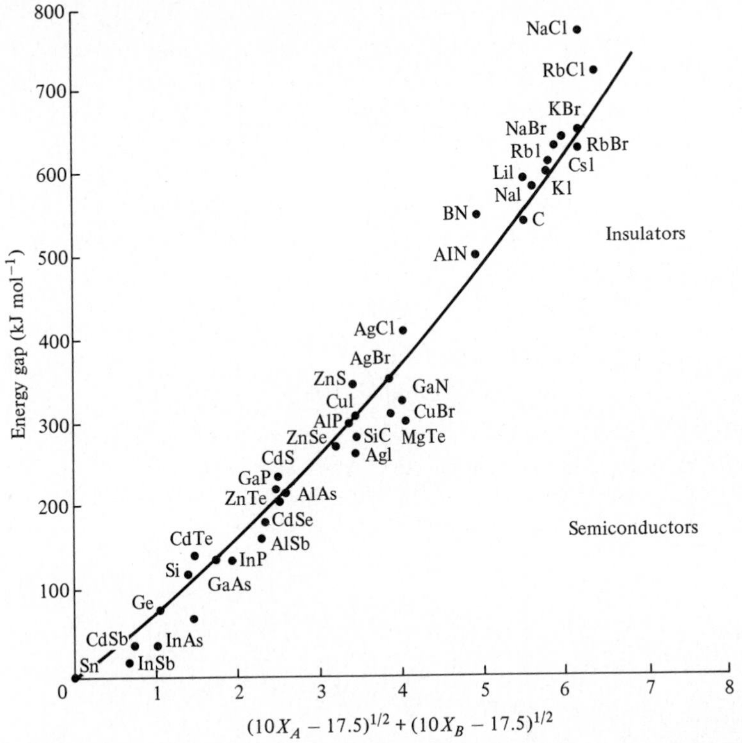

Fig. 4.22 Empirical relationship between energy gap and the electronegativities of the elements present. Note that substances made of a single, fairly electronegative atom (C, diamond) or from a very low-electronegative metal and high-electronegative nonmetal (NaCl) are good insulators. As the electronegativities approach 1.75, the electronegativity function rapidly approaches zero. [From N. B. Hannay, "Solid-state Chemistry," Prentice-Hall, Englewood Cliffs, N.J., **1967**. Reproduced with permission.]

Impurity and defect semiconductors

Consider a pure crystal of germanium. Like silicon it will have a low intrinsic conductivity at low temperatures. If we now dope some gallium atoms into this crystal, we shall have formed holes because each gallium atom contributes only three electrons rather than the requisite four to keep the band filled. These holes can conduct electricity by the process shown in Fig. 4.21b. By controlling the amount of gallium impurity, we can control the number of carriers.

Using Fig. 4.21b uncritically might indicate that there would be no energy gap in a gallium-doped germanium semiconductor. However, note that gallium lies to the left of germanium in the periodic table and is more electropositive; it thus tends to keep the positive hole. (Alternatively, germanium is more electronegative, and the electrons tend to stay on the germanium atoms rather than flow into the hole on the gallium atom.) This electronegativity effect creates an energy gap, as shown more graphically in Fig. 4.23. The

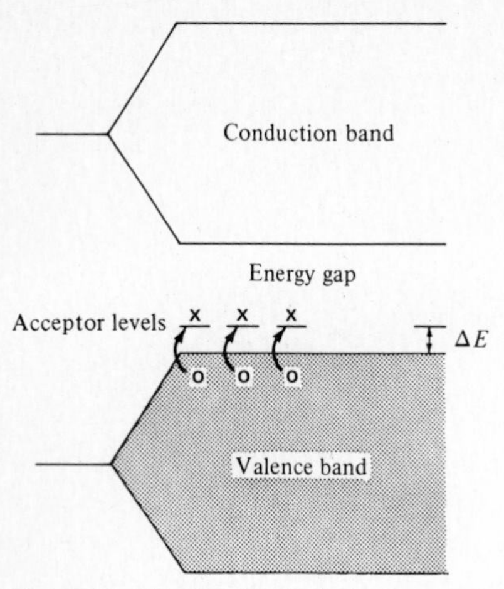

Fig. 4.23 Conduction by holes in an *acceptor* or *p*-type semiconductor.

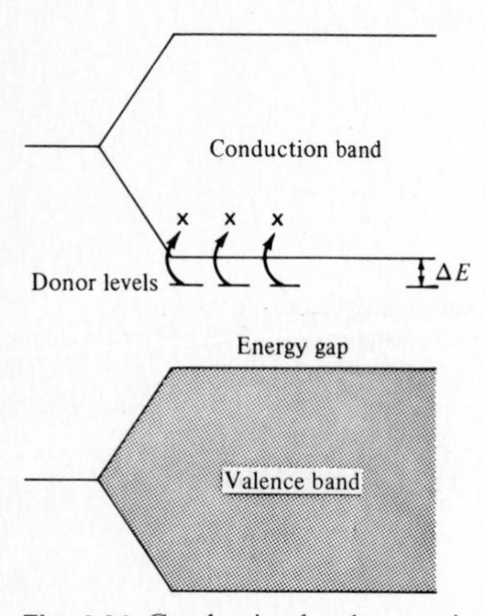

Fig. 4.24 Conduction by electrons in a *donor* or *n*-type semiconductor.

atomic energy level for gallium lies above that of germanium[17] and thus above the valence band. Providing a small ionization energy, ΔE, generates the holes for semiconduction. The resulting system is called an *acceptor* (since gallium can accept an electron) or *p*-type (p = positive holes) semiconductor.

In exactly the opposite manner, doping germanium with arsenic (five valence electrons) results in an excess of electrons and a *donor* (the arsenic donates the fifth electron) or *n*-type (n = negative electrons) semiconductor. The conduction can be viewed either by the simple picture (Fig. 4.21a) or the energy diagram in which the electrons can be ionized from the impurity arsenic atoms to the conduction band of the semiconductor (Fig. 4.24).

[17] The relation between energy levels and electronegativity was presented on pp. 149–152.

Various imperfections can lead to semiconductivity in analogous ways. For example, nickel(II) oxide may be doped by lithium oxide (see Fig. 4.7). The Ni^{+3} ions now behave as holes as they are reduced and produce new Ni^{+3} ions at adjacent sites. These holes can migrate under a potential (indicated by the signs on the extremes of the series of nickel ions):

$$
\begin{array}{c}
(+)\ Ni^{+2} \cdots\cdots Ni^{+3} \cdots\cdots Ni^{+2} \cdots\cdots Ni^{+2} \cdots\cdots Ni^{+2}\ (-) \\
\downarrow \\
(+)\ Ni^{+2} \cdots\cdots Ni^{+2} \cdots\cdots Ni^{+3} \cdots\cdots Ni^{+2} \cdots\cdots Ni^{+2}\ (-) \\
\downarrow \\
(+)\ Ni^{+2} \cdots\cdots Ni^{+2} \cdots\cdots Ni^{+2} \cdots\cdots Ni^{+3} \cdots\cdots Ni^{+2}\ (-)
\end{array}
$$

The range of possibilities for semiconduction is very great, and the applications to the operation of transistors and related devices has revolutionized the electronics industry, but an extensive discussion of these topics is beyond the scope of this text.[18] Note, however, that inorganic compounds are receiving intensive attention as the source of semiconductors, superconductors, and one-dimensional conductors (see Chapter 14).

PROBLEMS

4.1. Find the spinel exceptions to the structure field map in Fig. 4.2.

4.2. Predict the structures of
  a. $MgCr_2O_4$
  b. K_2MgF_4
  (i.e., to what mineral classes do they belong?).

4.3. Rationalize the fact that the Fluorite Field lies above and to the right of the Rutile Field from what you know about these structures. Does this insight enable you to predict anything about the silicon dioxide structure?

4.4. With regard to each of the following, does it make any difference whether one uses "correct" radii, such as empirically derived Shannon-Prewitt radii, or whether one uses theoretically sound but somewhat misassigned traditional radii?
  a. prediction of the interionic distance in a new compound, MX
  b. calculation of the radius ratio in M_2X
  c. calculation of the enthalpy of formation of a hypothetical compound, MX_2
  d. construction of a structure field map as shown in Figs. 4.1 and 4.2

4.5. Why is graphite a good conductor whereas diamond is not? (Both contain infinite lattices of covalently bound carbon atoms.)

4.6. It was stated casually (p. 201) that the energy levels of gallium are *above* those of germanium and, later, that those of arsenic lie *below* those of germanium. Can you provide any arguments, data, etc. to substantiate this?

---

[18] See "Semiconductors," N. B. Hannay, ed., American Chemical Society Monograph No. 140, Reinhold, New York, **1959**.

4.7. Cadmium sulfide is often used in the photometers of cameras. Suppose you were interested in infrared photography. Using Fig. 4.22, suggest some compounds that might be suitable for an infrared photocell.

4.8. Using Fig. 4.22, calculate the wavelength of light at which photoconduction will begin for a CdS light meter. If you are interested in black and white photography, can you tell why this wavelength is particularly appropriate?

4.9. A very important photographic reaction is the photolytic decomposition of silver bromide described approximately by the following equation:

$$AgBr_{(s)} \xrightarrow{h\nu} Ag_{(s)} + \tfrac{1}{2}Br_{2(l)} \qquad \textbf{(4.1)}$$

Assuming that the enthalpy of the reaction described in the equation can be equated with the energy of the photon, use a Born-Haber-type cycle to calculate the wavelength of light that is sufficiently energetic to effect the decomposition of silver bromide. What are some sources of error in your estimate?

5

The covalent bond: structure and reactivity

In Chapter 3 we have seen how the energetics of bond formation holds molecules together; in this chapter the forces that underlie molecular structure will be examined. Since these two topics are closely interwoven, it was necessary in the last chapter to anticipate some of the material to be presented here; similarly, it will be necessary to refer back continually to orbitals, overlap, energetics, repulsions, and the like. Although the discussion in this chapter applies broadly to all inorganic molecules, it will center on compounds of the nonmetals; the molecular structures of transition metal compounds will be discussed in Chapters 9 and 10.

STRUCTURE OF MOLECULES

In this section a few simple rules for predicting molecular structures will be investigated. We shall examine first a basically valence bond (VB) approach, followed by the *valence shell electron pair repulsion* (VSEPR) model, often couched in VB terms, though it need not be, and finally a purely molecular orbital (MO) treatment. It should be kept in mind that each molecule is a unique structure resulting from the interplay of several energy factors and that the following rules can only be a crude attempt to average the various forces.

1. First, from the electronic configuration of the elements, determine a reasonable Lewis-type structure. For example, in the carbon dioxide molecule, there will be a total of 16 valence electrons to distribute among three atoms:

$$\underset{(a)}{:\ddot{O}::C::\ddot{O}:} \qquad \text{or} \qquad \underset{(b)}{\begin{array}{l} :\ddot{O}::C \\ \phantom{:\ddot{O}::}\,\vdots\!\vdots \\ \phantom{:\ddot{O}:}:\ddot{O}: \end{array}}$$

Note that a Lewis structure says nothing about the bond angles in the molecule since both (a) and (b) meet all the criteria for a valid Lewis structure.

2. A structure should now be considered which lets all the electrons in the valence shell of the central atom(s) get as far away from each other as possible. In the usual σ–π treatment this usually means ignoring the π bonds temporarily since they will follow the σ bonds. In carbon dioxide there will be two σ bonds and no nonbonding electrons on the carbon atom, and so the preferred orientation is for the σ bonds to form on opposite sides of the carbon atom. This will require hybridization of the carbon $2s$ and $2p_z$ orbitals to form a digonal hybrid, with a bond angle of 180°.

3. Once the structure of the σ-bonded molecule has been determined, π bonds may be added as necessary to complete the molecule. In carbon dioxide, the p_x and p_y orbitals on the carbon atom were unused by the σ system and are available for the formation of π bonds. A complete structure for carbon dioxide would thus be as shown in Fig. 5.1a.

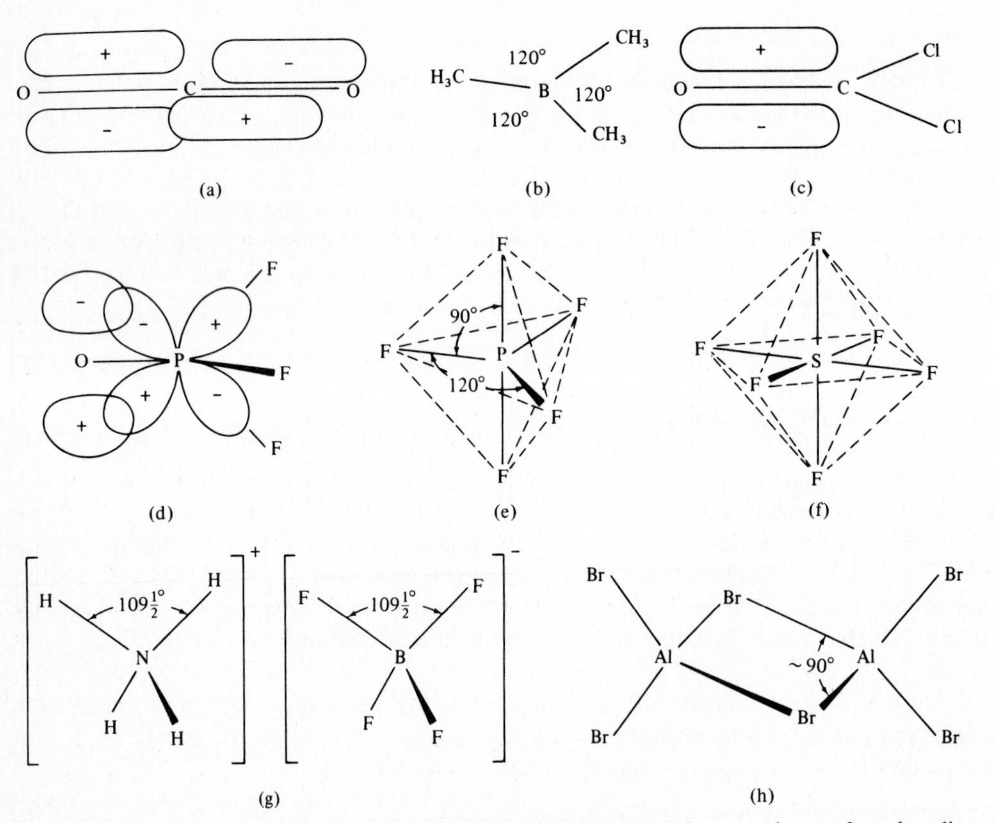

Fig. 5.1 Some simple molecular structures in which all electrons on the central atom form bonding pairs. (a) carbon dioxide, with two sp σ bonds (solid lines) and two π bonds; (b) trimethylborane, with three sp^2 σ bonds; (c) carbonyl chloride ("phosgene"), with three sp^2 σ bonds and one C—O π bond; (d) phosphorus oxyfluoride, with four approximately sp^3 σ bonds plus one p–d π bond; (e) phosphorus pentafluoride, with five sp^3d σ bonds; (f) sulfur hexafluoride, with six sp^3d^2 σ bonds; (g) ammonium tetrafluoroborate; each ion has four sp^3 σ bonds; (h) aluminum bromide dimer.

The following molecules illustrate the application of these rules to the prediction of structure of noncoordination compounds. The same factors work in coordination compounds as well, but an additional factor, the presence of an incompletely filled *d* subshell, must also be considered (see Chapters 9 and 10).

Trimethylborane. We may assume that the methyl groups will have their usual configuration found in organic compounds. The Lewis structure of $(CH_3)_3B$ will place six electrons in the valence shell of the boron atom, and in order that these electrons be as far apart as possible, the methyl groups should be located at the corners of an equilateral triangle. This results in sp^2, or trigonal, hybridization for the boron (Fig. 5.1b).

Phosgene. A Lewis structure for $OCCl_2$ has eight electrons about the carbon, but one pair forms the π bond of the double bond, so again an sp^2, or trigonal, hybridization will be the most stable (Fig. 5.1c).

Phosphoryl fluoride. Two Lewis-type structures can be drawn for the OPF_3 molecule:

```
     ..               ..
    :F:              :F:
     ..  ..       ..  ..  ..
 :O::P:F:        :O:P:F:
  ..  ..  ..      ..  ..  ..
    :F:              :F:
     ..               ..

    (a)              (b)
```

The relation between structures (a) and (b) will be discussed later, but for now either one suggests that the coordination number of the phosphorus atom will be 4. To a first approximation, the three fluorine atoms and the single oxygen atom will be bonded to the phosphorus atom with σ bonds formed from sp^3 tetrahedral orbitals. Now one of the five $3d$ orbitals on the phosphorus atom can overlap with a $2p$ orbital on the oxygen atom (Fig. 5.1d) and form a fifth bond, d_π—p_π, further stabilizing the molecule.

Phosphorus pentafluoride. A Lewis-type structure for the PF_5 molecule requires ten electrons in the valence shell of the phosphorus atom and the use of $3s$, $3p$, and $3d$ orbitals and five σ bonds. It is impossible to form five bonds such that they are all equidistant from one another, but the trigonal bipyramidal (Fig. 5.1e) and square pyramidal arrangements tend to minimize repulsions. Almost every 5-coordinate molecule (coordination compounds excepted) which has been carefully investigated has been found to have a trigonal bipyramidal structure. The structure of the PF_5 molecule is shown in Fig. 5.1e (sp^3d hybrid). The bonds are of two types: *axial*, the linear F—P—F system; and *equatorial*, the three P—F bonds forming a trigonal plane.

Sulfur hexafluoride. Six sulfur–fluorine σ bonds require twelve electrons in the valence shell. Six equivalent bonds require an octahedron and so sulfur will be hybridized sp^3d^2 as shown in Fig. 5.1f.

Ammonium tetrafluoroborate. Both the ammonium (NH_4^+) and tetrafluoroborate (BF_4^-) ions are isoelectronic with the methane molecule and we might therefore reasonably expect them to have similar structures. Indeed, the only difference among NH_4^+, CH_4, and BF_4^- (neglecting the difference between H and F) is the number of protons in the nucleus of the central atom. Thus the usual representation of the formation of these ions indicating that one bond is formed differently from the other three is misleading. All four bonds are equivalent, and since the electrons avoid each other as much as possible, the most stable arrangement is a tetrahedron (Fig. 5.1g).

Aluminum bromide. For the molecule $AlBr_3$, a structure similar to that of trimethylborane would be expected with 120° bond angles. Experimentally, however, it is found that aluminum bromide is a dimer, Al_2Br_6. This is readily explainable as a result of the tendency to maximize the number of bonds formed since Al_2Br_6 contains four bonds per aluminum atom. This is possible because the aluminum atom can accept an additional pair of electrons (Lewis acid, see Chapter 7) in its unused p orbital and rehybridize from sp^2 to sp^3. We should expect the bond angles about the aluminum to be approximately tetrahedral except for the strain involved in the Al—Br—Al—Br four-membered ring. Since the average bond angle within the ring must be 90°, we might expect both the aluminum and bromine atoms to use orbitals which are essentially purely p in character for the ring in order to reduce the strain. The structure of the Al_2Br_6 molecule is shown in Fig. 5.1h.

Valence shell electron pair repulsion theory

Although the discussions of the preceding molecules have been couched in valence bond terms (Lewis structures, hybridization, etc.) recall that the criterion for molecular shape (rule 2 above) was that the σ bonds of the central atom should be allowed to get as far from each other as possible: 2 at 180°, 3 at 120°, 4 at 109.5°, etc. This is the heart of the valence shell electron pair repulsion (VSEPR) method of predicting molecular structures, and is, indeed, independent of valence bond (VB) hybridization schemes, although it is most readily applied in a VB context.

The source of the repulsions that maximize bond angles is not completely clear. For molecules such as CO_2, $B(CH_3)_3$, or $O{=}PF_3$ we might suppose that van der Waals repulsions (analogous to the Born repulsions in ionic crystals) among, for example, the three methyl groups might open the bond angles to the maximum possible value of 120°. In the next section we shall see that nonbonding pairs of electrons ("lone pairs") are at least as effective as bonding pairs (or bonded groups) in repulsion, and so attention focuses on the electron pairs themselves. Although a number of theories have been advanced, the consensus seems to be that the physical force behind VSEPR is the Pauli force: *Two electrons of the same spin cannot occupy the same space.* However, it should be noted that there has been some disagreement over the matter. Nevertheless, as we shall see, the VSEPR model is an extremely powerful one for treating molecular structures.

Structures of molecules containing lone pairs of electrons

When we investigate molecules containing lone (unshared) electron pairs, we must take into account the differences between the bonding electrons and the nonbonding electrons. First, before considering hybridization and the energies implicit in the bonding rules (p. 227) let us consider the simplest possible viewpoint. Consider the water molecule in which the oxygen atom has a ground state electron configuration of $1s^2 2s^2 2p_z^2 2p_x^1 2p_y^1$.

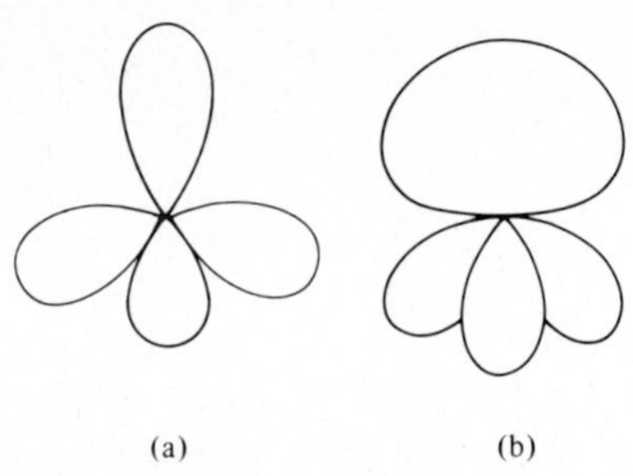

Fig. 5.2 (a) Four equivalent bonding electron pairs. (b) Three bonding electron pairs repelled by a nonbonding pair of electrons.

The unpaired electrons in the p_x and p_y orbitals may now be paired with electrons on two hydrogen atoms to give H_2O. Since the p_x and p_y orbitals lie at right angles to one another, maximum overlap is obtained with an H—O—H bond angle of 90°. The experimentally observed bond angle in water is, however, about $104\frac{1}{2}°$, much closer to a tetrahedral angle. Inclusion of repulsion of positive charges on the adjacent hydrogen atoms (resulting from the fact that the oxygen does not share the electrons equally with the hydrogens) might cause the bond angle to open up somewhat, but cannot account for the large deviation from 90°. Not only must the H—H repulsions be taken into consideration, but also every other energetic interaction in the molecule: all repulsions and all changes in bond energies as a function of angle and hybridization. It is impossible to treat this problem in a rigorous way, mainly as a result of our ignorance of the magnitude of the various energies involved; however, certain empirical rules have been formulated.[1]

First, as we have seen in examples on the previous pages, bond angles in molecules tend to open up as much as possible as a result of the repulsions between the electrons bonding the substituents to the central atoms. Repulsions between unshared electrons on the central atom and other unshared electrons or bonding electrons will affect the geometry. In fact, it is found that the repulsions between lone pair electrons are greater than those between the bonding electrons. The order of repulsive energies is lone pair–lone pair > long pair–bonding pair > bonding pair–bonding pair. This results from the absence of a second nucleus at the distal end of the lone pair which would tend to localize the electron cloud in the region between the nuclei. Since the lone pair does not have this second nucleus, it is attracted only by its own nucleus and tends to occupy a greater *angular* volume (Fig. 5.2).

The difference in spatial requirements between lone pairs and bonding pairs may perhaps be seen most clearly from the following example. Consider an atom or ion with a noble gas configuration such as C^{-4}, N^{-3}, O^{-2}, F^-, or Ne ($1s^22s^22p^6$). Assume that the eight electrons in the outer shell occupy four equivalent tetrahedral orbitals. Now let a proton interact with one pair of electrons to form an X—H bond (HC^{-3}, NH^{-2}, OH^-, HF, NeH^+). The proton will polarize the pair of electrons to which it attaches in the same

[1] R. J. Gillespie and R. S. Nyholm, *Quart. Rev. Chem. Soc.*, **1957**, *11*, 339; A. W. Searcy, *J. Chem. Phys.*, **1958**, *28*, 1237; **1959**, *31*, 1; R. J. Gillespie, *J. Am. Chem. Soc.*, **1960**, *82*, 5978. Best simple reviews are R. J. Gillespie, *J. Chem. Educ.*, **1963**, *40*, 295; **1970**, *47*, 18, and "Molecular Geometry," Van Nostrand–Reinhold, London, **1972**.

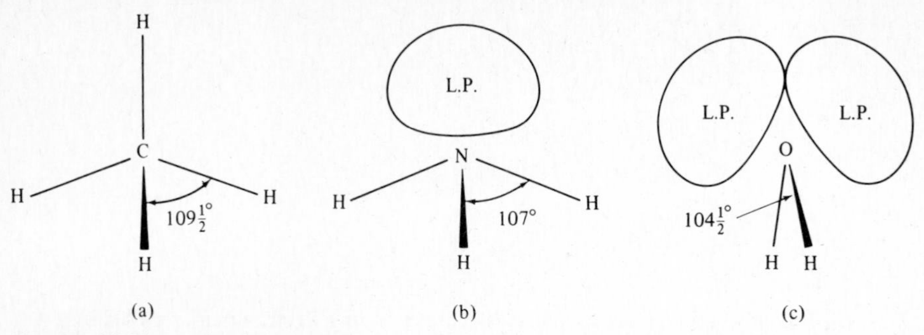

Fig. 5.3 (a) The molecular structure of methane. (b) The molecular structure of ammonia showing the reduction of bond angles. (c) The molecular structure of water showing the greater reduction of the bond angle by two lone pairs.

way that a proton or small, positive ion polarizes an anion (Fajans' rules, p. 129). Electron density will be removed from the vicinity of the nucleus of the first atom and attracted toward the hydrogen nucleus. The remaining, nonbonding pairs may thus expand at the expense of the bonding pair. Addition of a second proton produces two polarized, bonding pairs and two expanded lone pairs (H_2C^{-2}, N_2N^-, H_2O, H_2F^+). A third proton forms NH_3 with one expanded lone pair. A fourth proton produces CH_4 and NH_4^+ in which all four pairs of electrons have been polarized toward the hydrogen nuclei, are once more equivalent, and hence directed at tetrahedral angles.

From this point of view, the water molecule can be considered to be hybridized tetrahedrally to a first approximation. Since the two lone pairs will occupy a greater angular volume than the two bonding pairs, the angle between the latter two is reduced somewhat (from $109\frac{1}{2}°$ to $104\frac{1}{2}°$), allowing the angle between the lone pairs to open up slightly. The series methane, CH_4 (no lone pairs, bond angle = $109\frac{1}{2}°$); ammonia, NH_3 (one lone pair, bond angle = 107°); and water, H_2O (two lone pairs, bond angle = $104\frac{1}{2}°$) illustrates an isoelectronic series in which the increasing requirements of the nonbonding pairs reduce the bond angle (Fig. 5.3).

As a general rule, we can state that the lone pair will always occupy a greater angular volume than bonding electrons. Furthermore, if given a choice, the lone pair tends to go to that position in which it can expand most readily. Consider, for example, the following molecules, where if in each case we consider only the *bonding electrons*, we obtain wrong predictions concerning the geometry of the molecules. For example, BrF_3 would be trigonal, ICl_4^- tetrahedral, IF_5 trigonal bipyramidal, and SF_4 tetrahedral. In fact, none of these molecules has the structure just assigned to it. If, however, we include the lone pairs, we can predict not only the approximate molecular shape but also distortions which will take place.

Sulfur tetrafluoride. The molecule $:SF_4$ has ten electrons in the valence shell of sulfur, four bonding pairs and one nonbonding pair. In order to let each pair of electrons have as much room as possible, the approximate geometry will be a trigonal bipyramid, as in PF_5. However, the lone pair can be arranged in one of two possible ways, either equatorially (Fig. 5.4a) or axially (Fig. 5.4b). The experimentally derived structure is shown in Fig. 5.4c. The lone pair is in an equatorial position and tends to repel the bonding pairs and cause

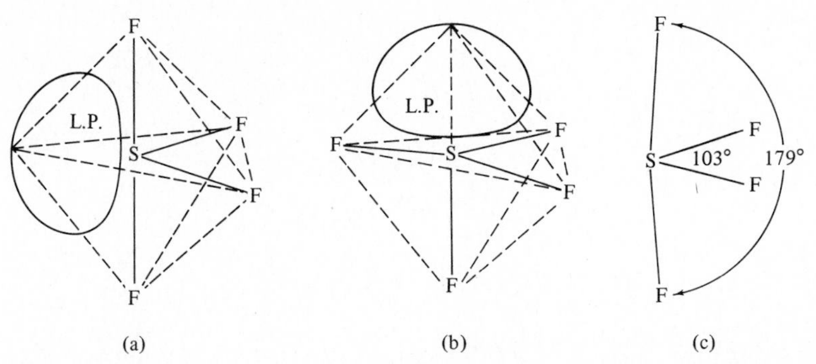

Fig. 5.4 Sulfur tetrafluoride. (a) Trigonal bipyramidal structure with *equatorial* lone pair. (b) Trigonal bipyramidal structure with *axial* lone pair. (c) Experimentally determined structure of sulfur tetrafluoride.

them to be bent back away from the position occupied in an undistorted trigonal bipyramid. We can rationalize the adoption of the equatorial position by the lone pair by noting that in this position it encounters only two 90° interactions (with the axial bonding pairs), whereas in the alternative structure it would encounter three 90° interactions (with the equatorial bonding pairs). Presumably the 120° interactions are sufficiently relaxed that they play no important role in determining the most stable arrangement. This is consistent with the fact that repulsive forces are important only at *very small distances*. In any event, lone pairs always adopt positions which minimize 90° interactions.

Bromine trifluoride. The $:\ddot{\mathrm{Br}}\mathrm{F}_3$ molecule also has ten electrons in the valence shell of the central atom, in this case three bonding pairs and two lone pairs. Again, the approximate structure is trigonal bipyramidal with the lone pairs occupying equatorial positions. The distortion from lone pair repulsion causes the axial fluorine atoms to be bent away from a linear arrangement so that the molecule is a slightly "bent T" with bond angles of $86\frac{1}{2}°$ (Fig. 5.5a).

Dichloroiodate (*I*) *anion.* The $:\underset{\cdot\cdot}{\ddot{\mathrm{I}}}\mathrm{Cl}_2^-$ anion has a linear structure as might have been supposed naively. However, note that three lone pairs are presumably still stereochemically

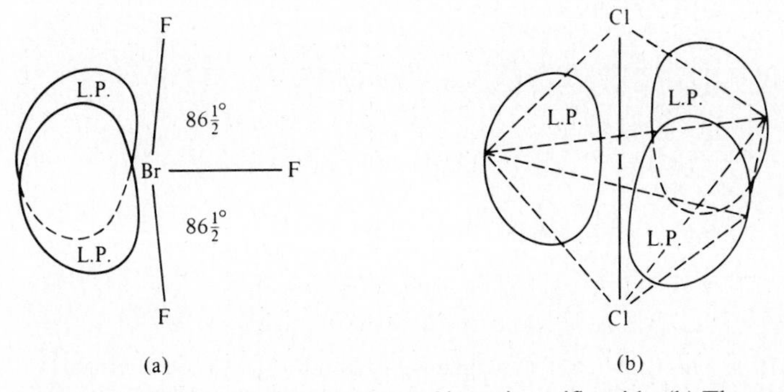

Fig. 5.5 (a) The molecular structure of bromine trifluoride. (b) The structure of the dichloroiodate(I) anion.

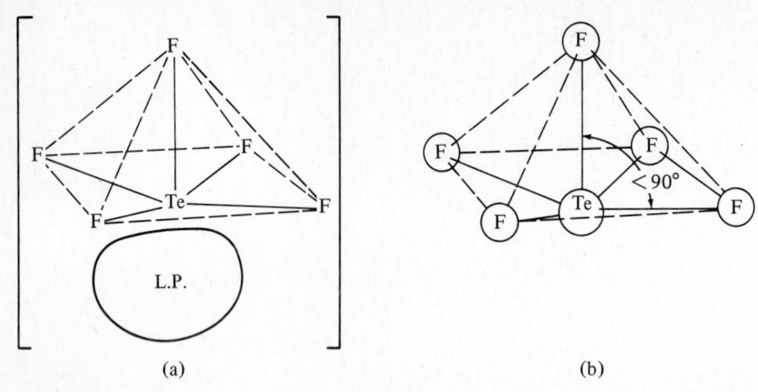

Fig. 5.6 (a) The pentafluorotellurate(IV) anion. Approximately octahedral arrangement of bonding and nonbonding electrons. (b) Experimentally determined structure. The tellurium atom is *below* the plane of the fluorine atoms. [From S. H. Martin et al., *Inorg. Chem.*, **1970**, *9*, 2100. Reproduced with permission.]

active, but by adopting the three equatorial positions they cause no distortion (Fig. 5.5b). [A note on bookkeeping for ions: Add 7 electrons (I) + 2 electrons (2Cl) + 1 electron (ionic charge) = 10 = 5 pairs.]

Pentafluorotellurate(IV) anion. In the $:TeF_5^-$ ion the tellurium atom has twelve electrons in its valence shell, five bonding pairs and one nonbonding. The most stable arrangement for six pairs of electrons is the octahedron which we should expect for a first approximation. Repulsion from the single lone pair should cause the adjacent fluorine atoms to move upward somewhat (Fig. 5.6a). The resulting structure is a square pyramid with the tellurium atom *below* the plane of the four fluorine atoms (Fig. 5.6b).

Tetrachloroiodate(III) anion. The $:\ddot{I}Cl_4^-$ anion is isoelectronic with the $[TeF_5]^-$ ion with respect to the central atom. In this case, however, there are four bonding pairs and two lone pairs. In an undistorted octahedron, all six points are equivalent, and the lone pairs could be adjacent, or *cis* (Fig. 5.7a), or opposite, or *trans* (Fig. 5.7b), to one another.

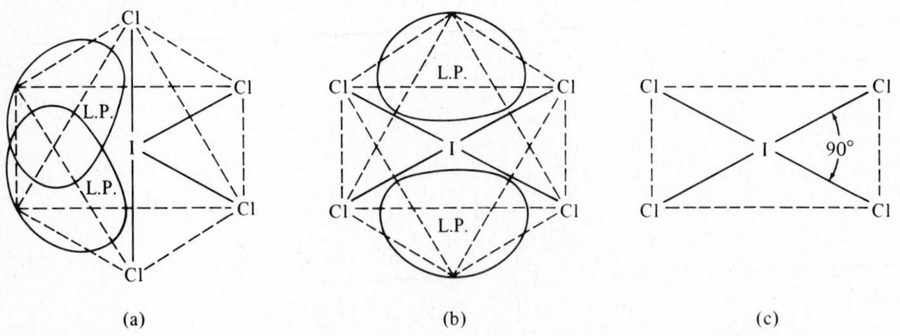

Fig. 5.7 The tetrachloroiodate(III) ion. (a) Octahedral arrangement of bonding and nonbonding electrons with lone pairs *cis* to each other. (b) Octahedral arrangement of bonding and nonbonding electrons with lone pairs *trans* to each other. (c) Experimentally determined structure.

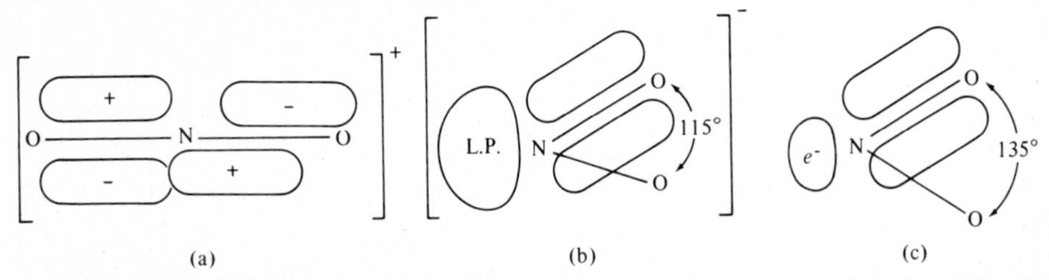

Fig. 5.8 (a) The linear nitryl ion, NO_2^+. (b) The effect of the lone pair in the nitrite ion, NO_2^-. Resonance has been omitted to simplify the discussion. (c) The effect of the unpaired electron, half of a "lone pair," in nitrogen dioxide.

In the *cis* arrangement the lone pairs will compete with each other for volume into which to expand, a less desirable arrangement than *trans*, in which they can expand at the expense of the bonding pairs. Since the lone pairs are not seen in a normal structural determination, the resulting arrangement of atoms is square planar (Fig. 5.7c).

Nitrogen dioxide, nitrite ion, and nitryl ion. The three species NO_2, NO_2^-, and NO_2^+ show the effect of steric repulsion of bonding and nonbonding electrons. The Lewis structures are

[:Ö::N::Ö:]$^+$:Ö::Ṅ:Ö: [:Ö::N̈:Ö:]$^-$

The nitryl ion, NO_2^+, is isoelectronic with carbon dioxide and will, like it, adopt a linear structure with two π bonds (Fig. 5.8a). The nitrite ion, NO_2^-, will have one π bond (stereochemically inactive), two σ bonds, and one lone pair. The resulting structure is therefore expected to be trigonal, with 120° sp^2 bonds to a first approximation. The lone pair should be expected to expand at the expense of the bonding pairs, however, and the bond angle is found to be 115° (Fig. 5.8b).

The nitrogen dioxide molecule is a free radical; i.e., it contains an unpaired electron. It may be considered to be a nitrite ion from which one electron has been removed from the least electronegative atom, nitrogen.[2] Instead of having a lone pair on the nitrogen, it has a single electron in an approximately trigonal orbital. Since a single electron would be expected to repel less than two, the bonding electrons can move so as to open up the bond angle and reduce the repulsion between them (Fig. 5.8c).

Nonmetal halides. The importance of *electron repulsions near the nucleus of the central atom* is nicely shown by the bond angles in phosphorus trihalide molecules: $PF_3 = 97.8°$, $PCl_3 = 100.3°$, $PBr_3 = 101.5°$, $PI_3 = 102°$.[3] The immediate inclination to ascribe the opening of the bond angle to van der Waals repulsions between the halogens must be rejected. Although the van der Waals radii increase F < Cl < Br < I, the *covalent* radii and hence the P—X bond length also increase in the same order. The two effects cancel each other (see Problem 5.14). The important factor appears to be the ionicity of the P—X bond. The more electronegative fluorine atom attracts the bonding electron pairs away

[2] This is an oversimplification since resonance can take place.

[3] Y. Morino et al. *Inorg. Chem.*, **1969**, *8*, 867.

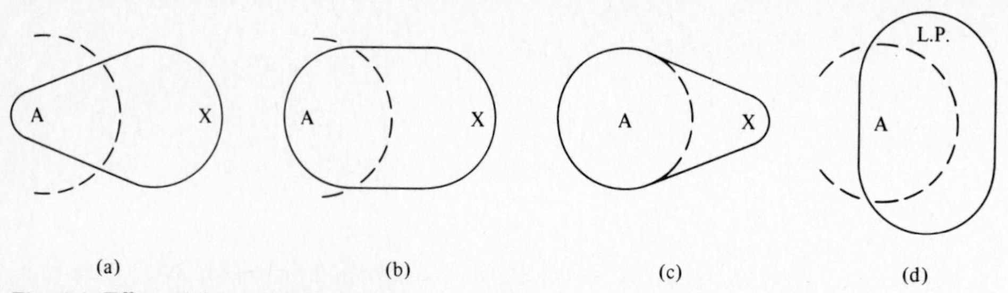

Fig. 5.9 Effect of decreasing electronegativity of X on the size of a bonding pair of electrons. (a) Electronegativity X > A; (b) electronegativity X = A; (c) electronegativity X < A; (d) X = lone pair of electrons, effective electronegativity of X is zero. [From R. J. Gillespie, *J. Chem. Educ.*, **1970**, *47*, 18. Reproduced with permission.]

from the phosphorus nucleus and allows the lone pair to expand while the F—P—F angle closes. Reduced bond angles in nonmetal fluorides are commonly observed. For the small atoms nitrogen and oxygen, where the VSEPR interactions seem to be especially important, *the fluorides have smaller bond angles than the hydrides* (NF_3 = 102.1°, NH_3 = 107.3°, OF_2 = 103.8°, OH_2 = 104.45°). Gillespie[1] has discussed the effect of substituent electronegativity and pointed out that the expansion of lone pairs relative to bonding pairs may be viewed simply as an example of the extreme effect when the nonexistent "substituent" on the lone pair has no electronegativity at all (see Fig. 5.9).

Carbonyl fluoride. Since fluorine and oxygen atoms are about the same size and similar in electronegativity, we might expect OCF_2 to have a rather symmetrical structure. There are no lone pairs on the carbon atom, so to a first approximation we might expect the molecule to be planar with approximately 120° bond angles (Fig. 5.10a). The molecule is indeed planar but distorted rather severely from a symmetrical trigonal arrangement (Fig. 5.10b). It is apparent that the oxygen atom requires considerably more room than the fluorine atoms. There are at least two steric reasons for this. First, the oxygen atom is doubly bonded to the carbon and the C=O bond length (120 pm) is somewhat less than that of C—F (135 pm); thus, the van der Waals repulsion of the oxygen atom will be greater. More important in the present case is the fact that the double bond contains *two* pairs of electrons, and whether viewed as a σ–π pair or twin bent bonds, it is reasonable to assume that they will require more space than a single bonding pair.

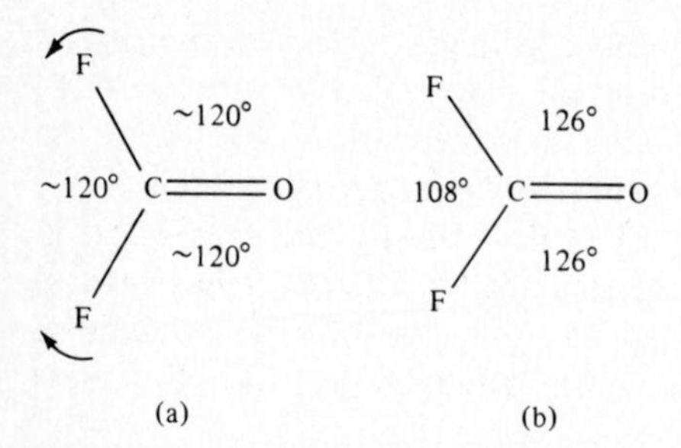

Fig. 5.10 (a) Possible structure of OCF_2. Arrows indicate *small* distortions resulting from electronegativity and size effects. (b) Actual molecular structure of OCF_2. Note very small FCF bond angles.

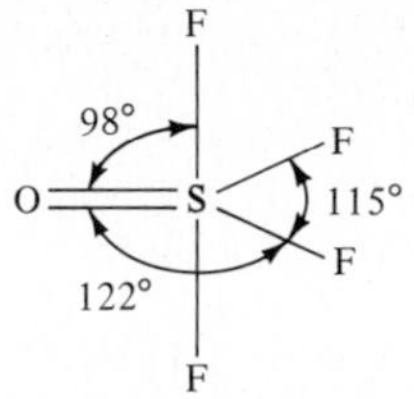

Fig. 5.11 Molecular structure of the OSF_4 molecule.

Table 5.1 Bond angles in molecules containing doubly bonded oxygen and electron lone pairs

Molecule	X—Y—X[a]	Molecule	X—Y—X[a]
$O{=}CF_2$	108°	$:GeF_2$	94 ± 4°
O_2SF_2	96°	$:SF_2$	98°
$O{=}PCl_3$	103.3°	$:PCl_3$	100.3°
$O{=}SF_4$	115,164°	$:SF_4$	101,173°
$O{=}IF_5$	<90°	$:IF_5$	81°

[a] Y = central atom, C, S, P, I; X = halogen atom, Cl, F.

This assumption is strengthened by other compounds with double bonds. In the OSF_4 molecule the doubly bonded oxygen atom seeks the more spacious equatorial position, and the fluorine atoms are bent away somewhat from the other two equatorial and the two axial positions (Fig. 5.11). Further examples are listed in Table 5.1. Note that the behavior of the doubly bonded oxygen atom is in several ways similar to that of a lone pair. Both require more room than a single bonding pair, both seek the equatorial position, and both repel adjacent bonds, thereby distorting the structure. For example, compare the structure of OSF_4 (Fig. 5.11) with that given previously for SF_4 (Fig. 5.4). However, Christe and Oberhammer cite one major difference: Although the lone pair in SF_4 is cylindrically symmetrical and exerts the same effect in all directions, the double bond, because of the π cloud, is more directional. Thus in $:SF_4$ the F—S—F bond angles are 101° in the equatorial plane and 173° in the axial plane for comparable "distortions" from perfect trigonal bipyramidal geometry of $-7°$ and $-9°$. In contrast, in $O{=}SF_4$ the same bond angles are 115° ($-5°$) in the equatorial plane and 164° ($-16°$) in the axial plane containing the S—O π bond. The greater distortion in the latter case is obvious. The interested reader is referred to the original paper for further discussion.[4]

No discussion of the VSEPR model of molecular structure would be complete without a brief discussion of some problems remaining. One of the most interesting current problems centers around the structure of XeF_6. The simplest MO treatment of this molecule predicted that the molecule would be perfectly octahedral.[5] In contrast, the VSEPR model considers the fact that there will be seven pairs of electrons in the valence shell (six bonding pairs

[4] K. O. Christe and H. Oberhammer, *Inorg. Chem.*, **1981**, *20*, 296.

[5] A discussion of MO orbital theory applied to noble gas compounds will be found in Chapter 15.

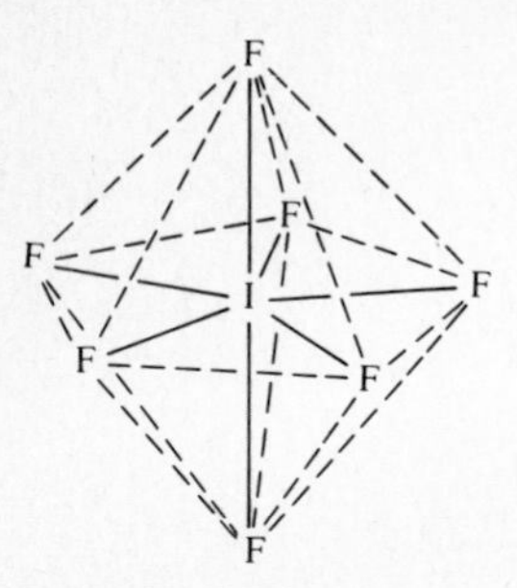

Fig. 5.12 Molecular structure of iodine heptafluoride.

and one lone pair) and predicts a structure based on 7-coordination.[6] Unfortunately, we have little to guide us in choosing the preferred arrangement. Gillespie suggested three possibilities for XeF_6: a distorted pentagonal bipyramid, a distorted octahedron, or a distorted trigonal prism. The lone pair should occupy a definite geometric position and a volume as great as or greater than a bonding pair. Unfortunately, only three neutral molecules with seven bonding pairs of electrons are known: IF_7, ReF_7, and OsF_7. The structures are known with varying degrees of certainty, but all three appear to have approximate D_{5h} symmetry, a distorted trigonal bipyramid (Fig. 5.12).[7] Unfortunately, knowing the MF_7 structures was of little help in studying XeF_6, for the pentagonal bipyramid was the first structure to be experimentally eliminated as a possibility.

Determining the exact structure of the gaseous XeF_6 molecule has proved to be unexpectedly difficult. It is known to be a slightly distorted octahedron. In contrast to the molecules discussed previously, however, *the lone pair appears to occupy less space than the bonding pairs.* The best model for the molecule appears to be a distorted octahedron in which the lone pair extends either through a face (C_{3v} symmetry) or through an edge (C_{2v} symmetry).[8,9] (See Fig. 5.13.) In the latter case the molecule would bear some resemblance to a distorted pentagonal bipyramid (cf. IF_7). The conformation of lowest energy appears to be that of C_{3v} symmetry. Part of the experimental difficulties stems from the fact that the molecule is highly dynamic and probably passes through several conformations. In either of the two models shown in Fig. 5.13, the Xe—F bonds near the lone pair appear to be somewhat lengthened and distorted away from the lone pair; however, *the distortion is less than would have been expected by the VSEPR model.* That the latter model correctly predicted a distortion at all at a time when others were predicting a purely octahedral molecule (all other hexafluorides such as SF_6 and UF_6 are octahedral) is a signal success, however.

[6] R. J. Gillespie, in "Noble-Gas Compounds," H. H. Hyman, ed., University of Chicago Press, Chicago, **1963**, p. 333.

[7] M. G. B. Drew, *Progr. Inorg. Chem.*, **1977**, *23*, 67.

[8] L. S. Bartell et al., *J. Chem. Phys.*, **1965**, *43*, 2547; R. M. Gavin, Jr., and L. S. Bartell, *J. Chem. Phys.*, **1968**, *48*, 2460, 2466.

[9] K. Seppelt and D. Lentz, *Progr. Inorg. Chem.*, **1982**, *29*, 172–179.

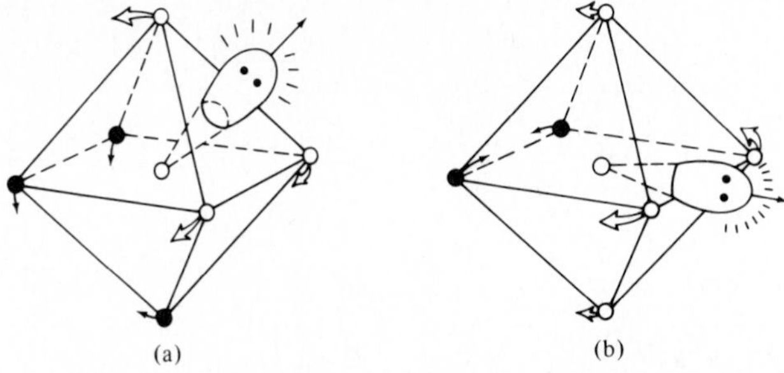

Fig. 5.13 Possible molecular structures of xenon hexafluoride: (a) lone pair emerging through face of the octahedron, C_{3v} symmetry; (b) lone pair emerging through edge of octahedron, C_{2v} symmetry. Arrows show movement of fluorine atoms to relieve strain from lone pair. [From R. M. Gavin, Jr., and L. S. Bartell, *J. Chem. Phys.*, **1968**, *48*, 2466. Reproduced with permission.]

The powerful technique of X-ray diffraction cannot be applied to the resolution of this question since solid XeF_6 polymerizes with a completely different structure (see below).

Finally, the isoelectronic compound $Xe(OTeF_5)_6$ crystallizes as a simple molecular solid, so that it may be studied by X-ray diffraction. Each molecule has C_{3v} symmetry (Fig. 5.13a), providing further support for that structure in XeF_6 as well.[9,10]

Even more puzzling are the structures of anions isoelectronic with XeF_6. Raman spectroscopy of the IF_6^- anion indicates that its symmetry, like that of XeF_6, is lower than octahedral.[11] Other anions such as SbX_6^{-3} and TeX_6^{-2} (X=Cl, Br, or I) have been assigned perfectly octahedral structures on the basis of X-ray crystallography.[12] More recent Raman and infrared spectroscopy has indicated, however, that these ions may be either nonoctahedral or extremely susceptible to deformation.[13] Complexes of Pb(II), As(III), Sb(III), and Bi(III) [electron configuration = $(n-2)f^{14}(n-1)d^{10}ns^2$] with polydentate ligands occupying six coordination sites have been found to have a stereochemically active lone pair.[14] Thus in Fig. 5.14 the six sulfur atoms bonded to the Bi^{+3} ion occupy six of the seven corners of a distorted pentagonal bipyramid. The seventh position, in the equatorial plane, is presumably occupied by the lone pair of the bismuth atom. All of these structures can be related to pentagonal bipyramids or distorted octahedra. Enough have now been characterized to indicate that it is indeed the lone pair that is stereochemically active, not constraints of the ligands.

The dichotomy of behavior of the heavier elements that have a lone pair is reflected in the crystal chemistry of Bi^{+3}. When forced into sites of high symmetry, the Bi^{+3} ion responds by assuming a spherical shape, but in crystals with lower symmetry the lone pair asserts itself and becomes stereochemically active.[14]

---

[10] E. Jacob et al., *Z. Anorg. Allg. Chem.*, **1981**, *427*, 7.

[11] K. O. Christe et al., *Inorg. Chem.*, **1968**, *7*, 626.

[12] S. L. Lawton and R. A. Jacobson, *Inorg. Chem.*, **1966**, *5*, 743, and references therein.

[13] C. J. Adams and A. J. Downs, *Chem. Commun.*, **1970**, 1699.

[14] M. C. Poore and D. R. Russell, *Chem. Commun.*, **1971**, 18; S. L. Lawton and G. T. Kokotailo, *Inorg. Chem.*, **1972**, *11*, 363; S. L. Lawton et al., ibid., **1974**, *13*, 135; R. D. Shannon, *Acta Crystallogr.*, **1976**, *A32*, 751; D. L. Kepert, *Progr. Inorg. Chem.*, **1977**, *23*, 1.

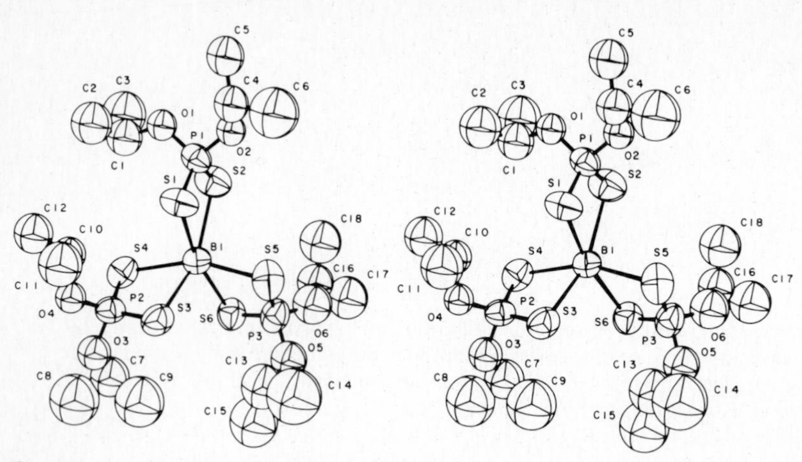

Fig. 5.14 A stereoview of the $Bi[(i\text{-}C_3H_7O)_2PS_2]_3$ molecule. Note that the line S_1-Bi-S_6 approximates an axis for a pentagonal bipyramid and that there is a position open (across from S_3 and S_4) for a stereochemically active lone pair. [From S. L. Lawton et al., *Inorg. Chem.*, **1974**, *13*, 135. Used with permission.]

There appears to be no simple "best" interpretation of the stereochemistry of species with 14 valence electrons. Rather, it should be noted that there appear to be several structures of comparable stability and small forces may tip the balance in favor of one or the other.

Finally, it should also be noted that the XeF_6 molecule exhibits a definite tendency to donate a fluoride ion and form the XeF_5^+ cation, which is isoelectronic and isostructural with IF_5 as expected by the VSEPR model.

The structure[15] of solid XeF_6 is complex, with 144 molecules of XeF_6 per unit cell; however, xenon hexafluoride does not exist as discrete XeF_6 molecules. The simplest way to view the solid is as being composed of pyramidal XeF_5^+ cations extensively bridged by "free" fluoride ions. Obviously, these bridges must contain considerable covalent character. They cause the xenon-containing fragments to cluster into tetrahedral and octahedral units (Fig. 5.15a,b). There are 24 tetrahedra and 8 octahedra per unit cell, packed very efficiently as pseudospheres into a Cu_3Au structure (Fig. 5.15c).[9,16] The structure thus provides us with no information about molecular XeF_6, but it does reinforce the idea that the VSEPR-correct, square pyramidal $:XeF_5^+$ is structurally stable.

Another problem arises with alkaline earth halide molecules, MX_2. These molecules exist only in the gas phase—the solids are ionic lattices (cf. CaF_2, Fig. 3.3). Most MX_2 molecules are linear, but some, such as SrF_2 and BaF_2, are bent.[17] If it is argued that the bonding in these molecules is principally ionic and therefore not covered by the VSEPR model, the problem remains. Electrostatic repulsion of the negative anions should also favor a 180° bond angle. At present, there is no simple explanation of these difficulties.

---

[15] There are actually four phases known of solid xenon hexafluoride.[9,16] All have several structural features in common and the phase described here, phase IV or the cubic phase, is the easiest to describe.

[16] R. D. Burbank and G. R. Jones, *Science*, **1970**, *168*, 248; **1971**, *171*, 485; *J. Am. Chem. Soc.*, **1974**, *96*, 43.

[17] A. Buchler et al., *J. Am. Chem. Soc.*, **1964**, *86*, 4544; V. Calder, et al., *J. Chem. Phys.*, **1969**, *51*, 2093.

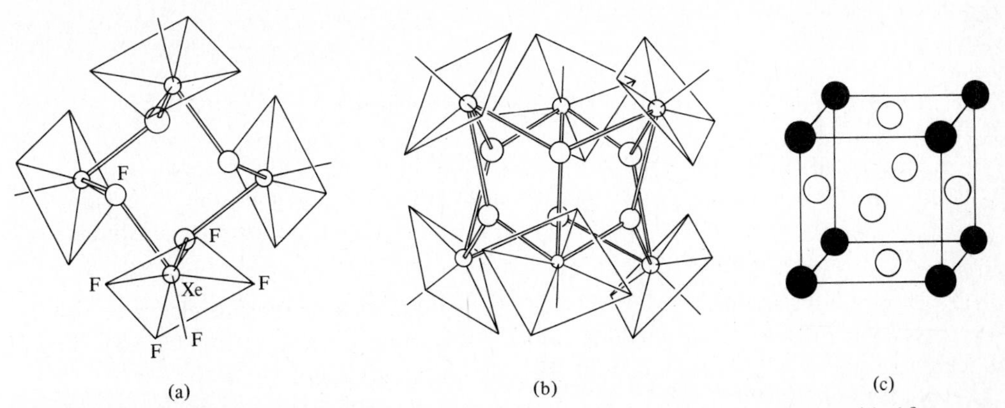

Fig. 5.15 The structure of solid XeF_6. Each Xe atom sits at the base of a square pyramid of five fluorine atoms. The bridging fluorine atoms are shown as larger circles. (a) The tetrameric unit with the Xe atoms forming a tetrahedron; (b) the hexameric unit with the Xe atoms forming an octahedron; (c) the Cu_3Au structure: The shaded circles represent the octahedral clusters and the open circles the tetrahedral clusters. [In part from R. Burbank and G. R. Jones, *J. Am. Chem. Soc.*, **1974**, *96*, 43. Used with permission.]

Summary of VSEPR rules

The preceding can be summed up in a few rules:

1. Electron pairs tend to minimize repulsions. Ideal geometries are:
   a. Coordination number 2 is linear.
   b. Coordination number 3 is trigonal.
   c. Coordination number 4 is tetrahedral.
   d. Coordination number 5 is trigonal bipyramidal.
   e. Coordination number 6 is octahedral.
2. Repulsions are of the order LP–LP > LP–BP > BP–BP.
   a. When lone pairs are present, the bond angles are smaller than predicted by rule 1.
   b. Lone pairs choose the largest site, e.g., equatorial in TBP.
   c. If all sites are equal, lone pairs will be *trans* to each other.
3. Double bonds occupy more space than single bonds.
4. Bonding pairs to electronegative substituents occupy less space than those to more electropositive substituents.

The VSEPR model has been well received by inorganic chemists, but the theoretical basis, although intuitively reasonable, has not received much support from chemists working in quantum mechanics.[18] There has been some criticism of the model as being unnecessary and no more accurate than a listing of empirical rules.[19] To include that discussion and the response to it[20] here would require too much space, but it does suggest one more rule that accounts for the very small bond angles in phosphines, arsines, hydrogen

[18] This situation is extremely reminiscent of the acceptance or not of the participation of *d* orbitals in bonding (see pp. 824–837).

[19] R. S. Drago, *J. Chem. Educ.*, **1973**, *50*, 244.

[20] R. J. Gillespie, *J. Chem. Educ.*, **1974**, *51*, 36.

sulfide, etc., and which is compatible with the energetics of hybridization discussed previously:

5. If the central atom is in the third row or below in the periodic chart, there are two possibilities:
   a. If the substituents are oxygen atoms or halogens, the above rules apply.
   b. If the substituents are less electronegative than halogens, the lone pair will occupy a nonbonding *s* orbital and the bonding will be through *p* orbitals and near 90° bond angles.

Thus phosphine has a bond angle of 93.3°. For substituents of borderline electronegativity it is reasonable to assume intermediate values. Thus Drago[19] suggests that Cl—S—S—Cl is such an example with bond angles of 104°.

Molecular orbitals and molecular structure

Because VB methods deal with easily visualized, localized orbitals, stereochemical arguments (such as the VSEPR model) have tended to be couched in VB terminology. Recently several workers[21,22] have attempted to modify simple LCAO-MO methods to improve their predictive power with respect to geometry and to reconcile some apparent conflicts between MO and VSEPR methods. Detailed consideration of their work is beyond the scope of the present discussion, but it may be noted that a number of approaches yield promising results and show interesting relationships to the VSEPR model. Although none has yet shown the simplicity of the latter, the following brief example will illustrate the general approach.[22] Consider the BeH_2 molecule discussed previously. In the preceding discussion only the filled, bonding orbitals were emphasized although other types of orbitals were mentioned. Figure 5.16 illustrates what happens to the energy of all of the orbitals in BeH_2—bonding, nonbonding, and antibonding—as the molecule is bent. Consider first the bonding σ_g orbital. It is constructed from atomic wave functions that are everywhere positive, and hence on bending there is an increase in overlap since the two hydrogen wave functions will overlap to a slightly greater extent (recall that the wave function of hydrogen never goes to zero despite our diagrammatic representation as a finite circle). The energy of the σ_g orbital (relabeled a_1) is lowered somewhat. In contrast, the energy of the σ_u orbital increases on bending. This is because the wave function changes in sign (as shown by the shading) and overlap of the terminal hydrogen wave function will be negative. In addition, the overlap of the hydrogen atoms with the linear *p* orbital must be poorer in the bent molecule, and so the σ_u orbital (relabeled b_2) will increase more rapidly than that of a_1 will decrease. BeH_2 has a molecular orbital electron configuration $\sigma_g^2\sigma_u^2$, or $a_1^2b_2^2$, and since b_2 loses more energy than a_1 gains, BeH_2 is linear, not bent.

Similar arguments can be applied to the nonbonding and antibonding orbitals (Fig. 5.16; see Problem 5.6). In the water molecule, H_2O, with eight valence electrons the MO configuration will be $\sigma_g^2\sigma_u^2\pi_{ux}^2\pi_{uy}^2$. Since the formerly nonbonding π_{ux} orbital is greatly stabilized (a_1) on bending, the water molecule is bent rather than linear.

---

[21] L. S. Bartell, *Inorg. Chem.*, **1966**, *5*, 1635; *J. Chem. Educ.*, **1968**, *45*, 754; H. B. Thompson, *Inorg. Chem.*, **1968**, *7*, 604; R. M. Gavin, Jr., *J. Chem. Educ.*, **1969**, *46*, 413; R. G. Pearson, *J. Am. Chem. Soc.*, **1969**, *91*, 4947.

[22] B. M. Gimarc, *J. Am. Chem. Soc.*, **1970**, *92*, 266; **1971**, *93*, 593; *Acc. Chem. Res.*, **1974**, *7*, 384. See also B. M. Deb, *J. Am. Chem. Soc.*, **1974**, *96*, 2030, 2044.

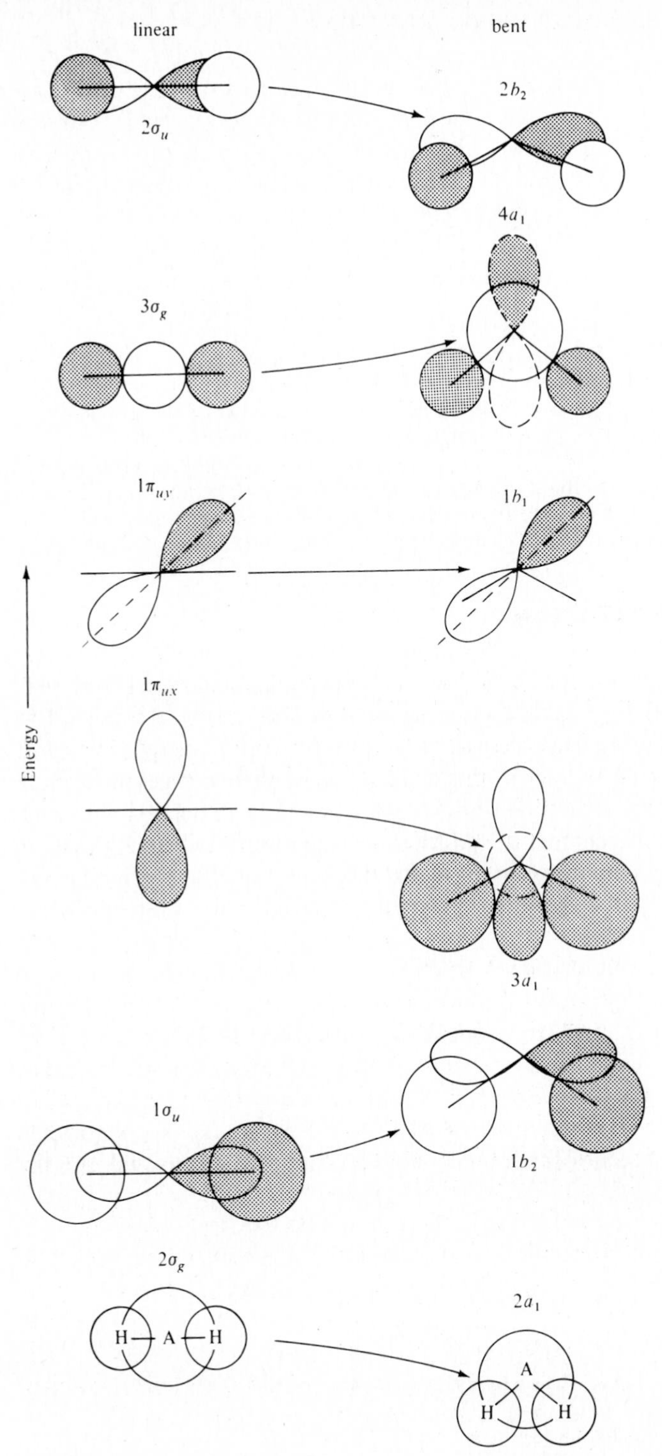

Fig. 5.16 Molecular orbital pictures and qualitative energies of linear and bent AH_2 molecules. Open and shaded areas represent differences in sign (+ or −) of the wave functions. Changes in shape which increase in-phase overlap lower the molecular orbital energy. [From B. M. Gimarc, *J. Am. Chem. Soc.*, **1971**, *93*, 593. Reproduced with permission.]

This brief discussion cannot do justice to the MO approach to stereochemistry, but it does illustrate the reduced importance of electron–electron repulsions (usually omitted in simple approximations) and the increased importance of overlap in this approach. Although the valence bond VSEPR approach and the LCAO-MO approach to stereochemistry appear on the surface to be very different, all valid theories of bonding, when carried sufficiently far, are in agreement that the most stable molecule will be the one which has the best compromise of (1) maximizing electron–nucleus attractions and (2) minimizing electron–electron repulsions.

Interest in accounting for molecular structure seems to be increasing, primarily because of the increase in amount of structural information available for systematization and partly because theoreticians remain unsatisfied with the theoretical foundations of the popular approaches. Methods have ranged from purely formal rules and definitions[23] to applications of second order or pseudo-Jahn-Teller theory,[24] electrostatic force theory,[25] and correlations with *ab initio* calculations.[26] These methods are beyond the scope of this text, but it may be noted that none has the simplicity of the VSEPR model and a few do not cover some geometries important to inorganic chemistry, such as trigonal bipyramidal and square pyramidal.

A CLOSER LOOK AT HYBRIDIZATION

As we have seen in Chapter 3, it is not correct to say that a particular structure is "caused" by a particular hybridization, though such factors as overlap and energy are related to hybridization. We have also seen the usefulness of viewing structures in terms of VSEPR. We shall encounter yet further factors later in this chapter. Nevertheless it is appropriate and useful to note here that certain structures and hybridizations are associated with each other. Some of the most common geometries and their corresponding hybridizations are shown in Fig. 5.17. In addition, there are many hybridizations possible for higher coordination numbers, but they are less frequently encountered and will be introduced as needed in later discussions.

The possible structures may be classified in terms of the coordination number of the central atom and the symmetry[27] of the resulting molecule (Fig. 5.17). Two groups about a central atom will form angular (p^2 orbitals, C_{2v} symmetry) or linear (sp hybrid, $D_{\infty h}$ symmetry) molecules; three will form pyramidal (p^3, C_{3v}) or trigonal planar (sp^2, D_{3h}) molecules; four will usually form tetrahedral (sp^3, T_d) or square planar (dsp^2, D_{4h}); five usually form a trigonal bipyramidal (D_{3h}), more rarely a square pyramidal (C_{4v}) molecule (both dsp^3 hybrids, but using different orbitals, see Table 5.2); and six groups will usually form an octahedral molecule (d^2sp^3, O_h).[28]

---

[23] J. F. Liebman, *J. Am. Chem. Soc.*, **1974**, *96*, 3053; J. F. Liebman and J. S. Vincent, ibid., **1975**, *97*, 1373; J. F. Liebman et al., ibid., **1976**, *98*, 5115.

[24] L. S. Bartell, *J. Chem. Educ.*, **1968**, *45*, 754; R. G. Pearson, *J. Am. Chem. Soc.*, **1969**, *91*, 1252, 4947; *J. Chem. Phys.*, **1970**, *52*, 2167; **1970**, *53*, 2986; *Chem. Phys. Lett.*, **1970**, *10*, 31.

[25] H. Nakatsuji, *J. Am. Chem. Soc.*, **1973**, *95*, 345, 354, 2084.

[26] G. W. Schnuelle and R. G. Parr, *J. Am. Chem. Soc.*, **1972**, *94*, 8974; C. A. Naleway and M. E. Schwartz, ibid., **1973**, *95*, 8235.

[27] See Appendix B for point groups and symmetry.

[28] These are obviously idealized structures with many modifying factors that affect the actual molecular geometry.

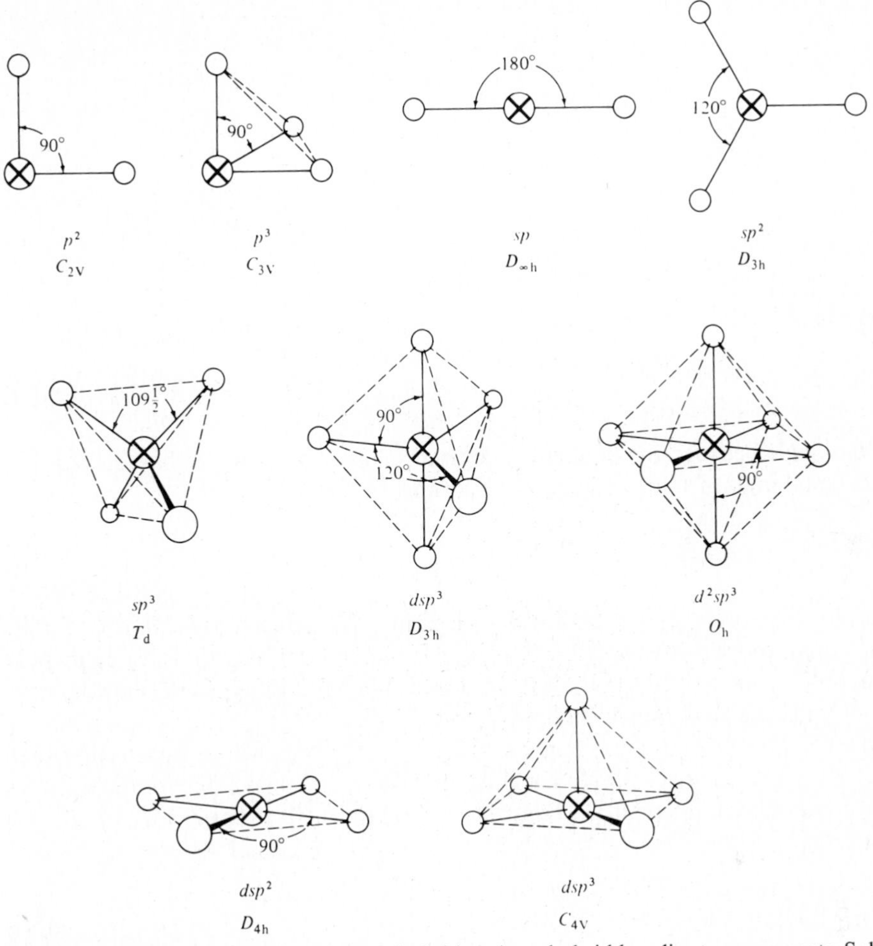

Fig. 5.17 Geometries of some common hybrid and nonhybrid bonding arrangements. Solid lines represent bonds formed from the orbitals on the central atom ⊗. The dashed lines are geometric lines added for perspective.

Most hybridizations result in *equivalent* hybrid orbitals; i.e., all the hybrid orbitals are identical in composition (% *s* and % *p* character) and in spatial orientation with respect to each other. They have very high symmetries, culminating in tetrahedral and octahedral symmetry. In the case of dsp^3 hybrid orbitals, the resulting orbitals are *not* equivalent. We shall see (p. 230) that the trigonal bipyramidal hybridization results in three strong equatorial bonds (sp^2 trigonal orbitals) and two weaker axial bonds (*dp* linear orbitals). The square pyramidal hybridization is approximately a square planar dsp^2 set plus the p_z orbital. As in the trigonal bipyramidal hybridization, the bond lengths and strengths are different (p. 472).

The relation between hybridization and bond angle is simple for *s*–*p* hybrids. For two or more equivalent orbitals, the percent *s* character (*S*) or percent *p* character (*P*) is given

Table 5.2 Component atomic orbitals involved in hybrid orbital formation

Hybridization	Atomic orbitals
sp, sp^2, sp^3	s + arbitrary p^n
dsp^2	$d_{x^2-y^2} + s + p_x + p_y$
dsp^3 (TBP)	$d_{z^2} + s + p_x + p_y + p_z$
dsp^3 (SP)	$d_{x^2-y^2} + s + p_x + p_y + p_z$
d^2sp^3	$d_{x^2-y^2} + d_{z^2} + p_x + p_y + p_z$

by the relationship:[29]

$$\cos\theta = \frac{S}{S-1} = \frac{P-1}{P} \tag{5.1}$$

where θ is the angle between the equivalent orbitals (°) and the s and p character are expressed as decimal fractions. In methane, for example,

$$\cos\theta = \frac{0.25}{-0.75} = -0.333; \qquad \theta = 109.5^\circ \tag{5.2}$$

In hybridizations involving nonequivalent hybrid orbitals, such as sp^3d, it is usually possible to resolve the set of hybrid orbitals into subsets of orbitals that are equivalent within the subset, as the sp^2 subset and the dp subset. We have seen (p. 118) that the nonequivalent hybrids may contain fractional s and p character, e.g., the water molecule which uses bonding orbitals midway between pure p and sp^3 hybrids. For molecules such as this, we can divide the four orbitals into the bonding subset (the bond angle is $104\frac{1}{2}^\circ$) and the nonbonding subset (angle unknown). We can then apply Eq. 5.1 to each subset of equivalent orbitals. In water, for example, the bond angle is $104\frac{1}{2}^\circ$, so:

$$\cos\theta = -0.250 = \frac{P-1}{P} \tag{5.3}$$

$$P = 0.80 = 0.80\%\ p \text{ character and } 20\%\ s \text{ character} \tag{5.4}$$

Now of course the total p character summed over all four orbitals on oxygen must be 3.00 ($= p_x + p_y + p_z$) and the total s character must be 1.00. If the bonding orbitals contain proportionately *more p* character, then the nonbonding orbitals (the two *line pairs*) must contain proportionately *less p* character, 70%: [0.80 + 0.80 + 0.70 + 0.70 = 3.00(p); 0.20 + 0.20 + 0.30 + 0.30 = 1.00(s)]. We can calculate the angle between the axes of the two lone pairs by the same method as used above:

$$\cos\phi = \frac{P-1}{P} = \frac{0.70 - 1.00}{0.70} = -0.429 \tag{5.5}$$

$$\phi = 115\tfrac{1}{2}^\circ \tag{5.6}$$

[29] Equation 5.1 is restricted to molecules such as water in which the angle is known between two equivalent orbitals (e.g., the orbitals bonding the hydrogen atoms). Equivalent and nonequivalent hybrids are discussed further on p. 227.

Although we have no method at present for accurately measuring the angle between the lone pairs, we have good reason from both experiment and VSEPR theory[30] to believe that they do indeed lie at a larger than tetrahedral angle. The opening of some bond angles and closing of others in nominally "tetrahedral" molecules is a common phenomenon. Usually the distortion is only a few degrees, but it should remind us that the terms "trigonal," "tetrahedral," etc., usually are only approximations. *Exactly* trigonal and tetrahedral hybridizations are probably exceedingly rare, restricted to molecules such as BCl_3 and CH_4, where all the substituents on the central atom are identical. We see this distortion epitomized in AL_5-type molecules. Unlike coordination numbers 2, 3, 4,[31] and 6, there is no unique, highly symmetrical set of equivalent orbitals that can be constructed for 5-coordination. Of the two hybridizations shown in Fig. 5.17, most compounds of non-metals favor the trigonal bipyramidal (TBP) structure.[32] Many coordination compounds are known, however, with square pyramidal (SP) structures (see pp. 472–476). More important for the present discussion, however, is the fact that there are many compounds that cannot be classified readily into TBP or SP geometries. Muetterties and Guggenberger[33] have shown that there is a continuous spectrum of compounds ranging from TBP to SP:

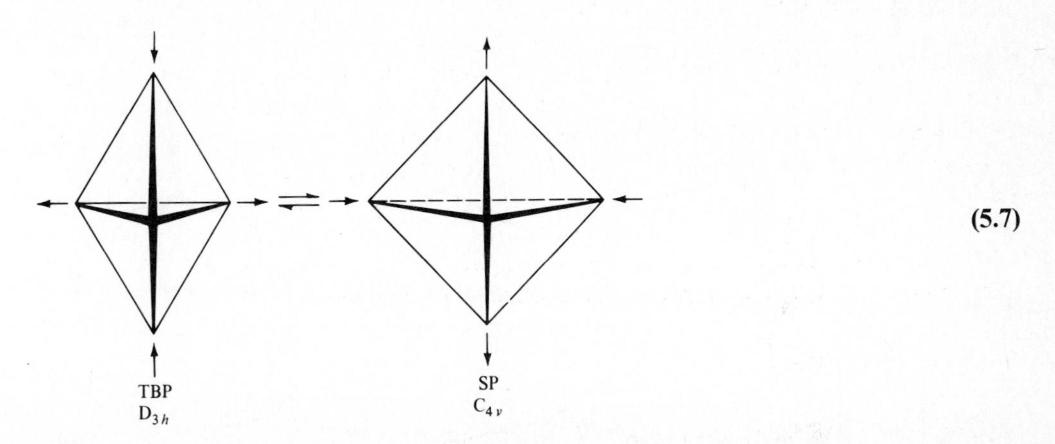

(5.7)

A list of these compounds showing the gradual change from trigonal bipyramidal to square pyramidal geometry is given in Table 5.3 (see Fig. 5.18). The gradual change can be quantified in terms of the dihedral angles between the faces of the polyhedra. For example, in the conversions shown above, when TBP → SP, the dihedral angle of the edge furthest

[30] The experimental method is careful X-ray diffraction similar to that described previously (pp. 72 and 175). [For example, see P. Coppens, *Angew. Chem. Int. Ed. (Eng.)*, **1977**, *16*, 32.] Since VSEPR predicts that lone pairs will occupy more space than bonding pairs, we should expect the angles between them to be larger.

[31] The apparent exception of *two* stable structures, of T_d and D_{4h} symmetry, for coordination number four can be misleading. The square planar structure is known only where there are special stabilizing energies (p. 409).

[32] The three exceptions seem to be pentaphenylantimony, pentacyclopropylantimony, and a bicyclic phosphorane (see Problem 5.15), which are square pyramidal. (P. J. Wheatley, *J. Chem. Soc.*, **1964**, 3718; A. L. Beauchamp et al., *J. Am. Chem. Soc.*, **1968**, *90*, 6675; A. H. Cowley et al., *J. Am. Chem. Soc.*, **1971**, *93*, 2150; J. A. Howard et al., *Chem. Commun.*, **1973**, 856.)

[33] E. L. Muetterties and L. J. Guggenberger, *J. Am. Chem. Soc.*, **1974**, *96*, 1748.

Table 5.3 Ideal and observed angles (deg) in ML_5 complexes

Compound	Dihedral angles (e_1 and e_2)	Dihedral angle (e_3)
Ideal trigonal bipyramid	53.1, 53.1	53.1
$CdCl_5^{-3}$	53.8, 53.8	53.8
$Ni[P(OC_2H_5)_3]_5^{+2}$	54.2, 57.3	50.8
$(C_6H_5)_5P$	51.8, 52.3	45.7
$Co(C_6H_7NO)_5^{+2}$	54.5, 58.5	37.8
$Ni(CN)_5^{-3}$	62.7, 68.6	32.2[a]
$Nb(NC_5H_{10})_5$(Nb2)	65.4, 67.0	23.2
$Nb(NC_5H_{10})_5$(Nb1)	68.6, 70.6	15.8
$Nb(NMe_2)_5$	70.2, 70.2	15.6
$(C_6H_5)_5Sb$	68.5, 69.2	14.4
$Ni(CN)_5^{-3}$	75.0, 79.4	0.3[a]
Ideal tetragonal pyramid	75.7, 75.7	0.0

[a] These two structures occur in different compounds. See "Coordination number 5," Chapter 10.

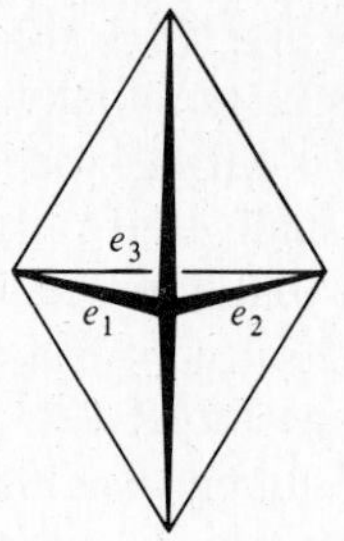

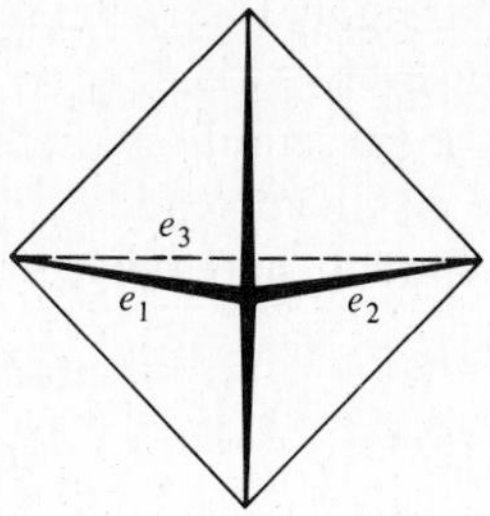

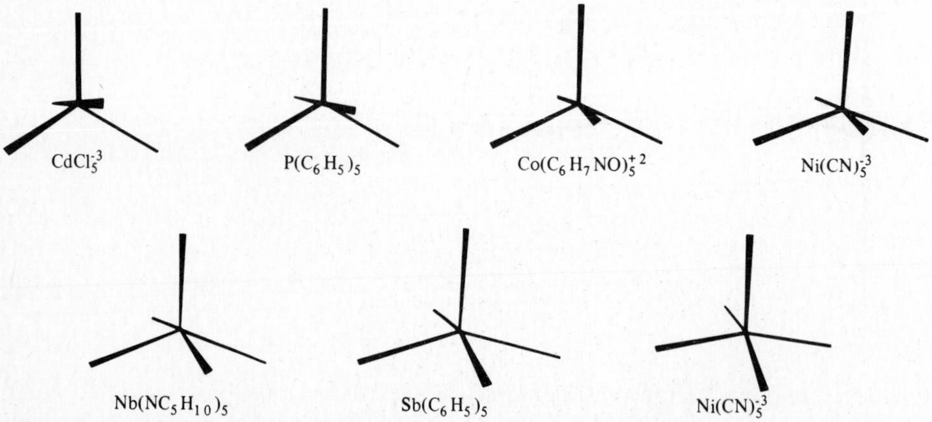

Fig. 5.18 Real five-coordinate molecular structures illustrating intermediates between TBP (D_{3h}) on left to SP (C_{4v}) on the right. [From E. L. Muetterties and L. J. Guggenberger, *J. Am. Chem. Soc.*, **1974**, *96*, 1748. Used with permission.]

from the viewer opens up gradually until it reaches 180°; i.e., the back face of the square pyramid becomes a plane. The angles of the edges (two of which are labeled e_1 and e_2 in the figures accompanying the table) are opened up, approaching right angles (76°) in the pyramid. Now as the reverse change takes place, SP → TBP, edge e_3 reappears as a "real" edge and e_1, e_2, and e_3 all close until they reach identical values in the idealized trigonal bipyramid. The gradual change in these angles as one progresses through the list of compounds in Table 5.3 indicates that just about every possible intermediate between the two limiting geometries is known. When the various substituents are different (and

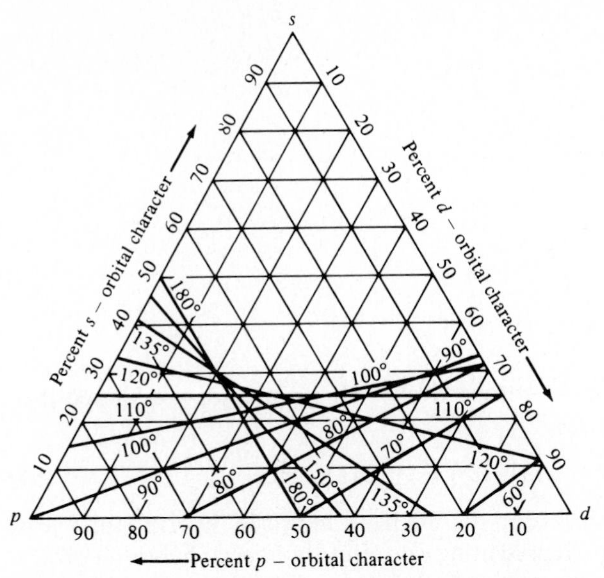

Fig. 5.19 Directional properties of hybrid orbitals from *s*, *p*, and *d* atomic orbitals. [From M. Kasha, adapted from G. Kimball, *Ann. Rev. Phys. Chem.*, **1951**, *2*, 177. Reproduced with permission.]

occasionally even when they are not—see pp. 471–479, these intermediate structures are the rule rather than the exception, and they should warn us to avoid overgeneralizing a structure to make it fit a preconceived pigeonhole. As Muetterties and Guggenberger[33] have said, describing a molecule as "a distorted trigonal bipyramid" conveys little information as to the extent of the distortion and the shape of the molecule. Methods are available for calculating the location of an intermediate on the TBP—SP spectrum, using the dihedral angles shown in Table 5.3. One can thus quantify the extent of distortion from either ideal structure towards the other. Thus, by using the dihedral angles in a particular 5-coordinate complex of zinc, it is possible to describe the intermediate structure as 60% TBP and 40% SP.[34]

The calculation of *s* and *p* character is more difficult if either *d* orbitals participate in the hybridization or if none of the orbitals is equivalent to another to form a subset. In the case of *d* orbitals, the relation between hybridization and bond angles is given in Fig. 5.19, although the accuracy is somewhat less than can be obtained from computation. For completely nonequivalent hybrid orbitals, simultaneous equations involving all of the bond angles and hybridizations may be solved.[35]

Energetics of hybridization

When a set of hybrid orbitals is constructed by a linear combination of atomic orbitals, the energy of the resulting hybrids is a weighted average of the energies of the participating atomic orbitals. For example, in the carbon atom the energies of the 2*s* and 2*p*

[34] W. S. Sheldrick et al., *Acta Crystallogr.*, **1980**, *B36*, 2316; M. C. Kerr et al., *J. Coord. Chem.*, **1981**, *11*, 111.

[35] S. O. Grim et al., *Phosphorus*, **1971**, *1*, 61.

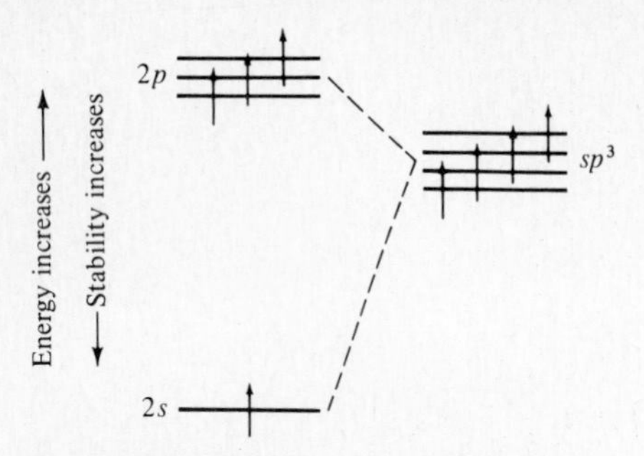

Fig. 5.20 Hybridization of a carbon atom (sp^3).

orbitals are −1878 and −1028 kJ mol^{-1}, respectively.[36] If sp^3 hybrids are formed their energy is:

$$E_{sp^3} = \tfrac{1}{4}(E_s + 3E_p) = \tfrac{1}{4}[-1878 + 3(-1028)] = -1241 \text{ kJ mol}^{-1} \quad (5.8)$$

This is shown graphically in Fig. 5.20. For the methane molecule, hybridization from $2s^12p^3$ to $(2sp^3)^4$ (or $2te^12te^12te^12te^1$, representing one electron in each hybrid *te* orbital) will cost no energy[37] since both configurations have the same total energy:

$$\sum E_{s+p^3} = [-1878 + 3(-1028)] = \sum E_{te} = [4(-1241)] = -4962 \text{ kJ mol}^{-1} \quad (5.9)$$

In contrast, if a phosphorus atom hybridizes, there will be an energy of hybridization involved because a phosphorus atom contains both half-filled and filled orbitals. This may be illustrated as follows. The energies of the phosphorus orbitals are $E_s = -1806$ kJ mol^{-1} and $E_p = -981$ kJ mol^{-1}. The energy of an sp^3 hybrid orbital is then:

$$E_{sp^3} = \tfrac{1}{4}[-1806 + 3(-981)] = -1187 \text{ kJ mol}^{-1} \quad (5.10)$$

Each electron in a particular orbital will have the energy associated with that orbital.[38] The energy of an unhybridized phosphorus atom ($3s^23p^3$) will therefore be:

$$\sum E_{s+p^3} = [2(-1806)] + [3(-981)] = -6555 \text{ kJ mol}^{-1} \quad (5.11)$$

For tetrahedrally hybridized phosphorus ($3te^23te^13te^13te^1$) the energy[39] will be:

$$\sum E_{te} = 5 \times (-1187) = -5935 \text{ kJ mol}^{-1} \quad (5.12)$$

In this case the hybridization has cost 620 kJ mol^{-1} of energy or roughly two "bonds' worth" of energy. This is shown graphically in Fig. 5.21. The energy differential between

[36] These energies represent calculated one-electron ionization energies. (A. Viste and H. Basch in H. B. Gray, "Electrons and Chemical Bonding," Benjamin, New York, 1964).

[37] Note that we are considering a carbon atom that has already been promoted from the ground state ($2s^22p^2$) and promotion energy does not enter here.

[38] This not quite true. The exact energy of an electron also depends upon whether there is a second electron in the same orbital or not. In the present case, however, we are not comparing states with different spin multiplicities.

[39] In both of these calculations only the valence electrons are considered since the energy of the electrons below the valence shell does not change on hybridization.

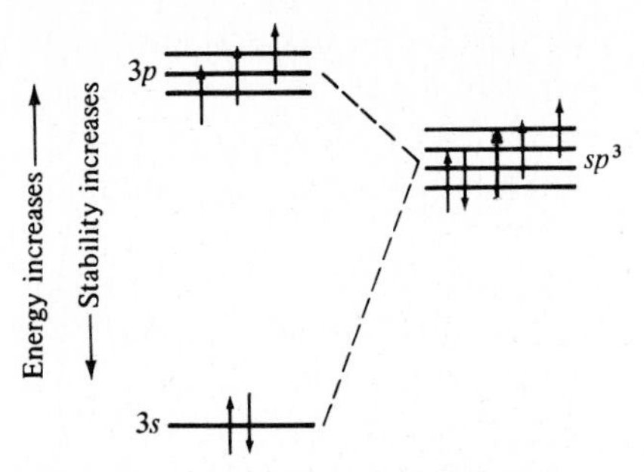

Fig. 5.21 Hybridization of a phosphorus atom (sp^3).

Table 5.4 Bond angles in the hydrides of Groups VA and VIA

NH_3 = 106° 47′	PH_3 = 93° 30′	AsH_3 = 92° 0′	SbH_3 = 91° 30′
H_2O = 104° 27′	H_2S = 92° 16′	H_2Se = 91° 0′	H_2Te = 89° 30′

the hybridized and unhybridized atom represents the increase in energy of the two electrons in the *filled* 3*s* orbital and the decrease in energy of the electrons in the *half-filled* 3*p* orbitals.

The energetics of hybridization, together with the principle of good overlap, are important in determining the electronic structure of molecules. Refer back to Fig. 3.54, which was presented without extensive comment as an example of a heteronuclear diatomic molecule (p. 141). Note first that upon hybridization of the oxygen atom, the more stable hybrid orbital accepts two electrons which become a lone pair on that atom because the orbital is too low in energy to match (and hence overlap efficiently) with any orbital on the carbon atom. In contrast, in the carbon atom the lower *sp* hybrid[40] matches the upper oxygen hybrid to form a strong σ bond. In order for this to happen, promotion must occur to place a single electron in the bonding atomic orbital on carbon. This promotion or hybridization energy is more than compensated by the very strong σ bond formed. A consequence of this hybridization scheme is that the lone pair on the oxygen atom is largely a low-energy, nondirectional *s* orbital, but that on the carbon atom is mostly a high-energy, directional *p* orbital. This has consequences for the chemical behavior of carbon monoxide (see Problem 3.34).

This energy of hybridization (a form of promotion energy) is occasionally important in determining the structure of molecules. For example, the hydrides of the Group VA and VIA elements are found to have bond angles considerably reduced as one progresses from the first element in each group to those that follow (see Table 5.4). An energy factor that favors reduction in bond angle in these compounds is the hybridization discussed above. It costs about 600 kJ mol^{-1} to hybridize the central phosphorus atom. From the standpoint of this energy factor alone, the most stable arrangement would be utilizing pure *p* orbitals in bonding and letting the lone pair "sink" into a pure *s* orbital. Opposing this tendency is the repulsion of electrons, both bonding and nonbonding (VSEPR). This

[40] The usage of *sp* here does not imply a perfect *di* hybrid with 50% *s* character, but simply some hybridization of *s* + *p* of this type. The lower orbital will have more *s* character, the upper one more *p* character.

favors an approximately tetrahedral arrangement. In the case of the elements N and O the steric effects are most pronounced because of the small size of atoms of these elements. In the larger atoms, such as those of P, As, Sb, S, Se, and Te, these effects are somewhat relaxed, allowing the reduced hybridization energy of more *p* character in the bonding orbitals to come into play.

Another factor which affects the most stable arrangement of the atom in a molecule is the variation of bond energy with hybridization. The directed lobes of *s–p* hybrid orbitals overlap more effectively than the undirected *s* orbitals, the two-lobed *p* orbitals, or the diffuse *d* orbitals. The increased overlap results in stronger bonds (p. 119). The molecule is thus sometimes forced to choose[41] between higher promotion energies and good overlap for an *s*-rich[42] hybrid or lower promotion energies and poor overlap for an *s*-poor hybrid.

A good example of the effect of differences in hybrid bond strength is the PCl_5 molecule. An sp^3d hybrid may be considered to be a combination of a $p_zd_z^2$ hybrid and an sp_xp_y hybrid. The former makes two linear hybrid orbitals bonding axially and the latter forms the trigonal, equatorial bonds (see p. 118). The sp^2 hybrid orbitals are capable of forming stronger bonds, and it is found that they are some 15 pm shorter than the longer, weaker axial bonds.[43] When the electronegativities of the substituents on the phosphorus atom differ, as in the mixed chlorofluorides, PCl_xF_{5-x}, and the alkylphosphorus fluorides, R_xPF_{5-x}, it is experimentally observed that the more electronegative substituent occupies the axial position and the less electronegative substituent is equatorially situated. This is an example of *Bent's rule*,[44] which states: *More electronegative substituents "prefer" hybrid orbitals having less s character and more electropositive substituents "prefer" hybrid orbitals having more s character.* Although proposed as an empirical rule, in the chlorofluorides of phosphorus, it is substantiated by molecular orbital calculations.[45]

A second example of Bent's rule is provided by the fluoromethanes. In CH_2F_2 the F—C—F bond angle is less than $109\frac{1}{2}°$, indicating less than 25% *s* character, but the H—C—H bond angle is larger and the C—H bond has more *s* character. The bond angles in the other fluoromethanes yield similar results (Problem 5.5).

The tendency of more electronegative substituents to seek out the low electronegativity $p_zd_{z^2}$ apical orbital in TBP structures is often termed "apicophilicity." It is well illustrated in a series of oxysulfuranes of the type

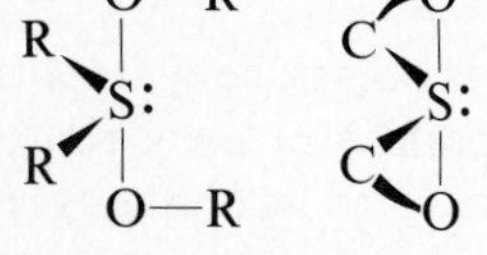

O—R
R
S:
R
O—R

O
C
S:
C
O

---

[41] The molecule, of course, does not "choose," but in order to understand its behavior, the *chemist* must choose the relative importance.

[42] Since most orbitals range from 0% to 50% *s* character, *s*-rich orbitals are those with more than 25% *s* character and *s*-poor are those with less than 25%.

[43] M. Roualt, *Ann. Phys.*, **1940**, *14*, 78.

[44] H. A. Bent, *J. Chem. Educ.*, **1960**, *37*, 616; *J. Chem. Phys.*, **1960**, *33*, 1258, 1259, 1260; *Chem. Rev.*, **1961**, *61*, 275.

[45] P. C. Van Der Voorn and R. S. Drago, *J. Am. Chem. Soc.*, **1966**, *88*, 3255.

prepared by Martin and co-workers.[46] These, as well as related phosphoranes, provide interesting insight into certain molecular rearrangements (see p. 247).

Bent's rule is also consistent with, and may provide alternative rationalization for, Gillespie's VSEPR model. Thus the Bent's-rule predication that highly electronegative substituents will "attract" *p* character and reduce bond angles is compatible with the reduction in angular volume of the bonding pair when held tightly by an electronegative substituent. Strong, *s*-rich covalent bonds require a larger volume in which to bond. Thus, doubly bonded oxygen, despite the high electronegativity of oxygen, seeks *s*-rich orbitals because of the shortness and better overlap of the double bond. Again, the explanation, whether in purely *s*-character terms (Bent's rule) or in larger angular volume for a double bond (VSEPR), predicts the correct structure.

It is sometimes philosophically unsettling to have multiple "explanations" when we are trying to understand what makes molecules behave as they do; it is only human "to want to know for sure" why things happen. On the other hand, alternative, non-conflicting hypotheses give us additional ways of remembering and predicting facts, and if we seem to find them in conflict, we have made either: (1) a mistake (which is good to catch) or (2) a discovery!

The mechanism operating behind Bent's rule is not completely clear. One factor favoring increased *p* character in electronegative substituents is the decreased bond angles of *p* orbitals and the decreased steric requirements of electronegative substituents (p. 213). There may also be an optimum "strategy" of bonding for a molecule in which *s* character (and hence improved overlap) is concentrated in those bonds in which the electronegativity difference is small and covalent bonding is important. The *p* character, if any, is then directed toward bonds to electronegative groups. The latter will result in greater ionic bonding in a situation in which covalent bonding would be low anyway (because of electronegativity differences).[47]

Some light may be shed on the workings of Bent's rule by observations of apparent exceptions to it.[48] The rare exceptions to broadly useful rules are unfortunate with respect to the universal application of those rules. They also have the annoying tendency to be confusing to someone who is encountering the rule for the first time. On the other hand, any such exception or apparent exception is a boon to the research scientist since it almost always provides insight into the mechanism operating behind the rule. Consider the cyclic bromophosphate ester:

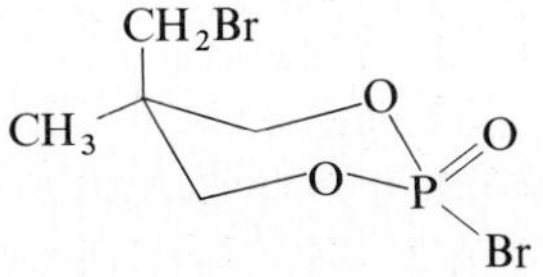

CH_2Br
CH_3 O O O P Br

The phosphorus atom is in an approximately tetrahedral environment using four σ bonds of approximately sp^3 character. We should expect the more electronegative oxygen atoms to bond to *s*-poor orbitals on the phosphorus and the two oxygen atoms in the ring do exhibit hybridizations of about 20%.[36] The most electropositive substituent on the phosphorus is

[46] G. W. Astrologes and J. C. Martin, *J. Am. Chem. Soc.*, **1976**, *98*, 2895, ibid., **1977**, *99*, 4390, 4400; J. C. Martin and E. F. Perozzi, *Science*, **1976**, *191*, 154.

[47] For a further discussion of ionic versus covalent bonding and the total bond energy resulting from the sum of the two, see pp. 142–143, 319–322.

[48] In addition to the example given here, see Problem 5.11.

the bromine atom and Bent's rule would predict an *s*-rich orbital, but instead it draws another *s*-poor orbital, slightly less than 20% *s*. The only *s*-rich orbital on the phosphorus atom is that involved in the σ bond to the exocyclic oxygen. This orbital has nearly 40% *s* character! This oxygen ought to be about as electronegative as the other two, so why the difference? The answer probably lies in the overlap aspect. (1) The large bromine atom has diffuse orbitals that overlap poorly with the relatively small phosphorus atom; thus, even though the bromine is less electronegative than is oxygen, it probably does not form as strong a covalent bond. (2) The presence of a π bond shortens the exocyclic double bond and increases the overlap of the σ orbitals. If molecules respond to increases in overlap by rehybridization in order to profit from it, the increased *s* character then becomes reasonable. From this point of view, Bent's rule might be reworded: *The p character tends to concentrate in orbitals with weak covalency (from either electronegativity or overlap considerations), and s character tends to concentrate in orbitals with strong covalency (matched electronegativities and good overlap).*

Some quantitative support for the above qualitative arguments comes from average bond energies of phosphorus, bromine, and oxygen (Appendix E):

P—Br	264 kJ mol^{-1}
P—O	335 kJ mol^{-1}
P=O	540 kJ mol^{-1}

Bent's rule is a useful tool in inorganic and organic chemistry. It is unfortunate that it hasn't been used more often and more critically. Recently it has been used to supplement the VSEPR interpretation of the structure of various nonmetal fluorides (see p. 215).[49]

Nonbonded repulsions and structure

For anyone who has encountered steric hindrance in organic chemistry, the emphasis thus far placed on electronic effects as almost the only determinant must seem puzzling. However, the preeminence of electronic over steric effects can be rationalized in terms of the following points: (1) Even the largest "inorganic atomic substituent," the iodine atom, is no larger than a methyl group (see Table 6.1 for van der Waals radii), to say nothing of *t*-butyl or di-orthosubstituted phenyl groups. (2) The wider range of hybridizations, the larger variety of atoms of varying electronegativities, and the greater importance of smaller molecules all combine to enhance electronic effects. So much so, in fact, that it is easy to forget nonbonded interactions in the discussions of electron-pair repulsions, overlap, etc., in the preceding sections. Bartell[50] has called attention to the importance of nonbonded repulsions[51] and to the situations in which they may be expected to be important. In general, the latter are apt to be molecules in which a small central atom is surrounded by large substituent atoms. Consider the water molecule, for example, with a bond angle of 104.5°. Replacing the hydrogen atoms with more electronegative halogen atoms should reduce the bond angle in terms of either Bent's rule or VSEPR-electronegativity models (rule 4, p. 219). Indeed, OF_2 has a slightly smaller bond angle, 103.2°.

[49] J. E. Huheey, *Inorg. Chem.*, **1981**, *20*, 4033.

[50] L. S. Bartell, *J. Chem. Educ.*, **1968**, *45*, 754.

[51] This is another example of the principle that if overlap is included, antibonding orbitals are more destabilizing than bonding orbitals are stabilizing. Nonbonded repulsions are merely another name for describing the forced overlap of two orbitals already filled with electrons and the destabilization that occurs as the antibonding orbital rises in energy faster than the bonding one lowers.

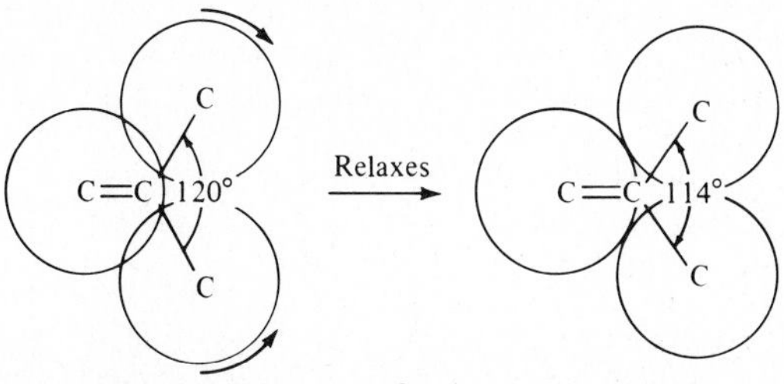

Fig. 5.22 Relaxation of nonbonded repulsions in a perfectly trigonal isobutylene molecule to form 114° bond angle observed experimentally. [From L. S. Bartell, *J. Chem. Educ.*, **1968**, *45*, 754. Used with permission.]

On the other hand, the bond angle in Cl_2O is larger than tetrahedral; it is 110.8°. Similarly, in the haloforms and methylene halides, HCX_3 and H_2CX_2, substitution of fluorine causes the bond angles to decrease (108.6°, 108.3°), but the corresponding chlorine compounds show larger bond angles (111.3°, 111.8°).

An interesting example of this effect is isobutylene, $CH_2{=}C(CH_3)_2$. Naively we might assume substituents on an ethylene to be bonded at 120°. We have seen that VSEPR predicts that the double bond will tend to close down the $CH_3{-}C{-}CH_3$ angle, but how much? Bartell has pointed out that if you assume that the two methyl groups and the methylene group can be portrayed by spheres representing the van der Waals radius of the carbon plus hydrogen substituents, then the short C=C double bond naturally causes repulsions at 120° that can be relaxed if the $CH_3{-}C{-}CH_3$ angle closes. Furthermore, it should close exactly enough to make all repulsions equal; thus the three substituent carbon atoms should lie very nearly on the corners of an equilateral triangle, which they do (Fig. 5.22).[52] Furthermore, it is possible to quantify these qualitative arguments and reproduce the bond angles in some representative hydrocarbons quite well (Fig. 5.23).

Bent bonds

Closely related to the subject of the structures of substituted double bonds is the model used to represent them. This brings us to the topic of "bent bonds." Such bonds are encountered in two contexts which are related but not exactly the same. The first relates to the ring strain of small cyclic molecules. Although rehybridization occurs as ring size decreases, placing more *p* character within the ring and more *s* character in exocyclic bonds, the minimum interorbital angle possible with only *s* and *p* orbitals is 90° (pure *p*). In three-membered rings the orbitals cannot follow the internuclear axis and the so-called σ bonds are not symmetrical about that axis but are distinctly bent. The bending of these bonds can actually be observed experimentally (Fig. 5.24).[53]

The second use of bent bonds is as an alternative interpretation of multiple bonding. There are two methods for interpreting the bonding in multiple-bonded compounds. We may for convenience distinguish the σ–π picture from the bent-bond picture. The former was developed in the context of MO theory, and the latter has been advocated by Pauling

[52] L. S. Bartell and R. A. Bonham, *J. Chem. Phys.*, **1960**, *32*, 824.

[53] P. Coppens, *Angew. Chem. Int. Ed. (Eng.)*, **1977**, *16*, 32.

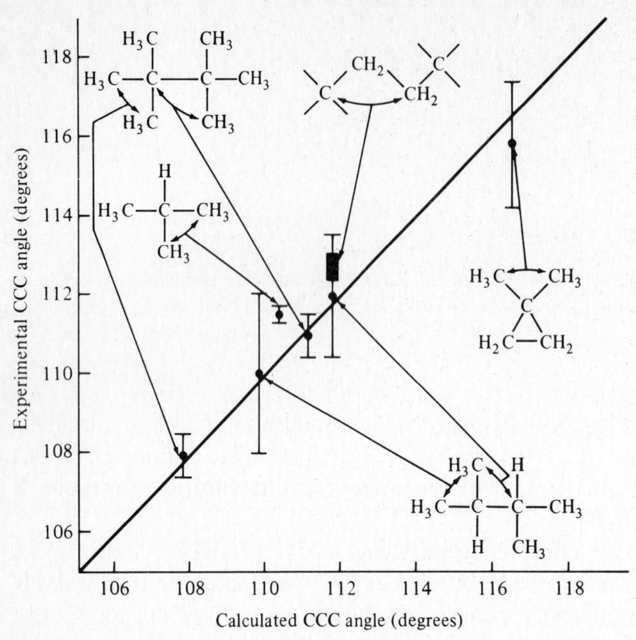

Fig. 5.23 Comparison of calculated C-C-C bond angles with experimental values. Calculations are based on nonbonded interactions. The solid bar represents a range of values found in *n*-alkanes. [From L. S. Bartell. *J. Chem. Educ.*, **1968**, *45*, 754. Used with permission.]

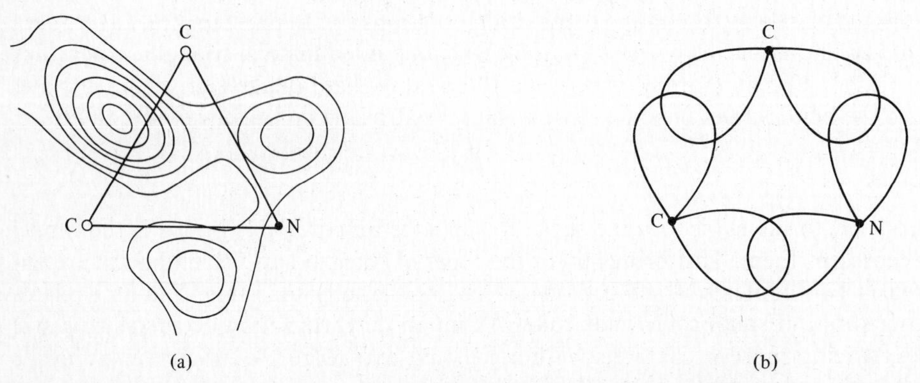

Fig. 5.24 Bonding in a three-membered ring: (a) Build-up of electron density upon bond formation. Note that the bulk of the electron density lies outside the triangle; (b) interpretation of (a) in terms of hybrid orbitals. [In part from T. Ito and T. Sakurai, *Acta Crystallogr.*, **1973**, *B29*, 1594. Reproduced with permission.]

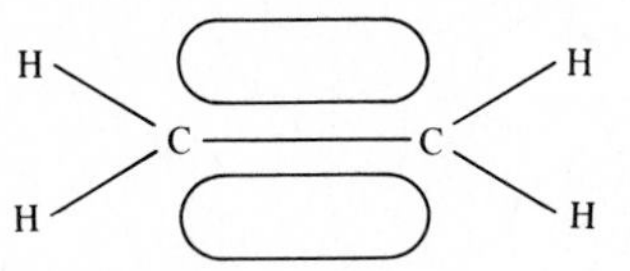

Fig. 5.25 Sigma–pi representation of the ethylene molecule.

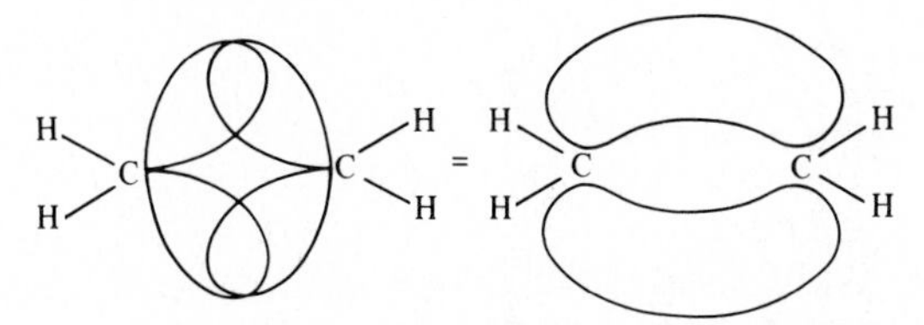

Fig. 5.26 Bent-bond representation of the ethylene molecule.

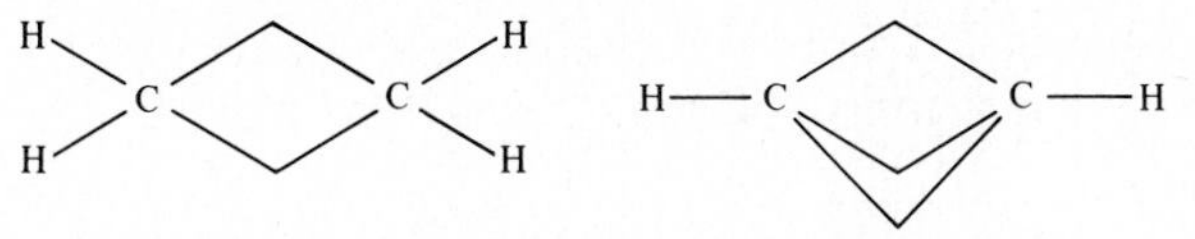

(a)

(b)

Fig. 5.27 Simplified representation of bent bonds in (a) ethylene and (b) acetylene molecules.

in a VB description, but neither is limited to either MO or VB theory.[54] The σ–π description is the one commonly encountered in the discussion of alkenes and alkynes. The double bond in ethylene, for example, may be considered to be a σ bond formed by the head-on overlap of trigonal hybrids on each carbon atom and a π bond from the sideways overlap of the remaining p orbital on each carbon atom (Fig. 5.25). These orbitals will resemble the MOs we have seen previously (Fig. 3.28). The two bonds are not equivalent since they differ in energy as well as symmetry. In accord with the principle of maximum overlap we should expect the σ bond to be stronger since the overlap is greater than in the π bond. If we assume that the σ bond in an ethylene molecule has the same strength as that in an ethane molecule (a good guess, but not exact), it would have a bond energy of about 347 kJ mol^{-1}. The experimental value for the dissociation energy of ethylene (breaking *both* σ and π) is 598 kJ mol^{-1}, indicating about 251 kJ mol^{-1} for the π bond. The difference of some 80 kJ mol^{-1} between the σ bond and the π bond can be attributed to the poorer overlap of the latter.

An alternative method of describing multiple bonds was suggested by Pauling.[55] In this approach the carbon atom is considered to be hybridized tetrahedrally to a first approximation in all compounds, saturated and unsaturated. Multiple bonds are considered to be formed from the oblique overlap of these orbitals, as shown in Figs. 5.26 and 5.27.

---

[54] In the usual qualitative description as presented here, the σ–π method contains aspects of both MO and VB theory.

[55] L. Pauling, "The Nature of the Chemical Bond," 3rd ed., Cornell University Press, Ithaca, N.Y., **1960**, pp. 136–142.

Table 5.5 Bond angles in substituted ethylene molecules and their interpretations in terms of σ–π versus bent bonding[a]

Molecule	Angle XCX	Hybridization				
		σ–π method			Bent-bond method	
		XCX orbitals	C—C orbital	Pi bond	XCX orbitals	C—C orbital
$F_2C{=}CH_2$	FCF = 109.3 ± .3°	2 × 74.8% *p*	1 × 50.4% *p*	100% *p*	2 × 74.8% *p*	2 × 75.2% *p*
$Cl_2C{=}CCl_2$	ClCCl = 113.3°	2 × 71% *p*	1 × 58% *p*	100% *p*	2 × 71% *p*	2 × 79% *p*
$H_2C{=}CH_2$	HCH = 117.6 ± .5°	2 × 68.5% *p*	1 × 63% *p*	100% *p*	2 × 68.5% *p*	2 × 81.5% *p*
$H_2C{=}CCl_2$	ClCCl = 122 ± 2°	2 × 65% *p*	1 × 70% *p*	100% *p*	2 × 65% *p*	2 × 85% *p*
$(H_3C)_2C{=}C(CH_3)_2$	H_3C—C—CH_3 = 137 ± 2°	2 × 57% *p*	1 × 86% *p*	100% *p*	2 × 57% *p*	2 × 93% *p*

[a] All structural data from "Tables of Interatomic Distances and Configurations in Molecules and Ions," L. E. Sutton, ed., Spec. Publ. No. 11 and Supplement, Spec. Publ. No. 18, The Chemical Society, London, **1958**, **1965**.

The resulting bonds are called either "bent bonds" or "banana bonds." Since the two bonds are equivalent, it is possible to assume that each bent bond has an energy of 299 kJ mol^{-1} (one-half the total bond energy). This is intermediate between the values assigned above for σ and π bonds, just as a bent bond is intermediate in overlap. The oblique overlap will be less than in a σ bond, but more than that in the parallel orbitals comprising a π bond.

It may have occurred to the reader that the HCH bond angles in the two models of ethylene will not be identical. The σ–π treatment assumes sp^2 hybridization and a bond angle of 120°. In contrast, the sp^3 bonds in the bent-bond approach would give HCH = $109\frac{1}{2}$°. Does this permit us to use the experimentally obtained bond angle to choose the better of the two models? Unfortunately, no. The apparent difference in predicted bond angles results from the oversimplified assumptions of integral hybrids. There is no compelling reason why the carbon orbitals bonding the hydrogen should have exactly 25% or 33.3% *s* character. Experimentally we find just about every value from 109° to 120° if we investigate a series of ethylene derivatives. Table 5.5 lists a few substituted ethylenes, the experimentally derived bond angles, and the interpretation of these bond angles in terms of the σ–π and the bent-bond picture. As long as fractional hybridization is permitted, both pictures can accommodate the experimental data. In fact, although the σ–π picture and the bent-bond picture seem to give widely differing pictures of double and triple bonds, they are *exactly* equivalent when treated rigorously by quantum-mechanical methods.[56] Their apparent differences result from the inadequacies of our simplified ways of picturing bonding.

The wide variation in bond angle in double-bonded compounds is but a further example of the fact that a molecule will adopt the most favorable geometry available to it and the hybridization will follow. Note that these compounds provide another facet of the structure/hybridization–cause/effect problem. We were not successful in mistakenly attempting to use assumptions about hybridization to predict bond angles in these compounds. However, *all* of the bond angles are consistent with Gillespie's, Bent's, and Bartell's ideas.

BOND LENGTHS

Bond lengths and multiple bonding were discussed in Chapter 3, and a comparison of various types of atomic radii will be discussed in the next chapter, but a short discussion of factors that affect the distance between two bonded atoms will be given here to complement the previous discussion of steric factors.

Bond multiplicity

One of the most obvious factors affecting the distance between two atoms is the bond multiplicity: Single bonds are longer than double bonds which are longer than triple bonds: C—C = 154 pm, C=C = 134 pm, C≡C = 120 pm; N—N = 145 pm, N=N = 125 pm, N≡N = 110 pm; O—O = 148 pm, O=O = 121 pm, etc. For carbon, Pauling[57]

[56] E. A. Walter, *J. Chem. Educ.*, **1966**, *43*, 134. See also F. A. Robinson and R. J. Gillespie, ibid., **1980**, *57*, 329; F. M. Akeroyd, **1982**, *59*, 371.

[57] L. Pauling, "The Nature of the Chemical Bond," 3rd ed., Cornell University Press, Ithaca, N.Y., **1960**, p. 239.

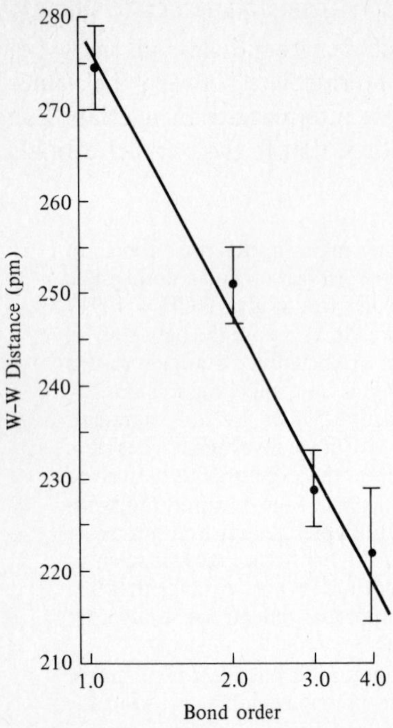

Fig. 5.28 Tungsten-tungsten bond lengths as a function of bond order. Each of the points represents the mean and range of several values which will be discussed at greater length in Chapter 14. Bond orders are plotted on a logarithmic scale in accord with Eq. 5.13.

has derived the following empirical relationship between bond length (D in pm) and bond order (n):

$$D_n = D_1 - 71 \log n \tag{5.13}$$

This relationship holds not only for integral bond orders but also for fractional ones (in molecules with resonance, etc.). One can thus assign variable bond order depending upon the length of the bond.

In view of the many factors, to be discussed below, affecting bond lengths, it does not seem wise to attempt to quantify the bond order–bond length relationship accurately, Nevertheless, bonds formed by elements other than carbon show similar trends (Fig. 5.28), and the general concept certainly is a valid one.

We have seen in Chapter 3 that the strength of a bond depends to a certain extent upon the hybridizations of the atoms forming the bond. We should therefore expect bond length to vary with hybridization. Bent[58] has shown that this variation is quite regular: C—C bond lengths are proportional to p character (Fig. 5.29) or, to say it another way, increasing s character increases overlap and bond strength and thus shortens bonds.

[58] H. A. Bent, *Chem. Rev.*, **1961**, *61*, 275.

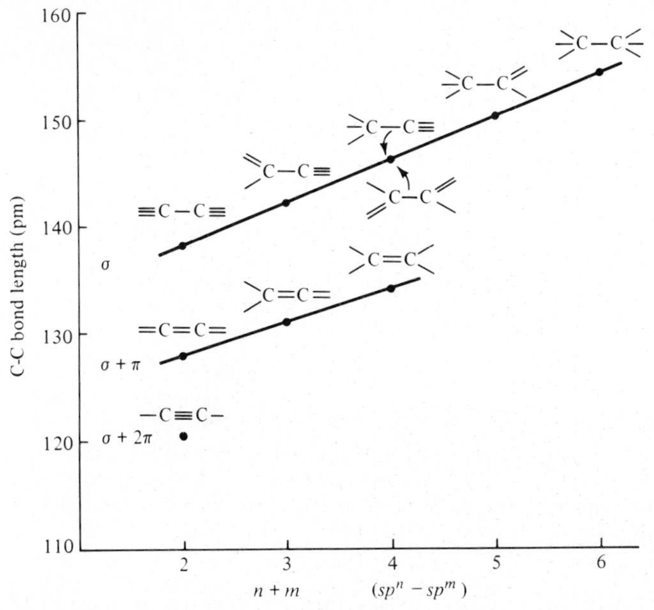

Fig. 5.29 Carbon-carbon bond lengths as a function of the hybridizations of the bonding atoms. [Data from H. A. Bent, *Chem. Rev.*, **1961**, *61*, 283.]

Another factor that affects bond length is electronegativity. Bonds tend to be shortened, relative to the expectations for nonpolar bonds, in proportion to the electronegativity difference of the component atoms. Thus the experimental bond length in HF is 91.8 pm versus an expected value of 108 pm. This topic will be discussed further in Chapter 6.

EXPERIMENTAL DETERMINATION OF MOLECULAR STRUCTURE

It is impossible to present the theory and practice of the various methods of determining molecular structure completely. No attempt will be made here to go into these methods in depth, but a general feeling of the importance of the different techniques can be gathered together with their strengths and shortcomings. For more material on these subjects, the reader is referred to texts on the application of physical methods to inorganic chemistry.[59]

[59] See R. S. Drago, "Physical Methods of Inorganic Chemistry," Van Nostrand–Reinhold, New York, **1965**; H. O. Hill and P. Day, "Physical Methods in Advanced Inorganic Chemistry," Wiley (Interscience), New York, **1968**; W. L. Jolly, "The Synthesis and Characterization of Inorganic Compounds," Prentice-Hall, Englewood Cliffs, N.J., **1970**; R. S. Drago, "Physical Methods in Chemistry," Saunders, Philadelphia, **1977**. See also Appendix G., pp. 1007–1009.

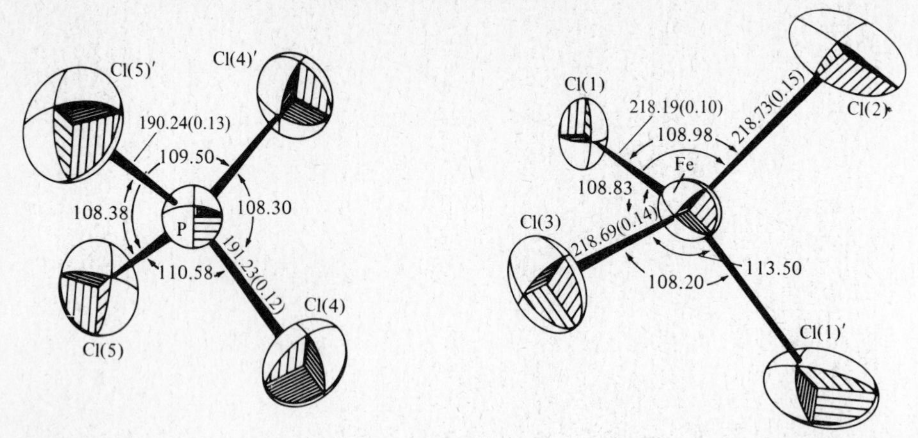

Fig. 5.30 Molecular structure of the ions in $PCl_4^+FeCl_4^-$ illustrating the thermal ellipsoids of motion. Note that the ellipsoids of the chlorine atoms tend to be larger than those of the phosphorus and iron atoms. All measurements in picometres. [From T. J. Kistenmacher and G. D. Stucky, *Inorg. Chem.* **1968**, *7*, 2150. Reproduced with permission.]

X-ray diffraction

X-ray diffraction has provided more structural information for the inorganic chemist than any other technique. It allows the precise measurement of bond angles and bond lengths. Unfortunately, in the past it was a time-consuming and difficult process, and molecular structures were solved only when there was reason to believe they would be worth the considerable effort involved. The advent of more efficient methods of gathering data and doing the computations has made it relatively easy to solve most structures.

In order to solve a structure by X-ray diffraction one generally needs a single crystal. Although powder data can provide "fingerprint" information and, in simple cases, considerable data, it is necessary to be able to grow crystals for more extensive analysis. X rays are diffracted by electrons, and so what are located are the centers of electron clouds, mainly the core electrons. This has two important consequences. First, if there is a great disparity in atomic number between the heavy and light atoms in a molecule, it may not be possible to locate the light atom (especially if it is hydrogen), or to locate it as accurately as the heavier atom. Secondly, there is a small but systematic tendency for the hydrogen atom to appear to be shifted 10–20 pm *toward* the atom to which it is bonded.[60] This is because hydrogen is unique in not having a core centered on the nucleus (which is what we are seeking) and the bonding electrons are concentrated toward the binding atom.

Since the location of an atom in a molecule as obtained by X rays is the time average of all positions it occupied while the structure was being determined, the resultant structure is often presented in terms of *thermal ellipsoids*, which are probability indicators of where the atoms are most likely to be found (see Fig. 5.30). Occasionally, from the size and orientation of an ellipsoid, something may be ventured on the bonding in a molecule. Obviously,

[60] P. Coppens in "Electronic Structure of Polymers and Molecular Crystals," J-M. André and J. Ladik, eds., Plenum, New York and London, **1974**, p. 232; *Angew. Chem. Int. Ed.* (*Eng.*), **1977**, *16*, 32.

if the ellipsoid is prolate (football shaped), the motion of the atom is mostly back and forth, and if oblate (curling-stone shaped), the motion is mostly wagging or wobbling about its bond axis.

Great improvements have been made in equipment, computer programs, and techniques in general. These have reduced the effort necessary for structure solution and improved the accuracy of the resultant structure. However, certain limitations remain which require that greater judgment be used in interpretation of the data than might appear obvious from their apparent accuracy. For example, consider $PCl_4^+FeCl_4^-$, illustrated in Fig. 5.30. The simplicity and symmetry of the ions, as well as the absence of light atoms such as hydrogen that scatter poorly, contribute to solution with an excellent *R*-factor[61] of 0.027. Nevertheless, assumptions made concerning the thermal motions of the atoms makes the uncertainty in bond lengths greater than the 0.1 pm standard deviation shown in Fig. 5.30.

Neutron diffraction is very similar in principle to X-ray diffraction. However, it differs in two important characteristics. Since neutrons are diffracted by the nuclei, one indeed locates the latter directly. Furthermore, the hydrogen nucleus is a good scatterer; thus the hydrogen atoms can be located easily and precisely. The chief drawback of neutron diffraction is that one must have a source of neutrons, and so the method is expensive and not readily available. The two methods may be used in tandem to obtain extremely useful results (cf. Fig. 10.33).

Methods based on molecular symmetry

Since a molecule with a center of symmetry (or certain other highly symmetrical arrangements) cannot have a dipole moment, no matter how polar the individual bonds, dipole moments have proved to be useful in distinguishing between two structures. Much of the classic chemistry of square planar coordination compounds of the MA_2B_2 type was elucidated on the basis of dipole moments present in the *cis* isomers and their absence in the *trans* isomers (see p. 470).[62]

Both infrared (IR) and Raman spectroscopy have selection rules based on the symmetry of the molecule. As we have seen, any molecular vibration that results in a change of dipole moment is infrared active. For a vibration to be Raman active, there must be a change of polarizability of the molecule as the transition occurs. It is thus possible to determine which modes will be either IR or Raman active, or both, from the symmetry of the molecule.[63] In general, these two modes of spectroscopy are complementary, and specifically, if a molecule has a center of symmetry, no IR-active vibration is also Raman active.

---

[61] In a least squares refinement, one minimizes the summation of the squares of the residuals between observed and calculated values. The crystallographer's *R*-factor is a normalized sum of residuals; $R \leq 0.05$ is usually quite acceptable.

[62] For further examples of the use of dipole moments in structure analysis, see G. J. Moody and J. D. R. Thomas, "Dipole Moments in Inorganic Chemistry," Edward Arnold, London, **1971**.

[63] For methods, see M. Orchin and H. H. Jaffé, "Symmetry, Orbitals, and Spectra," Wiley, New York, **1971**; W. L. Jolly, "The Synthesis and Characterization of Inorganic Compounds," Prentice-Hall, Englewood Cliffs, N.J., **1970**; R. S. Drago, "Physical Methods in Inorganic Chemistry," Van Nostrand–Reinhold, New York, **1965**.

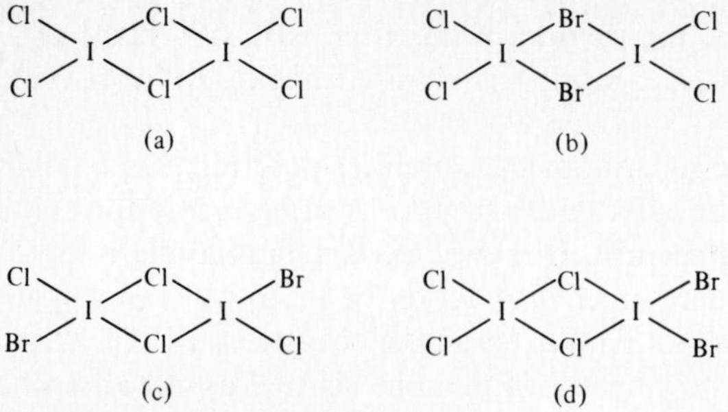

(a) (b)

(c) (d)

Fig. 5.31 Possible structures of iodine trihalides. Note that structure (d) has a different environment for each of the iodine atoms.

Methods using equivalence or nonequivalence of atoms

There are many methods that give spectroscopic shifts, usually called chemical shifts, depending upon the electronic environment of the atoms involved. For example, we saw in Chapter 3 that X-ray photoelectron spectroscopy indicates that there are four different types of carbon atoms in ethyl trifluoroacetate (Fig. 3.69) and three different nitrogen atoms in the cobalt complex (Fig. 3.70). In the same way, nuclear magnetic resonance (NMR) distinguishes two types of protons in ethyl trifluoroacetate, the three in the methyl group and two further downfield in the methylene group. In NMR, further structural information can be obtained from the fact that the methyl proton signal is split into a triplet and the methylene signal split into a quartet.[64] Other nuclei such as ^{11}B, ^{14}N, ^{19}F, ^{29}Si, and ^{31}P may be studied by NMR techniques. For example, ^{19}FNMR indicates that all of the fluorine atoms in PF_5 are magnetically equivalent. This means that: (1) All of the P—F bonds are identical, or (2) the fluorine atoms are interchanging positions faster than the NMR technique can follow. Since we have seen (p. 207) that it is structurally impossible to have a three-dimensional arrangement of five equivalent points in space, we are led to accept the second alternative. This process of interconversion will be discussed later in this chapter.

Mössbauer spectroscopy, simple infrared group frequencies, and other techniques may be applied to the resolution of molecular structure. Several examples of the various methods will be given at appropriate places in the text.[65] For now, the general method may be illustrated by the Mössbauer elucidation of the structure of $I_2Cl_4Br_2$. By analogy with I_2Cl_6 (Fig. 5.31a),[66] we might expect a bridged structure with either chloro or bromo bridges (Fig. 5.31b–d). The experimental result[67] that the two iodine atoms are in different

[64] This is analogous to part of the spectrum of C_2H_5OH or that of C_2H_5X, used almost universally to introduce this topic in organic chemistry. Since it is thoroughly covered in most organic courses, space will not be devoted to it here. See, for example, R. T. Morrison and R. N. Boyd, "Organic Chemistry," 3rd ed., Allyn & Bacon, Boston, **1973**, pp. 416–430.

[65] These are all listed under "Physical methods" in the index.

[66] Interhalogen compounds are discussed in Chapter 15, pp. 774–777.

[67] M. Pasternak and T. Sonnino, *J. Chem. Phys.*, **1968**, *48*, 1997.

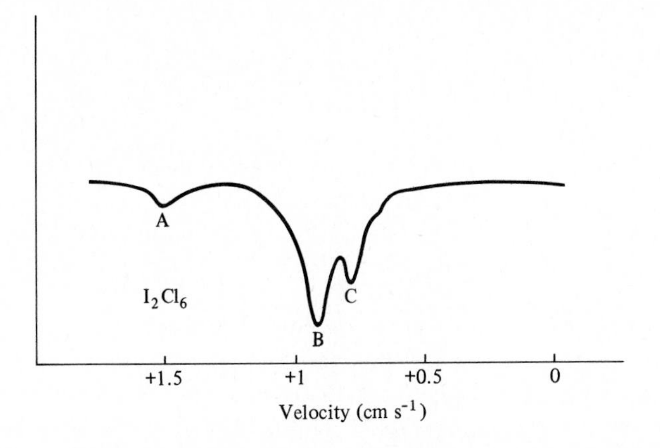

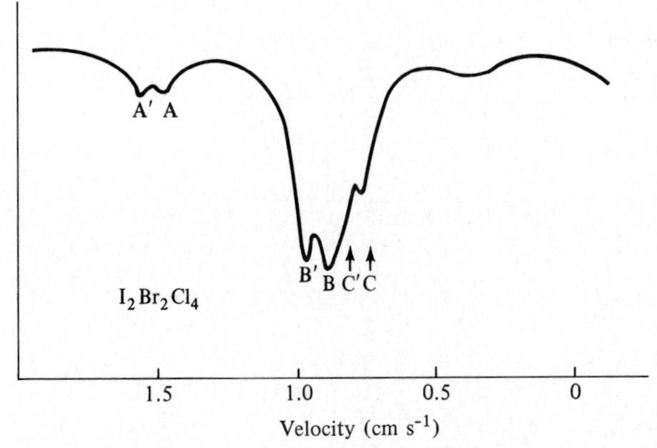

Fig. 5.32 Partial Mössbauer spectra of I_2Cl_6 (top) and $I_2Br_2Cl_4$ (bottom). Note splitting of peak A in I_2Cl_6 into A and A′ in $I_2Br_2Cl_4$. Presumably C′ is hidden under a shoulder of B. [From M. Pasternak and T. Sonino, *J. Chem. Phys.*, **1968**, *48*, 1997. Reproduced with permission.]

environments (Fig. 5.32) rules out the symmetrical structures shown in Fig. 5.31b,c and strongly suggests that the correct structure is Fig. 5.31d.

Summary on structural methods

This has been a brief survey of some of the methods available to the inorganic chemist for the determination of structure. Further examples will be encountered later in the text illustrating methods. A useful summary of some of the methods of structure determination has been provided by Beattie,[68] listing some of the characteristics that have been discussed above as well as some other features of their use (Table 5.6).

[68] I. R. Beattie, *Chem. Soc. Rev.*, **1975**, *4*, 107. The interested reader should also refer to "Physical Methods in Chemistry," by R. S. Drago, Saunders, Philadelphia, **1977**.

Table 5.6 Comparison of some physical techniques for structural studies

Technique	Nature of the effect	Information	Interaction time	Sensitivity	Comments
X-ray diffraction	Scattering, mainly by electrons, followed by interference ($\lambda = 0.01$–1 nm)	Electron density map of crystal	10^{-18} s but averaged over vibrational motion	crystal $\sim 10^{-3}$ cm^3	Location of light atoms or distinction between atoms of similar scattering factor difficult in presence of heavy atoms
Neutron diffraction	Scattering, mainly by nuclei, followed by interference ($\lambda = 0.1$ nm)	Vector internuclear distances	10^{-18} s but averaged over vibrational motion	crystal ~ 1 cm^3	Extensively used to locate hydrogen atoms. May give additional information due to spin $\frac{1}{2}$ on neutron leading to magnetic scattering
Electron diffraction	Diffraction (atom or molecule) mainly by nuclei, but also by electrons ($\lambda = 0.01$–0.1 nm)	Scalar distances due to random orientation	10^{-18} s but averaged over vibrational motion	100 Pa	Thermal motions cause blurring of distances. Preferably only one (small) species present. Heavy atoms easy to detect
Microwave	Absorption of radiation due to dipole change during rotation ($\lambda = 0.1$–30 cm; 300–1 GHz in frequency)	Mean value of r^{-2} terms; potential function	10^{-10} s	10^{-2} Pa	Mean value of r^{-2} does not occur at r_e even for harmonic motion. Dipole moment necessary. Only one component may be detected. Analysis difficult for large molecules of low symmetry
Vibrational infrared	Absorption of radiation due to dipole change during vibration ($\lambda = 10^{-1}$–10^{-4} cm)	Qualitative for large molecules	10^{-13} s	100 Pa	Useful for characterization. Some structural information from number of bands, position and possibly isotope effects. All states of matter

Vibrational Raman	Scattering of radiation with changed frequency due to polarizability change during a vibration (λ = visible usually)	Qualitative for large molecules	10^{-14} s	10^4 Pa (v^4 dependent)	Useful for characterization. Some structural information from number of bands, position, depolarization ratios, and possibly isotope effects. All states of matter
Electronic	Absorption of radiation due to dipole change during an electronic transition ($\lambda = 10$–10^2 nm)	Qualitative for large molecules	10^{-15} s	1 Pa	Useful for characterization. Some structural information from number of bands and position. All states of matter
Nuclear magnetic resonance	Interaction of radiation with a nuclear transition in a magnetic field ($\lambda = 10^2$–10^7 cm; 3 kHz to 300 MHz)	Number of magnetically equivalent nuclei in each environment	10^{-1}–10^{-9} s	10^3 Pa (^{1}H)	Applicable to solutions and gases. In conjunction with molecular weight measurements may be possible to choose one from several possible models
Mass spectrometry	Detection of fragments by charge/msss	Mass number, plus fragmentation patterns	—	10^{-9} Pa	Useful for characterization of species in a vapour, complicated by reactions in spectrometer. Does not differentiate isomers directly. Important for detecting hydrogen in a molecule
Extended X-ray absorption fine structure (EXAFS)	Back scattering of photoelectrons off ligands	Radial distances, number, and types of bonded atoms	10^{-18} s, but averaged over vibrational motion	Any state	Especially useful for metallobiomolecules and heterogeneously supported catalysts

SOURCE: Taken, in part, from I. R. Beattie, *Chem. Soc. Rev.*, **1975**, *4*, 107. Used with permission.

SOME SIMPLE REACTIONS OF COVALENTLY BONDED MOLECULES

One of the major differences between organic and inorganic chemistry is the relative emphasis placed on structure and reactivity. Structural organic chemistry is relatively simple, as it is based on digonal, trigonal, or tetrahedral carbon. Thus organic chemistry has turned to the various mechanisms of reaction as one of the more exciting aspects of the subject. In contrast, inorganic chemistry has a wide variety of structural types to consider, and even for a given element there are many factors to consider. Inorganic chemistry has been, and to a large extent still is, more concerned with the "static" structures of reactants or products than with the way in which they interconvert. This has also been largely a result of the paucity of unambiguous data on reaction mechanisms. However, this situation is changing. Interest is increasingly centering on how inorganic molecules change and react. Most of this work has been done on coordination chemistry, and much of it will be considered in Chapter 11, but a few simple reactions of covalent molecules will be discussed here.

Atomic inversion

The simplest reaction a molecule such as ammonia can undergo is the simple inversion of the hydrogen atoms about the nitrogen atom, analogous to the inversion of an umbrella in a high wind:

$$NH_3 \rightleftharpoons NH_3 \qquad (5.14)$$

One might argue that Eq. 5.14 does not represent a reaction because the "product" is identical to the "reactant" and no bonds were formed or broken in the process.[69] Semantics aside, the process illustrated in Eq. 5.14 is of chemical interest and worthy of chemical study. For example, consider the trisubstituted amines and phosphines shown in Fig. 5.33. Because these molecules are nonsuperimposable upon their mirror images (i.e., they are chiral) they are potentially optically active, and separation of the enantiomers is at least theoretically possible. Racemization of the optically active material can take place via the mechanism shown in Eq. 5.14. It is of interest to note that the energy barrier to inversion is strongly dependent on the nature of the central atom and that of the substituents. For example, the barrier to inversion of methylpropylphenylphosphine (Fig. 5.33b) is about 120 kJ mol^{-1}. This is sufficient to allow the separation of optical isomers, and their racemization may be followed by classical techniques. In contrast, the barrier to inversion in most amines is low ($\sim$40 kJ mol^{-1} in methylpropylphenylamine; about 25 kJ mol^{-1} in ammonia). With such low barriers to inversion, optical isomers cannot be separated because racemization takes place faster than the resolution can be effected. Since traditional chemical separations cannot effect the resolution of the racemic mixture, the chemist must turn

[69] Obviously, the same result can be obtained by dissociating a hydrogen atom from the nitrogen atom and allowing it to recombine to form the opposite configuration. For discussion of the various competing mechanisms that must be distinguished in studying Eq. 5.14, as well as values for barrier energies and methods for obtaining them, see J. B. Lambert, *Topics Stereochem.*, **1971**, *6*, 19.

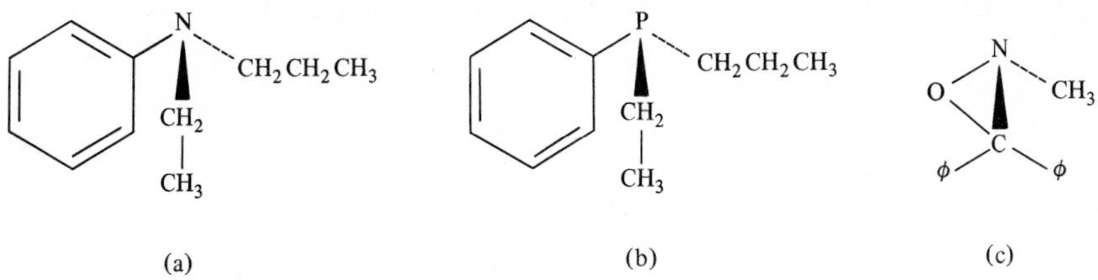

(a) (b) (c)

Fig. 5.33 Chiral amines and phosphines.

to spectroscopy to study the rate of interconversion of the enantiomers. The techniques involved are similar to those employed in the study of fluxional organometallic molecules (Chapter 13), and for now we may simply note that for inversion barriers of 20–100 kJ mol^{-1}, nuclear magnetic resonance is the tool of choice.

Since the transition state in the atomic inversion process of Eq. 5.14 involves a planar, sp^2-hybridized central atom, the barrier to inversion will be related to the ease with which the molecule can be converted from its pyramidal ground state. We should therefore expect that highly strained rings such as that shown in Fig. 5.33c would inhibit inversion, and this is found (145 kJ mol^{-1}). Furthermore, all of the effects we have seen previously affecting the bond angles in amines and phosphines should be parallel in the inversion phenomenon. For example, the smaller bond angles in phosphines require more energy to open up to the planar transition state than those of the corresponding amines; hence the optical stability of phosphines in contrast to the usual instability of most amines. In addition, the presence of electron-withdrawing substituents tends to increase the height of the barrier, but electron-donating groups can lower it. Just as in the case of the stereochemistry of pyramidal molecules, the results can be rationally accommodated by a variety of interpretations.[70]

Berry pseudorotation

We have seen previously that in PF_5 the fluorine atoms are indistinguishable by means of ^{19}F NMR (p. 242). This means that they are exchanging with each other faster than the NMR instrument can distinguish them. The mechanism for this exchange is closely related to the inversion reaction we have seen for amines and phosphines. The mechanism for this exchange is believed to take place through conversion of the ground state trigonal bipyramid (TBP) into a square pyramidal (SP) transition state and back to a new TBP structure (Fig. 5.34). This process results in complete scrambling of the fluorine atoms at the equatorial and axial positions in phosphorus pentafluoride, and if it occurs faster than the time scale of the NMR experiment (as it does), all of the fluorine atoms appear to be identical. Because it was first suggested by Berry,[71] and because, if all of the substituents are the same as in PF_5, the two TBP arrangements (Fig. 5.34) are related to each other by simple rotation, the entire process is called a *Berry pseudorotation.* Note that the process can take place very readily because of the similarity in energy between TBP and SP structures (p. 225). In fact, the series of 5-coordinate structures collected by Muetterties and

[70] R. D. Baechler et al., *J. Am. Chem. Soc.*, **1972**, *94*, 8060; C. C. Levin, ibid., **1975**, *97*, 5649; W. Cherry and N. Epiotis, ibid., **1976**, *98*, 1135; J. F. Liebman et al., ibid., **1976**, *98*, 5115; W. Cherry et al., *Acc. Chem. Res.*, **1977**, *10*, 167.

[71] R. S. Berry, *J. Chem. Phys.*, **1960**, *32*, 933; *Rev. Mod. Phys.*, **1960**, *32*, 447.

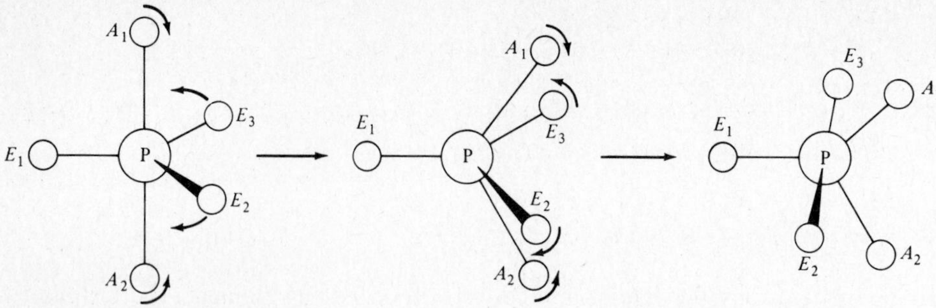

Fig. 5.34 Berry pseudorotation in a pentavalent phosphorus compound.

Guggenberger, which are intermediate between TBP and SP geometries (Table 5.3), effectively provides a reaction coordinate between the extreme structures in the Berry pseudorotation.

Substitution of alkyl groups on the phosphorus atom provides some interesting effects. If a single methyl group replaces a fluorine atom, it occupies one of the equatorial positions as expected (p. 230) and rapid exchange of the two axial and the two equatorial fluorine atoms is observed, as in PF_5. If two methyl groups are present, $(CH_3)_2PF_3$, the molecule becomes rigid and there is no observable exchange among the three remaining fluorine atoms. This dramatic change in behavior appears to be attributable to the intermediate formed, as shown in Fig. 5.34. In the pseudorotation of CH_3PF_4 the methyl group can remain at position E_1 and thus remain in an equatorial position both before and after the pseudorotation. In contrast, in $(CH_3)_2PF_3$ one of the methyl groups is forced to occupy either E_2 or E_3; therefore, after one pseudorotation it is forced to occupy the energetically unfavorable (for a substituent of low electronegativity) axial position. Apparently the difference in energy between equatorial and axial substitution of the methyl group is enough to inhibit the pseudorotation.

This difference in energy can be shown dramatically in the sulfurane

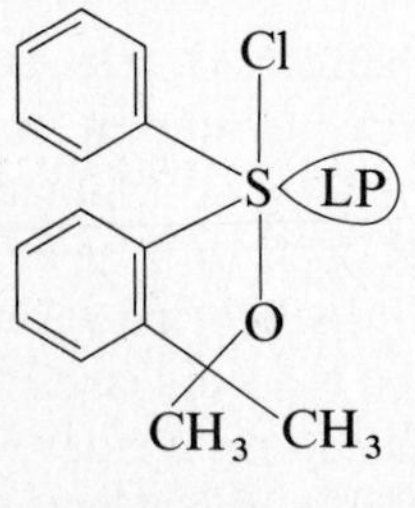

This molecule has an approximately TBP structure and is chiral. However, potentially it could racemize *via* a series of Berry pseudorotations.[72] That it does not do so readily, and is therefore the first optically active sulfurane to have been isolated, has been attributed to the fact that all racemization pathways must proceed through a TBP with an apical lone

[72] Of course, there are other potential mechanisms for racemization. If there were a trace of free Cl^- ion, an S_N2 displacement might be possible, or merely a simple S_N1 dissociation of the Cl atom and racemization. Neither of these reactions appears to take place either.

pair.[73] As we have seen in the preceding chapter, there is a very strong tendency for the lone pair to seek an equatorial site. The reluctance of the lone pair to occupy an apical site appears to be a sufficient barrier to allow the enantiomers to be isolated.

The question might be asked: Are there similar mechanisms for changing the configuration of molecules without breaking bonds in molecules with coordination numbers other than three and five? The answer is "yes." One of the most important series of inorganic compounds consists of 6-coordinate chelate compounds exemplified by the tris(ethylenediamine)cobalt(III) ion. Because of the presence of the three chelate rings, the ion is chiral and racemization can take place by a mechanism that is closely related to atomic inversion or Berry pseudorotation (the mechanism for 6-coordination is termed the "Bailar twist"; cf. p. 558).

Nucleophilic displacement

The crux of organic mechanistic stereochemistry may be the Walden inversion, the inversion of stereochemistry about a 4-coordinate carbon atom by nucleophilic attack of, for example, a hydroxide ion on an alkyl halide. Many reactions of inorganic molecules follow the same mechanism. In contrast, the dissociative mechanism of tertiary halides to form tertiary carbonium ion intermediates is essentially unknown among the nonmetallic elements silicon, germanium, phosphorus, etc. The reason for this is the generally lower stability of species with coordination numbers of less than 4, together with an increased stability of 5-coordinate intermediates. This difference is attributable to the presence of *d* orbitals in the heavier elements (Chapter 17).

The simplest reaction path for nucleophilic displacement may be illustrated by the solvolysis of a chlorodialkylphosphine oxide:

$$CH_3O^- + \mathrm{R(R')P(=O)Cl} \longrightarrow \left[\overset{\delta-}{CH_3O}\cdots \mathrm{P(=O)(R)(R')} \cdots \overset{\delta-}{Cl} \right] \longrightarrow CH_3O{-}\mathrm{P(=O)(R)(R')} \tag{5.15}$$

We would expect the reaction to proceed with inversion of configuration of the phosphorus atom. This is generally observed, especially when the entering and leaving groups are highly electronegative and are thus favorably disposed at the axial positions, and when the leaving group is one that is easily displaced. In contrast, in some cases when the leaving group is a poor one it appears as though front side attack takes place because there is a retention of configuration.[74] In either case, the common inversion or the less common retention, there is a contrast with the *loss* of stereochemistry associated with a carbonium ion mechanism.

The stability of 5-coordinate intermediates also makes possible the ready racemization of optically active silanes by catalytic amounts of base. The base can add readily to form a 5-coordinate intermediate. The latter can undergo Berry pseudorotation with

[73] J. C. Martin and T. M. Balthazor, *J. Am. Chem. Soc.*, **1977**, *99*, 152.

[74] For a discussion of the various possibilities, see M. L. Tobe, "Inorganic Reaction Mechanisms," Thomas Nelson & Sons, London, **1972**, pp. 25–37.

complete scrambling of substituents followed by loss of the base to yield the racemized silane.

Free radical mechanisms

Most of the reactions the inorganic chemist encounters in the laboratory involve ionic species such as the reactants and products in the reactions just discussed or those of coordination compounds (Chapter 11). However, in the atmosphere there are many *free radical* reactions initiated by sunlight. One of the most important and controversial sets of atmospheric reactions at present is that revolving around stratospheric ozone. The importance of ozone and the effect of ultraviolet (UV) radiation on life will be discussed in Chapter 18, but we may note briefly that only a small portion of the sun's spectrum reaches the surface of the earth and that parts of the UV portion that are largely screened can cause various ill effects to living systems.

The earth is screened from far-UV (extremely high energy) radiation by oxygen in the atmosphere. The UV radiation cleaves the oxygen molecule to form two free radicals (oxygen atoms):

$$O_2 + h\nu \text{ (below 242 nm)} \longrightarrow \cdot O \cdot + \cdot O \cdot \qquad \textbf{(5.16)}$$

The oxygen atoms can then attack oxygen molecules to form ozone:

$$\cdot O \cdot + O_2 + M \longrightarrow O_3 + M \qquad \textbf{(5.17)}$$

The neutral body M carries off some of the kinetic energy of the oxygen atoms. This reduces the energy of the system and allows the bond to form to make ozone. The net reaction is therefore:

$$3O_2 + h\nu \longrightarrow 2O_3 \qquad \textbf{(5.18)}$$

This process protects the earth from the very energetic, short-wavelength UV radiation and at the same time produces ozone, which absorbs somewhat longer-wavelength radiation (moderately high energy) by a similar process:

$$O_3 + h\nu \text{ (220–320 nm)} \longrightarrow O_2 + \cdot O \cdot \qquad \textbf{(5.19)}$$

The products of this reaction can recombine as in Eq. 5.17, in which case the ozone has been regenerated and the energy of the ultraviolet radiation has been degraded to thermal energy. Alternatively, the oxygen atoms can recombine to form oxygen molecules by the reverse of Eq. 5.16, thereby reducing the concentration of ozone. An equilibrium is set up between this destruction of ozone and its generation via Eq. 5.18 and so under normal conditions the concentration of ozone remains constant.

The controversy over supersonic transports (SSTs) of the Concorde type revolves around the production of nitrogen oxides whenever air containing oxygen and nitrogen passes through the very high temperatures of a jet engine. One of these products, nitric oxide, reacts directly with ozone, thereby reducing its concentration in the stratosphere:

$$NO + O_3 \longrightarrow NO_2 + O_2 \qquad \textbf{(5.20)}$$

Furthermore, nitrogen dioxide formed in Eq. 5.20 or directly in the combustion process can react to scavenge oxygen free radicals and prevent their possible recombination with molecular oxygen to regenerate ozone (Eq. 5.17):

$$NO_2 + \cdot O \cdot \longrightarrow NO + O_2 \tag{5.21}$$

Note that a combination of reactions 5.20 and 5.21 results in the net conversion of ozone to oxygen:

$$\cdot O \cdot + O_3 \longrightarrow 2O_2 \tag{5.22}$$

and that the nitrogen oxides, either NO or NO_2, continuously recycle and thus act as catalysts for the decomposition of ozone:

$$\begin{array}{ccc} O_3 & NO & O_2 \\ O_2 & NO_2 & \cdot O \cdot \end{array} \tag{5.23}$$

The current controversy revolves around the extent to which nitrogen oxides, NO_x, would be formed by SSTs and how much the ozone concentration would be affected.[75]

The ozone question is complicated by the fact that other chemicals are implicated in its destruction. Chlorofluorocarbons are widely used as propellants in spray cans and as refrigerants. They are extremely stable and long-lived in the environment. However, they too can undergo photolysis in the upper atmosphere:

$$F_3CCl + h\nu\ (190\text{–}220\ \text{nm}) \longrightarrow F_3C\cdot + Cl\cdot \tag{5.24}$$

The chlorine free radical can then interact with ozone in a manner analogous to the NO_x process:

$$Cl + O_3 \longrightarrow ClO + O_2 \tag{5.25}$$

$$ClO + O \longrightarrow Cl + O_2 \tag{5.26}$$

for a net reaction of:

$$O + O_3 \longrightarrow 2O_2 \tag{5.27}$$

with regeneration of the atomic chlorine. The chlorine thus acts as a catalyst and present evidence indicates that the ClO_x cycle may be three times more efficient in the destruction of ozone than the NO_x cycle.[76]

PROBLEMS

5.1. Draw Lewis structures for the following molecules and predict the molecular geometry:

a. BCl_3
b. BeH_2
c. $SnBr_4$
d. TeF_6

---

[75] H. S. Johnston, *Acc. Chem. Res.*, **1975**, *8*, 289; F. N. Alyea et al., *Science*, **1975**, *188*, 117.

[76] "Fluorocarbons and the Environment," Report of Federal Task Force on Inadvertent Modification of the Stratosphere, Council on Environmental Quality, Washington, D.C., **1975**.

e. AsF_5
f. XeO_4

5.2. Draw Lewis structures for the following molecules and predict the molecular geometry including expected distortions:
a. $TeCl_4$
b. ICl_2^+
c. ClF_3
d. SO_2
e. XeF_2
f. XeF_4
g. XeO_3

5.3. Use Eq. 5.1 to derive the bond angles in *sp*, sp^2, and sp^3 hybrid orbitals.

5.4. Assuming that the orbitals are directed along the internuclear axes (i.e., the bonds are not bent) use Eq. 5.1 to calculate the *p* character in the bonds of NH_3. The bond angle in NH_3 is 107.5°. What is the *p* character of the lone pair?

5.5. The bond angles in the fluoromethanes are:

Molecule	H—C—H	F—C—F
CH_3F	110–112°	—
CH_2F_2	111.9° ± 0.4°	108.3° ± 0.1°
CHF_3	—	108.8° ± 0.75°

a. Calculate the *s* character used by the carbon atom in the orbitals directed to the hydrogen and fluorine atoms.
b. Discuss the results in terms of Bent's Rule.

5.6. Show in a qualitative way why the energy levels of AH_2 in Fig. 5.16 increase or decrease in the way they do upon bending the molecule. Attempt to account for small and large changes.

5.7. Consider the free radicals CH_3 and CF_3. One is planar, the other pyramidal. Which is which? Why?

5.8. Group VIA tetrafluorides act as Lewis acids and form anions:

$$Cs^+F^- + SF_4 \longrightarrow Cs^+[SF_5]^-$$

Predict the structures of these anions. [K. O. Christe et al., *Inorg. Chem.*, **1972**, *11*, 1679.]

5.9. From Fig. 5.28, derive an equation for tungsten analogous to Eq. 5.13 for carbon.

5.10. Predict the carbon–carbon bond length in benzene.

5.11. Consider the molecule $CH_3C{\equiv}CH$. Applying Bent's rule in its classical form, predict whether the bond angles, H—C—H, are greater or less than $109\frac{1}{2}$°. Considering the arguments on overlap on p. 231, predict again. [The experimental result is given by C. C. Costain, *J. Chem. Phys.*, **1958**, *29*, 864.]

5.12. Consider the molecule ClF_3O_2 (with chlorine the central atom). How many isomers are possible? Which is the most stable? Consult Appendix B and assign point group designations to each of the isomers.

5.13. The structure for Al_2Br_6 (Fig. 5.1h) is assumed by both Al_2Br_6 and Al_2Cl_6 in the gas phase. In the solid, however, the structure can be described as a closest-packed array of halogen atoms (or ions) with aluminum atoms (or ions) in tetrahedral or octahedral holes. In solid aluminum bromide the aluminum atoms are found in pairs in adjacent tetrahedral holes. In solid aluminum chloride, atoms are found in one-third of the octahedral holes.

a. Discuss these two structures in terms of an ionic model for the solid. What factors favor or disfavor this interpretation?

b. Discuss these two structures in terms of covalent bonding in the solid. What factors favor or disfavor this interpretation?

5.14. Obtain the covalent and van der Waals radii of phosphorus and the halogens from Table 6.1.

a. Show that for an assumed bond angle of $109\frac{1}{2}°$ in the phosphorus trihalides there must be van der Waals contacts among the halogen atoms.

b. Show that because of the concomitant increase in both covalent and van der Waals radii, the repulsion between the halogens does not become worse as one progresses from PF_3 to PI_3.

5.15. One of the few phosphorus compounds that exhibit square pyramidal geometry is shown in Fig. 5.35. Rationalize the preferred geometry of SP over TBP in terms of the presence of the four- and five-membered rings. [R. R. Holmes, *J. Am. Chem. Soc.*, **1975**, *97*, 5379.]

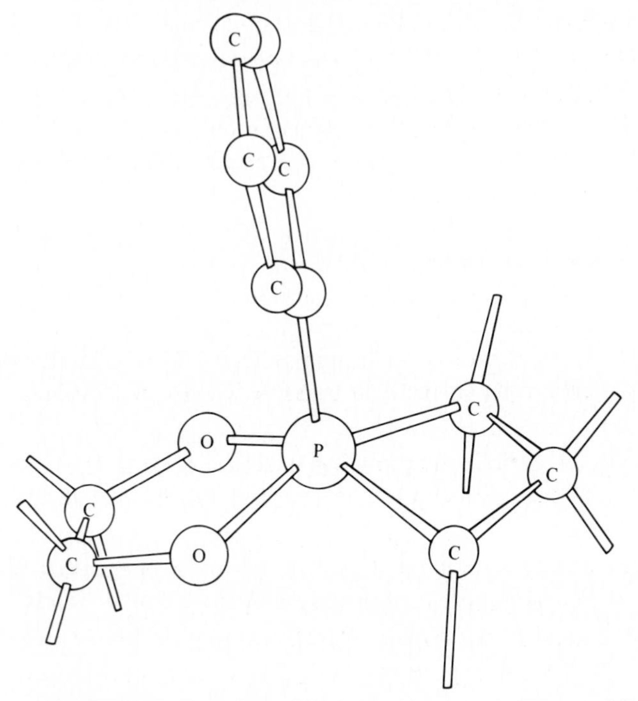

Fig. 5.35 Square pyramidal dioxophosphorane with five- and four-membered rings. [J. A. Howard et al., *Chem. Commun.*, **1973**, 856. Reproduced with permission.]

5.16. Consider the cyclic compounds I and II. In I the rapid exchange of the fluorine atoms is inhibited just as it is in $(CH_3)_2PF_3$. However, exchange in II is very rapid. Suggest a reason.

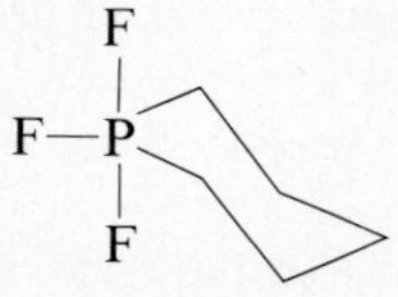

F
F—P
F

I

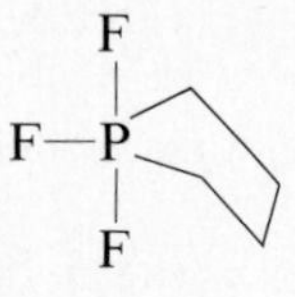

F
F—P
F

II

5.17. Suggest the most likely stereochemistry of the phosphinate ester resulting from ethanolysis of the following compound:

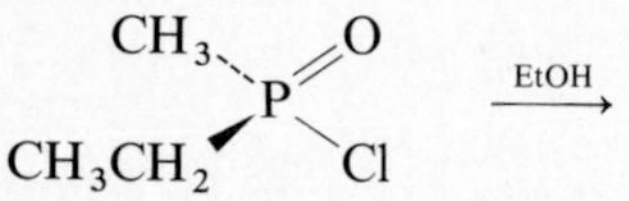

CH_3, O, P, CH_3CH_2, Cl $\xrightarrow{\text{EtOH}}$

5.18. Predict the geometries of $(CH_3)_2P(CF_3)_3$ and $(CH_3)_3P(CF_3)_2$. Do you expect these molecules to undergo pseudorotation? Explain. [K. I. The and R. G. Cavell, *Chem. Commun.*, **1975**, 716.]

5.19. In an sp^3d hybridized phosphorus atom in a TBP molecule, will the atom have a greater electronegativity when bonding through equatorial or axial orbitals? Explain.

5.20. Earlier (p. 215) it was stated that the repulsive effects of a lone pair and a doubly bonded oxygen atom in VSEPR theory were very similar. Discuss qualitative and quantitative differences that you feel should exist. [K. Christe and H. Oberhammer, *Inorg. Chem.*, **1981**, *20*, 296.]

5.21. Consult Table 5.5 for bond angle data in substituted ethylenes.

a. Rationalize the observed bond angles in terms of the various structural theories discussed in this chapter.

b. Calculate the *s* character in one of the bonds from the bond angle data.

c. Will the hydrogen atoms carry more positive charge in $H_2C{=}CH_2$ or in $H_2C{=}CCl_2$? This part can be answered quite correctly at two levels of sophistication—try both.

5.22. Predict and say as much as you can about the probable structure of solid InCl. *Be careful*! [*Hint:* Why do you think this problem was included in *this* chapter rather than in Chapters 3 or 4?]

5.23. If careful X-ray diffraction can distinguish the bending of bonds in three-membered rings, why can it not be used to distinguish bent double bonds from σ–π bonds?

5.24. Consider Fig. 5.36 which is an electron-density contour map of the sodium cyanide crystal. Interpret this diagram in terms of everything that you know about the structure of solid sodium cyanide.

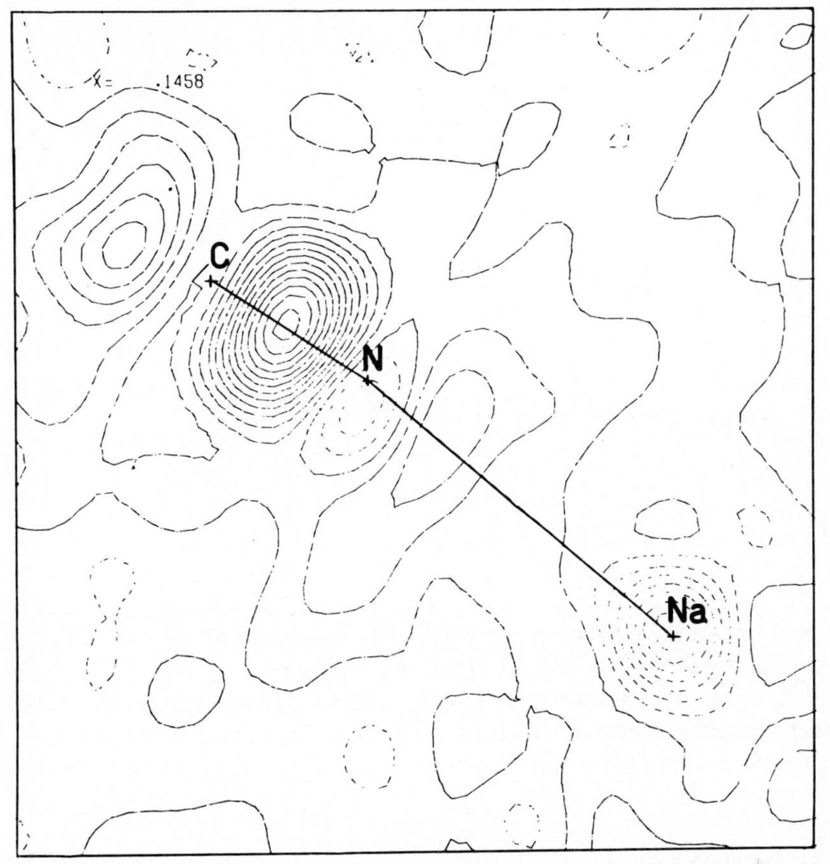

Fig. 5.36 Differential electron-density map of crystalline sodium cyanide, NaCN. Dashed contours indicate decreased electron density upon compound formation, solid contours indicate increased electron density. [From P. Coppens, *Angew. Chem.* (*Eng.*), **1977**, *16*, 32. Reproduced with permission.]

5.25. Read the section on symmetry in Appendix B and identify the symmetry elements and operations in the molecules and ions shown in the figures listed below. Determine the appropriate point group for each molecule and ion.

a. 5.1
b. 5.4
c. 5.5
d. 5.6
e. 5.7
f. 5.8
g. 5.10
h. 5.11
i. 5.12
j. 5.18

5.26. Calculate the hybridization of the carbon and nitrogen atoms in Fig. 5.24.

5.27. Considering the molecular orbital diagram of carbon monoxide (Fig. 3.54) and the discussion concerning hybridization and energy (pp. 227–229), predict which end of the carbon monoxide molecule will be the more basic (i.e., donate electrons most readily and form the strongest, direct covalent bond). [H. H. Jaffé and M. Orchin, *Tetrahedron*, **1960**, *10*, 212.]

6

Chemical forces

In the preceding chapters attention has been called to the importance of the forces between atoms and ions in determining chemical properties. In this chapter these forces will be examined more closely and a comparison made among them. The important aspects of each type of force are its relative strength, how rapidly it decreases with increasing distance, and whether it is directional or not. The last property is extremely important when considering the effects of a force in determining molecular and crystal structures. Since distance is an important factor in all interaction energies, a brief discussion of interatomic distances should preface any discussion of energies and forces.

INTERNUCLEAR DISTANCES AND ATOMIC RADII

It is valuable to be able to predict the internuclear distance of atoms within and between molecules, and so there has been much work done in attempting to set up tables of "atomic radii" such that the sum of two will reproduce the internuclear distances. Unfortunately there has been a proliferation of these tables and a bewildering array of terms including bonded, nonbonded, ionic, covalent, metallic, and van der Waals radii, as well as the vague term atomic radii. This plethora of radii is a reflection of the necessity of specifying what is being measured by an atomic radius. Nevertheless, it is possible to simplify the treatment of atomic radii without causing unwarranted errors.

Van der Waals radii

If two noble gas atoms are brought together with no kinetic energy tending to disrupt them, they will "stick" together. The forces holding them together are the weak London dispersion forces discussed in a later section (see pp. 266–267). The internuclear distance will be such that the weak attractive forces are exactly balanced by the Pauli repulsive forces of the closed shells. If the two noble gas atoms are identical, one-half of the internuclear distance

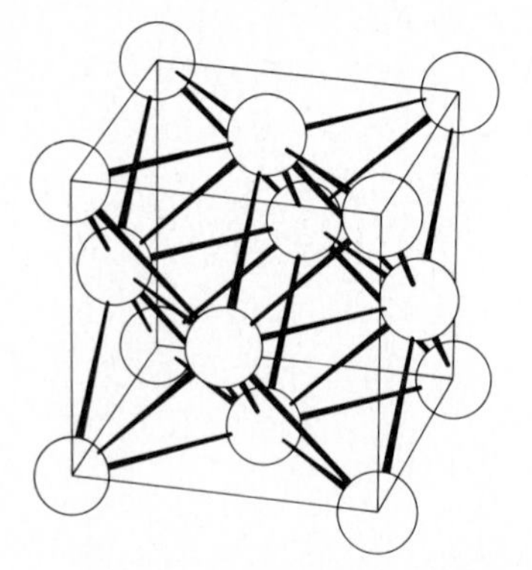
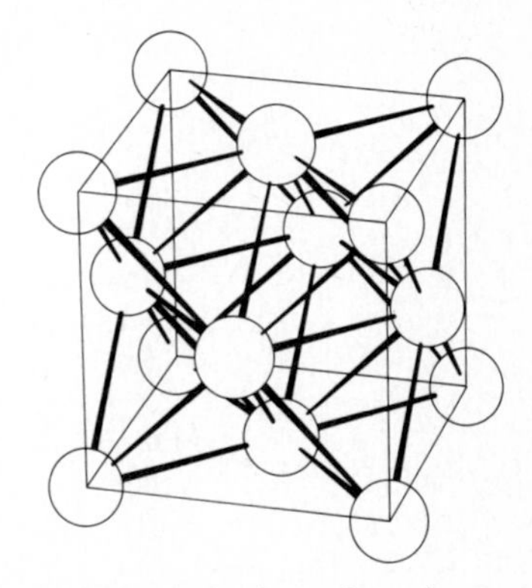

Fig. 6.1 Unit cell of argon. Note that the connecting lines are for geometric perspective only and do not represent bonds. [From M. F. C. Ladd, "Structure and Bonding in Inorganic Chemistry," Ellis Horwood, Chichester, **1979**. Reproduced with permission.]

may be assigned to each atom as its nonbonded or *van der Waals radius*. Solid argon (Fig. 6.1), for example, consists of argon atoms spaced at a distance of 380 pm yielding a van der Waals radius of 190 pm for argon.

Although the van der Waals radius of an atom might thus seem to be a simple, invariant quantity, such is not the case. The size of an atom depends upon how much it is compressed by external forces and upon substituent effects. For example, in XeF_4 the van der Waals radius of xenon appears to be closer to 170 pm than the accepted value of 218 pm obtained from solid xenon.[1] The explanation lies in the fact that the xenon is reduced in size because of electron density shifted to the more electronegative fluorine atom. In addition, the partial charges induced ($Xe^{\delta+}$, $F^{\delta-}$) may cause the xenon and fluorine atoms to attract each other and approach more closely.

Although we must therefore expect van der Waals radii to vary somewhat depending upon the environment of the atom, we can use them to estimate nonbonded distances with reasonable success. Table 6.1 lists the van der Waals radii of some atoms.

Ionic radii

Ionic radii are discussed thoroughly in Chapters 3 and 4 (also see Appendix I). For the present discussion it is only necessary to point out that the principal difference between ionic and van der Waals radii lies in the difference in the *attractive force*, not the difference in *repulsion*. The interionic distance in LiF, for example, represents the distance at which the repulsion of a He core (Li^+) and a Ne core (F^-) counterbalances the strong electrostatic or Madelung force. The attractive energy for Li^+F^- is considerably over 400 kJ mol^{-1}

[1] W. C. Hamilton and J. A. Ibers, in "Noble Gas Compounds," H. H. Hyman, ed., University of Chicago Press, Chicago, **1963**, p. 195; D. H. Templeton et al., ibid., p. 203; J. H. Burns, P. A. Agron, and H. Levy, ibid., p. 211. In XeF_4 the xenon atoms do not touch each other. The estimate of the van der Waals radius must be made by subtracting the van der Waals radius of fluorine from the shortest nonbonded (i.e., *between* molecules) xenon fluorine distance (320–330 pm).

Table 6.1 Atomic radii (pm)

Element	r_{VDW} [a]	r_{ion} [b]	r_{cov} [c]
1. H	120[a]–145[e]		37
2. He	180[f]		(32)
3. Li	180	90(+1)	134
4. Be		59(+2)	125
5. B		41(+3)	90
6. C	165[e]–170[a]		77
7. N	155		75
8. O	150	126(−2)	73
9. F	150–160	119(−1)	71
10. Ne	160[f]		(69)
11. Na	230	116(+1)	154
12. Mg	170	86(+2)	145
13. Al		68(+3)	130
14. Si	210		118
15. P	185		110
16. S	180	170(−2)	102
17. Cl	170–190	167(−1)	99
18. Ar	190[f]		(97)
19. K	280	152(+1)	196
20. Ca		114(+2)	
21. Sc		88(+3)	
22. Ti		74(+4)	
23. V			
24. Cr			
25. Mn			139[g]
26. Fe			125[h]
27. Co			126[h]
28. Ni	160		121(Td)[i] 116(Sq)[i]
29. Cu	140	91(+1)	
30. Zn	140	88(+2)	120
31. Ga	190	76(+3)	120
32. Ge			122
33. As			122
34. Se	190	184(−2)	117
35. Br	180–200	182(−1)	114
36. Kr	200[f]		110
37. Rb		166(+1)	
38. Sr		132(+2)	
39. Y		104(+3)	
40. Zr			
41. Nb			
42. Mo			
43. Tc			
44. Ru			
45. Rh			
46. Pd	160		
47. Ag	170	108(+1)	
48. Cd	160	109(+2)	
49. In	190	94(+3)	
50. Sn	220		140
51. Sb			143

Table 6.1 Atomic radii (pm)

Element	r_{VDW}[a]	r_{ion}[b]	r_{cov}[c]
52. Te	210	207(−2)	135
53. I	195–212	206(−1)	133
54. Xe	220[f]		130
55. Cs		181(+1)	
56. Ba		149(+2)	
57. La		117(+3)	
71. Lu		100(+3)	
72. Hf			
73. Ta			
74. W			
75. Re			
76. Os			
77. Ir			
78. Pt	170–180		
79. Au	170	151(+1)	
80. Hg	150	116(+2)	
81. Tl	200	102(+3)	
82. Pb	200		
83. Bi			
84. Po			
85. At			
86. Rn			(145)
92. U	190		
Organic groups			
CH_3	200[j]		
C_6H_5	170[j,k]		

[a] Values of van der Waals radii from A. Bondi, *J. Phys. Chem.*, **1964**, *68*, 441, unless otherwise noted.

[b] Ionic radii (C.N. = 6) are from Table 3.4 and are listed for comparative purposes only. For accurate work, see multiple values in Table 3.5.

[c] Covalent radii estimated from homonuclear bond lengths where available and from selected heteronuclear bonds otherwise. Bond lengths from L. Sutton, ed., "Tables of Interatomic Distances and Configuration in Molecules and Ions," Spec. Publ. No. 11 and 18, The Chemical Society, London, **1958**, **1965**, except where noted. Values in parentheses are for noble gases not known to form compounds and are extrapolated from the values of neighboring nonmetals: L. C. Allen and J. E. Huheey, *J. Inorg. Nucl. Chem.*, **1980**, *42*, 1523.

[d] J. C. Slater, *J. Chem. Phys.*, **1964**, *41*, 3199.

[e] N. L. Allinger et al., *J. Am. Chem. Soc.*, **1968**, *90*, 1199.

[f] G. A. Cook, "Argon, Helium and the Rare Gases," Wiley (Interscience), New York, **1961**, Vol. I, p. 13.

[g] F. A. Cotton and D. C. Richardson, *Inorg. Chem.*, **1966**, *5*, 1851.

[h] L. F. Dahl et al., *J. Am. Chem. Soc.*, **1969**, *91*, 1655.

[i] B. T. Kilbourn and H. M. Powell, *J. Chem. Soc.*, *A*, **1970**, 1688.

[j] L. Pauling, "The Nature of the Chemical Bond," 3rd ed., Cornell University Press, Ithaca, N.Y., **1960**.

[k] In direction perpendicular to ring.

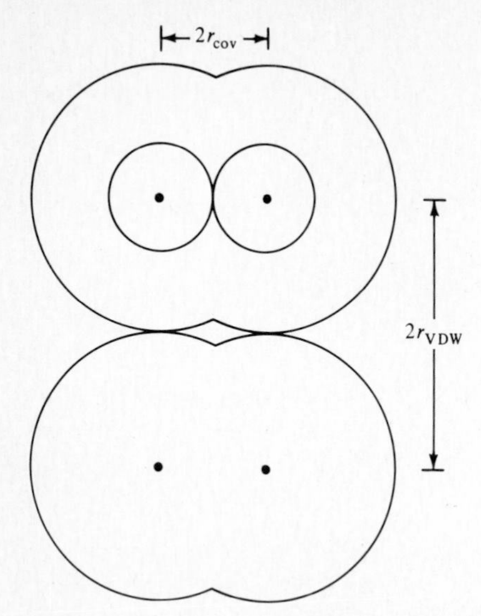

Fig. 6.2 Illustration of the difference between van der Waals and covalent radii in the F_2 molecule.

and the London energy of He–Ne is of the order of 4 kJ mol^{-1}. The forces in the LiF crystal are therefore considerably greater and the interionic distance (201 pm) is less than expected for the addition of He and Ne van der Waals radii (340 pm).

Covalent radii

The internuclear distance in the fluorine molecule is 142 pm, which is shorter than the sum of two van der Waals radii. The difference obviously comes from the fact that the electron clouds of the fluorine atoms overlap extensively in the formation of the F—F bond whereas little overlap of the van der Waals radii occurs between the molecules (Fig. 6.2) because of the rapidity with which repulsive energies increase with decreasing distance. Now it might be supposed that the equilibrium distance in the F_2 molecule is that at which the maximum overlap of the bonding orbitals occurs. However, if this were the sole criterion, the F_2 molecule would "collapse" until the two nuclei were superimposed. This would cause the orbital wave functions to have identical spatial distributions and the maximum possible overlap. Obviously this does not occur because of repulsions between the two positive nuclei and repulsions between the inner electron cores. If we consider (for *speculative* purposes only!) the F_2 molecule to be composed of F^+F^-, then we might picture it as in Fig. 6.3. The radius of the helium core can be estimated from Pauling's calculation for F^{+7}.[2] The F^+—F^- distance would be predicted to be $7 + 136 = 143$ pm, in remarkable agreement with the experimental value of 142! Corresponding values for the other halogens

[2] It might be argued that the radius of F^{+7} should not be used in the present hypothetical example with F^+. However, we are attempting to estimate the size of the He core, $1s^2$, which will be little affected by the presence or absence of the relatively nonpenetrating $2s$ and $2p$ electrons.

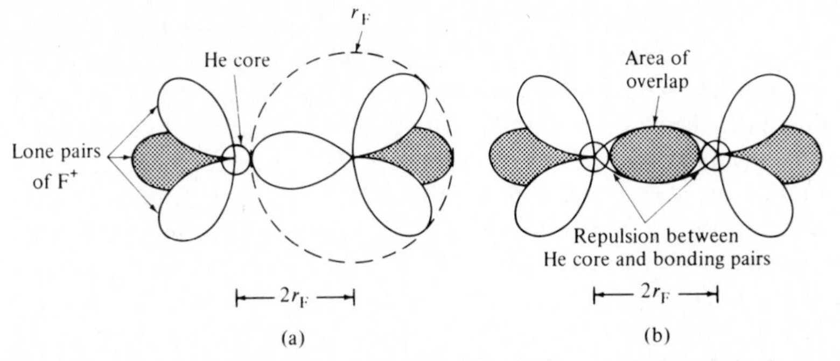

Fig. 6.3 (a) Hypothetical F^+F^- ion-pair molecule illustrating repulsion between the inner He core and the "lone pair" of the "F^- ion." (b) More realistic representation of repulsions between inner core and valence shell electrons. (The He core is not drawn to scale in either sketch.)

are 207 versus 199 (Cl), 234 versus 228 (Br), and 266 versus 267 (I). This is not meant to imply that the bonding in these halogen molecules is ionic; quite the contrary: The chief factor in determining the covalent radii of atoms is the size of the core electron cloud beneath the valence shell. This might be termed the "van der Waals" radius of the core.

Table 6.1 lists covalent radii obtained by dividing homonuclear bond distances by two. In many cases the appropriate homonuclear single bond has not been measured and the assigned covalent radius is obtained indirectly by subtracting the covalent radius of element B in a heteronuclear bond AB to obtain the radius of atom A.

The values in Table 6.1 are reasonably additive; i.e., the covalent bond distance in a molecule AB_n can be estimated reasonably well from $r_A + r_B$. Some typical values are listed in Table 6.2. The agreement is fairly good. In the case of molecules with several large substituent atoms around a small central atom such as CBr_4 and CI_4, the crowding apparently causes some lengthening of the bond. There are other cases in which the additivity

Table 6.2 Comparison of calculated and experimental radii (pm)

Molecule	Bond	$r_A + r_B$	r_{exp}
HF	HF	108	92
HCl	HCl	136	128
HBr	HBr	151	142
HI	HI	170	161
ClF	ClF	170	163
BrF	BrF	185	176
BrCl	BrCl	213	214
ICl	ICl	232	232
CH_4	CH	114	109
CF_4	CF	148	136
CCl_4	CCl	176	176
CBr_4	CBr	191	194
CI_4	CI	210	215

of the radii is rather poor. For example, the H—H and F—F bond distances are 74 and 142, respectively, yielding covalent radii of 37 and 71 pm. However, the bond length in the HF molecule is not 108 pm, but 92 pm. If we assume that the size of the fluorine atom is constant, then the radius of hydrogen in HF is 21. Alternatively, we can assume that the fluorine atom is somewhat smaller in the HF molecule than in the F_2 molecule, an extremely unlikely situation. Or more realistically, we can admit that the hydrogen atom is unique, that it has no inner repulsive core to determine its covalent radius but that in bonding the proton often partially penetrates the electron cloud of the other atom and that the bond distance is determined by a delicate balance of electron–nucleus attractions and nucleus–nucleus repulsions. However, this does not really solve our problem, for a widespread deviation from additivity results from the effect of differences in electronegativity between the bonding atoms. It is usually observed that the bond length between an electropositive atom and an electronegative atom is somewhat shorter than expected on the basis of their assigned covalent radii. Schomaker and Stevenson[3] have suggested the relation:

$$r_{AB} = r_A + r_B - 9\Delta_\chi \tag{6.1}$$

where Δ_χ is the difference in electronegativity between atoms A and B. Wells[4] has discussed deficiencies in Eq. 6.1, and Pauling[5] has suggested modifications to improve the accuracy. These details will not be discussed here, but the significance of the bond shortening in these molecules is reasonably clear. Heteropolar bonds are almost always stronger than expected on the basis of the corresponding homopolar bonds (see discussion of ionic resonance energy, Chapter 3). The atoms in the molecule AB are therefore held together more tightly and compressed somewhat relative to their situation in the molecules AA and BB, which are the basis of r_A and r_B. It is helpful to analyze the source of this stabilization somewhat more closely than merely labeling it "ionic resonance energy." To a first approximation, it results from the extra bonding energy ("ionic" or Madelung energy) resulting from the partial charges on the atoms:

$$H^{\delta+}\text{—}Cl^{\delta-} \qquad E = \frac{\delta^+\delta^-}{4\pi\varepsilon_0 r}$$

Now for a polyvalent atom the partial charge builds up every time another highly electronegative substituent is added. Thus the partial charge on the carbon atom in the carbon tetrafluoride molecule is considerably larger than it is in the methyl fluoride molecule, and so *all* of the carbon–fluorine bonds shrink proportionately more:

	C—F (pm)
CH_3F	139.1
CH_2F_2	135.8
CHF_3	133.2
CF_4	132.3

[3] V. Schomaker and D. P. Stevenson, *J. Amer. Chem. Soc.*, **1941**, *63*, 37.

[4] A. F. Wells, *J. Chem. Soc.*, **1949**, *55*.

[5] L. Pauling, "The Nature of the Chemical Bond," 3rd ed., Cornell University Press, Ithaca, N.Y., **1960**, pp. 228–230.

TYPES OF CHEMICAL FORCES

Covalent bonding

This topic has been discussed extensively in Chapters 3, 4, and 5, so only those aspects pertinent to comparison with other forces will be reviewed here. In general, the covalent bond is strongly directional as a result of the overlap criterion for maximum bond strength. We have seen previously the implications this has for determining molecular structures. In addition, the covalent bond is very strong. Some typical values[6] for purely covalent bonds are P—P, ~200 kJ mol^{-1}; C—C, 346 kJ mol^{-1}; and H—H, 432 kJ mol^{-1}. The smaller atoms can effect better overlap and hence have stronger bonds (for a discussion of this effect see p. 321). Bond polarity can *increase* bond strength (cf. Pauling's electronegativity calculations, p. 144), and so we find a few much stronger bonds such as Si—F (which probably includes some π bonding as well), 565 kJ mol^{-1}. Homopolar bonds between small atoms with repulsive lone pairs tend to be somewhat *weaker* than average, e.g., N—N, 167 kJ mol^{-1}, and F—F, 155 kJ mol^{-1}. Nevertheless, a good rule of thumb is that a typical covalent bond will have a strength of about 250–400 kJ mol^{-1}. As we shall see, this is stronger than all other chemical interactions with the exception of ionic bonds.

Because of the complexity of the forces operating in the covalent bond, it is not possible to write a simple potential energy function as for the electrostatic forces such as ion–ion and dipole–dipole. Nevertheless, it is possible to describe the covalent energy qualitatively as a fairly short-range force (as the atoms are forced apart, the overlap decreases).

Ionic bonding

The strength of a purely ionic bond between two ions can be obtained quite accurately by means of the Born equation (Chapter 3).[7] Neglecting repulsive forces, van der Waals forces, and other small contributions, we can estimate the energy of an ion pair simply as:

$$E = \frac{Z^+Z^-}{4\pi r \varepsilon_0} \tag{6.2}$$

For a pair of very small ions, such as Li^+ and F^-, we can estimate a bond energy of about 686 kJ mol^{-1}. The experimental values are 755.2 kJ mol^{-1} for dissociation to ions and 575.3 kJ mol^{-1} for dissociation to atoms. For a pair of larger ions, such as Cs^+ and I^-, the energy is correspondingly smaller or about half as much. It is evident that the strength of ionic bonds is of the same order of magnitude as covalent bonds. The common notion that ionic bonds are considerably stronger than covalent bonds probably results from mistaken interpretations of melting-point and boiling-point phenomena, which will be discussed further later (pp. 276–279).

Ionic bonding is nondirectional insofar as it is purely electrostatic. The attraction of one ion for another is completely independent of direction, but the sizes and numbers of

---

[6] Tables of bond strength values can be found in Appendix E.

[7] The "bond strength" thus obtained refers, of course, to the dissociation of the ion pair to the separated ions, $M^+X^- \rightarrow M^+ + X^-$. It is somewhat easier to dissociate an ion pair into the uncharged constituent atoms, $M^+X^- \rightarrow M + X$, because the ionization energy of the metal is greater than the electron affinity of the nonmetal.

ions determine crystal structures. Compared with the forces to follow, ionic bonding is relatively insensitive to distance. It is true that the force between two ions is inversely proportional to the square of the distance between them and hence decreases fairly rapidly with distance, but much less so than most other chemical forces.

Ion-dipole forces[8]

The various factors affecting the magnitude of the dipole moment in a polar molecule were discussed in the previous chapters. For the present discussion it is sufficient to picture a molecular dipole as two equal and opposite charges ($q^{\pm}$) separated by a distance r'. The dipole moment, μ, is given by:

$$\mu = qr' \tag{6.3}$$

When placed in an electric field, a dipole will attempt to orient and become aligned with the field gradient. If the field results from an ion, the dipole will orient itself so that the attractive end (the end with charge opposite to that of the ion) will be directed toward the ion and the other, repulsive end directed away. In this sense, ion–dipole forces may be thought of as "directional," in that they result in preferred orientations of molecules even though electrostatic forces are nondirectional.

The potential energy of an ion–dipole interaction is given as:

$$E = -\frac{|Z^{\pm}|\mu}{4\pi r^2 \varepsilon_0} \tag{6.4}$$

where $Z^{\pm}$ is the charge on the ion and r is the distance between the ion and the molecular dipole:

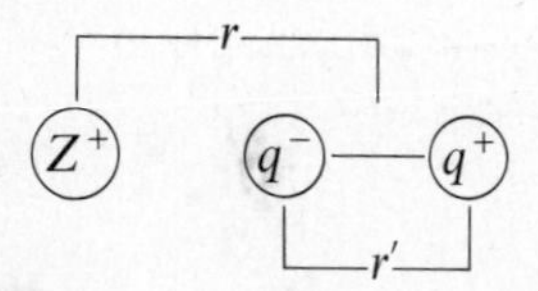

Ion–dipole interactions are similar to ion–ion interactions, except that they are more sensitive to distance ($1/r^2$ instead of $1/r$) and tend to be somewhat weaker since the charges (q^+, q^-) comprising the dipole are usually considerably less than a full electronic charge.

Ion–dipole forces are important in solutions of ionic compounds in polar solvents where solvated species such as $Na(OH_2)_x^+$ and $F(H_2O)_y^-$ (for solutions of NaF in H_2O) exist. In the case of some metal ions these solvated species can be sufficiently stable to be considered as discrete species, such as $[Co(NH_3)_6]^{+3}$. Complex ions such as the latter may thus be considered as electrostatic ion-dipole interactions, but this oversimplification

[8] For a short discussion of the various dipole interactions, see H. H. Jaffé, *J. Chem. Educ.*, **1963**, *40*, 649. For a more extensive discussion, see M. Davies, "Some Electrical and Optical Aspects of Molecular Behavior," Pergamon, Elmsford, N. Y., **1965**, pp. 158–176.

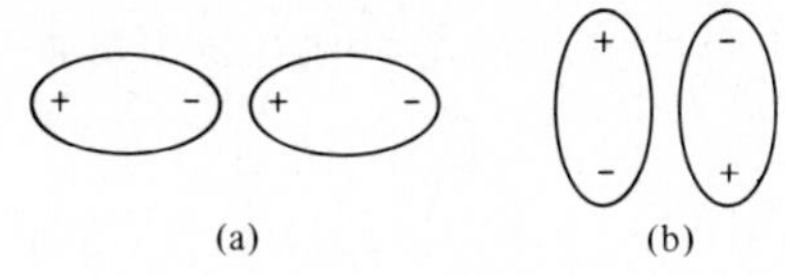

Fig. 6.4 (a) Head-to-tail arrangement of dipoles; (b) antiparallel arrangement of dipoles.

(Electrostatic Crystal Field Theory; see Chapter 9) is less accurate than are alternative viewpoints.

Dipole-dipole interactions

The energy of interaction of two dipoles[9] may be expressed as:

$$E = \frac{-2\mu_1\mu_2}{4\pi r^3 \varepsilon_0} \tag{6.5}$$

This energy corresponds to the "head-to-tail" arrangement shown in Fig. 6.4a. An alternative arrangement is the antiparallel arrangement in Fig. 6.4b. The second arrangement will be the more stable if the molecules are not too "fat." It can be shown that the energies of the two arrangements are equal if the long axis is 1.12 times as long as the short axis.[10] Both arrangements can exist only in situations in which the attractive energy is larger than thermal energies ($RT = 2.5$ kJ mol^{-1} at room temperature). In the solids and liquids in which we shall be interested, this will generally be true. At higher temperatures and in the gas phase there will be a tendency for thermal motion to randomize the orientation of the dipoles and the energy of interaction will be considerably reduced.

Dipole–dipole interactions tend to be even weaker than ion–dipole interactions and to fall off more rapidly with distance ($1/r^3$). Like ion–dipole forces, they are "directional" in the sense that there are certain preferred orientations and they are responsible for the association and structure of polar liquids such as water and hydrogen fluoride.

Induced-dipole interactions

If a charged particle, such as an ion, is introduced into the neighborhood of an uncharged, nonpolar molecule (e.g., an atom of a noble gas such as xenon), it will distort the electron cloud of the atom or molecule in much the same way as a charged cation can distort the electron cloud of a large, soft anion (Fajans' rules, pp. 129–131). The polarization of the neutral species will depend upon its inherent polarizability ("softness"), α, and on the polarizing field afforded by the charged ion, $Z^{\pm}$. The energy of such an interaction is:

$$E = -\frac{1}{2}\frac{Z^2\alpha}{r^4} \tag{6.6}$$

[9] Multiplying Eqs. 6.2, 6.4, 6.6, and 6.7 by Avogadro's number yields the correct energy for one mole of each species interacting. Since Eq. 6.5 involves *two molecules* of the polar species, multiplying by N yields the energy of *two moles* of dipoles.

[10] R. D. Gilardi, "The Analysis of the Vibrational Behavior of Bound Polar Complexes," Doctoral Dissertation, University of Maryland, College Park, **1966**, p. 18.

In a similar manner, a dipole can induce another dipole in an otherwise uncharged, nonpolar species. The energy of such an interaction is

$$E = \frac{-\mu^2\alpha}{r^6} \tag{6.7}$$

where μ is the moment of the inducing dipole.

Both of these interactions tend to be very weak since the polarizabilities of most species are not large. Because the energies vary inversely with high powers of r, they are effective only at very short distances. Their importance in chemistry is limited to situations such as solutions of ionic or polar compounds in nonpolar solvents.

Instantaneous dipole-induced dipole interactions[11]

Even in atoms in molecules which have no permanent dipole, instantaneous dipoles will arise as a result of momentary imbalance in electron distribution. Consider the helium atom, for example. It is extremely improbable that the two electrons in the 1*s* orbital of helium will be diametrically opposite each other at all times. Hence there will be instantaneous dipoles capable of inducing dipoles in adjacent atoms or molecules. Another way of looking at this phenomenon is to consider the electrons in two or more "nonpolar" molecules as synchronizing their movements (at least partially) to minimize electron–electron repulsion and maximize electron–nucleus attraction. Such attractions are extremely short-ranged and weak, as are dipole-induced dipole forces. The energy of such interactions may be expressed as:

$$E = \frac{-2\bar{\mu}\alpha}{r^6} \tag{6.8}$$

where $\bar{\mu}$ is the mean instantaneous dipole, or more conveniently as:

$$E = \frac{-3I\alpha^2}{4r^6} \tag{6.9}$$

where α is the polarizability and I is the ionization energy of the species.

London forces are extremely short-range in action (depending upon $1/r^6$) and the weakest of all attractive forces of interest to the chemist. As a result of the α^2 term, London forces increase rapidly with molecular weight, or more properly, with the *molecular volume* and the number of polarizable electrons.

It can readily be seen that molecular weight *per se* is not important in determining the magnitude of London forces as reflected by the boiling points of H_2, MW = 2, bp = 20 K; D_2, MW = 4 (a factor of *two* different), bp = 23 K; T_2, MW = 6, bp = 25 K—as well

[11] These are also sometimes referred to as *London dispersion forces* or *van der Waals forces*. The former name, although widely used (including this text!), is unfortunate inasmuch as it seems to imply that the forces tend to "disperse" the molecules, whereas they are always attractive. Usage of the term "van der Waals" forces varies: Some authors use it synonymously with London forces; others use it to mean all of the forces which cause deviation from ideal behavior by real gases. The latter would include not only London forces but also dipole interactions, etc.

as similar compounds, such as hydrocarbons containing different isotopes of hydrogen. Fluorocarbons have unusually low boiling points because tightly held electrons in the fluorine atoms have a small polarizability.

Repulsive forces

All of the interactions discussed thus far are inherently attractive and would become infinitely large at $r = 0$. Countering these attractive forces are repulsive forces resulting from nucleus–nucleus repulsion (important in the H_2 molecule) and, more important, the repulsion of inner or core electrons. At extremely short interatomic distances the inner electron clouds of the interacting atoms begin to overlap and Pauli repulsion (cf. pp. 29, 256) becomes extremely large. The repulsive energy is given by:

$$E = \frac{+k}{r^n} \tag{6.10}$$

where k is a constant and n may have various values, comparatively large. For ionic compounds, values of n ranging from 5 to 12 prove useful (p. 63), and the Lennard-Jones function, often used to describe the behavior of molecules, is sometimes referred to as the 6–12 function since it employs r^6 for the attractive energies (cf. Eq. 6.9) and r^{12} for repulsions. In any event, repulsive energies come into play only at extremely short distances.

Summary

Various forces acting on chemical species are summarized in Table 6.3. The forces are listed in order of decreasing strength from the ionic and covalent bonds to the very weak London forces. The application of a knowledge of these forces to interpretation of chemical phenomena requires a certain amount of practice and chemical intuition. In general, the importance of a particular force in affecting chemical and physical properties is related to its position in Table 6.3. For example, the boiling points of the noble gases are determined by London forces because no other forces are in operation. In a crystal of an ionic compound, however, although the London forces are still present they are dwarfed in comparison to the very strong ionic interactions and may be neglected to a first approximation (as was one in Chapter 3).

Table 6.3 Summary of chemical forces and interactions

Type of interaction	Strength	Energy–distance function
Covalent bond	Very strong	Complex, but comparatively long-range
Ionic bond	Very strong	$1/r$, comparatively long-range
Ion–dipole	Strong	$1/r^2$, short-range
Dipole–dipole	Moderately strong	$1/r^3$, short-range
Ion-induced dipole	Weak	$1/r^4$, very short-range
Dipole-induced dipole	Very weak	$1/r^6$, extremely short-range
London dispersion energy	Very weak[a]	$1/r^6$, extremely short-range

[a] Since London forces increase with increasing size and there is no limit to the size of molecules, these forces can become rather large. In general, however, they are very weak.

Table 6.4 Van der Waals distances and observed distances (pm) for some common hydrogen bonds

Bond type	A···B[a] (calc)	A···B (obs)	H···B (calc)	H···B (obs)
F—H—F	270	240	260	120
O—H···O	280	270	260	170
O—H···F	280	270	260	170
O—H···N	290	280	270	190
O—H···Cl	320	310	300	220
N—H···O	290	290	260	200
N—H···F	290	280	260	190
N—H···Cl	330	330	300	240
N—H···N	300	310	270	220
N—H···S	340	340	310	240
C—H···O	300	320	260	230

[a] The values in column 2 are not those to be obtained by the use of Table 6.1 because Hamilton and Ibers used van der Waals radii from Pauling.

SOURCE: W. C. Hamilton and J. C. Ibers, "Hydrogen Bonding in Solids," W. A. Benjamin, New York, **1968**, p. 16. Used with permission.

HYDROGEN BONDING

Although some would contend that hydrogen bonding is merely an extreme manifestation of dipole–dipole interactions, it appears to be sufficiently different to warrant a short, separate discussion. In addition, there is no universal agreement on the best description of the nature of the forces in the hydrogen bond.

We shall adopt an operational definition of the hydrogen bond: *A hydrogen bond exists when a hydrogen atom is bonded to two or more other atoms.*[12] This definition implies that the hydrogen bond cannot be an ordinary covalent bond since the hydrogen atom has only one orbital (1*s*) at sufficiently low energy to engage in covalent bonding.

Macroscopically the effects of hydrogen bonding are seen indirectly in the greatly increased melting and boiling points of such species as NH_3, H_2O, and HF. This phenomenon is well documented in freshman texts and need not be discussed further here. On the molecular level we can observe hydrogen bonding in the greatly reduced distances between atoms, distances that fall below that expected from van der Waals radii. Indeed this is a practical method of distinguishing between a true bonding situation and one in which a hydrogen atom is *close* to two atoms but bonded to only one. Table 6.4 lists

[12] W. C. Hamilton and J. A. Ibers, "Hydrogen Bonding in Solids," W. A. Benjamin, New York, **1968**, p. 13. Similar definitions are offered by G. C. Pimentel and A. L. McClellan, "The Hydrogen Bond," Freeman, San Francisco, **1969**; M. D. Joesten and L. J. Schaad, "Hydrogen Bonding," Dekker, New York, **1974**; P. Schuster et al., "The Hydrogen Bond," Vols. 1–3, North-Holland, Amsterdam, **1976**. These four books are excellent references for further reading on hydrogen bonding. There are three recent reviews: P. A. Kollman and L. C. Allen, *Chem. Rev.*, **1972**, *72*, 283; P. A. Kollman, *Appl. Electr. Struct. Theory*, **1977**, *4*, 109; and M. D. Joesten, *J. Chem. Educ.*, **1982**, *59*, 362.

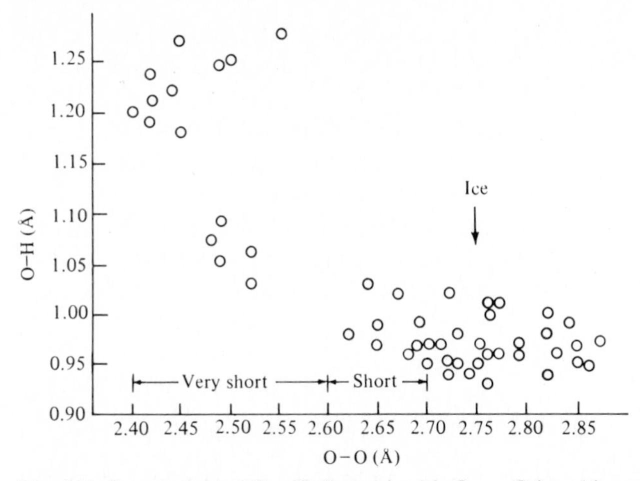

Fig. 6.5 Correlation of O—H distance with O · · · O bond length determined by neutron diffraction studies of a number of compounds containing O—H · · · O bonds. Note the short, symmetrical bonds clumped in the upper left hand corner. [From J. M. Williams and S. W. Peterson, *Spectr. Inorg. Chem.*, **1971**, *2*, 1. Reproduced with permission.]

some distances in hydrogen-bonded systems compared with the sum of the van der Waals radii for the species involved. In many hydrogen bonds, A—H · · · B, the atoms A and B are closer together than the sum of the van der Waals radii. Even more characteristic is that the hydrogen atom is considerably closer to atom B than predicted from the sum of the van der Waals radii, indicating penetration (or compression) of atom B's electron cloud by the hydrogen.

In the typical hydrogen bonding situation the hydrogen atom is attached to two very electronegative atoms. The system is usually linear and the hydrogen atom is nearer one nucleus than the other. Thus, for most of the systems in Table 6.4, the hydrogen atom is assumed to be attached to atom A by a short, "normal" covalent bond and attached to atom B by a longer, weaker "hydrogen bond." This situation is usually obtained even if both A and B are the same element. There are a few important exceptions, however. These are salts of the type $M^+HA_2^-$, where A may be the fluoride ion or the anions of certain organic acids such as acetic acid or benzoic acid. This latter type of hydrogen bonding is termed *symmetric* in contrast to the more common *unsymmetric* form. These hydrogen bonds are also characterized as strong and weak, since the short, strong hydrogen bonds, epitomized by FHF^- and some AHA^-, tend to be symmetrical, and the longer, weaker bonds are always unsymmetrical (see Fig. 6.5).

Although the subject of symmetric versus unsymmetric hydrogen bonding has received considerable attention, there is yet little understanding of the factors involved. Certainly, for the long, weak hydrogen bond we can approximate the situation by assuming the hydrogen atom to be covalently bonded to one atom and to be attracting the other. Obviously, it will be closer to the covalently bound atom than to the "dipole-attracted" atom. It is not so easy to see when or why the bond will become symmetrical, although if a resonance or delocalized molecular orbital model is invoked (see below), an analogy with the equivalent bond lengths in benzene can be appealed to. The situation is more complicated than that, however, for *unsymmetrical* FHF^- ions are known in some crystals.[13] Whether symmetrical hydrogen bonds are forced by a symmetrical environment or whether unsymmetrical bonding is induced by crystals of lower symmetry is, perhaps, a moot point.

Since hydrogen bonding occurs only when the hydrogen atom is bound to a highly electronegative atom,[14] the first suggestion concerning the nature of the hydrogen bond was that it consists of a dipole–ion or dipole–dipole interaction of the sort $A^{\delta-}—H^{\delta+} \cdots B^-$ or $A^{\delta-}—H^{\delta+} \cdots B^{\delta-}—R^{\delta+}$, where R is simply the remainder of a molecule containing the negative atom B. Support for this viewpoint comes from the fact that the strongest hydrogen bonds are formed in systems in which the hydrogen is bonded to the most electronegative elements:

$$F^- + HF \longrightarrow FHF^- \qquad \Delta H = -155^{15}\ \text{kJ mol}^{-1} \tag{6.11}$$

$$(CH_3)_2CO + HF \longrightarrow (CH_3)_2CO \cdots HF \qquad \Delta H = -46\ \text{kJ mol}^{-1} \tag{6.12}$$

$$H_2O + HOH \longrightarrow H_2O \cdots HOH\ (\text{ice}) \qquad \Delta H = -25\ \text{kJ mol}^{-1} \tag{6.13}$$

The simplistic electrostatic model qualitatively accounts for relative bond energies and the geometry (a linear arrangement maximizes the attractive forces and minimizes the repulsions). Nevertheless, there are reasons to believe that more is involved in hydrogen bonding than simply an exaggerated dipole–dipole or ion–dipole interaction. First, the shortness of hydrogen bonds indicates considerable overlap of van der Waals radii, and this should lead to considerable repulsive forces unless otherwise compensated. Secondly, symmetrical hydrogen bonds of the type F—H—F would not be expected if the hydrogen atom were covalently bound to one fluorine atom but weakly attracted by an ion–dipole force to the other. Of course, one can invoke resonance in this situation to account for the observed properties:

$$F—H \cdots F^- \longleftrightarrow F^- \cdots H—F$$

This implies delocalization of the covalent bond over both sides of the hydrogen atom. One might then ask whether there might not be a simpler molecular orbital treatment of the delocalization that would be more straightforward. The answer is yes. The mechanics will not be given here (see Chapter 15, the three-center–four-electron bond), but the results are that the covalent bond is "smeared" over all three atoms. In symmetric hydrogen bonds it is equal on both sides; in unsymmetric hydrogen bonds more electron

[13] J. M. Williams and L. F. Schneemeyer, *J. Am. Chem. Soc.*, **1973**, *95*, 5780.

[14] The least electronegative element forming true hydrogen bonds is carbon. These bonds are usually rather weak. However, there is enough known about them now, in contrast to a few years ago, to constitute a book: R. D. Green, "Hydrogen Bonding by C—H Groups," Wiley, New York, **1974**. See also U. T. Mueller-Westerhoff, *J. Am. Chem. Soc.*, **1981**, *103*, 7678. Other systems involving bridging hydrogen atoms will be discussed below.

[15] There has been considerable difficulty in determining the strength of the hydrogen bond in FHF^-, the strongest known. The value of -155 kJ mol^{-1} is a conservative one[16]; other estimates, both theoretical[17] and empirical[18] range upward, some as high as -252 kJ mol^{-1}.

[16] S. A. Harrell and D. H. McDaniel, *J. Am. Chem. Soc.*, **1964**, *86*, 4497.

[17] P. A. Kollman and L. C. Allen, *J. Am. Chem. Soc.*, **1970**, *92*, 6101; P. N. Noble and R. N. Kortzeborn, *J. Chem. Phys.*, **1970**, *52*, 5375.

[18] H. P. Dixon et al., *J. Chem. Phys.*, **1972**, *57*, 4388.

density is concentrated on the shorter link. Several workers[19–21] have attempted molecular orbital calculations of the energies involved in the hydrogen bond in O—H · · · O systems (all values expressed in kJ mol^{-1}):

	Coulson[19]	van Duijneveldt[20]	Morokuma[21]
Dipole–dipole ("electrostatic")	−25	−30.1	−33.4
Delocalization (resonance)	−33	−4.6	−34.1
Dispersion (London or VDW)	−13	−4.2	−1.0
Repulsion	+35	+20.9	+41.2
Total energy (theory)	−36	−18.0	−27.3
Experimental energy	−25		

From these numbers we can see some very rough agreement (in spite of the fact that not all of these workers defined their terms in exactly the same way) between the various calculations and between theory and experiment. The dipole–dipole, delocalization, and repulsion energies are generally about the same magnitude, with the dispersion energy being rather small. As a result, the last three terms tend to cancel; thus, ignoring everything except the first term gives a surprisingly good answer; hence the success of the simple electrostatic model. However, we should be cautious in using these numbers because a calculation that makes use of a different wave function gives a different partitioning.[22]

Allen[23] has presented a simple model for understanding the nature of the hydrogen bond and estimating the bond strength. The model involves three variables: (1) The bond moment, $\mu_{\mathrm{H-A}}$, which gives an estimate of the positive charge on the hydrogen atom, and therefore of its acceptor strength; (2) the ease of removing an electron from the donor[24] molecule, ΔI (We should expect the basicity of the donor to be related to the ease of removing an electron completely, the ionization energy.); (3) the distance over which the bond extends, R. (We have seen that short bonds are strong bonds and vice versa.) Using these variables, the energy of breaking a hydrogen bond can be estimated:

$$E = k\mu_{\mathrm{H-A}}(\Delta I)/R \quad k = 2.44 \times 10^{18}\ \mathrm{C}^{-1} \tag{6.14}$$

The model is probably more important for the insight it provides into hydrogen bonding than for calculational purposes. The reader is referred to the original paper[23] for further discussion of the model and the properties of hydrogen bonds. Since the hydrogen bond is so very important in systems ranging from inorganic solvents to DNA base-pairing, it

[19] C. A. Coulson, *Research*, **1957**, *10*, 149.

[20] J. G. C. M. van Duijneveldt-van de Ridjt and F. B. van Duijneveldt, *Theor. Chim. Acta*, **1970**, *19*, 83.

[21] K. Morokuma, *J. Chem. Phys.*, **1971**, *55*, 1236.

[22] P. A. Kollman and L. C. Allen, *Theor. Chim. Acta*, **1970**, *18*, 399.

[23] L. C. Allen, *J. Am. Chem. Soc.*, **1975**, *97*, 6921.

[24] There is confusion in the literature over the word "donor" in hydrogen-bonding systems. In this book the donor will always be an electron-pair donor or Lewis base, but there is some usage of "H-atom donor" (Brønsted acid) in the literature. The quantity ΔI is obtained as follows: The ability to form a hydrogen bond is assumed to be zero for a noble gas atom. The energy of a hydrogen bond therefore should be proportional to the difference in ionization energy of the donor atom (I_A) and the nearest noble gas atom (I_X). The energy of interest is thus $\Delta I = I_X - I_A$.

has received considerable attention.[25] That we still understand it only partially is a tribute to the complexity of hydrogen-bonding systems.

The requirement of linear geometry for a hydrogen bond is not so stringent as was once thought. In a sense, the hydrogen bond itself is nondirectional when looked at from the viewpoint of either an electrostatic attraction or overlap of the spherical 1*s* orbital. However, the repulsion from the other atoms will be minimized with a linear arrangement, and if the hydrogen bond is free to do so, it will adopt this geometry. If constraints of crystal packing or other geometrical factors prevent linearity, the hydrogen bond can still form providing the deviation from linearity is not too severe. In such cases there is a lengthening and weakening of the bond resulting from the increased repulsions.[26]

Bernal[27] has pointed out an interesting consequence of the fact that the hydrogen bond is stronger than the London dispersion forces but weaker than ionic bonds. In "soft" solids composed of covalent molecules loosely held by London forces, hydrogen bonding may strengthen and harden the structure. Examples are organic compounds containing hydrogen-bonding moieties such as the amide and carboxylic acid groups. In contrast, in ionic lattices hydrogen-bonding may weaken the structure. Examples are ionic salts, which become hydrated with a weakening of the ionic attractions (because the ions become separated by water molecules), and their replacement (in part) with weaker hydrogen-bonding links. The latter may serve as "weak links" and points of rupture.

The requirements of the stereochemistry of hydrogen bonding determine the structure of ice, e.g., the well-known fact that solid water is actually less dense than liquid water at the melting point. This is because the structure of solid water is actually rather open as a result of the requirements of hydrogen bonding (Fig. 6.6).

Finally, there are other systems such as B—H—B and W—H—W which formally meet the operational definition of hydrogen bonding given above. Such bonds can be very strong.[28] However, these systems differ fundamentally in the number of bonding electrons and in having *electropositive* atoms bonded to the hydrogen atom. To distinguish them from the electronegative, hydrogen-bonded systems, they are often termed *hydrogen-bridged* systems. (See Chapter 14.)

Hydrates and clathrates

The hydration of ions upon solution in water has been mentioned previously and its importance to solution chemistry discussed. In the solid crystalline hydrates, hydrogen bonding becomes important in addition to the ion–dipole attractions.[29] Often the water

[25] M. D. Joesten and L. J. Schaad, "Hydrogen Bonding," Dekker, New York, **1974**; A. Novak, *Struct. Bond.*, **1974**, *18*, 177; J. C. Speakman, ibid., **1972**, *12*, 141; P. A. Kollman and L. C. Allen, *Chem. Rev.*, **1972**, *72*, 283.

[26] W. C. Hamilton, *Ann. Rev. Phys. Chem.*, **1962**, *13*, 19; R. Chidambaram and S. K. Sikka, *Chem. Phys. Letters*, **1968**, *2*, 162.

[27] J. D. Bernal, in "Hydrogen Bonding," D. Hadzi, ed., Pergamon, Elmsford, N.Y., **1959**.

[28] See W. G. Evans et al., *Inorg. Chem.*, **1968**, *7*, 1746.

[29] W. C. Hamilton and J. A. Ibers, "Hydrogen Bonding in Solids," W. A. Benjamin, New York, **1968**, pp. 204–221.

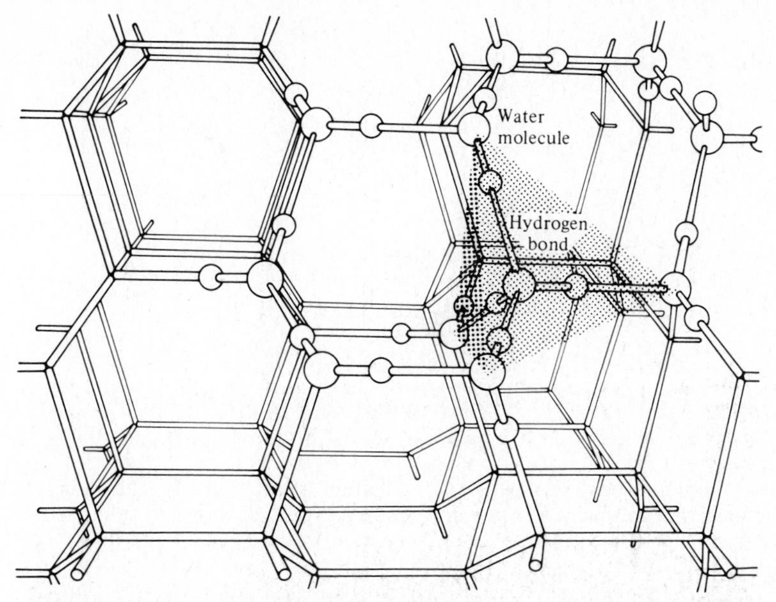

Fig. 6.6 The "open" structure of normal ice that results from the directionality of the hydrogen bonding. [From Richard E. Dickerson and Irving Geis, "Chemistry, Matter, and the Universe," copyright © 1976 by W. A. Benjamin, Inc., Menlo Park, California.]

molecules serve to fill in the interstices and bind together a structure which would otherwise be unstable because of disproportionate sizes of the cation and anion. For example, both $FeSiF_6 \cdot 6H_2O$ and $Na_4XeO_6 \cdot 8H_2O$ are well-defined, crystalline solids. The anhydrous materials are unknown. The large, highly charged anions presumably repel each other too much to form a stable lattice unless there are water molecules present. In general, some water molecules will be found coordinated directly to the cation and some will not. All the water molecules will be hydrogen-bonded, either to the anion or to another water molecule.

A specific example of these types of hydrates is $CuSO_4 \cdot 5H_2O$. Although there are five molecules of water for every Cu^{+2} ion, only *four* are coordinated to the cation, its 6-coodination being complete by coordination from SO_4^{-2} (Fig. 6.7a). The fifth water molecule is held in place by hydrogen bonds, O—H—O, between it and two coordinated water molecules and the coordinated sulfate anions (Fig. 6.7b). Dehydration to $CuSO_4 \cdot 3H_2O$, $CuSO_4 \cdot H_2O$, and eventually anhydrous $CuSO_4$ results in the water molecules coordinated to the copper being gradually replaced by oxygen atoms from the sulfate.[30]

In some cases hydrates form with neutral molecules. Examples of this type are compounds of the limiting formula $6X \cdot 46H_2O$ where X = Ar, Kr, Xe, Cl_2, CH_4, etc. The

[30] A. G. Wells, "Structural Inorganic Chemistry," 4th ed., Clarendon Press, Oxford, **1975**, pp. 557–558, 566, 897.

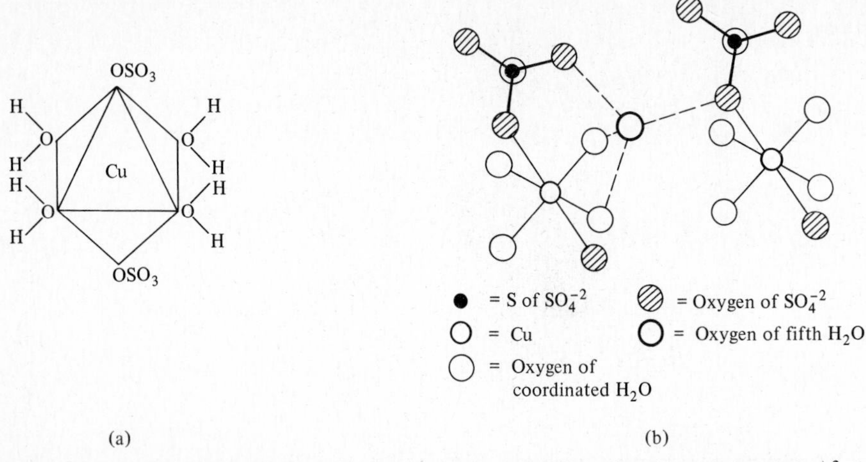

Fig. 6.7 Structure of copper(II) sulfate pentahydrate. (a) Coordination sphere of Cu^{+2}, four water molecules and two sulfate ions; (b) Position of fifth water molecule (oxygen shown by heavy circle). Normal covalent bonds depicted by solid lines, O—H—O hydrogen bonds depicted by dashed lines.

basic building block for this type of structure is a dodecahedron formed from 20 molecules of water (Fig. 6.8). Each water molecule is bonded to three others by hydrogen bonds. Half of the oxygen atoms have their fourth coordination position occupied by a hydrogen atom that can hydrogen bond to adjacent polyhedra, and the other half have a lone pair at the fourth position, which can accept a hydrogen bond from an adjacent polyhedron. When these dodecahedra pack together to form larger units relatively large voids are formed in the structure, and it is in these spaces that the *guest* molecules, X, reside. These gas hydrates in which the guest molecules are not bound chemically but are retained by the structure of the *host* are called *clathrates*. Since the structure can exist with incomplete

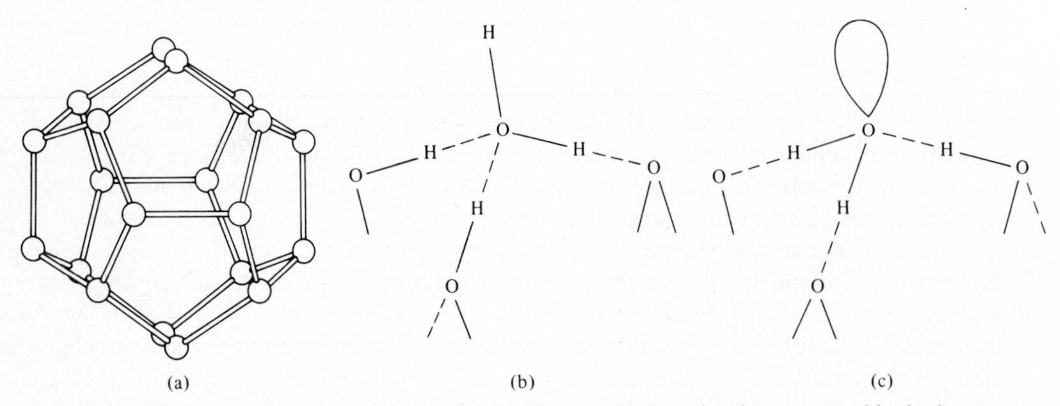

Fig. 6.8 (a) Pentagonal dodecahedron composed of twenty H_2O molecules connected by hydrogen bonds. (b) Apex of dodecahedron with external hydrogen atom capable of accepting a lone pair from an adjacent polyhedron to form linking hydrogen bond. (c) Apex of dodecahedron with external lone pair capable of hydrogen bonding to a hydrogen in an adjacent polyhedron.

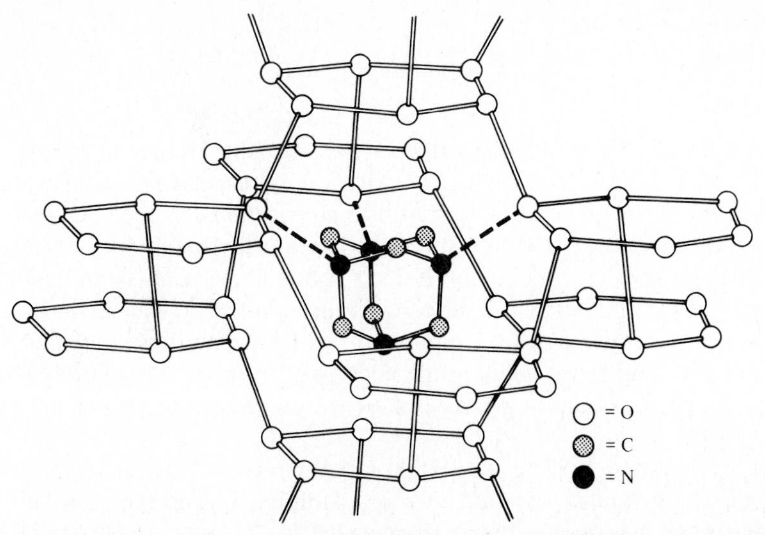

Fig. 6.9 Perspective drawing of the crystal structure of $(CH_2)_6N_4 \cdot 6H_2O$. The solid lines represent the O · · · H—O bonds in the clathrate cage and the dashed lines the N · · · H—O bonds. [From G. A. Jeffreys and T. C. W. Mak, *Science*, **1965**, *149*, 178. Copyright 1965 by the American Association for the Advancement of Science. Reproduced with permission.]

filling of the holes, formulas such as $6X \cdot 46H_2O$ must be considered limiting formulas that are not always realized in practice.[31]

Not all clathrates are hydrates. Other well-known examples have host lattices formed from hydrogen-bonded aggregates of hydroquinone, phenol, and similar organic compounds. Nonhydrogen-bonded host structures are also known. These consist of three-dimensional lattices composed of coordination polymers. In these the cage structures are formed by ligands, such as CN^- and SCN^-, which coordinate to metal ions at both ends (see Chapter 11). Perhaps the best known of this type of clathrate is the $Ni(CN)_2NH_3 \cdot M$ compound, where M is benzene, thiophene, furan, pyrrole, aniline, or phenol.[32]

Jeffrey and Mak[33] have described an interesting intermediate between true clathrates and hydrogen-bonded hydrates. In hexamethylenetetramine hexahydrate, $(CH_2)_6N_4 \cdot 6H_2O$, the hexamethylenetetramine molecule is trapped inside a cage composed of water molecules. Unlike the gas clathrates discussed above, three nitrogens in the trapped molecule are hydrogen-bonded to water molecules in the cage (Fig. 6.9).

Other clathrates are known in which ions participate in building the cage network, and as more structures are elucidated, it is likely that a complete transition series will be possible from the "pure," nonbonded guest–host relation in gas clathrates to the intimate association in crystalline ionic hydrates.

---

[31] For reviews of hydrate clathrates, see G. A. Jeffrey, *Acc. Chem. Res.*, **1969**, *2*, 344; G. A. Jeffrey and R. K. McMullan, *Progr. Inorg. Chem.*, **1967**, *8*, 43.

[32] M. M. Hagan, "Clathrate Inclusion Compounds," Van Nostrand-Reinhold, New York, **1962**, p. 67, lists over forty complexes which will serve as hosts in forming clathrate compounds.

[33] G. A. Jeffrey and T. C. W. Mak, *Science*, **1965**, *149*, 178.

EFFECTS OF CHEMICAL FORCES

Melting and boiling points

Fusion and vaporization result from supplying sufficient thermal energy to a crystal to overcome the potential energy holding it together. It should be noted that in most cases the melting and vaporization of a crystal does not result in *atomization*, i.e., the complete breaking of *all* chemical forces. In order to understand the relationship between chemical forces and physical properties such as melting and boiling points, it is necessary to compare the binding energies of the species in the vapor with those in the crystal. Only the *difference* between these two energies must be supplied in order to vaporize the solid. The following discussion will emphasize energy differences with respect to variation in melting and boiling points, but it should be realized that entropy effects can also be very important.

Crystals held together solely by London dispersion forces melt at comparatively low temperatures and the resulting liquids vaporize easily. Examples of this type are the noble gases which boil at temperatures ranging from −269 °C (He) to −62 °C (Rn). Many organic and inorganic molecules with zero dipole moments such as CH_4 (bp = −162 °C), BF_3 (bp = −101 °C), and SF_6 (sublimes at −64 °C) fall into this category. Because London forces increase greatly with polarizability, many larger molecules form liquids or even solids at room temperature despite having only this type of attraction between molecules. Examples are $Ni(CO)_4$ (bp = 43 °C), CCl_4 (bp = 77 °C), borazine, $B_3N_3H_6$ (bp = 53 °C), and trimeric phosphonitrilic chloride, $P_3N_3Cl_6$ (mp = 114 °C).

It should be noted that these compounds are a trivial illustration of the principle stated in the first paragraph. Although all of the molecules contain very strong *covalent bonds*, none is broken on melting or vaporization, and hence they play no part in determining the melting and boiling points.

Molecules in polar liquids such as water, liquid ammonia, sulfuric acid, and chloroform are held together by dipole–dipole and hydrogen–bonding interactions. For molecules of comparable size, these latter are stronger than London forces resulting in the familiar trends in boiling points of nonmetal hydrides. For the heavier molecules, such as H_2S, H_2Se, PH_3, and HI, dipole effects are not particularly important (the electronegativities of the nonmetals are very similar to that of hydrogen) and the boiling points are low and increase with increasing molecular weight. The first member of each series (H_2O, NH_3, HF) is strongly hydrogen bonded in the liquid state and has a higher boiling point.

Ionic compounds are characterized by very strong electrostatic forces holding the ions together. Vaporization results in ion pairs and other small clusters in the vapor phase. Although the stabilizing energies of these species are large, they are considerably less than those of the crystals. Assuming a hard-sphere model as a first approximation, the difference in electrostatic energies of an ion pair in the gas and the solid lattice would lie in their Madelung constants. For NaF, $A = 1.00$ for an ion pair, 1.75 for the lattice. We should thus expect that if crystalline sodium fluoride vaporized to form ion pairs, the bond energy would be slightly more than half ($1.00/1.75 = 0.57$) of the lattice energy. There are several factors that help stabilize the species in the gas phase and make their formation somewhat less costly. Polarization can occur more readily in a single ion pair than in the lattice. This results in a somewhat greater covalent contribution and shorter bond distances in the gas phase. Secondly, in addition to ion pairs there are small clusters of ions with a greater number of interactions and more attractive energy. It is not surprising to learn therefore, that vaporization costs only about one-fourth of the lattice energy, *not* almost one-half (Table 6.5). Nevertheless, since lattice energies are large, the energy necessary to vaporize an ionic compound is large and responsible for the high boiling points of ionic compounds.

Increasing the ionic charges will certainly increase the lattice energy of a crystal. For compounds which are predominantly ionic, increased ionic charges will result in increased melting and boiling points. Examples are NaF, mp = 997 °C, and MgO, mp = 2800 °C.

The situation is not always so simple as in the comparison of sodium fluoride and magnesium oxide. According to Fajans' rules, increasing charge results in increasing covalency, especially for small cations and large anions. Covalency *per se* does not necessarily favor either high or low melting and boiling points. For species which are strongly covalently bonded in the solid, but have weaker or fewer covalent bonds in the gas phase, melting and boiling points can be extremely high. Examples are carbon in the diamond

Table 6.5 Dissociation energies of the alkali halides for the solid and gas phases (kJ mol^{-1})[a]

Compound	$M{-}X_{(g)} \rightarrow M^+_{(g)} + X^-_{(g)}$	$MX_{(s)} \rightarrow M^+_{(g)} + X^-_{(g)}$	E_{subl}	Ratio
LiF	766	1033	268	0.26
LiCl	636	845	209	0.25
LiBr	615	799	184	0.23
LiI	573	741	167	0.23
NaF	644	916	272	0.30
NaCl	556	778	222	0.28
NaBr	536	741	205	0.28
NaI	506	690	184	0.27
KF	582	812	230	0.29
KCl	494	707	213	0.30
KBr	477	678	201	0.30
KI	448	640	192	0.30
RbF	565	778	213	0.27
RbCl	498	686	188	0.27
RbBr	464	661	197	0.30
RbI	439	623	184	0.30
CsF	548	749	201	0.27
CsCl	464	653	197	0.29
CsBr	448	632	184	0.29
CsI	414	602	188	0.31

[a] Gas-phase data are from the bond energies in Appendix E corrected to the ionic case by addition of the ionization energy and electron affinity. Lattice energies are from the best values in Table 3.3. Energies of sublimation (assuming ion pairing) are the difference between the energy of the lattice and that of the ion pairs. The *ratio* is that of E_{subl}/U which yields the fraction of the energy "lost" on sublimation.

Table 6.6 Melting points of potassium and silver halides

KF = 880 °C	AgF = 435 °C
KCl = 776 °C	AgCl = 455 °C
KBr = 730 °C	AgBr = 434 °C

and graphite forms (sublime about 3700 °C) and silicon dioxide (melts at 1710 °C, boils above 2200 °C). For example, in the latter compound the transition consists of changing four strong tetrahedral σ bonds in the solid polymer to two σ and two relatively weak π bonds in the isolated gas molecules:

$$\mathrm{Si(O-)_4} \longrightarrow \mathrm{O{=}Si{=}O} \qquad (6.15)$$

On the other hand, if the covalent bonds are almost as stable and as numerous in the gas phase molecules as in the solid, vaporization takes place readily. Examples are the depolymerization reactions that take place at a few hundred degrees. For example, red phosphorus sublimes and recondenses as white phosphorus:[34]

$$\left[\mathrm{P_4{-}P_4}\right]_n \longrightarrow 2n\ \mathrm{P_4} \qquad (6.16)$$

Thus increased covalent bonding resulting from Fajans-type phenomena can *lower* the transition temperatures. For example, both the potassium halides and silver halides (except AgI) crystallize in the NaCl structure. The potassium and silver ions closely resemble each other in size (K^+ = 133 pm, Ag^+ = 126 pm), yet the melting points of the halides are considerably different (Table 6.6). The greater covalent character of the silver halide bond compared with the potassium halide bond helps stabilize discrete AgX molecules in the liquid and thus makes the melting points of the silver compounds lower than those of the potassium compounds. A similar comparison can be made between the predominantly ionic species CsF and BaF_2 and the more covalent species KBr and $CaBr_2$ (Table 6.7). The change from 1:1 to 1:2 composition in the highly ionic fluorides produces the expected increase in lattice energy and corresponding increase in the transition temperatures. For the more covalent bromides, however, the molecular species $CaBr_2$ (in the gas phase and possibly to some extent in the liquid) has sufficient stability via its covalency so that the melting point is only slightly higher than that of KBr, and the boiling point is actually *lower*.

[34] The exact structure of red phosphorus is unknown, but this structure has been suggested. The exact structure is irrelevant to the argument here.

Table 6.7 Melting and boiling points of some alkali and alkaline earth halides

Melting point (°C)		Boiling point (°C)	
KBr = 730	$CaBr_2$ = 765	KBr = 1380	$CaBr_2$ = 812
CsF = 684	BaF_2 = 1280	CsF = 1250	BaF_2 = 2137

In extreme cases of Fajans' effects, as in BeI_2 and transition metal bromides and iodides, the stabilization resulting from covalency is very large. Distortion of the lattice occurs and direct comparison with ionic halides is difficult. For metal halides the boiling points of these compounds are comparatively low as expected: BeI_2 = 590 °C, ZnI_2 = 624 °C, $FeCl_3$ = 315 °C. The extreme of this trend is for the covalent forces to become so strong as to define discrete molecules even in the solid (e.g., Al_2Br_6, mp = 97 °C, bp = 263 °C). At this point we have come full circle and are back at the SF_6 and CCl_4 situation.

Solubility

Solubility and the behavior of solutes is a complicated subject,[35] and only a brief outline will be given here. A further discussion of solutions will be found in Chapter 8.

Solutions of nonpolar solutes in nonpolar solvents represent the simplest type of solution. The forces involved in solute–solvent and solvent–solvent interactions are all London dispersion forces and relatively weak. The presence of these forces resulting in a condensed phase is the only difference from the mixing of ideal gases. As in the latter case, the only driving force is the entropy ("randomness") of mixing. In an ideal solution ($\Delta H_{\text{mixing}} = 0$) at constant temperature the free energy change will be composed solely of the entropy term:

$$\Delta G = \Delta H - T\Delta S \qquad \textbf{(6.17)}$$

$$\Delta G = -T\Delta S \qquad (\text{for } \Delta H = 0) \qquad \textbf{(6.18)}$$

The change in entropy for the formation of a solution of this type is:[36]

$$\Delta S = -R(n_A \ln x_A + n_B \ln x_B) \qquad \textbf{(6.19)}$$

where x_A and x_B are the mole fractions of solute and solvent. For an equimolar mixture of "solute" and "solvent" the change in free energy upon solution at room temperature is rather small, only -1.7 kJ mol^{-1}.

[35] For more detailed discussions of solute behavior, see J. H. Hildebrand and R. L. Scott, "The Solubility of Non-electrolytes," Van Nostrand-Reinhold, New York, **1950**, and R. A. Robinson and R. H. Stokes, "Electrolyte Solutions," 2nd rev. ed., Butterworths, London, **1959**.

[36] For the origin of this expression and an excellent discussion of the thermodynamics of solution formation, see G. Barrow, "Physical Chemistry," McGraw-Hill, New York, **1961**.

At the other extreme from the ideal solutions of nonpolar substances are solutions of ionic compounds in a very polar solvent such as water. In order for an ionic compound to dissolve, the Madelung energy or electrostatic attraction between the ions in the lattice must be overcome. In a solution in which the ions are separated by molecules of a solvent with a high dielectric constant ($\varepsilon_{H_2O} = 81.7\varepsilon_0$) the attractive force will be considerably less (see p. 54).

The process of solution of an ionic compound in water may be considered by a Born-Haber type of cycle. The overall enthalpy of the process is the sum of two terms, the enthalpy of dissociating the ions from the lattice (the lattice energy) and the enthalpy of introducing the dissociated ions into the solvent (the solvation energy):

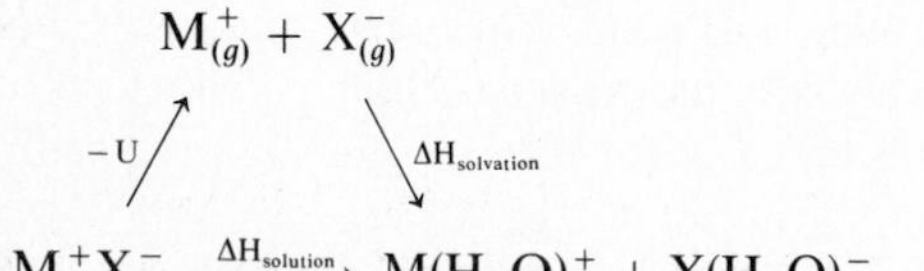

$$\begin{array}{ccc} & M^+_{(g)} + X^-_{(g)} & \\ {}^{-U}\nearrow & & \searrow^{\Delta H_{\text{solvation}}} \\ M^+X^-_{(s)} & \xrightarrow{\Delta H_{\text{solution}}} & M(H_2O)^+_x + X(H_2O)^-_y \end{array}$$

Two factors will contribute to the magnitude of the enthalpy of solvation. One is the inherent ability of the solvent to coordinate strongly to the ions involved. Polar solvents are able to coordinate well through the attraction of the solvent dipole to the solute ions. The second factor is the type of ion involved, particularly its size. The strength and number of interactions between solvent molecules and an ion will depend upon how large the latter is. The lattice energy of the solute also depends upon ionic size. The forces in the lattice are inherently stronger (ion–ion) than those holding the solvent molecules to the ion (ion–dipole), but there are several of the latter interactions for each ion. As a result, the enthalpy of solvation is roughly of the same order of magnitude as the lattice enthalpy, and so the total enthalpy of solution can be either positive or negative depending upon the particular compound. When the enthalpy of solution is negative, the free energy of solution is especially favorable since then the enthalpy and entropy of solution reinforce each other.

In many cases the enthalpy of solution for ionic compounds in water is positive. In these cases we find the solution cooling as the solute dissolves. The mixing tendency of entropy is *forcing* the solution to do work to pull the ions apart, and since in an adiabatic process such work can be done only at the expense of internal energy, the solution cools. If the enthalpy of solution is sufficiently positive, the entropy factor will not be able to overcome it and the compound will be insoluble. Thus some ionic compounds, such as $BaSO_4$, CaF_2, and Al_2O_3, are essentially insoluble in water.

The fact that the solubility of a salt depends critically upon the enthalpy of solution raises an interesting question concerning the magnitude of this quantity. Obviously, a large solvation enthalpy contributes toward a favorable enthalpy of solution. However, we find that the solvation enthalpy alone provides us with little predictive usefulness. Water soluble salts are known with both large (CaI_2, -2180 kJ mol^{-1}) and small (KI, -611 kJ mol^{-1}) hydration energies; insoluble salts are also known with large (CaF_2, -6782 kJ mol^{-1}) or small (LiF, -1004 kJ mol^{-1}) hydration energies. It is apparent that the hydration energies alone do not determine the solubility. Countering the hydration energies in these cases is the lattice energy. Both lattice energy (p. 59) and hydration energy (pp. 132, 264) are favored by large charge (Z) and small size (r). The difference lies in the nature of the dependence upon distance. The Born-Landé equation for the lattice energy (Eq. 3.14) may be written as a function of distance:

$$U = f_1\left(\frac{1}{r_+ + r_-}\right) \tag{6.20}$$

The simplest equation for the enthalpies of hydration of the cation and anion (Eq. 3.70) may be rewritten as:

$$\Delta H_{\text{hyd}} = f_2\left(\frac{1}{r_+}\right) + f_3\left(\frac{1}{r_-}\right) \tag{6.21}$$

Now the *lattice energy* is inversely proportional to the *sum of the radii*, whereas the *hydration enthalpy* is the *sum of two quantities* inversely proportional to the *individual radii*. Clearly the two functions will respond differently to variation in r_+ and r_-. Without delving into the details of the calculations, we may note that Eq. 6.20 is favored relative to Eq. 6.21 when $r_+ = r_-$ and the reverse is true for $r_- \ll r_+$ or $r_- \gg r_+$. To express it in terms of a physical picture, the lattice energy is favored when the ions are similar in size—the presence of either a much larger cation or a much larger anion can effectively reduce it. In contrast, the hydration enthalpy is the sum of the two individual ion enthalpies, and if just one of these is very large (from a single, small ion), the total may still be sizable even if the counterion is unfavorable (because it is large). The effects of this principle may be seen from the solubility of the alkali halides in water. Lithium fluoride is simultaneously the least soluble lithium halide and the least soluble alkali fluoride. Cesium iodide is the least soluble cesium halide and the least soluble alkali iodide. The most soluble salts in the series are those with the most disparate sizes, cesium fluoride and lithium iodide.[37]

The enthalpy of solution has been discussed somewhat more quantitatively by Morris.[38] He has pointed out the relation between the enthalpy of solution and the difference between the hydration enthalpy of the cation and that of the anion. This difference will be largest when the cation and anion differ most in size (Fig. 6.10). In these cases the enthalpy of solution tends to be large and negative and favors solution. When the hydration enthalpies (and the sizes) are more nearly alike, the crystal is favored. When entropy effects are added, a very nice correlation with the solubility and the free energy solution is found (Fig. 6.11).

There is a very practical consequence of the relation of solubility to size. It is often possible to make a large, complex ion from a metal and several ligands that is stable in solution but difficult to isolate without decomposition. Isolation of such large complex ions is facilitated by attempting to isolate them as salts of equally large counterions. This favors the stability of the crystalline state relative to solution and makes it easier to

[37] For a very thorough discussion of enthalpy, entropy, and the solubility of ionic compounds, see D. A. Johnson, "Some Thermodynamic Aspects of Inorganic Chemistry," Cambridge University Press, London, **1968**, Chap. 5.

[38] D. F. C. Morris, *Struct. Bonding*, **1968**, *4*, 63; **1968**, *6*, 157.

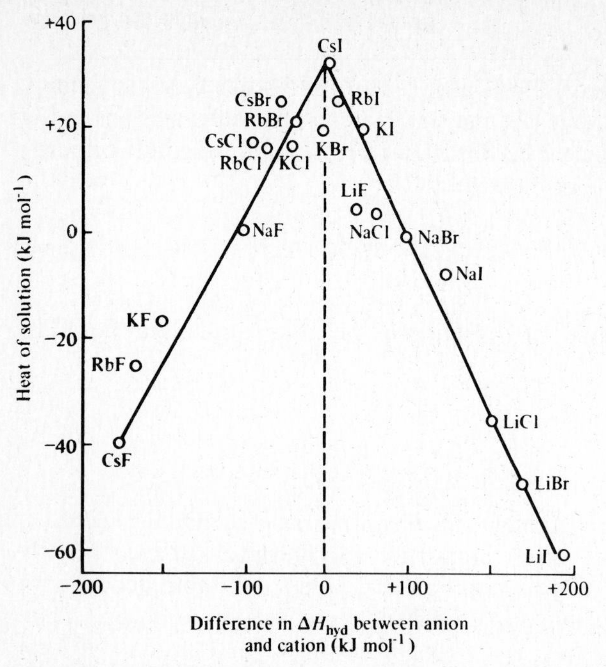

Fig. 6.10 Relation between the heat of solution of a salt and the individual heats of hydration of the component ions. [From D. F. C. Morris, *Struct. Bonding*, **1969**, *6*, 157. Reproduced with permission.]

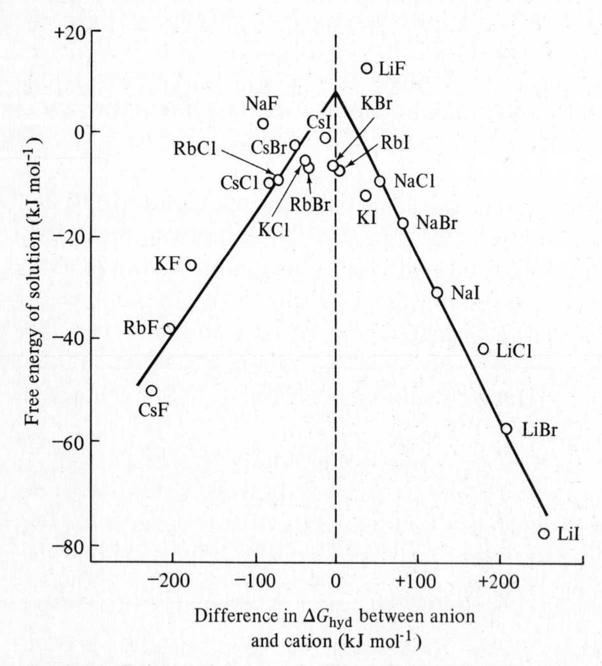

Fig. 6.11 Relation between the free energy of solution of a salt and the individual free energies of hydration of the component ions. [From D. F. C. Morris, *Struct. Bonding*, **1969**, *6*, 157. Reproduced with permission.]

obtain crystals of the desired complex. For example, the $[Ni(CN)_5]^{-3}$ ion was known to exist in solution but when solutions were evaporated, even in the presence of saturated KCN, only $K_2Ni(CN)_4 \cdot H_2O$ could be isolated. However, addition of large complex ions of chromium such as hexaamminechromium(III), $[Cr(NH_3)_6]^{+3}$, and tris(ethylenediamine)chromium(III), $[Cr(NH_2CH_2CH_2NH_2)_3]^{+3}$, allows the separation of $[Cr(NH_3)_6][Ni(CN)_5] \cdot 2H_2O$ and $[Cr(NH_2CH_2CH_2NH_2)_3][Ni(CN)_5] \cdot 1.5H_2O$, both of which are stable at room temperature.[39]

The insolubility of ionic compounds in nonpolar solvents is a similar phenomenon. The solvation energies are limited to those from ion-induced dipole forces, which are considerably weaker than ion–dipole forces and not large enough to overcome the very strong ion–ion forces of the lattice.

The reason for the insolubility of nonpolar solutes in some polar solvents such as water is less apparent. The forces holding the solute molecules to each other (i.e., the force tending to keep the crystal from dissolving) are very weak London forces. The interactions between water and the solute (dipole-induced dipole) are also weak but expected to be somewhat stronger than London forces. It might be supposed that this small solvation energy plus the entropy of mixing would be sufficient to cause a nonpolar solute to dissolve. In fact, it does not because any entropy resulting from the disordering of the hydrogen-bonded structure of the solvent water is more than offset by the loss of energy from the breaking of hydrogen bonds. Anthropomorphically we might say that the solute would willingly dissolve but that the water would rather associate with itself.

We can summarize the energetics of solution as follows. There will usually be an entropy driving force favoring solution. In cases where the enthalpy is negative, zero, or slightly positive, solution will take place. If the enthalpy change accompanying solution is too positive, solution will not occur. In qualitatively estimating the enthalpy effect, solute–solute, solvent–solvent, and solute–solvent interactions must be considered:

$$\Delta H_{\text{solution}} = \Delta H_{\text{solute–solvent}} - \Delta H_{\text{solute–solute}} - \Delta H_{\text{solvent–solvent}} \tag{6.22}$$

where the various energies result from ion–ion, ion–dipole, ion-induced dipole, dipole–dipole, and London forces.

PROBLEMS

6.1. From Fig. 6.1 describe the crystal structure of solid argon.

6.2. Confirm the statement made on p. 257 that the van der Waals radius of argon is 190 pm, using Fig. 6.1 and the knowledge that the unit cell is 535 pm on each edge.

6.3. Predict the internuclear distances in the following molecules and lattices by use of the appropriate van der Waals, ionic, covalent, and atomic radii. In those cases where two or more sets of values are applicable, determine which yield the values closest to the experimental values.

---

[39] For discussions of this subject, see F. Basolo, *Coord. Chem. Rev.*, **1968**, *3*, 213, and D. H. McDaniel, *Ann. Rep. Inorg. Gen. Synth.—1972*, 293, **1973**.

System	Distance	r (pm)
LiF molecule	Li—F	155
LiF crystal	Li—F	201
CsI molecule	Cs—I	332
CsI crystal	Cs—I	395
LiI molecule	Li—I	239
LiI crystal	Li—I	302
XeF_4 molecule	Xe—F	194
XeF_4 crystal	F—F (different molecules)	313
H_2O molecule	H—O	96
$SnCl_4$ molecule	Sn—Cl	233

6.4. At one time the melting points of the fluorides of the third-row elements were taken to indicate a discontinuity between ionic bonding (AlF_3) and covalent bonding (SiF_4). Explain the observed trend assuming that the bond polarity decreases uniformly from NaF to SF_6.

NaF = 997 °C AlF_3 = 1040 °C PF_5 = −94 °C

MgF_2 = 1396 °C SiF_4 = −90 °C SF_6 = −56 °C

6.5. List the following in order of increasing boiling point:

H_2O Xe LiF LiI H_2 BaO $SiCl_4$ SiO_2

6.6. The majority of clathrate compounds involve hydrogen bonding in the host cages. Discuss how the intermediate nature of the hydrogen bond (i.e., stronger than van der Waals forces, weaker than ionic forces) is related to the prevalence of hydrogen-bonded clathrates.

6.7. Two forms of boron nitride are known. The ordinary form is a slippery gray material. The second, formed artificially at high pressures, is the second hardest substance known. Both remain as solids at temperatures approaching 3000 °C. Suggest structures.

6.8. Predict the length of the following bonds: H_2O, HCl, NF_3, CF_4, H_2S, SF_2.

6.9. The Schomaker-Stevenson relationship states that heteropolar bonds are always stronger and shorter than hypothetical, purely covalent bonds between the same atoms. In an ionic crystal, would you expect some covalency to shorten or lengthen the bond? Explain. [R. D. Shannon and H. Vincent, *Struct. Bond.*, **1974**, *19*, 1.]

6.10. Find the melting points and boiling points of the elements or compounds listed. For each series, tabulate the data and explain the trends you observe in terms of the forces involved:

a. He, Ne, Ar, Kr, Xe
b. H_2O, H_2S, H_2Se, H_2Te
c. CH_4, CH_3Cl, CH_2Cl_2, $CHCl_3$, CCl_4
d. Carbon, nitrogen, oxygen, fluorine, neon

6.11. Consider the sizes of the isoelectronic species N^{-3}, O^{-2}, F^{-}, and Ne. Discuss the forces operating. *Caveat*! Be careful in choosing which numbers to use in your discussion.

6.12. The "stability" of noble gas configurations was discussed in Chapter 3, where it was pointed out that many ions are *not* stable; i.e., they are endothermic with respect to the corresponding atoms, but they are stabilized by the ionic lattice. However, some chemists argue that these ions *are* stable because they exist in solution as well as in lattices. Discuss.

6.13. Consider the ions $[\phi_3B{-}C{=}N{-}B\phi_3]^-$ and $[\phi_3P{=}N{=}P\phi_3]^+$.

a. Work out the electronic structures of these ions in detail including assigning formal charges.

b. Compare the geometries and other similarities or differences.

c. How should they prove useful in inorganic synthesis?

[C. M. Giandomenico et al., *J. Am. Chem. Soc.*, **1981**, *103*, 1407; S. W. Kirtley et al., ibid., **1974**, *96*, 7601.]

6.14. Water is well known to have an unusually high heat capacity. Not so well known is that liquid XeF_6 also has a high heat capacity compared to "normal" liquids such as argon, carbon tetrachloride, or sulfur dioxide. From your knowledge of the structures of the solids and the gaseous molecules of these materials (most of them are sketched in this text), explain the "anomalous" heat capacity of XeF_6.

6.15. Find the solubilities in water of the alkali halides. Calculate the molar concentration of a saturated solution for each and plot them in matrix form with columns headed F, Cl, Br, I, and horizontal rows labeled Li, Na, K, Rb, and Cs. Discuss any trends you notice.

6.16. In these first six chapters you have encountered many tables of atomic and molecular properties. They may be classified into two groups: (1) Radial wave functions, ionization energies, electron affinities, etc.; (2) ionic radii, covalent radii, electronegativities, etc.

a. What distinguishes and separates these two groups?

b. Leland Allen (pers. comm.) has suggested that it is the *problems* associated with Group 2 that makes chemistry a distinct and interesting science, not just a subbranch of chemical physics. Discuss.

6.17. Predict which of the following bonding interactions will be the stronger:

a. O=O or O—O

b. C—C or Si—Si

c. Ne—Ne or Xe—Xe

d. Li^+F^- or $Mg^{+2}O^{-2}$ (ion pair)

e. Li^+F^- or Ba^{+2} Te^{-2} (ion pair)

f. Li^+F^- or C—C (in diamond)

7

Acid-base chemistry

ACID–BASE CONCEPTS

The first point to be made concerning acids and bases is that so-called acid–base "theories" are in reality *definitions* of what an acid or base is; they are not theories in the sense of valence bond theory or molecular orbital theory. In a very real sense, we can make an acid be anything we wish—the differences between the various acid–base concepts are not concerned with which is "right" but which is *most convenient to use in a particular situation.* All of the current definitions of acid–base behavior are compatible with each other. In fact, one of the objects in the following presentation of *many* different definitions is to emphasize their basic parallelism and hence to direct the student toward a cosmopolitan attitude toward acids and bases which will stand him in good stead in dealing with various chemical situations, whether they be in aqueous solutions of ions, organic reactions, nonaqueous titrations, or other situations.

Brønsted-Lowry definition

In 1923 J. N. Brønsted and T. M. Lowry independently[1] suggested that *acids be defined as proton donors and bases as proton acceptors.* For aqueous solutions the Brønsted-Lowry definition does not differ appreciably from the Arrhenius definition of hydrogen ions (acids) and hydroxide ions (bases):

$$\underset{\text{Pure solvent}}{2H_2O} \rightleftharpoons \underset{\text{Acid}}{H_3O^+} + \underset{\text{Base}}{OH^-} \qquad \textbf{(7.1)}$$

The usefulness of the Brønsted-Lowry definition lies in its ability to handle any protonic

[1] J. N. Brønsted, *Recl. Trav. Chim. Pays-Bas*, **1923**, *42*, 718; T. M. Lowry, *Chem. Ind. (London)*, **1923**, *42*, 43.

solvent such as liquid ammonia or sulfuric acid:

$$\underset{\text{Acid}}{NH_4^+} + \underset{\text{Base}}{NH_2^-} \longrightarrow \underset{\text{Neutralization product}}{2NH_3} \tag{7.2}$$

$$\underset{\text{Acid}}{H_3SO_4^+} + \underset{\text{Base}}{HSO_4^-} \longrightarrow \underset{\text{Neutralization product}}{2H_2SO_4} \tag{7.3}$$

In addition, other proton-transfer reactions that would not normally be called neutralization reactions but which are obviously acid–base in character may be treated as readily:

$$\underset{\text{Acid}}{NH_4^+} + \underset{\text{Base}}{S^{-2}} \longrightarrow \underset{\text{Base}}{NH_3} + \underset{\text{Acid}}{HS^-} \tag{7.4}$$

Chemical species that differ from each other only to the extent of the transferred proton are termed *conjugates* (connected by brackets in Eq. 7.4). Reactions such as the above proceed in the direction of forming weaker species. The stronger acid and the stronger base of each conjugate pair react to form the weaker acids and bases. The emphasis which the Brønsted-Lowry definition places on competition for protons is one of the assets of working in this context, but it also limits the flexibility of the concept. However, as long as one is dealing with a protonic solvent system, the Brønsted-Lowry definition is as useful as any. The acid–base definitions given below were formulated in an attempt to extend acid–base concepts to systems not containing protons.

Lux-Flood definition

In contrast to the Brønsted-Lowry theory, which emphasizes the proton as the principal species in acid–base reactions, the definition proposed by Lux[2] and extended by Flood[2] describes acid–base behavior in terms of the oxide ion. This acid–base concept was advanced to treat nonprotonic systems which were not amenable to the Brønsted-Lowry definition. For example, in high temperature inorganic melts, reactions such as the following take place:

$$\underset{\text{Base}}{CaO} + \underset{\text{Acid}}{SiO_2} \longrightarrow CaSiO_3 \tag{7.5}$$

The base (CaO) is an *oxide donor* and the acid (SiO_2) is an *oxide acceptor*. The usefulness of the Lux-Flood definition is mostly limited to systems such as molten oxides.

[2] H. Lux, *Z. Elektrochem.*, **1939**, *45*, 303; H. Flood and T. Förland, *Acta Chem. Scand.*, **1947**, *1*, 592, 781; H. Flood, T. Förland, and B. Roald, ibid., **1947**, *1*, 790.

This approach emphasizes the acid- and basic-anhydride aspects of acid–base chemistry, certainly useful though often neglected. The Lux-Flood base is a basic anhydride:

$$Ca^{+2} + O^{-2} + H_2O \longrightarrow Ca^{+2} + 2OH^- \quad (7.6)$$

and the Lux-Flood acid is an acid anhydride:

$$SiO_2 + H_2O \longrightarrow H_2SiO_3 \quad (7.7)$$

(This latter reaction is very slow as written and is of more importance in the reverse, dehydration reaction.) The characterization of these metal and nonmetal oxides as acids and bases is of help in rationalizing the workings of, e.g., a basic Bessemer converter in steelmaking. The identification of these acidic and basic species will also prove useful in developing a general definition of acid–base behavior.

Solvent system definition

Many solvents autoionize with the formation of a cationic and an anionic species as does water:

$$2H_2O \rightleftharpoons H_3O^+ + OH^- \quad (7.8)$$

$$2NH_3 \rightleftharpoons NH_4^+ + NH_2^- \quad (7.9)$$

$$2H_2SO_4 \rightleftharpoons H_3SO_4^+ + HSO_4^- \quad (7.10)$$

$$2OPCl_3 \rightleftharpoons OPCl_2^+ + OPCl_4^- \quad (7.11)$$

For the treatment of acid–base reactions, especially neutralizations, it is often convenient to define an acid as *a species that increases the concentration of the characteristic cation of the solvent* and a base as *a species that increases the concentration of the characteristic anion*. The advantages of this approach are principally those of convenience. One may treat nonaqueous solvents by analogy with water. For example:

$$K_w = [H_3O^+][OH^-] = 10^{-14} \quad (7.12)$$

$$K_{AB} = [A^+][B^-] \quad (7.13)$$

where $[A^+]$ and $[B^-]$ are the concentrations of the cationic and anionic species characteristic of a particular solvent. Similarly, scales analogous to the pH scale of water may be constructed with the neutral point equal to $-\frac{1}{2}\log K_{AB}$, although, in practice, little work of this type has actually been done. Some examples of data of this type for nonaqueous solvents are listed in Table 7.1. The "leveling" effect follows quite naturally from this viewpoint, however. All acids and bases *stronger* than the characteristic cation and anion of the solvent will be "leveled" to the latter. Acids and bases *weaker* than those of the solvent system will remain in equilibrium with them. For example:

$$\mathbf{H_2O + HClO_4 \longrightarrow H_3O^+ + ClO_4^-} \quad (7.14)$$

Table 7.1 Ionic products, pH ranges, and neutral points of some solvents

Solvent	Ionic product	pH range	Neutral point
H_2SO_4	10^{-4}	0–4	2
CH_3COOH	10^{-13}	0–13	6.5
H_2O	10^{-14}	0–14	7
C_2H_5OH	10^{-20}	0–20	10
NH_3	10^{-29}	0–29	14.5

SOURCE: Data from J. Jander and C. Lafrenze, "Ionizing Solvents," Verlag Chemie Gmbh., Weinheim, **1970**.

but

$$\mathbf{H_2O} + CH_3C\begin{matrix}\diagup OH \\ \diagdown\!\!\diagdown O\end{matrix} \rightleftharpoons \mathbf{H_3O^+} + CH_3C\begin{matrix}\diagup O^- \\ \diagdown\!\!\diagdown O\end{matrix} \qquad \mathbf{(7.15)}$$

Similarly,

$$\mathbf{NH_3} + HClO_4 \longrightarrow \mathbf{NH_4^+} + ClO_4^- \qquad \mathbf{(7.16)}$$

and

$$\mathbf{NH_3} + HC_2H_3O_2 \longrightarrow \mathbf{NH_4^+} + C_2H_3O_2^- \qquad \mathbf{(7.17)}$$

but

$$\mathbf{NH_3} + NH_2CONH_2 \rightleftharpoons \mathbf{NH_4^+} + NH_2CONH^- \qquad \mathbf{(7.18)}$$

The solvent system concept has been used extensively as a method of classifying solvolysis reactions. For example, one can compare the hydrolysis of nonmetal halides with their solvolysis by nonaqueous solvents:

$$3H_2O + OPCl_3 \longrightarrow OP(OH)_3 + 3HCl\uparrow \qquad \mathbf{(7.19)}$$

$$3ROH + OPCl_3 \longrightarrow OP(OR)_3 + 3HCl\uparrow \qquad \mathbf{(7.20)}$$

$$6NH_3 + OPCl_3 \longrightarrow OP(NH_2)_3 + 3NH_4Cl \qquad \mathbf{(7.21)}[3]$$

Considerable use has been made of these analogies, especially with reference to nitrogen compounds and their relation to liquid ammonia as a solvent.[4]

One criticism of the solvent system concept is that it concentrates too heavily on ionic reactions in solution and on the *chemical properties* of the solvent to the neglect of the

[3] Although this reaction *appears* to be basically different from the others in stoichiometry and products, the difference lies merely in the relative basicity of H_2O, ROH, and NH_3 and the stability of their conjugate acids toward dissociation: $BH^+Cl^- \rightarrow B + HCl\uparrow$.

[4] L. F. Audrieth and J. Kleinberg, "Non-aqueous Solvents," Wiley, New York, **1953**; E. C. Franklin, "The Nitrogen System of Compounds," Van Nostrand-Reinhold, New York, **1935**.

physical properties. For example, reactions in phosphorus oxychloride (= phosphoryl chloride) have been systematized in terms of the hypothetical autoionization:

$$OPCl_3 \rightleftharpoons OPCl_2^+ + Cl^- \quad (7.22)$$

or

$$2OPCl_3 \rightleftharpoons OPCl_2^+ + OPCl_4^- \quad (7.23)$$

Substances which increase the chloride ion concentration may be considered bases and substances which strip chloride ion away from the solvent with the formation of the dichlorophosphoryl ion may be considered acids:

$$OPCl_3 + PCl_5 \rightleftharpoons OPCl_2^+ + PCl_6^- \quad (7.24)$$

Extensive studies of reactions between chloride ion donors (bases) and chloride ion acceptors (acids) have been conducted by Gutmann,[5] who interpreted them in terms of the above equilibria. An example is the reaction between tetramethylammonium chloride and iron(III) chloride, which may be carried out as a titration and followed conductometrically:

$$(CH_3)_4N^+Cl^- + FeCl_3 \xrightarrow[OPCl_3]{} (CH_3)_4N^+FeCl_4^- \quad (7.25)$$

which was interpreted by Gutmann in terms of:

$$(CH_3)_4N^+Cl^- \xrightarrow[OPCl_3]{\text{dissolve in}} (CH_3)_4N^+ + \mathbf{Cl^-} \quad (7.26)$$

$$FeCl_3 + OPCl_3 \rightleftharpoons \mathbf{OPCl_2^+} + FeCl_4^- \quad (7.27)$$

$$\mathbf{OPCl_2^+ + Cl^-} \longrightarrow OPCl_3 \quad (7.28)$$

Meek and Drago[6] showed that the reaction between tetramethylammonium chloride and iron(III) chloride can take place just as readily in triethyl phosphate, $OP(OEt)_3$, as in phosphorus oxychloride, $OPCl_3$. They suggested that the similarities in physical properties of the two solvents, principally the dielectric constant, were more important in this reaction than the difference in chemical properties, namely, the presence or absence of autoionization to form chloride ions.[7]

One of the chief difficulties with the solvent system concept is that in the absence of data, one is tempted to push it further than can be justified. For example, the reaction of thionyl halides with sulfites in liquid sulfur dioxide may be supposed to occur as follows, assuming that autoionization occurs:

$$2SO_2 \rightleftharpoons SO^{+2} + SO_3^{-2} \quad (7.29)$$

---

[5] V. Gutmann, *Z. Anorg. Allgem. Chem.*, **1952**, *270*, 179; *Monatsh. Chem.*, **1954**, *85*, 1077; *J. Phys. Chem.*, **1959**, *63*, 378; V. Gutmann and M. Baaz, *Monatsh. Chem.*, **1959**, *90*, 729; M. Baaz et al., ibid., **1960**, *91*, 548.

[6] D. W. Meek and R. S. Drago, *J. Am. Chem. Soc.*, **1961**, *83*, 4322. For a complete discussion of this point of view and critique of the solvent system approach, see R. S. Drago and K. F. Purcell, *Progr. Inorg. Chem.*, **1964**, *6*, 271.

[7] See p. 344 for further discussion of this point.

Accordingly, sulfite salts may be considered bases since they increase the sulfite ion concentration. It may then be supposed that thionyl halides behave as acids because of dissociation to form thionyl and halide ions:

$$SOCl_2 \rightleftharpoons SO^{+2} + 2Cl^- \quad (7.30)$$

The reaction between cesium sulfite and thionyl chloride may now be considered to be a neutralization reaction in which the thionyl ions and the sulfite ion combine to form solvent molecules:

$$SO^{+2} + SO_3^{-2} \longrightarrow 2SO_2 \quad (7.31)$$

Indeed, solutions of cesium sulfite and thionyl chloride in liquid sulfur dioxide yield the expected products:

$$Cs_2SO_3 + SOCl_2 \longrightarrow 2CsCl + 2SO_2 \quad (7.32)$$

Furthermore, the amphoteric behavior of the aluminum ion can be shown in sulfur dioxide as readily as in water. Just as $Al(OH)_3$ is insoluble in water but dissolves readily in either strong acid or basic solution, $Al_2(SO_3)_3$ is insoluble in liquid sulfur dioxide. Addition of either base (SO_3^{-2}) or acid ($SOCl_2$) causes the aluminum sulfite to dissolve, and it may be reprecipitated upon neutralization.

The application of the solvent system concept to liquid sulfur dioxide chemistry stimulated the elucidation of reactions such as those of aluminum sulfite. Unfortunately, there is no direct evidence at all for the formation of SO^{+2} in solutions of thionyl halides. In fact, there is evidence to the contrary. When solutions of thionyl bromide or thionyl chloride are prepared in ^{35}S-labeled sulfur dioxide, almost no exchange takes place. The half-life for the exchange is about two years or more. If ionization took place:

$$2S^*O_2 \rightleftharpoons S^*O^{+2} + S^*O_3^{-2} \quad (7.33)$$

$$SOCl_2 \rightleftharpoons SO^{+2} + 2Cl^- \quad (7.34)$$

one would expect rapid scrambling of the tagged and untagged sulfur in the two compounds. The lack of such a rapid exchange indicates that either Eq. 7.33 or 7.34 (or both) is incorrect.

The fact that labeled thionyl bromide exchanges with thionyl chloride indicates that perhaps the ionization shown in Eq. 7.34 actually occurs as:[8]

$$SOCl_2 \rightleftharpoons SOCl^+ + Cl^- \quad (7.35)$$

In a solvent with a permittivity as low as sulfur dioxide ($\varepsilon = 15.6\varepsilon_0$ at 0 °C) the formation of highly charged ions such as SO^{+2} is energetically unfavorable.

When the ionic species formed in solution are known, the solvent system approach can be useful. In solvents that are not conducive to ion formation and for which little or nothing is known of the nature or even the existence of ions, one must be cautious. Our

[8] R. E. Johnson, T. H. Norris, and J. L. Huston, *J. Am. Chem. Soc.*, **1951**, *73*, 3052; L. F. Johnson and T. H. Norris, ibid., **1957**, *79*, 1584; T. H. Norris, *J. Phys. Chem.*, **1959**, *63*, 3831.

familiarity with aqueous solutions of high permittivity ($\varepsilon_{H_2O} = 88.0\varepsilon_0$) characterized by ionic reactions tends to prejudice us toward parallels in other solvents and thus tempts us to overextend the solvent system concept.

Lewis definition

In 1923 G. N. Lewis[9] proposed a definition of acid–base behavior in terms of electron-pair donation and acceptance. The Lewis definition is perhaps the most widely used of all because of its simplicity and wide applicability, especially in the field of organic reactions. Lewis defined a base as *an electron-pair donor* and an acid as *an electron-pair acceptor.* In addition to all of the reactions discussed above, the Lewis definition includes reactions in which no ions are formed and no hydrogen ions or other ions are transferred:

$$R_3N + BF_3 \longrightarrow R_3N \rightarrow BF_3 \quad (7.36)$$

$$4CO + Ni \longrightarrow Ni(CO)_4 \quad (7.37)$$

$$2L + SnCl_4 \longrightarrow SnCl_4L_2 \quad (7.38)$$ [10]

$$2NH_3 + Ag^+ \longrightarrow Ag(NH_3)_2^+ \quad (7.39)$$

The Lewis definition thus encompasses all reactions entailing hydrogen ion, oxide ion, or solvent interactions, as well as the formation of acid–base adducts such as R_3NBF_3 and all coordination compounds.

Usanovich definition

The Usanovich definition[11] of acids and bases has not been widely used, probably because of (1) the relative inaccessibility of the original to non-Russian-reading chemists and (2) the awkwardness and circularity of Usanovich's original definition. The Usanovich definition includes all reactions of Lewis acids and bases and extends the latter concept by removing the restriction that the donation or acceptance of electrons be as shared pairs. The complete definition is as follows: *An acid is any chemical species which reacts with bases, gives up cations, or accepts anions or electrons, and, conversely, a base is any chemical species which reacts with acids, gives up anions or electrons, or combines with cations.* Although perhaps unnecessarily complicated, this definition simply includes all Lewis acid–base reactions plus redox reactions, which may consist of complete transfer of one or more electrons. Usanovich also stressed unsaturation involved in certain acid–base reactions:

$$OH^- + O{=}C{=}O \longrightarrow HOCO_2^- \quad (7.40)$$

---

[9] G. N. Lewis, "Valence and the Structure of Molecules," The Chemical Catalogue Co., New York, **1923**. See also W. F. Luder and S. Zuffanti, "The Electronic Theory of Acids and Bases," 2nd rev. ed., Dover, New York, **1961**, and R. S. Drago and N. A. Matwiyoff, "Acids and Bases," Heath, Lexington, Mass., **1968**.

[10] L = electron pair donating ligand such as acetone, various amines, or halide ion.

[11] M. Usanovich, *Zhur. Obschei Khim.*, **1939**, *9*, 182; H. Gehlen, *Z. Phys. Chem.*, **1954**, *203*, 125; H. L. Finston and A. C. Rychtman, "A New View of Current Acid-Base Series," Wiley, New York, **1982**.

Unfortunately the Usanovich definition of acids and bases is often dismissed casually with the statement that it includes "almost all of chemistry and the term 'acid–base reaction' is no longer necessary—the term 'reaction' is sufficient." If some chemical reactions were called "acid–base" reactions simply to distinguish them from other, non-acid–base reactions, this might be a valid criticism. However, most workers who like to talk in terms of one or more acid–base definitions do so because of the great systematizing power which they provide. As an example, Pearson has shown that the inclusion of many species, even organic compounds not normally considered acidic or basic, in his principle of hard and soft acids and bases helps the understanding of the nature of chemical reactions (pp. 312–313). It is unfortunate that a good deal of faddism and provincialism has been shown by chemists in this area. As each new concept came along, it was opposed by those who felt ill at ease with the new definitions. For example, when the solvent system was first proposed, some chemists refused to call the species involved acids and bases, but insisted that they were "acid analogues" and "base analogues"! This is semantics, not chemistry. A similar controversy took place when the Lewis definition became widely used and later when the Usanovich concept was popularized. Since the latter included redox reactions, the criticism that it included too much was especially vehement. That the dividing line between electron-pair donation–acceptance (Lewis definition) and oxidation–reduction (Usanovich definition) is not a sharp one may be seen from the following example. The compound C_5H_5NO, pyridine oxide, can be formed by the oxidation of pyridine. Now this may be considered to be a Lewis adduct of pyridine and atomic oxygen:

$$C_5H_5N: + :\ddot{\underset{..}{O}}: \longrightarrow C_5H_5N \rightarrow \ddot{\underset{..}{O}}: \tag{7.41}$$

Yet no one would deny that this is a redox reaction, even though no electron transfer has occurred between ionic species.

An example of the different points of view and different tastes in the matter of acid–base definitions was provided to the author in graduate school while attending lectures on acid–base chemistry from two professors. One felt that the solvent system was very useful, but that the Lewis concept went too far because it included coordination chemistry. The second used Lewis concepts in all of his work, but felt uncomfortable with the Usanovich definition because it included redox chemistry! To the latter's credit, however, he realized that the separation was an artificial one, and he suggested the pyridine oxide example given above.

In the presence of such a plethora of definitions, one can well ask which is the "best" one. Each concept, properly used, has its strong points and its weaknesses. One can do no better than to quote the concluding remarks of one of the best discussions of acid–base concepts:[12] "Actually each approach is correct as far as it goes, and knowledge of the fundamentals of all is essential."

[12] T. Moeller, "Inorganic Chemistry," Wiley, New York, **1952**, p. 330. For another excellent discussion of acid–base concepts, see T. Moeller, "Inorganic Chemistry, A Modern Introduction," Wiley, New York, **1982**, pp. 585–603.

A generalized acid–base concept

One justification for discussing a large number of acid–base definitions, including a few that are little used today, is to illustrate their fundamental similarity. All define the *acid* in terms of *donating a positive species* (a hydrogen ion or solvent cation) or *accepting a negative species* (an oxide ion, a pair of electrons, etc.). A *base* is defined as *donating a negative species* (a pair of electrons, an oxide ion, a solvent anion) or *accepting a positive species* (hydrogen ion). We can generalize all these definitions by defining *acidity as a positive character of a chemical species which is decreased by reaction with a base*; similarly *basicity is a negative character of a chemical species which is decreased by reaction with an acid.* The advantages of such a generalization are twofold: (1) it incorporates the information content of the various other acid–base definitions; (2) it provides a useful criterion for correlating acid–base strength with electron density and molecular structure. Some examples may be useful in illustrating this approach. It should be kept in mind that acid–base concepts do not *explain* the observed properties; these lie in the principles of structure and bonding. Acid–base concepts help *correlate* empirical observations.

1. *Basicity of metal oxides.* In a given periodic group, basicity of oxides tends to increase as one progresses down the periodic chart. For example, in Group IIA, BeO is amphoteric but the heavier oxides (MgO, CaO, SrO, BaO) are basic. In this case the charge on the metal ion is the same in each species, but in the Be^{+2} ion it is packed into a much smaller volume, hence its effect is more pronounced. As a result BeO is more acidic or less basic than the oxides of the larger metals. In this case the "positiveness" is a matter of the size and charge of the cation. This is closely related, of course, to the Fajans' polarizing ability (pp. 129–131). The same effect is seen in Group III oxides: B_2O_3 is acidic, Al_2O_3 is amphoteric, and Sc_2O_3 is quite basic.

2. *Hydration and hydrolysis reactions.* We have seen (Chapter 6) that large charge-to-size ratio for cations results in an increase in hydration energy. Closely related to hydration and, in fact, inseparable from it except in degree is the phenomenon of hydrolysis. In general, we speak of hydration if no reaction beyond coordination of water molecules to the cation occurs:

$$Na^+ + nH_2O \longrightarrow [Na(H_2O)_n]^+ \qquad \textbf{(7.42)}$$

In the case of hydrolysis reactions, the acidity (charge-to-size ratio) of the cation is so great as to cause rupture of H—O bonds with ionization of the hydrate to yield hydronium ions:

$$Al^{+3} + 6H_2O \longrightarrow [Al(H_2O)_6]^{+3} \xrightarrow{H_2O} H_3O^+[Al(H_2O)_5OH]^{+2} \qquad (7.43)$$

Cations that hydrolyze are those which are either small (e.g., Be^{+2}) or are highly charged (e.g., Fe^{+3}, Sn^{+4}), i.e., or both, and have a high charge/size density. Values of pK_h (negative log of the hydrolysis constant) are compared to the $(\text{charge})^2/(\text{size})$ ratio[13] in Table 7.2. The correlation is good for the main group elements and La^{+3} but less so for the transition metals, especially the heavier ones. The reason for the apparently anomalous behavior of metal ions such as Hg^{+2}, Sn^{+2}, and Pb^{+2} is not completely clear, but it may be related to the "softness" of the latter (see p. 312).

---

[13] Z^2/r has been used here, but any of the Z^n/r^m functions would give similar results. See p. 130.

Table 7.2 Hydrolysis constants and charge-size function

Z^2/r		pK_h^a		
		Main group elements and lanthanides	Lighter transition and posttransition metals	Heavier transition and posttransition metals
2.2[b]	0.87[c]	$Na^+ = 14.48$		$Ag^+ = 6.9$
3.5	1.35	$Li^+ = 13.82$		
7.6	2.94	$Ba^{+2} = 13.82$		
8.4	3.28			$Sn^{+2} = 4.30$
8.7	3.39			$Pb^{+2} = 7.78$
8.8	3.45	$Sr^{+2} = 13.18$		
10.1	3.92			$Hg^{+2} = 3.70$
10.3	4.00	$Ca^{+2} = 12.70$		
10.8	4.21		$Cd^{+2} = 11.70$	
12.5	4.89		$Mn^{+2} = 10.70$	
13.3	5.19		$Fe^{+2} = 10.1$	
13.7	5.33		$Zn^{+2} = 9.60$	
13.9	5.40		$Co^{+2} = 9.6$	
14.1	5.48		$Cu^{+2} = 7.53$	
14.3	5.56	$Mg^{+2} = 11.42$		
14.7	5.71		$Ni^{+2} = 9.40$	
21.8	8.49	$La^{+3} = 10.70$		
		⋮		$Pu^{+3} = 6.95$
22.6	8.82	⋮ Lanthanides		$Bi^{+3} = 1.58$
26.3	10.23	⋮		$Tl^{+3} = 1.15$
27.2	10.59	$Lu^{+3} = 6.6$		
29.2	11.39			$In^{+3} = 3.70$
31.6	12.33		$Sc^{+3} = 4.6$	
33.1	12.90	$Be^{+2} = 6.50$		
35.5	13.85		$Fe^{+3} = 2.19$	
36.1	14.06		$V^{+3} = 2.92$	
			$Cr^{+3} = 4.01$	
37.3	14.52		$Ga^{+3} = 3.40$	
38.7	15.09			$Th^{+3} = 3.89$
41.1	16.00			$U^{+3} = 1.50$
43.6	16.98	$Al^{+3} = 5.14$		
51.3	20.00			$Pu^{+4} = 1.6$
57.0	22.22			$Zr^{+4} = 0.22$
57.8	22.54			$Hf^{+4} = 0.12$

Increasing tendency to hydrolyze because of charge-size function ↓

Increasing tendency to hydrolyze because of electronic structure →

[a] Values of pK_h from K. B. Yatsimirksii and V. P. Vasil'ev, "Instability Constants of Complex Compounds," Pergamon, Elmsford, N.Y., **1960**, except for Bi, Hf, Lu, Pu, Sc, and Tl, which are from J. Bjerrum, G. Schwarzenbach, and L. G. Sillen, eds., "Stability Constants of Metal–Ion Complexes: Part II, Inorganic Ligands," The Chemical Society, London, **1958**. For many elements there is considerable uncertainty in the hydrolysis constants not only as a result of experimental errors but also because some have not been corrected to infinite dilution. Z^2/r values were calculated from ionic radii in Table 3.4.

[b] $C^2\ m^{-1} \times 10^{28}$.

[c] $e^2\ m^{-10}$.

The concept of hydrolysis may also be extended to the closely related phenomenon of the reaction of nonmetal halides with water:

$$PCl_3 + 6H_2O \longrightarrow H_3PO_3 + 3H_3O^+ + 3Cl^- \tag{7.44}$$

In this case the water attacks and hydrolyzes not a cation but a small, highly charged center (the trivalent phosphorus atom) resulting from the inductive effect of the chlorine atoms.

3. *Basicity of substituted amines.* In water, ammonia is a weak base, but nitrogen trifluoride shows no basicity whatsoever. In the NH_3 molecule, the nitrogen atom is partially charged negatively from the inductive effects of the hydrogen atoms, but the reverse is true in the NF_3 molecule. Replacement of a hydrogen atom in the ammonia molecule with an electron-withdrawing group such as —OH or —NH_2 also results in decreased basicity. Since alkyl groups are normally electron-donating (more so than hydrogen) toward electronegative elements, we might expect that replacement of a hydrogen atom by a methyl group would increase the basicity of the nitrogen atom. This effect is shown in Table 7.3.

As expected, substitution of an alkyl group for a hydrogen atom in the ammonia molecule results in increased electron density on the nitrogen atom and increased basicity. Substitution of a second alkyl group also increases the basicity, although less than might have been expected from the previous substitutional effect. The trialkylamines do *not* continue this trend and surprisingly are as weak as or weaker than the monoalkylamines. Although the explanation of this apparent anomaly is fairly simple, it does not depend upon electron density and so will be postponed to the next section.

4. *Acidity of oxyacids.* The strength of an oxyacid is dependent upon several factors that relate to the inductive effect of the central atom on the hydroxyl group: (a) *The inherent electronegativity of the central atom:* Perchloric acid, $HClO_4$, and nitric acid, HNO_3, are among the strongest acids known; sulfuric acid, H_2SO_4 is only slightly weaker. In contrast, phosphoric acid, H_3PO_4, is considerably weaker and carbonic acid, H_2CO_3, and boric acid, H_3BO_3, are rather weak. (b) *The inductive effect of substituents:* Although acetic acid, CH_3COOH, is rather weak, successive substitution of chlorine atoms on the methyl group increases the dissociation of the proton until trichloroacetic acid is considerably stronger than phosphoric acid, for example.

More important for inorganic oxyacids is the number of oxygen atoms surrounding the central atom. Thus in the series of chlorine oxyacids, acid strength increases in the order $HOCl < HOClO < HOClO_2 < HOClO_3$. The trends in acidity of oxyacids,

Table 7.3 Basicity of ammonia and amines

Electron-withdrawing substitution ←		pK_b $NH_3 = 4.74$	Electron-donating substitution →	
$NH_2OH = 7.97$	$NH_2NH_2 = 5.77$	$MeNH_2 = 3.36$	$Me_2NH = 3.29$	$Me_3N = 4.28$
		$EtNH_2 = 3.25$	$Et_2NH = 2.90$	$Et_3N = 3.25$
		$i\text{-}PrNH_2 = 3.28$	$i\text{-}Pr_2NH = 1.95$	
		$i\text{-}BuNH_2 = 3.51$	$i\text{-}Bu_2NH = 3.32$	$i\text{-}Bu_3N = 3.58$

H_aXO_b, can be correlated with a set of rules:[14]

1. The successive pK_a values for an oxyacid differ by five units: $pK_2 = pK_1 + 5$; $pK_3 = pK_2 + 5$, etc.
2. For the various values of a and b, the value of pK_1 may be estimated as follows:

If $a = b$, $pK_1 \cong 7$, very weak acid, e.g., HClO, $pK_1 = 7.48$; H_3BO_3, $pK_1 = 9.24$.[15]
If $a = b - 1$, $pK_1 = 2$, weak acid, e.g., $HClO_2$, $pK_1 = 2.02$; H_2SO_3, $pK_1 = 1.997$; H_3PO_4, $pK_1 = 2.12$.
If $a = b - 2$, $pK_1 = -3$, strong acid, e.g., H_2SO_4, $pK_1 = -3$, $pK_2 = 1.9$; HNO_3, $pK_1 = -3$, $pK_2 = 1.4$.
If $a = b - 3$, $pK_1 = -8$, very strong acid, e.g., $HClO_4$.

5. "*Ultimate acids and bases.*" Familiarity with the idea that acidity and basicity are related to electron density at reacting sites and charge-to-size ratio might lead one to ask if there exists a single strongest acid or base species. A little reflection would suggest the bare proton as having the highest positive charge-to-size ratio. Of course the proton never occurs uncoordinated or unsolvated in chemical systems, but attaches itself to any chemical species containing electrons. It is too strong an acid to coexist with any base without reacting. Even a noble gas atom, not normally considered to be a base, will combine with the exceedingly acidic proton (see p. 301). The choice of the proton as the "characteristic" exchanged species in the Brønsted-Lowry concept was not fortuitous.

Concerning the "ultimate" base, one might choose various small, highly charged ions such as H^-, F^-, or O^{-2}, all of which are indeed quite basic. However, the electron appears to be the complement of the proton. It might be objected that the isolated electron has even less justification as a chemical entity than the proton, but solutions are known in which electrons (solvated of course) are the anionic species! And interestingly, such solutions are very basic. This topic will be discussed in further detail in the section on liquid ammonia chemistry in the next chapter.

MEASURES OF ACID–BASE STRENGTH

There is no universally satisfactory way of measuring the strength of acids and bases that will apply to all systems. In protonic solvents such as water, acid and base strengths can be adequately treated by ionization constants in a Brønsted-Lowry context. One of the advantages of this approach is its emphasis on the competitive nature of acid–base equilibria in protonic solvents. Since the solvated hydrogen ion is the strongest acid that can exist in these solvents (see Chapter 8), the conjugate base of each acid competes for it. The stronger base reacts with the hydrogen ion to form the weaker, undissociated acid. In addition, as we have seen, the pK_a and pK_b values of acids and bases can often be correlated with electronegativity and inductive effects, especially if the compounds being compared are structurally similar. However, solvent effects can seriously bias conclusions

[14] L. Pauling, "General Chemistry," 3rd ed., W. H. Freeman, San Francisco, **1970**, pp. 499–502.

[15] J. J. Christensen et al., "Handbook of Proton Ionization Heats and Related Thermodynamic Quantities," Wiley, New York, **1970**.

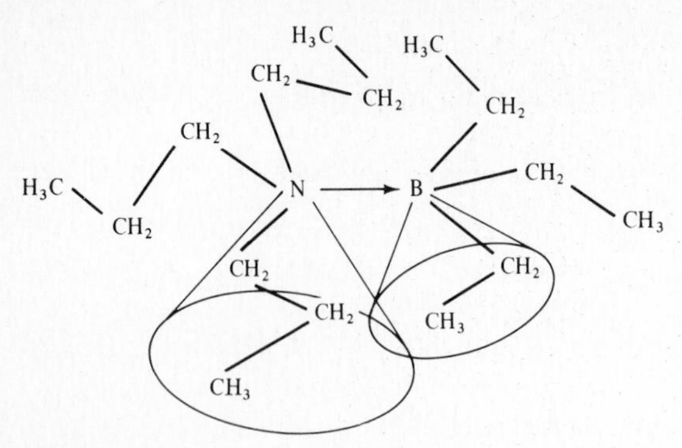

Fig. 7.1 Tripropylamine—triethylborane adduct illustrating steric hindrance between the bulky substituents on the nitrogen and the boron.

based on such solution data. If, however, one is more interested in the nature of the reaction than other factors, proper use of pK_a values will enable the user to predict results.

Direct comparison of pK_a values provides only a qualitative estimate of relative acid strength for weak acids and does not provide a comparison with strong acids. Recently, a quantitative scale of acid and base strengths for any given solvent system has been suggested.[16] A comparison between two acids or bases is made by comparing the fractions, α, of the electrolyte ionized, summed over molar concentrations from 0 to 1. This is accomplished by taking the integral (S) of the Ostwald dilution law in this range. This function equals unity for a strong acid or base (completely ionized in solution) and approaches zero for a very weak acid or base. For example, a comparison of hydrochloric acid and acetic acid (HOAc) may be made as follows. In water $S_{\text{HCl}} = 1$ and $S_{\text{HOAc}} = 8.4 \times 10^{-3}$, so $S_{\text{HCl}}/S_{\text{HOAc}} = 110$ and hydrochloric acid is estimated to be one hundred times stronger than acetic acid. For weak acids $S_{\text{A}}/S_{\text{B}}$ is nearly equivalent to $\sqrt{K_{\text{A}}/K_{\text{B}}}$.

Steric effects

In reactions between Lewis acids and bases such as amines and boranes or boron halides, bulky substituents on one or both species can affect the stability of the acid–base adduct. Perhaps the most straightforward type of effect is simple steric hindrance between substituents on the nitrogen atom and similar large substituents on the boron atom. Figure 7.1 is a diagrammatic sketch of the adduct between molecules of tripropylamine and triethylborane. This phenomenon is known as *front* or "F-strain" and can have a considerable influence on the stability of the adduct since the alkyl groups tend to sweep out large volumes as they rotate randomly.

A second, similar effect is known as *back* or "B-strain." It results from the structural necessity for the nitrogen atom in amines to be approximately tetrahedral (sp^3) in order to bond effectively through its lone pair. If the alkyl groups on the nitrogen atom are sufficiently bulky, presumably they can force the bond angles of the amine to open up, causing more s character to be used in these bonds and more p character to be left in

[16] L. S. Levitt and H. F. Widing, *Chem. Ind.*, **1974**, 781.

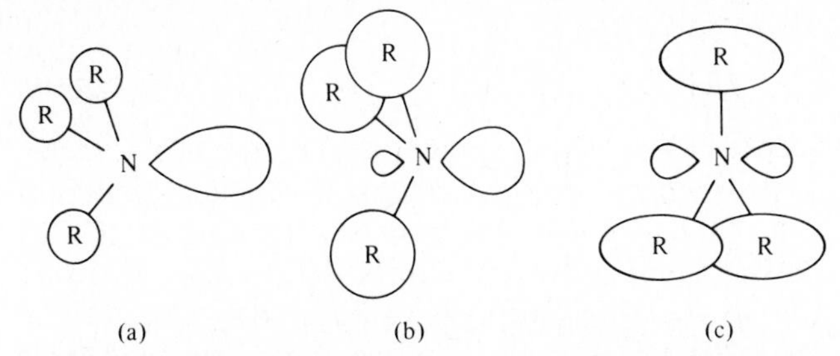

Fig. 7.2 B-strain in substituted amines. (a) Small substituents, no strain, good base; (b) moderate strain from intermediate-sized substituents, some rehybridization; (c) extreme bulkiness of substituents, nitrogen atom forced into planar, $sp^2 + p$ hybridization, weak base.

the lone pair. The extreme result of this would be the formation of a planar, trigonal molecule with a lone pair in a pure *p* orbital, poorly suited for donation to an acid (Fig. 7.2).

Related to B-strain, but less well understood, is "I-strain" (for *internal* strain). In cyclic amines and ethers, such as $(CH_2)_nO$, the basicity varies with ring size. In such compounds the hybridization (and hence the overlapping ability and electronegativity) of not only the basic center (N, O, etc.), but also of the carbon atoms in the ring will vary with ring size, and there are no simple rules for predicting the results.

When the basic center is exocyclic, as in lactams, lactones, etc., the results can be interpreted in a straightforward way analogous to the argument given previously (p. 150) for biphenylene. Consider the series of lactams:

CH_3–N (O) > CH_3–N (O) > CH_3–N (O)

As the ring size is reduced, the internal bond angles must reduce, and the hybridization of the cyclic atoms must have less *s* character and lower electronegativity. Towards the exocyclic oxygen atom, the basic center,[17] the cyclic carbon atom must in turn exhibit *greater s* character and a *higher* electronegativity. The carbonyl groups in small ring compounds are therefore less basic.[18]

Solvation effects and acid–base "anomalies"[19]

One might consider either F-strain or B-strain, or both, as being responsible for the unusual behavior of reduced basicity in trialkylamines. The addition of *three* alkyl groups might pose steric problems on a small nitrogen atom. F-strain can *not* be responsible for the reduced basicity of tertiary amines in Table 7.3 since the pK_b values are measured

[17] See Problem 7.13.
[18] C. A. L. Filgueiras and J. E. Huheey, *J. Org. Chem.*, **1976**, *41*, 49.
[19] E. M. Arnett, *Acc. Chem. Res.*, **1983**, *16*, in press.

in aqueous solution and represent the equilibrium:

$$R_3N + H_2O \rightleftharpoons R_3NH^+ + OH^- \tag{7.45}$$

and the proton being added to the base has no steric requirements because it is so small. Nevertheless B-strain could be involved since it results from intramolecular strain in a particular molecule, and indeed this explanation has been offered to account for the decreased basicity of trialkyl amines. Of interest in this regard is the fact that if the basicities of methylamines are measured in the *gas phase*, they increase regularly $NH_3 < MeNH_2 < Me_2NH < Me_3N$ (see Table 7.4). Therefore the "anomaly" of the basicity of trimethylamine must lie in some solution effect. Solvation through hydrogen-bonding will tend to increase the apparent strength of all amines because the positively charged ammonium ions will be more extensively solvated (ΔH ten to one hundred times larger) than the uncharged amine. Hence the basicity of the amines is enhanced

$$R_3N + 2H_2O \longrightarrow OH^- + R_3N^+\text{—H}\cdots OH_2 \tag{7.46}$$

$$RNH_2 + 4H_2O \longrightarrow OH^- + R\text{—}N^+(\text{—H}\cdots OH_2)_3 \tag{7.47}$$

in proportion to the extent of solvation of the conjugate ammonium ion, and the energies of solvation are $RNH_3^+ > R_2NH_2^+ > R_3NH^+$. This is the reverse order of increase in basicity that results from electronic (inductive) effects. Two opposing, nonlinear trends will give a maximum or a minimum. Therefore it is not surprising to find a maximum in basicity (as measured in aqueous solution) for the dialkylamines.

Many of the "anomalies" of recent findings are historical artifacts: Accurate experimental data for species in solution had been accumulated for decades, and corresponding theories had been proposed long before the first gas phase data were collected. For example, it has been found that the acidity of water and alcohols goes in the order $H_2O > R(1^\circ)OH > R(2^\circ)OH > R(3^\circ)OH$ with the "explanation" being that the electron-releasing alkyl groups force electron density onto the oxygen of the conjugate base making it more basic. But note that the electronegativities of branched and unbranched alkyl groups are practically identical, and if there *is* any trend, those groups having more carbon atoms are slightly more electronegative: Me = 2.30, Et = 2.32, *i*-Pr = 2.33, *t*-Bu = 2.34 (see Table 3.13). These groups differ significantly *only* in their *charge capacity* (p. 155). Thus highly branched groups are both *better donors* (when attached to electronegative centers) and *better*

Table 7.4 Estimated proton affinities (kJ mol^{-1})[a]

Trinegative ions	Dinegative ions	Uninegative ions[b]	Neutral molecules	
		H^- = 1674	H_2 = 423	Group IA
		CH_3^- = 1742	CH_4 = 536	Group IVA
		$C_6H_5CH_2^-$ = 1582	$C_6H_5CH_3$ = 808	
		$C_6H_5^-$ = 1664	C_6H_6 = 777	
		CN^- = 1461	CO = 582	
			$(CH_3)_3N$ = 938	Group VA
			$(CH_3)_2NH$ = 923	
			C_5H_5N = 922	
			$C_2H_5NH_2$ = 908	
			CH_3NH_2 = 896	
			NH_2OH = 870[d]	
N^{-3} = 3084[b]	NH^{-2} = 2565[c]	NH_2^- = 1671	NH_3 = 858	
		NO^- = 1423	CH_3CN = 799	
		N_3^- = 1360	HCN = 715	
			NCl_3 = 795[d]	
			NF_3 = 556[d]	
			N_2 = 475	
			$(CH_3)_3P$ = 944	
			$(CH_3)_2PH$ = 908	
			CH_3PH_2 = 857	
		PH_2^- = 1541	PH_3 = 800	
			$(CH_3)_3As$ = 894	
		AsH_2^- = 1501	AsH_3 = 768	
			$CH_3C(O)NH_2$ = 1000[d]	Group VIA
			CH_3NO_2 = 874[d]	
			t-C_4H_9OH = 816	
			C_6H_5OH = 824	
			$(CH_3)_2CO$ = 825	
			$(CH_3)_2O$ = 808	
			$CH_3C(O)H$ = 790	
			CH_3OH = 734	
			HC(O)OH = 765	
			H_2CO = 741	
		OH^- = 1671	H_2O = 724	
			CO_2 = 530	
			O_2 = 423	
	O^{-2} = 2318[c], 2548[b]	OOH^- = 1381[b]	CH_3SH = 789	
	S^{-2} = 2300[c]	SH^- = 1471	H_2S = 739	
	Se^{-2} = 2200[b]	SeH^- = 1417	H_2Se = 742	
		F^- = 1553	HF = 469	Group VIIA
		Cl^- = 1393	HCl = 586[e]	
		Br^- = 1351	HBr = 590[e]	
		I^- = 1314	HI = 607[e]	
			Kr = 424	Group VIIIA
			Xe = 478	

[a] Proton affinity = $-\Delta H$ of reaction 7.49. Unless otherwise indicated, all data from D. H. Aue and M. T. Bowers, in "Gas Phase Chemistry," Vol. 2, M. T. Bowers, ed., Academic, New York, **1979**, Chapter 9.

[b] Calculated from bond energies of HX molecules and the electron affinities of X. For example, see J. E. Bartmess and R. T. McIver, in "Gas Phase Chemistry," Vol. 2, M. T. Bowers, ed., Academic Press, New York, **1979**, Chapter 11.

[c] T. C. Waddington, *Adv. Inorg. Chem. Radiochem.*, **1959**, *1*, 157.

[d] W. L. Jolly and D. N. Hendrickson, *J. Am. Chem. Soc.*, **1970**, *92*, 1863.

[e] D. A. Dixon et al., *Inorg. Chem.*, **1972**, *11*, 1972.

acceptors (when attached to electropositive centers). Seemingly paradoxically (but refer again to Figs. 2.15 and 3.61) O^- is *electropositive: The oxygen atom will be stabilized if the anionic charge is delocalized.* This can best be accomplished by groups with larger charge capacities. Relative to the hydrogen atom (1.0), the charge capacities of groups are Me = 2.8, Et = 3.4, *i*-Pr = 3.9, and *t*-Bu = 4.2. The net result is that in gas-phase reactions with no complicating solvation energies, the order of basicity is $OH^- > R(1°)O^- > R(2°)O^- > R(3°)O^-$.[20]

So why the reversal of basicity as one proceeds from the gas phase to solution? Once again, solvation effects overcome inherent electronic effects. As in the case of amines, hydrogen bonding is the predominant factor, and as the organic portion of the ion grows, it becomes increasingly like a ball of wax. The anion loses the special solvation stability normally enjoyed over neutral molecules and thus more readily accepts a proton. The enhanced basicity of the *t*-butoxide ion arises not because the electron density on the oxygen is higher (it is lower),[21] but because the anion lacks stabilizing solvation.

Despite these somewhat jarring conflicts with "rules" learned in preceding courses, it is obvious that the best measure of the inherent basicity of a species is its gas-phase *proton affinity*. It is defined as the enthalpy of the reaction:

$$B + H^+ \longrightarrow BH^+ \tag{7.48}$$

It may be obtained by a technique known as *ion cyclotron resonance spectroscopy* and related methods, which measure the competition by two bases for a proton in the gas phase. It is thus unaffected by solvation effects. Until recently, few data of this type were available, but the technique has now been used successfully on a wide variety of compounds.[22] Some typical values are listed in Table 7.4. In general, these values are in qualitative agreement with those expected on the basis of inductive effects and charge, e.g., $NH_3 > H_2O > HF$; $O^{-2} > OH^- > H_2O$.

Lewis interactions in nonpolar solvents

Of great current interest are the attempts to evaluate and correlate strengths of Lewis acids and bases. Many of these have been measured in the gas phase and many more in aprotic, nonpolar solvents in which, it is hoped, the solvent effects will be minimized. There are several methods available for evaluating acid and base strengths. It is common to equate the extent of interaction of an acid and a base with the enthalpy of reaction. In some cases this enthalpy has been measured by direct calorimetry: ΔH equals q for an adiabatic process at constant pressure.

Often the enthalpy of reaction is obtained by measuring the equilibrium constant of an acid–base reaction over a range of temperatures. If ln K is plotted versus $1/T$, the

[20] J. I. Brauman and L. K. Blair, *J. Am. Chem. Soc.*, **1968**, *90*, 6561; **1970**, *92*, 5986.

[21] N. C. Baird, *Can. J. Chem.*, **1969**, *47*, 2306; T. P. Lewis, *Tetrahedron*, **1969**, *25*, 4117; J. E. Huheey, *J. Org. Chem.*, **1971**, *36*, 204.

[22] J. L. Beauchamp, *Ann. Rev. Phys. Chem.*, **1971**, *22*, 527; *idem*, in "Interactions Between Ions and Molecules," P. Ausloos, ed., Plenum, New York, **1975**, p. 413. H. Hartmann, K.-H. Lebert, and K.-P. Wancsek, *Top. Curr. Chem.*, **1973**, *43*, 57; D. H. Aue and M. T. Bowers in "Gas Phase Ion Chemistry," Vol. 2, M. T. Bowers, ed., Academic Press, New York, **1979**, Chapter 9.

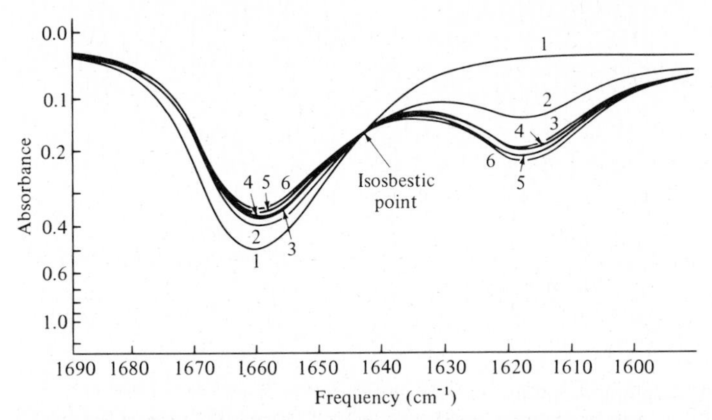

Fig. 7.3 Infrared absorption spectra of dimethylacetamide-iodine system: (1) dimethylacetamide only; (2–6) increasing concentrations of iodine. Peak at 1662 cm^{-1} is from free dimethylacetamide, that at 1619 cm^{-} is from the DMA · I_2 adduct. [From C. D. Schmulbach and R. S. Drago, *J. Am. Chem. Soc.*, **1960**, *82*, 4484. Reproduced with permission.]

slope will be equal to $\Delta H/R$. Thus various experimental methods have been devised to measure the equilibrium constant by spectrophotometric methods. Any absorption which differs between one of the reactants (either acid or base) and the acid–base adduct is a potential source of information on the magnitude of the equilibrium constant since it gives the concentration of two of the three species involved in the equilibrium directly and the third indirectly from a knowledge of the stoichiometry of the reaction. For example, consider the extensively studied reaction between organic carbonyl compounds and iodine. The infrared carbonyl absorption frequency is shifted in frequency in the adduct with respect to the free carbonyl compound (for the reason, see p. 168). Thus the equilibrium mixture exhibits two absorption bands in the carbonyl region of the spectrum (Fig. 7.3) and the relative concentrations of the free carbonyl and the adduct can be obtained.[23] Alternatively, one can observe the absorption of the iodine molecule, I_2, in the 300–600 nm portion of the visible spectrum. Again, the adduct absorbs at a different frequency from the free iodine and the two absorption maxima provide information on the relative concentrations of the species present.[24]

Two complications can prevent a simple determination of the concentration of each species from a measurement of absorbance at a chosen frequency. Although most of the acid–base reactions of interest result in a one-to-one stoichiometry, one cannot assume this *a priori*, and two-to-one and three-to-one adducts might also be present. Fortunately, this is usually an easy point to resolve. The presence of an *isosbestic point* or point of constant absorbance (see Fig. 7.3) is usually a reliable criterion that only two absorbing species (the free acid or base and a single adduct) are present.[25]

---

[23] C. D. Schmulbach and R. S. Drago, *J. Am. Chem. Soc.*, **1960**, *82*, 4484.

[24] R. S. Drago, R. L. Carlson, N. J. Rose, and D. A. Wenze, *J. Am. Chem. Soc.*, **1961**, *83*, 3572.

[25] The present discussion of isosbestic points is oversimplified; the reader is warned that one can get into difficulties with such oversimplifications. For further discussion of the use of isosbestic points, see M. D. Cohen and E. Fischer, *J. Chem. Soc.*, **1962**, 3044; R. G. Mayer and R. S. Drago, *Inorg. Chem.*, **1976**, *15*, 2010; T. Nowicka-Jankowska, *J. Inorg. Nucl. Chem.*, **1971**, *33*, 2043; D. V. Styne, *Inorg. Chem.*, **1975**, *14*, 453.

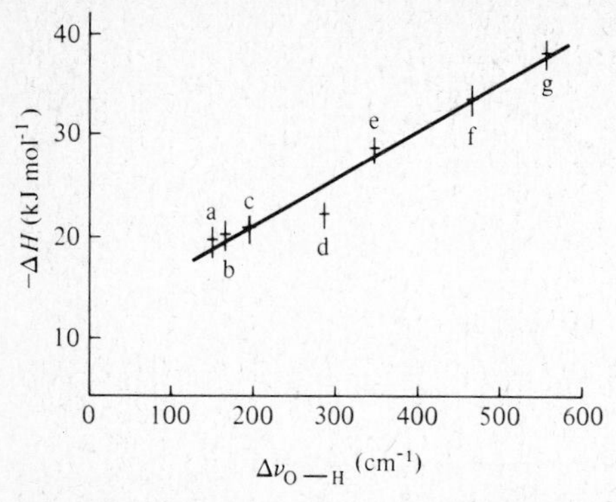

Fig. 7.4 Relation between the enthalpy of formation of base–phenol adducts and the stretching frequencies of the O—H bond in the phenol. Bases: (a) acetonitrile, (b) ethyl acetate, (c) acetone, (d) tetrahydrofuran, (e) dimethylacetamide, (f) pyridine, (g) triethylamine. [From T. D. Epley and R. S. Drago, *J. Am. Chem. Soc.*, **1967**, *89*, 5770. Reproduced with permission.]

The second problem is somewhat more troublesome. The separation between the absorption maximum of the adduct and that of the free acid (or base) is seldom large and so there is considerable overlapping of bands (see Fig. 7.3). If the absorptivities of each of the species at each frequency were known, it would be a simple matter to ascribe a proportion of the total absorbance at a given frequency to each species. It is usually a relatively simple matter to measure the absorptivity of the free acid (or base) over the entire working range. Since it is often impossible to prepare the pure adduct (in the absence of equilibrium concentrations of free acid and base) its absorptivity cannot be measured. However, if the equilibrium is studied at two different concentrations of acid (or base), it is possible to set up two simultaneous equations in terms of the two unknowns K and a and solve for both.[26]

Alternative methods of measuring the enthalpy of acid–base reactions involve measuring some physical property which depends upon the strength of the interaction. In general, such methods must be calibrated against one of the previous types of measurement but once this is done they may often be extended to reactions that prove difficult to measure by other means. One example is the study of phenol as a Lewis acid.[27] Phenol forms strong hydrogen bonds to Lewis bases, especially those that have a donor atom with a large negative charge. The formation of the hydrogen bond alters the electron density in the —O—H group of the phenol and the hydrogen stretching frequency in the infrared. Once the frequencies of a series of known phenol–base adducts have been used for calibration (Fig. 7.4), it is possible to estimate the enthalpy of adduct formation of bases with similar functional groups directly from IR spectra.

A second method involves the relation between *s* character and NMR coupling constants (see pp. 167–168). Drago and co-workers[28] have shown that there is a good

[26] N. J. Rose and R. S. Drago, *J. Am. Chem. Soc.*, **1959**, *81*, 6138.

[27] M. D. Joesten and R. S. Drago, *J. Am. Chem. Soc.*, **1962**, *84*, 3817; T. D. Epley and R. S. Drago, ibid., **1967**, *89*, 5770; **1969**, *91*, 2883; G. C. Vogel and R. S. Drago, ibid., **1970**, *92*, 5347.

[28] T. F. Bolles and R. S. Drago, *J. Am. Chem. Soc.*, **1966**, *88*, 5730.

correlation between the ^{119}Sn—H coupling constant in chlorotrimethylstannane–base adducts and the strength of the base–tin bond. It has been suggested[28] that the stronger bases force the tin to rehybridize to a greater extent (sp^3 in free Me_3SnCl, sp^3d in the limit of strong base adduct) than weaker ones and the change in *s* character of the Sn—C bonds is reflected in the coupling constants.

Systematics of Lewis acid–base interactions

Drago and co-workers have proposed a number of ways of expressing enthalpies of reactions in terms of contributing parameters of acids and bases. The first was[29]

$$-\Delta H = E_A E_B + C_A C_B \tag{7.49}$$

where ΔH is the enthalpy of formation of a Lewis acid–base adduct, E_A and C_A are parameters characteristic of the acid, and E_B and C_B are parameters characteristic of the base. The E parameters are interpreted as the susceptibility of the species to undergo electrostatic ("ionic" or dipole–dipole) interaction and the C parameters are the susceptibility to form covalent bonds. From this we expect those acids which bond well electrostatically (E_A is large) to form the most stable adducts with bases that bond electrostatically (since the product $E_A E_B$ will then be large). Conversely, acids that bond well covalently will tend to form their most stable adducts with bases that bond well covalently. Equation 7.49 performs satisfactorily in predicting the enthalpy of reaction of many acids and bases, and it is in fact, the preferred equation when dealing with neutral acids and bases. Some typical values of E_A, E_B, C_A, and C_B are listed in Tables 7.5A,B. The original papers[29] should be consulted for details of estimating these parameters. The application of Eq. 7.49 may be illustrated with the reaction between pyridine ($E = 1.17$, $C = 6.40$) and iodine ($E = 1.00$, $C = 1.00$):[30]

$$\begin{aligned} -\Delta H_{calc} &= E_A E_B + C_A C_B = 1.17 + 6.40 = 7.57 \text{ kcal mol}^{-1} \quad (31.7 \text{ kJ mol}^{-1}) \\ -\Delta H_{exp} &= 7.8 \text{ kcal mol}^{-1} \quad (33 \text{ kJ mol}^{-1}) \end{aligned} \tag{7.50}$$

If one or both species are charged, the formation of the adduct will be accomplished with considerable electron density transferred from the negative to the positive species. The electron density transfer will be accompanied by a release of energy associated with ionization energy–electron affinity terms. Equation 7.49 thus no longer gives satisfactory results. In order to overcome these difficulties, Kroeger and Drago[31] added a third parameter, t, to evaluate this transfer energy and thus obtained the equation:

$$-\Delta H = e_A e_B + c_A c_B + t_A t_B \tag{7.51}$$

---

[29] R. S. Drago and B. B. Wayland, *J. Am. Chem. Soc.*, **1965**, *87*, 3571; R. S. Drago et al., *J. Am. Chem. Soc.*, **1971**, *93*, 6014; R. S. Drago, *Struc. Bond.*, **1973**, *15*, 73; R. S. Drago, L. B. Parr, and C. S. Chamberlain, *J. Am. Chem. Soc.*, **1977**, *99*, 3203.

[30] Equations 7.49 and 7.51 and the values presented in Table 7.4A,B are predicated on the use of kcal mol^{-1}. These values can be made compatible with SI usage by incorporating 4.184 kJ $kcal^{-1}$ into either the acid *or* the base sets of E–C terms, or by incorporating $(4.184 \text{ kJ kcal}^{-1})^{1/2}$ into all E, C, e, c, and t factors for *both* acids and bases.

[31] M. K. Kroeger and R. S. Drago, *J. Am. Chem. Soc.*, **1981**, *103*, 3250.

Table 7.5A Acid parameters

Acid	E	C	e	c	t
Iodine	1.00	1.00	0.231	1.200	0.122
Sulfur dioxide	0.920	0.808	3.777	0.721	0.010
Phenol	4.33	0.62	4.561	0.274	0.315
Iodine monochloride	5.10	0.83			
Hexafluoroisopropyl alcohol	5.93	0.623	5.214	0.426	0.475
Antimony pentachloride	7.38	5.13			
Boron trifluoride	9.88	1.62	2.000	1.668	0.914
Methyl cation, CH_3^+			20.326	3.592	6.627
Proton, H^+			8.654	8.554	15.040

Table 7.5B Base parameters

Base	E	C	e	c	t
Pyridine	1.17	6.40	0.629	5.112	11.486
Methylamine	1.30	5.88	0.786	4.655	11.281
Ammonia	1.36	3.46	0.694	2.713	11.587
Trimethylphosphine	0.84	6.55			
Chloride ion			5.111	2.730	17.508
Hydroxide ion			6.599	34.817	2.394

Note: More values of E, C, e, c, and t may be found in Appendix E.

where the e and c parameters have essentially the same significance as in Eq. 7.49, but lowercase is used to distinguish the two equations. In addition, the $t_A t_B$ term adds the energy of the electron density transfer effect. Values of e_A, e_B, c_A, c_B, t_A, and t_B for some common acids and bases are also listed in Tables 7.5A,B.

The importance of the E–C or e–c–t parameters is manifold. First, they enable predictions to be made of the enthalpies of reactions that have not been studied. Thus the parameters in Tables 7.5A,B and Appendix E were obtained from a few hundred reactions of acids and bases, but they can be used to predict the enthalpies of thousands of reactions. For example, accurate values for the enthalpy of reaction can be obtained for reactions such as (kcal mol^{-1} values; multiply by 4.184 kJ kcal^{-1} for kJ mol^{-1}):

$$C_6H_6 + SO_2 \longrightarrow [C_6H_6SO_2] \qquad \Delta H_{exp} = -1.0,\ \Delta H_{calc} = -1.2 \tag{7.52}$$

$$O(CH_2CH_2)_2O + ICl \longrightarrow [O(CH_2CH_2)_2O \rightarrow ICl] \qquad \Delta H_{exp} = -7.5,\ \Delta H_{calc} = -7.5 \tag{7.53}$$

$$H^+ + NH_3 \longrightarrow NH_4^+ \qquad \Delta H_{exp} = -203.5,\ \Delta H_{calc} = -202.3 \tag{7.54}$$

$$CH_3^+ + F^- \longrightarrow CH_3F \qquad \Delta H_{exp} = -257.3,\ \Delta H_{calc} = -254.0 \tag{7.55}$$

Table 7.6 Examples of the use of acid and base parameters[a]

Acid	Base	Calculated enthalpy (kcal mol^{-1})	Experimental enthalpy (kcal mol^{-1})
Proton, H^+	Ammonia	−203.5	−202.3
	Methylamine	−216.3	−211.3
	Fluoride ion	−367.3	−371.3
	Chloride ion	−330.9	−333.3
	Hydride ion	−398.9	−400.4
Methyl cation, CH_3^+	Ammonia	−100.6	−99.3
	Methylamine	−107.5	−110.3
	Fluoride ion	−257.3	−254.0
	Chloride ion	−229.7	−227.0
	Hydride ion	−311.1	−310.0
Phenol, ϕOH	Ammonia	−7.4	−7.8
	Methylamine	−8.2	−8.6
	Benzene	−1.1	[b]
Iodine, I_2	Ammonia	−4.8	−4.8
	Methylamine	−7.2	−7.2
	Benzene	−1.5	−1.3

[a] Data are for reactions in the absence of solvation effects. Strictly speaking, this means for gas-phase reactions only. In practice, however, the acids and bases may be dissolved in nonpolar solvents, and these parameters applied to solution studies.

[b] No experimental data available.

Typical calculated values obtained by combining acid and base parameters are listed together with experimental values in Table 7.6. In some cases it appears that we can predict values for certain systems more accurately than they could be measured (because of experimental difficulties, etc.).

The second item of importance with respect to systems of this sort is that it enables us to obtain some insight into the nature of the bonding in various systems. Thus, if we compare the C_A and E_A parameters of iodine and phenol, we find that I_2 is twice[32] as good a "covalent-bonder" as phenol, but that the latter is about five times as effective through electrostatic attractions. This is not unexpected inasmuch as phenol, C_6H_5OH, is a very strong hydrogen-bonding species. In contrast, iodine has no dipole moment but must react with a Lewis base by expanding its octet and accepting electrons to form a covalent bond.

A similar effect can be observed in the bases. The E_B value of acetonitrile is much larger than that of *p*-xylene, $(CH_3)_2C_6H_4$, corresponding to the large dipole moment of acetonitrile ($\mu = \sim 13$ C m) compared with the latter ($\mu = 0.0$ C m). On the other hand, the C_B values are reversed, corresponding to the enhanced ability of the aromatic ring to donate π electron density to an acid.

[32] One must be careful in making comparisons using these numbers. A comparison of E_{I_2} to $E_{\phi OH}$ is valid, but one cannot compare the E and C parameters directly for a single species because of the artificial assumption $E_{I_2} = C_{I_2} = 1.00$.

A CLOSER LOOK AT BOND ENERGIES AND BOND LENGTHS IN ACID–BASE COMPOUNDS

Bond energies

There are two ways to approach the formation of a polar bond $X^+—Y^+$. We have already encountered both. One is to consider the formation of a nonpolar molecule, X—Y, followed by an electronegativity-controlled shift of electron density from X towards Y. Alternatively, one can form the ions X^+ and Y^-, followed by their interaction. We can view the latter as the Fajans' polarization of X^- by Y^+, or basic attack of $:Y^-$ upon X^+. Whatever model is used, there occurs, in one form or another, three contributing energies:

1. The *covalent energy*, E_C, arising from electron sharing. It is a maximum in a homopolar bond and decreases with ionicity.
2. The *Madelung energy*, E_M, arising from the coulombic attraction of the partial charges:

$$X^{\delta+}—Y^{\delta-} \qquad E_M = \frac{\delta^+\delta^-}{4\pi r} \tag{7.56}$$

This energy is termed the Madelung energy since it represents a "lattice energy" internal to the molecule with a Madelung constant, of course, equal to 1.00. It is a maximum in a purely ionic bond ($\delta^+ = Z^+$) and decreases to whatever extent the charges on X and Y decrease.

3. The *electronegativity energy*, E_χ, or *IE–EA energy* arising from ionization energy–electron affinity terms in the total energy sum. It is a more complex function than E_M and E_C, but it will be clarified by some examples below.[33]

Consider the *ionic resonance energy* of a principally covalent bond with a little ionic character (see p. 144). To a first approximation, the ionic resonance energy may be equated with Madelung energy, E_M, and the electronegativity energy, E_χ, that stabilizes the XY molecule more than the small loss of covalent energy.

The simplest example of these three terms has already been encountered in the form of the Born-Haber cycle and the Born-Landé Equation (taken only to the point of isolated gas-phase molecules; see p. 59 (Eq. 3.4)). The equation in that form shows that the gas-phase ion pairs ("ionic molecules") are stabilized by the Madelung energy holding them together and are destabilized by the ionization energy–electron affinity energy that had to be paid to form the ions. Note that if we brought two ions together (acid–base reaction) and they were not *completely* ionic (no compounds are), electron density would flow from the anion (base) to the cation (acid); there would be some loss of Madelung energy (the charges decrease) but a stabilization of the formerly unfavorable E_χ as the metal flows down the energetically steep part of curve Fig. 3.63b (A^+) and the nonmetal flows up the relatively mild slope of Fig. 3.63a (B^-). To say it another way Fig. 3.64 will be approximated, no matter whether we approach equilibrium from $A\cdot + \cdot B$ or $A^+ + :B^-$. Finally,

[33] See also R. S. Evans and J. E. Huheey, *J. Inorg. Nucl. Chem.*, **1970**, *32*, 373, 383, 777; *Chem. Phys. Lett.*, **1973**, *19*, 144.

inherent in the bond A—B is the covalent energy term arising from the overlap of orbitals, whether it be from the covalent bond picture or polarization of ionic species.

We can now examine in further detail the *e–c–t* parameters of Drago's systems. We have seen that $E_A E_B$ and $C_A C_B$ terms indicate tendencies to form electrostatic or covalent acid–base interactions. Likewise, the $e_A e_B$ acid–base term in Eq. 7.51 provides us a measure of E_M and the $c_A c_B$ term provides a measure of E_C. That leaves only the $t_A t_B$ term to be a measure of E_χ, which it is. It can be clearly shown to be related to an IE–EA energy.[31] Note, however, that in contrast to the ion-pair example given above, where the E_χ was destabilizing, in covalent, acid–base reactions of ionic species, the E_χ term will be strongly *stabilizing*, in accordance with the electronegativity argument given above, and the $t_A t_B$ term will be a major contributor to the stabilization of the donor-acceptor bond.

An example of the importance of the energy associated with this transfer of electron density is given by Kroeger and Drago.[31] Because of the very large value of t_{H^+} (resulting from the high ionization energy of hydrogen), the energy associated with the gas-phase attachment of H^+ to bases (proton affinity) is unique; other acids, including the aqueous hydronium ion, are different:

$$B\!: + H^+ \longrightarrow B\!:\!H^+ \qquad \text{gas phase} \tag{7.57}$$

$$B\!: + H_3O^+ \longrightarrow B\!:\!H^+ + H_2O \qquad \text{aqueous phase} \tag{7.58}$$

This is because the bare proton has an extremely high charge/size ratio and releases an extremely large amount of E_χ upon adduct formation ($t_{H^+} = 15.040$) compared to all other acids that have several atoms over which to delocalize the cationic charge. This is analogous to the *b* parameter of the Mulliken-Jaffé electronegativities (see p. 153). which measures the capacity of a group to "soak up" charge and stabilize it. Thus methyl groups can stabilize charge (have a low electronegativity *b* term) for exactly the same reason that alkyl groups as ions have lower *t* values than the proton (Me = 6.627; Et = 3.106; Pr = 1.697; Bu = 0.442).

Bond lengths

Although chemists have spent considerable time and energy in determining ionic and covalent radii, much less effort has been expended on determining how bond lengths change as the molecular environment changes. Gutmann[34] has provided a useful set of three rules describing the behavior of bonds in acid–base adducts:

Rule 1. The stronger and shorter the bond formed between a donor atom (Lewis base) *and an acceptor atom* (Lewis acid), *the greater will be the lengthening of the adjacent bonds in both the donor and acceptor molecules*:

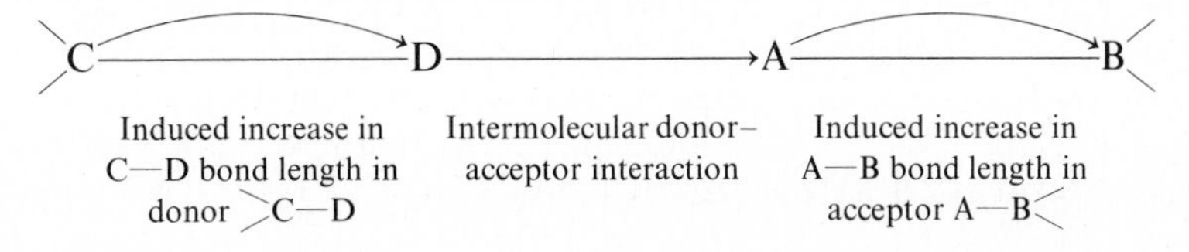

[34] V. Gutmann, *Coordin. Chem. Rev.*, **1975**, *15*, 207; **1976**, *18*, 225; *Electrochim. Acta*, **1976**, *21*, 661; "The Donor-Acceptor Approach to Molecular Interactions," Plenum, New York, **1978**, pp. 4–16.

Thus the donor–acceptor bond, D—A, forms somewhat at the expense of bonds C—D and B—A. There may be one or more C—D or B—A bonds, and either may be single or multiple. One of the useful aspects of Gutmann's Rules is their generality in this regard. Data illustrating *Rule 1* include the following for bond lengthening in boron trifluoride upon adduct formation:

Free acid or acid adduct	N—B (pm)	B—F (pm)
BF_3	—	130
CH_3CNBF_3	163	133
H_3NBF_3	160	138
$(CH_3)_3NBF_3$	158	139

The effect of *Rule 1* can be seen in the behavior of the carbonyl donor group as it interacts with a Lewis acid such as iodine:

C═O—I—I

C═O	O—I	I—I
Decrease in bond energy	Increase in bond energy	Decrease in bond energy
Bond lengthens	Bond shortens	Bond lengthens
Force constant and ν_{CO} decreases		

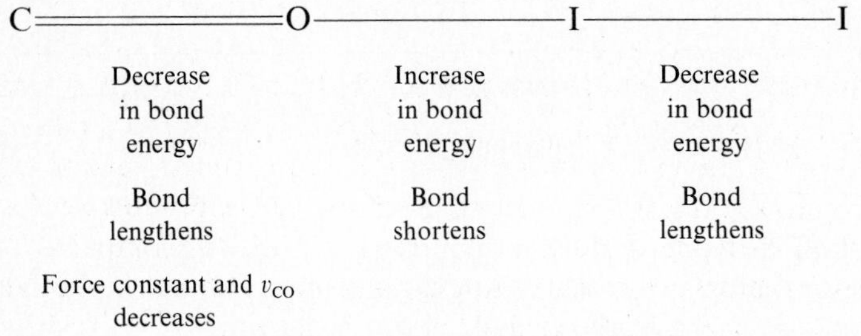

As a result, the interaction between a carbonyl base and a Lewis acid such as iodine causes a distinct shift in the frequency of the carbonyl group, a fact that has been used extensively to study the basicity of organic carbonyl compounds (see Fig. 7.3).

Rule 2. A sigma bond is lengthened when, as a result of the interaction, the electron shift occurs from a nucleus carrying a positive fractional charge, whereas the sigma bond is shortened when the electron shift is in the opposite direction. This is one of those strange rules that is easier to apply than to state briefly. Consider the adduct of tetrachloroethylene carbonate. Note especially the partial charges in the reactants *before* formation of the O—Sb bond:

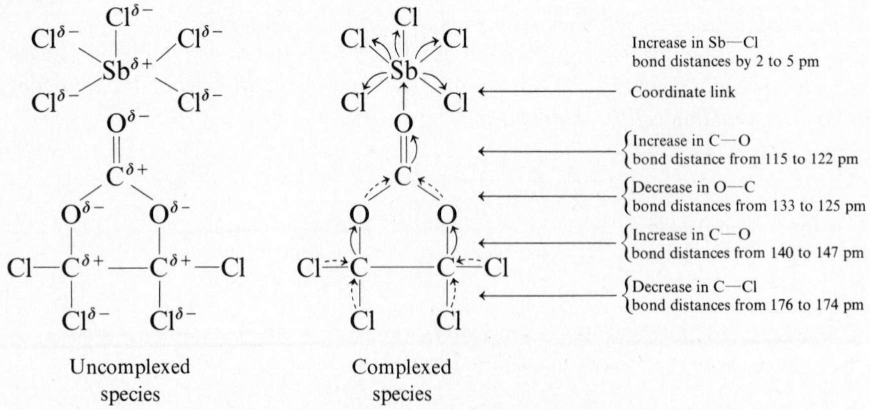

Uncomplexed species: $Sb^{\delta+}Cl^{\delta-}_5$; $O^{\delta-}$═$C^{\delta+}$($O^{\delta-}$)$_2$; Cl—$C^{\delta+}$—$C^{\delta+}$—Cl with $Cl^{\delta-}$ substituents

Complexed species: $SbCl_5$←O═C(O)$_2$ ring with C_2Cl_4

Increase in Sb—Cl bond distances by 2 to 5 pm

Coordinate link

Increase in C—O bond distance from 115 to 122 pm

Decrease in O—C bond distances from 133 to 125 pm

Increase in C—O bond distances from 140 to 147 pm

Decrease in C—Cl bond distances from 176 to 174 pm

Uncomplexed species

Complexed species

1. The Sb—Cl and C=O bonds are lengthened on complex formation in accord with *Rule 1*.

2. The C—O (ester) bonds are weakened and lengthened as more electron density is shifted from the positive carbon atom to the negative oxygen.

3. The Cl—C and C—O (carbonate) bonds are strengthened and shortened because electron density is shifted from the more electronegative, negatively charged atoms (Cl, O) to the positive carbon.

Rule 3. As the coordination number of an atom increases, so do the lengths of the bonds from that atom. This is illustrated by:

$$\underset{r_{Si\text{—}F} = 157\ \text{pm}}{SiF_4} + 2F^- \longrightarrow \underset{171\ \text{pm}}{SiF_6^{-2}} \quad (7.59)$$

This is but a variant of *Rule 1* since the Si—F bonds (like the B—F bonds above are expected to lengthen upon adduct formation. However the phenomenon is widespread and worthy of separate consideration (Table 7.7).

Gutmann has intentionally avoided interpreting his three Bond Length Rules in terms of a specific model of bonding (rehybridization, electronegativity, etc.) so that their success as empirical rules would not be tied to the success of a particular theory. Several factors suggest themselves as important, and so the following discussion is not meant to be exclusive, but intends merely to provide a rationale.

Consider once more the drawing associated with *Rule 1*, noting that a base is a molecule that has built up electron density that can be donated to an atom with relative little

Table 7.7 Bond lengths, M—X, in acceptor molecules and in their donor–acceptor complexes[a]

Acceptor	M—X (pm)	Complex ion	M—X (pm)
$CdCl_2$	223.5	$CdCl_6^{4-}$	253
SiF_4	154	SiF_6^{2-}	171
$TiCl_4$	218–221	$TiCl_6^{2-}$	235
$ZrCl_4$	233	$ZrCl_6^{2-}$	245
$GeCl_4$	208–210	$GeCl_6^{2-}$	235
GeF_4	167	GeF_6^{2-}	177
$SnBr_4$	244	$SnBr_6^{2-}$	259–264
$SnCl_4$	230–233	$SnCl_6^{2-}$	241–245
SnI_4	264	SnI_6^{2-}	285
$PbCl_4$	243	$PbCl_6^{2-}$	248–250
PF_5	154–157	PF_6^-	173
$SbCl_5$	231	$SbCl_6^-$	247
SO_2	143	SO_3^{2-}	150
SeO_2	161	SeO_3^{2-}	174
ICl	230	ICl_2^-	236
I_2	266	I_3^-	283

[a] From V. Gutmann, "The Donor-Acceptor Approach to Molecular Interactions," Plenum, New York, **1978**. Reproduced with permission.

electron density (see pp. 294–297); thus before interaction, the CDAB system will appear as:

$$C^{\delta+}\text{———}D^{\delta-} \qquad A^{\delta+}\text{———}B^{\delta-}$$

Bond shortened by Madelung energy　　　Bond shortened by Madelung energy

The C—D and A—B bonds will be shortened as a result of their polarity (Schomaker-Stevenson Rule). However, upon formation of a D—A bond, the charges will change:

$$C^{\delta\delta+}\text{———}D^{\delta\delta-}\text{———}A^{\delta\delta+}\text{———}B^{\delta\delta-}$$

More +　　　Less −　　　Less +　　　More −

While there is a net gain in bonding energy in the acid–base region, there is some loss within both the acid molecule and the base molecule, leading to reduction in Madelung energy and Schomaker-Stevenson shortening in these bonds; hence the lengthening of *Rule 1*.

HARD AND SOFT ACIDS AND BASES

For some time coordination chemists have been aware of certain trends in the stability of metal complexes. One of the earliest correlations was the *Irving-Williams series of stability*.[35] For a given ligand, the stability of complexes with dipositive metal ions follows the order: $Ba^{+2} < Sr^{+2} < Ca^{+2} < Mg^{+2} < Mn^{+2} < Fe^{+2} < Co^{+2} < Ni^{+2} < Cu^{+2} > Zn^{+2}$. This order arises in part from a decrease in size across the series and in part from ligand field effects (Chapter 9). A second observation is that certain ligands form their most stable complexes with metal ions such as Ag^{+}, Hg^{+2}, and Pt^{+2}, but other ligands seem to prefer ions such as Al^{+3}, Ti^{+4}, and Co^{+3}.[36] Ligands and metal ions were classified[37] as belonging to type (*a*) or (*b*)[38] according to their preferential bonding. Class (*a*) metal ions include those of alkali metals, alkaline earth metals, and lighter transition metals in higher oxidation states such as Ti^{+4}, Cr^{+3}, Fe^{+3}, Co^{+3} and the hydrogen ion, H^{+}. Class (*b*) metal ions include those of the heavier transition metals, and those in lower oxidation states such as Cu^{+}, Ag^{+}, Hg^{+}, Hg^{+2}, Pd^{+2}, and Pt^{+2}.[39] According to their preferences toward either class (*a*) or class (*b*) metal ions, ligands may be classified as

[35] H. Irving and R. J. P. Williams, *Nature*, **1948**, *162*, 746; *J. Chem. Soc.*, **1953**, 3192.

[36] The existence of isolated ions of high charge such as Ti^{+4} in chemical systems is energetically unfavorable. Nevertheless, complexes exist with these elements in high *formal* oxidation states.

[37] S. Ahrland, J. Chatt, and N. R. Davies, *Quart. Rev. Chem. Soc.*, **1958**, *12*, 265. See also G. Schwarzenbach, *Experientia Suppl.*, **1956**, *5*, 162.

[38] The (*a*) and (*b*) symbolism is arbitrary and should not be confused with the A and B subgroups of the periodic table.

[39] Only a limited selection of examples of class (*a*) and (*b*) metal ions is given here for the purpose of illustration. A complete listing is provided in Tables 7.8 and 7.9.

type (*a*) or (*b*), respectively. Stability of these complexes may be summarized as follows:

Tendency to complex with class (*a*) metal ions	Tendency to complex with class (*b*) metal ions
N >> P > As > Sb	N << P > As > Sb
O >> S > Se > Te	O << S < Se ~ Te
F > Cl > Br > I	F < Cl < Br < I

For example, phosphines (R_3P) and thioethers (R_2S) have a much greater tendency to coordinate with Hg^{+2}, Pd^{+2}, and Pt^{+2}, but ammonia, amines (R_3N), water, and fluoride ion prefer Be^{+2}, Ti^{+4}, and Co^{+3}. Such a classification has proved very useful in accounting for and predicting the stability of coordination compounds.

Pearson[40] has suggested the terms "hard" and "soft" to describe the members of class (*a*) and (*b*). Thus a hard acid is a type (*a*) metal ion and a hard base is a ligand such as ammonia or the fluoride ion. Conversely, a soft acid is a type (*b*) metal ion and a soft base is a ligand such as a phosphine or the iodide ion. A thorough discussion of the factors operating in hard and soft interactions will be postponed temporarily, but it may be noted now that the hard species, both acids and bases, tend to be small, slightly polarizable species and that soft acids and bases tend to be larger and more polarizable. Pearson has suggested a simple rule (sometimes called Pearson's principle) for predicting the stability of complexes formed between acids and bases: *Hard acids prefer to bind to hard bases and soft acids prefer to bind to soft bases.* It should be noted that this statement is not an explanation or a theory, but a simple rule of thumb which enables the user to predict qualitatively the relative stability of acid–base adducts.

Classification of acids and bases as hard or soft

In addition to the (*a*) and (*b*) species discussed above that provide the nucleus for a set of hard and soft acids and bases, it is possible to classify any given acid or base as hard or soft by its apparent preference for hard or soft reactants. For example, a given base, B, may be classified as hard or soft by the behavior of the following equilibrium:[41]

$$BH^+ + CH_3Hg^+ \rightleftharpoons CH_3HgB^+ + H^+ \quad (7.60)$$

In this competition between a hard acid (H^+) and a soft acid (CH_3Hg^+), a hard base will cause the reaction to go to the left, but a soft base will cause the reaction to proceed to

[40] R. G. Pearson, *J. Am. Chem. Soc.*, **1963**, *85*, 3533. Recent summaries have been provided by Pearson [*J. Chem. Educ.*, **1968**, *45*, 581, 643; *Surv. Progr. Chem.*, **1969**, *1*, 1, A. Scott, ed., Academic Press, New York.] For further reading on this topic, see *Struc. Bonding*, **1966**, *1*, C. K. Jørgensen et al., which contains papers from a symposium on this subject. For the interesting application to organic chemistry, see R. G. Pearson and J. Songstad, *J. Am. Chem. Soc.*, **1967**, *89*, 1827, and T-L. Ho, *Chem. Rev.*, **1975**, *75*, 1, and "Hard and Soft Acids and Bases Principle in Organic Chemistry," Academic Press, New York, **1977**.

[41] If this equilibrium is studied in aqueous solution as is usually the case, all species will be hydrated, and, specifically, the acids will occur as $CH_3Hg(H_2O)^+$ and H_3O^+. For data on equilibria of this type, see G. Schwarzenbach and M. Schellenberg, *Helv. Chim. Acta*, **1965**, *48*, 28.

Table 7.8 Classification of hard and soft acids

Hard acids
H^+, Li^+, Na^+, K^+(Rb^+, Cs^+)
Be^{+2}, $Be(CH_3)_2$, Mg^{+2}, Ca^{+2}, Sr^{+2}(Ba^{+2})
Sc^{+3}, La^{+3}, Ce^{+4}, Gd^{+3}, Lu^{+3}, Th^{+4}, U^{+4}, UO_2^{+2}, Pu^{+4}
Ti^{+4}, Zr^{+4}, Hf^{+4}, VO^{+2}, Cr^{+3}, Cr^{+6}, MoO^{+3}, WO^{+4}, Mn^{+2}, Mn^{+7}, Fe^{+3}, Co^{+3}
BF_3, BCl_3, $B(OR)_3$, Al^{+3}, $Al(CH_3)_3$, $AlCl_3$, AlH_3, Ga^{+3}, In^{+3}
CO_2, RCO^+, NC^+, Si^{+4}, Sn^{+4}, CH_3Sn^{+3}, $(CH_3)_2Sn^{+2}$
N^{+3}, RPO_2^+, $ROPO_2^+$, As^{+3}
SO_3, RSO_2^+, $ROSO_2^+$
Cl^{+3}, Cl^{+7}, I^{+5}, I^{+7}
HX (hydrogen-bonding molecules)
Borderline acids
Fe^{+2}, Co^{+2}, Ni^{+2}, Cu^{+2}, Zn^{+2}
Rh^{+3}, Ir^{+3}, Ru^{+3}, Os^{+2}
$B(CH_3)_3$, GaH_3
R_3C^+, $C_6H_5^+$, Sn^{+2}, Pb^{+2}
NO^+, Sb^{+3}, Bi^{+3}
SO_2
Soft acids
$Co(CN)_5^{-3}$, Pd^{+2}, Pt^{+2}, Pt^{+4}
Cu^+, Ag^+, Au^+, Cd^{+2}, Hg^+, Hg^{+2}, CH_3Hg^+
BH_3, $Ga(CH_3)_3$, $GaCl_3$, $GaBr_3$, GaI_3, Tl^+, $Tl(CH_3)_3$
CH_2, carbenes
Pi-acceptors: trinitrobenzene, chloroanil, quinones, tetracyanoethylene, etc.
HO^+, RO^+, RS^+, RSe^+, Te^{+4}, RTe^+
Br_2, Br^+, I_2, I^+, ICN, etc.
O, Cl, Br, I, N, $RO\cdot$, $RO_2\cdot$
M^0 (metal atoms) and bulk metals

the right.[42] The methylmercury cation is convenient to use because it is a typical soft acid and, being monovalent like the proton, simplifies the treatment of the equilibria. Complete listings of hard and soft acids and bases are given in Tables 7.8 and 7.9.

An important point to remember in considering the information in these tables is that the terms hard and soft are relative with no sharp dividing line between them. This is illustrated in part by the third category, "borderline," for both acids and bases. But even within a group of hard or soft, not all will have equivalent hardness or softness. Thus, although all alkali metal ions are hard, the larger, more polarizable cesium ion will be somewhat softer than the lithium ion. Similarly, although nitrogen is usually hard because of its small size, the presence of polarizable substituents can affect its behavior. Pyridine, for example, is sufficiently softer than ammonia to be considered borderline.

[42] An interesting historical sidelight on this type of soft–soft interaction is the origin of the name "mercaptan," a *mer*cury *capt*urer: $Hg^{+2} + 2RSH = Hg(SR)_2 + 2H^+$

Table 7.9 Classification of hard and soft bases

Hard bases
NH_3, RNH_2, N_2H_4
H_2O, OH^-, O^{-2}, ROH, RO^-, R_2O
CH_3COO^-, CO_3^{-2}, NO_3^-, PO_4^{-3}, SO_4^{-2}, ClO_4^-
$F^-(Cl^-)$
Borderline bases
$C_6H_5NH_2$, C_5H_5N, N_3^-, N_2
NO_2^-, SO_3^{-2}
Br^-
Soft bases
H^-
R^-, C_2H_4, C_6H_6, CN^-, RNC, CO
SCN^-, R_3P, $(RO)_3P$, R_3As
R_2S, RSH, RS^-, $S_2O_3^{-2}$
I^-

Acid–base strength and hardness and softness

Hardness and softness refer to special stability of hard–hard and soft–soft interactions and should be carefully distinguished from inherent acid or base strength. For example, both OH^- and F^- are hard bases; yet the basicity of the hydroxide ion is about 10^{13} times that of the fluoride ion. Similarly, both SO_3^{-2} and Et_3P may be considered soft bases; however, the latter is 10^7 times as strong (toward CH_3Hg^+). It is possible for a *strong* acid or base to displace a weaker one, even though this *appears* to violate the principle of hard and soft acids and bases. For example, the *stronger*, *softer* base, the sulfite ion, can displace the *weak*, *hard base*, fluoride ion, from the *hard acid*, the proton, H^+:

$$SO_3^{-2} + HF \longrightarrow HSO_3^- + F^- \qquad K_{eq} = 10^4 \tag{7.61}$$

Likewise the *very strong*, *hard base*, hydroxide ion, can displace the weaker *soft base*, sulfite ion, from the *soft acid*, methylmercury cation:

$$OH^- + CH_3HgSO_3^- \longrightarrow CH_3HgOH + SO_3^{-2} \qquad K_{eq} = 10 \tag{7.62}$$

In these cases the *strengths* of the bases ($SO_3^{-2} > F^-$, Eq. 7.61; $OH^- > SO_3^{-2}$, Eq. 7.62), are sufficient to force these reactions to the right in spite of hard–soft considerations. Nevertheless, if a competitive situation is set up in which both strength *and* hardness–softness are considered, the hard–soft rule works:

$$\underset{\text{Soft–hard}}{CH_3HgF} + \underset{\text{Hard–soft}}{HSO_3^-} \longrightarrow \underset{\text{Soft–soft}}{CH_3HgSO_3^-} + \underset{\text{Hard–hard}}{HF} \qquad K_{eq} \sim 10^3 \tag{7.63}$$

$$CH_3HgOH + HSO_3^- \longrightarrow CH_3HgSO_3^- + HOH \qquad K_{eq} > 10^7 \tag{7.64}$$

Table 7.10 Basicity toward the proton and methylmercury cation

Base	Linking atom	pK_s^a, (CH_3Hg^+)	pK_h^b, (H^+)
F^-	F	1.50	2.85
Cl^-	Cl	5.25	−7.0
Br^-	Br	6.62	−9.0
I^-	I	8.60	−9.5
OH^-	O	9.37	15.7
HPO_4^-	O	5.03	6.79
S^{-2}	S	21.2	14.2
$HOC_2H_4S^-$	S	16.12	9.52
SCN^-	S	6.05	~4
SO_3^{-2}	S	8.11	6.79
$S_2O_3^{-2}$	S	10.90	negative
NH_3	N	7.60	9.42
$NH_2C_6H_4SO_3^-(p)$	N	2.60	3.06
$\phi_2PC_6H_4SO_3^-$	P	9.15	~0
$Et_2PC_2H_4OH$	P	14.6	8.1
Et_3P	P	15.0	8.8
CN^-	C	14.1	9.14

[a] $pK_s = \log[CH_3HgB]/[CH_3Hg^+][B]$.
[b] $pK_h = \log[HB]/[H^+][B]$.

In considering acid–base interactions, it is necessary to consider both strength *and* hardness–softness. Table 7.10 lists the strengths of various bases toward the proton (H^+) and the methylmercury cation (CH_3Hg^+). Bases such as the sulfide ion (S^{-2}) and triethylphosphine (Et_3P) are very strong toward both the methylmercury ion and the proton, but about a million times better toward the former; hence they are considered *soft*. The hydroxide ion is a strong base toward both acids, but in this case about a million times better toward the proton; hence it is hard. The fluoride ion, F^-, is not a particularly good base toward either acid but slightly better toward the proton as expected from its hard character.

The importances of both inherent acidity and a second hard–soft factor is well shown by the Irving-Williams series and some oxygen, nitrogen, and sulfur chelates (Fig. 7.5). The Irving-Williams series of increasing stability from Ba^{+2} to Cu^{+2} is a measure of increasing inherent acidity of the metal (largely due to decreasing size). Superimposed upon this is a hardness–softness factor in which the softer species coming later in the series (greater number of *d* electrons, see p. 319) favor ligands S > N > O. The harder alkaline earth and early transition metals ions (few or no *d* electrons) preferentially bind in the order O > N > S.

Symbiosis

As noted above, the hardness or softness of an acidic or basic site is not an inherent property of the particular atom at that site but can be influenced by the substituent atoms. The addition of soft, polarizable substituents can soften an otherwise hard center and the presence of electron-withdrawing substituents can reduce the softness of a site. The acidic

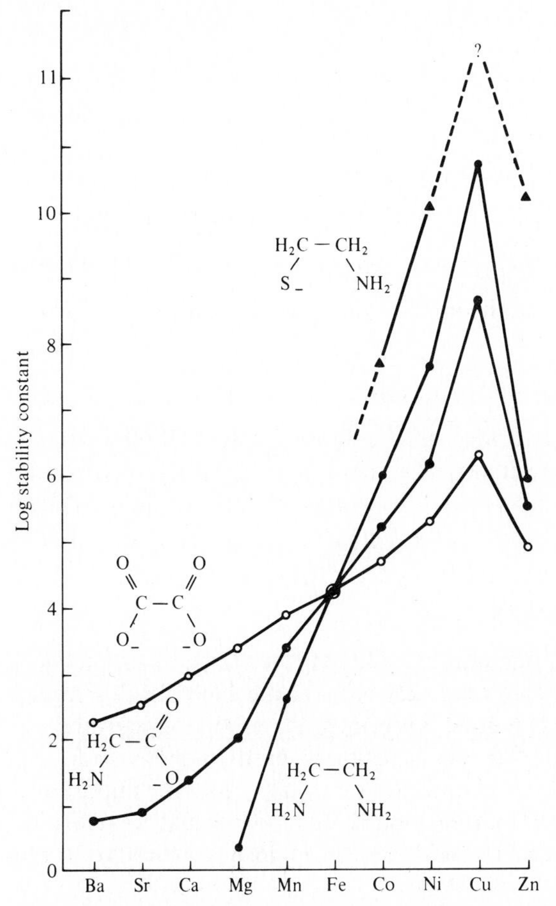

Fig. 7.5 The Irving-Williams effect: The stability increases in the series Ba–Cu, decreases with Zn. [From H. Sigel and D. B. McCormick, *Acc. Chem. Res.*, **1970**, *3*, 201. Reproduced with permission.]

boron atom ("B^{+3}") is borderline between hard and soft. Addition of three hard, electronegative fluorine atoms hardens the boron and makes it a hard Lewis acid. Conversely, addition of three, soft, electropositive hydrogens[43] softens the boron and makes it a soft Lewis acid. Examples of the difference in hardness of these two boron acids are:

$$R_2SBF_3 + R_2O \longrightarrow R_2OBF_3 + R_2S \tag{7.65}$$

$$R_2OBH_3 + R_2S \longrightarrow R_2SBH_3 + R_2O \tag{7.66}$$

[43] In a manner analogous to the usual treatment of balancing redox equations, it is necessary here to do some careful "bookkeeping." Although this is merely a formalism, it is necessary to make certain that the proper comparison is being made. In the present example, the formation of BF_3 is formally considered to be $B^{+3} + 3F^-$. The three F^- ions harden the B^{+3}. The analogous comparison is $B^{+3} + 3H^- = BH_3$. In this case the *soft* hydride ions soften the B^{+3}. One must be careful to distinguish between the small, hard proton (H^+) and the large ($r = 208$ pm), soft hydride ion (H^-).

In a similar manner, the hard BF_3 molecule will prefer to bond to another fluoride ion, but the soft BH_3 acid will prefer a softer hydride ion:

$$BF_3 + F^- \longrightarrow BF_4^- \qquad \textbf{(7.67)}$$

$$B_2H_6 + 2H^- \longrightarrow 2BH_4^- \qquad \textbf{(7.68)}$$[44]

In a competitive reaction, therefore, the following reaction will proceed to the right:

$$BF_3H^- + BH_3F^- \longrightarrow BF_4^- + BH_4^- \qquad \textbf{(7.69)}$$

The fluorinated methanes isoelectronic with the above behave in a similar manner:[45]

$$CF_3H + CH_3F \longrightarrow CF_4 + CH_4 \qquad \textbf{(7.70)}$$

Jørgensen[46] has referred to this tendency of fluoride ions to favor further coordination by a fourth fluoride (the same is true for hydrides) as "symbiosis." Although other factors can work to oppose the symbiotic tendency, it has widespread effect in inorganic chemistry and helps to explain the tendency for compounds to be symmetrically substituted rather than to have mixed substituents.

Theoretical basis of hardness and softness

Although the hard–soft rule is basically a pragmatic one allowing the prediction of chemical properties, it is of interest to investigate the theoretical basis of the effect. In this regard there is no complete unanimity among chemists concerning the relative importance of the various possible factors that might affect the strength of hard–hard and soft–soft interactions. Indeed, it is probable that the various factors may have differing importance depending upon the particular situation. The following discussions should therefore be considered as preliminary ideas in a relatively new area of interest; further work will probably clarify the picture somewhat.

A simple explanation for hard–hard interactions would be to consider them to be primarily electrostatic or ionic interactions. Most of the typical hard acids and bases are those that we might suppose to form ionic bonds such as Li^+, Na^+, K^+, F^-, and OH^-. Since the electrostatic or Madelung energy of an ion pair is inversely proportional to the interatomic distance, the smaller the ions involved, the greater is the attraction between the hard acid and base. Since an electrostatic explanation cannot account for the apparent stability of soft–soft interactions (the Madelung energy of a pair of large ions should be relatively small), it has been suggested that the predominant factor here is a covalent one. This would correlate well for transition metals, Ag, Hg, etc., since it is usually assumed that bonds such as Ag—Cl are considerably more covalent than the corresponding ones

[44] The simple BH_3 molecule does not exist in appreciable quantities, but always dimerizes to B_2H_6. See Chapter 14.

[45] Equation 7.70 is not meant to imply that a mixture of trifluoromethane and fluoromethane would react spontaneously to form tetrafluoromethane and methane, although the reaction would be *exothermic* if it occurred. In this case, as in many others in chemistry, kinetic considerations (lack of a suitable mechanism) override favorable thermodynamics.

[46] C. K. Jørgensen, *Inorg. Chem.*, **1964**, *3*, 1201.

of the alkali metals. In this regard the polarizing power and the polarizability of *d* electrons becomes important. It has been pointed out that all really soft acids are transition metals with six or more *d* electrons, with the d^{10} configuration (Ag^+, Hg^{+2}) being extremely good.[47] From this point of view the polarization effects in soft–soft interactions resemble in some ways the ideas of Fajans (pp. 129–131), although there are notable differences.

Π bonding has been suggested[48] as possibly contributing to soft–soft interactions. Π bonding occurs most readily in those metal ions that have low oxidation states and large numbers of *d* electrons (see Chapter 9). Class (*b*) metal ions (soft Lewis acids) satisfy this criterion. Furthermore, the important π-bonding ligands such as carbon monoxide, phosphines, phosphites, and heavier halogens are all soft bases. The presence of *d* orbitals on the ligand in each case except carbon monoxide[49] enhances the π bonding. Thus the second-row elements N, O, and F are precluded from entering into this type of interaction.

Finally, it should be pointed out that London dispersion energies increase with increasing size and polarizability (see pp. 266–267) and might thus stabilize a bond between two large, polarizable (soft) atoms.[50]

Electronegativity and hardness and softness

In general, species having relatively high electronegativities are hard and those having low electronegativities are soft. In this regard it should be recalled that we are considering *ions* and that although Li, for example, has a low electronegativity, the Li^+ ion has a relatively high electronegativity resulting from the extremely high second ionization potential. In contrast, transition metals in low oxidation states (Cu^+, Ag^+, etc.) have relatively low ionization energies and low electronegativities.[51] The same may be said of hard and soft bases. This relation between hardness and electronegativity helps explain the fact that the trifluoromethyl group is considerably harder than the methyl group and boron trifluoride is harder than borane.

Pearson[52] has called attention to an interesting anomaly between the rule of hard and soft acids and bases and Pauling's original method of defining electronegativity. According to the latter, the ionic resonance energy is proportional to the square of the difference in electronegativities of the constituent atoms (pp. 144–145). Taken literally, this would imply that the most stabilization would occur when bonds are formed between elements furthest apart in electronegativity, such as cesium and fluorine. One might then predict the following reaction based on the presumed dominating stability resulting

---

[47] S. Ahrland, *Struct. Bonding*, **1966**, *1*, 207.

[48] J. Chatt, *Nature (London)*, **1956**, *177*, 852; *J. Inorg. Nucl. Chem.*, **1958**, *8*, 515; R. S. Mulliken, *J. Am. Chem. Soc.*, **1955**, *77*, 885.

[49] In the case of carbon monoxide, π^* orbitals resulting from the triple bond can serve the same function. See p. 366.

[50] K. S. Pitzer, *J. Chem. Phys.*, **1955**, *23*, 1735.

[51] It requires less energy to oxidize $Cu^0 \rightarrow Cu^{+3}$ than $Li^+ \rightarrow Li^{+2}$! Strictly speaking, it appears that, instead of electronegativity, hardness and softness may be determined by a related quantity, the energies of the *frontier orbitals*. The latter are the *highest occupied* and the *lowest unoccupied orbitals* in the interacting atoms (G. Klopman, *J. Am. Chem. Soc.*, **1968**, *90*, 223).

[52] R. G. Pearson, *Chem. Commun.*, **1968**, 65.

from the ionic resonance energy of Cs—F:

$$CsI + LiF \longrightarrow LiI + CsF \tag{7.71}$$

Experimentally, it is found that the reverse is true—that the reaction actually proceeds:

$$LiI + CsF \longrightarrow LiF + CsI \qquad \Delta H = \sim -63 \text{ kJ mol}^{-1} \tag{7.72}$$

Note, however, that the hard–soft acid–base rule works: The two *harder* species (Li^+, F^-) prefer each other and the two *softer* species (Cs^+, I^-) also seem to prefer each other.

Analysis of this apparent paradox can be approached at two levels. At the simplest level we can look at the heats of atomization of the four compounds in Eq. 7.71: LiF = +573, CsF = +502, LiI = +347, CsI = +335 (kJ mol^{-1}). Although the hard–hard interaction (LiF) forms the strongest bond, as expected, the soft–soft interaction is the least stable! It appears that the driving force for Eq. 7.72 is the hard–hard interaction and that this causes the reaction to proceed to the right *in spite of* the weak soft–soft interaction rather than being abetted by it as might have been supposed. Lest it be objected that Eq. 7.72 represents predominantly ionic species in which "covalent" soft–soft interactions cannot take place, the same results can be obtained from a consideration of reactions involving more typical soft–soft species:

$$\underset{\text{Soft–hard}}{HgF_2} + \underset{\text{Hard–soft}}{BeI_2} \longrightarrow \underset{\text{Hard–hard}}{BeF_2} + \underset{\text{Soft–soft}}{HgI_2} \qquad \Delta H = -397 \text{ kJ mol}^{-1} \tag{7.73}$$

The heats of the atomization of the species involved are: BeF_2 = +1264; HgF_2 = +536; BeI_2 = +577; HgI_2 = +293 (kJ mol^{-1}). The driving force for the reaction illustrated in Eq. 7.73 is almost entirely the very high bond energies in BeF_2. We are therefore led to believe that, at least in these examples, the presumably electrostatic energy of the hard–hard interaction is the major driving force.

A more thorough examination of the situation involves attempting to partition the bonding energy of a molecule among the various types of energies: "covalent," "ionic," etc. Obviously, this approach is somewhat artificial, but since as chemists we are accustomed to thinking in terms of these energies, such a treatment may help to illustrate the problem posed by the "Pearson-Pauling paradox."

As a first step, we may consider bonds that are completely covalent (in the sense that both atoms have zero partial charge). The bond energies of some common homopolar bonds are plotted versus bond distances in Fig. 7.6. The most striking feature of these data is the remarkable correlation between bond energy and bond distance. In general, small atoms form strong covalent bonds and large atoms form weaker covalent bonds. Apparent exceptions to this trend are the weak O—O, N—N, and F—F bonds. In these molecules lone pairs on the adjacent small atoms cause strong repulsions. If it were not for these repulsions, the bond energies for these bonds would be about 350–400 kJ mol^{-1}.[53] As evidence that the relative instability of these bonds results from lone-pair repulsions and not from some inherent weakness in the bonding ability of these atoms, it may be

[53] See L. Pauling, "The Nature of the Chemical Bond," 3rd ed., Cornell University Press, Ithaca, N.Y., **1960**, pp. 142–144; K. S. Pitzer, *J. Am. Chem. Soc.*, **1947**, *70*, 2140; P. Politzer, ibid., **1969**, *91*, 6235; *Inorg. Chem.*, **1977**, *16*, 3350.

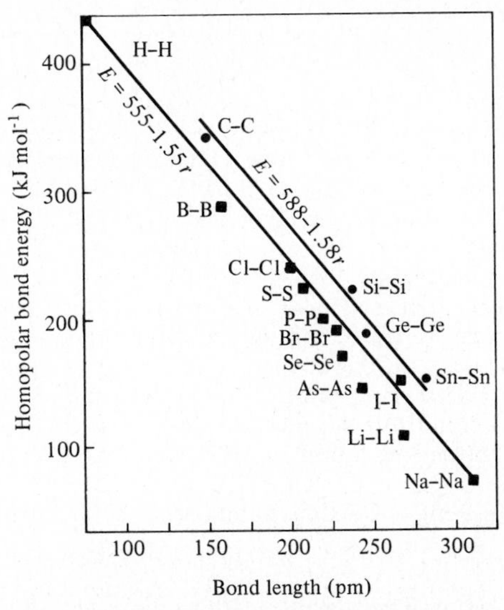

Fig. 7.6 Relation between bond length and bond energy for homopolar bonds. Upper line is for the Group IVA elements only, lower line for the entire set. [From J. E. Huheey and R. S. Evans, *J. Inorg. Nucl. Chem.*, **1970**, *32*, 383. Reproduced with permission.]

noted that the N≡N triple bond is stronger (941 kJ mol^{-1}) and shorter (109.8 pm) than the C≡C triple bond (837 kJ mol^{-1}; 120 pm) as expected. It may be noted that even bonds between larger atoms containing lone pairs (Cl, S, Br, etc.) are somewhat weaker than between comparable atoms without lone-pair repulsions (Si, Ge, Sn), although presumably the repulsions are somewhat relaxed with increasing distance between the bonding atoms.

The data in Fig. 7.6 indicate that interpreting hard–hard interactions (bonds between small atoms) as purely electrostatic and those between larger atoms (soft–soft interactions) as strong covalent bonds is not accurate. Covalent bonding is strongest between small atoms that can achieve very good overlap (the strongest homopolar bond is H—H, 431 kJ mol^{-1}).

In estimating the contributions to the bond energy in a molecule (or ion pair) such as LiF, we must therefore consider the possibility that these two atoms can form a strong *covalent* bond as well as the possibility of forming a strong Madelung attraction (in an Li^+F^- ion pair). This should not be taken to imply that lithium fluoride can form a covalent bond *in addition* to, an ionic bond, but *instead* of, if it proves to be energetically more feasible, or more realistically a bond with part ionic and part covalent character to maximize the bond energy. Attempts to calculate the relative importance of these two methods of bonding are only approximate, but they indicate that each contributes to an appreciable extent in LiF. Of the total bond energy of the LiF bond (573 kJ mol^{-1}), roughly one-fourth comes from covalent bonding, one-half from a Madelung (electrostatic) attraction between the partial charges on the Li and F atoms (approximately two-thirds of an electronic charge), and about one-fourth from the transfer of partial charge from the more electropositive lithium atom to the more electronegative fluorine atom (this latter corresponds roughly to Pauling's ionic resonance energy).[54]

Both the HSAB principle and the E_AE_B–C_AC_B system were proposed and developed in the 1960s. Insofar as the HSAB principle employs ideas of electrostatic and covalent bonding to account for hardness and softness, it was natural to attempt a correlation with the E and C parameters.[55] The early 1970s showed repeated attempts to correlate the two ideas, prove one superior to the other, or to improve their theoretical bases. For example, both Drago[56] and Pearson[57] have discussed the possible quantification of the HSAB principle along the lines of the *E–C* system, but have come to diametrically opposed conclusions. Drago and co-workers[58] have even suggested that the HSAB model is no longer tenable.

Part of the difficulties encountered in comparing these two approaches results from the different ways in which they are used. The *E–C* approach treats the interaction of only two species at a time; to the extent that the nonpolar solvents used in these studies minimize solvation effects, the results are comparable to gas-phase proton affinities. In contrast, the HSAB principle is usually applied to exchange or competition reactions of the sort:

$$A_1B_1 + A_2B_2 \rightleftharpoons A_1B_2 + A_2B_1 \qquad \textbf{(7.74)}$$

We have already seen that in the gas phase the stability of all metal halides follows the order $F^- > Cl^- > Br^- > I^-$, contrary to the simplest possible interpretation of the HSAB rule. Perhaps the latter should be restated as follows: Soft acids prefer to bond to soft bases *when* hard acids are preferentially bonding to hard bases. Although the HSAB rule works in the gas phase,[52] by far its greatest usefulness lies in the interpretation of complexes in aqueous solution. These ions will always be hydrated though this may not be explicitly stated. Under these circumstances, it is somewhat surprising that the HSAB rule works as well as it does. In a recent analysis of both HSAB and *E–C* approaches[59] it was concluded that the HSAB principle works reasonably well when applied to metal-ion complexation, but not to the Drago-Wayland set of acids and bases.

McDaniel and co-workers[59] have presented a graphical means of portraying some of the ideas discussed in this chapter. For the reaction of hard and soft acids and bases:

$$A_hB_s + A_sB_h \longrightarrow A_hB_h + A_sB_s \qquad \textbf{(7.75)}$$

it can be shown that the enthalpy change for this reaction, ΔH_r, can be related to the affinities of the bases for the two acids as shown in Fig. 7.7. If the affinities for a hard acid (e.g., H^+) and a softer acid (e.g., CH_3^+) are plotted and lines of unit slope are drawn through them, ΔH_r for the reaction can be measured by the distance between the lines in either the *x* or *y* direction. Furthermore, if two bases were to fall on the same line in Fig. 7.7, they would be equally soft. If the line for a given base lies above that for another, the first base is softer and the second is harder. Finally, since strength is related to the magnitude of acid–base interactions, the further a given base lies from the origin, the stronger it is.

[54] J. E. Huheey and R. S. Evans, *Chem. Commun.*, **1969**, 968; *J. Inorg. Nucl. Chem.*, **1970**, *32*, 373, 383, 777; see also p. 144.

[55] R. G. Pearson, *Science*, **1966**, *151*, 172.

[56] R. S. Drago and R. A. Kabler, *Inorg. Chem.*, **1972**, *11*, 3144; R. S. Drago, ibid., **1973**, *12*, 2211.

[57] R. G. Pearson, *Inorg. Chem.*, **1972**, *11*, 3146.

[58] R. S. Drago et al., *J. Am. Chem. Soc.*, **1971**, *92*, 6014; R. S. Drago, *J. Chem. Educ.*, **1974**, *51*, 300.

[59] D. H. McDaniel, personal communication. See also, S. V. Lucas, Doctoral Dissertation, University of Cincinnati, Cincinnati, 1975.

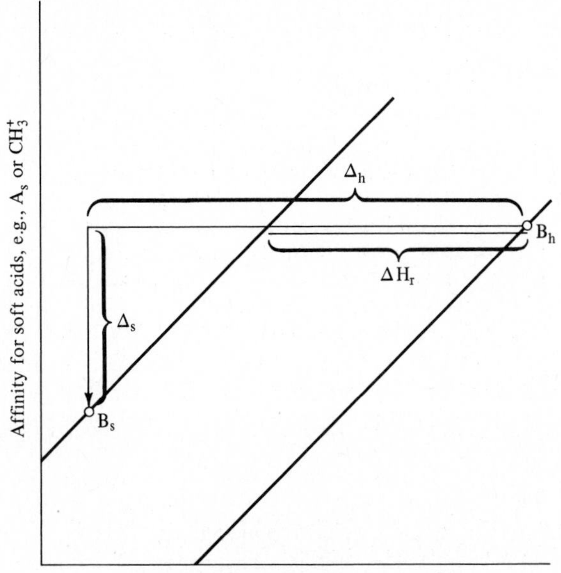

Fig. 7.7 McDaniel diagram illustrating HSAB parameters. Vertical axis, affinity for soft acids; horizontal axis, affinity for hard acids. Δ_s is the difference in affinity of two bases, B_s and B_h for the soft acid, A. Δ_h is the difference in affinity of two bases, B_s and B_h for the hard acid, A_h. The reaction enthalpy of Eq. 7.75 is given by the horizontal distance (or vertical distance) between the two lines of unit slope. [Courtesy of D. H. McDaniel.]

Some typical bases are plotted with their proton affinities and their methyl cation affinities in Fig. 7.8. The reader is urged to find analogous pairs, such as I^- and F^-, PH_3 and NH_3, CO and H_2O, and to interpret their positions on this graph in terms of inherent strength, hardness and softness, etc.

Recently, Staley and co-workers have provided direct measurement of HSAB effects in gas-phase dissociation energies between transition metals (where the principle has always proved most useful) and various ligands, both hard and soft. As expected, Cu^+ (d^{10}) is significantly softer than Co^+ (d^8), which in turn is softer than Mn^+ (d^6).[60] Other transition metals such as iron and nickel give similar results.[61] A comparison of the results for Co^+ and Mn^+ is given in Fig. 7.9. There is the expected correlation of dissociation energies for a large series of oxygen bases, the variation along the line resulting from differences in substituents, hybridization, electronegativity, etc. However, as soon as the hard oxygen bases are replaced by softer bases such as MeSH, EtSH, Me_2S, and HCN, a new line is generated with the softer Co^+ ion showing ~7 kcal mol^{-1} (29 kJ mol^{-1}) greater dissociation energies. The interested reader is referred to the original articles that contain far more data and interesting figures than can be presented here. In summary, all of these data are consistent with the HSAB effect acting in the absence of complicating solvent effects to stabilize either hard–hard interactions, or soft–soft interactions, or both.

[60] R. W. Jones and R. H. Staley, *J. Am. Chem. Soc.*, **1982**, *104*, 2296; *J. Phys. Chem.*, **1982**, *86*, 1387.
[61] M. M. Kappes and R. H. Staley, *J. Am. Chem. Soc.*, **1982**, *104*, 1813, 1819.

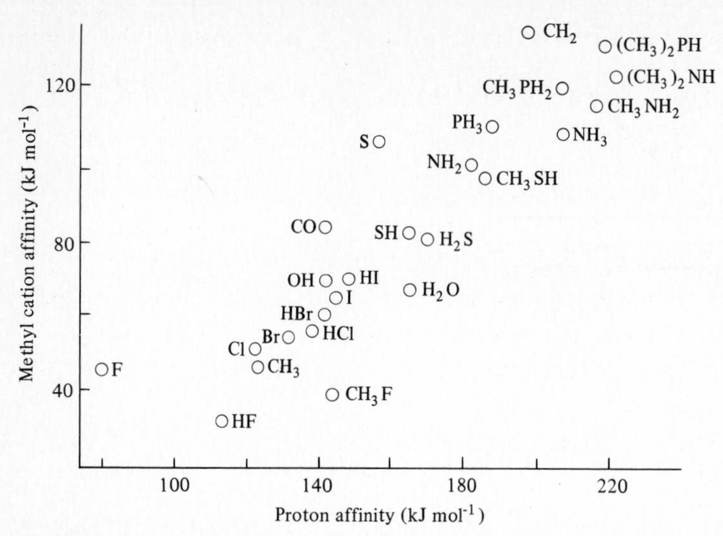

Fig. 7.8 Metal cation affinity versus proton affinity for neutral species. The methyl cation can be considered a softer acid than the proton. [Courtesy of D. H. McDaniel.]

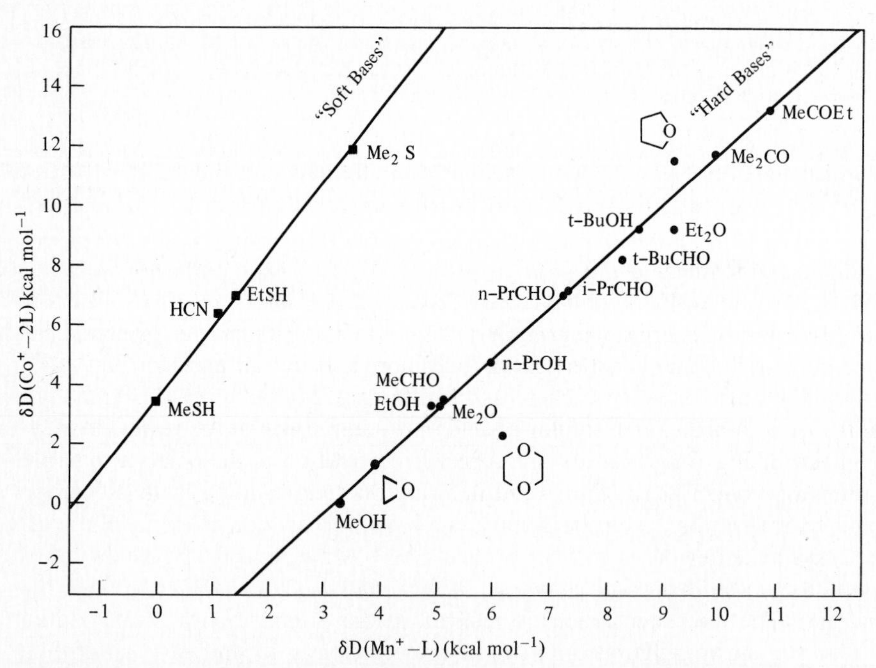

Fig. 7.9 Relative ligand dissociation energies for Mn^+ and Co^+. Zero points for the scales have been arbitrarily chosen. Note that for the *soft* ligands, MeSH, HCN, EtSH, and Me_2SH, the data points lie above and to the left of those for the oxygen bases. This indicates relatively stronger hard–hard bonding (O—Mn^+) or soft–soft bonding (S—Co^+), or both. [Modified from R. W. Jones and R. H. Staley, *J. Phys. Chem.*, **1982**, *86*, 1387. Reproduced with permission.]

In summary, acid–base chemistry is conceptually rather simple, but the multiplicity of factors involved makes its treatment somewhat involved. Until more unifying concepts are developed, as they undoubtedly will be, it will be necessary to apply to each problem that is encountered the ideas, rules, and (when available) the parameters available to it.

PROBLEMS

7.1. Which would you expect to be a better Lewis acid, BCl_3 or $B(CH_3)_3$? Explain.

7.2. The order of acidity of boron halides is $BF_3 < BCl_3 < BBr_3$. Is there anything unexpected in this order? Suggest possible explanations.

7.3. B-strain can occur in amines to lower their basicity. Will B-strain inhibit or enhance the acidic behavior of boranes?

7.4. Predict which way the following reactions will go (left or right):

$$HI + NaF = HF + NaI$$

$$AlI_3 + 3NaF = AlF_3 + 3NaI$$

$$CaS + H_2O = CaO + H_2S$$

$$CuI_2 + 2CuF = CuF_2 + 2CuI$$

$$TiF_4 + 2TiI_2 = TiI_4 + 2TiF_2$$

$$COF_2 + HgBr_2 = COBr_2 + HgF_2$$

7.5. Calculate the values for the proton affinities of the halide anions shown in Table 7.4 from a Born-Haber thermochemical cycle and values for ionization energies, electron affinities, and bond energies.

7.6. When the Drago-Wayland relationship was first set up, the following adducts had not been studied quantitatively:
a. $CH_3CN{\cdot}(CH_3)_3SnCl$; $(CH_3)_2N{-}C{\equiv}N{\cdot}ICl$; $(CH_3)_2N{-}C{\equiv}N{\cdot}p\text{-}ClC_6H_4OH$
b. $HCCl_3$ adducts with the bases $(CH_3)_2CO$, $(CH_3)_2NH$, $(C_2H_5)_2NH$, and $(CH_2)_4O$.
Predict the enthalpy of formation of these adducts. [(a) H. F. Henneike, Jr., and R. S. Drago, *Inorg. Chem.*, **1968**, *7*, 1908; (b) L. J. Sachs, R. S. Drago, and D. P. Eyman, *Inorg. Chem.*, **1968**, *7*, 1484.]

7.7. Predict the enthalpies of adduct formation for the following acid–base pairs and compare with the experimental values shown in parentheses.
a. $NH_3{\cdot}B(CH_3)_3$ (13.75)
b. $CH_3NH_2{\cdot}B(CH_3)_3$ (17.64)
c. $(CH_3)_2NH{\cdot}B(CH_3)_3$ (19.26)
d. $(CH_3)_3N{\cdot}B(CH_3)_3$ (17.62).
Explain any discrepancies observed.

7.8. In general, the best data for correlating acid–base phenomena are obtained in gas-phase experiments rather than in solution. Discuss factors present in solution, especially in polar solvents, that make solution data suspect.

7.9. In contrast to the generalization made in Problem 7.8, there is reason to believe the solution data for $CH_3O(CH_2)_nNH_2$ may be more indicative of inherent basicity than the gas-phase work. Can you suggest a reason? [*Hint:* Consider the possibilities for hydrogen bonding (P. Love, R. B. Cohen, and R. W. Taft, *J. Am. Chem. Soc.*, **1968**, *90*, 2455).]

7.10. **a.** Estimate the approximate pK_a of phosphoric and arsenic acids.
b. Refine your answer by deciding whether H_3PO_4 or H_3AsO_4 is stronger.

7.11. Phosphorous acid can exist as either of two tautomers,

```
  OH           O
 /             ||
P—OH   or   H—P—OH
 \             |
  OH           OH
```

From the pK_a of phosphorous acid, 1.8, assign a structure to the form of phosphorous acid in aqueous solution. The pK_a of hypophosphorous acid, H_3PO_2, is 2.00. Assign a reasonable structure. [See p. 825.]

7.12. In Fig. 7.3, could you have assigned the peaks if the legend had not? [*Hint:* See pp. 168–169.]

7.13. The discussions of basicity of amides on pp. 299 and 303 are based upon the carbonyl oxygen being the basic site. It is also possible that the amide nitrogen atom could act as a base. What experimental work could you suggest to determine which atom is the most basic site?

7.14. If you did not answer Problem 3.41 when you read Chapter 3, do so now.

7.15. Using Fig. 7.8, explain why H_2S is considered softer than H_2O even though it binds more tightly to the hard acid H^+.

7.16. Balance the following equations, identifying the acid and the base:
a. $SO_3 + K_2O$
b. $MgO + Al_2O_3$
c. $Al_2O_3 + Na_2O$
d. $CaO + P_4O_{10}$
e. $SiO_2 + K_2O + Al_2O_3 + MgO$

7.17. If you did not answer Problem 3.34 when you read Chapter 3, do so now.

7.18. Potassium metal reacts with graphite to form an intercalation compound approximating C_8K. Will this material act as an acidic or basic catalyst? [*Chem. Commun.*, **1976**, 913.]

7.19. Predict the order of proton affinities for the following bases: NR_3, S^{-2}, NF_3, O^{-2}, NH_3, OH^-, NCl_3, N^{-3}. Pick any pair of bases from this series and explain why you decided that one was stronger than the other.

7.20. The enthalpy of hydration for 1 mol of protons *and* 1 mol of hydroxide ions is 1552 ± 21 kJ mol^{-1}. Using this value and other appropriate thermodynamic data, calculate the enthalpy of neutralization of strong acids and strong bases in aqueous solution.

7.21. Calculate the enthalpies for all of the possible 1:1 reactions between the acids H^+, CH_3^+, BF_3, and $SbCl_5$, and the bases OH^-, H^-, $(CH_3)_3P$, and Cl^-. Use both Eq. 7.49 and 7.51 where possible. You may check some of your answers against Table 7.5. How accurate are your calculations? Do you notice any systematic relationships between E and e, C and c?

7.22. The N—H bond length in ammonia is 101.5 pm but 103.1 pm in the ammonium ion. Explain.

7.23. Examine Fig. 7.4 and provide a rationale for the relationship therein.

7.24. Most frequently the change in frequency of the carbonyl group upon coordination to a Lewis acid is stated in terms of bond order rather than as explained on p. 310. Develop such an argument.

7.25. If you did not answer Problem 6.17 when you read Chapter 6, do so now.

7.26. Plot t values for:
a. H^+, CH_3^+, $C_2H_5^+$, and $C_3H_7^+$ (p. 306) versus Mulliken-Jaffé b values (Table 3.12)
b. H^-, F^-, Cl^-, Br^-, and I^- (Table E.3) versus Mulliken-Jaffé b values
What can you conclude from your graphs?

7.27. **a.** We learn in organic chemistry that $C_6H_5NH_2$ and C_5H_5N are weaker bases than NH_3, but Table 7.4 indicates otherwise. Discuss, including the important molecular attributes of each molecule. [*Hint:* See RO^-, p. 302.]
b. Water is a weak acid, but most hydrocarbons are usually considered to have virtually no acidity whatsoever (this will be discussed further in Chapter 12). However, in the gas phase $C_6H_5CH_3$ is 10^{12} stronger as an acid than H_2O. Discuss the particular molecular properties that cause the gas-phase values to be different from solution data and to differ so much between these two species.

7.28. Reconcile the values of the proton affinities of pyridine (922 kJ mol^{-1}) and ammonia (858 kJ mol^{-1}) with the argument on p. 300 concerning the relationship between pK_b and electronegativity. The latter argument seems to go with the "conventional wisdom" rather than the discussion in this chapter. Criticize.

7.29. Using a Born-Haber-type cycle, clearly show all of the terms that one should evaluate in considering the energetics involved in transferring the competition (as given by the enthalpy of reaction):

$$BH^+_{(g)} + B'_{(g)} \xrightarrow{\Delta H} B_{(g)} + B'H^+_{(g)} \tag{7.76}$$

into solution:

$$BH^+_{(s)} + B'_{(s)} \xrightarrow{\Delta H} B_{(s)} + B'H^+_{(s)} \tag{7.77}$$

Once you have your Born-Haber diagram drawn, return to the preceding questions and see if it helps to clarify your answers. [E. M. Arnett, *Acc. Chem. Res.*, **1983**, *16*, in press.]

8

Chemistry in aqueous and nonaqueous solutions

Almost all of the reactions that the practicing inorganic chemist observes in the laboratory take place in solution. Although water is the best-known solvent, it is not the only one of importance to the chemist. The organic chemist often uses nonpolar solvents such as carbon tetrachloride and benzene to dissolve nonpolar compounds. These are also of interest to the inorganic chemist and, in addition, polar solvents such as liquid ammonia, sulfuric acid, glacial acetic acid, sulfur dioxide, and various nonmetal halides have been studied extensively. The study of solution chemistry is intimately connected with acid–base theory, and the separation of this material into a separate chapter is merely a matter of convenience. For example, nonaqueous solvents are often interpreted in terms of the solvent system concept, the formation of solvates involve acid–base interactions, and even redox reactions may be included within the Usanovich definition of acid–base reactions.

There are several physical properties of a solvent that are of importance in determining its behavior. Two of the most important from a pragmatic point of view are the melting and boiling points. These determine the liquid range and hence the potential range of chemical operations. More fundamental is the dielectric constant. A high dielectric constant is necessary if solutions of ionic substances are to form readily. Coulombic attractions between ions are inversely proportional to the permittivity (dielectric constant) of the medium:

$$E = \frac{q^+ q^-}{4\pi r \varepsilon} \quad \textbf{(8.1)}$$

where ε is the permittivity. In water, for example, the attraction between two ions is only slightly greater than 1% of the attraction between the same two ions in the absence of the solvent:

$$\varepsilon_{H_2O} = 81.7\varepsilon_0 \quad \textbf{(8.2)}$$

where ε_0 is the permittivity of a vacuum. Solvents with high permittivities (dielectric constants) will tend to be water-like in their ability to dissolve salts.

Table 8.1 Physical properties of water

Boiling point	100 °C
Freezing point	0 °C
Density	1.00 g cm^{-3} (4 °C)
Permittivity (dielectric constant)	81.7 ε_0 (18 °C)
Specific conductance	$4 \times 10^{-6}\ \Omega^{-1}\ m^{-1}$ (18 °C)
Viscosity	0.00101 Pa s (20 °C)
Ion product constant	$1.008 \times 10^{-14}\ mol^2\ dm^{-6}$ (25 °C)

WATER

Water will be discussed only briefly here but a summary of its physical properties is given in Table 8.1 for comparison with the nonaqueous solvents to follow. One notable property is the very high permittivity which makes it a good solvent for ionic and polar compounds. The solvating properties of water and some of the effects have been discussed in Chapter 6. Electrochemical reactions in water are discussed on pp. 350–353.

NONAQUEOUS SOLVENTS

Although many nonaqueous solvent systems have been studied,[1] the discussion here will be limited to a few representative solvents: ammonia,[2] a basic solvent; sulfuric acid, an acidic solvent; and bromine trifluoride, an aprotic solvent. In addition a short discussion of the chemistry taking place in solutions of molten salts is included.

Ammonia

Ammonia has probably been studied more extensively than any other nonaqueous solvent. Its physical properties resemble those of water except that the dielectric constant is considerably smaller (Table 8.2). The lower dielectric constant results in a generally decreased ability to dissolve ionic compounds, especially those containing highly charged ions (e.g., carbonates, sulfates, and phosphates are practically insoluble). In some cases the solubility is higher than might be expected on the basis of the dielectric constant alone. In these cases there is a stabilizing interaction between the solute and the ammonia. One type of interaction is between certain metal ions such as Ni^{+2}, Cu^{+2}, and Zn^{+2} and the ammonia molecule, which acts as a ligand to form stable ammine complexes. A second type

[1] The interested reader is referred to the following books on nonaqueous solvents: L. F. Audrieth and J. Kleinberg, "Non-Aqueous Solvents," Wiley, New York, **1953**; T. C. Waddington, "Non-Aqueous Solvents," Appleton, New York, **1969**; J. Jander and C. Lafrenze, "Ionizing Solvents," Wiley, New York, **1970**. For a synopsis of over 5000 papers on reactions in nonaqueous solvents, including melts, see G. Charlot and B. Tremillon, "Chemical Reactions in Solvents and Melts," P. J. J. Harvey, transl., Pergamon, Elmsford, N.Y., **1969**.

[2] It is customary to refer to substances which are gases at room temperature as *liquid* ammonia, *liquid* sulfur dioxide, *liquid* hydrogen chloride, etc., when these materials are liquefied for use as solvents. This is redundant—all solvents, including water and sulfuric acid, are liquids under the usual conditions for solution study.

Table 8.2 Physical properties of ammonia

Boiling point	$-33.38\ °C$
Freezing point	$-77.7\ °C$
Density	$0.725\ g\ cm^{-3}\ (-70\ °C)$
Permittivity (dielectric constant)	$26.7\varepsilon_0\ (-60\ °C)$
Specific conductance	$1 \times 10^{-9}\ \Omega^{-1}\ m^{-1}$
Viscosity	$0.00254\ Pa\ s\ (-33\ °C)$
Ion product constant	$5.1 \times 10^{-27}\ mol^2\ dm^{-6}$

is between the mutually polarizing and polarizable ammonia molecule and polarizable solute molecules or ions. Ammonia may thus be a better solvent than water toward nonpolar molecules. Ionic compounds containing large, polarizable ions such as iodide and thiocyanate are quite soluble.

Precipitation reactions take place in ammonia just as they do in water. Because of the differences in solubility between the two solvents, the results may be considerably different. As an example, consider the precipitation of silver chloride in aqueous solution:

$$KCl + AgNO_3 \longrightarrow AgCl\downarrow + KNO_3 \tag{8.3}$$

In ammonia solutions the direction of the reaction is reversed so that:

$$AgCl + KNO_3 \longrightarrow KCl\downarrow + AgNO_3 \tag{8.4}$$

Ammonia undergoes autoionization with the formation of ammonium and amide ions:

$$2NH_3 \rightleftharpoons NH_4^+ + NH_2^- \tag{8.5}$$

Neutralization reactions can be run that parallel those in water:

$$KNH_2 + NH_4I \longrightarrow KI + 2NH_3 \tag{8.6}$$

Furthermore, amphoteric behavior resulting from complex formation with excess amide also parallels that in water:

$$Zn^{+2} + 2OH^- \longrightarrow Zn(OH)_2\downarrow \xrightarrow{\text{excess } OH^-} Zn(OH)_4^{-2} \tag{8.7}$$

$$Zn^{+2} + 2NH_2^- \longrightarrow Zn(NH_2)_2\downarrow \xrightarrow{\text{excess } NH_2^-} Zn(NH_2)_4^{-2} \tag{8.8}$$

All acids that behave as strong acids in water react completely with ammonia (are "leveled") to form ammonium ions:

$$HClO_4 + NH_3 \longrightarrow NH_4^+ + ClO_4^- \tag{8.9}$$

$$HNO_3 + NH_3 \longrightarrow NH_4^+ + NO_3^- \tag{8.10}$$

In addition, some acids which behave as weak acids in water (with pK_a up to about 12) react completely with ammonia and hence are strong acids in this solvent:

$$HC_2H_3O_2 + NH_3 \longrightarrow NH_4^+ + C_2H_3O_2^- \tag{8.11}$$

Furthermore, molecules which show no acidic behavior at all in water may behave as weak acids in ammonia:

$$NH_2C(O)NH_2 + NH_3 \rightleftharpoons NH_4^+ + NH_2C(O)NH^- \qquad \textbf{(8.12)}$$

The basic solvent ammonia levels all species showing significant acidic tendencies and enhances the acidity of very weakly acidic species.

Most species that would be considered bases in water are either insoluble or behave as weak bases in ammonia. Extremely strong bases, however, may be leveled to the amide ion and behave as strong bases:

$$H^- + NH_3 \longrightarrow NH_2^- + H_2\uparrow \qquad \textbf{(8.13)}$$

$$O^{-2} + NH_3 \longrightarrow NH_2^- + OH^- \qquad \textbf{(8.14)}$$

Solvolysis reactions are well known in ammonia, and again many reactions parallel those in water. For example, the solvolysis of halogens may be illustrated by:

$$Cl_2 + 2H_2O \longrightarrow HOCl + H_3O^+ + Cl^- \qquad \textbf{(8.15)}$$

$$Cl_2 + 2NH_3 \longrightarrow NH_2Cl + NH_4^+ + Cl^- \qquad \textbf{(8.16)}$$

Many nonmetal halides behave as acid halides in solvolysis reactions:

$$OPCl_3 + 6H_2O \longrightarrow OP(OH)_3 + 3H_3O^+ + 3Cl^- \qquad \textbf{(8.17)}$$

$$OPCl_3 + 6NH_3 \longrightarrow OP(NH_2)_3 + 3NH_4^+ + 3Cl^- \qquad \textbf{(8.18)}$$

The resemblance of these two reactions and the structural resemblance between phosphoric acid ($OP(OH)_3$) and phosphoramide ($OP(NH_2)_3$) has led some people to use the term "ammono acid" to describe the latter.

In a manner analogous to that used in water, a pH scale can be set up for ammonia: pH = 0(1 mol dm^{-3} NH_4^+); pH = 13.5([NH_4^+] = [NH_2^-]), neutrality; pH = 27 (1 mol dm^{-3} NH_2^-). Likewise oxidation–reduction potentials may be obtained, based on the hydrogen electrode (see p. 353):

$$NH_4^+ + e^- = NH_3 + \tfrac{1}{2}H_2 \qquad \mathscr{E}^0 = 0 \qquad \textbf{(8.19)}$$

In summary, the chemistry of ammonia solutions is remarkably parallel to that of aqueous solutions. The principal differences are in the increased basicity of ammonia and its reduced dielectric constant. The latter not only reduces the solubility of ionic materials, it promotes the formation of ion pairs and ion clusters. Hence even "strong" acids, bases, and salts are highly associated.

Solutions of metals in ammonia

If a small piece of an alkali metal is dropped into a Dewar flask containing liquefied ammonia, the latter immediately assumes an intense deep blue color. If more alkali metal is dissolved in the ammonia, eventually a point is reached where a *bronze-colored* phase separates and floats on the blue solution.[3] Further addition of alkali metal results in the gradual conversion of blue solution to bronze solution until the former disappears. Evaporation of the ammonia from the bronze solution allows one to recover the alkali metal unchanged.[4] This unusual behavior has fascinated chemists since its discovery in 1864. Complete agreement on the theoretical interpretation of experimental observations made on these solutions has not been achieved, but the following somewhat simplified discussion will indicate the most popular interpretations.[5]

The blue solution is characterized by (1) its color, which is independent of the metal involved; (2) its density, which is very similar to that of pure ammonia; (3) its conductivity, which is in the range of electrolytes dissolved in ammonia; and (4) its paramagnetism, indicating unpaired electrons, and its electron paramagnetic resonance "g-factor," which is very close to that of the free electron. This has been interpreted as indicating that in dilute solution, alkali metals dissociate to form alkali metal cations and solvated electrons:

$$\mathrm{M} \xrightarrow{\text{dissolve in NH}_3} \mathrm{M}^+ + [e(\mathrm{NH}_3)_x]^- \qquad \textbf{(8.20)}$$

The dissociation into cation and anion accounts for the electrolytic conductivity. The solution contains a very large number of unpaired electrons, hence the paramagnetism, and the g-factor value indicates that the interaction between solvent and electrons is rather weak. It is common to talk of the electron existing in a cavity in the ammonia, loosely solvated by the surrounding molecules. The blue color is a result of an absorption peak that has a maximum of about 1500 nm. This peak results from an absorption of photons by the electron as it is excited to a higher energy level, but not all workers are in agreement as to the nature of the excited state.

The very dilute solutions of alkali metals in ammonia thus come close to presenting the chemist with the hypothetical "ultimate" base, the free electron (p. 297). As might be expected, such solutions are metastable, and when catalyzed, the electron is "leveled" to the amide ion:

$$[e(\mathrm{NH}_3)_x]^- \xrightarrow{\mathrm{Fe_2O_3}} \mathrm{NH}_2^- + \tfrac{1}{2}\mathrm{H}_2 + (x-1)\mathrm{NH}_3 \qquad \textbf{(8.21)}$$

The bronze solutions have the following characteristics: (1) a bronze color with a definite metallic luster; (2) very low densities; (3) conductivities in the range of metals; and (4) magnetic susceptibilities similar to those of pure metals. All of these properties are consistent with a model describing the solution as a "dilute metal" or an "alloy" in which the electrons behave essentially as in a metal, but the metal atoms have been moved apart (compared with the pure metal) by interspersed molecules of ammonia.

The nature of these two phases helps to throw light on the metal–nonmetal transition. For example there has been much speculation that hydrogen molecules at sufficiently

[3] Cesium appears to be an exception. Although the solution changes from blue to bronze with increasing concentration, a two-phase system is never obtained.

[4] One must be very careful to exclude water and other materials which might react with the alkali metal and thus prevent the reversibility of the solution.

[5] The interested reader is referred to W. L. Jolly, *Progr. Inorg. Chem.*, **1959**, *1*, 235; W. L. Jolly, "Metal-Ammonia Solutions," Dowden, Hutchinson, and Ross, Stroudsburg, Pa., **1972**; P. P. Edwards, *Adv. Inorg. Chem. Radiochem.*, **1982**, *25*, 135.

high pressures, such as those occurring on the planet Jupiter, might undergo a transition to an "alkali metal." The fundamental transition is one of a dramatic change of the van der Waals interactions of H_2 molecules into metallic cohesion.[6]

Solutions of alkali metals in ammonia have been the best studied, but other metals and other solvents give similar results. The alkaline earth metals except beryllium form similar solutions readily, but upon evaporation a solid "ammoniate," $M(NH_3)_x$, is formed. Lanthanide elements with stable +2 oxidation states (europium, ytterbium) also form solutions. Cathodic reduction of solutions of aluminum iodide, beryllium chloride, and tetraalkylammonium halides yields blue solutions, presumably containing Al^{+3}, $3e^-$; Be^{+2}, $2e^-$; and R_4N^+, e^-, respectively. Other solvents such as various amines, ethers, and hexamethylphosphoramide have been investigated and show some propensity to form this type of solution. Although none do so as readily as ammonia, stabilization of the cation by complexation (see p. 531) results in typical blue solutions in ethers.[7] The solvated electron is known even in aqueous solution, but it has a very short ($\sim 10^{-3}$ s) lifetime.[8]

These solutions of electrons are not mere laboratory curiosities. In addition to being strong bases, they are also good one-electron reducing agents. For example, pure samples of alkali metal superoxides may be readily prepared in these solutions:

$$M^+ + e^- + O_2 \longrightarrow M^+ + O_2^- \tag{8.22}$$

The superoxide ion is further reducible to peroxide:

$$M^+ + e^- + O_2^- \longrightarrow 2M^+ + O_2^{-2} \tag{8.23}$$

Some metal complexes may also be forced into unusual oxidation states:

$$[Pt(NH_3)_4]^{+2} + 2M^+ + 2e^- \longrightarrow [Pt(NH_3)_4]^0 + 2M^+ \tag{8.24}$$

$$Mo(CO)_6 + 4Na^+ + 4e^- \longrightarrow Na_4[Mo(CO)_4] + 2Na_2C_2O_2 \tag{8.25}$$[9]

$$Au + M^+ + e^- \longrightarrow M^+ + Au^- \tag{8.26}$$[9]

The chemistry of metal electrides has been extensively studied and although the formulation M^+e^- is undoubtedly the best, most chemists have the all-too-human emotion of feeling more secure in their science if they have something more tangible then solutions and equations on paper. Therefore the very recent isolation of a material believed to be cesium electride, $[Cs(ligand)]^+e^-$, as single crystals was welcome, indeed.[9a] The crystals are dark blue with a single absorption maximum at 1500 nm, have no likely anions present (the empirical formula is 1:1 Cs:ligand with a trace of lithium impurity, an artifact of the synthetic technique, and are most readily formulated as a complex of cesium electride.

[6] P. P. Edwards and M. J. Sienko, *J. Am. Chem. Soc.*, **1981**, *103*, 2967.

[7] J. L. Dye, M. G. DeBacker, and V. A. Nicely, *J. Am. Chem. Soc.*, **1970**, *92*, 5226.

[8] E. J. Hart, *Acc. Chem. Res.*, **1969**, *2*, 161, in "Survey of Progress in Chemistry," A. F. Scott, ed., Academic Press, New York, **1969**, Vol. 5, pp. 129–184.

[9] See Chapter 13, "Carbonylate anions."

[9a] D. Issa and J. L. Dye, *J. Am. Chem. Soc.*, **1982**, *104*, 3781. For further discussion of this compound, see p. 53.

Table 8.3 Physical properties of sulfuric acid

Boiling point	300 °C (with decomposition)
Freezing point	10.371 °C
Density	1.83 g cm^{-3} (25 °C)
Permittivity (dielectric constant)	$110\varepsilon_0$ (20 °C)
Specific conductance	1.04 Ω^{-1} m^{-1} (25 °C)
Viscosity	0.2454 Pa s (20 °C)
Ion product constant	2.7×10^{-4} mol^2 dm^{-6} (25 °C)

Sulfuric acid

The physical properties of sulfuric acid are listed in Table 8.3. The dielectric constant is even higher than that of water, making it a good solvent for ionic substances and leading to extensive autoionization. The high viscosity, some 25 times that of water, introduces experimental difficulties: Solutes are slow to dissolve and slow to crystallize. It is also difficult to remove adhering solvent from crystallized materials. Furthermore, solvent that has not drained from prepared crystals is not readily removed by evaporation because of the very low vapor pressure of sulfuric acid.

Autoionization of sulfuric acid results in the formation of the hydrogen sulfate (bisulfate) ion and a solvated proton:

$$2H_2SO_4 \rightleftharpoons H_3SO_4^+ + HSO_4^- \qquad \textbf{(8.27)}$$

As expected, a solution of potassium hydrogen sulfate is a strong base and may be titrated with a solute forming $H_3SO_4^+$ ions. Such a titration may readily be followed conductometrically with a minimum in conductivity at the neutralization point.[10] The nature and amount of solution products may be determined by titration with standard acid or base. In this way it may be determined if a solute dissolves to form acidic or basic solutions and how many equivalents are formed per mole.

Another method that has proved extremely useful in obtaining information about the nature of solutes in sulfuric acid solution is measurement of freezing point depressions. The freezing point constant (k) for sulfuric acid is 6.12 kg °C mol^{-1}. For ideal solutions, the depression of the freezing point is given by:

$$\Delta T = kmv \qquad \textbf{(8.28)}$$

where m is the stoichiometric molality (moles solute per kilogram solvent) and v is the number of particles formed when one molecule of solute dissolves in the solvent. For example, ethanol reacts with sulfuric acid as follows:

$$C_2H_5OH + 2H_2SO_4 \longrightarrow C_2H_5HSO_4 + HSO_4^- + H_3O^+ \qquad v = 3 \qquad \textbf{(8.29)}$$

[10] This statement is not quite true. The concentration of ions is at a minimum at the neutralization point, but since conductivity depends on viscosity as well (which changes with composition), the absolute minimum conductivity does not occur exactly when $[H_3SO_4^+] = [HSO_4^-]$. The slight difference is not important in practice, however.

It is found that all species that are basic in water are also basic in sulfuric acid:

$$OH^- + 2H_2SO_4 \longrightarrow 2HSO_4^- + H_3O^+ \quad \nu = 3 \tag{8.30}$$

$$NH_3 + H_2SO_4 \longrightarrow HSO_4^- + NH_4^+ \quad \nu = 2 \tag{8.31}$$

Likewise, water behaves as a base in sulfuric acid:

$$H_2O + H_2SO_4 \longrightarrow HSO_4^- + H_3O^+ \quad \nu = 2 \tag{8.32}$$

Amides, such as urea, which are nonelectrolytes in water and acids in ammonia accept protons from sulfuric acid:

$$NH_2C(O)NH_2 + H_2SO_4 \longrightarrow HSO_4^- + NH_2C(O)NH_3^+ \quad \nu = 2 \tag{8.33}$$

Acetic acid is a weak acid in aqueous solution and nitric acid a strong acid, but *both* behave as bases in sulfuric acid!

$$CH_3C(=O)OH + H_2SO_4 \longrightarrow HSO_4^- + CH_3C(OH)_2^+ \quad \nu = 2 \tag{8.34}$$

$$HNO_3 + 2H_2SO_4 \longrightarrow 2HSO_4^- + NO_2^+ + H_3O^+ \quad \nu = 4 \tag{8.35}$$

Sulfuric acid is a very acidic medium, and so almost all chemical species which react upon solution do so with the formation of hydrogen sulfate ions and are bases. Because of the extreme tendency of the H_2SO_4 molecule to donate protons, molecules exhibiting basic tendencies will be leveled to HSO_4^-.

Perchloric acid is one of the strongest acids known, but in sulfuric acid it is practically a nonelectrolyte, behaving as a very weak acid:

$$HClO_4 + H_2SO_4 \rightleftharpoons H_3SO_4^+ + ClO_4^- \tag{8.36}$$

One of the few substances found to behave as an acid in sulfuric acid is disulfuric ("pyrosulfuric") acid. It is formed from sulfur trioxide and sulfuric acid:

$$SO_3 + H_2SO_4 \longrightarrow H_2S_2O_7 \tag{8.37}$$

$$H_2S_2O_7 + H_2SO_4 \rightleftharpoons H_3SO_4^+ + HS_2O_7^- \tag{8.38}$$

The only truly strong acid in sulfuric acid is hydrogen tetrakis(hydrogen sulfato)borate, $HB(HSO_4)_4$.[11] The compound has not been prepared and isolated in pure form, but solutions of it may be prepared in sulfuric acid:

$$H_3BO_3 + 6H_2SO_4 \longrightarrow B(HSO_4)_4^- + 3H_3O^+ + 2HSO_4^- \quad \nu = 6 \tag{8.39}$$

---

[11] For a discussion of the chemistry of the sulfuric acid solvent system, see R. J. Gillespie and E. A. Robinson, *Adv. Inorg. Chem. Radiochem.*, **1959**, *1*, 385; R. J. Gillespie and E. A. Robinson, in "Non-Aqueous Solvent Systems," T. C. Waddington, ed., Academic Press, New York, **1965**; W. H. Lee, in "The Chemistry of Non-Aqueous Solvents," J. J. Lagowski, ed., Academic Press, New York, **1967**.

Addition of SO_3 removes the H_3O^+ and HSO_4^- ions:

$$B(HSO_4)_4^- + 3H_3O^+ + 2HSO_4^- + 3SO_3 \longrightarrow H_3SO_4^+ + B(HSO_4)_4^- + 4H_2SO_4 \tag{8.40}$$

Solutions of strong acids in very acidic solvents are extremely effective protonating agents and have been termed "*superacids*."[12] Among the more interesting reactions of superacids prepared in fluorosulfonic acid and hydrogen fluoride solvent systems is the apparent protonation of alkanes:

$$\left.\begin{array}{l} SbF_5 + 2HF \longrightarrow SbF_6^- + H_2F^+ \quad (8.41) \\ SbF_5 + 2HSO_3F \longrightarrow FSO_3SbF_5^- + H_2SO_3F^+ \quad (8.42) \end{array}\right\} \text{superacids}$$

$$(CH_3)_3C{-}CH_3 + \text{superacid} \longrightarrow \left[(CH_3)_3C{-}CH_4^+ \right] \longrightarrow (CH_3)_3C^+ + CH_4 \tag{8.43}$$

Even such unlikely bases as Xe, H_2, Cl_2, Br_2, and CO_2 have been shown to accept H^+ ions from superacids, though perhaps only to a small extent. There is no evidence that Ar, O_2, or N_2 become protonated.[12]

Summary of protonic solvents

Despite certain differences, the three protonic solvents discussed above (water, ammonia, and sulfuric acid) share a similarity in their acid–base behavior. All are autoionizing, with the ionization taking place through the transfer of a proton from one molecule of solvent to another with the formation of a solvated proton (Brønsted acid, solvent system acid) and a deprotonated anion (Brønsted and Lewis base, solvent system base). The inherent acidities and basicities of these three solvents differ, however, and so their tendencies to protonate or deprotonate solutes differ. It is possible to list solvents in order of their inherent acidity or basicity.[13] Water is obviously less acidic than sulfuric acid but more so than ammonia. Glacial acetic acid lies between water and sulfuric acid in acidity. Figure 8.1 graphically illustrates the relative acidities and basicities of four solvents, together with various acid–base conjugate pairs. They are listed in order of the pK_a in water. In an ideal aqueous solution the pH of an equimolar mixture of conjugates is given by the pK_a, and similar acidity scales may be used in other solvents.[14] The pK_a is

[12] R. J. Gillespie, *Acc. Chem. Res.*, **1968**, *1*, 202; G. A. Olah and R. H. Schlosbert, *J. Am. Chem. Soc.*, **1968**, *90*, 2726; R. J. Gillespie and G. P. Pez, *Inorg. Chem.*, **1969**, *8*, 1233; G. A. Olah and J. Shen, *J. Am. Chem. Soc.*, **1973**, *95*, 3582; J.-Y. Calves and R. J. Gillespie, *J. Am. Chem. Soc.*, **1977**, *99*, 1788, and references therein.

[13] In the following discussion and in considering Fig. 8.1, it must be remembered that the ordering of acids and bases is only an approximate process. Differing solvation effects and similar phenomena prevent a precise ordering of solutes, and the acidity of a species such as H_2O will differ depending upon whether it is a solute surrounded by another solvent or is situated in the hydrogen-bonded structure of pure water.

[14] If, instead of an equimolar mixture of the acid and base conjugates, pure acid or pure base is added, the resulting solution will be somewhat more acid or basic, respectively.

thus a rough estimate of acidity in solvents other than water. Any given acid is stronger than the acids listed above it and, conversely, any base is stronger than the bases below it. All species that lie within the extremes of a particular solvent behave as weak electrolytes in that solvent and form weakly acidic or weakly basic solutions. All species that lie beyond the enclosed range are leveled by the solvent.

An example may serve to illustrate the information that may be obtained from Fig. 8.1. Consider acetic acid. In water, acetic acid behaves as an acid or, to be more precise, an equimolar mixture of acetic acid and an acetate salt will have a pH of 4.7. If acetic acid is added to sulfuric acid, it will behave as a base and be leveled to $CH_3C(OH)_2^+$, the acetic acidium ion, and HSO_4^- (cf. Eq. 8.34; note the equilibrium lying at about -9 on the scale in Fig. 8.1).

If dissolved in ammonia, acetic acid will behave as a strong acid and be leveled to NH_4^+ and CH_3COO^- (cf. Eq. 8.11; note equilibrium lying at 4.7 on the scale). The different behavior of acetic acid as a strong acid (ammonia) or a weak acid (water) depends upon the basicity of the solvent.

The "boxes" for the solvents (Fig. 8.1) indicate the range over which differentiation occurs; outside the range of a particular solvent, all species are leveled. For example, water can differentiate species (i.e., they are weak acids and bases) with pK_a's from about 0 to 14 (such as acetic acid). Ammonia, on the other hand, behaves the same toward acetic acid and sulfuric acid because both lie below the differentiating limit of ~ 12. The extent of these ranges is determined by the autoionization constant of the solvent (e.g., ~ 14 units for water). The acid–base behavior of several species discussed previously may be seen to correlate with Fig. 8.1.

A complete discussion of relative acidities and basicities would be too extensive to be covered here. Nevertheless it is possible to summarize the behavior of acids and bases as involving (1) the inherent acidity–basicity of the solvent, (2) the inherent acidity–basicity of the solute, and (3) the interaction of solute and solvent to form an equilibrium (the "weak" case, differentiation) or alternatively to go essentially to completion (the "strong" case, leveled). Finally, it must be recalled that only solvents of high dielectric constants can support electrolytic solutions. Solvents of low dielectric constants will result in "weak" electrolytes irrespective of acidity or basicity arguments.

Aprotic solvents

Thus far, the solvents discussed have had one feature in common with water, namely, the presence of a transferable hydrogen and the formation of onium ions. In this section we shall look briefly at solvents which do not ionize in this way. These may be conveniently classified into three groups. The first group consists of solvents such as carbon tetrachloride and cyclohexane which are nonpolar, essentially nonsolvating, and do not undergo autoionization. These are useful when it is desired that the solvent play a *minimum role* in the chemistry being studied, e.g., in the determination of E and C parameters discussed in the previous chapter.

The second group consists of those solvents that are highly polar (usually), yet do not ionize to an appreciable extent. Some examples of solvents of this type are acetonitrile, $CH_3C{\equiv}N$; dimethylacetamide (DMA), $CH_3C(O)N(CH_3)_2$; dimethyl sulfoxide (DMSO),

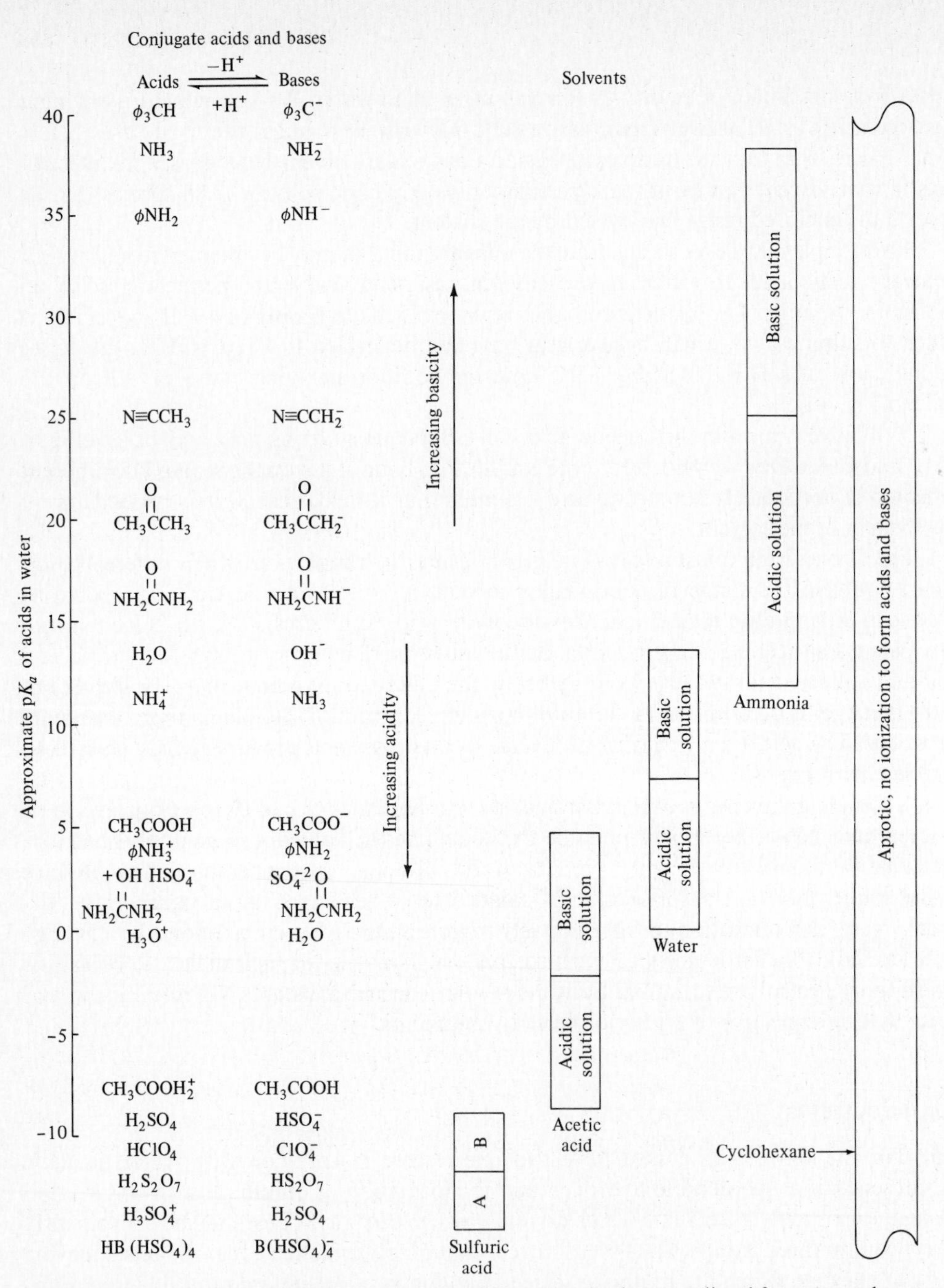

Fig. 8.1 Relative acidity and basicity of solvents. Solvents and solutes are listed from top to bottom in order of decreasing basicity and increasing acidity. Solutes are listed in order of decreasing pK_a as determined in water. Some values of pK_a are estimated. In ideal aqueous solutions, equimolar mixtures of an acid and its conjugate base will have a pH equal to the pK_a. The range of acidity and basicity over which a particular solvent is differentiating is shown at the right. All acids lying below and all bases lying above the enclosed box will be leveled to the characteristic cation and anion of the solvent.

$(CH_3)_2S{=}O$; and sulfur oxide, SO_2. Although these solvents do not ionize to a significant extent, they are good coordinating solvents because of their polarity. The polarity ranges from low (SO_2) to extremely high (DMSO). Most are basic solvents tending to coordinate strongly with cations and other acidic centers:

$$CoBr_2 + 6DMSO \longrightarrow [Co(DMSO)_6]^{+2} + 2Br^- \qquad \textbf{(8.44)}$$

$$SbCl_5 + CH_3C{\equiv}N \longrightarrow [CH_3C{\equiv}N \rightarrow SbCl_5] \qquad \textbf{(8.45)}$$

A few, the nonmetal oxides and halides, can behave as acceptor solvents, reacting with anions and other basic centers:

$$\phi_3CCl + SO_2 \longrightarrow \phi_3C^+ + SO_2Cl^- \qquad \textbf{(8.46)}$$

This group of solvents ranges from the limiting case of a nonpolar solvent (Group I) to an autoionizing solvent (Group III, see below). Within this range a wide variety of reactivity is obtained. Gutmann[15] has defined the *donor number* (DN) as a measure of the basicity or donor ability of a solvent. It is defined as the negative enthalpy of reaction of a base with the Lewis acid antimony pentachloride, $SbCl_5$:

$$B + SbCl_5 \longrightarrow B \rightarrow SbCl_5 \qquad DN_{SbCl_5} \equiv -\Delta H_r \qquad \textbf{(8.47)}$$

These donor numbers provide an interesting comparison of the relative donor abilities of the various solvents (Table 8.4), ranging from the practically nonpolar 1,2-dichloroethane to the highly polar hexamethylphosphoramide, $[(CH_3)_2N]_3PO$. Note, however, that there is no exact correlation between donor number and dielectric constant. Some solvents with relatively high dielectric constants such as nitromethane and propylene carbonate ($\varepsilon/\varepsilon_0$ = 38.6 and 65.1) may be very poor donors (DN = 2.7 and 15.1). Conversely, the best donors do not always have high dielectric constants: pyridine (DN = 33.1, $\varepsilon/\varepsilon_0 = 12.3$) and diethyl ether (DN = 19.2, $\varepsilon/\varepsilon_0 = 4.3$). This should serve to remind us that solubility is not merely an electrostatic interaction but that solvation also involves the ability to form covalent donor bonds. Note that pyridine and tributyl phosphate may be considered to be relatively "soft" bases (pp. 312–325). Drago[16] has criticized the donor-number concept because it does not go far enough in accounting for differences in hardness and softness (or electrostatic and covalent differences). By limiting the evaluation of donor numbers to a single acid ($SbCl_5$), the donor-number system in effect presents only half of the information available from the $E_AE_B + C_AC_B$ four-parameter equation. More recently, Gutmann has extended the concept to include an *acceptor number* (AN), that measures the electrophilic behavior of a solvent (Table 8.4).[17]

The third group of solvents consists of those that are highly polar and autoionizing. They are usually highly reactive and are difficult to keep pure since they react with traces

[15] V. Gutmann and E. Wychera, *Inorg. Nucl. Chem. Letters*, **1966**, *2*, 257; V. Gutmann, "Coordination Chemistry in Non-Aqueous Solutions," Springer, New York, **1968**; U. Mayer and V. Gutmann, *Adv. Inorg. Chem. Radiochem.*, **1975**, *17*, 189. See also V. Gutmann, "The Donor-Acceptor Approach to Molecular Interactions," Plenum, New York, **1978**.

[16] R. S. Drago et al., *J. Am. Chem. Soc.*, **1971**, *93*, 6014.

[17] V. Gutmann, *Electrochim. Acta*, **1976**, *21*, 661; *Chimia*, **1977**, *31*, 1.

Table 8.4 Donor number (DN), acceptor number (AN), and relative permittivity (dielectric constant, $\varepsilon/\varepsilon_0$) of selected solvents[a]

Solvent	DN	AN	$\varepsilon/\varepsilon_0$
Acetic acid	—	52.9	6.2
Acetic anhydride	10.5	—	—
Acetone	17.0	12.5	20.7
Acetonitrile	14.1	19.3	36
Acetyl chloride	0.7	—	15.8
Antimony pentachloride	—	100.0	—
Benzene	0.1	8.2	2.3
Benzonitrile	11.9	15.5	—
Benzoyl chloride	2.3	—	23
Benzoyl fluoride	2.3	—	—
Benzyl cyanide	15.1	—	—
iso-Butyronitrile	15.4	—	—
n-Butyronitrile	16.6	—	—
Carbon tetrachloride	—	8.6	2.2
Chloroform	—	23.1	4.8
Dichloromethane	—	20.4	—
N,N-Diethylacetamide (DEA)	32.2	—	—
Dichloroethylene carbonate	3.2	16.7	—
Diethyl ether	19.2	3.9	4.3
N,N-Diethylformamide (DEF)	30.9	—	—
Diglyme	24.0	10.2	—
Dimethylacetamide (DMA)	27.3	13.6	37.8
Dimethylformamide (DMF)	24.0	16.0	36.7
Dimethylsulfoxide (DMSO)	29.8	19.3	45
Dioxane	14.8	10.8	2.2
Diphenylphosphoric chloride	22.4	—	—
Ethanol	19.0	37.1	24.3
Ethyl acetate	17.1	—	—
Ethylene carbonate	16.4	—	—
Ethylene sulfite	15.3	—	—
Formamide	—	39.8	—
Hexamethylphosphorotriamide (HMPA)	38.8	10.6	—
Methanol	20.0	41.3	32.6
Methyl acetate	16.5	—	—
N-Methylpyrolidinone	27.3	13.3	—
Methylsulfonic acid	—	126.1	—
Nitrobenzene	4.4	14.8	34.8
Nitromethane	2.7	20.5	38.6
Phenylphosphoric dichloride	18.5	—	—
Phenylphosphoric difluoride	16.4	—	—
Phosphorus oxychloride	11.7	—	—
Propanol	18.0	33.5	20.1
Propionitrile	16.1	—	—
Propylene carbonate	15.1	18.3	65.1
Pyridine (py)	33.1	14.2	12.3
Selenium oxychloride	12.2	—	—
Tetrachloroethylene carbonate	0.8	—	—
Tetrahydrofuran	20.0	8.0	7.3
Tetramethylenesulfone	14.8	—	—

Table 8.4 *(Continued)*

Solvent	DN	AN	$\varepsilon/\varepsilon_0$
Thionyl chloride	0.4	—	—
Tributyl phosphate	23.7	—	—
Trifluoroacetic acid	—	105.3	—
Trifluorosulfonic acid	—	129.1	—
Trimethyl phosphate	23.0	—	—
Water	18	54.8	81.7

[a] The ratio $\varepsilon/\varepsilon_0$ is more convenient to use than the value of the permittivity in absolute units.

SOURCE: V. Gutmann, *Chimia*, **1977**, *31*, 1; "The Donor-Acceptor Approach to Molecular Interactions," Plenum, New York, **1978**. Used with permission.

of moisture and other contaminants. Some even react slowly with silica containers or dissolve electrodes of gold and platinum. An example of one of the more reactive of these solvents is bromine trifluoride. Nonfluoride salts, such as oxides, carbonates, nitrates, iodates, and other halides, are fluorinated:

$$Sb_2O_5 \xrightarrow{BrF_3} [BrF_2^+][SbF_6^-] \quad \textbf{(8.48)}$$

$$GeO_2 \xrightarrow{BrF_3} [BrF_2^+]_2[GeF_6^{-2}] \quad \textbf{(8.49)}$$

$$PBr_5 \xrightarrow{BrF_3} [BrF_2^+][PF_6^-] \quad \textbf{(8.50)}$$

$$NOCl \xrightarrow{BrF_3} [NO^+][BrF_4^-] \quad \textbf{(8.51)}$$

Fluoride salts dissolve unchanged except for fluoride ion transfer to form conducting solutions:

$$KF \xrightarrow{BrF_3} K^+, [BrF_4^-] \quad \textbf{(8.52)}$$

$$AgF \xrightarrow{BrF_3} Ag^+, [BrF_4^-] \quad \textbf{(8.53)}$$

$$SbF_5 \xrightarrow{BrF_3} [BrF_2^+], [SbF_6^-] \quad \textbf{(8.54)}$$

$$SnF_4 \xrightarrow{BrF_3} 2[BrF_2^+], [SnF_6^{-2}] \quad \textbf{(8.55)}$$

These solutions can be considered acids or bases by analogy to the presumed autoionization[18] of BrF_3:

$$2BrF_3 \rightleftharpoons [BrF_2^+] + [BrF_4^-] \quad \textbf{(8.56)}$$

[18] This expression for the autoionization of BrF_3 is based on the conductivity of pure BrF_3 and the characterization of the BrF_4^- ion in salts such as $KBrF_4$. The evidence for BrF_2^+ is weaker. Further support for this formulation is obtained from the ICl_3 system where X-ray evidence for ICl_2^+ and ICl_4^- ions has been obtained. See A. G. Sharpe, in "Non-Aqueous Solvent Systems," T. C. Waddington, ed., Academic Press, New York, **1965**, pp. 292–295.

Reactions 8.48 to 8.50, 8.54, and 8.55 above may be considered to form acid solutions (BrF_2^+ ions formed) and reactions 8.51 to 8.53 may be considered to form basic solutions (BrF_4^- ions formed). Acid solutions may be readily titrated by bases:

$$[BrF_2^+][SbF_6^-] + Ag^+[BrF_4^-] \longrightarrow Ag^+[SbF_6^-] + 2BrF_3 \quad (8.57)$$

Such reactions may be followed conveniently by measuring the conductivity of the solutions—a minimum occurs at the 1:1 endpoint. Solutions of SnF_4 behave as a dibasic acid:

$$[BrF_2^+]_2[SnF_6^{-2}] + 2KBrF_4 \longrightarrow 2K^+[SnF_6^{-2}] + 4BrF_3 \quad (8.58)$$

with minimum conductivities corresponding to 1:2 mole ratios.

A similar, although less reactive, aprotic solvent is phosphorus oxychloride (phosphoryl chloride). A tremendous amount of work on the properties of this solvent has been done by Gutmann and co-workers.[19] They have interpreted their results in solvent system terms based on the supposed autoionization:

$$OPCl_3 \rightleftharpoons OPCl_2^+ + Cl^- \quad (8.59)$$

or the more general solvated forms:

$$(m + n)OPCl_3 \rightleftharpoons [OPCl_2^+(OPCl_3)_{n-1}] + [Cl^-(OPCl_3)_m] \quad (8.60)$$

It is extremely difficult to measure this autoionization since contamination with traces of water yields conducting solutions which may be described approximately as:

$$3H_2O + 2OPCl_3 \longrightarrow 2[H_3O^+]Cl^- + Cl_2P(O)OP(O)Cl_2 \quad (8.61)$$

If autoionization does occur, the ion product, $(c_{OPCl_2^+})(c_{Cl^-})$, is equal to or less than 5×10^{-14}.

Salts which dissolve in phosphorus oxychloride to yield solutions with high chloride ion concentrations are considered bases:

$$K^+Cl^- \xrightarrow{OPCl_3} K^+ + Cl^- \quad \text{strong base} \quad (8.62)$$

$$Et_3N \overset{OPCl_3}{\rightleftharpoons} [Et_3NP(O)Cl_2^+] + Cl^- \quad \text{weak base} \quad (8.63)$$

Most molecular chlorides behave as acids:

$$FeCl_3 \xrightarrow{OPCl_3} [OPCl_2^+] + [FeCl_4^-] \quad (8.64)$$

$$SbCl_5 \xrightarrow{OPCl_3} [OPCl_2^+] + [SbCl_6^-] \quad (8.65)$$

As might be expected, basic solutions may be titrated with acidic solutions and the neutralization followed by conductometric, potentiometric, photometric, and similar

---

[19] For complete lists of references to this work, see V. Gutmann, "Coordination Chemistry in Non-Aqueous Solvents," Springer, New York, **1968**; V. Gutmann, "Halogen Chemistry," Academic Press, New York, **1967**, Vol. II, pp. 399–488; and D. S. Payne, "Non-Aqueous Solvent Systems," T. C. Waddington, ed., Academic Press, New York, **1965**, pp. 301–352.

Table 8.5 Relative chloride ion donor and acceptor abilities

Chloride ion donors (↑ Increased basicity)	Chloride ion acceptors (↓ Increased acidity)
$[R_4N]Cl$	
KCl	
$AlCl_3$	$AlCl_3$
$TiCl_4$	$ZnCl_2$
PCl_5	PCl_5
$ZnCl_2$	$TiCl_4$
	$HgCl_2$
BCl_3	BCl_3
	BF_3
	$InCl_3$
$SnCl_4$	$SnCl_4$
$AlCl_2^+$	
$HgCl_2$	$SbCl_5$
$SbCl_3$	$FeCl_3$

methods. Some metal and nonmetal chlorides are amphoteric in phosphorus oxychloride:

$$K^+ + Cl^- + AlCl_3 \xrightarrow{OPCl_3} K^+ + [AlCl_4^-] \quad \textbf{(8.66)}$$

$$SbCl_5 + AlCl_3 \xrightarrow{OPCl_3} [AlCl_2^+] + [SbCl_6^-] \quad \textbf{(8.67)}$$

A table of relative chloride ion donor and acceptor abilities can be established[20] from equilibrium and displacement reactions (Table 8.5). As expected, good donors are generally poor acceptors and vice versa with but few exceptions (e.g., $HgCl_2$).

There has been some controversy in the literature over the proper interpretation of reactions in solvents such as phosphorus oxychloride. Drago and co-workers[21] have suggested the "coordination model" as an alternative to the solvent system approach. They have stressed the errors incurred when the solvent system concept has been pushed further than warranted by the facts.[22] In addition, they have pointed out that iron(III) chloride dissolves in triethyl phosphate with the formation of tetrachloroferrate(III) ions, $FeCl_4^-$, just as in phosphorus oxychloride. In triethyl phosphate, however, the solvent cannot behave as a chloride ion donor and so a reaction such as Eq. 8.64 is not applicable. In triethyl phosphate the chloride ion transfer *must* take place from one $FeCl_3$ molecule to another with the formation of a cationic iron(III) species:

$$2FeCl_3 \xrightarrow{OP(OEt)_3} [FeCl_2^+(OP(OEt)_3)_n] + [FeCl_4^-] \quad \textbf{(8.68)}$$

[20] The ordering of this list is not invariant—some of the compounds listed have very similar donor and acceptor abilities and exchange places depending upon the nature of other ions in solution. This is to be expected in a solvent of relatively low dielectric constant ($\varepsilon/\varepsilon_0 = 13.9$) where ion-pair formation will be important and the nature of the counterion can affect the stability of a chloride adduct.

[21] D. W. Meek and R. S. Drago, *J. Am. Chem. Soc.*, **1961**, *83*, 4322; R. S. Drago and K. F. Purcell, *Progr. Inorg. Chem.*, **1964**, *6*, 271; R. S. Drago and K. F. Purcell, in "Non-Aqueous Solvent Systems," T. C. Waddington, ed., Academic Press, New York, **1965**, pp. 211–251.

[22] Compare the discussion of the supposed autoionization of the sulfur dioxide solvent system, p. 291.

By analogy, Drago and co-workers argue that in view of the similarity in physical and chemical properties between phosphorus oxychloride, $OPCl_3$, and triethyl phosphate, $OP(OEt)_3$, it is probable that the formation of $FeCl_4^-$ in phosphorus oxychloride proceeds by a reaction similar to Eq. 8.68:

$$2FeCl_3 \xrightarrow{OPCl_3} [FeCl_2^+(OPCl_3)_n] + [FeCl_4^-] \quad \textbf{(8.69)}$$

They argue that the similar coordinating ability of these phosphoryl ($\geqslant$P=O) solvents (and to a lesser extent their dielectric constants) is more important than their chemical differences (supposed autoionization and chloride ion transfer in phosphorus oxychloride).

Gutmann[23] has rejoined that the dichloroiron(III) ion, $[FeCl_2^+$ (solvent)$]$, is not found in *dilute* solutions in phosphorus oxychloride but only in concentrated solutions or those to which a strong acid such as $SbCl_5$ has been added. In such cases the chloride donor ability of the solvent has been exceeded and chloride ions are abstracted from the iron(III) chloride. This point was made earlier[24] by the observation that the controversy is at least partially a semantic one. The only "characteristic property" of the solvo-cations and solvo-anions in the solvent system autoionization is that *they are the strongest acids and bases that can exist in that particular solvent without being leveled.* In triethyl phosphate (a nonleveling solvent) the dichloroiron(III) ion is perfectly stable. In phosphorus oxychloride a mechanism for leveling exists, namely:

$$FeCl_2^+ + OPCl_3 \rightleftharpoons OPCl_2^+ + FeCl_3 \quad \textbf{(8.70)}$$

This equilibrium will lie to the right if the dichloroiron(III) ion is stronger than the dichlorophosphoryl ion and to the left if the acid strengths are reversed. The important point is that neither the solvent system approach nor the coordination model can, *a priori*, predict the nature of the equilibrium in Eq. 8.70. To make this prediction, one must turn to the generalized acid–base definition given above together with some knowledge of the relative electron densities on the central atoms in $FeCl_2^+$ and $OPCl_2^+$ (presently unavailable). The essence of the acidity of iron(III) chloride lies in its tripositive ion of rather small radius and high charge, which is compensated only in part by three coordinated chloride ions and which seeks elsewhere for electron density to reduce its positive character. It is thus an acid irrespective of the solvent chosen and will accept the strongest base available to it. If the basicity of the phosphoryl group is sufficient (as it must of necessity be in triethyl phosphate or in phosphorus oxychloride if the chloride ion concentration is too low), then the iron(III) chloride is less acidic than if it can abstract a chloride ion (possible only in phosphorus oxychloride).

MOLTEN SALTS

The chemistry of molten salts as nonaqueous solvent systems is one that has blossomed in recent years and only a brief survey can be given here. Fortunately, there are several

---

[23] V. Gutmann, "Coordination Chemistry in Non-Aqueous Solutions," Springer, New York, **1968**, and references cited therein.

[24] J. E. Huheey, *J. Inorg. Nucl. Chem.*, **1962**, *24*, 1011.

good discussions available of chemistry in high-temperature melts.[25] The most obvious differences when compared with the chemistry of aqueous solutions are the strongly bonded and stable nature of the solvent, a concomitant resistance to destruction of the solvent by vigorous reactions, and higher concentrations of various species, particularly coordinating anions, than can be obtained in saturated solutions in water.

Solvent properties

On the basis of the structure of the liquid, molten salts can be conveniently classified into two groups although there is no distinct boundary between the two. The first consists of compounds such as the alkali halides that are bonded chiefly by ionic forces. On melting, very little change takes place in these materials. The coordination of the ions tends to drop from six in the crystal to about four in the melt and the long-range order found in the crystal is destroyed, but a local order, each cation surrounded by anions, etc., is still present. These fused salts are all very good electrolytes because of the presence of large numbers of ions. They behave normally with respect to cryoscopy and this is a useful means of study. The number of ions, ν, may be determined in these systems just as in the sulfuric acid system (p. 334). For example, if sodium chloride is the solvent, $\nu_{KF} = 2$, $\nu_{BaF_2} = 3$, etc. One interesting point is that a salt with a common ion behaves somewhat anomalously in that the common ion does not behave as a "foreign particle" and ν is correspondingly lower. In sodium chloride solutions, $\nu_{NaF} = 1$.

The second group consists of compounds in which covalent bonding is important. These compounds tend to melt with the formation of discrete molecules although autoionization may occur. For example, the mercury(II) halides ionize as follows:

$$2HgX_2 \rightleftharpoons HgX^+ + HgX_3^- \tag{8.71}$$

This is analogous to the aprotic halide solvents discussed in the previous section. Acidic solutions may be prepared by increasing the concentration of HgX^+ and basic solutions by increasing the concentration of HgX_3^-:

$$Hg(ClO_4)_2 + HgX_2 \longrightarrow 2HgX^+ + 2ClO_4^- \tag{8.72}$$

$$KX + HgX_2 \longrightarrow K^+ + HgX_3^- \tag{8.73}$$

and neutralization reactions occur on mixing the two:

$$HgX^+ + ClO_4^- + K^+ + HgX_3^- \longrightarrow 2HgX_2 + K^+ + ClO_4^- \tag{8.74}$$

A similar autoionization of the nitrate ion has been postulated[26] to occur at high temperatures:

$$NO_3^- \rightleftharpoons NO_2^+ + O^{-2} \tag{8.75}$$

[25] "Molten Salt Chemistry," M. Blander, ed., Wiley (Interscience), New York, **1964**; H. Bloom, "The Chemistry of Molten Salts," Benjamin, New York, **1967**; H. Bloom and J. W. Hastie, in "Non-Aqueous Solvent Systems," T. C. Waddington, ed., Academic Press, New York, **1965**, Chap. IX; A. K. Holliday and A. G. Massey, "Inorganic Chemistry in Non-Aqueous Solvents," Pergamon, Elmsford, N.Y., **1965**, Chap. VI; "Fused Salts," B. R. Sundheim, ed., McGraw-Hill, New York, **1964**.

[26] R. N. Kust and F. R. Duke, *J. Am. Chem. Soc.*, **1963**, *85*, 3338. See also F. R. Duke, in "Fused Salts," B. R. Sundheim, ed., McGraw-Hill, New York, **1964**, p. 409.

The nitryl (or nitronium) ion, NO_2^+, may in some ways be considered an analogue of the hydronium ion in water, and acidic solutions would be formed when oxide acceptors are added to the melt:

$$NO_3^- + S_2O_7^{-2} \longrightarrow NO_2^+ + 2SO_4^{-2} \qquad (8.76)$$

Just as a hydrogen electrode can measure the activity of hydronium ions in water, an electrode made by passing gaseous nitrogen dioxide over platinum should be capable of measuring the concentration of NO_2^+:

$$H_2O + \tfrac{1}{2}H_2 \rightleftharpoons H_3O^+ + e^- \qquad \mathscr{E} = -(RT/\mathscr{F})\ln[H_3O^+] \qquad (8.77)$$

$$NO_2 \rightleftharpoons NO_2^+ + e^- \qquad \mathscr{E} = -(RT/\mathscr{F})\ln[NO_2^+] \qquad (8.78)$$

Unfortunately, the situation is complicated by the fact that the nitryl ion and the nitrate ion can also react with each other, though slowly:

$$NO_2^+ + NO_3^- \longrightarrow 2NO_2\uparrow + \tfrac{1}{2}O_2\uparrow \qquad (8.79)$$

Thus acidic solutions in nitrate melts decompose with evolution of nitrogen dioxide and oxygen. It should be kept in mind that as in the case for most solvent systems models, the reactions proposed (Eqs. 8.75 and 8.76) are those that are *consistent* with the behavior of the system but are not *proved* to exist. Indeed, it has been suggested[27] that acids react directly with the nitrate ion to form nitrogen dioxide and oxygen:

$$2NO_3^- + S_2O_7^{-2} \longrightarrow 2NO_2\uparrow + \tfrac{1}{2}O_2\uparrow + 2SO_4^{-2} \qquad (8.80)$$

without benefit of a nitryl ion intermediate. Later evidence, however, seems to strengthen the case for nitryl ion formation.[28] More startling is work[29] indicating that the oxide ion itself can exist in nitrate melts only at very low concentrations. The limiting factor is the oxidation of the oxide ion by nitrate to peroxide and superoxide:

$$O^{-2} + NO_3^- \longrightarrow O_2^{-2} + NO_2^- \qquad (8.81)$$

$$O_2^{-2} + 2NO_3^- \longrightarrow 2O_2^- + 2NO_2^- \qquad (8.82)$$

Ionization processes in nitrate melts thus appear to be more complicated than implied by Eq. 8.75.

Solutions of metals

One of the most interesting aspects of molten salt chemistry is the readiness with which metals dissolve. For example, the alkali halides dissolve large amounts of the corresponding alkali metal, and some systems (e.g., cesium in cesium halides) are completely

[27] L. E. Topol et al., *J. Phys. Chem.*, **1966**, *70*, 2857.
[28] B. J. Brough et al., *J. Inorg. Nucl. Chem.*, **1968**, *30*, 2870.
[29] P. G. Zambonin and J. Jordan, *J. Am. Chem. Soc.*, **1969**, *91*, 2225.

miscible at all temperatures above the melting point. On the other hand, the halides of zinc, lead, and tin dissolve such small amounts of the corresponding free metal that special analytical techniques must be devised in order to estimate the concentration accurately.

At one time solutions of metals in their molten salts were thought to be colloidal in nature, but this has been shown not to be true. However, no completely satisfactory theory has been advanced to account for all the properties of these solutions. One hypothesis involves reduction of the cation of the molten salt to a lower oxidation state. For example, the solution of mercury in mercuric chloride undoubtedly involves reduction:

$$Hg + HgCl_2 \longrightarrow Hg_2Cl_2 \tag{8.83}$$

and mercury(I) chloride remains when the melt is allowed to solidify. For most transition and posttransition metals the evidence for the formation of "subhalides" is considerably weaker. The Cd_2^{+2} ion is believed to exist in solutions of cadmium in molten cadmium chloride but can be isolated only through the addition of aluminum chloride:

$$Cd + CdCl_2 \longrightarrow [Cd_2Cl_2] \xrightarrow{Al_2Cl_6} Cd_2^{+2}[AlCl_4^-]_2 \tag{8.84}$$

In many cases although the presence of reduced species is suspected, it is impossible to isolate them. On solidification the melts disproportionate to solid metal and solid cadmium(II) salt.

In solutions of alkali metals in alkali halides, reduction of the cation, at least in the sense of forming discrete species such as M_2^+, is untenable. Although no universally acceptable theory has been formulated, it is probable that in these salts ionization takes place upon solution:

$$M \longrightarrow M^+ + e^- \tag{8.85}$$

The presence of "free" electrons thus bears a certain similarity to solutions of these same metals in liquid ammonia. If the electrons are thought to be trapped in anion vacancies in the melt, an analogy to F-centers (see p. 189) may be made. Undoubtedly the situation is considerably more complex with the possibility of the electron being delocalized in energy levels or "bands" characteristic of several atoms, but a thorough discussion of this problem is beyond the scope of this book.[30]

Complex formation

Molten salts provide a medium in which the concentration of anionic ligands can be much higher than is possible in aqueous solutions. The concentration of the chloride ion in concentrated aqueous hydrochloric acid is about 12 mol dm^{-3} for example. In contrast, the concentration of the chloride ion in molten lithium chloride is about 35 mol dm^{-3}. Furthermore, there are no other competing ligands (such as H_2O) present to interfere. As a result, it is possible to form not only complex ions that are well known in aqueous solution:

$$CoCl_2 + 2Cl^- \longrightarrow CoCl_4^{-2} \tag{8.86}$$

[30] J. D. Corbett, in "Fused Salts," B. R. Sundheim, ed., McGraw-Hill, New York, **1964**, p. 341.

but also those that cannot exist in aqueous solution because of their susceptibility to hydrolysis:

$$FeCl_2 + 2Cl^- \longrightarrow FeCl_4^{-2} \tag{8.87}$$

$$CrCl_3 + 3Cl^- \longrightarrow CrCl_6^{-3} \tag{8.88}$$

$$TiCl_3 + 3Cl^- \longrightarrow TiCl_6^{-3} \tag{8.89}$$

Some of these complexes are discussed further in Chapter 9.

Unreactivity of molten salts

Many reactions that cannot take place in aqueous solutions because of the reactivity of water may be performed readily in molten salts. Both chlorine and fluorine react with water (the latter vigorously), and so the use of these oxidizing agents in aqueous solution produces hydrogen halides, etc., in addition to the desired oxidation products. The use of the appropriate molten halide obviates this difficulty. Even more important is the use of molten halides in the preparation of these halogens:

$$KHF_2 \xrightarrow{\text{electrolysis}} \tfrac{1}{2}F_2 + \tfrac{1}{2}H_2 + KF \tag{8.90}$$

$$NaCl \xrightarrow{\text{electrolysis}} \tfrac{1}{2}Cl_2 + Na \tag{8.91}$$

The latter reaction is also important in the commercial production of sodium, which, like the halogens, is too reactive to coexist with water.

The reactions in Eqs. 8.90 and 8.91 are typical of the many important industrial processes involving high-temperature molten salts. Other examples are the production of magnesium and aluminum and the removal of silica impurities (in a blast furnace, for example) by a high-temperature acid–base reaction:

$$\underset{\text{Gangue}}{SiO_2} + \underset{\text{Flux}}{CaO} \longrightarrow \underset{\text{Slag}}{CaSiO_3} \tag{8.92}$$

Although high-temperature industrial reactions have been of great importance for many years, it is only recently that these systems have been investigated and better understood.[31]

Low-temperature molten salts

Although the term "molten salts" conjures up images of very high-temperature fused systems, some salts are liquid at room temperature. For example, if one mixes the crystalline solids triethylammonium chloride and copper(I) chloride, an endothermic reaction takes place to form a light green oil.[32] The most reasonable reaction is the coordination of a

[31] For further reading on this subject see T. Førland et al., "Selected Topics in High Temperature Chemistry," Universitatetsforlaget, Oslo, **1966**.

[32] J. T. Yoke, *Inorg. Chem.*, **1963**, *2*, 1210.

second chloride ion to the copper(I) ion:

$$Et_3NHCl + CuCl \longrightarrow Et_3NHCuCl_2 \tag{8.93}$$

to form the dichlorocuprate ion, $CuCl_2^-$. It is tempting to assume that the low lattice energy of this salt with a large cation and large anion is so low that the compound does not crystallize, even at room temperature. However, triethylammonium tetraphenylborate and triethylammonium dibromocuprate are both solids, even though they contain even *larger* anions. The source of the low melting point seems rather to be the following equilibria:

$$CuCl_2^- + CuCl \rightleftharpoons Cu_2Cl_3^- \tag{8.94}$$

$$2CuCl_2^- \rightleftharpoons Cu_2Cl_3^- + Cl^- \tag{8.95}$$

$$CuCl_2^- + Cl^- \rightleftharpoons CuCl_3^{-2} \tag{8.96}$$

Evidence for these equilibria comes from the Raman spectra, which show an absorption peak (or unresolved peaks), probably attributable to $Cu_2Cl_3^-$. Addition of CuCl or Cl^- causes this peak to increase or decrease as expected by the above equilibria. The system thus probably contains at least four anionic species, and the "impurities" account for the depression of the melting point. In accordance with this interpretation is the fact that the material is oily and never forms a crystalline solid with a true freezing point, but congeals to a glass ~0 °C.

The existence of a molten salt system at room temperature has proved to be of interest for a variety of reasons. First, some of the chemistry performed in high-temperature chloride melts (p. 344) can be done at room temperature to check for similarities and differences. Much of the chemistry is the same, but a few of the complexes formed are different:

$$Fe^{+2} + 6Cl^- \xrightarrow{Et_3NHCuCl_2} FeCl_6^{-4} \tag{8.97}$$

Of perhaps greater potential interest and more related to the other topics of this chapter is the use of this system as both solvent and reactant in a voltaic cell.[33] If two platinum gauze electrodes are immersed in liquid chlorocuprates and a potential is applied, the cell begins charging. At less than 1% of full charge, the potential stabilizes at 0.85 V and remains at that value until the cell is fully charged. The half-reactions for charging are:

$$CuCl_2^- + e^- \longrightarrow Cu + 2Cl^- \tag{8.98}$$

$$CuCl_2^- \longrightarrow CuCl_2 + e^- \tag{8.99}$$

Allowing the reactions to proceed spontaneously (reverse of Eqs. 8.98 and 8.99) produces 0.85 V with low current flow. The chief difficulty with the cell is the fact that $CuCl_2$ is soluble in the melt. It thus diffuses and allows the cell to decay through direct reaction

[33] W. W. Porterfield and J. T. Yoke, "Inorganic Compounds with Unusual Properties," *Advances in Chemistry Series No. 150*, R. B. King, ed., American Chemical Society, Washington, D.C., **1976**.

of the electrode materials:

$$CuCl_2 + Cu + 2Cl^- \longrightarrow 2CuCl_2^- \qquad \textbf{(8.100)}$$

The fact that the solvent can be both oxidized and reduced is an asset in the above reactions, but it is a handicap when the system is used merely as a solvent. For example, the chlorocuprate solvent must be handled in the absence of air to prevent oxidation. Some solutes cannot be studied. Even so gentle an oxidizing agent as $FeCl_3$ oxidizes the solvent:

$$FeCl_3 + Cl^- + CuCl_2^- \longrightarrow FeCl_4^{-2} + CuCl_2 \qquad \textbf{(8.101)}$$

Solid acid and base catalysts

While they are not solvents and solutions in the usual sense of the word, it is convenient here to introduce the concept of solid acids and bases. For example, consider the class of compounds known as zeolites. These are aluminosilicate structures with variable amounts of Al(III), Si(IV), metal cations, and water (see Chapter 14).

Zeolites may behave as Lewis acids at Al^{+3} sites, or as Brønsted-Lowry acids by means of adsorbed H^+ ions. Because they have relatively open structures, a variety of small molecules may be accommodated within the —O—Al—O—Si— framework. The latter may then be catalyzed to react by the acidic centers. Coordinatively unsaturated oxide ions can act as basic sites, and in some catalytic reactions both types of centers are believed to be important. Catalysis by zeolites is discussed further in Chapter 13.

ELECTRODE POTENTIALS AND ELECTROMOTIVE FORCES

As we have seen, acidity and basicity are intimately connected with electron transfer. When the electron transfer involves an integral number of electrons it is customary to refer to the process as a redox reaction. When occurring in solution these reactions comprise the subject of electrochemistry. This is not the place for a thorough discussion of the thermodynamics of electrochemistry; that may be found in any good textbook of physical chemistry. Rather, we shall investigate the applications of electromotive force (*emf*) of interest to the inorganic chemist. Nevertheless, a very brief review of the conventions and thermodynamics of electrode potentials and half-reactions will be presented.

1. The standard hydrogen electrode ($a_{H^+} = 1.00$; $f_{H_2} = 1.00$) *is arbitrarily assigned an electrode potential of* 0.00 V.

2. If we construct a cell composed of a hydrogen electrode and a second electrode (M/M^{+n}) of metal M immersed in a solution of M^{+n} of unit activity, we can measure the potential between the electrodes of the cell. Since the hydrogen electrode was assigned a potential of 0.00 V, the potential of the electrode, M/M^{+n}, is by definition the same as the measured potential of the cell. If the metal electrode is *positively charged* with respect to the hydrogen electrode (e.g., Cu/Cu^{+2}) the electrode potential of the metal is assigned a positive sign ($V_{Cu/Cu^{+2}} = +0.337$ V). If the metal tends to lose electrons more readily than hydrogen and thus becomes *negatively charged* (e.g., Zn/Zn^{+2}), the electrode is assigned a *negative sign* ($V_{Zn/Zn^{+2}} = -0.763$ V). This assignment is often called the European convention because of its supposed popularity in Europe. Actually, the "European" vs.

"American" preferences are more closely related to specializations: The electrochemist and analytical chemist often prefer the practicality of this convention. It is often convenient because it represents the actual, experimentally observed *electrostatic potential*, independent of speculations concerning the direction of the reaction.

This convention is convenient in that it results in a single, invariant quantity for the electrode potential for each electrode (the zinc electrode is always electrostatically negative whether the reaction under consideration occurs in a galvanic cell or an electrolytic cell). However, most physical and inorganic chemists are more interested in the *thermodynamics of half-reactions* rather than the *electrostatic potential* that obtains in conjunction with the standard hydrogen electrode. The convention related to thermodynamics, popularized by Latimer,[34] is often known as the "American convention." More properly, it might be termed the *thermodynamic convention*. This convention assigns to the electromotive force ($\mathscr{E}$) a sign such that

$$\Delta G = -n\mathscr{F}\mathscr{E} \tag{8.102}$$

where ΔG is the change in Gibbs free energy, n is the number of moles of electrons involved, and $\mathscr{F}$ is Faraday's constant, 96 500 coulombs per mol e^-. It is necessary to specify the direction in which a reaction is proceeding. Thus if we consider the reaction

$$Zn + 2H^+ \longrightarrow Zn^{+2} + H_2 \tag{8.103}$$

and find that for the reaction, as written, $\Delta G < 0$, then (since H^+/H_2 is defined as 0.00 V):

$$Zn \longrightarrow Zn^{+2} + 2e^- \qquad \mathscr{E} > 0 \tag{8.104}$$

For the nonspontaneous reaction:

$$H_2 + Zn^{+2} \longrightarrow Zn + 2H^+ \tag{8.105}$$

$\Delta G > 0$, and so for

$$Zn^{+2} + 2e^- \longrightarrow Zn \qquad \mathscr{E} < 0 \tag{8.106}$$

Accordingly, the sign of the *emf* of either a half-reaction ("electrode") or the overall redox reaction depends upon the direction in which the equation for the reaction is written (as is true for any thermodynamic quantity such as enthalpy, entropy, or free energy).

The sign of the reduction *emf* is always algebraically the same as that of the "European" electrostatic potential although the two refer to different concepts.[35] Since we shall be

---

[34] W. Latimer, "The Oxidation Potentials of the Elements and Their Values in Aqueous Solution," Prentice-Hall, Englewood Cliffs, N. J., **1952**. This convention was suggested by G. N. Lewis and M. Randall, "Thermodynamics," McGraw-Hill, New York, **1923**.

[35] F. C. Anson, *J. Chem. Educ.*, **1959**, *36*, 394, in an interesting illustration has called attention to the difference in meaning between *plus* and *minus* as applied *physically* (i.e., with regard to Benjamin Franklin's assignment of plus and minus to charges of electricity) and as applied *algebraically* (i.e., with respect to free energy changes). For further discussion of this point, see M. C. Day and J. Selbin, "Theoretical Inorganic Chemistry," 2nd ed., Van Nostrand-Reinhold, New York, **1968**, pp. 342–343, and T. S. Licht and A. J. de Bethune, *J. Chem. Educ.*, **1957**, *34*, 433.

In an effort to reduce confusion in this matter and to emphasize the difference in meaning of the two conventions, the International Union of Pure and Applied Chemistry has adopted the "Stockholm convention": (a) *Electrode potentials* shall refer only to the electrostatic potential, V; (b) the term *emf* shall be applied to the thermodynamic quantity, $\mathscr{E}$, where $\Delta G = -n\mathscr{F}\mathscr{E}$.

more interested in thermodynamics than in electrostatic potentials, we shall deal exclusively with *emfs* in this book (see Table G.1).

3. *The Nernst equation* applies to the *emfs* of both half-reactions and total redox reactions:

$$\mathscr{E} = \mathscr{E}^0 - \frac{RT}{n\mathscr{F}} \ln \frac{\prod \text{products}}{\prod \text{reactants}} \tag{8.107}$$

where $\mathscr{E}^0$ represents the *emf* with all species at unit activity and $\prod$ represents the usual products of the activities raised to the power corresponding to the stoichiometry of the reaction (as in K_{sp} and K_{eq}, for example).

4. *Reactions resulting in a decrease in free energy* ($\Delta G < 0$) *are spontaneous.* This is a requirement of the second law of thermodynamics. Concomitantly, redox reactions in which $\mathscr{E} > 0$ are therefore spontaneous.

5. In aqueous solutions two half-reactions are of special importance: (a) *the reduction of hydrogen in water or hydronium ions*:

$$1\ \text{mol dm}^{-3}\ \text{acid}\ H_3O^+ + e^- \longrightarrow H_2O + \tfrac{1}{2}H_2 \qquad \mathscr{E}^0 = 0.00\ \text{V} \tag{8.108}$$

$$\text{Neutral solution}\ H_2O + e^- \longrightarrow OH^- + \tfrac{1}{2}H_2 \qquad \mathscr{E} = -0.414\ \text{V} \tag{8.109}$$

$$1\ \text{mol dm}^{-3}\ \text{base}\ H_2O + e^- \longrightarrow OH^- + \tfrac{1}{2}H_2 \qquad \mathscr{E}^0 = -0.828\ \text{V} \tag{8.110}$$

and (b) *the oxidation of oxygen in water or hydroxide ions*:

$$1\ \text{mol dm}^{-3}\ \text{acid}\ H_2O \longrightarrow \tfrac{1}{2}O_2 + 2H^+ + 2e^- \qquad \mathscr{E}^0 = -1.229\ \text{V} \tag{8.111}$$

$$\text{Neutral solution}\ H_2O \longrightarrow \tfrac{1}{2}O_2 + 2H^+ + 2e^- \qquad \mathscr{E} = -0.815\ \text{V} \tag{8.112}$$

$$1\ \text{mol dm}^{-3}\ \text{base}\ 2OH^- \longrightarrow \tfrac{1}{2}O_2 + H_2O + 2e^- \qquad \mathscr{E}^0 = -0.401\ \text{V} \tag{8.113}$$

These reactions limit the *thermodynamic stability*[36] of species in aqueous solution.

6. In calculating the "*skip-step emf*" for a multivalent species it is necessary to take into account the total change in free energy. Suppose we know the *emfs* for the oxidation of Fe to Fe^{+2} and Fe^{+2} to Fe^{+3} and wish to calculate the skip-step *emf* for Fe to Fe^{+3}:

$$Fe \longrightarrow Fe^{+2} \qquad \mathscr{E}^0 = 0.44 \qquad \Delta G = -2 \times 0.44 \times \mathscr{F} \tag{8.114}$$

$$Fe^{+2} \longrightarrow Fe^{+3} \qquad \mathscr{E}^0 = -0.77 \qquad \Delta G = -1 \times -0.77 \times \mathscr{F} \tag{8.115}$$

$$Fe \longrightarrow Fe^{+3} \qquad \Delta G = -0.11\mathscr{F} \tag{8.116}$$

$$\mathscr{E}^0 = -\frac{\Delta G}{n\mathscr{F}} = 0.11/3 = 0.04\ \text{V} \tag{8.117}$$

[36] As is always the case when dealing with thermodynamic stabilities, it must be borne in mind that a species may possibly be *thermodynamically unstable* yet *kinetically stable*; i.e., no mechanism of low activation energy may exist for its decay.

Although the *emf*s are not additive, the free energies *are*, allowing simple calculation of the overall *emf* for the three-electron change.

7. *Emf diagrams* are useful for summarizing a considerable amount of a thermodynamic information about the oxidation states of an element in a convenient way. For example, the following half-reactions may be taken from Table G.1:

$$Mn^{+2} + 2e^- \longrightarrow Mn \qquad \mathscr{E}^0 = -1.18 \tag{8.118}$$

$$Mn^{+3} + e^- \longrightarrow Mn^{+2} \qquad \mathscr{E}^0 = +1.51 \tag{8.119}$$

$$MnO_2 + 4H^+ + e^- \longrightarrow Mn^{+3} + 2H_2O \qquad \mathscr{E}^0 = +0.95 \tag{8.120}$$

$$MnO_4^{-2} + 4H^+ + 2e^- \longrightarrow MnO_2 + 2H_2O \qquad \mathscr{E}^0 = +2.26 \tag{8.121}$$

$$MnO_4^- + e^- \longrightarrow MnO_4^{-2} \qquad \mathscr{E}^0 = +0.56 \tag{8.122}$$

$$MnO_2 + 4H^+ + 2e^- \longrightarrow Mn^{+2} + 2H_2O \qquad \mathscr{E}^0 = +1.23 \tag{8.123}$$

$$MnO_4^- + 4H^+ + 3e^- \longrightarrow MnO_2 + 2H_2O \qquad \mathscr{E}^0 = +1.70 \tag{8.124}$$

$$MnO_4^- + 8H^+ + 5e^- \longrightarrow Mn^{+2} + 4H_2O \qquad \mathscr{E}^0 = +1.51 \tag{8.125}$$

By omitting species such as H_2O, H^+, and OH^-, all of the above information can be summarized as:

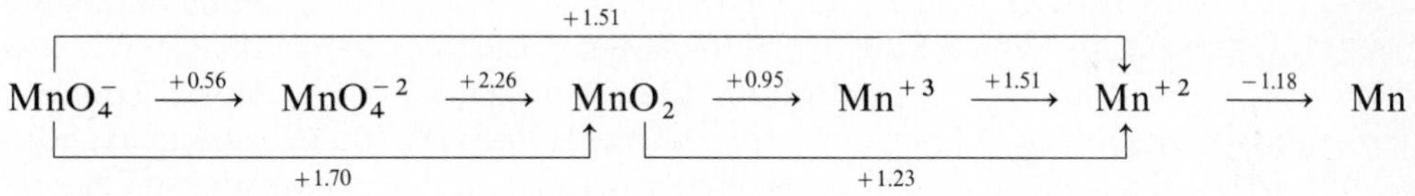

+1.51

$$MnO_4^- \xrightarrow{+0.56} MnO_4^{-2} \xrightarrow{+2.26} MnO_2 \xrightarrow{+0.95} Mn^{+3} \xrightarrow{+1.51} Mn^{+2} \xrightarrow{-1.18} Mn$$

+1.70 +1.23

The highest oxidation state is listed on the left and the reduction *emf*s are listed between each species and the next reduced form, with the lowest oxidation state appearing on the right.[37]

Electrochemistry in nonaqueous solutions

Although the entire discussion of electrochemistry thus far has been in terms of aqueous solutions, the same principles apply equally well to nonaqueous solvents. As a result of differences in solvation energies *emf* values may vary considerably from those found in aqueous solution. In addition the reduction and oxidation *emf*s characteristic of the solvent (see Appendix F for values) vary with the chemical behavior of the solvent. As a result of these two effects, it is often possible to carry out reactions in a nonaqueous

[37] This convention originated with Latimer and is widespread in the inorganic chemistry literature. Unfortunately, Latimer used oxidation *emfs* and so his diagram (a mirror image of the one above) is incompatible with the current emphasis on reduction *emfs*. This has resulted in a wide variety of modified "Latimer" diagrams, often with no indication as to whether oxidation or reduction *emfs* are involved, or which way the reaction is proceeding. To avoid possible confusion, arrows (not present in the original Latimer diagrams) have been added. Further discussion and use of Latimer diagrams may be found in Chapter 12.

solvent that would be impossible in water. For example, both sodium and beryllium are too reactive to be electroplated from aqueous solution, but beryllium can be electroplated from liquid ammonia and sodium from solutions in pyridine.[38] Unfortunately, the thermodynamic data necessary to construct complete tables of *emf* values are lacking for most solvents other than water. Jolly[39] has compiled such a table for liquid ammonia (Table F.2). The hydrogen electrode is used as the reference point to establish the scale as in water:

$$NH_4^+ + e^- \longrightarrow \tfrac{1}{2}H_2 + NH_3 \qquad \mathscr{E}^0 = 0.000 \tag{8.126}$$

A single example of the application of *emf* values to chemistry in ammonia will suffice. The *emf* diagram for mercury in acidic solution is

$$Hg^{+2} \xrightarrow{-0.2} Hg_2^{+2} \xrightarrow{+1.5} Hg \qquad (Hg^{+2} \xrightarrow{+0.67} Hg)$$

and for the insoluble mercury(I) iodide the diagram is:

$$Hg^{+2} \xrightarrow{+0.66} Hg_2I_2 \xrightarrow{+0.68} Hg$$

It may readily be seen that the mercurous ion (whether free or in Hg_2I_2) is thermodynamically unstable with respect to disproportionation in ammonia, in contrast to its stability in water.

Electrochemistry in nonaqueous solvents is not merely a laboratory curiosity. We have already seen batteries made with solid electrolytes (sodium beta alumina) that are certainly "nonaqueous." In looking for high-efficiency cells one desires the cathode and anode to be highly reactive (large positive emf) and to have a low mass per mole of electrons transferred. In these terms, lithium appears to be highly desirable. Its very reactivity, however, precludes its use in aqueous systems or even liquid ammonia. One successful battery utilizing lithium has been developed using sulfur dioxide or thionyl chloride ($OSCl_2$) as solvent *and* oxidant. Others involve mass-efficient lithium metal with other oxidants and solvents.[40] Highly efficient batteries of this sort are widely used in specialized applications where low mass and long life are important.

Hydrometallurgy[41]

Traditionally the winning of metals from their ores has been achieved by pyrometallurgy—the reduction of relatively concentrated metallic ores at high temperatures. The reactions of the blast furnace form a typical example (see also p. 287):

$$Fe_2O_3 + 3CO \longrightarrow 2Fe + 3CO_2 \tag{8.127}$$

$$CO_2 + C \longrightarrow 2CO \tag{8.128}$$

---

[38] R. W. Parry and E. H. Lyons, Jr., in "The Chemistry of Coordination Compounds," J. C. Bailar, Jr., ed., Van Nostrand-Reinhold, New York, **1956**, pp. 669–671.

[39] W. L. Jolly, *J. Chem. Educ.*, **1956**, *33*, 512.

[40] K. J. Jones and E. S. Hatch, Jr., *Indust. Res. Dev.*, Feb. **1982**, *24*, 182; Mar. **1982**, *24*, 89.

[41] For a recent review, see F. Habashi, *Chem. Eng. News*, Feb. 8, **1982**, 46.

Table 8.6 A comparison of pyrometallurgy and hydrometallurgy[a]

	Pyrometallurgy	Hydrometallurgy
Energy consumption	Because high temperatures (about 1500 °C) are involved, reaction rates are high but much energy is consumed. Heat recovery systems are needed to make the process economical. Heat can be recovered readily from hot gases (although the equipment needed is bulky and expensive), but is rarely recovered from molten slag or metal, so that a great deal of energy is lost.	Because low temperatures are involved in dissolution processes, they require little energy, although reaction rates are slow. However, a requirement for electrowinning or for cleaning effluents and recovering reagents may more than offset this energy advantage.
Dust	Most processes emit large amounts of dust, which must be recovered to abate pollution or because the dust itself contains valuable metals; equipment for dust recovery is bulky and expensive.	No problem, because materials handled usually are wet.
Toxic gases	Many processes generate toxic gases, so that reactors must be gas-tight and the gases removed by scrubbers or other systems; this is expensive, especially when the gases are hot and corrosive.	Many processes do not generate gases, and if they do, reactors can be made gas-tight easily.
Solid residues	Many residues, such as slags, are coarse and harmless, so that they can be stored in exposed piles without danger of dissolution, although the piles may be esthetically unacceptable.	Most residues are finely divided solids that, when dry, create dust problems and, when wet, gradually release metal ions in solution that may contaminate the environment.
Treatment of sulfide ores	Sulfur dioxide is generated, which in high concentrations must be converted to sulfuric acid (for which a market must be found) and in low concentrations must be disposed in other ways (available but expensive).	Ores can be treated without generating sulfur dioxide, eliminating the need to make and market sulfuric acid; sulfide sulfur can be recovered in the elemental form.
Treatment of complex ores	Unsuitable because separation is difficult	Suitable
Treatment of low-grade ores	Unsuitable because large amounts of energy are required to melt gangue materials	Suitable if a selective leaching agent can be used
Economics	Best suited for large-scale operations requiring a large capital investment	Can be used for small-scale operations requiring a low capital investment

[a] From F. Habashi, *Chem. Eng. News*, Feb. 8, **1982**, *60*, 46. Used with permission.

Carbon monoxide for the reduction of the iron is formed not only from the recycling of carbon dioxide (Eq. 8.146) but also from the direct oxidation of the coke in the charge by hot air:

$$2C + O_2 \longrightarrow 2CO \qquad \textbf{(8.129)}$$

The energy released by the combustion is sufficient to raise the temperature well above the melting point of iron, 1535 °C. One of the incentives for development of alternative methods of producing metals is the hope of finding a less energy-intensive process.

Hydrometallurgy is not new—it has been used for almost a century in the separation of gold from low-grade ores. This process is typical of the methods used. Gold is normally a very unreactive metal:

$$Au \longrightarrow Au^+ + e^- \qquad \mathscr{E}^0 = -1.69\ V \qquad \textbf{(8.130)}$$

With such a negative oxidation *emf*, it is too noble to react with either O_2 ($\mathscr{E}^0 = +1.185\ V$) or Cl_2 ($\mathscr{E}^0 = +1.36\ V$). By complexation of the Au(I) ion, however, the *emf* can be shifted until it is much more favorable:

$$Au + 2CN^- \longrightarrow Au(CN)_2^- + e^- \qquad \mathscr{E}^0 = +0.60\ V \qquad \textbf{(8.131)}$$

Oxygen in the air is now a sufficiently strong (and cheap!) oxidizing agent to effect the solution of the gold. It may then be reduced and precipitated by an active metal such as zinc powder ($\mathscr{E}^0 = -0.763\ V$). Such hydrometallurgical processes offer definite advantages:

1. Low-grade ores may be leached, with complexing agents if necessary, and profitably exploited.
2. Complex ores may be successfully treated and multiple metals separated under more carefully controlled processes.
3. Since the reactions are carried out at room temperature, potential energy savings are possible.
4. Because no stack gases are involved, air pollution does not represent the problem faced by pyrometallurgy.

These aspects do not form an unmixed blessing, however. If the metal must be reduced by electrolysis, the process may become energy intensive. Thus attractive solutions to this problem are reduction of more valuable gold by less expensive zinc or of more valuable copper by scrap iron. Finally, in view of the large amounts of waste water formed as by-product, one may be trading an air pollution problem for a water pollution problem. A comparison of the two types of processes is given in Table 8.6.

PROBLEMS

8.1. Suggest the specific chemical and physical interactions responsible for the reversal of Eqs. 8.3 and 8.4 in water and ammonia solutions.

8.2. Using a Born-Haber type cycle employing the various energies contributing to the formation of $M^+, e(NH_3)_x^-$ species in ammonia solutions, explain why such solutions form only with the most active metals.

8.3. When 1 mol N_2O_5 is dissolved in sulfuric acid, it requires 3 mol $H_3SO_4^+$ to titrate the resulting solution to a minimum-conducting system. Conductivity studies also indicate that $v = 6$ for N_2O_5. Propose an equation representing the solvolysis of N_2O_5 by sulfuric acid.

8.4. What is the strongest acid listed in Fig. 8.1? The strongest base?

8.5. From Fig. 8.1 determine how the following solutes will react with the solvents, and how the equilibria will lie; i.e., will the solute be completely leveled or in equilibrium? State whether the solution formed in each case will be more acidic or more basic then the pure solvent.

Solute	Solvent
H_2SO_4	Acetic acid
H_2SO_4	Water
H_2SO_4	Ammonia
$CH_3C(O)CH_3$	Ammonia
$CH_3C(O)CH_3$	Water
ϕNH^-	Ammonia
ϕNH^-	Water
ϕNH^-	Acetic acid
ϕNH^-	Sulfuric acid

8.6. Construct the *emf* diagram for manganese in basic solution (from values in Table 8.1), and predict which oxidation states will be stable. Explain the source of instability for each unstable species.

8.7. Construct the *emf* diagram for uranium in acid solution, and predict which oxidation states will be stable. Explain the source of the instability for each unstable species.

8.8. Examination of the *emf*s of plutonium in acid solution indicates that Pu^{IV} is *stable* with respect to disproportionation but Pu^{V} is not. Note, however, that the *differences* in the appropriate *emf* values are not large. Using the appropriate thermodynamic expressions, calculate the equilibrium concentrations of Pu^{III}, Pu^{IV}, Pu^{V}, and Pu^{VI} when $Pu(ClO_4)_4$ is dissolved in water to a concentration of 0.1 mol dm^{-3}. [See D. A. Johnson, "Some Thermodynamic Aspects of Inorganic Chemistry," 2nd ed., Cambridge University Press, London, **1982**, p. 85.]

8.9. With equations and words describe what happens:

a. when metallic potassium is dissolved in ammonia to form a dilute solution
b. when more potassium is added to form concentrated solutions
c. when solutions (a) or (b) are evaporated carefully *in vacuo*
d. when (a) is treated with Fe_2O_3

How can (d) be considered a leveling reaction?

8.10. Consider each of the following solvents individually: (1) ammonia, (2) acetic acid, (3) sulfuric acid.

a. Give equations for autoionization of the pure solvent.
b. Discuss what will happen if CH_3COOH is dissolved in each of the solvents, i.e., what ions will form. Give appropriate equations. Will the solution be acidic or basic

with respect to the pure solvent? Will the solute act as a weak or a strong acid (base)?

c. Give an example of a strong base, a strong acid, and a neutralization reaction.

8.11. As a working hypothesis, assume that you accept the solvent system picture of $OPCl_3$ and a value of 5×10^{-14} mol^2 dm^{-6} for the ion product. Set up a pCl scale for $OPCl_3$, draw the equivalent of Fig. 8.1 for it, and discuss how you would go about obtaining data for compounds to complete your diagram.

8.12. The stability constant, K, for $Au(CN)_2^-$ is defined as $[Au^+][CN^-]^2$.

a. From the $\mathscr{E}^0$ of $+0.60\ V$ for Eq. 8.149, estimate K.

b. Qualitatively describe why this complex is so stable.

8.13. Correlate the behavior of various solutes in "superacids" with their gas-phase proton affinities. What factors besides proton affinities affect their solution chemistry? Predict what species will be present when XH_3 (Group VA), H_2X (Group VIA), and HX (Group VIIA) are dissolved in "superacids." [G. Olah and J. Shen, *J. Am. Chem. Soc.*, **1973**, *95*, 3582.]

8.14. Single-crystal "cesium electride" is almost entirely diamagnetic. Reconcile this with the formulation $[Cs(ligand)]^+e^-$. Is there a paradox here?

8.15. On p. 343, $HgCl_2$ is mentioned as an exception to the obviously intuitive rule that "good acceptors should not be good donors, and *vice versa.*" Can you suggest a reason why Hg(II) might be paradoxical?

9

Coordination chemistry: theory

Coordination compounds have always been a challenge to the inorganic chemist. In the early days of chemistry they seemed unusual (hence the name "complex" ions) and seemed to defy the usual rules of valence. Today they comprise a large body of current inorganic research. A survey of articles in recent issues of the journal *Inorganic Chemistry* indicates that perhaps 70% could be considered to deal with coordination compounds. Although the usual bonding theories can be extended to accommodate these compounds, they still provide stimulating problems to be resolved. In synthetic work they continue to provide a challenge in the laboratory. The rapidly developing field of bioinorganic chemistry is centered on the presence of coordination compounds in living systems.

The modern study of coordination compounds begins with two men, Alfred Werner and Sophus Mads Jørgensen. Both men were astute chemists, not only in the synthetic or laboratory aspects but also in the area of interpretation and theory. As it turned out, they differed fundamentally in their interpretation of the phenomena they observed and thus served as protagonists, each spurring the other to perform further experiments to augment the evidence for his point of view. From our viewpoint a half century later, we can conclude that Werner was "right" and Jørgensen was "wrong" in the interpretation of the experimental evidence they had. Indeed, Werner was the first inorganic chemist to be awarded the Nobel Prize in Chemistry (1913).[1] Nevertheless, Jørgensen's contributions

[1] Alfred Werner won the Prize in 1913. It long appeared that he might be the first, last, and *only* inorganic chemist to win the Prize. Then, sixty years later, Geoffrey Wilkinson and E. O. Fischer shared it in 1973 for their work on ferrocene. (See Chapter 13. For a personal account from the American side of the Altantic, see G. Wilkinson, *J. Organometal. Chem.*, **1975**, *100*, 273.) However, there has been a flurry of Prizes in recent years on work related closely to inorganic chemistry: Linus Pauling, 1954, for development of valence bond theory; Karl Ziegler and Giulio Natta for organometallic catalysts, 1963 (see Chapter 13); Robert S. Mulliken, 1966, for development of molecular orbital theory; William Lipscomb, 1976, and H. C. Brown, 1979, for their theoretical and synthetic work in borane chemistry (see Chapter 14; more recently Professor Lipscomb has done excellent work in the area of enzymes, see Chapter 18); and Roald Hoffmann, 1981, for his theoretical work in organometallic chemistry (see "Building Bridges Between Inorganic and Organic Chemistry," R. Hoffmann, *Angew. Chem.* **1982**, in press).

should not be slighted—as an experimentalist he was second to none, and had he not been prejudiced by some of the theories of valence current in his day, he might well have achieved the same results and fame as Werner.

This is a textbook neither of historical chemistry nor of the philosophy of scientific discovery, so an extensive discussion of the Werner-Jørgensen controversy would be out of place here,[2] but we may briefly outline the problem faced by chemists in the latter part of the nineteenth century. Many elements had fixed "valences"[3] such as Na = +1 and O = −2, and some exhibited two or three stable "valences," e.g., Cu = +1 and +2 and P = −3, +3, and +5. Some metals, however, exhibited combining powers that were hard to reconcile with this simple picture. The standard valence of chromium was +3 and those of platinum +2 and +4. Yet chlorides of these metals will react with ammonia (in which the valences of nitrogen and hydrogen are already satisfied):

$$CrCl_3 + 6NH_3 \longrightarrow CrCl_3 \cdot 6NH_3 \tag{9.1}$$

$$PtCl_2 + 4NH_3 \longrightarrow PtCl_2 \cdot 4NH_3 \tag{9.2}$$

Jørgensen attempted to formulate these compounds by analogy to organic compounds, such as[4]

$$Pt\begin{matrix} \diagup NH_3-NH_3-Cl \\ \diagdown NH_3-NH_3-Cl \end{matrix}$$

Werner, in formulating his ideas about the structure of coordination compounds, had before him facts such as the following. Four ammonia complexes of cobalt(III) chloride had been discovered and named according to their colors:

Complex	Color	Early name
$CoCl_3 \cdot 6NH_3$	Yellow	*Luteo* complex
$CoCl_3 \cdot 5NH_3$	Purple	*Purpureo* complex
$CoCl_3 \cdot 4NH_3$	Green	*Praseo* complex
$CoCl_3 \cdot 4NH_3$	Violet	*Violeo* complex

One of the more interesting facts about this series is the presence of two compounds of identical empirical formula, $CoCl_3 \cdot 4NH_3$, but having distinct properties, the most noticeable being the difference in color. Furthermore, Werner noted that the reactivities of the chloride ions in these four compounds differed considerably. Addition of silver nitrate

[2] For a complete discussion of the earliest work in coordination chemistry, see G. B. Kauffman, *J. Chem. Educ.*, **1959**, *36*, 521; "Classics in Coordination Chemistry," Part 1: The Selected Papers of Alfred Werner; Part 2: Selected Papers (1798–1899), Dover, New York, **1968**, **1976**.

[3] The "valence" of earlier chemists is most directly related to oxidation state as presently used.

[4] Modern valence theory prohibits the existence of pentacovalent nitrogen, so these structures look strange at the present time. The lack of distinction between covalence and oxidation state in Jørgensen's day allowed many chemists to accept these formulations, however.

resulted in different amounts of precipitated silver chloride:

$$CoCl_3 \cdot 6NH_3 + \text{excess } Ag^+ \longrightarrow 3AgCl\downarrow \quad \textbf{(9.3)}$$

$$CoCl_3 \cdot 5NH_3 + \text{excess } Ag^+ \longrightarrow 2AgCl\downarrow \quad \textbf{(9.4)}$$

$$CoCl_3 \cdot 4NH_3 + \text{excess } Ag^+ \longrightarrow 1AgCl\downarrow \quad \text{For either } \textit{praseo} \text{ or } \textit{violeo} \text{ salt} \quad \textbf{(9.5)}$$

The correlation between the number of ammonia molecules present and the number of equivalents of silver chloride precipitated led Werner to the following conclusion:

> We can thus make the general statement: Compounds $M(NH_3)_5X_3$ [M = Cr, Co; X = Cl, Br, etc.] are derived from compounds $M(NH_3)_6X_3$ by loss of one ammonia molecule.
> With this loss of an ammonia molecule, however, a simultaneous change in function of one acid residue X [= chloride ion] occurs. . . . [In] $Co(NH_3)_5Cl_3$ . . . two chlorine atoms behave as ions and are precipitated by silver nitrate at room temperature, while the third behaves completely analogously to chlorine in chloro-ethane, that is, it no longer acts as an ion.[5]

From this conclusion Werner postulated perhaps the most important part of his theory: that in this series of compounds cobalt exhibits a constant coordination number of 6, and that as ammonia molecules are removed, they are replaced by chloride ions which now act as though they are covalently bound to the cobalt, rather than as free chloride ions. To describe the complex chemistry of cobalt, one must therefore consider not only the oxidation state of the metal but also its coordination number.[6] Werner thus formulated these four salts as $[Co(NH_3)_6]Cl_3$, $[Co(NH_3)_5Cl]Cl_2$, and $[Co(NH_3)_4Cl_2]Cl$.

Realizing that these formulations implied a precise statement of the number of ions formed in solution, one of Werner's first experimental studies was the measurement of the conductivities of a large number of coordination compounds.[7] Some of the results of this work are listed in Table 9.1 together with values for simple ionic compounds for comparison.

The second important contribution that Werner made to the study of coordination chemistry was the postulate that the bonds to the ligands were fixed in space and therefore could be treated by the application of structural principles. By means of the numbers and properties of the isomers obtained, Werner was able to assign the correct geometric structures to many coordination compounds long before any direct experimental method was available for structure determination. Werner's method was that used previously by organic chemists[8] to elucidate the structures of substituted benzenes, namely isomer counting. Werner postulated that the six ligands in an ion such as the $[Co(NH_3)_6]^{+3}$ ion were situated in some symmetrical fashion with each NH_3 group equidistant from the central cobalt

---

[5] A. Werner, *Z. Anorg. Chem.*, **1893**, *3*, 267. For translation, see footnote 2. All bracketed material and ellipses are mine.

[6] Werner's terminology and symbolism differed in small, relatively unimportant ways from that used today. For example, Werner referred to oxidation state as "primary valence" (*Hauptvalenz*) and coordination number as "secondary valence" (*Nebenvalenz*). His use of brackets was, e.g., $\{Co_{Cl}^{(NH_3)_5}\}Cl_2$, instead of $[Co(NH_3)_5Cl]Cl_2$.

[7] A. Werner and A. Miolati, *Z. Phys. Chem.*, **1893**, *12*, 35; **1894**, *14*, 506.

[8] W. Körner, *Gazz. Chim. Ital.*, **1874**, *4*, 305.

Table 9.1 Conductivities of coordination compounds

Empirical formula	Conductivity[a]	Werner formulation
	Nonelectrolytes	
$PtCl_4 \cdot 2NH_3$	3.52[b]	$[Pt(NH_3)_2Cl_4]$ (*trans*)
$PtCl_4 \cdot 2NH_3$	6.99[b]	$[Pt(NH_3)_2Cl_4]$ (*cis*)
	1:1 Electrolytes	
NaCl	123.7	—
$PtCl_4 \cdot 3NH_3$	96.8	$[Pt(NH_3)_3Cl_3]Cl$
$PtCl_4 \cdot NH_3 \cdot KCl$	106.8	$K[Pt(NH_3)Cl_5]$
	1:2 and 2:1 Electrolytes	
$CaCl_2$	260.8	—
$CoCl_3 \cdot 5NH_3$	261.3	$[Co(NH_3)_5Cl]Cl_2$
$CoBr_3 \cdot 5NH_3$	257.6	$[Co(NH_3)_5Br]Br_2$
$CrCl_3 \cdot 5NH_3$	260.2	$[Cr(NH_3)_5Cl]Cl_2$
$CrBr_3 \cdot 5NH_3$	280.1	$[Cr(NH_3)_5Br]Br_2$
$PtCl_4 \cdot 4NH_3$	228.9	$[Pt(NH_3)_4Cl_2]Cl_2$
$PtCl_4 \cdot 2KCl$	256.8	$K_2[PtCl_6]$
	1:3 and 3:1 Electrolytes	
$LaCl_3$	393.5	—
$CoCl_3 \cdot 6NH_3$	431.6	$[Co(NH_3)_6]Cl_3$
$CoBr_3 \cdot 6NH_3$	426.9	$[Co(NH_3)_6]Br_3$
$CrCl_3 \cdot 6NH_3$	441.7	$[Cr(NH_3)_6]Cl_3$
$PtCl_4 \cdot 5NH_3$	404	$[Pt(NH_3)_5Cl]Cl_3$
	1:4 Electrolytes	
$PtCl_4 \cdot 6NH_3$	522.9	$[Pt(NH_3)_6]Cl_4$

[a] This is the molar conductivity measured at a concentration of 0.001 mol dm^{-3}. Values are from Werner and Miolati, except for $[Pt(NH_3)_5Cl]Cl_3$, which is from L. A. Chugaev and N. Vladimirov, *Compt. Rend.*, **1915**, *160*, 840.

[b] The theoretical value is, of course, zero, but impurities or a reaction with the solvent water could produce a small concentration of ions.

atom. Three such arrangements come to mind: a planar hexagon, similar to the benzene ring, and two solid polyhedra, the trigonal prism and the octahedron. The latter is closely related to the former, being formed by a 60° rotation of one of the trigonal faces (in fact, the octahedron can be considered to be a trigonal *anti*-prism). For a "disubstituted" complex, MA_4B_2, the planar arrangement gives three isomers, the familiar *ortho*, *meta*, and *para* arrangements of organic chemistry. The trigonal prism yields three isomers also, but there are only *two* octahedral arrangements for this formulation. The total number of isomers expected for each geometrical arrangement together with the experimental results for various compositions is listed in Table 9.2.

In every case Werner investigated, the number of isomers found was equal to that expected for an octahedral complex. For $[Co(NH_3)_4Cl_2]Cl$, for example, a green isomer

Table 9.2 Numbers of isomers expected and found for C.N. = 6

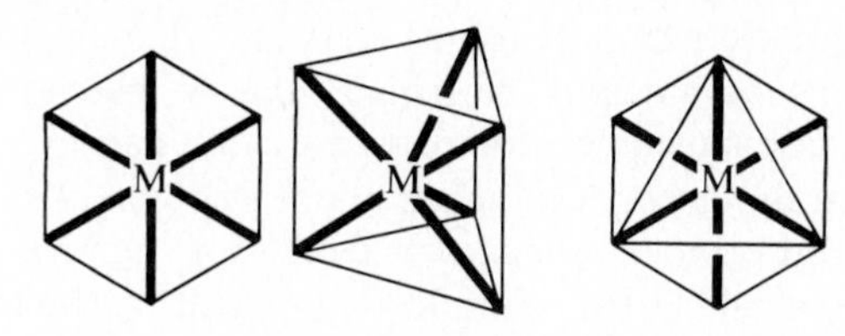

Formula	Planar	Trigonal prism	Octahedral	Experimental
MA_5B	1	1	1	1
MA_4B_2	3	3	2	2
MA_3B_3	3	3	2	2

and a violet isomer were known to Werner. Although the correlation is perfect, it must be borne in mind that the presence of two instead of three known isomers for this compound and others constitutes negative evidence concerning the structure of these complexes. Although Werner worked carefully and examined many systems, there was always the possibility, admittedly small, that the third isomer had escaped his detection. The failure to synthesize a compound, to observe a particular property, or to effect a particular reaction can never be positive proof of the nonexistence of that compound, property, or reaction. It may simply reflect some failure in technique on the part of the chemist. One well-known example of the failure of negative evidence is the overthrow of the dogma of the chemical inertness of the noble gases (see Chapter 15).

Werner was correct, however, in his conclusions concerning the octahedral geometry of coordination number 6 for cobalt and platinum(IV). He was also correct, and on a firmer logical footing, in his assignment of square planar geometries to the complexes of palladium and platinum from the fact that two isomers have been isolated from compounds of the formula MA_2B_2. The most likely alternative structure, the tetrahedron, would produce only one isomer for this composition.

VALENCE BOND THEORY

The first successful application of bonding theory to coordination compounds was made by Linus Pauling. It is usually referred to as the *valence bond theory* of coordination compounds.[9] It is closely related to the hybridization and geometry of noncomplex compounds as discussed in Chapter 5. From the valence bond point of view the formation of a complex is a reaction between a Lewis base (ligand) and a Lewis acid (metal or metal ion) with the

[9] The best account of valence bond theory as applied to coordination compounds is probably Pauling's own book, "The Nature of the Chemical Bond," 3rd ed., Cornell University Press, Ithaca, N.Y., **1960**.

formation of a coordinate covalent (or dative) bond between the ligand and the metal. For example, the ions Ni^{+2}, Pd^{+2}, and Pt^{+2} have d^8 configurations. Complexes of Pd^{II} and Pt^{II} are customarily 4-coordinate, square planar, and diamagnetic, and this arrangement is often found in Ni^{II} complexes as well. Inasmuch as the ground state of these ions is paramagnetic (^3F), the ligands in these diamagnetic complexes must cause pairing of the two odd electrons. Pauling suggested that this occurs via the use of one *d* orbital by the ligands. It is customary to represent the metal orbitals by boxes or circles for the purpose of accounting for the distribution of the metal electrons and those donated by the ligand:

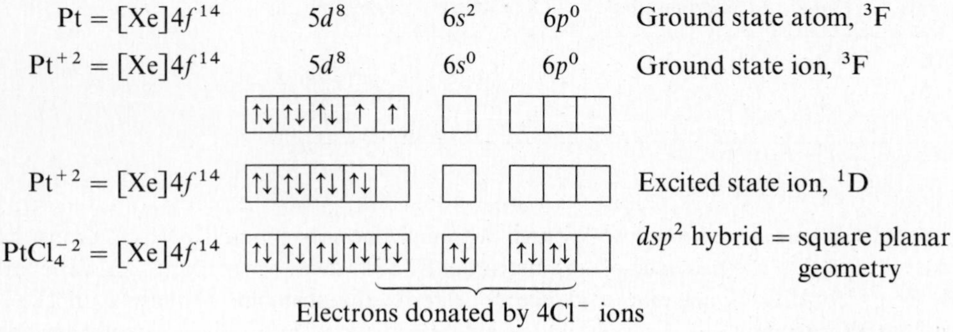

$Pt = [Xe]4f^{14}$	$5d^8$	$6s^2$	$6p^0$	Ground state atom, ^3F
$Pt^{+2} = [Xe]4f^{14}$	$5d^8$	$6s^0$	$6p^0$	Ground state ion, ^3F
	[↑↓][↑↓][↑↓][↑][↑]	[ ]	[ ][ ][ ]	
$Pt^{+2} = [Xe]4f^{14}$	[↑↓][↑↓][↑↓][↑↓][ ]	[ ]	[ ][ ][ ]	Excited state ion, ^1D
$PtCl_4^{-2} = [Xe]4f^{14}$	[↑↓][↑↓][↑↓][↑↓][↑↓]	[↑↓]	[↑↓][↑↓][ ]	dsp^2 hybrid = square planar geometry

Electrons donated by $4Cl^-$ ions

In those cases (limited to Ni^{+2}) in which paramagnetic 4-coordinate complexes are known, all five *d* orbitals of the metal must be occupied by the d^8 electrons, eliminating the possibility of involvement of a low-lying *d* orbital (3*d* in Ni) in the ligand–metal bonds. Pauling suggested that these complexes would involve sp^3 hybrids:

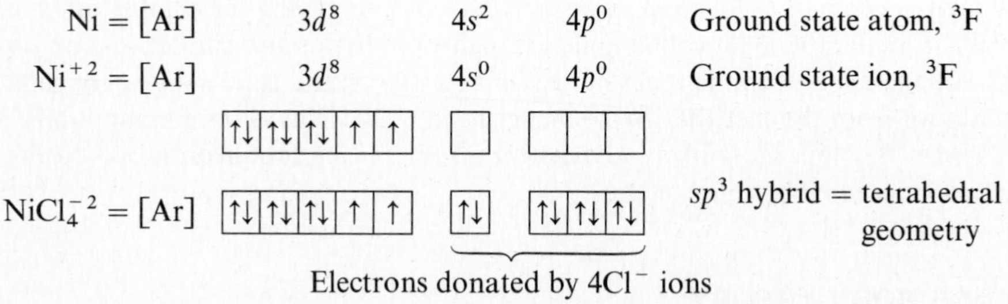

$Ni = [Ar]$	$3d^8$	$4s^2$	$4p^0$	Ground state atom, ^3F
$Ni^{+2} = [Ar]$	$3d^8$	$4s^0$	$4p^0$	Ground state ion, ^3F
	[↑↓][↑↓][↑↓][↑][↑]	[ ]	[ ][ ][ ]	
$NiCl_4^{-2} = [Ar]$	[↑↓][↑↓][↑↓][↑][↑]	[↑↓]	[↑↓][↑↓][↑↓]	sp^3 hybrid = tetrahedral geometry

Electrons donated by $4Cl^-$ ions

According to this interpretation, it should be possible to predict the geometry of a d^8 complex if the magnetic susceptibility is known, namely: diamagnetic = square planar; paramagnetic = tetrahedral. This rule, the "magnetic criterion of bond type," proved exceedingly useful for a number of years, but we now know that it can be misleading, which indicates one of the weaknesses of the simplified VB interpretation of bonding in complexes.

In a similar manner, two possibilities were postulated for 6-coordinate complexes of d^6 ions, such as Fe^{+2}, Co^{+3}. Consider Co^{III} complexes, for example. Werner synthesized a very large number of these complexes. All of Werner's Co^{III} complexes as well as all others synthesized prior to the formulation of Pauling's VB treatment of coordination were diamagnetic. They were thus formulated as follows:

$Co = [Ar]$	$3d^7$	$4s^2$	$4p^0$	Ground state atom, ^4F
$Co^{+3} = [Ar]$	$3d^6$	$4s^0$	$4p^0$	Ground state ion, ^5D
	[↑↓][↑][↑][↑][↑]	[ ]	[ ][ ][ ]	
$Co^{+3} = [Ar]$	[↑↓][↑↓][↑↓][ ][ ]	[ ]	[ ][ ][ ]	Excited state ion, ^1I
$Co(NH_3)_6^{+3} = [Ar]$	[↑↓][↑↓][↑↓][↑↓][↑↓]	[↑↓]	[↑↓][↑↓][↑↓]	d^2sp^3 hybrid = octahedral geometry

Electrons donated by $6NH_3$ molecules

The resulting VB description adequately accounts for the properties of the complex, i.e., no unpaired electrons and a perfectly octahedral structure. The discovery that a paramagnetic complex of Co^{III} could be formed—CoF_6^{-3}, having four unpaired electrons—required an adjustment of the theory. Pauling suggested that the fluoride ion could bond through the "outer" $4d$ orbitals. Such a hybrid, $3sp^34d^2$, would have identical symmetry with the $3d^2sp^3$ hybrids invoked above. The hexafluorocobaltate(III) ion can thus be formulated as:

	$3d^6$	$4s$	$4p$	$4d$	
$CoF_6^{-3} = [Ar]$	[↑↓][↑][↑][↑][↑]	[↑↓]	[↑↓][↑↓][↑↓]	[↑↓][↑↓][ ][ ][ ]	sp^3d^2 hybrid = octahedral geometry

Electrons donated by $6F^-$ ions

Electroneutrality principle and back bonding

One apparent difficulty with the VB assumption of the donation of electrons from Lewis base ligands to Lewis acid metal ions is the buildup of formal negative charge on the metal. Since this is a problem that arises, in one form or another, in all complete treatments of coordination compounds, the following discussion is appropriate to all of the following methods.

Consider a complex of Co^{II} such as $[CoL_6]^{+2}$. The six ligands share twelve electrons with the metal atom, and since all of these electrons come from the ligands, this process results in a formal charge buildup of -6 on the metal which is only partially canceled by the ionic charge of $+2$ on the combining metal ion. From a formal charge point of view the cobalt would become negative; however, Pauling pointed out two reasons why metals would not in fact exist with such an unfavorable negative charge. First, because donor atoms on ligands are, in general, highly electronegative atoms such as the halogens, N and O, the electrons will not be shared equally between the metal and the ligands, thus inducing positive charges on the metal to help offset the unfavorable formal charge. Pauling suggested that complexes would be most stable when the electronegativity of the ligand was such that the metal achieved an essentially neutral condition. This rule of thumb is known as the *electroneutrality principle.* Pauling and co-workers[10] have made semiquantitative calculations correlating the stability of complexes with the charges on the central metal atom. Inasmuch as these calculations involve very rough approximations, they are not apt to be very accurate and the method of calculation need not concern us. The values do indicate *qualitatively* how the buildup of excessive negative charge on a metal can destabilize a complex. Some typical charges are:

$[Be(H_2O)_4]^{+2}$	$[Be(H_2O)_6]^{+2}$	$[Al(H_2O)_6]^{+3}$	$[Al(NH_3)_6]^{+3}$
Be = −0.08	Be = −1.12	Al = −0.12	Al = −1.08
4O = −0.24	6O = −0.36	6O = −0.36	6N = 1.20
8H = 2.32	12H = 3.48	12H = 3.48	18H = 2.88
Total = +2.00	Total = +2.00	Total = +3.00	Total = +3.00

Although the above values should not be taken too literally, they do indicate some qualitative principles in the formation of coordination compounds. First, four water molecules effectively neutralize the +2 ionic charge of beryllium, but six water molecules donate too much electron density. In contrast, Al^{+3} can adequately accommodate six water molecules. Hence $[Be(H_2O)_4]^{+2}$ and $[Al(H_2O)_6]^{+3}$ are stable, but $[Be(H_2O)_6]^{+2}$ is not. Similarly, $[Al(NH_3)_6]^{+3}$ is unstable because the nitrogen atom of the ammonia ligands is not sufficiently electronegative to remove the excess buildup of electron density.

The methods used to estimate these charges are only approximate, and the experimental methods used to measure charge density are not very accurate, but they confirm the essential correctness of the electroneutrality principle. In six cobalt(III) complexes the absolute value of the charge on the cobalt atom was experimentally determined (see p. 175)[10a] to average 0.6 electron.

The electroneutrality principle seems to apply most readily to complexes which are unstable because of excessive buildup of electron density on the metal. Although the ionization energy of a metal atom should also discourage the formation of complexes with large positive charges on the metal, such charges lead to increased Madelung ("ionic") bond energy which may compensate and help stabilize the complex. It is thus possible to find metals in high oxidation states in oxide and fluoride complexes such as CoF_6^{-2}, MnF_6^{-2}, RuF_6^-, CrO_3F^-, MnO_4^-.

In carbonyl, nitrosyl, and related compounds the metal usually exists in a low oxidation state and is bonded to an element of fairly low electronegativity. The σ bonds formed by donation of electrons from the ligand to the metal atom thus greatly increase the electron density on the metal atom. These complexes could not be stable if the metal did not have a second way of decreasing its electron density. Pauling suggested that such a mechanism for delocalization of metal electron density could be found in "back bonding" or partial

[10] L. Pauling, ibid., pp. 172–174; E. S. Gould, "Inorganic Reactions and Structure," Holt, Rinehart & Winston, New York, **1956**, pp. 140–141.

[10a] Y. Saito, "Inorganic Molecular Dissymmetry," Springer-Verlag, Berlin, **1979**.

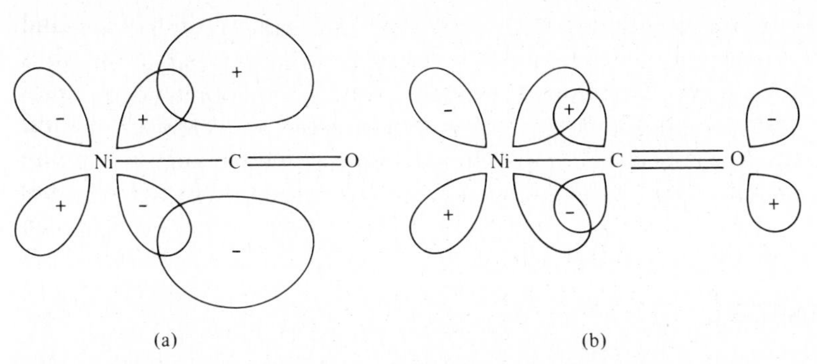

Fig. 9.1 Effect of metal → ligand π bonding on increasing the bond order of the Ni—C bond and decreasing the bond order of the C—O bond. (a) VB viewpoint: π bond between d orbital on Ni atom and p orbital on C atom. (b) MO viewpoint: π bond between d orbital on Ni atom and antibonding orbital (π^*) on the carbonyl group.

double bond resonance:

$$\underset{\text{(I)}}{\text{Ni—C}\equiv\text{O}} \longleftrightarrow \underset{\text{(II)}}{\text{Ni}^+\text{=C=O}^-}$$

To whatever extent canonical form (II) contributes to the resonance hybrid, electron density will be shifted from the nickel atom to the oxygen atom. A more precise examination of this process indicates that the delocalization of the electron density occurs via the overlap of d orbitals on the metal with orbitals of the carbonyl ligand. Valence bond theory would require the use of one of the p orbitals of the carbon atom (thus making it unavailable for a π bond to the oxygen) whereas the molecular orbital theory would speak in terms of overlap of the metal d orbital with the π^* orbital[11] of carbon monoxide (see Fig. 3.53). In either representation (Fig. 9.1) the electron density is shifted via the π bond from the metal atom to the carbonyl group and the bond order of the C—O bond is reduced. The metal, already in a low oxidation state, is prevented from becoming excessively negative. The formation of π bonds of this type will be discussed at greater length later in this chapter.

Strengths and weaknesses of the valence bond approach

Until about 30 years ago, the valence bond theory was almost the only one applied to coordination compounds by chemists. In the following decade a revolution occurred, and today few inorganic chemists use simple valence bond theory in discussions of these complexes. It is not because valence bond theory is wrong—it isn't. It has simply proved

[11] Confusion should not be allowed to occur simply because of the semantics involved here—namely, the formation of a *bond* by means of an *antibonding orbital*. The term "antibonding" is merely a convenient way of saying that in the isolated carbon monoxide molecule this orbital lies higher in energy than the original atomic orbitals in carbon and oxygen. Nevertheless, it may still be sufficiently low in energy to combine with a d orbital on the metal and form a bond.

to be more convenient to explain the properties of these molecules in different terms. The valence bond theory of complex compounds is still alive.[12] As we shall see (p. 745), Pauling is currently interpreting some unusual properties of coordination compounds in valence bond terms. The student should therefore be aware of the technique and the terminology, although the present-day inorganic chemist needs a much broader theoretical viewpoint in order to interpret adequately the experimental data presented by coordination chemistry.

CRYSTAL FIELD THEORY

The crystal field theory (CFT)[13] was developed by Bethe[14] and Van Vleck.[15] Its development was contemporary with that of Pauling's VB approach. Although used to some extent by physicists, it remained largely unknown to chemists until the 1950s. For a 20-year period, all coordination chemistry was interpreted in terms of valence bond theory. The popularity which the latter enjoyed over CFT was due in large part to the fact that the valence bond theory answered the questions which the chemists of that period were asking about the geometry and magnetic susceptibility of isolated complex ions. It is unfortunate that chemists did not make more use of the pioneering work of Bethe, Van Vleck, and others and develop coordination chemistry from both points of view. As Van Vleck pointed out as early as 1935[15] and as will be discussed later, both valence bond and crystal field theory are special cases of the more general molecular orbital treatment.

Just as crystal field theory largely replaced valence bond theory in treating coordination compounds, it in turn is being replaced by molecular orbital theory. However, *and this is the important point*, CFT is conceptually so simple, its symmetry arguments identical with and its energetics and predictions essentially so much the same as molecular orbital theory that it is far and away the best way for a newcomer to approach the modern theory of coordination chemistry. We shall proceed to more recent treatments of theory in turn.

In order to understand clearly the interactions that are responsible for crystal field effects, it is necessary to have a firm grasp of the geometrical relationships of the d orbitals. There is no unique way of representing the five d orbitals, but the most convenient representations are shown in Figs. 9.2 and 9.3. In fact, there are *six* wave functions that can be written for orbitals having the typical four-lobed form (d_{xy}, for example). Inasmuch as there can be only five d orbitals having any physical reality, the fifth d orbital (d_{z^2}) must be a linear combination of two of the other "orbitals," $d_{z^2-y^2}$ and $d_{z^2-x^2}$. Thus the latter

[12] A prepublication reviewer of this book has suggested that this sentence should be amended to include: ". . . but is currently in the intensive care unit!" In contrast, Pauling might claim that alternative theories are unnecessary. Rather than trying to adopt the exact viewpoint of Pauling, Huheey, or a reviewer, the reader is advised to develop a wide understanding of all of these ideas. Once an eclectic, universal view of chemistry has been achieved, any nonuseful and unused ideas will automatically be discarded, a much easier process than filling unappreciated lacunae later.

[13] As used here, *crystal field theory* will refer to a purely electrostatic interaction between ligands and the central metal ion; i.e., covalency will be excluded. As we shall see later, it is possible to modify the theory to include some covalency (this modification is usually called *ligand field theory*).

[14] H. Bethe, *Ann. Physik*, **1929**, [5], *3*, 135.

[15] J. H. Van Vleck, *Phys. Rev.*, **1932**, *41*, 208; *J. Chem. Phys.*, **1935**, *3*, 803, 807.

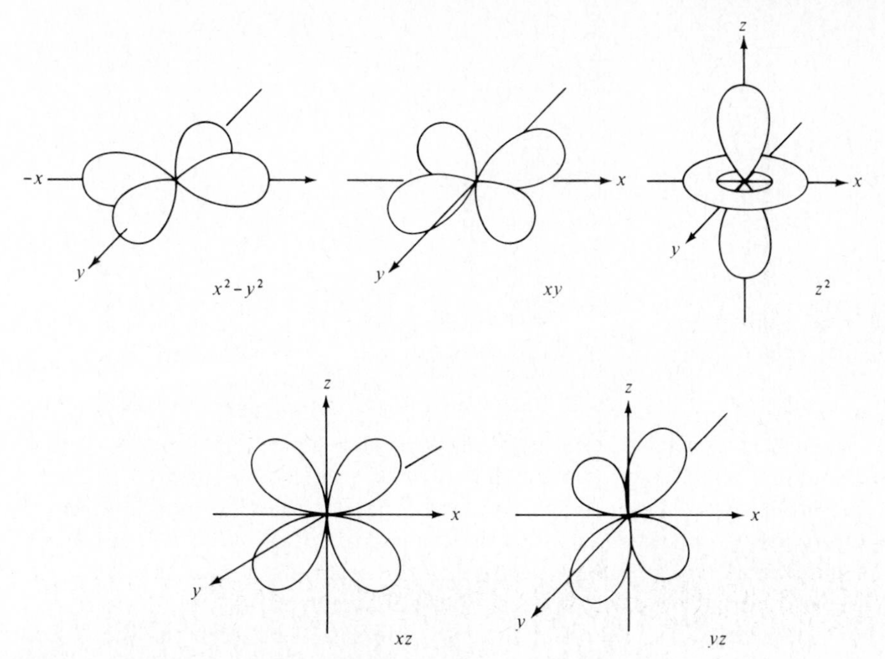

Fig. 9.2 Spatial arrangement of the five d orbitals.

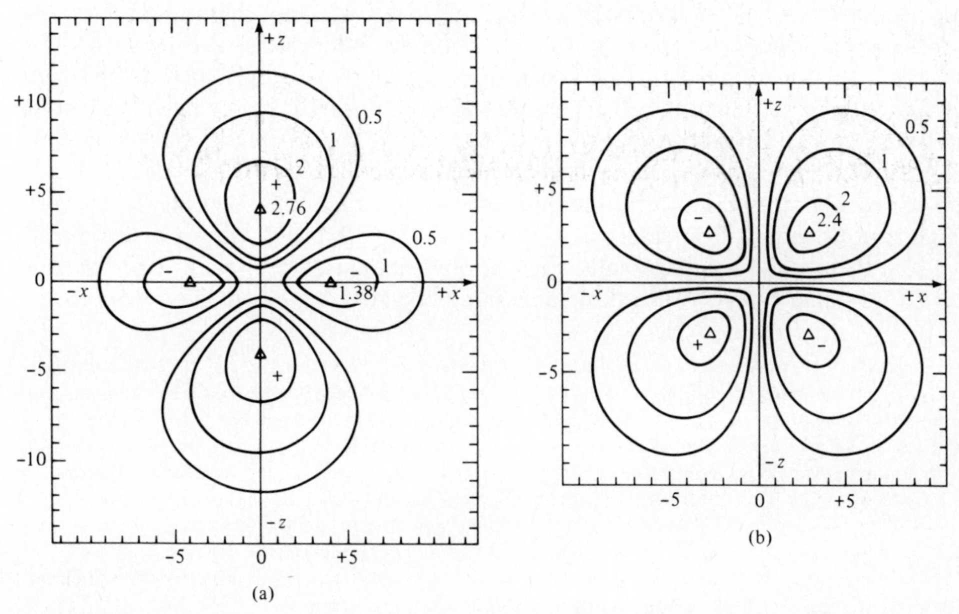

Fig. 9.3 Electron density contours for (a) d_{z^2} and (b) d_{xz} orbitals. The triangles represent the points of maximum electron density. [From B. Perlmutter-Hayman, *J. Chem. Educ.*, **1969**, *46*, 428. Reproduced with permission.]

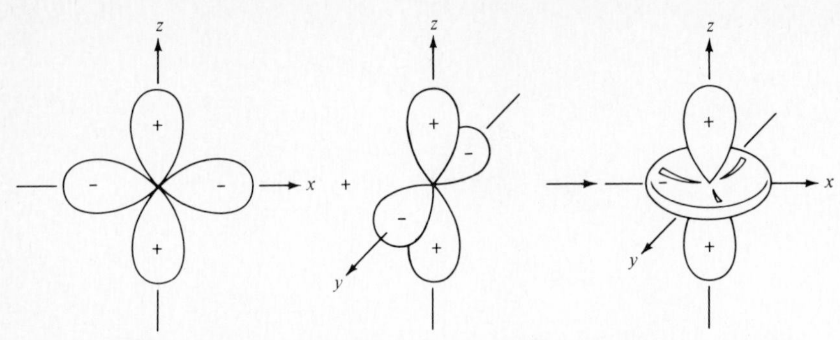

Fig. 9.4 Illustration of the d_{z^2} orbital as a linear combination of the $d_{z^2-x^2}$ and $d_{z^2-y^2}$ orbitals.

two "orbitals" have no independent existence, but it is often convenient to think of the d_{z^2} orbital as a combination of the two, having the "average" properties of the two (Fig. 9.4). Therefore, since both have high electron density along the z axis, the d_{z^2} orbital has a large amount of electron density concentrated along the axis. Since one of the component wave functions is maximized along the x axis ($d_{z^2-x^2}$) and the other along the y axis ($d_{z^2-y^2}$), the resultant d_{z^2} has a torus[16] of electron density in the xy plane.

Pure crystal field theory assumes that the only interaction between the metal ion and the ligands is an electrostatic ("ionic") one. The five d orbitals in an isolated, gaseous metal ion are degenerate. If a spherically symmetric field of negative charges is placed around the metal ion all of the orbitals will be raised in energy as a result of the repulsion between the negative field and the negative electrons in the orbitals,[17] but the five d orbitals will still remain degenerate. If the field results from the influence of real ligands (either from negative anionic ligands or the negative end of dipolar ligands such as NH_3) the symmetry *must* be less than spherical because of the finite number of ligands involved (usually four or six). Let us consider the case of six ligands approaching to form an octahedral complex. For convenience, we may consider the six ligands to enter along the axes of the coordinate system, i.e., from the directions z, $-z$, x, $-x$, y, $-y$ (Fig. 9.5). Under these conditions they will interact strongly with the e_g orbitals[18] lying along the x, y, and z axes, the d_{z^2} and $d_{x^2-y^2}$ orbitals. Furthermore, these two orbitals will be raised in energy to the same extent.[19] The three remaining orbitals, d_{xy}, d_{yz}, and d_{xz}, collectively

[16] This is more often referred to as a "doughnut" or, less elegantly, as a "belly band." It is often neglected, especially in figures in which it is difficult to portray adequately all of the lobes of all five d orbitals simultaneously. Nevertheless, it is important to remember that the d_{z^2} has a component in the xy plane.

[17] Because the electrons of all of the orbitals are repelled, it has been suggested that this be shown on the orbital splitting diagrams (such as Fig. 9.6) by tie lines showing the increased orbital energy upon approach of the ligands. On the other hand, the attraction between the metal ion and the ligand ions or dipoles will result in a lowering of energy (as it must if the complex is to be stable) and it seems unnecessary to include all of these effects in the simple orbital splitting diagram as long as one is aware of their existence.

[18] For present purposes the terms e_g and t_{2g} may be taken merely as convenient labels to differentiate the two groups of orbitals. The labels have their origin in group theory (see Appendix B).

[19] It might appear as though the $d_{x^2-y^2}$ orbital would be repelled more since it is confronted with four ligands in contrast to the (apparent) two ligands of the d_{z^2} orbital. Recall, however, that the d_{z^2} orbital has electron density in the xy plane (or that it is a linear combination of a $z^2 - y^2$ and a $z^2 - x^2$ wave function).

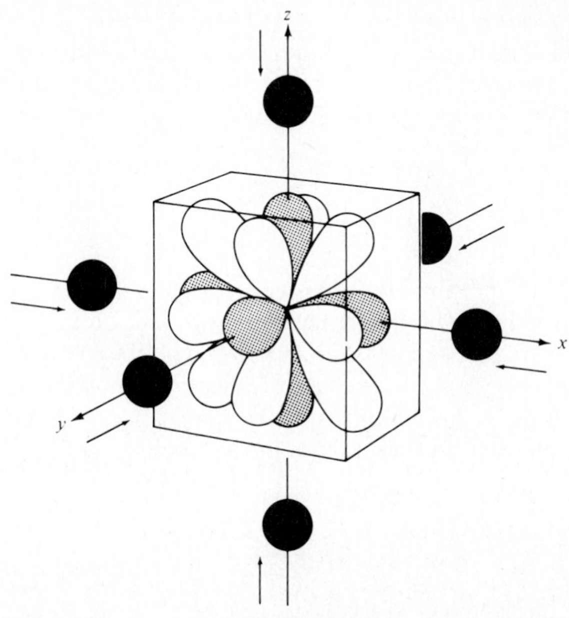

Fig. 9.5 Complete set of d orbitals in an octahedral field. The e_g orbitals are shaded and the t_{2g} orbitals are unshaded. The torus of the d_{z^2} orbital has been omitted for clarity.

known as the t_{2g} orbitals, will be repelled to a lesser extent inasmuch as they are directed *between* the approaching ligands (Fig. 9.5). The extent to which the e_g and the t_{2g} orbitals are separated is denoted by the quantity $10Dq$ or Δ (Fig. 9.6). As we shall see, this quantity may be a large energy or a small one but is in any event $10Dq$ *by definition.* Let us consider the approach of the ligands as a two-step process. In the first the ligands approach but produce a hypothetical spherical field which repels all of the orbitals to the same extent, and in the second the octahedral field splits the orbital degeneracy. We find that the barycenter, or "center of gravity," of the orbitals remains constant during the second step. That is to say the energy of all the orbitals may be raised by the repulsion of the advancing ligands in step one, but merely rearranging the ligands from a (hypothetical) spherical

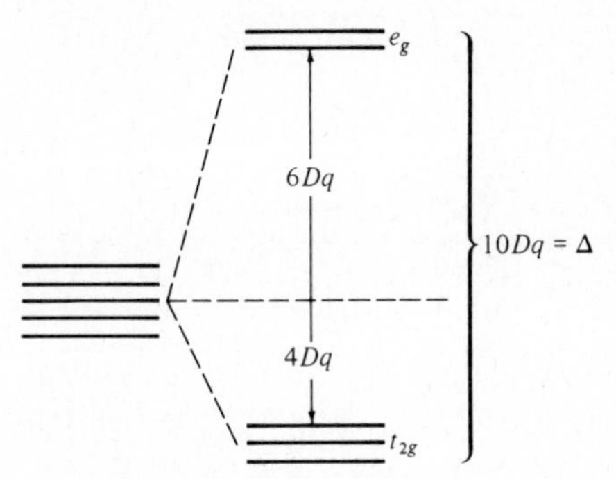

Fig. 9.6 Splitting of the degeneracy of the five d orbitals by an octahedral ligand field.

field to an octahedral field (or cubic or tetrahedral) does not alter the *average* energy of the set of d orbitals. Hence to maintain the barycenter constant it is necessary for the *two* e_g orbitals to be repelled $6Dq$ to balance the stabilization of the *three* t_{2g} orbitals to the extent of $4Dq$. This constancy of the barycenter of the d orbitals may be likened to a conservation of energy of the orbitals.

Measurement of 10*Dq*

Before further consequences of crystal field splitting are discussed, it may be helpful to provide some insight into the magnitude of $10Dq$ and how it may be measured. Consider, for example, the $[Ti(H_2O)_6]^{+3}$ ion. Ti^{+3} has a d^1 electron configuration with the electron occupying the lowest d orbital available to it. In the case of an octahedral complex this will be one of the three degenerate t_{2g} orbitals. Solutions of the Ti^{+3} ion appear violet colored as a result of the absorption of photons and promotion of the electron with the transition

$$t_{2g}^1 e_g^0 \longrightarrow t_{2g}^0 e_g^1$$

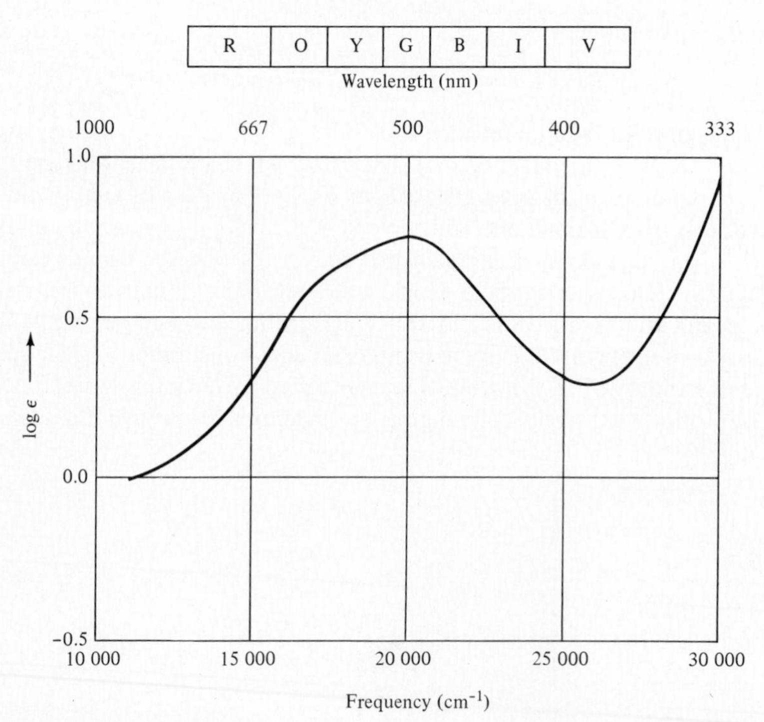

Fig. 9.7 Visible spectrum of 0.1 mol dm^{-3} aqueous solution of the $[Ti(H_2O)_6]^{+3}$ ion. The letters at the top indicate the colors of the portions of the visible spectrum. [From H. Hartman et al., *Z. anorg. allg. Chem.*, **1956,** *284*, 153. Reproduced with permission.]

This transition is found[20] to occur with a maximum at 20 300 cm^{-1} (see Fig. 9.7). Its energy is thus:[21]

$$20\,300\ \text{cm}^{-1} \times \frac{1\ \text{kJ mol}^{-1}}{83.6\ \text{cm}^{-1}} = 243\ \text{kJ mol}^{-1} \qquad \textbf{(9.6)}$$

By way of comparison, the absorption maximum of ReF_6 (also a d^1 species) is 32 500 cm^{-1} corresponding to an energy of 388 kJ mol^{-1} for $10Dq$.[22] These are typical values for $10Dq$ and are thus of the same order of magnitude as a typical chemical bond (p. 263). As we shall see, complexes can be stabilized by more or less than $10Dq$, but we can expect stabilizations that are energetically comparable to the formation of one or more bonds.

The d^1 situation is the simplest possible in that the observed transition reflects the actual energy levels of the e_g and t_{2g} levels. For the more general d^n situation electron–electron interactions must be taken into account and the calculation becomes somewhat more involved. The appropriate methods are discussed on pp. 442ff.

Crystal field stabilization energy: weak field case

As we have seen, the energy difference between the t_{2g} and e_g orbitals is defined as $10Dq$ and the energy of the t_{2g} level relative to the barycenter of unperturbed d orbitals is $-4Dq$. In the d^1 case discussed above, the *crystal field stabilization energy* (CFSE) is $-4Dq$.[23] For d^2, the CFSE $= -8Dq$ and for d^3, CFSE $= -12Dq$. In these configurations the electrons obey Hund's rule; that is, the electrons remain unpaired and enter the different degenerate orbitals (see p. 33). Upon reaching the d^3 configuration, however, the t_{2g} level becomes half-filled and there are no further orbitals of this energy to accept electrons without pairing. When one more electron is added to form the d^4 case, two possibilities arise. In the *weak field* limit the splitting of the orbitals is small with respect to the energy necessary to cause electron pairing in a single orbital (the *pairing energy*, P). Since $P > 10Dq$, the fourth electron enters one of the e_g levels rather than "pay the price" of pairing with one of the electrons in the t_{2g} orbitals. Instead, it loses somewhat less energy (in the form of CFSE) by entering the destabilized e_g level. The net CFSE energy is thus:

$$\text{CFSE} = (3 \times -4Dq) + (1 \times +6Dq) = -6Dq \qquad \textbf{(9.7)}$$

The electron configuration for d^4 may be written $t_{2g}^3e_g^1$.

---

[20] H. Hartman and H. L. Schläfer, *Z. Physik. Chem.* (*Leipzig*), **1951**, *197*, 116; H. Hartman et al., *Z. Anorg. Allg. Chem.*, **1956**, *284*, 153.

[21] It is suggested that the reader verify the value of the constant 83.6 cm^{-1} = 1 kJ mol^{-1} by means of the conversion factors involving the speed of light, Avogadro's number, joules, etc.

[22] W. Moffitt et al., *Mol. Phys.*, **1959**, *2*, 109.

[23] Inasmuch as CFSE energy always *stabilizes* a complex (unless it happens to be equal to zero), it should be expressed as a *negative* quantity: $-4Dq$, $-8Dq$, $-12Dq$, etc. In practice, however, usage is somewhat loose—crystal field stabilization energies of $4Dq$, $8Dq$, and $12Dq$ are often listed. If it is remembered that CFSE $\leqslant 0$, no confusion can occur.

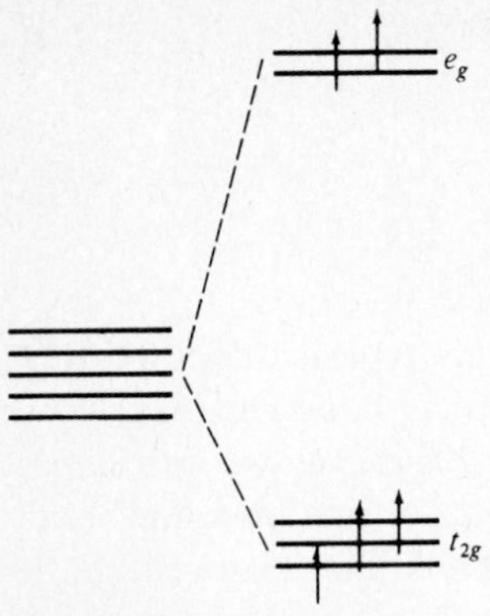

Fig. 9.8 Electron configuration of a d^5 ion in a weak octahedral field.

The addition of a fifth electron results in a half-filled d subshell and an electron configuration $t_{2g}^3 e_g^2$. The crystal field stabilization energy is zero. The presence of two electrons in the unfavorable e_g level exactly balances the stabilization resulting from the three electrons in the t_{2g} level (see Fig. 9.8). In other words, the half-filled subshell (d^5) is spherically symmetrical and no stabilization can occur through the application of an external ligand field, octahedral or otherwise.

In the same way, the configurations for d^6 to d^{10} may be obtained. The configurations and their resultant CFSEs are listed in Table 9.3, together with the number of unpaired electrons expected for each configuration.

Table 9.3 Crystal field effects for weak and strong octahedral fields[a]

Weak field				Strong field			
d^n	Configuration	Unpaired electrons	CFSE	d^n	Configuration	Unpaired electrons	CFSE
d^1	t_{2g}^1	1	$-4Dq$	d^1	t_{2g}^1	1	$-4Dq$
d^2	t_{2g}^2	2	$-8Dq$	d^2	t_{2g}^2	2	$-8Dq$
d^3	t_{2g}^3	3	$-12Dq$	d^3	t_{2g}^3	3	$-12Dq$
d^4	$t_{2g}^3 e_g^1$	4	$-6Dq$	d^4	t_{2g}^4	2	$-16Dq + P$
d^5	$t_{2g}^3 e_g^2$	5	$0Dq$	d^5	t_{2g}^5	1	$-20Dq + 2P$
d^6	$t_{2g}^4 e_g^2$	4	$-4Dq$	d^6	t_{2g}^6	0	$-24Dq + 2P$
d^7	$t_{2g}^5 e_g^2$	3	$-8Dq$	d^7	$t_{2g}^6 e_g^1$	1	$-18Dq + P$
d^8	$t_{2g}^6 e_g^2$	2	$-12Dq$	d^8	$t_{2g}^6 e_g^2$	2	$-12Dq$
d^9	$t_{2g}^6 e_g^3$	1	$-6Dq$	d^9	$t_{2g}^6 e_g^3$	1	$-6Dq$
d^{10}	$t_{2g}^6 e_g^4$	0	$0Dq$	d^{10}	$t_{2g}^6 e_g^4$	0	$0Dq$

[a] This table is somewhat oversimplified, being based on one-electron energy levels. In other words, configuration interaction or electron–electron effects have been neglected.

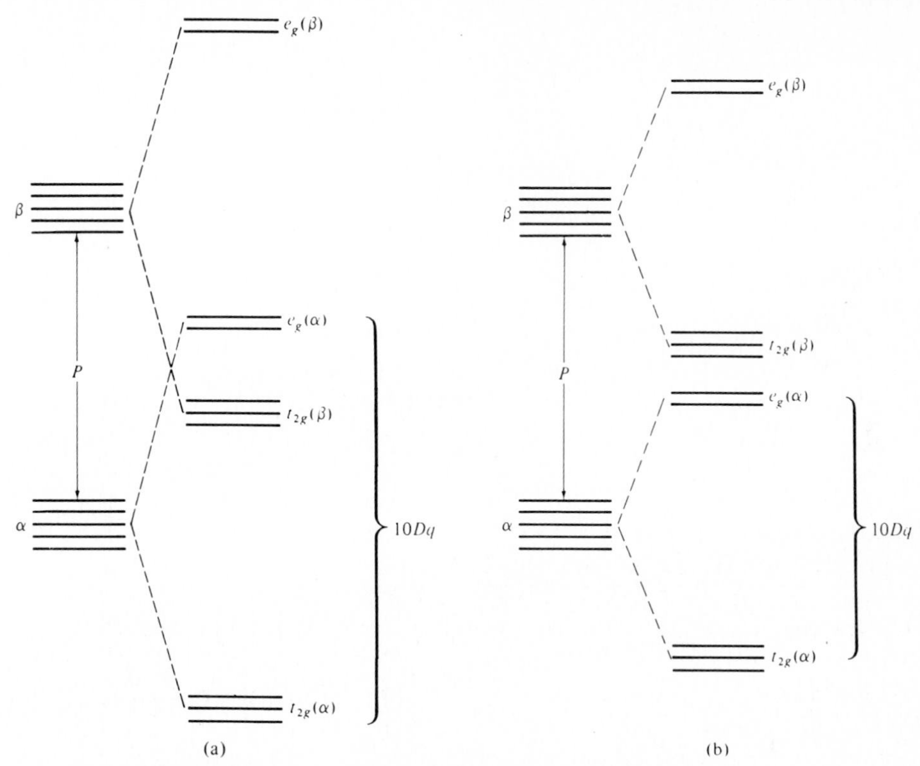

Fig. 9.9 Comparison of (a) strong field and (b) weak field cases of crystal field splittings. Note that each line does *not* represent an orbital but the *complete one-electron wave function* including the spin (α or β).

Crystal field stabilization: strong field case

If the splitting of the d orbitals is large with respect to the pairing energy ($10Dq > P$) it is more favorable for the electrons to pair up in the t_{2g} level than to enter the strongly unfavorable e_g level. This situation is known as the *strong field* case.[24] A comparison of energy levels for the weak field case and the strong field case is shown in Fig. 9.9. If the pairing energy is larger than $10Dq$ (because of large P's or small Dq's) the electrons remain unpaired as long as possible and no electron pairing occurs until the d^6 configuration is reached (Fig. 9.10). In the strong field case, the first three electrons remain unpaired in the degenerate t_{2g} orbitals, but the *fourth* pairs in order to remain in the t_{2g} level rather than enter the strongly unfavorable e_g level (Fig. 9.10). As a result, the crystal field stabilization energy of configurations with more than three electrons will in general be greater for the strong field case. For d^4, the configuration is t_{2g}^4 and the CFSE is $-16Dq + P$;

[24] *Strong field* and *weak field* are, of course, interpretive and correspond to the experimental observations of *low* and *high spin*.

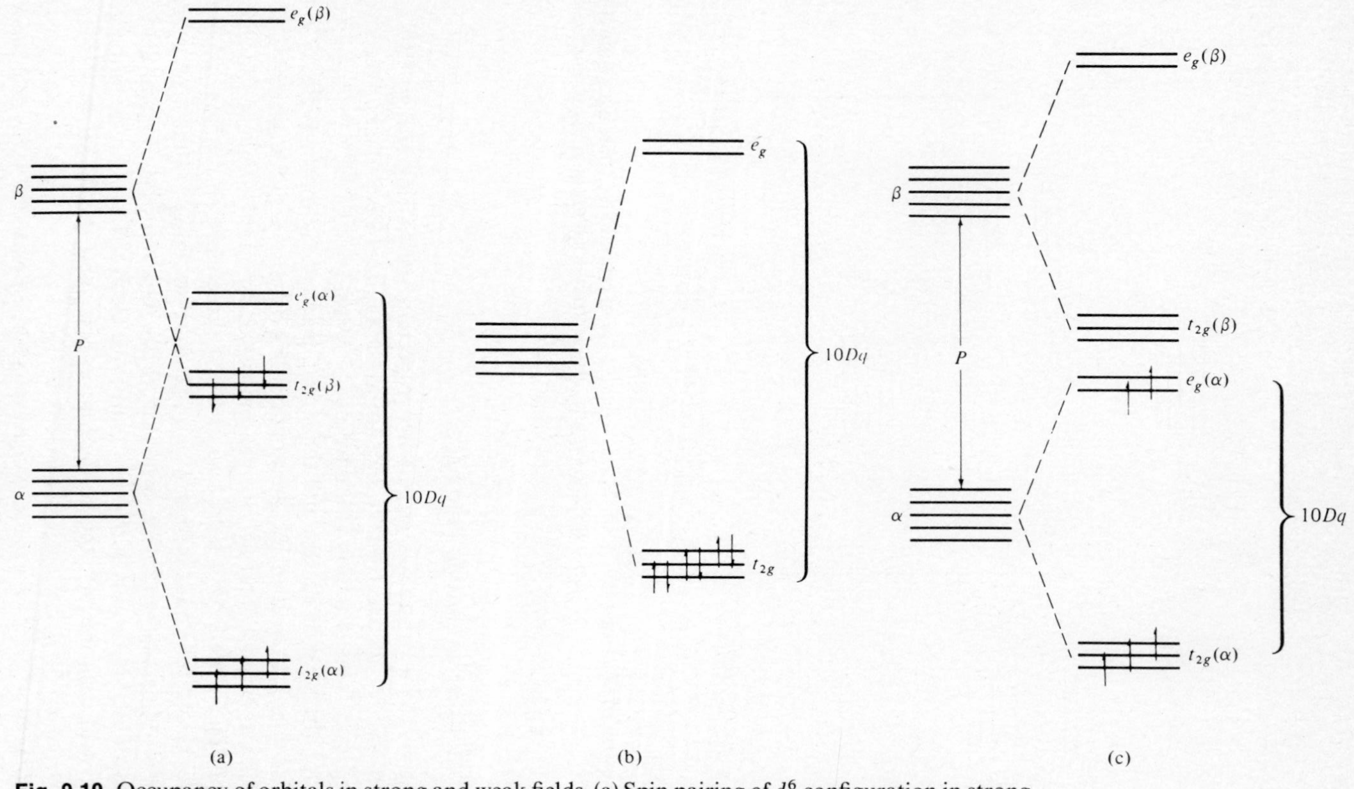

Fig. 9.10 Occupancy of orbitals in strong and weak fields. (a) Spin pairing of d^6 configuration in strong octahedral field. (b) Usual representation of spin pairing of d^6; the energy of pairing not explicitly shown. (c) Nonpairing of d^5 configuration in weak octahedral field. The more common representation that does not explicitly show the pairing energy is given in Fig. 9.8.

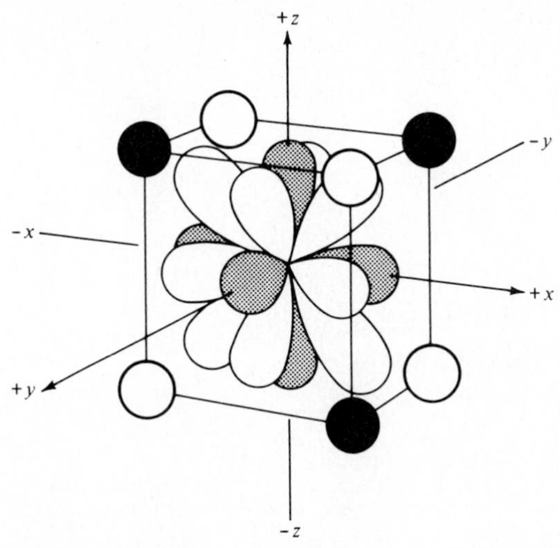

Fig. 9.11 Complete set of d orbitals in a cubic field. All eight ligands produce a field $\frac{8}{9}$ as strong as a corresponding octahedral field (see Fig. 9.5). Either set of four tetrahedral ligands (○ or ●) produces a field $\frac{4}{9}$ as strong as the octahedral field.

for d^5, CFSE $= -20Dq + 2P$; and for d^6, CFSE $= -24Dq + 2P$.[25] The seventh electron enters the destabilizing e_g level and the CFSE decreases:

$$\text{CFSE} = (6 \times -4Dq) + (1 \times 6Dq) + P = -18Dq + P \tag{9.8}$$

A summary of configurations, Crystal Field Stabilization Energies and number of unpaired electrons for d^n, where $n = 1$ to 10 is given in Table 9.3.

Crystal field effects: tetrahedral symmetry

Two of the most common geometries for 4-coordinate complexes are the tetrahedral and square planar arrangements. The square planar arrangement is a special case of the more general D_{4h} symmetry involving tetragonal distortion of octahedral complexes, discussed later. Tetrahedral coordination is closely related to cubic coordination. Consider eight ligands approaching the central metal atom aligned on the corners of the cube shown in Fig. 9.11. In this arrangement, the ligands do not directly approach any of the metal d orbitals, but they come closer to the t_{2g} orbitals directed to the *edges* of the cube than to the e_g levels directed to the *centers of the faces* of the cube. Hence the t_{2g} levels are raised in energy and the e_g levels stabilized. Furthermore, since the "center of gravity"

[25] It is common to refer to the CFSEs in these cases merely as $-16Dq$, $-20Dq$, and $-24Dq$, respectively, It is understood that all configurations with $n > 3$ will have pairing energies involved if they are strong field or low spin. If the *pairing energies*, P, appear confusing, it is because they are calculated not from the absolute number of electron pairs but, rather, from the *change* in pairing going from the weak field to the strong field situation. See p. 374.

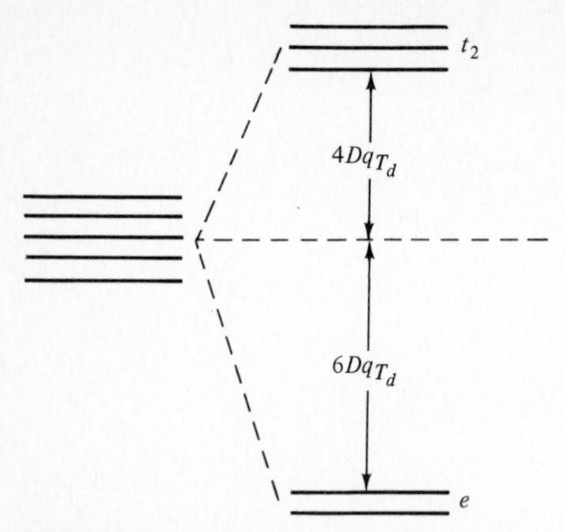

Fig. 9.12 Splitting of d orbitals in a tetrahedral field.

rule holds, the t_{2g} levels are raised $4Dq$ and the e_g levels are lowered $6Dq$ from the barycenter (Fig. 9.12). *The energy level scheme for cubic symmetry is exactly the inverse of that for octahedral symmetry.*

If four ligands are removed from alternate corners of the cube in Fig. 9.11, the remaining ligands form a tetrahedron about the metal. The energy level scheme for tetrahedral symmetry is qualitatively similar to that for cubic symmetry, but the splitting ($10Dq$) is only half as large. *Only the weak field case need be considered for tetrahedral complexes* and this greatly simplifies the treatment of electron configurations and crystal field stabilizing energies.[26] The pairing energy is larger than $10Dq$ so the electrons enter the five orbitals remaining unpaired until the sixth electron forces pairing. For example, the d^4 case results in an $e^2t_2^2$ configuration with the CFSE $= -4Dq$:

$$\text{CFSE} = (2 \times -6Dq) + (2 \times 4Dq) = -4Dq \tag{9.9}$$

The number of unpaired electrons, the configurations, and the crystal field stabilization energies are given for $d^1 - d^{10}$ in Table 9.4.

Because spin-pairing does not occur in tetrahedral complexes the large crystal field stabilization energies ($-16Dq$ to $-24Dq$) observed in low-spin octahedral complexes are not seen here. Further, since the absolute value of $10Dq$ (in kJ mol^{-1}) is less in tetrahedral complexes than in octahedral ones (because of the indirect effect of the ligands and the smaller number), the total crystal field stabilization is much less important. We should therefore expect to find crystal field effects *minimized* in tetrahedral complexes as compared with other geometries.

[26] One might wonder whether it would not be possible under exceptional circumstances (high field strength from the ligands, low-pairing energy of the metal, etc.) for a tetrahedral complex to be low spin. Experimentally none has been observed, and we shall see later that under the circumstances favoring strong fields other geometries are favored over tetrahedral structures.

Table 9.4 Crystal field effects for weak cubic and tetrahedral fields[a]

d^n	Configuration	Unpaired electrons	CFSE
d^1	e_g^1	1	$-6Dq$
d^2	e_g^2	2	$-12Dq$
d^3	$e_g^2t_{2g}^1$	3	$-8Dq$
d^4	$e_g^2t_{2g}^2$	4	$-4Dq$
d^5	$e_g^2t_{2g}^3$	5	$0Dq$
d^6	$e_g^3t_{2g}^3$	4	$-6Dq$
d^7	$e_g^4t_{2g}^3$	3	$-12Dq$
d^8	$e_g^4t_{2g}^4$	2	$-8Dq$
d^9	$e_g^4t_{2g}^5$	1	$-4Dq$
d^{10}	$e_g^4t_{2g}^6$	0	$0Dq$

[a] Since tetrahedral fields do not have a center of symmetry, the $_g$ subscripts are inappropriate for that symmetry.

Pairing energies

The difference in energy between a low-spin configuration and a high-spin configuration (the "pairing energy") is composed of two terms. One is the inherent repulsion which must be overcome when forcing two electrons to occupy the same orbital. One might expect this to be fairly constant from one element to another and reasonably independent of other factors. The larger, more diffuse d orbitals in the heavier transition metals ($5d$) might more readily accommodate two negative charges than the smaller $3d$ orbitals, but otherwise little variation is expected. The second factor of importance is the loss of exchange energy (the basis of Hund's rule, p. 33) which occurs as electrons with parallel spins are forced to have antiparallel spins. The exchange energy is proportional to the number of pairs[27] of electrons of the same spin that can be arranged from n parallel electrons:

$$E_{ex} = \frac{n(n-1)}{2}K \quad \textbf{(9.10)}$$

The greatest loss of exchange energy occurs when the d^5 configuration is forced to pair—hence the apparent stability of the half-filled d subshell.

The electron–electron interactions are commonly expressed in terms of the *Racah parameters* (see p. 445) and may be obtained from the spectra of the free, gaseous ions. Some typical values are listed in Table 9.5.

The transition from high-spin to low-spin states may be portrayed graphically as in Fig. 9.13. Increasing field strength results in increasing stabilization for all configurations

[27] The word "pair" as used here simply means a set of two electrons, *not* two electrons with their *spins paired*.

Table 9.5 Pairing energies for some 3*d* metal ions[a]

	Ion	P_{coul}	P_{ex}	P_T
d^4	Cr^{+2}	71.2 (5 950)	173.1 (14 475)	244.3 (20 425)
	Mn^{+3}	87.9 (7 350)	213.7 (17 865)	301.6 (25 215)
d^5	Cr^{+}	67.3 (5 625)	144.3 (12 062)	211.6 (17 687)
	Mn^{+2}	91.0 (7 610)	194.0 (16 215)	285.0 (23 825)
	Fe^{+3}	120.2 (10 050)	237.1 (19 825)	357.4 (29 875)
d^6	Mn^{+}	73.5 (6 145)	100.6 (8 418)	174.2 (14 563)
	Fe^{+2}	89.2 (7 460)	139.8 (11 690)	229.1 (19 150)
	Co^{+3}	113.0 (9 450)	169.6 (14 175)	282.6 (23 625)
d^7	Fe^{+}	87.9 (7 350)	123.6 (10 330)	211.5 (17 680)
	Co^{+2}	100 (8 400)	150 (12 400)	250 (20 800)

[a] Pairing energies, in kJ mol^{-1} (and cm^{-1}) calculated from formulas and data given by L. E. Orgel, *J. Chem. Phys.*, **1955**, *23*, 1819, and J. S. Griffith, *J. Inorg. Nucl. Chem.*, **1956**, *2*, 1, 229. The values pertain to the free ion and may be expected to be from 15% to 30% smaller for the complexed ion as a result of the nephelauxetic effect. P_{coul}, P_{ex}, and P_T refer to the coulombic, exchange, and total energy opposing pairing of electrons.

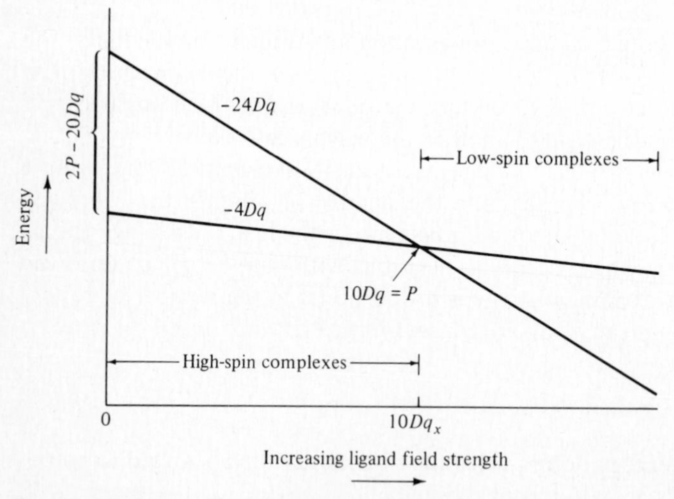

Fig. 9.13 Relation of pairing energy ($2P$) of d^6 complexes to the crystal field stabilization energy. Complexes in the vicinity of $10Dq_x$ may exist as a thermal equilibrium of high- and low-spin forms.

d^1 to d^9 (except weak field d^5). However, the slope of the energy lines is determined by the CFSE which is always greater for the strong field case than for the weak field case (see Table 9.3). At some value of the field strength ($10Dq_x$) the energies of the two states become equal. For weaker fields, left of the crossover point, the high-spin state is more stable, and to the right the low-spin state is favored. The two states are in equilibrium with one another at $10Dq_x$. On either side of the crossover point there will still be a thermal equilibrium established to the extent that the difference in energies is of the order of kT. This means that both species will exist in ratios determined by the Boltzmann distribution. The exact description of such situations can get rather complicated,[28] but the following oversimplification is not too misleading: Both species occur, and so the measured magnetic susceptibility is a weighted average of the two species. The first example discovered was that of iron(III) *N,N*-dialkyldithiocarbamates, $[Fe(S_2CNR'R'')_3]$.[29] The iron(III) ion may be either high-spin ($S = \frac{5}{2}$) or low-spin ($S = \frac{1}{2}$) depending upon the ligands. When the iron(III) dithiocarbamates are cooled to the neighborhood of absolute zero, the magnetic susceptibility corresponds to that expected for $S = \frac{1}{2}$. Upon warming, the magnetic susceptibility rises slowly until at temperatures above 350 K it *approaches* that of a completely high-spin complex, indicating a change in equilibrium from predominantly spin paired to predominantly spin free.

Since in the high-spin complexes the electrons in the e_g orbitals are directed at the ligands, we should expect the latter to "back off" a bit from the iron atom. In the low-spin complexes the ligands should be allowed to come somewhat closer because the e_g orbitals are empty. Experimentally, this is exactly what we find: The Fe—S distance in a group of high-spin complexes ($\mu_{eff} = 5.9$ BM) averages about 240 pm; in a group of low-spin complexes $\mu_{eff} = 2.3$ BM) it averages about 230 pm.

Factors affecting the magnitude of 10*Dq*

There are several factors that affect the extent of splitting of the *d* orbitals by ligands. Several values for $10Dq$ for aqua complexes of the first transition cations are listed in Table 9.6. From this table, two factors become apparent.

Oxidation state of metal ion. The ionic charge on the metal ion has a direct effect upon the magnitude of Dq. This is to be expected in terms of the electrostatic crystal field model: The increased charge on the metal ion will draw the ligands in more closely, hence they will have a greater effect in perturbing the metal *d* orbitals. Theoretically the change from +2 to +3 ionic charge would result in a corresponding increase in Dq of 50%. We see that in some cases (Cr^{+2}, Mn^{+3}) this expectation is borne out, but in others the increase is somewhat less as a result of increased repulsions.

Number and geometry of ligands. The splitting in an octahedral field is somewhat more than twice as strong as for a tetrahedral field for the same metal ion and the same ligands. There are two factors involved here: Four ligands instead of six would result in a 33% decrease in field, other factors being equal. Furthermore, in the tetrahedral complex the ligands are directed much less efficiently than in the octahedral (O_h) complex. In the latter the ligands exert maximum influence on the e_g level and as little as possible on the t_{2g}

[28] For an excellent discussion of the problem of high-spin–low-spin "crossovers," see R. L. Martin and A. H. White, *Trans. Metal Chem.*, **1968**, *4*, 113, and R. L. Carlin and A. J. van Duyneveldt, "Magnetic Properties of Transition Metal Compounds," Springer-Verlag, New York, **1977**, pp. 222–226. The latter is an excellent treatment of the entire topic of the magnetochemistry of coordination compounds.

[29] L. Cambi and A. Cagnasso, *Atti Accad. Naz. Lincei*, **1931**, *13*, 809; L. Cambi and L. Szegö, *Chem. Ber.*, **1931**, *64*, 2591; L. Cambi et al., *Atti Accad. Naz. Lincei*, **1932**, *15*, 266, 329. See footnote 28.

Table 9.6 Crystal field theory data for metal ions of the first transition series in aqua complexes[a]

Number of d electrons	Ion	Free ion ground state	Octahedral field ground state	Tetrahedral field ground state	Dq (cm^{-1}) oct.	Dq (cm^{-1}) tetr.	Stabilization ($kJ\ mol^{-1}$) oct.	Stabilization ($kJ\ mol^{-1}$) tetr.	Oct. site preference energy ($kJ\ mol^{-1}$)
1	Ti^{+3}	^2D	t_{2g}^1	e^1	2030	900	96.6	64.4	32.3
2	V^{+3}	^3F	t_{2g}^2	e^2	1800	840	174.5	120.0	54.5
3	V^{+2}	^4F	t_{2g}^3	$e^2t_2^1$	1180	520	168.0	36.4	131.6
	Cr^{+3}	^4F	t_{2g}^3	$e^2t_2^1$	1760	780	250.8	55.6	195.2
4	Cr^{+2}	^5D	$t_{2g}^3e_g^1$	$e^2t_2^2$	1400	620	100.3	29.3	71.0
	Mn^{+3}	^5D	$t_{2g}^3e_g^1$	$e^2t_2^2$	2100	930	150.1	44.3	105.8
5	Mn^{+2}	^6S	$t_{2g}^3e_g^2$	$e^2t_2^3$	750	330	0	0	0
	Fe^{+3}	^6S	$t_{2g}^3e_g^2$	$e^2t_2^3$	1400	620	0	0	0
6	Fe^{+2}	^5D	$t_{2g}^4e_g^2$	$e^3t_2^3$	1000	440	47.6	31.4	16.3
	Co^{+3}	^5D	t_{2g}^6	$e^3t_2^3$	—	780	188	107	81[b]
7	Co^{+2}	^4F	$t_{2g}^5e_g^2$	$e^4t_2^3$	1000	440	71.5	62.7	8.8
8	Ni^{+2}	^3F	$t_{2g}^6e_g^2$	$e^4t_2^4$	860	380	122.4	27.2	95.2
9	Cu^{+2}	^2D	$t_{2g}^6e_g^3$	$e^4t_2^5$	1300	580	92.8	27.6	65.2
10	Zn^{+2}	^1S	$t_{2g}^6e_g^4$	$e^4t_2^6$	0	0	0	0	0

[a] The data given are:
1. Number of *d* electrons
2. Transition metal ions
3. Free ion Russell-Saunders ground term (spin-orbit coupling is neglected in the term designation)
4. Electron configuration of octahedral ground state
5. Electron configuration of tetrahedral ground state
6. *Dq* values for octahedral hydrates of the ions
7. *Dq* calculated for tetrahedral coordination

8, 9. The thermodynamic stabilization in octahedral or tetrahedral fields

10. The octahedral site preference, or the difference between columns 8 and 9

[b] The octahedral site stabilization of Co^{+3} was estimated from the heat of hydration increment caused by the crystal field, and the tetrahedral site stabilization was taken to be the same as for Cr^{+3}.

SOURCE: T. M. Dunn, D. S. McClure, and R. G. Pearson, "Some Aspects of Crystal Field Theory," Harper & Row, New York, **1965**, p. 82. Used with permission.

level. In the tetrahedral (T_d) complex the ligands are not aimed directly at any of the orbitals but exert somewhat more influence on the t_{2g} orbitals than on the e_g orbitals (see Fig. 9.11). It can be shown that for a point-charge model:[30]

$$10Dq_{T_d} = \tfrac{4}{9}[10Dq_{O_h}] \tag{9.11}$$

In real molecules the conversion factor is not exactly $\frac{4}{9}$, but this is a good rough approximation for converting from one type to the other.

Nature of the ligands. Different ligands cause different degrees of splitting. This is shown in Fig. 9.14 illustrating the absorption spectra of CrL_6 species. There are two

[30] C. J. Ballhausen, "Introduction to Ligand Field Theory," McGraw-Hill, New York, **1962**, pp. 108–110.

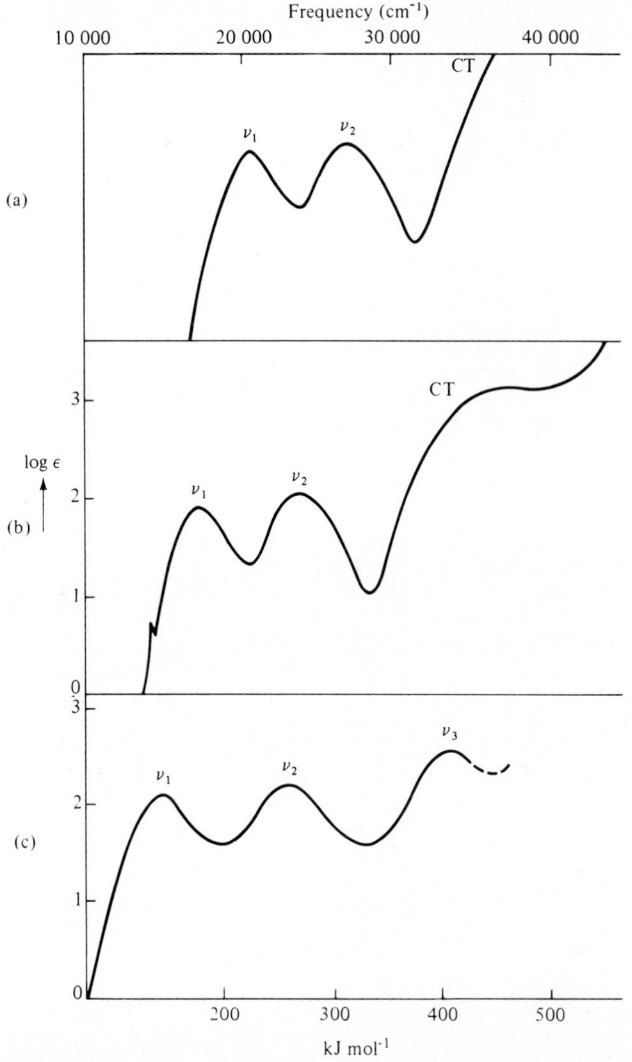

Fig. 9.14 Comparative spectra of some chromium(III) complexes: (a) $[Cr(en)_3]^{+3}$; (b) $[Cr(ox)_3]^{-3}$; (c) $[CrF_6]^{-3}$; v_1 corresponds to $10Dq$; CT = charge transfer band (see p. 456). [Spectra a and b from H. L. Schläfer, *Z. Physik. Chem.* (*Frankfurt am Main*), **1957**, *11*, 65. Reproduced with permission. Spectrum c sketched from data given by C. K. Jørgensen, "Absorption Spectra and Chemical Bonding in Complexes," Pergamon, Elmsford, N. Y., **1962**.]

absorption maxima since there are two transitions possible for $t_{2g}^3e_g^0 \rightarrow t_{2g}^2e_g^1$, differing in the extent of electron–electron repulsion in the excited state (see pp. 442–448). Note that there is a steady progression in frequency of the absorption maxima as the ligands change with the linking atoms progressing Cl → S → O → N → C, corresponding to a progressive increase in the value of $10Dq$ (Fig. 9.15). A more complete list of typical values of $10Dq$ with various ligands is given in Table 9.7. In general, it is possible to list ligands in order of

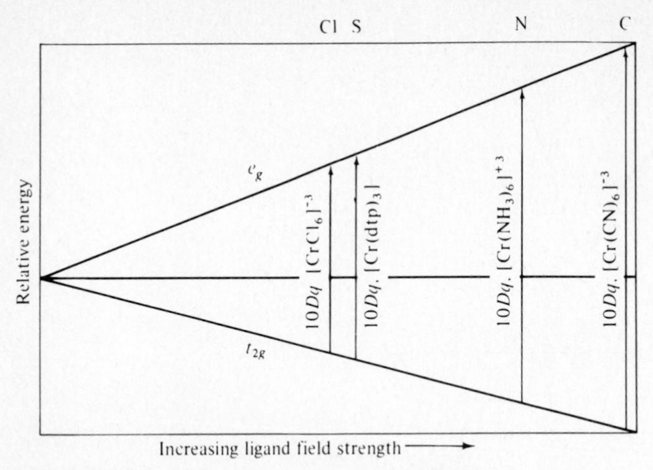

Fig. 9.15 Effect of ligand on magnitude of 10*Dq*.

Table 9.7 Some values of 10*Dq*, in kJ mol^{-1} (and cm^{-1}), for transition metal complexes[a]

Complex	10*Dq*		Complex	10*Dq*		Complex	10*Dq*	
$[CrCl_6]^{-3}$	158	(13 200)	$[MoCl_6]^{-3}$	230	(19 200)	$[WCl_6]^{-3}$	[b]	[b]
$[Cr(dtp)_3]$	172	(14 400)	$[Mo(dtp)_3]$	[b]	[b]	$[W(dtp)_3]$	[b]	[b]
$[CrF_6]^{-3}$	182	(15 200)	$[MoF_6]^{-3}$	[b]	[b]	$[WF_6]^{-3}$	[b]	[b]
$[Cr(H_2O)_6]^{+3}$	208	(17 400)	$[Mo(H_2O)_6]^{+3}$	[b]	[b]	$[W(H_2O)_6]^{+3}$	[b]	[b]
$[Cr(NH_3)_6]^{+3}$	258	(21 600)	$[Mo(NH_3)_6]^{+3}$	[b]	[b]	$[W(NH_3)_6]^{+3}$	[b]	[b]
$[Cr(en)_3]^{+3}$	262	(21 900)	$[Mo(en)_3]^{+3}$	[b]	[b]	$[W(en)_3]^{+3}$	[b]	[b]
$[CoCl_6]^{-3}$	[b]	[b]	$[RhCl_6]^{-3}$	243	(20 300)	$[IrCl_6]^{-3}$	299	(25 000)
$[Co(dtp)_3]$	170	(14 200)	$[Rh(dtp)_3]$	263	(22 000)	$[Ir(dtp)_3]$	318	(26 600)
$[Co(H_2O)_6]^{+3}$	218	(18 200)	$[Rh(H_2O)_6]^{+3}$	323	(27 000)	$[Ir(H_2O)_6]^{+3}$	[b]	[b]
$[Co(NH_3)_6]^{+3}$	274	(22 900)	$[Rh(NH_3)_6]^{+3}$	408	(34 100)	$[Ir(NH_3)_6]^{+3}$	490	(41 000)
$[Co(en)_3]^{+3}$	278	(23 200)	$[Rh(en)_3]^{+3}$	414	(34 600)	$[Ir(en)_3]^{+3}$	495	(41 400)
$[Co(CN)_6]^{-3}$	401	(33 500)	$[Rh(CN)_6]^{-3}$	544	(45 500)	$[Ir(CN)_6]^{-3}$	[b]	[b]

[a] Abbreviations: dtp = diethyldithiophosphate, en = ethylenediamine.
[b] These complexes are either unknown or the 10*Dq* values are not available.

increasing field strength in a *spectrochemical series.*[31] Although it is not possible to form a complete spectrochemical series of all ligands with one metal ion, it is possible to construct one from overlapping sequences, each illustrating a portion of the series:[32]

$$I^- < Br^- < S^{-2} < SCN^- < Cl^- < NO_3^- < F^- < OH^- < ox^{-2} < H_2O < NCS^-$$
$$< CH_3CN < NH_3 < en < bipy < phen < NO_2^- < phosph < CN^- < CO$$

[31] K. Fajans, *Naturwissenschaften*, **1923**, *11*, 165; R. Tsuchida, *Bull. Chem. Soc. Japan*, **1938**, *13*, 388, 436, 471; Y. Shimura and R. Tsuchida, *Bull. Chem. Soc. Japan*, **1956**, *29*, 311; C. K. Jørgensen, "Absorption Spectra and Chemical Bonding in Complexes," Pergamon, Elmsford, N.Y., **1962**.

[32] Abbreviations are: *ox* = oxalate, *en* = ethylenediamine, *bipy* = bipyridine, *phen* = *o*-phenanthroline, *phosph* = 4-methyl-2,6,7-trioxa-l-phosphabicyclo[2.2.2]octane.

Table 9.8 Values of f factors for various ligands

Ligand	f	Ligand	f
$\mathbf{Br}^{-\,a}$	0.72	$p\text{-}CH_3C_6H_4NH_2$	1.15
SCN^-	0.73	NC^-	1.15
$\mathbf{Cl}^-$	0.78	CH_3NH_2	1.17
$dsep^- = (C_2H_5O)_2P\mathbf{Se}_2^-$	0.8	$gly^- = NH_2CH_2CO_2^-$	1.18
$\mathbf{N}_3^-$	0.83	CH_3CN	1.22
$dtp^- = (C_2H_5O)_2P\mathbf{S}_2^-$	0.83	$Py = C_5H_5N$	1.23
$\mathbf{F}^-$	0.9	NH_3	1.25
$dtc^- = (C_2H_5)_2NC\mathbf{S}_2^-$	0.90	$en = NH_2CH_2CH_2NH_2$	1.28
$dmso = (CH_3)_2S\mathbf{O}$	0.91	$dien = NH(CH_2CH_2NH_2)_2$	1.29
$urea = (NH_2)_2C\mathbf{O}$	0.92	NH_2OH	1.30
CH_3COOH	0.94		
$C_2H_5\mathbf{O}H$	0.97	bipy = (2,2′-bipyridine structure)	1.33
$dmf = (CH_3)_2NCH\mathbf{O}$	0.98		
$ox^{-2} = C_2\mathbf{O}_4^{-2}$	0.99	phen = (1,10-phenanthroline structure)	1.34
$H_2\mathbf{O}$	1.00		
NCS^-	1.02	CN^-	~1.7

[a] The ligating atoms are shown in boldface type.
SOURCE: C. K. Jørgensen, "Oxidation Numbers and Oxidation States," Springer, New York, **1969**, pp. 84–85. Used with permission.

Although the spectrochemical series allows one to rationalize differences in spectra to a large extent and even permits some predictability, it presents serious difficulties in interpretation for the pure (i.e., ionic) crystal field theory. If the splitting of the d orbitals resulted simply from the effect of point charges (whether from ions or dipoles), one should expect that anionic ligands would exert the greatest effect. To the contrary, most of the anionic ligands lie on the low end of the spectrochemical series. Furthermore, OH^- lies below the neutral H_2O molecule. Again, NH_3 produces a greater field than H_2O, although the dipole moments are the reverse ($\mu_{NH_3} = 4.90 \times 10^{-30}$ C m; $\mu_{H_2O^-} = 6.17 \times 10^{-30}$ C m). Such results cause us to doubt the original assumption of purely electrostatic interactions between ligands and central metal ions. Later we shall explore this point somewhat further.

Jørgensen[33] has attempted to quantify the spectrochemical series relative to water as a standard ligand with a field factor (f) of 1.00. Values range from about 0.7 for weak-field bromide ions to about 1.7 for the very strong-field cyanide ion (Table 9.8).

A fourth factor in determining the size of the crystal field splitting is the nature of the metal ion involved. Within a given transition series the differences are not great, but noticeable changes occur between the congeners in a given group in progressing $3d \rightarrow 4d \rightarrow 5d$. This may be readily seen from the values in Table 9.7. In progressing from Cr to Mo or Co to Rh the value of Dq increases about 50%. Likewise, the values for Ir complexes are some 25% greater than for Rh complexes. This is a general trend for all transition elements and the most important result is that complexes of the second and third transition series

[33] C. K. Jørgensen, "Absorption Spectra and Chemical Bonding in Complexes," Pergamon, Elmsford, N.Y., **1962**, and "Oxidation Numbers and Oxidation States," Springer, New York, **1969**.

Table 9.9 Values of *g* factors for various metal ions[a]

$3d^5$	Mn(II)	8.0	$4d^6$	Ru(II)	20
$3d^8$	Ni(II)	8.7	$3d^3$	Mn(IV)	23
$3d^7$	Co(II)	9	$4d^3$	Mo(III)	24.6
$3d^3$	V(II)	12.0	$4d^6$	Rh(III)	27.0
$3d^5$	Fe(III)	14.0	$4d^3$	Tc(IV)	30
$3d^3$	Cr(III)	17.4	$5d^6$	Ir(III)	32
$3d^6$	Co(III)	18.2	$5d^6$	Pt(IV)	36

[a] In units of kK ($=1000\ cm^{-1}$).

SOURCE: C. K. Jørgensen, "Oxidation Numbers and Oxidation States," Springer, New York, **1969**, pp. 84–85. Used with permission.

(4*d* and 5*d*) are almost exclusively low spin compared to large numbers of both high- and low-spin complexes for the first transition series.[34]

Jørgensen[33] has provided a measure of the tendency of a metal ion to form a low-spin complex. These *g* factors (Table 9.9) provide an estimate of the value of $10Dq$ for an octahedral complex when combined with the *f* value for the appropriate ligands:

$$10Dq = f_{\text{ligand}} \times g_{\text{ion}} \tag{9.12}$$

By combining an estimate of $10Dq$ from Eq. 9.12 with the pairing energy given in Table 9.5 it is possible to rationalize the high- or low-spin state of various metal complexes as well as predict the magnetic behavior of new complexes.

Evidence for crystal field stabilization

The ability to account for magnetic and spectral properties of complexes is strong evidence for the splitting of the degenerate *d* orbitals of transition metal ions. The correlation of the magnitude of $10Dq$ obtained from spectral data,[35] the pairing energy, and the spin state of the complex presents compelling evidence for the advantages of a crystal field model or some modification of that model. In this section and the one to follow further evidence obtained from the structures and energetics of complexes will be presented.

The first indication that crystal field stabilization energy (CFSE) might be of importance in transition metal compounds was obtained from the computation of the lattice energies. As we have seen in Chapter 3, although the prediction of lattice energies is not highly accurate, it was found that such predictions could be made much better for ions such as Na^+, K^+, Ca^{+2}, Mn^{+2}, and Zn^{+2} than for Cr^{+2}, Fe^{+2}, Co^{+2}, Ni^{+2}, and Cu^{+2}. Wherever a serious discrepancy is found, it may be attributable to the CFSE for the metal ion. Those ions that do not show such discrepancies are those with d^0, d^5 (weak field), and

[34] In any system like this in which there are many variables, it is impossible to assign sole responsibility for a phenomenon to one factor. Other factors of some importance in leading to low-spin complexes for 4*d* and 5*d* transition elements are the lower spin-pairing energy, higher coordination numbers, and tendency toward higher oxidation states in the latter.

[35] The relation between $10Dq$ and the spectrum of $[Ti(H_2O)_6]^{+3}$ has been discussed on p. 372. The treatment of spectra of the more general case, d^n, is discussed on p. 442.

d^{10} configurations, which are not split by a crystal field. Consider the lattice energies of the halides from CaX_2 to ZnX_2. Inasmuch as we might expect a *gradual decrease* in ionic radius from Ca^{+2} to Zn^{+2} (p. 40), we should also expect a *gradual and smooth increase* in lattice energy based on the Born-Landé equation (p. 62). However, as shown by Fig. 9.16, the expected smooth curve is not observed. The ions Ca^{+2}, Mn^{+2}, and Zn^{+2} lie on a curve that is nearly a straight line, but deviations from this curve are maximized in two places: in the region of V^{+2} and the region of Ni^{+2}. Table 9.3 indicates that for a weak octahedral field (recall that the halide ions are on the weak end of the spectrochemical series) $V^{+2}(d^3)$ and $Ni^{+2}(d^8)$ have the greatest CFSE ($12Dq$). The d^2, d^4, d^7, and d^9 ions have somewhat less (6 and $8Dq$) and the d^0, d^5, and d^{10} have $0Dq$ CFSE, qualitatively confirming the shape of the curve within the unfortunately large experimental errors.

Somewhat better data are available for the enthalpies of hydration of transition metal ions. Although this enthalpy is measured at (or more properly, extrapolated to) infinite dilution, only six water molecules enter the coordination sphere to form an octahedral aqua complex. The enthalpies of hydration are thus closely related to the enthalpy of formation of the hexaqua complex. If the values of ΔH_H for the $+2$ and $+3$ ions of the

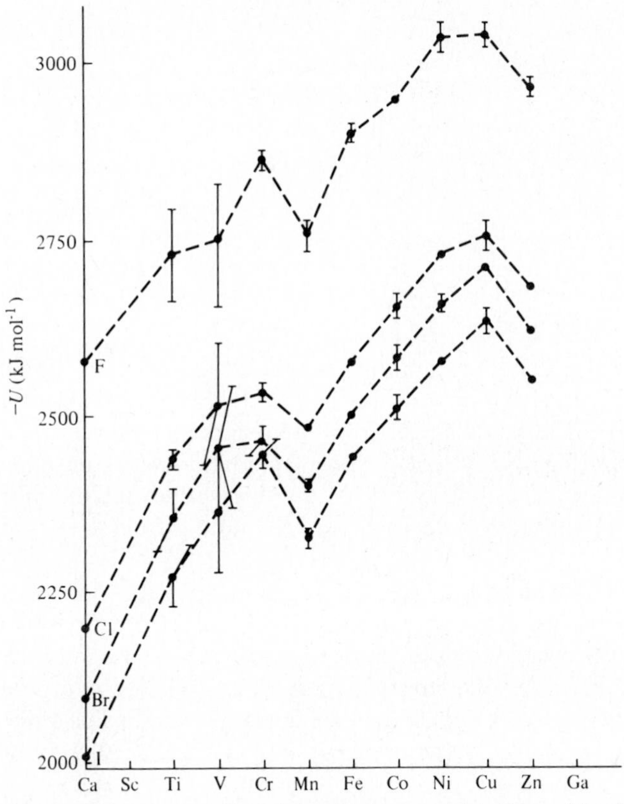

Fig. 9.16 Lattice energies of the divalent metal halides of the first transition series. Vertical bars indicate uncertainties in experimental values. [Modified from P. George and D. S. McClure, *Progr. Inorg. Chem.*, **1959**, *1*, 381. Reproduced with permission.]

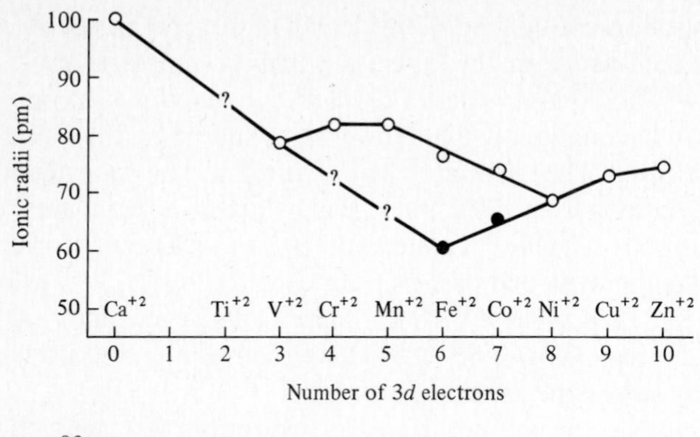

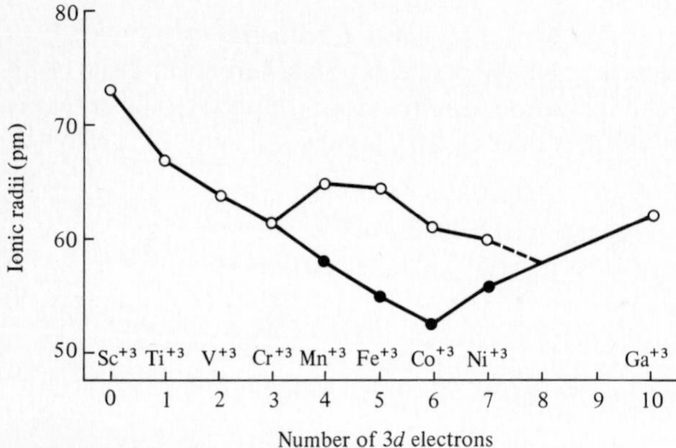

Fig. 9.17 Radii of the divalent ions Ca^{+2} to Zn^{+2} (above) and the trivalent ions Sc^{+3} to Ga^{+3} as a function of the number of *d* electrons. Low-spin ions are indicated by solid circles. [From R. D. Shannon and C. T. Prewitt, *Acta Crystallogr.*, **1970**, *B26*, 1076. Reproduced with permission.]

first transition elements (except Sc^{+2} which is unstable) are plotted as a function of atomic number, curves much like those shown in Fig. 9.16 are obtained. If one subtracts the predicted CFSE from the experimental enthalpies, the resulting points lie very nearly on a straight line from Ca^{+2} to Zn^{+2} and from Sc^{+3} to Fe^{+3} (the +3 oxidation state is unstable in water for the remainder of the first transition series). Many thermodynamic data for coordination compounds follow this pattern.[36] In each case the double-humped curve can be ascribed to CFSE.

A slightly different form of the typical two-humped curve is shown by the ionic radii of the 3*d* divalent metals. These are plotted in Fig. 9.17 (from Table 3.4). For both dipositive and tripositive ions there is a steady decrease in radius for the strong field case until the t_{2g}^6 configuration is reached. At this point the next electron enters the e_g level, into an orbital directed at the ligands, repelling them and causing an increase in the effective radius of the metal. In the case of high-spin ions the increase in radius occurs with the $t_{2g}^3 e_g^1$ configuration for the same reason.

---

[36] For a general discussion of the consequences of splitting of *d* orbitals, see P. George and D. S. McClure, *Progr. Inorg. Chem.*, **1959**, *1*, 381.

Although Fig. 9.17 was drawn from the composite data of many compounds, the same effect can be seen from a single series of complexes formed with the ligand $N(CH_2CH_2N{=}CHC_5H_4N)_3$. The geometry of the complexes formed with this ligand is discussed in some detail in Chapter 10, but for now we may note that six of the nitrogen atoms coordinate to the metal in an approximately octahedral arrangement and are thus directed toward the e_g orbitals. The seventh nitrogen atom is located in the center of an octahedral face ("axial coordination") and thus interacts most directly with a t_{2g} orbital. Figure 9.18 illustrates the metal–nitrogen bond distances in a series of complexes of this type. The shorter distances represent the lengths of the "octahedral" bonds. The six bonds of this type in a complex are not identical in length (hence the two values shown for each compound), but the slight difference is not important for the present discussion and may be ignored. The longer, axial "bond" in each complex represents the distance to the seventh nitrogen atom. There are several effects to be seen in this series: (1) There is a general decrease in size progressing across the series, to be expected from the usual decrease in size across a series. (2) An increase in electrons in orbitals directed at the nitrogen atoms causes an increase in bond length, both for the e_g orbitals (octahedral nitrogen atoms) and the t_{2g} orbitals (axial nitrogen atom). (3) Specifically, in the case of the Fe^{+2} complex, the only low-spin complex in the series, there is a dramatic decrease in one distance and a concomitant increase in the other. The low-spin t_{2g}^6 configuration maximizes electron density toward the axial nitrogen atom while minimizing it in the direction of the octahedral nitrogen atoms.

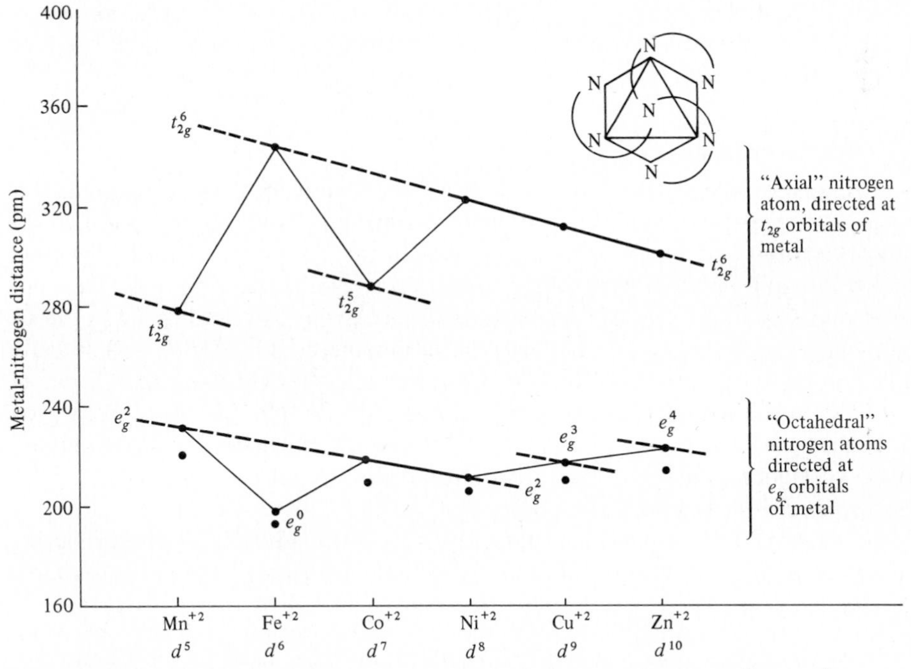

Fig. 9.18 Effect of orbital occupancy on metal–ligand distance. Upper set: distance from metal to apical nitrogen atom. Lower set: distance to remaining six "octahedral" nitrogen atoms.

Further evidence of the importance of CFSE comes from the stability of particular oxidation states. In aqueous solution Co(III) is unstable with respect to reduction by water to form Co(II). Although there are several energy terms involved, this may be viewed as a reflection of the high third ionization energy of cobalt. If various moderate to strong field ligands are present in the solution, however, the Co(III) ion is perfectly stable; in fact, in some cases it is difficult or impossible to prevent the spontaneous transformation of Co(II) to Co(III).

For example, the appropriate oxidation *emf*s are:

$$[Co(H_2O)_6]^{+2} \rightleftharpoons [Co(H_2O)_6]^{+3} + e^- \qquad \mathscr{E}^0 = -1.84 \qquad \textbf{(9.13)}$$

$$[Co(EDTA)]^{-2} \rightleftharpoons [Co(EDTA)_6]^{-} + e^- \qquad \mathscr{E}^0 = -0.60 \qquad \textbf{(9.14)}$$

$$[Co(ox)_3]^{-4} \rightleftharpoons [Co(ox)_3]^{-3} + e^- \qquad \mathscr{E}^0 = -0.57 \qquad \textbf{(9.15)}$$

$$[Co(phen)_3]^{+2} \rightleftharpoons [Co(phen)_3]^{+3} + e^- \qquad \mathscr{E}^0 = -0.42 \qquad \textbf{(9.16)}$$

$$[Co(NH_3)_6]^{+2} \rightleftharpoons [Co(NH_3)_6]^{+3} + e^- \qquad \mathscr{E}^0 = -0.10 \qquad \textbf{(9.17)}$$

$$[Co(en)_3]^{+2} \rightleftharpoons [Co(en)_3]^{+3} + e^- \qquad \mathscr{E}^0 = +0.26 \qquad \textbf{(9.18)}$$

$$[Co(CN)_5]^{-3} + CN^- \rightleftharpoons [Co(CN)_6]^{-3} + e^- \qquad \mathscr{E}^0 = +0.83 \qquad \textbf{(9.19)}$$

Note that the order of ligands in Eqs. 9.13–9.19 is approximately that of the spectrochemical series and, hence, that of increasing crystal field stabilization energies. The oxidation of Co(II) to Co(III) results in a change from high to low spin. We think of the oxidation as taking place in two steps, the first being the rearrangement of electrons to the low-spin state and the second the removal of the seventh electron to produce Co(III):

$$Co^{II}(t_{2g}^5e_g^2) \longrightarrow Co^{II}(t_{2g}^6e_g^1) \xrightarrow{-e} Co^{III}(t_{2g}^6e_g^0) \qquad \textbf{(9.20)}$$

This is not to imply that the reaction actually occurs in this manner but that we may consider the thermodynamics as the sum of this hypothetical sequence. Now the first step involves the pairing of electrons and the pairing energy will be in part compensated for by the extra CFSE of the low-spin configuration (18*Dq* versus 8*Dq*). Strong field ligands favor this step. The second step is the removal of the electron occupying the antibonding e_g level. This step is endothermic because of the high ionization energy (Co^{+2} to Co^{+3}), but the CFSE will favor the ionization (18*Dq* to 24*Dq*). The stronger the field, the larger the value of *Dq*. It should be pointed out that CFSE is only one of a number of factors affecting the *emf*. In particular, entropy effects associated with chelate rings can be important and are largely responsible for the fact that the order of ligands in Eqs. 9.13–9.19 is different from that of the spectrochemical series.[37]

Further evidence will be presented later to indicate the advantages of considering coordination compounds in terms of splitting of *d* orbital degeneracy, but one more simple

---

[37] P. A. Rock, *Inorg. Chem.*, **1968**, *7*, 837, has shown how both the entropy effects and the stabilization from the ligand field may be estimated and correlated nicely with the known *emf* values for Co(II)/Co(III) and Fe(II)/Fe(III) systems. Note, however, that difficulties are occasionally encountered (J. J. Kim and P. A. Rock, *Inorg. Chem.*, **1969**, *8*, 563).

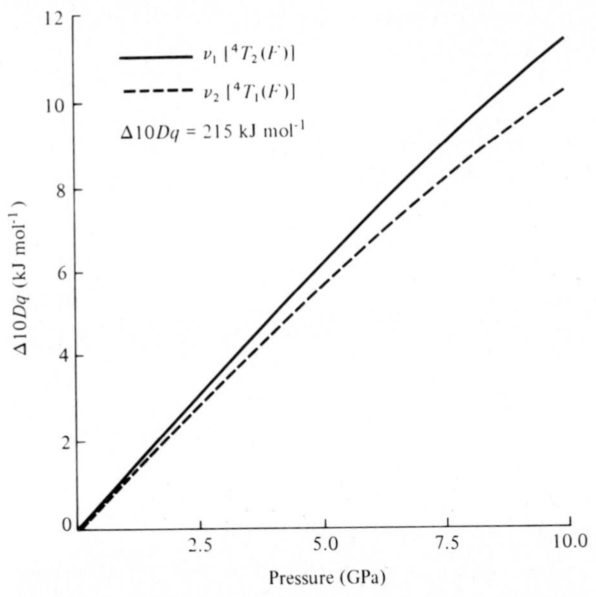

Fig. 9.19 Effect of pressure on electronic transitions of Cr(III) in Al_2O_3 (ruby). The spectrum resembles that of $[Cr(ox)_3]^{-3}$ shown in Fig. 9.14 and v_1 and v_2 refer to the absorption maxima in that spectrum. [From H. G. Drickamer, *Solid State Physics*, **1965**, *17*, 1. Reproduced with permission.]

example will be presented here to conclude this discussion. It is possible to study the spectra and magnetic properties of coordination compounds under pressures of as much as 1.5×10^{10} Pa.[38] Although the strong repulsive forces of the atoms prevent much compression, the results are exactly what we would expect as a result of the ligands being pushed closer to the metal and interacting more strongly with the *d* orbitals of the latter. The spectra are thus altered (Fig. 9.19) with the absorption maxima shifted toward the blue or high-energy end of the spectra.[39] The magnitude of separation of the energy levels, $10Dq$, is correspondingly increased. The increase in $10Dq$, the rate of increase with distance (proportional to r^{-5}), and the increase in integrated intensity (proportional to r^{-8}) are all in reasonably good agreement with a crystal field perturbation of the *d* orbitals.[38]

Octahedral versus tetrahedral coordination

Thus far we have discussed three types of complexes: high-spin octahedral, low-spin octahedral, and high-spin tetrahedral. For the first transition series large numbers of each type of complex are known, together with some complexes of other geometries to be considered later. There are several factors that influence the adoption of tetrahedral or octahedral coordination, and occasionally the balance between opposing factors is a delicate one.

[38] H. G. Drickamer, *Solid State Physics*, **1965**, *17*, 1.

[39] As we shall see later in the discussion of the interpretation of spectra, the Racah parameters become important in spectra of ions having more than one *d* electron and so exceptions to this oversimplification are found.

Table 9.10 Relative crystal field stabilization energies (CFSE) and resulting octahedral site stabilization energies (OSSE) for various electron configurations, d^1–d^9

Configuration	CFSE, tetrahedral complex[a]		CFSE, octahedral high-spin (Dq_{O_h})	OSSE, high-spin (Dq_{O_h})	CFSE, octahedral low-spin (Dq_{O_h})	OSSE, low-spin (Dq_{O_h})
	Units = Dq_{T_d}	Units = Dq_{O_h}				
d^1	6	2.67	4	1.33	4	1.33
d^2	12	5.33	8	2.67	8	2.67
d^3	8	3.55	12	8.45	12	8.45
d^4	4	1.78	6	4.22	$16 - p$	$14.22 - p$
d^5	0	0	0	0	$20 - 2p$	$20.00 - 2p$
d^6	6	2.67	4	1.33	$24 - 2p$	$21.33 - 2p$
d^7	12	5.33	8	2.67	$18 - p$	$12.67 - p$
d^8	8	3.55	12	8.45	12	8.45
d^9	4	1.78	6	4.22	6	4.22

[a] Tetrahedral stabilization energies calculated in Dq_{O_h} units by the assumption that $Dq_{T_d} = \frac{4}{9}Dq_{O_h}$.

From a purely electrostatic viewpoint, octahedral coordination is favored simply on the basis of six versus four ligands. This conclusion is not dependent on the electrostatic model, however, for a covalent model would predict that, in general, six bonds would be stronger than four.

Opposing octahedral coordination is ligand–ligand repulsion. Steric requirements favor the formation of four tetrahedral bonds rather than six octahedral bonds if the ligands are bulky. This situation is very similar to the choice of coordination number in ionic crystals on the basis of the radius ratio of the cation and anion (see p. 83).

Tetrahedral complexes are always high spin. As a result, the maximum CFSE can only be $12Dq$, which, converted to octahedral field equivalents by the $\frac{4}{9}$ conversion factor, is only about $5Dq$. The CFSE is therefore only rarely as important in tetrahedral complexes as it is in octahedral ones. In fact, when comparing a given ion in a tetrahedral field with the same ion in an equivalent (i.e., same ligands) octahedral hole, the ion is always at least as stable—usually more so—in the octahedral hole. The difference in energy, which always favors the octahedral case, is termed the *octahedral site stabilization energy* (OSSE), since the term was originally applied to cationic preference for octahedral holes in anionic lattices. For some configurations, such as d^1, d^2, d^5, d^6, and d^7, the advantage of the octahedral high-spin arrangement is little or nothing. Others, such as d^3 and d^8, are strongly favored to be octahedral (see Table 9.10). Some estimates of OSSE in kJ mol^{-1} may be obtained from Table 9.6.

High cationic charges tend to increase the value of Dq and thus favor octahedral coordination.[40] Furthermore, high charges tend to increase the possibility of spin pairing in the complex. Certain spin-paired configurations such as d^5 and d^6 yield extremely large amounts of CFSE and strongly favor octahedral coordination.

[40] An exception to this and the next rule lies in extremely low oxidation state complexes with π-bonding ligands such as cyanides and carbonyls. For these complexes the crystal field approach provides an extremely poor approximation and they are better discussed in terms of molecular orbital theory (see p. 412).

Some examples of the operation of these factors in determining the nature of complex compounds may be illuminating. Consider Mn^{+2} and Fe^{+3}, both d^5 metal ions. As a result of the large exchange stabilization of the half-filled d subshell, the pairing energies are large and high-spin complexes are strongly favored. For Mn^{+2}, the $+2$ charge on the ion does not help produce large ligand fields and almost all of the complexes known are high spin. A few low-spin complexes are known—those with the very strongest field ligands such as the cyanide ion. With ordinary ligands such as water, high-spin octahedral complexes result. These provide no CFSE, and so it is not surprising that if the ligands are even slightly bulky, like the chloride ion, for example, tetrahedral complexes, MnX_4^{-2}, are formed.

Iron(III) resembles Mn(II) in its tendency to resist spin pairing, but the increased ionic charge tends to favor octahedral coordination and to increase the effective field of the ligands. It is not surprising, therefore, to find some tetrahedral species such as $FeCl_4^-$ but a large number of octahedral complexes and a larger fraction of the latter to be low spin.

Changing the electron configuration by one electron to Co^{+3}, d^6, causes a rather abrupt change in properties. In contrast to Fe(III) *all* complexes of Co(III) are octahedral and almost all are low spin. The reasons are the more favorable spin-pairing energy and the more favorable CFSE for the octahedral case, *especially in the strong field case.* Diamagnetic complexes such as $[Co(NH_3)_6]^{+3}$ thus typify the coordination chemistry of Co(III). The few high-spin octahedral complexes known, such as $[CoF_6]^{-3}$ and $[Co(H_2O)_3F_3]$, have ligands from the very weak field end of the spectrochemical series.

Addition of one electron, $Co^{+3} + e^- \rightarrow Co^{+2}$, results in completely different complexes. The OSSE for low-spin complexes is almost cut in half (see Table 9.10). Again, the lowered charge on Co^{+2} does not favor strong fields and spin pairing. Co(II) complexes are therefore typically high-spin[41] octahedral with ligands such as NH_3 and H_2O and tend to be unstable with respect to oxidation if the ligands are strong field (see p. 390). High-spin octahedral complexes are not particularly stabilized by CFSE compared with tetrahedral coordination, so it is not surprising to find that halides and pseudohalides form complexes such as $[CoCl_4]^{-2}$, $[CoBr_4]^{-2}$, and $[Co(NCS)_4]^{-2}$.

Site selection in spinels and other systems

Further evidence concerning the relative stability of octahedral versus tetrahedral coordination can be found by examining site preference in certain crystals, such as the spinels. Spinels have the formula $A^{II}B_2^{III}O_4$, where A^{II} can be a Group IIA metal or a transition metal in the $+2$ oxidation state and B^{III} is a Group IIIA metal or a transition metal in the $+3$ oxidation state. The oxide ions form a close-packed cubic lattice with eight tetrahedral holes and four octahedral holes per "molecule" of AB_2O_4 (see p. 79). In a so-called *normal* spinel such as $MgAl_2O_4$, the Mg^{+2} ions occupy one-eighth of the tetrahedral holes and the Al^{+3} ions occupy half of the octahedral holes. This is the most stable arrangement inasmuch as it yields a coordination number of 4 for the divalent ion and 6 for the trivalent

[41] The hexacyano complex might be expected to be spin paired. All evidence indicates that the species present in solution is the pentacyano complex, $[Co(CN)_5]^{-3}$ or $[Co(CN)_5H_2O]^{-3}$, and that no hexacoordinate low-spin species are known. In either tetrahedral or octahedral complexes, Co(II) is expected to form distorted complexes as a result of the Jahn-Teller effect (see p. 396).

Table 9.11 Theoretical and experimental cation distributions in spinels $A^{II}B^{III}_2O_4$[a]

	B														
	Al^{+3}		Ga^{+3}		Fe^{+3}		Cr^{+3}		Mn^{+3}		V^{+3}		Co^{+3}	
A	Exp.	Th.	Exp.	Th.	Exp.	Th.	Exp.	Th.	Exp.	Th.	Exp.	Th.	Exp.	Th.
Mg^{+2}	0.88I	O	I	O	I	O	N	N		N	N	N		N
Zn^{+2}	N	O	N	O	N	O	N	N	N, T	N	N	N		N
Cd^{+2}	N	O	N	O	N	O	N	N	N	N		N		N
Mn^{+2}	N	O		O	I	O	N	N	N	N	N	N		N
Fe^{+2}	N	I		I	I	I	N	N		N	N	I + N		N
Co^{+2}	N	I		I	I	I	N	N		N		I + N	N	N
Ni^{+2}	$\frac{3}{4}$I + $\frac{1}{4}$N	I	I	I	I	I	N	N		I + N		I		I
Cu^{+2}	I	I		I	0.86I, T	I	N	N		N		I		N

[a] Abbreviations: N = normal; I = inverse; O = no prediction made by OSSE; T = tetragonal distortion from Jahn-Teller effect.

SOURCE: T. M. Dunn, D. S. McClure, and R. G. Pearson, "Some Aspects of Crystal Field Theory," Haper & Row. New York, **1965**, p. 86, and A. F. Wells, "Structural Inorganic Chemistry," 3rd ed., Oxford University Press, London, **1962**, p. 489. Used with permission.

ion [cf. $Be(H_2O)_4^{+2}$, and $Al(H_2O)_6^{+3}$, p. 366]. Most interesting, therefore, are those spinels with the so-called *inverse* structure in which the A^{II} ions and *one-half* of the B^{III} ions have exchanged places; i.e., the A^{II} ions occupy octahedral holes along with one-half of the B^{III} ions while half of the B^{III} ions are in tetrahedral holes. An example of an inverse spinel is $NiFe_2O_4$. The oxide ions provide a moderately weak crystalline field, and so both the Ni^{+2}, d^8, and Fe^{+3}, d^5, remain high spin. For the d^5 Fe^{+3} ion the CFSE is zero for both tetrahedral and octahedral sites but the d^8 Ni^{+2} ion has an OSSE of $8.45Dq$ (see Table 9.10) or approximately 96 kJ mol^{-1} (based on the assumption that O^{-2} and H_2O provide similar fields; see Table 9.6, last column). This CFSE advantage for the Ni^{+2} ion in the octahedral holes is sufficient to invert the structure.

A second example of an inverse spinel is magnetite, Fe_3O_4. Although both A and B ions in this case are iron, some are ferrous and others ferric: $Fe^{II}Fe^{III}_2O_4$. Again, the d^5 Fe^{III} ion provides no CFSE in either tetrahedral or octahedral weak fields. The d^6 Fe^{II} ion, however, is stabilized to the extent of $4Dq_{O_h}$ versus $6Dq_{T_d}$ for a new OSSE of $1.33Dq_{O_h}$. This provides only a few kJ mol^{-1} of stabilization for the inverse structure but appears to be sufficient.

Not all transition metal spinels have the inverse structure. All the chromium spinels, $A^{II}Cr^{III}_2O_4$, have the normal structure as a result of the strong octahedral site preference of Cr^{III} (ca. 188 kJ mol^{-1}; see Table 9.6). Two other metal oxides, Mn_3O_4 and Co_3O_4, also have the normal structure. In the former the Mn^{III} ion provides an OSSE of $4.22Dq$.[42] In the case of Co_3O_4 the Co^{III} is *spin paired* resulting in the extra stabilization of d^6 in a strong field. As a result the CFSE favors Co^{III} in the octahedral holes. The difference between Co^{II} in tetrahedral versus octahedral holes is not significant, and so the normal structure is preferred.

Further examples of spinel structures are listed in Table 9.11. It will be noted that the correlation between experimental results and predictions by crystal field theory is remarkable especially when it is remembered that the CFSE contributes only about 5–10% to the total bonding energy of the system. This is a point that is easy to forget. There are many

[42] In this as well as earlier examples, the reader should verify these results to his own satisfaction.

factors other than *d* orbital splittings that can affect the total energy and the preferred structure of a transition metal compound and it is accurate to use CFSE only when these other factors are reasonably constant. This is not always the case, and for the spinels that seemingly violate the *d* orbital splitting prediction, it can be shown that a closer examination of all the energies removes the apparent discrepancy.[43] (See Problem 9.27.)

A final example of the importance of *d* orbital splittings on the selection of tetrahedral versus octahedral coordination is found in the examination of solutions of transition metal ions in molten salts. These solutions may be considered as disordered crystal lattices in which long-range order has been destroyed, but local order prevails. Transition metals can then occupy (or create) either tetrahedral or octahedral holes. Alternatively, they may be viewed as analogous to aqueous solutions but with much larger concentrations of ligating anions than could ever be produced in the latter.

Extensive studies of the absorption spectra of transition metal ions in molten salts have been undertaken by Gruen and co-workers.[44] Their findings are readily interpretable in terms of stabilization from the field splitting of the metal *d* orbitals by the ligands. The chloride ion provides a rather weak field, so all of the metal ions are high-spin. Some metal ions show a preference for 6 coordination and others for 4 coordination, and some exist as equilibria between 6- and 4-coordinate complexes. Since the concentration of chloride ions is constant and size effects are unlikely to be too important (the metal ions do not differ too much in size), the main factors governing the selection of coordination number 4 or 6 lie in the size of the octahedral site stabilization energy. Chromium(III) exists only as the 6-coordinate, octahedral $[CrCl_6]^{-3}$ ion as might be expected from the very great OSSE (Table 9.10) and the $+3$ ionic charge. Two other tripositive ions, Ti^{+3} and V^{+3}, also form octahedral species, but since the OSSE is somewhat smaller, tetrahedral species are also known and equilibria are established:

$$[TiCl_6]^{-3} \rightleftharpoons [TiCl_4]^{-} + 2Cl^{-} \tag{9.21}$$

As expected, the divalent species tend to favor tetrahedral coordination. Only V^{+2} with the largest possible OSSE is able to form an octahedral complex and that only in equilibrium with a tetrahedral species:

$$[VCl_6]^{-4} \rightleftharpoons [VCl_4]^{-2} + 2Cl^{-} \tag{9.22}$$

For all the other divalent ions of the first transition series, only 4-coordinate complexes such as $[CoCl_4]^{-2}$ and $[MnCl_4]^{-2}$ are known.

In the case of both the spinels and the molten chloride complexes, further strong support for crystal field theory comes from the distortions from perfect octahedral or tetrahedral symmetry about the metal ion. The factor responsible for these distortions is discussed in the next section.

[43] See A. Navrotsky and O. J. Kleppa, *J. Inorg. Nucl. Chem.*, **1967**, *29*, 2701; **1968**, *30*, 479; T. M. Dunn et al., "Some Aspects of Crystal Field Theory," Harper & Row, New York, **1965**, p. 85, and references cited therein. For the "classic" papers on spinels see D. S. McClure, *J. Phys. Chem. Solids*, **1957**, *3*, 311, and J. D. Dunitz and L. E. Orgel, *J. Phys. Chem. Solids*, **1957**, *3*, 318. Also see J. D. Dunitz and L. E. Orgel, *Adv. Chem. Inorg. Radiochem.*, **1960**, *3*, 1.

[44] See D. M. Gruen, *Pure Appl. Chem.*, **1963**, *6*, 23, or D. M. Gruen, in "Fused Salts," B. R. Sundheim, ed., McGraw-Hill, New York, **1964**, pp. 301–339.

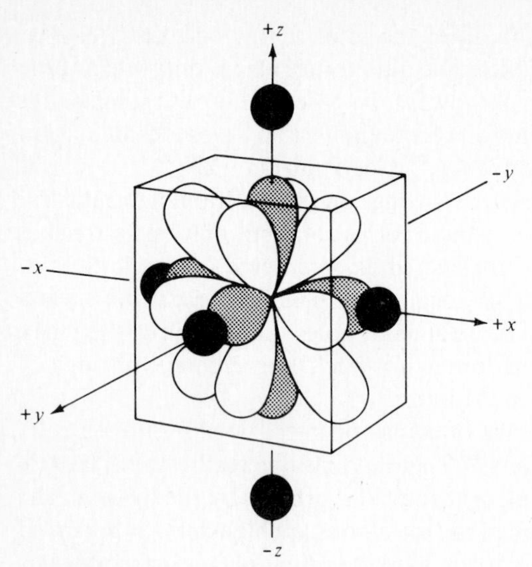

Fig. 9.20 Geometric arrangement of ligands of tetragonal symmetry (z-out).

Tetragonal distortions from octahedral symmetry

If two *trans* ligands in an octahedral complex (for example, those along the z axis) are moved either toward or away from the metal ion, the resulting complex is said to be tetragonally distorted. Ordinarily such distortions are not favored since they only result in a loss of bonding energy. In certain situations, however, the presence of a Jahn-Teller effect favors such a distortion.

The Jahn-Teller theorem[45] states that for a nonlinear molecule in an electronically degenerate state, distortion must occur to lower the symmetry, remove the degeneracy, and lower the energy. Consider Fig. 9.20 in which the ligands on the z axis have moved out. As a result, they interact less with those orbitals having a z component, i.e., the d_{z^2}, d_{xz}, and d_{yz}, and these orbitals are stabilized (Fig. 9.21). As a result of the "center of gravity" rule, those orbitals without a z component, $d_{x^2-y^2}$ and d_{xy}, will be raised a corresponding amount. It is not possible, *a priori*, to predict the extent of these splittings since it is not possible to predict the amount of distortion that will take place. We can say that the splitting of the upper, e_g, orbitals (δ_1) will be somewhat larger than that of the lower, t_{2g}, orbitals (δ_2) and that both will be relatively small with respect to $10Dq$.

Distortion can proceed only to a limited extent before being balanced by other energy factors. If we assume a hypothetical undistorted complex subject to Jahn-Teller distortion (which can, of course, have no real existence since the distorted form will always lie lower in energy) the atoms will be placed at equilibrium positions such that bonding energies (exclusive of Jahn-Teller effects) are maximized and repulsions minimized. If distortion is now allowed to occur, the complex will *gain stabilizing energy* from the Jahn-Teller splitting of the degener-

[45] H. A. Jahn and E. Teller, *Proc. Roy. Soc.*, **1937**, *A161*, 220; H. A. Jahn, *Proc. Roy. Soc.*, **1938**, *A164*, 117. See L. E. Orgel, "An Introduction to Transition-Metal Chemistry," 2nd ed., Wiley, New York, **1966**, pp. 57–61. C. J. Ballhausen, "Introduction to Ligand Field Theory," McGraw-Hill, New York, **1962**, p. 193.

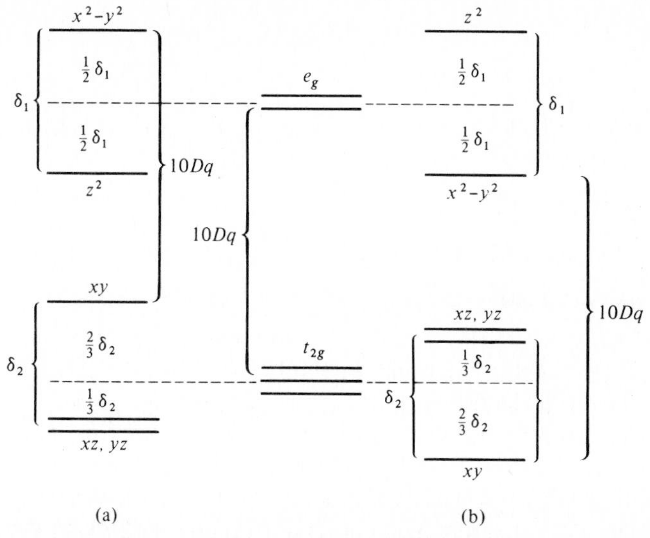

Fig. 9.21 Orbital energy diagram for a tetragonal field: (a) z ligands out; (b) z ligands in. $10Dq \gg \delta_1 > \delta_2$; the drawing is not to scale.

ate levels, but there will be a tendency for the other energy terms to be *less* favorable. Hence the condition of minimum energy is achieved when the rate of gain in energy from the Jahn-Teller distortion is balanced by the loss from other factors.

The Jahn-Teller theorem *per se* does not predict which type of distortion will take place other than that the center of symmetry will remain. The z ligands can move out as in the example discussed above or the z ligands can move in. In the latter case the splitting is similar but the secondary splitting within the e_g and t_{2g} levels is inverted (see Fig. 9.21).

It is possible to view the Jahn-Teller effect in terms of a physical picture that may be intuitively more acceptable than an abstract theorem.[46] Note that an electron configuration susceptible to Jahn-Teller distortion,[47] such as d^1, is not spherically symmetrical. Let us consider the odd electron to be present in the d_{z^2} orbital. In the approach of six ligands to form a quasioctahedral complex, the ligands approaching along the z axis will be somewhat repelled by the presence of the odd electron. Viewed somewhat differently, the electronegativity or attraction of the metal ions for the electrons of the ligands will be less in the z direction than in the x and y directions. This anisotropic electronegativity will result in stronger bonds to the x and y ligands and weaker bonds to the z ligands. The former are therefore expected to be shorter and the latter somewhat longer ("z-out"). Conversely, if we initially had placed the odd electron in the $d_{x^2-y^2}$ orbital, the anisotropic electronegativity would have acted in the reverse manner and the shortening would have occurred along the z axis ("z-in"). Although this qualitative picture has the advantages of being somewhat easier to visualize, it fails to give the more quantitative results of Fig. 9.21.

[46] For a more nearly complete discussion along these lines, see J. D. Dunitz and L. E. Orgel, *Adv. Inorg. Chem. Radiochem.*, **1960**, *2*, 1.

[47] A complete listing of such configurations is given later.

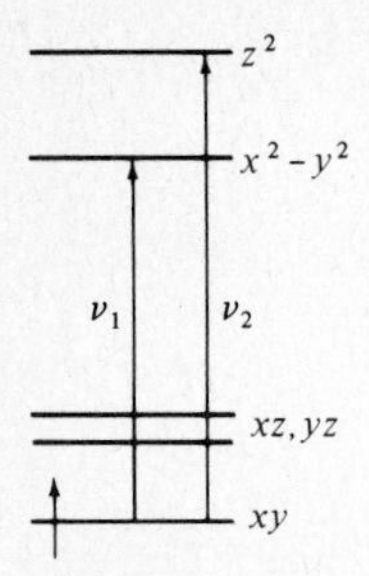

Fig. 9.22 Energy levels in $[Ti(H_2O)_6]^{+3}$ and transitions to excited states.

Consider an ion subject to Jahn-Teller distortion, $[Ti(H_2O)_6]^{+3}$. The Ti(III) ion is a d^1 species and in an octahedral field the configuration is t_{2g}^1. The t_{2g} level is triply degenerate, however, and the Jahn-Teller theorem forbids it to be occupied by a single electron without undergoing distortion. (Note that the Cr(III) ion, d^3, would be acceptable since no degeneracy occurs.[48]) If distortion occurs, the odd electron can occupy a lower energy level. Note, however, that the Jahn-Teller theorem *per se* does not predict which way the distortion will occur. In this case, however, it is easy to see that the stabilization of the odd electron in the "z-in" distortion is twice that of the "z-out" distortion,[49] and we would therefore expect the $[Ti(H_2O)_6]^{+3}$ ion to be a distorted octahedron, compressed along the z axis. The CFSE energy will be $\frac{2}{3}\delta_2$ larger than it would have been in a regular octahedron. Although the upper e_g levels are split as well, there is no energetic effect since they are unoccupied in the ground state. The excited state, $t_{2g}^0e_g^1$, will also be subject to distortion of the same type. The spectrum of the ion (Fig. 9.7) shows the result of this splitting. Instead of a single, Gaussian curve the absorption peak shows a shoulder resulting from the superposition of two peaks: one from the ground state to the lower of the two e_g levels[50] and the other to the upper (Fig. 9.22). In some species, such as the $[CoF_6]^{-3}$ ion, the splitting of the two bands is sufficient to cause two distinct peaks to be found in the spectrum (Fig. 9.23).

The best evidence for the presence of Jahn-Teller effects in transition metal compounds comes from structural studies of solids containing Cu(II). The latter is a d^9 configuration, and in an octahedral field the ninth electron has the option of entering the d_{z^2} or the $d_{x^2-y^2}$ orbital. The octahedral complex is therefore degenerate and not expected to exist in an undistorted form. Alternatively, the d^9 system may be viewed by means of the "hole formalism," which would describe Cu(II) as the spherically symmetrical d^{10} system with a hole, or missing electron. From this point of view, the hole behaves exactly the way the electron does, but instead of finding the *lowest* orbital available to it, the hole tends to

---

[48] It is necessary to distinguish between two slightly different usages of the word "degenerate." In both cases the unsplit t_{2g} level is triply degenerate. In the case of the Ti^{+3} ion, however, there are three different allocations of the single electron: (a) $xy^1xz^0yz^0$; (b) $xy^0xz^1yz^0$; (c) $xy^0xz^0yz^1$. For Cr^{+3} there is a single configuration: $xy^1xz^1yz^1$.

[49] Note, incidentally, that the elongation or "z-out" distortion does not completely remove the degeneracy since the lowest level is still twofold degenerate (Fig. 9.21), and so some other types of distortion would still be necessary to satisfy the Jahn-Teller theorem.

[50] It is not really proper to speak of the energy levels in a tetragonal complex as t_{2g} or e_g since strictly these refer to the upper and lower levels derived from the t_{2g} and e_g levels of the undistorted octahedron.

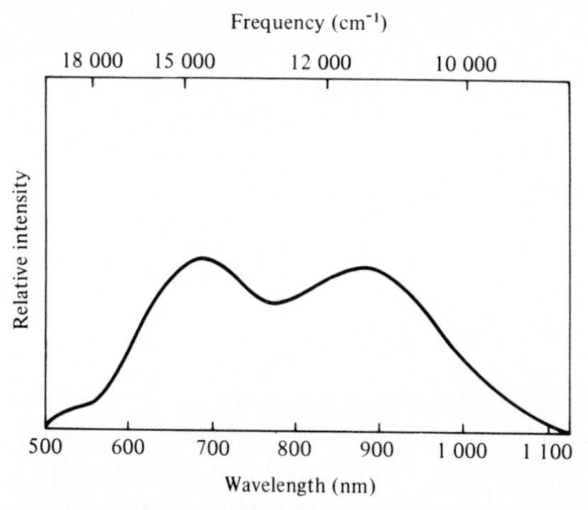

Fig. 9.23 Absorption spectrum of K_3CoF_6 illustrating transitions from the ground state to the Jahn-Teller split excited state. [From F. A. Cotton and M. D. Meyers, *J. Am. Chem. Soc.*, **1960**, *82*, 5023. Reproduced with permission.]

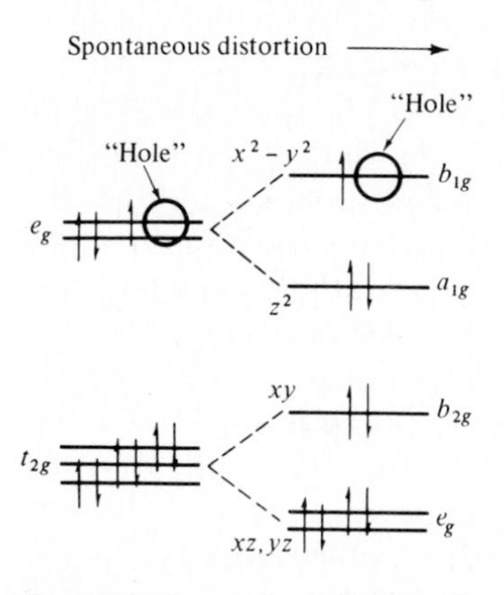

Fig. 9.24 Energy level diagram for d^9 configuration in an octahedral and a tetragonal field illustrating the tendency of the "hole" to "float."

"float" to the top. Jahn-Teller distortion thus lets it float somewhat higher than it would in the undistorted complex (Fig. 9.24). An analogous scheme for viewing such systems is to consider a d^n system as an *inverted* d^{10-n} system as shown in Fig. 9.25. All these schemes are useful, and they will be encountered again in the following discussions.

In the case of Cu(II), distortion via either elongation or compression will lead to a stabilization of $\frac{1}{2}\delta_1$. Again, the Jahn-Teller theorem provides us with no clue as to which way the distortion will occur, but simply says that such distortion will lower the energy of the system. Undoubtedly there has in the past been too much stress placed on the Jahn-Teller effect as an abstract theorem without looking into these energies, their source, and their magnitude. Indeed, some recent workers have explained these distortions in quite different ways.[51]

---

[51] J. K. Burdett, *Inorg. Chem.*, **1981**, *20*, 1959.

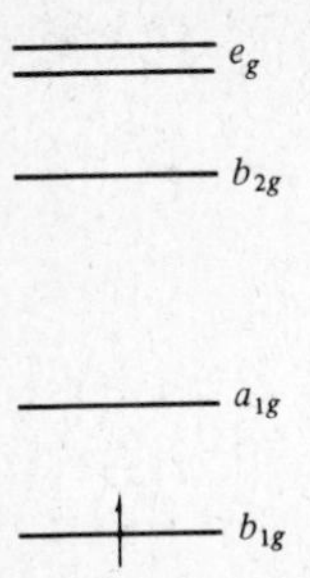

Fig. 9.25 The d^9 configuration of Fig. 9.24 represented as an inverted d^1 case.

Experimentally we find that in the Cu(II) series of compounds the distortion is usually elongation along the z axis. Table 9.12 lists some distances found in crystals containing hexacoordinate Cu(II) ions. Each compound has both "short" and "long" bonds. It is of interest to note that the "short" bonds represent a nearly constant radius for the Cu^{+2} ion, whereas the "long" bonds show no such constancy. This might be expected since the distortion or lengthening of the long bonds can proceed to various degrees.

Orbital degeneracy will occur only when the e_g or t_{2g} level is partially filled.[52] Since the splitting of the e_g level is somewhat greater than the splitting of the t_{2g} level, we should therefore expect weak-field d^4 and d^9 and strong-field d^7 and d^9 to show the greatest effects. Cu(II) is an excellent example of a d^9 species undergoing Jahn-Teller distortion. We have fewer data to support Jahn-Teller distortion in high-spin d^4 or low-spin d^7. Cr(II) and Mn(III) are d^4, and both have been found to be distorted in some compounds (see Table 9.13). Furthermore, extensive studies of the spectra of Mn(III) compounds are readily interpreted in terms of elongation along the z axis.[53] The d^7 configuration of CO^{+2} is less satisfying. Only the cyanide ion can provide a sufficiently strong field to induce spin pairing. The expected cyano complex, $[Co(CN)_6]^{-4}$, is not found, however, but instead the principal species in solution has five cyano groups per cobalt, probably $[Co(CN)_5H_2O]^{-3}$.[54] One might suggest that this is the "ultimate" in Jahn-Teller distortion, namely complete dissociation of one cyanide (from the hypothetical $[Co(CN)_6]^{-4}$ ion), but this is probably stretching a tenuous analogy to an extreme.

Degeneracy arising from odd electrons in the t_{2g} level should also produce a Jahn-Teller distortion, but the effects should be somewhat smaller because the t_{2g} orbitals are less influenced by the ligands. Configurations that might lead to observable Jahn-Teller distortions are d^1, d^2, high-spin d^6, d^7, and low-spin d^4 and d^5. Evidence for Jahn-Teller distortions in these configurations is considerably weaker than for the configurations involving splitting of the e_g level.

No discussion of the Jahn-Teller effect in coordination compounds would be complete without inclusion of the problems of chelate compounds (see Chapter 10 for a more

[52] Note that a t_{2g}^4 configuration is degenerate in exactly the same way as a t_{2g}^1 configuration, since there are three ways of writing the precise configuration: $xy^2xz^1yz^1$, etc.

[53] T. S. Davis et al., *Inorg. Chem.*, **1968**, *7*, 1994.

[54] The exact nature of the pentacyano species in solution is not completely clear. The solutions are difficult to study since they are readily oxidized to Co(III). The $[Co(CN)_5H_2O]^{-3}$ formulation has been suggested by J. M. Pratt and R. J. P. Williams, *J. Chem. Soc.*, A, **1967**, 1291.

Table 9.12 Metal–ligand distances in Cu(II) compounds[a,b]

Compound	Short distances	r_{Cu}[c]	Long distances	r_{Cu}
CuF_2	4F at 193	122	2F at 227	156
Na_2CuF_4	4F at 191	120	2F at 237	166
K_2CuF_4	2F at 195	124	4F at 208	147
$KCuF_3$	2F at 196	125	4F at 207	146
$CuF_2 \cdot 2H_2O$	2F at 189	118	2F at 247	176
	$2H_2O$ at 193			
$CuCl_2$	4Cl at 230	131	2Cl at 295	196
$CsCuCl_3$	4Cl at 230	131	2Cl at 265	166
$CuCl_2 \cdot 2H_2O$	2Cl at 228	129	2Cl at 295	196
	$2H_2O$ at 193			
$CuCl_2 \cdot 2C_5H_5N$	2N at 202	127	2Cl at 305	206
	2Cl at 228	128		
$CuBr_2$	4Br at 240	126	2Br at 318	204
α-$Cu(NH_3)_2Br_2$	$2NH_3$ at 193	118	2Br at 308	194
	2Br at 254	140		
β-$Cu(NH_3)_2Br_2$	$2NH_3$ at 203	128	4Br at 288	174
$Cu(HCOO)_2 \cdot 4H_2O$	4HCOO at 200	127	$2H_2O$ at 236	163
$Cu(proline)_2 \cdot 2H_2O$	2N at 199	124	$2H_2O$ at 252	179
	2O at 203	130		
$Cu(DMG)_2$	4N at 194	119	2O at 243	170
$Cu(NH_3)_2Cl_2$	$2NH_3$ at 195	120	4Cl at 276	177
$Cu(NH_3)_4SO_4 \cdot H_2O$	$4NH_3$ at 205	130	$1H_2O$ at 259	186
			$1H_2O$ at 337	264
$Cu(NH_3)_6^{+2}$[d]	$4NH_3$ at 207	132	$2NH_3$ at 262	187

[a] Data, unless noted otherwise, from A. F. Wells, "Structural Inorganic Chemistry," 3rd ed., Oxford University Press, London, **1962**, and J. D. Dunitz and L. E. Orgel, *Adv. Inorg. Chem. Radiochem.*, **1960**, *2*, 1.
[b] All distances are in picometres.
[c] The radius of Cu(r_{Cu}) was obtained by subtracting the covalent radius of the ligating atom (Table 6.1) from the Cu—X distance.
[d] T. Tistler and P. A. Vaughan, *Inorg. Chem.*, **1967**, *6*, 126.

Table 9.13 Metal–ligand distances in Cr(II) and Mn(III) compounds[a,b]

Compound	Short distance	r_M[c]	Long distance	r_M
CrF_2	4F at 200	129	2F at 243	172
$KCrF_3$	2F at 200	129	4F at 214	143
MnF_3	2F at 179	108	2F at 191	120
			2F at 209	138
γ-MnO(OH)	4O at 188	115	2O at 230	157

[a] Data from A. F. Wells, "Structural Inorganic Chemistry," 3rd ed. Oxford University Press, London, **1962**.
[b] All distances are in picometres.
[c] The metal radii (r_M) were obtained by subtracting the covalent radius of the ligating atom (Table 6.1) from the M—X distance.

thorough discussion of chelates). The very nature of the chelate ring tends to restrict the distortion of a complex from a perfect octahedron[55] since the chelate will have a preferred "bite," or distance between the coordinating atoms:

$$H_2N\!-\!CH_2\!-\!CH_2\!-\!NH_2$$

bite

An example of the conflict between stabilization from the Jahn-Teller effect and chelate geometrical requirements is found in the ethylenediamine complexes of Cu(II). Most divalent transition metal ions form complexes with ethylenediamine (en) by stepwise replacement of water:

$$[M(H_2O)_6]^{+2} + en \longrightarrow [M(H_2O)_4en]^{+2} + 2H_2O \quad \textbf{(9.23)}$$

$$[M(H_2O)_4en]^{+2} + en \longrightarrow [M(H_2O)_2(en)_2]^{+2} + 2H_2O \quad \textbf{(9.24)}$$

$$[M(H_2O)_2(en)_2]^{+2} + en \longrightarrow [M(en)_3]^{+2} + 2H_2O \quad \textbf{(9.25)}$$

The stepwise stability constants, K_1, K_2, K_3, measure the tendency of ethylenediamine to displace two, four, and six molecules of water with the formation of the mono-, bis-, and tris-(ethylenediamine) complexes. The values for these constants for the ions Mn^{+2} to Zn^{+2} show a rather uniform trend of slightly increasing stability of all the ethylenediamine complexes toward the end of the series (the Irving-Williams order). Cu(II) provides a striking exception, however. The tris(ethylenediamine)copper(II) complex, $[Cu(en)_3]^{+2}$, is remarkably unstable. At one time its very existence was questioned[56] and although it has subsequently been prepared,[57] the value of K_3 (a measure of the tendency to add the third ethylenediamine ligand) is the lowest of the ions discussed even though the K_1 and K_2 values are the highest (Fig. 9.26). This lack of stability can be traced directly to the distortion necessary in a d^9 Cu^{+2} ion. The bis(ethylenediamine) complex, $[Cu(en)_2(H_2O)_2]^{+2}$, can distort readily by letting the two *trans* water molecules move out from the copper with the two ethylenediamine rings relatively unchanged. In contrast, the tris(ethylenediamine) complex cannot distort tetragonally without straining at least two of the chelate rings:

$$[Cu(en)_2(H_2O)_2]^{+2} + en \longrightarrow [Cu(en)_3]^{+2} \text{ (strain)} + 2H_2O \quad \textbf{(9.26)}$$

[55] In some cases, chelate rings could favor the distortion of an octahedral complex toward trigonal prismatic structures (see drawing associated with Table 9.2), but they will always inhibit tetragonal distortion since there is a limited range over which the chelate can span two coordination sites.

[56] J. Bjerrum and E. J. Nielson, *Acta Chem. Scand.*, **1948**, *2*, 297.

[57] G. Gordon and R. K. Birdwhistell, *J. Am. Chem. Soc.*, **1959**, *81*, 3567.

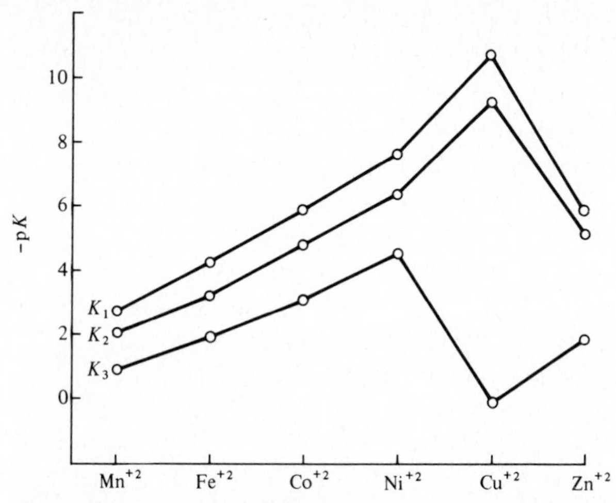

Fig. 9.26 Step-wise stability constants, K_1, K_2, and K_3, of ethylenediamine complexes in aqueous solution at 25 °C. [From C. S. G. Phillips and R. J. P. Williams, "Inorganic Chemistry," Oxford University Press, Oxford, **1965**. Reproduced with permission.]

Alternatively, it is possible that the constraint of the chelate ring system can prevent tetragonal distortion and form an undistorted octahedron, but the resulting complex would lack the stabilization inherent in Jahn-Teller distortion:

(9.27)

Both types of chelates are known. For example, [Cu(bipy)(hfacac)$_2$] is[58] known to be tetragonally distorted.[59] Both bipyridine and hexafluoroacetylacetone are chelating ligands bonding through nitrogen and oxygen atoms, respectively. The structure of this interesting molecule is shown in Fig. 9.27. The two nitrogen atoms form the bipyridine bond to the copper at a distance of 200 pm, consistent with the short bonds from other nitrogen ligands (Table 9.12). One oxygen from each hexafluoroacetylacetone ligand bonds at a distance of 197 pm, again consistent with short Cu—O bonds. The two remaining

[58] bipy = bipyridine,

hfacac = hexafluoroacetylacetone, $CF_3C(O)CH_2C(O)CF_3$(= 1,1,1,5,5,5-hexafluoropentanedione-2,4).

[59] M. V. Veidis, G. H. Schreiber, T. E. Gough, and G. J. Palenik, *J. Am. Chem. Soc.*, **1969**, *91*, 1859.

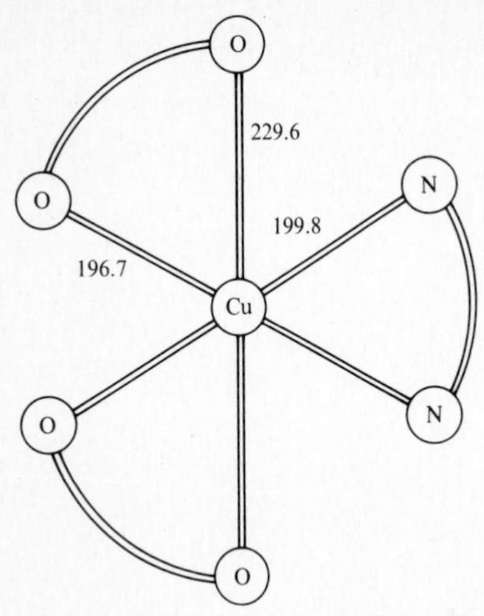

Fig. 9.27 Structure of [Cu(dipy)(hfacac)$_2$] together with Cu—O and Cu—N bond lengths (in pm). [Modified from M. V. Veidis et al., *J. Am. Chem. Soc.*, **1969**, *91*, 1859. Reproduced with permission.]

oxygen atoms form Cu—O bonds which are some 33 pm *longer*, indicative of severe Jahn-Teller distortion *despite* the presumably restraining effects of the chelate rings.

Unfortunately, not all chelates subject to the Jahn-Teller effect provide us with the simple picture afforded by [Cu(bipy)(hfacac)$_2$]. There are a few that appear to exist *without any detectable distortion.* Examples are tris(ethylenediamine)copper(II), [Cu(en)$_3$]$^{+2}$, tris(octamethylpyrophosphoramide)copper(II), [Cu(OMPA)$_3$]$^{+2}$, both d^9, and tris-(2,4-pentanedionato)manganese(III), [Mn(acac)$_3$], d^4.[60] In each of these the central metal lies at the center of a perfect octahedron of oxygen or nitrogen atoms within experimental error. Figure 9.28 illustrates the [Cu(OMPA)$_3$]$^{+2}$ ion, representative of the group with six Cu—O bonds of 206.5 pm.

There is no simple answer to the question as to why these complexes do not distort, nor is it presently possible to predict which complexes will distort and which will not. Certainly a chelate ring system tends to inhibit distortion, but we have seen that some chelates do indeed distort. It may be argued that the distortion is small (the Jahn-Teller theorem does not require a *large* distortion, merely *some* distortion) and less than the experimental error, but this begs the question. The answer seems to lie in the fact that these systems are not static but dynamic; one must consider the dynamic-static Jahn-Teller effect.[61] Simply stated, this interprets the apparently high symmetry of complexes such as in Figure 9.28 as a result of oscillation of the complex among three tetragonal structures. Thus at any instant the complex is distorted but the time-averaged, observable structure appears undistorted. This might seem to be an explanation that evades the

[60] M. Cola et al., *Atti Accad. Sci. Torino*, **1962**, *96*, 381; D. L. Cullen and E. C. Lingafelter, *Inorg. Chem.*, **1970**, *9*, 1858; M. D. Joesten et al., ibid., **1970**, *9*, 151 (OMPA = [(CH$_3$)$_2$N]$_2$P(O)OP(O)[N(CH$_3$)$_2$]$_2$); B. Morosin and J. R. Brathovde, *Acta Crystallogr.*, **1964**, *17*, 705; T. S. Davis et al., *Inorg. Chem.*, **1968**, *7*, 1994.

[61] A. Abragam and M. H. L. Pryce, *Proc. Roy. Soc.*, **1951**, *A206*, 164.

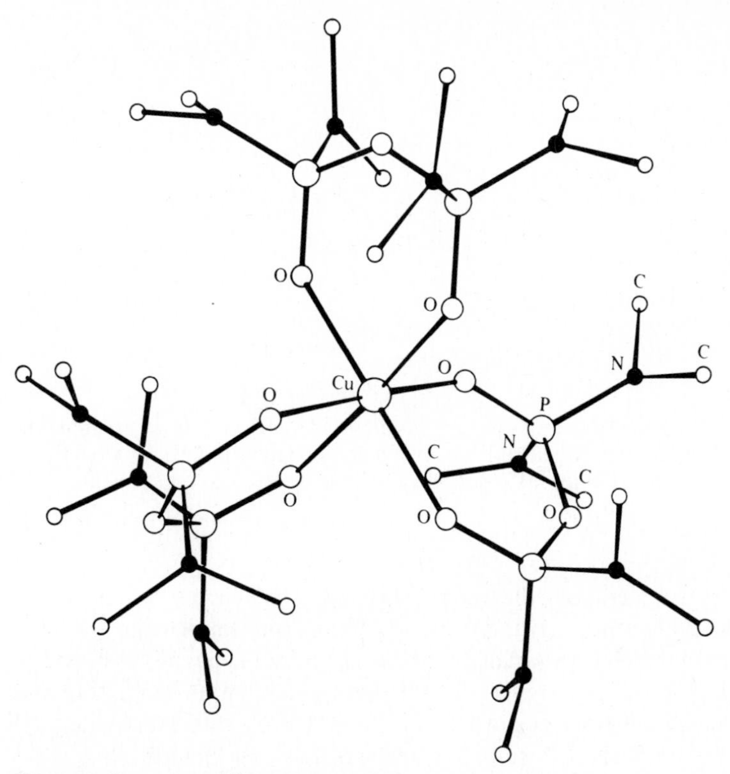

Fig 9.28 Structure of $[Cu(OMPA)_3]^{+2}$. All Cu—O bond lengths are 206.5 pm. The hydrogen atoms have been omitted for clarity. [From M. D. Joesten et al., *J. Am. Chem. Soc.*, **1968**, *90*, 5623. Reproduced with permission.]

question rather than answering it, but fortunately we have experimental evidence to support it. If the $[Cu(OMPA)_3]^{+2}$ complex is cooled to 175 K, the lattice changes from one of very high symmetry to one of considerably less symmetry, indicating that perhaps the shape of the complex has changed. Furthermore, the electron spin resonance measurements at 90 K are typical of tetragonally distorted copper(II) complexes.[62] Although the crystal structure of this complex could not be resolved because of difficulties attending work at low temperatures, fortunately an excellent series of compounds containing the hexanitrocuprate(II) ion, $[Cu(NO_2)_6]^{-4}$, is available.[63] Some of these, such as $K_2BaCu(NO_2)_6$, $K_2CaCu(NO_2)_6$, $Rb_2PbCu(NO_2)_6$, and $K_2SrCu(NO_2)_6$, are distorted at room temperature (298 K), but others, such as $K_2PbCu(NO_2)_6$ and $Tl_2PbCu(NO_2)_6$, are in the dynamic phase, i.e., they appear undistorted. However, if these latter compounds are cooled a few degrees, the dynamic Jahn-Teller effect is "frozen" out. For example, at 276 K, $K_2PbCu(NO_2)_6$ has obvious distortion (Cu—N distances of 205.8, 215.3, and 216.6 pm).

---

[62] R. C. Koch et al., *J. Chem. Phys.*, **1973**, *59*, 6312.

[63] See S. Takagi et al., *Acta Crystallogr.*, **1975**, *B31*, 596; **1976**, *B32*, 326; M. D. Joesten et al., *Inorg. Chem.*, **1977**, *11*, 2680.

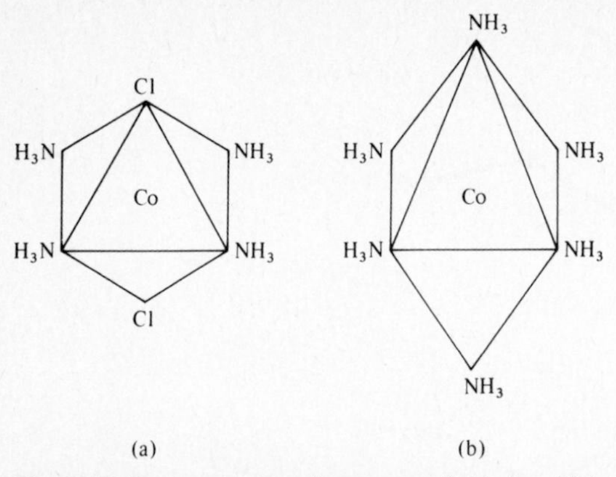

Fig. 9.29 Two examples of tetragonal fields in which the fields along the *z* axis are weaker than in the *xy* plane. (a) Tetraamminedichlorocobalt(III) ion with weak-field chloride ions on the *z* axis. (b) Hypothetical distorted hexaamminecobalt(III) ion with weakened field along the *z* axis resulting from movement of ammonia ligands away from metal.

Although we do not yet understand the Jahn-Teller effect sufficiently well to make accurate predictions concerning these dynamic-static effects, our knowledge of these systems is constantly improving and they no longer appear as puzzling as they once did.[64]

Although it is not always possible to predict in advance the exact nature of tetragonal distortions, the study of such systems is nevertheless of considerable interest. As implied above, tetragonal fields may be produced via the presence of different ligands along the *z* axis from those along the *x* and *y* axes. If the *z* ligands are lower in the spectrochemical series than the *x* and *y* ligands, they will exert less of a field on the d_{z^2} orbital and the effect will be the same as though the *z* ligands were moved out. Consider the *trans*-tetraammine-dichlorocobalt(III) ion, $[Co(NH_3)_4Cl_2]^+$ (the *praseo* complex on pp. 360 and 361). Even though there is no distortion in this molecule (it is not Jahn-Teller degenerate) and the bond lengths are those to be expected on the basis of addition of radii, the field experienced by the Co(III) ion is nevertheless tetragonal. This is because the chloride ion is much lower in the spectrochemical series than ammonia and exerts a less powerful effect along the *z* axis. Figure 9.29 illustrates this idea by comparing the *trans*-dichloro complex with a *hypothetical* hexaamminecobalt(III) complex, $[Co(NH_3)_6]^{+3}$, in which we have (*by thought processes only*!) moved two ammonia ligands out along the *z* axis to produce the same tetragonal field.

An interesting example of the effect of varying degrees of tetragonal distortion has been provided by Belford and co-workers.[65] They have studied the spectra of $[Cu(acac)_2]$ in a variety of solvents. Copper acetylacetonate has square planar coordination[66] which may be regarded as an extreme tetragonal distortion. In basic solvents such as ethers, alcohols,

---

[64] For reviews, see F. S. Ham in "Electron Paramagnetic Resonance," S. Geschwind, ed., Plenum, New York, **1972**, Chapter 1; R. Englman, "The Jahn-Teller Effect in Molecules and Crystals," Wiley (Interscience), London, **1972**; I. B. Bersuker, *Coord. Chem. Rev.*, **1975**, *14*, 357.

[65] R. L. Belford et al., *J. Chem. Phys.*, **1957**, *26*, 1165.

[66] H. Koyama et al., *J. Inst. Polytech. Osaka City Univ.*, **1953**, *4*, 43.

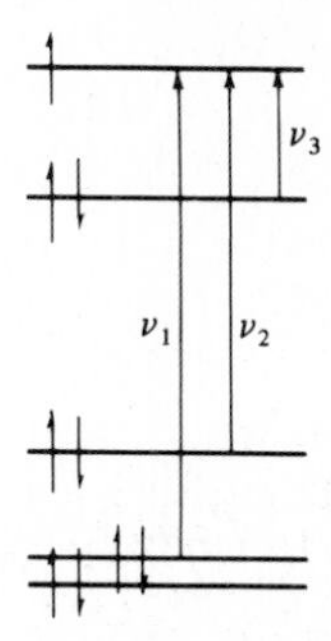

Fig. 9.30 Transition for Cu(II), d^9, in a tetragonal field.

and amines, two solvent molecules coordinate along the z axis:

$$\text{Cu(O}\frown\text{O)}_2 \xrightarrow[\text{solvent B}]{\text{dissolve in}} \text{B–Cu(O}\frown\text{O)}_2\text{–B} \qquad \textbf{(9.28)}$$

These solvent molecules exert a z-component field proportional to their position in the spectrochemical series. Some, such as chloroform, exert little or no effect since their basicity is extremely low. Others, such as the amines, are considerably more basic and coordinate more strongly. Repelling the advance of the solvent molecules are two electrons in the d_{z^2} orbital, so only the most basic species can force their way in to form strong bonds. There are three d–d electronic transitions possible in a tetragonal complex (Fig. 9.30): $d_{xz}(d_{yz}) \rightarrow d_{x^2-y^2}(\nu_1)$; $d_{xy} \rightarrow d_{x^2-y^2}(\nu_2)$; $d_{z^2} \rightarrow d_{x^2-y^2}(\nu_3)$.[67] The frequency of these transitions will be strongly dependent upon the field experienced by the various d orbitals. Frequently ν_2 should remain constant at $10Dq_{O_h}$ since neither the d_{xy} nor the $d_{x^2-y^2}$ orbital has a z component. They should thus be completely unaffected by any maneuvering of ligands on the z axis and the separation corresponds to the splitting $10Dq_{O_h}$ in a hypothetical purely octahedral complex having six identical ligands. As the tetragonal field approaches the octahedral configuration (as the z field strength approaches that of x and y) ν_1 should diminish and approach ν_2 (see Fig. 9.31). The third transition, ν_3, should also decrease as z ligand basicity increases eventually disappearing in the octahedral limit as $x^2 - y^2$ and z^2 collapse to form the degenerate e_g orbitals. The spectra of copper acetylacetonate in various solvents are shown in Fig. 9.32. We note immediately that the observed curves change from one with two maxima ($HCCl_3$) toward one with a distinct single major peak with a shoulder on the low-energy side. Belford and co-workers resolved the experimental spectra into their component bands. They noted that band 2 remained unchanged in frequency (14 800 to 15 200 cm^{-1}) and thus assigned it to transition ν_2. Band 1 decreases in frequency (18 800 to 15 100 cm^{-1}) and approaches ν_2, so it was assigned to transition ν_1. Band 3 decreases steadily in frequency corresponding to ν_3. All of the experimental values correlate very well with the expectations based on the expected splittings of the d orbitals in a tetragonal field. The extremely weak base, chloroform, provides the extreme tetragonal field approaching

[67] Belford et al., (footnote 65) labeled the xy and $x^2 - y^2$ orbitals in a different sense from that given here. This is of importance only if reference is made to their paper in conjunction with the present discussion. It corresponds to rotation of the coordinate system 45° about the z axis.

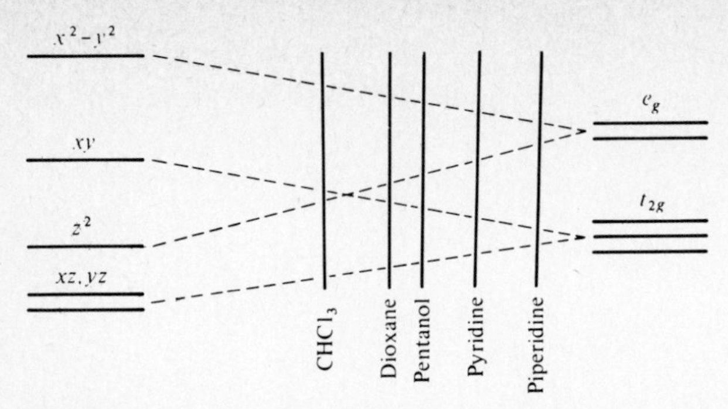

Fig. 9.31 Gradual transition between square planar symmetry (left) and purely octahedral symmetry (right) with increasing strength of the axial field (z ligands). The exact position of the d_{z^2} orbital is uncertain.

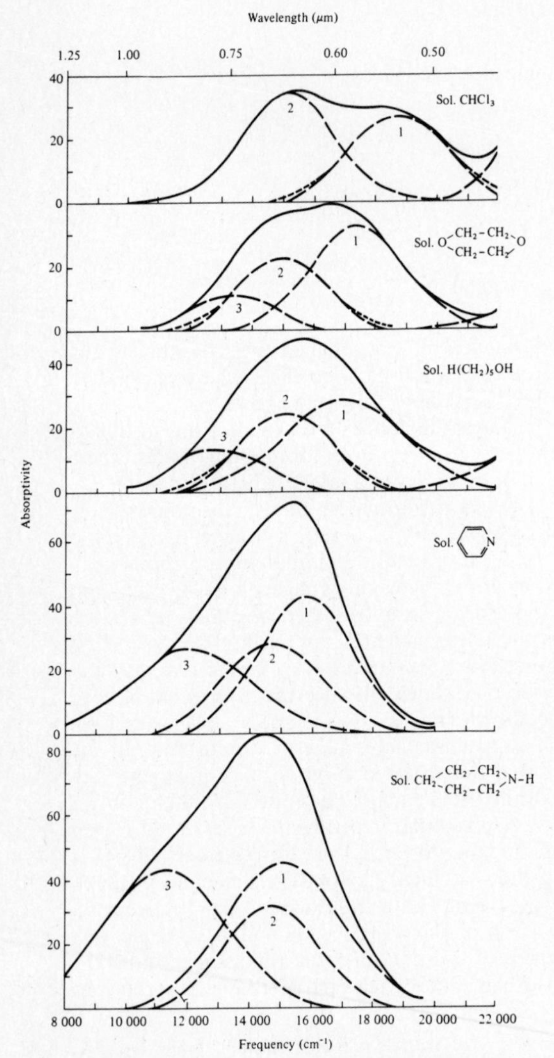

Fig. 9.32 Spectra of copper acetylacetonate dissolved in various bases. See text for explanation. [From R. L. Belford et al., *J. Chem. Phys.*, **1957**, *26*, 1165. Reproduced with permission.]

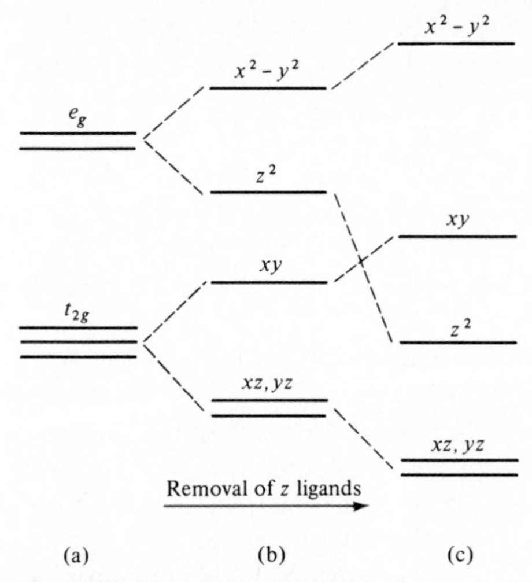

Fig. 9.33 Distortion of octahedral complex (a) through tetragonal symmetry (b) to square planar limit (c). The z^2 orbital may lie below the xz and yz orbital in the square planar complex.

square planar splitting with a large disparity between the xy ligands and the z ligands. The other extreme is piperidine, $C_5H_{10}NH$, which provides an almost purely octahedral field. The ready interpretation of such effects was a major driving force in the displacing of valence bond theory by crystal field theory and its successors.

It is an unfortunate fact that often Nature does not deign to behave in a simple fashion merely for the sake of pedagogical elegance. In the present case, of which the above is a somewhat simplified description, coordination can occur through either *one* or *two* ligands along the z axis. This does not cause any serious problems directly since the tetragonal distortion from one ligand is exactly half that of two (see Table 9.14 for square pyramidal effects). Unfortunately, there is evidence that in the case of the monoadduct there is distortion of the acetylacetone groups away from the entering base; this would affect the energy levels differently. Nevertheless, the above analysis is essentially accurate in portraying the effects of axial ligation on Cu(II) spectra. For more recent and detailed papers on the subject, see footnote 68.

Square planar coordination

If tetragonal distortion progresses sufficiently (in the case of "z-out") that the ligands on the z axis are removed to infinity, a square planar complex is obtained. As mentioned above, crystal field theory therefore does not consider square planar complexes as a new type of coordination but merely the limiting case of extreme tetragonal distortion. Figure 9.33 illustrates this relationship. Metal ions having a d^8 configuration and ligands high in the spectrochemical series favor square planar complexes. This combination will result in low-spin complexes with the d^8 electrons occupying the low-energy yz, xz, z^2, and xy orbitals while the high-energy $x^2 - y^2$ orbital remains unoccupied. The stronger the field applied on the $x^2 - y^2$ orbital, the higher it will be raised. Since it is unoccupied

---

[68] R. L. Belford and J. W. Carmichael, Jr., *J. Chem. Phys.*, **1967**, *46*, 4515; L. L. Funck and T. R. Ortolano, *Inorg. Chem.*, **1968**, *7*, 567; B. B. Wayland and A. F. Garito, *Inorg. Chem.*, **1969**, *8*, 182; B. L. Libutti et al., *Inorg. Chem.*, **1969**, *8*, 1510; A. B. P. Lever and E. Mantovani, *Inorg. Chem.*, **1971**, *10*, 817.

(in the low-spin case), such an increase in energy will do no harm and since, as a result of the barycenter rule, the four filled orbitals must be lowered a corresponding amount, the complex will be further stabilized. Note that the transition of lowest energy, $xy \rightarrow x^2 - y^2$, corresponds to $10Dq_{O_h}$. Typical low-spin square planar complexes are $[Ni(CN)_4]^{-2}$, $[PdCl_4]^{-2}$, $[Pt(NH_3)_4]^{+2}$, $[PtCl_4]^{-2}$, and $[AuCl_4]^-$, all d^8 species. In the first transition series, only the extremely strong field ligands such as cyanide can effect the spin pairing necessary for stabilization of the square planar arrangement, but heavier metals that are more prone to form low-spin complexes exist as square planar complexes even with halide ligands.

The relation between tetragonal fields and square planar fields in conjunction with the effect of increased effectiveness of ligand fields is of value in rationalizing the chemistry of the coinage metals, Group IB. For copper, the Cu(II) ion is the most stable, existing, as we have seen, predominantly in tetragonal complexes. In contrast, gold is known almost exclusively as Au(I) and Au(III). Since gold is a 5*d* element, it experiences a splitting of the *d* orbitals some 80% greater than copper does. A d^9 Au(II) complex would be strongly distorted tetragonally and the ninth electron would occupy the highly unfavorable $x^2 - y^2$ orbital, which will be raised correspondingly higher in energy (Fig. 9.34). It is therefore much easier to ionize than the odd electron in Cu(II). As a result, Au(II) disproportionates into Au(I) and Au(III) species.

With respect to low-spin square planar complexes the valence bond theory and the crystal field theory agree in assigning the nonbonding d^8 electrons on the metal to the

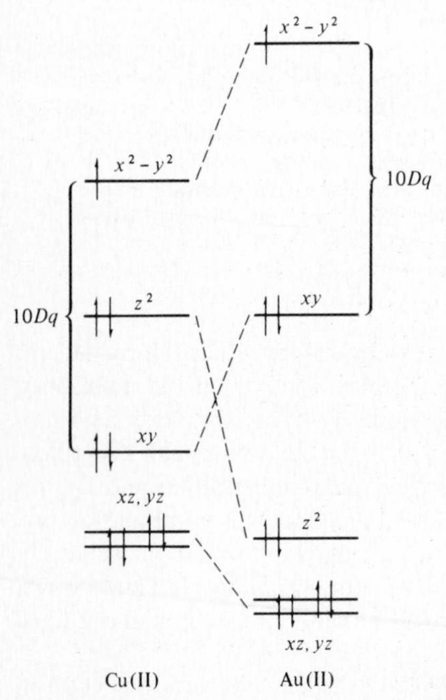

Fig. 9.34 Comparison of Cu(II) and Au(II) in tetragonal and square planar fields to illustrate the instability of Au(II) toward the removal (oxidation) of the odd electron.

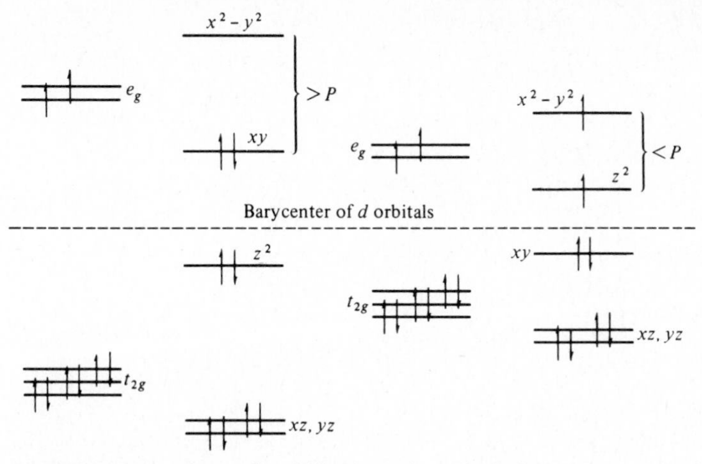

Fig. 9.35 Comparison of stabilization of octahedral vs. tetragonal symmetry for strong field (left) and weak field (right) ligands. Note that no stabilization is afforded in the weak field in contrast to the strong field case.

d_{xy}, d_{yz}, d_{xz}, and d_{z^2} orbitals and using the $d_{x^2-y^2}$ orbital in a different manner (as part of a hybridization, dsp^2, in the valence bond treatment, and unoccupied, directed at the ligands in crystal field theory). Note, however, that VBT allows only *low-spin* square planar complexes since the $d_{x^2-y^2}$ orbital *must* be involved in the dsp^2 hybrid and is therefore unavailable for metal electrons. In contrast, CFT simply says that this orbital is *highly unfavorable*, but if the pairing energy is greater than the crystal field splitting, *high-spin complexes are not forbidden* and could exist. Experimentally, no high-spin square planar complexes have been observed as yet. In the absence of strong splitting, there is no driving force to convert an octahedral complex into a square planar one (see Fig. 9.35).

Although no square planar complexes with a high-spin ground state are known, as we have seen they are not prohibited by simple crystal field theory. For some time it was thought that tetrabutylammonium bis(toluene-3,4-dithiolato)cobaltate(I) had a high-spin ground state because it is paramagnetic at room temperature. Very careful studies at low temperature[69] have shown that the ground state is actually a singlet (spin-paired) but that a low-lying triplet state exists only 0.10 kJ mol^{-1} higher in energy. Because of the Boltzmann distribution the compound exists as a mixture of low- and high-spin species at room temperature. Although this finding allows us still to say that no high-spin (ground state) square planar complexes are known, it is obvious that in this case the difference in energy between low- and high-spin configurations is practically nonexistent.

Orbital splittings in fields of other symmetries

Although most complexes can be accommodated under the theory for tetrahedral complexes or that for octahedral complexes (including distortion to square planar), there are many stable complexes which exist in other configurations, and some of these unusual symmetries are thought to be important as intermediates in the reactions of coordination

[69] C. R. Ollis et al., *J. Am. Chem. Soc.*, **1971**, *93*, 547.

Table 9.14 The energy levels of *d* orbitals in crystal fields of different symmetries[a]

C.N.	Structure	d_{z^2}	$d_{x^2-y^2}$	d_{xy}	d_{xz}	d_{yz}
1	Linear[b]	5.14	−3.14	−3.14	0.57	0.57
2	Linear[b]	10.28	−6.28	−6.28	1.14	1.14
3	Trigonal[c]	−3.21	5.46	5.46	−3.86	−3.86
4	Tetrahedral	−2.67	−2.67	1.78	1.78	1.78
4	Square planar[c]	−4.28	12.28	2.28	−5.14	−5.14
5	Trigonal bipyramid[d]	7.07	−0.82	−0.82	−2.72	−2.72
5	Square pyramid[d]	0.86	9.14	−0.86	−4.57	−4.57
6	Octahedron	6.00	6.00	−4.00	−4.00	−4.00
6	Trigonal prism	0.96	−5.84	−5.84	5.36	5.36
7	Pentagonal bipyramid	4.93	2.82	2.82	−5.28	−5.28
8	Cube	−5.34	−5.34	3.56	3.56	3.56
8	Square antiprism	−5.34	−0.89	−0.89	3.56	3.56
9	ReH_9 structure (see Fig. 10.56)	−2.25	−0.38	−0.38	1.51	1.51
12	Icosahedron	0.00	0.00	0.00	0.00	0.00

[a] All energies are in *Dq* units.
[b] Ligands lie along *z* axis.
[c] Ligands lie in *xy* plane.
[d] Pyramid base in *xy* plane.
SOURCE: J. J. Zuckerman, *J. Chem. Educ.*, **1965**, *42*, 315, and R. Krishnamurthy and W. B. Schaap, ibid., **1969**, *46*, 799. Used with permission.

compounds as well. These configurations will not be discussed in detail, but the values for the various energy levels for these as well as the more familiar symmetries are given in Table 9.14. From these values orbital energy level diagrams can be constructed similarly to those given previously for octahedral, tetrahedral, and square planar structures.

MOLECULAR ORBITAL THEORY

Although the crystal field theory adequately accounts for a surprisingly large amount of data, it has serious defects. It was soon realized that the assumption of a point–charge model was a drastic oversimplification.[70] There are several experimental and semi-theoretical arguments that can be presented against the assumption that the splitting of the *d* orbitals is a result solely of electrostatic effects and that the bonding is "ionic" with no covalent character. We have seen that interpretation of the spectrochemical series in terms of a point–charge model is extremely tenuous. The ligand exerting the strongest field and thus lying at the head of the series is the carbonyl group, CO, with no ionic charge and almost no dipole moment ($\mu = 0.374 \times 10^{-30}$ C m).

On the basis of theory alone we might expect the crystal field theory to be deficient. If we examine the radial wave functions for metal and ligand atoms, we note that there must be some overlap and hence some covalent bonding involved, although the fact that

[70] In one of his earliest papers Van Vleck pointed out that both VBT and CFT could be considered as simplified approaches to the more general molecular orbital treatment.

Table 9.15 The nephelauxetic series of ligands and metal ions[a]

Ligand	*h*	Metal	*k*
F^-	0.8	Mn(II)	0.07
H_2O	1.0	V(II)	0.1
dmf	1.2	Ni(II)	0.12
urea	1.2	Mo(III)	0.15
NH_3	1.4	Cr(III)	0.20
en	1.5	Fe(III)	0.24
ox^{-2}	1.5	Rh(III)	0.28
Cl^-	2.0	Ir(III)	0.28
CN^-	2.1	Tc(IV)	0.3
Br^-	2.3	Co(III)	0.33
N_3^-	2.4	Mn(IV)	0.5
I^-	2.7	Pt(IV)	0.6
dtp^-	2.8	Pd(IV)	0.7
$dsep^-$	3.0	Ni(IV)	0.8

[a] The total nephelauxetic effect in a complex MX_n is proportional to the product $h_X \cdot k_M$.

SOURCE: C. K. Jørgensen, "Oxidation Numbers and Oxidation States," Springer, New York, **1969**, p. 106. Used with permission.

we do not have exact wave functions for the heavier elements leads to some uncertainty as to the extent of overlap.

Indirect evidence for sharing of electrons between ligands and the central metal ion comes from the *nephelauxetic effect*. It is found that the electron–electron repulsion in complexes is somewhat less than in the free ion. A nephelauxetic series may be set up for both various metal ions and various ligands indicating the order of increased nephelauxetic effect (Table 9.15). The decreased electron–electron repulsion may be attributed to an increased distance between electrons and, hence, to an effective increase in the size of the orbitals (*nephelauxetic* means "cloud expanding"). This increase apparently results from the combination of orbitals on the metal and ligand to form larger *molecular orbitals* through which the electrons can move. Ligands which can delocalize the metal electrons over a large space (those containing larger atoms with *d* orbitals for π bonding) are most effective in this manner.

Perhaps the best direct experimental evidence for sharing of electrons between metal and ligand comes from electron spin resonance. Unpaired electrons behave as magnets as a result of their spin[71] and line up parallel or antiparallel to an applied magnetic field. These two alignments will have slightly different energies, and transitions from one to the other can be detected by applying the energy of transition in the form of radio frequency electromagnetic radiation. An electron on an isolated metal ion shows a single absorption for this transition. In many complexes, however, more complicated spectra are observed

[71] *All* electrons have a spin, but those which are paired with other electrons are effectively "canceled out."

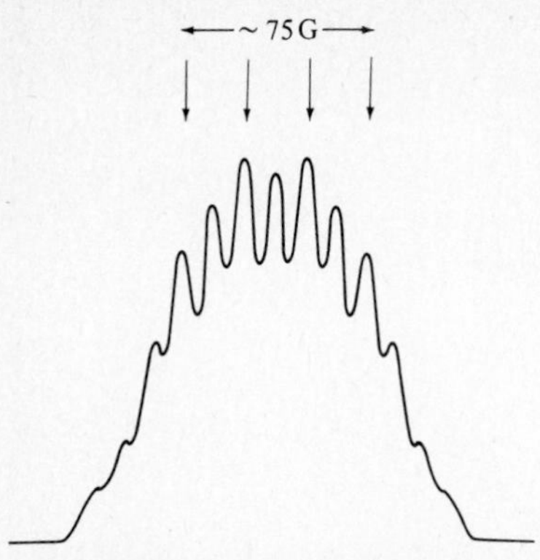

Fig. 9.36 The electron spin resonance spectrum of K_2IrCl_6 in K_2PtCl_6. [From J. Owen, *Disc. Faraday Soc.*, **1955**, *19*, 132. Reproduced with permission.]

(Fig. 9.36). The hyperfine splitting results from the effect of the magnetic moment of the ligand nuclei on the odd electron. This indicates that at least part of the time the odd electron is occupying an orbital on the ligand, or, more properly, the orbital occupied by the electron is a molecular orbital formed from atomic orbitals of the metal *and* the ligand.[72]

Beryllium hydride as a coordination compound

Consider a complex compound formed as follows:

$$Be^{+2} + 2H^- \longrightarrow BeH_2 \qquad \textbf{(9.29)}$$

This reaction may not be the most typical reaction possible for a coordination compound, but it has the advantage of being one of the simplest. Be(II) acts as an acceptor and the hydride ions can behave as ligands. We can now construct appropriate molecular orbitals for this complex from the 2*s* and 2*p* wave functions of the beryllium atom and the 1*s* wave functions of the two hydride ions. When this is done, we obtain two bonding orbitals, two nonbonding orbitals, and two antibonding orbitals. The four electrons from the hydride ions are placed in the two bonding orbitals. Not unexpectedly, the result is identical to that obtained from the reaction:

$$Be + 2H \longrightarrow BeH_2 \qquad \textbf{(9.30)}$$

which was discussed previously (Fig. 3.38, p. 120).

This examination of beryllium hydride has added little to the understanding of BeH_2, but it does focus attention on the fact that all neutral molecules AB_n may be considered as coordination compounds formed from A^{+n} and nB^-, and, conversely, coordination compounds may be treated by the same methods as used previously for other neutral molecules. For ionic complexes the methods are the same but a greater or lesser number of electrons is involved.

---

[72] This description of ESR spectra has been extremely simplistic to indicate the relation between covalent bonding and hyperfine splitting. For a more thorough treatment, see A. Carrington and A. D. MacLachlan, "Introduction to Magnetic Resonance," Harper & Row, New York, **1967**, pp. 167–171.

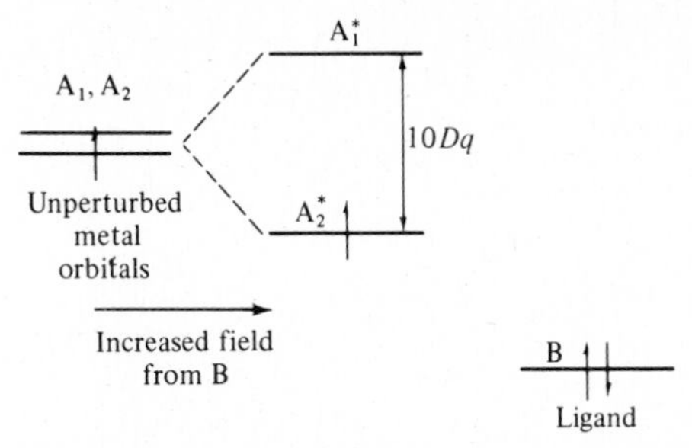

Fig. 9.37 Pure crystal field model. See text for explanation.

A simplified comparison of crystal field theory and molecular orbital theory approaches to bonding

Before delving into the problem of a complete treatment of a coordination compound which may have many interactions between orbitals and several electrons to be filled in, a brief comparison of molecular orbital results with the already familiar crystal field theory may be of interest. Let us consider one of the simplest possible systems: a Lewis base, B, having one orbital containing a pair of electrons to be donated, and a Lewis acid (metal ion), A^+, having two *sp* orbitals available for bonding and a single odd electron:

$$B\!\!\subset\!\!\supset + (\subset\!\!\supset A \subset\!\!\supset)^+ \longrightarrow [AB]^+ \tag{9.31}$$

In the isolated A^+ ion the two *sp* hybrids (A_1, A_2) are initially degenerate, but under the approach of the ligand lone pair they will be split into a higher energy orbital ($A_1^\dagger$) and a lower energy orbital ($A_2^\dagger$). (The † denotes that the energy has been perturbed by the approach of B.) The CFT interpretation is thus very simple: The odd electron on A^+ will occupy orbital $A_2^\dagger$ since it is repelled least by the ligand. The transition $A_2^\dagger \rightarrow A_1^\dagger$ represents $10Dq$ (Fig. 9.37).

The molecular orbital theory appears at first glance to be considerably different, but the results are identical if both are carried to the same degree of preciseness. Upon the approach of B, we assume that the orbital on A facing B (orbital A_1) will overlap and mix with B. We thus have a *bonding* and an *antibonding* orbital formed:

$$\Psi_b = A_1 + B \tag{9.32}$$

$$\Psi_a = A_1 - B \tag{9.33}$$

The second orbital on A, A_2, faces away from B so that it does not overlap[73] and hence does not change in energy; it thus remains as a *nonbonding orbital*. The energy of the system will be minimized if the three electrons (two from B, one from A) are distributed $\Psi_b^2\Psi_n^1\Psi_a^0$. The lowest energy transition will then be $\Psi_n \rightarrow \Psi_a$, and it is assigned the value $10Dq$ (see Fig. 9.38).

[73] This statement is true only if one ignores the small region of electron density on the "back side" of a hybrid orbital. Since this smaller lobe is very much smaller than the main lobe, we will, for purposes of simplification, ignore it.

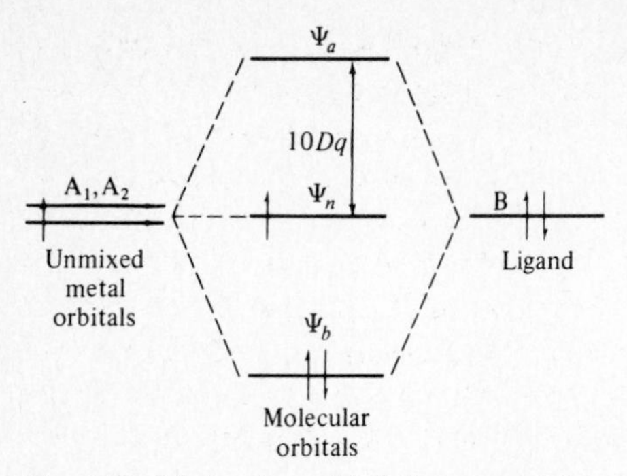

Fig. 9.38 Molecular orbital theory model, covalent extreme. See text for explanation.

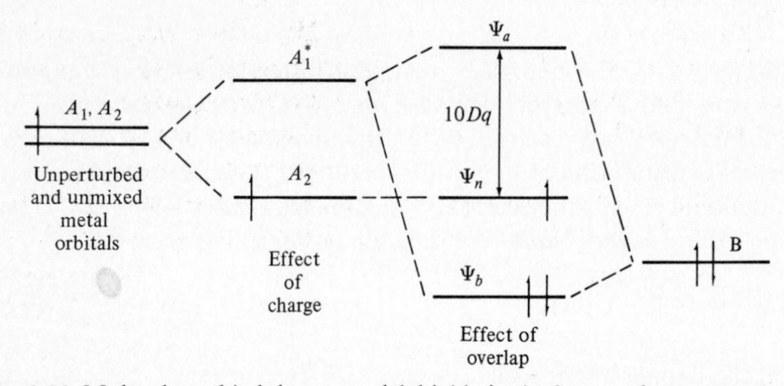

Fig. 9.39 Molecular orbital theory model, highly ionic. See text for explanation.

The results of the two treatments are thus qualitatively similar: (1) The odd electron occupies an orbital on A directed away from B. (2) Addition of $10Dq$ of energy will promote the electron to an orbital directed toward B. (3) The presence of the odd electron in this latter orbital will *destabilize* the molecule. (4) The energy of transition ($10Dq$) will be proportional to the extent of interaction between A and B. The differences which arise because of the relative energies of A_1, A_2, B, and the other orbitals are a result of our original assumptions: The crystal field model assumes no covalent bonding, so the energy of the electron on B must be much lower than on A. Molecular orbital theory assumes (to a first approximation) that the orbital energies are similar. As the electronegativity of B is increased, it will drop in energy and the CFT diagram (Fig. 9.37) may be obtained as the limit when the bonding is completely ionic (Fig. 9.39).[74] We shall return to this point later (see p. 428).

Use of orbital symmetry

The construction of molecular orbitals for an octahedral complex such as $[Co(NH_3)_6]^{+3}$ differs from the previous simple case of BeH_2 only in the fact that there is a greater number of overlapping orbitals to consider: $3d$, $4s$, $4p$ on the Co^{+3} and six approximately sp^3

[74] In this regard, the reader may wish to review the relation between electronegativity and molecular orbital formation, pp. 137–143.

hybrid (lone pair) orbitals on the ammonia ligands. For π-bonding ligands, further orbitals must be considered. In addition, the number of electrons is greater: six from Co^{+3}, twelve from the ligands. Although it might at first appear that finding the proper combination of nine metal orbitals with six ligand orbitals would be a hopeless task, the use of symmetry properties of the orbitals greatly simplifies the problem. We have seen (Figs. 3.25 and 3.28) that certain atomic orbitals can overlap in certain ways to have positive overlap integrals and the formation of bonding molecular orbitals. Changing the orientation of the orbitals or changing the symmetry type can result in zero or negative overlap. It is therefore possible to simplify the problem by consideration of overlap possibilities. This can be done rather elegantly by means of group theory,[75] but we shall follow the more easily visualized, although less elegant, method of pictorial representation.

Before proceeding with transition metal complexes involving d orbitals, consider once more the simple BeH_2 molecule as a complex of $Be^{+2} + 2H^-$. The molecular orbitals in BeH_2 can be obtained by inspection without need to resort to *ligand group orbitals* and symmetry properties, but for this very reason they provide a useful comparison.

There are two orbitals available on the beryllium for bonding: the $2s$ (with A_{1g} symmetry) and the $2p_z$ (with A_{2u} symmetry). Since the resulting molecular orbitals will be linear combinations of the atomic orbitals of metal and ligand with the same symmetry, it is appropriate to construct linear combinations of the ligand orbitals to match the symmetry of the metal orbitals. An a_{1g} *ligand group orbital* (LGO) can be constructed by *adding* the $1s$ wave functions of the hydrogen atoms:

$$\Psi_{a_{1g}} = \Psi_{\mathrm{H}} + \Psi_{\mathrm{H'}} \tag{9.34}$$

This LGO has the required symmetry since, like the beryllium $2s$ orbital, the wave function is everywhere positive.[76] The second LGO can be constructed with the *ungerade* symmetry, A_{2u}:

$$\Psi_{a_{2u}} = \Psi_{\mathrm{H}} - \Psi_{\mathrm{H'}} \tag{9.35}$$

The appropriate molecular orbitals may now be written:

$$\sigma_g = \Psi_{2s} + \Psi_{a_{1g}} \tag{9.36}$$

$$\sigma_u = \Psi_{2p_z} + \Psi_{a_{2u}} \tag{9.37}$$

It can be seen that this method is a somewhat formalized way of obtaining the same molecular orbitals that were written down by inspection previously (pp. 119–121). The ligand group method has the advantage of systematizing the process of forming molecular orbitals.

Consider the five $3d$ orbitals, for example. Two of these, the d_{z^2} and $d_{x^2-y^2}$, are directed toward the ligands providing positive overlap. The three remaining, the d_{xy}, d_{xz}, and d_{yz}, are directed between the ligands and the net overlap is zero (see Fig. 9.40). Note that these

[75] H. Bethe, *Ann. Physik*, **1929**, *3*, 133. For discussions in modern terminology see F. A. Cotton, "Chemical Applications of Group Theory," 2nd ed., Wiley (Interscience), New York, **1970**, Chap. 8; B. N. Figgis, "Introduction to Ligand Fields," Wiley (Interscience), New York, **1966**, Chap. 8; S. F. A. Kettle, *J. Chem. Educ.*, **1966**, *43*, 652.

[76] This ignores the small infranodal region of the $2s$ orbital having a negative sign. See p. 23.

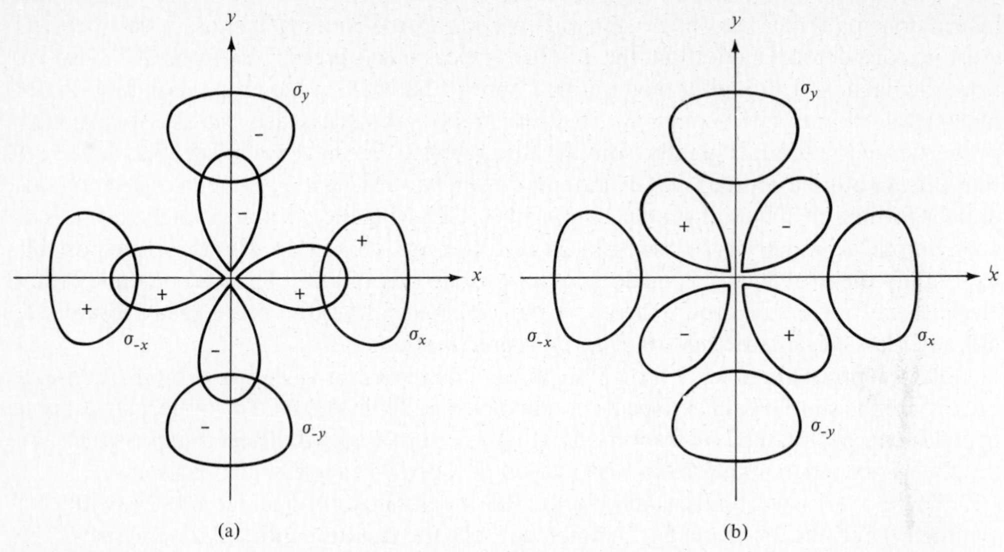

Fig. 9.40 Overlap of ligands in xy plane with $d_{x^2-y^2}$ (a) and d_{xy} (b) orbitals. Note that an appropriate choice of sign for the ligand orbitals provides positive overlap for the $d_{x^2-y^2}$ orbital. No single sign for the ligand wave function produces positive overlap with the d_{xy} orbital, however.

two groups of d orbitals are the same as we have seen previously in crystal field theory, namely, t_{2g} and e_g.

An LGO can now be constructed on the basis of Fig. 9.40:

$$\Psi_{\mathrm{LGO},x^2-y^2} = \tfrac{1}{2}(\Psi_{\sigma_x} + \Psi_{\sigma_{-x}} - \Psi_{\sigma_y} - \Psi_{\sigma_{-y}}) \quad \textbf{(9.38)}$$

or expressed in a somewhat simpler symbolism:

$$\sum\nolimits_{x^2-y^2} = \tfrac{1}{2}(\sigma_x + \sigma_{-x} - \sigma_y - \sigma_{-y}) \quad \textbf{(9.39)}$$

where $\sum$ and σ represent the wave functions for the ligand group orbital and the contributing atomic orbitals, respectively. The second e_g orbital[77] of the metal is the d_{z^2} orbital and the appropriate $\sum_{z^2}$ LGO can be written:[78]

$$\sum\nolimits_{z^2} = \frac{1}{2\sqrt{3}}(2\sigma_z + 2\sigma_{-z} - \sigma_x - \sigma_{-x} - \sigma_y - \sigma_{-y}) \quad \textbf{(9.40)}$$

The pictorial representation of the e_g orbitals is given in Fig. 9.41b.

---

[77] There is no universal agreement concerning the usage of e_g versus E_g, t_{2g} versus T_{2g}, etc. Both refer to the symmetry properties of the orbitals. In general, when speaking of the orbitals of an isolated (or hypothetically isolated, as in pure CFT) metal ion, it is customary to use lower case, e_g, t_{2g}, etc. When speaking of the symmetry of AOs, LGOs, or resulting MOs, upper case symbols, E_g, T_{2g}, etc., are often used. Since upper case symbols are also used to refer to spectroscopic states of complexes (see next section), in order to prevent possible confusion, the use of upper case symbols will be restricted to spectroscopic states and symmetry symbols. Orbitals (whether atomic or molecular) will be labeled with lower case equivalents. For example, d^1 in an octahedral complex has T_{2g} symmetry, has an electron configuration t_{2g}^1, and is a T_{2g} spectroscopic state.

[78] The coefficients 1/2 and $1/2\sqrt{3}$ before the entire wave functions for the LGOs as well as the coefficients for individual AOs (e.g., $2\sigma_x$) result from the necessity to normalize the wave functions. Note that the "z-lobe" of the d_{z^2} orbital is twice as large as the "xy-band" of that orbital (p. 370; Fig. 9.4).

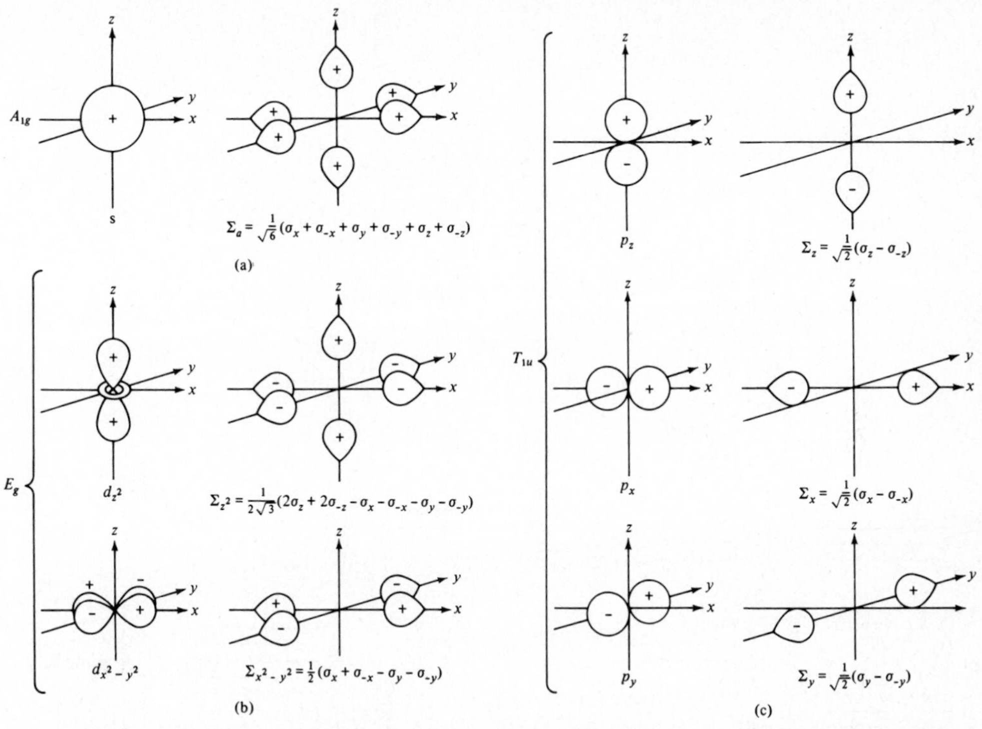

Fig. 9.41 Ligand Group Orbitals (LGOs) and matching Atomic Orbitals (AOs) of the same symmetry.

The metal t_{2g} orbitals cannot form σ bonds (see Fig. 9.40) and the discussion of them will be deferred to the section on π bonding. In the meantime they may be considered nonbonding orbitals. The spherical 4*s* orbital has A_{1g} symmetry and a ligand group orbital of A_{1g} symmetry can readily be constructed (Fig. 9.41a). In a similar manner, LGOs can be written for the 4*p* orbitals (T_{1u} symmetry, Fig. 9.41c).

It is possible to set up an energy level diagram for an octahedral complex such as $[Co(NH_3)_6]^{+3}$. Note that there are many approximations involved and that the resulting diagram is only qualitatively accurate. Even the ordering of the energy levels may be a matter of uncertainty. Although this is unfortunate, it does not seriously detract from the usefulness of the MO diagram. It is certain that the overlap of the 4*s* and 4*p* orbitals with the ligands is considerably better than that of the 3*d* orbitals.[79] As a result, the a_{1g} and t_{1u} molecular orbitals are the lowest in energy and the corresponding a_{1g}^* and t_{1u}^* antibonding orbitals the highest in energy. The e_g and e_g^* orbitals arising from the 3*d* orbitals are displaced less from their barycenter because of poorer overlap. The t_{2g} orbitals are nonbonding (in a σ-only system) and not displaced. The resulting energy level diagram appears in Fig. 9.42.

Electrons may now be added to the molecular orbitals of the complex in order of increasing energy. In a complex such as $[Co(NH_3)_6]^{+3}$ there will be a total of eighteen

[79] In general, *d* orbitals tend to be large and diffuse and as a result overlap of *d* orbitals may be *quantitatively* poor even when *qualitatively* favorable. This problem is discussed in Chapter 17.

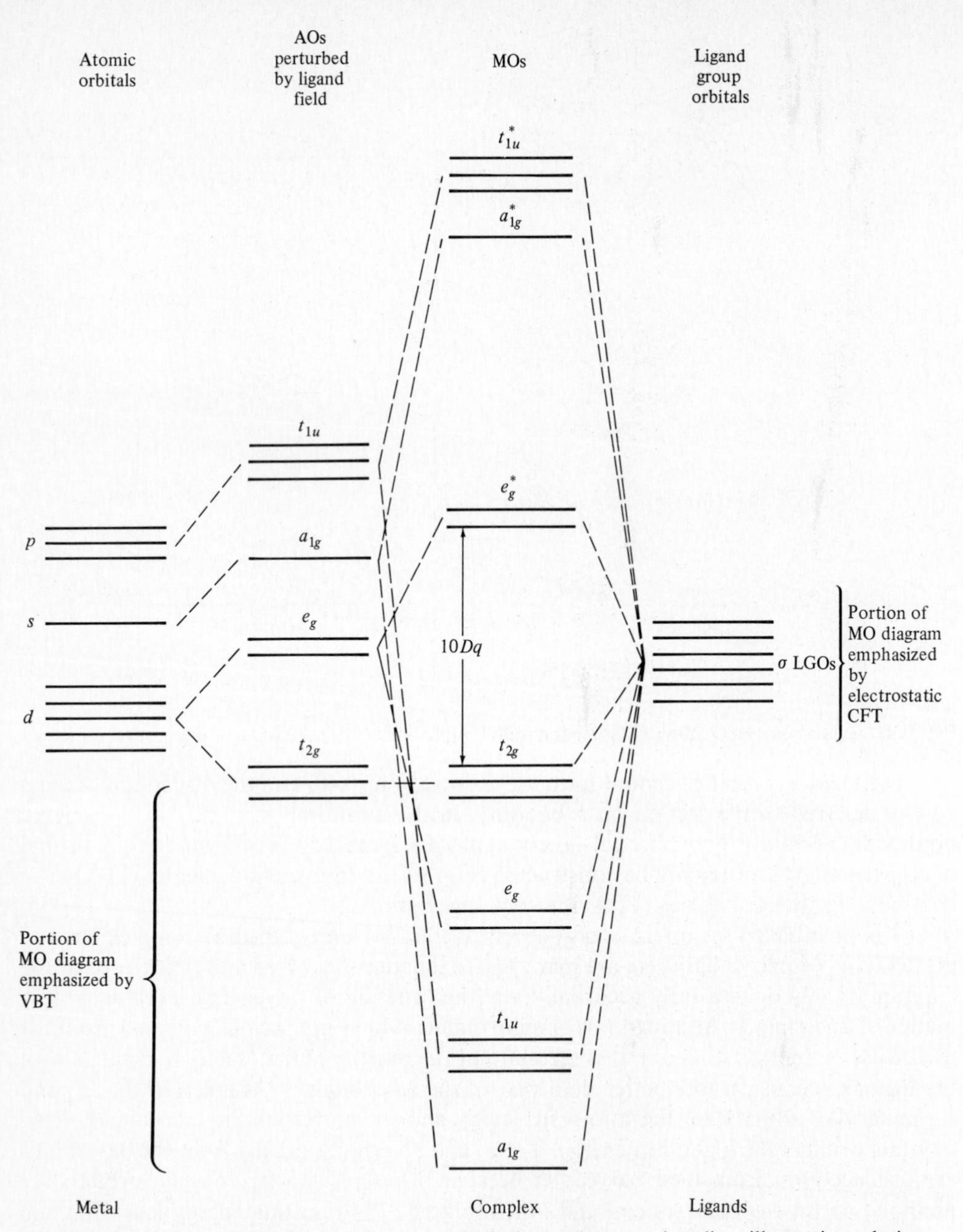

Fig. 9.42 Molecular orbital diagram for an octahedral complex, no π bonding, illustrating relation between MOT, CFT, and VBT. The perturbing influence of the ligand field on the atomic orbitals (*à la* CFT) is unnecessary in MOT but has been added to illustrate the difference between the e_g and t_{2g} orbitals.

electrons, twelve from lone pairs on the nitrogen atom and six from the $3d^6$ configuration of the Co^{+3} ion. The electron configuration will then be:[80] $a_{1g}^2 t_{1u}^6 e_g^4 t_{2g}^6$. (See Problem 9.33.) Note that the complex is diamagnetic because the electrons pair in the t_{2g} level rather than enter the higher energy e_g^* level.[81] If this energy difference is small, as in $[CoF_6]^{-3}$, the electrons will be distributed $t_{2g}^4 e_g^{*2}$. Thus both molecular orbital theory and crystal field theory account for the magnetic and spectral properties of octahedral complex ions by supposing the existence of two sets of orbitals separated by an energy gap, $10Dq$. If this energy is greater than the pairing energy, low-spin complexes will be formed, but if the energy necessary to pair the electrons is greater than $10Dq$, high-spin complexes will result.[82]

THE ANGULAR OVERLAP METHOD[83]

Although a diagram such as Fig. 9.42 "explains everything," in a certain sense, it is intellectually unsatisfying in that it does not provide a complete understanding of the physical rationale for the molecular orbital theory of complexes or a method to quantify the energies involved. There are many ways of approaching the problem, none perfect and all involving many assumptions, so the following qualitative discussion is meant to illustrate a general method rather than provide details.

We have seen (Fig. 3.56) that the interaction of two atomic orbitals having symmetry appropriate for positive overlap will give a bonding and an antibonding orbital with energies[84] determined by the overlap and the energies of the original orbitals:

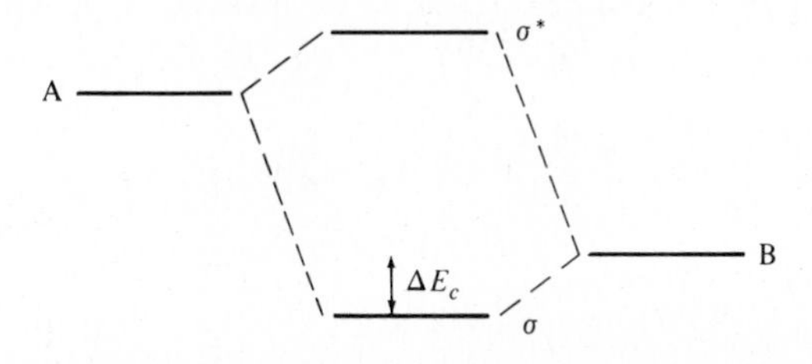

[80] It is not common to write out the complete molecular orbital configuration in this way but rather to abbreviate, as in CFT, to t_{2g}^6.

[81] Note that in CFT the equivalent notation would be $t_{2g}^4 e_g^2$ because the latter theory ignores the distinction between bonding, nonbonding, and antibonding orbitals.

[82] Note that Fig. 9.42 corresponds to Fig. 9.6 for CFT. To be precise, the energies of the two spin sets, α and β, should be shown as in Figs. 9.9 and 9.10. Such diagrams would be extremely difficult to follow because of the number of levels involved, and so the representation of Fig. 9.42 is usually followed. It is well to keep in mind the representation of Fig. 9.9 when discussing the populations of electrons in t_{2g} and e_g^* levels.

[83] For more thorough discussions of AOM see J. K. Burdett, *Adv. Inorg. Chem. Radiochem.*, **1978**, *21*, 113; "Molecular Shapes," Wiley, New York, **1980**, pp. 142–195; R. L. DeKock and H. B. Gray, "Chemical Structure and Bonding," Benjamin/Cummings, Menlo Park, Calif., **1980**, pp. 391–405. Some authors refer to the "angular orbital model" if they are stressing the theory, rather than to the "angular orbital method," which stresses its application. Both are abbreviated AOM.

[84] In Fig. 3.56 and other figures dealing with simple diatomic molecules, energy was denoted by ΔE; it is more common in AOM to use E, e, or ε.

It is also not difficult to see that for orbitals other than s orbitals the extent of overlap will be critically dependent upon the angle between the orbitals:

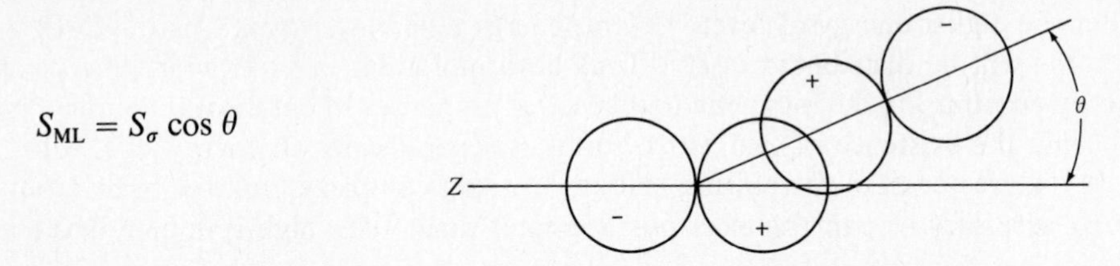

This is the basis of the *angular overlap model* (AOM) for treating coordination compounds by molecular orbital theory. The energy of interaction, E, is taken as:

$$E = \beta S_{ML}^2 \tag{9.41}$$

where S_{ML} is the overlap integral between orbitals on atoms M and L and β is a constant inversely proportional to the difference in energy between the original orbitals. Because the radial portions of the wave functions (see p. 16) will be constant for a given metal ion and set of ligands, only the angular dependence of the overlap integral need be considered. For the simple case of a p_z orbital given above, this is merely $\cos\theta$. Similar functions for d orbitals are given in Table 9.16. We may then write Eq. 9.41 as:

$$E = \beta S_\sigma^2 f(\theta, \phi) \tag{9.42}$$

where S_σ is the overlap of the d_z^2 orbital with a ligand lying on the z axis ($\theta = 0°$).

The use of Eq. 9.42 and Table 9.16 may be briefly illustrated by considering the two e_g orbitals of an octahedral set. The two LGOs have been given previously (Eqs. 9.39 and 9.40). By means of Fig. 9.43 we can evaluate S_{ML} for Eq. 9.39:

$$\theta = 90° \tag{9.43}$$

$$\phi = 0°, 90°, 180°, 270° \tag{9.44}$$

$$S_{ML} = \tfrac{1}{2} \cdot 4 \frac{\sqrt{3}}{2} S_\sigma \tag{9.45}$$

Table 9.16 Overlap integrals of metal atom d orbitals as a function of the ligand positions in polar coordinates (θ, ϕ)[a]

d_{z^2}	$\frac{1}{2}(3\cos^2\theta - 1)S_\sigma$
$d_{x^2-y^2}$	$(\sqrt{\frac{3}{2}})\cos^2\phi\sin^2\theta\ S_\sigma$
d_{xz}	$(\sqrt{\frac{3}{2}})\cos\phi\sin^2\theta\ S_\sigma$
d_{yz}	$(\sqrt{\frac{3}{2}})\sin\phi\sin^2\theta\ S_\sigma$
d_{xy}	$(\sqrt{\frac{3}{2}})\sin^2\phi\sin^2\theta\ S_\sigma$

[a] See Fig. 9.43. From J. K. Burdett, *Adv. Inorg. Chem. Radiochem.*, **1978**, *21*, 113. Used with permission.

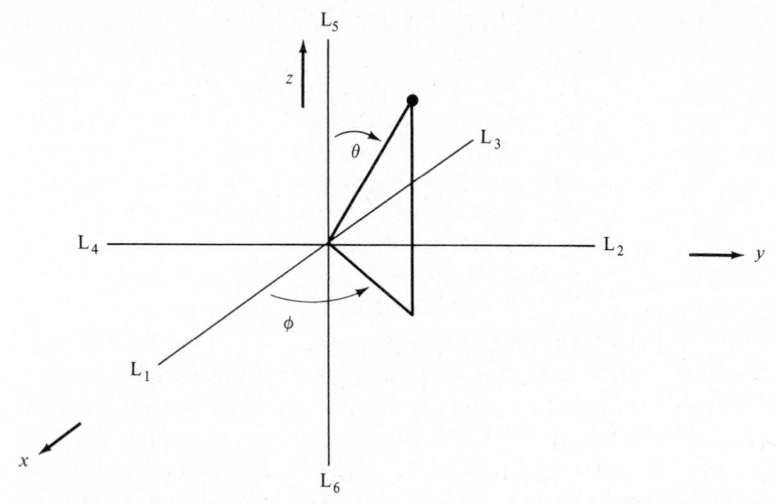

Fig. 9.43 Polar coordinates and ligand σ-orbital labeling in the octahedron. [From J. K. Burdett, *Adv. Inorg. Chem. Radiochem.*, **1978**, *21*, 113. Reproduced with permission.]

for an energy of:

$$E = 3\beta S_\sigma^2 \tag{9.46}$$

The energy for the second e_g orbital may be likewise evaluated from Eq. 9.40 and is also $3\beta S_\sigma^2$ (as it *must* be from symmetry considerations), and so the destabilization by electrons in e_g orbitals in octahedral complexes will occur in units of $3\beta S_\sigma^2$.

A comparison of the results from the AOM and crystal field theory is shown in Fig. 9.44. Note that by the definitions of the two methods, $10Dq = 3\beta S_\sigma^2$. The source of the splitting

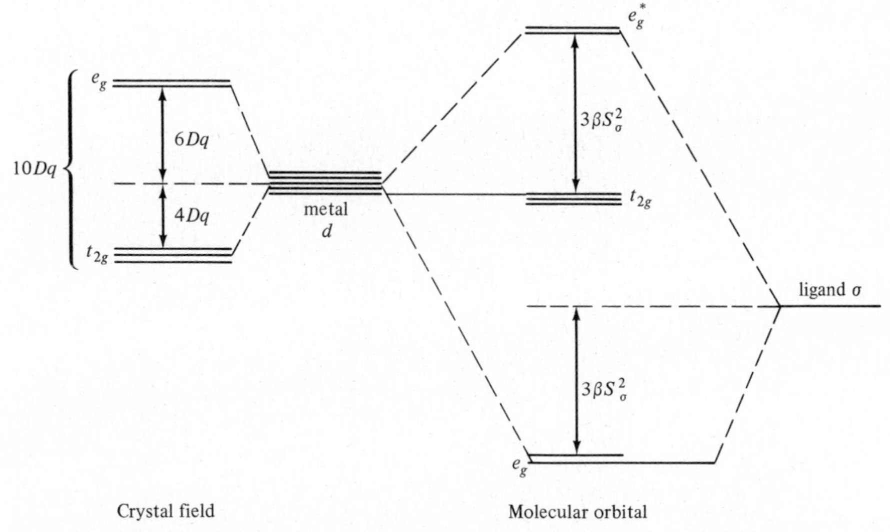

Fig. 9.44 Crystal field and molecular orbital energy levels for an octahedral complex.

of the t_{2g} and $e_g(e_g^*)$ orbitals is ascribed to different causes, and the origins of the scales on which the energies are measured differ. We have seen that the most common reference point for crystal field energies is the barycenter of the unperturbed d orbitals. The AOM generally uses the barycenter of the metal d orbitals and the ligand σ donor orbitals. Thus an octahedral complex gains $12\beta S_\sigma^2$ of *molecular orbital stabilization energy* (MOSE) immediately on formation from the four "ligand" electrons occupying the e_g level, each stabilized to an extent of $3\beta S_\sigma^2$, without regard to d electrons. Now addition of the first three electrons (d^1–d^3) will cause no change in the MOSE because in a σ-only system the t_{2g} level is strictly nonbonding. However, high-spin d^4 will be less stabilized, $(12 - 3)\beta S_\sigma^2$, just as in the weak ligand field case. It is not difficult to work out the energies shown in Table 9.17.

If the MOSE and crystal field stabilization energy (CFSE) numbers do not seem comparable, it is because although in a certain sense they purport to measure the same quantities, in a very real sense they do not. A comparison will illustrate this difference as well as indicate how the AOM treats ligand effects. Recall that the "two-humped" curve of lattice energy as a function of d-orbital filling, Ca^{+2} to Zn^{+2}, was presented as evidence for ligand field effects (p. 387). A similar curve of hydration enthalpies of transition metal ions is shown in Fig. 9.45a. Note that the dashed line representing zero CFSE passes through weak-field d^0–d^5–d^{10}, Ca^{+2}–Mn^{+2}–Zn^{+2}. The deviations of the experimental data from the dashed line are directly proportional to the various CFSEs. The CFSE is taken as the difference between the actual enthalpy and that expected if the metal ion had no d electrons.

Now compare Fig. 9.45b with Table 9.17. The heavy dashed line here represents a hypothetical energy curve, Ca^{+2} through Zn^{+2}, were there no MOSE. This line has a fairly sharp upward slope because of the *decrease* in ionic radius, Ca^{+2} (114 pm) through Zn^{+2} (88 pm), which enhances favorable contributions from *both* the covalent (s and p

Table 9.17 Comparison of stabilization energies of high-spin and low-spin octahedral complexes[a]

	MOSE[b]		CFSE[c]	
Configuration	High-spin	Low-spin	High-spin	Low-spin
d^0	12	12	0	0
d^1	12	12	4	4
d^2	12	12	8	8
d^3	12	12	12	12
d^4	9	12	6	16
d^5	6	12	0	20
d^6	6	12	4	24
d^7	6	9	8	18
d^8	6	6	12	12
d^9	3	3	6	6
d^{10}	0	0	0	0

[a] Pairing energies, P, have been omitted for simplicity.
[b] Molecular orbital stabilization energies in units of βS_σ^2.
[c] Crystal field stabilization energies in units of Dq.

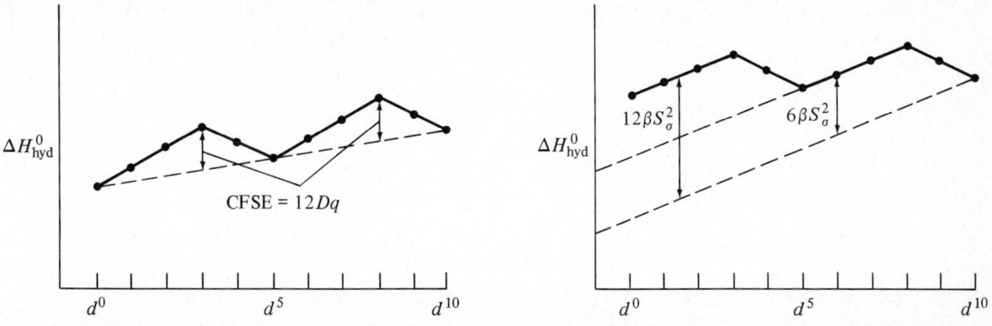

Fig. 9.45 Enthalpies of hydration of M^{+2} transition metal ions. (a) Crystal field interpretation of variation in energy by weak-field CFSE model. (b) Molecular orbital description by high-spin MOSE model. [From J. K. Burdett, *Adv. Inorg. Chem. Radiochem.*, **1978**, *21*, 113. Reproduced with permission.]

overlap) *and* the Madelung interactions to the bonding energy (see p. 322). Over and above this, we then have the MOSE adding a constant 12 units for the first four elements. The d^4 configuration (high-spin) sees a reduction in MOSE of 3 units resulting from one antibonding electron, and the d^5 configuration another reduction of 3 units. This results in a net stabilization of 6 units for the following elements until another antibonding electron appears at d^9, followed by the last at d^{10}. Note that Figs. 9.45a and 9.45b are identical as far as the experimental data are concerned (as they *must* be, of course!), but they differ in how the *base line* is drawn. The MOSE is the difference between the actual experimental value and that expected from a metal ion with *no* d-orbital-type occupancy, *including* e_g *electrons donated by ligands.*

Different ligand geometries

Using Table 9.16, the reader could work through all of the bond angles for the various geometries (and a little practice *can't* hurt!), but this has already been done by others, and so the information in Table 9.18 can be used to determine S_{ML} for several common geometries. Repeating the above type of calculation, one obtains:

$$E_{z^2} = (\tfrac{1}{4} + \tfrac{1}{4} + \tfrac{1}{4} + \tfrac{1}{4})\beta S_\sigma^2 \qquad \textbf{(9.47)}$$

$$E_{x^2-y^2} = (\tfrac{3}{4} + \tfrac{3}{4} + \tfrac{3}{4} + \tfrac{3}{4})\beta S_\sigma^2 \qquad \textbf{(9.48)}$$

for βS_σ^2 and $3\beta S_\sigma^2$ of MOSE for a square planar complex and thus destabilizations of βS_σ^2 for occupation of the $\sigma_{z^2}^*$ antibonding orbital and of $3\beta S_\sigma^2$ for occupation of the $\sigma_{x^2-y^2}^*$ orbital. Similarly, it is not difficult to compute the energies of the t_2 and e levels in tetrahedral symmetry:

$$E_{xz} = E_{xy} = E_{yz} = \tfrac{4}{3}\beta S_\sigma^2 \qquad \textbf{(9.49)}$$

$$E_{x^2-y^2} = E_{z^2} = 0 \qquad \textbf{(9.50)}$$

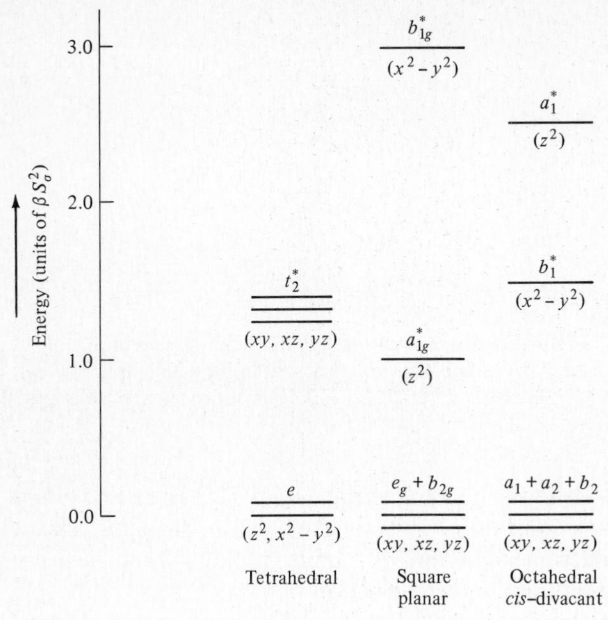

Fig. 9.46 Relative energies in some 4-coordinate geometries as estimated by the angular overlap method (AOM). Only σ orbitals are considered.

These results are of considerable interest in that they provide insight into the crystal field relationship (Eq. 9.11): $Dq_{T_d} = \frac{4}{9}Dq_{O_h}$. Consider our results for tetrahedral and octahedral symmetry:

$$(\tfrac{4}{3})\beta S_\sigma^2 \text{ (tetrahedral)} \div 3\beta S_\sigma^2 \text{ (octahedral)} = \tfrac{4}{9} \qquad \textbf{(9.51)}$$

Thus all of the "weak field" expectations of crystal field theory will likewise be anticipated in a *low-E*, high-spin AOM view of tetrahedral complexes. The energy levels for tetrahedral and square planar complexes are shown in Fig. 9.46. Just as we have seen previously in terms of the CFT model, the square planar arrangement has four relatively stable orbitals and one *very* unstable orbital. Hence low-spin d^8 is particularly suitable.

Also shown in Fig. 9.46 are the energy levels of "octahedral *cis*-divacant" complexes, a geometry we have not encountered previously. Inspection of Fig. 9.46 indicates why it is rare. For most occupancy numbers it is inferior in bonding energy to alternative arrangements, and it has some steric problems as well. Octahedral *cis*-divacant consists of a 4-coordinate complex formed from the hypothetical removal of two *cis* ligands from an octahedral complex. "Square planar" could be considered "octahedral *trans*-divacant" (see Chapter 10, p. 470 for *cis–trans* isomerism in octahedral complexes). The interest in this geometry is that it should occur at all—in $Cr(CO)_4$, for example. It serves as a warning that one must not forget the simple assumptions made above leading to Table 9.18 and Fig. 9.46. The geometry of $Cr(CO)_4$ can be rationalized,[83] but it requires theory beyond the scope of this book.

Table 9.18 Values of S_{ML}^2 (in units of S_σ^2) for d orbitals interacting with ligands forming complexes of various geometries. To find the total energy, add values for all rows numbered according to each ligand position in the diagrams. Square brackets indicate geometries formed by combination of other geometries.

Ligand position	z^2	$x^2 - y^2$	xz	xy	yz
1	$\frac{1}{4}$	$\frac{3}{4}$	0	0	0
2	$\frac{1}{4}$	$\frac{3}{4}$	0	0	0
3	$\frac{1}{4}$	$\frac{3}{4}$	0	0	0
4	$\frac{1}{4}$	$\frac{3}{4}$	0	0	0
5	1	0	0	0	0
6	1	0	0	0	0
7	$\frac{1}{4}$	$\frac{3}{16}$	0	0	$\frac{9}{16}$
8	$\frac{1}{4}$	$\frac{3}{16}$	0	0	$\frac{9}{16}$
9	0	0	$\frac{1}{3}$	$\frac{1}{3}$	$\frac{1}{3}$
10	0	0	$\frac{1}{3}$	$\frac{1}{3}$	$\frac{1}{3}$
11	0	0	$\frac{1}{3}$	$\frac{1}{3}$	$\frac{1}{3}$
12	0	0	$\frac{1}{3}$	$\frac{1}{3}$	$\frac{1}{3}$

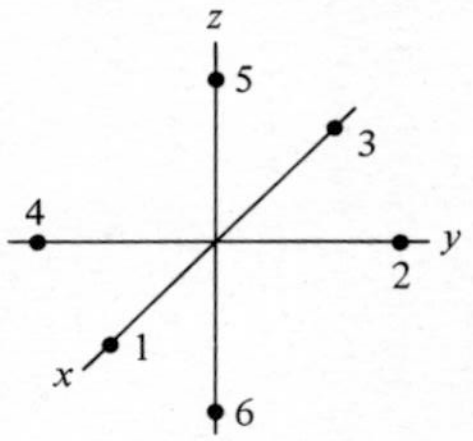

(Column headings z^2 to yz: Metal atomic orbitals)

Octahedral (positions 1–6): Square planar = 1 + 2 + 3 + 4 → (+5) → [Square pyramidal = 1 + 2 + 3 + 4 + 5]

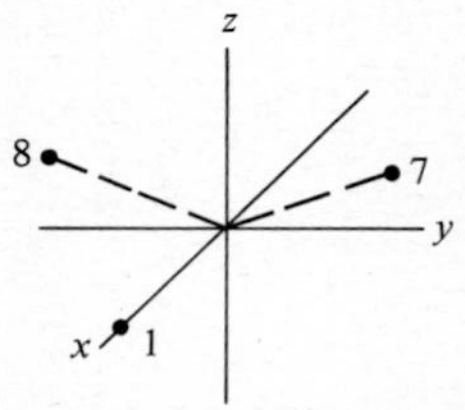

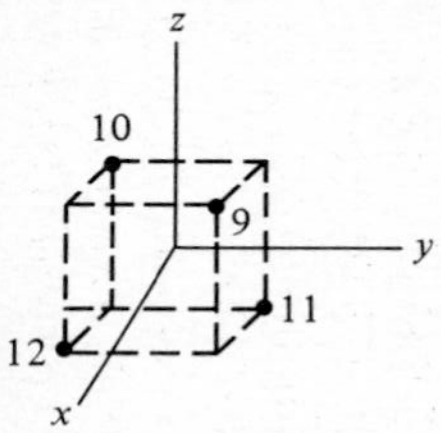

Linear = 5 + 6

Trigonal planar = 1 + 7 + 8 → (with Linear) → [Trigonal bipyramidal = 1 + 5 + 6 + 7 + 8]

Tetrahedral (positions 9–12): Cubic = 2 × (9 + 10 + 11 + 12)

In summary, the AOM gives us a direct means of evaluating the overlap of ligand and metal orbitals and the effect of orbital occupancy on molecular orbital stabilization energies.

Comparison of the molecular orbital, crystal field, and valence bond theories

Refer back to Fig. 9.48. Both molecular orbital theory (MOT) and crystal field theory (CFT) account for the magnetic and spectral properties of octahedral complex ions by supposing the existence of two sets of orbitals separated by an energy gap, $10Dq\ (= 3\beta S\sigma^2)$. If the energy necessary to pair the electrons is greater than this, high-spin complexes will form; otherwise, low-spin complexes form. Likewise, the visible spectra of complexes are attributed to electronic transitions such as $t_{2g} \rightarrow e_g^*$. The qualitative results of crystal field theory and molecular orbital theory are quite similar although the fundamental assumptions—purely electrostatic perturbations versus orbital mixing—seem considerably different. One might say that crystal field theory gives the "right" answer for "wrong" reasons. Cotton[85] has applied the wit and wisdom of G. B. Shaw to the characterization of the two theories: Crystal field theory is "too good to be true"; i.e., it is easy to understand and apply, but its physical reality is poor—a point-charge model does *not* represent a complex. On the other hand, molecular orbital theory is "too true to be good"; i.e., it corresponds so closely to reality that the simplicity of the crystal field theory is largely lost. This points to the crux of the difficulty of the molecular orbital theory: computational problems. The estimation of the appropriate overlap integrals in the absence of accurate wave functions presents many difficulties. Often in practice one uses simplifying assumptions, of which the angular overlap model (AOM) discussed above is typical. For many purposes a qualitative molecular orbital approach as described above is adequate, and we may use it without too much concern for the computational problems.[86]

To return to the comparison of molecular orbital theory with those discussed previously, it should be noted that despite the differences in basic assumption, both CFT and MOT describe complexes in terms of interactions between the metal orbitals and the ligands, and *the greater this interaction, the greater will be* $10Dq$. Since the six ligands always will donate twelve electrons to fill the a_{1g}, t_{1u}, and e_g orbitals, the d electrons will occupy the t_{2g} and e_g^* levels, subject to the influence of the size of $10Dq$, and so the distribution of electrons is identical to that predicted by simple CFT (cf. Fig. 9.42). The magnitude of $10Dq$ now results from the strength of the ligand–metal bond rather than from electrostatic effects. For strongly bonding ligands the a_{1g}, t_{1u}, and e_g levels will be lowered more, and hence the *antibonding* orbitals e_g^*, t_{1u}^*, and a_{1g}^* are correspondingly raised. As the ligands are moved away from the metal, the overlap of orbitals decreases, the bonding is weakened, and the orbitals move closer to their barycenter. In the extreme, with the ligands removed completely, we obtain the degenerate d orbitals as in CFT.

The valence bond theory in its usual form concentrates on the formation of a d^2sp^3 hybrid from the metal orbitals. Note that the a_{1g}, t_{1u}, and e_g molecular orbitals are formed from the single 4s, three 4p, and two of the five 3d orbitals of the metal (cf. Fig. 9.42). As is commonly the case with simple VB descriptions, the excited states are ignored.

---

[85] F. A. Cotton, *J. Chem. Educ.*, **1964**, *41*, 466.

[86] For a recent review of the status of molecular orbital theory in inorganic and organometallic chemistry, see R. F. Fenske, *Progr. Inorg. Chem.*, **1976**, *21*, 179. See also M. Gerloch et al., *Struct. Bond.*, **1981**, *46*, 1.

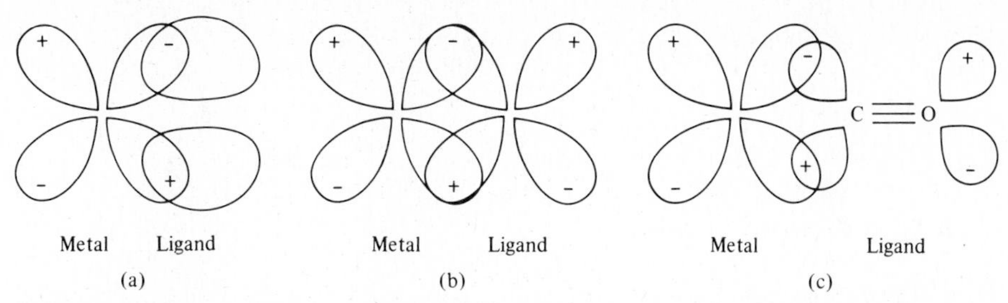

Fig. 9.47 Π bonds between a metal *d* orbital and ligand (a) *p* orbitals, (b) *d* orbitals, and (c) π^* antibonding orbitals.

Molecular orbital theory thus contains the best aspects of both valence bond theory and crystal field theory, and as shown by Van Vleck almost 50 years ago,[87] the latter two are but special cases of the more general molecular orbital theory.

π BONDING AND MOLECULAR ORBITAL THEORY

We have seen previously (pp. 366–367) that π bonding is useful in explaining by means of valence bond (VB) theory the stability of certain complexes. The angular overlap method (AOM) approach toward π bonding is very similar to that of σ bonding. Although the energy of the π interactions can be evaluated,[83] the values of E cannot be checked so readily against experimental data, and so the following discussion will be qualitative. Three types of ligand orbitals are available for π bonding with a metal *d* orbital: (1) a *p* orbital perpendicular to the σ-bond axis, (2) a *d* orbital lying in a plane that includes the metal atom, and (3) a π^* orbital lying in a plane that includes the metal atom (Fig. 9.47). The metal atom could conceivably use both its $t_{1u}(4p)$ and $t_{2g}(3d)$ orbitals. The former, however, are directed toward the ligands and used to form strong σ bonds. The t_{2g} metal orbitals are directed *between* the ligands and are thus nonbonding in a σ-only system (Fig. 9.40) but can readily π bond to LGOs having the appropriate T_{2g} symmetry (Fig. 9.48). In forming t_{2g} group orbitals from the twelve appropriate atomic orbitals (p_x and p_y on each ligand atom, for example), three other sets of group orbitals are also obtained: t_{1u}, t_{2u}, t_{1g}.[88] The t_{1u} LGOs have appropriate symmetry to overlap with the metal t_{1u} orbitals, but, as explained above, π bonding using these orbitals would tend to weaken the σ system. The t_{2u} and t_{1g} orbitals *must* remain nonbonding for the simple reason that there are no t_{2u} or t_{1g} orbitals on the metal since only those orbitals having the same symmetry can interact, mix, or combine. π bonding is thus restricted to the orbitals of T_{2g} symmetry.

[87] J. H. Van Vleck, *J. Chem. Phys.*, **1935**, *3*, 803, 807; J. H. Van Vleck and A. Sherman, *Rev. Modern Phys.*, **1935**, *7*, 167.

[88] The derivation of these sets will not be given here. See footnote 75. Note, however, that the t_{2g} LGO in Fig. 9.48 does indeed have the same symmetry with respect to rotation, reflection, etc., as the metal t_{2g} orbital. Also note that there *must* be four sets of triply degenerate (T) LGOs to account for the twelve atomic orbitals.

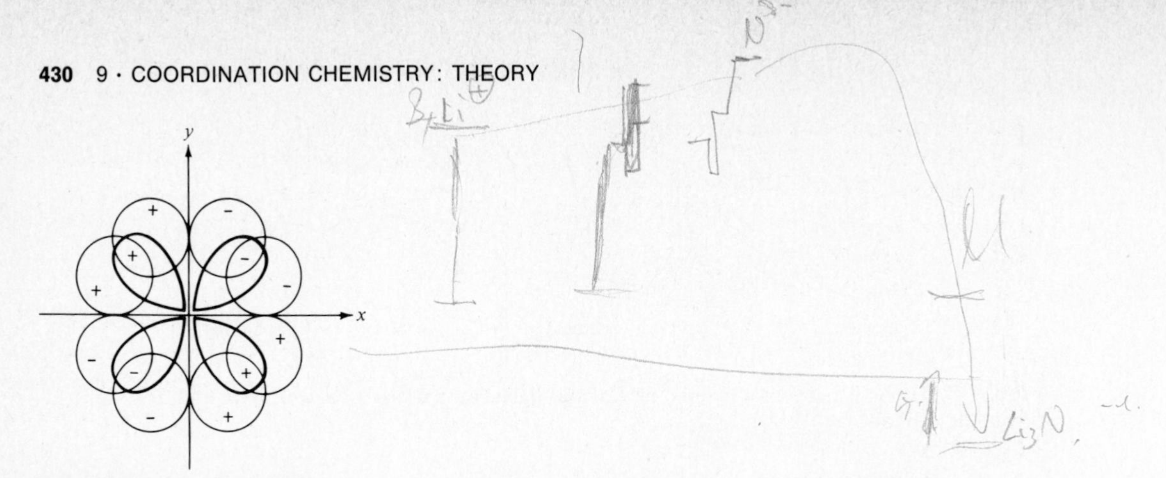

Fig. 9.48 Overlap of metal t_{2g} orbital (d_{xy}) with $t_{2g}(p)$ LGOs. There are two more sets mutually perpendicular.

We shall first consider the simplest possible case of π bonding in octahedral complexes, although one that is relatively unimportant. Consider the complex $[CoF_6]^{-3}$. The σ system will be similar to Fig. 9.42. The t_{2g} orbitals of the metal can interact with t_{2g} LGOs from the fluorine $2p$ orbitals. Since fluorine is more electronegative than cobalt, the fluorine $2p$ orbitals lie at a lower energy than the corresponding metal $3d$ orbitals. Under the circumstances, the bonding π orbital will resemble the fluorine orbital more than the metal orbital, and conversely the π^* orbital will more closely resemble the metal orbital (see pp. 139–141). The energy level diagram for the π system in $[CoF_6]^{-3}$ is shown in Fig. 9.49. Since the $2p$ orbitals on the fluorine are filled, these electrons will fill the resultant molecular t_{2g} π orbitals. The electrons from the $3d(t_{2g})$ orbitals of the cobalt are therefore in π^* orbitals at a higher energy than they would be if π bonding had not taken place. Since the level of the e_g^* orbitals is unaffected by the π interaction, $10Dq(e_g^* - t_{2g}^*$ or $e_g^* - t_{2g})$ is reduced as a result of the π bonding. It is felt that this is the source of the

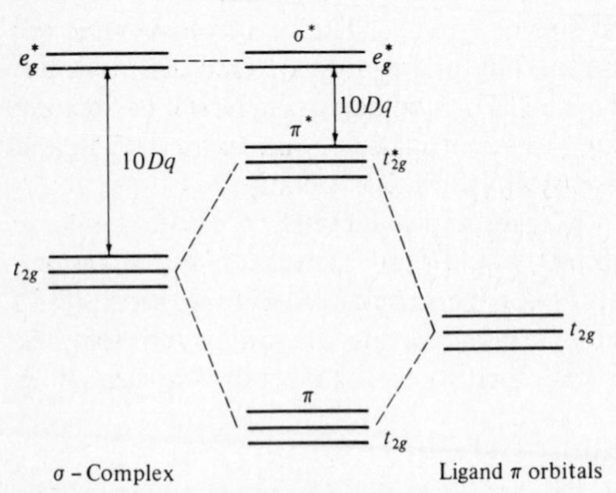

Fig. 9.49 Energy level diagram for π system of $[CoF_6]^{-3}$. Left: Unperturbed MOs for a σ-only octahedral complex. Right: t_{2g} LGO from fluorine $2p$ orbitals perpendicular to the σ bonds. Center: Resulting MOs *after* π interaction. Note that $10Dq$ is diminished by the π interaction and that there is little or no net gain in bonding energy.

position of fluoride (and other halides) at the weak-field extreme in the spectrochemical series (weaker than most σ-only ligands). Note also that the gain in bond energy is slight—the filled t_{2g} orbitals are lowered in energy somewhat, but the nearly filled t_{2g}^* orbitals are raised an equal quantity. Thus the only net stabilizing energy is that derived from the slightly different populations of the two sets of orbitals.

A more important case of π bonding occurs with ligands such as R_3P and R_2S. In these molecules the ligating atom can σ bond to the metal through an approximately sp^3 hybrid orbital in a manner similar to NH_3. Both phosphorus and sulfur, however, have empty $3d$ orbitals which can receive electron density from the metal. These orbitals have fairly low electronegativity (compared with a *positively* charged metal), and so the t_{2g} LGOs will lie at a *higher energy* than the corresponding metal orbitals. The resulting energy level diagram is shown in Fig. 9.50. Although the t_{2g} is lowered and the t_{2g}^* raised in a manner almost identical to that of the previous case, the fact that the ligand t_{2g} orbitals are *empty* allows the t_{2g}^* orbitals to rise with *no cost of energy* while the bonding t_{2g} orbitals are *stabilized*. Π bonding of this type (identical to the "back bonding" of valence bond theory) thus can stabilize a complex by increasing the bond energy. In addition, since the resulting t_{2g} pi orbital is delocalized over both metal and ligand (as opposed to the localized *metal* t_{2g} orbitals in which the electrons would have been without π bonding), electron density is removed from the metal. This will not be particularly desirable in a complex containing a metal in a high formal oxidation state since it will already carry a positive partial charge. In low oxidation states, on the other hand, electron density that tends to be built up via the σ system can be dispersed through the π system. A synergistic effect can cause the two systems to help each other. The more electron density that the π system can transfer from the metal to the ligand, the more the metal can accept via the σ system. In turn, the more the σ system removes from the ligand, the more readily the ligand can accept electron density through the π system. Up to a certain point, then, each system can augment the bonding possibilities of the other.

Π bonding by ligands such as those containing phosphorus and sulfur provides a simple explanation for the *raison d'etre* of strong field ligands that proved so vexing to a purely electrostatic interpretation of crystal field theory (pp. 381, 412). If we examine the strong-field end of the spectrochemical series (p. 384), we find ligands such as the nitrite ion, the cyanide ion, carbon monoxide, and a phosphorus-bearing ligand. The latter owes its position in the series to its ability to π bond as described above and *increase* the value of $10Dq$ relative to what it would be in a σ-only system (cf. Fig. 9.50). The other three ligands π bond in a very similar fashion except that the acceptor orbital is a π^* orbital (cf. Fig. 9.47c).[89] The result in each case is identical to that found for the phosphorus example: The bonding t_{2g} level is lowered so that the quantity $10Dq$ is *increased*. The increase in $10Dq$ is so substantial that in many cases (metal carbonyls, for example) the resulting complexes are colorless resulting from a blue shift of the absorption maximum completely out of the visible region into the UV.

Halide ions such as Cl^-, Br^-, and I^- present difficulties. Like the fluoride ion, they have *filled p* orbitals, but unlike the fluoride ion, they have *empty d* orbitals. At present

[89] See footnote 11 (p. 367) concerning the avoidance of semantic confusion over "bonding" by means of "antibonding" orbitals.

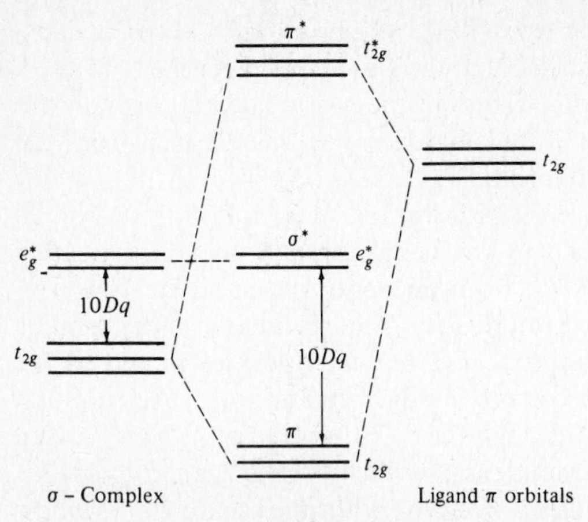

Fig. 9.50 Energy level diagram for a π system with an acceptor ligand such as P, S, or CO. Such a ligand provides an *empty* π orbital at a *higher energy* (lower electronegativity) than the metal orbitals. Note that π bonding in this case increases $10Dq$ and also increases the total bond energy of the system.

there is no *a priori* method of determining which set of t_{2g} LGOs (filled p or empty d) will interact more strongly with the t_{2g} orbitals of the metal, although with the advent of increasingly accurate wave functions, this should someday be possible. Empirically, we observe that all of the halide ions lie at the weak-field end of the spectrochemical series, and this would indicate that the p-orbital interaction is more important than the d-orbital interaction. On the other hand, the fact that the chloride ion acts as a very weak *trans*-director (see p. 542) indicates that in complexes such as $[PtCl_4]^{-2}$ at least there is a modicum of d–d π bonding.

Finally, there is the possibility that the d orbitals of the metal are empty and there are filled orbitals on the ligands capable of donating electrons to form π bonds. This situation is not likely to be important in most complexes since the metal will normally have as much or more electron density than optimal simply from the σ system. Only in complexes having extremely high formal charges would we expect to find any of this type of bonding. Fluoride complexes in which the metal is in a very high oxidation state (MnF_6^-, TaF_8^{-3}, etc.) might possibly have some π bonding of this type, although there is no evidence that it is important. On the other hand, the oxyanions of transition metals in high oxidation states such as CrO_4^{-2}, MnO_4^-, and FeO_4^{-2} probably contain appreciable π bonding. In principle this π bonding can be treated as above but since the complexes are tetrahedral, the problem of π bonding is somewhat more complex and will not be discussed here.[90]

MEASUREMENT OF π-BONDING EFFECTS

The extent and importance of π bonding in coordination compounds is currently stimulating considerable discussion and experimentation. We have seen that π bonding provides a reasonable rationale for much of the spectrochemical series. In Chapter 11 we shall find

[90] See B. N. Figgis, "Introduction to Ligand Fields," Wiley (Interscience), New York, **1966**, pp. 196–201.

that π bonding appears to be important in determining patterns of ligand substitution reactions. In this section various experimental methods of evaluating π bonding will be compared.

Infrared spectroscopy

One of the most widely used experimental methods is the study of infrared spectra. As we shall soon see (p. 592), it is possible to differentiate between bridging carbonyl groups and terminal carbonyl groups by means of the frequency of the C—O absorption in the infrared. The frequency (or more properly, the force constant, k) is a measure of the resistance of the bond to displacement of the atoms and hence of the bond strength. Since triple bonds are stronger than double bonds their infrared absorption occurs at a higher frequency. This method may also be used to estimate qualitative differences in bond strength. For example, consider Table 9.19, which lists IR data for two isoelectronic series of metal carbonyls. On the basis of the absorption maxima we can say that the C—O bond energies in these two series decrease in the order $[Mn(CO)_6]^+ > [Cr(CO)_6] > [V(CO)_6]^-$ and $[Ni(CO)_4] > [Co(CO)_4]^- > [Fe(CO)_4]^{-2}$. These qualitative results are consistent with the π-bonding theory discussed above: The greater the positive charge on the central metal atom, the less readily the metal can back bond electrons into the π^* orbitals of the carbon monoxide ligands. In contrast, in the carbonylate anions the metal has a greater electron density to be dispersed and hence π bonding is enhanced. Note that the IR frequencies relate to the C—O bond and hence do not provide *direct* measurement of M—C bonds. However, the greater the amount of M—C π bonding, the less π bonding can occur between carbon and oxygen. This can be explained in molecular orbital terms as the competition of both the metal and the oxygen for the appropriate orbitals of the carbon, but it is expressed perhaps more succinctly in terms of resonance in VB theory:

$$\underset{\text{(I)}}{M^- — C{\equiv}O^+} \longleftrightarrow \underset{\text{(II)}}{M{=}C{=}O} \qquad \textbf{(9.52)}$$

Canonical form I tends to build up electron density on the metal (as a result of the dative σ bond) and is thus favored in complexes such as $[Mn(CO)_6]^+$. The C—O bond is thus

Table 9.19 Infrared absorptions of some metal carbonyls

Compound	Frequency (cm^{-1})
$[Mn(CO)_6]^+$	2090
$[Cr(CO)_6]$	2000
$[V(CO)_6]^-$	1860
$[Ni(CO)_4]$	2060
$[Co(CO)_4]^-$	1890
$[Fe(CO)_4]^{-2}$	1790

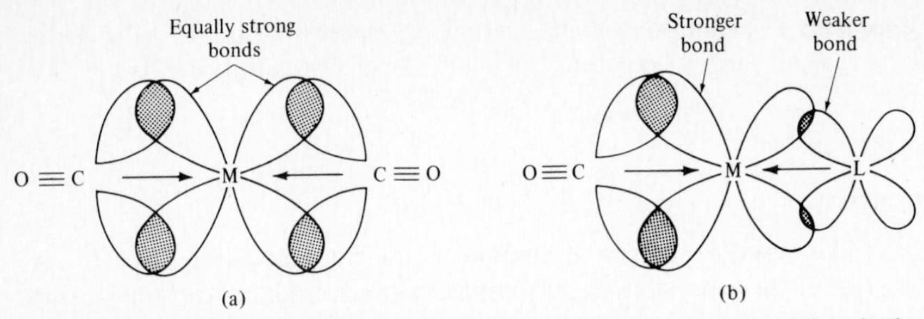

Fig. 9.51 Competition for π-bonding d orbitals of a central metal atom. (a) Equal π bonding resulting from equal competition. (b) Stronger OC—M π bonding resulting from lessened competition of weak π-bonding ligand, L. The "d orbital" left-right symmetry has been lost due to polarization resulting from unequal overlap by CO and L.

strengthened. Canonical form II is favored by negatively charged species since it prevents excessive buildup of negative charge by dispersing it back onto the ligands via the π system.

Competition for π electrons can be observed in a similar manner. Consider the two hypothetical molecules shown in Fig. 9.51. Although these molecules are hypothetical, they can be considered to represent the three perpendicular OC—Mo—CO or OC—Mo—L arrays in the octahedral molecules $[Mo(CO)_6]$ and $[Mo(CO)_3L_3]$ (Fig. 9.52). In the OC—Mo—CO system both carbonyls compete equally for electron density from the metal, and so the π bonds are equivalent. In contrast, if ligand L is a poor π-bonding ligand (or in the extreme, a σ-only ligand), it will not be able to compete successfully with the extremely good π-bonding ligand CO. Hence the electron density will tend to drain off the metal toward the carbon monoxide, making a very strong π bond to it and a very weak π bond (or none at all) to L. Again, this can be illustrated by a resonance picture:

$$\underset{\text{(I)}}{\text{OC—M=L}} \longleftrightarrow \underset{\text{(II)}}{\text{OC=M—L}} \qquad \textbf{(9.53)}$$

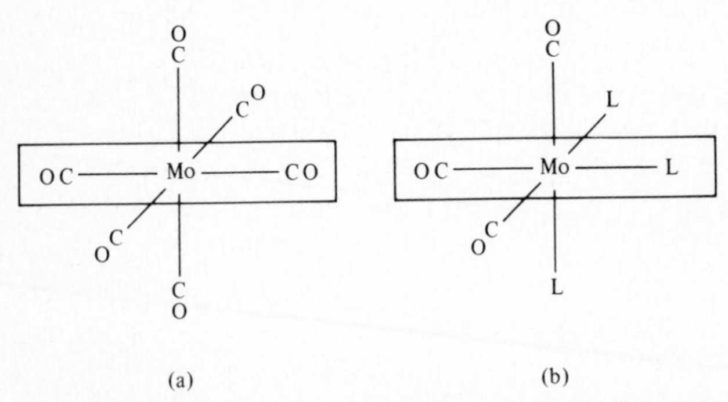

Fig. 9.52 Comparison of π bonding in the molecules $[Mo(CO)_6]$ and $[Mo(CO)_3L_3]$. Note that the portion enclosed in the box is the same as that illustrated in Fig. 9.51.

Table 9.20 IR frequencies of molybdenum carbonyls

Compound[a]	Frequency (cm^{-1})
$(PCl_3)_3Mo(CO)_3$, L = PCl_3	1989, 2041
$(\phi PCl_2)_3Mo(CO)_3$, L = ϕPCl_2	1943, 2016
$(\phi_2PCl)_3Mo(CO_3)$, L = ϕ_2PCl	1885, 1977
$(\phi_3P)_3Mo(CO)_3$, L = ϕ_3P	1835, 1949
$Py_3Mo(CO)_3$, L = Py	1746, 1888
dien$Mo(CO)_3$, L = dien/3	1723, 1883

[a] ϕ = C_6H_5; Py = C_5H_5N; dien = $H_2NCH_2CH_2NHCH_2CH_2NH_2$.

SOURCE: E. W. Abel et al., *J. Chem. Soc.*, **1959**, 2325. Used with permission.

For the case where L = CO, canonical forms I and II will contribute equally to the hybrid wave function, and we can say that there is half a π bond to each ligand. If L is a poorer π bonder than CO, canonical form II will contribute more to the hybrid and the metal–carbon π bond will be strengthened. This is the commonly found situation since carbon monoxide is probably the best π-bonding ligand, although there is a possibility that the nitrosyl group or phosphorus trifluoride may sometimes be a better π acceptor. In this case of better π bonding by L, canonical form I would contribute more to the hybrid than canonical form II.

Table 9.20 lists some infrared data for variously substituted molybdenum carbonyls. The decrease in IR frequencies on progressing down the table corresponds to decreasing C—O π bonding and increasing C—Mo π bonding. This indicates that the carbon monoxide becomes increasingly better at competing with the opposite ligand as we progress from phosphorus trichloride to diethylenetriamine (dien). For the phosphorus ligands this corresponds to replacement of the chlorine atoms by less electronegative phenyl groups, making the phosphorus less willing to accept negative charge. If pyridine can π bond at all, it will certainly do so to a limited extent. Finally, diethylenetriamine, like ammonia, should be a σ-only ligand.

Although IR frequencies provide an interesting *qualitative* picture of π bonding, in order to obtain quantitative data the corresponding force constants must be estimated. This is commonly done by means of the Cotton-Kraihanzel force-field technique. This procedure makes certain simplifying assumptions concerning the nature of the force constants in order to provide a practical solution to a problem which would be extremely difficult to solve rigorously.[91] The results of this type of calculation provide a means of

[91] The Cotton-Kraihanzel technique is beyond the scope of this book. The interested reader is referred to F. A. Cotton and C. S. Kraihanzel, *J. Am. Chem. Soc.*, **1962**, *84*, 4432; W. D. Horrocks, Jr., and R. C. Taylor, *Inorg. Chem.*, **1963**, *2*, 723; F. A. Cotton, *Inorg. Chem.*, **1964**, *3*, 702; F. A. Cotton et al., *Inorg. Chem.*, **1967**, *6*, 1357; J. K. Burdett et al., *Inorg. Chem.*, **1978**, *17*, 948; for a qualitative discussion, see F. A. Cotton and G. W. Wilkinson, "Advanced Inorganic Chemistry," 4th ed., Wiley, New York, **1980**, p. 1078.

setting up a π-acceptor series:[92]

$$NO > CO \sim RNC \sim PF_3 > PCl_3 > PCl_2OR > PCl_2R > PBr_2R > PCl(OR)_2 > PClR_2 > P(OR)_3 > PR_3 \sim SR_2 > RCN > o\text{-phenanthroline} > \text{alkyl amines, ethers, and alcohols}$$

This series shows many of the trends that might have been expected on the basis of electronegativity, especially the phosphorus-bearing ligands: $PF_3 > PCl_3 >$ halophosphites and halophosphines > phosphines. The similarity of phosphites and phosphines is more than might have been predicted on electronegativity arguments. This indicates that there may be significant O—P π bonding in the phosphites and competition for the phosphorus *d* orbitals. Alkyl amines, ethers, and alcohols have no empty low-lying orbitals and hence form the weak end of the π-acceptor system.

Graham[93] has extended this type of calculation still further and assigned numbers to various ligands in an effort to quantify their relative σ donor(−)/acceptor(+) and π donor(−)/acceptor(+) abilities:

Ligand	σ	π
Phosphorus trifluoride	−0.09	+0.79
Carbon monoxide	−0.06	+0.74
Phosphorus trichloride	−0.09	+0.71
Trimethyl phosphite	−0.36	+0.58
Tri-*n*-butylphosphine	−0.48	+0.48
Cyclohexylamine	0.00	0.00

As an example of the significance of these numbers, phosphorus trifluoride can be seen to be as good a π acceptor (+0.79) as or perhaps slightly better than carbon monoxide (+0.74), but neither is a particularly good σ donor. On the other hand, tributylphosphine, in line with expectations from the three electron-donating alkyl groups, is a fairly good σ donor (−0.48) but a comparatively poor π acceptor (+0.48). Note that cyclohexylamine has been arbitrarily assigned values of 0.00 rather than assuming that $\sigma = \pi = 0$. However, π bonding should be impossible in cyclohexylamine.

Recently another method of analysis has permitted the reproduction of experimental carbonyl stretching frequencies with remarkable accuracy.[94,95] The procedure involves group contributions which depend not only upon the group and the metal it is substituted

[92] There is no universal agreement on the ordering of this series since many of the ligands have π-accepting tendencies which are virtually indistinguishable. Arsenic and antimony compounds fall into the series alongside the corresponding phosphorus compound. Although there is some uncertainty as to whether these compounds are slightly better or slightly poorer acceptors than their phosphorus analogues, the accepting ability is probably in the order P > As > Sb. In the above series R can represent either phenyl or alkyl groups with the phenyl-substituted ligands exhibiting better acceptor ability.

[93] W. A. G. Graham, *Inorg. Chem.*, **1968**, *7*, 315.

[94] J. A. Timney, *Inorg. Chem.*, **1979**, *18*, 2502.

[95] Because of interactions among the carbonyl groups, the actual force constants vary about 2–4%. For information on this and on the process for converting force constants to frequencies and *vice versa*, see footnotes 91 and 94.

upon but also upon the position of substitution. For example, for a given carbonyl group the stretching force constant, k_{CO}, is given by

$$k_{CO} = k_d + n_\theta e_\theta + e_c \tag{9.54}$$

where k_d is the stretching force constant of the isolated monocarbonyl, M—C≡O, having a certain number (d) of electrons in the nd orbitals of the metal; n_θ is the number of other substituent groups at angle θ having substituent effects, e_θ; and e_c is a correction for charge effects, i.e., if the metal carbonyl is a cation or anion. Both k_d and e_θ have been tabulated (Tables 9.21 and 9.22) and may be used directly in Eq. 9.54. The value of e_c is $+197\ \mathrm{N\ m^{-1}}$, added for cations and subtracted for anions.

Application of Eq. 9.54 may be shown by a calculation for $V(CO)_6^-$. The molecule is octahedral, so $\theta = 90°$ (*cis* ligands) and 180° (the *trans* ligand).

$$k_{CO} = k_d + 4e_{cis} + e_{trans} + e_c \tag{9.55}$$

$$k_{CO} = 1387 + 4(33.5) + 1(126.1) + 197(-1) = 1450\ \mathrm{N\ m^{-1}}$$

Table 9.21 Values of k_d for equation 9.54 ($\mathrm{N\ m^{-1}}$)

Orbitals	k_5	k_6	k_7	k_8	k_9	k_{10}
3*d*	1372	1387	1444	1498	1554	1610
4*d*	—	1389	—	1506	—	1636?
5*d*	1353	1381	1445	1498	—	1613

SOURCE: J. A. Timney, *Inorg. Chem.*, **1979**, *18*, 2502. Used with permission.

Table 9.22 Some typical values of ligand effect constants ($\mathrm{N\ m^{-1}}$)

Ligand	$e_{90(cis)}$	$e_{180(trans)}$	$e_{109(td)}$	$e_{90(ax,eq)}$	$e_{120(eq,eq)}$
CO	33.5	126.1	37.3	25.5	51.4
NO	42	160	65		
N_2	14.0	52.0	45	22	
Cl	143	106	145		
Br	134	101	141		
I	112	104	125		
H	75	129		70	
CH_3	71	92	71		
PF_3	33.2	141.6	16.0	44.6	
PCl_3	30.6	109.3	35.3	21	
PPh_3	−21	39	−31.7	−52	
$P(OMe)_3$	−15.2	29.8	−11.2	−30	
en	−60	−54			
py	−29	−43			
Et_2O	−38	−5			

SOURCE: J. A. Timney, *Inorg. Chem.*, **1979**, *18*, 2502, in which additional values may be found. Reproduced with permission.

The details of the calculation are not so important for our purposes as are the general insights: (1) The increased and decreased frequencies of charged species that we have seen before (Table 9.19) are reflected in the increased and decreased (± 197 N m^{-1}) force constants of related species, such as $M(CO)_n$. (2) Once again, species such as NO, CS, and PF_3 most strongly affect the force constant and hence the stretching frequency of the carbonyl group, (3) *but effectively only when* in the *trans position.* Note the differences between the *cis* and *trans* values for the good π-bonding ligands, CO, CS, NO, and PF_3: The ratio averages around 1:4. In the same way, the tetrahedral constants are depressed by a large factor compared with the *trans* values, again consistent with the π-bonding hypothesis. However, we should note the perhaps somewhat overcautionary words of the designer of the above scheme: "Indeed, it is not obvious why the method works at all. The field demands considerable theoretical discussion."[94]

Nuclear magnetic resonance

Further evidence for the π-bonding model can be obtained through the use of nuclear magnetic resonance to study the bonding. We have seen in Chapter 2 that the electronic spin is quantized, $m_s = \pm\frac{1}{2}$. Likewise, the particles comprising the nucleus have quantized spins, and so the nucleus may have a net magnetic moment. The coupling constant J_{AB} is a measure of the interaction of the magnetic moments of the nuclei A and B. The coupling of the nuclear spins occurs indirectly via the bonding electrons. The relation between the magnitude of the coupling constant and the bond A—B may be pictured as follows:[96] If an electron is in the vicinity of nucleus A, its spin can interact with the nuclear spin. If this electron also interacts with the nucleus B on the other end of the A—B bond, it can cause a coupling of the nuclei A and B. The magnitude of the coupling will be proportional to the "time" an electron spends near each nucleus, i.e., to the bonding electron density at each nucleus. Briefly, three factors will be involved: (1) *hybridization* because only *s* electrons have finite electron density at the nucleus (Figs. 2.3, 2.6–2.11; note that all orbitals except *s* orbitals have a node at the nucleus); (2) *charges* on atoms A and/or B because positive charges tend to attract the electrons and increase coupling but negative charges tend to decrease it; and (3) the *strength* of the covalent bond between A and B. For example, if the bond were completely ionic, no exchange of electrons could take place, and hence no "information" could be conveyed from one nucleus to the other. Obviously, the interpretation of NMR data is difficult since many variables are involved. If factors 2 and 3 are relatively constant, coupling constants may provide information on the hybridization or *s* character of the orbitals involved in the bonding.[97] On the other hand, if the hybridization is assumed to be constant, estimates of the partial charges on the atoms may be obtained.[98]

---

[96] Space does not permit an extensive discussion of the principles of nuclear magnetic resonance, rather merely its applications to inorganic chemistry. For a thorough discussion, see A. Carrington and A. D. McLachlan, "Introduction to Magnetic Resonance," Harper & Row, New York, **1967**. For more recent papers discussing the relation between coupling constants, hybridization, and other factors, see A. C. Blizzard and D. P. Santry, *Chem. Commun.*, **1970**, 87, and references therein.

[97] C. Juan and H. S. Gutowsky, *J. Chem. Phys.*, **1962**, *37*, 2198; N. Muller and D. E. Pritchard, *J. Chem. Phys.*, **1959**, *31*, 768, 1471.

[98] D. M. Grant and W. M. Litchman, *J. Am. Chem. Soc.*, **1965**, *87*, 3994; J. E. Huheey, *J. Chem. Phys.*, **1966**, *45*, 405.

Table 9.23 Π bonding and NMR coupling constants (Hz)[a]

Metal	Phosphine	$R_3P—M—Cl$	$R_3P—M—PR_3$	$R_3P—M—CO$
Tungsten	Tributylphosphine		265	225
Platinum	Triethylphosphine	3520	2400	
Rhodium	Tributylphosphine	114	84	

[a] See footnotes 99–101 for additional data on compounds of this type.

Grim and co-workers have measured the coupling constants between ^{31}P in ligands and various complexed metals such as platinum(II),[99] tungsten(0),[100] and rhodium(I, III).[101] They found that the coupling constant between phosphorus and the central metal atom was consistently larger in those complexes in which P—M π bonding is expected to be highest and smaller in complexes in which π bonding is thought to be less important (see Table 9.23). Inasmuch as the hybridization of these complexes is not expected to deviate significantly from "ideal" dsp^2 [Pt(II), Rh(I)] and d^2sp^3 [W(0), Rh(III)] hybrids,[102] it was concluded that the differences in coupling resulted from stronger bonding[103] when the phosphorus was *trans* to a poor π bonder. Note that for a given metal, the coupling is greatest when the phosphorus is *trans* to the weakly π-bonding chloride ion, less when *trans* to the intermediate π bonder, phosphorus, and least when the phosphorus is *trans* to a carbonyl group.

Recently there has been increasing skepticism of the simple picture of π bonding and its effect either in the form of arguments favoring "σ and π effects" or even complete denial of the importance of π bonding in affecting the properties studied. For example, Angelici[104] has shown the IR frequencies discussed above can be related to the basicity (and hence σ-bonding ability) of the ligand. For example, amines are stronger bases than phosphines, and this corresponds to an increased carbonyl stretching frequency, presumably resulting from a weakening of the metal–carbonyl σ bond and strengthening of the carbonyl triple bond. If basicity of ligands is plotted versus stretching frequency, a linear relationship is found. Angelici interprets this as indicating that σ effects are important and π effects unimportant.

Other authors have argued that on the basis of theoretical analysis of IR frequencies[105] or study of IR force constants as the electron density on the metal is varied[106] that *both* σ and π effects must be considered and that there has been an overemphasis on the π system.

[99] S. O. Grim et al., *Inorg. Chem.*, **1967**, *6*, 1133. See also A. Pidcock et al., *Proc. Chem. Soc.*, **1962**, 184.

[100] S. O. Grim and D. A. Wheatland, *Inorg. Chem.*, **1969**, *8*, 1716.

[101] S. O. Grim and R. A. Ference, *Inorg. Nucl. Chem. Letters*, **1966**, *2*, 205.

[102] For a contrary view, see L. M. Venanzi, *Chem. Brit.*, **1968**, 162.

[103] Note that the π bond cannot be *directly* responsible for the coupling since it has a node at the nucleus. A synergistic effect from the π bond could strengthen the σ bond, however.

[104] R. J. Angelici, *J. Inorg. Nucl. Chem.*, **1966**, *28*, 2627; R. J. Angelici and M. D. Malone, *Inorg. Chem.*, **1967**, *6*, 1731; R. J. Angelici and C. M. Ingemanson, *Inorg. Chem.*, **1969**, *8*, 83.

[105] T. L. Brown and D. J. Darensbourg, *Inorg. Chem.*, **1968**, *7*, 959.

[106] R. E. Dessy and L. Wieczorek, *J. Am. Chem. Soc.*, **1969**, *91*, 4963.

Photoelectron spectroscopy

Further support for a σ-donor–π-acceptor interpretation of ligand-metal bonding comes from ESCA data.[107] In a large series of complexes of Fe, Co, Ni, Cu, Mo, Ru, Rh, Pd, W, Re, Os, Ir, and Pt, the metal binding energies become smaller (less positive) when σ-only ligands are added, but become larger for π-acceptor ligands such as: CS (+0.2 eV) < N_2 < CO < RN_2^+ < NO^+ (1.8 eV). Furthermore, there is some evidence that when more than one π acceptor is present they inhibit (compete with) each other. All of this is consistent with a model of ligands donating electron density to the metal through σ bonds and removing it (where possible) through π bonds.

Crystallography

The most convincing evidence, pro or con, in the π-bonding controversy comes from crystallographic determinations of bond lengths. Extensive π bonding should result in a shortening of the M—C bond distance and a lengthening of the C—O bond distance (see Eq. 9.52). Two problems beset the direct interpretation of carbonyl bond lengths. First, the bond length in carbon monoxide is relatively insensitive to bond order. The difference in length between a triple bond (113 pm) and a double bond (~120 pm) is small. Thus it is difficult to establish that the observed differences in bond lengths are meaningful since in most metal carbonyls the value is 114–115 pm, and unless the accuracy is great, this difference of 1–2 pm may not be statistically significant. The differences in bond lengths of metal–carbon bonds are greater, but now it is difficult to establish the proper covalent radius of the metal for a bond in the absence of π bonding.[108] However, Cotton and Wing[109] have accounted for most of the difficulties inherent in this type of calculation and shown that it is reasonable to attribute the bond lengths in carbonyl compounds to π bonding.

The best evidence for π bonding and *trans* competition for the metal *d* orbitals comes from the determination of bond lengths in phosphine and phosphite complexes of chromium[110] (Fig. 9.53). The bond lengths in pentacarbonyltriphenylphosphinechromium, pentacarbonyl(triphenyl phosphite)chromium, and *trans*-tetracarbonylbis(triphenyl phosphite)chromium are listed in Table 9.24. On the basis of infrared and other evidence, triphenyl phosphite is expected to be a better π acceptor than triphenylphosphine. All the bond lengths in Table 9.24 support this view. The Cr—P bond is *shorter* in the phosphite complex, as expected if the phosphite is better able to compete with the carbonyl group for electron density in the π bond. The *trans* (to P) Cr—C bond is longer and the *trans* C—O bond is *shorter* in the phosphite, again in support of the hypothesis that the phosphite is better able to compete with the carbonyl group. Furthermore, the comparison of the bond lengths in the carbonyl groups *trans* to each other (and hence *cis* to the phosphorus ligand) versus the carbonyl groups *trans* to the phosphorus ligand also confirms the π-bonding picture. Carbon monoxide ligands competing against each other should be able

[107] R. D. Feltham and P. Brant, *J. Am. Chem. Soc.*, **1982**, *104*, 641.

[108] Note that there are no simple compounds with single M—M bonds (see Chapter 6) and the *metallic radii* represent the metal in a totally different environment.

[109] F. A. Cotton and R. M. Wing, *Inorg. Chem.*, **1965**, *4*, 314.

[110] H. J. Plastas et al., *J. Am. Chem. Soc.*, **1969**, *91*, 4324; H. S. Preston et al., *Inorg. Chem.*, **1972**, *11*, 161.

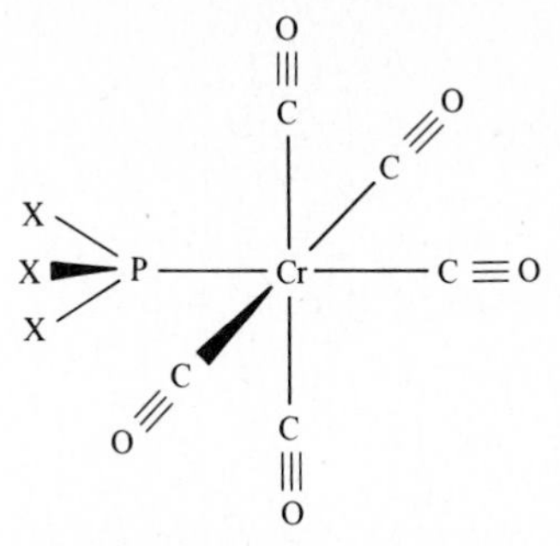

Fig. 9.53 Structure of the phosphine (X = phenyl) and phosphite (X = phenoxy) complexes of chromium carbonyl.

Table 9.24 Bond lengths (pm) in chromium carbonyl complexes with phosphorus ligands

Bond	$[Cr(P(\phi)_3)(CO)_5]$	$[Cr(P(O\phi)_3)(CO)_5]$	*trans*-$[Cr(P(O\phi)_3)_2(CO)_4]$
Cr—P	242.2	230.9	225.2
Cr—C (*trans* to P)	184.4	186.1	—
Cr—C (*trans* to CO)	188.0	189.6	187.8
C—O (*trans* to P)	115.4	113.6	—
C—O (*trans* to CO)	114.7	113.1	114.0

to π bond less efficiently than when competing against a less efficient phosphorus ligand. Hence it is not surprising that the C—O bond lengths are longer and Cr—C bond lengths are shorter for the carbonyl groups *trans* to the phosphorus than when *trans* to another carbon monoxide molecule.

The bond lengths in the *trans*-bis(triphenyl phosphite) complex also illustrate the expected trend. Here the two phosphite ligands competing with each other fare better in the π-acceptor competition than against a carbonyl in the mono(phosphite) complex and the Cr—P bonds are consequently shortened.

In view of the conflicting experimental evidence and opposing views, what conclusion can we draw concerning the importance of π bonding in metal complexes? As in any field in which data are rapidly accumulating, we cannot come to a definite conclusion. However, in the opinion of the author it is possible at this time to reject the σ-only viewpoint; it is incompatible with the evidence currently available. The *extent* of π bonding is still an open question. When π bonding has been pushed to an extreme, to the exclusion of all other considerations, the resulting model is apt to be distorted, as is the case in any description of molecules which considers only one variable. It is probable that we shall see a modified theory of balanced σ and π effects in complexes of this type.

SUMMARY OF MOLECULAR ORBITAL THEORY

Molecular orbital theory undoubtedly offers us the best current interpretation of the properties of coordination compounds. As we have seen, the valence bond theory in its usual form and the crystal field theory are but simplified versions of the molecular orbital theory in which certain aspects of the latter have been conveniently ignored. Often these omissions are necessary in order that the problem be tractable. It is quite legitimate to use either valence bond or crystal field theory to predict the qualitative properties of complexes as long as one realizes the inherent limitations. With increasingly accurate wave functions and increased use of high-speed computers to carry out the tedious calculations, it is certain that the molecular orbital theory will assume a preeminent position in the theoretical interpretation of metal complexes.

A CLOSER LOOK AT SPECTRA

Although it was indicated previously that electron–electron interactions must be considered if accurate energies are to be obtained for crystal field splittings, all of the energy diagrams used thus far have been the so-called "one-electron case" even when many *d* electrons were included. This disregard of electron–electron interaction does not cause serious problems as long as the discussion is qualitative, but the quantitative interpretation of spectra requires a more careful approach.

First let us return to the d^1 case discussed much earlier. As we have seen in Chapter 2, in the absence of external forces the set of five *d* orbitals is degenerate, the ground state of a d^1 species is ^2D, and the electron has an equal probability of being in any of the five *d* orbitals.[111] Under the influence of an octahedral field, either the electrostatic field of pure crystal field theory or the more complicated interaction with octahedral ligand group orbitals, the degenerate orbitals are split into t_{2g} and e_g (or e_g^*) orbitals and, likewise, the ^2D state is split into a $^2T_{2g}$ state (representing the $t_{2g}^1e_g^0$ configuration) and a ^2E_g state (representing the $t_{2g}^0e_g^1$ configuration) at higher energy. Before proceeding with the investigation of d^n states, let us first look at the interaction of an octahedral field with s^1, p^1, and f^1 configurations.

An *s* orbital is completely symmetrical and also nondegenerate; hence it is unaffected by all fields, including an octahedral field. The *p* orbitals are not split by an octahedral field since all interact equally, but fields of lower symmetry[112] can cause a splitting of the *p* orbitals. On the other hand, the set of *f* orbitals[113] is split by an octahedral field into three levels, a triply degenerate level $6Dq$ below the barycenter (t_1), a triply degenerate level $2Dq$ above the barycenter (t_2), and a single level $12Dq$ above the barycenter (a_2). We can thus say that these orbitals and their states transform in an octahedral field as follows:

State	Transforms in octahedral field as	States
^2S	$\longrightarrow$	$^2S(A_{1g})$
^2P	$\longrightarrow$	$^2P(^2T_{1g})$
^2D	$\longrightarrow$	$^2T_{2g}$, ^2E_g
^2F	$\longrightarrow$	$^2T_{1g}$, $^2T_{2g}$, $^2A_{2g}$

A complete set of transformations for *S*, *P*, *D*, *F*, *G*, *H*, and *I* states is given in Table 9.25.

[111] Actually, in the absence of an external reference the five *d* orbitals remain undefined.

[112] For example, consider the effect of a tetragonal field (either *z*-in or *z*-out) on the degeneracy of the *p* orbitals.

[113] A representation of the *f* orbitals will be found in Chapter 16.

Table 9.25 Splitting of d^n terms in an octahedral field

Term		Components in an octahedral field
S	$\longrightarrow$	A_{1g}
P	$\longrightarrow$	T_{1g}
D	$\longrightarrow$	$E_g + T_{2g}$
F	$\longrightarrow$	$A_{2g} + T_{1g} + T_{2g}$
G	$\longrightarrow$	$A_{1g} + E_g + T_{1g} + T_{2g}$
H	$\longrightarrow$	$E_g + T_{1g} + T_{1g} + T_{2g}$
I	$\longrightarrow$	$A_{1g} + A_{2g} + E_g + T_{1g} + T_{2g} + T_{2g}$

Next, let us return to the problem of d^n configurations, namely the d^2. In the absence of an external field this produces two states, a lower energy ^3F and higher energy ^3P.[114] We have seen that the ground state consists of two electrons in different orbitals. If an octahedral field is imposed, the two orbitals must be two of the t_{2g} orbitals. Now if we excite one of these electrons to the higher energy e_g orbitals, we find that there are two possibilities. If the electron being promoted comes from the xz or yz orbitals it will cost less energy to promote it to the d_{z^2} orbital than to the $d_{x^2-y^2}$ orbital. The reason for this is that the electron–electron repulsion causes the two resulting configurations $(xy)^1(x^2-y^2)^1$ and $(xy)^1(z^2)^1$ to be at different energies. The source of this energy difference is simply that the $(xy)^1(x^2-y^2)^1$ configuration has both electrons confined to the vicinity of the xy plane; i.e., neither orbital has a z component. The alternative configuration allows the electrons to separate somewhat and spread along all three coordinate axes. Finally, if two electrons are promoted (configuration e_g^2), we obtain a fourth, high-energy state.

To return to the subject of the transformation of orbitals in an octahedral field, we recall that p orbitals remain unchanged but that f orbitals are split into three levels: T_{1g}, T_{2g}, and A_{2g}. Now recall that the two triplet states arising from the d^2 configuration are the ^3F and ^3P. These states behave in an octahedral field exactly as the F and P states arising from the f^1 and p^1 configurations discussed above; i.e., the ^3F state is split into $^3T_{1g}(F)$, $^3T_{2g}$, and $^3A_{2g}$ states and the ^3P is unsplit and becomes the $^3T_{1g}(P)$ state.[115] The splitting of $d^1(^2D)$ and $d^2(^3F, {}^3P)$ configurations is shown in Fig. 9.54. Note from Table B.1, p. 954 that the low-energy states for all d^n configurations are S, P, D, or F states: d^1 and d^9, ^2D; d^2 and d^8, ^3F and ^3P; d^3 and d^7, ^4F and ^4P; d^4 and d^6, ^5D; and d^5, ^6S. The S and P states are not split in an octahedral field and the D and F states split as shown in Fig. 9.54.

By means of the hole formalism all the splittings can be interpreted in terms of Fig. 9.54. It was mentioned earlier that the d^9 configuration of Cu^{+2} could be treated as an inverted d^1 configuration; i.e., the single "hole" which tends to "float" to the top in a $t_{2g}^6e_g^3$ configuration is the equivalent of an electron in the inverted $e_g^1t_{2g}^0$ configuration. So the ^2D term of Cu^{+2} may be treated by Fig. 9.54a *inverted*; i.e., the ground state in an octahedral field is ^2E_g with an energy lying $10Dq$ below the first excited state $^2T_{2g}$. The d^8 configuration of Ni^{+2} (^3F and ^3P) can be treated in a similar manner as an inverted d^2.

[114] This disregards the higher-energy singlet states not favored by Hund's rules: ^1S, ^1D, and ^1F.

[115] The parenthetical F and P refer to the original ^3F and ^3P state since the states are otherwise the same, i.e., have the same symmetry.

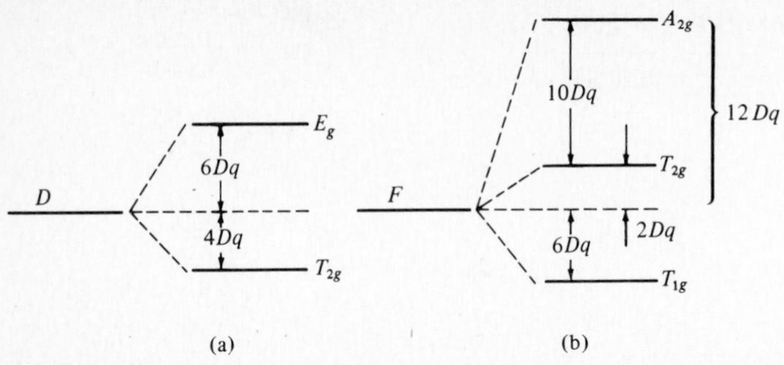

Fig. 9.54 Splitting of terms arising from (a) d^1 and (b) d^2 electron configurations.

Consideration of the splitting of F and P states in Fig. 9.54 provides us with the ^3F split into $^3A_{2g}$ lying $10Dq$ below the $^3T_{2g}$ which lies $8Dq$ below the $^3T_{1g}$. Note that the inversion applies only to the F states and that the F state is always (whether from d^2 or d^8) *lower* in energy than the ^3P.

In a similar manner, by considering a d^3 configuration to be a spherically symmetrical d^5 with two holes and the d^4 to be a d^5 with one hole, these configurations can be determined. Finally, the d^6 and d^7 configurations can be related to the d^5 configuration. These will not be discussed here since the reader can work them out for himself. The results are shown in Table 9.26.

This procedure may be applied to the spectra of complexes discussed previously. The spectra of several complexes of Cr(III) were illustrated. Each spectrum shows at least two well-defined absorption bands in the visible region. A third band is sometimes discernible although it is often obscured by the "charge transfer band" at a higher energy. The latter corresponds to the transfer of an electron from a molecular orbital having primarily ligand character (a ligand orbital in the crystal field approximation) to one having primarily metal character (metal orbital in CFT approximation). It thus corresponds

Table 9.26 Splitting of terms for d^n configurations in weak fields

Configuration of free ion	Ground state of free ion	Configuration of complexed ion	Energy level diagram (Fig. 9.54)
d^1	^2D	t_{2g}^1	(a)
d^2	^3F	t_{2g}^2	(b)
d^3	^4F	t_{2g}^3	Inverted (b)
d^4	^5D	$t_{2g}^3e_g^1$	Inverted (a)
d^5	^6S	$t_{2g}^3e_g^2$	No splitting
d^6	^5D	$t_{2g}^4e_g^2$	(a)
d^7	^4F	$t_{2g}^5e_g^2$	(b)
d^8	^3F	$t_{2g}^6e_g^2$	Inverted (b)
d^9	^2D	$t_{2g}^6e_g^3$	Inverted (a)
d^{10}	^1S	$t_{2g}^6e_g^4$	No splitting

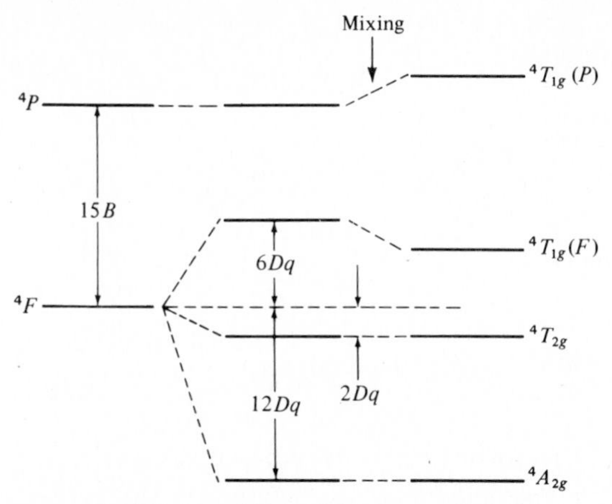

Fig. 9.55 Splitting of ^4F and ^4P terms of a $3d^3$ ion in octahedral fields illustrating the inversion of Fig. 9.54b and the mixing of the T_{1g} terms (right).

essentially to the ionization of the ligand in that considerable energy must be involved to force the electron from the stable ligand orbital to the higher energy metal antibonding orbital. It is thus a redox process and occurs most easily (i.e., at lowest frequencies) when the metal is in a high oxidation state and the ligand is relatively easy to oxidize (see p. 456).

The other two bands can be assigned on the basis of Table 9.26. We note that Cr^{+3} is a d^3 ion with a ^4F ground state. Under the influence of an octahedral field the latter will split into a $^4A_{2g}$ ground state and two excited states, $^4T_{2g}$ and $^4T_{1g}$. The transitions are thus $^4A_{2g} \rightarrow {}^4T_{2g}(\nu_1)$ and $^4A_{2g} \rightarrow {}^4T_{1g}(\nu_2)$. The first transition is seen to be $10Dq$ from Fig. 9.54. The second transition would be $18Dq$ were it not for the interaction of the $^4T_{1g}(F)$ and $^4T_{1g}(P)$ terms. Since the wave functions for these two states have identical symmetry they will "mix," the amount of mixing being inversely proportional to the difference in energy between the ^4F and ^4P levels. This mixing is analogous to molecular orbital formation in which two orbitals of appropriate symmetry mix (form linear combinations) to yield two new orbitals, one at a lower energy and one at a higher energy than the contributing orbitals. As a result of the mixing, the $^4T_{1g}(F)$ will lie somewhat lower in energy and the $^4T_{1g}(P)$ will lie somewhat higher in energy than they would if mixing had not occurred (Fig. 9.55).

The interelectronic repulsion responsible for the differences in energy between the ^4F, ^4P, and other terms is expressed by means of the Racah parameters B and C.[116] These are linear combinations of certain coulomb and exchange integrals. Although they could be obtained by accurate evaluation of these integrals, in general this is not feasible and they are treated as empirical parameters instead, obtained from the spectra of free ions. The parameter B is usually sufficient to evaluate the difference in energy between states of the same spin multiplicity (e.g., the difference between the ^4F and ^4P states in the free ion is $15B$), while both parameters are necessary for terms of different spin multiplicity (the

[116] Occasionally the Slater-Condon-Shortley parameters F_2 and F_4 are used instead. Their relation to the Racah parameters is $B = F_2 - 5F_4$ and $C = 35F_4$.

Table 9.27 Spectral correlations for octahedral chromium(III) complexes[a]

Energy level diagram	Predicted (Fig. 9.55)	$[CrF_6]^{-3}$ I	$[CrF_6]^{-3}$ II	$[CrF_6]^{-3}$ Exp.	$[Cr(ox)_3]^{-3}$ I	$[Cr(ox)_3]^{-3}$ II	$[Cr(ox)_3]^{-3}$ Exp.	$[Cr(en)_3]^{+3}$ I	$[Cr(en)_3]^{+3}$ II	$[Cr(en)_3]^{+3}$ Exp.
$^4T_{1g}(P)$	$12Dq+15B+x$	$30.7+x$	34.8	34.4	$34.8+x$	38.1	[b]	$40.0+x$	46.5	[b]
$^4T_{1g}(F)$	$18Dq-x$	$26.8-x$	22.4	22.7	$31.5-x$	24.0	23.9	$39.3-x$	28.7	28.5
$^4T_{2g}$	$10Dq$	14.9	14.9	14.9	17.5	17.5	17.5	21.85	21.85	21.85
$^4A_{2g}$ (ν_1 ν_2 ν_3)	0	0	0	0	0	0	0	0	0	0
		$Dq=1.49$			$Dq=1.75$			$Dq=2.185$		
		$B'=0.827$[c]			$B'=0.64$[d]			$B'=0.64$[d]		

[a] All energies are expressed as $\bar{\nu}/1000\ cm^{-1}$.
[b] This transition is not experimentally observed since it is masked by the charge transfer spectrum.
[c] Obtained from the three observed transitions and Eq. 9.58.
[d] Obtained from the value for the free ion, values of h and k from Table 9.15 and Eqs. 9.56 and 9.57.

difference between the ^4F ground state and the lowest lying doublet, ^2G, is $4B + 3C$). We shall not present here the derivation of the differences between the various terms as functions of the Racah parameters,[117] but we shall simply present the results for Cr(III). For most transition metal ions B can be estimated as approximately 1000 cm^{-1} and $C \sim 4B$ (for Cr^{+3} the exact values are 918 and 4133 cm^{-1}, respectively) for the free ion.

We can compare the experimental results from spectra (such as Fig. 9.14) with those expected from theory. In general, the observed transitions will be $^4A_{2g} \rightarrow {}^4T_{2g}$, $^4A_{2g} \rightarrow {}^4T_{1g}(F)$, and $^4A_{2g} \rightarrow {}^4T_{1g}(P)$. The transitions arising from the ^2G state are either not observed or are very weak since they are spin forbidden.[118] When we compare the experimental results for $[CrF_6]^{-3}$, $[Cr(ox)_3]^{-3}$, and $[Cr(en)_3]^{+3}$ with the theoretical expectations, we obtain some interesting results. The $^4A_{2g} \rightarrow {}^4T_{2g}$ transition (Band I, equal to $10Dq$) increases in progressing from L = F^- to L = en, as expected on the basis of the spectrochemical series. From Fig. 9.54, the $^4A_{2g} \rightarrow {}^4T_{1g}(F)$ transition is expected to be $18Dq$, or 80% greater than the $^4A_{2g} \rightarrow {}^4T_{2g}$. The transition $^4A_{2g} \rightarrow {}^4T_{1g}(P)$ is expected, on the basis of the free ion, to be $15B$, or $15 \times 918\ cm^{-1} = 13\,770\ cm^{-1}$. It can be seen that these expectations (approximation I, Table 9.27) are not realized very closely by the experimental results. Two corrections must be made in order to improve the interpretation and correlation of the spectra. First, the extent of the mixing of the F and P terms (T_{1g} in Fig. 9.55) was not included in approximation I. This will be discussed presently, but we must first note that even if this is known exactly, the experimental results cannot be duplicated using the free-ion value of B, even if it, too, is known exactly. The apparent value of B in complexes is always smaller than that of the free ion. This phenomenon is known as the nephelauxetic effect (p. 413) and is attributed to delocalization of the metal electrons over molecular orbitals that encompass not only the metal but the ligands as well. As a result of this delocalization or "cloud expanding" the average interelectronic

[117] These may be found in C. J. Ballhausen, "Introduction to Ligand Field Theory," McGraw-Hill, New York, **1962**, in terms of the Slater-Condon-Shortley parameters.
[118] One of the selection rules for permitted transitions is $\Delta S = 0$; i.e., the spin multiplicity must not change for an allowed transition. See p. 453.

Table 9.28 Values of B for transition metal ions (cm^{-1})

Metal	M^{+2}	M^{+3}
Ti	695	—
V	755	861
Cr	810	918
Mn	860	965
Fe	917	1015
Co	971	1065
Ni	1030	1115

SOURCE: D. Sutton, "Electronic Spectra of Transition Metal Complexes," McGraw-Hill, New York, **1968**. Used with permission.

repulsion is *reduced* and B' (representing B in the complex) is smaller. The nephelauxetic ratio, β, is given by:

$$\beta = \frac{B'}{B} \tag{9.56}$$

It is always less than one and decreases with increasing delocalization. Estimates[119] of β may be obtained from the nephelauxetic parameters h_X for the ligand and k_M for the metal:

$$(1 - \beta) = h_X \cdot k_M \tag{9.57}$$

If all three transitions are observed, it is a simple matter to assign a value to B' since the following equation must hold:

$$15B' = \nu_3 + \nu_2 - 3\nu_1 \tag{9.58}$$

where ν_1 is the absorption occurring at the lowest frequency. For example, the value of B' in the fluoro complex is:

$$B' = \frac{34\,400 + 22\,700 - 3(14\,900)}{15} = 827 \text{ cm}^{-1} \tag{9.59}$$

If only two transitions are observed (for example, ν_3 may be obscured by a charge transfer band, p. 444) it is still possible to calculate B' but the methods are beyond the scope of this book.[120] However, sufficient spectra have been analyzed for the more common ligands and metal ions that β may be estimated from Eq. 9.57 and Table 9.15, and in turn used to estimate B' from the free-ion value B (Table 9.28).

If $10Dq$ can be measured directly as in the case of Cr(III) spectra ($10Dq = \nu_1$) and B' evaluated by means of Eq. 9.58 or Eqs. 9.56 and 9.57, it is quite simple to estimate all of

[119] Values for h_X and k_M from Jørgensen are given in Table 9.15. For very large values of $h_X \cdot k_M$ Eq. 9.57 breaks down because the delocalization becomes so extensive that the bonding becomes somewhat metallic in character.

[120] A. B. P. Lever, *J. Chem. Educ.*, **1968**, *45*, 711.

the transitions. For high-spin octahedral d^3 and d^8 and tetrahedral d^2 and d^7 species the appropriate equations are:

$$\nu_1 = A_{2g} \rightarrow T_{2g} = 10Dq \qquad \textbf{(9.60)}$$

$$\nu_2 = A_{2g} \rightarrow T_{1g}(F) = 7.5B' + 15Dq - \tfrac{1}{2}(225B'^2 + 100Dq^2 - 180B'Dq)^{1/2} \qquad \textbf{(9.61)}$$

$$\nu_3 = A_{2g} \rightarrow T_{1g}(P) = 7.5B' + 15Dq + \tfrac{1}{2}(225B'^2 + 100Dq^2 - 180B'Dq)^{1/2} \qquad \textbf{(9.62)}$$

Using these equations, more accurate estimates (Table 9.27, column II) can be made of the transitions and the spectrum fitted quite suitably.[121]

The use of adjusted B and C parameters is the basis of the *ligand field theory* (LFT) or *adjusted crystal field theory* (ACFT) in which B' and C' are treated as empirical, adjustable parameters which are fitted to experimentally obtained spectral data. The above examples have provided an introduction to the methods of treating spectral data.[122] The spectra of transition metal compounds have been studied extensively by these methods, and the resulting data are the source of the generalizations that have been made previously on field strength, the nephelauxetic effect, etc.

In many cases only two absorption bands are observed, and so the number of variable parameters (two: Dq and B') is as great as the number of experimental values to be correlated. Since it is always possible to fit n observations with n variables, it is of some concern to decide how valuable such endeavors are.[123] In the example above we were able to reproduce three experimental data with two parameters, and in more complicated spectra it is possible to account for a larger number of absorption peaks. This substantiates our faith that we are truly "explaining" something and not merely juggling numbers.

Orgel diagrams

It is possible to use Fig. 9.54 in conjunction with the appropriate combinations of B and C to obtain the values such as those in Table 9.27, and if adjusted B' and C' values are used, the resulting energy level diagrams will be fairly accurate. In general, however, this is too bothersome for routine use and the chemist turns to pictorial representation of this information, including the effect of variable field strength. Such representations are known as *Orgel diagrams*,[124] and we have seen the simplest possible one, that for d^1, previously (Fig. 9.15).[125] More complicated (and more informative) Orgel diagrams are shown in Figs. 9.56 and 9.57. The former illustrates the splitting of the quartet terms arising from the d^3 configuration of the Cr^{+3} ion. If the mixing of the ${}^4T_{1g}(F)$ and ${}^4T_{1g}(P)$ is omitted (dashed lines), the figure is the same as Fig. 9.54b inverted. Terms of the same symmetry (in this case T_{1g}) cannot cross inasmuch as the closer they approach each other the more mixing occurs, raising the upper and lowering the lower (see pp. 109–110).

[121] For further examples of the treatment of the spectra of d^3 ions and a discussion of the accuracy of the methods, see E. König, *Inorg. Chem.*, **1971**, *10*, 2632.

[122] See J. Ferguson, *Progr. Inorg. Chem.*, **1970**, *12*, 159.

[123] For a discussion of this subject see T. M. Dunn, *Pure Appl. Chem.*, **1963**, *6*, 1.

[124] L. E. Orgel, *J. Chem. Phys.*, **1955**, *23*, 1004.

[125] Figure 9.15 is labeled with orbital configurations t_{2g}^1 and e_g^1 which for d^1 are equivalent to ${}^2T_{2g}$ and ${}^2E_g$, respectively, for the splitting of ${}^2D$.

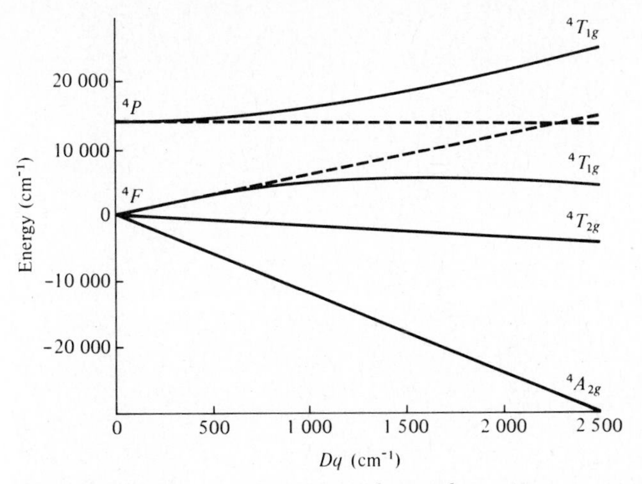

Fig. 9.56 Orgel diagram for the Cr^{+3} ion (d^3) in an octahedral field. The dashed lines represent the energies of the $^4T_{1g}$ prior to mixing. [From L. E. Orgel, *J. Chem. Phys.*, **1955**, *23*, 1004. Reproduced with permission.]

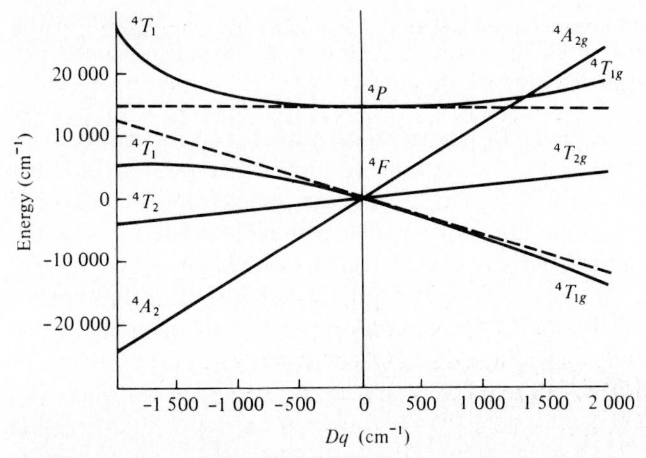

Fig. 9.57 Orgel diagram for the Co^{+2} ion in tetrahedral (left) and octahedral (right) fields. The dashed lines represent the $^4T_1$ terms before mixing. [From L. E. Orgel, *J. Chem. Phys.*, **1955**, *23*, 1004. Reproduced with permission.]

Figure 9.57 illustrates the Orgel diagram for the d^7 configuration of Co^{+2}. Again momentarily ignoring the mixing of the T_1 terms, we note that the tetrahedral splitting (left) is exactly the inverse of that for the octahedral case (right) recalling the formalism employed earlier (p. 378) that a tetrahedral field may be considered a *negative* octahedral field; i.e., the octahedral splitting is the "normal" one from Fig. 9.54b, but the tetrahedral splitting is the "inverted" one. Further, we note that mixing takes place for the T_1 levels[126]

[126] Since a tetrahedron has no center of symmetry, the *gerade* and *ungerade* labels are inappropriate for tetrahedral complexes.

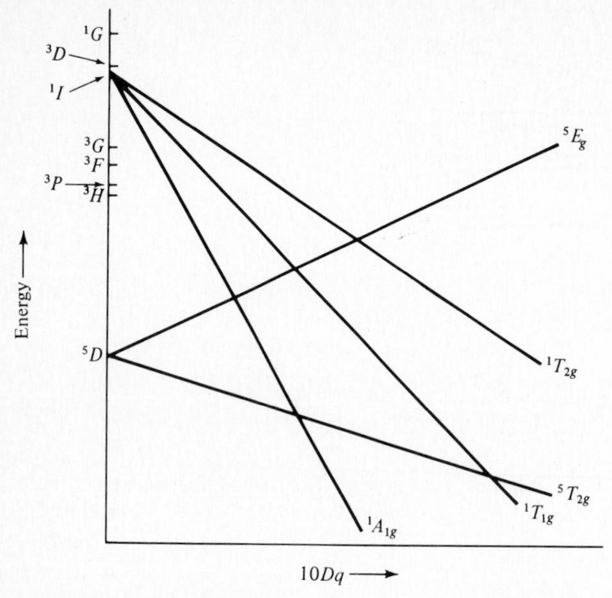

Fig. 9.58 Modified Orgel diagram to illustrate the stabilization of low-spin states with increasing field strength for Co^{+3} (d^6). [From F. A. Cotton and G. Wilkinson, "Advanced Inorganic Chemistry," 2nd ed., Wiley (Interscience), New York, **1966**. Reproduced with permission.]

and more so for the tetrahedral case than the octahedral case since with increasing fields the F and P terms approach each other in the former while getting further apart in the latter.

Orgel diagrams treat only the weak-field or high-spin case, and so the ^2G state is not included. Low-spin cases are not discussed *per se*, although it is possible to add the low-spin states to an Orgel diagram and show clearly that with increasing field strength, configurations arising from d^4, d^5, d^6, and d^7 potentially can change from high spin to low spin as we have already seen qualitatively. Figure 9.58 thus represents the Orgel diagram for the ^5D term for Co^{+3} to which have been added lines for the singlet states. The latter are favored by strong fields, and so it is not surprising that eventually the ground state becomes $^1A_{1g}$ instead of $^5T_{2g}$.[127]

Tanabe-Sugano diagrams

In order to treat fully the problem of interpretation of spectra, including both weak and strong fields, it is common to use diagrams provided by Tanabe and Sugano.[128] These are related to Orgel diagrams except that: (1) low-spin terms are included; (2) the ground state is always taken as the abscissa and the energy of the other states plotted relative to it; and (3) although the diagram is basically one of energy as a function of Dq, the units are E/B and $10Dq/B$, respectively. Furthermore, in order to represent the energy levels at all accurately it is necessary to make some assumptions about the relative value of C/B.

[127] For the derivation of the terms for the strong field limit, see B. N. Figgis, "Introduction to Ligand Fields," Wiley (Interscience), New York, **1966**, Chap. 7. This is one of the few easily readable discussions of the origin of term splitting in weak and strong fields.

[128] Y. Tanabe and S. Sugano, *J. Phys. Soc. Japan*, **1954**, 753, 766.

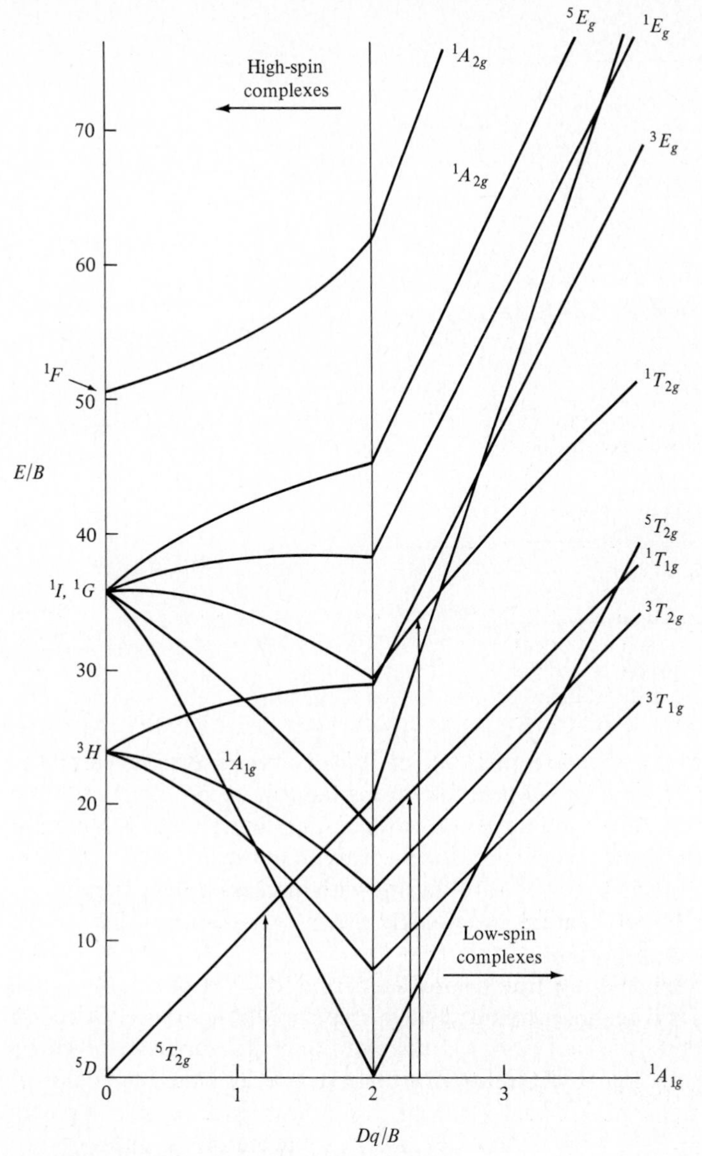

Fig. 9.59 Modified Tanabe-Sugano diagram showing only the ^5D, ^3H, ^1F, ^1G, and ^1I terms. Arrows represent transitions in $[CoF_6]^{-3}$ and $[Co(en)_3]^{+3}$.

A simplified version of the Tanabe-Sugano diagram for Co^{+3} is shown in Fig. 9.59.[129] Only a few terms are shown. As expected from the Orgel diagram, the free-ion ground state, ^5D, is split by the increasing octahedral field into a $^5T_{2g}$ ground state and an excited ^5E_g state. The singlet, ^1I, which lies at a very high energy in the free ion, is split by the

[129] A complete set of Tanabe-Sugano diagrams is given in Appendix D.

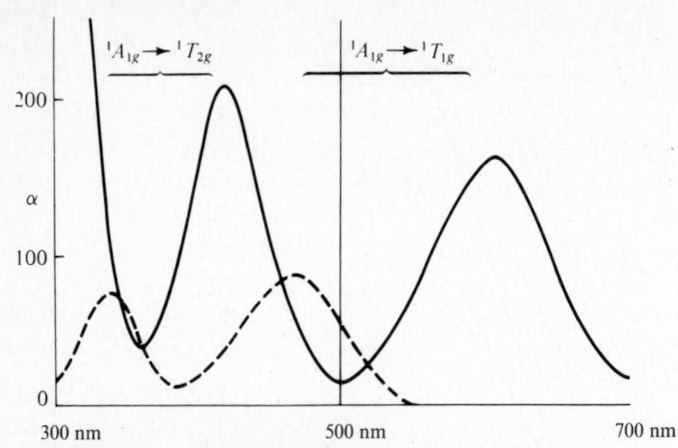

Fig. 9.60 Spectra of $[Co(en)_3]^{+3}$ (--------) and $[Co(ox)_3]^{-3}$ (———). [From A. Mead, *Trans. Faraday Soc.*, **1934**, *30*, 1052. Reproduced with permission.]

Table 9.29 Absorption maxima for cobalt complexes

Complex	ν_1 (cm^{-1})	ν_2 (cm^{-1})	$\nu_1 - \nu_2$ (cm^{-1})
$K_3[Co(C_2O_4)_3]$	16 500	23 800	7300
$[Co(en)_3]Cl_3$	21 400	29 500	8100

application of the ligand field into several terms, only one of which is important. This term, the $^1A_{1g}$, is greatly stabilized by the ligand field and drops rapidly, becoming the ground state at $10Dq/B = 20$. At this point spin pairing takes place, and hence there is a discontinuity in the diagram, shown by the vertical line. Beyond this point the low-spin $^1A_{1g}$ term is the ground state. The ^5E_g and $^5T_{2g}$ continue to diverge with increasing field strength, as might be supposed from the Orgel diagram, but quickly rise in energy with respect to the $^1A_{1g}$ state and soon become unimportant.

The spectra of Co(III) complexes can now be predicted from Fig. 9.59. We shall assume that the transitions observed will be those without change in spin multiplicity (see "Selection Rules"). For high-spin complexes such as $[CoF_6]^{-3}$, therefore, only the quintet states will be important and one transition, $^5T_{2g} \rightarrow {}^5E_g$, should be observed. Indeed the blue color of this complex results from a single peak[130] at 13 000 cm^{-1}. For low-spin Co(III) complexes we would expect transitions $^1A_{1g} \rightarrow {}^1T_{1g}$ and $^1A_{1g} \rightarrow {}^1T_{2g}$. Furthermore, although we should expect both of these transitions to increase in energy as the field increases, we might expect the latter to increase slightly more rapidly than the former (compare the slopes of T_{1g} and T_{2g}). The spectra of Co(III) complexes are therefore expected to show two absorption peaks and these peaks should appear more widely spaced at large values of Dq. The spectra of the yellow $[Co(en)_3]^{+3}$ and the green $[Co(ox)_3]^{-3}$ shown in Fig. 9.60 confirm these expectations (see also Table 9.29).

[130] Actually the excited E_g state is split by a Jahn-Teller effect resulting in two peaks. See Fig. 9.23. F. A. Cotton and M. D. Meyers, *J. Am. Chem. Soc.*, **1960**, *82*, 5023.

Selection rules

The above calculation of the energy differences between the various states that give rise to spectral absorptions is all well and good, but it does not address the more fundamental issue of whether a transition will take place at all. There are a number of limitations, called *selection rules*, that reflect the restricted ability of an atom to change states. Any transition in violation of the selection rules is said to be "forbidden," though as we shall see, to paraphrase George Orwell, "some transitions are more forbidden than others." In this section we shall not pursue the theoretical basis of the rules in any detail, but outline simple tests for their application. The first selection rule states that *any transition in which $\Delta S \neq 0$ is forbidden*; i.e., an allowed transition must be one in which the spin state of the complex does not change. A low-spin complex of Co^{+3} such as tris(ethylenediamine)cobalt(III) or tris(oxalato)cobalt(III), t_{2g}^6, should therefore be expected to undergo *d–d* transitions to another singlet state: ${}^1A_{1g} \rightarrow {}^1T_{1g}$ and ${}^1A_{1g} \rightarrow {}^1T_{2g}$ (Fig. 9.60). This is why in the Tanabe-Sugano diagram of Fig. 9.59 there is no transition from the ground state to the triplet states ${}^3T_{1g}$ and ${}^3T_{2g}$. The transition to the ${}^5E_g$ state is even more unlikely, since it is not only spin-forbidden but also involves an improbable simultaneous two-electron transition.

In the same way any high-spin complex of a d^5 ion will have *all* of its transitions spin-forbidden since its ground state, ${}^6A_{1g}$ cannot undergo any *d–d* transition without spin-pairing as is obvious either from inspection of a simple splitting diagram

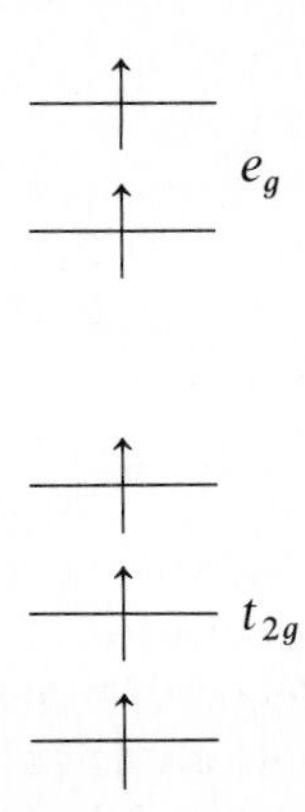

or of the high-spin states in the Tanabe-Sugano diagram for a d^5 ion (Appendix D). An example of this effect is the hexafluoroferrate(III) ion, FeF_6^{-3}, which is colorless despite having five *d* electrons potentially capable of undergoing transitions.

The second rule is the *Laporte selection rule*: *In complexes with a center of symmetry the only allowed transitions are those with a change of parity*[131]; i.e., *gerade to ungerade* and $u \rightarrow g$ are allowed, but not $g \rightarrow g$ and $u \rightarrow u$. Since all *d* orbitals have *gerade* symmetry, this means that *all d–d* transitions in centrosymmetric molecules (octahedral symmetry being the most important example) are formally forbidden.

Tetrahedral complexes do not have a center of symmetry. Furthermore, *ungerade p* orbitals can mix with the *d* orbitals to remove some of the "forbiddenness" from the Laporte

[131] P. W. Atkins, "Physical Chemistry," 2nd ed., Freeman, San Francisco, **1982**, pp. 608–609.

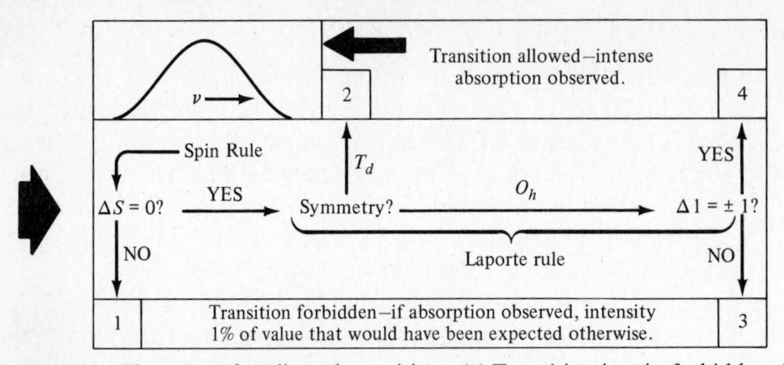

Fig. 9.61 Flow chart for allowed transitions. (1) Transition is spin-forbidden. (2) Transitions in noncentrosymmetric molecules are allowed by admixture of *p* character. (3) Transitions in centrosymmetric molecules are forbidden if no change in parity is involved; this includes pure *d–d* transitions. (4) Transitions are permitted if change in parity is involved, as in charge-transfer spectra (see pp. 456–458).

rule. This may be viewed in two seemingly different, but very similar ways. First, a set of sd^3 hybrids has identical (tetrahedral) symmetry to that of a set of sp^3 hybrids, so that any tetrahedral molecule can range between 100% sd^3 to 100% sp^3, neither *extreme* being likely, or we can note that d_{xy}, d_{xz}, and d_{yz} in tetrahedral symmetry transform as t_2, as do p_x, p_y, p_z; thus the $t_2(d)$ and $t_2(p)$ orbital sets can mix.

Finally, we should note that, as in atomic spectra, transitions are allowed only for $\Delta l = \pm 1$. Thus in the spectrum of the sodium atom we observe only $s \rightarrow p$, $p \rightarrow d$, and like transitions, but no $s \rightarrow s$, $p \rightarrow p$, or $d \rightarrow d$ transitions.[132] This point will become important when we examine charge transfer transitions that are normally *fully* allowed and the extinction coefficients should thus be very high. A flow diagram providing a simplified synopsis of these rules is given in Fig. 9.61. Start on the left and work through until a decision can be made as to whether a transition is allowed or not.

With all of the restrictions on allowed transitions it might seem, contrary to what you already know to be true, that visible and UV spectroscopy of transition metals would be virtually nonexistent. For example, there would be absolutely *no d–d* transitions observed for octahedral complexes. The fact that optical spectroscopy is responsible for many of the facts and theories in this chapter is due to the relaxation of the selection rules through various mechanisms, thus allowing transitions to occur at low intensities. They may thus be observed in concentrated solutions or in crystals.

For example, as a result of the coupling of the spin vector of the electron and the angular momentum vector of the orbitals, the $\Delta S = 0$ rule may be relaxed giving observable absorptions. Although such *spin-orbit coupling* permits the observation of the transitions,

[132] G. Herzberg, "Atomic Spectra and Atomic Structure," Dover, New York, **1944**, Chapter 5 (Li); M. C. Day, Jr., and J. Selbin, "Theoretical Inorganic Chemistry," 2nd ed., Reinhold, New York, **1969**, pp. 13ff. (Na); M. Karplus and R. N. Porter, "Atoms and Molecules," Benjamin, New York, **1970**, pp. 139ff. (H). Each of these has a figure indicating the allowed transitions for the element noted in parentheses.

the extinction coefficients are low. Typically, spin-forbidden absorptions have extinction coefficients only 1% as large as spin-allowed transitions.

Various unsymmetrical vibrations of an octahedral complex can destroy its center of symmetry and allow transitions that would otherwise be Laporte-forbidden. Such *vibronic* (*vib*rational/electr*onic*) transitions will be observable, though weak. (The number of molecules in an unsymmetrical conformation at any instant will be a small fraction of the total.) Typically, extinction coefficients of octahedral complexes are in the range of 5–100 $dm^3\ mol^{-1}\ cm^{-1}$. Noncentrosymmetric tetrahedral complexes will have the highest extinction coefficients, in the range of 500–5000 $dm^3\ mol^{-1}\ cm^{-1}$. We can thus use as a *very rough* rule of thumb a 10^{-2} factor for each degree of "forbiddenness."

One further factor that does not deal with selection rules but, like intensities, is of considerable interest when interpreting spectra is the line width of an absorption. Consider again the first spectrum we encountered in this chapter (Fig. 9.7), that of $Ti(H_2O)_6^{+3}$ in aqueous solution. Note how broad the absorption peak is. The transition involved is $t_{2g}^1 \rightarrow e_g^{*1}$. Whether viewed from crystal field theory or molecular orbital theory, the t_{2g} orbitals are relatively unaffected by the ligands, whereas the e_g^* levels will be very sensitive to the M—L bond length. As the ligand approaches (in a molecular vibration), the energy of e_g^* will rise; as the ligand retreats, it will drop. The energy of the t_{2g} level will remain relatively constant. The breadth of the absorption peak in Fig. 9.7 is a result of the fact that it is integrated over a collection of molecules with all possible molecular structures and many, many values for $10Dq$. Such ligand motions will be exaggerated through molecular collisions in solution. It is generally expected that spectra will sharpen in the gas phase.

In contrast, if a transition occurs *within* the t_{2g} level, its magnitude will be unaffected to a first approximation by ligand vibration. Hence the broadening effects will be minimal and the absorptions sharp.

A good example of a substance exhibiting several features discussed in this section is the ruby, composed of α-Al_2O_3 in which Cr^{+3} ions replace some Al^{+3} ions in approximately octahedral holes. The ground state of the free, d^3, Cr^{+3} ion is ^4F, which becomes $^4A_{1g}$ in ruby. Spin-allowed transitions (see the d^3 Tanabe-Sugano diagram in Appendix D) will be limited to $^4A_{1g} \rightarrow {}^4T_{1g}$ and $^4A_{1g} \rightarrow {}^4T_{2g}$. These will be broad peaks (because they involve the e_g^* level—they are $t_{2g} \rightarrow e_g^*$) of relatively high absorbance (because they are spin-allowed). The absorption maxima occur at 18 000 cm^{-1} and 25 000 cm^{-1}. Hence all of the visible light except the extreme red ("ruby red!") will be absorbed (cf. Fig. 9.7 for the relation between color and frequency). The other possible transitions to low-lying states (^2E_g, $^2T_{1g}$, and $^2T_{2g}$) will all be quartet-to-doublet, violating the $\Delta S \neq 0$ rule, and thus they must have considerably smaller extinction coefficients. However, because they involve transitions within only the nonligated t_{2g} orbitals, they are very sharp. These transitions form the basis of the ruby laser in which electrons are pumped to these excited doublets. Since the transition back to the ground state is also spin-forbidden, it does not occur readily, and it is possible to build up the population of molecules in the excited state without their decay until one obtains in-phase, stimulated emission of light that is essentially monochromatic because of the very narrow band width. Thus we can see how several features of the behavior of electronic transitions (some of which are not touched upon here, such as the pumping procedure) facilitate the operations of the ruby laser.

Charge-transfer spectra

Although the majority of visible spectra studied by the inorganic chemist in evaluating coordination compounds has been of the *d–d* type, perhaps the most important, at least from an applied chemistry point of view, have been those involving *charge-transfer transitions*. These result from an ionization energy–electron affinity process that is the exact reverse of that which took place when the compound originally formed. Consider a crystal of sodium chloride. Imagine ionizing an electron from a chloride ion (= negative of electron affinity) and transferring it to the sodium ion (= negative of ionization energy). It can be seen that the sum of these plus related energies could be supplied by a high-energy photon that would effect this reverse ionization process. Indeed such photons exist, but their energy is so high that they belong to the ultraviolet portion of the spectrum. Hence, as you know, sodium chloride does not absorb visible light—it is colorless.

Now consider a slightly different compound, one in which the ionization energy of the metal is higher—such as a transition or posttransition metal, especially in higher oxidation states—and so in which the reverse process of returning an electron to the metal is energetically more favorable. Also consider a nonmental, either as anion or ligand, in which the electron affinity is lower—a chalcogen or heavier halogen—so that the cost of removal of the electron from the anion is less. What is the result? The energies associated with the reverse ionizations described above will be considerably smaller and may well be associated with a photon in the visible or near-ultraviolet region. Consider the intensely colored permanganate ion, MnO_4^-. The color *cannot* arise from *d–d* transitions (see Problem 9.31). Note, however, that the complex is ideally designed for the process described above: Manganese is in a formal oxidation state of +7 and combined with four oxide ions (−2). One can almost say that there is an internal redox reaction poised to take place—a charge transfer from an oxygen atom to the manganese—needing only the energy of a photon to activate it. In the same way one can formulate the charge transfer spectrum of orange CrO_4^{-2} ion as being facilitated by the high oxidation state of chromium(VI).

As in the case of crystal field theory, one must be careful not to take the ionic terminology too literally. The charge transfer that actually takes place is from a low-lying, filled molecular orbital having *mostly ligand* character to a higher-energy, empty molecular orbital having *mostly metal* (often antibonding) character. Thus although it is convenient to label the transitions in terms of the predominant atomic orbitals involved (as we *always* do in CFT and *often* do in MOT) we should not lose sight of the fact that the orbitals involved are linear combinations of orbitals from both metal and ligand. Thus, in Fig. 9.62, although the observed transitions may include the e_L (ligand nonbonding orbital if π bonding is ignored) $\rightarrow e_M$ and $t_2 \rightarrow e$ transitions (if e is not filled) as well as the $t_1(e) \rightarrow t_2^*$ transition, the shorthand notation ($L_{2p(nb)} \rightarrow M_{d(e)}$, $L_{2p(b)} \rightarrow M_{d(t_2)}$, and $L_{2p(nb)} \rightarrow M_{d(t_2)}$, or even simply $L_{2p} \rightarrow M_{3d}$) is permissible as long as no one takes it literally.

Some examples of "ionic" solids exhibiting charge-transfer spectra are listed in Table 9.30. The purpose of Table 9.30 is manifold: (1) to illustrate the appropriate conditions for charge-transfer spectra—metals such as Cd, Hg, As, Pb, and Fe; nonmetals such as O, S, or Se (not shown); Br and I are also important; (2) to call attention to the high extinction coefficients of these transitions arising from the fact that they are fully allowed, in contrast to *d–d* transitions; (3) to illustrate a few of the many pigments that owe the nature and

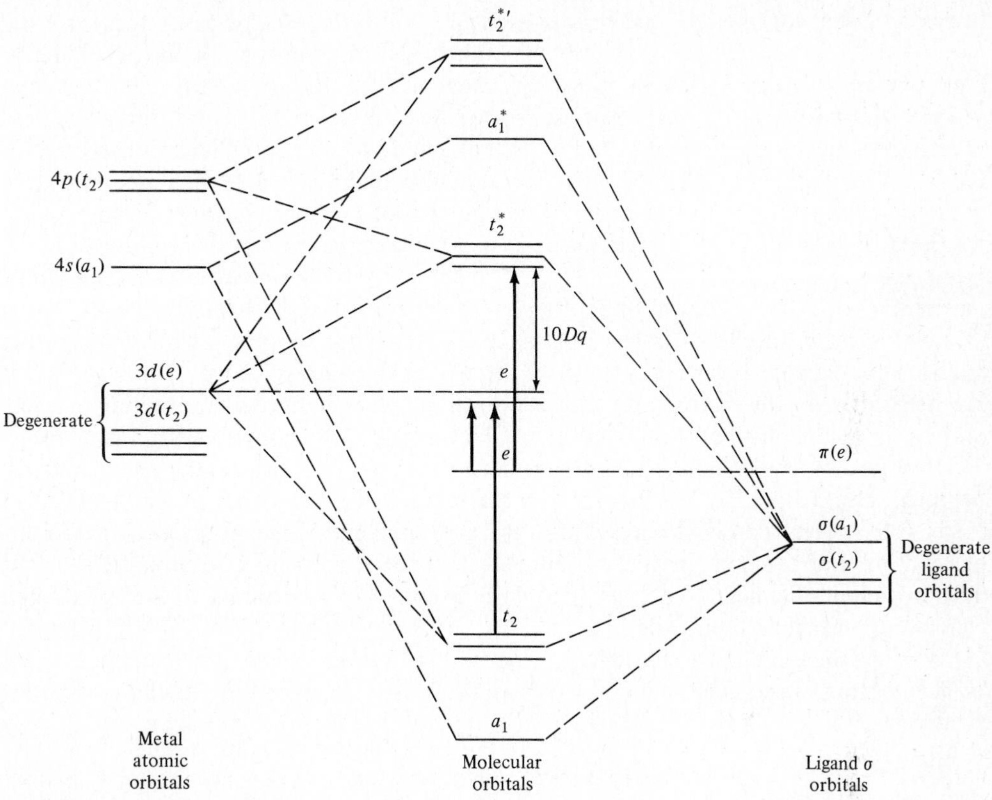

Fig. 9.62 Molecular orbital energy level diagram for tetrahedral complexes. Π bonding has not been included, though the ligand lone pairs of π symmetry are shown. Possible charge-transfer transitions are shown as heavy arrows.

Table 9.30 Pigments in which color is produced by charge-transfer transitions from an ionic ground state

Pigment	Charge transfer reaction	Primary orbitals involved[a]
Cadmium yellow (CdS)	$Cd^{+2}S^{-2} \xrightarrow{h\nu} Cd^{+}S^{-}$	Ligand $p_\pi \rightarrow$ metal 5*s*
Vermilion (HgS)	$Hg^{+2}S^{-2} \xrightarrow{h\nu} Hg^{+}S^{-}$	Ligand $p_\pi \rightarrow$ metal 6*s*
Orpiment and realgar (As_2S_3, As_4S_4)	$As^{+3}S^{-2} \xrightarrow{h\nu} As^{+2}S^{-}$	Ligand $p_\pi \rightarrow$ metal 4*s* or 4*p*
Naples yellow [$Pb_3(SbO_4)_2$]	$Sb^{+5}O^{-2} \xrightarrow{h\nu} Sb^{+4}O^{-}$	Ligand $p_\pi \rightarrow$ metal 5*s* or 5*p*
Massicot (PbO)	$Pb^{+2}O^{-2} \xrightarrow{h\nu} Pb^{+}O^{-}$	Ligand $p_\pi \rightarrow$ metal 6*s*
Chrome yellow ($PbCrO_4$)	$Cr^{+6}O^{-2} \xrightarrow{h\nu} Cr^{+5}O^{-}$	Ligand $p_\pi \rightarrow$ metal 3*d*
Red and yellow ochres (iron oxides)	$Fe^{+3}O^{-2} \xrightarrow{h\nu} Fe^{+2}O^{-}$	Ligand $p_\pi \rightarrow$ metal 3*d*

[a] Simplified notation. See text.

SOURCE: T. B. Brill, "Light, Its Interaction with Art and Antiquities," Plenum Press, New York, **1980**. Used with permission.

intensity of their colors to charge-transfer spectra. Some of these have been long known and used by man in the attempt to beautify his immediate environment: The use of ochres as pigments dating from prehistoric times followed their natural abundance [the reds and yellows of the deserts and some other soils are caused by iron(III) oxide]. Pigments used in antiquity included orpiment found in Tutankamen's tomb near Thebes and lead antimonate (later dubbed "Naples yellow") important in Babylonian glazes.[133] Later, vermilion became an important pigment in the Venetian school of painting, and chrome yellow continues to be an important pigment in several contexts because of its brightness. For a more nearly complete discussion of inorganic pigments and their relation to art, see Brill's interesting book on the subject of light and art.[134]

Note that most charge-transfer pigments are reds and yellows. This is because the main charge-transfer band tends to lie in the UV as mentioned above. If it is shifted to lower energies, the absorption may trail over into the blue-violet region, resulting in a red, orange, or yellow pigment. And recall that CT bands are fully allowed and thus intense.

The phenomena of charge-transfer processes are not limited to ligand–metal systems. Although electrostatically less favorable, metal–ligand charge-transfer processes can also occur with an electron being ionized from the metal atom to a vacant orbital on the nonmetal atom. For example, in an octahedral complex, electrons may be promoted from a filled orbital on the metal (e.g., the t_{2g}) to various antibonding orbitals on the ligand (like π^* orbitals on carbon monoxide).

Finally, there are more subtle factors beyond the simple picture portrayed here, such as heavy atom effects and reduction of symmetry that enhance simple charge-transfer transitions.[135]

PROBLEMS

9.1. The complex $[Ni(CN)_4]^{-2}$ is diamagnetic but $[NiCl_4]^{-2}$ is paramagnetic with two unpaired electrons. Likewise, $[Fe(CN)_6]^{-3}$ has only one unpaired electron, but $[Fe(H_2O)_6]^{+3}$ has five.

a. Explain these experimental observations using valence bond theory.

b. Explain these experimental observations using simple crystal field theory.

9.2. Diamagnetic complexes of cobalt(III) such as $[Co(NH_3)_6]^{+3}$, $[Co(en)_3]^{+3}$, and $[Co(NO_2)_6]^{-3}$ are orange-yellow. In contrast, the paramagnetic complexes $[CoF_6]^{-3}$ and $[Co(H_2O)_3F_3]$ are blue. Explain *qualitatively* this difference in color.

9.3. D. M. Gruen[44] has given a general rule for spectra of 3*d* metal ions dissolved in LiCl–KCl eutectic: ν_{max} in LiCl–KCl = 0.75 ν_{max} in water. Explain.

9.4. Complexes of cobalt(III) having more than one type of ligand, such as $[Co(NH_3)_4Cl_2]^+$, come in a wide variety of colors: red, green, purple, etc. (see p. 360,

[133] C. Singer et al., "A History of Technology," Vol. I, "From Early Times to Fall of Ancient Empires," Oxford University Press, London, **1954**.

[134] T. B. Brill, "Light, Its Interaction with Art and Antiquities," Plenum, New York, **1980**.

[135] S. P. McGlynn et al., *Chem. Rev.*, **1981**, *81*, 475.

for example) rather than the simple yellow versus blue of Problem 9.2. Suggest a source of this variability.

9.5. Use Eq. 9.12 to estimate values of $10Dq$ for the complexes listed in Table 9.7, and compare your estimates with the experimental values.

9.6. Show that in the purely electrostatic approximation $\delta_2 = 0.75\delta_1$ for small tetragonal splittings.

9.7. $[ReF_8]^{-2}$ has a square antiprismatic structure. Why is this more stable than cubic coordination for rhenium(VI)?

9.8. Show qualitatively that the approach of octahedrally situated ligands to the $4f$ orbitals in Fig. 16.5 results in energy levels of three, three, and one orbital each.

9.9. Consider the following data for nickel(II) complexes [from R. S. Drago, "Physical Methods in Inorganic Chemistry," Van Nostrand-Reinhold, New York, **1965**]:

Complex	ν_1 (cm^{-1})	ν_2 (cm^{-1})	ν_3 (cm^{-1})
$[Ni(H_2O)_6]^{+2}$	8 500	15 400	26 000
$[Ni(NH_3)_6]^{+2}$	10 750	17 500	28 200
$[Ni(OS(CH_3)_2)_6]^{+2}$	7 728	12 970	24 038
$[Ni(dma)_6]^{+2}$	7 576	12 738	23 809

Calculate appropriate values of B' and Dq for these complexes.

9.10. The following absorption bands are found in the spectrum of $[Cr(CN)_6]^{-3}$: 264 nm, 310 nm, and 378 nm. Calculate appropriate values of B' and Dq. [*Caution:* This problem is not as straightforward as it might at first appear.

9.11. The Ag^+ ion is larger in a square planar environment than it is in a tetrahedral one (Table 3.4). For Ni^{+2} the reverse is true. Explain.

9.12. Explain why for a given series of ions, M^{+3}, one might expect the radii to decrease in the order $d^2 > d^3 > d^4 > d^5 > d^6$. Explain why, in fact, a quite different ordering is often found.

9.13. Both nickel and platinum belong to the same family of the periodic table, but the complexes $NiCl_4^{-2}$ and $PtCl_4^{-2}$ differ considerably in geometry, color, and magnetism. Using any theory of your choice, discuss the magnetic, geometric, and spectroscopic properties of these complexes.

9.14. Copper does not form any carbonyls that are stable at room temperature. However, it is possible to isolate molecules that contain both copper atoms and carbonyl groups by matrix isolation in frozen carbon monoxide at 10–15 K. What should be the formula of the most stable copper carbonyl? [H. Huber et al., *J. Am. Chem. Soc.*, **1975**, *97*, 2097.]

9.15. Consult Appendix J, Section 7, and name the following coordination compounds:

$[Co(NH_3)_6][Cr(CN)_6]$	$Na[Co(CO)_4]$
$[Co(NH_3)_5(H_2O)]Cl_3$	$[Co(NH_3)_5Cl]Br_2$
$[Co(NH_3)_6]Cl_2$	$[Cr(en)_3][FeCl_4]_3$

9.16. Draw the structures of the following coordination compounds:

a. sodium tetracyanonickelate(II)

b. potassium tetrachloronickelate(II)

c. hexaamminechromium(III) tetrachloromanganate(II)

d. hexaquatitanium(III) chloride

e. tetrakis(pyridine)platinum(II) tetraphenylborate(III)

f. di-μ-carbonylbis(tricarbonylcobalt).

9.17. a. From your chemical "intuition," predict the spin state of the metal ion in the following complexes.

b. Using data from Tables 9.5, 9.8, and 9.9, quantify your predictions.

$[Co(NH_3)_6]^{+3}$ $[Fe(H_2O)_6]^{+3}$ $[Fe(CN)_6]^{-3}$

9.18. Discuss the bond lengths found in the hexanitrocuprate(II) anion (Fig. 9.63).

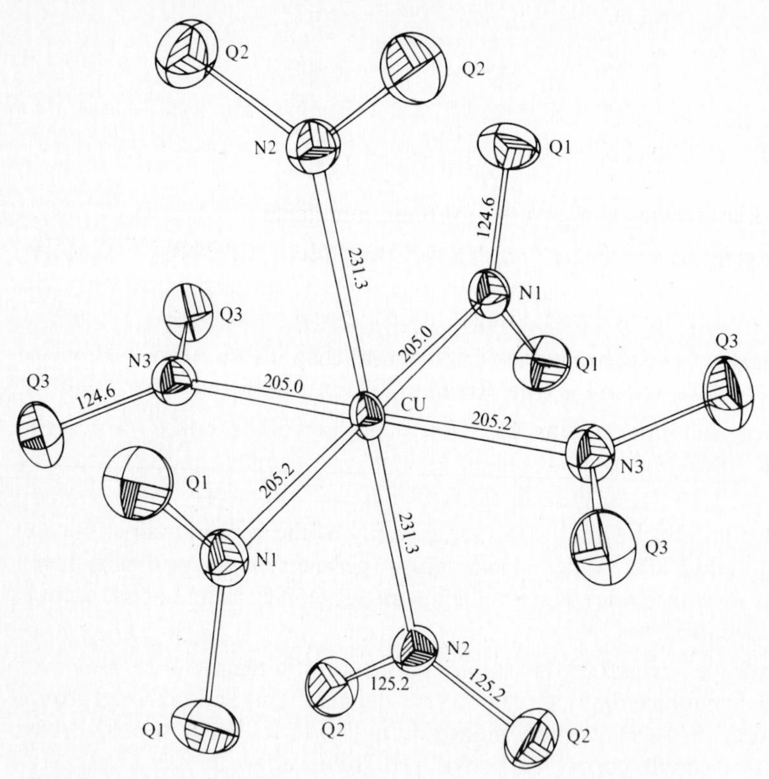

Fig. 9.63 Structure of the hexanitrocuprate(II) anion. Distances in pm. [From S. Takagi et al., *J. Am. Chem. Soc.*, **1974**, *96*, 6606. Reproduced with permission.]

9.19. Use Eq. 2.14 to estimate the spin-only effective magnetic moments (μ_{eff}) of the iron complexes discussed on p. 381.

9.20. The lengthening of the Fe—S bonds in high-spin complexes was discussed in CFT terms on p. 381. Explain this lengthening in equivalent MOT terms.

9.21. Do a radius ratio calculation to rationalize that CoF_6^{-4} and CoX_4^{-2} are the stoichiometries of these complexes rather than CoF_4^{-2} and CoX_6^{-4} (X = Cl, Br, and I).

9.22. On p. 402 the "bite" of a bidentate ligand was defined as:

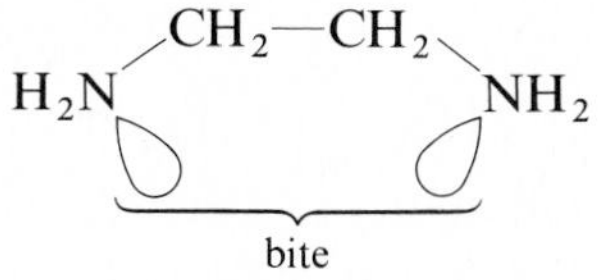

$$H_2N\text{—}CH_2\text{—}CH_2\text{—}NH_2$$

bite

rather than a seemingly more logical distance between the lone-pair "jaws":

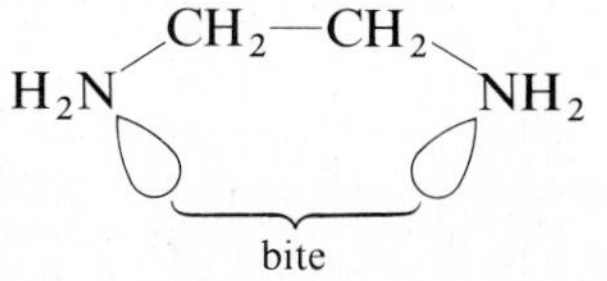

$$H_2N\text{—}CH_2\text{—}CH_2\text{—}NH_2$$

bite

Why? Discuss. What is the relationship to the radius of the metal?

9.23. On p. 390 the ease of loss of an electron (as shown by the *emf*) was related to the electron being repelled by a strong-field ligand (CFT). Explain this effect in equivalent MOT terms.

9.24. Figure 9.19 illustrated the effect of pressure on spectra and was explained in CFT terms. Explain Fig. 9.19 in AOM terms.

9.25. Although we are accustomed to thinking of complexes of transition metals as brightly colored, there are several that are not. For each of the following exceptions, present a reason for its lack of intense color:

a. $Cu(NH_3)_4^+$ is a completely colorless complex in contrast to its sister complex, $Cu(NH_3)_4^{+2}$, which is intensely blue.

b. $Co(H_2O)_6^{+2}$ is very pale pink, although $CoCl_4^{-2}$ is deep blue.

c. $Au(CN)_4^-$ and $Co(CN)_6^{-3}$ complexes form white crystals when combined with colorless cations.

d. Unlike Co^{+2}, Mn^{+2} forms pale pink complexes with *both* water and chloride ions as ligands: $Mn(H_2O)_6^{+2}$ and $MnCl_4^{-2}$.

9.26. There are several *non*transition metal ions that are *colorless* when isolated, i.e., in solution: Cd^{+2}, Hg^{+2}, and Pb^{+2}. Nevertheless, when combined with certain anions that are also colorless in solution, the final compounds are strongly colored. Explain.

9.27. Burdett[136] has shown that the distribution of "normal" versus "inverse" spinels (Table 9.11) fits a structure field map quite well, seemingly in complete contradiction to the arguments presented on pp. 393–395 (see Fig. 9.64). Can you reconcile these two viewpoints by common factors?

[136] J. K. Burdett et al., *J. Am. Chem. Soc.*, **1982**, *104*, 92.

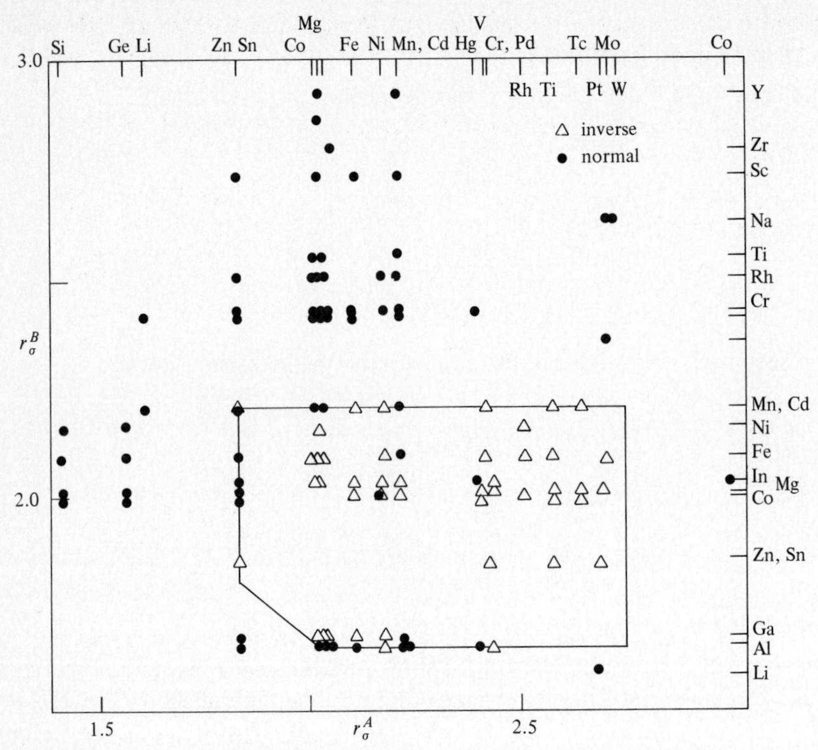

Fig. 9.64 Structure sorting map for normal and inverse spinels. Note that all inverse spinels occur within the enclosed area; few normal ones do. [From J. K. Burdett et al., *J. Am. Chem. Soc.*, **1982**, *104*, 92. Reproduced with permission.]

9.28. If you did not answer Problem 8.12 when you read Chapter 8, do so now.

9.29. Using *only* Table 9.18, confirm that the splitting in a tetrahedral complex is only $\frac{4}{9}$ of that in an octahedral complex assuming the same metal and ligands.

9.30. Interpret Tables 9.14 and 9.15 in terms of the angular overlap method.

9.31. Consider Fig. 9.62.

a. Show that the purple color of the permanganate ion cannot arise from d–d transitions.

b. Suggest possible reasons for the color.

c. If the frequencies of the observed transitions are 1.85×10^4 cm^{-1}, 3.22×10^4 cm^{-1}, and 4.44×10^4 cm^{-1}, show how we might be able to estimate the value of $10Dq$, assuming we can assign the transitions correctly, even though $10Dq$ cannot be observed directly.

9.32. Estimate the frequency of the charge-transfer spectrum of NaCl mentioned in the text (p. 456).

9.33. **a.** Using a photocopy of Fig. 9.42, fill in the electrons in hexaamminecobalt(III) and confirm that the molecular orbital electron configuration is that stated on p. 421.
b. How many bonds are shown in Fig. 9.42?

9.34. Why was sodium chosen to illustrate experimentally the $\Delta l = \pm 1$ rule, rather than the simpler hydrogen atom?

9.35. After reading Appendix B, find the symmetry elements and assign the appropriate point group to the hexanitrocuprate anion (Fig. 9.63).

10

Coordination chemistry: structure

Although a modicum of structural information on coordination compounds was introduced earlier to illustrate the work of Alfred Werner and to provide the context for the discussion of Jahn-Teller distortions, the previous chapter was largely restricted to the principles of formation of coordination compounds as illustrated in terms of bonding theory. In this chapter, structure will be examined more closely, particularly with reference to the existence of various coordination numbers and molecular structures in the various metal ion complexes and the effect of these structural aspects on the physical and chemical properties of these complexes.

The coordination numbers of metal ions range from 1, as in ion pairs such as Na^+Cl^- in the vapor phase, to 12 in some mixed metal oxides. The lower limit, 1, is hardly within the realm of coordination chemistry, since the Na^+Cl^- ion pair would not usually be considered a coordination compound although this is a matter of convenience rather than anything distinctly different about the bonding. A more likely candidate is the *vanadyl ion*, VO^{+2} in which the vanadium(IV) species appears to have a coordination number of 1. The existence of a free VO^{+2} ion is transient at best since it readily coordinates with four or five more ligating atoms to form species such as $[VO(acac)_2]$ and $[VO(H_2O)_5]^{+2}$. The upper limit of 12 is equally unimportant since the definition of solid crystal lattices such as perovskite and hexagonal $BaTiO_3$ as coordination compounds[1] is not particularly rewarding. The lowest and highest coordination numbers found in "typical" coordination compounds are 2 and 9, respectively, with the intermediate number 6 being the most important.

We have seen in Chapter 3 that the coordination number of ions in lattices is related to the ratio of the radii of the ions. The same general principles apply to coordination compounds, especially when a single coordination number, such as 4, has two common geometries—tetrahedral and square planar. An extended list of radius ratios is given in Table 10.1.

---

[1] See A. F. Wells, "Structural Inorganic Chemistry," 4th ed., Oxford University Press, London, **1975**, pp. 483–487.

Table 10.1 Radius ratios

C.N.	Minimum radius ratio	Coordination polyhedron
4	0.225	Tetrahedron
6	0.414	Octahedron
	0.528	Trigonal prism
7	0.592	Capped octahedron
8	0.645	Square antiprism
	0.668	Dodecahedron (bisdisphenoid)
	0.732	Cube
9	0.732	Tricapped trigonal prism
12	0.902	Icosahedron
	1.000	Cuboctahedron

COORDINATION NUMBER 2

Few complex ions are known with a coordination number of 2. They are generally limited to the +1 ions of the Group IB metals and the closely related Hg(II) species. Examples are $[Cu(NH_3)_2]^+$, $[Ag(NH_3)_2]^+$, $[CuCl_2]^-$, $[AgCl_2]^-$, $[AuCl_2]^-$, $[Ag(CN)_2]^-$, $[Au(CN)_2]^-$, and $[Hg(CN)_2]$. Even these may react with additional ligands to form higher-coordinate complexes such as:

$$[Ag(NH_3)_2]^+ + 2NH_3 \longrightarrow [Ag(NH_3)_4]^+ \qquad \textbf{(10.1)}$$

$$[Hg(CN)_2] + 2CN^- \longrightarrow [Hg(CN)_4]^{-2} \qquad \textbf{(10.2)}$$

The low stability of 2-coordinate complexes with respect to other possible structures is well illustrated by the cyano complexes. Although silver(I) and gold(I) form discrete bis(cyano) complexes, solid $KCu(CN)_2$ possesses a chain structure in which the coordination number of the copper(I) is 3.

The geometry of coordination number 2 would be expected to be linear (but there are exceptions—see p. 218), either from the point of view of simple electrostatics or from the use of *sp* hybrids by the metal (note that all of the above are d^{10} ions). Orgel[2] has suggested that since the $(n-1)d$ orbitals in these metals are of nearly the same energy as the *ns* and *np* orbitals, the d_{z^2} orbital can enter into this hybridization to remove electron density from the region of the ligands, thus stabilizing the complex to a certain extent. The first stage would be the hybridization of the *s* and d_{z^2} orbitals to form a hybrid orbital, Ψ_1, having maximum electron density in the *xy* plane, and a second hybrid, Ψ_2, having maximum electron density along the *z* axis (Fig. 10.1). The pair of electrons from the d_{z^2} orbital can be placed in the hybrid Ψ_1:

$$\text{Hybridization: } \Psi_{d_{z^2}} \pm \Psi_s \rightarrow \Psi_1 + \Psi_2$$
$$\text{Electron configuration: } ((n-1)d_{z^2})^2(ns)^0 \rightarrow \Psi_1^2 + \Psi_2^0 \qquad \textbf{(10.3)}$$

[2] L. E. Orgel, *J. Chem. Soc.*, **1958**, 4186.

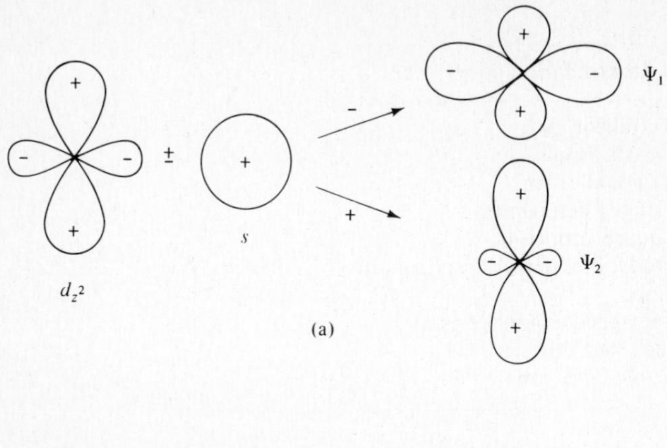

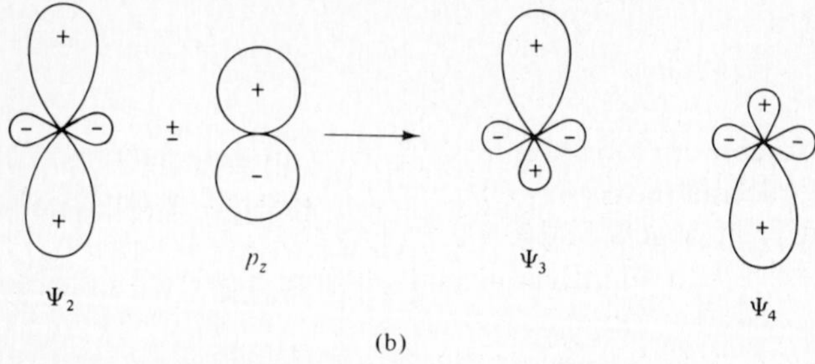

Fig. 10.1 Hybridization of $(n-1)d_{z^2}$, ns, and np_z orbitals: (a) formation of hybrid orbitals Ψ_1 and Ψ_2 removing electron density (into Ψ_1) *away* from the z axis; (b) formation of linear-bonding hybrid orbitals from Ψ_2 and np_z orbitals.

The hybrid Ψ_2 can then hybridize with the p_z orbital to form two hybrid orbitals, Ψ_3 and Ψ_4, at 180° to each other (as the simple *sp* hybrids would be) and lying in the region evacuated of electron density from the first hybridization step:

$$\text{Hybridization: } \Psi_2 \pm \Psi_{p_z} \rightarrow \Psi_3 + \Psi_4 \tag{10.4}$$

The hybrid orbitals Ψ_3 and Ψ_4 are thus available to accept electron density from ligands approaching along the z axis. The above stepwise procedure is, of course, artificial. The actual complex formation would involve the mixing of the metal $(n-1)d_{z^2}$, ns, and np_z orbitals with the ligand orbitals σ_z and σ_{-z} to form five molecular orbitals, two bonding (occupied), one nonbonding (occupied), and two antibonding (unoccupied).

Fisher and Drago[3] have extended Orgel's scheme to explain why Hg^{+2} has a strong tendency towards coordination number 2 as opposed to its congeners, Zn^{+2} and Cd^{+2}, which form stable tetrahedral complexes. The nearness in energy of the 5*d* and 6*s* energy levels allows the *d–s* hybridization to be important, and the latter, in turn, makes the Hg^{+2} ion quite soft (high C parameter) and it bonds preferentially along the z axis to strong, soft ligands, such as mercaptans (recall that CH_3Hg^+ can be used as a reference soft acid,

[3] K. J. Fisher and R. S. Drago, *Inorg. Chem.*, **1975**, *14*, 2804.

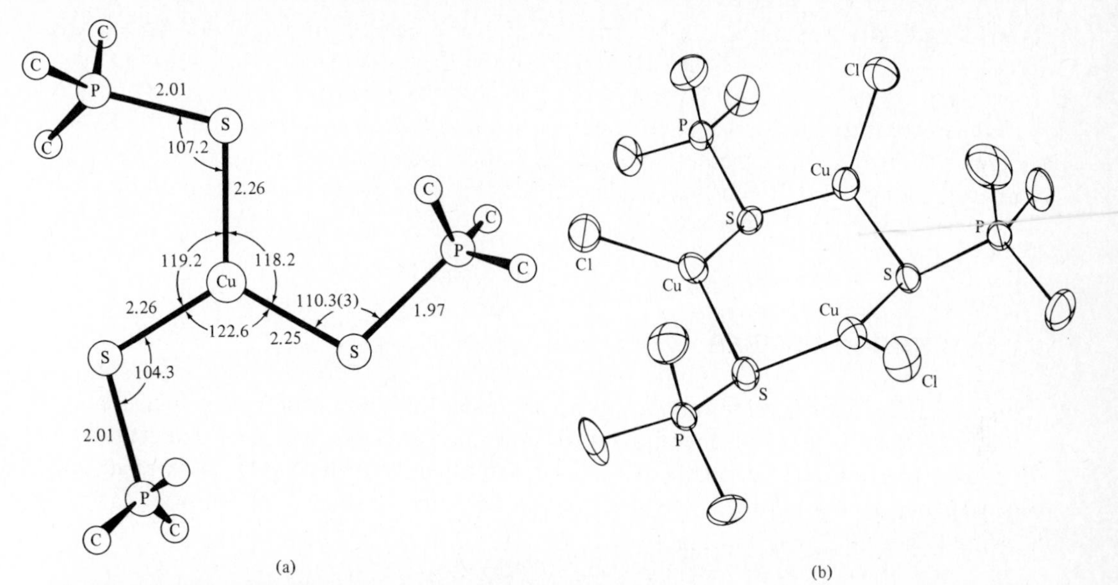

Fig. 10.2 Three-coordinate copper: (a) $Cu(SP(CH_3)_3)$; (b) $Cu(SP(CH_3)_3)Cl_3$. [From P. G. Eller and P. W. R. Corfield, *Chem. Commun.*, **1971**, 105; J. A. Tiethof et al., *Inorg. Chem.*, **1973**, *12*, 1170. Reproduced with permission.]

p. 313). In contrast, the lighter congeners have a larger separation of their $(n-1)d$ and ns levels, and d orbital participation in the hybridization is discouraged. These ions tend to be harder and tend to form more electrostatic complexes, such as $ZnCl_4^{-2}$ and $Zn(H_2O)_4^{+2}$, with the hybridization restricted to sp^3.

COORDINATION NUMBER 3

This is a rare coordination number. Many compounds which might appear to be 3-coordinate as judged from their stoichiometry are found upon examination to have higher coordination numbers. Examples are $CsCuCl_3$ (infinite chains, —Cl—$CuCl_2$—Cl—, C.N. = 4 at 230 pm; two more Cl^- from adjacent segments at 265 pm), $KCuCl_3$ (infinite double chains, Cl_4—(Cu_2Cl_2)—Cl_4, C.N. = 6, distorted octahedron), and NH_4CdCl_3 (infinite chains, C.N. = 6).

The $KCu(CN)_2$ chain described above (—CN—Cu(CN)—CN—Cu(CN)—CN—) is an example of true 3-coordination. Four other examples of 3-coordination that have been verified by X-ray studies are tris(trimethylphosphine sulfide)copper(I) perchlorate, $[Cu(SPMe_3)_3]^+[ClO_4]^-$ (Fig. 10.2a), *cyclo*-tris(chloro-μ-(trimethylphosphine sulfide)copper(I)) (Fig. 10.2b), the triiodomercurate(II) anion, $[HgI_3]^-$, and tris(triphenylphosphine)platinum(0),[4] $[(\phi_3P)_3Pt]^0$. In all four examples the geometry approximates an equilateral triangle with the metal atom at the center as expected for sp^2 hybridization.

[4] J. A. Tiethof, *Inorg. Chem.*, **1973**, *12*, 1170, and **1974**, *13*, 2505; R. H. Fenn et al., *Nature*, **1963**, *198*, 381; V. Albano et al., *Chem. Commun.*, **1966**, 507.

Some d orbital participation can be expected as in the case of linear hybrids since a trigonal sd^2 hybrid is also possible. C.N. = 3 is also enhanced by steric considerations, and since electronic factors do *not* favor it, the former *must* be dominant.

There has been an upsurge of interest in coordination number 3, and the above examples illustrate some recent findings. Whether 3-coordinate complexes are more common than previously believed or whether they have attracted undue attention because of their novelty remains to be seen.[5]

COORDINATION NUMBER 4

This is the first coordination number to be discussed that has an important place in coordination chemistry. It is also the first for which isomerism is to be expected. The structures formed with coordination number 4 can be conveniently divided into *tetrahedral* and *square planar* forms although intermediate and distorted structures are known.[6]

Tetrahedral complexes

Tetrahedral complexes are favored by steric requirements, either simple electrostatic repulsions of charged ligands or van der Waals repulsions of large ones. A valence bond (VB) point of view ascribes tetrahedral structures to sp^3 hybridization. From a crystal field (CF) or molecular orbital (MO) viewpoint we have seen that, in general, tetrahedral structures are not stabilized by large CFSE. Tetrahedral complexes are thus favored by large ligands like Cl^-, Br^-, and I^- and small metal ions of three types: (1) those with a noble gas configuration, ns^2np^6, such as Be^{+2}; (2) those with a pseudo-noble gas configuration, ns^2np^6 and $(n-1)d^{10}$, such as Zn^{+2} and Ga^{+3}; and (3) those transition metal ions which do not strongly favor other structures by virtue of the CFSE, such as Co^{+2}, d^7.

Tetrahedral complexes do not exhibit geometrical isomerism. However, they are potentially optically active just as is tetrahedral carbon. The simple form of optical isomerism exhibited by most optically active organic chemistry, namely four different substituents, is rarely observed because substituents in tetrahedral complexes are usually too labile[7] for the complex to be resolved, i.e., they racemize rapidly. However, an interesting series of cyclopentadienyliron phosphine carbonyl compounds (see Chapter 13 for further discussion of organometallic compounds) has been synthesized and characterized.[8] A

[5] P. G. Eller et al., *Coordin. Chem. Rev.*, **1977**, *24*, 1.

[6] M. C. Favas and D. L. Kepert, *Progr. Inorg. Chem.*, **1980**, *27*, 325.

[7] *Labile* refers not to thermodynamic stability *per se* but, rather, to the ease of substitution by other ligands. In addition to bond strength, the accessibility of a suitable mechanism also contributes to the stability or lability of a complex (see p. 547). Labile is one of those words the pronunciation of which American chemists seem unable to agree upon. Some follow the dictionary and rhyme it more or less with its antonym, *stable*. Others rhyme with it M*obile*, and a third group shows an English or Australian bent and rhymes it with *hay-stile*.

[8] G. M. Reisner et al., *Inorg. Chem.*, **1978**, *17*, 783; J. D. Korp and I. Bernal, *J. Organometal. Chem.*, **1981**, *220*, 355.

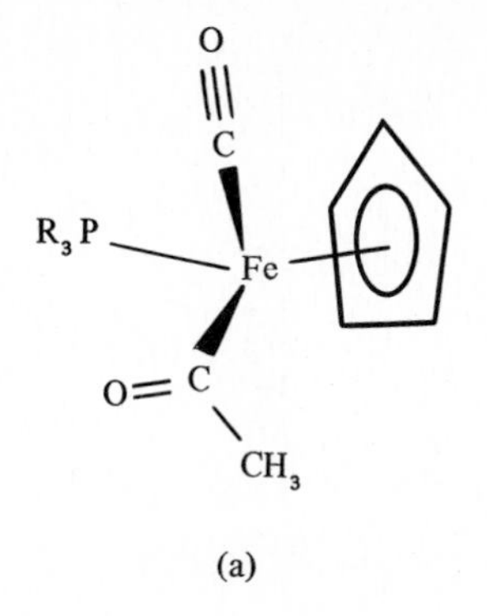

(a)

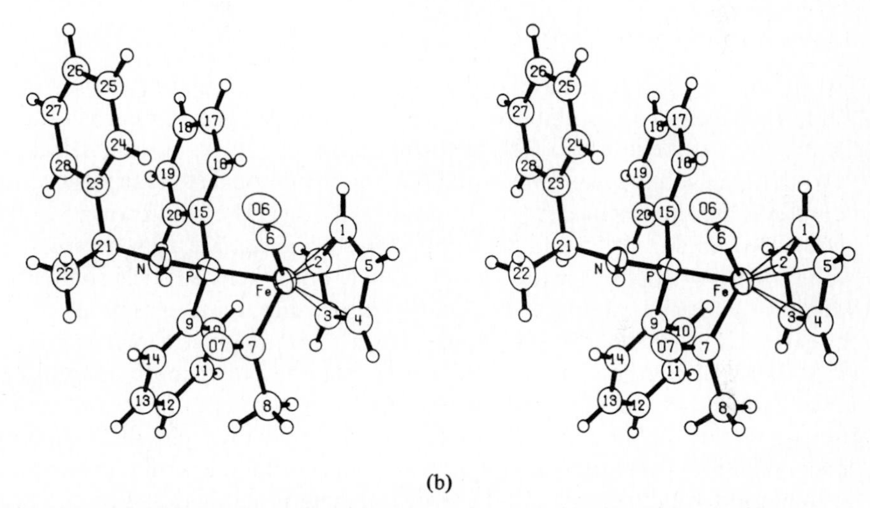

(b)

Fig. 10.3 (a) Line drawing of the inner coordination sphere of an acetyl(carbonyl)chloro(phosphine)iron complex. (b) Stereoview of the same molecule. [From J. D. Korp and I. Bernal, *J. Organometal. Chem.*, **1981**, *220*, 355. Reproduced with permission.]

line drawing and stereoview of one of these is shown in Fig. 10.3. Note that the large C_5H_5 ring forces the other ligands until the bond angles are essentially 90° rather than $109\frac{1}{2}$°. Indeed an argument could be made for considering the complex to be 8-coordinate, though little is gained by such a view. The chirality of the molecule is the important feature to be noted.

A second form of optical isomerism analogous to that shown by the organic spiranes has been demonstrated. Any molecule will be optically active if it does not contain a mirror plane of symmetry or center of inversion. The two optical isomers of bis(benzoylacetonato)-beryllium are illustrated in Fig. 10.4. In order for isomers to be resolvable, the chelate must be unsymmetrical (with respect to the two ends of the chelating molecule—*not* necessarily asymmetric, that is, optically active itself); $[Be(acac)_2]$ is not optically active.

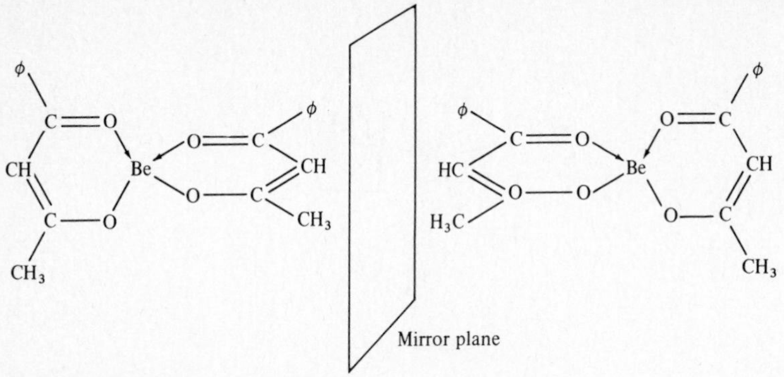

Mirror plane

Fig. 10.4 Enantiomers of bis(benzoylacetonato)beryllium.

Square planar complexes

Square planar complexes are less favored sterically than tetrahedral complexes (see Table 10.1) and so are prohibitively crowded by large ligands. On the other hand, if the ligands are small enough to form a square planar complex, an octahedral complex with two additional σ bonds can form with little or no additional steric repulsion. Square planar complexes are thus formed by only a few metal ions. The best known are the d^8 species such as Ni^{+2}, Pd^{+2}, Pt^{+2}, and Au^{+3} (pp. 409–411). There are a few complexes of $Co^{+2}(d^7)$ with bidentate ligands that are square planar, but otherwise such complexes are rather scarce. Chlorophyll and other biocomplexes are important exceptions to this rule, but the geometry is dictated by the rigid porphyrin structure (see Chapter 18). The prerequisite for stability of square planar complexes is thus the presence of nonbulky, strong field ligands which π-bond sufficiently well to make up through this means the energy "lost" through 4- rather than 6-coordination. For Ni^{+2}, for example, the cyanide ion forms a square complex, whereas ammonia and water form 6-coordinate octahedral species and chloride, bromide, and iodide form tetrahedral complexes. For the heavier metals the steric requirements are relaxed and the effective field strength of all ligands is increased. Under these conditions, even the tetrachloropalladate(II), tetrachloroplatinate(II), and tetrachloroaurate(III) anions are square planar.

Square planar complexes of the formula $[MA_2B_2]$ exhibit *cis–trans* isomerism:

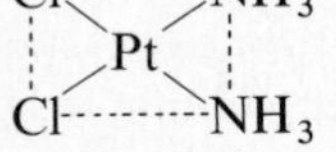

$\mu \neq 0$ 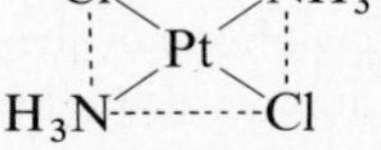 $\mu = 0$

cis-diamminedichloroplatinum (II) *trans*-diamminedichloroplatinum(II)

If such complexes are neutral molecules as in the above example, they may be readily distinguished (and often separated as well) by the presence of a dipole moment (μ) in the *cis* isomer but none in the *trans* isomer. Only in the unlikely event that the M—A and M—B bond moments were identical could the *cis* isomer have a zero dipole moment. (See Appendix B for further discussion of symmetry and dipole moments.)

Square planar complexes rarely show optical isomerism. The plane formed by the four ligating atoms and the metal ion will ordinarily be a mirror plane and prevent the

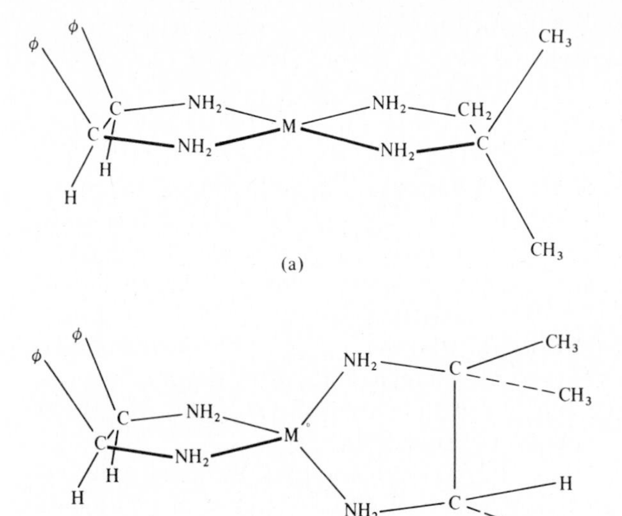

Fig. 10.5 Possible structures of (*meso*-stilbenediamine) (*iso*-butylenediamine)palladium(II) and platinum(II) complexes: (a) planar structure, optically active; (b) tetrahedral structure, optically inactive.

possibility of optical asymmetry. An unusual exception to this general rule was used in an ingenious experiment to prove that platinum(II) and palladium(II) complexes were *not* tetrahedral.[9] Certain complexes (Fig. 10.5a) with square planar structures have no mirror plane and hence are optically active. If these complexes were tetrahedral (Fig. 10.5b) there would be a mirror plane (defined by the metal and two nitrogen atoms from isobutylenediamine) reflecting the phenyl groups, methyl groups, etc. Inasmuch as optical activity *is found experimentally*, these complexes *must* be square planar.

COORDINATION NUMBER 5

In the past a coordination number of 5 was considered almost as rare as that of coordination number 3. Again, many of the compounds which might appear to be 5-coordinate on the basis of stoichiometry are found to have other coordination numbers upon close examination. Thus Cs_3CoCl_5 and $(NH_4)_3ZnCl_5$ contain discrete MCl_4^{-2} tetrahedral anions and free chloride ions. Thallium pentafluoroaluminate is composed of infinite chains, —F—AlF_4—F—, in which the coordination number of the aluminum is 6. The complex of cobalt(II) chloride and diethylenetriamine, $NH_2CH_2CH_2NHCH_2CH_2NH_2$, of empirical formula $[CoCl_2dien]$ is not a 5-coordinate molecule but a salt, $[Co(dien)_2]^{+2}$ $[CoCl_4]^{-2}$ containing octahedral cations and tetrahedral anions.[10]

[9] W. H. Mills and T. H. H. Quibell, *J. Chem. Soc.*, **1935**, 839; A. G. Lidstone and W. H. Mills, *J. Chem. Soc.*, **1939**, 1754.

[10] M. Ciampolini and G. P. Speroni, *Inorg. Chem.*, **1966**, 5, 45.

Nyholm and Tobe[11] have shown that if electrostatic forces were the only forces operating in bonding, 5-coordinate compounds would always disproportionate into 4- and 6-coordinate species:

$$2[MX_5]^{n-5} \longrightarrow [MX_4]^{n-4} + [MX_6]^{n-6} \qquad (10.5)$$

Since covalent bonding is obviously of great importance in coordination compounds, we should not expect the calculation of Nyholm and Tobe to hold strictly, but it is true that there is a delicate balance of forces in 5-coordinate complexes and their stability with respect to other possible structures is not great. For example, the compound $[Ni(PNP)X_2]$ (where $PNP = \phi_2PCH_2CH_2NRCH_2CH_2P\phi_2$) is a true 5-coordinate species, but on warming slightly it converts to $[Ni(PNP)X]_2[NiX_4]$, which contains both square planar and tetrahedral species.[12] Another example is the pair of compounds of empirical formula $MX_2(Et_4dien)$. The cobalt complex is 5-coordinate, but the corresponding nickel compound is 4-coordinate, $[NiX(Et_4dien)]^+X^-$.[13]

Although 5-coordinate compounds are still less common than those of either coordination number 4 or 6, recently there has been considerable interest in this subject and the number of known compounds has increased rapidly. There is now a sufficiently large number of these compounds that some systematization can be accomplished although their chemistry is still rather incompletely understood.[14]

Furlani[15] has suggested the following classification for the stereochemical arrangements of coordination number 5:

1. *Regular trigonal bipyramid* (*TBP*). All five ligands must be the same with no distortions present. An example of a noncoordination compound with this structure is PF_5. The pentachlorocuprate(II) anion, $[CuCl_5]^{-3}$, is a good example of a complex ion with this geometry (Fig. 10.6).[16] In contrast to the bond lengths in PF_5 (see pp. 207, 230), the *equatorial* bond is longer than the *axial* bond in the copper complex. This had been predicted by the observation that the d_{z^2} orbital, directed at the axial ligands, contains only one electron (the odd electron in the d^9 configuration; see Fig. 10.10 or Table 9.14) in contrast to two electrons in the other orbitals.[17] The model would further predict that in a TBP complex of a d^{10} species, the symmetric, filled d subshell would remove this effect and the axial bonds should be about as long as the equatorial bonds. Unfortunately, attempts to isolate $[ZnCl_5]^{-3}$ led to double salts of the type[18] $[Co(NH_3)_6][ZnCl_4]Cl$, but $[Co(NH_3)_6][CdCl_5]$ has been prepared and found to have axial and equatorial bond lengths within about 1% of each other.[19]

---

[11] R. S. Nyholm and L. M. Tobe, in "Essays in Coordination Chemistry," W. Schneider et al., eds., Birkhäuser Verlag, Basel, **1964**, p. 112.

[12] L. Sacconi, *Pure Appl. Chem.*, **1968**, *17*, 95.

[13] Z. Dori and H. B. Gray, *J. Am. Chem. Soc.*, **1966**, *88*, 1394.

[14] For reviews, in addition to refs. 6, 11, 12, and 15, see P. L. Orioli, *Coordin. Chem. Rev.*, **1971**, *6*, 285; J. S. Wood, *Progr. Inorg. Chem.*, **1972**, *16*, 227; R. Morassi et al., *Coord. Chem. Rev.*, **1973**, *11*, 343.

[15] C. Furlani, *Coordin. Chem. Rev.*, **1968**, *3*, 141.

[16] K. N. Raymond et al., *Inorg. Chem.*, **1968**, *7*, 1111. See also S. A. Goldfield and K. N. Raymond, ibid., **1971**, *10*, 2604.

[17] R. J. Gillespie, *J. Chem. Soc.*, **1963**, 4672.

[18] D. W. Meek and J. A. Ibers, *Inorg. Chem.*, **1970**, *9*, 465.

[19] T. V. Long II et al., *Inorg. Chem.*, **1970**, *9*, 459.

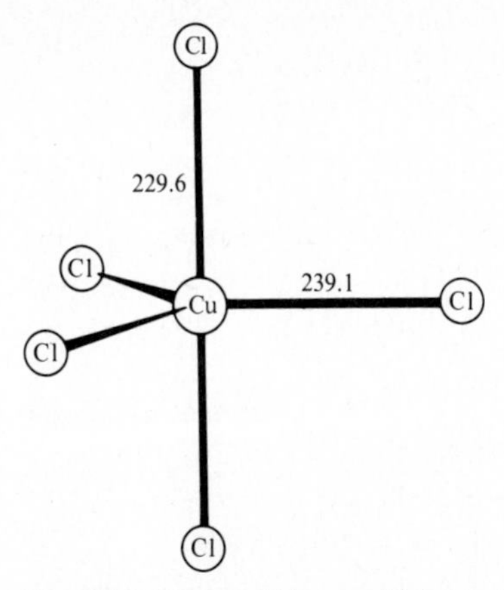

Fig. 10.6 Trigonal bipyramidal structure (type I) of the pentachlorocuprate(II) anion in the compound $[Cr(NH_3)_6][CuCl_5]$. Note difference in bond lengths. [From K. N. Raymond et al., *Inorg. Chem.*, **1968**, *7*, 1111. Reproduced with permission.] Distances in pm.

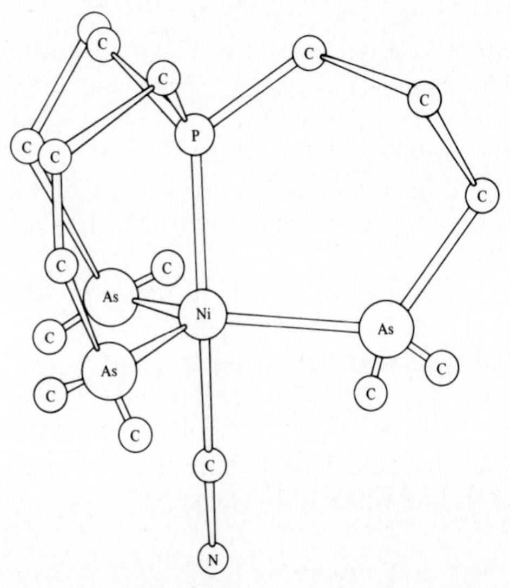

Fig. 10.7 Distorted trigonal bipyramidal structure (type II) of the cyanotris(3-dimethylarsinopropyl)-phosphinenickel(II) cation in $[Ni(TAP)CN][ClO_4]$. [From D. L. Stevenson and L. F. Dahl, *J. Am. Chem. Soc.*, **1967**, *89*, 3424. Reproduced with permission.]

2. *Slightly distorted trigonal bipyramid.* This arrangement often occurs with tetradentate ligands in one axial position and the three equatorial positions retaining the threefold rotational symmetry, as in cyanotris(3-dimethylarsinopropyl)phosphinenickel(II) perchlorate (Fig. 10.7).[20] The metal atom may or may not be in the trigonal plane, but for most purposes complexes of this type may be regarded as TBP structures.

[20] D. L. Stevenson and L. F. Dahl, *J. Am. Chem. Soc.*, **1967**, *89*, 3424.

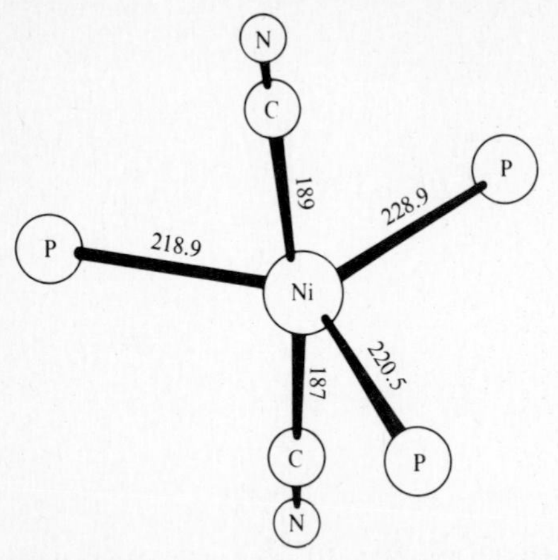

Fig. 10.8 Intermediate structure (type III) of [$Ni(CN)_2(P\phi(OEt)_2)_3$]. Only the ligating atoms are shown. [From J. K. Stalnick and J. A. Ibers, *Inorg. Chem.*, **1969**, *8*, 1084. Reproduced with permission.] Distances in pm.

3. *Highly distorted structures.* These may be viewed as either highly distorted trigonal bipyramidal or highly distorted square pyramidal structures. In fact, just about every possible distortion between TBP and SP has been observed. An example is dicyanotris(phenyldiethoxyphosphine)nickel(II), [$Ni(CN)_2(\phi P(OEt)_2)_3$] (Fig. 10.8).[21]

4. "*Regular*" *square pyramid.* This may be considered as a modification of the following structure in which the metal atom lies in the plane of the basal ligands. It is not an important structure.

5. "*Distorted*" *square pyramid* (*SP*). This geometry is "distorted" in the sense that the metal atom lies *above* the plane of the four basal ligands. In another sense, it is "regular" in that if all four ligands are the same, as in the pentacyanonickelate(II) anion, [$Ni(CN)_5$]$^{-3}$ (Fig. 10.9);[22,23] arguments based on either hybridization schemes or on minimizing steric repulsions favor the square pyramid with the metal atom above the basal plane. In this sense, SP is as "regular" as the corresponding TBP structure. Fortunately, little confusion is apt to occur since there are probably no compounds in which the metal atom lies *exactly* in the basal plane.

The above five classes form a convenient method of discussing the geometry of coordination number 5, but not all compounds fit the "pigeonholes" neatly. The differences between the various structures are often slight and the energy barriers tending to prevent interconversion are also small. Hence structures are found ranging from perfect TBP to perfect SP with various degrees of intermediacy (see p. 225). The mechanisms for interconversion of the two limiting structures are of interest since they provide some insight into the reaction possibilities. It requires only slight movements of the ligands to cause

[21] J. K. Stalick and J. A. Ibers, *Inorg. Chem.*, **1969**, *8*, 1084.
[22] A. Terzis et al., *Inorg. Chem.*, **1970**, *9*, 2415.
[23] K. N. Raymond et al., *Inorg. Chem.*, **1968**, *7*, 1362.

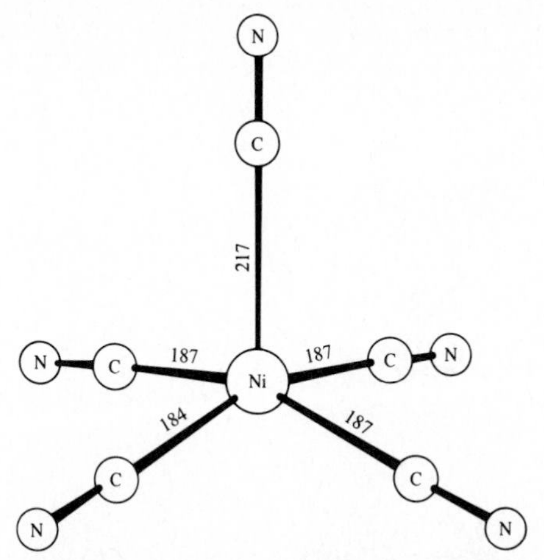

Fig. 10.9 Square pyramidal structure (type V) of the pentacyanonickelate(II) anion in $[Cr(en)_3]$ $[Ni(CN)_5] \cdot 1.5H_2O$. Note the difference in bond lengths. See text for complete discussion of this compound. [From K. N. Raymond and P. W. R. Corfield, *Inorg. Chem.*, **1968**, *7*, 1362. Reproduced with permission.] Distances in pm.

the conversion of one structure into the other.[24] Of particular interest in regard to the delicate balance between the forces favoring TBP versus SP structures are two pentacyanonickelate(II) salts with different but very similar cations. Tris(1,3-propanediamine)-chromium(III) pentacyanonickelate, $[Cr(tn)_3][Ni(CN)_5]$, contains square pyramidal anions similar to Fig. 10.9.[22] In contrast, crystalline tris(ethylenediamine)chromium(III) pentacyanonickelate(II) sesquihydrate, $[Cr(en)_3][Ni(CN)_5] \cdot 1.5H_2O$ contains both square pyramidal anions (Fig. 10.9) and slightly distorted trigonal bipyramidal anions.[23] The IR and Raman spectra of this solid exhibit two sets of bands, one of which (the TBP set) disappears when the sesquihydrate is dehydrated.[22] In aqueous solution the structure is apparently also square pyramidal. It would appear that the SP structure is inherently more stable but by such a slight margin that forces arising in the hydrated crystal can stabilize a TBP structure.

The forces favoring each of the limiting structures are not completely understood but the following generalizations can be made. On the basis of ligand repulsions alone, whether they be considered naively as purely electrostatic or as Pauli repulsions from the bonding pairs (see p. 207), the trigonal bipyramid is favored.[17] For this reason almost every[25] 5-coordinate compound with a nonmetallic central element (such as PF_5) has the TBP structure (unless there are lone pairs), since effects arising from incompletely filled d orbitals are not present. Likewise, we should expect d^0 and d^{10} to favor the TBP structure. Comparing the relative energies of the orbitals in TBP (D_{3h}) versus SP (C_{4v}) geometry (Fig. 10.10), d^1 and d^2 also appear to favor TBP geometry for the same reasons. The d^3

[24] R. R. Holmes et al., *Inorg. Chem.*, **1969**, *8*, 2612, have analyzed these conversion mechanisms in terms of IR- and Raman-active vibrations in the molecule.

[25] The exceptions are discussed on pp. 225–226.

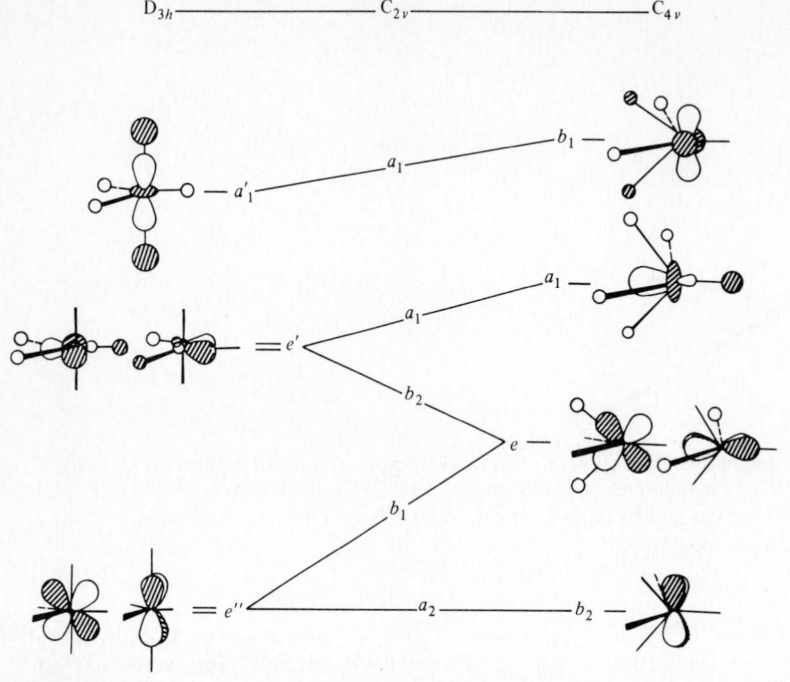

Fig. 10.10 Wave function and energy changes along a Berry pseudorotation coordinate. (Note relative energies of the $e'(D_{3h})$ and a_1 and $e(C_{4v})$ levels. [From A. R. Rossi and R. Hoffmann, *Inorg. Chem.*, **1975**, *14*, 365. Reproduced with permission.]

and d^4 configurations should also favor TBP versus SP even more since the e'' orbitals of D_{3h} are more stable than the e of SP. In contrast, low-spin d^6 should favor the SP configuration since the e orbitals of the latter are lower in energy than are the e' orbitals of a trigonal bipyramidal complex. For d^8 the order of stability again switches for the same sort of reason and continues through d^9 and d^{10}. Unfortunately, there are few data available to test these predictions, but there are a number of d^8 5-coordinate complexes (exemplified by $Fe(CO)_5$), most of which are TBP. The low-spin d^7 complex $Co(dpe)_2Cl^+$ crystallizes in two forms—a red solid that contains SP ions and a green form that contains TBP ions (Fig. 10.11).[26] Apparently, the slight ligand field stabilization energy favoring the SP arrangement balances the inherent superiority of the TBP arrangement and allows the isolation of both isomers. In solution, the two forms interconvert readily, either by a Berry pseudorotation or through dissociation and recombination (see p. 247).

We have seen (p. 230) that with nonmetallic central atoms (d^0), more electronegative elements prefer the axial positions of a TBP structure. A molecular orbital analysis of metal complexes[27] indicates that most d^n configurations follow this same pattern. A notable exception is d^8, which favors electropositive substituents at apical sites and electronegative

[26] J. K. Stalick et al., *Inorg. Chem.*, **1973**, *12*, 1668.
[27] A. R. Rossi and R. Hoffmann, *Inorg. Chem.*, **1975**, *14*, 365.

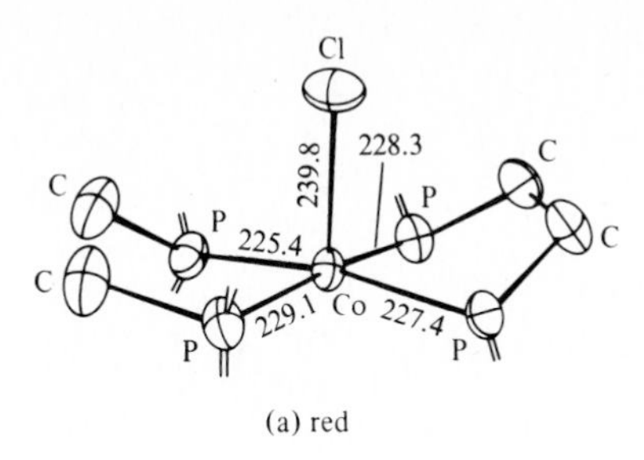

(a) red

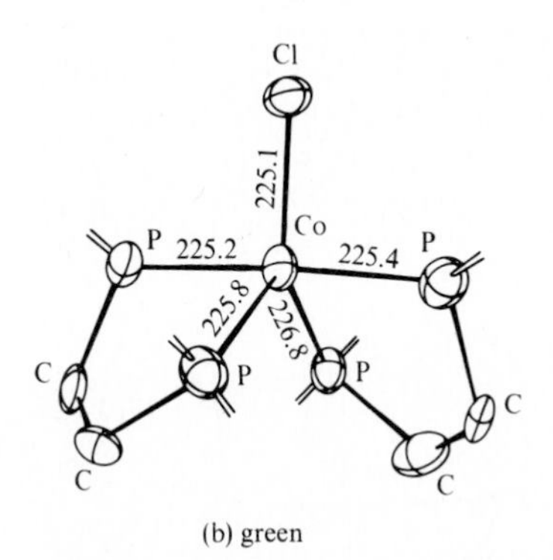

(b) green

Fig. 10.11 The structures of the red, square pyramidal isomer (a) and the green, trigonal bipyramidal isomer (b) of the chlorobis[1,2-bis(diphenylphosphino)ethane]cobalt(II) cation. Phenyl groups and other substituents have been removed for clarity. [From J. K. Stalnick et al., *Inorg. Chem.*, **1973**, *12*, 1668. Reproduced with permission.] Distances in pm.

substituents at equatorial sites. In the same way, the normally weak bonding of axial substituents is reversed with the d^8 configuration. Thus we find the methyl group in the axial position in the d^8 Ir(I) complex shown in Fig. 10.12a in contrast to its universal equatorial position in phosphoranes (p. 231). Also, in contrast to the phosphoranes, the axial bonds are shorter in $Fe(CO)_5$ than are the equatorial bonds (Fig. 10.12b); however, it must be stressed that there are exceptions to this behavior (Table 10.2). In addition, there are complications in seemingly straightforward complexes. For example, it was noted above that the bond lengths are very nearly the same in $CdCl_5^{-2}$. Since this is d^{10} we should expect the axial bonds to be somewhat longer than the equatorial; however, the reverse is actually observed, although the difference is not great.

The same type of analysis[28] predicts that in d^8 complexes good π-accepting ligands will prefer the equatorial position. The series of compounds shown in Fig. 10.13 allows us to test this. Note that most of the ligands occur in both axial and equatorial positions depending upon what other ligands are present. If we assume that the best π-acceptors

[28] There is insufficient space here to go through the complete derivation, but it may be noted that the method is not unlike that given previously for octahedral complexes (p. 416). For the complete method, see footnote 27.

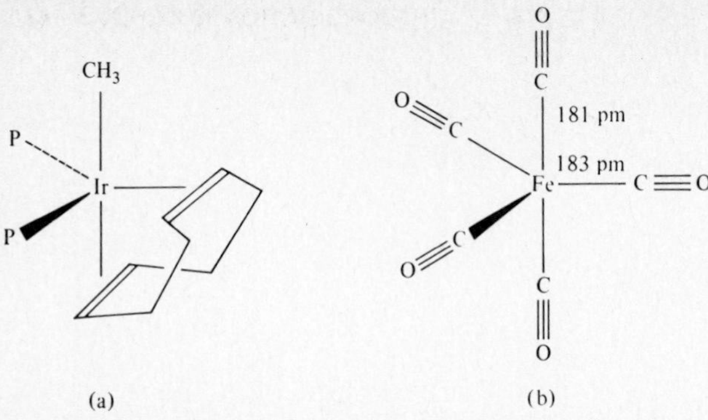

Fig. 10.12 Complexes showing *apparent* "exceptions" to the rules for trigonal bipyramidal bonding. (a) The methyl ligand seeks the axial position and allows the strong π-acceptors to occupy equatorial positions. Substituent groups on the phosphine ligands omitted for clarity. [From A. R. Rossi and R. Hoffmann, *Inorg. Chem.*, **1975**, *14*, 365. Reproduced with permission.] (b) The equatorial bonds in $Fe(CO)_5$ are slightly longer than the axial bonds.

Table 10.2 Bond lengths in some d^8 TBP complexes

Molecule	M—L (pm)	
	Axial	Equatorial
$Ni(phosph)_5^{+2}$	214	219
$Fe(CO)_5$	181	183
$Co(CNCH_3)_5^+$	184	188
$Pt(SnCl_3)_5^{-3}$	240	243
$Mn(CO)_5^-$	182	180

SOURCE: A. R. Rossi and R. Hoffmann, *Inorg. Chem.*, **1975**, *14*, 365. Reproduced with permission.

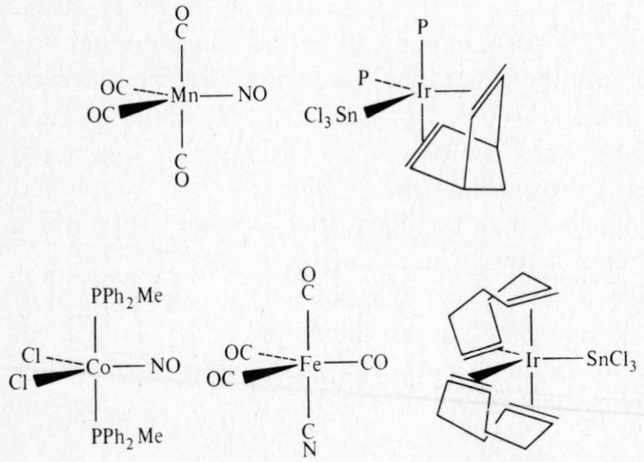

Fig. 10.13 A series of trigonal bipyramidal complexes which allow a π-acceptor series to be arranged by noting equatorial vs. axial site preference. [From A. R. Rossi and R. Hoffmann, *Inorg. Chem.*, **1975**, *14*, 365. Reproduced with permission.]

Table 10.3 Bond lengths in some SP complexes

	M—L (pm)		
Molecule	Apical	Basal	d^n
$Nb(NMe_2)_5$	198	204	d^0
$MnCl_5^{-2}$	258	230	d^4
$Ni(CN)_5^{-2}$	217	186	d^8
$InCl_5^{-2}$	242	246	d^{10}
$Sb\phi_5$	212	222	d^{10}

SOURCE: A. R. Rossi and R. Hoffmann, *Inorg. Chem.*, **1975**, *14*, 365. Reproduced with permission.

will always choose an equatorial position, we can arrange them in the following order:

$$NO^+ > CO > CN^-(?)^{29}SnCl_3^- > Cl > P > C{=}C > CH_3$$

This series may be compared with that given on p. 436 derived from completely different assumptions. The general concurrence is reassuring.

The square pyramidal geometry is complicated by the possibility that the central atom may be in the plane of the basal ligands ("regular") or above it to varying degrees ("distorted") and the molecular orbital predictions depend upon the extent of "distortion." The following discussion assumes that the metal atom is lying above the basal plane as is commonly found. Under these conditions the "normal" situation (d^0–d^6, d^{10}) is for the apical bond to be the strongest with weaker basal bonds. As in the TBP case the d^8 configuration is reversed with stronger basal bonds and a weak apical bond. Likewise, good donors usually (d^0–d^6, d^{10}) seek the apical position, but in d^8 complexes electronegative ligands should prefer the apical position. The bond lengths shown in Table 10.3 generally support these conclusions although, as before, there are some puzzling exceptions.

If the 5-coordinate complex is a result of the addition of a fifth, weakly bound ligand to a strongly π-bonded, square planar complex:

$$[Pd(diars)_2]^{+2} + X^- \longrightarrow [Pd(diars)_2X]^+ \qquad \textbf{(10.6)}$$

then the π-bonding requirements of the former ligands (*diars* = *o*-phenylenebis(dimethylarsine)) require that they remain coplanar or nearly so with the metal atom; hence the square pyramidal arrangement is strongly favored.

Finally, polydentate ligands can affect the geometry of a complex merely as a result of their own steric requirements. For example, we find some tetradentate ligands such as tris(2-dimethylaminoethyl)amine, ("Me_6tren" = $((CH_3)_2NCH_2CH_2)_3N$), form only 5-coordinate complexes (Fig. 10.14), apparently because the polydentate ligand cannot span a 4-coordinate tetrahedral or square planar complex and cannot conform ("fold") to fit a portion of an octahedral coordination sphere.[30] This view is strengthened by studies of a

[29] Since CN^- and $SnCl_3^-$ do not occur in the same complex in this series, the inequality here is uncertain.

[30] M. Ciampolini and N. Nardi, *Inorg. Chem.*, **1966**, *5*, 41, 1150; M. Di Vaira and P. L. Orioli, *Inorg. Chem.*, **1967**, *6*, 955.

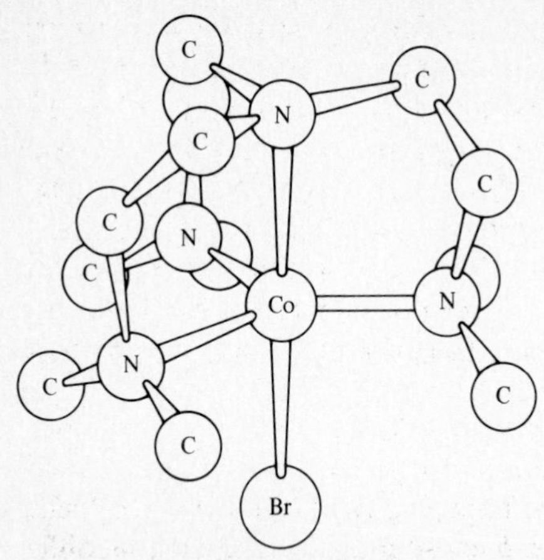

Fig. 10.14 Molecular structure of bromotris(2-dimethylaminoethyl)aminecobalt(II) cation in [CoBr(Me_6tren)]Br. [From M. DiVaira and P. L. Orioli, *Inorg. Chem.*, **1967**, *6*, 955. Reproduced with permission.]

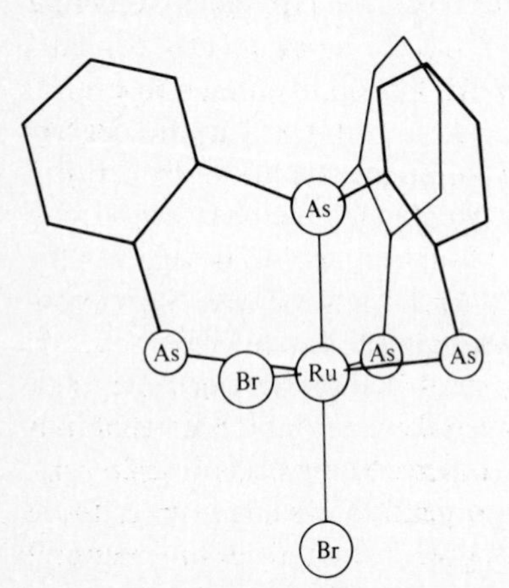

Fig. 10.15 Molecular structure of dibromotris(*o*-diphenylarsinophenyl)arsineruthenium(II). Note the distortion from octahedral symmetry. [From P. B. H. Mais et al., *Chem. Ind.* (*London*), **1963**, 1204. Reproduced with permission.]

few octahedral complexes which do form with related tetradentate ligands, such as tris(*o*-diphenylarsinophenyl)arsine ("QAS"). Figure 10.15 illustrates the structure of dibromotris(*o*-phenylarsinophenyl)arsineruthenium(II).[31] The constraints of the ligand chelate rings require that the resulting octahedron be distorted. Some polydentate ligands can favor TBP geometries, others the SP arrangement.[32] For example, zinc(II) forms TBP

[31] R. H. B. Mais et al., *Chem. Ind.* (*London*), **1963**, 1204.

[32] For further discussions of this point see L. M. Venanzi, *Angew. Chem. Int. Ed. Engl.*, **1964**, *3*, 453; L. Sacconi, *Pure Appl. Chem.*, **1968**, *17*, 95; C. Furlani, *Coordin. Chem. Rev.*, **1968**, *3*, 141.

complexes (Furlani type 2, p. 473) with tris(2-aminoethyl)amine (tren) occupying one axial and three equatorial positions and the fifth, axial ligand isothiocyanate[33] or chloride.[34]

Square pyramidal complexes are also known with chelate rings such as *trans*–*bis*(hydrazinecarboxylato-N′,O)zinc(II).[35] In addition, there are examples of the whole spectrum between SP and TBP (see pp. 225–227). For example, *trans*-aquabis(8-hydroxyquinoline)zinc(II) is roughly halfway between the two limiting geometries.[36] Since Zn(II) has a d^{10} configuration, stabilization effects from partially filled sets of orbitals cannot be responsible for the choice of geometry about the central metal atom. Probably the ligand geometry (chelate "bite") and the electronic effects mentioned above[27] provide the necessary stabilization of these complexes.

Five-coordinate complexes with d^5, d^6, d^7, and d^8 may be either high or low spin. The magnetic susceptibility of the low-spin complexes is that expected if one of the *d* orbitals is unavailable for occupancy by the metal *d* electrons. Thus *S* equals $0(d^8)$, $\frac{1}{2}(d^7)$, $1(d^6)$, and $\frac{3}{2}(d^5)$. The magnetic susceptibilities of the low-spin 5-coordinate complexes thus differ significantly from corresponding low-spin octahedral complexes. The unavailability of the fifth *d* orbital can be rationalized in terms of inner orbital dsp^3 bonding. The TBP structure results from $d_{z^2}sp^3$ hybridization and SP from $d_{x^2-y^2}sp^3$ hybridization.[37] In this sense valence bond theory is in qualitative agreement with simple crystal field theory or more elaborate molecular orbital schemes. The latter two methods, however, also account for the energy levels of the other *d* orbitals as well as assign the difference between high- and low-spin complexes to the relative energies of the d_{z^2} and $d_{x^2-y^2}$ orbitals (see Fig. 10.10). One complex is known in which high- and low-spin forms are in equilibrium (see Fig. 9.16).[38] Valence bond theory can resort to "inner" and "outer" orbital complexes to account for high-spin species, but the results are somewhat less aesthetic. The situation is very similar to that of tetragonal distortion discussed in Chapter 9 in that there are several energy levels involved (as opposed to t_2 and e in octahedral and tetrahedral complexes), and hence the placement of these energy levels can be reasonably interpreted by means of spectral measurements. In some low-spin complexes the interelectron repulsions can be neglected to a first approximation and the spectrum interpreted solely on the basis of the simple one-electron energy level diagrams (Fig. 10.16). In the more general case, however, these electron–electron effects must be treated in a manner analogous to that given in Chapter 9 for octahedral complexes.[39]

Although it has long been known that various geometric and optical isomers are possible for coordination number 5, examples have been few and mostly very recently

---

[33] G. D. Andretti et al., *J. Am. Chem. Soc.*, **1969**, *91*, 4112.

[34] R. J. Sime et al., *Inorg. Chem.*, **1971**, *10*, 537.

[35] F. Bigoli et al., *Chem. Commun.*, **1970**, 120.

[36] C. S. Kerr et al., *J. Coord. Chem.* **1981**, *11*, 111.

[37] These are the simplest hybridizations to visualize. It is possible to substitute more *d* character for *p* character and arrive at the same symmetry. For example, d^3sp also forms TBP and d^4s and d^4p also form SP. The actual %*s*, %*p*, and %*d* character will depend upon energetic factors such as promotion energy and quality of overlap of the resulting hybrids.

[38] S. M. Nelson and W. S. Kelly, *Chem. Commun.*, **1968**, 436.

[39] The complete discussion will not be given here. The interested reader is referred to C. Furlani, *Coordin. Chem. Rev.*, **1968**, *3*, 141; M. Ciampolini, *Struct. Bonding*, **1969**, *6*, 52.

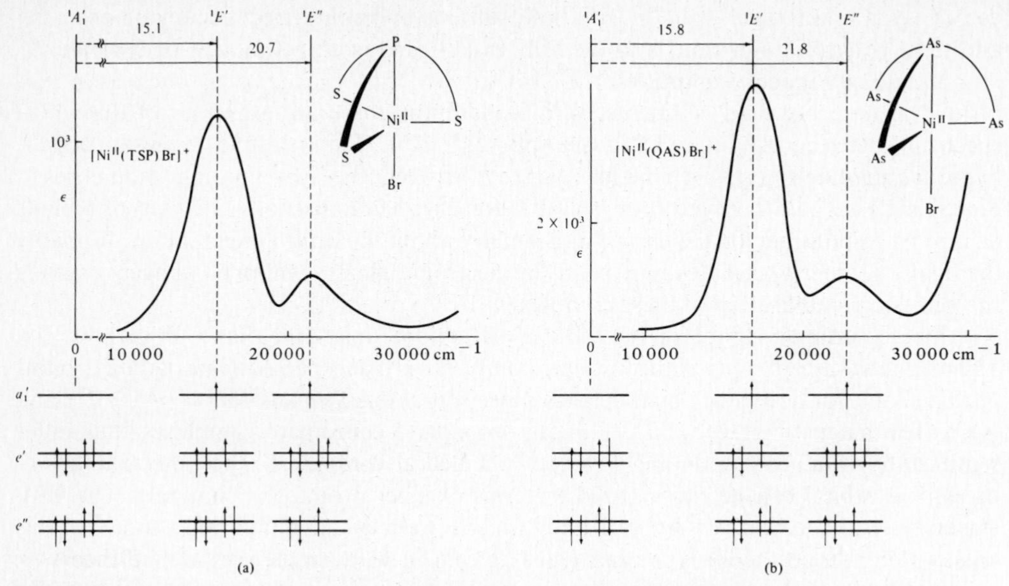

Fig. 10.16 Spectra of (a) bromotris(methylmercapto-*o*-phenyl)phosphinenickel(II) and (b) bromotris(dimethylarsino-*o*-phenyl)arsinenickel(II) cations. [From C. Furlani, *Coordin. Chem. Rev.*, **1968**, *3*, 141. Reproduced with permission.]

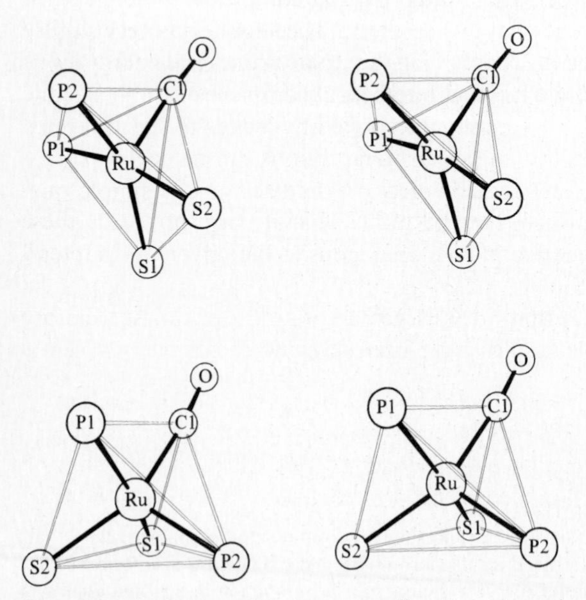

Fig. 10.17 Stereoview of the inner coordination sphere about the central ruthenium atom in the orange (top) and violet (bottom) isomers of $(\phi_3P)_2[(CF_3)_2C_2S_2]Ru(CO)$. [From I. Bernal et al., *J. Cryst. Mol. Struct.*, **1974**, *4*, 43. Reproduced with permission.]

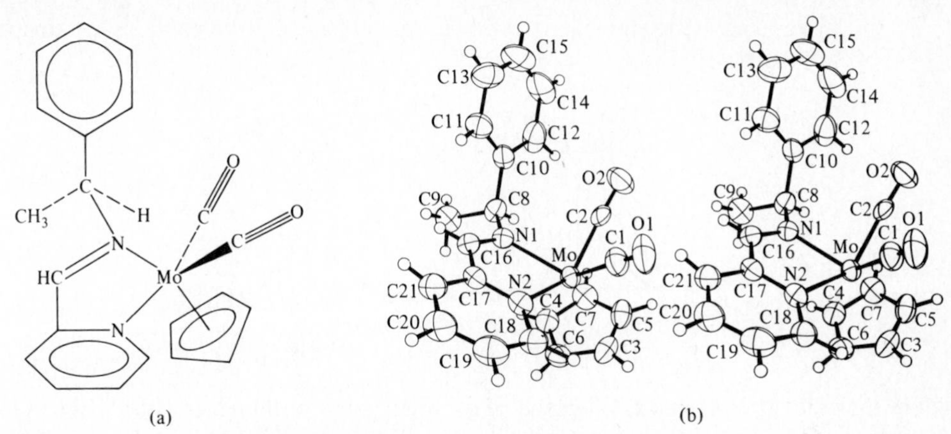

Fig. 10.18 The Schiff base complex of dicarbonylcyclopentadienylmolybdenum(II) cation. (a) Conventional drawing. (b) Stereoview. [From I. Bernal et al., *Inorg. Chem.*, **1978**, *17*, 382. Reproduced with permission.]

discovered. We have already seen examples of TBP-SP isomerism in a Ni(II) complex (p. 475) and a Co(II) complex (p. 476). Another example is $(\phi_3P)_3[(CF_3)_2C_2S_2]Ru(CO)$. It exists as two isomers, one orange and one violet, which coexist in solution and which may by crystallized as pure materials. Both isomers are SP, but the orange isomer has the carbon monoxide ligand in the apical position and the violet isomer has a basal carbon monoxide with one of the phosphorus atoms at the apex of the pyramid (Fig. 10.17).[40] A related type of geometric isomerism is found in the organometallic complex dibromodicarbonylcyclopentadienylrhenium(III). Both isomers have the cyclopentadienyl ring at the apex of a square pyramid with the basal ligands in either a *cis* or a *trans* arrangement:[41]

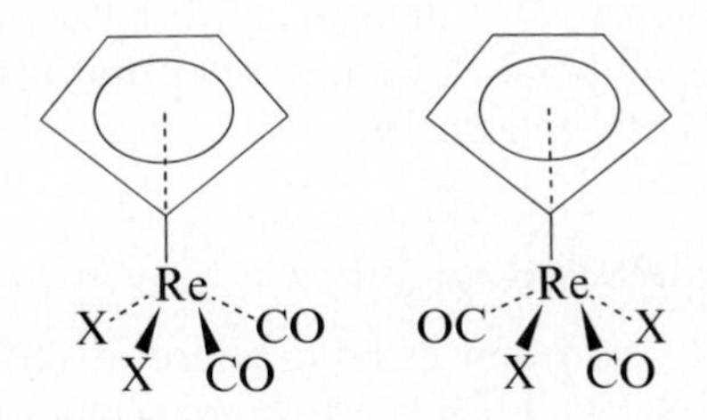

Finally, optical isomerism is even more rare. The first example[42] of the determination of such a complex is shown in Fig. 10.18. Note that if the bidentate ligand had been ethylenediamine, dipyridine, or the oxalate ion, there would have been a mirror plane and no optical activity. There is a large group of bioinorganic compounds with coordination number 5 and optical activity (see Problem 18.26).

[40] I. Bernal et al., *Chem. Commun.*, **1973**, 39; I. Bernal et al., *J. Cryst. Mol. Struct.*, **1974**, *4*, 43; A. Clearfield et al., *J. Coordin. Chem.*, **1977**, *6*, 227.

[41] R. B. King and R. H. Reimann, *Inorg. Chem.*, **1976**, *15*, 179.

[42] S. J. La Placa et al., *Angew. Chem.* (*Int.*), **1975**, *14*, 353. See also M. G. Reisner et al., *Angew. Chem.* (*Int.*), **1976**, *15*, 776; M. G. Reisner et al., *J. Organometal. Chem.*, **1977**, *137*, 329.

Another interesting example that combines geometric and optical activity consists of complexes of the type[43]

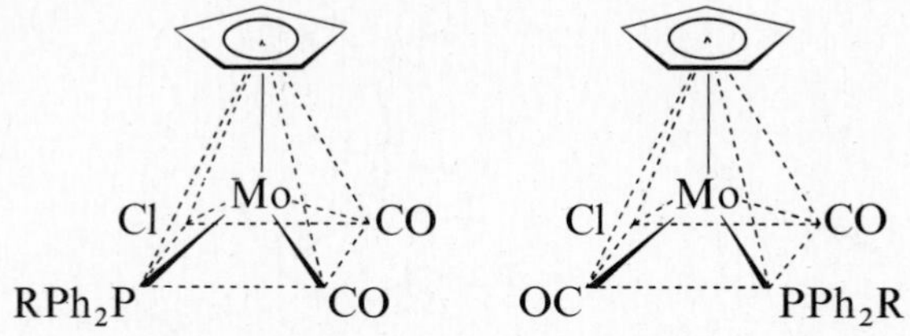

Cl, Mo, CO, RPh_2P, CO

cis (chiral)

Cl, Mo, CO, OC, PPh_2R

trans (achiral)

Note that the *cis* isomer lacks a plane of symmetry and is therefore chiral, but that the *trans* isomer has such a plane of symmetry and will be achiral in absence of an asymmetric carbon in the phosphine ligand.[44] As in the case of the previously encountered (p. 483) cyclopentadienyl complex, it can be argued whether the coordination number is 5 or 9. In either semantic interpretation these compounds are of considerable interest since isomerism in 9-coordinate complexes is even less well documented than in 5-coordinate.

COORDINATION NUMBER 6[45]

This is by far the most common coordination number. With certain ions 6-coordinate complexes are formed almost exclusively. For example, Cr(III) and Co(III) are almost exclusively octahedral in their complexes.[46] It was this large series of octahedral Cr(III) and Co(III) complexes which led Werner to formulate his theories of coordination chemistry and formed the basis for almost all of the classic work on complex compounds. Before discussing the various isomeric possibilities for octahedral complexes it is convenient to dispose of the few nonoctahedral geometries.

Distortions from perfect octahedral symmetry

Two forms of distortion of octahedral complexes are of some importance. The first is tetragonal distortion, either elongation or compression along one of the fourfold rotational axes of the octahedron (Fig. 10.19a). This type of distortion has been discussed previously in connection with the Jahn-Teller effect. Another possibility is elongation or compression along one of the four threefold rotational axes of the octahedron that pass through the

[43] G. M. Reisner et al., *Chem. Commun.*, **1978**, 691.

[44] In point of fact, chiral ligands are often used in synthesizing molecules of this type, both for observing steric effects and for aid in solving the X-ray crystal structures. See the discussion of absolute configuration of complexes for C.N. = 6 (p. 495).

[45] D. L. Kepert, *Progr. Inorg. Chem.*, **1977**, *23*, 1.

[46] A series of trigonal bipyramidal complexes of the general formula $[CoX_3(PR_3)_2]$ has been reported by K. A. Jensen et al., *Acta Chem. Scand.*, **1963**, *17*, 1126. The only nonoctahedral complex of chromium(III) appears to be $[CrCl_3(NMe_3)_2]$, G. W. A. Fowles and P. T. Greene, *Chem. Commun.*, **1966**, 784.

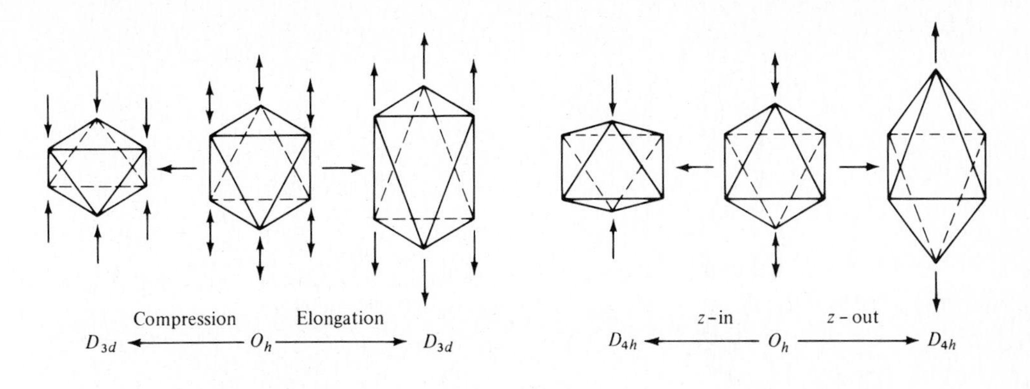

Fig. 10.19 Trigonal and tetragonal distortion of an octahedral complex. Either may occur via elongation or compression.

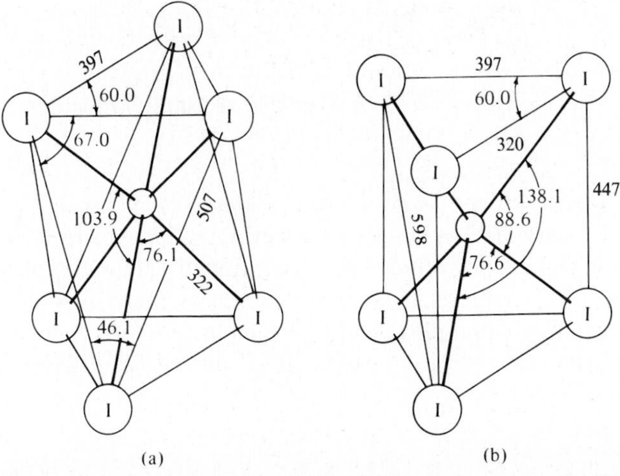

Fig. 10.20 Partial crystal structure of ThI_2 showing (a) trigonal antiprismatic coordination of half of the thorium atoms and (b) trigonal prismatic coordination of the other half. [From L. J. Guggenberger and R. A. Jacobson, *Inorg. Chem.*, **1968**, *7*, 2257. Reproduced with permission.] Distances in pm.

centers of the faces (Fig. 10.19b), resulting in a trigonal antiprism. Trigonal antiprismatic coordination[47] is found for half of the thorium atoms in ThI_2 and trigonal prismatic coordination for the other half (Fig. 10.20), but it is not generally important in discrete molecular complexes.[48]

Another configuration that is not really a distortion but involves a reduction of symmetry may be mentioned here. It consists of the replacement of six unidentate ligands

[47] L. J. Guggenberger and R. A. Jacobson, *Inorg. Chem.*, **1968**, *7*, 2257.

[48] Trigonal fields superimposed upon octahedral fields may be important in the interpretation of predominantly ionic solids such as ruby (see C. J. Ballhausen, "An Introduction to Ligand Field Theory," McGraw-Hill, New York, **1962**).

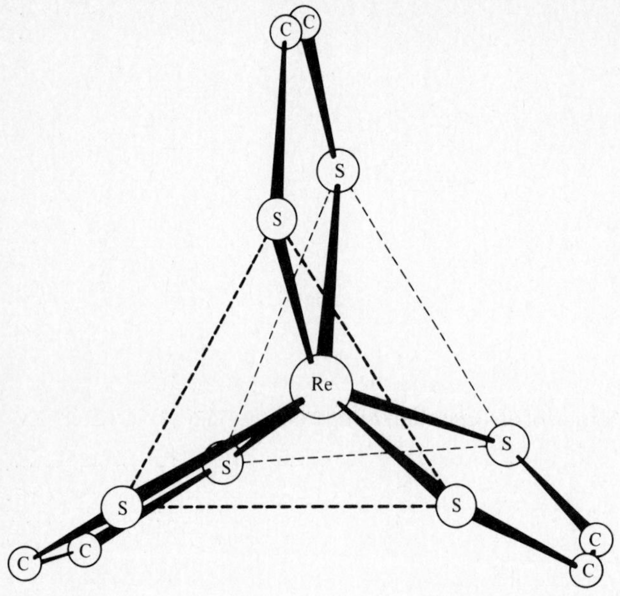

Fig. 10.21 Structure of $[Re(S_2C_2\phi_2)_3]$. The phenyl rings have been omitted for clarity. [From R. Eisenberg and J. A. Ibers, *Inorg. Chem.*, **1966**, *5*, 411. Reproduced with permission.]

in a complex such as $[Co(NH_3)_6]^{+3}$ with chelate rings such as ethylenediamine to form $[Co(en)_3]^{+3}$. The latter complex has no plane of symmetry and the symmetry of the complex has been reduced from O_h to D_3. For most purposes this reduction in symmetry has little effect (the spectrum of the ethylenediamine complex is virtually identical to that of the hexaammine) except to make it possible to resolve optically active isomers (see p. 491).

Trigonal prism

Although by far the greatest number of 6-coordinate complexes may be derived from the octahedron, a few interesting complexes have the geometry of the trigonal prism (Figs. 10.20b and 10.21). For many years the only examples of trigonal prismatic coordination were in crystal lattices such as the sulfides of heavy metals (MoS_2 and WS_2, for example).[49] The first example of this geometry in a discrete molecular complex was tris(*cis*-1,2-diphenylethene-1,2-dithiolato)rhenium, $[Re(S_2C_2\phi_2)_3]$ (Fig. 10.21).[50] Since then a significant series of trigonal prismatic complexes of ligands of the type $R_2C_2S_2$ has been fully characterized with rhenium, molybdenum, tungsten, vanadium, zirconium, and niobium and suggested for other metals.[51]

---

[49] A. F. Wells, "Structural Inorganic Chemistry," 4th ed., Oxford University Press, London, **1975**, p. 612.

[50] R. Eisenberg and J. A. Ibers, *J. Am. Chem. Soc.*, **1965**, *87*, 3776; *Inorg. Chem.*, **1966**, *5*, 411.

[51] A. E. Smith et al., *J. Am. Chem. Soc.*, **1965**, *87*, 5798; R. Eisenberg et al., ibid., **1966**, *88*, 2874; R. Eisenberg and H. B. Gray, *Inorg. Chem.*, **1967**, *6*, 1849; E. I. Stiefel et al., *J. Am. Chem. Soc.*, **1967**, *89*, 3353; M. J. Bennett et al., ibid., **1973**, *95*, 7504.

Table 10.4 Map of twist angles (θ) in tris(dithiolato)metal complexes. Trigonal prism, $\theta = 0°$; octahedron, $\theta = 60°$.

Number of electrons					
n		$V\ (\theta = 0°)$			
$n + 1$	$Zr^{2-}\ (\theta \sim 20°)$	$Nb^{-}\ (\theta = 0°)$ $Ta^{-}\ (\theta \sim 16°)$	$Mo\ (\theta = 0°)$		
$n + 2$		$V^{2}\ (\theta = 17°)$		$Re\ (\theta = 0°)$	
$n + 3$			$Mo^{2-}\ (\theta = 14°)$ $W^{2-}\ (\theta = 14°)$		
$n + 4$					
$n + 5$					$Fe^{2-}\ (\theta = 25°)$

SOURCE: D. L. Kepert, *Progr. Inorg. Chem.*, **1977**, *23*, 1. Reproduced with permission.

There is considerable ambiguity concerning the charge on this type of ligand. This is because it may be formulated either as a neutral dithioketone ($n = 0$) or the dianion of an unsaturated dithiol ($n = 2$). The difference of two electrons can be represented formally as

$$\underset{R}{\overset{S}{\|}}C{-}\underset{R}{\overset{S}{\|}}C \xrightarrow{2e^-} \underset{R}{\overset{^-S}{|}}C{=}\underset{R}{\overset{S^-}{|}}C \xrightarrow{2H^+} \underset{R}{\overset{HS}{|}}C{=}\underset{R}{\overset{SH}{|}}C$$

Since the electrons involved are delocalized over molecular orbitals not only of the ligand but of the metal as well, it is impossible and not very meaningful to attempt to assign a formal charge to the ligand or the metal.

In addition to the neutral complexes, it is possible to add 1, 2, or 3 electrons to form reduced species of the type $[M(S_2C_2R_2)_2]^{-p}$. Present evidence is that the reduced species tend to retain the trigonal prismatic coordination with some distortion toward a regular octahedron with increasing addition of electrons (see Table 10.4).

Molecular orbital diagrams have been offered in explanation of the spectral and polarographic properties of these complexes.[52] One of the most interesting features of the 1,2-ethenedithiolate or 1,2-dithiolene complexes is the short distance between the two sulfur atoms within a chelate ring. This distance is remarkably constant at about 305 pm, some 60 pm less than the sum of the van der Waals radii (Table 6.1), indicating the strong possibility of some S—S bonding that may stabilize the trigonal structure.[53] One way in which this might come about is by pulling the sulfur atoms toward each other, reducing

---

[52] E. I. Stiefel et al., *J. Am. Chem. Soc.*, **1966**, *88*, 2956; G. N. Schrauzer and V. P. Mayweg, ibid., **1966**, *88*, 3235.

[53] For reviews of complexes containing these ligands, see J. A. McCleverty, *Progr. Inorg. Chem.*, **1968**, *10*, 49; G. N. Schrauzer, *Trans. Metal Chem.*, **1968**, *4*, 299; R. Eisenberg, *Progr. Inorg. Chem.*, **1970**, *12*, 295; D. L. Kepert, ibid., **1977**, *23*, 1. Related ligands are reviewed by D. Coucouvanis, ibid., **1980**, *27*, 223.

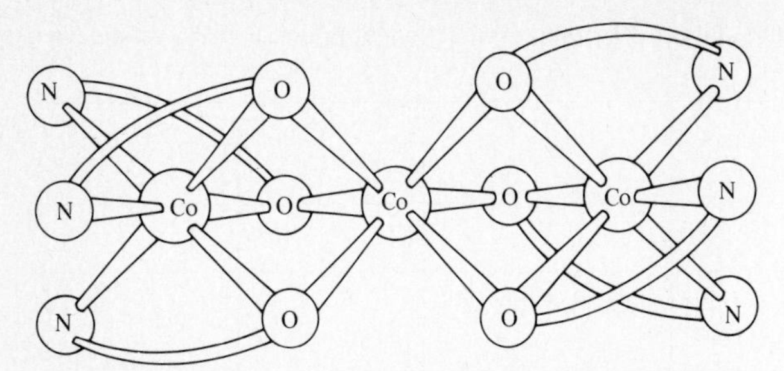

Fig. 10.22 Schematic drawing of $[Co(Co(OCH_2CH_2NH_2)_3)_2]^{+2}$. The central cobalt atom has trigonal prismatic coordination. [From J. A. Bertrand et al., *J. Am. Chem. Soc.*, **1969**, *91*, 2394. Reproduced with permission.]

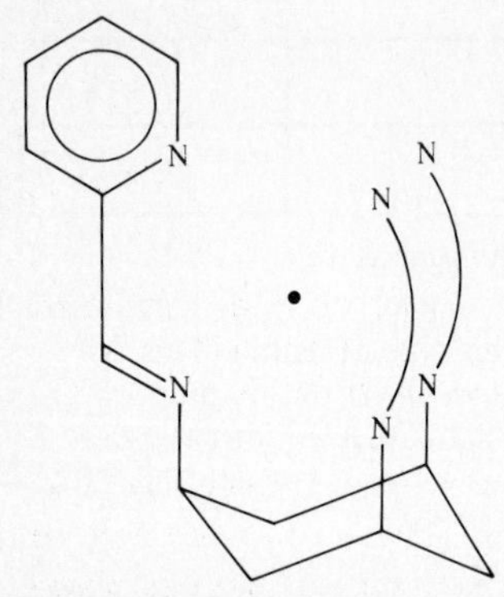

Fig. 10.23 Structure of *cis, cis*-1,3,5-tris(pyridine-2-aldimino)cyclohexane with the substituents in the axial positions. The dot marks the location of the metal atom in the tris(chelate).

the bite angle (it is about 81° in the complex in Fig. 10.21). Perfect octahedral coordination requires 90°. Another way to look at it is to imagine Fig. 10.21 undergoing at 60° twist of one of the S_3 triangles to form an octahedron. If the other dimensions remain the same, the sulfur atoms would have to move away from each other.

One complex is known with a trigonal prismatic coordination sphere of oxygen atoms. The bis[tris(amidoethoxido)cobalt(III)]cobalt(II) cation has two Co(III) ions octahedrally coordinated and a Co(II) ion with trigonal prismatic coordination (Fig. 10.22).[54] In this complex it is believed that the trigonal prism is favored over a third octahedron because of interligand repulsions in the latter.

Another way one can induce trigonal prismatic coordination is to tailor the ligand to be rigid and to favor this geometry. For example, consider the ligand shown in Fig. 10.23. If the three pyridinealdimino groups occupy axial positions on the cyclohexane ring (as they *must* if all six nitrogen atoms are to coordinate to the same metal) the N–N′ chelate rings will be vertical ("parallel"). If a Zn^{+2} is coordinated to this ligand, the resulting complex has almost perfect trigonal prismatic geometry.[55] However, in general trigonal prismatic geometry offers less ligand field stabilization energy (LFSE) than octahedral

[54] J. A. Bertrand et al., *J. Am. Chem. Soc.*, **1969**, *91*, 2394.
[55] W. O. Gillum et al., *Chem. Commun.*, **1969**, 843.

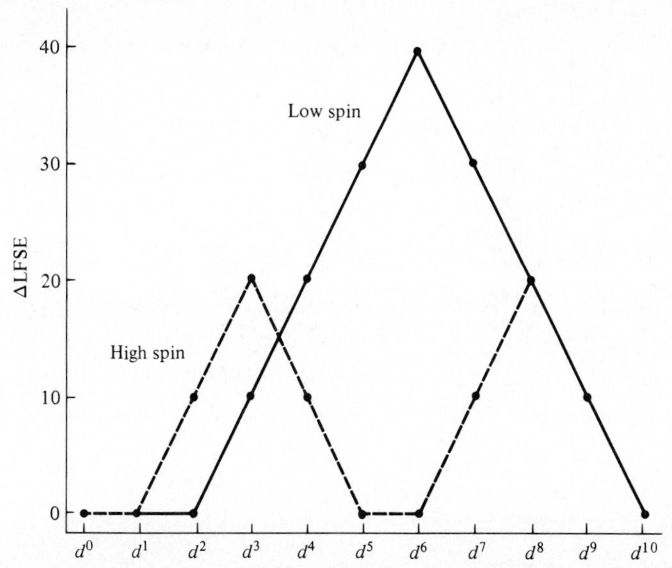

Fig. 10.24 Relative stability (LFSE) of octahedral coordination vs. trigonal prismatic coordination as a function of *d* orbital occupancy. Note maximum for low-spin d^6 and minima for high-spin d^5 and d^{10}. [From W. O. Gillum et al., *Inorg. Chem.*, **1970**, 9, 1825. Reproduced with permission.]

geometry, never more (Fig. 10.24).[56] The $Zn^{+2}(d^{10})$, $Mn^{+2}(d^5)$, and $Co^{+2}(d^7)$ compounds have virtually indistinguishable powder patterns, indicating negligible distortion from TP geometry. In contrast, the complexes of Fe^{+2} and Ni^{+2} differ from the first three and from each other, suggesting that they have different geometries.[57] The complete X-ray structure has been obtained from the Ni^{+2} complex,[58] and it has been found to be almost exactly halfway between trigonal prismatic and octahedral. On the basis of Fig. 10.24 we should expect the extra LFSE of the octahedral complexes to be in the order: low-spin Co^{+3} > low-spin Fe^{+2} > Ni^{+2} > high-spin Co^{+2} > high-spin $Mn^{+2} = Zn^{+2}$. This accounts for the TP geometry of the Co(II), Mn(II), and Zn(II) complexes and the distortion of the Ni(II) complex, and predicts that the Fe(II) complex should be even more distorted towards the octahedral geometry.

Geometric isomerism in octahedral complexes[59]

There are two simple types of *cis–trans* isomerism possible for octahedral complexes. The first exists for complexes of the type MA_2B_4 in which the A ligands may be either next to each other (Fig. 10.25a) or on opposite apexes of the octahedron (10.25b). Complexes

[56] For a discussion of this effect, as well as other factors that affect trigonal prismatic coordination, see R. A. D. Wentworth, *Coord. Chem. Rev.*, **1972/73**, *9*, 171.

[57] W. O. Gillum et al., *Inorg. Chem.*, **1970**, *9*, 1825.

[58] E. B. Fleischer et al., *Inorg. Chem.*, **1972**, *11*, 2775. For a more recent discussion of these effects as well as other coordinate geometries, see R. Hoffmann et al., *J. Am. Chem. Soc.*, **1976**, *98*, 2484.

[59] For a fuller discussion of geometric isomerism, see Y. Saito, "Inorganic Molecular Dissymmetry," Springer-Verlag, Berlin, **1979**, pp. 6–7, 73–88.

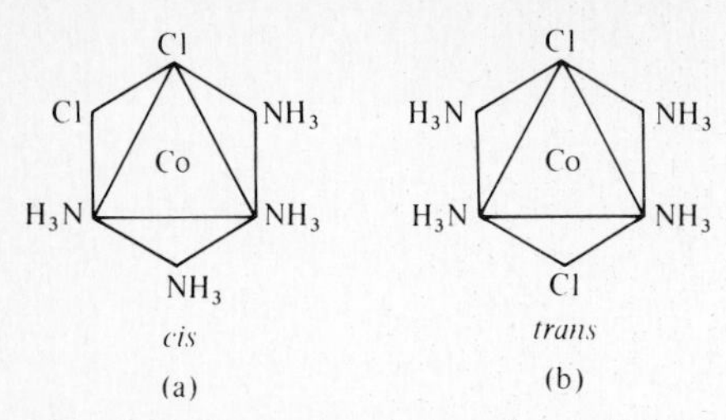

Fig. 10.25 Examples of *cis* and *trans* isomers.

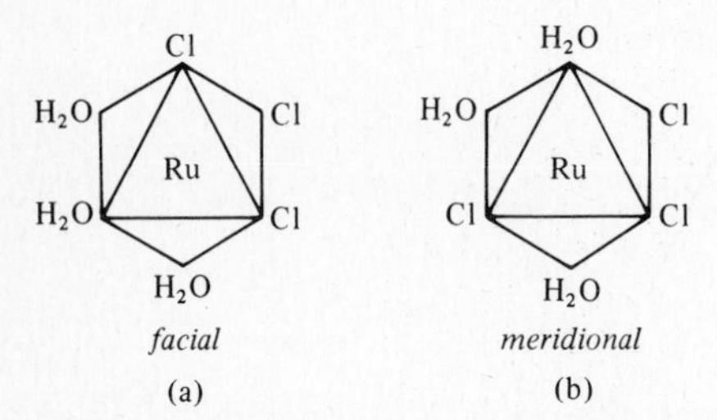

Fig. 10.26 Examples of *facial* and *meridional* isomers.

of this type were studied by Werner, who showed that the *praseo* and *violeo* complexes of cobalt(III) were of this type (see p. 360). A very large number of these complexes is known, and classically they provided a fertile area for the study of structural effects.[60] More recently there has been renewed interest in them as indicators of the effects of lowered symmetry on electronic transition spectra.

Two isomers are also possible for complexes of the type MA_3B_3: (1) The ligands of one type form an equilateral triangle on one of the faces (the *facial* isomer, Fig. 10.26a) or (2) they may span three positions such that two are opposite, or *trans*, to each other (the *meridional* isomer, Fig. 10.26b).[61] In contrast to *cis–trans* isomers of MA_2B_4, which number in the hundreds, only about a half dozen *fac–mer* isomers have been characterized: $[Ru(H_2O)_3Cl_3]$, $[Pt(NH_3)_3Br_3]^+$, $[Pt(NH_3)_3I_3]^+$, $[Ir(H_2O)_3Cl_3]$, $[Rh(CH_3CN)_3Cl_3]$, and $[Co(NH_3)_3(NO_2)_3]$.[62]

If the number of different kinds of ligands is increased or if polydentate ligands are involved, more cases of isomerism can occur, but they may be related to the *cis–trans* isomers above.[63] In place of the terms *cis–trans* or *facial–meridional*, a numbering or lettering system can be used (Fig. 10.27). The *facial* isomer is thus the *a, b, c* or *1, 2, 3* isomer and the meridional isomer the *a, b, f* or *1, 2, 6* isomer.

There have been a number of discussions concerning the best method of "counting isomers" for the more complicated complexes such as $[Pt(py)(NH_3)(NO_2)(Cl)(Br)(I)]$ (fifteen geometric isomers, each of which consists of a DL-pair!). The interested reader is

---

[60] See F. Basolo, in "The Chemistry of the Coordination Compounds," J. C. Bailar, Jr., ed., Van Nostrand-Reinhold, New York, **1956**, pp. 277–308.

[61] These are sometimes called *cis–trans* isomers but the *fac–mer* nomenclature is preferable.

[62] See R. B. Hagel and L. F. Druding, *Inorg. Chem.*, **1970**, *9*, 1496, and references therein.

[63] For a short discussion of other types of isomerism, see J. V. Quagliano and L. M. Vallerino, "Coordination Chemistry," Heath, Lexington, Mass., **1969**, Chapter 5.

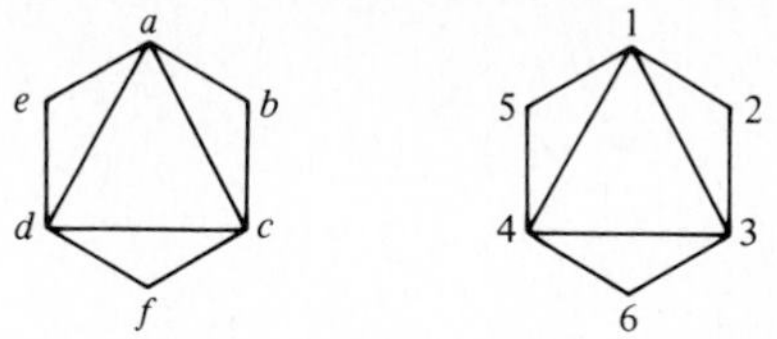

Fig. 10.27 Methods of designating positions on an octahedral complex.

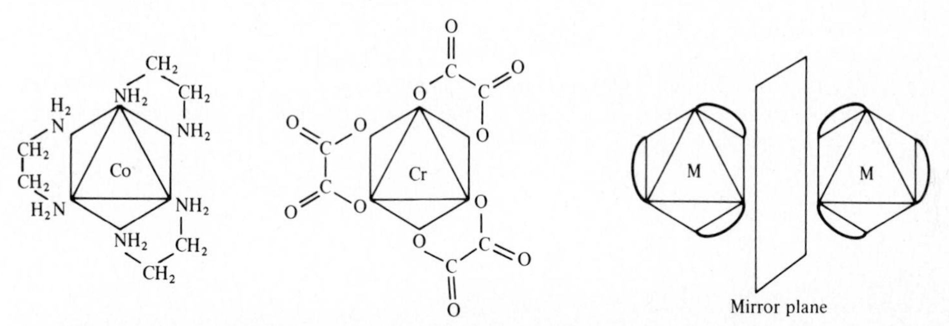

Fig. 10.28 Structures of the optically active complexes $[Co(en)_3]^{+3}$ and $[Cr(ox)_3]^{-3}$ (one isomer of each) and a stylized drawing of the two optical isomers of any tris(chelate) complex.

referred to the original method of Bailar[64] and a modification for computer usage.[65] Douglas and McDaniel[66] have presented a scheme of drawing out the isomers utilizing symmetry principles as an aid in preventing duplication. Finally, papers utilizing Pòlya's theorem and group theory have been presented.[67]

Optical isomerism of octahedral complexes

It was mentioned above that tris(chelate) complexes of the type $[Co(en)_3]^{+3}$ lack a plane of symmetry. As a result, such complexes can exist in either of two optically active forms (or a racemic mixture of the two). Figure 10.28 illustrates the complex ions $[Co(en)_3]^{+3}$ and $[Cr(ox)_3]^{-3}$, each of which exhibits optical isomerism. An operational criterion for lack of a plane of symmetry (and hence for optical isomerism) is the nonsuperimposability of the optical isomers. Although with practice this may be determined by inspection of sketches on paper, the reader is urged to handle three-dimensional models to improve the depth perception and orientation necessary to interpret drawings with facility. (Appendix H provides directions for the construction of simple paper models for this use.)

It is not necessary to have three chelate rings present. An example of a resolvable complex analogous to the tetraamminedichlorocobalt(III) cation discussed is shown in Fig. 10.29. Two geometrical isomers exist, *cis* and *trans*. The latter has three internal planes

---

[64] J. C. Bailar, Jr., *J. Chem. Educ.*, **1957**, *34*, 334, 626; S. A. Mayper, ibid., **1957**, *34*, 623.

[65] W. E. Bennett, *Inorg. Chem.*, **1969**, *8*, 1325.

[66] B. E. Douglas and D. H. McDaniel, "Concepts and Models of Inorganic Chemistry," Blaisdell, Waltham, Mass., **1965**, pp. 365–368.

[67] B. A. Kennedy et al., *Inorg. Chem.*, **1964**, *3*, 265; I. V. Krivoshei, *Zh. Strukt. Khim.*, **1965**, *6*, 322 [*J. Struct. Chem.* (*Engl. transl.*) **1965**, *6*, 304].

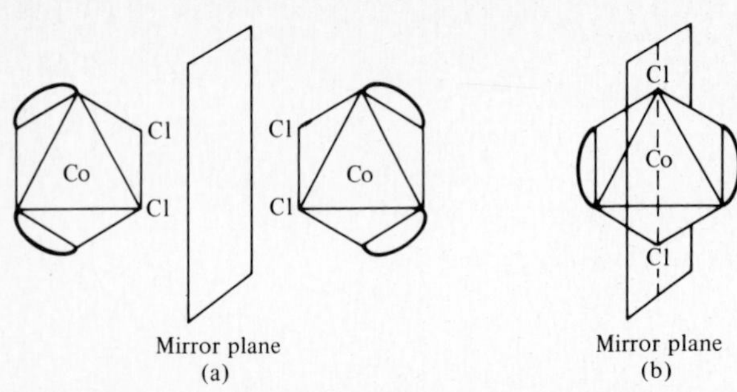

Fig. 10.29 (a) Optical isomers of *cis*-dichlorobis(ethylenediamine)cobalt(III) ion ("*violeo*" salt). (b) *trans*-Dichlorobis(ethylenediamine)cobalt(III) ion ("*praseo*" salt) showing one internal plane of symmetry (there are two others perpendicular to the one shown).

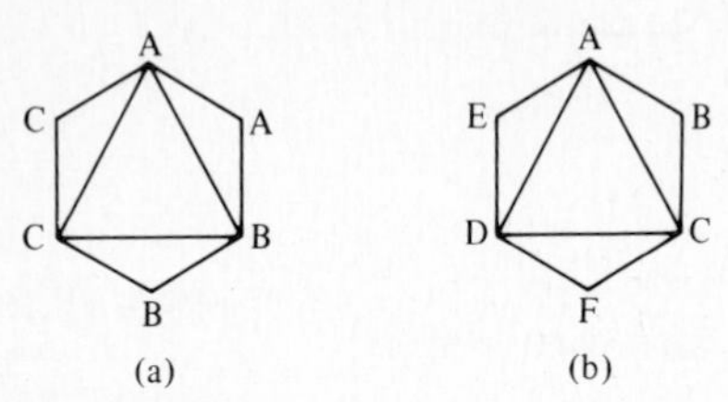

Fig. 10.30 Examples of optically active arrangements of ligands not containing chelate rings.

of symmetry and hence is optically inactive. The former is optically active, however. Since the two chloride ions replace two nitrogen atoms from an ethylenediamine ring without disturbing the remaining geometry, the optical activity is preserved.[68] For purposes of optical dissymmetry, the change of two nitrogen atoms to two chlorides is insignificant. In principle all of the ethylenediamine rings could be replaced by unidentate ligands as long as the "screw" arrangement of the ligands is retained (Fig. 10.30a). Finally, if all six ligands attached to the central metal atom were different (Fig. 10.30b), the complex would obviously be optically active. There are many other possibilities for optical isomerism with varying numbers of different groups, with and without chelate rings. In practice, resolution of optical isomers is limited to complexes possessing chelate rings for two reasons: (1) The presence of the chelate rings provides the complex with additional stability (see "The Chelate Effect," p. 527); (2) the nonchelate complexes are difficult to synthesize, and it is difficult to separate the large number of geometric isomers.

Werner synthesized and resolved optical isomers as strong corroborative evidence for his theory of octahedral coordination. Although he did not discuss the subject in his first paper, the idea appears to have come to him about four or five years later,[69] and he was able to resolve amminechlorobis(ethylenediamine)cobalt(III) cation (*cf*. Fig. 10.29 with one chloro group replaced by an ammine) about ten years after he first became

[68] The "replacement" of ethylenediamine by chloride ions cited here is a "paper and pencil reaction" and refers to the formal change. It does not imply that in solution two chloride ions could attack a $[Co(en)_3]^{+3}$ cation with retention of configuration.

[69] Werner mentioned the possibility in a letter to his co-worker Arturo Miolati in 1897. See "Classics in Coordination Chemistry," G. B. Kauffman, ed., Dover, New York, **1968**, pp. 155–158.

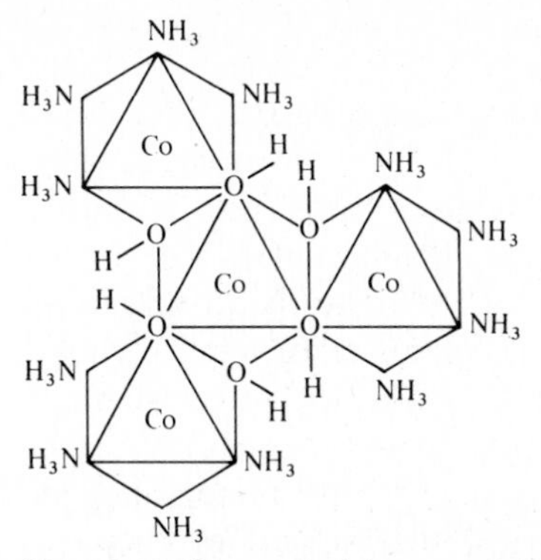

Fig. 10.31 One optical isomer of the tris[tetraammine-μ-dihydroxocobalt(III)]cobalt(III) cation.

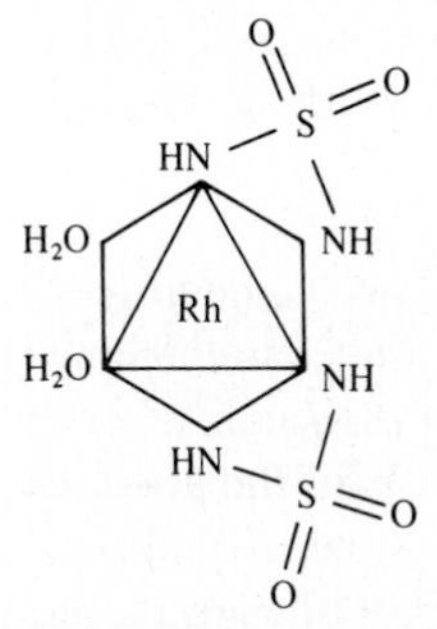

Fig. 10.32 One optical isomer of the *cis*-diaqubis(sulfamido)rhodate(III) anion.

interested in this aspect.[70] Werner realized that the presence of optical isomers in the *cis* compound but not the *trans* compound was incompatible with alternative formulations of these compounds. He succeeded in resolving a large series of complexes including Co(III), Cr(III), Fe(II), and Rh(III) ions, and ethylenediamine, oxalate, and dipyridyl as chelating ligands. Nevertheless, a few of his critics pointed out that all his ligands contained carbon. By associating optical activity somehow with "organic" versus "inorganic" compounds and ignoring all symmetry arguments they proceeded to discount his results. In order to silence these specious arguments, Werner synthesized and resolved a polynuclear complex containing no carbon atoms, tris[tetraammine-μ-dihydroxocobalt(III)]cobalt(III) (Fig. 10.31).[71] This laid to rest the distinction between carbon and the other elements that should have died in 1828 with Wöhler's work. It is interesting to note that in all of the work since Werner only one other completely carbon-free complex has been resolved, sodium *cis*-diaquabis(sulfamido)rhodate(III) (Fig. 10.32),[72] chiefly because of the difficulty of preparing noncarbon chelating agents.

---

[70] A. Werner and V. L. King, *Chem. Ber.*, **1911**, *44*, 1887. A translation may be found in "Classics in Coordination Chemistry," G. B. Kauffman, ed., Dover, New York, **1968**, pp. 159–173.

[71] A. Werner, *Chem. Ber.*, **1914**, *47*, 3087. A translation may be found in "Classics in Coordination Chemistry," G. B. Kauffman, ed., Dover, New York, **1968**, pp. 177–184.

[72] F. G. Mann, *J. Chem. Soc.*, **1933**, 412.

Resolution of optically active complexes

Few inorganic chemists have been as lucky as Pasteur in having their optically active compounds crystallize as recognizable, hemihedral crystals of the two enantiomers which may be separated by visual inspection. Various chemical methods have been devised to effect the resolution of coordination compounds.[73] They all involve interaction of the optically active species with some other optically active species. For optically active cations, interaction with optically active tartrate, α-bromocamphor-π-sulfonate, or metal complex anions will result in differentially soluble diastereoisomers, the shifting of equilibria in solution, or various other changes in physical properties that allow the separation to be effected. For the resolution of an anion, optically active bases such as strychnine or brucine (protonated to form cations) may be used. Neutral species present difficulties since it is impossible to form salts. Differential physical properties toward asymmetric substrates may be employed such as preferential extraction into an asymmetric solvent, preferential adsorption from solution upon quartz or sugar, or preferential adsorption from the gas phase using standard vapor phase chromatography on a column packed with substrate.[74]

Absolute configuration of complexes

The determination of the absolute spatial relationship (the *chirality* or "handedness") of the atoms in a dissymmetric coordination compound is a problem that has intrigued inorganic chemists from the days of Werner. The latter had none of the physical methods now available for such determinations. Note that it is not possible to assign the absolute configuration simply on the basis of the direction of rotation of the plane of polarized light,[75] although we shall see that through analysis of the rotatory properties of enantiomers strong clues can be provided as to the configuration. Werner suggested a solubility rule for relating configurations to each other: *Given two optically active compounds providing four enantiomers, A, A′, B, and B′*[76] *which form diastereoisomers with an optically active resolving agent, C, four compounds will form, AC, A′C, BC, and B′C. The two least soluble diastereoisomers will have related configurations.* That is, if A′C and B′C are the least soluble, it indicates that the configurations of A′ and B′ are similar and the reduced solubility results from the more effective packing of these ions in the presence of the optically active counter ion, C.[77] Although Werner was able to apply this rule to the relation of one complex to another, he had no way of determining the absolute configuration of *any* complex.

[73] For surveys of resolution and other experimental methods, see F. Basolo, in "Coordination Chemistry," J. C. Bailar, Jr., ed., Van Nostrand-Reinhold, New York, **1956**, Chapter 8, and S. Kirchner, in "Preparative Inorganic Reactions," W. M. Jolly, ed., Wiley (Interscience), New York, **1964**, Vol. I, Chapter 2.

[74] R. E. Sievers, R. W. Moshier, and M. L. Morris, *Inorg. Chem.*, **1962**, *1*, 966; R. E. Sievers, in "Coordination Chemistry," S. Kirschner, ed., Plenum Press, New York, **1969**, pp. 270–288.

[75] For example, it is known that the enantiomers of $[Co(en)_3]^{+3}$ and $[Rh(en)_3]^{+3}$, which rotate sodium D light in the *same* direction, have the *opposite* absolute configuration; i.e., they are mirror images of each other (ignoring the difference between Co and Rh).

[76] A and A′ refer to the two enantiomers of compound No. 1, and B and B′ refer to the two enantiomers of compound No. 2.

[77] As discussed in Chapters 3 and 6, there are many factors affecting lattice energy and solubility, and so any rule such as this can only be approximately true. A restatement to improve the reliability has been given: "*If the less soluble diastereoisomers formed by certain enantiomers of two ions with a given resolving counter ion are isomorphous, then those enantiomers have related configurations.*" [K. Garbett and R. D. Gillard, *J. Chem. Soc., A.*, **1966**, 802.]

Before discussing the methods of experimentally determining absolute configurations let us briefly discuss means of denoting such configurations. As was the case in organic chemistry, these rules grew up before much was known about the absolute configurations, so they leave much to be desired in terms of logical interrelationships. The simplest method, already mentioned, is simply the experimental notation of the direction of rotation of polarized light, + or *d*, − or *l*. This serves to identify enantiomers but little else, although we shall see that by using certain techniques and assumptions, strong clues can be provided with regard to configuration. If proof be needed that the direction of rotation does *not* relate to absolute configuration, consider $[Co(en)_3]^{+3}$ and $[Rh(en)_3]^{+3}$. The two enantiomers that rotate sodium D light in the *same* direction are known to have the *opposite* absolute configuration; i.e., they are mirror images of each other (ignoring the difference between cobalt and rhodium).

Next, as in Emil Fischer's system for D-glyceraldehyde and sugars, we can arbitrarily assign the D configuration to $(+)$-$[Co(en)_3]^{+3}$ and compare all known configurations with it. This now immediately tells us the configuration of a D-isomer (by comparing it with a figure of D-$[Co(en)_3]^{+3}$), but the D symbol is an arbitary one that could have as readily been ZZZ versus WWW. Furthermore, it can even seduce the unwary into thinking that D has something to do with being "dextrorotatory" or having "right-handedness."

More recently, the suggestion has been made[78] that the best way to view a tris(chelate), the most common species of optically active complexes, is down the threefold rotation axis. If the helix thus viewed is right-handed, the isomer is the Δ-isomer, and its mirror image is the Λ-isomer. The D-, L-, Δ-, and Λ-isomers may thus be portrayed as follows:

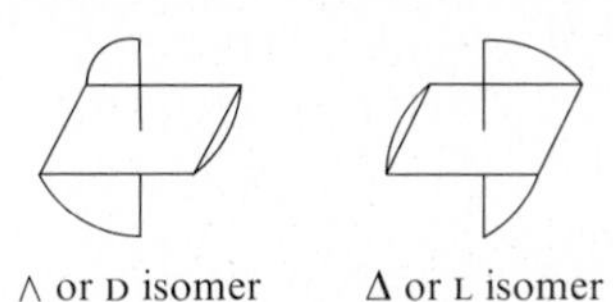

Λ or D isomer   Δ or L isomer

Note that it is an unfortunate result of these systems that Δ ≠ D and Λ ≠ L, furthering possible confusion.

In ordinary X-ray diffraction work both enantiomers give the same diffraction pattern, and thus this method gives no information on the absolute configuration about the metal atom. However, absolute configurations of coordination compounds can be directly determined by means of the anomalous dispersion of X rays, called *Bijvoet analysis.*[79] The method has not been widely applied, but as in the related problem in organic chemistry of the absolute configuration of D-glyceraldehyde, once *one* absolute configuration is known, there are methods to correlate others. The absolute configuration of D-$(+)[Co(en)_3]^{+3}$ has been determined as the chloride and bromide salts, and L-$(-)[Co(-pn)]^{+3}$ as the bromide (en = ethylenediamine; −pn = *l*- or R-propylenediamine).

If one of the ligands is optically active and its absolute configuration is known, one can determine the absolute configuration of the metal atom by X-ray (or neutron) diffraction. Of the two enantiomeric structures consistent with the X-ray data, the one having

[78] T. S. Piper, *J. Am. Chem. Soc.*, **1961**, *83*, 3908. Other systems have also been proposed.

[79] J. M. Bijvoet, *Endeavour*, **1955**, *14*, 71. See also Y. Saito, "Inorganic Molecular Dissymmetry," Springer-Verlag, Berlin, **1979**, Chapter 2.

the correct configuration about the *known* asymmetric atom is chosen. For example, consider tris(propylenediamine)cobalt(III). This was synthesized

$$[Co(CO_3)_3]^{-3} + 3(-)pn \longrightarrow [Co(-pn)_3]^{+3} + 3CO_3^{-2} \tag{10.7}$$

Surprisingly, the synthesis is stereospecific and we obtain only one isomer (see p. 505). It is then identified as the Δ- or L-isomer as follows. Since we know the absolute configuration of the asymmetric carbon in R-propylenediamine, we choose the enantiomeric solution of the data that provides the correct configuration for that carbon, thus automatically fixing the correct configuration about the cobalt (see Fig. 10.33). This method is

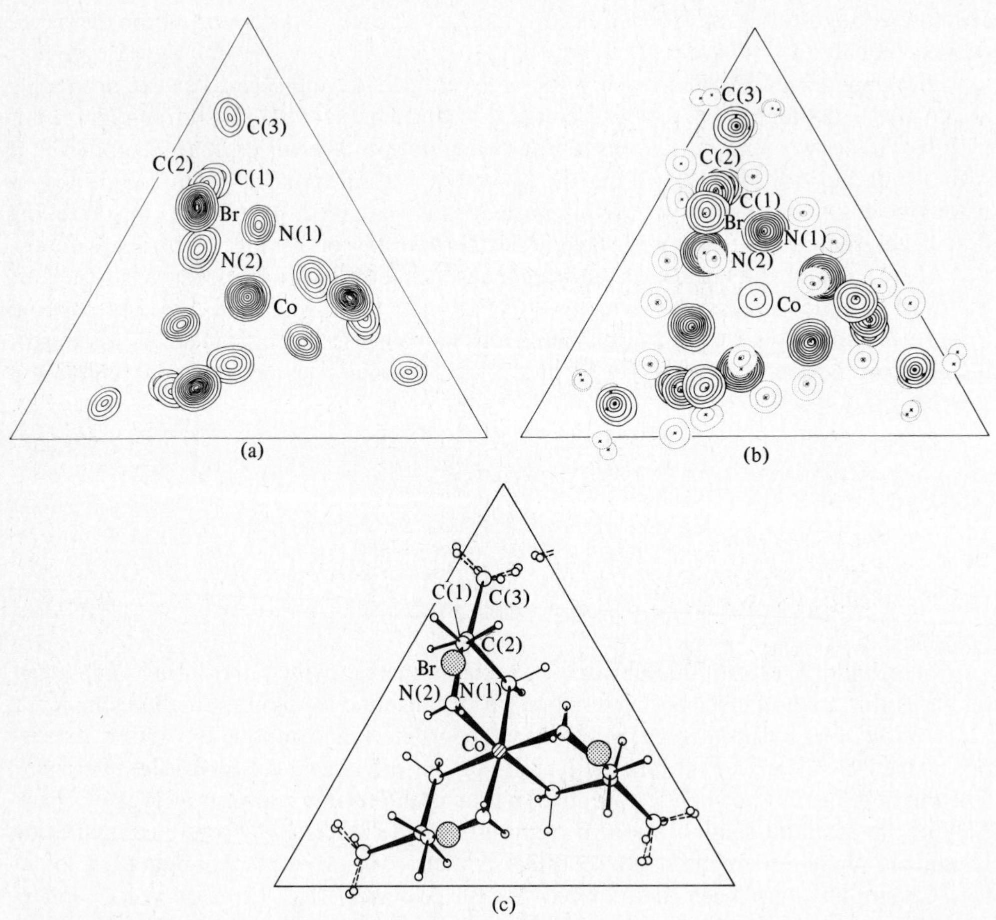

Fig. 10.33 Determination of the absolute configuration of $(-)_{589}$-tris(R-propylenediamine)cobalt(III) bromide. (a) Electron density map from X-ray diffraction data. Contours around cobalt and bromine are drawn at intervals of $10e^- \text{Å}^{-3}$; those for lighter atoms at $2e^- \text{Å}^{-3}$. Note that scattering increases with increasing atomic number: C < N < Co < Br. The H atoms do not show. (b) Neutron scattering density. The scattering is no longer proportional to Z (note weak scattering of cobalt). Hydrogen atoms are now locatable. (c) Final structure of compound. The multiple hydrogen locations in the corners of the triangle represent rotational disorder. [From Y. Saito, "Inorganic Molecular Dissymmetry," Springer-Verlag, Berlin, **1979**. Reproduced with permission.]

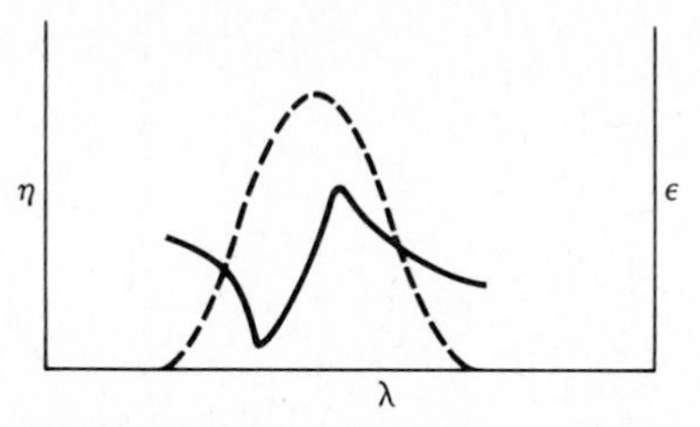

Fig. 10.34 Change in refractive index of a compound in the region of absorption: (solid line) refractive index; (dashed line) absorption band. [From R. D. Gillard *Progr. Inorg. Chem.*, **1966**, *7*, 215. Reproduced with permission.]

especially useful when studying complexes of naturally occurring L-amino acids such as alanine and glutamate.

Two related methods of optical analysis, optical rotatory dispersion (ORD) and circular dichroism (CD), have been used extensively in assigning absolute configurations. In order to discuss these two methods it is necessary to say a word or two about the interaction between electromagnetic radiation and matter.[80] The refraction of light (as measured by the refractive index, η) results from the induction of dipoles in the medium (from the interaction between the electric vector of the light and the electrons of matter). If all of the molecules are symmetric and nonabsorbing, the only effect will be that the light is slowed (relative to a vacuum) when passing through the medium.

In that part of the spectrum over which a particular compound is transparent, the index of refraction is relatively constant. In the neighborhood of an absorption band, however, dramatic changes take place (Fig. 10.34). The absorption bands result from excitation of electrons from lower to higher levels, and under these conditions the interaction between electromagnetic radiation and the electrons is at a maximum.

If plane-polarized light is passed through a solution containing optically active molecules of one of the two enantiomeric configurations, it will be rotated either right (+ or dextrorotatory) or left (− or levorotatory) as seen by the observer. The specific rotation at wavelength λ is defined as

$$[\alpha]_\lambda = \frac{\alpha}{lc} \tag{10.8}$$

where α is the rotation measured in degrees, l is the length of the light path (in decimetres) through the optically active medium, and c is the concentration in grams per cubic centimetre of the solution. The molecular rotation, $[\phi]$, may be defined as

$$[\phi] = \frac{M[\alpha]}{100} \tag{10.9}$$

where M is the molecular weight of the optically active compound.

---

[80] For more nearly complete discussions of these topics see R. D. Gillard, *Progr. Inorg. Chem.*, **1966**, *7*, 215, and R. D. Gillard, in "Physical Methods of Advanced Inorganic Chemistry," H. A. O. Hill and P. Day, eds., Wiley (Interscience), New York, **1968**, p. 167; S. Kirschner and N. Ahmad, in "Coordination Chemistry," S. Kirschner, ed., Plenum Press, New York, **1969**, pp. 42–63; R. S. Drago, "Physical Methods in Chemistry." Saunders, Philadelphia, **1977**, pp. 123–125.

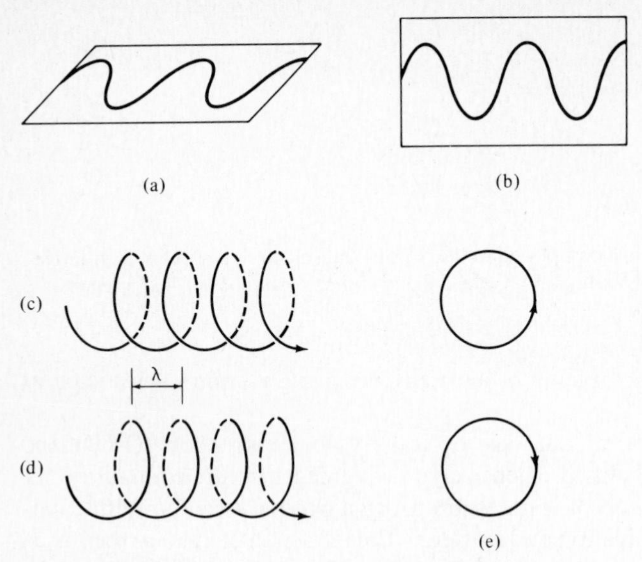

Fig. 10.35 (a) Plane polarized light, horizontal plane; (b) plane polarized light, vertical plane; (c) left circularly polarized light; (d) right circularly polarized light; (e) left (top) and right (bottom) circularly polarized light seen by an observer.

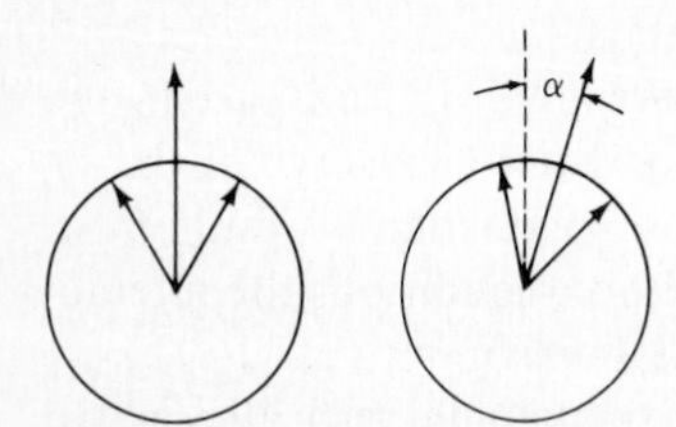

Fig. 10.36 Dextrorotation of plane polarized light resulting from unequal rotation of left and right circularly polarized light: (left) light entering sample; (right) light emerging from sample.

The optical rotation arises from unequal refraction of left and right circularly polarized light. Plane-polarized light may be considered to be light in which the electric vector vibrates only in one plane (Fig. 10.35a,b). It may be decomposed into two circularly polarized components of equal intensities, rotating in opposite directions with the electric vector passing through one revolution for each wavelength (Fig. 10.35c,d). The rotation of plane-polarized light arises from the fact that asymmetric molecules have different refractive indices, η_l and η_d, for left and right circularly polarized light. The two components interact differently with the medium and emerge out of phase. If recombined, the plane of polarization has been rotated (Fig. 10.36).

If the rotation is measured in a region far from absorption bands, the difference in refractive index is the only factor involved. Nevertheless, the specific rotation depends upon the wavelength of the light used and this should always be stated. Thus if the sodium D line ($\lambda = 589$ nm) is used, as is common, the rotation can be reported as $[\alpha]_D$ or $[\alpha]_{589}$.[81]

[81] Since the D line is used so often, the subscript D is often omitted, but it should *always* be stated for other wavelengths.

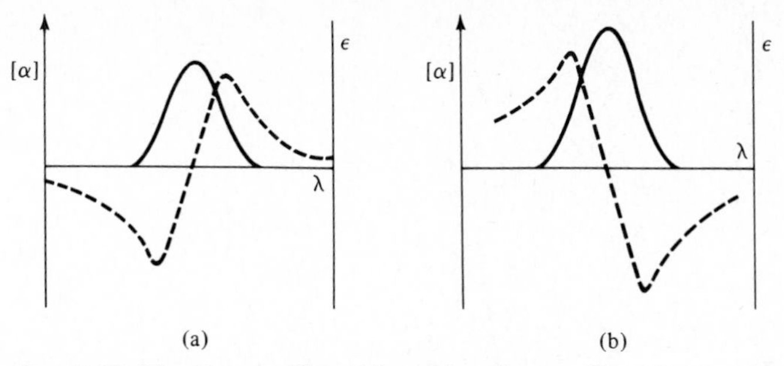

Fig. 10.37 The Cotton effect: (a) positive Cotton effect; (b) negative Cotton effect; (solid line) absorption band; (dashed line) ORD curve. [From R. D. Gillard, *Progr. Inorg. Chem.*, **1966**, *7*, 215. Reproduced with permission.]

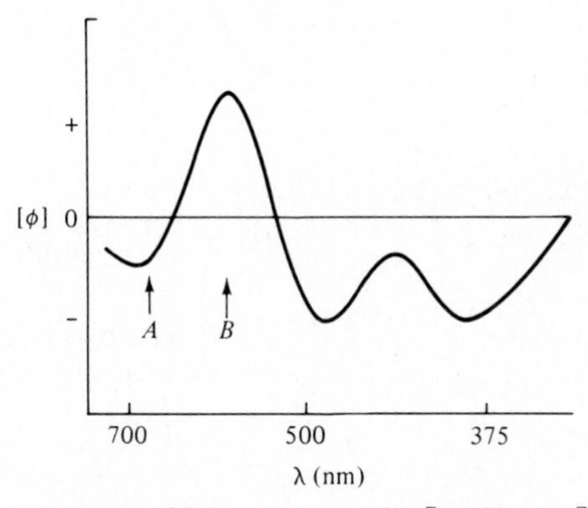

Fig. 10.38 ORD spectrum of D-$[CoCl_2(en)_2]^+$. Wavelength A (644 nm) is the red Cd line used by Werner yielding (−) rotation; B (589 nm) was used by subsequent workers who obtained (+) rotation. [From R. D. Gillard, *Progr. Inorg. Chem.*, **1966**, *7*, 215. Reproduced with permission.]

The variation in specific rotation increases as an absorption band is approached. The reason is that the refractive indices η_l and η_d change rapidly with change in wavelength just as for nonpolarized light (Fig. 10.34). As a result, the specific rotation drops to a minimum, passes through zero at maximum absorption, and rises quickly to a maximum. The measurement of the variation of rotation with wavelength is called optical rotatory dispersion (ORD) and the abrupt reversal of rotation in the vicinity of an absorption band is called the *Cotton effect*[82] (Fig. 10.37a). If the complex was initially dextrorotatory, the effect is reversed with the ORD curve rising to a maximum, falling sharply to a minimum, and then slowly rising (Fig. 10.37b). As a result of the change in sign of the rotation as a function of wavelength, confusion has arisen in the literature when the wavelength was not specified (Fig. 10.38).

ORD curves such as shown in Fig. 10.37 are useful in the assignment of absolute configurations. For example, the configurations of the enantiomers of tris(ethylenediamine)cobalt(III), tris(alaninato)cobalt(III), and bis(ethylenediamine)glutamatocobalt(III)

[82] Named after Aime Cotton, a French physicist.

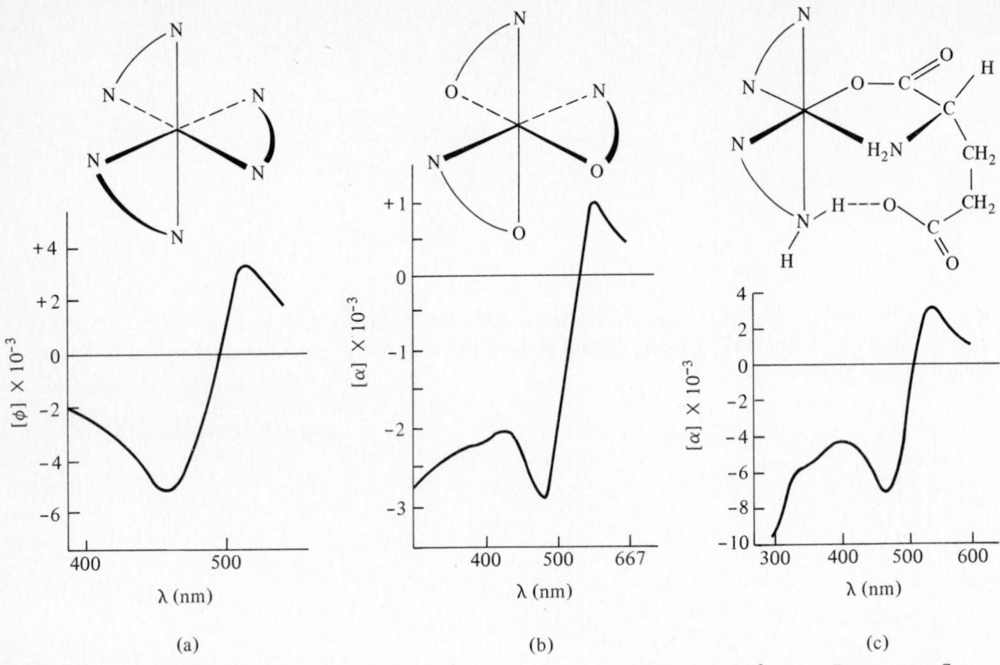

Fig. 10.39 The absolute configurations and ORD spectra of (a) D-$[Co(en)_3]^{+3}$; (b) D-$[Co(L\text{-}ala)_3]$ (L-ala = L-alanine); (c) D-$[Co(en)_2(glu)]^+$ (glu = the dianion of L-glutamic acid). All of these complexes have the Λ configuration.

are known from X-ray investigations and it is found that the three D-enantiomers of these complexes have similar ORD spectra (Fig. 10.39).

A general rule may be stated: *If, in analogous compounds, corresponding electronic transitions show Cotton effects of the same sign, the compounds have the same optical configuration.*[83] On this basis the D-configuration could have been assigned to any of these in the absence of X-ray data simply on the basis of the similarity of the ORD spectra to *one of known configuration.* The chief difficulty is the requirement of "corresponding electronic transitions." Often it is difficult or impossible to resolve spectra with overlapping Cotton effects (see Fig. 10.38 for an example of the complexity of some ORD spectra).

In addition to the rapid change in refractive indices in the neighborhood of absorption peaks, there will be *differential absorption* of the two circularly polarized components. The difference between the extinction coefficients of left and right polarized light ($\varepsilon_l - \varepsilon_d$) is called the circular dichroism (CD) (see Fig. 10.40). It is intimately related to the ORD and the absorption spectrum of the complex. In the past ORD has often been of more value than CD, but this has been a result of instrumentation since usually CD can give at least as much information as ORD. Since CD occurs only within an absorption band it is somewhat simpler to interpret than ORD, which may be complicated by the overlapping of effects from several bands, some far removed from the area of the spectrum being studied. Complete reviews and comparisons of the two methods are available.[80]

---

[83] R. D. Gillard, *Progr. Inorg. Chem.*, **1966**, *7*, 215.

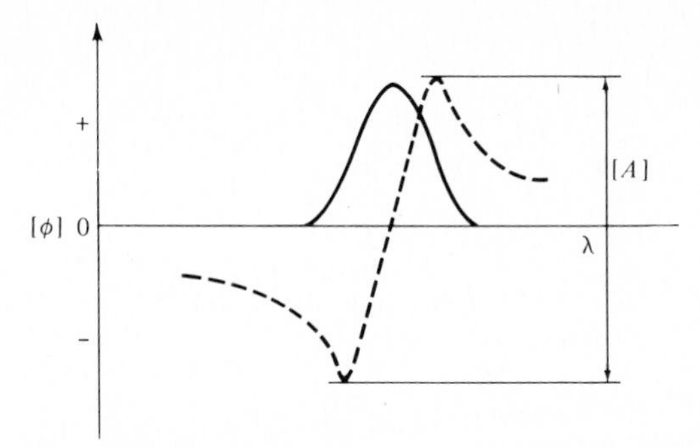

Fig. 10.40 Optical rotatory dispersion (dashed line) and circular dichroism (solid line) for an isolated asymmetric chromophore. Relationships are $[A] = 4028(\varepsilon_l - \varepsilon_r)$ and $\Delta v = 0.925(v_{max} - v_{min})$ where Δv is the width of the CD band at half-height and v_{max} and v_{min} are the frequencies of the maximum and minimum in the ORD spectrum. [From R. D. Gillard, *Progr. Inorg. Chem.*, **1966**, *7*, 215. Reproduced with permission.]

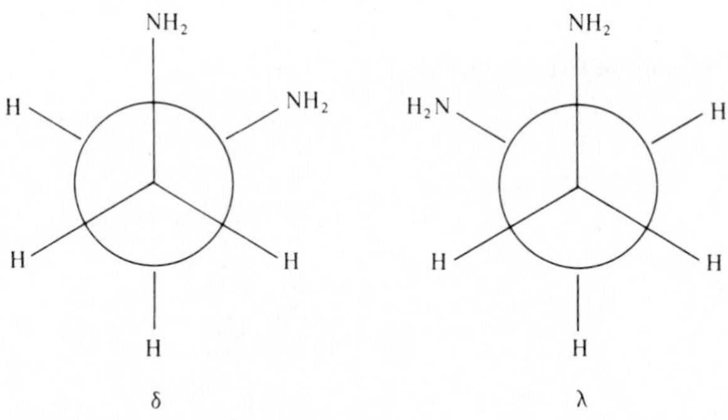

Fig. 10.41 Enantiomeric conformations of *gauche* ethylenediamine (1,2-diaminoethane). Note that δ represents a "right-handed helix" and λ a "left-handed helix."

Stereoselectivity and the conformation of chelate rings

In addition to the asymmetry generated by the tris(chelate) structure of octahedral complexes, it is possible to have asymmetry in the ligand as well. For example, the *gauche* configuration that ethylenediamine assumes when bonding to a metal is inherently asymmetrical (Fig. 10.41) and could in principle be resolved were it not for the almost complete absence of an energy barrier preventing racemization. Attachment of the chelate to the metal retains the asymmetry of the *gauche* form, but the two enantiomers can still interconvert through a planar conformation at a very low energy, similar to the interconversions of organic ring systems (Fig. 10.42). Thus, although it is possible in principle to describe two optical isomers of a complex such as $[Co(NH_3)_4(en)]^{+3}$ (Fig. 10.43), in practice it proves to be impossible to isolate them.[84]

If two or more rings are present in one complex, they can interact with each other and certain conformations might be expected to be stabilized as a result of possible reductions in interatomic repulsions. For example, consider a square planar complex containing

[84] Y. Shimura, *Bull. Chem. Soc. (Japan)*, **1958**, *31*, 311.

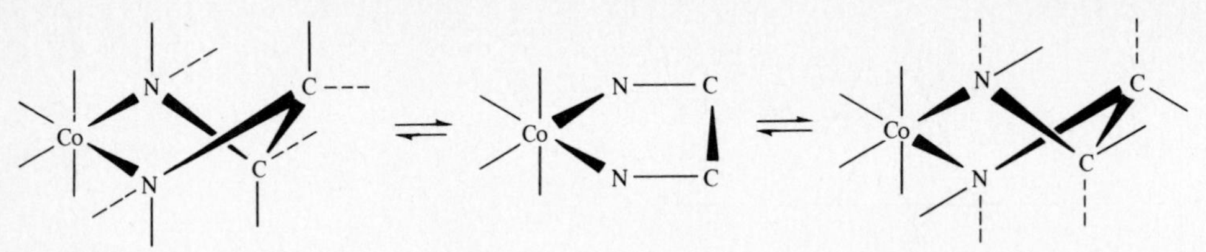

Fig. 10.42 Enantiomeric conformations of the five-membered ethylenediamine–metal chelate ring and the intermediate planar conformation responsible for interconversion.

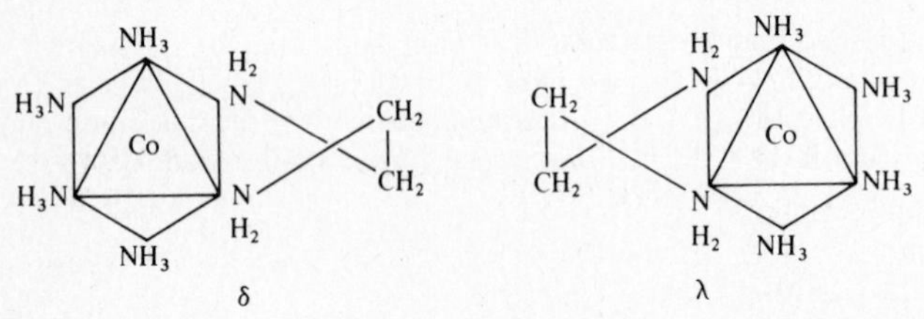

Fig. 10.43 Enantiomeric conformations of tetraammine(ethylenediamine)cobalt(III) ion.

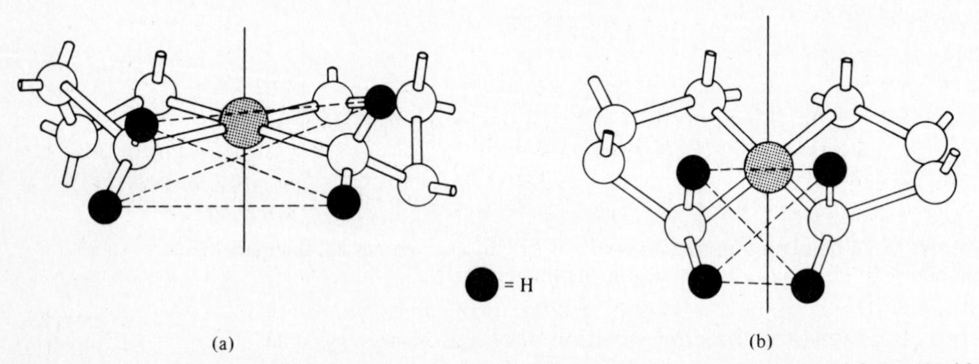

Fig. 10.44 Conformational interactions in bis(chelate) square planar complexes: (a) $\lambda\lambda$ form; (b) $\lambda\delta$ form. All hydrogen atoms except four have been omitted for greater clarity. Dashed lines represent inter-ring H—H repulsions. [Modified from E. J. Corey and J. C. Bailar, *J. Am. Chem. Soc.*, **1959**, *81*, 2620. Reproduced with permission.]

two chelated rings of ethylenediamine. From a purely statistical point of view we might expect to find three structures, which may be formulated M$\delta\delta$, M$\lambda\lambda$, and M$\delta\lambda$ (which is identical to M$\lambda\delta$). The first two molecules lack a plane of symmetry, but M$\delta\lambda$ is a *meso* form. Corey and Bailar[85] were the first to show that the M$\delta\delta$ and M$\lambda\lambda$ forms should predominate over the *meso* form since the latter has unfavorable H—H interactions of the axial–axial and equatorial–equatorial type between the two rings (Fig. 10.44). The

[85] E. J. Corey and J. C. Bailar, Jr., *J. Am. Chem. Soc.*, **1959**, *81*, 2620. This is the classic paper in the field upon which all of the subsequent work has been based. For more recent analyses of conformational effects, see J. R. Gollogly et al., *Inorg. Chem.* **1971**, *10*, 317; J. K. Beattie, *Acc. Chem. Res.* **1971**, *4*, 253; and Y. Saito, "Inorganic Molecular Dissymmetry," Springer-Verlag, Berlin, **1979**, Chapter 3.

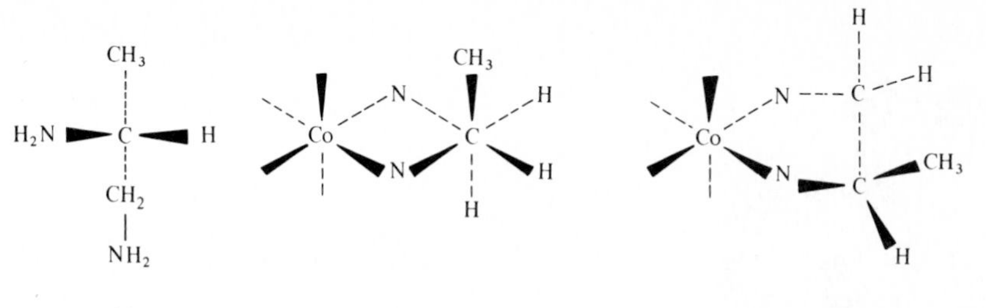

Fig. 10.45 (a) The absolute configuration of R-(−)-propylenediamine; (b) a propylenediamine chelate ring with a δ conformation and an axial methyl group (this is like a Newman projection, there is a second carbon atom behind the one shown); (c) λ conformation resulting in an equatorial methyl group.

enantiomeric $M\delta\delta$ and $M\lambda\lambda$ forms are expected to be about 4 kJ mol^{-1} more stable, other factors being equal.[86]

More important consequences result for octahedral tris(chelate) complexes. Again, from purely statistical arguments, we might expect to find $M\delta\delta\delta$, $M\delta\delta\lambda$, $M\delta\lambda\lambda$, and $M\lambda\lambda\lambda$ forms. In addition, these will all be optically active from the tris(chelate) structure as well, so there are expected to be *eight* distinct isomers formed. In general, a much smaller number is found, usually only two. This stereoselectivity is most easily followed by using an optically active ligand such as propylenediamine, $CH_3CH(NH_2)CH_2NH_2$. The five-membered chelate ring will give rise to two types of substituent positions, those that are essentially axial and those that are essentially equatorial.[87] All substituents larger than hydrogen will cause the ring to adopt a conformation in which the substituent is in an equatorial position (Fig. 10.45). As a result of this strong conformational propensity, (−)-propylenediamine bonds preferentially as a λ chelate and (+)-propylenediamine bonds as a δ chelate. This reduces the number of expected isomers to four: Λ-$M\delta\delta\delta$ (= D-$M\delta\delta\delta$), Λ-$M\lambda\lambda\lambda$ (= D-$M\lambda\lambda\lambda$), Δ-$M\delta\delta\delta$ (= L-$M\delta\delta\delta$), and Δ-$M\lambda\lambda\lambda$ (= L-$M\lambda\lambda\lambda$) where Λ, Δ, D, and L refer to the absolute configuration about the metal related to Λ-$(+)_{589}$-$[Co(en)_3]^{+3}$ (= the D enantiomer; see Fig. 10.39). In a typical reaction such as the oxidation of cobalt(II) chloride in the presence of racemic propylenediamine, only two isomers were isolated.[88]

$$[Co(H_2O)_6]^{+2} + (+, -)\text{pn} \xrightarrow{[o]} \text{D-}[Co(+\text{pn})_3]^{+3} + \text{L-}[Co(-\text{pn})_3]^{+3} \quad \mathbf{(10.10)}$$

The difference in stability between the various isomers has been related to preferred packing arrangements of chelate rings about the central metal atom. Thus, for (+)-propylenediamine (which forms a δ chelate) the most efficient method of fitting around

[86] The *trans*-dichlorobis(ethylenediamine)cobalt(III) ion, $[Co(en)_2Cl_2]^+$, has a square planar arrangement of the ethylenediamine rings as shown in Fig. 10.44. In the crystalline $[Co(en)_2Cl_2]Cl \cdot 2H_2O$, the *meso* isomer has been found [A. Nakahura et al., *Bull. Chem. Soc.* (*Japan*), **1952**, *25*, 331]. This has been attributed[85] to crystal effects overcoming the slight conformational stability of the $\delta\delta$ and $\lambda\lambda$ isomers. Indeed, in the (−)-propylenediamine analogue the greater steric requirements of the methyl group presumably dominate since the predictions of Corey and Bailar are confirmed [H. Iwasaki and Y. Saito, *Bull. Chem. Soc.* (*Japan*), **1962**, *35*, 1131.]

[87] The difference between "axial" and "equatorial" positions in a five-membered ring are less extreme than in the well-known chair form of cyclohexane, but the energetic differences are nevertheless clear-cut.

[88] A. P. Smirnoff, *Helv. Chim. Acta*, **1920**, *3*, 177.

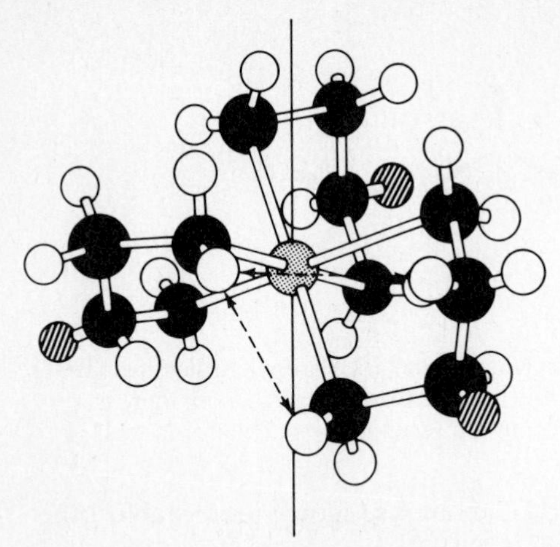

Fig. 10.46 The *lel* conformer of the D or Λ isomer of tris(diamine) metal complexes. The hatched circles represent the positions of the methyl group in the propylenediamine complex. For propylenediamine, this represents the Dδδδ or Λδδδ isomer. [Modified from E. J. Corey and J. C. Bailar, *J. Am. Chem. Soc.*, **1959**, *81*, 2620. Reproduced with permission.]

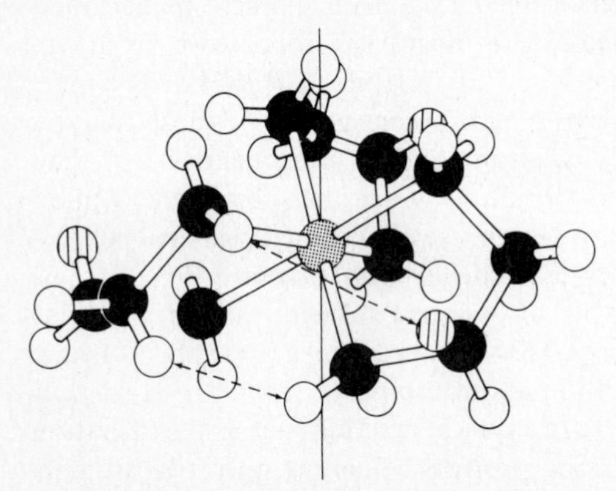

Fig. 10.47 The *ob* conformer of the L or Δ isomer of tris(diamine) metal complexes. The hatched circles represent the positions of the methyl group in the propylenediamine complex. For propylenediamine this represents the Lδδδ or Δδδδ isomer. [Modified from E. J. Corey and J. C. Bailar, *J. Am. Chem. Soc.*, **1959**, *81*, 2620. Reproduced with permission.]

a metal will be in the form of a left-handed helix. This arrangement minimizes the various repulsions. It has been termed the *lel* isomer since the C—C bonds are parallel to the threefold axis of the complex (Figs. 10.46, 10.48a). The alternative isomer, in which the ligands form a right-handed helix about the metal, is known as the *ob* isomer since the C—C bonds are oblique to the threefold axis (Figs. 10.47, 10.48b). The interactions between the hydrogen atoms of the various rings stabilize the *lel* isomer by a few kilojoules per mole.

Dwyer and co-workers[89] have studied these systems extensively, and by careful work have been able to show the existence of the unstable *ob* isomers as well as the *lel* isomers.

[89] F. P. Dwyer et al., *J. Am. Chem. Soc.*, **1964**, *86*, 590.

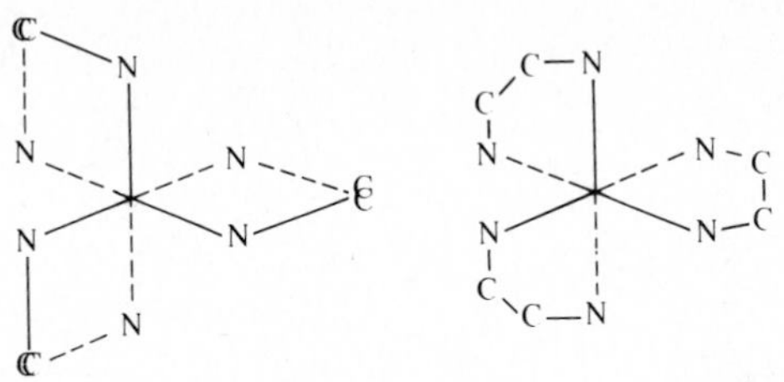

Fig. 10.48 Schematic portrayal of the *lel* (left) and the *ob* (right) conformers of the D or Λ isomer of a tris(diamine) complex such as D$\delta\delta\delta$ and $\Lambda\delta\delta\delta$ (left) and D$\lambda\lambda\lambda$ or $\Lambda\lambda\lambda\lambda$ (right).

The equilibrium constant for the reaction

$$\text{L}\delta\delta\delta \rightleftharpoons \text{D}\delta\delta\delta \qquad (10.11)$$

is 14.6, which corresponds to a ΔG of -6.7 kJ mol^{-1}.[90] Note that this greater stability of the one conformation is the source of the stereospecific synthesis of Δ-tris(R-propylenediamine)cobalt(III) seen previously (p. 496).

Asymmetric synthesis catalyzed by coordination compounds[91]

There has been interest in the stereospecific synthesis of organic compounds using optically active coordination compounds. This would allow the specific synthesis of essential L-amino acids such as lysine, for example, which is notably deficient in many staple foods, such as corn (maize), preventing the latter from being a well-balanced food. Coordination of achiral glycine to a transition metal activates the α carbon, allowing it to be sustituted. If the transition metal is part of an asymmetric molecule, the preferred conformations just discussed can give a kinetically controlled, preferred configuration to the product. In general, little success has yet been achieved, but the field is young. One example of partial success is the reaction of Λ-bis(ethylenediamine)glycinatocobalt(III) with acetaldehyde under basic conditions:

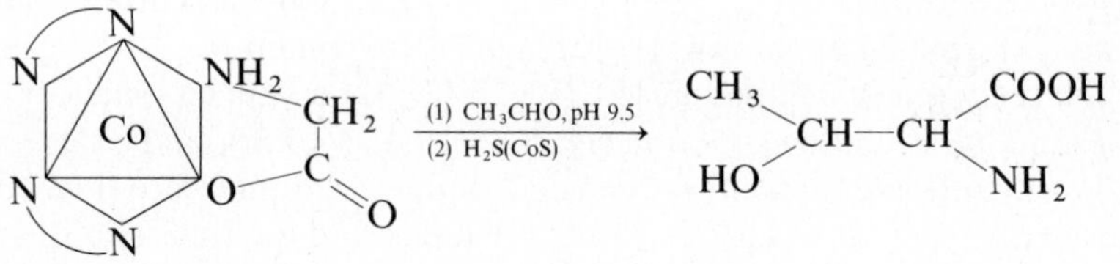

$$[\text{Co(en)}_2(\text{NH}_2\text{CH}_2\text{C(O)O})] \xrightarrow[\text{(2) H}_2\text{S(CoS)}]{\text{(1) CH}_3\text{CHO, pH 9.5}} \text{CH}_3\text{CH(OH)}\text{—CH(NH}_2\text{)COOH} \qquad (10.12)$$

Under these conditions threonine and allothreonine are synthesized with 58% and 67%, respectively, of the product being the S- or L-isomer.

[90] For more nearly complete discussions of stereoselectivity and conformational effects, see J. H. Dunlop and R. D. Gillard, *Adv. Inorg. Chem. Radiochem.*, **1966**, *9*, 185; A. M. Sargeson, *Transition Metal Chem.*, **1966**, *3*, 303; C. J. Hawkins, "Absolute Configuration of Metal Complexes," Wiley, New York, **1971**, and the works cited in footnote 85.

[91] D. A. Phipps, *J. Mol. Catal.*, **1979**, *5*, 81; B. Bosnich and M. D. Fryzuk, "Topics in Inorganic and Organometallic Stereochemistry," G. L. Geoffroy, ed., "Topics in Stereochemistry," Vol. 12, Wiley, New York, **1981** pp. 119–154.

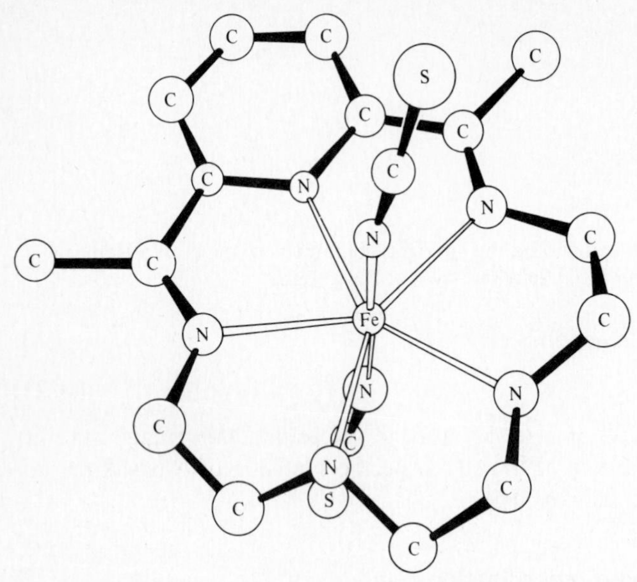

Fig. 10.49 Molecular structure of the 7-coordinate complex involving 2,13-dimethyl-3,6,9,12,18-pentaazabicyclo[12.3.1]-octadeca-1(18),2,12,14,16-pentaene and two axial ligands (SCN^-). Hydrogen atoms omitted. [From E. Fleischer and S. Hawkinson, *J. Am. Chem. Soc.*, **1967**, *89*, 720. Reproduced with permission.]

COORDINATION NUMBER 7

Coordination number 7 cannot be considered at all common. The relative instability of these species can be attributed to the fact that the additional bond energy of the seventh bond is offset by (1) increased ligand–ligand repulsion, (2) weaker bonds, and (3) generally reduced crystal field stabilization energy (CFSE) as a result of nonoctahedral geometry. There are three distinct geometries known: (1) pentagonal bipyramid (Fig. 10.49), which is also found in the main-group compound IF_7 (Fig. 5.12); (2) a capped octahedron in which a seventh ligand has been added to a *triangular* face (Fig. 10.50); and (3) a capped trigonal prism in which a seventh ligand has been added to a *rectangular* face (Fig. 10.51). In addition, there are many intermediate cases and transitions, and the situation is reminiscent of 5-coordinate geometries.[92]

In many of these complexes the requirements of polydentate ligands favor coordination number 7. Thus it is not difficult to see the effects of five macrocyclic and co-planar nitrogen atoms in Fig. 10.49 on the resulting pentagonal bipyramidal structure. In some cases, even unfavorable interactions may be forced by the ligand geometry. For example, in one type of "7-coordinate" complex, it appears as though the seventh coordination, forced by the geometry of the other six coordinating atoms, might better be considered an "antibond" rather than a bond. One example is illustrated in Fig. 10.52. The series of

[92] M. G. B. Drew, *Progr. Inorg. Chem.*, **1977**, *23*, 67; D. L. Kepert, ibid., **1979**, *25*, 41.

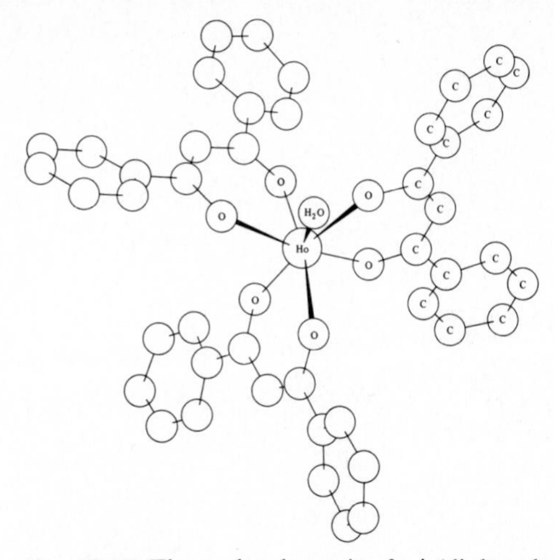

Fig. 10.50 The molecular unit of tris(diphenylpropanedionato)aquaholmium projected down the threefold axis. The water molecule is directly above Ho but has been displaced slightly in this drawing to show the structure better. [From A. Zalkin et al., *Inorg. Chem.*, **1969**, 8, 2680. Reproduced with permission.]

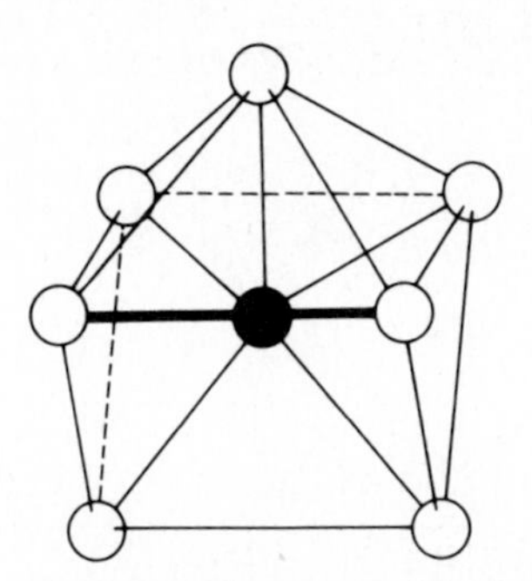

Fig. 10.51 Structure of the heptafluoroniobate(V) anion. [From J. L. Hoard, *J. Am. Chem. Soc.*, **1939**, *61*, 1252. Reproduced with permission.]

$[M(trenpy)]^{+2}$ complexes ($M^{+2} = Mn^{+2}$, Fe^{+2}, Co^{+2}, Ni^{+2}, Cu^{+2}, Zn^{+2}; trenpy = $(C_5H_4NCH{=}NCH_2CH_2)_3N$) was discussed previously (p. 389; Fig. 9.18) as an example of ligand field effects. The amine nitrogen atom (N_7) capping the octahedron is directed at the t_{2g} orbitals, which already contain from three to six electrons of the metal. A repulsion is set up between these electrons and the lone pair of this nitrogen atom. Although the M–N distance (280–340 pm) may be less than the sum of the van der Waals radii ($r_{X_{VDW}} + r_{Y_{VDW}}$ = 320 pm (an X–Y distance less than the van der Waals sum is often the accepted criterion for bonding), it is considerably longer than the other six M–N bond distances (230 pm) and the constraints of the polydentate system restrict the movement of N_7 *away* from the metal.

At the same time the bond angles (C—N—C) at this position vary from 112° ($\sim sp^3$ as expected for an amine ligand) in the manganese complex (where repulsion is least) up

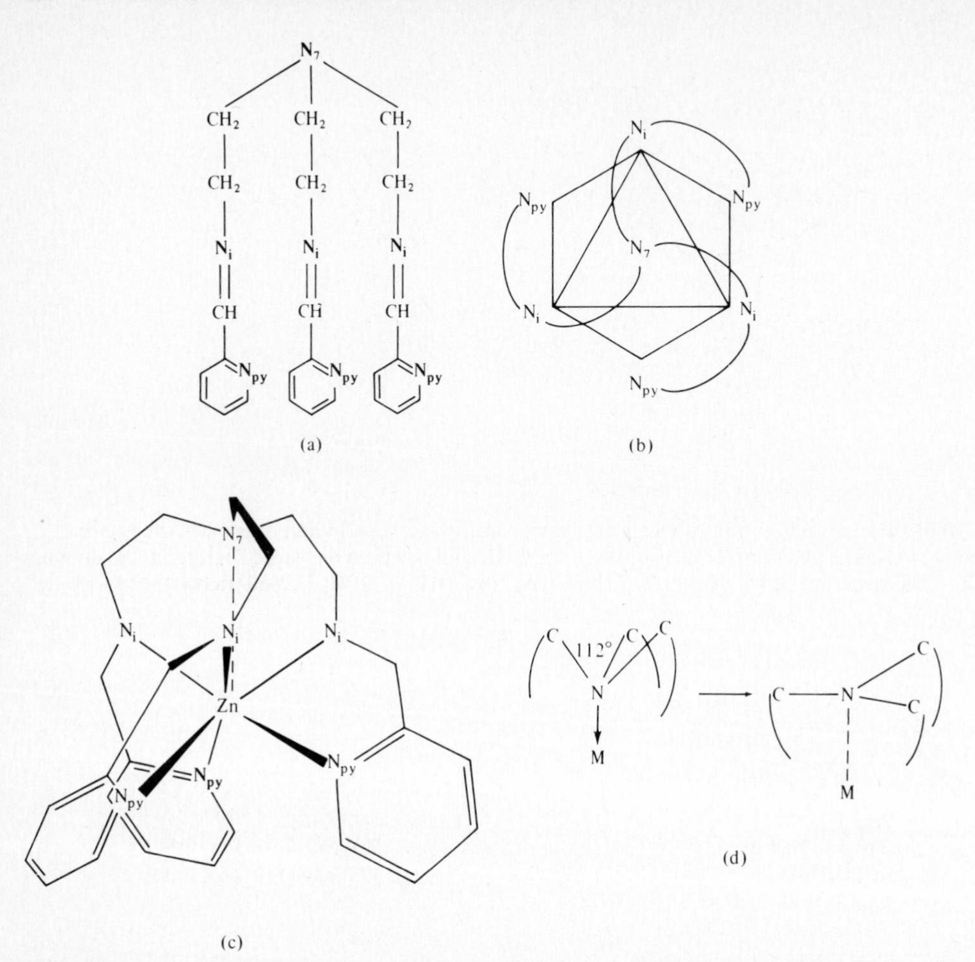

Fig. 10.52 $[M(trenpy)]^{++}$ complexes: (a) The "trenpy" ligand: $N(CH_2CH_2N{=}CHC_5H_5N)_3$; the nitrogen atoms are labeled N_{py}, the "pyridine nitrogen," N_i, the "imide nitrogen," and N_7, the "seventh' or unique nitrogen atom; (b) a diagrammatic representation of the $[M(trenpy)]^{+2}$ complex, viewed down the threefold axis; (c) the molecular structure of the zinc(II) complex as viewed perpendicularly to the threefold axis; the long Zn—N "bond" is shown by the dashed line; (d) the change in bond angles at the N_7 atom as the t_{2g} levels fill with electrons. [Courtesy of E. C. Lingafelter.]

to a maximum value of 120° in the t_{2g}^6 iron complex with maximum repulsions.[93] The amine nitrogen atom corresponds to a three-ribbed umbrella that has been inverted by the wind (the handle is the lone pair directed at the metal). As the t_{2g} level fills, the repulsions increase, and the metal–nitrogen distance increases and the "umbrella" begins to flatten (Fig. 10.52d).

Although isomerism in 7-coordinate complexes is possible in principle, no examples are known. Note, for example, that the trenpy complexes must be optically active (see Fig. 10.52), subject of course to kinetic stability with respect to racemization.

[93] C. Mealli and E. C. Lingafelter, *Chem. Commun.*, **1970**, 885; E. C. Lingafelter, personal communication.

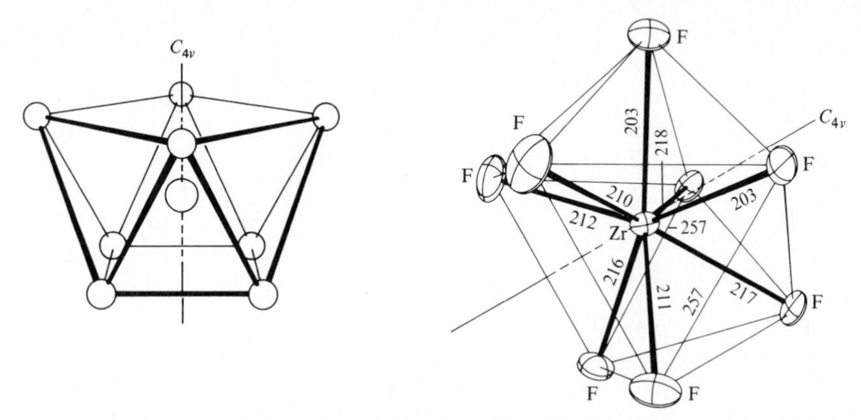

Fig. 10.53 Square antiprismatic coordination of fluoride ligands about a central zirconium(IV) atom $Na_7Zr_6F_{31}$. Bond distances in picometres. The mean F—F distance is 256 pm. [From J. H. Burns et al., *Acta Crystallogr.*, **1968**, *B24*, 230. Reproduced with permission.]

COORDINATION NUMBER 8

Coordination number 8 cannot be considered to be common; yet the number of known compounds has increased rapidly in recent years, and it is exceeded only by 4- and 6-coordination. The factors important in this increase in knowledge of 8-coordination can be traced largely to improved three-dimensional X-ray techniques and to increased interest in the coordination chemistry of lanthanide and actinide elements (see Chapter 16).

Two factors are important in favoring 8-coordination. One is the size of the metal cation—it must be sufficiently large to accommodate eight ligands without undue crowding. Relatively few 8-coordinated complexes are known for the first transition series. The largest numbers of this type of complex are found for the lanthanides and actinides, and it is fairly common for zirconium, hafnium, niobium, tantalum, molybdenum, and tungsten. As a corollary is the requirement that the ligands be relatively small. The commonest ligating atoms are carbon, nitrogen, oxygen, and fluorine. The second requirement is that the metal be in a high formal oxidation state. This requirement arises out of the electroneutrality principle. The formation of eight dative σ bonds to a metal in a low oxidation state would result in excess electron density on the metal. The common oxidation states are thus $+3$ or greater, resulting in electron configurations with few remaining electrons such as d^0, d^1, and d^2.[94]

There are several coordination polyhedra available for 8-coordination. The most regular, the cube, is not found in discrete complexes but occurs only in lattices such as CsCl. The two common structures are the square antiprism (Fig. 10.53) and the

[94] The picture of the formation of, for example, $[Mo(CN)_8]^{-4}$, as a combination of $Mo^{+4}(d^2) + 8CN^-$ has no physical meaning and is merely a bookkeeping device, but hardly more so than $Co^{+3} + 6NH_3$ discussed extensively earlier.

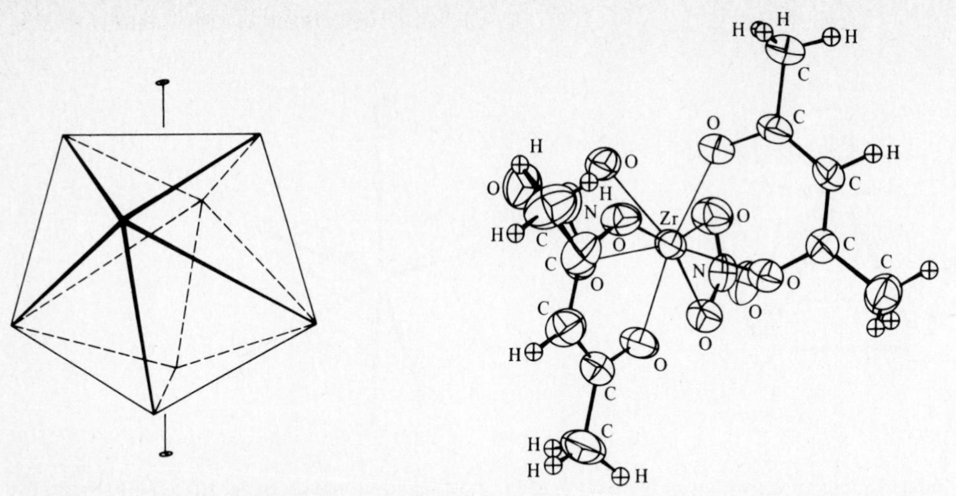

Fig. 10.54 Dodecahedral coordination in tetrakis(acetylacetonato)zirconium. [From V. W. Day and R. C. Fay, *J. Am. Chem. Soc.*, **1975**, *97*, 5136. Reproduced with permission.]

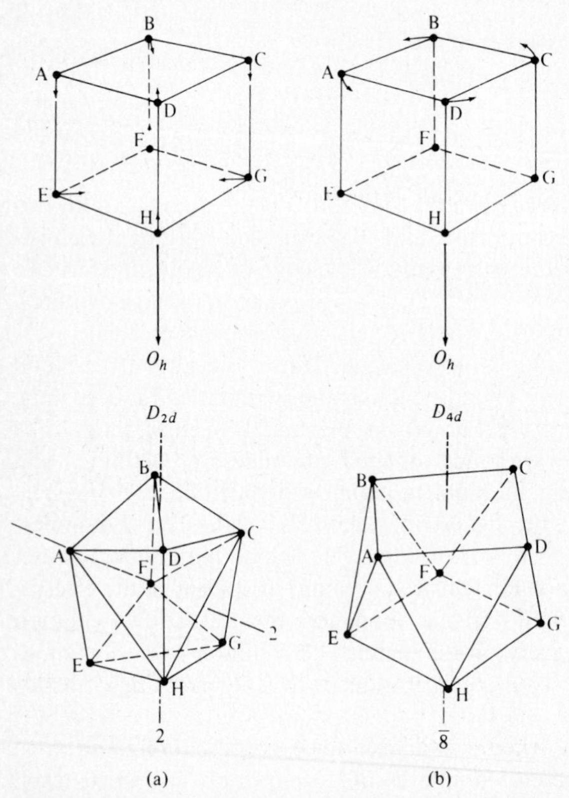

Fig. 10.55 Distortions of the cube to form: (a) the dodecahedron; (b) the square antiprism. Note the similarity of the dodecahedron, viewed down the puckered CDHG face (2-axis) and the antiprism, viewed down the ABCD face (8-axis). [From S. J. Lippard, *Progr. Inorg. Chem.*, **1968**, *8*, 109. Reproduced with permission.] Related to Fig. 10.54a by 90° rotation about the vertical twofold axis.

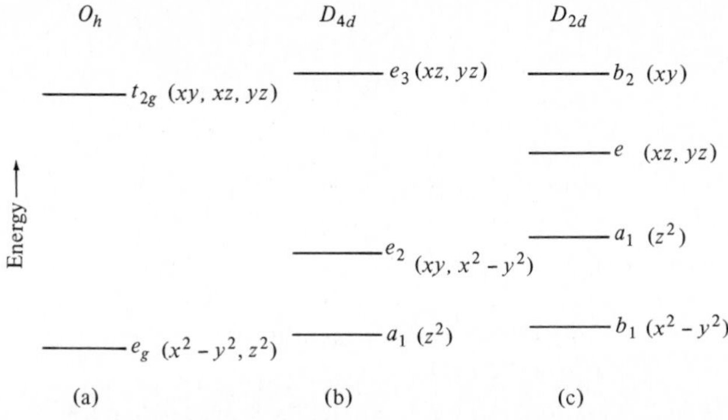

Fig. 10.56 Energy level diagrams for 8-coordination: (a) cubic (O_h); (b) square antiprismatic (D_{4d}); (c) dodecahedral (D_{2d}); (d) trigonally bicapped trigonal prism (D_{3h}); (e) rectangularly bicapped trigonal prism (C_{2v}). [From J. K. Burdett et al., *Inorg. Chem.*, **1978**, *17*, 2553. Reproduced with permission.]

dodecahedron (Fig. 10.54). Both may be considered to be distortions of the simple cube resulting in reduced ligand–ligand repulsions (Fig. 10.55). Other geometries are known, but these two are the most important.

From a valence bond point of view, the formation of a dodecahedron can arise from sp^3d^4 hybridization. The square antiprism can form from either sp^3d^4 or p^3d^5. The necessity of using four or five *d* orbitals in the hybridization for the ligand bonding rationalizes the common occurrence of d^0, d^1, and d^2 configurations from a valence bond model. Crystal field theory and molecular orbital theory give a similar picture in that both the square antiprism and dodecahedron give a nondegenerate lower level (to accept the one or two *d* electrons) and the remaining four *d* orbitals (or molecular orbitals derived principally from metal *d* orbitals) at higher levels (Fig. 10.56). The ligand field stabilization energies (LFSEs) of both structures are comparable, and the choice between the two is a delicate balance of forces, as was the case for 5-coordination. At present there is no satisfactory treatment of the problem of proper selection of geometry for 8-coordination.

There are extensive possibilities for the formation of geometric and/or optical isomers in 8-coordinate complexes. Thus far, none has been characterized. The problem is less one of lack of stability than the fact that complete X-ray analysis is essentially the only tool available for their study. The preparation, isolation, and characterization of such isomers remains an "unanswered challenge."[95]

HIGHER COORDINATION NUMBERS[96]

There are few structures known with coordination numbers larger than 8. The existence of coordination number 12 in some crystal lattices was mentioned above. Discrete 9-coordinate structures are known for complexes such as $[Ln(H_2O)_9]^{+3}$ (where Ln is an

[95] S. J. Lippard, *Progr. Inorg. Chem.*, **1968**, *8*, 109. See also S. J. Lippard, ibid., **1975**, *21*, 91; D. L. Kepert, ibid., **1978**, *24*, 179; J. K. Burdett et al., *Inorg. Chem.*, **1978**, *17*, 2553.

[96] See B. E. Robertson, *Inorg. Chem.*, **1977**, *16*, 2735; M. C. Favas and D. L. Kepert, *Progr. Inorg. Chem.*, **1981**, *28*, 309.

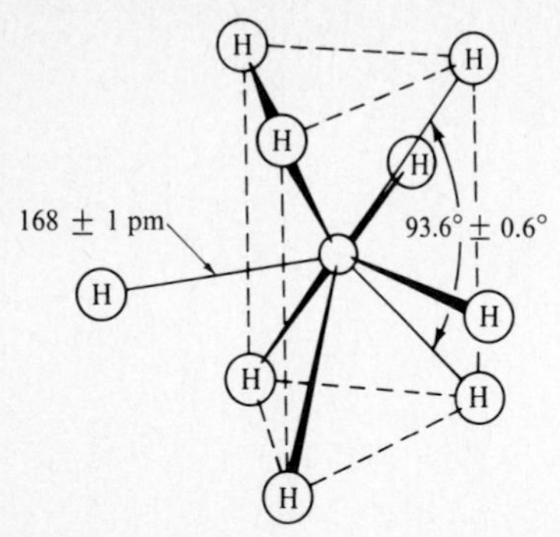

Fig. 10.57 Molecular structure of the $[ReH_9]^{-2}$ anion. [From S. C. Abrahams et al., *Inorg. Chem.*, **1964**, *3*, 558. Reproduced with permission.]

early member of the lanthanide series) and for the hydride complexes $[MH_9]^{-2}$ (where M = Tc or Re). These structures are formed by adding a ligand to each of the rectangular faces of a trigonal prism (Fig. 10.57). Few 10-coordinate structures have been described, but structures consisting of bicapped square antiprisms (Fig. 10.58) and "double" trigonal bipyramids (Fig. 10.59) are known. Nine- and 10-coordinate species have been suggested as intermediates in reactions of 8-coordinate species. Coordination numbers as high as 12 are also known (Fig. 10.60).

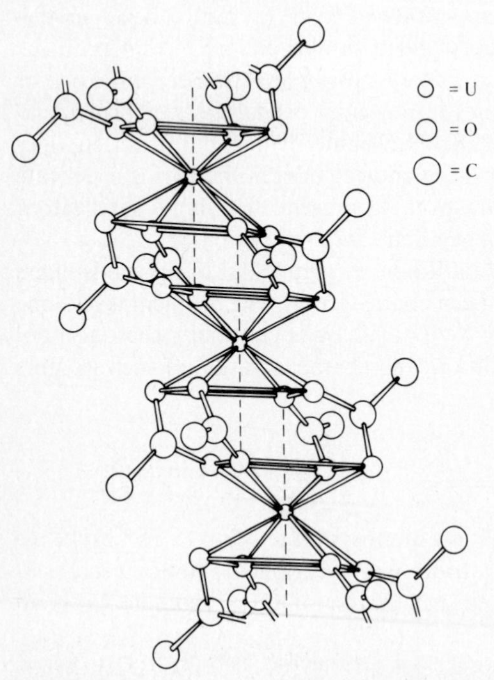

Fig. 10.58 Structure of uranium(IV)acetate polymer revealing effective 10-coordination for uranium. The ninth and tenth "long bonds" are shown by dashed lines. [From I. Jelenic et al., *Acta Crystallogr.*, **1964**, *17*, 758. Reproduced with permission.]

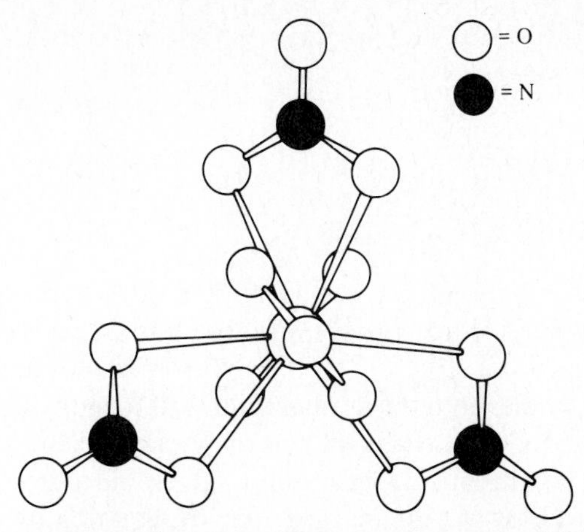

Fig. 10.59 Structure of the pentanitratocerate(III) anion. Each nitrate ion may be thought of as occupying an apex of a trigonal bipyramid. The resulting coordination number is 10 since each nitrate ion is bidentate. The view is down the principal axis of the trigonal bipyramid and the axial nitrogen atoms as well as the central cerium atom are partially obscured. [From A. R. Al-Karaghouli and J. S. Wood, *Chem. Commun.*, **1970**, 135. Reproduced with permission.]

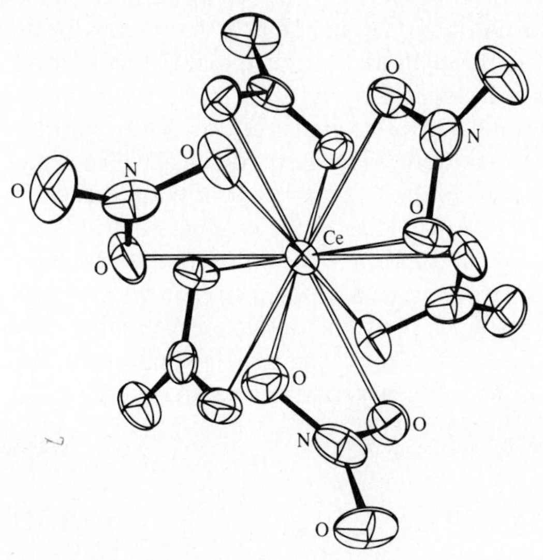

Fig. 10.60 Structure of the hexanitratocerate(III) anion. Each nitrate ligand is bidentate to give a coordination number of 12. [From T. A. Beineke and J. Delgaudio, *Inorg. Chem.*, **1968**, *1*, 715. Reproduced with permission.]

LINKAGE ISOMERISM

In addition to the geometric and optical isomerism discussed previously, there is another type of isomerism that is becoming increasingly important in inorganic chemistry. It deals with ligands that are capable of bonding through one type of donor atom in one situation but a different atom in another complex. The first example of this type of isomerism was provided by Jørgensen, Werner's contemporary. His method of preparation was as follows:

$$[Co(NH_3)_5Cl]Cl_2 \xrightarrow{NH_3} \xrightarrow{HCl} \xrightarrow{NaNO_2} \text{“solution A”} \quad (10.13)$$

$$\text{“Solution A”} \xrightarrow{\text{let stand in cold}} [(NH_3)_5CoONO]Cl_2 \quad \text{red} \quad (10.14)$$

$$\text{“Solution A”} \xrightarrow{\text{heat}} \xrightarrow[\text{conc. HCl}]{\text{cool}} [(NH_3)_5CoNO_2]Cl_2 \quad \text{yellow} \quad (10.15)$$

Jørgensen and Werner agreed that the difference between the two isomers resides in the linkage of the NO_2 group to the cobalt. The N-bonded (or "nitro") structure was assigned to the yellow isomer and the O-bonded (or "nitrito") structure to the red isomer on the basis of the color of similar compounds. For example, both the hexaammine and tris(ethylenediamine) complexes of cobalt (assuredly N-bonded) are yellow, and the aquapentaammine and nitratopentaammine complexes, containing one oxygen atom and five nitrogen atoms in the coordination sphere, are red. Thus long before the electronic explanation of spectra was evolved, the correct assignment of structure was made on the basis of color.

In the following years, these compounds were the subject of considerable controversy. A brief history of the disputing claims is given here both because it indicates some of the methods applicable in such studies, and also because it indicates the errors that can be perpetuated if reports in the literature are accepted uncritically. The red form is less stable than the yellow and is slowly converted to the latter on standing or more rapidly by heating or addition of hydrochloric acid to a solution. Piutti[97] claimed that the absorption spectra of the two forms were identical. This was disputed by Shibata,[98] who claimed that they had quite different spectra! Lecompte and Duval[99] compared the X-ray powder patterns[100] of the two forms and found that they were "rigorously identical." They suggested that the red color in the supposed nitrito complex was a result of the presence of some unreacted starting material, namely $[Co(NH_3)_5Cl]Cl_2$, in the product.

Adell[101] measured the rate of conversion of the red form to the yellow photometrically and found it to be a first-order reaction. This is to be expected if the conversion is an intramolecular rearrangement involving no other species (with the possible exception of the solvent). On the other hand, if the red isomer is actually unreacted starting material in the form of $[Co(NH_3)_5Cl]ClNO_2$, the reaction might be expected to be second order:[102]

$$Co(NH_3)_5Cl^{+2} + NO_2^- \longrightarrow (NH_3)_5CoNO_2^{+2} + Cl^- \quad (10.16)$$

$$\frac{-d[Co(NH_3)_5Cl]}{dt} = k[Co(NH_3)_5Cl^{+2}][NO_2^-] \quad (10.17)$$[103]

[97] A. Piutti, *Ber. Deut. Chem. Ges.*, **1912**, *45*, 1832.

[98] Y. Shibata, *J. Coll. Sci. Imp. Univ. Tokyo*, **1915**, *37*, 15.

[99] J. Lecompte and C. Duval, *Bull. Soc. Chim.*, **1945**, *12*, 678.

[100] Powder patterns are determined by the type of crystal lattice and by the spacings in the lattice. They are useful as "fingerprinting" devices for the identification of crystals. In the case of simple cubic systems they are sufficient to solve the structure of the compound, but less symmetrical lattices require more elaborate analysis.

[101] B. Adell, *Z. Anorg. Chem.*, **1944**, *252*, 272.

[102] First-order kinetics is to be expected for an intramolecular reaction, but its presence is not *proof* that the reaction takes place by such a mechanism.

[103] The brackets in this equation represent the concentrations of the various species in mol dm^{-3} rather than indications of structural moieties.

Murmann and Taube[104] have shown that the formation of the *nitrito* complex occurs *without* the rupture of the Co—O bond. They used ^{18}O-labeled $[Co(NH_3)_5OH]^{+2}$ as a starting material and found that all of the ^{18}O remained in the complex. This argues in favor of reaction 10.18 in preference to 10.19:

$$Co(NH_3)_5{}^{18}OH^{+2} + N_2O_3 \longrightarrow (NH_3)_5Co^{18}O\cdots H^{+2} \quad (O{-}N\cdots ONO) \longrightarrow (NH_3)_5Co^{18}ONO^{+2} \qquad \textbf{(10.18)}^{105}$$

$$Co(NH_3)_5{}^{18}OH^{+2} + NO_2^- \longrightarrow (NH_3)_5CoNO_2^{+2} + {}^{18}OH^- \qquad \textbf{(10.19)}^{105}$$

The labeled nitrite complex may be caused to rearrange by heating. In this process no loss of ^{18}O is found even in the presence of excess nitrite, confirming Adell's hypothesis that the reaction is an intramolecular rearrangement:

$$[(NH_3)_5Co^{18}ONO]^{+2} \longrightarrow \left[(NH_3)_5Co\begin{matrix}\cdot\cdot{}^{18}O \\ | \\ \cdot\cdot N{-}O\end{matrix}\right]^{+2} \longrightarrow (NH_3)_5CoNO^{18}O^{+2} \qquad \textbf{(10.20)}$$

Finally, the ^{18}O can be quantitatively removed by the basic hydrolysis of the nitro isomer:

$$[(NH_3)_5CoNOO^{18}]^{+2} + OH^- \longrightarrow [(NH_3)_5CoOH^{+2}] + {}^{18}ONO^- \qquad \textbf{(10.21)}$$

All these experiments are consistent with the original hypothesis of Jørgensen and Werner of linkage isomerism. It is difficult to rationalize the "rigorous" contrary evidence of some of the early workers except by the general phenomenon that it is deceptively easy to obtain the experimental results that one expects and desires.

Werner knew of two other examples of linkage isomerism, both nitro–nitrito isomers, and they underwent the same period of skepticism and confirmation as the compounds discussed above although considerably less work was done with them. A period of more than 50 years passed before Basolo and co-workers attacked the problem with rather amazing results. Linkage isomerism, once relegated to a few lines as an "exceptional" situation in discussions of isomerism, now boasts an extensive chemistry which continues to develop.[106] The first new linkage isomers prepared were nitro–nitrito isomers of Cr(III), Rh(III), Ir(III), and Pt(IV).[107] In all cases except Cr(III), the nitrito isomer converts readily to the more stable nitro isomer.

---

[104] R. K. Murmann and H. Taube, *J. Am. Chem. Soc.*, **1956**, *78*, 4886.

[105] The differences in these reactions of N_2O_3 versus NO_2^-, the presence or absence of OH^- as a product, etc., are more apparent than real since these species will interact with each other to form an equilibrium mixture. The general argument does not depend upon the exact nature of the reactants and products.

[106] J. L. Burmeister, *Coord. Chem. Rev.*, **1968**, *3*, 225. See also footnote 125.

[107] F. Basolo and G. S. Hammaker, *J. Am. Chem. Soc.*, **1964**, *82*, 1001; *Inorg. Chem.*, **1962**, *1*, 1.

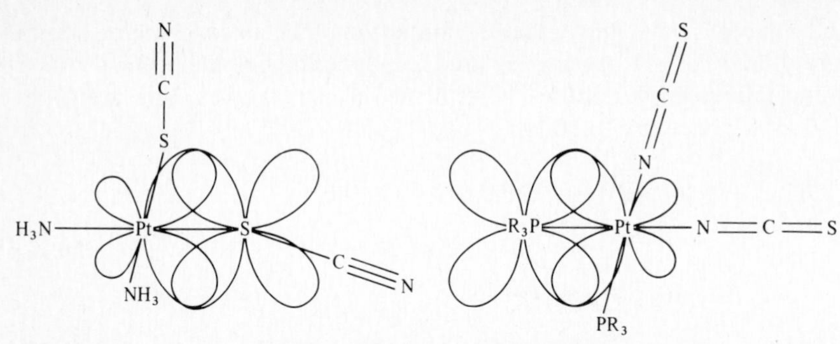

Fig. 10.61 Structures of $[Pt(SCN)_2(NH_3)_2]$ and $[Pt(NCS)_2(PR_3)_2]$ illustrating the competition for π-bonding *d* orbitals of the metal. One set of π bonds has been omitted for clarity. The "*d* orbital" left-right symmetry has been lost due to polarization Cf. Fig. 9.51.

The first thiocyanate linkage isomers were isolated after it was noted[108] that the structures of *cis* complexes containing thiocyanate and either ammonia or phosphine were S- or N-linked, respectively (Fig. 10.61). The hypothesis provided was that these isomers were more stable than the alternatives (i.e., S-bonded in the phosphine complex, N-bonded in the ammine complex) because of the competition for π-bonding orbitals on the metal. The phosphine forms the best π bonds and hence tends to monopolize the π-bonding *d* orbitals of the platinum, reducing the stability of the weaker sulfur π bond, hence the thiocyanate ion bonds through the nitrogen atom. In the absence of competition for π orbitals (ammonia cannot form a π bond), the sulfur atom is preferentially bonded. Using this hypothesis as a basis, Basolo and co-workers[109] attempted to find complexes in which the π-bonding tendencies were balanced allowing the isolation of both isomers. Examples of the complexes thus isolated are $[(\phi_3As)_2Pd(SCN)_2]$, $[(\phi_3As)_2Pd(NCS)_2]$, and $[(dipy)Pd(SCN)_2]$, $[(dipy)Pd(NCS)_2]$. In both cases, on warming, the S-bonded isomer is converted to the N-bonded isomer, which is presumably slightly more stable.

The competition for π bonding is indicated in the behavior of the selenocyanate group, $SeCN^-$. This group readily bonds to the heavier group VIIIB metals via the selenium atom to form complexes such as $[Pd(SeCN)_4]^{-2}$ and *trans*-$[Rh(P\phi_3)_2(SeCN)_2]$. However, in a closely related complex, *trans*-$[Rh(P\phi_3)_2(CO)(NCSe)]$, the presence of a *trans* carbonyl group apparently favors coordination via the non-π-bonding nitrogen atom.[110]

Another example of apparent electronic (i.e., π bonding) control of linkage isomerism comes from bidentate chelates having one strong and one weak donor atom (Fig. 10.62). The presence of an S-bonded thiocyanato group *trans* to the non-π-bonding nitrogen atom, but an N-bonded isothiocyanato group *trans* to the π-bonding phosphine donor is indicative of π competition in this complex.[111]

---

[108] A. Turco and C. Pecile, *Nature*, **1961**, *191*, 66.

[109] F. Basolo et al., *J. Am. Chem. Soc.*, **1963**, *85*, 1700; J. L. Burmeister and F. Basolo, *Inorg. Chem.*, **1964**, *3*, 1587.

[110] J. L. Burmeister and N. J. DeStefano, *Chem. Commun.*, **1970**, 1698.

[111] D. W. Meek et al., *J. Am. Chem. Soc.*, **1970**, *92*, 5351; G. R. Clark and G. J. Palenik, *Inorg. Chem.*, **1970**, *9*, 2754.

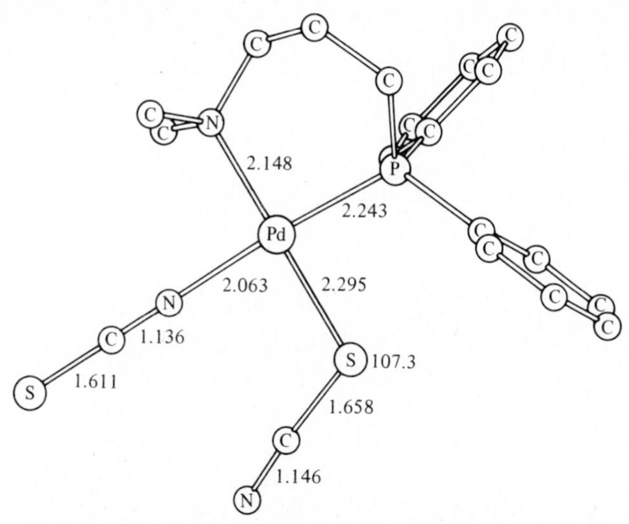

Fig. 10.62 The molecular structure of isothiocyanatothiocyanato(1-diphenylphosphino-3-dimethylaminopropane)palladium(II). *Note:* (1) *trans* arrangement of P—Pd—N and N—Pd—S bonds: (2) linear vs. bent arrangement of the NCS group. [From D. W. Meek et al., *J. Am. Chem. Soc.*, **1970**, *92*, 5351. Reproduced with permission.]

In the nitro–nitrito,[112] thiocyanato–isothiocyanato,[113] and similar selenocyanato–isoselenocyanato[114] cases, steric factors may play an important role in determining the relative stability of the variously linked isomers. For example, the nitro group is sterically more hindered than the nitrito group (Fig. 10.63a). Likewise, the N-bonded isomer of the latter two is less hindered than the S- or Se-bonded isomer (Fig. 10.63b). In the ammonia versus phosphine problem of Turco and Pecile, the steric requirements of the larger substituted phosphines can augment the electronic effects of π bonding.

Burmeister and co-workers[115,116] have investigated the steric problems involved in linkage isomerism. For example, if three selenocyanate groups are displaced from $[Pd(SeCN)_4]^{-2}$ by 1,1,7,7-tetraethyldiethylenetriamine (Et_4dien), and the product is isolated at low temperatures, the Se-bonded isomer is obtained. When this product is dissolved in a polar solvent it slowly isomerizes to the N-bonded isomer:

$$[Pd(HN(C_2H_4N(C_2H_5)_2)_2)(SeCN)] \longrightarrow [Pd(HN(C_2H_4N(C_2H_5)_2)_2)(NCSe)] \qquad (10.22)$$

[112] D. M. L. Goodgame and M. A. Hitchman, *Inorg. Chem.*, **1964**, *3*, 1389.

[113] F. Basolo et al., *Inorg. Chem.*, **1964**, *3*, 1202.

[114] J. L. Burmeister and H. J. Gysling, *Chem. Commun.*, **1967**, 543.

[115] J. L. Burmeister et al., *J. Am. Chem. Soc.*, **1969**, *91*, 44.

[116] J. L. Burmeister and N. J. DeStefano, *Inorg. Chem.*, **1969**, *8*, 1546; J. L. Burmeister and J. C. Lim, *Chem. Commun.*, **1969**, 1154.

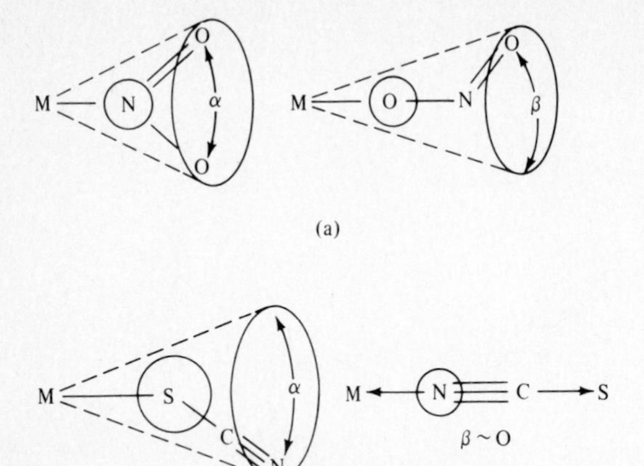

Fig. 10.63 Steric requirements of ambidentate ligands. Note that the angles marked α are larger than β. In addition, the van der Waals radii of S and Se are larger than that of N. (In computing the angles α and β quantitatively, the van der Waals radii of the terminal O and N would have to be included.)

The isomerization process can be followed by observing changes in the visible–ultraviolet spectrum (Fig. 10.64). The presence of two well-defined isosbestic points (see p. 303) indicates that the reaction proceeds without side reactions (such as hydrolysis, etc.). The correlation between the rate of isomerization and the dielectric constant of the solvent is indicative of a dissociative process, namely ionization of the $SeCN^-$ ion and reassociation via an N-linkage with a reduction in steric repulsions from the ethyl groups. In the complex formed with the related ligand *without* ethyl groups (dien), the Se-bonded isomer is stable and no isomerization takes place.[115]

Although steric factors are usually assumed to favor the linear isothiocyanato group, it has been suggested that the presence of the phenyl rings in the compound illustrated in Fig. 10.62 favors the S-linkage in the adjacent thiocyanato group since it can bend away.[117]

Further evidence of the importance of steric effects comes from isomerizations in the solid state in which the presence of a particular counterion (large versus small) may determine which linkage isomer is the most stable.[115,118] Clearly, the factors involved in this type of linkage isomerism are only partially understood and further work is necessary to clarify the situation.

One or more factors may be operating simultaneously to provide a delicate balance of counterpoising effects. Palenik and co-workers[119] have prepared an interesting series of compounds (Fig. 10.65a–c) that illustrates the steric effects in linkage isomers of square planar palladium(II) complexes. The six-membered chelate ring in Fig. 10.65c allows an

[117] See D. W. Meek et al., *J. Am. Chem. Soc.*, **1970**, *92*, 5351, for a complete analysis of electronic and steric factors in compounds of this type.

[118] J. L. Burmeister and J. C. Lim, *Chem. Commun.*, **1968**, 1346.

[119] G. J. Palenik et al., *Inorg. Nucl. Chem. Lett.*, **1974**, *10*, 125.

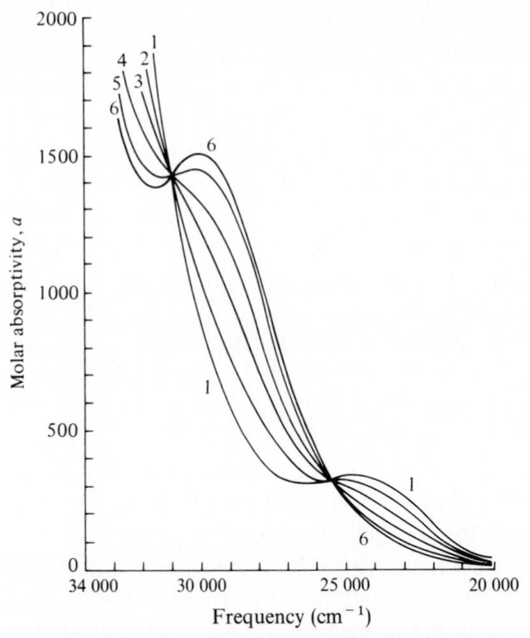

Fig. 10.64 Changes in the absorption spectrum of a solution of [Pd(Et_4dien)SeCN][$B\phi_4$] at 25°: (1) after 2.5 min; (2) 32 min; (3) 1.2 h; (4) 2 h; (5) 3.7 h; (6) 11.5 h (isomerization complete). Note presence of two isosbestic points. [From J. L. Burmeister et al., *J. Am. Chem. Soc.*, **1969**, *91*, 44. Reproduced with permission.]

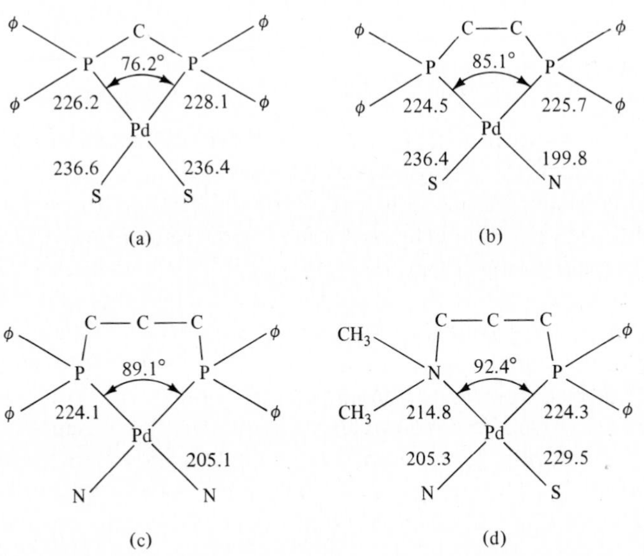

Fig. 10.65 Structure of four palladium complexes illustrating combined steric and electronic effects on bonding of the thiocyanate ligand. [From G. J. Palenik et al., *Inorg. Nucl. Chem. Letters*, **1974**, *10*, 125 Reproduced with permission.] Distances in pm.

essentially unstrained angle of 89.1° at the palladium atom. The aryl-alkyl-substituted phosphines are only weakly π-bonding, but the expected N-bonded isomer obtains. As the chelate ring is contracted to five atoms (Fig. 10.65b) and then to four atoms (Fig. 10.65a) the electronic environment on the phosphorus is essentially constant, but the steric constraints are relaxed as shown by the decreasing P—Pd—P bond angle. First one (Fig. 10.65b), then both (Fig. 10.65a) thiocyanate groups rearrange as the large sulfur atom is allowed more room around the palladium atom. However, the same effect can be accomplished by holding the geometry essentially constant (Fig. 10.65d), if one of the phosphorus atoms is replaced by a smaller, non-π-bonding nitrogen atom. Rearrangement of one of the thiocyanate ligands occurs. Significantly, it is the one *trans* to the nitrogen atom. Furthermore, the ubiquitous presence of the *trans influence*[120] in these complexes indicates the pervasive consequences of electronic effects. We may therefore conclude that if either the electronic *or* the steric factor in a series of complexes is held constant, it is possible for the *other* factor to determine the nature of the resulting linkage isomer.[121]

Jørgensen[122] has proposed the principle of *symbiosis* with respect to hard and soft acid–base behavior. This rule of thumb states that hard species will tend to make the atom to which they are bound harder and increase the tendency to attract *more* hard species. Conversely, the presence of some soft ligands enhances the ability of the central atom to accept other soft ligands. In terms of the electrostatic versus covalent picture of Pearson's "hard and soft" or Drago's E_AE_B and C_AC_B parameters, the best "strategy" of a complex is to "put all its eggs in one basket," i.e., form all hard ("electrostatic") or all soft ("covalent") bonds to ligands. There are many examples that could be given to illustrate this tendency in metal complexes:

All ligands "hard"	All ligands "soft"
$[Co(NH_3)_5NCS]^{+2}$	$[Co(CN)_5SCN]^{-3}$
$[Rh(NH_3)_5NCS]^{+2}$	$[Rh(SCN)_6]^{-3}$
$[Fe(NCSe)_4]^{-2}$	$[cpFe(CO)_2(SeCN)]$

In the first example the hard ammonia ligands tend to "harden" the cobalt and so the thiocyanate bonds preferentially through the nitrogen atom. Conversely, the soft cyanide ligands "soften" the cobalt, making it bond to the "soft" end of the thiocyanate (the sulfur atom). Similarly, in the case of Rh(III) five ammonia ligands result in preference for nitrogen at the sixth position; if all six soft sulfur atoms can ligate, they will. Iron(II) appears to prefer the hard nitrogen atom unless softened by the presence of carbonyl groups.

The symbiotic theory adequately covers most of the linkage isomerism and linkage preferences for octahedral complexes. Unfortunately, it contradicts *exactly* the π-bonding theory applied above to square planar complexes. Π bonding can be equated with softness, σ bonding with hardness. In the case of octahedral complexes we say that the presence of soft, π-bonding ligands favors the addition of more soft, π-bonding ligands (symbiotic theory), but in the case of square planar complexes we say that soft, π-bonding ligands discourage the presence of other π bonders and favor the addition of hard, σ-only ligands

[120] See pp. 542–545 for a discussion of the electronic factors operative in the *trans* influence.
[121] J. E. Huheey and S. O. Grim, *Inorg. Nucl. Chem. Lett.*, **1974**, *10*, 973.
[122] C. K. Jørgensen *Inorg. Chem.*, **1964**, *3*, 1201.

(π-competition theory). Obviously the situation is somewhat less than perfect if the two theories are applicable only in limited areas and appear to contradict each other in their basic *raisons d'être*. Nevertheless they have heuristic value and serve to emphasize that there are many factors involved, both electronic and steric, in determining which of the possible isomers will be preferred. At present the number of cases known is insufficient to sort out all of the factors but it may be hoped that as more data accumulate a more comprehensive theory will emerge.[123]

Pearson has elaborated upon the above ideas, distilling the essence of the π-competition theory to: Two soft ligands in mutual *trans* positions will have a destabilizing influence on each other when attached to class b (soft) metal atoms. He also provided additional examples of the rule that symbiosis prevails in octahedral complexes, antisymbiosis in square planar complexes. He predicted that tetrahedral complexes would show antisymbiosis but on a much reduced scale compared with the square planar complexes.[124]

Linkage isomerism is but a special case of the phenomenon of *ambidentate*[125] behavior of ligands. In addition to those ligands such as SCN^-, $SeCN^-$, and CN^- for which linkage isomers have been isolated, there are closely related ligands such as CO and NCO^- which can bond through either of two ligating atoms, although no examples have been produced in which the two types of molecules are isomers of each other. For example, CO acts as a bridging molecule[126] in $[cp(OC)FeCOAlEt_3]_2$ and $[cpFeCOAlEt_3]_4$ in which the Lewis acid triethylalane has accepted a pair of electrons from the cyclopentadienylironcarbonyl group (Fig. 10.66).[127] The cyanate ion normally bonds through the nitrogen atom as in $[cp_2TiNCO]$, but the closely related complex $[cp_2Ti(OCN)_2]$ is thought to involve O-coordination.[128] Other examples of ambidentate ligands are dimethylsulfoxide, urea, thiourea, the sulfite ion, and the cyanide ion. The latter provides a good example of ambidentate behavior.[129] In discrete complexes it almost always bonds through the carbon atom although it has been reported to form a few linkage isomers such as *cis*-$[Co(trien)(CN)_2]^+$ and *cis*-$[Co(trien)(NC)_2]^+$. In addition a large number of polymeric complexes is known which contain ambidentate cyanide bridging groups and which may be related to the "Prussian blue" structure. The latter is formed by the addition of ferric salts to ferrocyanides:

$$K^+ + Fe^{+3} + [Fe(CN)_6]^{-4} \longrightarrow [KFe(CN)_6Fe]_x\downarrow \qquad \textbf{(10.23)}$$

[123] See N. J. DeStefano and J. L. Burmeister, *Inorg. Chem.*, **1971**, *10*, 998, for further discussion of this problem. It should also be noted that although the *phenomenon* of symbiosis is very real, the choice of the word symbiosis to describe it is unfortunate. As DeStefano and Burmeister point out, *symbiosis* in biology refers to the "flocking together" of *different species*, rather than the same species, in intimate association. Nevertheless inorganic chemists will continue to use the term in its current sense.

[124] R. G. Pearson, *Inorg. Chem.*, **1973**, *12*, 712.

[125] A. H. Norbury and A. I. P. Sinha, *Quart. Rev. Chem. Soc.*, **1970**, *24*, 69; A. H. Norbury, *Adv. Inorg. Chem. Radiochem.*, **1975**, *17*, 232; J. L. Burmeister, "The Chemistry and Biochemistry of Thiocyanic Acid and Its Derivatives," A. A. Newman, ed., Academic Press, New York, 1975.

[126] Note the difference between the bridging carbonyl here consisting of $-C\equiv O\rightarrow$ and the bridging carbonyl groups found in polynuclear carbonyls, $>C=O$, see Chapter 13.

[127] N. J. Nelson et al., *J. Am. Chem. Soc.*, **1969**, *91*, 5173.

[128] J. L. Burmeister et al., *Inorg. Chem.*, **1969**, *8*, 170.

[129] D. F. Shriver, *Struct. Bonding*, **1966**, *1*, 32.

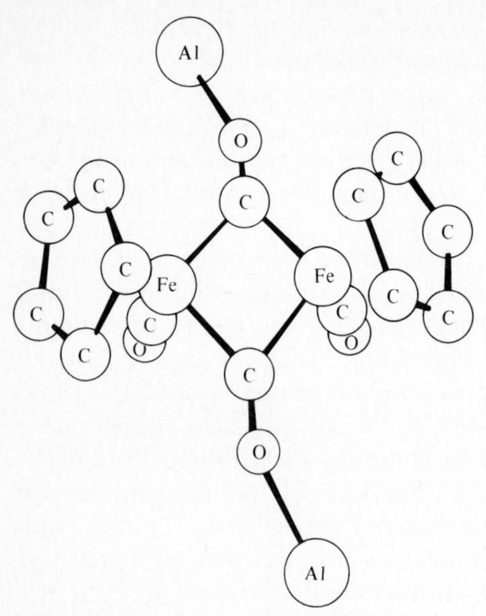

Fig. 10.66 Structure of $[cp(OC)FeCOAlEt_3]_2$. The ethyl groups and hydrogen atoms have been omitted for clarity. [From N. J. Nelson et al., *J. Am. Chem. Soc.*, **1969**, *91*, 5173. Reproduced with permission.]

Addition of ferrous salts to ferricyanides produces "Turnbull's blue":

$$K^+ + Fe^{+2} + [Fe(CN)_6]^{-3} \longrightarrow [KFe(CN)_6Fe]_x\downarrow \qquad \textbf{(10.24)}$$

It has been shown that the iron–cyanide framework is the same in Prussian blue, Turnbull's blue, and other related polymeric cyanide complexes (Fig. 10.67), differing only in the number of potassium ions (for example) necessary to maintain electrical neutrality. These cations may thus be present or absent from the large cubic holes. Various quantities of water molecules may also be present in these sites. Prussian blue, itself, has an empirical formula $Fe_4[Fe(CN)_6]_3$ corresponding to a structure having hexacoordinate, low-spin Fe(II) bonded through the carbon atoms and hexacoordinate, high-spin Fe(III) bonded through the nitrogen atoms of the cyanide. To achieve this stoichiometry, one fourth of the Fe(II) sites are occupied by water molecules. This reduces the number of bridging cyanide groups (Fe(II)—C≡N—Fe(III)) somewhat, and water molecules occupy the otherwise empty ligand positions thus created. There is also one water molecule in each cubic site.[130] Turnbull's blue is identical.[131] Although X-ray and magnetic data support this identity, the best evidence comes from the fact that the Mössbauer spectra of Prussian blue and Turnbull's blue are identical.[132] Since Mössbauer spectra are extremely sensitive to the

[130] H. J. Buser et al., *Inorg. Chem.*, **1977**, *16*, 2704.

[131] The method of preparation and precipitation may affect the nature of the precipitate (colloidal, etc.) and so varying amounts of water and cations may be incorporated in the structure.

[132] J. F. Duncan and P. W. R. Wigly, *J. Chem. Soc.*, **1963**, 1120; E. Fluck et al., *Angew. Chem. Int. Ed. Engl.*, **1963**, *2*, 277; A. K. Bonnette, Jr., and J. F. Allen, *Inorg. Chem.*, **1971**, *10*, 1613.

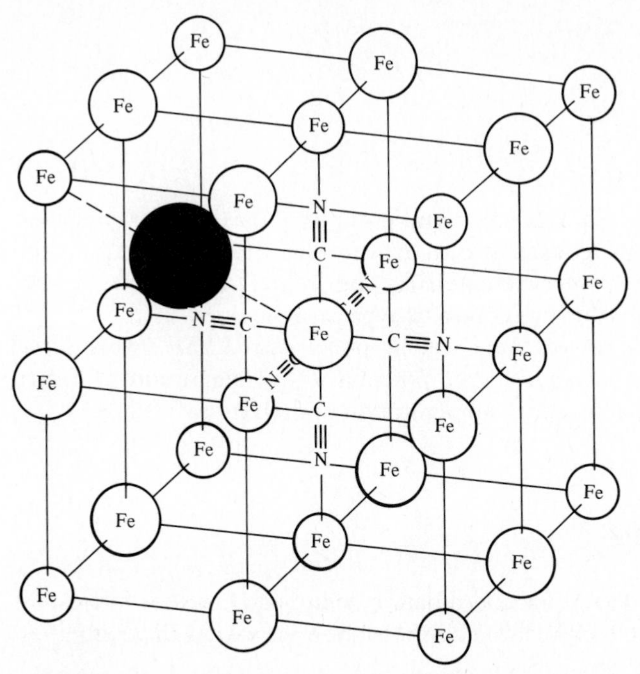

Fig. 10.67 Portion of the crystal structure of Prussian blue showing the bridging by ambidentate cyanide ions. Circles represent iron(II) (◯), iron(III) (○), and oxygen in water (●). The remaining interstitial or "zeolitic" water in the cubic sites has been omitted for clarity, as have most of the cyanide ions. In addition, some of the cyanide ions are replaced by water molecules coordinated to iron(III), and there are also vacancies in the structure. [Modified from H. J. Buser et al., *Inorg. Chem.*, **1977**, *11*, 2704. Reproduced with permission.]

electron density and microenvironment about the iron atoms (see pp. 165, 242) this confirms the identity. Ferric ferricyanide (Berlin green) consists of Fe^{III} at all iron sites, and the white "ferrous ferrocyanide" (actually $[K_2Fe(CN)_6Fe]$) consists of Fe^{II} at all iron sites and potassium ions in all of the interstices.

If one pyrolyzes Prussian blue gently in vacuum[133] or, better, precipitates ferrous ferricyanide in the presence of a reducing agent such as potassium iodide or sucrose,[134] the compound formed is truly iron(II) hexacyanoferrate(III). However, it reverts to Prussian blue upon warming with dilute hydrochloric acid.

A particularly interesting example of linkage isomerism has been reported by Shriver and co-workers.[135] Mixing solutions of iron(II) salts and potassium hexacyanochromate(III) results in a brick-red precipitate which turns dark green on heating:

$$Fe^{+2} + K^{+} + [Cr(CN)_6]^{-3} \longrightarrow KFe[Cr(CN)_6] \quad \text{brick red} \qquad \textbf{(10.25)}$$

$$KFe[Cr(CN)_6] \xrightarrow{100^{\circ}C} KCr[Fe(CN)_6] \quad \text{dark green} \qquad \textbf{(10.26)}$$

---

[133] J. G. Cosgrove et al., *J. Am. Chem. Soc.*, **1973**, *95*, 1083.

[134] R. Robinette and R. L. Collins, *J. Coord. Chem.*, **1974**, *3*, 333.

[135] D. F. Shriver et al., *Inorg. Chem.*, **1965**, *4*, 725.

This has been interpreted in terms of linkage isomers of the type:

$$\underset{\text{brick red}}{\text{—Fe—NC—Cr—CN—Fe—NC—Cr—}} \longrightarrow \underset{\text{dark green}}{\text{—Fe—CN—Cr—NC—Fe—CN—Cr—}} \tag{10.27}$$

in which the linear arrays shown in Eq. 10.27 represent portions of the cubic arrays shown in Fig. 10.67. The initial product is C-coordinated to the chromium(III) since that was the arrangement in the original hexacyanochromate(III). The iron(II) coordinates to the available nitrogen atoms to form the Prussian-blue-type structure. As in the case of Prussian blue discussed above, however, there will be preferential LFSE favoring the coordination of the strong field C-linkage to the potential t_{2g}^6 configuration of Fe(II), approximately twice as great as that of the t_{2g}^3 configuration of Cr(III) (see Table 9.3).

OTHER TYPES OF ISOMERISM

In general the other types of isomerism for coordination compounds are less interesting than those discussed previously, but will be listed briefly to show the variety of possibilities.

Ligand isomerism

Since many ligands are organic compounds and the latter have a large number of possibilities for isomerism, the resulting complexes can show isomerism from this source. Examples of isomeric ligands are 1,2-diaminopropane ("propylenediamine," pn) and 1,3-diaminopropane ("trimethylenediamine," tn) or *ortho*, *meta*, and *para* toluidine ($CH_3C_6H_4NH_2$).

Ionization isomerism

The compounds $[Co(NH_3)_5Br]^{+2}SO_4^{-2}$ and $[Co(NH_3)_5SO_4]^+Br^-$ dissolve in water to yield different ions and thus react differently to various reagents:

$$[Co(NH_3)_5Br]SO_4 \xrightarrow{Ba^{+2}} BaSO_4\downarrow \tag{10.28}$$

$$[Co(NH_3)_5SO_4]Br \xrightarrow{Ba^{+2}} \text{No reaction} \tag{10.29}$$

$$[Co(NH_3)_5SO_4]Br \xrightarrow{Ag^+} AgBr\downarrow \tag{10.30}$$

$$[Co(NH_3)_5Br]SO_4 \xrightarrow{Ag^+} \text{No reaction} \tag{10.31}$$

Solvate isomerism

This is a somewhat special case of the above interchange of ligands involving neutral solvate molecules. The best known example involves isomers of "chromic chloride hydrates," of which three are known: $[Cr(H_2O)_6]Cl_3$, $[Cr(H_2O)_5Cl]Cl_2 \cdot H_2O$, and

$[Cr(H_2O)_4Cl_2]Cl \cdot 2H_2O$. These differ in their reactions:

$$[Cr(H_2O)_6]Cl_3 \xrightarrow{\text{dehydr. over } H_2SO_4} [Cr(H_2O)_6]Cl_3 \quad \text{(no change)} \tag{10.32}$$

$$[Cr(H_2O)_5Cl]Cl_2 \cdot H_2O \xrightarrow{\text{dehydr. over } H_2SO_4} [Cr(H_2O)_5Cl]Cl_2 \tag{10.33}$$

$$[Cr(H_2O)_4Cl_2]Cl \cdot 2H_2O \xrightarrow{\text{dehydr. over } H_2SO_4} Cr[(H_2O)_4Cl_2]Cl \tag{10.34}$$

$$[Cr(H_2O)_6]Cl_3 \xrightarrow{Ag^+} [Cr(H_2O)_6]^{+3} + 3\,AgCl \tag{10.35}$$

$$[Cr(H_2O)_5Cl]Cl_2 \cdot H_2O \xrightarrow{Ag^+} [Cr(H_2O)_5Cl]^{+2} + 2\,AgCl \tag{10.36}$$

$$[Cr(H_2O)_4Cl_2]Cl \cdot 2H_2O \xrightarrow{Ag^+} [Cr(H_2O)_4Cl_2]^+ + AgCl \tag{10.37}$$

Coordination isomerism

Salts that contain complex cations and anions may exhibit isomerism through the interchange of ligands from cation and anion. For example, both hexaamminecobalt(III) hexacyanochromate(III), $[Co(NH_3)_6][Cr(CN)_6]$, and its coordination isomer, $[Cr(NH_3)_6][Co(CN)_6]$, are known. Another example is $[Cu(NH_3)_4][PtCl_4]$ and $[Pt(NH_3)_4][CuCl_4]$ in which the isomers differ in color (as a result of the d^9 Cu^{+2} chromophore), being violet and green, respectively. There is a very large number of cases of this type of isomerism.

A special case of coordination isomerism has sometimes been given the name "polymerization isomerism" since the various isomers differ in formula weight from one another. However, the term is unfortunate since polymerization is normally used to refer to the reaction in which a monomeric unit builds a larger structure consisting of repeating units. The isomers in question are represented by compounds such as $[Co(NH_3)_4(NO_2)_2][Co(NH_3)_2(NO_2)_4]$, $[Co(NH_3)_6][Co(NO_2)_6]$, $[Co(NH_3)_5NO_2][Co(NH_3)_2(NO_2)_4]_2$, $[Co(NH_3)_6][Co(NH_3)_2(NO_2)_4]_3$, $[Co(NH_3)_4(NO_2)_2]_3[Co(NO_2)_6]$, and $[Co(NH_3)_5NO_2]_3[Co(NO_2)_6]_2$. These all have the empirical formula $Co(NH_3)_3(NO_2)_3$, but they have formula weights that are 2, 2, 3, 4, 4, and 5 times this, respectively.

STRUCTURAL EQUILIBRIA OF COMPLEXES

We have seen in Chapter 9 that the coordination number of chloro complexes of transition metal ions in molten salts may be either 4 or 6 and that in many cases an equilibrium exists between the two. This is a common phenomenon wherever the various factors that favor one type of geometry over another (LFSE, steric repulsions, π bonding, etc.) are balanced so as to neutralize each other. These equilibria were not amenable to extensive study by chemical means, but with the advent of physical methods (e.g., spectroscopy) of estimating the concentrations of the species present in solution their study has increased.

One type of equilibrium is between species of different coordination numbers as the octahedral–tetrahedral equilibria found in molten salts:

$$[MX_4]^{m-4} + 2X^- \rightleftharpoons [MX_6]^{m-6} \tag{10.38}$$

Five-coordinate species have been studied which form equilibria with 4- or 6-coordinate species:

$$[Co(PR_3)_3(NCS)_2] \rightleftharpoons [Co(PR_3)_2(NCS)_2] + PR_3 \qquad (10.39)$$

$$[OV(acac)_2] + NH_3 \rightleftharpoons [OV(acac)_2NH_3] \qquad (10.40)$$

An extremely important biological compound of this sort is oxyhemoglobin, $Hb \cdot O_2$, a 5-coordinate species in equilibrium with O_2 + Hb, a 4-coordinate species (see Chapter 18).

Within a given coordination number there are interesting isomeric equilibria. For example, Chatt and Wilkins[136] measured *cis–trans* equilibria in square planar platinum(II) complexes such as:

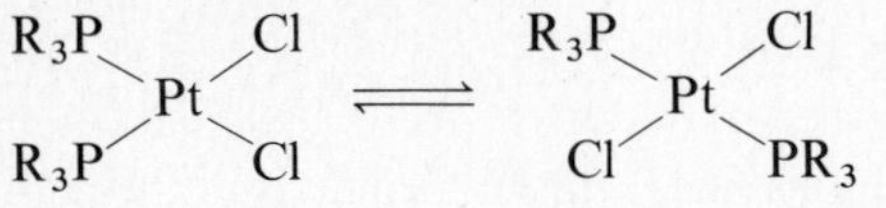

$$\begin{matrix} R_3P & & Cl \\ & Pt & \\ R_3P & & Cl \end{matrix} \rightleftharpoons \begin{matrix} R_3P & & Cl \\ & Pt & \\ Cl & & PR_3 \end{matrix} \qquad (10.41)$$

They were able to follow the equilibrium by measuring the dipole moment of the mixture and relating it to the concentration of the *cis* isomer. Their results are of interest because they found that: (1) the concentration of the *trans* isomer is greater than that of the *cis* isomer; (2) the position of the equilibrium favoring the *trans* isomer results from the strong influence of the *entropy* effect (*ca.* 40 J K^{-1}) arising from the release of solvent molecules accompanying the transformation from the polar *cis* isomer ($\mu = 33 \times 10^{-30}$ C m) to the nonpolar *trans* isomer; (3) the *bond energy* favors the *cis* isomer as might be expected from simple π-bonding arguments; and (4) increasing steric hindrance from either increasing the size of the alkyl groups on the phosphorus or changing the chloride ion to iodide favors the *trans* isomer.

An interesting example of isomerism arises for Ni(II) complexes. These are capable of existing as tetrahedral high-spin species or as square planar low-spin species. If the ligands have the appropriate field strength and steric requirements, it is possible to have a system in which the tetrahedral and square planar species have comparable energies and will exist in equilibrium with one another:

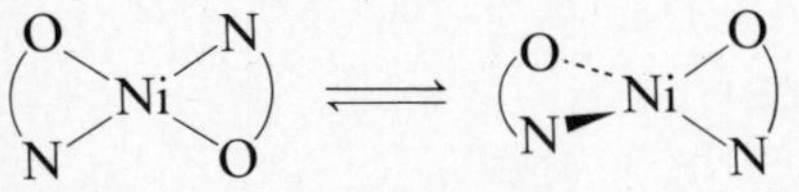

$$\begin{matrix} O & & N \\ & Ni & \\ N & & O \end{matrix} \rightleftharpoons \begin{matrix} O & & O \\ & Ni & \\ N & & N \end{matrix} \qquad (10.42)$$

The position of the equilibrium can be measured by a variety of means: (1) magnetic susceptibility (square planar, $S = 0$; tetrahedral, $S = 1$); (2) optical spectra (square planar and tetrahedral species have characteristic spectra that may be used to estimate their concentrations); and (3) isotropic NMR shifts.[137] These provide convenient methods for following the equilibria in these systems, which may then be interpreted in terms of the steric and electronic effects of the substituents. For example, bulky groups tend to favor the

[136] J. Chatt and R. G. Wilkins, *J. Chem. Soc.*, **1952**, *273*, 4300; **1956**, 525.

[137] R. H. Holm, *Acc. Chem. Res.*, **1969**, *2*, 307.

tetrahedral isomers as a result of reduced steric hindrance in the latter. On the other hand, substituents that are capable of π bonding can be stabilized in the planar configuration. It is possible to balance these effects, one against the other, and obtain tetrahedral and square planar complexes of essentially equal stability. There are several nickel(II) complexes of the type $[NiX_2(PR_3)_2]$ which occur in two crystalline varieties: red and diamagnetic (presumably square planar) and blue or green and paramagnetic (presumably tetrahedral).[138] In at least one case the balance between the two forms is so delicate that *both* forms exist *in a single crystal.*[139] Bis(benzyldiphenylphosphine)dibromonickel(II) crystallizes with one square planar molecule and two tetrahedral molecules per unit cell. The magnetic susceptibility corresponds to one diamagnetic and two paramagnetic nickel ions, corresponding to the structural results.

A similar balance of forces and resulting interconversions and equilibria were discussed previously for 5-coordinate molecules and ions (see pp. 225–227). Fluxional organometallic molecules have been studied extensively. They are discussed in Chapter 13.

THE CHELATE EFFECT

Reference has been made previously to the enhanced stability of complexes containing chelate rings. This extra stability is termed the *chelate effect*. The chief effect is one of entropy common to all chelate systems, but some chelates may have additional stabilizing effects as well. The entropy factor may be viewed from two points of view which are equivalent in that they are both statistical and probabilistic in nature and therefore relate to the entropy of the system, but they look at the problem from somewhat different aspects. One is simply to consider the difference in dissociation between ethylenediamine complexes and ammonia complexes, for example, in terms of the effect of the ethylenediamine ring (the electronic effects of ethylenediamine and ammonia are practically identical). If a molecule of ammonia dissociates from the complex, it is quickly swept off into the solution and the probability of its returning to its former site is remote. On the other hand, if one of the amino groups of ethylenediamine dissociates from a complex it is retained by the other end still attached to the metal. The nitrogen atom can move only a few hundred picometres away and can swing back and attach to the metal again. The complex has a smaller probability of dissociating and is therefore experimentally found to be more stable toward dissociation.

A more sophisticated explanation would be to consider the equilibrium:

$$[Co(NH_3)_6]^{+3} + 3en \rightleftharpoons [Co(en)_3]^{+3} + 6NH_3 \qquad \textbf{(10.43)}$$

in terms of the enthalpy and entropy. Since the bonding of ammonia and ethylenediamine is very similar[140] we expect ΔH for this reaction to be near zero. To a first approximation the change in entropy of this reaction will be proportional to the difference in the number

[138] M. C. Browning et al., *J. Chem. Soc.*, **1962**, 693.

[139] B. T. Kilbourn and H. M. Powell, *J. Chem. Soc., A*, **1970**, 1688.

[140] Ethylenediamine and ammonia are almost identical in the field strengths (their f factors differ by less than 3%, p. 385), and so the LFSE for complexes with the same metal ion will be almost identical.

of particles present in the system. The reaction proceeds to the right with an increase in number of particles and hence the entropy factor favors the production of the chelate system instead of the hexaammine. In the replacement of water molecules by chelates, the increase in number of molecules in solution causes an increase of entropy given by $\Delta S° = xR \ln 55.5 = 33.4x$ J K^{-1} mol^{-1}, where x is the number of chelate rings, contributing 10.0 kJ mol^{-1} to the free energy of the complex at 300 K for each chelate ring formed. One might therefore assume that it is easy to treat the chelate effect quantitatively in terms of enthalpy and entropy, but such is not the case. While the above discussion is basically correct, a careful analysis of all entropy and enthalpy terms reveals that there are enough unexpected variations in the contributing terms that if one actually finds the *ideal* entropies or enthalpies or both, it is "largely coincidental."[141]

Although "the chelate effect" normally refers to the results of the above entropy effects, there are additional, secondary effects that tend to favor the stabilization of complexes by certain bidentate ligands. For example, King[142] has reported interesting differences between "phosphene" chelates of the type $(R_2P)HC{=}CH(PR_2)$ and their saturated diphosphine analogues, $R_2PCH_2{-}CH_2PR_2$. The latter can form chelate rings but instead often act as bridging ligands to form binuclear complexes, $L_5MoPR_2CH_2CH_2PR_2MoL_5$. In contrast, the ethylene analogues form chelates of the type $(OC)_2Mo(R_2PCH{=}CHPR_2)_2$. Two effects are important. First, the phosphene ligands are sterically restricted from effective bridging between two metals, whereas the more flexible saturated diphosphine can effectively bridge. Furthermore, even in the absence of a competing bridging reaction, the phosphene will enjoy a stability over the less rigid ligand since the latter has a greater rotational freedom upon partial dissociation and hence an entropy effect favoring dissociation of the flexible ligand over the rigid phosphene (Fig. 10.68).

Finally, chelating ligands such as acetylacetone enjoy resonance stabilization as a result of the formation of six-membered rings having some aromatic character. Acetylacetone (2,4-pentanedione) acts as a ligand through the acidic enol form

$$\underset{\text{diketone}}{CH_3{-}\overset{\overset{\displaystyle O}{\|}}{C}{-}CH_2{-}\overset{\overset{\displaystyle O}{\|}}{C}{-}CH_3} \rightleftharpoons \underset{\text{enol}}{CH_3{-}\overset{\overset{\displaystyle O}{\|}}{C}{-}CH{=}\overset{\overset{\displaystyle OH}{|}}{C}{-}CH_3} \xrightarrow{-H^+}$$

$$\underset{\text{resonance forms of the enolate anion}}{CH_3{-}\overset{\overset{\displaystyle O}{\|}}{C}{-}CH{=}\overset{\overset{\displaystyle O^-}{|}}{C}{-}CH_3 \longleftrightarrow CH_3{-}\overset{\overset{\displaystyle O^-}{|}}{C}{=}CH{-}\overset{\overset{\displaystyle O}{\|}}{C}{-}CH_3} \tag{10.44}$$

Acetylacetone thus forms neutral complexes with trivalent metals such as $[Ti(acac)_3]$, $[Cr(acac)_3]$, and $[Co(acac)_3]$. In these complexes the rings are symmetrical with the two M—O bonds equal in length, the two C—O bonds equal, and the two C—C bonds

[141] R. T. Myers, *Inorg. Chem.*, **1978**, *17*, 952.

[142] R. B. King and C. A. Eggert, *Inorg. Chim. Acta*, **1968**, *2*, 33; R. B. King and P. N. Kapoor, *Inorg. Chem.*, **1969**, *8*, 1792; R. B. King, *J. Coordin. Chem.*, **1971**, *1*, 67.

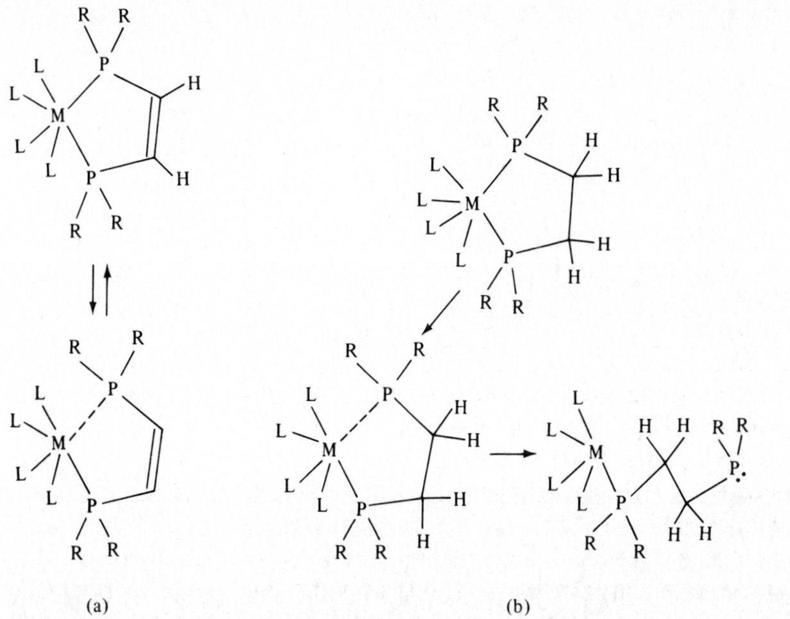

(a) (b)

Fig. 10.68 Differing steric effects resulting from the rigidity of "phosphene" ligands compared with analogous saturated diphosphine ligands: (a) rigid phosphene ligand, complexed and partially dissociated; (b) flexible diphosphine ligand complexed and partially dissociated.

equal as a result of the resonance equivalence (only one ring shown):

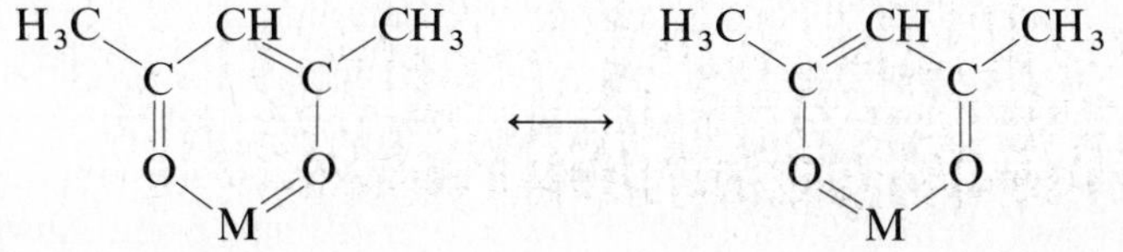

(10.45)

As a result of π bonding, the delocalization of electrons is greater in the chelated ring than in the free enolate and some resonance stabilization takes place.

An interesting example of the destruction of resonance (at least partially) resulting from Jahn-Teller distortion is given by bipyridinebis(hexafluoroacetylacetone)copper(II) (see Fig. 9.27 and accompanying discussion of the Jahn-Teller effect, pp. 396–409). As a result of this distortion, the two Cu—O distances are no longer equivalent (197 versus 230 pm) and presumably the π bonding is no longer equivalent. Therefore one resonance form is favored over the other, and as a result there is an alternation of bond lengths through the ring (Fig. 10.69).[143]

The chelate effect is amplified in the case of polydentate ligands forming several rings with a single metal atom. The extreme of this form of stabilization is found in

[143] M. V. Veidis et al., *J. Am. Chem. Soc.*, **1969**, *91*, 1859.

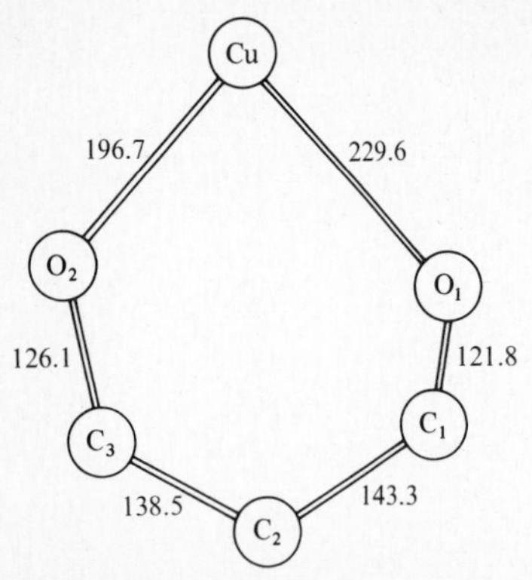

Fig. 10.69 Bond lengths in the acetylacetonate ring of [Cu(py)(hfacac)$_2$]. [From M. V. Veidis et al., *J. Am. Chem. Soc.*, **1969**, *91*, 1859. Reproduced with permission.] Distances in pm.

hexadentate ligands such as ethylenediaminetetraacetic acid (EDTA), $(HOOCCH_2)_2$-$NCH_2CH_2N(CH_2COOH)_2$, the anion of which has six ligating atoms.[144]

We have seen that chelate rings obey much the same type of steric requirements with respect to conformations as do organic rings. Unlike organic ring systems, maximum stability usually arises from five-membered rings since the metal atom is larger than a carbon atom and the bond angles at the metal ($\angle$L—M—L) will be 90° in square planar or octahedral complexes in contrast to an optimum angle of $109\frac{1}{2}°$ for carbon. For rings exhibiting significant resonance effects, such as acetylacetone, six-membered rings are quite stable. Larger and smaller chelate rings are known, but they are not nearly so stable as the five- and six-membered species.

An area of particular research interest in recent years has been the construction of planar, macrocyclic ligands. These are special types of polydentate ligands in which the ligating atoms are constrained in a large planar (or nearly planar) ring encircling the metal atom. For example, polyethers[145] are known in which the ether oxygen atoms lie in a nearly planar arrangement about the central metal atom (Fig. 10.70) and the remainder of the molecule lies in a "crown" arrangement. All of the oxygen atoms "point" inward toward the metal atom, and these macrocycles have the unusual property of forming stable complexes with alkali metals. This exceptional stability seems to arise from the close fitting of the alkali metal ion into the hole in the center of the ligand.[146] For example,

[144] The potentially hexadentate ligancy of EDTA may not always be realized if the size of the metal ion prevents the fitting of the polydentate ligand around it.

[145] C. J. Pedersen, *J. Am. Chem. Soc.*, **1967**, *89*, 7017; **1970**, *92*, 391; D. Bright and M. R. Truter, *Nature*, **1970**, *225*, 176. The nomenclature of the organic ring systems is complex (the ligand in Fig. 10.70 is 2,3,11. 12-dibenzo-1,4,7,10,13,16-hexaoxacyclooctadeca-2,11-diene), and Pedersen has condensed this to "dibenzo-18-crown-6," in which the number 18 refers to the macrocyclic ring size, "crown" is a trivial generic name for the class, and 6 refers to the number of oxygen atoms. This name may be further abbreviated to 18C6. For purposes of the present discussion these compounds may be referred to as "crown-4," "crown-5," and "crown-6," with the implication that the ring size increases with the number of oxygen atoms. The cryptands, to be discussed shortly, may be abbreviated to the point of C222, standing for "cryptand with three pairs of oxygen atoms on the 'seams of the football'" (see p. 533).

[146] See H. K. Frensdorff, *J. Am. Chem. Soc.*, **1971,** *93*, 600.

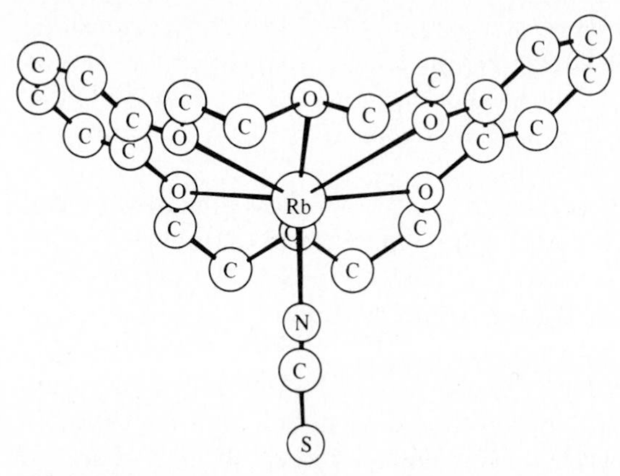

Fig. 10.70 Structure of a "crown" macrocyclic complex of rubidium thiocyanate with dibenzo-18-crown-6. [From D. Bright and M. R. Truter, *Nature*, **1970**, *225*, 176. Reproduced with permission.]

crown-4 selectively complexes Li^+, crown-5 complexes best with Na^+, and K^+ is favored by crown-6. These have the unusual ability to promote the solubility of alkali salts in organic solvents as a result of the large hydrophobic organic ring. For example, alkali metals do not normally dissolve in ethers as they do in ammonia (see pp. 331–334), but will do so if crown ligands are present:[147]

$$K + \text{crown-6} \xrightarrow[\text{or THF}]{Et_2O} [K(\text{crown-6})]^+ + [e(\text{solvent})]^- \qquad \textbf{(10.46)}$$

The ability to complex and stabilize alkali metal ions has been exploited several times to effect syntheses that might otherwise be difficult or impossible. Consider the hypothetical compound $Cs^+ Au^-$. It might at first appear to be an improbable compound since it is an *ionic* compound between two *metals* rather than between a metal and a *nonmetal*. Yet inspection of the ionization energy of cesium and the electron affinity of gold (or merely comparing the Pauling electronegativity of gold, 2.54, with that of iodine, 2.66) sparks curiosity—it should be investigated. Indeed, based on the suggestion of the late Sir Ronald Nyholm, the first edition of this book (1972), as does this one (footnote 3, p. 55), mentioned the possibility of such a bond. The experimental stumbling block is obvious, however: If you mix two metals, how do you determine that you indeed *do* have an ionic compound and not an alloy? Cesium was alloyed to react with gold and the conductivity in the melt was characteristic of an ionic compound. Further evidence was desirable, however. Two approaches were used. One was to take advantage of the solubility of alkali metals in liquid ammonia, the strong reducing power of the free electron, and the stability of reduced species in this medium. (Note that Au^- should be a powerful reductant.) The second was to stabilize metals, such as sodium, that would otherwise not react, with macrocyclic polyethers to form $Na(\text{macrocycle})^+Au^-$ salts. A combination of these techniques with physical methods such as nonaqueous electrochemistry and, particularly, the anomalously low ESCA binding energy (see Fig. 1.7) of the -1 gold.[148]

---

[147] J. L. Dye et al., *J. Am. Chem. Soc.*, **1970**, *92*, 5226.

[148] J. Knecht et al., *Chem. Comm.*, **1978**, 905; W. J. Peer and J. J. Lagowski, *J. Am. Chem. Soc.*, **1978**, *100*, 6280; A. J. Bond et al., ibid., **1978**, *100*, 7768; B. Busse and K. G. Weil, *Angew. Chem. Int. Engl. Ed.*, **1979**, *18*, 629.

In a similar manner the so-called Zintl salts composed of alkali metal cations and clusters of metals as anions (see p. 752) were known in liquid ammonia solution but proved to be impossible to isolate—upon removal of the solvent they reverted to alloys. Stabilization of the cations by complexation with macrocyclic ligands allowed the isolation and determination of the structures of these compounds.

This general trait of crown ethers and cryptands (see below) stabilizing alkali metal salts has been extended to even more improbable compounds, the *alkalides*, which exist as complexed alkali metal cations and *alkalide anions*. Thus crystalline $Na(macrocycle)^+$ Na^- consists approximately of closest-packed, large complex cations with *sodide anions* in the octahedral holes.[149]

Finally, from the arguments presented in Chapter 7, the fluoride ion, F^-, should be a strong base and nucleophile. Normally, however, it does not show these expected properties because its very basicity attracts it to its countercation so strongly that it is ion-paired in solution and not free to react. Addition of, for example, crown-6 ether to a solution of potassium fluoride in benzene increases the solubility tenfold and also increases the nucleophilicity of the fluoride ion. We shall encounter this phenomenon again in Chapter 13 on organometallic chemistry.[150]

The formation of macrocyclic rings is favored by the presence of a cation of appropriate size that can serve to hold the partially formed ligand in position as the remainder of the ring is synthesized. This process is called the "template effect." For example, the polyether crown compounds are synthesized in higher yields in the presence of an alkali metal ion. Busch and co-workers[151] have used template syntheses to form macrocyclic complexes of transition metals. The study of the properties and syntheses of macrocyclic complexes may have important consequences for biochemistry inasmuch as many important compounds in living systems such as chlorophyll, hemoglobin, and cytochromes contain macrocyclic porphyrin rings attached to metal atoms. The biosynthesis of these molecules probably involves some form of template synthesis. In addition to their direct structural relationship to biological molecules, macrocycles such as the polyethers may provide clues to the discrimination shown by biological tissues toward various ions. This selectivity provides the so-called sodium pump necessary for the proper Na^+/K^+ ionic balance responsible for electrical gradients and potentials in muscle action.[152]

If a macrocycle is sufficiently large and flexible, it is possible for it to wrap around six octahedral positions of a metal ion (Fig. 10.71).[153] The ultimate in encirclement of metal ions by the ligand is shown by encapsulation reactions in which the ligand forms a three-dimensional cage about the metal. For example, the hydroxyl groups in tris-(dimethylglyoximato)cobalt(III) can be esterified to form a monofluoroborate ester group

---

[149] See J. L. Dye et al., *J. Am. Chem. Soc.*, **1974**, *96*, 608, 7203; B. van Eck et al., *Inorg. Chem.*, **1982**, *21*, 1966.

[150] J. H. Clark, *Chem. Rev.*, **1980**, *80*, 429. See also W. P. Weber and G. W. Gokel, "Phase Transfer Catalysis in Organic Synthesis," Springer Verlag, Berlin, **1977**, pp. 117 ff., 253, 257.

[151] L. T. Taylor, S. C. Vergez, and D. H. Busch, *J. Am. Chem. Soc.*, **1966**, *88*, 3170; L. T. Taylor and D. H. Busch, *Inorg. Chem.*, **1969**, *8*, 1366; V. Katovic, L. T. Taylor, and D. H. Busch, *J. Am. Chem. Soc.*, **1969**, *91*, 2122, and references therein.

[152] A. L. Lehninger, "Biochemistry," 2nd ed., Worth, New York, **1975**, pp. 789–793; G. Weissmann and R. Claiborne, eds., "Cell Membranes: Biochemistry, Cell Biology & Pathology," HP Publishing, New York, **1975**, pp. 98–103.

[153] See P. A. Tasker and E. B. Fleischer, *J. Am. Chem. Soc.*, **1970**, *92*, 7072, and references therein.

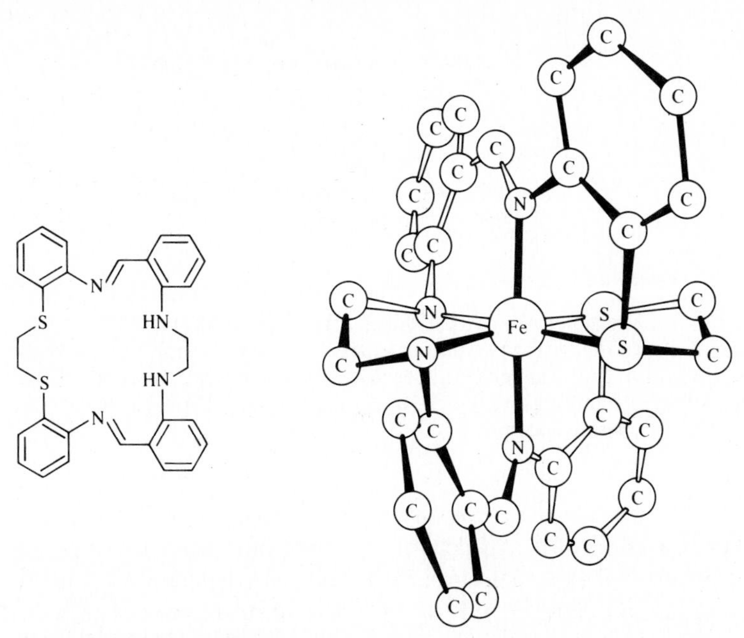

Fig. 10.71 Complex between Fe^{+2} and a macrocyclic ligand illustrating the "wrapping effect" of the macrocyclic ligand. [From P. A. Tasker and E. B. Fleischer, *J. Am. Chem. Soc.*, **1970**, *92*, 7072. Reproduced with permission.]

at each end of the complex, completely sealing in the metal:[154]

$$3\,\mathrm{HO{-}N{=}C(CH_3){-}C(CH_3){=}N{-}OH}\ (\mathrm{Co}/3) \xrightarrow{\mathrm{Et_2OBF_3}} \left[\mathrm{F{-}B}\left(\mathrm{O{-}N{=}C(CH_3){-}C(CH_3){=}N{-}O}\ (\mathrm{Co}/3)\right)_3\mathrm{B{-}F}\right]^{+} \tag{10.47}$$

The resulting case (Fig. 10.72) is called a *clathro-chelate* or a *cryptate*.[155] One class of cryptate-forming ligands of the type $N(CH_2CH_2OCH_2CH_2OCH_2CH_2)_3N$ has been called "football ligands"[156] because the polyether bridges between the two nitrogen atoms resemble the seams of a football. Ligands of this type form exceptionally stable complexes with alkali metals and show high selectivity when the size of the "football" is adjusted to fit the desired cation.[157]

---

[154] D. R. Boston and N. J. Rose, *J. Am. Chem. Soc.*, **1968**, *90*, 6859; J. E. Parks et al., *J. Am. Chem. Soc.*, **1970**, *92*, 3500; G. A. Zakrewski et al., *J. Am. Chem. Soc.*, **1971**, *93*, 4411; S. C. Jackels and N. J. Rose, *Inorg. Chem.*, **1973**, *12*, 1232.

[155] D. H. Busch, *Rec. Chem. Progr.*, **1964**, *25*, 107; J. M. Lehn et al., *J. Am. Chem. Soc.*, **1970**, *92*, 2916; *Chem. Commun.*, **1970**, 1055. The free ligand is often called a *cryptand.*

[156] P. B. Chock and E. O. Titus, *Progr. Inorg. Chem.* **1973**, *18*, 287.

[157] J. M. Lehn and J. P. Sauvage, *J. Am. Chem. Soc.*, **1975**, *97*, 6700. For a review of alkali metal transport, see footnote 156, and for a short review of several aspects of crown ether chemistry, see A. C. Knipe, *J. Chem. Educ.*, **1976**, *53*, 618; R. M. Izatt and J. J. Christensen, eds., "Progress in Macrocyclic Chemistry," Vol. 2, Wiley, New York, **1981**; G. W. Gokel and S. H. Korzeniowski, "Macrocyclic Polyether Syntheses," Springer Verlag, Heidelberg, **1982**.

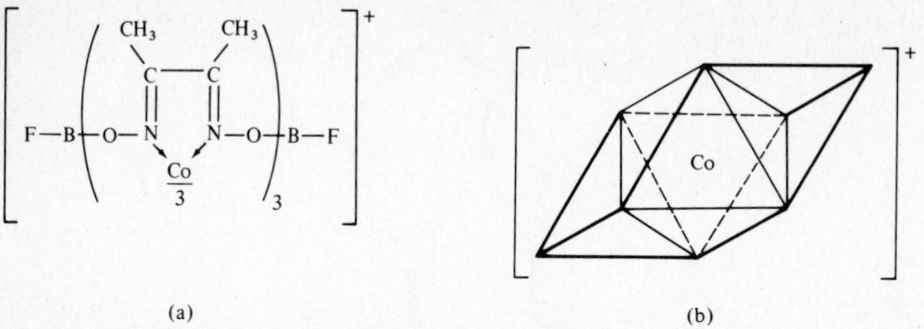

(a) (b)

Fig. 10.72 Clathro-chelate derived from dimethylglyoxime, boron trifluoride, and cobalt(III): (a) formula; (b) geometry of clathro-chelate; boron atoms form apexes on lower left and upper right; coordinate positions on the octahedron are occupied by nitrogen atoms. The heavier lines represent edges of the polyhedron spanned by chelate rings. [From D. R. Boston and N. J. Rose, *J. Am. Chem. Soc.*, **1968**, *90*, 6860. Reproduced with permission.]

Closely related to the "football ligands" are the so-called *sepulchrate* ligands. One can be formed by the condensation of formaldehyde and ammonia onto the nitrogen atoms of tris(ethylenediamine)cobalt(III). This results in tris(methylene)amino caps on opposite faces of the coordination octahedron. If the synthesis utilizes one of the (Δ, Λ)-enantiomers, the chirality of the complex is retained. Furthermore, the complex may be reduced to the corresponding cobalt(II) cation and reoxidized to cobalt(III) without loss of chirality. This is particularly unusual in that, as we shall see in the following chapter (Fig. 11.5), cobalt(II) complexes are quite labile in contrast to the stability of cobalt(III) complexes.[158] Once again the extra stability of polydentate complexes is demonstrated.

In contrast, Bernal[159] has isolated a self-resolving complex (see p. 494) whose chirality depends only on the conformation of an acyl bridge:

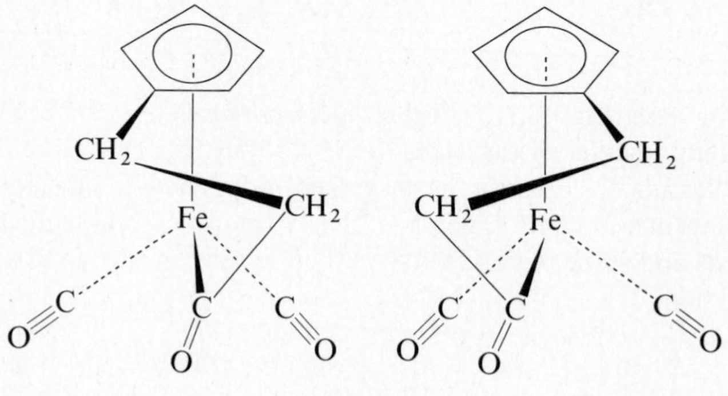

Single crystals consist of only one enantiomer and so in the solid state the enantiomers do not racemize despite the seeming lack of barriers to rotation of the rings (see p. 627).

[158] I. I. Creaser et al., *J. Am. Chem. Soc.*, **1977**, *99*, 3181; Y. Saito, "Inorganic Molecular Dissymmetry," Springer-Verlag, Heidelberg, **1979**, p. 88.

[159] I. Bernal, personal communication.

Presumably crystal packing forces "lock in" the chirality. Immediately upon solution, the complex racemizes, indicating the fragility of the forces stabilizing it.

PROBLEMS

10.1. A complex of nickel(II), $[NiCl_2(P\phi_3)_2]$, is paramagnetic. The analogous complex of palladium(II) is diamagnetic. Predict the number of isomers that will exist for each of these formulations.

10.2. **a.** Discuss the possibilities of identifying the *cis–trans* isomers of compounds MA_2B_4 by dipole moments.
b. Discuss similar possibilities for *fac–mer* isomers of MA_3B_3. Are there any problems arising in octahedral complexes that make this application less certain than in square planar complexes?

10.3. When attempts were made to isolate the 5-coordinate complexes $[CuCl_5]^{-3}$, $[ZnCl_5]^{-3}$, and $[CdCl_5]^{-3}$ (p. 472), the anions were isolated as salts of $[Co(NH_3)_6]^{+3}$ and $[Cr(NH_3)_6]^{+3}$. Why?

10.4. Five-coordinate complexes of tris(2-dimethylaminoethyl)amine (Me_6tren) are even more stable in the trigonal bipyramidal configuration (compared with square pyramidal) than tren complexes. Suggest an explanation. [R. Sime et al., *Inorg. Chem.*, **1971**, *10*, 537.]

10.5. Draw out all the isomers, geometric and optical, of the following:

$[Co(en)_2Cl_2]^+$ $\quad$ $[Co(en)_2NH_3Cl]^{+2}$ $\quad$ $[Co(en)(NH_3)_2Cl_2]^+$

10.6. Consider a compound having the formula $[(NH_3)(RNH_2)M(R)(CO)_2]^+X^-$ and having square pyramidal geometry with the R group apical (M is any metal; R is an organic group). Draw out all the isomers and discuss whether they are geometric or optical isomers.

10.7. There is an optically active complex (C.N. = 6) illustrated in this chapter that is not so labeled. Find it and determine if the isomer shown is Λ or Δ.

10.8. Which of the following is the most likely structure for pentacyanocobalt(III)-*μ*-cyanopentaamminecobalt(III)?

$[(NH_3)_5Co{-}CN{-}Co(CN)_5]$ or $[(NH_3)_5Co{-}NC{-}Co(CN)_5]$

Why? [B.-C. Wang et al., *Inorg. Chem.*, **1971**, *10*, 1492.]

10.9. Draw the most likely structure of pentaamminecobalt(III)-*μ*-thiocyanatopentacyanocobalt(III). [F. R. Fronczek and W. P. Schaefer, *Inorg. Chem.*, **1975**, *14*, 2066.]

10.10. Which of the two isomers—$Co(DH)_2(SCN)py$ or $Co(DH)_2(NCS)py$—would you predict to be thermodynamically the most stable? DH^- represents the monoanion dimethylglyoximate, $HON{=}C(CH_3)C(CH_3){=}NO^-$. [A. H. Norbury and S. Raghunathan, *J. Inorg. Nucl. Chem.*, **1975**, *37*, 2133.]

10.11. Draw the molecular structure of the following complexes:
a. *cis*-dichlorotetracyanochromate(III)
b. *mer*-triamminetrichlorocobalt(III)

c. *trans*-dichlorobis(trimethylphosphine)palladium(II)
d. *fac*-triaquatrinitrocobalt(III)

10.12. Why does iron(II) hexacyanoferrate(III) spontaneously isomerize to Prussian blue?

10.13. With the aid of the table of ligand abbreviations given in Appendix J find the names of each of the ligands listed below. Sketch the structure of each ligand and classify it as monodentate, bidentate, tridentate, etc. Sketch the mode of attachment of the ligand to a metal ion.

a. acac	**e.** fod	**i.** tap
b. chxn	**f.** Hedta	**j.** pc
c. dtp	**g.** ox	**k.** pn
d. edta	**h.** phen	**l.** dmf

10.14. If you did not answer Problem 9.22 when you read Chapter 9, do so now.

10.15. Figure 10.73 illustrates two forms of the pentanitrocuprate(II) ion, $[Cu(NO_2)_5]^{-3}$. Discuss all of the types of isomerism exhibited in these ions. [K. A. Klanderman et al., *Inorg. Chim. Acta*, **1977**, *23*, 117.]

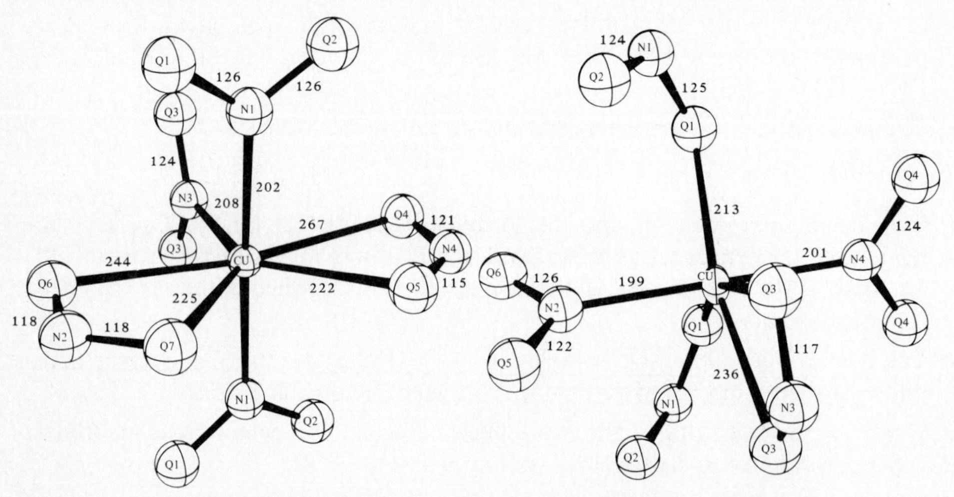

Fig. 10.73 Two forms of the $[Cu(NO_2)_5]^{-3}$ ion. All bond distances in picometres. [From K. A. Klanderman et al., *Inorg. Chim. Acta*, **1977**, *23*, 117. Reproduced with permission.]

10.16. The molecule shown in Fig. 10.2b appears to be "left-handed." Does this molecule have chirality? Explain.

10.17. If you learned the Cahn-Ingold-Prelog rules in organic chemistry, test your recall by assigning the appropriate *R,S* notation to the molecule shown in Fig. 10.3. Assume that the C_5H_5 ligand is a "single atom" of mass 60 ($= 5 \times 12$).[160]

---

[160] No complete unanimity on exact rules for treating the so-called *polyhapto* ligands has been reached, though most people favor some variant of the above scheme. See K. Stanley and M. C. Baird, *J. Am. Chem. Soc.*, **1975**, *97*, 6599.

10.18. Read the section on symmetry in Appendix B and identify the symmetry elements and operations in the molecules and ions shown in the figures listed below. Determine the appropriate point group for each molecule and ion.

a. 9.29
b. 10.4
c. 10.5
d. 10.6
e. 10.7
f. 10.9
g. 10.21
h. 10.25
i. 10.26
j. 10.28
k. 10.51
l. 10.52c
m. 10.57

10.19. Occasionally, in the preparation of the artwork for a research article or a textbook, the photographic negative taken from the original line drawing of the artist is inserted "up-side-down" (or *reverse*, front-to-back) and the resulting image is reversed. Does this make any difference? Discuss. Are there any exceptions to the general rule? Illustrate your argument with sketches.

11

Coordination chemistry: reactions, kinetics, and mechanisms

Despite extensive study, inorganic chemistry has yet to achieve the understanding of reaction mechanisms enjoyed by organic chemistry. This fact, alluded to previously (Chapter 5), results from the task of trying to handle more than one hundred elements with a single scheme. Unfortunately, even attempts to predict from one element in a group to another in the same group are not always successful. The lack of understanding of mechanisms carries over into synthesis; the methods used in synthesizing coordination compounds often include some combination of redox chemistry and displacement of ligands (often water) by the desired ligands. Thus the well-known synthesis of hexaammine-cobalt(III) cation is typical. Starting with stable and common cobalt(II) salts (such as the nitrate or carbonate) the desired ligand, ammonia, is added in high concentrations to replace those present (usually water or chloride ion), and an oxidizing agent (air or hydrogen peroxide, with a charcoal catalyst) effects the change in oxidation state:

$$[CoCl_4]^{-2} \xrightarrow[-4Cl^-]{+6NH_3} [Co(NH_3)_6]^{+2} \xrightarrow{[O]} [Co(NH_3)_6]^{+3} \qquad \textbf{(11.1)}$$

Occasionally, synthetic "tricks" are discovered that allow the synthesis of a desired compound or isomer without interference from an undesired one. Thus, quite early, Werner discovered that certain *cis* isomers could be prepared by using a carbonato compound that would react with acid to evolve carbon dioxide and insert two anions of the acid as ligands:

$$[Co(en)_2CO_3]^+ + 2HCl \longrightarrow \textit{cis}\text{-}[Co(en)_2Cl_2]^+ + H_2O + CO_2\uparrow \qquad \textbf{(11.2)}$$

More rarely, reactions are known in which predictions can be made about a wide variety of complexes involving different metals and different ligands and for which the mechanism is fairly well understood. A good example of this type of reaction is that of 4-coordinate, square complexes in which the *trans effect* operates and allows systematization to be made.

THE *TRANS* EFFECT

The presence of large deposits of platinum ores in Russia caused an intensive study of the coordination compounds of platinum early in the development of coordination chemistry. As a result of these studies by the Russian school, the first stereospecific displacement reaction was discovered. Consider two means of forming diamminedichloroplatinum(II): (1) displacement of Cl^- ions from $[PtCl_4]^{-2}$ by NH_3; (2) displacement of NH_3 from $[Pt(NH_3)_4]^{+2}$ by Cl^- ions. It is found that two different isomers are formed:

$$[PtCl_4]^{-2} \xrightarrow[-Cl^-]{+NH_3} [PtCl_3(NH_3)]^{-} \xrightarrow[-Cl^-]{+NH_3} PtCl_2(NH_3)_2$$

trans-diamminedichloro-platinum(II) [Not found in this reaction]

cis-diamminedichloro-platinum(II) [Exclusive product]

(11.3)

$$[Pt(NH_3)_4]^{+2} \xrightarrow[-NH_3]{+Cl^-} [Pt(NH_3)_3Cl]^{+} \xrightarrow[-NH_3]{+Cl^-} PtCl_2(NH_3)_2$$

trans-diamminedichloro-platinum(II) [Exclusive product]

cis-diamminedichloro-platinum(II) [Not found in this reaction]

(11.4)

Reactions in Eqs. 11.3 and 11.4 can be rationalized as follows: (1) Step one is a simple displacement, and since all four groups present (either NH_3 or Cl) are identical, only one compound is formed; (2) in step two, two products can potentially be formed in either reaction, but in practice only one is found and it differs between the two reactions. In both reactions, however, the isomer that is found is that which forms by substitution of a ligand *trans* to a chloride ion. The ligands *trans* to chloride ions have been circled in Eqs. 11.3 and 11.4 to emphasize this fact. The *trans effect* may be defined as *the labilization of ligands trans to other, trans-directing ligands.* By comparison of a large number of reactions, it is possible to set up a *trans-directing series*:

$$\begin{array}{c}A\quad Cl\\ Pt\\ B\quad Cl\end{array} \xrightarrow[-Cl^-]{+NH_3} \begin{cases} \begin{array}{c}A\quad Cl\\ Pt\\ B\quad NH_3\end{array} & A > B \\ \text{OR} & \\ \begin{array}{c}A\quad NH_3\\ Pt\\ B\quad Cl\end{array} & B > A \end{cases} \tag{11.5}$$

The ordering of ligands in this series is as follows: $CN^- \sim CO \sim NO \sim H^- > CH_3^- \sim SC(NH_2)_2 \sim SR_2 \sim PR_3 > SO_3H^- > NO_2^- \sim I^- \sim SCN^- > Br^- > Cl^- > py > RNH_2 \sim NH_3 > OH^- > H_2O$.

The *trans* effect in synthesis

By ordering the sequence of addition of substituents, the *trans* effect may be utilized to provide the desired isomer in an otherwise complicated system. For example, consider the problem of the synthesis of the three geometrical isomers of amminebromochloro-(pyridine)platinum(II). By utilizing a knowledge of the *trans* effect, the following reactions were suggested and carried out:[1]

$$\left[\begin{array}{c}Cl\quad Cl\\ Pt\\ Cl\quad Cl\end{array}\right]^{-2} \xrightarrow[-Cl^-]{+NH_3} \left[\begin{array}{c}Cl\quad \textcircled{Cl}\\ Pt\\ \textcircled{Cl}\quad NH_3\end{array}\right]^{-} \xrightarrow[-Cl^-]{+Br^-} \left[\begin{array}{c}Cl\quad Br\\ Pt\\ \textcircled{Cl}\quad NH_3\end{array}\right]^{-} \xrightarrow[-Cl^-]{+py} \begin{array}{c}Cl\quad Br\\ Pt\\ py\quad NH_3\end{array} \tag{11.6}$$

In addition to the *trans* effect, use is made of the general principle that, other things being equal, a metal–halogen bond is more labile than a metal–nitrogen bond. In this case "other things equal" means that in step 2 the chloride ion *trans* to another chloride ion will be replaced more readily than a nitrogen atom *trans* to a chloride ion. The group *trans* to the most influential *trans* director is again circled. Note that the results of the last reaction indicate the bromide ion to be a better *trans* director than chloride.

The synthesis of the second isomer follows a similar pattern:

$$\left[\begin{array}{c}Cl\quad Cl\\ Pt\\ Cl\quad Cl\end{array}\right]^{-2} \xrightarrow[-Cl^-]{+py} \left[\begin{array}{c}Cl\quad \textcircled{Cl}\\ Pt\\ \textcircled{Cl}\quad py\end{array}\right]^{-} \xrightarrow[-Cl^-]{+Br^-} \left[\begin{array}{c}Cl\quad Br\\ Pt\\ \textcircled{Cl}\quad py\end{array}\right]^{-} \xrightarrow[-Cl^-]{NH_3} \begin{array}{c}Cl\quad Br\\ Pt\\ H_3N\quad py\end{array} \tag{11.7}$$

[1] A. D. Hel'man, E. F. Karandashova, and L. N. Essen, *Dokl. Akad. Nauk. S.S.S.R.*, **1948**, *63*, 37.

Here, too, the lability of a metal–chlorine bond results in preferential replacement of chloride ion instead of pyridine in the second step and the *trans* influence of the bromide ion determines the final geometry.

The third isomer may be formed as follows:

$$\left[\begin{matrix}Cl & & Cl\\ & Pt & \\ Cl & & Cl\end{matrix}\right]^{-2} \xrightarrow[-2Cl^-]{+2NH_3} \begin{matrix}Cl & & NH_3\\ & Pt & \\ Cl & & NH_3\end{matrix} \xrightarrow[-Cl^-]{+py} \left[\begin{matrix}Cl & & NH_3\\ & Pt & \\ py & & \boxed{NH_3}\end{matrix}\right]^{+} \xrightarrow[-NH_3]{+Br^-} \begin{matrix}Cl & & NH_3\\ & Pt & \\ py & & Br\end{matrix} \tag{11.8}$$

In this synthesis the *trans* effect predicts the formation of the *cis* isomer in the first step and the replacement of the ammonia molecule *trans* to the chloride ion (rather than the one *trans* to pyridine) in the final step. The inherent lability of the platinum–chlorine bond directs the second step. In the third step this inherent lability runs counter to the labilizing *trans* effect. Hence the fact that the entering bromide ion displaces an ammonia molecule rather than the chloride ion can only be set down as an empirical observation. This is a good example of the fact that the *trans* effect provides us with qualitative information concerning which ligands will be *more* labile (than they would be in the absence of a *trans* director) but no information on the *absolute* lability of the various ligands. Nevertheless, considerable ordering is provided by application of the *trans* effect and empirical observations (such as the inherent lability of M—Cl bonds). If on occasion the two rules run counter to one another and we are unable to choose with certainty, we should not be too discouraged—imagine the chaos resulting from reactions 11.6 to 11.8 if they were completely random!

An interesting application of the *trans* effect is in distinguishing between *cis* and *trans* isomers of complexes of the type $[PtA_2X_2]$ where A = ammine and X = halide (Kurnakov test). Addition of thiourea (tu) to the *cis* complex results in complete replacement of the former ligands:

$$\begin{matrix}Cl & & NH_3\\ & Pt & \\ Cl & & NH_3\end{matrix} \longrightarrow \left[\begin{matrix}Cl & & NH_3\\ & Pt & \\ tu & & NH_3\end{matrix}\right]^{+} \longrightarrow \left[\begin{matrix}Cl & & tu\\ & Pt & \\ tu & & NH_3\end{matrix}\right]^{+} \longrightarrow \left[\begin{matrix}tu & & tu\\ & Pt & \\ tu & & NH_3\end{matrix}\right]^{+2} \longrightarrow \left[\begin{matrix}tu & & tu\\ & Pt & \\ tu & & tu\end{matrix}\right]^{+2} \tag{11.9}$$

but in the *trans* isomer the replacement stops after the two halide ions have been replaced since the *trans* ammonia molecules do not labilize each other:

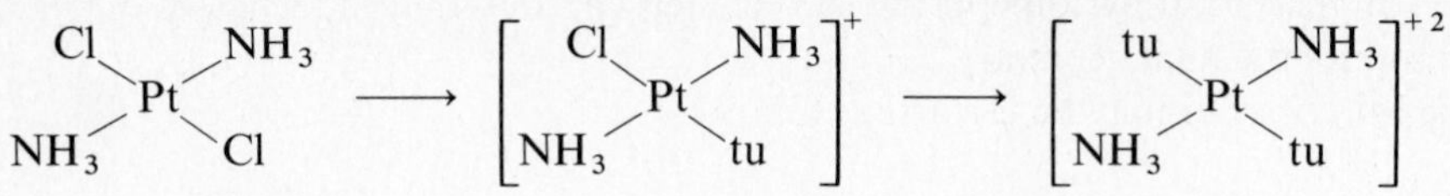

$$\begin{array}{ccc} Cl & & NH_3 \\ & Pt & \\ NH_3 & & Cl \end{array} \longrightarrow \left[\begin{array}{ccc} Cl & & NH_3 \\ & Pt & \\ NH_3 & & tu \end{array}\right]^{+} \longrightarrow \left[\begin{array}{ccc} tu & & NH_3 \\ & Pt & \\ NH_3 & & tu \end{array}\right]^{+2} \tag{11.10}$$

Mechanism of the *trans* effect

The *trans* effect must be kinetically controlled since the thermodynamically most stable isomer is not always produced. This is obvious since it is possible to form two different isomers (Eqs. 11.3 and 11.4) or three different isomers (Eqs. 11.6, 11.7, and 11.8) depending upon the reaction sequence and only one of the isomers can be the most stable in a thermodynamic sense. Furthermore, we observe that the best *trans* directors (T) are often those that are the best π acceptors. Hence in the formation of bis isomers of the sort $[PtCl_2T_2]$ from $[PtCl_4]^{-2}$ the *trans* isomer is always observed:

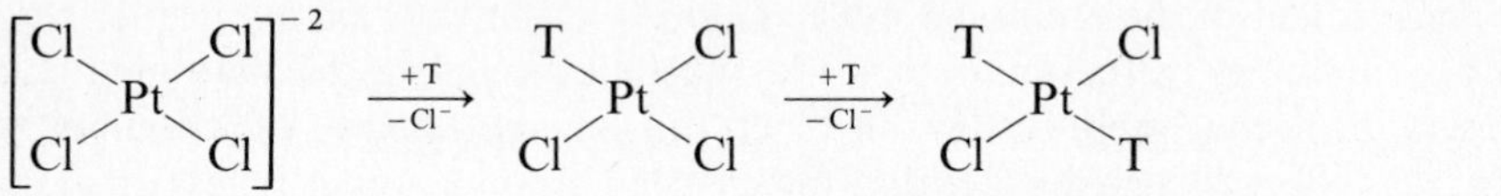

$$\left[\begin{array}{ccc} Cl & & Cl \\ & Pt & \\ Cl & & Cl \end{array}\right]^{-2} \xrightarrow[-Cl^-]{+T} \begin{array}{ccc} T & & Cl \\ & Pt & \\ Cl & & Cl \end{array} \xrightarrow[-Cl^-]{+T} \begin{array}{ccc} T & & Cl \\ & Pt & \\ Cl & & T \end{array} \tag{11.11}$$

even though, in general, the *cis* isomers are favored thermodynamically (see p. 526). Such kinetically controlled reactions are common in both organic and inorganic chemistry and represent examples of reactions in which the energy of activation of the activated complex is more important in determining the course of reaction than the energy of products.

Two viewpoints have been advanced with respect to the mechanism of the *trans* effect. The first is essentially a static one emphasizing a weakening of the *trans* bond, the second is the lowering of the activation energy of *trans* replacement. The earlier literature made no attempt to distinguish between the two effects, but in view of the above discussion it is important to separate them.[2] The "*trans effect*" may be defined[3] as "the effect of a coordinated group [T] upon the *rate* of substitution reactions of ligands opposite to it." In contrast, the "*trans influence*" may be defined[4] as "the extent to which that ligand weakens the bond *trans* to itself in the equilibrium state of a substrate."

The earliest theory that still has current application is the polarization theory of Grinberg.[5] The latter suggested that in a completely symmetrical complex such as $[PtCl_4]^{-2}$ the bond dipoles to the various ligands will all be identical and cancel (Fig. 11.1a).

[2] Whether the static *trans* influence is operative in the dynamic *trans* effect is a moot point. Certainly it is reasonable to assume that a weakening of a *trans* bond may be *one* of the important factors in the latter mechanism.

[3] F. Basolo and R. G. Pearson, *Progr. Inorg. Chem.*, **1962**, *4*, 381. The symbol T has been substituted for the symbol A used by Basolo and Pearson to make the definition consistent with the symbolism used in the following discussion.

[4] A. Pidcock et al., *J. Chem. Soc., A*, **1966**, 1707.

[5] A. A. Grinberg, *Acta Physiochim, U.R.S.S.*, **1935**, *3*, 573 [*C.A.*: *30*, 4074^{6}]; B. V. Nekrasov, *Zh. Obshch. Khim.*, **1937**, *7*, 1594; for an English translation see A. A. Grinberg, in "An Introduction to the Chemistry of Complex Compounds," 2nd ed., J. R. Leach (transl.), D. H. Busch and R. F. Trimble, Jr., eds., Pergamon, Elmsford, N.Y., **1962**.

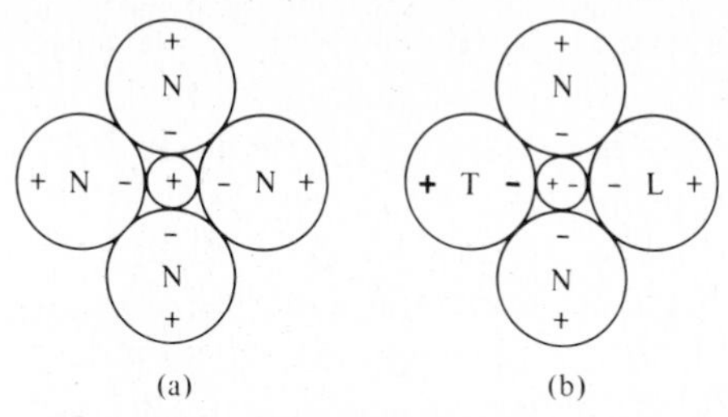

Fig. 11.1 *Trans*-directing in a square planar complex according to the polarization theory. T = *trans*-director, N = non-*trans*-director, L = labilized leaving group. (a) No *trans* effect; (b) *trans* effect operating.

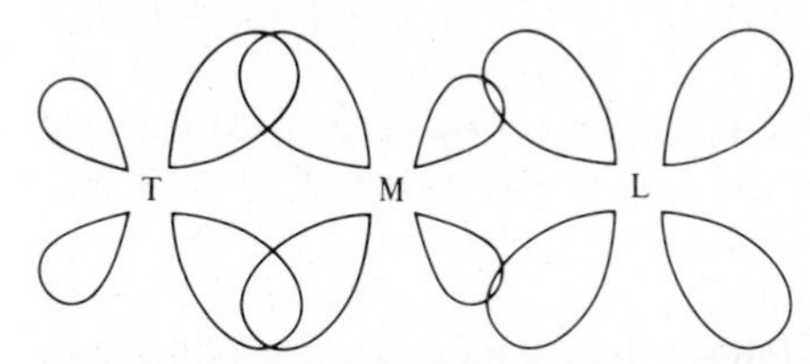

Fig. 11.2 Weakening of π bonding resulting in a *trans* influence. The two non-*trans*-directing ligands have been omitted for clarity. The "*d* orbital" left-right symmetry has been lost due to polarization. Cf. Fig. 9.51.

In contrast, if a more polarizable and polarizing ligand, T, is introduced (such as an I^- ion), its polarizability will induce an additional, uncompensated dipole in the metal (cf. the mutual polarization of the Fajans effect, p. 129). This induced dipole in the metal will oppose the natural dipole of the ligand *trans* to the polarizing ligand (Fig. 11.1b). The polarization theory is supported by the following facts: (1) It should be most important when the central metal atom is large and polarizable, and indeed the ordering of the *trans* effect is $Pt^{II} > Pd^{II} > Ni^{II}$; (2) the *trans* series listed above should also be a polarizability series and, in general, the *trans*-directing groups are more polarizable either because they are large (iodide) or multiply-bonded (ethylene, cyanide, carbon monoxide). Nevertheless, the emphasis on *weakening* of the *trans* bond limits its application to the *trans* influence.

A second approach that involves the weakening of the *trans* bond is the static π-bonding theory. According to this viewpoint, two π-bonding ligands vying for the *d* orbitals of the metal atom will tend to labilize each other (as opposed to the more stable *cis* isomer where they do not compete), and the stronger π bonder will dominate, weakening the bonding of the group *trans* to it (Fig. 11.2). Such a theory suffers from a number of difficulties. The emphasis on *weakening* of the *trans* bond presents the same ambiguities between *trans* influence and *trans* effect as Grinberg's polarization theory. In addition, it is difficult to see how the superior π bonding of cyanide ion or a phosphine molecule could labilize an ammonia molecule *trans* to it inasmuch as *the ammonia is not stabilized by π bonding either in the presence or absence of the trans director*. Thus π-bonding olefins exhibit a *trans* influence on halogens but not on nitrogen ligands.[6] Furthermore, ligands such as the

[6] J. A. Wunderlich and D. P. Mellor, *Acta Crystallogr.*, **1954**, *7*, 130; **1955**, *8*, 57; P. R. H. Alderman et al., *Acta Crystallogr.*, **1960**, *13*, 149.

hydride ion and alkyl groups which are exceptionally good *trans* directors obviously cannot π bond. Nevertheless they show a strong *trans* influence toward all ligands.[7]

Two recent attempts have been made to revive a non-π-bonding explanation for the *trans* influence. They are, in effect, a restatement of Grinberg's polarization theory in terms of modern molecular orbital theory. Langford and Gray[8] have pointed out that ligands *trans* to each other effectively compete for the *p* orbital of the metal ion that lies along the T—M—L axis. A ligand that competes more efficiently can "monopolize" this *p* orbital to the detriment of the bonding of the *trans* ligand, L. Similar explanations of the *trans* influence and *trans* effect in terms of preferential use of low-energy *ns* and/or $(n-1)d$ orbitals have been offered.[9] The details of these explanations are beyond the scope of this discussion, but the basic premise is the same in all: The formation of an extraordinarily strong σ bond by T will cause the hybrid orbital directed to T to be optimized to the detriment of the group *trans* to it.

All static theories suffer from the fact that the stress on the *weakening* of the *trans* bond implies (but does not necessitate) a *dissociative mechanism*[10] which is not found. In contrast, we might expect from what we have seen previously of 5-coordinate complexes that the most ready transition state would be one in which a square pyramid or trigonal bipyramid was formed (*associative mechanism*)[10] followed by elimination of the leaving group. Experimentally this has been verified in that those reactions that have been studied are second order, depending upon the concentration of both the complex and the attacking ligand:

$$[PtX_4]^{-2} + E \xrightarrow{\text{slow}} [PtX_4E]^{-2} \xrightarrow{\text{fast}} [PtX_3E]^{-} + X^{-} \qquad \textbf{(11.12)}$$

Cardwell[11] has suggested that by consideration of the activated complex,[12] the *trans* effect can be explained in terms of the electronegativity of the substituents. This approach is related to the polarization theory of Grinberg, which is basically electrostatic, but is a dynamic or kinetic explanation. Cardwell suggested that the activated complex consisted of a trigonal bipyramid (Fig. 11.3). The most electronegative substituents will tend to assume axial positions in such a trigonal bipyramid (see p. 230). Loss of the leaving group, L, from the transition state will re-form the square complex with the entering group, E,

[7] F. Basolo et al., *J. Chem. Soc.*, **1961**, 2207; A. Pidcock et al., *J. Chem. Soc.*, *A*, **1966**, 1707.

[8] C. H. Langford and H. B. Gray, "Ligand Substitution Processes," Benjamin, New York, **1965**, p. 25.

[9] Y. K. Syrkin, *Izv. Akad. Nauk S.S.S.R. Otdel. Khim. Nauk*, **1948**, 69; A. Pidcock et al., *J. Chem. Soc.*, *A*, **1966**, 1707; L. M. Venanzi, *Chem. Brit.*, **1968**, *4*, 162; R. S. Tobias, *Inorg. Chem.*, **1970**, *9*, 1296. The views in these papers are not all the same but represent a spectrum of "σ-only" approaches to the problem.

[10] Although the terms *dissociative* ($\approx S_{N1}$) and *associative* ($\approx S_{N2}$) are used for simplicity, it should be realized that most often the mechanisms are not so clear-cut. The terms represent two extremes and the actual mechanisms may lie somewhere between, in effect resulting in a *concerted* reaction. See discussion, p. 552.

[11] H. M. E. Cardwell, *Chem. Ind.* (*London*), **1955**, 422.

[12] The *activated complex* refers to the configuration of the reactants and products at the peak of the reaction profile energy curve (transition state). The term *intermediate* implies that the species has some stability (although perhaps slight) with respect to an activated complex leading from the reactants and an activated complex leading toward the products. In the present discussion the distinction is rather finer than warranted since it is not known whether the proposed species occur as pictured, much less whether they occur at the peak of the curve or in a minor minimum.

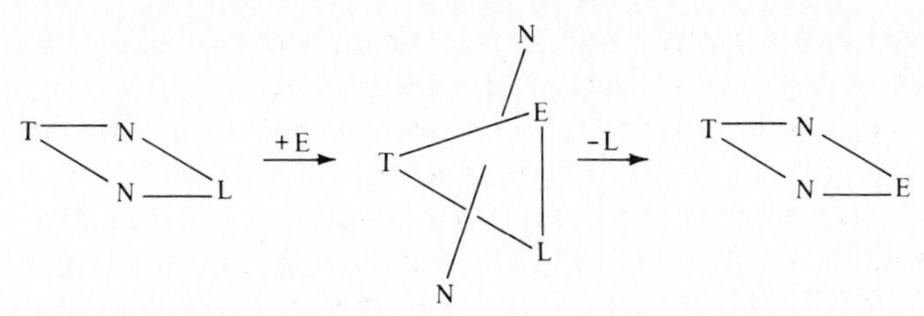

Fig. 11.3 Trigonal bipyramidal transition state containing a *trans*-directing group (T) of low electronegativity; two non-*trans*-directing groups (N) of high electronegativity; and the entering (E) and leaving (L) groups.

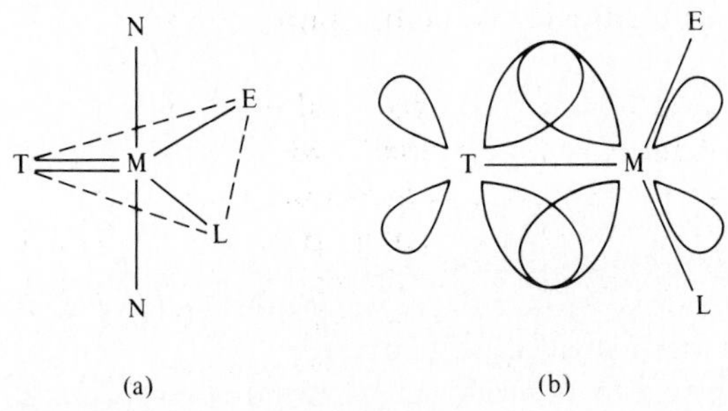

Fig. 11.4 (a) The π-bonding transition state according to Orgel. (b) The π-bonding acceptance of T removes electron density from the region of E and L (viewed down the N—M—N axis).

trans to the least electronegative group, T, in place of the former group L. Substituents of high electronegativity such as F, O, or N will tend to assume the axial positions preferentially and will not act as *trans* directors. It should be pointed out that the relative position of the chloride ion is consistent with the high electronegativity of chlorine if it is recalled that the ligand in question is the chloride anion, the electronegativity of which is considerably lower than that of the neutral atom (pp. 153–155). However, the estimation of the effective electronegativities of various ligands in complexes is difficult, and this reduces the usefulness of this approach. This factor, together with the evidence that other factors must be involved, has tended to reduce the popularity of this theory.

Chatt[13] and Orgel[14] independently pointed out that consideration ot the activated complex improves the π-bonding theory. These workers assumed a trigonal bipyramidal intermediate or activated complex and proposed that a π-bonding ligand would stabilize such an intermediate. Chatt and co-workers suggested that a strong π-bonding ligand would remove electron density from the metal through back-bonding, and hence the more positive metal could more readily accept the entering, fifth ligand. Orgel pointed out that the *d* orbital involved in π bonding on the metal will lie in the trigonal plane formed by T, E, and L (Fig. 11.4). We have already seen (p. 477) that π-bonding (and hence *trans*-directing) ligands will preferentially assume equatorial positions in a trigonal bipyramidal

[13] J. Chatt et al., *J. Chem. Soc.*, **1955**, 4456.
[14] L. E. Orgel, *J. Inorg. Nucl. Chem.*, **1956**, *2*, 137.

molecule and stabilize the 5-coordinate intermediate. The correlation of the *trans*-directing series with the amount of π bonding expected on other grounds is generally quite good. The best *trans* directors are the same ones encountered in other cases where π bonding is thought to be important: CN^-, CO, NO, and to a lesser extent, phosphines and sulfides.

Currently, the best interpretation of the *trans* effect is uncertain. The presence of the hydride ion and methanide ion high in the *trans*-directing series even though they are incapable of π bonding indicates that in some cases, at least, a σ-bonding mechanism must be important. Some workers would support a σ-only theory, whereas others believe π bonding to be important, especially in ligands such as the cyanide ion. The entire controversy is reminiscent of the π-bonding arguments presented earlier (pp. 429–441). As in that argument, the σ- versus π-bonding viewpoints are not mutually exclusive since in ligands such as cyanide, π bonding could act synergistically to enhance a σ effect.[15]

Kinetics of square planar substitution reactions

As we might expect, the rate of a substitution reaction in a square planar complex depends upon the nature of the *trans* ligand, and good *trans* directors behave as activating groups and increase the rate of reaction. Although this is perhaps the most obvious factor in these reactions, there are several other variables that will affect reaction rate.

The nature of the entering group and the nature of the leaving group are two properties which will affect the rate of reaction. Both can be rationalized in terms of an intermediate of the structure shown in Fig. 11.4 and the same π-bonding arguments that were discussed above. A general ordering of entering groups that enhance the reaction is $R_3P > \text{tu} > CN^- > SCN^- > I^- > Br^- > N_3^- > OH^-$. An entering group that can come in and assume a strongly π-bonding equatorial position (E) can stabilize the intermediate and promote the reaction. The leaving groups can act in the same way to retard the reaction. Although strongly π-bonding ligands can stabilize the intermediate, they will be poor leaving groups (L) and thus will retard the reaction unless the sum of σ and π bonding is not great. The important thing here is not type of bonding but bond strength. Thus CN^- and SCN^- are poor leaving groups because they form strong σ and π bonds, whereas the halides form weaker bonds and are good leaving groups in complexes as they are in organic compounds.

A closer look at mechanisms

In the above discussion the simplifying equalities S_{N2} = "associative" and S_{N1} = "dissociative" mechanisms were made. Many chemists prefer to avoid the labels S_{N1} and S_{N2} and to speak of a spectrum of possibilities of associative (*A*) mechanisms ranging from an S_{N2} limit through a variety of possibilities to an interchange (*I*) mechanism consisting of a concerted process with simultaneous action of entering and leaving groups through a variety of dissociative (*D*) mechanisms to a "pure" S_{N1} limit. For the limited discussion of mechanisms in this book we need not worry unduly over these distinctions, but it should

[15] See F. R. Hartley, *Chem. Soc. Rev.*, **1973**, *2*, 163, and R. G. Wilkins, "The Study of Kinetics and Mechanism of Reactions of Transition Metal Complexes," Allyn and Bacon, Boston, **1974**, pp. 223–233.

be kept in mind that the classification of reaction mechanisms is not a clear-cut, "pigeon-hole" process and that no two reactions probably proceed with exactly the same interplay of these mechanisms.[16]

Lability, inertness, stability, and instability

The *trans* effect illustrates the importance of the study of the mechanisms of complex substitution reactions. Before continuing with a discussion of these, the differentiation of the thermodynamic terms *stable* and *unstable* from the kinetic terms *labile* and *inert* should be made. As indicated, stability refers to *thermodynamic stability*. Consider the following cyano complexes: $[Ni(CN)_4]^{-2}$, $[Mn(CN)_6]^{-3}$, and $[Cr(CN)_6]^{-3}$. All of these complexes are extremely stable from a thermodynamic point of view,[17] yet kinetically they are quite different. If the rate of exchange of radiocarbon-labeled cyanide is measured, we find that despite the thermodynamic stability, one of these complexes is extremely *labile* kinetically, a second is moderately so, and only $[Cr(CN)_6]^{+3}$ can be considered to be *inert*:

$$[Ni(CN)_4]^{-2} + 4^{14}CN^- \longrightarrow [Ni(^{14}CN)_4]^{-2} + 4CN^- \quad t_{1/2} = ca.\ 30\ s \quad \mathbf{(11.13)}$$

$$[Mn(CN)_6]^{-3} + 6^{14}CN^- \longrightarrow [Mn(^{14}CN)_6]^{-3} + 6CN^- \quad t_{1/2} = ca.\ 1\ h \quad \mathbf{(11.14)}$$

$$[Cr(CN)_6]^{-3} + 6^{14}CN^- \longrightarrow [Cr(^{14}CN)_6]^{-3} + 6CN^- \quad t_{1/2} = ca.\ 24\ d \quad \mathbf{(11.15)}$$

The terms *labile* and *inert* are obviously relative, and two chemists might not use them in identical ways. Taube[18] has suggested that those complexes which react completely within about *one minute* at 25°C should be considered *labile* and those that take longer be considered inert.

The tetracyanonickelate ion is a good example of a thermodynamically stable complex which is kinetically labile. The classic example of the opposite case, i.e., the kinetically inert complex which is thermodynamically unstable, is the hexaamminecobalt(III) cation in acid solution. One might expect it to decompose:

$$[Co(NH_3)_6]^{+3} + 6H_3O^+ \longrightarrow [Co(H_2O)_6]^{+3} + 6NH_4^+ \quad \mathbf{(11.16)}$$

The tremendous thermodynamic driving force of *six* basic ammonia molecules attaching to six protons results in an equilibrium constant for the reaction (Eq. 11.16) of $\sim 10^{25}$. Nevertheless, acidification of a solution of hexaamminecobalt(III) results in no noticeable change and several days are required (at room temperature) for degradation of the complex despite the driving force of favorable thermodynamics. The inertness of the complex

---

[16] For a more nearly complete discussion, see F. Basolo and R. G. Pearson, "Mechanisms of Inorganic Reactions," 2nd ed., Wiley, New York, **1967**; C. H. Langford and H. B. Gray, "Ligand Substitution Processes," Benjamin, New York, **1965**; R. G. Wilkins, "The Study of Kinetics and Mechanism of Reactions of Transition Metal Complexes," Allyn and Bacon, Boston, **1974**.

[17] Note that for the one complex for which there are reasonably accurate data, the equilibrium constant corresponds to less than *one free* Ni^{+2} *ion per cubic decimetre* for a solution of 0.01 mol dm^{-3} Ni^{+2} in 1 mol dm^{-3} NaCN! The other complexes are probably even more stable.

[18] H. Taube, *Chem. Rev.*, **1952**, *50*, 69.

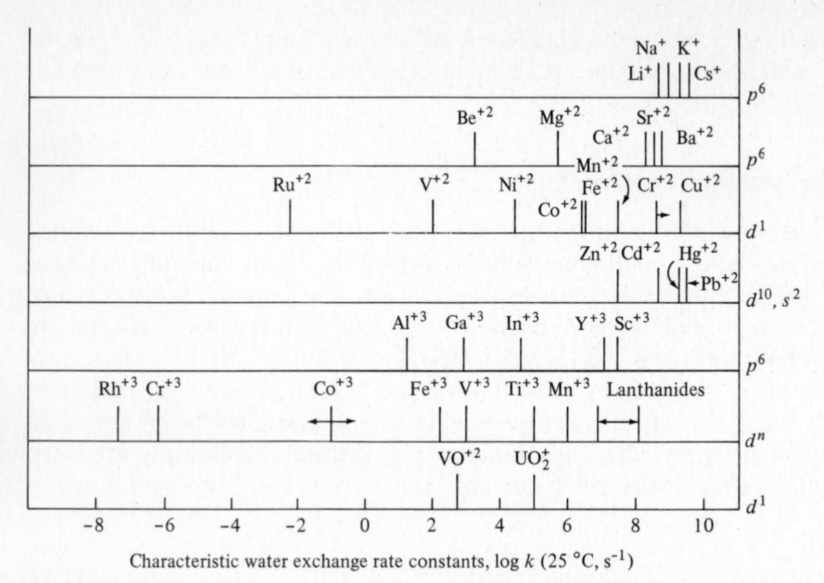

Fig. 11.5 Comparison of the rates of exchange for various metal aqua complexes. [From D. W. Margerum et al., in "Coordination Chemistry," Vol. 2, A. E. Martell, ed., *ACS Monograph No. 174*, Washington, D.C., **1978**. Reproduced with permission.]

results from the absence of a suitable low-energy pathway for the acidolysis reaction. The difference between stable and inert can be expressed succinctly: Stable complexes have large positive free energies of reaction, ΔG; inert complexes merely have large positive free energies of activation, ΔG^*. To anticipate the following discussion somewhat, the lability of Ni^{+2} complexes can be associated with the ready ability of Ni^{+2} to form 5- or 6-coordinate complexes. The additional bond energy of the fifth (or fifth and sixth) bond in part compensates for the loss of ligand field stabilizing energy. In contrast, the reaction for $[Co(NH_3)_6]^{+3}$ must involve either an extremely unstable 7-coordinate species (note the reluctance of 7-coordinate complexes to form) or the formation of a 5-coordinate species with concomitant loss of bond energy *and* LFSE.

KINETICS AND REACTION RATES OF OCTAHEDRAL SUBSTITUTION

Although it is probable that if the reaction rates were known for all possible complexes a continuous series could be formed, it is still convenient to classify metal ions in four categories based on the rate of exchange of coordinated water. Gray and Langford[19] have suggested the following classification (compare also Fig. 11.5).

Class I. The exchange of water is extremely fast. In general, even the fastest kinetic techniques cannot follow the reaction. First-order rate constants are of the order of $10^8\ s^{-1}$. The complexes are bound by essentially purely electrostatic forces and include

[19] H. B. Gray and C. H. Langford, *Chem. Eng. News*, April 1, **1968**, p. 68.

the complexes of the alkali metals and larger alkaline earth metals. The Z^2/r ratio for these ions ranges up to about 10 C^2 m^{-1} (see Table 7.1).

Class II. The exchange of water is fast. First-order rate constants for water exchange range from 10^5 to 10^8 s^{-1}. Such reactions can be studied by the fastest kinetic techniques such as relaxation techniques. In these methods, the system at equilibrium is perturbed by a fast variation of a physical parameter such as pressure (ultrasonic or "P-jump" methods) or temperature ("T-jump" methods) and the response of the system used to estimate rates of reaction.[20] Metal ions belonging to this group are the dipositive transition metals, Mg^{+2}, and tripositive lanthanides. These may be considered as ions in which the bonding is somewhat stronger than in Class I, but LFSEs are relatively small. The Z^2/r for these ions ranges from about 10 to 30 C^2 m^{-1}.

Class III. The exchange of water is relatively slow compared with Classes I and II, although fast on an absolute scale, with first-order rate constants of 1 to 10^4 s^{-1}. The reaction can be followed, however, by more or less conventional kinetic techniques if they are augmented by flow techniques so that the fast reaction times can be related to the rate of flow. The metal ions of this group are most of the tripositive transition metal ions, stabilized to some extent by LFSE, and two very small ions Be^{+2} and Al^{+3}: The Z^2/r ratios are greater than about 30 C^2 m^{-1}.

Class IV. The exchange of water is slow. These are the only inert complexes. First-order rate constants may range from 10^{-1} to 10^{-9} s^{-1}. These ions are comparable in size to Class III ions and exhibit considerable LFSE: Cr^{+3} (d^3), Co^{+3} (low-spin d^6), Pt^{+2} (low-spin d^8).

Ligand field effects and reaction rates

The complexes of metal ions in Class IV are typically stabilized to a great extent by LFSE: Co^{+3} (24*Dq*, low-spin) and Cr^{+3} (12*Dq*). Of course it is not the absolute LFSE that prevents reaction but the loss upon formation of the activated complex. The difficulty here is, of course, to assign the LFSE of the activated complex without exact knowledge of its structure. In the absence of such knowledge, approximations can be made on the basis of likely structures. Basolo and Pearson[21] have presented values for strong and weak fields for square pyramidal (C.N. = 5) and pentagonal bipyramidal (C.N. = 7) intermediates. The values are listed in Table 11.1. The experimental results, in general, correlate rather well with this simple picture. For tripositive metal ions, lability is expected to increase in the order Co(III) < Cr(III) < Mn(III) < Fe(III) < Ti(III) < Ga(III) < Sc(III). The comparison of dipositive with tripositive species is difficult since the rate is also affected by the charge on the central metal ion which strengthens the metal–ligand bonds. For example, in the nontransition metal series $[AlF_6]^{-3} > [SiF_6]^{-2} > [PF_6]^{-} > SF_6$, the lability decreases in the order shown with SF_6 being exceptionally inert.

As a result of the generally greater lability of dipositive metals, octahedral Ni(II) is considerably more labile than corresponding tripositive ions in line with the exchange

[20] K. Kustin, ed., "Methods in Enzymology," Vol. XVI (Fast Reactions), Academic Press, New York, **1969**; K. Kustin and J. Swinehart, *Progr. Inorg. Chem.*, **1970**, *13*, 107; R. G. Wilkins, "The Study of Kinetics and Mechanism of Reactions of Transition Metal Complexes," Allyn and Bacon, Boston, **1974**, pp. 122–177.

[21] F. Basolo and R. G. Pearson, "Mechanisms of Inorganic Reactions," 2nd ed., Wiley, New York, **1967**. See also footnote 35.

Table 11.1 Change in LFSE (units *Dq*)[a] upon changing a 6-coordinate complex to a 5-coordinate (square pyramidal) or a 7-coordinate (pentagonal bipyramidal) species

	High spin		Low spin	
System	C.N. = 5	C.N. = 7	C.N. = 5	C.N. = 7
d^0	0	0	0	0
d^1	+0.57	+1.28	+0.57	+1.28
d^2	+1.14	+2.56	+1.14	+2.56
d^3	−2.00	−4.26	−2.00	−4.26
d^4	+3.14	−1.07	−1.43	−2.98
d^5	0	0	−0.86	−1.70
d^6	+0.57	+1.28	−4.00	−8.52
d^7	+1.14	+2.56	+1.14	−5.34
d^8	−2.00	−4.26	−2.00	−4.26
d^9	+3.14	−1.07	+3.14	−1.07
d^{10}	0	0	0	0

[a] The common convention is used: Negative quantities refer to *loss* of LFSE and destabilization of the complex.

SOURCE: Modified from F. Basolo and R. G. Pearson, "Mechanisms of Inorganic Reactions," 2nd ed., Wiley, New York, **1967**. Used with permission.

rates shown in Eqs. 11.13 to 11.15. When considering dipositive metals only, however, it is found that other ions such as Mn(II), Fe(II), Co(II), and Cu(II) that possess less LFSE are even more labile.

The difference in lability between Ni(II) complexes and Pt(II) complexes can be related to effects of ligand fields as well. Ni(II) is a Class II metal, whereas Pt(II) belongs to Class IV. The explanation parallels that provided for Cu(II) versus Au(I)–Au(III) systems (see p. 410). In contrast to odd electrons being oxidized in the latter, the Ni(II) versus Pt(II) comparison involves the availability of the antibonding d_{z^2} orbital for occupancy by a fifth, attacking group in an associative mechanism. Pt(II) will lose proportionately more energy in such a transition since the LFSE is considerably greater for the heavier metal (see pp. 385–386).

Mechanisms of substitution reactions

The study of substitution mechanisms has occupied many workers for a long time and continues to be an active area of investigation. Only a brief outline of the results of such work can be given here.[22] Consider a reaction of the type:

$$[N_nML] + E \longrightarrow [N_nME] + L \qquad \textbf{(11.17)}$$

[22] For more nearly complete discussions of reaction mechanisms, see D. Benson, "Mechanisms of Inorganic Reactions in Solution," McGraw-Hill, New York, **1968**; F. Basolo and R. G. Pearson, "Mechanisms of Inorganic Reactions," 2nd ed., Wiley, New York, **1967**; C. H. Langford and H. B. Gray, "Ligand Substitution Processes," Benjamin, New York, **1965**; R. G. Wilkins, "The Study of Kinetics and Mechanisms of Reactions of Transition Metal Complexes," Allyn and Bacon, Boston, **1974**.

where $n = 3$ for a square planar complex, $n = 5$ for an octahedral complex, N = a "noninvolved" ligand, L = the leaving ligand, and E = the entering ligand. We have seen that for square planar complexes the reaction has an associative (A or S_{N^2}) mechanism:

$$[N_3PtL] + E \longrightarrow [N_3PtLE] \longrightarrow [N_3PtE] + L \tag{11.18}$$

Octahedral complexes conceivably can react via either an associative, dissociative (D or S_{N^1}), or intermediate (I) mechanism:

$$[N_5CoL] + E \longrightarrow [N_5CoLE] \longrightarrow [N_5CoE] + L \tag{11.19}$$

$$[N_5CoL] \xrightarrow{-L} [N_5Co] \xrightarrow{+E} [N_5CoE] \tag{11.20}$$

involving 7-coordinate and 5-coordinate intermediates, respectively.

It is likely that most reactions will be found not to fall cleanly into the extreme mechanism of the first two types, but will involve to a certain degree a concerted mechanism.

Substitution in octahedral complexes

One might expect the *trans* effect to be operable in octahedral complexes as well as in square planar complexes. Evidence for its presence in octahedral complexes is not abundant, however. There is definite evidence for a *trans influence* on the ground state of cobalt(III) complexes of the type $[Co(NH_3)_5X]^{+2}$. Significant lengthening is found in the Co—N bond *trans* to the nitrosyl group and shortening *trans* to a chloride ion.[23] In metal carbonyls the *trans* effect seems to be operative to the extent that substitution proceeds readily as long as there are carbonyl groups *trans* to each other (and presumably activating each other):

$$Mo(CO)_6 \xrightarrow[-3CO]{+3R_3P} fac\text{-}Mo(CO)_3(PR_3)_3 \tag{11.21}$$

$$MnBr(CO)_5 \xrightarrow[-2CO]{+2R_3P} MnBr(CO)_3(PR_3)_2 \tag{11.22}$$

These products are almost invariably *facial* if trisubstituted or *cis* if the reaction is stopped at the disubstituted product. If steric hindrance exists between the *cis* substituents, isomerization to a *trans* product may occur. Otherwise greater stability is found with the carbonyl groups *trans* to the weaker π-bonding phosphines.[24]

Some octahedral complexes show both a definite *trans* influence and a *trans* effect although it is difficult to quantify the relationship between the two. Ions such as the *trans*-aquabis(ethylenediamine)-S-thiosulfatocobalt(III) and the corresponding S-sulfito- and

[23] For a complete discussion of *cis* and *trans* effects in cobalt(III) complexes, see J. M. Pratt and R. G. Thorp, *Adv. Inorg. Chem. Radiochem.*, **1969**, *12*, 375.

[24] R. B. King, "Transition-Metal Organometallic Chemistry," Academic Press, New York, **1969**, pp. 73–75.

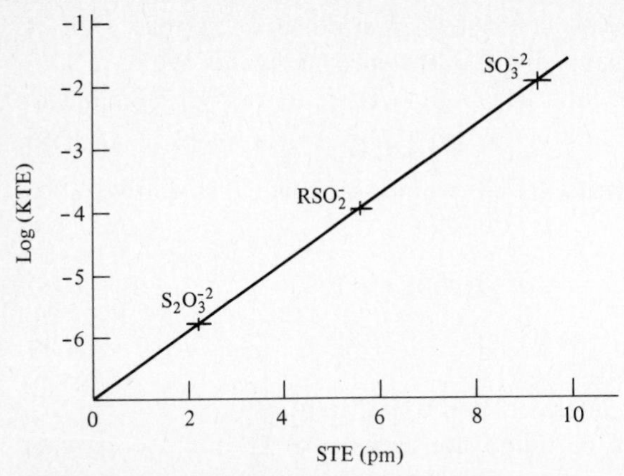

Fig. 11.6 Plot of log (KTE) vs. STE (pm) for the S-bonded ligands SO_3^{-2}, RSO_2^-, and $S_2O_3^{-2}$. [From J. N. Cooper et al., *Inorg. Chem.*, **1980**, *19*, 2265. Reproduced with permission.]

S-alkylsulfinato-complexes show a lengthening of the bond *trans* to the sulfur atom and a labilization of the *trans* ligand. Furthermore, there seems to be a linear relation between the *trans influence* (sometimes termed the "static *trans* effect," or STE) and the usual *trans effect* (sometimes termed the "kinetic *trans* effect," or KTE). See Fig. 11.6.[25]

Substitution in octahedral cobalt(III) complexes

The best-studied systems have been those involving Co(III) because of the stability of the complexes and their availability. As we shall see, other systems sometimes parallel those of Co(III) and sometimes not. For example, in common with most octahedral complexes, Co(III) usually does not show a *trans* effect. Apparently this is because substitution reactions at an octahedral Co(III) center occur via a dissociative mechanism (D or S_{N1}):

$$[N_5CoL]^{+n} \xrightarrow[\text{slow}]{-L^-} [CoN_5]^{+n+1} \xrightarrow[\text{fast}]{+E^-} [N_5CoE]^{+n} \tag{11.23}$$

or a concerted interchange mechanism tending toward the dissociative side (I_d or $S_{N1}IP$) with ion pair formation prior to reaction so that the entering group can enter as soon as the leaving group departs:[26]

$$[N_5CoL]^{+n} \xrightarrow[\text{fast}]{+E^-} [N_5CoL]^{+n}E \xrightarrow[\text{slow}]{} [N_5CoE]^{+n}L \xrightarrow[\text{slow}]{-L^-} [N_5CoE]^{+n} \tag{11.24}$$

There is considerable evidence for mechanisms of this sort, but only a few examples will be given. First, the rates of such reactions correlate fairly well (inversely) with the thermodynamic bond strength of the Co—L bond, indicating that the Co—L bond is broken

[25] R. C. Elder et al., *Inorg. Chem.*, **1978**, *17*, 431; J. N. Cooper et al., ibid., **1980**, *19*, 2265.

[26] The distinction between these mechanisms is a rather fine one and need not concern us here. Basically, it revolves around whether there is a 5-coordinate intermediate with a finite lifetime that can discriminate among potential entering groups.

Table 11.2 Rate constants for the acid hydrolysis of $[Co(en)_2(X\text{-}py)Cl]^{+2}$ at 50 °C

X-py	pK	Rate constant, k (s^{-1})
Pyridine	8.82	1.1×10^{-5}
3-Methylpyridine	8.19	1.3×10^{-5}
4-Methylpyridine	7.92	1.4×10^{-5}
4-Methoxypyridine	7.53	1.5×10^{-5}

Table 11.3 Rate constants and isomeric products for the acid hydrolysis of chlorobis-(ethylenediamine)(ligand)cobalt(III) complexes at 25 °C

	Cis isomer		*Trans* isomer		
Ligand, C	Rate constant, *cis* isomer, k_c (s^{-1})	% *cis* isomer in product	Rate constant, *trans* isomer (s^{-1})	% *trans* isomer in product	k_c/k_t
NCS^-	1.1×10^{-5}	100	5.0×10^{-8}	30–50	220
OH^-	1.3×10^{-2}	100	1.4×10^{-3}	25	9.2
Cl^-	2.4×10^{-4}	100	3.2×10^{-5}	65	7.5
N_3^-	2.5×10^{-4}	100	2.4×10^{-4}	80	1.04
NH_3	5.0×10^{-7}	84	4.0×10^{-7}	17	1.25
NO_2^-	1.1×10^{-4}	100	1.0×10^{-3}	100	0.11

initially. We might expect that this bond between the metal and an anion would be broken more readily if electron density were increased on the metal. Electron-donating ligands should therefore assist the replacement of L. This proves to be true as shown by the data for the acid hydrolysis of chloride complexes in which various pyridines are present:

$$[Co(en)_2(X\text{-}py)Cl]^{+2} + H_2O \longrightarrow [Co(en)_2(X\text{-}py)H_2O]^{+3} + Cl^- \qquad \textbf{(11.25)}$$

It is found that increasing the basicity of the pyridine (and hence increasing the electron density on the metal) *does* indeed increase the rate of the reaction (Table 11.2).[27]

Increasing the total negative charge on the complex facilitates the reaction by more readily allowing the departure of the leaving ligand. If the reaction were an associative one, increasing the negative charge would be expected to *repel* the incoming ligand.

Further evidence for the proposed mechanism comes through operation of the *cis* effect in certain reactions. The nature of the *cis* effect can be seen from Table 11.3. Certain ligands such as the thiocyanate and hydroxide ions greatly accelerate the hydrolysis of a complex when *cis* to the leaving group as compared to the analogous reaction in which the leaving group is in the *trans* position. For example, consider the reaction:

$$[Co(en)_2CCl]^{+n} + H_2O \longrightarrow [Co(en)_2C(H_2O)]^{+n+1} + Cl^- \qquad \textbf{(11.26)}$$

[27] F. Basolo et al., *J. Am. Chem. Soc.*, **1956**, *78*, 2676.

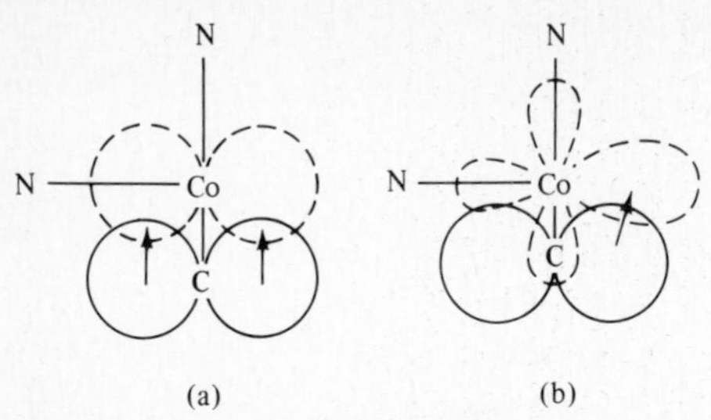

Fig. 11.7 Overlap of filled *p* orbital of C with vacant (a) *p* orbital or (b) d^2sp^3 hybrid orbital of cobalt in a 5-coordinated square pyramidal activated complex resulting from the dissociation of X from *cis*-[CoN_4CL]. The two N ligands not shown are above and below the plane of the paper. [From F. Basolo and R. G. Pearson, *Progr. Inorg. Chem.*, **1962**, *4*, 381. Reproduced with permission.]

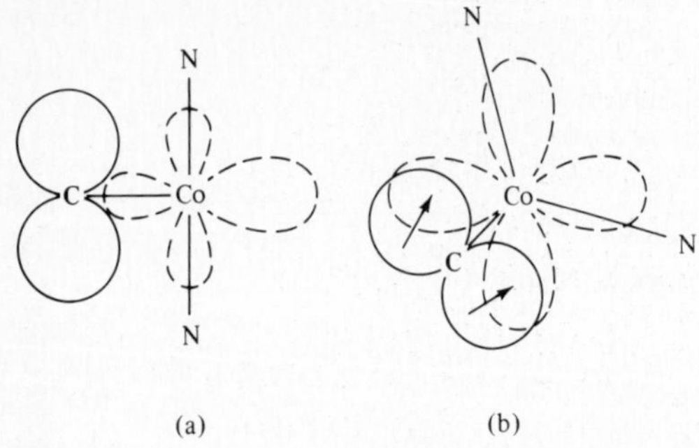

Fig. 11.8 (a) No overlap of filled *p* orbital of C with vacant d^2sp^3 hybrid orbital of cobalt in the square pyramidal activated complex resulting from the dissociation of X from *trans*-[CoN_4CX]. (b) Efficient overlap with vacant $d_{x^2-y^2}$ orbital if there is rearrangement of a trigonal bipyramidal structure. The two N ligands not shown are above and below the plane of the paper. [From F. Basolo and R. G. Pearson, *Progr. Inorg. Chem.*, **1962**, *4*, 381. Reproduced with permission.]

where C represents a *cis*-activating ligand such as hydroxide ion. When the hydroxide ion is *cis* to the leaving chloride ion, the reaction rate is about ten times as great as when it is in the *trans* position. The reaction rates for the acid hydrolysis of *cis* and *trans* isomers are listed in Table 11.3 together with the ratio of the reaction rates k_c/k_t. Those ligands that exhibit a strong *cis* effect have unshared pairs of electrons (in addition to the pair used in the σ dative bond). Pearson and Basolo[28] have suggested that this electron pair is available to be donated to the metal in a p–d π bond (Fig. 11.7). Such π bonding by a *cis*-substituent can stabilize a square pyramidal activated complex by lowering the positive charge on the metal. This also allows the reaction to proceed without extensive rearrangement and the product is 100% *cis* isomer. When these same ligands are in the *trans* position there is no orbital available for overlap unless the complex rearranges to a trigonal bipyramidal structure (Fig. 11.8). This raises the energy of activation and makes substitution reactions more difficult and results in mixtures of isomers in the product. Those ligands which fail to exhibit a positive *cis* effect are of two types: (1) those that lack another lone pair to donate to the metal such as ammonia, and (2) those that are π *acceptors*, such

[28] R. G. Pearson and F. Basolo, *J. Am. Chem. Soc.*, **1956**, *78*, 4878 (1956).

as the nitro group.[29] The latter tend to withdraw electrons via π bonding and hence cannot operate through the mechanism outlined above.

In acid hydrolysis the reaction rate is independent of the nature or the concentration of the entering group, which is further evidence for a dissociative mechanism. The only anion that has been shown to have an appreciable effect upon the rate of hydrolysis is the hydroxide ion. The rate constant for hydrolysis in basic solution is often a million times that found for acidic solutions. Furthermore, the reaction rate is found to be second order and dependent upon the hydroxide ion concentration:[30]

$$\frac{-d[\text{complex}]}{dt} = k_B[\text{complex}][\text{OH}^-] \quad \textbf{(11.27)}$$

The dependence upon hydroxide ion concentration can be taken as evidence of an associative (S_{N2}) mechanism:

$$[(\text{NH}_3)_5\text{CoCl}]^{+2} + \text{OH}^- \longrightarrow [(\text{NH}_3)_5\text{CoCl(OH)}]^+ \longrightarrow [(\text{NH}_3)_5\text{CoOH}]^{+2} + \text{Cl}^- \quad \textbf{(11.28)}$$

Although this accounts adequately for the kinetics of the reaction, the prevailing opinion is that the reaction takes place via an alternative mechanism. First, we have seen that, in general, 7-coordinate complexes are not very stable compared to the relative stability of 5-coordinate complexes. An alternative mechanism involving coordination number 5 can be suggested in which the hydroxide ion promotes the reaction through proton abstraction:

$$[(\text{NH}_3)_5\text{CoCl}]^{+2} + \text{OH}^- \overset{\text{fast}}{\rightleftharpoons} [(\text{NH}_3)_4\text{Co(NH}_2)\text{Cl}]^+ + \text{H}_2\text{O} \quad \textbf{(11.29)}$$

$$[(\text{NH}_3)_4\text{Co(NH}_2)\text{Cl}]^+ \xrightarrow{\text{slow}} [(\text{NH}_3)_4\text{CoNH}_2]^{+2} + \text{Cl}^- \quad \textbf{(11.30)}$$

$$[(\text{NH}_3)_4\text{CoNH}_2]^{+2} + \text{H}_2\text{O} \xrightarrow{\text{fast}} [(\text{NH}_3)_5\text{CoOH}]^{+2} \quad \textbf{(11.31)}$$

According to this viewpoint, the hydroxide ion would rapidly set up an equilibrium with the amidocobalt complex. The rate-determining step would be the dissociation of this complex as in the acid hydrolysis discussed above. The concentration of the amido complex would be dependent, however, on the hydroxide ion concentration through equilibrium (Eq. 11.29), hence the reaction rate would be proportional to the hydroxide ion concentration.

This mechanism, assigned the symbolism $S_{N1}CB$ for first-order reaction acting on the conjugate base of the complex, is supported by a number of observations. It rationalizes the fact that the hydroxide ion is unique in its millionfold increase in rate over acid hydrolysis; other anions which are incapable of abstracting protons from the complex but which would otherwise be expected to be good nucleophiles in an S_{N2} reaction do *not* show this increase. Furthermore, the $S_{N1}CB$ mechanism can apply only to complexes in

---

[29] The azide ion, N_3^-, can potentially behave as either a π donor or a π acceptor. It does not appear to act strongly in either capacity.

[30] Equation 11.27 is not precisely accurate. The complex continues to hydrolyze by the dissociative (or "acid") mechanism even at high pH values, so a more accurate expression would be $-d[\text{complex}]/dt = k_A[\text{complex}] + k_B[\text{complex}][\text{OH}^-]$. Since k_B is from 10^5 to 10^8 times as great as k_A, Eq. 11.27 is adequate if the $[\text{OH}^-]$ is sufficiently high.

which one or more ligands have ionizable hydrogen atoms. Thus complexes such as $[Co(py)_4Cl_2]^+$ and $[Co(CN)_5Cl]^{-3}$ would not be expected to exhibit typical base hydrolysis and indeed they do not. The hydrolysis proceeds slowly and without dependence upon the hydroxide ion.[31]

If the hydroxide ion accelerates the reaction only via proton abstraction rather than by direct attack, it might be supposed that it would be possible to trap the 5-coordinate activated complex by addition of large amounts of an anion other than hydroxide:

$$[CoA_5Cl]^{+2} + OH^- \rightleftharpoons [CoA_4(A-H)Cl]^+ + H_2O \tag{11.32}$$

$$[CoA_4(A-H)Cl]^+ \longrightarrow [CoA_4(A-H)]^{+2} + Cl^- \tag{11.33}$$

$$[CoA_4(A-H)]^{+2} + X^- \longrightarrow [CoA_4(A-H)X]^+ \tag{11.34}$$

$$[CoA_4(A-H)X]^+ + H_2O \longrightarrow [CoA_5X]^{+2} + OH^- \tag{11.35}$$

where A represents an amino group from ammonia or an amine, (A − H) is the corresponding deprotonated conjugate base, and X^- is a competing nucleophilic anion. In aqueous solution it is found that the only product isolated is the hydrolyzed complex, presumably because $[H_2O]$ is so much greater than can possibly be obtained for $[X^-]$ that reaction 11.31 proceeds faster than 11.34. By allowing the reaction to take place in a nonaqueous solvent (dimethylsulfoxide) it was shown that reactions 11.32 to 11.35 could be made to take place.[32] Under these circumstances it was found that the reaction is catalyzed by small amounts of hydroxide ion although no hydroxide appears in the product if large concentrations of a competing nucleophile such as nitrite, azide, or thiocyanate ions are present. It was even found possible to replace hydroxide by piperidine and observe the same base-catalyzing replaçement of chloride by X. All of these observations are consistent with an $S_{N1}CB$ mechanism but not with an S_{N2} mechanism.

An elegant experiment has been performed by Green and Taube[33] providing strong support for the $S_{N1}CB$ mechanism for the base hydrolysis of halopentaamminecobalt(III) complexes. Naturally occurring oxygen contains about 0.2% of the heavy isotope ^{18}O. It is found that hydroxide ions in aqueous solution have a smaller abundance of ^{18}O than the surrounding water molecules. This isotope effect does not involve the past history of the hydroxide ions inasmuch as very rapid exchange takes place and so the system is always at thermodynamic equilibrium. Rather it involves a slight preference of ^{18}O for water molecules instead of hydroxide ions. It provides an effective label for OH^- and H_2O since by measuring the $^{18}O/^{16}O$ ratio the source of the oxygen (i.e., whether from OH^- or H_2O) can be ascertained. In aqueous solution, the *fractionation factor* is given by:

$$f_{OH^-} = \frac{[H_2{}^{18}O][^{16}OH^-]}{[H_2{}^{16}O][^{18}OH^-]} = 1.040 \tag{11.36}$$

[31] This statement, like most others made about mechanisms, may have exceptions. Busch and co-workers suggested that the base hydrolysis of $[Co(EDTA)]^-$ takes place via an S_{N2} mechanism and a 7-coordinate intermediate. (See D. H. Busch et al., in "Advances in the Chemistry of the Coordination Compounds," S. Kirschner, ed., Macmillan, New York, **1961**.)

[32] R. G. Pearson et al., *J. Am. Chem. Soc.*, **1960**, *82*, 4434. These workers used *trans*-chloronitrobis(ethylenediamine)cobalt(III) salts.

[33] M. Green and H. Taube, *Inorg. Chem.*, **1963**, *2*, 948.

Green and Taube hydrolyzed halopentaamminecobalt(III) complexes in basic solutions of various concentrations. They isolated the resulting hydroxo complex and measured the fractionation factor for it compared with water:

$$f_{\text{complex}} = \frac{[Co(NH_3)_5{}^{16}OH][H_2{}^{18}O]}{[Co(NH_3)_5{}^{18}OH][H_2{}^{16}O]} \tag{11.37}$$

The observed fractionation factor for chloro, bromo, and iodo complexes was 1.0056 ± 0.0001. If the hydroxide had entered the coordination sphere directly, the value should have been 1.040. On the other hand, if a 5-coordinate intermediate was involved which then reacted with water (Eq. 11.31), the fractionation factor should be 1.000. The proximity of the value of f_{complex} to unity rules out the possibility that the hydroxide ion can be involved in direct attack on the metal.[34]

Although the mass of evidence now indicates that the normal mode of reaction for Co(III) complexes involves a dissociative mechanism, the question is not completely settled, and there is some evidence, both pro and con, that Cr(III) complexes proceed via an S_{N2} mechanism. A tentative prediction has been made on the basis of the effect of LFSE on the activation energy that Co(III) complexes will react via a dissociative mechanism but that Cr(III) will react via an associative mechanism involving a capped trigonal prism (C.N. = 7).[35]

The preceding discussion of substitution mechanisms barely scratches the surface of a field that has occupied the attention of many of the world's best coordination chemists. It is an area in which a seeming infinity of problems and methods of attack exists. It is unfortunate that it is not possible to present a more comprehensive theory of substitution mechanisms. In fact, the above discussion errs on the side of omission of fine points and controversial interpretations rather than otherwise.[36] For every experiment designed to confirm a mechanism, an alternative explanation can usually be found. As one noted researcher once said: "[The members of the other school of thought] are extremely ingenious at coming up with alternative explanations for all of the conclusive experiments that we seem to do."[37] This should serve to remind us of the truism that it is not possible to prove that a particular mechanism is the correct one; it is only possible sometimes to prove that an alternative mechanism is *not* correct. To this might be added a corollary: Often it is extremely difficult to prove that the alternative *is* impossible.

Other octahedral mechanisms

Although it is impossible to examine all of the many reactions that have been studied, it is useful to look at a couple of reactions that yield different results. Surprisingly, we need go no further than the congeners of cobalt and examine some of the chemistry of Rh(III) and Ir(III). Evidence that complexes such as $[Rh(NH_3)_5H_2O]^{+3}$ and $[Ir(NH_3)_5H_2O]^{+3}$

[34] The small discrepancy between 1.0000 and 1.0056 was explained as resulting from small kinetic isotope effects as the reaction proceeded.

[35] S. T. Spees, Jr., et al., *J. Am. Chem. Soc.*, **1968**, *90*, 6626.

[36] For a review of some of these problems associated with base hydrolysis in octahedral complexes, see M. L. Tobe, *Acc. Chem. Res.*, **1970**, *3*, 377.

[37] R. G. Pearson in "Mechanisms of Inorganic Reactions," Advances in Chemistry Series, No. 49, J. Kleinberg, Symposium Chairman, American Chemical Society, Washington, D.C., **1965**, p. 25.

react via associative mechanisms is not as extensive as data for Co(III) but seems clear. First, reaction of these complexes with anions such as Cl^- and Br^- proceed several times more rapidly than exchange of water, indicating that the incoming anion plays an active role in the transition, presumably in the form of a 7-coordinate intermediate. Further evidence comes from the volume of activation, $\Delta V^{\ddagger}$. Cobalt(III) complexes react with a positive $\Delta V^{\ddagger}$ as expected for dissociation in the transition state. In contrast, exchange of water in $[Rh(NH_3)_5H_2O]^{+3}$ and $[Ir(NH_3)_5H_2O]^{+3}$ proceeds with a *negative* $\Delta V^{\ddagger}$, indicating an associative mechanism.

Racemization and isomerization

Another set of reactions that has received considerable attention is that in which optically active complexes, especially tris(chelate) compounds, racemize:

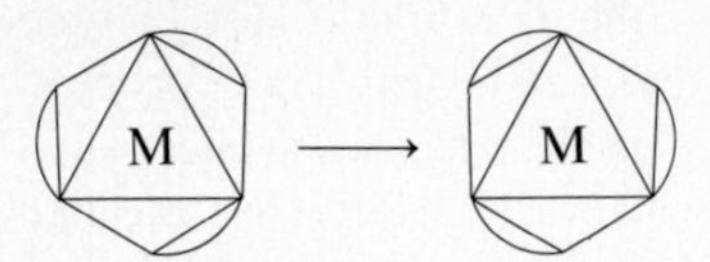

There are several mechanisms that are possible for such an inversion, some of which can be eliminated by appropriate experiments. For example, one mechanism would be complete dissociation of one chelate ring with formation of a square planar complex or a *trans* diaqua complex:

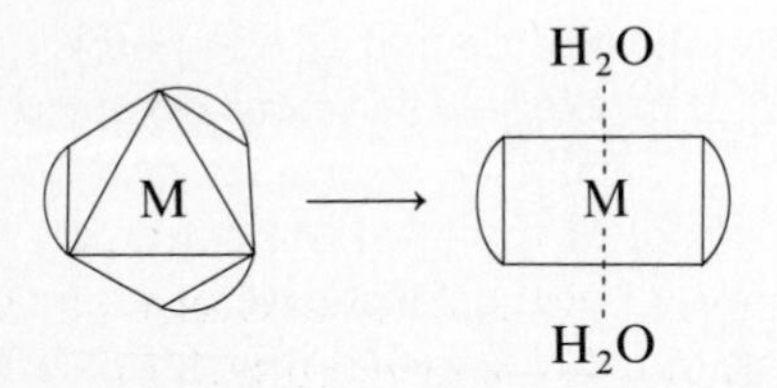

The asymmetry has been lost, and if the chelate ring is reformed, it has a 50-50 chance of forming either the Δ or Λ isomer. Since the rate-determining step in the racemization is the dissociation of the ligand, the rate of racemization (k_r) must equal the rate of dissociation (k_d). For example, tris(phenanthroline)nickel(II) racemizes at the same rate ($k_r = 1.5 \times 10^{-4}\ s^{-1}$) as it dissociates ($k_d = 1.6 \times 10^{-4}\ s^{-1}$). If racemization takes place faster than dissociation (e.g., tris(phenanthroline)iron(II), $k_r = 6.7 \times 10^{-4}$; $k_d = 0.70 \times 10^{-4}\ s^{-1}$), this mechanism can be eliminated.

More reasonable is that only one end of the chelate ring detaches with formation of a 5-coordinate complex. This complex could undergo a Berry pseudorotation (see p. 247) with scrambling of the positions. Reforming of the chelate ring would then form a racemic mixture of Δ and Λ isomers.

"Twist" mechanisms that do not require bond rupture have been proposed to account for racemization. The earliest mechanism of this sort was proposed by Rây and Dutt[38] and is known as the rhombic twist (Fig. 11.9a). More recently, Bailar suggested a trigonal

[38] P. C. Rây and N. K. Dutt, *J. Indian Chem. Soc.*, **1943**, *20*, 81.

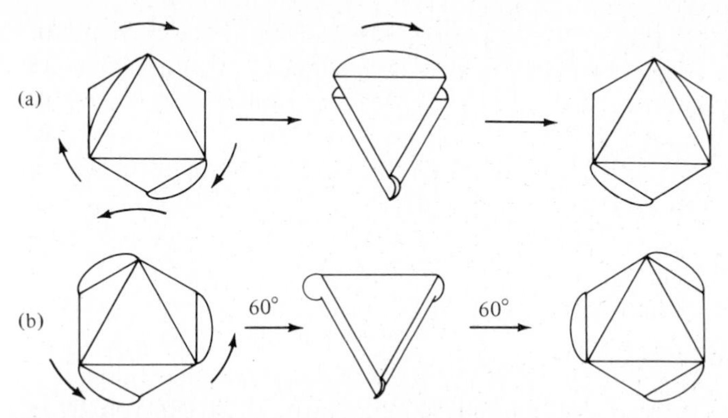

Fig. 11.9 (a) Rhombic or Rây-Dutt twist; (b) Trigonal or Bailar twist. Note that the difference does not lie in the direction of the twist but in the twist axis. An octahedron has eight C_3 axes (normal to the eight trigonal faces). Addition of three chelate rings reduces this number to two. A 120° twist about a C_3 axis results in a Bailar twist (b). If the twist is about one of the six remaining pseudo-C_3 axes, it is a Rây-Dutt twist (a). (The third ligand in (a) is hidden behind the octahedron.)

twist mechanism (Fig. 11.9b).[39] It is a commentary on the experimental difficulties encountered that so many years have passed with so little firm experimental evidence. It appears that rigid chelates and those with small bite angles favor a trigonal twist. Rigidity in the chelate ring inhibits one-ended dissociation. Small bite angles are known to stabilize trigonal prismatic coordination (see p. 488) and thus might be expected to reduce the energy barrier to the twist.[40]

MECHANISMS OF REDOX REACTIONS[41,42]

It might be thought that there would be little to study in the mechanism of electron transfer—that the reducing agent and the oxidizing agent would simply "bump" into each other and electron transfer would take place. Reactions in solution are complicated, however, by the fact that the oxidized and reduced species are often metal ions surrounded by shields of ligands and/or solvating molecules.

Outer sphere mechanisms

In this type of reaction the coordination spheres of the metal ions are not altered. Consider the reaction:

$$[Fe(CN)_6]^{-4} + [Mo(CN)_8]^{-3} \longrightarrow [Fe(CN)_6]^{-3} + [Mo(CN)_8]^{-4} \quad \textbf{(11.38)}$$

[39] J. C. Bailar, Jr., *J. Inorg. Nucl. Chem.*, **1958**, *8*, 165.

[40] For further discussion of the factors affecting twist mechanisms, see D. L. Kepert, *Prog. Inorg. Chem.*, **1977**, *23*, 1.

[41] H. Taube, *J. Chem. Educ.*, **1968**, *45*, 452; "Electron Transfer Reactions of Complex Ions in Solution," Academic, New York, **1970**; A. G. Sykes, *Essays Chem.*, **1970**, *1*, 25; A. Haim, *Acc. Chem. Res.*, **1975**, *8*, 264.

[42] D. E. Pennington, in "Coordination Chemistry," Vol. 2, A. E. Martell, ed., *ACS Monograph No. 174*, Washington, D. C., **1978**, pp. 476–590.

Such a reaction may be considered to approximate the simple collision model expressed above. In general, the rate of the redox reaction is faster than that of the exchange of the ligands, and so we may consider the reaction to be a simple electron transfer from one stable complex to another.

However, consider the following reaction, which might seem to be the ultimate example of two ions merely "bumping together":

$$[Fe(H_2O)_6]^{+3} + [Fe^*(H_2O)_6]^{+2} \longrightarrow [Fe(H_2O)_6]^{+2} + [Fe^*(H_2O)_6]^{+3} \quad \mathbf{(11.39)}$$

Yet when we examine the reaction more closely, we find that $\Delta G^* \simeq 33$ kJ mol^{-1}. Why not zero?

The bond lengths in Fe^{+2} and Fe^{+3} complexes are not the same (see Table 3.4). If the electron transfer could take place without input of energy, we would obtain as products the Fe(II) complex with bond lengths typical of Fe^{+3} and the Fe(III) complex with bond lengths typical of Fe^{+2}; both could then relax with the release of energy. This clearly violates the First Law of Thermodynamics. The electron transfer creates a system with the potential energy of a coiled spring and thus requires work. The actual process occurs with shortening of the bonds in the Fe(II) complex, lengthening of the bonds in the Fe(III) complex, followed by transfer of the electron. There has been considerable interest in correlating the rate of electron transfer with the structural properties of the reacting complexes. This topic, which goes under the general rubric of *Marcus theory*, is beyond the scope of this book, but fortunately there are good reviews.[42,43]

Inner sphere mechanisms

In this type of reaction a ligand is intimately involved in the transfer of the electron from one metal to another. The first example of this type of mechanism was provided by Taube and co-workers.[44] The system involved the reduction of cobalt(III) (in $[Co(NH_3)_5Cl]^{+2}$) by chromium(II) (in $[Cr(H_2O)_6]^{+2}$) and was specifically chosen because (1) both Co(III) and Cr(III) form inert complexes and (2) the complexes of Co(II) and Cr(II) are labile (see p. 548). Under these circumstances the chlorine atom, firmly attached to the inert Co(III) ion, can displace a water molecule from the labile Cr(II) complex to form a bridged intermediate:

$$[Co(NH_3)_5Cl]^{+2} + [Cr(H_2O)_6]^{+2} \longrightarrow [(NH_3)_5Co{-}Cl{-}Cr(H_2O)_5]^{+4} + H_2O \quad \mathbf{(11.40)}$$

The redox reaction now takes place within this binuclear complex with formation of reduced Co(II) and oxidized Cr(III). The latter species forms an inert chloroaqua complex, but the cobalt(II) has been labilized so the intermediate dissociates with the chlorine atom remaining with the chromium:

$$[(NH_3)_5Co{-}Cl{-}Cr(H_2O)_5]^{+4} \longrightarrow [(NH_3)_5Co]^{+2} + [Cr(H_2O)_5Cl]^{+2} \quad \mathbf{(11.41)}$$

[43] R. A. Marcus, *Ann. Rev. Phys. Chem.*, **1964**, *15*, 155.
[44] H. Taube et al., *J. Am. Chem. Soc.*, **1953**, *75*, 4118.

The 5-coordinate cobalt(II) species presumably immediately picks up a water molecule to fill its sixth coordination position, and it is so labile that hydrolysis to $[Co(H_2O)_6]^{+2}$ occurs rapidly.

Formally, such an inner sphere reaction consists of the transfer of a chlorine *atom* from cobalt to chromium, decreasing the oxidation state of the former but increasing that of the latter. In addition to the self-consistency of the above model (inert and labile species) and the observed formation of a chlorochromium complex, further evidence for this mechanism can be obtained by running the reaction in the presence of free radioisotopically labeled chloride ions in the solution. Very little of this chloride is ever found in the product, indicating that the chloride transfer has indeed been through the bridge rather than indirectly through free chloride.

Inner sphere mechanisms of this type have several consequences. An obvious one is the transfer of a ligand from one coordination sphere to another. A second is that for a reaction such as described above, the rate can be no faster than the rate of exchange of the ligand in the absence of a redox reaction, since exchange of the ligand is an intimate part of the process. A third, less obvious consequence is that the reaction is zero order in one important reactant. If the rate-determining step is a dissociation of a ligand from one complex (to form a site for bridging), the reaction will be first order in that species but zero order in the second. If the rate-determining step, however, is the attack of the second species, the reaction will be first order in it but zero order in the first complex.[45]

If the bridging ligand contains more than one atom, the thermodynamically favored isomer may *not* be the product obtained. The geometry of the bridge may result in a linkage isomer (*remote* attack) of the most stable product:

$$[(NH_3)_5Co(CN)]^{+2} + [Co(CN)_5]^{-3} \longrightarrow [\ldots Co{-}C{\equiv}N{-}Co \ldots] \longrightarrow [(NH_3)_5Co]^{+2} + [CN{-}Co(CN)_5]^{-3} \quad \textbf{(11.42)}$$

In at least one case,[46] *both* linkage isomers form, either through *remote* attack, or through *adjacent* attack:

$$[(NH_3)_5CoSCN]^{+2} + [Cr(H_2O)_6]^{+2} \longrightarrow [\ldots Co{-}S{-}C{\equiv}N^{+}{-}Cr \ldots]^{+4} \longrightarrow [(NH_3)_5Co)]^{+2} + [S{=}C{=}N{-}Cr(H_2O)_5]^{+2} \quad \text{71\% remote}$$

$$[(NH_3)_5CoSCN]^{+2} + [Cr(H_2O)_6]^{+2} \longrightarrow \left[\ldots Co{-}\overset{\overset{\displaystyle N^{-}}{\|}\atop{\overset{\displaystyle C}{\|}}}{S^{+2}}{-}Cr \ldots \right]^{+4} \longrightarrow [(NH_3)_5Co]^{+2} + [N{\equiv}C{-}S{-}Cr(H_2O)_5]^{+2} \quad \text{29\% adjacent}$$

(11.43)

All of the above is predicated upon a mechanism that involves transfer of the bridging ligand. Although such transfer is commonly observed, there are a few instances in which

[45] For a more thorough discussion of the kinetics of these systems, see footnote 42 and E. Chaffee and J. O. Edwards, *Progr. Inorg. Chem.*, **1970**, *13*, 205.

[46] C. Shea and A. Haim, *J. Am. Chem. Soc.*, **1971**, *93*, 3055.

no ligand exchange takes place.[47] If the bridging ligand stabilizes its former complex more than it would the newly formed complex, it would not be surprising if it failed to transfer. For example:

$$[Fe(CN)_6]^{-3} + [Co(CN)_5]^{-3} \longrightarrow [Fe(CN)_6]^{-4} + [Co(CN)_5]^{-2} \quad \textbf{(11.44)}$$

Presumably the C-cyano group stabilizes the d^6 configuration of Fe(II) more than the N-cyano group would stabilize d^6 Co(III).

PHOTOCHEMISTRY OF COORDINATION COMPOUNDS[48]

Photochemistry is attracting increased interest in inorganic chemistry, both as an area of study for better understanding of the properties and mechanisms of coordination compounds and for the potential applications of light-capturing reactions. Space limitations forbid extensive discussion of photochemistry,[49] but the following discussion will illustrate some of the more salient features.

Gray and co-workers[50] have synthesized an interesting bridged dimer of rhodium(I), $[Rh(CNCH_2CH_2CH_2NC)_4Rh]^{+2}$ (abbreviated $[Rh_2(bridge)_4]^{+2}$). Since Rh(I) is d^8, the environment about each rhodium atom is square planar. This, combined with the constraints of the short trimethylene bridges, results in a Rh–Rh distance of only 326 pm, longer than a Rh—Rh bond, but probably somewhat shorter than the sum of the van der Waals radii, suggesting the possibility of interaction of the two filled d_{z^2} orbitals on the rhodium atoms. Evidence for the interaction of the two orbitals comes from a shift in the frequency of the transition ${}^1A_{1g}(d_{z^2}) \rightarrow {}^1A_{2u}$.

If $[Rh_2(bridge)_4]^{+2}$ is dissolved in 12 mol dm^{-3} HCl, a blue solution results. Upon irradiation at 546 nm, the blue color disappears and is replaced by a yellow color. Hydrogen gas is simultaneously released. The absorption spectrum of the yellow solution is identical to that of the product of the oxidation of $[Rh_2(bridge)_4]^{+2}$ by chlorine:

$$[Rh_2(bridge)_4]^{+2} + Cl_2 \longrightarrow [Cl{-}Rh(bridge)_4Rh{-}Cl]^{+2} \quad \textbf{(11.45)}$$

The photoredox reaction may thus be formulated:

$$\underset{\text{blue } (\lambda_{max}\ 578\ \text{nm})}{[(Cl^-){-}Rh(bridge)_4{-}(H^+)]^{+3}} \xrightarrow[12\ \text{mol dm}^{-3}\ \text{HCl}]{546\ \text{nm}}$$

$$[(Cl^-){-}Rh^{I\frac{1}{2}}(bridge)_4Rh^{I\frac{1}{2}}{-}(H\cdot)]^{+2*} \xrightarrow[\text{fast}]{H^+}$$

$$\underset{\text{yellow } (\lambda_{max}\ 338\ \text{nm})}{[Cl{-}Rh^{II}(bridge)_4Rh^{II}{-}Cl]^{+2}} + H_2 \quad \textbf{(11.46)}$$

---

[47] B. Grossman and A. Haim, *J. Am. Chem. Soc.*, **1970**, *92*, 4835.

[48] One of the pioneers in this area has written a useful book on the subject: A. W. Adamson and P. D. Fleischauer, "Concepts of Inorganic Photochemistry," Wiley, New York, **1975**.

[49] In addition to ref. 48, see M. S. Wrighton, ed., "Inorganic and Organometallic Chemistry," *ACS Adv. Chem. Ser. No. 168*, Washington, D. C., **1978**.

[50] H. B. Gray et al., Chapter 3 in ref. 49.

Applications of photochemistry

Reactions such as that shown in Eq. 11.46 are of considerable interest with regard to a possible solar-driven photochemical splitting of water:

$$H_2O \xrightarrow{h\nu} H_2 + \tfrac{1}{2}O_2 \qquad \textbf{(11.47)}$$

The production of hydrogen has already been effected, and if some means could be found in which the Rh(II) complex could oxidize water to oxygen gas, either directly or indirectly, the catalytic cycle would be completed and a useful source of fuel obtained directly from solar energy.

There has been considerable interest in the use of photosemiconductors such as cadmium sulfide in the photolysis of water. The incoming photon promotes an electron from the valence band (VB) to the conducting band (CD) (see Fig. 11.10). To be feasible with solar energy, the band gap must be equal to $h\nu$ for visible, not UV, light. The promoted electrons migrate through the conduction band to the surface where they reduce water to hydrogen. The holes migrate through the valence band to the surface where they oxidize water to oxygen. Unfortunately, in cadmium sulfide the holes can also oxidize sulfide ions to free sulfur, destroying the system:

$$CdS \xrightarrow{h\nu} Cd^{+2} + S\downarrow + 2e^- \qquad \textbf{(11.48)}$$

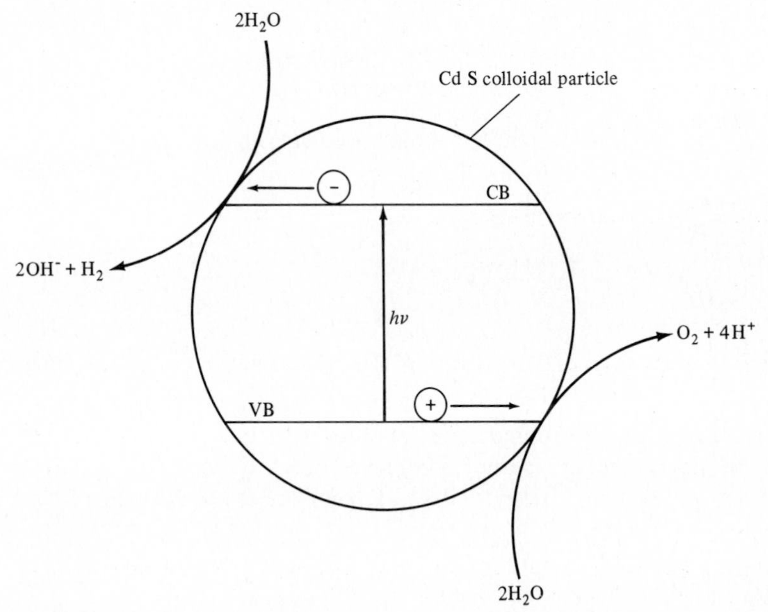

Fig. 11.10 Schematic of electron promotion by photon ($h\nu$) from valence band (VB) to conduction band (CB) of a colloidal particle of cadmium sufide. The electron reduces water to molecular hydrogen and the hole oxidizes water to molecular oxygen.

It has been found[51] that addition of sulfide or polysulfide anions to the electrolyte can protect the CdS from oxidation of this type. Alternatively, addition of ruthenium dioxide to the system catalyzes the oxidation of water preferentially, thus protecting the system. A direct conversion of solar energy to split water into hydrogen and oxygen may thus become practical and many systems are under current study.[52] As yet, however, none rivals the efficiency of catalytic chlorophyll in photosynthesis (see Chapter 18).

A modification of the system shown in Fig. 11.10 may provide a useful means of cleaning up gaseous effluents of oil refineries and other industries. A cadmium sulfide/ruthenium dioxide catalyst is used to split hydrogen sulfide into hydrogen and sulfur:

$$H_2S \longrightarrow H_2 + S \qquad \textbf{(11.49)}$$

Thus the objectional and polluting hydrogen sulfide is converted into two chemicals, hydrogen and sulfur, of considerable industrial value.[53]

Summary

Considerable work has been expended in an effort to systematize the mechanisms of reactions of coordination compounds. As in so many other areas of inorganic chemistry, much of the complexity of the subject stems from a large variety of elements and the various oxidation states of those elements. Even simple expectations based on the periodic chart often are unwarranted: The mechanistic chemistry of Rh(III) does not follow what would be expected on the basis of that of its lighter congener, Co(III). Nevertheless, more and more systems are being understood better, and the principles of reactivity are becoming useful in synthesis and other applications.

PROBLEMS

11.1. Predict the products of the following reactions (one mole of each reactant):

$$[Pt(CO)Cl_3]^- + NH_3$$

$$[Pt(NH_3)Br_3]^- + NH_3$$

$$[(C_2H_4)PtCl_3]^- + NH_3$$

11.2. Why does the reaction of two moles of R_3P with K_2PtCl_4 produce a different isomer than the reaction of two moles of R_3N?

11.3. Why does the existence of a well-ordered series of entering groups (p. 546) provide evidence that substitution in square planar complexes proceeds via an associative (S_{N2}) mechanism? [R. G. Wilkins, "The Study of Kinetics and Mechanism of Re-

---

[51] A. B. Ellis, S. W. Kaiser, and M. S. Wrighton, *J. Am. Chem. Soc.*, **1976**, *98*, 6855; A. B. Ellis et al., ibid, **1977**, *99*, 2839, 2848.

[52] M. Grätzel, *Acc. Chem. Res.*, **1981**, *14*, 376; J. Kiwi, K. Kalyanasundaram, and M. Grätzel, *Struct. Bond.*, **1982**, *49*, 37; D. Huonghong et al., *J. Am. Chem. Soc.*, **1982**, *104*, 2977.

[53] M. Grätzel, in D. O'Sullivan, *Chem. Eng. News*, July 27, **1981**, p. 40.

actions of Transition Metal Complexes," Allyn and Bacon, Boston, 1974, pp. 229–230.

11.4. Consider a square planar complex in which the ligands provide considerable "shrubbery" tending to extend up and down from the plane of the ligating atoms. Such a complex could be $[M(Et_4dien)SCN]^+$ where $Et_4dien = Et_2NCH_2CH_2NHCH_2CH_2NEt_2$ and M = Pd(II), Pt(II), or Au(III). If you found that replacement of SCN^- by X^- proceeded by an essentially first order reaction, how could you reconcile that with your knowledge of the reaction mechanisms of square planar and octahedral complexes? [R. G. Wilkins, ibid., pp. 231, 235–238.]

11.5. Based upon the discussion on pp. 559–562, predict whether the following reaction will proceed via an inner-sphere or outer-sphere mechanism:

$$[Co(NH_3)_6]^{+3} + [Cr(H_2O)_6]^{+2} \rightarrow$$

11.6. Do you expect the shift cited for the ${}^1A_{1g} \rightarrow {}^1A_{2u}$ transition (p. 562) to be a red shift or a blue shift? Explain.

12

Some descriptive chemistry of the transition metals

In the preceding chapters the principles guiding the structure and the reactions of complexes have been examined. The present chapter will concentrate on the properties of the individual metals in various oxidation states. The stability of various oxidation states will be examined and the similarities and differences compared. The material in this chapter and much of the next may be characterized as the "descriptive chemistry" of the transition metals. Unfortunately, descriptive chemistry has not been especially popular with students and teachers, particularly in recent years. Admittedly, complete mastery of all the properties and all the reactions of the compounds of one element would be an impossible task, to say nothing of attempting it with 106. Furthermore, new reactions and properties are constantly being discovered that require the continual revision of one's knowledge. Nevertheless, it is impossible to ignore descriptive chemistry and try only for the mastery of the theoretical side of chemistry. Theory can only be built upon and checked against facts. Actually, in reading this book you have already encountered a vast body of descriptive chemistry, perhaps without consciously being aware of it. As each theory and model has been presented, an appeal has been made to the real world to support or modify that concept. Much of this "descriptive" chemistry goes unnoticed, but consider that almost all metal carbonyls are diamagnetic (p. 591), that magnetite has an inverse spinel structure (p. 394), and that potassium permanganate is purple (p. 456) or that it is a strong oxidizing agent (p. 579). Furthermore

$$[PtCl_4]^{-2} + 2NH_3 \longrightarrow cis\text{-}PtCl_2(NH_3)_2 + 2Cl^- \qquad (12.1)$$

When it comes to what is "important" descriptive chemistry, *chacun à son goût*![1]

---

[1] This is why any single volume, even the so-called "descriptive" ones, must pick and choose which "facts" are to be presented. The reader should become familiar with two comprehensive sets of volumes on inorganic chemistry: "Comprehensive Inorganic Chemistry," J. C. Bailar, Jr., et al., eds., Pergamon Press, Elmsford, N.Y., **1973**, and "MTP International Review of Science: Inorganic Chemistry, Series One," H. J. Emeléus, ed., Butterworths, London, **1972**. In general, either is capable of telling you more than you are apt to want to know about a particular element. In addition, consult Appendix G for sources of information.

Even so, it sometimes seems as though *no* facts are retained. Each of us has a "silver chloride is a pale green gas" story. My local one concerns the organic graduate student who reported that he couldn't get bromine to dissolve in carbon tetrachloride (!). He even tried *grinding* it with a mortar and pestle (!!). When his professor investigated, it turned out that the *bromine* was still in an unopened container in the bottom of the packing can, and the material in question, being subjected to grinding and solubilization tests, was the vermiculite packing material (!!!). Truly,

> A *little* learning is a dangerous thing;
> Drink deep, or taste not the Pierian spring.[2]

In this chapter the theories developed previously will be used to help correlate the important facts of transition metal chemistry. Much of the chemistry of these elements has already been included in the chapters on coordination chemistry (Chapters 9 and 10). More will be discussed in the chapters on organometallic chemistry (Chapter 13), lanthanides and actinides (Chapter 16), and the descriptive biological chemistry of the transition metals (Chapter 18). The present chapter will thus concentrate on the trends within the series (Sc to Zn, Y to Cd, and La to Hg),[3] the differences between series ($Ti \rightarrow Zr \rightarrow Hf$; $Cu \rightarrow Ag \rightarrow Au$), and the stable oxidation states of the various metals.

General periodic trends

As the effective atomic number increases across a series of transition metals, the size decreases from poor shielding by the d electrons. For transition metal ions the results of ligand field effects override a smooth decrease and so minima in the ionic radii curves are found for d^6 low-spin ions, etc. (Fig. 9.17). The decrease in ionic radii favors the formation of stable complexes, and this, together with ligand field stabilization energies (LFSEs) arising from incompletely filled d orbitals, is responsible for the general order of stability of complexes. The increasing availability of d electrons for back bonding via π orbitals (especially in low oxidation states) increases the "softness" of the metal ions toward the ends of the series.

The differences between one series and another are discussed later in this chapter but we may note some seeming paradoxes: The heavier metals tend to be somewhat less reactive as elements, yet they are more easily oxidized to higher oxidation states; the first and second series appear to be more closely related to each other (than to the third) on the basis of ionization energies, yet it is common to group the second and third series together on the basis of chemical properties and to differentiate them from the first series.

The first ionization energies of the transition metals are listed in Table 12.1. The first two series do not differ significantly from each other—one time an element of one series is higher, sometimes the other. Beginning with cesium the third series has a noticeably lower ionization energy as we might expect on the basis of larger size ($Cs > Rb$, $Ba > Sr$, $La > Y$).

---

[2] A. Pope, "A Essay on Criticism," Part II, 1. 15; see J. Bartlett, "Familiar Quotations," E. M. Beck, ed., Little, Brown Boston, **1968**, p. 402b; B. Evans, "Dictionary of Quotations," Bonanza Books, New York, **1968**, p. 381:10.

[3] The lanthanide and actinide elements are not included in this chapter but are discussed in Chapter 16.

Table 12.1 Ground state ionization energies of the transition metals in kJ mol^{-1} (eV mol^{-1})

Group number	IA	IIA	IIIB	IVB	VB	VIB	VIIB	VIIIB			IB	IIB
4th period	419	590	632	658	651	654	710	760	759	738	747	907
(K–Zn)	(4.34)	(6.11)	(6.54)	(6.82)	(6.74)	(6.77)	(7.44)	(7.87)	(7.86)	(7.64)	(7.73)	(9.39)
5th period	404	551	616	661	665	686	703	712	721	806	732	869
(Rb–Cd)	(4.18)	(5.70)	(6.38)	(6.84)	(6.88)	(7.10)	(7.28)	(7.37)	(7.46)	(8.34)	(7.58)	(8.99)
6th period	376	503	539	680	762	771	761	840	880	870	891	1099
(Cs–Hg)	(3.89)	(5.21)	(5.58)—(lanthanides)—	(7.0)	(7.89)	(7.98)	(7.88)	(8.7)	(9.1)	(9.0)	(9.22)	(10.44)

The addition of fourteen poorly shielding $4f$ electrons results in increasing ionization energies and decreasing size (the "lanthanide contraction"). This specific phenomenon and others related to it are discussed at some length in the chapter on periodicity (Chapter 17), but for now we may note that atoms of postlanthanide metals are (1) smaller than would otherwise have been expected—so much so that they are often essentially the same size as their lighter congeners; and (2) more difficult to oxidize. This latter phenomenon proceeds with ever increasing ionization energies until the most noble metals are reached: iridium, platinum, and gold. As a result of the relatively small size, heavy nuclei, and tightly held electrons of their atoms, the elements in this series are also the densest elements (d_{Os} = 22.5 g cm^{-3}; d_{Ir} = 22.4 g cm^{-3}).

Although in many respects these effects of the lanthanide contraction serve to make these elements less reactive than would otherwise be the case, in one respect their bonding ability is increased. The relatively high effective nuclear charge appears to enhance the ability of these metals to use their d orbitals in π bonding.[4] The most pronounced effects that have been ascribed to π bonding are found in these elements, e.g., square planar platinum(II) and octahedral tungsten(0), and appear to result from effects of the lanthanide contraction.[5]

CHEMISTRY OF THE VARIOUS OXIDATION STATES OF TRANSITION METALS

Low and negative oxidation states

The entire question of oxidation state[6] is an arbitrary one and the assignment of appropriate oxidation states is often merely a matter of convenience (or inconvenience!). The concept of oxidation state is best defined in compounds between elements of considerably different electronegativity in which the resulting molecular orbitals are clearly more closely related to the atomic orbitals of one atom than another. In those cases in which the differences in electronegativity are small and especially those in which there are extensive delocalized molecular orbitals that are nonbonding, weakly bonding, or antibonding, the situation becomes difficult. The former situation is found with ligands containing halogen, oxygen, and nitrogen σ bonding ligands. The latter type of complex is often encountered among organometallic species for which often no attempt is made to assign oxidation states. Other ligands stabilizing low (if imprecise) oxidation states are cyanide and phosphorus trifluoride, both excellent π-bonding ligands. Hence it is possible to prepare zero-valent nickel complexes with these ligands:

$$Ni(CO)_4 + 4PF_3 \longrightarrow Ni(PF_3)_4 + 4CO \qquad \textbf{(12.2)}$$

$$Ni^{+2} \xrightarrow{KCN} 2K^+[Ni(CN)_4]^{-2} \xrightarrow{K} 4K^+[Ni(CN)_4]^{-4} \qquad \textbf{(12.3)}$$

[4] See pp. 582–583 for a discussion of the properties of d orbitals and the effect of varying charge.

[5] R. B. King, *Inorg. Nucl. Chem. Letters*, **1969**, *5*, 905.

[6] For a book devoted entirely to the problems related to the difficulties in defining oxidation states, see C. K. Jørgensen, "Oxidation Numbers and Oxidation States," Springer, New York, **1969**.

Table 12.2 Some oxidation states of the metals of the first transition series

Oxidation number	Group				
	IIIB	IVB	VB	VIB	VIIB
VII					$[MnO_4]^-$
VI				$[CrO_4]^{-2}$	$[MnO_4]^{-2}$
V			$[VO_4]^{-3}$	$[CrO_4]^{-3}$	$[MnO_4]^{-3}$
IV		TiO_2	$[VOF_4]^{-2}$	$[CrO_4]^{-4}$	MnO_2
III	$[Sc(H_2O)_6]^{+3}$	$[Ti(H_2O)_6]^{+3}$	$[V(H_2O)_6]^{+3}$	$[Cr(H_2O)_6]^{+3}$	$[Mn(CN)_6]^{-3}$
II		$TiCl_2$	$[V(H_2O)_6]^{+2}$	$[Cr(H_2O)_6]^{+2}$	$[Mn(H_2O)_6]^{+2}$
I				$[Cr(dipy)_3]^+$	$[Mn(CN)_6]^{-5}$
0		$[Ti(dipy)_3]$	$[V(dipy)_3]$	$[Cr(dipy)_3]$	$[Mn(dipy)_3]$
−I			$[V(dipy)_3]^-$	$[Cr(dipy)_3]^-$	$[Mn(dipy)_3]^-$
−II				$[Cr(dipy)_3]^{-2}$	
−III			$[V(CO)_5]^{-3}$	$[Cr(dipy)_3]^{-3}$	$[Mn(CO)_4]^{-3}$
−IV				[a]	

[a] Both Mo and W form −IV complexes, $[M(CO)_4]^{-4}$.

Ligands with extensively delocalized molecular orbitals that are essentially nonbonding can make the assignment of precise oxidation states difficult or impossible. For example, we have already seen this in the thiolene-thiolate ligands (see pp. 486–488). A similar ligand is dipyridine, which forms complexes that may formally be classified as containing +1, 0, or even −1 oxidation states for the metal. In the lower oxidation states much electron density on the metal can be delocalized over the π system. Other examples of extensively delocalized systems containing metals in low oxidation states (see pp. 853–868) are often encountered in biological systems.

Range of oxidation states

Although the definition of low oxidation states is somewhat subjective, it is still possible to discuss the range of oxidation states exhibited by various metals. When the oxidation states for the metals of the first transition series are examined, the results shown in Table 12.2 are found. There is a general trend between a minimum number of oxidation states (one or two) at each end of the series (Sc^{+3} and Zn^{+2}) to a maximum number in the middle (manganese, −III to VII). The paucity of oxidation states at the extremes stems from either too few electrons to lose or share (Sc, Ti) or too many *d* electrons (and hence fewer orbitals open in which to share electrons with ligands) for high valency (Cu, Zn). A second factor tending to reduce the stability of high oxidation states toward the end of the transition series is the steady increase in effective atomic number. This acts to lower the energy of the *d* orbitals and draw them into the core of electrons not readily available for bonding.[7] Thus, early in the series it is difficult to form species that do not utilize the *d* electrons. Scandium(II) is virtually unknown and Ti^{IV} is more stable than Ti^{III}, which is much more

[7] See C. S. G. Phillips and R. J. P. Williams, "Inorganic Chemistry," Oxford University Press, London, **1966**, Vol. II, pp. 157–166, and G. E. Coates et al., "Principles of Organometallic Chemistry," Methuen, London, **1968**, pp. 153–157.

Table 12.2 *(Continued)*

Group				
VIIIB			IB	IIB
$[FeO_4]^{-2}$				
$[FeO_4]^{-3}$				
$[FeO_4]^{-4}$	$[CoF_6]^{-2}$	$[NiF_6]^{-2}$	$[CuF_6]^{-2}$	
$[Fe(H_2O)_6]^{+3}$	$[Co(CN)_6]^{-3}$	$[NiF_6]^{-3}$	$[CuF_6]^{-3}$	
$[Fe(H_2O)_6]^{+2}$	$[Co(H_2O)_6]^{+2}$	$[Ni(H_2O)_6]^{+2}$	$[Cu(H_2O)_6]^{+2}$	$[Zn(H_2O)_6]^{+2}$
	$[Co(dipy)_3]^{+}$	$[Ni\{N(SiMe_3)_3\}(P\phi_3)_2]^{+}$	$[Cu(CN)_2]^{-}$	
$[Fe(dipy)_3]$	$[Co(dipy)_3]$	$[Ni(dipy)_2]$		$[Zn(dipy)_3]$
$[Fe(dipy)_3]^{-}$	$[Co(CO)_4]^{-}$			
$[Fe(CO)_4]^{-2}$				

stable than Ti^{II}. At the other extreme, the only oxidation state for zinc is +2 (no *d* electrons are involved) and for nickel, Ni^{II} is much more stable than Ni^{III}. As a result, maximum oxidation states of reasonable stability occur equal in value to the sum of the *s* and *d* electrons through manganese ($Ti^{IV}O_2$, $V^{V}O_2^{+}$, $Cr^{VI}O_4^{-2}$, $Mn^{VII}O_4^{-}$), followed by a rather abrupt decrease in stability for higher oxidation states, so that the typical species to follow are $Fe^{II,III}$, $Co^{II,III}$, Ni^{II}, $Cu^{I,II}$, Zn^{II}.

Comparison of properties by oxidation state

There are certain resemblances among metal ions that can best be discussed in terms of oxidation state and are relatively independent of electron configuration. These are the ones which relate principally to size and charge phenomena. For example, the ordinary alums, $KAl(SO_4)_2 \cdot 12H_2O$, are isomorphous with the chrome alums, $KCr(SO_4)_2 \cdot 12H_2O$, and mixed crystals of any composition between the two extremes may be prepared by substitution of Cr^{+3} for Al^{+3}. In this case the two cations have the same charge and similar radii ($r_{Al^{+3}} = 53$ pm; $r_{Cr^{+3}} = 62$ pm). There are examples of resemblance between Mg^{+2} ($r = 72$ pm), Mn^{+2} ($r = 82$ pm), and Zn^{+2} ($r = 75$ pm),[8] despite the fact that one has a noble gas configuration (s^2p^6) and the others do not (d^5 and d^{10}).[9] The remarkable resemblances of the lanthanides to each other (see Chapter 16) bear witness to the overwhelming influence of identical charge and similar size in these species.

The resemblances that depend more on charge than on electron configuration might be termed "physical" ones. They relate to crystal structure and hence to solubilities and

[8] The similarity of the names of magnesium and manganese results from the confusion of these two elements by early chemists, an error which has persisted among neophyte chemists to this day. Both names derive ultimately from Magnesia, an ancient city in Asia Minor.

[9] Note that Mg^{+2} most closely resembles Zn^{+2} and high-spin Mn^{+2}. The resemblance of Mg^{+2} to the other +2 transition metal cations is less because of ligand field stabilization energies in complexes of the latter.

tendencies to precipitate. Coprecipitation is often more closely related to oxidation state than to family relationships. Thus carriers for radioactive tracers need not be of the same chemical family as the radioisotope. Technetium(VII) may be carried not only by perrhenate but also by perchlorate, periodate, and tetrafluoroborate. Lead(II) has the same solubility characteristics as the heavier alkaline earth metals. Thallium(I) ($r = 150$ pm) often resembles potassium ($r = 138$ pm), especially with respect to oxygen and other highly electronegative elements. It thus forms a soluble nitrate, carbonate, phosphate, sulfate, and fluoride. It is also incorporated into many potassium enzymes and is exceedingly poisonous. The electronic structure of cations also affects their properties, especially with respect to polarization of anions (see pp. 130–132) and thus we should not be surprised that with respect to the heavier halogens, Tl^+ resembles Ag^+ more closely than it does K^+.

Finally, one chemical property that depends upon the cationic charge is the coordination number. Although this is greatly influenced by size (see p. 83) there is also a tendency for cations with larger charges to have larger coordination numbers, e.g., Co^{+2} (C.N. = 4 and 6) versus Co^{+3} (C.N. = 6 only); Mn^{+2} (C.N. = 4 in $MnCl_4^{-2}$) versus Mn^{+4} (C.N. = 6 in MnF_6^{-2}). This is a consequence of the Pauling electroneutrality principle (see pp. 365–366). On the other hand, in extremely high oxidation states (Cr^{VI}, Mn^{VII}, Os^{VIII}) there is a tendency to form metal–oxygen double bonds (considerable π bonding from the oxygen to the ligand), and tetrahedral species (CrO_4^{-2}, MnO_4^-, OsO_4) result.

THE CHEMISTRY OF ELEMENTS POTASSIUM–ZINC: COMPARISON BY ELECTRON CONFIGURATION

Although there are resemblances that depend only upon the charge or oxidation state, transition metal chemistry is more often governed by the electron configuration of the metal ions. Thus, although there is a natural tendency for lower oxidation states to be reducing in character and higher ones to be oxidizing, the electron configuration may well make the divalent, trivalent, or even higher oxidation state the most stable one for a particular metal. In this section the properties of the metals of the first transition series are briefly examined in terms of electron configuration.[10]

There are a few properties of the elements of the first transition series that are sufficiently constant to be worth mentioning prior to the investigation of the individual cases. The magnetic moments of most species are given well by the "spin-only" value; i.e., the orbital contribution to the magnetic moment is effectively quenched. Hence the magnetic moment may be approximated[11] by the equation:

$$\mu = \sqrt{n(n+2)} \qquad \textbf{(12.4)}$$

[10] It is often helpful to view the descriptive chemistry of the transition metals from two different points of view in a comparative study. For a thorough review of transition metal chemistry in an element-by-element approach, see F. A. Cotton and G. Wilkinson, "Advanced Inorganic Chemistry," 4th ed., Wiley (Interscience), New York, **1980**, Chapters 20-22. For the chemistry of individual elements, see Appendix G.

[11] See p. 29. For ions with S ground states (e.g., Mn^{+2} and Fe^{+3}) there is no orbital angular momentum and the observed magnetic moment is expected to be given exactly by Eq. 12.4. For other ions of the first transition series the experimental values usually differ only slightly from the spin-only values. See B. N. Figgis and J. Lewis, *Progr. Inorg. Chem.*, **1964**, *6*, 37, for a complete discussion of the magnetochemistry of transition metal complexes.

where n is the number of unpaired electrons and the magnetic moment is measured in Bohr magnetons.

The most common coordination numbers for the first-row transition metals are 4 and 6, with other values being relatively rare (see pp. 465–513). Both high-spin and low-spin complexes are commonly observed depending upon the properties of the metal and the ligands (see pp. 381–386).

The d^0 configuration

This configuration occurs for simple ions such as K^+, Ca^{+2}, and Sc^{+3} and for the formal oxidation states equal to the group numbers for many of the transition metals. This holds true as far as Mn^{VII}, but Fe^{VIII} is unknown.

All metal ions with d^0 configurations are hard acids, and with the exception of the higher oxidation states there is little tendency to form complexes. There is no tendency to behave as reducing agents (there are no electrons to lose) and little tendency to behave as oxidizing agents until species such as CrO_4^{-2} and MnO_4^- are reached. In general, therefore, the aqueous chemistry may be described simply: The lower charged species (K^+, Ca^{+2}, Sc^{+3}) behave as simple, uncomplexed (other than by water) free ions in aqueous solution. The unusual "crown" complexes of the alkali metals have been discussed previously (see p. 530) and Ca^{+2} may be complexed by polydentate ligands such as EDTA. The higher oxidation states (Cr^{VI}, Mn^{VII}) tend to form oxyanions, which are good oxidizing agents, especially in acidic solution; the oxides of the intermediate species tend to be insoluble (TiO_2) and/or amphoteric (VO_2^+, VO_4^{-3}).

A quick survey of the metals with the d^0 configuration yields the following descriptive information. For potassium, calcium, and scandium it is the only stable electron configuration. It is by far the most stable for titanium, occurring in the halides, TiX_4, the oxide, TiO_2, and a fluoro complex, TiF_6^{-2}. Vanadium(V) occurs in the vanadate ion, VO_4^{-3}, and in polyvanadate species (see p. 695). Vanadium(V) is a mild oxidizing agent, and hence several other configurations are important for vanadium. At very low pH's the VO_2^+ ion is the principal species present. The strongly oxidizing species CrO_4^{-2} ($\mathscr{E}^0$ to Cr^{+3} is $+1.33$ V) and MnO_4^- ($\mathscr{E}^0$ to Mn^{+2} is $+1.55$ V) are unstable relative to lower oxidation states.

The d^1 configuration

This does not tend to be a stable configuration. It is completely unknown for scandium and strongly reducing in Ti^{III}. The later members of the series tend to disproportionate to more stable configurations:

$$3[CrO_4]^{-3} + 8H^+ \longrightarrow 2[CrO_4]^{-2} + Cr^{+3} + 4H_2O \tag{12.5}$$

$$3[MnO_4]^{-2} + 4H^+ \longrightarrow 2[MnO_4]^- + MnO_2 + 2H_2O \tag{12.6}$$

The only d^1 species of importance is the vanadyl ion, VO^{+2}, which is the most stable form of vanadium in aqueous solution.

The d^2 configuration

This configuration ranges from Ti^{II}, very strongly reducing, to Fe^{VI}, very strongly oxidizing. It is not a particularly stable configuration. Both Ti^{II} and V^{III} are reducing agents and Cr^{IV} and Mn^{V} are relatively unimportant. The ferrate(VI) ion, FeO_4^{-2}, is of some interest. It is formed by vigorous oxidation of iron or iron compounds. It is reasonably stable in basic solutions but becomes a powerful oxidizing agent as the pH is lowered.

The d^3 configuration

This is usually not a stable configuration, being strongly reducing in V^{+2} and strongly oxidizing in Mn^{IV}.[12] However, Cr^{+3} is the most stable form of chromium in aqueous solution. In octahedral coordination it benefits from the LFSE of a half-filled sublevel, t_{2g}^3.

The d^4 configuration

Although there are no really stable species with this configuration, there are still interesting aspects to its chemistry. The chromous ion, Cr^{+2}, is strongly reducing yet may be prepared readily:

$$Cr^{+3} \xrightarrow{Zn(Hg)} Cr^{+2} \tag{12.7}$$

$$Cr + 2H^+ \longrightarrow Cr^{+2} + H_2 \tag{12.8}$$

For the latter reaction very pure chromium is required to prevent formation of Cr^{+3}, but the reaction with a zinc amalgam is "clean" and a convenient source of a strongly reducing species. Addition of these solutions to saturated solutions of sodium acetate precipitates chromium(II) acetate:

$$2Cr^{+2} + 4CH_3COO^- + 2H_2O \longrightarrow Cr_2(OOCCH_3)_4(H_2O)_2 \tag{12.9}$$

The latter is a sparingly soluble red crystalline precipitate, stable with respect to oxidation when perfectly dry. It contains a metal–metal bond (see pp. 747–748).

Manganese(III) is a strong oxidizing agent and is subject to disproportionation as well. Complexes of Mn^{III} are also relatively unstable with the exception of $[Mn(CN)_6]^{-3}$, which forms readily upon exposure of a solution of manganese(II) and cyanide to air. A few iron(IV) compounds are known.

The d^4 configuration is the first for which we can expect both high- and low-spin octahedral species. Most seem to be high spin,[13] although the cyanide complexes of Cr^{+2} and Mn^{+3} are low spin.

---

[12] The relative stability of MnO_2 results from its insolubility. Other Mn^{IV} species are strong oxidizing agents.

[13] The known complexes of Fe^{IV}, $[Fe(diars)_2X_2]^{+2}$, are low spin. This indicates a strong spin-pairing tendency from the +4 cationic charge since it must overcome the opposing tendency of the disparate ligands (diars = strong, halide = weak) to form a strongly distorted tetragonal field. The latter might favor a high-spin d^4 configuration (see Fig. 9.21a).

The d^5 configuration

There are two important species having this configuration, Mn^{+2} and Fe^{+3}. Both are quite stable, although the latter may be reduced to Fe^{+2} with suitable reducing agents. The exchange energy favors the high-spin configuration and almost all of the known complexes are high spin. Some of the few exceptions are $[Mn(CN)_6]^{-4}$, $[Fe(CN)_6]^{-3}$, and $[Fe(dipy)_3]^{+3}$. One very interesting and somewhat unexpected, low-spin d^5 complex is $[CoF_6]^{-2}$, in which the $+4$ cationic charge is able to overcome the loss of exchange energy of the d^5 configuration and the weak field of the fluoride ion.[14] As expected from the spin-forbidden nature of *d–d* transitions in high-spin d^5 octahedral complexes, those of Mn^{+2} and Fe^{+3} are colorless or nearly so (see pp. 453 and 461).

The d^6 configuration

Octahedral complexes with strong field ligands provide the greatest possible amount of LFSE with d^6 configurations. Iron(II) is relatively stable although a mild reducing agent. Cobalt(III) is extremely stable in the presence of strong-field ligands but a strong oxidizing agent in their absence. Nickel(IV) is strongly oxidizing.

The influence of charge on spin-pairing tendencies is well documented in this series. Most Fe^{II} complexes are high spin, exceptions being ferrocyanide, $[Fe(CN)_6]^{-4}$, "nitroprusside," $[Fe(CN)_5NO]^{-2}$, and the orthophenanthroline complex, $[Fe(phen)_3]^{+2}$. Cobalt(III) complexes, in contrast, tend to be low spin except in the presence of the weakest ligands, $[CoF_6]^{-3}$ and $[Co(H_2O)_3F_3]$. Finally, with Ni^{IV} even the weakest-field ligands form low-spin complexes (e.g., $[NiF_6]^{-2}$ is diamagnetic).

The d^7 configuration

The only important species with this configuration is Co^{+2}, although Ni^{III} species are known. Cobalt(II) occurs in tetrahedral ($[CoCl_4]^{-2}$), square planar ($[Co(DMG)_2]$), square pyramidal ($[Co(ClO_4)(OAsMe\phi_2)_4]^{+}$), trigonal bipyramidal ($[CoBr(Me_6tren)]^{+}$), and octahedral ($Co(NH_3)_6^{+2}$) complexes. Although there is some tendency for tetrahedral complexes to predominate, especially with larger ligands, the lack of large ligand field stabilization for any single geometry results in the proliferation of types of complexes.

Cobalt(II) is stable in aqueous solution, but in the presence of strong field ligands it is easily oxidized to form Co(III) complexes. The isoelectronic Ni^{III} species are strong oxidizing agents.

The d^8 configuration

This configuration is ideal for the formation of low-spin square planar complexes with strong field ligands. They are typically red or yellow, although other colors are sometimes found. Tetrahedral, high-spin complexes are typically formed with bulky ligands such as triphenylphosphine oxide or the halides. An interesting exception to this rule is

[14] G. C. Allen and K. D. Warren, *Inorg. Chem.*, **1969**, *8*, 1902.

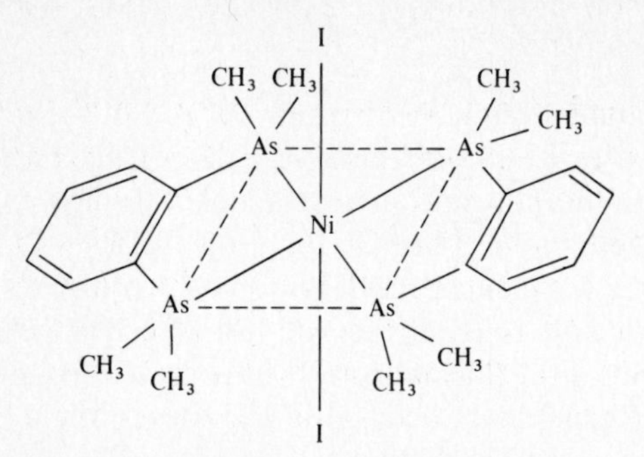

Fig. 12.1 Molecular structure of diiodobis(*o*-phenylenebisdimethylarsine)nickel(II).

$[Ni(SPMe_2NPMe_2S)_2]$, in which the nickel atom is surrounded by a tetrahedral array of sulfur atoms, although there appears to be no steric reason preventing the square planar arrangement found in related complexes such as $[Ni(NHP\phi_2NP\phi_2NH)]$.[15] Tetrahedral nickel(II) complexes are typically a deep blue or green-blue resulting from an intense absorption band from the $T_1(F) \rightarrow T_1(P)$ transition.

Five-coordinate Ni^{II} complexes may be either high or low spin depending upon the nature of the ligands involved. With soft ligating atoms such as sulfur, phosphorus, and arsenic the complexes tend to be low spin and are high spin with similar complexes with nitrogen-containing ligands.[16] Both trigonal bipyramidal and square pyramidal complexes are known (see pp. 472–479).

Six-coordinate Ni^{II} complexes may either have six identical ligands, as in $[Ni(H_2O)_6]^{+2}$, $[Ni(NH_3)_6]^{+2}$, and $[Ni(en)_3]^{+2}$, or have axial ligands which are different from the remaining four, $[NiL_4L'_2]^{+2}$. The former are paramagnetic with two unpaired electrons as expected on the basis of simple ligand field splittings. The latter form as products of the reaction of square planar complexes with two additional ligands (which may be solvent molecules). The resulting tetragonal field is usually sufficiently near to an undistorted octahedral field so that the complex becomes paramagnetic like the octahedral complexes. If there is sufficient disparity between the positions in the spectrochemical series of the two ligands [e.g., diiodobis(*o*-phenylenebisdimethylarsine)nickel(II); Fig. 12.1], the resulting adduct is diamagnetic. The complex may be viewed as a square planar, diamagnetic complex that has not been sufficiently perturbed by the tetragonal field of the weak iodo ligands to cause unpairing of the electrons.[17]

Some copper(III) complexes, such as CuO_2^-, are known and behave as might be expected from their analogy to Ni(II) forming square planar structures. However, unlike Ni(II), all purely inorganic Cu(III) species are very strong oxidizing agents. Recently, relatively stable Cu(III) has been found in biological systems.

A few Co^+ complexes are known, and although they must be considered exceptional, vitamin B_{12} depends upon this oxidation state for its action (see p. 878).

---

[15] M. R. Churchill et al., *J. Am. Chem. Soc.*, **1969**, *91*, 6518.

[16] See C. Furlani, *Coordin. Chem. Rev.*, **1968**, *3*, 141.

[17] See pp. 396–400. For a comprehensive review of the electronic structure and bonding of Ni(II) complexes, see L. Sacconi, *Transition Metal Chem.*, **1968**, *4*, 199.

The d^9 configuration

This configuration is found in copper(II) compounds but is otherwise unimportant. It has neither the closed-subshell stability of d^{10} nor the LFSE possible for d^8. Copper(II) may be fairly easily reduced to copper(I) compounds. (See Eq. 12.10.)

Six-coordinate complexes are expected to be distorted from pure octahedral symmetry by the Jahn-Teller effect and this distortion is generally observed (see pp. 396–400). A number of 5-coordinate complexes are known, both square pyramidal and trigonal bipyramidal. Four-coordination is exemplified by square planar and tetrahedral species as well as intermediate configurations.

The d^{10} configuration

For the first transition series this configuration is limited to Cu^+ and Zn^{+2}, but it is also exhibited by the posttransition metals in their highest oxidation states (Ga^{III}, Ge^{IV}). The copper(I) complexes are good reducing agents, being oxidized to Cu^{II}. They may be stabilized by precipitation with appropriate counterions to the extent that Cu^I may form to the exclusion of Cu^{II}:

$$Cu^{+2} + 2I^- \longrightarrow CuI\downarrow + \tfrac{1}{2}I_2 \qquad \textbf{(12.10)}$$

$$Cu^{+2} + 2CN^- \longrightarrow CuCN\downarrow + \tfrac{1}{2}(CN)_2 \qquad \textbf{(12.11)}$$

Zinc(II), gallium(III), and germanium(IV) are the most stable oxidation states for these elements, but the later nonmetals (arsenic, selenium, and bromine) show a reluctance to assume their highest oxidation state.

The spherically symmetrical d^{10} configuration affords no LFSE, so the preferred coordination is determined by other factors. For Cu^I the preferred coordination appears to be linear (*sp*), 2-coordination, although a few 3-coordinate complexes are known as well as several tetrahedral complexes. Zinc(II) is typically either tetrahedral (e.g., $[ZnCl_4]^{-2}$) or octahedral ($[Zn(H_2O)_6]^{-2}$), but both trigonal bipyramidal and square pyramidal 5-coordinate complexes are known (see p. 480). The posttransition metals form tetrahedral (e.g., $[GaCl_4]^-$) and octahedral (e.g., $[Ge(acac)_3]^+$, $[GeCl_6]^{-2}$, and $[AsF_6]^-$) complexes.

THE CHEMISTRY OF THE HEAVIER TRANSITION METALS

A detailed account of the descriptive chemistry of the heavier transition metals is beyond the scope of this book.[18] Many aspects of the chemistry of these elements such as metal–metal multiple bonds, metal clusters, organometallic chemistry, and coordination chemistry are discussed under those topics. The present discussion will be limited to a comparison of the similarities and differences of the heavier metals and their lighter congeners.

In general, the coordination numbers of the elements of the second and third transition series tend to be greater than for the first series because the ionic radii are larger by about

[18] For an extensive discussion on an element-by-element basis, see F. A. Cotton and G. Wilkinson, "Advanced Inorganic Chemistry," 4th ed., Wiley (Interscience), New York, **1980**, Chapters 21–24, and references in Appendix G.

15–20 pm for corresponding species.[19] Thus tetrahedral coordination is considerably less frequent although present in species such as $[ReO_4]^-$. Square planar coordination is found in d^8 species such as Pd^{II}, Pt^{II}, and Au^{III} which are especially stabilized by LFSE. Octahedral species are quite common, and the presence of coordination numbers 7, 8, 9, and 10 is not particularly uncommon.

The heavier congeners show a pronounced tendency toward higher oxidation states. Whereas the +2 state is known for all elements of the first transition series except scandium, it is relatively unimportant for the heavier metals. Cadmium is the only one to be restricted to the +2 oxidation state and Hg^{II}, Pd^{II}, and Pt^{II} are the only other important dipositive species. Although cobalt is known as both Co^{II} and Co^{III}, its congeners rhodium and iridium are essentially limited to the +3 oxidation state or higher. Chromium(III) is the most stable oxidation state of chromium, but both molybdenum and tungsten are strongly reducing in that oxidation state, and the +6 oxidation state is much more important. In general, the stability of the highest oxidation state (i.e., the group number oxidation state) is considerably greater in the heavier metals. Thus $[ReO_4]^-$ is not a strong oxidizing agent like $[MnO_4]^-$. The trend is extended further along the series as well, culminating in ruthenium tetroxide and osmium tetroxide, two of the few cases of a valid +8 oxidation state.[20] Further examples are the stability of Pd^{IV}, Pt^{IV}, and Au^{III} relative to their lighter congeners. Gold is even able to achieve the unexpected oxidation state of +5 (see pp. 3, 772).

As discussed in Chapter 9, there is a much greater tendency toward spin pairing in the heavier transition metals and the presence of high-spin complexes is much less common. Thus in contrast to Ni^{II}, which is known in tetrahedral, square planar, square pyramidal, trigonal bipyramidal, and octahedral complexes, Pd^{II} and Pt^{II} complexes are almost universally low spin, square planar. A few weakly bonded 5-coordinate adducts are known (see p. 479) and there is one high-spin Pd^{II} complex which may be tetrahedral,[21] but these are certainly exceptional situations. The effect of ligand field strength on the instability of the d^9 configuration in silver(II) and gold(II) has been discussed previously (see p. 410).

There is also an important difference among heavier transition metals in magnetic properties. Because of extensive spin-orbit coupling the spin-only approximation of Eq. 12.4 is no longer valid. As there is extensive temperature-dependent paramagnetism, the simple interpretation of the magnetic moment in terms of the number of unpaired electrons cannot be extended from the first transition series to their heavier congeners.[22]

Finally, the heavier posttransition metals have group number oxidation states corresponding to d^{10} configurations: indium(III), thallium(III), tin(IV), lead(IV), antimony(V), bismuth(V), etc. However, there is an increasing tendency, termed the "inert pair effect," for the metals to employ *p* electrons only and thus to exhibit oxidation states two less than those listed (see pp. 843–845).

---

[19] Differences in coordination number and in spin state complicate a direct comparison. The above range was taken from comparison of 6-coordinate Sc^{+3} and Y^{+3}, $\Delta r = 15$ pm and Zn^{+2} and Cd^{+2}, $\Delta r = 20$ pm. Because of the lanthanide contraction the radii of the third series are very similar to those of the second series. See p. 797.

[20] Earlier claims for the synthesis of OsF_8 have been discredited, but the synthesis of OsF_7 has been reported [O. Glemser et al., *Chem. Ber.*, **1966**, *99*, 2652].

[21] A. L. Lott and P. G. Rasmussen, *J. Am. Chem. Soc.*, **1969**, *91*, 6502.

[22] See B. N. Figgis and J. Lewis, *Progr. Inorg. Chem.*, **1964**, *6*, 37.

OXIDATION STATES AND *EMF*s OF THE TRANSITION METALS

Having compared in general terms the properties of transition metals both on the basis of the d electron configuration and the properties of the light versus heavier metals, we shall now look more specifically at the stability of the various oxidation states of each element in aqueous solution. Every oxidation state will not be examined in detail, but the *emf* data to make such an evaluation will be presented in the form of Latimer diagrams.

If you are not throughly familiar with the principles of electrochemistry, before considering the following discussion for determining the stability of oxidation states, review pp. 350–353 and the Latimer diagram derived there:

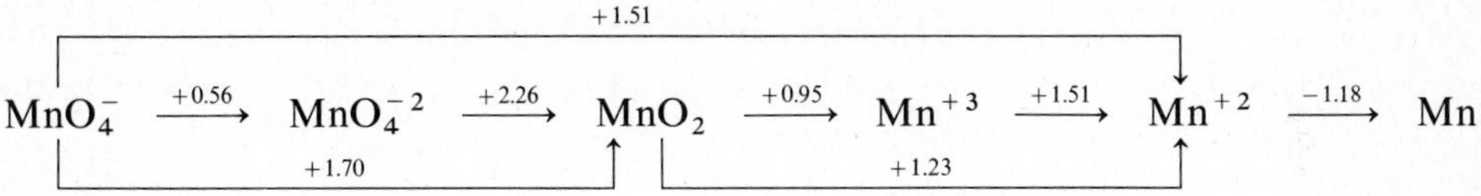

$$MnO_4^- \xrightarrow{+0.56} MnO_4^{-2} \xrightarrow{+2.26} MnO_2 \xrightarrow{+0.95} Mn^{+3} \xrightarrow{+1.51} Mn^{+2} \xrightarrow{-1.18} Mn$$

$MnO_4^- \rightarrow Mn^{+2}$: $+1.51$

$MnO_4^- \rightarrow MnO_2$: $+1.70$

$MnO_2 \rightarrow Mn^{+2}$: $+1.23$

Stability of oxidation states

There are three sources of thermodynamic instability for a particular oxidation state of an element in aqueous solution: (1) The element may reduce the hydrogen in water or hydronium ions; (2) it may oxidize the oxygen in water or hydroxide ions; and (3) it may disproportionate.

The *emf* values for reduction of hydrogen are given in Eqs. 8.108 to 8.110. These determine the minimum *oxidation emf* necessary for a species to effect reduction of the hydrogen: 1 mol dm^{-3} acid, $\mathscr{E}^0 > 0.00$; neutral solution, $\mathscr{E}^0 > +0.414$; 1 mol dm^{-3} base, $\mathscr{E}^0 > +0.828$. For manganese the only oxidation state that is unstable in this way is Mn^0:

$$Mn \longrightarrow Mn^{+2} + 2e^- \qquad \mathscr{E}^0 = +1.18 \tag{12.12}$$

$$2H^+ + 2e^- \longrightarrow H_2 \qquad \mathscr{E}^0 = 0.00 \tag{12.13}$$

$$Mn + 2H^+ \longrightarrow Mn^{+2} + H_2 \qquad \mathscr{E}^0 = +1.18 \tag{12.14}$$

The *emf* values for oxidation of the oxygen in water are given in Eqs. 8.111 to 8.113. These determine the minimum *reduction emf* necessary for a species to effect oxidation of the oxygen: 1 mol dm^{-3} acid, $\mathscr{E}^0 > +1.229$; neutral solution, $\mathscr{E}^0 > +0.815$; 1 mol dm^{-3} base, $\mathscr{E}^0 > +0.401$. There are several oxidation states of manganese that are reduced by water, but the manganate ion is typical:

$$MnO_4^{-2} + 4H^+ + 2e^- \longrightarrow MnO_2 + 2H_2O \qquad \mathscr{E}^0 = +2.26 \tag{12.15}$$

$$H_2O \longrightarrow \tfrac{1}{2}O_2 + 2H^+ + 2e^- \qquad \mathscr{E}^0 = -1.23 \tag{12.16}$$

$$MnO_4^{-2} + 4H^+ + H_2O \longrightarrow MnO_2 + \tfrac{1}{2}O_2 + 2H^+ \qquad \mathscr{E}^0 = +1.03 \tag{12.17}$$

Species that reduce or oxidize water can be spotted rapidly in *emf* diagrams such as the one given above for manganese. For example, in acid solution all negative *emf*s result in reduction of H^+ ion by the species to the *right* of that *emf* value. All values more positive than $+1.229$ result in oxidation of water by the species to the *left* of that value. Examination of the manganese diagram for acid solution reveals the following species are unstable: Mn^0 (oxidized to Mn^{+2}), Mn^{+3} (reduced to Mn^{+2}), and MnO_4^{-2} (reduced to MnO_2). One should also examine the skip-step *emf* values for possible reactions leading to instability. Thus, although water will not reduce MnO_4^- to MnO_4^{-2}, the skip-step *emf* for MnO_4^- to MnO_2 ($+1.70$) is sufficiently large to make the reaction proceed:

$$2MnO_4^- + 2e^- \longrightarrow 2MnO_4^{-2} \qquad \mathscr{E}^0 = +0.56 \tag{12.18}$$

$$H_2O \longrightarrow \tfrac{1}{2}O_2 + 2H^+ + 2e^- \qquad \mathscr{E}^0 = -1.23 \tag{12.19}$$

$$2MnO_4^- + H_2O \longrightarrow 2MnO_4^{-2} + \tfrac{1}{2}O_2 + 2H^+ \qquad \mathscr{E}^0 = -0.67 \tag{12.20}$$

$$2MnO_4^- + 8H^+ + 6e^- \longrightarrow 2MnO_2 + 4H_2O \qquad \mathscr{E}^0 = +1.70 \tag{12.21}$$

$$3H_2O \longrightarrow \tfrac{3}{2}O_2 + 6H^+ + 6e^- \qquad \mathscr{E}^0 = -1.23 \tag{12.22}$$

$$2MnO_4^- + 2H^+ \longrightarrow 2MnO_2 + \tfrac{3}{2}O_2 + H_2O \qquad \mathscr{E}^0 = +0.47 \tag{12.23}$$

In this example the MnO_4^{-2} ion may be considered a species high in free energy acting as a barrier to the reduction of MnO_4^-. Since the three-electron reduction leads to an overall decrease in free energy, the MnO_4^{-2} is bypassed.

Disproportionation occurs when a species is both a good reducing agent and a good oxidizing agent. In basic solution, for example, Cl_2 disproportionates to Cl^- and ClO^- ions:

$$\tfrac{1}{2}Cl_2 + e^- \longrightarrow Cl^- \qquad \mathscr{E}^0 = +1.36 \tag{12.24}$$

$$\tfrac{1}{2}Cl_2 + 2OH^- \longrightarrow ClO^- + H_2O + e^- \qquad \mathscr{E}^0 = -0.42 \tag{12.25}$$

$$Cl_2 + 2OH^- \longrightarrow Cl^- + ClO^- + H_2O \qquad \mathscr{E}^0 = +0.94 \tag{12.26}$$

Species susceptible to disproportionation are readily picked out from an *emf* diagram such as that given for manganese. The "normal" behavior of an element (i.e., when uncomplicated by disproportionation) is for the *emf* values to decrease steadily from left to right. The good reducing agents are on the right, the good oxidizing agents are on the left, and the stable species are toward the middle. Whenever this gradual change from more positive to more negative is broken, disproportionation will occur. For manganese in acid solution such "breaks" occur at two species: Mn^{+3} and MnO_4^{-2}. As it turns out, both ions are also unstable because they are reduced by water, but even if they were stable in this regard they would be unstable as a result of disproportionation reactions:

$$Mn^{+3} + e^- \longrightarrow Mn^{+2} \qquad \mathscr{E}^0 = +1.51 \tag{12.27}$$

$$Mn^{+3} + 2H_2O \longrightarrow MnO_2 + 4H^+ + e^- \qquad \mathscr{E}^0 = -0.95 \tag{12.28}$$

$$2Mn^{+3} + 2H_2O \longrightarrow Mn^{+2} + MnO_2 + 4H^+ \qquad \mathscr{E}^0 = +0.56 \tag{12.29}$$

Other applications of *emf*s include the prediction of thermodynamically possible redox reactions (e.g., Will Sn^{IV} oxidize Fe^{II} to Fe^{III}?) and the stabilization of oxidation states through the formation of complexes. The former is a straightforward application of thermodynamics and will not be discussed further here. The second is of great importance. It was introduced in Chapter 9 and will be discussed further below.

The effect of concentration on stability

The Nernst equation was given before (p. 352), and more recently (p. 579) the effect of pH on the reduction potential of the proton was mentioned, but the effect in general should be emphasized. There are several types of reactions where concentrations of the reactants and products affect the stability of various oxidation states. This can be understood through application of the Nernst equation. The reduction potential of hydrogen will obviously vary with the concentration of the hydrogen ion—hence the commonly known fact that many reasonably active metals dissolve in acid but not in base.

Perhaps even more important is the effect of hydrogen ion concentration on the *emf* of a half-reaction of a particular species. Consider the permanganate ion as an oxidizing agent in acid solution (as it often is). From the Latimer diagram above we can readily see that the reduction *emf* is 1.51 V when all species have an activity of 1 mol dm^{-3}. What is *not* shown is the complete equation:

$$MnO_4^- + 8H^+ + 5e^- \longrightarrow Mn^{+2} + 4H_2O \qquad \textbf{(12.30)}$$

which makes it clear that the concentration of the hydrogen ion enters the Nernst equation to the eighth power—the oxidizing power of the permanganate ion is strongly pH-dependent. If the hydrogen ion concentration is reduced to 10^{-14} mol dm^{-3} (1 mol dm^{-3} OH^-), a completely different set of values is obtained:

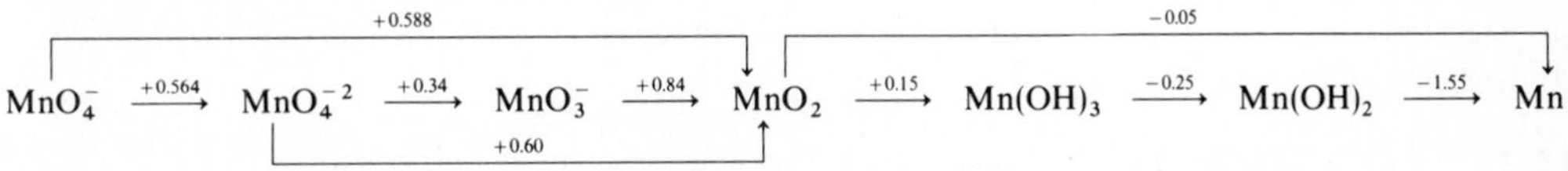

$$MnO_4^- \xrightarrow{+0.564} MnO_4^{-2} \xrightarrow{+0.34} MnO_3^- \xrightarrow{+0.84} MnO_2 \xrightarrow{+0.15} Mn(OH)_3 \xrightarrow{-0.25} Mn(OH)_2 \xrightarrow{-1.55} Mn$$

($MnO_4^- \to MnO_2$: $+0.588$; $MnO_4^{-2} \to MnO_2$: $+0.60$; $MnO_2 \to Mn$: -0.05)

Thus the oxidation states of manganese that are stable in 1 mol dm^{-3} base are different from those in acid (see Problem 12.13).

The subtlety of concentration effects may well be illustrated by the puzzlement once occasioned by the inclusion of Dry Ice® in a list of ingredient for the preparation of potassium permanganate—all the more so because it was obvious that the Dry Ice® was a true *reagent*, not a *coolant*.

The preparation takes advantage of the fact that the oxidation *emf*s of manganese are more favorable in basic than in acid solution: One can oxidize the readily available manganese dioxide to the green manganate ion, MnO_4^{-2}, with an *emf* of only -0.60 V to overcome. Since this half reaction

$$2H_2O + MnO_{2(s)} \longrightarrow MnO_4^- + 4H^+ + 2e^- \qquad \mathscr{E}^0 = -0.60V \qquad \textbf{(12.31)}$$

is so highly hydrogen ion sensitive,

$$\mathscr{E} = -0.60 - \frac{0.059}{2}\log\,[MnO_4^{-2}][H^+]^4 \qquad \textbf{(12.32)}$$

we can force the reaction even more by increasing the hydroxide ion concentration above 1 mol dm^{-3}, say, by using *fused* potassium hydroxide (an example of a nonaqueous, fused-salt reaction, see p. 344). Now, how can we get the manganate(VI) oxidized the remainder of the way to permanganate [=manganate(VII)]? By increasing the hydrogen ion concentration and gradually shifting over from the basic towards the acidic Latimer diagram. As shown by the latter, manganate(VI) disproportionates in acidic solution, forming 2 moles of permanganate for every one of manganese dioxide. But how can this "acidification" be effected without adding large amounts of strong acid (recall that permanganate is unstable in concentrated acid)? Simple: Dissolve the potassium manganate(VI)/potassium hydroxide mixture in water, throw in a few chunks of Dry Ice®, and in the "witches cauldron" effect, watch the solution turn from green to deep purple!

Group IB

The elements copper, silver, and gold show such anomalies that there sometimes appears to be little congruence as a family, with that member which is least reactive as a metal (Au) being the only one that has an appreciable chemistry in the +3 oxidation state and the only one to reach the −1 and +5 oxidation states (AuF_5), although both copper and silver may be oxidized to +4. The members of the family more or less routinely (silver infrequently) violate the very useful rule of thumb you may have learned in General Chemistry: *The maximum oxidation state of an element is equal to or less than its group number*. Thus we have $CuSO_4$, AgF_2, $AuCl_4^-$. Each member of the family has a different preferential oxidation state (Cu, +2; Ag, +1; Au, +3). The one property they *do* have in common is that none has a positive *emf*, $M \rightarrow M^{+n}$; therefore, the free metals are unaffected by simple acids, nor are they readily oxidized otherwise, leading to their use in materials intended to last. This, together with their value, has led to the term "coinage metals" for the members of this family.

$$Cu^{+3} \xrightarrow{+1.8} Cu^{+2} \xrightarrow{+0.153} Cu^{+} \xrightarrow{+0.521} Cu$$

$$Cu^{+2} \xrightarrow{+0.337} Cu$$

$$Cu^{+2} \longrightarrow CuX \longrightarrow Cu$$

X = Cl, +0.538; +0.137
X = Br, +0.640; +0.033
X = I, +0.860; −0.1852
X = CN, ~1.2; ~ −0.43

$$AgO^{+} \xrightarrow{\sim +2.1} Ag^{+2} \xrightarrow{+1.980} Ag^{+} \xrightarrow{+0.7991} Ag$$

$$Ag(S_2O_3)_2^{-3} \xrightarrow{+0.017} Ag$$

$$AgX \longrightarrow Ag$$

X = OAc, +0.643
X = Cl, +0.2222
X = Br, +0.0713

$$Au^{+3} \xrightarrow{<+1.29} Au^{+2} \xrightarrow{>+1.29} Au^{+} \xrightarrow{+1.691} Au$$

$$Au^{+3} \xrightarrow{+1.498} Au$$

$$AuCl_4^- \xrightarrow{+1.00} Au$$

$$AuBr_4^- \xrightarrow{+0.87} AuBr_2^- \xrightarrow{+0.956} Au$$

Although copper forms compounds in any of four different oxidation states, only the +2 state enjoys much stability. The +3 state is generally too strong an oxidizing agent, though recently Cu(III) has been found in biological systems. Complexation by peptides can lower the reduction *emf* to the range 0.45–1.05 V.[23] The free +1 ion will spontaneously disproportionate (+0.521 > +0.153). Copper(I) compounds are known, however, in the form of complexes such as $[Cu(CN)_2]^-$, or as the sparingly soluble halides.

Silver forms stable compounds only in the +1 oxidation state, all higher states being strong oxidizing agents. Even silver(I) is not overly stable as shown by the large (+0.8 V) reduction potential. The photosensitized reduction of silver halides is, of course, the basis of photography.

None of the oxidation states occurring in gold compounds can really be said to be thermodynamically stable. Gold(II) and gold(I) are subject to disproportionation. The reduction potential of gold(III) to gold(I) is marginally above that necessary to oxidize water, but the presence of complexing agents can stabilize both +1 and +3, with the latter usually being the more stable.

Group IIB

The $d^{10}s^2$ configuration of this family is not conducive to an extensive redox chemistry. The overwhelming tendency is to lose the *s* electrons to become a stable +2 cation; indeed, this essentially describes the entire redox chemistry of zinc and cadmium.

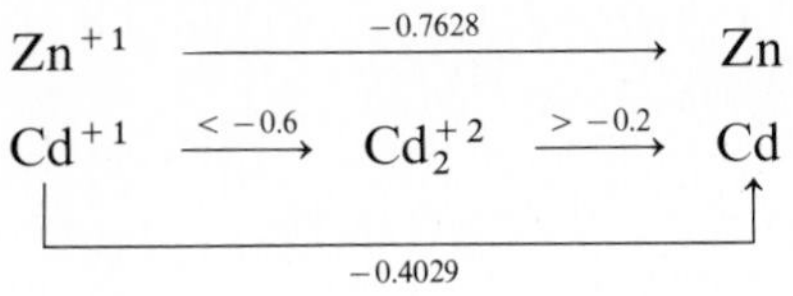

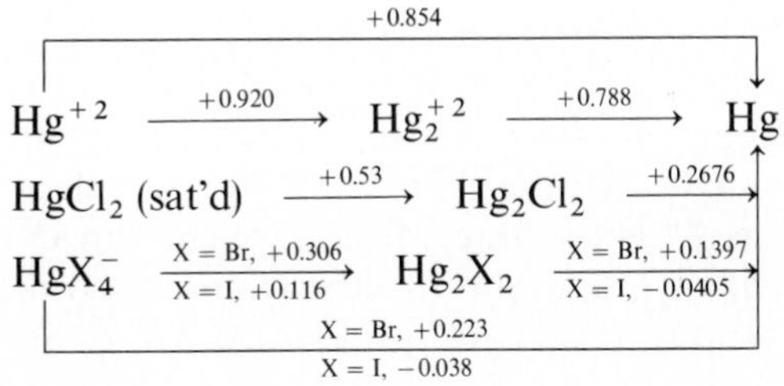

$$Zn^{+1} \xrightarrow{-0.7628} Zn$$

$$Cd^{+1} \xrightarrow{<-0.6} Cd_2^{+2} \xrightarrow{>-0.2} Cd$$

$$Cd^{+1} \xrightarrow{-0.4029} Cd$$

$$Hg^{+2} \xrightarrow{+0.920} Hg_2^{+2} \xrightarrow{+0.788} Hg$$

$$Hg^{+2} \xrightarrow{+0.854} Hg$$

$$HgCl_2\ (\text{sat'd}) \xrightarrow{+0.53} Hg_2Cl_2 \xrightarrow{+0.2676} Hg$$

$$HgX_4^- \xrightarrow[X = I,\ +0.116]{X = Br,\ +0.306} Hg_2X_2 \xrightarrow[X = I,\ -0.0405]{X = Br,\ +0.1397} Hg$$

$$HgX_4^- \xrightarrow[X = I,\ -0.038]{X = Br,\ +0.223} Hg$$

The ability of mercury to form a Hg—Hg bond plus a greater tendency to form coordination compounds increases the complexity of its Latimer diagram somewhat, but not much. The electrochemistry is straightforward, with both mercury(I) and mercury(II) being stable in aqueous solution.

Group IIIB and the Lanthanides

These metals are all quite active, resembling the alkaline earth metals (IIA) to a certain degree. The +3 oxidation state is stable for all, as are an occasional +2 and +4 species. See Chapter 16.

[23] D. W. Margerum and G. D. Owens, *Metal Ions Biol. Syst.*, **1981**, *12*, 75.

Group IVB

The redox chemistry of this family is straightforward and not particularly exciting. All have a stable +4 state. In addition, titanium has two lower states: Ti(II), which is unstable because it reduces H^+ except at low acidities; and Ti(III), which is a reducing agent but reasonably stable (see p. 372).

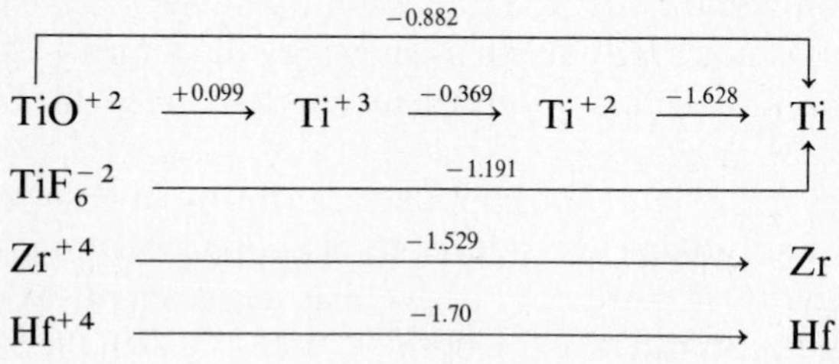

$$TiO^{+2} \xrightarrow{+0.099} Ti^{+3} \xrightarrow{-0.369} Ti^{+2} \xrightarrow{-1.628} Ti \qquad (TiO^{+2} \xrightarrow{-0.882} Ti)$$

$$TiF_6^{-2} \xrightarrow{-1.191} Ti$$

$$Zr^{+4} \xrightarrow{-1.529} Zr$$

$$Hf^{+4} \xrightarrow{-1.70} Hf$$

Group VB

As was the case in Group IVB, only the first member of the family, vanadium, has an extensive redox chemistry.

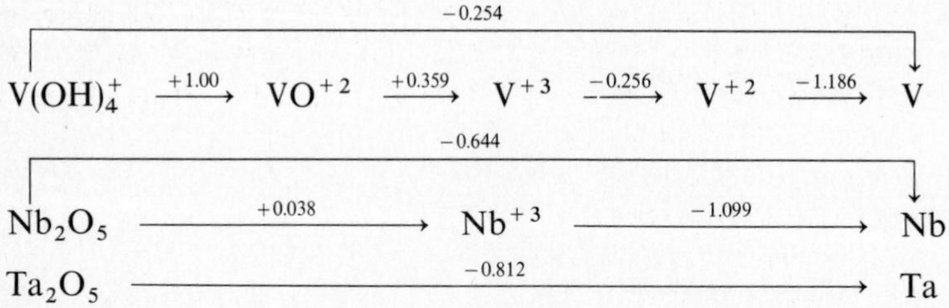

$$V(OH)_4^+ \xrightarrow{+1.00} VO^{+2} \xrightarrow{+0.359} V^{+3} \xrightarrow{-0.256} V^{+2} \xrightarrow{-1.186} V \qquad (V(OH)_4^+ \xrightarrow{-0.254} V)$$

$$Nb_2O_5 \xrightarrow{+0.038} Nb^{+3} \xrightarrow{-1.099} Nb \qquad (Nb_2O_5 \xrightarrow{-0.644} Nb)$$

$$Ta_2O_5 \xrightarrow{-0.812} Ta$$

All of the oxidation states shown for vanadium are known, though vanadium(II) is strongly reducing and unstable with respect to the hydrogen ion. Vanadium(III), though stable, is a fairly strong reducing agent. Vanadium(V) is a strong oxidizing agent, but only in concentrated acid. The dependency upon the hydrogen ion concentration is such that in the neutral solutions the reduction potential ($\mathscr{E} = 1.00$ V, $[H^+] = 1$) is lowered to such an extent that it becomes difficult to reduce V(V).

Because of the general insolubility of the oxides and the lack of stable lower oxidation states, there is little solution redox chemistry of niobium and tantalum. Niobium(III) *does* appear to form upon the reduction of niobium(V) with zinc, and it seems stable in the cold in the absence of air, but if the solution is heated, decomposition occurs with precipitation of mixed oxides.

Group VIB

Chromium continues the "normal" pattern we have seen for vanadium: The highest oxidation state is strongly oxidizing, the lower ones are strongly reducing, and there are unusual values leading to disproportionation. The chromium(II) ion can readily be prepared, but it is too strong a reducing agent to exist except as a complex, insoluble salt, etc. Chromium(III) is the most stable form. Chromium(VI) is a powerful oxidizing agent in acid solution (dichromate), but because of the hydrogen ion dependence, it is much less so in basic solution (chromate).

$$Cr_2O_7^{-2} \xrightarrow{+1.33} Cr^{+3} \xrightarrow{-0.406} Cr^{+2} \xrightarrow{-0.913} Cr$$

$$Cr_2O_7^{-2} \xrightarrow{+0.293} Cr \qquad Cr^{+3} \xrightarrow{-0.744} Cr$$

$$H_2MoO_4 \xrightarrow{+0.4} MoO_2^{+} \xrightarrow{+0.0} Mo^{+3} \xrightarrow{-0.20} Mo$$

$$H_2MoO_4 \xrightarrow{0.0} Mo$$

$$WO_3 \xrightarrow{-0.03} W_2O_5 \xrightarrow{-0.04} WO_2 \xrightarrow{-0.15} W^{+3} \xrightarrow{-0.11} W$$

$$WO_3 \xrightarrow{-0.090} W \qquad WO_2 \xrightarrow{-0.12} W$$

The heavier congeners, molybdenum and tungsten, have a less interesting redox chemistry. The *emf*s are small and the differences relatively unimportant. The chemistry of these elements in iso- and hetero-poly acids, multiple bonds, etc. is generally of more interest (see Chapter 14).

Group VIIB

The first member of this family, manganese, exhibits one of the most interesting redox chemistries known; thus it has already been discussed in detail above. Technetium has a "normal" distribution with modest *emf* values, as does rhenium, except that two species unstable with respect to disproportionation, rhenium trioxide and the rhenium(III) cation, are known.

$$TcO_4^- \xrightarrow{+0.7} TcO_2 \xrightarrow{+0.6} Tc^{+2} \xrightarrow{+0.4} Tc$$

$$TcO_4^- \xrightarrow{+0.6} Tc \qquad TcO_2 \xrightarrow{+0.5} Tc$$

$$ReO_4^- \xrightarrow{(+0.4)} ReO_3 \xrightarrow{(+0.6)} ReO_2 \xrightarrow{(+0.2)} Re^{+3} \xrightarrow{(+0.3)} Re$$

$$ReO_4^- \xrightarrow{+0.362} Re \qquad ReO_4^- \xrightarrow{+0.510} ReO_2 \qquad ReO_2 \xrightarrow{+0.2513} Re$$

$$ReO_4^- \xrightarrow{(+0.5)} ReCl_6^{-2} \xrightarrow{(+0.3)} ReCl_4^- \xrightarrow{(+0.15)} Re$$

Group VIIIB

The Group VIIIB metals form a heterogeneous assortment of elements combined into a single family more from a desire not to have any group number exceed eight, a "magic number" in chemistry even before Lewis formalized it in his octet theory, than from any compelling logic. This, of course, ignores the fact that the set of five *d* orbitals have a capacity of ten electrons, and thus there should be *ten* families of transition metals. Indeed, Sanderson[24] has advocated a revision of the periodic chart that includes this change among others, but it has not been generally accepted by inorganic chemists. As long as one realizes that the chemistry of iron is no more similar to nickel than it is to chromium, no problem exists.

$$FeO_4^{-2} \xrightarrow{+2.20} Fe^{+3} \xrightarrow{+0.771} Fe^{+2} \xrightarrow{-0.4402} Fe$$

$$Fe(CN)_6^{-3} \xrightarrow{+0.36} Fe(CN)_6^{-4} \xrightarrow{\text{ca, } -1.5} Fe$$

$$RuO_4 \xrightarrow{(+0.9)} RuO_4^{-} \xrightarrow{(+0.6)} RuO_4^{-2} \xrightarrow{(+0.4)} RuO_2 \xrightarrow{(+0.1)} Ru_2O_3 \xrightarrow{(-0.1)} Ru$$

$$RuO_4 \xrightarrow{+0.3} Ru \qquad RuO_4 \xrightarrow{(+0.75)} RuO_4^{-2} \qquad RuO_4 \xrightarrow{+0.58} RuO_2 \qquad RuO_2 \xrightarrow{-0.04} Ru$$

$$RuCl_5^{=} \xrightarrow{+0.601} Ru$$

$$OsO_4 \xrightarrow{+1.0} OsCl_6^{-2} \xrightarrow{+0.85} OsCl_6^{-3} \xrightarrow{(+0.4)} Os^{+2} \xrightarrow{(+0.85)} Os$$

$$OsO_4 \xrightarrow{+0.85} Os \qquad OsO_4 \xrightarrow{(+0.97)} OsCl_6^{-3} \qquad OsCl_6^{-3} \xrightarrow{(+0.71)} Os$$

$$CoO_2 \xrightarrow{>+1.8} Co^{+3} \xrightarrow{+1.808} Co^{+2} \xrightarrow{-0.277} Co$$

$$RhO_4^{-2} \xrightarrow{+1.5} RhO^{+2} \xrightarrow{+1.4} Rh^{+3} \xrightarrow{+1.2} Rh^{+2} \xrightarrow{+0.6} Rh^{+} \xrightarrow{+0.6} Rh$$

$$Rh^{+3} \xrightarrow{+0.8} Rh$$

$$RhCl_6^{-2} \xrightarrow{\sim +1.2} RhCl_6^{-3} \xrightarrow{+0.431} Rh$$

$$IrO_4^{-2} \xrightarrow{<+1.3} IrO_2 \xrightarrow{+0.7} Ir^{+3} \xrightarrow{+1.15} Ir$$

$$Ir^{+3} \xrightarrow{<+1.1} Ir^{+2} \xrightarrow{>+1.1} Ir \qquad IrO_2 \xrightarrow{+0.93} Ir$$

$$IrCl_6^{-2} \xrightarrow{+1.017} IrCl_6^{-3} \xrightarrow{+0.77} Ir$$

$$IrBr_6^{-3} \xrightarrow{+0.99} IrBr_6^{-4}$$

[24] R. T. Sanderson, "Inorganic Chemistry," Van Nostrand-Reinhold, New York, **1967**. See also a discussion of the history of group numbering, together with some editorial opinion, by W. C. Fernelius and W. H. Powell, *J. Chem. Educ.*, **1982**, *59*, 504.

$$NiO^{=} \xrightarrow{>1.8} NiO_2 \xrightarrow{+1.678} Ni^{+2} \xrightarrow{-0.250} Ni$$

$$Ni(NH_3)_6^{+2} \xrightarrow{-0.49} Ni$$

$$PdO_3 \xrightarrow{\sim +2} Pd^{+4} \xrightarrow{+1.6} Pd^{+2} \xrightarrow{+0.987} Pd$$

$$PdCl_6^{-2} \xrightarrow{+1.288} PdCl_4^{-2} \xrightarrow{+0.62} Pd$$

$$PdBr_4^{-2} \xrightarrow{+0.60} Pd$$

$$PtO_3 \xrightarrow{> +2.0} PtO_2 \xrightarrow{\sim +1.1} Pt^{+2} \xrightarrow{\sim +1.2} Pt$$

$$PtX_6^{-2} \xrightarrow[X = Br,\ +0.59]{X = Cl,\ +0.68} PtX_4^{-2} \xrightarrow[+0.581]{+0.73} Pt$$

The pattern we have seen in the immediately preceding elements continues with iron and its congeners—the metal and +2 oxidation state are reducing, the higher oxidation states are oxidizing species. Members of the cobalt and nickel families, however, tend to be stable only in the +2 state unless stabilized by complexation. The reader may readily apply the methods illustrated previously to examine the relative stability of the individual oxidation states.

The Group VIIIB metals illustrate well the point made previously that heavier congeners more readily assume higher oxidation states. Thus iron, cobalt, and nickel are effectively limited to +2 and +3 oxidation states, but all of their congeners have reasonably stable higher oxidation states.

[*Note:* For reasons of space only a representative number of *emf* was included in the Latimer diagrams. A complete list of values is given in Appendix F.]

PROBLEMS

12.1. Prepare lists of similarities and differences of

a. manganese and rhenium

b. scandium and zinc

c. copper, silver, and gold

d. iron(II) and tin(II)

12.2. Pick a transition metal family (e.g., VIB) and look up all of the oxidation *emf*s from $M \rightarrow M^{+2}$ to $M^{+n-1} \rightarrow M^{+n}$ (where n = group number). Discuss these values and their effects on the chemistry of these elements.

12.3. Of the d^4 species discussed, chromium(II) is strongly reducing but manganese(III) is strongly oxidizing. Explain the differences in properties of these two species.

12.4. Consult Appendix G, "The Literature of Inorganic Chemistry," and select a transition metal for which a recent review has been written. Read it carefully and summarize the descriptive chemistry of this metal, especially with regard to the various oxidation states. [*Note:* This can be a time-consuming job, but it gives a "feel" for the principles of transition metal chemistry that cannot be obtained by reading a short chapter in a textbook.]

12.5. Hexadentate ligands of the general type **N N N N N N**, where **N** represents amine, imine, or hydroxylamine functions connected by two-carbon bridges (⌒), readily form complexes with Ni^{+2}, e.g., $[NiN_6]^{+2}(NO_3^-)_2$. These compounds dissolve in concentrated nitric acid, "nitrous fumes are evolved, and a deep red solution results." In some cases, crystals can be obtained. They are diamagnetic and quantitatively oxidize two moles of Fe^{+2} to Fe^{+3}. Describe the electronic state of the nickel.

12.6. For elements with rather low maximum oxidation states such as Cu, Ag, and Au, the highest oxidation state is found in the fluorides. For elements with high maximum oxidation states (Fe, Mn, Ru, Rh, Os) the maximum oxidation state is found in oxides or oxyanions. Explain.

12.7. Explain the *emf*s of the silver halides in terms of their solubilities. What about AgOAc, which is *soluble*?

12.8. **a.** How can one verify, just by looking at the Latimer diagram of silver, that sodium thiosulfate ("hypo") is useful in photographic processes that require the removal of excess, unreacted silver halide?

b. Is this process ("fixing") *actually* a redox reaction? Explain.

12.9. When citing the Sandmeyer reaction, organic textbooks frequently write the needed copper(I) halide as Cu_2Cl_2, Cu_2Br_2, etc. Comment.

12.10. Consider the complex ions dibromoaurate(I) and tetrabromoaurate(III). Which is more stable in aqueous solution? Explain.

12.11. Explain in terms of redox chemistry how the formation of the chloro complexes stabilizes rhenium(III).

12.12. In the following equations predict whether the reaction will proceed to the left or the right:

a. Fe(III) + Sn(II) = Fe(II) + Sn(IV)

b. $2Cu^{+2} + 4I^- = 2CuI + I_2$

c. $2Cu^{+2} + 4Br^- = 2CuBr + Br_2$

d. $TiO^{+2} + Cu^+ = Ti^{+3} + Cu^{+2}$

12.13. Compare the Latimer diagrams for manganese in acidic and in basic solution. Interpret the diagrams in terms of which oxidation states are stable.

12.14. Show that the seemingly disparate Eqs. 2.14 and 12.4 are, in fact, identical.

12.15. If you did not do Problem 9.19 when you read Chapter 9, do so now.

13

Organometallic chemistry

Organometallic compounds may be simply characterized as forming the organic chemistry of metals. Yet the borderlines are sometimes unclear. All would characterize nickel tetracarbonyl, $Ni(CO)_4$, as an organometallic compound; yet carbon monoxide is hardly a typical organic compound. Likewise organophosphorus, organosulfur, and organoselenium compounds form a border of organometallic chemistry even though phosphorus, sulfur, and selenium are borderline nonmetals.

Organometallic chemistry provides interesting insights into bonding, structure, and reactivity of molecules, some of which are unique to organometallic chemistry. On the practical side, organometallic catalysts become increasingly important in an age when temperature (and hence fuel) needs to be minimized in chemical processes.

The effective atomic number rule and the chemistry of organometallic compounds

Valence theory had not progressed sufficiently far in Werner's time to permit him to say much about the nature of the bonding in the compounds he studied. Nevertheless, his distinction between ionizable and nonionizable groups (such as Cl^-) led the way to the development of bonding theory in terms of ionic versus covalent bonding.

The first attempt to account for the bonding in complexes was by Sidgwick,[1] who extended the octet theory of G. N. Lewis to coordination compounds. Ligands were considered to be Lewis bases which donated electrons (usually one electron pair per ligand) to the metal ion acting as a Lewis acid. Stability was assumed to be attendant to a noble gas configuration. The sum of the electrons on the metal plus the electrons donated from the ligands was called the *effective atomic number* (EAN), and when it was 36 (Kr), 54 (Xe), or 86 (Rn), the EAN Rule was said to be obeyed. Alternatively, one can state that when

[1] N. V. Sidgwick, "The Electronic Theory of Valency," Cornell University Press, Ithaca, N.Y., **1927**.

the electron configuration is ns^2 $(n-1)d^{10}np^6$, there will be 18 electrons in the valence orbitals to make a closed configuration. Indeed, with increasing frequency this rule of thumb is referred to as the *18-electron Rule*, which has the advantage of being the same for all rows of the periodic chart and one does not have to refer to the different EAN for each noble gas. Furthermore, the number is an easy one to remember since it is merely the capacity of nine orbitals, one set each of $s(1)$, $p(3)$, and $d(5)$ orbitals. The rule is obeyed with rather high frequency by organometallic compounds, especially the metal carbonyls and nitrosyls. It has considerable usefulness as a predictive rule only when applied to a member of these groups of organometallic compounds that have been found to obey it. Examples will be given below to illustrate how the stability of certain organometallic compounds can be correlated with the EAN.

Molecular orbital theory and the 18-electron Rule

The importance of π bonding in stabilizing complexes was discussed in Chapter 9. For an octahedral complex the inclusion of π bonding results in an increase in $10Dq$ and therefore an increase in the total bonding energy (Fig. 9.50). The most stable bonding arrangement will be that in which *all* of the bonding orbitals are occupied and *none* of the antibonding orbitals is occupied. This will require 18 electrons, which when added to the noble gas configuration of the inner core provide an effective atomic number equal to the next noble gas. Viewed in another way, it means that *all of the orbitals of the valence shell* as correlated by their filling in the 18 elements across the periodic table ($4s$ = K, Ca; $3d$ = Sc–Zn; $4p$ = Ga–Kr, for the first transition) *are employed in the bonding and the resulting bonding molecular orbitals are filled.*

The picture is somewhat more complicated with regard to the other possible geometries, but in every case the use of all nine of the valence shell orbitals results in the need for 18 electrons in bonding or approximately nonbonding orbitals. In the tetrahedral nickel carbonyl the four σ bonds from the carbonyl groups result in four strongly bonding molecular orbitals (a_1 and t_2), accommodating eight electrons; the remaining ten electrons must occupy the e and t_2^* orbitals. The latter are formally antibonding in a σ-only system (Fig. 9.54), and this might at first appear to be contradictory to the discussion above. It should be noted, however, that π bonding may lower the energy of both of these sets of orbitals in a manner analogous to that in octahedral complexes.

The quantitative treatment of the molecular orbitals in carbonyls is hampered by our lack of knowledge of the relative energies of the various orbitals. Nevertheless we can see that in all of the carbonyl complexes, and to a lesser extent in other organometallic compounds, strong σ bonding enhanced by the synergistic effect of π bonding results in a wide separation of the bonding and antibonding orbitals. Under these circumstances there is only one configuration of maximum stability, usually corresponding to a total of 18 electrons. This is in contrast to situations described in Chapter 12 in which the differences in energy levels are small, allowing a variety of stable oxidation states and electron configurations.[2]

[2] P. R. Mitchell and R. V. Parish, *J. Chem. Educ.*, **1969**, *46*, 811.

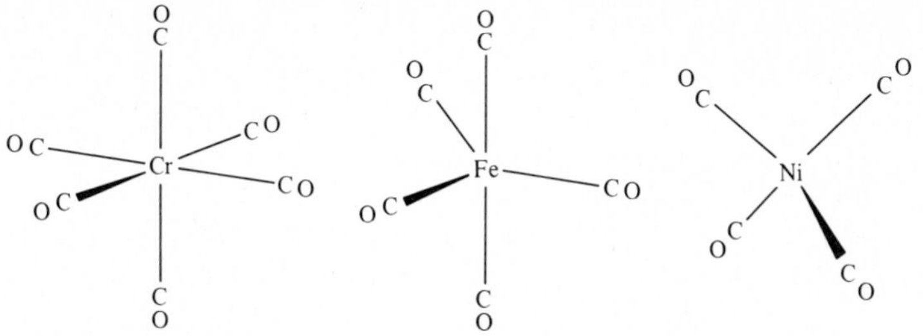

Fig. 13.1 Structures of simple carbonyls of chromium, iron, and nickel.

METAL CARBONYLS

Carbonyls

Almost all of the transition metals form compounds in which carbon monoxide acts as a ligand. There are three points of interest with respect to these compounds: (1) Carbon monoxide is not ordinarily considered a very strong Lewis base, and yet it forms strong bonds to the metals in these complexes; (2) the metals are always in a low oxidation state, most often formally in an oxidation state of zero, but also in low positive and negative oxidation states; and (3) the 18-electron Rule is obeyed with remarkable frequency, perhaps 99% of the time. Examples of the application of the rule to some simple metal carbonyls are

Cr	$= 24e^-$	Fe	$= 26e^-$	Ni	$= 28e^-$
6CO	$= 12e^-$	5CO	$= 10e^-$	4CO	$= 8e^-$
$Cr(CO)_6$	$= 36e^-$	$Fe(CO)_5$	$= 36e^-$	$Ni(CO)_4$	$= 36e^-$

Alternatively, we can write $Cr = 4s^2 3d^4 4p^0$. This totals six electrons, which when supplemented by six pairs from the carbon monoxide equals 18 ($\Sigma = 6 + 12 = 18$). Thus counting only electrons in the $3d$, $4s$, and $4p$ orbitals the count becomes:[3]

Cr	$= 6e^-$	Fe	$= 8e^-$	Ni	$= 10e^-$
6CO	$= 12e^-$	5CO	$= 10e^-$	4CO	$= 8e^-$
$Cr(CO)_6$	$= 18e^-$	$Fe(CO)^5$	$= 18e^-$	$Ni(CO)_4$	$= 18e^-$

The molecular structures of these compounds are shown in Fig. 13.1. They are compatible with both valence shell-electron pair repulsion (VSEPR) and valence bond ideas of directed valence.

Metals with odd atomic numbers cannot satisfy the 18-electron Rule by simple addition of carbon monoxide ligands since the resultant moiety will have an odd number of electrons (and hence not 18) no matter how many carbonyls are added. There are several options

[3] Both the EAN Rule and 18-electron Rule have been discussed here, to show their relationship to each other and because they are both found with considerable frequency in the literature. The student is recommended to choose *one* system and use it consistently, as mixing systems and numbers (18, 32, 36, 54, and 86) can easily lead to confusion at worst and careless errors at best.

Table 13.1 The 18-electron Rule and metals with odd atomic numbers

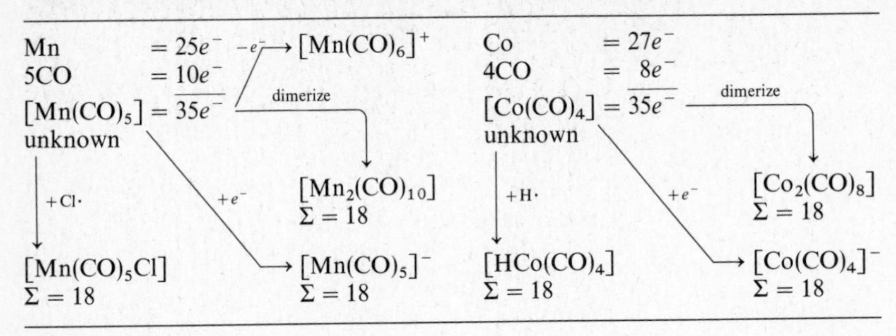

open to these metals by which the 18-electron Rule can be satisfied. The simplest is the addition of an electron by a reducing agent to form an anion such as $[M(CO)_n]^-$. Alternatively, the electron deficient moiety can bond covalently with an atom or group that also has a single unpaired electron available, e.g., hydrogen or chlorine: $HM(CO)_n$ or $M(CO)_nCl$. Finally, if no other species are available with which to react, two moieties each with an odd electron can dimerize with resultant pairing of the odd electrons. The various options for metals of the first transition series with odd atomic numbers are listed in Table 13.1. This table should be considered principally a bookkeeping device, and it does not imply any real existence of the monomers $[Mn(CO)_5]$ and $[Co(CO)_4]$. The preparative reactions leading to the carbonyl hydrides and halides will be discussed later.

The pairing of the odd electron on each metal atom in the dimer (the dimers are diamagnetic) implies the existence of a metal–metal bond in $[Mn_2(CO)_{10}]$ and $[Co_2(CO)_8]$. The molecular structure of the former (Fig. 13.2b) clearly reveals the presence of Mn—Mn bond, whereas Fe—Fe and Co—Co bonds may be inferred in the latter (Fig. 13.2d,f). They represent a large number of polynuclear carbonyls containing *bridging carbonyl groups* (>C=O) in addition to the *terminal carbonyl groups* (—C≡O) found in all metal carbonyls. The latter represents a carbon monoxide moiety *relatively* unchanged from free carbon monoxide; i.e., the bond between the carbon and oxygen can best be approximated as a *triple bond*. The bond order is reflected in the carbon–oxygen stretching frequencies, 2143 cm^{-1} for the carbon monoxide and 2000 ± 100 cm^{-1} for terminal carbonyl groups. Bridging carbonyl groups are, on the other hand, electronically much closer to the carbonyl group of organic chemistry as found in ketones, etc., with a carbon–oxygen bond order of approximately *two*. Again, the bond order may be inferred from the carbon–oxygen stretching frequency which is 1800 ± 75 cm^{-1} in bridging carbonyl groups compared with 1715 ± 10 cm^{-1} for saturated ketones, for example. The infrared spectrum of a metal carbonyl thus provides important information on the nature of the carbonyl groups present and may allow one to distinguish between a structure with only terminal carbonyls, such as Fig. 13.2b, and one containing one or more bridging carbonyl groups such as Fig. 13.2d. Note that alternative structures with bridging groups in dimanganese decarbonyl and none in dicobalt octacarbonyl may be drawn, such as those in Fig. 13.3. Although the bridged manganese carbonyl is unknown, there is infrared evidence that in solution structure 13.3b exists in equilibrium with 13.2f. At very low temperatures the bridged

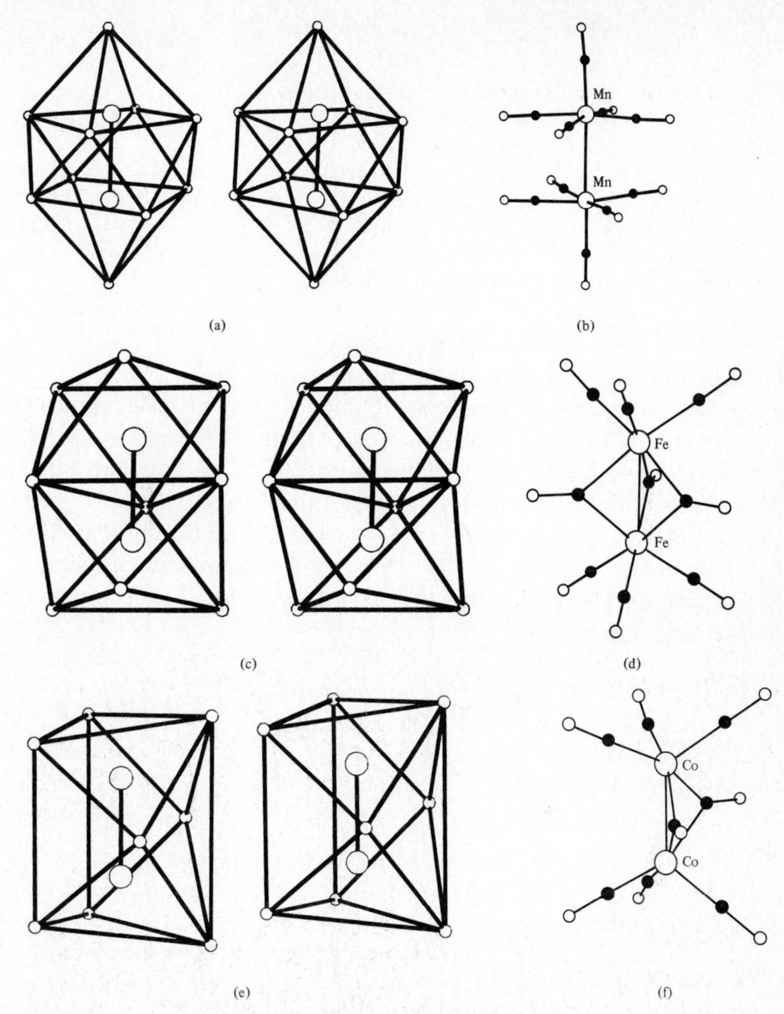

Fig. 13.2 Stereoviews (left) and conventional computer plots (right) of dimeric metal carbonyls: (a,b) $Mn_2(CO)_{10}$; (c,d) $Fe_2(CO)_9$; (e,f) $Co_2(CO)_8$. [From B. F. G. Johnson and R. E. Benfield, "Topics in Inorganic and Organometallic Stereochemistry," G. L. Geoffroy, ed., Wiley, New York, **1981**. Reproduced with permission.]

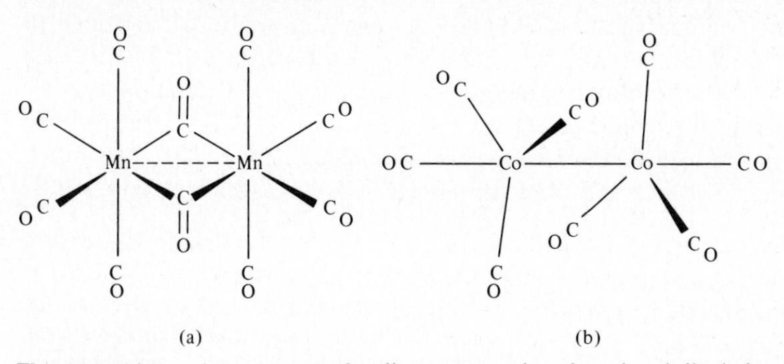

Fig. 13.3 Alternative structures for dimanganese decarbonyl and dicobalt octacarbonyl. Structure (a) is unknown, but there is infrared evidence for the existence of (b) in solution.

Table 13.2 Metal carbonyls of the first transition series

Monomer	$V(CO)_6$[a]	$Cr(CO)_6$	[a]	$Fe(CO)_5$	[a]	$Ni(CO)_4$
Dimer	$V_2(CO)_{12}$		$Mn_2(CO)_{10}$	$Fe_2(CO)_9$	$Co_2(CO)_8$	
Trimer			[a]	$Fe_3(CO)_{12}$	[a]	
Tetramer					$Co_4(CO)_{12}$	

[a] These formulations forbidden by the EAN Rule.

form predominates, and as the temperature is raised, the second isomer appears. If the temperature is increased to over 100 °C, a third form, of uncertain structure, predominates.[4]

In addition to the dimers required of metals with odd atomic numbers, dimers may also form which may be formally considered to be "carbon monoxide deficient." For example, in addition to the iron pentacarbonyl described above, there are two additional iron carbonyls: the so-called "enneacarbonyl" (diiron nonacarbonyl), $Fe_2(CO)_9$, and triiron dodecacarbonyl, $Fe_3(CO)_{12}$. A rather similar polymeric cobalt carbonyl, $Co_4(CO)_{12}$, is also known. These compounds obey the EAN Rule if metal–metal bonds are included in the computation:

$$\begin{array}{lll}
2\text{Fe} = 52e^- & 3\text{Fe} = 78e^- & 4\text{Co} = 108e^- \\
9\text{CO} = 18e^- & 12\text{CO} = 24e^- & 12\text{CO} = 24e^- \\
\hline
[\text{Fe}_2(\text{CO})_9] = 70e^- & [\text{Fe}_3(\text{CO})_{12}] = 102e^- & [\text{Co}_4(\text{CO})_{12}] = 132e^- \\
\text{EAN} = 35e^-/\text{Fe} & \text{EAN} = 34e^-/\text{Fe} & \text{EAN} = 33e^-/\text{Co} \\
\Sigma = 17e^-/\text{Fe} & \Sigma = 16e^-/\text{Fe} & \Sigma = 15e^-/\text{Co}
\end{array}$$

The 18-electron Rule requires that the enneacarbonyl be a *dimer* with a metal–metal bond. Since the rule requires that a 16-electron configuration must form *two* metal–metal bonds on each atom, the result will be a *trimer* formed from a triangle of iron atoms. Finally, if the metal atom is three electrons short, three metal–metal bonds must be formed on each cobalt directed towards three other cobalt atoms. This forms a tetrameric cluster of metal atoms arranged in the form of a tetrahedron. Larger clusters are known, but the rule breaks down and other methods must be used.[5]

Table 13.2 lists the formulas of metal carbonyls of the first transition series. Certain formulations violate the 18-electron Rule and should not be expected to be stable. The only exception to the predictions based on the EAN is vanadium carbonyl, expected to be the dimer, but existing as the paramagnetic monomer, $V(CO)_6$, instead. In this case the problem may relate to steric hindrance. The dimer would have a coordination number of 7. The repulsion of the ligands could overcome a weak bond energy or might provide a kinetic barrier to dimerization. Once $V(CO)_6$ has formed, it may require too much energy to rearrange the octahedral molecule to allow a V—V bond to form. To test this

[4] G. Bor and K. Noack, *J. Organometal. Chem.*, **1974**, *64*, 367; G. Bor et al., *Chem. Commun.*, **1976**, 914.

[5] See B. F. G. Johnson and R. E. Benfield, in "Topics in Inorganic and Organometallic Stereochemistry," G. L. Geoffroy, ed., Wiley, New York, **1981**, pp. 253–335, and K. Wade, *Adv. Inorg. Chem. Radiochem.*, **1976**, *18*, 1. A thorough explanation of these methods as applied to boranes and a short account as applied to metal clusters may be found in Chapter 14.

hypothesis, vanadium carbonyls were synthesized by co-condensing carbon monoxide and vanadium atoms at low temperatures with a system relatively rich in vanadium. The initial fragments would thus be expected to have fewer than six carbonyl groups and could dimerize before becoming "saturated." The resulting infrared spectrum shows four peaks not present in the monomer, indicating at least one new compound. One of the IR bands is at 1850 cm^{-1}, suggestive of a bridging carbonyl, so on this evidence and other arguments, the dimer has been tentatively formulated as the bridged dimer:[6]

$$2V + 12CO \xrightarrow[\text{gas matrix}]{\text{CO or rare}} (OC)_5V(\mu\text{-}C{=}O)_2V(CO)_5 \tag{13.1}$$

In any event, $V(CO)_6$ shows less stability than carbonyls obeying the 18-electron Rule, decomposing at 70°. If given the opportunity, it readily accepts an electron to form the EAN-allowed anion:

$$Na + V(CO)_6 \longrightarrow Na^+ + [V(CO)_6]^- \tag{13.2}$$

Finally, although the foregoing discussion has dealt almost exclusively with metals of the first transition series, analogous compounds are known for the heavier transition metals as well. These also obey the EAN Rule, in general, and thus tend to resemble the compounds discussed above: $Mo(CO)_6$ and $W(CO)_6$ versus $Cr(CO)_6$, $Tc_2(CO)_{10}$ and $Re_2(CO)_{10}$ versus $Mn_2(CO)_{10}$, etc. To be sure, there are some differences between the carbonyls of the heavier elements and the first-row transition elements (no one has succeeded in synthesizing enneacarbonyls of ruthenium and osmium, for example), but the resemblance is probably greater for the carbonyls than for other classes of coordination compounds.

Preparation and properties of metal carbonyls[7]

Some metal carbonyls can be made by direct interaction of the finely divided metal and carbon monoxide:

$$Ni + 4CO \longrightarrow Ni(CO)_4 \tag{13.3}$$

$$Fe + 5CO \xrightarrow[\text{high pressure}]{200^\circ C} Fe(CO)_5 \tag{13.4}$$

[6] T. A. Ford et al., *Inorg. Chem.*, **1976**, *15*, 1666. Unfortunately we know little about the strength of metal–metal bonds. One estimate (F. A. Cotton and R. R. Monchamp, *J. Chem. Soc.*, **1960**, 533) of the Mn—Mn bond strength in $Mn_2(CO)_{10}$ is 140 ± 55 kJ mol^{-1}, which may be somewhat high. More recently D. V. Korol'kov, Kh. Missner, and K. V. Ovchinnikov (*Zh. Strukt. Khim.*, **1972**, *14*, 717; *J. Struct. Chem. U.S.S.R.*, **1972**, *14*, 616) have estimated the Re=Re double bond at 350–425 kJ mol^{-1}.

[7] See J. C. Hileman, *Prep. Inorg. Reactions*, **1964**, *1*, 77.

In most cases, however, the metal must be reduced in the presence of carbon monoxide:

$$VCl_3 + 3Na + 6CO \longrightarrow V(CO)_6 + 3NaCl \quad \textbf{(13.5)}$$

$$Re_2O_7 + 17CO \longrightarrow Re_2(CO)_{10} + 7CO_2 \quad \textbf{(13.6)}$$

In the last reaction the carbon monoxide itself acts as a reducing agent.

The structures of the simple carbonyls have been shown previously (Fig. 13.1) and may be summarized briefly: Octahedral (Group VIB carbonyls and $V(CO)_6$), trigonal bipyramidal (Fe, Ru, and Os pentacarbonyls), and tetrahedral ($Ni(CO)_4$) species are known.

The reactions of carbonyl compounds are many and varied and will be discussed separately as methods of preparing other organometallic compounds. For now we may note that they may be considered to be of two types:

1. Simple replacement reactions in which a carbon monoxide molecule is replaced by another two-electron donor:

$$[Ni(CO)_4] + PX_3 \longrightarrow [Ni(CO)_3PX_3] + CO \quad \textbf{(13.7)}$$

or a many-electron donor that may replace several carbon monoxide ligands:

$$[Cr(CO)_6] + C_6H_6 \longrightarrow [Cr(C_6H_6)(CO)_3] + 3CO \quad \textbf{(13.8)}$$

2. Redox reactions (including the formation or cleavage of metal–metal bonds):

$$[Co_2(CO)_8] + 2Na \longrightarrow 2Na^+[Co(CO)_4]^- \quad \textbf{(13.9)}$$

$$[Mn(CO)_5Br] + [Mn(CO)_5]^- \longrightarrow [Mn_2(CO)_{10}] + Br^- \quad \textbf{(13.10)}$$

Because of their ready synthesis and wide reactivity, metal carbonyls are useful starting materials for many other organometallic compounds.

Polynuclear carbonyls without bridging groups

Above it was noted that carbonyls of metals with odd atomic numbers must be dimeric in order to obey the EAN Rule. In addition there are polynuclear carbonyls which may be regarded as "carbon monoxide deficient" in the sense that they are related to simple metal carbonyls but have metal–metal bonds in place of one or more carbonyl groups. Polynuclear carbonyls without bridging groups are a restricted class consisting of binuclear species such as $Mn_2(CO)_{10}$, the trinuclear species $Ru_3(CO)_{12}$ and $Os_3(CO)_{12}$ (Fig. 13.4b), tetranuclear $Ir_4(CO)_{12}$, pentanuclear $Os_5(CO)_{16}$ (Fig. 13.4d), and hexanuclear $Os_6(CO)_{18}$. The triruthenium and triosmium compounds are isostructural and contain a triangular array of metal atoms, the iridium compound has a tetrahedral cluster metal atoms, and the $Os_5(CO)_{16}$ osmium cluster is trigonal bipyramidal. When dodecacarbonyltriosmium(0) is pyrolyzed, new polynuclear compounds are formed having compositions corresponding to $Os_4(CO)_{13}$, $Os_5(CO)_{16}$, $Os_6(CO)_{18}$, $Os_7(CO)_{21}$, $Os_8(CO)_{23}$, and $Os_5(CO)_{15}C$, none of which show infrared bands characteristic of bridging carbonyl groups.[8] A reasonable guess

[8] C. R. Eady et al., *J. Organometal. Chem.*, **1972**, *37*, C39.

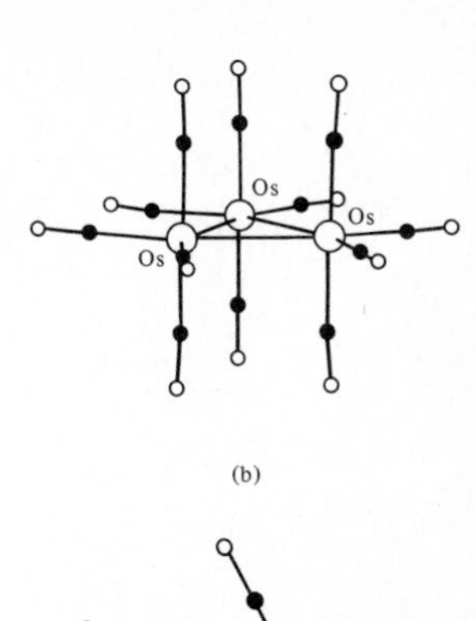

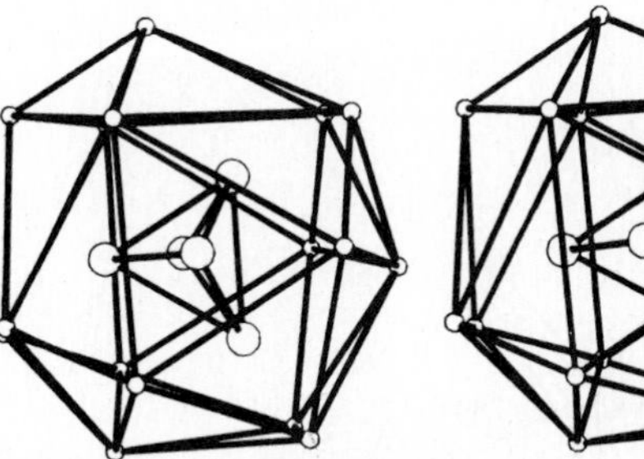

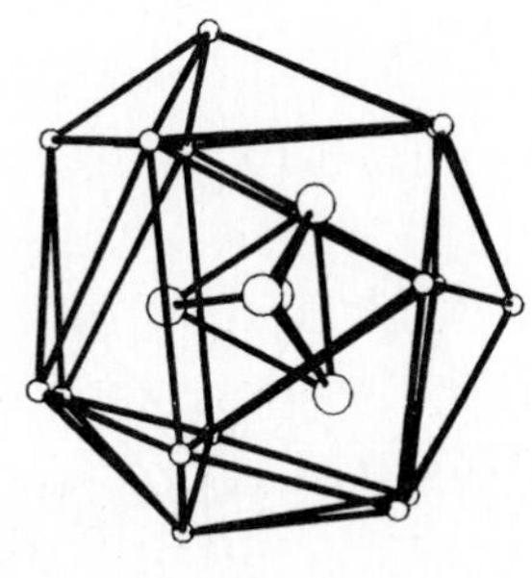

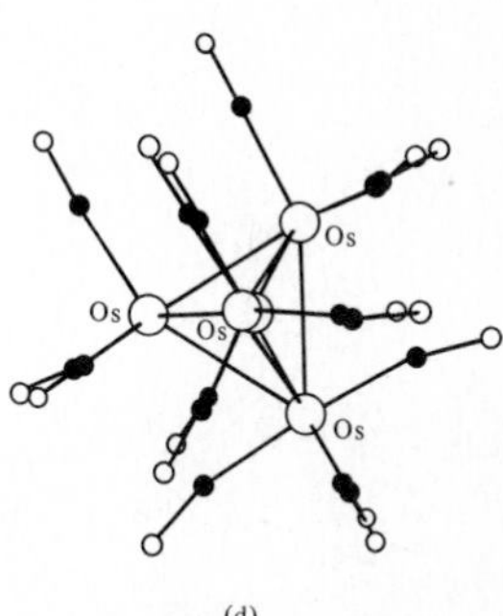

Fig. 13.4 Stereoviews (left) and conventional computer plots (right) of polynuclear metal carbonyls *without* bridging ligands: (a,b) $Os_3(CO)_{12}$; (c,d) $Os_5(CO)_{12}$. [From B. F. G. Johnson and R. E. Benfield, "Topics in Inorganic and Organometallic Stereochemistry," G. L. Geoffroy, ed., Wiley, New York, **1981**. Reproduced with permission.]

of the structure of $Os_6(CO)_{18}$ based on that of the triosmium compound and the noble gas rule would be an octahedron of osmium atoms, each of which has three terminal carbonyl groups. Surprisingly, the structure (Fig. 13.5) is a bicapped tetrahedron in which the osmium atoms are bonded to three, four, or five other osmium atoms as well as three terminal carbonyl groups.[9] The dianion, $[Os_6(CO)_{18}]^{-2}$, which differs by having only two more electrons, has a structure based on an octahedron of osmium atoms, however.[10] It may be noted that both of these results are consonant with the polyhedron skeleton electron pair theory (PSEPT), also known as Wade's Rules, which treats metal clusters analogously to polyhedral boranes.[5]

Polynuclear carbonyls with bridging carbonyl groups

The elucidation of the structures of some of these polynuclear carbonyls (Fig. 13.6) has been the subject of much recent research. The presence of bridging carbonyl groups gives rise to characteristic infrared absorptions. However, if the structure is complex, the presence of absorptions in the carbonyl region cannot be taken as unequivocal proof of the presence of bridging carbonyls since they may possibly arise as overtones or combination bands. Interpretations of infrared spectra must therefore be accepted with caution. They

[9] R. Mason et al., *J. Am. Chem. Soc.*, **1973**, *95*, 3802.

[10] M. McPartlin et al., *Chem. Commun.*, **1976**, 883.

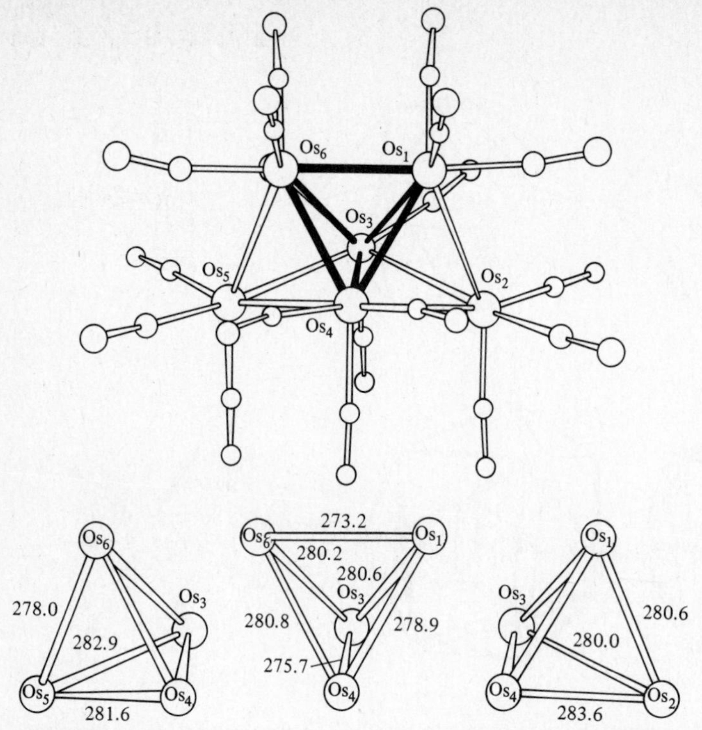

Fig. 13.5 Molecular structure of $Os_6(CO)_{18}$. The structure is termed a bicapped tetrahedron. The central tetrahedron is outlined in black for clarity. Distances in pm. [From R. Mason, K. M. Thomas, and D. M. Mingos, *J. Am. Chem. Soc.*, **1973**, *95*, 3802. Reproduced with permission.]

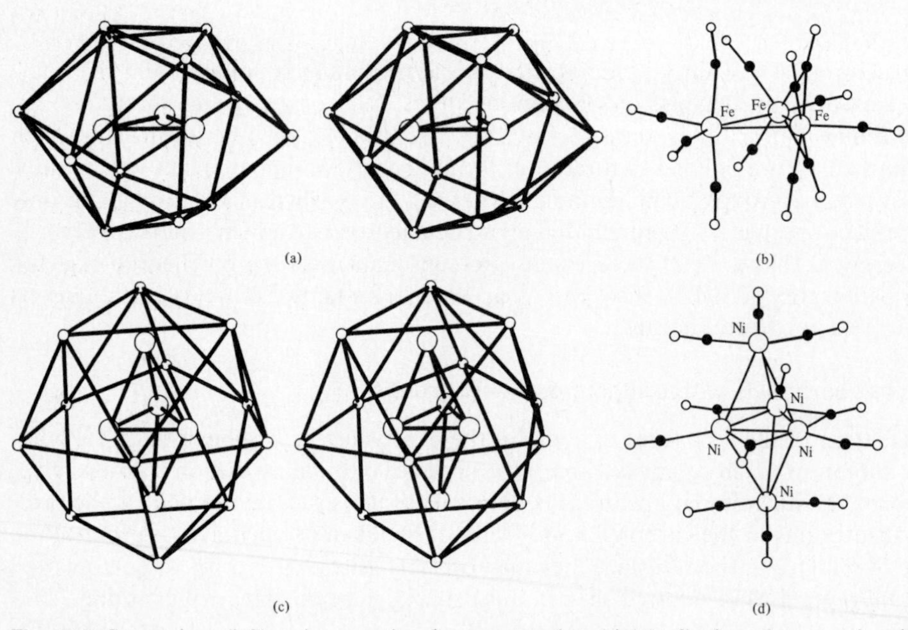

Fig. 13.6 Stereoviews (left) and conventional computer plots (right) of polynuclear metal carbonyls *with* bridging ligands: (a,b) $Fe_3(CO)_{12}$; (c,d) $Ni_5(CO)_{16}^{-2}$. [From B. F. G. Johnson and R. E. Benfield, "Topics in Inorganic and Organometallic Stereochemistry," G. L. Geoffroy, ed., Wiley, New York, **1981**. Reproduced with permission.]

do have one advantage over methods such as X-ray crystallography in the ability to study structures in solution, which may differ from those found in the solid.[11] X-ray crystallography provides unambiguous information on structure but occasionally has been beset with very difficult problems. The classic example is triiron dodecacarbonyl, for which disorder in the solid prevented a complete solution for many years.[12] The available X-ray data were compatible with a triangular array of iron atoms, and a structure similar to that of the isoelectronic $Os_3(CO)_{12}$ (Fig. 13.4b) was suggested by the X-ray evidence and the weak IR bands at about 1800 cm^{-1}. (The latter were assigned to overtones and/or combinations rather than as fundamental C—O stretches.) This simple model was overthrown by Mössbauer evidence[13] which proved that the three iron atoms are *not* in identical environments (see pp. 164, 242). The final solution[12] of the X-ray structure (Fig. 13.6b) showed that the iron atoms do indeed form a nearly equilateral triangle, but that one of the iron atoms is bonded only by Fe—Fe bonds without carbonyl bridges.

Reduction of nickel tetracarbonyl by alkali metals in tetrahydrofuran results in a mixture of carbonylate anions (see next section) from which salts of $[Ni_5(CO)_{12}]^{-2}$ can be isolated. X-ray diffraction provided the structure shown in Fig. 13.6d, a trigonal bipyramidal cluster of nickel atoms with both terminal and bridging carbonyl groups.[14]

The structure of hexarhodium hexadecacarbonyl (Fig. 13.7) is of interest because in addition to 12 terminal carbonyl groups it contains carbonyl groups bridging *three* metal atoms and lying on alternate triangular faces of the octahedral Rh_6 cluster.[15] The corresponding cobalt compound is thought to have a similar structure.[16]

A most unusual polynuclear species is $[Rh_{13}(CO)_{24}H_3]^{-2}$; see Fig. 13.8.[17] Note that the central metal atom is coordinated to nothing except the other 12 metal atoms. Furthermore, compare the arrangement of the 13 rhodium atoms with a set of BAB layers in Fig. 3.15a. The metal atoms in this cluster are hexagonally closest packed. In a sense, the cluster is a microscopic "chunk" of rhodium with carbon monoxide coördinated to its "surface." There has been considerable interest in compounds of this type with respect to catalysis. These clusters have been described as "little pieces of metal with chemisorbed species on the periphery"[18]—an allusion to the resemblance of the cluster to the bulk metal, of carbon monoxide ligands to adsorbed carbon monoxide, and the use of these compounds as models for catalysis.

The factors causing carbonyl groups to bridge or not to bridge are not at all well understood. Steric factors are probably important, for just as vanadium hexacarbonyl

---

[11] See F. A. Cotton, *Inorg. Chem.*, **1966**, *5*, 1083; B. Haas and R. K. Sheline, *J. Inorg. Nucl. Chem.*, **1967**. *29*, 693; E. A. C. Lucken et al., *J. Chem. Soc.*, *A*, **1967**, 148.

[12] See C. H. Wei and L. F. Dahl, *J. Am. Chem. Soc.*, **1969**, *91*, 1351, for a discussion of the history of this problem and the final X-ray structure. See also F. A. Cotton, *Progr. Inorg. Chem.*, **1975**, *21*, 1, for a more accurate determination and a discussion of the structural chemistry of this and related polynuclear carbonyls.

[13] M. Kalvius et al., *Z. Naturforsch. A*, **1962**, *17*, 494; E. Fluck et al., *Angew. Chem. Int. Ed. Engl.*, **1963**, *2*, 277; R. H. Herber et al., *Inorg. Chem.*, **1963**, *2*, 153.

[14] G. Longoni et al., *J. Am. Chem. Soc.*, **1975**, *97*, 5034.

[15] E. R. Corey et al., *J. Am. Chem. Soc.*, **1963**, *85*, 1202.

[16] P. Chini, *Inorg. Chem.*, **1969**, *8*, 1206.

[17] V. G. Albano, *Chem. Commun.*, **1975**, 859.

[18] E. L. Muetterties, *Science*, **1977**, *196*, 839. This is an excellent review not only of metal cluster chemistry but also of some of the important considerations of catalysis in general.

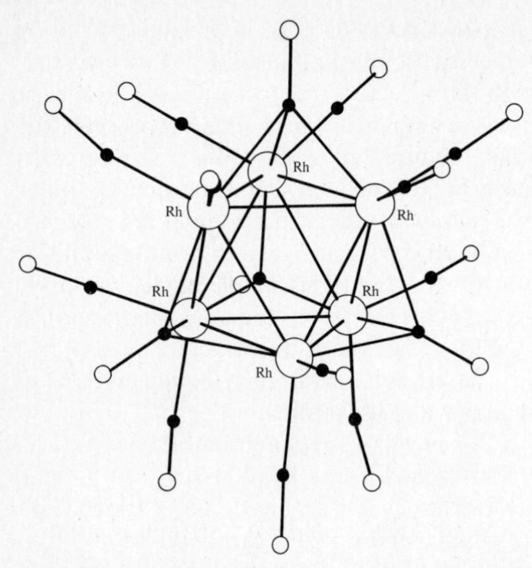

Fig. 13.7 The $Rh_6(CO)_{16}$ molecule. Note the presence of carbonyl groups on octahedral faces bridging *three* rhodium atoms. [From B. F. G. Johnson and R. E. Benfield, "Topics in Inorganic and Organometallic Stereochemistry," G. L. Geoffroy, ed., Wiley, New York, **1981**. Reproduced with permission.]

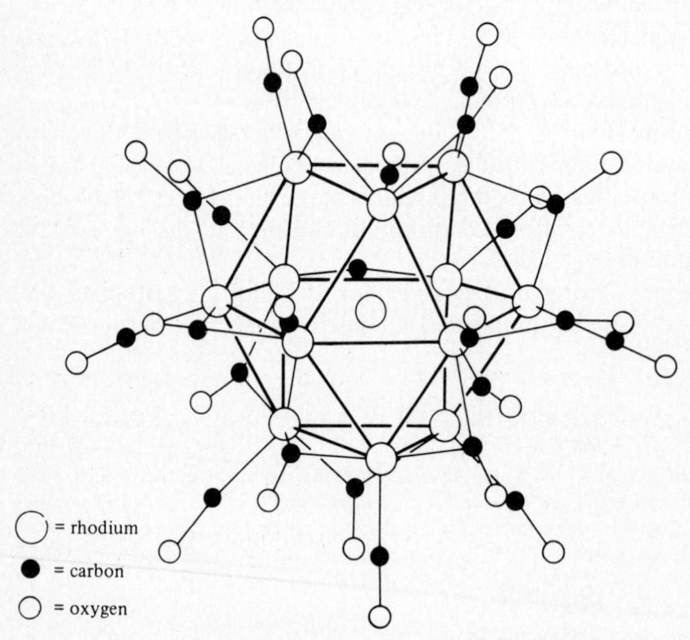

Fig. 13.8 The molecular structure of $Rh_{13}(CO)_{24}H_3^{-2}$ anion. Note central rhodium atom closest packed in contact with other twelve rhodium atoms. [From V. Albano, *Chem. Commun.*, **1975**, 860. Used with permission.]

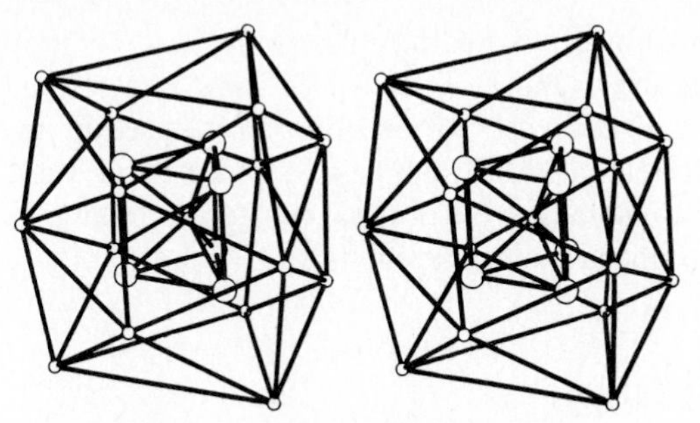

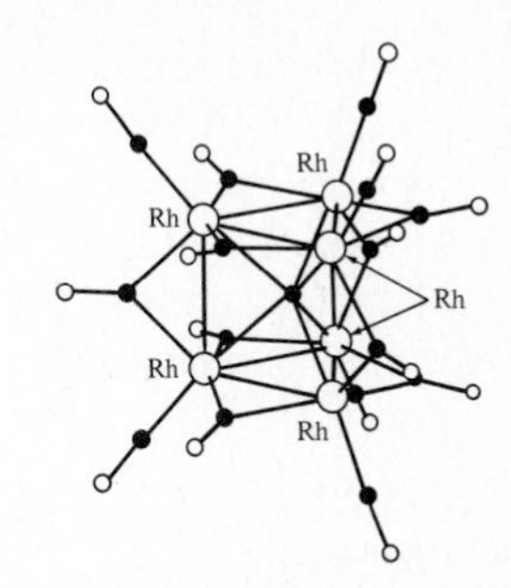

Fig. 13.9 Stereoview (left) and conventional computer plot (right) of a polynuclear metal carbonyl carbide, $Rh_6(CO)_{15}C$. Note carbon lying at the center of a trigonal prism of rhodium atoms. [From B. F. G. Johnson and R. E. Benfield, "Topics in Inorganic and Organometallic Stereochemistry," G. L. Geoffroy, ed., Wiley, New York, **1981**. Reproduced with permission.]

refuses to dimerize (it would become 7-coordinate), a bridged dimanganese decacarbonyl would have a coordination number of 7 or more (depending upon the number of bridges) rather than 6. Not even a trace of a bridged dimanganese species has been observed in equilibrium with the nonbridged species, in contrast to cobalt (see above) or diiron. However, this does not account for a common observation: Within a given family the heavier congeners tend to be less bridged than the lighter ones. Thus we have seen differences between cobalt and iridium or iron and osmium, and other examples could be cited. This cannot be steric hindrance but may be a size effect of a different sort. As the M—M bond lengthens (Co → Ir) the M—C bond must lengthen *more* or the M—C—M angle must open up, or both. Both effects could destabilize the structure, though it is not clear that this is the explanation.

A closely related question deals with the general distribution of carbonyl ligands about a metal cluster. Until recently this problem could not be attacked for lack of data, but as was pointed out recently[19] the number of known structures is now growing exponentially. Several workers have noted that the ligands tend to lie on the vertices of a regular polyhedron. This suggests an approach to this question similar to that of Bartell for substituted ethylenes (p. 232) or Gillespie (VSEPR, except that nonbonding electrons are here delocalized). The regular polyhedra represent minimum ligand–ligand repulsion (tetrahedron, octahedron, dodecahedron, icosahedron), and so an effort was made to calculate the other polyhedra of minimum repulsion and compare them to the experimental results.[20] The results are encouraging, especially if one notes that some of the distortions may come from the ligands attempting to fit a large polyhedron while attached to a smaller metal polyhedron of different symmetry. The stereoviews in Figs. 13.4–13.6 and 13.9 illustrate these polyhedra.

Finally, an unusual type of carbonyl compound is exemplified by the carbonyl carbides,[21] $[Fe_5(CO)_{15}C]$, $[Ru_6(CO)_{17}C]$, and $[Rh_6(CO)_{15}C]$ (Fig. 13.9), which contain

[19] F. A. Cotton and G. Wilkinson, "Advanced Inorganic Chemistry," 4th ed., Wiley, New York, **1980**, p. 1057.

[20] B. F. G. Johnson and R. E. Benfield, in "Topics in Inorganic and Organometallic Stereochemistry," G. L. Geoffroy, ed., Wiley, New York, **1981**, 253.

[21] M. Tachikawa and E. L. Muetterties, *Progr. Inorg. Chem.*, **1981**, *28*, 203.

5- and 6-coordinate carbon atoms. The polyhedral clusters in these compounds are a square pyramid, an octahedron, and a trigonal prism, respectively. Recently a number of similar compounds have been synthesized with nitrogen, phosphorus, sulfur, and arsenic[22] at the centers of large clusters of metal atoms. In all cases the nonmetal atoms have coordination numbers higher than those based on normal covalent bonding. Instead, the electrons presumably exist in delocalized molecular orbitals.

Carbonylate anions

Elements with odd atomic numbers may still satisfy the EAN Rule by accepting electrons:

$$[V(CO)_6] + Na \longrightarrow Na^+[V(CO)_6]^- \tag{13.11}$$

$$[Mn_2(CO)_{10}] + 2Na \longrightarrow 2Na^+[Mn(CO)_5]^- \tag{13.12}$$

Elements with even atomic numbers may also form carbonylate anions by accepting *two* electrons:

$$[Fe_3(CO)_{12}] + 6Na \longrightarrow 6Na^+ + 3[Fe(CO)_4]^{-2} \tag{13.13}$$

Under exceptional conditions, it is even possible to force *three* electrons onto manganese and rhenium:[23]

$$Re_2(CO)_{10} + 6Na \xrightarrow[\text{hexamethylphosphoramide}]{} 2Na_3[Re(CO)_4] \tag{13.14}$$

The elements molybdenum and tungsten, having even atomic numbers, would be expected to form dianions, and indeed their salts can be isolated in low yield from liquid ammonia:[24]

$$2M(CO)_6 + 6Na \longrightarrow 2Na_2[M(CO)_5] + Na^+ \ {}^-OC{\equiv}CO^- \ {}^+Na \qquad (M = Mo, W) \tag{13.15}$$

where most of displaced carbon monoxide forms the sparingly soluble disodium ethynediolate. The low yield of the dianion results from the further reduction to the -4 anion:[25]

$$2Na_2[M(CO)_5] + 6Na \longrightarrow Na_4[M(CO)_4] + Na^+ \ {}^-OC{\equiv}CO^- \ {}^+Na \qquad (M = Mo, W) \tag{13.16}$$

Other preparations involve reduction of the metal by carbon monoxide already present in the metal carbonyl or disproportionation of the metal. The first synthesis of a metal carbonylate ion involved the former procedure:

$$Fe(CO)_5 + 4OH^- \longrightarrow [Fe(CO)_4]^{-2} + CO_3^{-2} + 2H_2O \tag{13.17}$$

---

[22] See J. L. Vidal, *Inorg. Chem.*, **1981**, *20*, 243.

[23] J. E. Ellis and R. A. Faltynek, *Chem. Commun.*, **1975**, 966.

[24] See H. Behrens, *Adv. Organometal. Chem.*, **1980**, *18*, 1, for a review of metal carbonyl chemistry in liquid ammonia.

[25] J. E. Ellis et al., *J. Am. Chem. Soc.*, **1978**, *100*, 3605.

Often the use of Lewis bases effects the disproportionation of the metal. For example:

$$3Mn_2(CO)_{10} + 12py \longrightarrow 2[Mn(py)_6]^{+2} + 4[Mn(CO)_5]^- + 10CO \qquad \textbf{(13.18)}$$

(py = pyridine)

Although such syntheses are convenient, they suffer from an obvious disadvantage—they are wasteful of the difficultly obtained dimanganese dodecacarbonyl. One-third of the manganese originally present in the metal carbonyl is lost as the cationic manganese(II) complex.

In the case of some metals, mild reducing conditions may allow the reduction to be carried out stepwise:

$$2Cr(CO)_6 + 2NaBH_4 \xrightarrow[NH_3]{liq.} Na_2[Cr_2(CO)_{10}] + 2CO + H_2 + \text{boron hydrides} \qquad \textbf{(13.19)}$$

$$Na_2[Cr_2(CO)_{10}] + 2Na \xrightarrow[NH_3]{liq.} 2Na_2[Cr(CO)_5] \qquad \textbf{(13.20)}$$

These are but a few examples of a large class of anions synthesized from metal carbonyls. Without exception, these compounds obey the 18-electron Rule and are of interest because of the information they provide regarding bonding and structure as well as their usefulness in the synthesis of other carbonyl derivatives. The structures are often related to those of neutral metal carbonyls, but not necessarily those of the neutral metal carbonyl of the same element. For example $Fe(CO)_4^{-2}$ and $Co(CO)_4^-$ are isoelectronic and isostructural with $Ni(CO)_4$ (Fig. 13.1c). Likewise, $Fe_2(CO)_8^{-2}$ is isoelectronic and isostructural with the Co—Co bonded form of $Co_2(CO)_8$ (Fig. 13.2f) and quite unlike the "parent" iron enneacarbonyl (Fig. 13.2d) since the former does not have the bridging carbonyl groups of the latter.

Carbonylate ions are useful in the preparation of other carbonyl derivatives. Typical reactions involve nucleophilic attack of the anion on a positive center:

$$RX + Mn(CO)_5^- \longrightarrow RMn(CO)_5 + X^- \qquad \textbf{(13.21)}$$

$$RCOX + Co(CO)_4^- \longrightarrow RCOCo(CO)_4 + X^- \qquad \textbf{(13.22)}$$

$$Mn(CO)_5Br + Mn(CO)_5^- \longrightarrow Mn_2(CO)_{10} + Br^- \qquad \textbf{(13.23)}$$

Although Eq. 13.23 is of little importance in the manufacture of $Mn_2(CO)_{10}$ (the reactants are synthesized via the metal carbonyl itself), it illustrates a general and useful method of synthesizing metal–metal bonds in which the metals may be *different*:

$$Mn(CO)_5Br + Re(CO)_5^- \longrightarrow (CO)_5MnRe(CO)_5 + Br^- \qquad \textbf{(13.24)}$$

$$HgSO_4 + 2Mn(CO)_5^- \longrightarrow (CO)_5Mn{-}Hg{-}Mn(CO)_5 + SO_4^{-2} \qquad \textbf{(13.25)}$$

$$\phi SnCl_3 + 3Co(CO)_4^- \longrightarrow \phi Sn(Co(CO)_4)_3 + 3Cl^- \qquad \textbf{(13.26)}$$

$$(\phi_3P)_2NiCl_2 + 2Co(CO)_4^- \longrightarrow (OC)_4CoNi(\phi_3P)_2Co(CO)_4 + 2Cl^- \qquad \textbf{(13.27)}$$

The nucleophilicity of carbonylate anions varies with the nature of the metal and the substituents. When some of the substituents present are poorer π acceptors than carbon monoxide (which is the case for almost all ligands), there is a greater buildup of electron density on the metal atom and the nucleophilicity is increased. In addition, the nucleophilicity depends upon the relative stability of the free anion versus the new adduct with increased coordination number. The most stable coordination numbers for metal carbonyls and carbonylate anions appear to be 4 (tetrahedral) and 6 (octahedral). Coordination number 5 is less favorable. Thus the $[Mn(CO)_5]^-$ anion is quite nucleophilic (coordination number changes from 5 to 6 on reaction), but $[Co(CO)_4]^-$ is not (coordination number changes from 4 to 5). The variability in nucleophilicity and the variety of possible reactions make carbonylate anions useful synthetic reagents.[26]

An example of a carbonylate anion useful in the synthesis of organic compounds is tetracarbonylferrate(-II), isolable as a dioxane complex of its sodium salt, $Na_2Fe(CO)_4 \cdot 1\frac{1}{2}(CH_2)_4O_2$. Reaction with alkyl halides yields alkyliron carbonylate:

$$RX + Na_2Fe(CO)_4 \xrightarrow{THF} Na[RFe(CO)_4]^- \qquad \textbf{(13.28)}$$

Treatment with carbon monoxide under low pressure gives the corresponding acyl compound:

$$Na[RFe(CO)_4] + CO \longrightarrow Na[R\overset{\overset{O}{\|}}{C}Fe(CO)_4] \qquad \textbf{(13.29)}$$

Although these intermediates have been isolated and characterized,[27] this is not necessary, and the acyl derivative may be treated directly to form the desired products:

$$Na[R\overset{\overset{O}{\|}}{C}Fe(CO)_4] \xrightarrow{H^+} RC\overset{\displaystyle /\!\!/ O}{\underset{\displaystyle \backslash H}{}} \qquad \textbf{(13.30)}^{28}$$

$$Na[R\overset{\overset{O}{\|}}{C}Fe(CO)_4] \xrightarrow{R'X} R\overset{\overset{O}{\|}}{C}R' \qquad \textbf{(13.31)}^{29}$$

Carboxylic acids and their esters and amides can be made by oxidation of either the alkyl or acyl iron compound (presumably to form an acyl chloride) and treatment with the appropriate reagent:

$$Na[RFe(CO)_4] \xrightarrow{Cl_2} \xrightarrow{R'OH} R\overset{\overset{O}{\|}}{C}OR' \qquad \textbf{(13.32)}^{29}$$

[26] See R. B. King, *Acc. Chem. Res.*, **1970**, *3*, 417.

[27] W. D. Siegl and J. P. Collman, *J. Am. Chem. Soc.*, **1972**, *94*, 2516.

[28] M. P. Cooke, *J. Am. Chem. Soc.*, **1970**, *92*, 6080.

[29] J. P. Collman et al., *J. Am. Chem. Soc.*, **1972**, *94*, 1788. J. P. Collman et al., *J. Am. Chem. Soc.*, **1973**, *95*, 249; J. P. Collman and N. W. Hoffman, *ibid.*, **1973**, *95*, 2689.

The same type of reaction can be used to produce iron carbonyl complexes containing unsymmetrical phosphine or arsine ligands:

$$Na_2[Fe(CO)_4] \xrightarrow{\phi_2PCl} Na[\phi_2PFe(CO)_4] \xrightarrow{RX} R\phi_2PFe(CO)_4 \quad \textbf{(13.33)}$$[30]

Carbonyl hydrides

Acidification of carbonylate anions often results in the formation of the metal hydrides which may be considered the conjugate acids of the anions:

$$Co(CO)_4^- + H_3O^+ \longrightarrow HCo(CO)_4 + H_2O \quad \textbf{(13.34)}$$

$$Re(CO)_5^- + H_2O \longrightarrow HRe(CO)_5 + OH^- \quad \textbf{(13.35)}$$

$$Fe(CO)_4^{-2} + H_3O^+ \longrightarrow HFe(CO)_4^- + H_2O \quad \textbf{(13.36)}$$

$$HFe(CO)_4^- + H_3O^+ \longrightarrow H_2Fe(CO)_4 + H_2O \quad \textbf{(13.37)}$$

The hydrides may also be synthesized from the appropriate metal carbonyl or directly from the elemental metals, hydrogen, and carbon monoxide:

$$Mn_2(CO)_{10} + H_2 \longrightarrow HMn(CO)_5 \quad \textbf{(13.38)}$$

$$Co + 4CO + \tfrac{1}{2}H_2 \longrightarrow HCo(CO)_4 \quad \textbf{(13.39)}$$

When acting as acids, these compounds vary from cobalt tetracarbonyl hydride, which behaves as a very strong acid, to rhenium pentacarbonyl hydride, which is such a weak acid that the carbonylate anion may be readily hydrolyzed by water as shown in Eq. 13.35. Iron tetracarbonyl dihydride is an unstable gas at room temperature which behaves as a dibasic acid ($pK_1 \simeq 4.4$; $pK_2 \simeq 14$). The large difference in ionization constants for the two protons provided the first evidence that the hydrogen atoms in $H_2Fe(CO)_4$ were both bonded to the same atom (and hence the iron atom). This has subsequently been confirmed, and it is believed that a metal–hydrogen bond occurs in all of these compounds. They all exhibit very high-field NMR chemical shifts which are useful in characterizing hydride complexes.

The very large δ values (as far upfield as $\delta = -50$ ppm compared with typical values of 0 to 15 for hydrogen atoms in most organic compounds)[31] are consistent with an environment of high electron density. For this reason the proton is sometimes spoken of as being "buried" in the electron clouds of the metal. However, it is incorrect to view the proton vaguely as occupying some indefinite position inside the complex. Although the location of hydrogen atoms by X-ray crystallography is notoriously difficult because of the low scattering of the X-rays, this limitation does not apply to neutron diffraction studies. The latter have found the hydrogen atom to occupy a definite stereochemical position in the coordination shell.[32] The bond lengths are compatible with the addition of simple covalent radii. Even X-ray studies that do

[30] J. P. Collman et al., *J. Am. Chem. Soc.*, **1973**, *95*, 2389.

[31] The -50 δ hydrogen atom is found in complexes of the type $[Ir(PR_3)_2Cl_2H]$ (C. Masters et al., *Chem. Commun.*, **1971**, 209). See also Footnote 33. Typical values for organic compounds may be found in R. T. Morrison and R. N. Boyd, "Organic Chemistry," 3rd ed., Allyn & Bacon, Boston, **1973**, p. 421. This book may also be consulted for a general discussion of the meaning and significance of τ values.

[32] S. J. La Placa et al., *Inorg. Chem.*, **1969**, *8*, 1928. See also Footnote 33 and Fig. 10.5.

not locate the hydrogen atom provide strong evidence for stereochemical activity; the remaining ligands are arranged symmetrically only if it is assumed that a hydrogen atom is occupying an equivalent position in the coordination sphere.[33]

The bond lengths in both carbonyl and noncarbonyl hydride complexes provide interesting information on the operation of a *trans* influence by the hydride ion (see pp. 542–544). In *trans*-$[PtHBr(PEt_3)_2]$ the Pt—Br distance is 256 pm compared with 250 pm when the bromine is *trans* to another bromine. The lengthening is even greater in $[OsHBr(CO)(\phi_3P)_3]$ in which the Os—P distance is 256 pm when *trans* to the hydrogen atom compared with 234 pm for two phosphorus atoms *trans* to each other. In contrast, the hydrogen atom has a *weaker trans* influence than the carbonyl group: In $[HMn(CO)_5]$ the Mn—C distance is *shorter* for the group *trans* to the hydrogen atom than when two carbonyl groups are *trans* to each other.

A very basic and reactive carbonylate anion is $Cr(CO)_5^{-2}$. It readily abstracts a hydrogen ion from water:

$$[Cr(CO)_5]^{-2} + H_2O \longrightarrow [HCr(CO)_5]^- + OH^- \quad \textbf{(13.40)}$$

to form the hydridopentacarbonylchromate(0) anion. The latter may be prepared more easily by reduction of the hexacarbonyl:

$$[Cr(CO)_6]^+ \xrightarrow{NaBH_4} [HCr(CO)_5]^- \quad \textbf{(13.41)}$$

or by a form of the water–gas shift reaction:

$$OH^- + [Cr(CO)_6] \longrightarrow (OC)_5Cr\text{:::}C\begin{matrix}\cdot\cdot O^- \\ OH\end{matrix} \longrightarrow [HCr(CO)_5]^- + CO_2 \quad \textbf{(13.42)}$$

However, unless it is precipitated by a large counterion (see p. 281), it will react further with water:

$$[HCr(CO)_5]^- + H^+ \longrightarrow [Cr(CO)_5]^0 + H_2 \quad \textbf{(13.43)}$$

The unstable $[Cr(CO)_5]^0$ (note that the 18-electron Rule is not obeyed) reacts immediately with more of the anion:

$$[Cr(CO)_5]^0 + [HCr(CO)_5]^- \longrightarrow [(OC)_5Cr\text{—}H\text{—}Cr(CO)_5]^- \quad \textbf{(13.44)}$$

The resulting dichromium anion is bridged through the hydride ion. Because of the strong basicity of the hydridochromate anion, synthesis of analogous, mixed-metal, bridged hydrides occurs readily:[34]

$$(pip)Mo(CO)_5 + HCr(CO)_5 \longrightarrow (OC)_5Mo\text{—}H\text{—}Cr(CO)_5 + pip \quad \textbf{(13.45)}$$

$$(pip)W(CO)_5 + HCr(CO)_5 \longrightarrow (OC)_5W\text{—}H\text{—}Cr(CO)_5 + pip \quad \textbf{(13.46)}$$

(where "pip" is the easily replaced ligand piperidine).

In some ways the hydrogen bridge in these carbonylates resembles a hydrogen bond, but in this case it is between two electron-rich metals rather than between two nonmetallic

[33] M. L. H. Green and D. J. Jones, *Adv. Inorg. Chem. Radiochem.*, **1965**, *7*, 115; P. W. Atkins, J. C. Green, and M. L. H. Green, *J. Chem. Soc., A*, **1968**, 2275.

[34] M. Y. Darensbourg and J. C. Deaton, *Inorg. Chem.*, **1981**, *20*, 1644.

anions or dipoles. The hydrogen bridge may be either linear, as in $[HCr_2(CO)_{10}]^-$, or nonlinear, as in $HW_2(CO)_9NO$.[35] The differences in bonding are not clear, but in the tungsten complex it appears as though there is appreciable tungsten–tungsten bonding with the hydrogen overlapping tangentially.[36] Similar B—H—B bridges will be discussed in Chapter 14.

Although in many ways metal carbonyl hydrides may be regarded as acids, they also behave in some ways similarly to the basic hydrides of the principal group metals (e.g., NaH, $LiAlH_4$). For example, they act as reducing agents toward many organic compounds and are capable of hydrogenation of alkynes and alkenes. They are thus intermediate between the strictly "hydridic" hydrogens in the saline hydrides and the "protonic" hydrogens in compounds with nonmetals (e.g., HCl, NH_3).

There are other ways in which carbonyl hydrides and carbonylate anions are atypical. Although we have seen that the metal atom has high basicity (see also p. 613) and readily accepts protons, not all the chemistry of carbonylate anions follows this pattern. For example, when carbonylate anions form ion pairs with alkali metal cations, the terminal oxygen atom behaves as the basic site

$$(OC)_3CoCO^-Na^+$$

in every case studied except one.[37] Furthermore, because of this balance of basic sites in the molecule, the course of a reaction occasionally may be modified simply by changing the solvent:[37]

$$(cp)Fe(CO)_2^- + RCl \xrightarrow{\text{hexane}} (cp)Fe(CO)_2R + Cl^- \quad \textbf{(13.47)}$$

$$(cp)Fe(CO)_2^- + RCl \xrightarrow{\text{THF}} (cp)Fe(CO)_2Cl + RH \quad \textbf{(13.48)}$$

Parallels with nonmetal chemistry: isolobal fragments

Many of the reactions of metal carbonyls parallel closely those of certain nonmetals. For example, the $\cdot Mn(CO)_5$ fragment has 17 valence electrons, one short of the 18 necessary to fulfill the EAN Rule. It is analogous to the chlorine atom with seven valence electrons, one short of a noble gas configuration. The compounds and reactions of this fragment may thus be related to similar ones for chlorine or the methyl group. All three are formally free radicals and their chemistry consists of pairing the odd electron: It normally exists as a dimer, $Mn_2(CO)_{10}$ (cf. Cl_2, C_2H_6), but it may be reduced to the anion, $Mn(CO)_5^-$ (cf. Cl^-, negative methyl in CH_3MgX), which is the conjugate base of an acid, $HMn(CO)_5$ (cf. HCl, CH_4?); furthermore, it will combine with other species having a single unpaired electron, R·, I·, to form neutral molecules, $RMn(CO)_5$ (cf. RCl, RCH_3) and $Mn(CO)_5I$ (cf. ICl, CH_3I). The $Mn(CO)_5$ fragment may be considered an *electronically equivalent group*[38] of the chlorine atom and methyl group or an *isolobal* fragment.[39] The concept is a simple outgrowth of equating the Lewis octet of organic and main-group chemistry with

[35] L. B. Handy et al., *J. Am. Chem. Soc.*, **1970**, *92*, 7312.

[36] R. A. Love, *J. Am. Chem. Soc.*, **1976**, *98*, 4491.

[37] K. H. Pannell and D. Jackson, *J. Am. Chem. Soc.*, **1976**, *98*, 4443.

[38] A. S. Foust, M. S. Foster, and L. F. Dahl, *J. Am. Chem. Soc.*, **1969**, *91*, 5631; J. E. Ellis, *J. Chem. Educ.*, **1976**, *53*, 1.

[39] R. Hoffmann, *Science*, **1981**, *211*, 995; *Angew. Chem.*, in press.

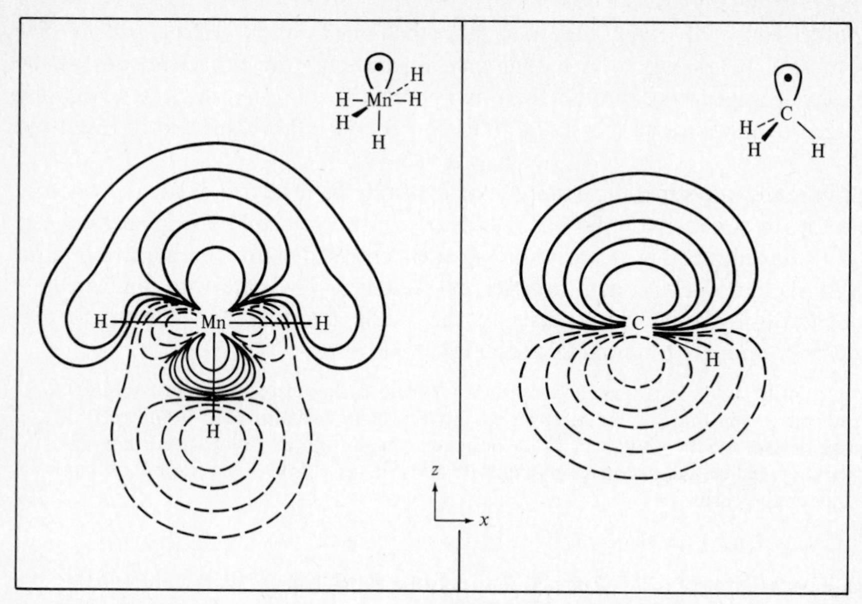

Fig. 13.10 Calculated contour diagram of the isolobal a_1 orbitals of MnH_5^- (left) and CH_3 (right). The contours of Ψ are plotted in a plane passing through manganese and three hydrogen atoms and through carbon and one hydrogen. [From R. Hoffmann, *Angew. Chem. Int. Ed.*, in press. Reproduced with permission of the Royal Academy of Science, Sweden.]

the 18-electron Rule of transition metal organometallic chemistry.[40] Of course one should not push these ideas too far. For example, unlike hydrochloric acid, manganese carbonyl hydride is a weak acid. But then, hydrogen cyanide is also a weak acid, and methane is normally not considered to be an acid at all; hence the question mark above. In many ways the isolobal $Mn(CO)_5$ may be considered a pseudohalogen (see p. 789). In any event the isolobal formalism is more concerned with isoelectronic[42] comparisons than with topics like polarity, etc.

Hoffmann[39] and co-workers have shown that isolobal fragments also have symmetry and relationships that go beyond simple electron counting. Thus the calculated electron density of the $\cdot MnH_5$ group (isolobal with $Mn(CO)_5$ but easier to calculate) may be compared with that of the methyl radical, $CH_3\cdot$ (Fig. 13.10). When one calculates the overlap integrals of these two isolobal fragments with respect to an incoming probe such as a hydrogen atom, the results are remarkably similar (Fig. 13.11). The manganese atom

[40] The isolobal concept, like most of Sherlock Holmes's explanations, seems "obvious" in hindsight (see footnote 12, Chapter 15), but it was first proposed in its present form by Halpern[41] as recently as 1968, elaborated by Dahl[38] and Ellis,[38] and extensively developed by Hoffmann.[39] It was the latter theoretical work, much of which is beyond the scope of this book, that formed the basis of Hoffmann's Nobel laureate address; see p. 359.

[41] J. Halpern, *Adv. Chem.*, Ser. No. 70, American Chemical Society, Washington, DC, **1968**, pp. 18–19.

[42] Care should be exercised with the use of the word "isoelectronic" here. These isolobal fragments are certainly not isostructural, and only $CH_3\cdot$ and $Cl\cdot$ are isoelectronic *internally*. However, the three, as do the isolobal fragments to be discussed, behave in an isoelectronic manner *externally*, as towards the hydrogen atom in Fig. 13.11.

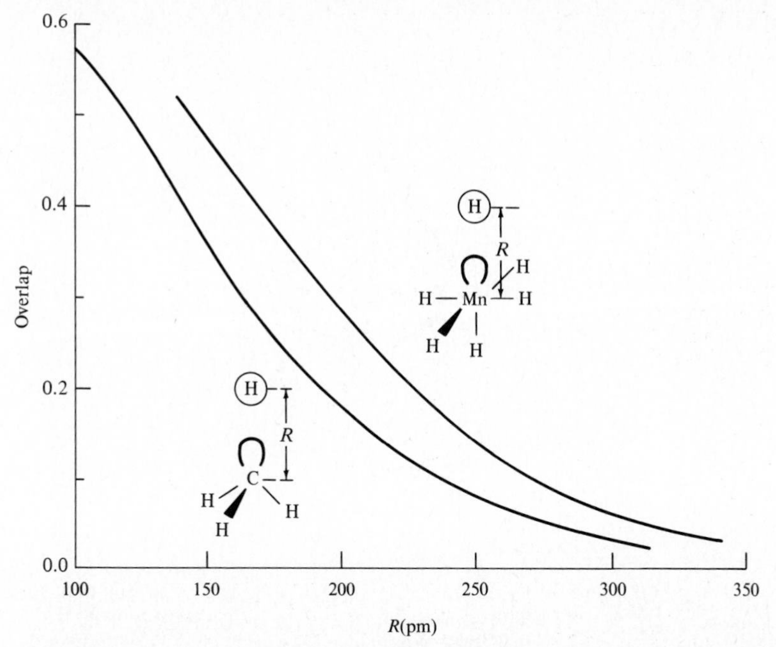

Fig. 13.11 Overlap intergrals between the a_1 frontier orbital of MnH_5^- and CH_3 and a H 1*s* orbital at a distance *R* from the Mn or C. [From R. Hoffmann, *Angew. Chem. Int. Ed.*, in press. Reproduced with permission of the Royal Academy of Science, Sweden.]

always has a somewhat greater overlap, but the dependence on distance is essentially identical.

Isolobal fragments with only 16 valence electrons will behave as pseudochalcogens. Thus $Fe(CO)_4$ may form $H_2Fe(CO)_4$ (cf. H_2S) and $Fe(CO)_4^{-2}$ (cf. S^{-2}). Fifteen-electron fragments such as $Ir(CO)_3$ are isolobal with the Group VA elements, such as phosphorus. Thus tetrameric iridium carbonyl (Fig. 13.12) is isostructural with the P_4 molecule.

While the isolobal concept is most useful as a structural tool, some chemical reactions can be systematized with it. For example the disproportionation of metal carbonyls (Eq. 13.18) parallels that of halogens and pseudohalogens (p. 790). Metal carbonyls may also undergo halogenation and hydrogenation. Finally, there is one case in which the metal carbonyl adds across a double bond:

$$Co_2(CO)_8 + F_2C{=}CF_2 \longrightarrow (OC)_4CoCF_2CF_2Co(CO)_4 \longrightarrow$$

$$(OC)_3Co\underset{\underset{\underset{O}{\|}}{C}}{\overset{\overset{CF_3}{|}}{CF}}Co(CO)_3 \xrightarrow{Co_2(CO)_8} CF_3CCo_3(CO)_9 \quad \mathbf{(13.49)}$$

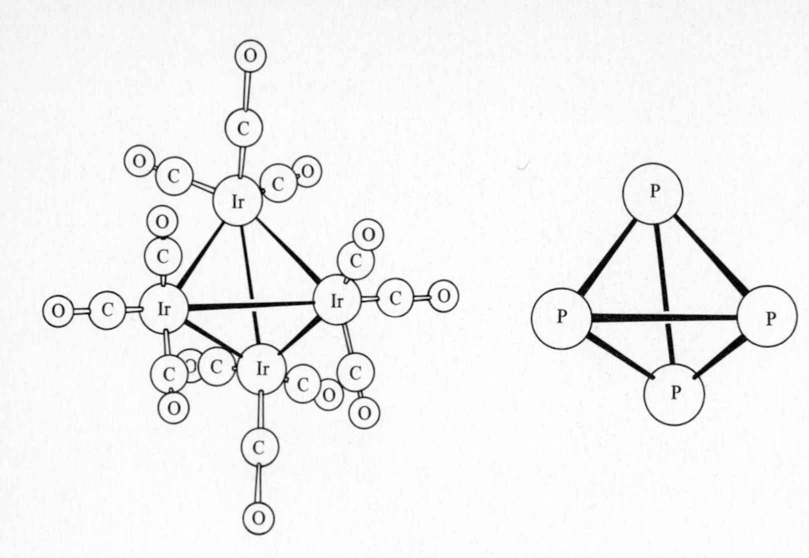

Fig. 13.12 Comparison of (a) $Ir_4(CO)_{12}$ and (b) P_4. Both are tetramers, composed of the isolobal fragments $Ir(CO)_3$ and P, respectively, each of which is trivalent. [Structure (a) from G. R. Wilkes and L. F. Dahl, in B. R. Penfold, *Perspectives in Structural Chemistry*, **1968**, *2*, 71. Reproduced with permission.]

We shall encounter further examples of isolobal fragments later in this chapter, but for now we can sum up the aspects that have been discussed as follows: *Two fragments are isolobal if the number, symmetry properties, approximate energies, and shapes of the frontier orbitals and the number of electrons in them are the same.*[39] It is obvious that the concept has developed far beyond mere bookkeeping.

METAL NITROSYLS

The nitric oxide molecule readily ionizes to form the nitrosyl cation, NO^+ (Eq. 3.50), which is isoelectronic with carbon monoxide. As a ligand, the nitrosyl group, NO, may ordinarily be considered to donate *three* electrons. Few complexes containing only nitrosyl ligands are well characterized, but many mixed carbonyl–nitrosyls are known. They may be formed readily by replacement of carbon monoxide with nitric oxide:

$$Fe(CO)_5 + 2NO \longrightarrow Fe(CO)_2(NO)_2 + 3CO \quad \textbf{(13.50)}$$

$$Co_2(CO)_8 + 2NO \longrightarrow 2Co(CO)_3NO + 2CO \quad \textbf{(13.51)}$$

Metal nitrosyls of this type generally obey the 18-electron Rule as illustrated by the isoelectronic series: $Ni(CO)_4$, $Co(CO)_3NO$, $Fe(CO)_2(NO)_2$, $MnCO(NO)_3$, $Cr(NO)_4$. The chemistry of metal nitrosyls is less well developed than that of the metal carbonyls. Typically, nitrosyl compounds form with the incorporation of nitric oxide into a complex to satisfy a need for an odd number of electrons to complete a noble gas configuration:

$$Ni(C_5H_5)_2 \xrightarrow{NO} (C_5H_5)NiNO \tag{13.52}$$

$$(C_5H_5)Mo(CO)_3H \xrightarrow{NO} (C_5H_5)Mo(CO)_2NO \tag{13.53}$$

$$CrCl_3 + Na^+C_5H_5^- \xrightarrow{NO} (C_5H_5)Cr(NO)_2Cl \tag{13.54}$$

Although the nitrosyl group generally occurs as a terminal group, bridging nitrosyl groups are also known:

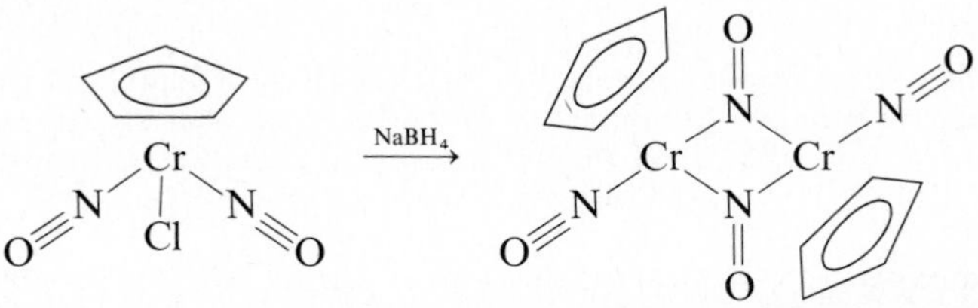

$$\xrightarrow{NaBH_4} \tag{13.55}$$

As in the case of the corresponding carbonyls, the infrared stretching frequencies[43] in this compound are diagnostic: Terminal NO = 1672 cm^{-1} and bridging NO = 1505 cm^{-1}.

Since the nitrosyl cation is isoelectronic with the carbonyl group, it is not surprising that there is a great similarity in the behavior of the two ligands. However, in one respect the nitrosyl group behaves in a manner unobserved for carbon monoxide. Although most nitrosyl groups appear to be linear (as expected for *sp*, —N≡O bonding), a few cases of distinctly bent species are known. The first well-characterized example was the nitrosyl tetrafluoroborate oxidative addition product of Vaska's complex:

$$(\phi_3P)_2Ir(CO)Cl + NO^+BF_4^- \longrightarrow [(\phi_3P)_2Ir(CO)Cl(NO)]^+BF_4^- \tag{13.56}$$

The structure of the product (Fig. 13.13a) is square pyramidal with a bent nitrosyl ligand ($\angle$Ir—N—O = 124°) at the apical position.[45] Since then several other complexes with similar nitrosyl groups have been found, including the remarkable complex $[(\phi_3P)_2Ru(NO)_2Cl]^+[PF_6]^-$ containing both a linear and a bent nitrosyl group[46] (Fig. 13.13b). The latter compound provides interesting data to test ideas concerning the nature of the bonding in the two types of groups.

Although the reason for the effect seems reasonably clear, there is the possibility of some confusion in the terminology since the discussion can be couched in terms from different points of view, representing different semantics but not different bonding.

Consider the nitrosyl ion, NO^+. Isoelectronic with carbon monoxide and molecular nitrogen, it has three bonding pairs between the atoms and a lone pair on both

[43] The stretching frequency of the nitrosyl group is sensitive to other bonding parameters in the molecule and ranges from 1550 to 1950 cm^{-1} for terminal nitrosyls and from 1330 to 1500 cm^{-1} (and beyond?) in bridging groups.[44]

[44] W. P. Griffith, *Adv. Organometal. Chem.*, **1968**, *7*, 211.

[45] D. J. Hodgson et al., *J. Am. Chem. Soc.*, **1968**, *90*, 4486.

[46] C. G. Pierpont et al., *J. Am. Chem. Soc.*, **1970**, *92*, 4760.

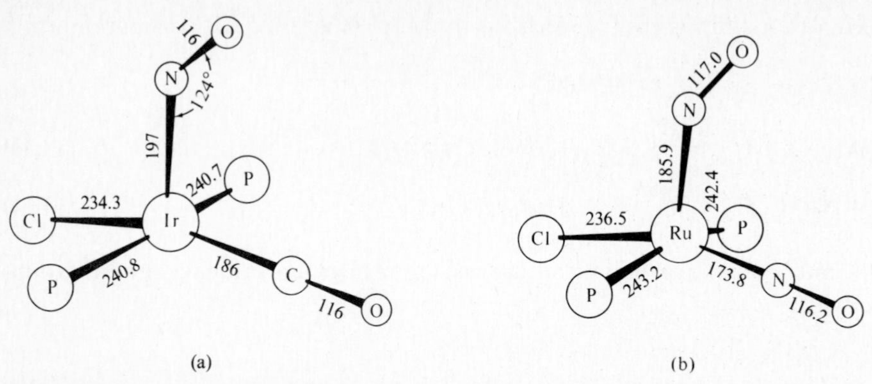

Fig. 13.13 Complexes containing bent nitrosyl groups: (a) $[(\phi_3P)_2Ir(CO)(NO)Cl]^+$; (b) $[(\phi_3P)_2Ru(NO)_2Cl]^+$. Phenyl groups have been omitted for clarity. Distances in picometres. [Structure (a) from D. J. Hodgsen et al., *J. Am. Chem. Soc.*, **1968**, *90*, 4486; structure (b) from C. G. Pierpont et al., *J. Am. Chem. Soc.*, **1970**, *92*, 4760. Reproduced with permission.]

the nitrogen atom and the oxygen atom. Both atoms are hybridized *sp*. Both atoms are potential donors, but the nitrogen coordinates preferentially (cf. carbon monoxide), avoiding a large formal positive charge on the more electronegative oxygen atom. Hence the nitrosyl ion can be characterized as a σ donor, donating the *sp* hybridized lone pair on the nitrogen. The resulting O≡N–metal system will be linear.

When bonding in this way, nitric oxide will always be a three-electron donor ($NO \rightarrow NO^+$ provides one; the :O≡N: lone pair provides two), and many examples can be given of its taking the place of carbon monoxide when an additional electron is needed. Thus the series of simple metal carbonyls discussed above has a parallel nitrosyl/carbonyl series: $Ni(CO)_4$, $Co(CO)_3NO$, $Fe(CO)_2(NO)_2$, $MnCO(NO)_3$, $Cr(NO)_4$.

A bent nitrosyl ligand is an analogue of an organic nitroso group, or the NO group in ClNO, consisting of a doubly bonded NO group, a single σ bond between the nitrogen and its substituent (R, Cl, Ru, etc.), and a lone pair on the nitrogen atom. It is this lone pair that causes the nitrosyl group to be bent (Fig. 13.14). In this case the nitrosyl ligand is acting like a one-electron donor (but see below).

The question as to whether a nitrosyl ligand will be linear or bent resolves itself into whether the pair of electrons in question will be forced to reside in an atomic orbital on the nitrogen atom (bent group) or whether there is a low-lying metal-based molecular orbital available to it.

If there are available nonbonding MOs on the metal (an *electron-poor* system),[47] the pair can reside there and allow the nitrogen to form an *sp* hybrid with concomitant π back bonding (Fig. 13.14a). On the other hand, if all the low-lying orbitals on the metal are already filled (an *electron-rich* system), the pair of electrons must occupy a nonbonding orbital[48] on the nitrogen requiring trigonal hybridization and a bent system (Fig. 13.14b).

[47] This is a relative term. All metals in carbonyl–nitrosyl systems are relatively electron-rich compared with, for example, a fluoride complex.

[48] It is common practice to refer to lone pairs as nonbonding electrons, but in the present instance these electrons have some antibonding character since the N—O bond order is reduced from three to two (to a first approximation) in going from the linear to the bent system.

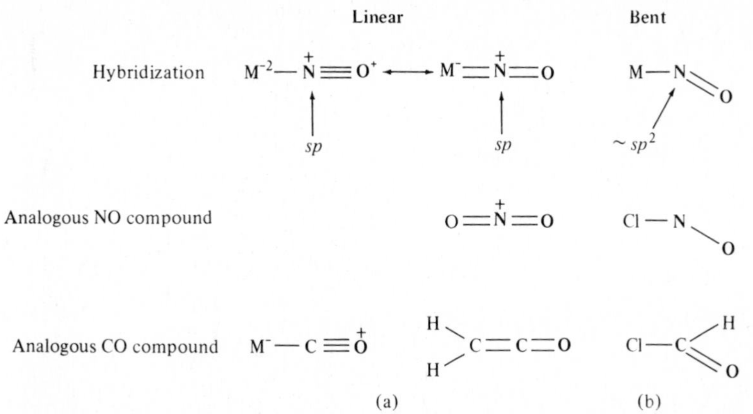

	Linear		Bent
Hybridization	$M^{-2}—\overset{+}{N}\equiv O^{+} \longleftrightarrow M^{-}=\overset{+}{N}=O$		$M—N=O$
	sp	sp	$\sim sp^2$
Analogous NO compound		$O=\overset{+}{N}=O$	$Cl—N=O$
Analogous CO compound	$M^{-}—C\equiv\overset{+}{O}$	$H_2C=C=O$	$Cl—C(H)=O$
	(a)		(b)

Fig. 13.14 Relative geometry, linear *vs.* bent, of nitrosyl ligands correlated with the hybridization of the nitrogen atom and the geometry of analogous, noncoordination compounds containing NO and CO groups.

The bond lengths in the ruthenium complex (Fig. 13.13b) containing both types of bonding are in accord with this view. In the "normal," linear system there is a short metal–nitrogen bond indicative of good π bonding (as in carbonyls, for example) *and* a short nitrogen–oxygen bond.[49] The latter is nominally a triple bond weakened by extensive metal–nitrogen π back bonding into the antibonding π orbitals of the nitrogen–oxygen system. The bent system, in contrast, is very similar to the nitroso group, —N=O, of organic chemistry with a double bond between nitrogen and oxygen and a relatively long, essentially σ-only nitrogen–metal bond. Π bonding will be reduced in the plane of the bend, and so the metal nitrogen bond is some 12 pm longer towards the apical nitrosyl than towards the NO group in the basal plane (see Fig. 13.13b).

We can actually see the process of electron pair shift with a resultant change in structure in the complex ion $Co(diars)_2NO^{+2}$ (where *diars* is a bidentate diarsine ligand). If NO is a three-electron donor, the 18-electron Rule is obeyed and we should expect the nitrosyl group to be linear, as indeed it is (Fig. 13.15). Reaction of this complex with the thiocyanate ion would violate the 18-electron Rule (since thiocyanate ion contributes a pair of its own) unless a pair can be shifted from a molecular orbital of largely metal character to the atomic orbital on nitrogen. This indeed happens and a "stereochemical control of valence" results.[50]

Metal basicity

Equation 13.56 indicates that NO^+ ion attacks a pair of electrons on the iridium atom to form the bond (and incidently gets a second pair for an essentially nonbonding orbital).

[49] Note that this is somewhat unusual. Quite often when a bond is strengthened and shortened it does so at the expense of an adjacent bond which is weakened and lengthened.

[50] J. H. Enemark and R. D. Feltham, *Proc. Nat. Acad. Sci., U.S.*, **1972**, *69*, 3534. See also J. H. Enemark and R. D. Feltham, *J. Am. Chem. Soc.*, **1974**, *96*, 5002, 5004; *Coordin. Chem. Rev.*, **1974**, *13*, 339. R. D. Feltham and J. H. Enemark, in "Topics in Inorganic and Organometallic Stereochemistry," G. L. Geoffroy, ed., Wiley, New York, **1981**, pp. 155–215.

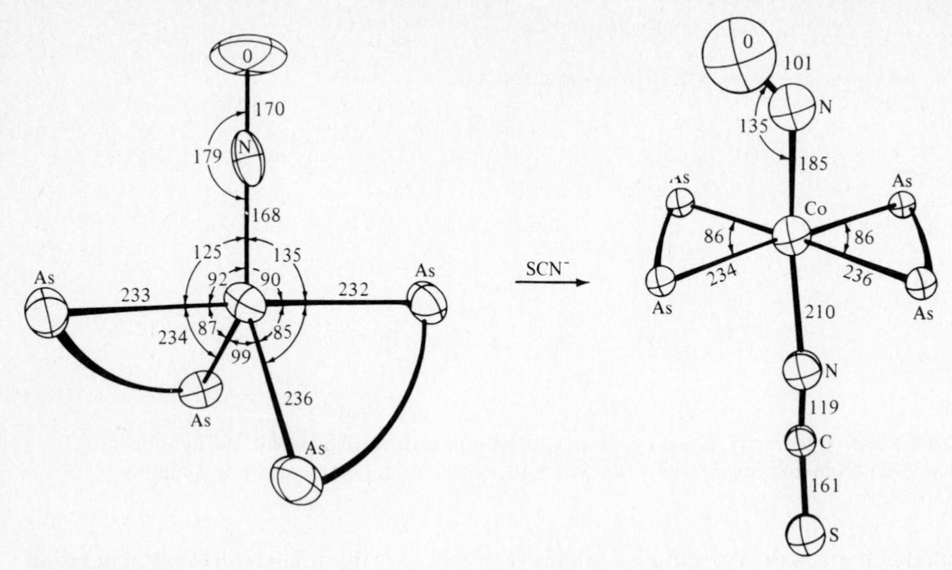

Fig. 13.15 Stereochemical control of valence. Note localization of lone pair on the nitrogen atom and bending of nitrosyl group upon addition of thiocyanate ion to coordination sphere of the cobalt ion. [From J. H. Enemark and R. D. Feltham, *Proc. Natl. Acad. Sci., U.S.* **1972**, *69*, 3524. Used with permission.] Distances in pm.

The nitrosyl atom is therefore a Lewis acid and the metal a Lewis base. This type of reaction might seem unusual. The nitrosyl cation is certainly a typical Lewis acid, but we are not accustomed to thinking of metals as bases. In fact, metal cations are typical acid species. However, in low oxidation states metals are good reducing agents and thus are sources of electron density. Although the concept of transition metal basicity is relatively recent,[51] some examples have been known for several years. For example, many carbonyl hydrides are acidic with the hydrogen readily dissociating:

$$HCo(CO)_4 \longrightarrow H^+ + [Co(CO)_4]^- \qquad \textbf{(13.57)}$$

$$HMn(CO)_5 \longrightarrow H^+ + [Mn(CO)_5]^- \qquad \textbf{(13.58)}$$

The hydrogen is bonded directly to the metal and hence the metal is the basic site in the conjugate base anion. The latter are strongly nucleophilic, displacing less basic species in several useful synthetic reactions (see also p. 603).

$$HgCl_2 + 2[Co(CO)_4]^- \longrightarrow Hg(Co(CO)_4)_2 + 2Cl^- \qquad \textbf{(13.59)}$$

$$Cr(CO)_6 + [Mn(CO)_5]^- \longrightarrow [(OC)_5CrMn(CO)_5]^- + CO \qquad \textbf{(13.60)}$$

$$B_5H_8Br + [Re(CO)_5]^- \longrightarrow B_5H_8Re(CO)_5 + Br^- \qquad \textbf{(13.61)}$$

Recently trinitrosylcobalt, $[Co(NO)_3]$, has been reported to be an extremely strong Lewis base, forming adducts with acids such as trimethylborane which are more stable

[51] D. F. Shriver, *Acc. Chem. Res.*, **1970**, *3*, 231.

than corresponding adducts of ammonia or trimethylamine.[52] Furthermore, the gas-phase proton affinity of iron pentacarbonyl has been found to be 855 kJ mol^{-1}, only slightly less than that of ammonia.[53] It is known that in solution the proton bonds directly to the metal.[54] Thus the basicity of transition metals in low oxidation states is a widespread, if little appreciated, phenomenon. Recently, the proton affinity of the uranium atom has been estimated as 995 kJ mol^{-1} and it has been dubbed a "superbase."[55]

DINITROGEN COMPLEXES

It has been pointed out previously that molecular nitrogen, N_2, is isoelectronic with both carbon monoxide and the nitrosyl cation. Despite the many complexes of the latter ligands, for many years it proved to be impossible to form complexes of dinitrogen. This difference in behavior was usually ascribed to the lack of polarity of N_2 and a resultant inability to behave as a π acceptor.[56]

The first dinitrogen complex was characterized in 1965 from the reduction of commercial ruthenium trichloride (containing some Ru^{IV}) by hydrazine hydrate. The pentaammine(dinitrogen)ruthenium(II) cation was formed and could be isolated in a variety of salts.[57] Soon other methods were found to synthesize the complex, such as the formation of the pentaammineazido complex, $[Ru(NH_3)_5N_3]^{+2}$, which slowly decomposes to form the dinitrogen complex. The latter may even be formed directly from nitrogen gas:

$$[Ru(NH_3)_5Cl]^{+2} \xrightarrow[H_2O]{Zn(Hg)} [Ru(NH_3)_5H_2O]^{+2} \quad \textbf{(13.62)}$$

$$[Ru(NH_3)_5H_2O]^{+2} + N_2 \longrightarrow [Ru(NH_3)_5N_2]^{+2} + H_2O \quad \textbf{(13.63)}$$

The unexpectedly large nucleophilicity of the dinitrogen group shown by the displacement of water in Eq. 13.63 is also shown in the formation of a bridge complex:[58]

$$[Ru(NH_3)_5N_2]^{+2} + [Ru(NH_3)_5H_2O]^{+2} \longrightarrow [Ru(NH_3)_5N_2Ru(NH_3)_5]^{+4} + H_2O \quad \textbf{(13.64)}$$

A variety of experimental methods has been brought to bear to elucidate the nature of the dinitrogen ligand. An X-ray study of the original ruthenium complex[59] indicated that the Ru—N—N linkage was linear, but disorder in the crystal prevented accurate determination of bond lengths. However, this was accomplished in the related complex $[Co(\phi_3P)_3N_2H]$ (see Fig. 13.16).[60] The results indicated that N_2 greatly resembles CO

[52] I. H. Sabherwal and A. B. Burg, *Chem. Commun.*, **1970**, 1001.

[53] M. S. Foster and J. L. Beauchamp, *J. Am. Chem. Soc.*, **1975**, *97*, 4808.

[54] A. Davison et al., *J. Chem. Soc.*, **1962**, 3653.

[55] P. Armentrout et al., *J. Am. Chem. Soc.*, **1977**, *99*, 3162.

[56] Note that the dipole moment of carbon monoxide is extremely low, however: 0.375×10^{-30} C m.

[57] A. D. Allen and C. V. Senoff, *Chem. Commun.*, **1965**, 621.

[58] For summaries of the chemistry of dinitrogen complexes of this type see A. D. Allen and F. Bottomley, *Acc. Chem. Res.*, **1968**, *1*, 360; R. Murray and D. C. Smith, *Coordin. Chem. Rev.*, **1968**, *3*, 429.

[59] F. Bottomley and S. C. Nyburg, *Chem. Commun.*, **1966**, 897; *Acta Crystallogr.*, **1968**, *B24*, 1289.

[60] B. R. Davis et al., *Inorg. Chem.*, **1969**, *8*, 2719.

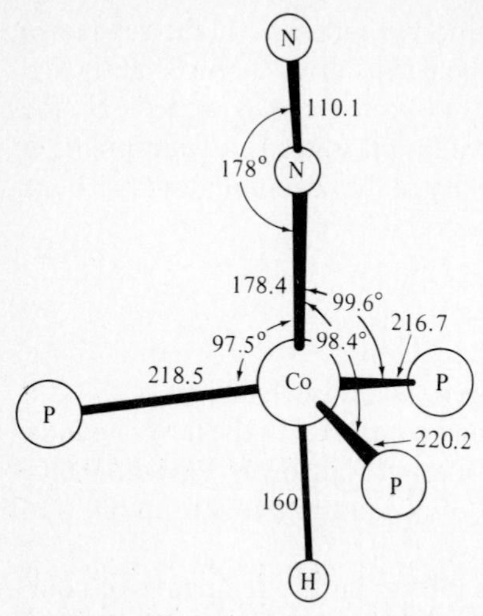

Fig. 13.16 Molecular structure of $[(\phi_3P)_3Co(N_2)H]$. Distances in pm. [From B. R. Davis et al., *Inorg. Chem.*, **1969**, *8*, 2719. Reproduced with permission.]

in its bonding to metals. Back bonding of electron density from the metal into π antibonding orbitals is apparent from the short Co—N bond (180.7 pm), which is closer to a Co—CO bond (175.3 pm) than the Co—NH_3 bond (195 pm) in ammine complexes. The N—N bond length (111.2 pm) is longer than that found in molecular nitrogen (109.8 pm) indicating a weakening of the nitrogen–nitrogen triple bond from the electron density in the antibonding orbitals of the nitrogen (cf. the value of 111.6 pm for bond order $2\frac{1}{2}$ in N_2^+).[61] Support for this interpretation also comes from an interpretation of the IR spectra of dinitrogen and carbonyl complexes of iridium(I)[62] and from Mössbauer spectroscopy.[63] It appears that the bonding in carbonyl and dinitrogen complexes is similar but that it is somewhat weaker in the dinitrogen complexes. Carbon monoxide is a better σ donor and/or π acceptor. This is to be expected both on the basis of whatever polarity exists in the CO molecule (see Fig. 3.51) and the fact that the π antibonding orbital is concentrated on the carbon atom (see Fig. 3.53) and overlap with the metal orbital is favored.

Although most dinitrogen complexes are of the "end on" type just discussed, one class of "side on" compounds has been discovered.[64] The compounds are unfortunately quite complicated, containing diphenylnickel moieties, "free" alkali metals, and ether and ethoxide groups. The portion of interest is shown in Fig. 13.17. It consists of a tetraphenyldinickel molecule with a nickel–nickel single bond of 275 pm (cf. Table 6.1). A dinitrogen molecule is coordinated "side on" to both nickel atoms. Each nitrogen atom also appears to be coordinated to the center of a "normal"[65] lithium–lithium bond. The most interesting

[61] Two papers further substantiate this interpretation: B. R. Davis and J. A. Ibers, *Inorg. Chem.*, **1970**, *9*, 2768; **1971**, *10*, 578.

[62] D. J. Darensbourg and C. L. Hyde, *Inorg. Chem.*, **1971**, *10*, 431; D. J. Darensbourg, ibid., **1971**, *10*, 2399.

[63] G. M. Bancroft et al., *Chem. Commun.*, **1969**, 585.

[64] K. Jonas, *Angew, Chem. Int. Engl. Ed.*, **1973**, *12*, 997; C. Krüger and Y.-H. Tsay, ibid., **1973**, *12*, 998.

[65] The bond distance between the lithium atoms is 258 pm compared to 267 pm in the gaseous dilithium molecule.

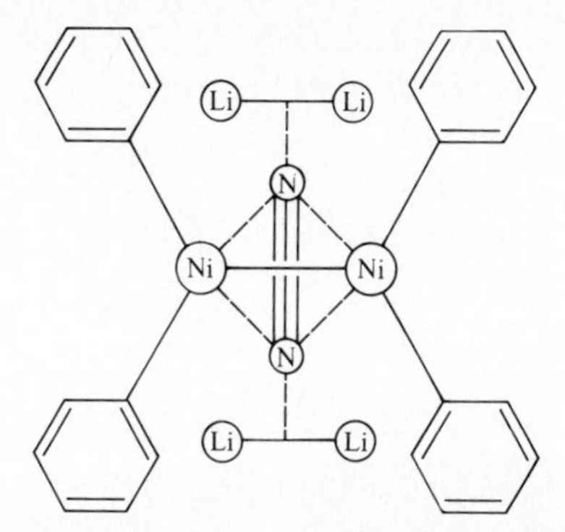

Fig. 13.17 Side-on coordination of the dinitrogen molecule to tetraphenyldinickel molecule. [From K. Jonas et al., *J. Am. Chem. Soc.*, **1976**, *98*, 74. Reproduced with permission.]

aspect of the structure is the nitrogen-nitrogen bond distance of 136 pm,[66] which is about midway between that for a single and a double bond (cf. the discussion of back bonding and bond lengthening in alkenes on pp. 641–643).

The ability to synthesize complexes containing the dinitrogen group, especially those with considerable alteration of the electronic state of nitrogen, opens up possibilities of direct fixation of nitrogen from the atmosphere, a long-standing challenge to the chemist. It also provides insight into the closely related process of biological fixation of nitrogen and the enzyme systems involved (see pp. 890–895).

METALLOCENES

The detailed chemistry of the metallocenes and related compounds will be encountered shortly, but first the application of the 18-electron Rule to these compounds and other hydrocarbon derivatives will be discussed briefly, to complete that aspect of the methodology of treating organometallic compounds. In addition to the coordinate covalent bond formed between carbon monoxide and metals, two other types of carbon–metal bonds may form. An alkyl group may form a σ bond, pairing an odd electron on a metal atom. In this type of bond the alkyl group is "isoelectronic" (as far as the 18-electron Rule is concerned) with the hydrogen atom and, like the latter, donates a single electron to the EAN of the metal.

Unsaturated hydrocarbons are capable of bonding to a metal atom through donation to the metal of the π electrons in the multiple bond. In these π complexes, each coordinated double bond of the ligand(s) is assumed to donate two electrons to the metal atom. Some examples of π complexes and the 18-electron Rule are:

Mn	= 25	Fe	= 26	Cr	= 24
Mn^+	= 24	$CH{=}CH_2$		$2C_6H_6$	= 12
C_2H_4	= 2	|	= 4	$[Cr(C_6H_6)_2]$	= 36
5CO	= 10	$CH{=}CH_2$			
$[Mn(CO)_5C_2H_4]^+$	= 36	3CO	= 6		
		$[Fe(C_4H_6)(CO)_3]$	= 36		

[66] K. Jonas et al., *J. Am. Chem. Soc.*, **1976**, *98*, 74.

A special case of π complexes consists of those that involve moieties that may be considered formally as free radicals. In many ways these are reminiscent of nitrosyl complexes inasmuch as an odd number of electrons is involved in the bonding. The most important examples of ligands of this type are the allyl radical, $CH_2{=}CH{-}CH_2\cdot$, a three-electron donor, and the cyclopentadienyl radical, $C_5H_5\cdot$, a five-electron donor. Examples of the application of the 18-electron Rule are:

Mn	= 25	Fe	= 26	Ni	= 28
$CH_2{=}CHCH_2\cdot$	= 3	$2C_5H_5\cdot$	= 10	$C_5H_5\cdot$	= 5
4CO	= 8	$[Fe(C_5H_5)_2]$	= 36	NO	= 3
$[Mn(C_3H_5)(CO)_4]$	= 36			$[Ni(C_5H_5)NO]$	= 36

The 18-electron Rule is not obeyed as consistently by these types of organometallic compounds as by the carbonyls, nitrosyls, and their derivatives. For example, in addition to ferrocene, $[Fe(C_5H_5)_2]$, the first metallocene to be discovered, similar compounds, $[M(C_5H_5)_2]$, are known for several other elements of the first transition series [M = Cr, Co, Ni] in which the 18-electron Rule *cannot be obeyed*. However, only ferrocene shows exceptional thermal stability (stable to 500 °C) and is not air sensitive. Furthermore, cobaltocene with an EAN = 37 is readily oxidized to $[Co(C_5H_5)_2]^+$, which then reflects much of the thermal stability of ferrocene. Mixed cyclopentadienyl–carbonyls are known for all the metals from vanadium to nickel: $[(C_5H_5)V(CO)_4]$, $[(C_5H_5)Cr(CO)_3]_2$, $[(C_5H_5)Mn(CO)_3]$, $[(C_5H_5)Fe(CO)_2]_2$, $[(C_5H_5)Co(CO)_2]$, and $[(C_5H_5)Ni(CO)]_2$. Of interest is the fact that these compounds, obeying the 18-electron Rule, require that the odd-atomic-number elements, V, Mn, and Co, form monomers and the even-atomic-number elements, Cr, Fe, and Ni, form dimers in contrast to the simple carbonyls (pp. 591–595).

A review of the 18-electron Rule

Table 13.3 reviews the number of electrons donated by various ligands for application of the 18-electron Rule. Judicious use of this rule can be of considerable help in predicting the correct formula of a hypothetical compound, the stability of such a compound, or even a likely structure. For example it is possible to prepare a compound of empirical formula $Fe(C_5H_5)_2(CO)_2$. If this compound is assumed to be a ferrocene-like molecule, it clearly violates the 18-electron Rule. It is possible to draw a structure, however, which

Table 13.3 Ligand contributions to the 18-electron Rule

Hydrogen	1
Alkyl, acyl groups	1
Carbonyl group	2
Nitrosyl group, linear	3
(Nitrosyl group, bent)	(1)
Simple Lewis bases, Cl^-, PR_3, etc.	2
Alkenes	2/double bond
Allyl, $CH_2{=}CH{-}CH_2\cdot$	3
Cyclopentadienyl, $C_5H_5\cdot$	5
Cycloheptatrienyl, $C_7H_7\cdot$	7

obeys the rule perfectly (see Problem 13.2). The proper use of the EAN formalism may thus suggest appropriate molecular structures and experiments to elucidate the structure and properties of a compound. Use of the rule as dogma *in place of experiment*, however, has no place in chemistry.

Molecular orbitals of metallocenes

Ferrocene-type molecules obey the general rules developed previously for the carbonyls, but it is worthwhile to investigate the nature of the molecular orbitals in some detail. The properties of the molecular orbitals of the π system of cyclic polyenes may be briefly summarized as follows.[67] There is a single orbital at lowest energy that consists of an unbroken, i.e., nodeless,[68] "doughnut" of electron density above and below the plane of the ring. At slightly higher energy there is a doubly degenerate set of orbitals each of which has *one* nodal plane containing the principal axis. This is followed by another doubly degenerate set with *two* nodal planes and yet higher energy. This pattern continues with doubly degenerate orbitals of increasing energy and increasing number of nodal planes until the number of molecular orbitals is equal to the number of atomic p orbitals, i.e., the number of carbons in the ring. If this number is odd, the highest antibonding orbital is doubly degenerate; if the number is even, the highest antibonding orbital is nondegenerate:

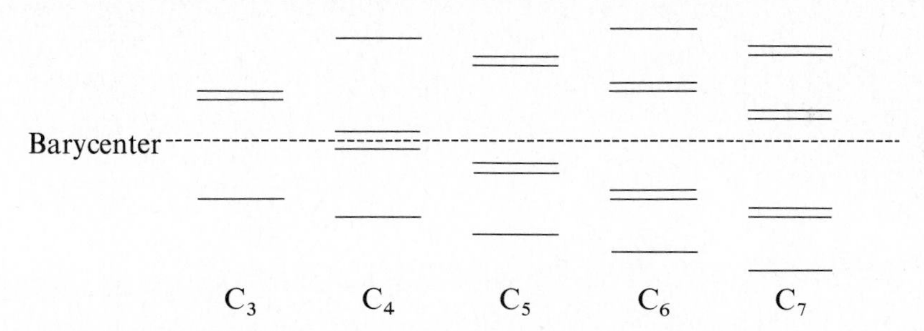

The increasing number of nodes will result in molecular orbitals with symmetries (as viewed down the ring–metal–ring axis in the metallocenes) of σ (complete cylindrical symmetry), π (one nodal plane), δ (two nodal planes), etc. These can now be combined with atomic orbitals of the same symmetry on the metal. For example, consider the lowest energy bonding orbital. This orbital has no nodes, and if the wave functions for this orbital on ring no. 1 and on ring no. 2 are written with the same sign, this produces a *gerade* ligand group orbital of the same symmetry (a_{1g}) as an atomic s orbital. On the other hand, if the two wave functions are written with opposite signs, an *ungerade* LGO of the same symmetry (a_{2u}) as the atomic p orbital is obtained. In the same manner other LGOs can be constructed by either adding or subtracting the higher molecular orbitals of the two rings. The resulting combinations are shown in Fig. 13.18.

---

[67] This is not the place to delve into the nature of organic ring systems. For a discussion of these see D. J. Royer, "Bonding Theory," McGraw-Hill, New York, **1968**, pp. 153–163, and L. E. Orgel, "An Introduction to Transition Metal Chemistry," 2nd ed., Wiley, New York, **1966**, Chap. 10.

[68] This refers to nodal planes perpendicular to the plane of the ring. The latter *must* be a nodal plane since the π system is constructed from atomic p orbitals.

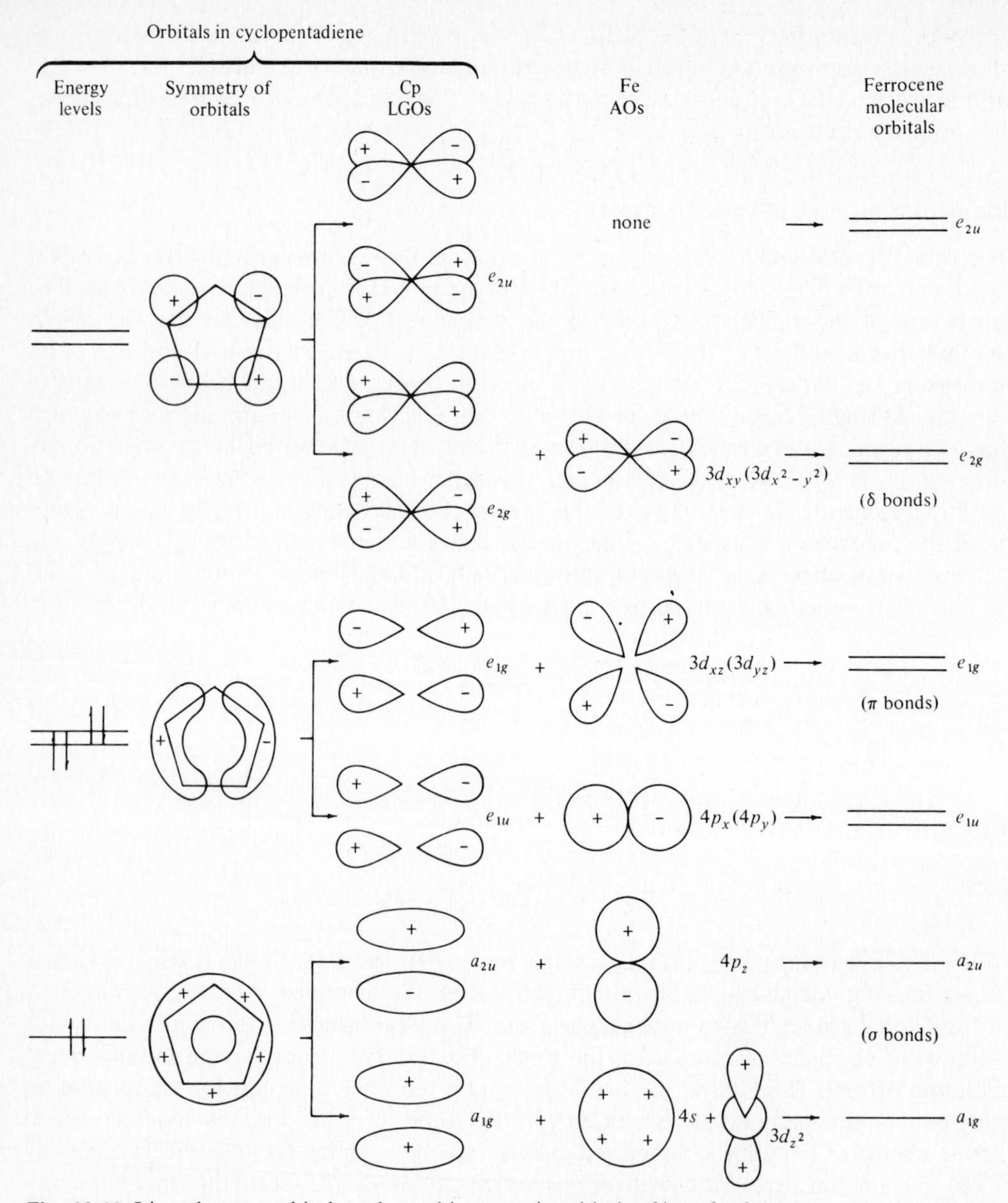

Fig. 13.18 Ligand group orbitals and matching atomic orbitals of iron for ferrocene.

Although symmetry considerations allow us to decide what molecular orbitals are possible, knowledge of relative energies and overlap integrals is necessary in order to estimate the nature of the resulting energy levels. There is still an uncertainty concerning several of these values,[69] but the energy level diagram shown in Fig. 13.19 is typical for

[69] See J. W. Lauher and R. Hoffmann, *J. Am. Chem. Soc.*, **1976**, *98*, 1729, and references therein.

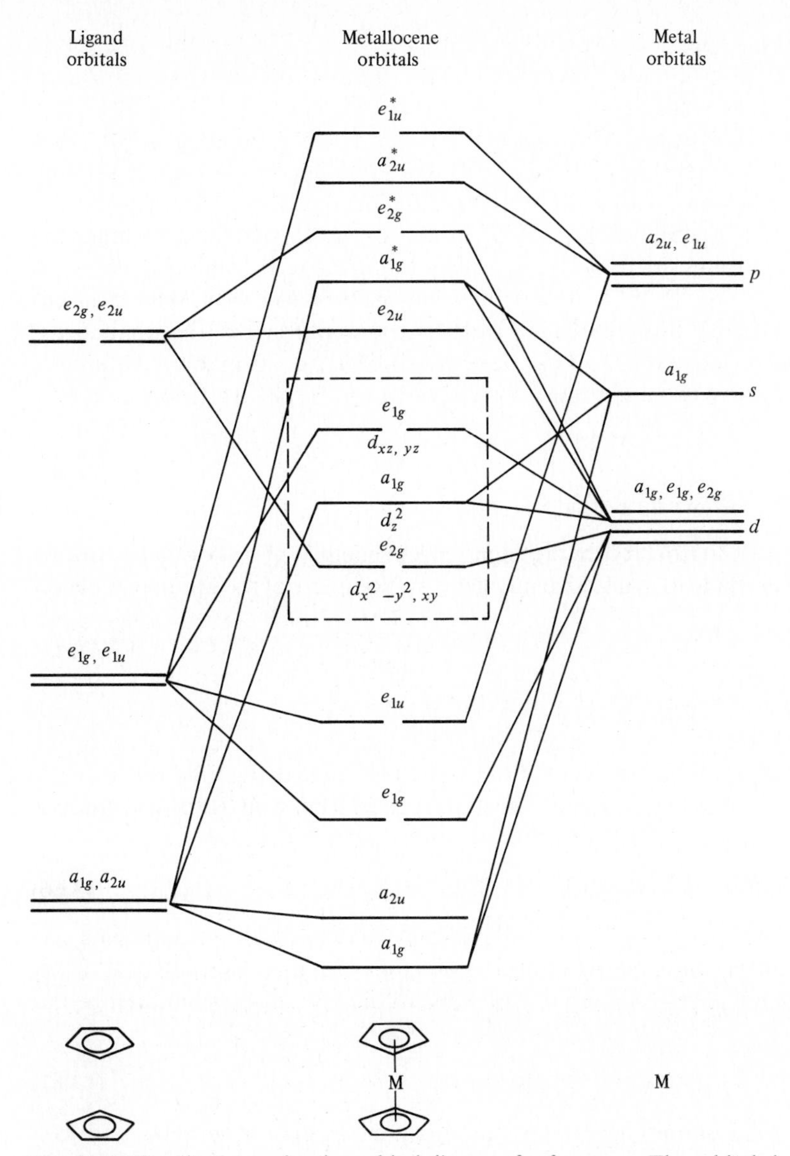

Fig. 13.19 Qualitative molecular orbital diagram for ferrocene. The orbitals being filled in metallocenes are enclosed in the box, e.g., ferrocene is $e^4_{2g}\,a'^2_{1g}$. [From J. W. Lauher and R. Hoffmann, *J. Am. Chem. Soc.*, **1976**, *98*, 1729. Reproduced with permission.]

ferrocene. The $a_{1g}(\sigma)$ orbitals of the cyclopentadiene are so stable relative to the metal orbitals that they interact but little. On the other hand, the $e_2(\pi)$ orbitals are so high in energy that they interact very little. The a_{1g} "bonding" orbitals and the e_2 "antibonding" orbitals are thus essentially localized on the cyclopentadienyl rings although there is a little stabilization from mixing. The 4*p* orbitals on the iron are at a high energy, and so the e_{1u} and a_{2u} orbitals do not contribute much to the bonding. The only orbitals that are well matched are the e_{1g} ring orbitals and the corresponding $3d$ orbitals of the metal which form two strong π bonds. These bonds are thought to supply most of the stabilization that holds the ferrocene molecule together.

If we supply enough electrons to fill all the bonding and nonbonding orbitals but none of the antibonding orbitals, nine pairs will be required. Once again we see that the 18-electron Rule is a reflection of filling strongly stabilized MOs.

The molecular orbital diagram in Fig. 13.19 also allows us to rationalize the magnetic properties of the metallocenes. Although it is probable that there is some change in the relative energies of the molecular orbitals in going from titanium to cobalt, self-consistent results can be obtained with this qualitative ordering. It appears that the a'_{1g} and e_{2g} energy levels converge and may even cross as one proceeds back from iron toward titanium. Thus Cp_2V^+ and Cp_2V have two and three unpaired electrons, respectively, with probable electron configurations of e_{2g}^2 and $e_{2g}^2 a'^1_{1g}$.[70]

Synthesis of cylopentadienyl compounds

The first metallocene was discovered by accident independently by two groups. In one group, an attempt[71] was made to synthesize fulvalene by oxidation of the cyclopentadienyl Grignard reagent:

$$C_5H_5{-}Mg{-}X \xrightarrow{Fe^{+3}} C_5H_4{=}C_5H_4 \qquad (13.65)$$

The synthesis of fulvalene was unsuccessful,[72] but a stable orange compound was isolated which was subsequently characterized and named ferrocene. The iron(III) is first reduced by the Grignard reagent to iron(II) which then reacts to form ferrocene:

$$Fe^{+3} \xrightarrow{CpMgX} Fe^{+2} \xrightarrow{2CpMgX} Cp_2Fe \qquad (13.66)$$

More recently, the use of sodium cyclopentadienide or thallium cyclopentadienide has proved to provide a more versatile synthesis of ferrocene and other metallocenes. In addition to the Grignard reagent, $CpMgX{\cdot}nEt_2O$, it is possible to prepare magnesium cyclopentadienide directly:

$$2C_5H_6 + Mg \xrightarrow{500-600\ ^\circ C} Mg^{+2} + 2C_5H_5^- + H_2 \qquad (13.67)$$

The latter is a white solid, stable indefinitely under dry nitrogen, and may be used as a source of cyclopentadienide ions.

Ferrocene, the most stable of the metallocenes, may be synthesized by methods not available to most of the other metallocenes. Thus iron will react directly with cyclo-

---

[70] Note that it is not necessary that the a'_{1g} and e_{2g} levels have identical energies in Cp_2V so long as the difference in energy is less than the spin-pairing energy.

[71] T. J. Kealy and P. L. Pauson, *Nature*, **1951**, *168*, 1039.

[72] Fulvalene has not yet been successfully synthesized and isolated although dilute solutions are known. See M. R. Churchill and J. Wormald, *Inorg. Chem.*, **1969**, *8*, 1970, and references therein.

pentadiene at high temperatures (it was discovered independently by this method[73] as well as by the Grignard synthesis). Use of amines facilitates the removal of the acidic hydrogen of cyclopentadiene and is sufficient to allow the formation of ferrocene from iron at lower temperatures:

$$Fe + 2R_3NH^+Cl^- \longrightarrow FeCl_2 + 2R_3N + H_2 \quad (13.68)$$

$$FeCl_2 + 2C_5H_6 + 2R_3N \longrightarrow Fe(C_5H_5)_2 + 2R_3NH^+Cl^- \quad (13.69)$$

$$\text{Net reaction: } Fe + 2C_5H_6 \longrightarrow Fe(C_5H_5)_2 + H_2 \quad (13.70)$$

Reactions of ferrocene-like molecules

The cyclopentadienyl rings in ferrocene are aromatic and undergo many of the same reactions as benzene. In general, all metallocenes are more reactive toward electrophilic reagents than benzene, indicating that electrons are more readily available. An example of this ease of reaction is the acylation of ferrocene by acetic anhydride containing phosphoric acid as catalyst:

$$Fe(C_5H_5)_2 + (CH_3CO)_2O \xrightarrow{H_3PO_4} (CH_3COC_5H_4)Fe(C_5H_5) + CH_3COOH \quad (13.71)$$

A second example of the reactivity of the ferrocene rings is their condensation with formaldehyde and amines (Mannich condensation):

$$Fe(C_5H_5)_2 + CH_2O + HNMe_2 \xrightarrow[H_3PO_4]{CH_3COOH} (C_5H_5)Fe(C_5H_4CH_2NMe_2) + H_2O \quad (13.72)$$

Ferrocene thus resembles the more reactive thiophene and phenol rather than benzene which does not undergo Mannich condensation.

Acetylation of ferrocene deactivates the molecule. In other to obtain the diacetyl compound, conditions similar to the usual acylation of aromatic rings are required:

$$(C_5H_5)Fe(C_5H_4COCH_3) + (CH_3CO)_2O \xrightarrow{AlCl_3} Fe(C_5H_4COCH_3)_2 \quad (13.73)$$

Almost all of the product is the 1,1′-isomer (acetyl groups on different rings); the 1,2-isomer is a minor by-product. Deactivation by acyl groups is expected by analogy with other organic ring systems, and so the reluctance to form the 1,2-isomer is not unexpected. The point of interest, however, is that since more rigorous conditions are necessary to effect substitution even *on the second ring*, some electronic effects must be passed from ring to ring through the iron atom. A second point of interest with regard to $Fe(C_5H_4COCH_3)_2$ is the existence of only one isomer, indicating free rotation of the rings with respect to each other.[74]

---

[73] S. A. Miller et al., *J. Chem. Soc.*, **1952**, 632.

[74] See p. 627 for further discussion of the rotation of these rings.

Acylation reactions provide a means of bridging the two rings in ferrocene:

$$\text{Fe}(C_5H_5)_2 \xrightarrow{\text{EtOCCH}_2\text{CCl (O, O)}} C_5H_5\text{Fe}C_5H_4\text{C(=O)CH}_2\text{C(=O)OEt} \xrightarrow{(1)\ H_2/PtO_2;\ (2)\ \text{hydrolyze};\ (3)\ PCl_3} C_5H_5\text{Fe}C_5H_4CH_2CH_2C(=O)Cl \xrightarrow{AlCl_3} \text{Fe}(C_5H_4CH_2CH_2C(=O)C_5H_4) \xrightarrow[HCl]{Zn/Hg} \text{Fe}(C_5H_4CH_2CH_2CH_2C_5H_4) \quad (13.74)$$

Compounds of this type are not only structurally interesting in their own right, but have proved useful in elucidating information on the bonding and structures of metallocene systems, for example, the transmission of electronic effects from one ring to the other.[75] Also, it should be noted that although the bridged compounds are inherently chiral, if the enantiomers *could* be separated, they would racemize immediately in solution from the limited "rocking" motions of the rings (see p. 534).

Other reactions typical of aromatic systems, such as nitration and bromination, are not feasible with metallocenes because of their sensitivity to oxidation (discussed below). Many of the derivatives of this type can be made indirectly by means of another type of reaction typical of aromatic systems: metalation.[76] Just as phenyllithium can be obtained from benzene, analogous ferrocene compounds can be prepared:

$$C_6H_6 + LiC_4H_9 \longrightarrow LiC_6H_5 + C_4H_{10} \quad (13.75)$$

$$\text{Fe}(C_5H_5)_2 \xrightarrow{n\text{-BuLi}} (C_5H_4Li)\text{Fe}(C_5H_5) \xrightarrow{n\text{-BuLi}} \text{Fe}(C_5H_4Li)_2 \quad (13.76)$$

[75] H. L. Lentzner and W. E. Watts, *Chem. Commun.*, **1970**, 906.

[76] For a review of the metalation of metallocenes, see D. W. Slocum et al., *J. Chem. Educ.*, **1969**, *46*, 144.

The monolithio and dilithio derivatives are useful intermediates in the synthesis of various ferrocenyl derivatives. Some typical reactions are:

Fe –Li $\xrightarrow[(2)\ H_2O]{(1)\ CO_2}$ Fe –C(=O)OH (13.77)

Fe –Li $\xrightarrow{N_2O_4}$ Fe –NO_2 $\xrightarrow[HCl]{Fe}$ Fe –NH_2

Fe –Li $\xrightarrow[(2)\ O]{(1)\ B(OR)_3}$ Fe –$B(OH)_2$ $\xrightarrow{HgCl_2}$ Fe –HgCl $\xrightarrow{Br_2}$ Fe –Br

Fe –$B(OH)_2$ $\xrightarrow{CuBr_2}$ Fe –Br (13.78)

Fe (–Li)(–Li) $\xrightarrow[(2)\ H_2O]{(1)\ CO_2}$ Fe (–C(=O)OH)(–C(=O)OH)

Fe (–Li)(–Li) $\xrightarrow{CH_2(SiMe_2Cl)_2}$ Fe (–Si(Me)(Me)–CH_2–Si(Me)(Me)–) (13.79)

It is possible to oxidize several of the metallocenes. Ferrocene loses an electron rather reluctantly since a noble gas configuration is being disrupted. On the other hand, cobaltocene is readily oxidized to the cobaltocinium ion, losing the thirty-seventh and *antibonding* electron. Nickelocene, with two antibonding electrons, may also be oxidized, although only one electron can be removed and the nickelocinium ion is not very stable.

Cobaltocene can undergo an alternative reaction to achieve a noble gas configuration. Since cobalt needs only 36 electrons to achieve a krypton configuration, the last, superfluous electron behaves almost as a free radical. Cobaltocene reacts with halogenated hydrocarbons to use this electron in the formation of a σ bond on the *cyclopentadiene* ring (four-electron donor; *not* cyclopentadienyl, a five-electron donor):

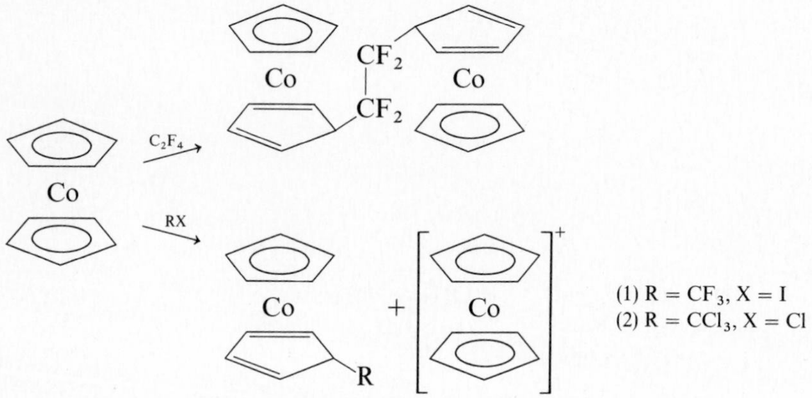

Co $\xrightarrow{C_2F_4}$ Co CF_2–CF_2 Co

Co $\xrightarrow{RX}$ Co(R) + $[Co]^+$ **(13.80)**

(1) $R = CF_3, X = I$
(2) $R = CCl_3, X = Cl$

Rhodocene is so much more reactive than cobaltocene that it can be formed and isolated only by reduction in the absence of solvents. Otherwise the cyclopentadiene complex is formed:

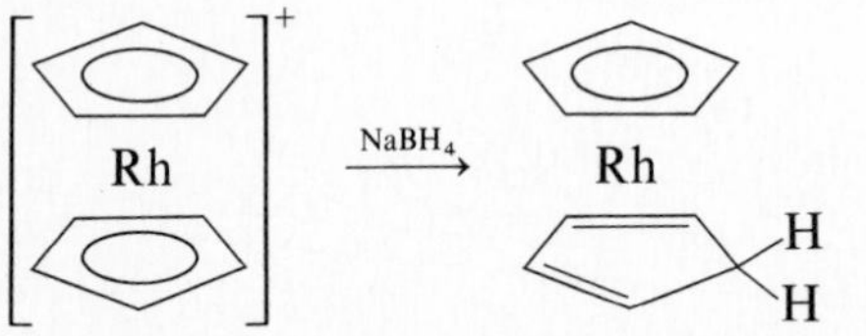

$[Rh]^+$ $\xrightarrow{NaBH_4}$ Rh(H)(H) **(13.81)**

Even when isolated by special techniques, rhodocene is unstable, behaving as a free radical and dimerizing spontaneously on warming to room temperature:

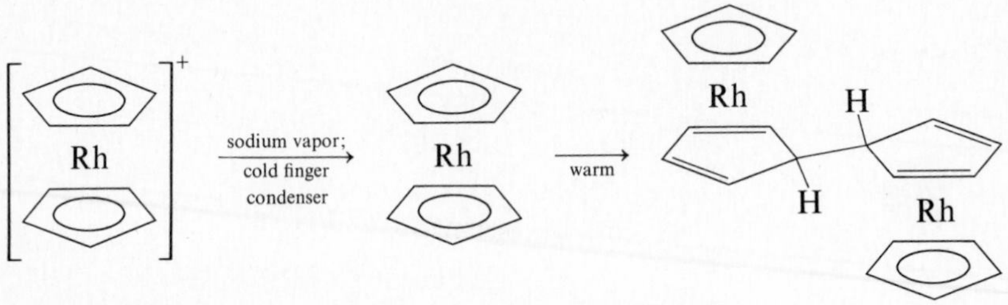

$[Rh]^+$ $\xrightarrow[\text{cold finger condenser}]{\text{sodium vapor;}}$ Rh $\xrightarrow{\text{warm}}$ Rh H H Rh **(13.82)**

Nickelocene contains two excess electrons and can react in a similar manner to change

Table 13.4 Properties of the bis-π-cyclopentadienyl complexes of first row transition metals

Cp_2V	Cp_2Cr	Cp_2Mn	Cp_2Fe	Cp_2Co	Cp_2Ni
Purple mp = 167–168°C	Scarlet mp = 172–173°C	Amber mp = 172–173°C	Orange mp = 173°C	Purple mp = 173–174°C	Green mp = 173–174°C

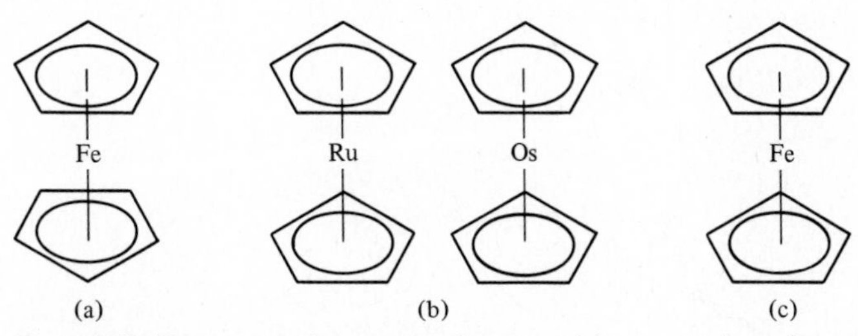

Fig. 13.20 Structures of some metallocenes: (a) staggered configuration of rings in ferrocene in the crystal: (b) eclipsed configurations of ruthenocene and osmocene in the crystals; (c) eclipsed configuration of ferrocene in the gas phase.

one cyclopentadienyl ring to a three-electron donor:

Ni $\xrightarrow{C_2F_4}$ Ni H H CF_2 CF_2 **(13.83)**

The three-electron π system thus formed is similar to that in allyl complexes (see p. 645).

Structures of cyclopentadienyl compounds

The bis-π-cyclopentadienyl compounds of the elements of the first transition series are isomorphous and have melting points which are remarkably constant at or near 173°C (Table 13.4). The structure of ferrocene in the solid state has been found to have the staggered configuration, but the heavier analogues ruthenocene and osmocene have the eclipsed configuration in the solid (Fig. 13.20). At one time this was interpreted as carbon–carbon or hydrogen–hydrogen (or both) repulsions between the two rings forcing the smaller ferrocene molecule into the staggered configuration. However, in the gas phase ferrocene has been found to have an eclipsed configuration and a barrier to rotation of only 4 ± 1 kJ mol^{-1}. Hence the rings are essentially free to rotate and the configuration adopted in the crystal is susceptible to packing forces.[77] Indeed some ferrocene derivatives

[77] For a discussion of this problem as well as references to the various structures, see M. R. Churchill and J. Wormald, *Inorg. Chem.*, **1969**, *8*, 716.

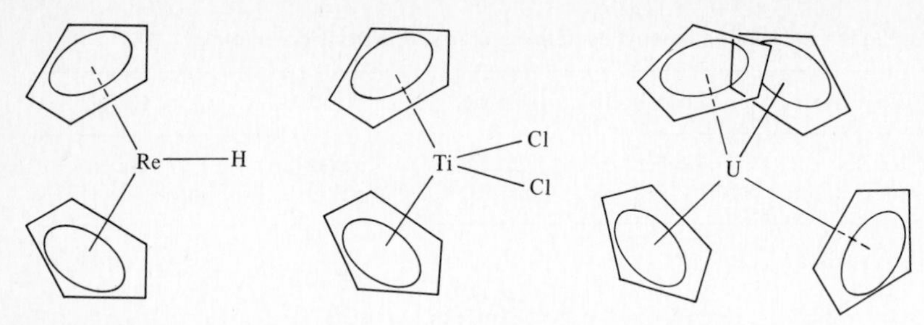

Fig. 13.21 Molecular structure of the $Ni_2(Cp)_3^+$ cation.

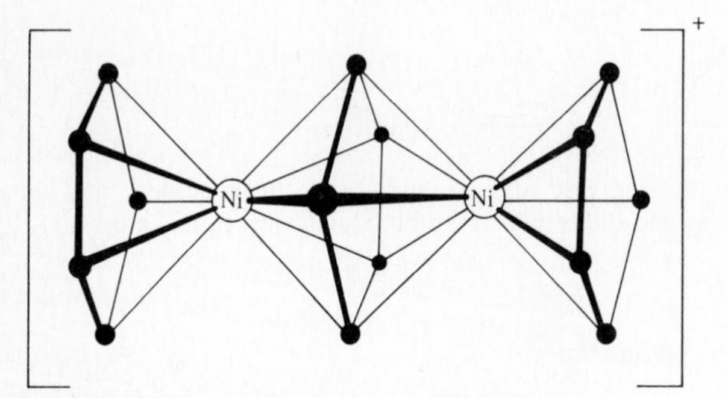

Fig. 13.22 Molecular structure of the $Ni_2(Cp)_3^+$ cation.

occur in the eclipsed form in the solid. For some, such as ferrocenedicarboxylic acid, $Fe(C_5H_4COOH)_2$, the eclipsed form is strongly favored by intermolecular forces (in this case, hydrogen bonds),[78] but more subtle forces must be important in others.

In typical metallocene compounds, Cp_2M, all of the C—C bonds are of the same length[79] and the rings are parallel. There are several cyclopentadienyl compounds in which the rings are tilted with respect to one another. Examples are Cp_2ReH and Cp_2TiCl_2 (Fig. 13.21), in which the steric requirements of additional ligands prevent parallel rings. In addition, lone pair requirements in Sn(II) and Pb(II) result in similar tilting of the rings in Cp_2Sn and Cp_2Pb. Finally, there are compounds with more than two cyclopentadienyl rings. Examples of compounds having several cyclopentadienyl rings attached to a single metal atom are tris(cyclopentadienyl)titanium and tetrakis(cyclopentadienyl)uranium (Fig. 13.21). A different type of structure is the layered arrangement of nickel atoms and cyclopentadienyl rings in $Cp_3Ni_2^+$, as shown in Fig. 13.22.[80] A molecular orbital study of the latter type of compound has been done,[81] but we shall not pursue the matter here.

[78] G. J. Palenik, *Inorg. Chem.*, **1969**, *8*, 2744.

[79] See discussion below on the structure of ϕ_3PCuCp for a more detailed analysis of the cyclopentadienyl ring.

[80] H. Werner and A. Salzer, *Syn. Inorg. Metal-Org. Chem.*, **1972**, *2*, 239; E. Dubler et al., *Angew. Chem., Int. Eng. Ed.*, **1974**, *13*, 135.

[81] J. W. Lauher et al., *J. Am. Chem. Soc.*, **1976**, *98*, 3219.

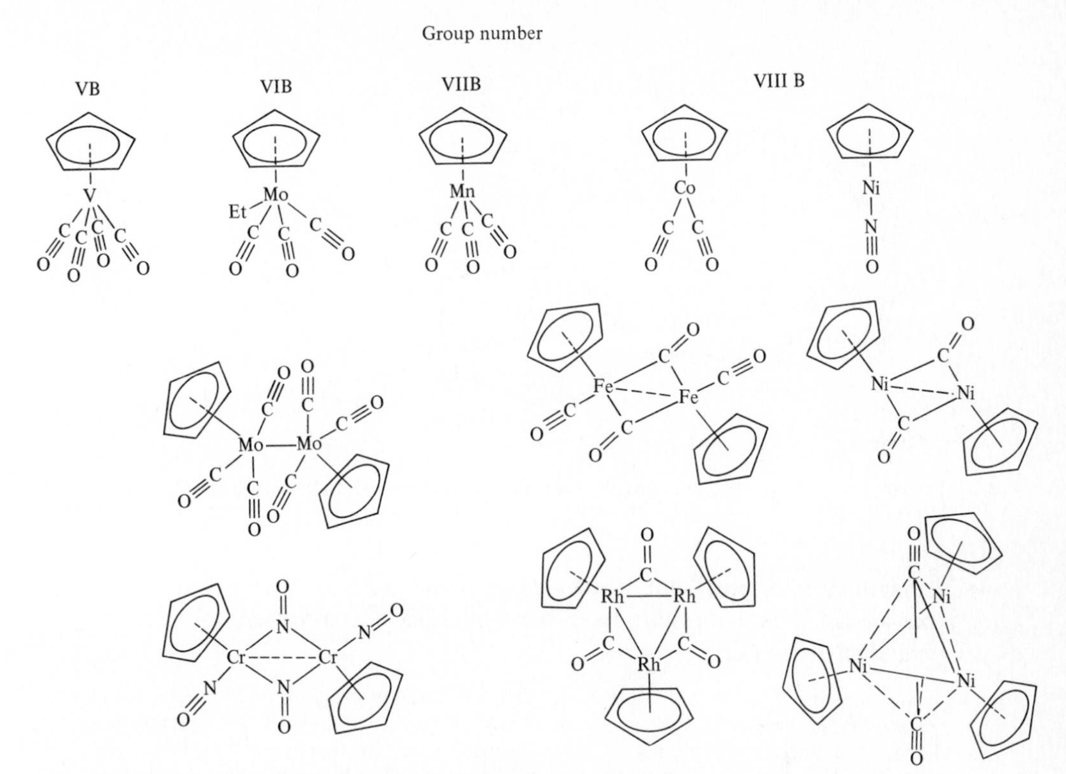

Fig. 13.23 Some monocyclopentadienyl complexes containing carbonyl and nitrosyl groups.

There are a few compounds known having only one cyclopentadienyl ring per metal atom. Sodium cyclopentadienide, cyclopentadienylthallium(I) (vapor), and cyclopentadienylindium(I) (vapor) have structures that may be described as "open-faced sandwiches." In the solid, cyclopentadienylindium(I) polymerizes to form an infinite sandwich structure of alternating cyclopentadienyl rings and indium atoms.

In addition to the few monocyclopentadienyl compounds just described, there is a considerably larger number of compounds containing a single ring and additional ligands such as carbon monoxide to complete the coordination sphere. These are monomers or dimers as necessary to obey the effective atomic number rule (see p. 591). Some examples are shown in Fig. 13.23.

One compound of this latter type, (cyclopentadienyl)triphenylphosphinecopper(I), is of special interest since its structure has been determined with greater precision than that of any other cyclopentadienyl compound[82] and it provides us with interesting structural information on the nature of the C_5H_5 ring. The C—C bonds are nearly constant at 139.9 pm, with a standard deviation of 0.6 pm. The positions of the hydrogen atoms, extremely difficult to obtain in molecules containing heavy atoms, were located, and it

[82] F. A. Cotton and J. Takats, *J. Am. Chem. Soc.*, **1970**, *92*, 2353.

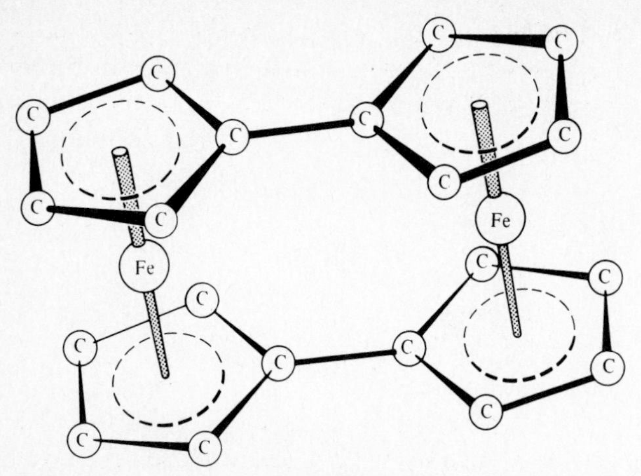

Fig. 13.24 Molecular structure of bis(fulvalene)diiron (=1,1′-biferrocenyl). [From M. R. Churchill and J. Wormald, *Inorg. Chem.*, **1969**, *8*, 1970. Reproduced with permission.]

appears that they lie slightly out of the plane of the carbons on the side of the ring *toward the metal.*[83] Similar results have been obtained for dibenzenechromium[84] and dicyclopentadienylchromium.[85]

There has probably been too much emphasis on the D_{5h} (or C_{5v} if the hydrogen atoms are out of plane) symmetry of the cyclopentadienyl rings in metallocenes. This because the first structural question asked after it was suggested that the ligands are π-bonded was: Do the rings have the D_{5h} symmetry of the cyclopentadienyl radical and anion, or do they have the C_{2v} symmetry of cyclopentadiene? The former implied delocalized bonding; the latter was consistent with more localized alkene–metal bonds.

Recently workers[86] have noted that the bond lengths in $CpM(CO)_n$ are not all of equal length. The deviations, though small, are statistically significant and probably arise from effects analogous to the *trans* influence noted previously (p. 549).

There has been considerable interest in polycyclic ferrocene derivatives. Two examples will be given. Although fulvalene, the object of the work that led to ferrocene, has never been isolated,[87] a bis(fulvalene)diiron complex, $(C_{10}H_8)_2Fe_2$, is known. Its structure has been determined by X-ray diffraction[88] and consists of two ferrocenyl moieties attached by single bonds (Fig. 13.24). Such stabilization of organic molecules is not uncommon. Formally, fulvalene behaves as a dicyclopentadienyl diradical, stabilized in the complex by the formation of two aromatic π systems in each of the ligands.

---

[83] The uncertainty in the positions of the hydrogen atoms is an order of magnitude greater than that of the carbon atoms, which in turn is an order of magnitude greater than that of the phosphorus and copper atoms. The apparent positions of the C_5H_5 hydrogen atoms thus lie at the threshold of statistical significance.

[84] E. Keulen and F. Jellinek, *J. Organometal. Chem.*, **1966**, *5*, 490.

[85] A. Haaland et al., *Chem. Commun.*, **1974**, 54.

[86] L. B. Byers and L. F. Dahl, *Inorg. Chem.*, **1980**, *19*, 277; F. J. Fitzpatrick et al., ibid. **1981**, *20*, 2852.

[87] See p. 622. Phenyl- and halogen-substituted fulvalenes have been prepared.

[88] M. R. Churchill and J. Wormald, *Inorg. Chem.*, **1969**, *8*, 1970.

A second, related molecule is bis(azulene)iron. Azulene is an isomer of naphthalene with a cyclopentadiene ring fused to a cycloheptatriene ring. The complex which it forms with iron, $(C_{10}H_8)_2Fe$, was first formulated as a metal–diene, metal–triene complex:

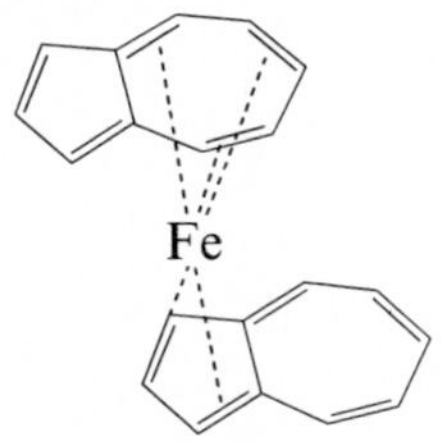

Fe

This was done on the basis of the need of ten electrons by iron (26 + 10 = EAN of 36) and hydrogenation which required 5 moles of hydrogen. When the structure was determined by X-ray diffraction,[89] it was found that the complex was a substituted ferrocene utilizing the five-membered rings of the azulene molecules with coupling between the seven-membered rings:

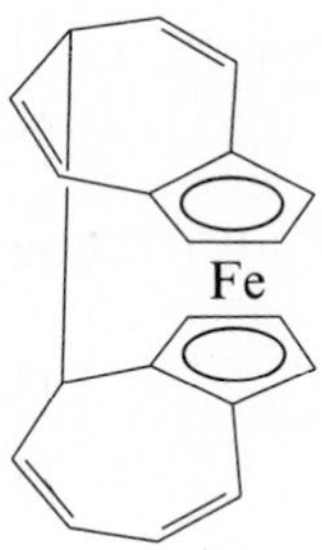

Fe

As in the bis(fulvalene) complex the stability of the ferrocene moiety is the driving force, in this case resulting in the formation of a carbon–carbon bond. This structure satisfies the effective atomic number rule as well as the one postulated earlier. The misleading data on hydrogenation result from the consumption of 4 mol of hydrogen by the remaining four double bonds and one mole in the hydrogenolysis of the interring C—C bond.

Covalent versus ionic bonding

Metallocenes such as Cp_2Cr, Cp_2Fe, and Cp_2Co are considered to have strong covalent bonding between the metal and the rings. Although not all of these metallocenes are stable with respect to oxidation, etc., all have strong bonding with respect to dissociation of the rings from the metal atom (Table 13.5). The bonds between the metal and rings certainly have some polarity, but these compounds do not react like polar organometallic

[89] M. R. Churchill and J. Wormald, *Inorg. Chem.*, **1969**, *8*, 716. For further discussion of complexes of this type, see M. R. Churchill, *Progr. Inorg. Chem.*, **1970**, *11*, 53.

Table 13.5 Dissociation energies of metallocene cations[a]

Reaction			ΔH (kJ mol^{-1})
Cp_2Mg^+	$\longrightarrow$	$CpMg^+ + Cp$	310
Cp_2V^+	$\longrightarrow$	$CpV^+ + Cp$	515
Cp_2Cr^+	$\longrightarrow$	$CpCr^+ + Cp$	633
Cp_2Mn^+	$\longrightarrow$	$CpMn^+ + Cp$	364
Cp_2Fe^+	$\longrightarrow$	$CpFe^+ + Cp$	641
Cp_2Co^+	$\longrightarrow$	$CpCo^+ + Cp$	754
Cp_2Ni^+	$\longrightarrow$	$CpNi^+ + Cp$	524

[a] The dissociation energies are for the +1 cations rather than the neutral molecules since the data were obtained from the appearance potentials shown in the mass spectra of these compounds.

SOURCE: M. Cais and M. S. Lupin, in *Adv. Organometal. Chem.*, **1970**, *8*, 211. Used with permission.

compounds (as exemplified by the well-known Grignard reagent):

$$RMgX + H_2O \longrightarrow RH + MgXOH \tag{13.84}$$

$$Cp_2Fe + H_2O \longrightarrow \text{No reaction} \tag{13.85}$$

In contrast, several compounds are known which contain a very reactive C_5H_5 group:

$$Cp^-Na^+ + H_2O \longrightarrow C_5H_6 + Na^+OH^- \tag{13.86}$$

$$2Cp^-Mg^{+2} + 2H_2O \longrightarrow 2C_5H_6 + Mg(OH)_2 \tag{13.87}$$

$$3Cp^-Ln^{+3} + 3H_2O \longrightarrow 3C_5H_6 + Ln(OH)_3 \tag{13.88}$$

These compounds are considered to have a salt-like nature and are usually referred to as metal cyclopentadienides rather than as cyclopentadienyl complexes, although the distinction is sometimes rather arbitrary. As is the case with all polar bonds, there is no sharp distinction between covalent and ionic bonding. Thus, although the lanthanide compounds are usually referred to as ionic, there may be a substantial amount of covalent character present.[90]

The beryllium and magnesium compounds are of interest with regard to the problem of covalent versus ionic bonding. Although magnesium cyclopentadienide is structurally almost identical to ferrocene, it is thought to be essentially ionic. The sandwich structure should be the most stable one not only for covalent complexes utilizing *d* orbitals, but also from an electrostatic viewpoint for a cation and two negatively charged rings. The structure of the beryllium compound is unusual and still somewhat uncertain. The first

[90] In the present discussion cyclopentadienide has been applied to compounds M^+Cp^- and cyclopentadienyl to complexes such as Cp_2M. Some chemists use the -yl form for all compounds, e.g., cyclopentadienylsodium by analogy with common usage for related compounds such as methylsodium (more properly named sodium methanide).

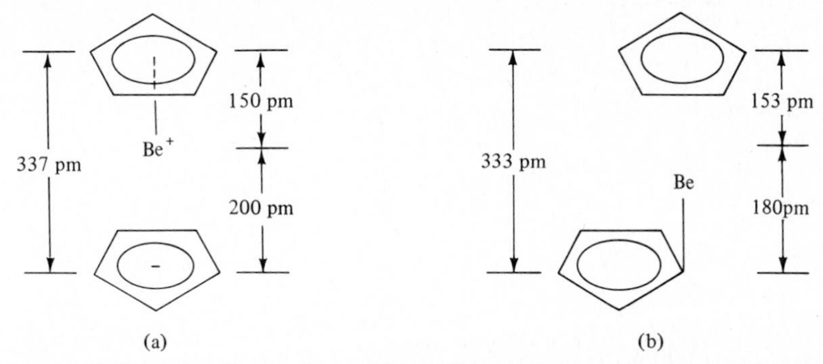

Fig. 13.25 Suggested structures, of $Be(C_5H_5)$: (a) Unsymmetrical sandwich proposed from electron diffraction data; (b) "Slipped sandwich" σ–π molecule determined by X-ray diffraction.

structure (Fig. 13.25a) was based on interpretation of electron diffraction data in the gas phase. The beryllium atom is closer to one ring than the other. The distance between the two rings (337 pm) is thought to be determined by the nonbonded repulsions between the rings and is that expected from the van der Waals radii of carbon (r_{VDW} = 165–170 pm). In this view the small Be^{+2} ion polarizes the π cloud of one ring, and an energetically more favorable situation is obtained from a strong, shorter "covalent" bond and a longer, weaker ionic bond than from two bonds of intermediate length. However, an X-ray diffraction study of solid beryllocene indicates a "slipped sandwich" molecule with a σ or monohapto (see below) bond to one ring and a "π" or pentahapto linkage to the other (Fig. 13.25b).[91]

Manganocene, Cp_2Mn, is thought to be largely ionic. One form consists of infinite chains of CpMn fragments bridged by cyclopentadienyl rings in the solid.[92] Upon warming to 159 °C the color changes from brown to orange, and the compound becomes isomorphous with ferrocene. The evidence for the ionic bonding is of three types: (1) Manganocene reacts instantaneously with ferrous chloride in tetrahydrofuran to form ferrocene and is hydrolyzed immediately by water; (2) the dissociation energy (Table 13.5) is closer to that of magnesium cyclopentadienide than to those of the other transition metal metallocenes; and (3) the magnetic moment of manganocene is 5.86 Bohr magnetons, corresponding to five unpaired electrons. All these data are consistent with the presence of a d^5Mn^{+2} ion. The evidence is not unequivocal, however. Other metallocenes such as chromocene react with iron(II) chloride to yield ferrocene as well, although admittedly not so rapidly. Assignment of "ionic bonding" to manganocene on the basis of the presence of high-spin Mn(II) is reminiscent of similar assignments to other high-spin complexes with other ligands (e.g., $[Fe(H_2O)_6]^{+2}$). The presence of high-spin manganese and a lower dissociation energy for manganocene are indicative of the absence of *strong* covalent bonding (large ligand field stabilization energy in ligand field terminology) but do not prevent the possibility of some covalent bonding. In any event, the stability of the half-filled subshell, d^5 configuration is responsible for the "anomaly" of manganocene.

[91] A. Almenningen et al., *J. Chem. Phys.*, **1964**, *40*, 3434; C. H. Wong et al., *Acta Crystallogr.*, **1972**, *B26*, 1662. See also N.-S. Chiu and L. Schläfer, *J. Am. Chem. Soc.*, **1978**, *100*, 2604.

[92] W. Bünder and E. Weiss, *Z. Naturforsch., B*, **1978**, *33*, 1235.

Σ-bonded cyclopentadienyl compounds

In a few cases the cyclopentadienyl ligand behaves as a one-electron donor, σ-bonded organic group. Examples are Cp_2Hg and Cp_4Sn in which the bonding is linear and tetrahedral, respectively:

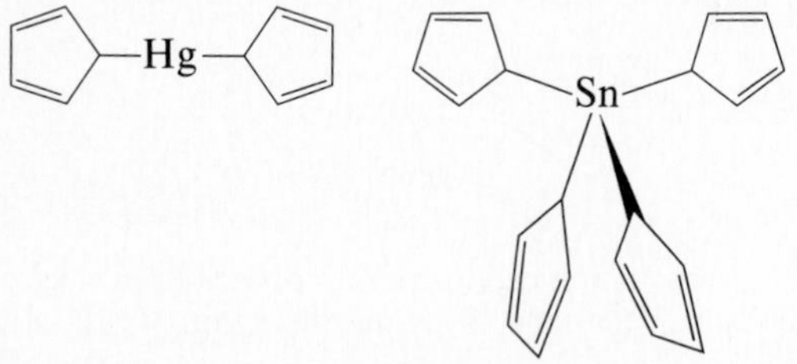

Hg Sn

Other examples are known in which a cyclopentadienyl group adopts a σ-bonded arrangement induced by the 18-electron Rule. For example, the compound $Fe(C_5H_5)_2(CO)_2$ can obey the 18-electron Rule only if one cyclopentadienyl ring is σ bonded and one is π bonded. The 18-electron Rule thus suggested a structure for this compound containing two types of cyclopentadienyl rings, and this was confirmed by experiment:

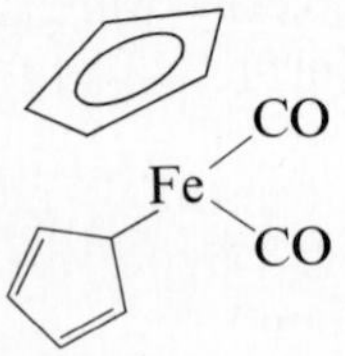

CO Fe CO

OTHER AROMATIC CYCLOPOLYENES

Although the cyclopentadienyl group is the best-known aromatic ligand, there are several others of considerable importance. None is as stable as the most stable metallocenes, however, and their chemistry is more severely limited.

Σ–π and hapto nomenclature

Two types of covalent bonding by C_5H_5, typified by ferrocene and dicyclopentadienylmercury, have been discussed. It is possible to distinguish these as σ (Cp_2Hg) and π (Cp_2Fe) complexes, but this terminology is not completely satisfactory. Semantic ambiguities arise. Is manganocene a π complex or not? Other problems exist. Cyclooctatetraene contains four π bonds which are all potential donors to transition metals. There is thus the possibility that one, two, three, or four π bonds could act as donors to a single metal atom. All would be "π complexes"; yet it is necessary to distinguish among them.

Cotton[93] has suggested the *hapto* system (Gr. ηαπτειν, to fasten) based on the number of atoms of the ligand attached[94] to the metal atom. Dicyclopentadienylmercury is a *monohapto* species.[95] Ferrocene is a *pentahapto* complex since all five of the carbon atoms of the ring are attached to the iron atom. A third type of cyclopentadienyl–metal bond exists, a three-electron donor (see p. 645). An example of a compound containing two types of bonding is $Cp_2W(CO)_2$:

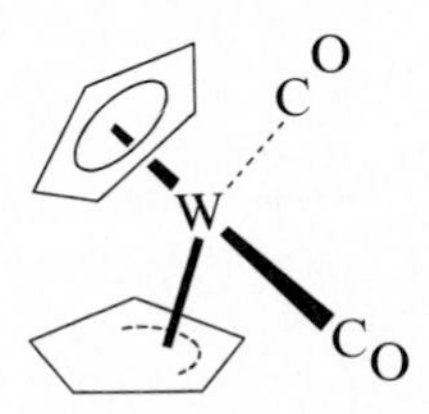

This compound could be named (η^3-cyclopentadienyl)(η^5-cyclopentadienyl)dicarbonyltungsten(II) (pronounced trihaptocyclopentadienylpentahaptocyclopentadienyl . . .) or formulated $(\eta^3\text{-}C_5H_5)(\eta^5\text{-}C_5H_5)W(CO)_2$. The system of naming is quite general and further examples will be given in the following discussions of other types of complexes. (See also Appendix J.)

Arene complexes

Benzene and substituted benzenes can act as six-electron donors. The best known example is dibenzenechromium, synthesized early in this century but not characterized until 1954. It was first synthesized via a Grignard synthesis:

$$\phi MgBr + CrCl_3 \xrightarrow[Et_2O]{} \xrightarrow[H_2O]{} Cr(\phi H)_2^+ + \text{related products} \qquad \textbf{(13.89)}$$

$$Cr(\phi H)_2^+ \xrightarrow{Na_2S_2O_4} Cr(\phi H)_2 \qquad \textbf{(13.90)}$$

Careful work has shown that the reaction takes place via *monohapto* species which rearrange to form the *hexahapto* or π complex. Unlike the cyclopentadienyl compounds, the two ligands are not isomers but differ by a hydrogen atom (phenyl versus benzene). By carrying out the reaction at $-20°C$ it is possible to isolate triphenylchromium (as the tetrahydrofuran solvate), which can then be converted via an assumed radical mechanism to arene complexes (Fig. 13.26). Arene complexes may also be synthesized by the Fischer-Hafner adaptation of the Friedel-Crafts reaction. In this reaction aluminum is used to

[93] F. A. Cotton, *J. Am. Chem. Soc.*, **1968**, *90*, 6230.

[94] The word "attached" is used in its widest sense. It does not imply covalent bonds. As such it is closely related to the idea of coordination number, which may indicate either the number of covalent bonds or the number of nearest neighbors. Even if sodium cyclopentadienide is considered to be completely ionic it may be named as a *pentahapto* species. This is one distinct advantage of the present system.

[95] It is true that the bond from the metal to the ring migrates around the ring rather rapidly. This topic is discussed on p. 673 under "Fluxional Molecules." At any instant, however, the metal is attached to *one* carbon atom rather than all five.

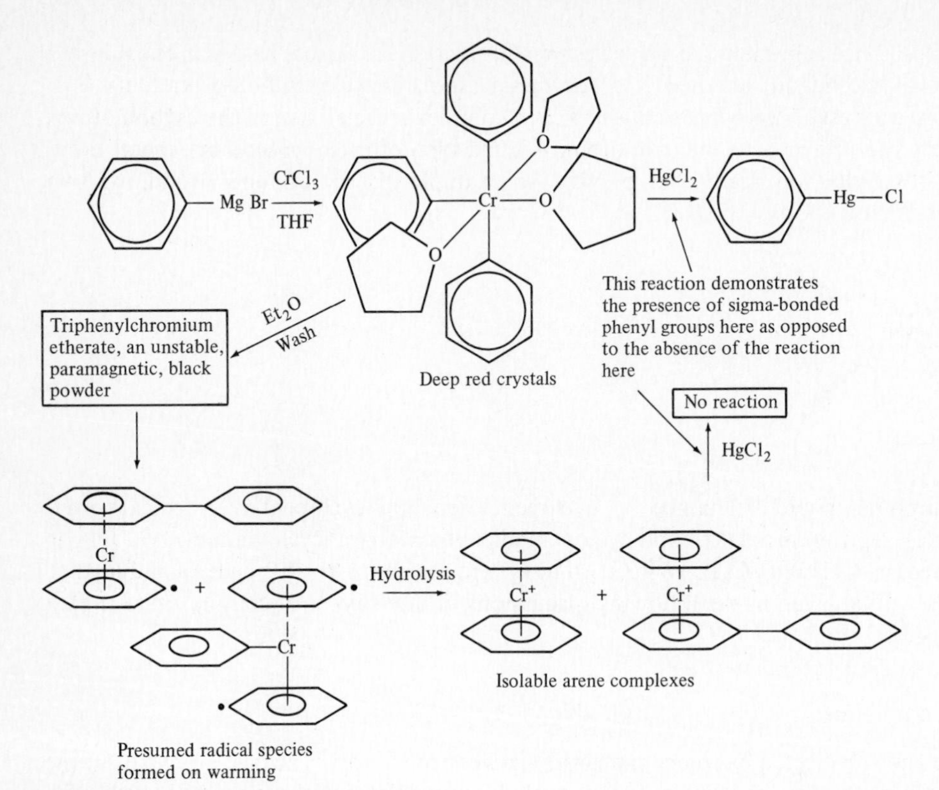

Fig. 13.26 Some known and assumed species in the formation of arene chromium compounds from chromium(III) chloride and phenyl Grignard reagent. [Modified from G. E. Coates et al., "Principles of Organometallic Chemistry," Methuen, London, **1968**. Reproduced with permission.]

reduce the metal salt to a lower oxidation state which is then added to the aromatic system via aluminum chloride catalyst:

$$3CrCl_3 + 2Al + AlCl_3 + 6C_6H_6 \longrightarrow 3[(C_6H_6)_2Cr]^+[AlCl_4]^- \qquad \textbf{(13.91)}$$

The structure of the dibenzenechromium molecule consists of two parallel planar benzene rings. It is possible to prepare compounds containing only one ring:

$$Cr(CO)_6 + Cr(C_6H_6)_2 \longrightarrow 2\,(C_6H_6)Cr(CO)_3 \qquad \textbf{(13.92)}$$

The benzene rings can be removed completely by the action of a more active ligand:

$$Cr(C_6H_6)_2 + 6PF_3 \longrightarrow Cr(PF_3)_6 + 2C_6H_6 \qquad \textbf{(13.93)}$$

Arene complexes exhibit aromaticity and undergo reactions expected for such compounds, but in contrast to the metallocenes they are less reactive than benzene. This has been interpreted in terms of lowered electron density on the rings, which is borne out by the susceptibility to nucleophilic attack:

$$(C_6H_5OH)Cr(C_6H_6) + MeO^- \longrightarrow (C_6H_5OMe)Cr(C_6H_6) + OH^- \quad \textbf{(13.94)}$$

Perhaps as a result of the transfer of electron density from the rings to the metal, arene complexes are more susceptible to oxidation. Even $Cr(C_6H_6)_2(EAN = 36)$ is oxidized to the cation on exposure to air.

Cycloheptatriene and tropylium complexes

Cycloheptatriene is a six-electron donor that can form complexes similar to those of benzene differing in the localization of the π electron in C_7H_8. The alternation in bond length in the free cycloheptatriene is retained in the complexes (Fig. 13.27). Furthermore, in $C_7H_8Mo(CO)_3$ the double bonds are located *trans* to the carbonyl groups, providing an essentially octahedral environment for the metal atom.

Cycloheptatriene complexes can be oxidized (hydride ion abstraction) to form tropylium complexes:

$$C_7H_8Mo(CO)_3 \xrightarrow{\phi_3C^+BF_4^-} [C_7H_7Mo(CO)_3]^+ + BF_4^- + \phi_3CH \quad \textbf{(13.95)}$$

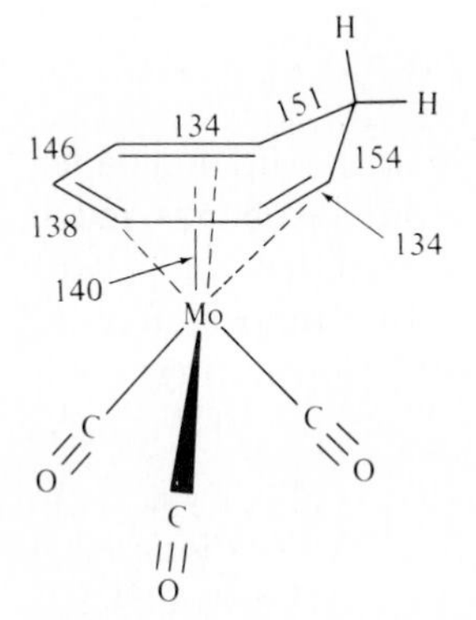

Fig. 13.27 Molecular structure of cycloheptatrienemolybdenum tricarbonyl illustrating alternation in bond lengths and location of double bonds *trans* to the carbonyl groups. Distances in pm.

The tropylium ring is planar with equal C—C distances. Like benzene and the cyclopentadienide anion, the tropylium cation is an aromatic, six-electron π system. Since one electron is lost to form a cation, the C_7H_7 group is a seven-electron donor which may be considered either as a tropylium ion donating six π electrons or a cycloheptatrienyl (tropyl) radical donating seven electrons.[96]

Since the cycloheptatrienyl ring carries a formal positive charge, it is particularly susceptible to nucleophilic attack by anions, effectively reversing the process of Eq. 13.95:

$$[(C_7H_7)Mo(CO)_3]^+ + X^- \longrightarrow (C_7H_7X)Mo(CO)_3 \qquad (13.96)$$

$X^- = CN^-, MeO^-$

Cyclobutadiene, cyclooctatetraene, and related systems

In accord with the Hückel rule of $4n + 2$ electrons, both cyclobutadiene and cyclooctatetraene are nonaromatic. In the case of cyclooctatetraene we can study the free molecule and find that it contains alternating bond lengths consistent with the structure:

This nonplanar molecule (see p. 709) becomes planar on reaction with an active metal to produce the cyclooctatetraenide anion:

$$2K + C_8H_8 \longrightarrow 2K^+ + C_8H_8^{-2} \qquad (13.97)$$

The anion has ten π electrons, conforming to Hückel's rule.

Cyclooctatetraene is relatively unimportant as a delocalized π ligand although a few complexes are known. More often it behaves as a polyolefin with localized bonds and as such it will be discussed below. The reaction of titanium alkoxide, $Ti(OC_4H_9)_4$, with cyclooctatetraene and triethylaluminum results in a compound of empirical formula $Ti_2(C_8H_8)_3$, which has been shown to contain two planar cyclooctatetraene rings and a

[96] The same problem arises with cyclopentadienyl which may be considered (formally) as either C_5H_5 (five-electron donor) or $C_5H_5^-$ (six-electron donor). As far as formal application of the EAN Rule is concerned, either representation is permissible since the number of electrons on the metal (i.e., the oxidation state) differs in the two cases.

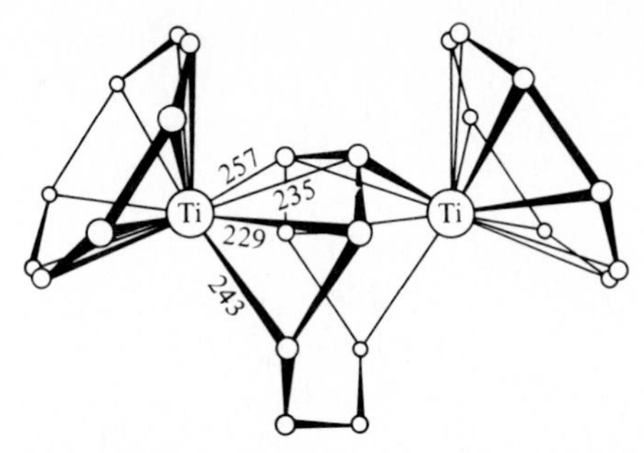

Fig. 13.28 Molecular structure of $[Ti_2(C_8H_8)_3]$. [From H. Dietrich and H. Dierks, *Angew. Chem. Inter. Engl. Ed.*, **1966**, *5*, 899. Reproduced with permission.] Distances in pm.

third, bridging ring behaving as a six-electron donor (Fig. 13.28).[97] Cyclooctatetraene also forms sandwich compounds with actinides and lanthanides (see pp. 809–812).

In the case of cyclobutadiene we are less fortunate—the free molecule has not been available for study. It is another of the many organic molecules that are nonexistent or unstable when free but are known in complexes. Theoretical studies[98] have suggested that if cyclobutadiene is square it will exist as a triplet:

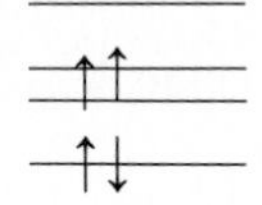

but that a rectangular species:

```
H           H
 \         /
  C-------C
  ||      ||
  C-------C
 /         \
H           H
```

should exist as a (thermodynamically) stable singlet. More recently, theoretical calculations including electron–electron repulsions, as well as the available experimental evidence (IR spectrum in an argon matrix at 8 K), indicate that the ground state of cyclobutadiene is a square singlet.[99]

The preceding discussion anticipates the experimental facts somewhat since no one had isolated cyclobutadiene and its existence appeared improbable prior to 1956. At that time it was predicted that the unstable molecule would be stabilized by bonding to a

[97] H. Dietrich and H. Dierks, *Angew. Chem. Int. Engl. Ed.*, **1966**, *5*, 899.

[98] M. J. S. Dewar and G. J. Gleicher, *J. Am. Chem. Soc.*, **1965**, *87*, 3255.

[99] W. T. Borden, *J. Am. Chem. Soc.*, **1975**, *97*, 5968, and references therein. See also E. A. Halevi et al., *J. Am. Chem. Soc.*, **1976**, *98*, 7088; G. Maier et al., *Angew. Chem. Int. Engl. Ed.*, **1976**, *15*, 226.

transition metal, and three years later the first complex containing the tetramethylcyclobutadiene ring was reported:

$$2\ \text{Me}_4\text{C}_4\text{Cl}_2 + 2\text{Ni(CO)}_4 \longrightarrow [(\text{Me}_4\text{C}_4)\text{NiCl}_2]_2 + 8\text{CO} \qquad \textbf{(13.98)}$$

The unsubstituted cyclobutadiene ring has also been synthesized in complexes:

$$2\ \text{C}_4\text{H}_2\text{Cl}_2 + \text{Fe}_2(\text{CO})_9 \longrightarrow 2\ (\text{C}_4\text{H}_4)\text{Fe(CO)}_3 + 3\text{CO} + 2\text{Cl}_2 \qquad \textbf{(13.99)}$$

Several methods of preparation and subsequent modes of reaction have been investigated.[100] The complexed cyclobutadiene ring behaves as a reactive aromatic system undergoing Friedel-Crafts, Mannich, and metalation reactions:

$$\text{C}_4\text{H}_4\text{–Fe(CO)}_3 \xrightarrow[\text{AlCl}_3]{\text{CH}_3\text{COCl}} \text{CH}_3\text{C(O)–C}_4\text{H}_3\text{–Fe(CO)}_3 \qquad \textbf{(13.100)}$$

$$\text{C}_4\text{H}_4\text{–Fe(CO)}_3 \xrightarrow[\text{Me}_2\text{NH}]{\text{CH}_2\text{O}} \text{Me}_2\text{NCH}_2\text{–C}_4\text{H}_3\text{–Fe(CO)}_3 \qquad \textbf{(13.101)}$$

$$\text{C}_4\text{H}_4\text{–Fe(CO)}_3 \xrightarrow[\text{NaCl}]{\text{Hg(OAc)}_2} \text{Cl–Hg–C}_4\text{H}_3\text{–Fe(CO)}_3 \qquad \textbf{(13.102)}$$

Finally, the aromatic cyclopropenyl ring has been found to form complexes as a three-electron donor (the cyclopropenyl cation is an $n = 0$, two-electron π system). The triphenylcyclopropenyl group behaves as a *trihapto* ligand but sterically resembles an extremely sterically hindered σ donor (Fig. 13.29), occupying one coordination site on the tetrahedron surrounding nickel.

[100] For a review of cyclobutadiene complexes see P. M. Maitlis, *Adv. Organometal. Chem.*, **1966**, *4*, 95.

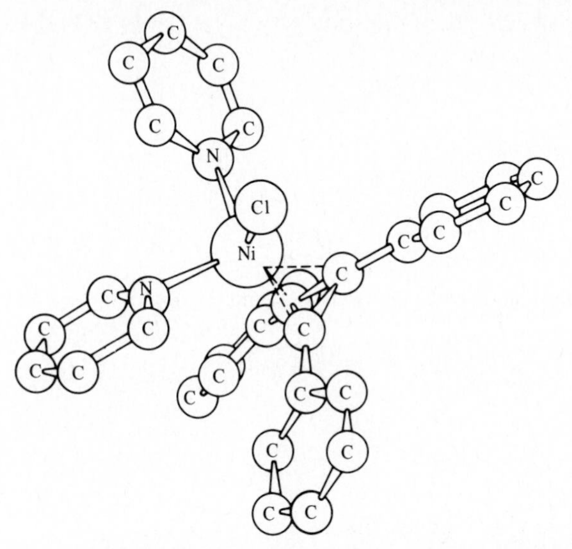

Fig. 13.29 Molecular structure of $[(\pi\text{-}\phi_3C_3)NiCl(py)_2]$. [From D. L. Weaver and R. M. Tuggle, *J. Am. Chem. Soc.*, **1969**, *91*, 6506. Reproduced with permission.]

NONAROMATIC OLEFIN AND ACETYLENE COMPLEXES

Ethylene complexes

Complexes between metal salts and olefins have been known since 1827 but not understood until comparatively recently. For example, Zeise isolated stable yellow crystals when he refluxed an alcoholic solution of potassium tetrachloroplatinate:[101]

$$[PtCl_4]^{-2} + C_2H_5OH \xrightarrow[\text{heat}]{\text{EtOH}} [Pt(C_2H_4)Cl_3]^- + Cl^- + H_2O \qquad \textbf{(13.103)}$$

Silver ions form similar olefinic complexes which are soluble in aqueous solution and may be used to effect the separation of unsaturated hydrocarbons from alkanes. Catalysts for the polymerization of olefins presumably are capable of forming metal–olefin complexes which lead to the polymerized product.

The structure of the anion in Zeise's salt is known (Fig. 13.30). The ethylene occupies the fourth coordination site of the square planar complex with the C—C axis perpendicular to the platinum–ligand plane.

The bond between the ethylene molecule and the metal ion may be considered as a dative σ bond to a suitable hybrid orbital on the metal (dsp^2 on Pt^{+2}, s or sp on Ag^+). The bonding is analogous to that in carbon monoxide with a dative carbon–metal σ[102]

[101] For a translation of Zeise's original paper, see Paper 2 in "Classics in Coordination Chemistry," Part 2, G. B. Kauffman, ed., Dover, New York, **1976**, pp. 21–37.

[102] The definition of a σ bond requires cylindrical symmetry about the bond axis. This is clearly impossible if the donated pair comes from an elongated π cloud on the ethylene. Therefore this modified "σ" bond is sometimes called a mu (μ) bond.

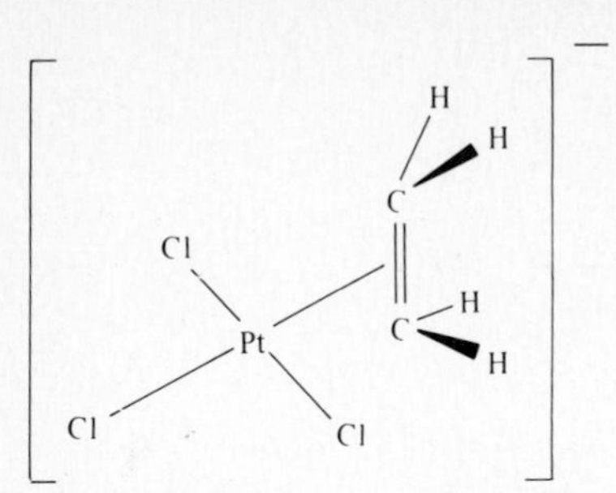

Fig. 13.30 The structure of the anion in Zeise's salt, trichloro(ethylene)platinate(II) ion.

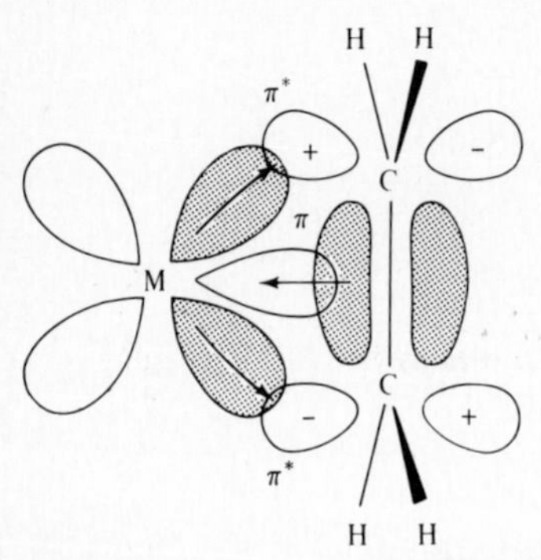

Fig. 13.31 Molecular orbital picture of bonding between ethylene and a metal ion.

bond and a reciprocal metal–carbon (antibonding orbital) π bond (Fig. 13.31). The extent of back bonding is undetermined and varies depending upon the substituents on the ethylene and the other ligands on the metal. For example, in the complexes of the type $[LRh(C_2H_4)(C_2F_4)]$ (where L = acac or cp), the tetrafluoroethylene molecule bonds more strongly and at a shorter distance (Rh—C = 201–202 pm) than does the unsubstituted ethylene (Rh—C = 217–219 pm).[103] This indicates that the π-acceptor ability ($C_2F_4 > C_2H_4$) is the most important factor in determining the bond lengths in this compound.

In the extreme case where both the "σ" dative bond ($C_2X_4 \rightarrow M$) and the back bond ($C_2X_4 \leftarrow M$) each have an order of one, the bonding can be represented by a simple valence bond structure (Eq. 13.104b) involving only σ bonds. We may therefore think of the true structure of these compounds as a resonance hybrid of:

$$M \leftarrow \|(C_2X_4) \longleftrightarrow M\langle(C_2X_4) \qquad \textbf{(13.104)}$$

(a) (b)

[103] J. A. Evans and D. R. Russell, *Chem. Commun.*, **1971**, 197; L. J. Guggenberger and R. Cramer, *J. Am. Chem. Soc.*, **1972**, *94*, 3779.

in which (13.104a) represents an ethylene → metal dative bond only and (13.104b) represents the extreme of "complete" back bonding and mixing of the "σ" and "π" orbitals to form two σ bonds between the metal and the ethylene. This viewpoint is useful since it emphasizes two structural variables in these complexes: (1) the increase in C—C bond length and (2) the bending of the substituents X away from the metal as back bonding becomes more important. The two effects operate together as the electronegativity of the X group increases, and the bending of substituents away ($\Delta\beta$) can be related directly to the lengthening of the C—C bond (Δl, pm):[104]

$$\Delta\beta = 2.09\,\Delta l \qquad \textbf{(13.105)}$$

Finally, there is the remarkable compound

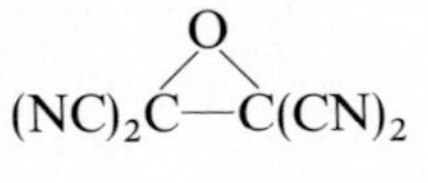

$$(NC)_2\overset{\displaystyle O}{C—C}(CN)_2 \quad \textbf{(1)}^{105}$$

which is structurally quite close to a nickel complex,

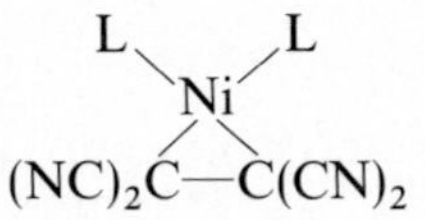

$$\begin{array}{c} L\quad L \\ Ni \\ (NC)_2C—C(CN)_2 \end{array} \quad \textbf{(2)}$$

Both the C—C bond lengths (**1** = 149.7 pm; **2** = 147.6 pm) and the bending of the substituents ($\Delta\beta = 32.2°$ (**1**) and 38.4° (**2**)) are nearly the same. Yet, in contrast to the nickel complex, the σ model is the *only* one applicable to tetracyanoethylene oxide. In view of these similarities we may feel justified in using it as an approximation in the complexes as well.[106]

Dienes

Olefins with more than one double linkage form many interesting complexes, only a few of which will be mentioned here. The simplest are those of butadiene:

$$C_4H_6 + Fe(CO)_5 \longrightarrow \begin{array}{c} CH—CH \\ H_2C \qquad CH_2 \\ Fe \\ C\ C\ C \\ O\ O\ O \end{array} \qquad \textbf{(13.106)}$$

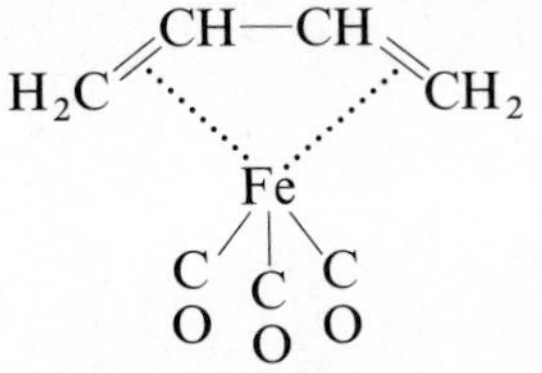

[104] J. K. Stalick and J. A. Ibers, *J. Am. Chem. Soc.*, **1970**, *92*, 5333. See Fig. 2 of this paper for exact definition of $\Delta\beta$.

[105] D. A. Matthews et al., *J. Am. Chem. Soc.*, **1971**, *93*, 5945.

[106] For a different interpretation of the bonding in these complexes, see J. H. Nelson et al., *J. Am. Chem. Soc.*, **1969**, *91*, 7005; K. S. Wheelock et al., ibid., **1970**, *92*, 5110.

The iron atom is equidistant from the four carbon atoms, and the plane of the latter is not quite parallel to that defined by the three carbonyl groups.

The preference of the cyclopentadienylcobalt moiety (EAN = 32) for a diene ligand (four-electron donor) rather than a cyclopentadienyl ligand (five-electron donor) has been discussed previously (see p. 626). In the case of iron the driving force is in the opposite direction, and so we find

$$C_5H_6 + Fe(CO)_5 \xrightarrow{-2CO} (C_5H_6)Fe(CO)_3 \xrightarrow{-CO} (C_5H_5)Fe(CO)_2H \xrightarrow{-H_2} [(C_5H_5)Fe(CO)_2]_2 \quad \textbf{(13.107)}$$

Many polyolefins use only part of their potential ligancy, for example, cyclooctatetraene which often behaves as a diene or even as two dienes bonded to two different metals:

$$C_8H_8 + Ru_3(CO)_{12} \xrightarrow{\text{heptane}} C_8H_8Ru(CO)_3 \text{ and } C_8H_8[Ru(CO)_3]_2 \quad \textbf{(13.108)}$$

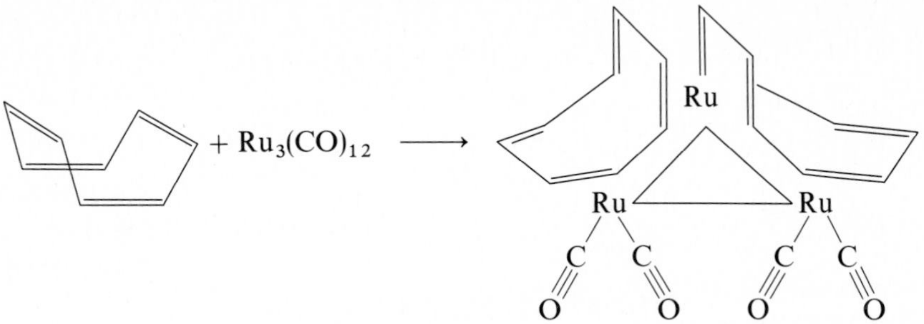

$$+ Ru_3(CO)_{12} \longrightarrow \quad \textbf{(13.109)}$$

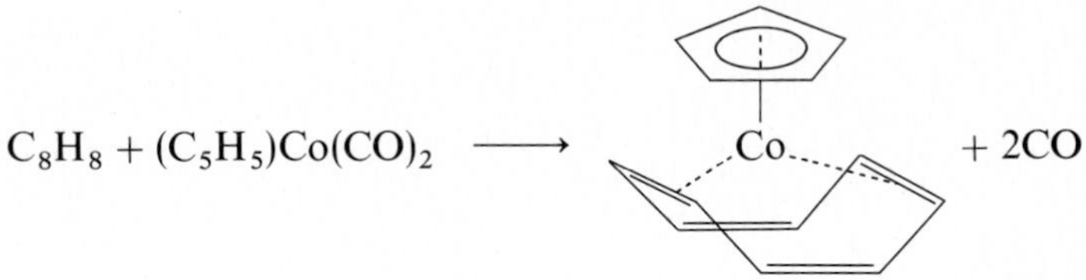

$$C_8H_8 + (C_5H_5)Co(CO)_2 \longrightarrow \quad + 2CO \quad \textbf{(13.110)}$$

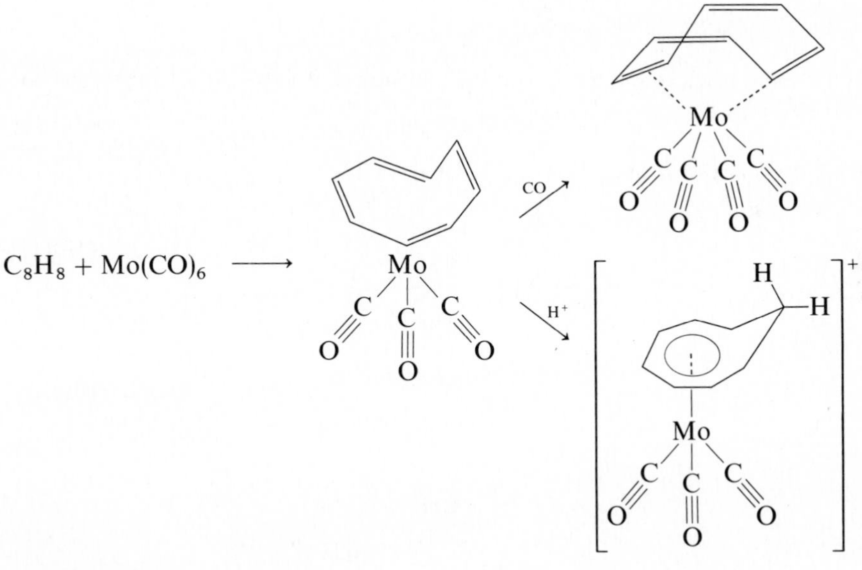

$$C_8H_8 + Mo(CO)_6 \longrightarrow \quad \textbf{(13.111)}$$

The last reaction is of especial interest as a result of the formation of a homotropylium ring.

Three-electron donors

The allyl group, $CH_2{=}CH{-}CH_2{-}$, can behave as a σ ligand (*monohapto* ligand) or as a σ plus π ligand (*trihapto* ligand). Examples of this behavior are

$$Na[Mn(CO)_5] + CH_2{=}CHCH_2Br \longrightarrow (OC)_5Mn{-}CH_2CH{=}CH_2 + NaBr \quad \textbf{(13.112)}$$

$$(OC)_5Mn{-}CH_2CH{=}CH_2 \longrightarrow (OC)_4Mn{-}\left(\begin{matrix}CH_2\\ \vdots\\ CH_2\end{matrix}\right)CH + CO \qquad (13.113)$$

$$Na^+[C_5H_5Mo(CO)_3]^- + C_3H_5Cl \xrightarrow{80^\circ} (C_5H_5)Mo(CO)_3{-}CH_2{-}CH{=}CH_2 + NaCl$$

$$\swarrow \qquad (13.114)$$

$$(C_5H_5)Mo(CO)_2{-}\left(\begin{matrix}CH_2\\ \vdots\\ CH_2\end{matrix}\right)CH$$

In some cases only the *trihapto* complex is known, although it is thought to form via the *monohapto* intermediate:

$$Na^+[Co(CO)_4]^- + C_3H_5Br \longrightarrow [(OC)_4Co{-}CH_2{-}CH{=}CH_2]$$

$$\swarrow \qquad (13.115)$$

$$CO + (OC)_3Co{-}\left(\begin{matrix}CH_2\\ \vdots\\ CH_2\end{matrix}\right)CH + NaBr$$

$$NiCl_2 + 2CH_2{=}CH{-}CH_2MgCl \longrightarrow Ni(\eta^3\text{-}CH_2CHCH_2)_2 + 2MgCl_2 \qquad (13.116)$$

X-ray photoelectron studies[107] indicate that η^1-allyl groups acquire a negative partial charge when bonded to an electropositive metal such as manganese. It thus behaves like

[107] A. J. Ricco et al., *Organometallics*, **1982**, *1*, 94.

other alkyl groups such as methyl and propyl. However, η^3-allyl groups show much less electron density because of the π donation to the metal. Incidentally, these same data indicate that the η^5-cyclopentadienyl group is probably positively charged in $CpMn(CO)_3$.

Other systems can act as three electron donors even though potentially they have more π electrons. The possibility that the cyclopentadienyl group utilizes only three electrons in some compounds has been encountered previously, and it has been suggested that cycloheptatriene, a potential seven-electron donor, utilizes only three electrons in $(C_7H_8)Co(CO)_3$. Protonation of butadiene complexes can convert the diene to a three-electron donor:

$$(C_4H_6)Fe(CO)_3 + HCl \longrightarrow (\eta^3\text{-}CH_3CHCHCH_2)Fe(CO)_3Cl \qquad (13.117)$$

The same type of conversion is also known with the addition of a fluoride ion to a diene:

$$(C_6F_8)Fe(CO)_3 + Cs^+F^- \longrightarrow Cs^+[(C_6F_9)Fe(CO)_3]^- \qquad (13.118)$$

Acetylene and other alkyne complexes

The chemistry of complexes between acetylenes and metals is somewhat more complicated than that of ethylene complexes because of the greater possibilities for bonding by acetylenes and the tendency of some of the complexes to act as intermediates in the formation of other organometallic compounds.

The simplest acetylene complexes resemble those of ethylene. Thus there are analogues of Zeise's salt containing platinum(II) (d^8) and an acetylene molecule which occupies a position like that of ethylene in Zeise's salt.[108] Of greater interest are complexes of the type $L_2Pt(RC{\equiv}CR)$, which may formally be considered 3-coordinate complexes of platinum(0). The structures of two complexes of this type have been determined[109] and they both have structures in which the acetylene lies almost in the plane of PtL_2, forming an approximately square planar complex if each carbon atom in the acetylene molecule is assumed to occupy a coordinate position (Fig. 13.32).

Alternative bonding schemes can be proposed for acetylene complexes similar to those discussed previously for ethylene. As in the previous case, either a μ-bonding model

[108] G. R. Davies et al., *J. Chem. Soc., A*, **1970**, 1873.

[109] J. O. Glanville et al., *J. Organometal. Chem.*, **1967**, *7*, P9; C. Panattoni et al., *J. Am. Chem. Soc.*, **1968**, *90*, 798.

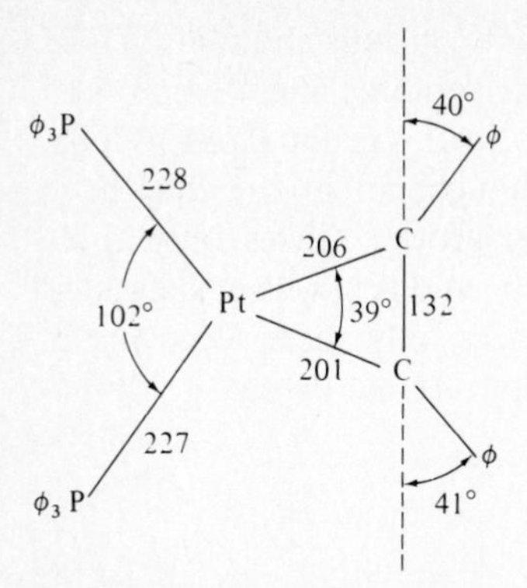

Fig. 13.32 Molecular structure of bis(triphenylphosphine)diphenylacetyleneplatinum. [From J. O. Glanville et al., *J. Organometal. Chem.*, **1967**, 7, P9. Reproduced with permission.] Distances in pm.

with more or less back bonding into antibonding π orbitals or a double σ-bonding model (one σ bond from each carbon) can be used to rationalize the bonding. In addition, acetylenes have an additional π bond and empty antibonding π orbital present.

Although the exact description of the bonding in these complexes can become complicated, it should be pointed out that the following simple model correctly predicts the structures of the compounds known to date. Assume that the back bonding is so extensive that the double σ-bonding model is a reasonable approximation. It can be confusing to speak of d^n configurations in these complexes because of the ambiguities of oxidation state, but the total number of electrons is distinct. For complexes of the type $PtL_2(RC{\equiv}CR)$ the total number of electrons in the valence shell molecular orbitals will be:

$$8(d_{Pt}) + 2(s_{Pt}) + 2(RC{\equiv}CR) + 4(2L) = 16 \tag{13.119}$$

These complexes thus fall two electrons short of that needed for the 18-electron Rule (EAN = 84; Rn = 86), are isoelectronic with complexes of the type $[Pt^{II}X_4]^{-2}$, and like the latter are square planar and diamagnetic.

Complexes of the type $[Pt(R_2C{=}CR_2)X_3]^-$ or $[Pt(RC{\equiv}CR)X_3]^-$ can most simply be described as a platinum(II) ion receiving a pair of electrons into a dsp^2 hybrid orbital, but there is some advantage to treating all of these complexes by the same terminology. In addition, the double σ model rationalizes the fact that the C—C axis is perpendicular to the PtX_3 plane; this model views the platinum as pentacoordinate ($3X + 2\sigma_{RC\equiv CR}$), and the choice lies between pentagonal planar (C—C axis coplanar) and trigonal bipyramidal (C—C axis perpendicular).

The double σ-bonding model is obviously a formalism that interprets the bonding solely in terms of resonance of the type shown by Eq. 13.104b rather than the intermediate situation of mixed "σ" and "π" bonding. However, as long as it is not taken literally it is useful since it successfully rationalizes (1) the arrangement of the C—C axis; (2) the bending of the substituents R and X away from the metal; and (3) the lengthening of the C—C bond. [Note that in Fig. 13.32 the C—C bond length is 132 pm, almost identical to that found in ethylene, 134 pm, substantiating the view that the platinum has "added across" one of the π bonds in acetylene, lowering the bond order to two.]

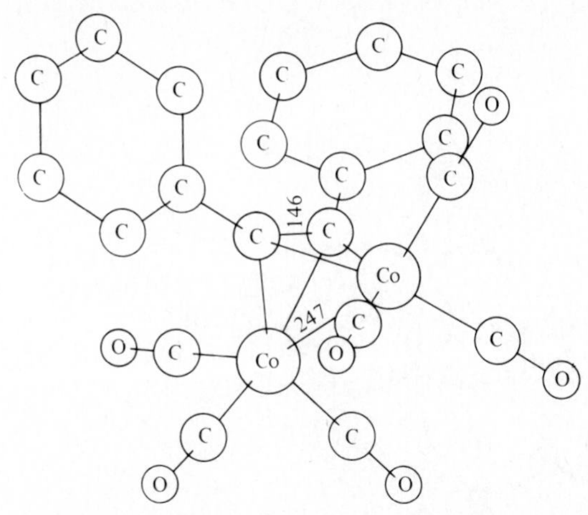

Fig. 13.33 Molecular structure of diphenylacetylenehexacarbonyldicobalt. Average cobalt–carbon (acetylene) bond distance is 196 pm. [From W. G. Sly, *J. Am. Chem. Soc.*, **1959**, *81*, 18. Reproduced with permission.] Distances in pm.

Acetylenes potentially have two pairs of π electrons to donate, but since the two π clouds lie in planes perpendicular to each other, it is impossible to get good overlap with metal orbitals by both at the same time in a mononuclear complex. However, if acetylene is allowed to react with dicobalt octacarbonyl, two moles of carbon monoxide are eliminated:

$$Co_2(CO)_8 + HC{\equiv}CH \longrightarrow H_2C_2Co_2(CO)_6 + 2CO \qquad \textbf{(13.120)}$$

The elimination of 2 mol of carbon monoxide and the effective atomic number rule lead one to suspect that the acetylene molecule is acting as a four-electron donor. This is confirmed by the crystallographic determination of the structure of the diphenylacetylene complex[110] (Fig. 13.33). The position of the two cobalt atoms is such as to allow overlap from both π orbitals in the carbon–carbon triple bond or, perhaps more realistically, approximately sp^3 hybridization[111] for the carbon and σ bonds to the phenyl group, the second "acetylenic" carbon atom, and the two cobalt atoms. The length of the carbon–cobalt bonds is consistent with single-bond radii for these elements, and the carbon–carbon distance indicates a bond order of about 1.3.[112] Analogous acetylene–dirhodium complexes are known (Fig. 13.34).[113]

Acetylene–cobalt carbonyl complexes of this type, in which at least one carbon substituent is a hydrogen atom, react with hydrochloric acid to produce a third type of complex:

$$3(RC{\equiv}C{-}H)Co_2(CO)_6 \xrightarrow[\text{MeOH}]{\text{HCl}} 2(RC_2H_2)Co_3(CO)_9 + \text{other products} \qquad \textbf{(13.121)}$$

[110] W. G. Sly, *J. Am. Chem. Soc.*, **1959**, *81*, 18.

[111] This is not meant to imply 25% *s* character in all four bonds—the bond angles are such as to prohibit this—but simply the formation of four σ (bonding orbitals each of which has some *s–p* hybridization).

[112] L. Pauling, "The Nature of the Chemical Bond," 3rd ed., Cornell University Press, Ithaca, N.Y., **1960**, p. 239.

[113] M. A. Bennett et al., *Inorg. Chem.*, **1976**, *15*, 97.

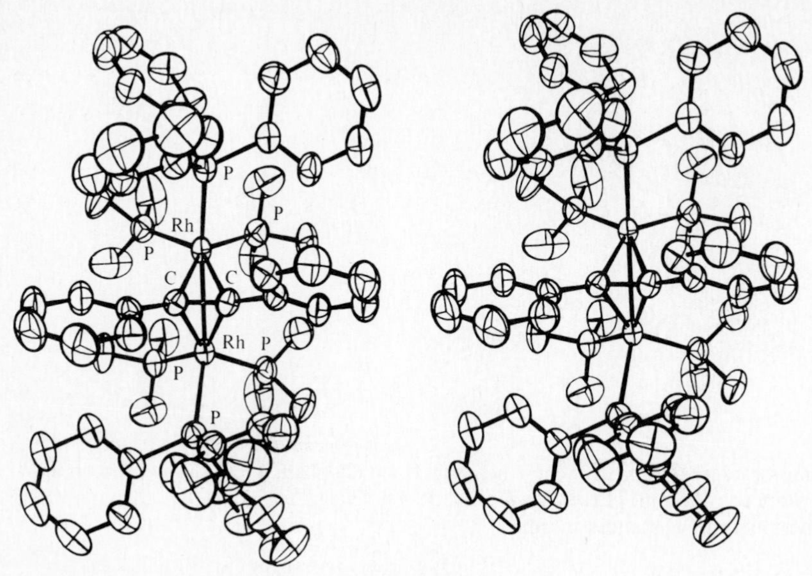

Fig. 13.34 Stereoview of $Rh_2(\phi C{\equiv}C\phi)(PF_3)_4(P\phi_3)_2$. [From M. A. Bennett et al., *Inorg. Chem.*, **1976**, *15*, 97. Reproduced with permission.]

Bromination of this product indicates that a rearrangement has occurred, so that only one carbon is bound to cobalt:

$$\mathrm{RCH_2C{-}Co_3(CO)_9} \xrightarrow{Br_2} \mathrm{RCH_2CBr_3} + \text{other products} \qquad \textbf{(13.122)}$$

This structure is further supported by the fact that the same compound can be synthesized from a trihaloalkane:

$$\mathrm{RCH_2CBr_3} + 3\mathrm{Co(CO)_4^-} \longrightarrow \mathrm{RCH_2C{-}Co_3(CO)_9} + 3\mathrm{Br^-} + 3\mathrm{CO}$$

(13.123)

An X-ray diffraction study of several compounds of this type confirms this structure.[114]

It is worth noting that these structures drawn as simple tetrahedra composed of σ bonds are somewhat oversimplified. There is delocalization in the metal cluster, and if the carbon is replaced by oxygen or nitrogen, there is considerable π-bonding to the cluster. Even with carbon it is possible to detect some π interaction with the cluster.[115]

The alkylidene group (RCH_2—C�) is isolobal with the $(OC)_3CO$⋜group. Thus alkylidene tricobalt nonacarbonyl is isostructural with tetracobalt dodecacarbonyl as far as the skeletal tetrahedron is concerned.[116] It is in the same way isostructural with the purely organic molecule tetrahedrane, C_4H_4. In fact, the complete series $Co_4(CO)_{12}$, $(\phi CH_2C)_1Co_3(CO)_9$, $(\phi CH_2C)_2Co_2(CO)_6$, $(\phi CH_2C)_3Co_1(CO)_3$, and $(\phi CH_2C)_4$ is known.

A related reaction provides similar ketone derivatives

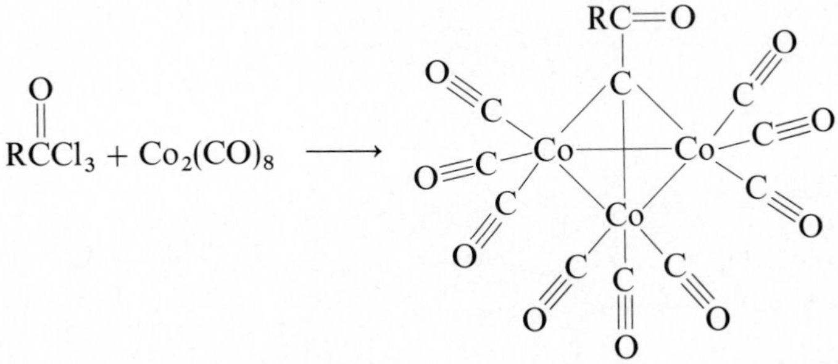

$$RC(=O)Cl_3 + Co_2(CO)_8 \longrightarrow (RC{=}O)CCo_3(CO)_9 \quad (13.124)$$

which have some unusual properties. One of the most interesting of these is that the ketonic CO group undergoes facile decarbonylation at relatively low temperatures (e.g., refluxing benzene) in sharp contrast to ordinary organic systems. The following mechanism has been suggested as accounting for the ready elimination of carbon monoxide:[117]

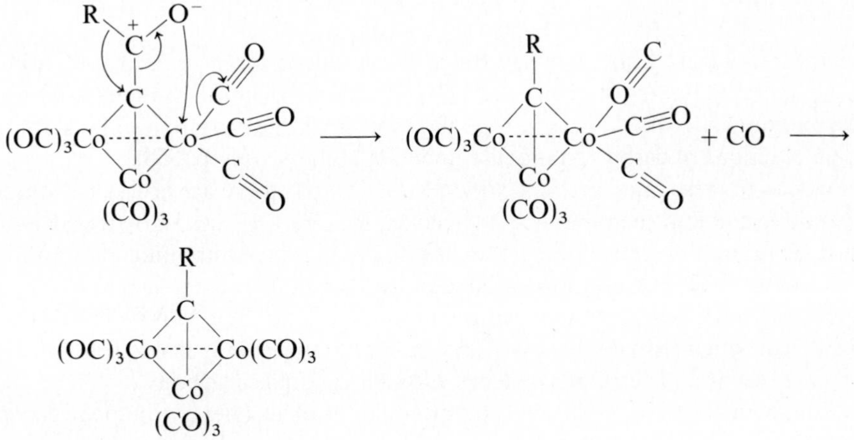

$$(OC)_3Co \cdots Co(CO)_3 \longrightarrow (OC)_3Co\cdots Co(CO)_3 + CO \longrightarrow (OC)_3Co\cdots Co(CO)_3 \quad (13.125)$$

[114] D. Seyferth, *Adv. Organometal. Chem.*, **1976**, *14*, 97.

[115] D. C. Miller and T. B. Brill, *Inorg. Chem.*, **1978**, *17*, 240; G. Granozzi et al., *J. Organometal. Chem.*, **1981**, *208*, C6; *Inorg. Chem.*, **1982**, *21*, 1081; S. F. Xiang et al., *Organometallics*, **1982**, *1*, 699.

[116] The Co_4 cluster has three bridging carbonyl groups, the Co_3C cluster none.

[117] D. Seyferth and M. O. Nestle, *J. Am. Chem. Soc.*, **1981**, *103*, 3320.

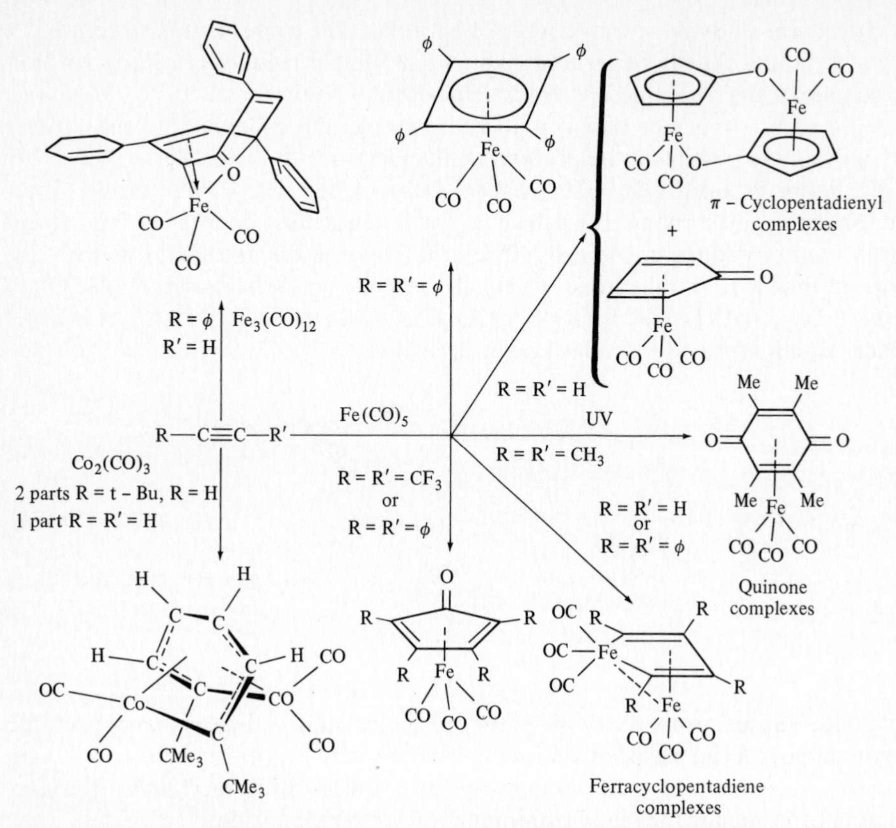

Fig. 13.35 Products of the reaction of acetylenes with metal carbonyls. [From G. E. Coates et al., "Principles of Organometallic Chemistry," Methuen, London, **1968**. Reproduced with permission.]

Although the O-bonded carbonyl looks strange and is, indeed, unknown in stable carbonyls, we have encountered linking carbonyls (p. 521) and ion pairs at the oxygen (p. 607), so an O-bonded intermediate should be possible. We shall see further examples of insertion and elimination reactions of carbon monoxide under "Catalysis" (see p. 658).

In addition to these dinuclear and trinuclear carbonyls there are acetylene complexes containing three and four metal atoms, such as $Fe_3(RCCR)(CO)_9$ and $Co_4(RCCR)(CO)_{10}$, in which the concept of localized bonds breaks down and the compounds resemble metal cluster compounds. The carbon atoms have more than four nearest neighbors and any bonding scheme based solely on conventional σ and/or π bonding will be inadequate.[118] The unusual carbonyl–carbide discussed earlier (pp. 5 and 601) is isolated in very small yield from the reaction of certain acetylenes with triiron dodecacarbonyl.

Some of the unexpected products from the reactions of acetylene with metal carbonyls are shown in Fig. 13.35. The carbonyl group may react to form cyclic ketones and quinones. Even iron is incorporated into ring structures to form ferracyclopentadiene rings (FeC_4R_4, to be distinguished from the FeC_5H_5 moiety). It is generally believed that these reactions

[118] For a discussion of some of these unusual acetylene complexes, see G. E. Coates et al., "Principles of Organometallic Chemistry," Methuen, London, **1968**, pp. 224–232.

go via initial formation of an acetylene adduct, but in only one case has it been possible to isolate such a compound: $(Me_3SiC{\equiv}CSiMe_3)Fe(CO)_4$, probably because of the bulkiness of the trimethylsilyl groups.[119]

The reaction with $Co_2(CO)_8$ results in trimerization of the acetylene to form an unusual C_6 bridging chain which behaves as an allyl group to *both* cobalt atoms. Coupling of the proximal carbon atoms attached to the cobalt atoms via σ bonds can result in the final trimerization to form substituted benzenes. This is an example of the catalyzation of the reaction:

$$3RC{\equiv}CR \longrightarrow C_6R_6 \qquad (13.126)$$

The conversion of acetylene into benzenes in the absence of catalysts occurs only at high temperatures and in poor yields. Formation of metal–acetylene complexes allows the reaction to proceed more readily. Another example is the previously discussed triphenylchromium complex, p. 635, which reacts as follows:

$$(THF)_3Cr(C_6H_5)_3 + 3CH_3C{\equiv}CCH_3 \longrightarrow (C_6H_5)_3Cr(CH_3C{\equiv}CCH_3)_3 \xrightarrow{CH_3C{\equiv}CCH_3} [C_6(CH_3)_6]_2Cr \qquad (13.127)$$

[119] K. H. Pannell and G. M. Crawford, *J. Coord. Chem.*, **1973**, *2*, 251.

Since chromium arene complexes are rather stable they may be isolated from reaction 13.127. If a metal is used which forms less stable arene complexes, the cyclotrimerization occurs with the metal acting as a catalyst. Examples are titanium alkyls (Ziegler catalysts) and nickel complexes (Reppe catalysts). The importance of metal catalysts has been one of the motivating forces in the study of organometallic compounds and is of sufficient importance to warrant separate discussion.

CATALYSIS BY ORGANOMETALLIC COMPOUNDS[120]

Coordinative unsaturation

A vacant coordination site has been suggested[121] as the single most important property of a homogeneous catalyst. The catalyst must have carefully balanced bonding properties such that it has the ability to accept the desired ligand and yet does not accept an alternative but nonreactive ligand instead. The action of carbon monoxide, phosphines, and sulfur compounds as "poisons"[122] in various catalytic systems can be attributed to the tenacity with which these ligands bind to the heavy metals present in the catalysts. One method by which a complex may remain stable although coordinatively unsaturated is the ability to have stable complexes with different coordination numbers. The complex may remain stable and relatively free from unwanted reaction until the appropriate species appears and then expand its coordination shell to receive it. If there is a formal loss of electrons accompanying this change in coordination, the reaction is classified as an *oxidative addition*:

$$\underset{d^8}{[PtCl_4]^{-2}} \xrightarrow{Cl_2} \underset{d^6}{[PtCl_6]^{-2}} \qquad (13.128)$$

The reverse reaction in which the coordination number decreases while there is a formal gain in electrons is termed *reductive elimination.*

The 16,18-electron Rule and oxidative addition reactions

We have seen that of all the complexes examined in Chapter 9, the only geometry that did *not* have maximum stability in an 18-electron configuration was square planar with a d^8 configuration and $\Sigma = 16$. The widest possibilities for oxidative addition lie with these square planar species. They are coordinatively unsaturated with but 16 electrons in the valence shell. They can become 5-coordinate through simple addition of a ligand and

[120] For more thorough discussions of this topic, see J. P. Collman and L. S. Hegedus, "Principles and Applications of Organotransition Metal Chemistry," University Science Books, Mill Valley, Calif., **1980**; G. W. Parshall, "Homogeneous Catalysis," Wiley, New York, **1980**; C. Masters, "Homogeneous Transition-Metal Catalysis," Chapman and Hall, London, **1981**.

[121] J. P. Collman, *Acc. Chem. Res.*, **1968**, *1*, 136.

[122] Note that these chemicals are poisons in biological systems as well, and for much the same reason. See pp. 914, 935.

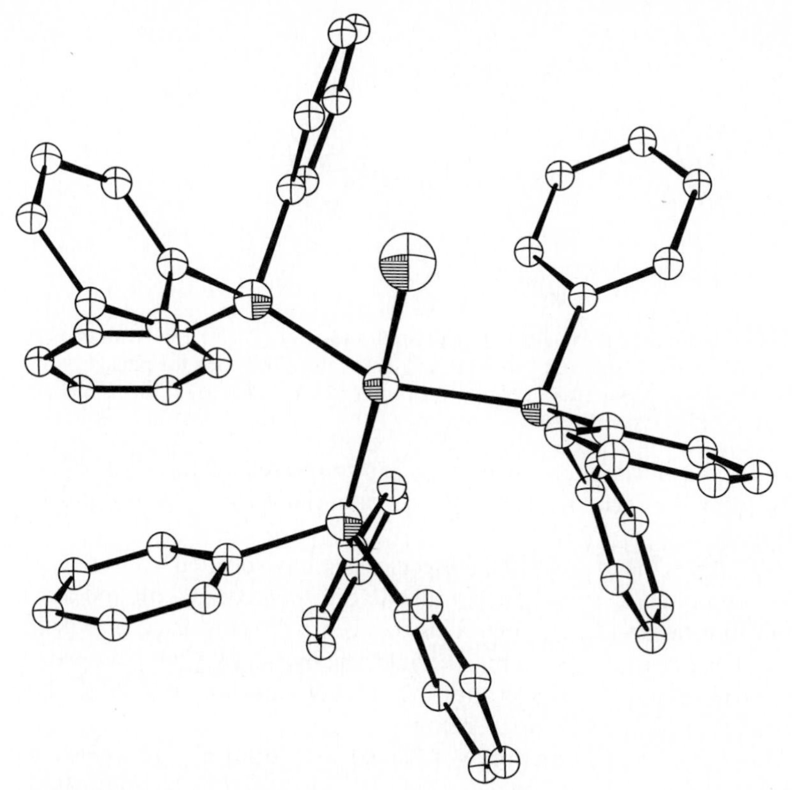

Fig. 13.36 Structure of Wilkinson's catalyst, $(\phi_3P)_3RhCl$. [From F. H. Jardine, *Progr. Inorg. Chem.*, **1981**, *28*, 63. Reproduced with permission.]

6-coordinate through addition combined with oxidation. In either case they become isoelectronic with the next noble gas, i.e., they will then have 18 electrons in the valence shell. A catalyst can undergo oxidative addition to bring two molecules together enhancing their reactivity. Thus if we can get an alkene and active hydrogen as ligands on the same metal atom, perhaps we can effect hydrogenation of the alkene to an alkane. Subsequent reductive elimination of the product would regenerate the original catalyst. For example, Wilkinson's catalyst,[123] chlorotris(triphenylphosphine)rhodium(I), $(\phi_3P)_3RhCl$ (see Fig. 13.36), is thought to behave as a hydrogenation catalyst as follows: In solution one of the phosphine ligands dissociates, leaving $(\phi_3P)_2RhCl$. The latter is too reactive to have been isolated as yet, but the closely related $(\phi_3P)_3Rh^+$ ion which could form from the dissociation of a chloride ion from Wilkinson's catalyst has been studied and has an unusual structure (Fig. 13.37). Unlike most 3-coordinate complexes (see p. 467), it is more T-shaped than triangular. The evidence for the dissociation to form $(\phi_3P)_2RhCl$ is indirect, but persuasive: (1) In less sterically hindered phosphine complexes the catalytic effect disappears—apparently steric repulsion forcing dissociation is necessary; and (2) in the corresponding

[123] F. H. Jardine, J. A. Osborn, and G. Wilkinson, *J. Chem. Soc., A*, **1966**, 1711; **1967**, 1574; F. H. Jardine, *Progr. Inorg. Chem.*, **1981**, *28*, 63.

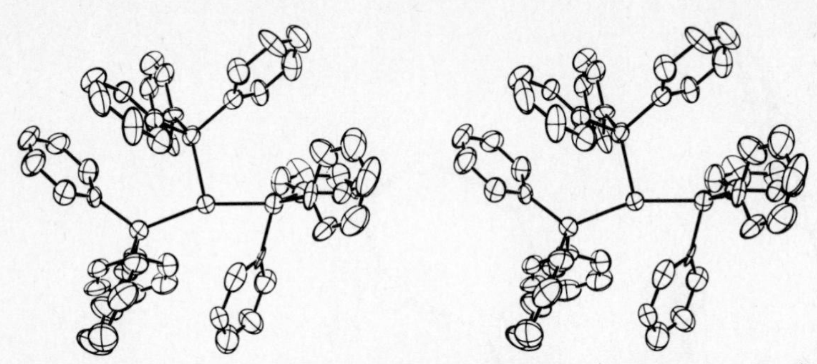

Fig. 13.37 Stereoview of the structure of the $(\phi_3P)_3Rh^+$ cation, showing the planar, approximately T-shaped coordination about the rhodium atom. Note the unusual manner in which the phenyl ring at the lower right is drawn toward the rhodium atom. [From Y. W. Yared et al., *J. Am. Chem. Soc.*, **1977**, *99*, 7076. Reproduced with permission.]

iridium complex in which the metal–phosphorus bond is stronger (cf. the increased LFSE of the third transition over the second) apparently no dissociation takes place and no catalysis is observed.

To return to the catalysis, the $(\phi_3P)_2RhCl$ molecule, possibly solvated, can undergo oxidative addition of a molecule of hydrogen. An alkene can then coordinate and react with a hydrogen atom to form an alkyl group. This reaction will result from a hydride ion migrating from the metal to a carbon atom. Although the hydrogen atom does essentially all of the moving, reactions of this sort are often called "insertion reactions." We shall see another example, "carbonyl insertion," below.

The overall reactions from hydrogenation with Wilkinson's catalyst thus appear to be as follows ($L = \phi_3P$):

$$\text{L—Rh(L)(Cl)—L} + \text{S} \longrightarrow \text{L—Rh(S)(Cl)—L} + \text{L} \qquad (13.129)$$

$$\text{L—Rh(S)(Cl)—L} + H_2 \longrightarrow \text{L—Rh(H)(H)(Cl)(S)—L} \qquad (13.130)$$

$$\text{L—Rh(H)(H)(Cl)(S)—L} + H_2C{=}CH_2 \longrightarrow \text{L—Rh(H)(H)(Cl)(}H_2C{=}CH_2\text{)—L} + \text{S} \qquad (13.131)$$

$$\text{L—Rh(H)(H)(Cl)(}H_2C{=}CH_2\text{)—L} \longrightarrow \text{Cl—Rh(H)(L)(L)(}CH_2CH_3\text{)} \qquad (13.132)$$

$$Cl{-}Rh(H)(L)_2(CH_2CH_3) + S \longrightarrow L{-}Rh(Cl)(H)(L)(S)(CH_2CH_3) \qquad \textbf{(13.133)}$$

$$L{-}Rh(Cl)(H)(L)(S)(CH_2CH_3) \longrightarrow L{-}Rh(Cl)(L)(S) + CH_3CH_3 \qquad \textbf{(13.134)}$$

Tolman[124] catalytic loops

A reaction involving a true catalyst can always be represented by a closed loop. Thus we may combine Eqs. 13.129–13.134 into a continuous cycle with the various catalytic species forming the main body of the loop with the reactants and products entering and leaving the loop at appropriate places:

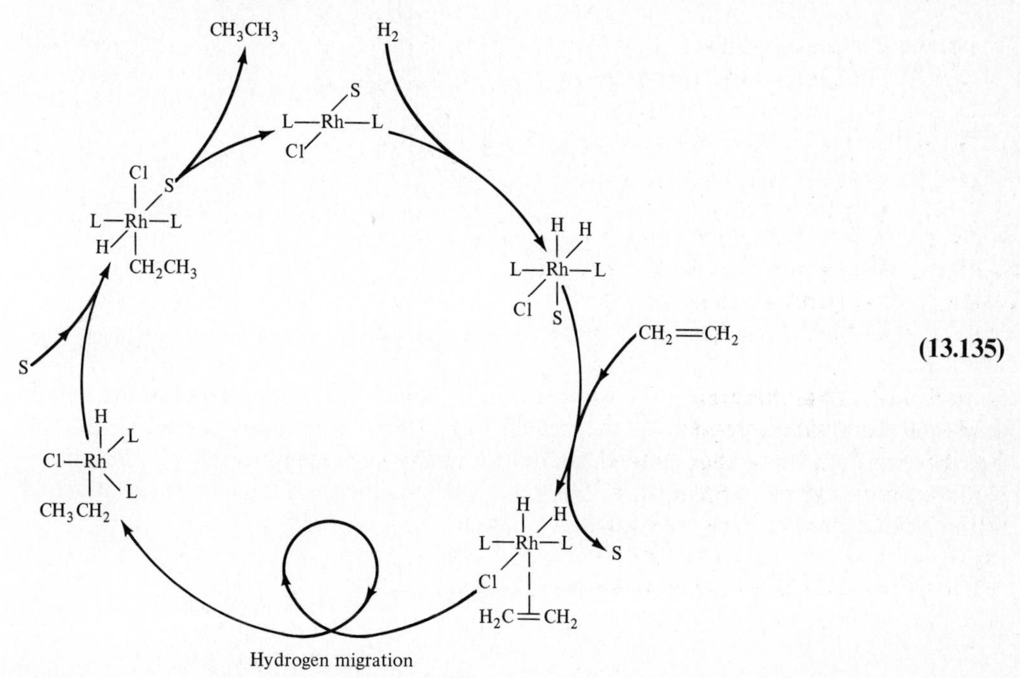

(13.135)

Insertion reactions

One of the most important types of reactions catalyzed by transition metal compounds is that in which a group, Y, is "inserted" into an X—Z bond:

$$Y + X{-}Z \longrightarrow X{-}Y{-}Z \qquad \textbf{(13.136)}$$

[124] C. A. Tolman, *Chem. Soc. Rev.*, **1972**, *1*, 337.

An insertion reaction with dicobalt octacarbonyl as catalyst hydroformylates alkenes is shown in Eq. 13.137. This hydroformylation method is sometimes called the *oxo* process, especially in the industrial literature. It involves both the hydrogenation of an alkene and the insertion of a molecule of carbon monoxide to form an aldehyde functional group. About five million tons of aldehydes and aldehyde derivatives (mostly alcohols) are produced annually, making the process the most important industrial synthesis using a metal carbonyl as catalyst.[125]

(13.137)

In Eq. 13.137 a carbon monoxide ligand from the cobalt is inserted between the cobalt atom and the alkyl group—this is the critical step in the formation of the aldehyde. Reactions of this type have been studied, and we know that although incoming carbon monoxide promotes the reaction, the insertion is by a carbon monoxide ligand already attached to the metal. Consider the equilibrium:

(13.138)

[125] M. Orchin, *Acc. Chem. Res.*, **1981**, *14*, **259**; R. L. Pruett, *Adv. Organometal. Chem.*, **1979**, *17*, 1.

The incoming carbon monoxide reacts with the 5-coordinate species and forces the equilibrium to the right. Other ligands such as triphenylphosphine can fill the vacant coordination site and do the same thing.

The Monsanto Company has built a plant capable of producing 140 000 metric tons of acetic acid per year using a rhodium catalyst to insert carbon monoxide into methanol. The mechanism is believed to be as follows (start with CH_3OH, lower right, and end with CH_3COOH, lower left):[126]

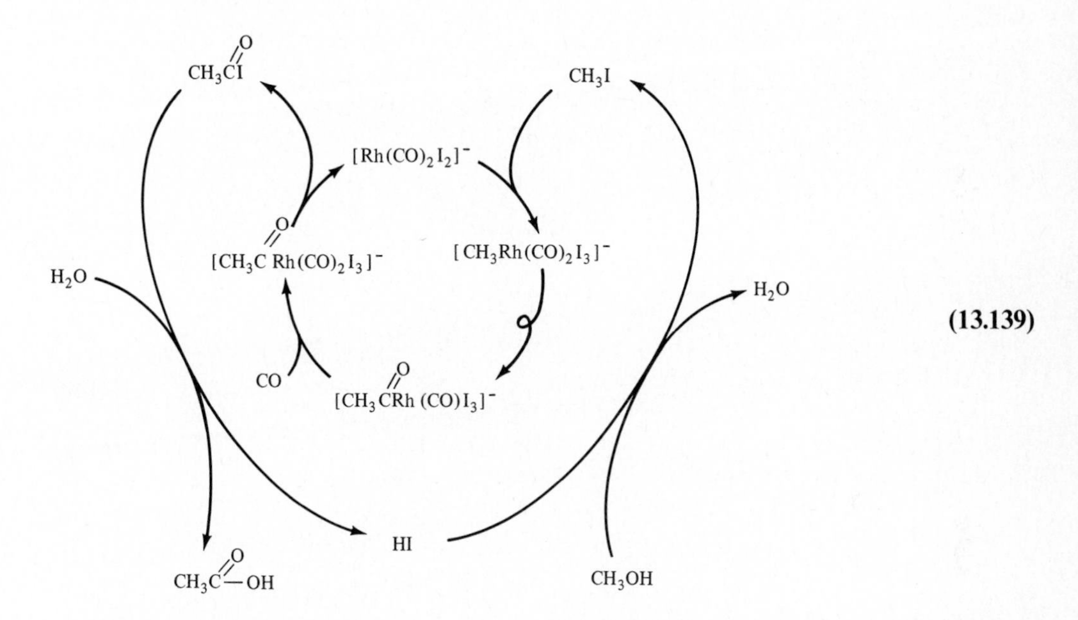

(13.139)

Note that the catalytic loop involves oxidative addition, insertion, and reductive elimination, with a net production of acetic acid from the insertion of carbon monoxide into methanol. The rhodium shuttles between the +1 and +3 oxidation states. The catalyst is so efficient that the reaction will proceed at atmospheric pressure, though in practice the system is pressurized to increase the rate of the reaction. This technology has been licensed worldwide for sufficient plant capacity to produce a million metric tons of acetic acid annually, which is equivalent to 40% of the world's current production.

Historical perspective

The history of organic industrial chemistry can be organized around the relative reactivities of the organic feedstocks.[127] Thus the years 1910–1950 can be characterized as the "acetylene period" since readily available, highly reactive, but rather expensive acetylene was employed. From 1950 to the present alkenes have predominated. As we shall see below,

[126] J. F. Roth, *Plat. Met. Rev.*, **1975**, *19*, 12; D. Forster, *J. Am. Chem. Soc.* **1976**, *98*, 846; *Adv. Organometal. Chem.*, **1979**, *17*, 255.

[127] G. W. Parshall, "Homogeneous Catalysis," Wiley, New York, **1980**, p. 223.

the alkenes in turn will probably be replaced by synthesis gas as the raw material of choice, and we shall enter the era of one-carbon feedstocks.

One of the first processes to foretell the importance of alkenes following World War II was the Wacker process developed in Germany. It has been said that "the invention of the Wacker process was a triumph of common sense."[128] The key to the process (Eq. 13.140) is the oxidation of the palladium catalyst by copper(II), which is in turn regenerated by oxygen from air:

$\frac{1}{2}O_2 + 2H^+$

H_2O

$2\,Cu^+$

$2\,Cu^{+2}$

$PdCl_4^{-2}$

$CH_2{=}CH_2$

Cl^-

$Cl_3Pd^- \cdots \| (CH_2{=}CH_2)$

H_2O

Cl^-

$H_2\ddot{O}{:}\ \ Cl{-}Pd(Cl)\cdots(CH_2{=}CH_2)$

$H_2O^+{-}CH_2{-}Pd^-(Cl)_2$

H^+

Cl^-

$Cl_3Pd^{=}{-}CH_2{-}C(H)_2\ \ \ddot{O}H_2$

$PdCl_4^{-4}$

$HC(=O)CH_3 + H^+$

(13.140)[129]

The net reaction is the oxidation of alkenes to aldehydes.

Catalysis of reactions of synthesis gas

It was mentioned above that the evolution of homogeneous catalysis for industrial uses has been extended to ever cheaper and less reactive feedstocks. Currently the organic chemicals industry seems to be moving ever further into syntheses from *synthesis gas*, or *syngas*. The latter is a more or less broad term to cover various mixtures of carbon monoxide and hydrogen. The first to be of commercial importance was *water gas*, formed by the

[128] G. W. Parshall, "Homogeneous Catalysis," Wiley, New York, **1980**, p. 102.

[129] There is some disagreement over the nature of the species involved. This loop depicts one possibility.

action of steam on red-hot coke:

$$H_2O + C \longrightarrow CO + H_2 \qquad \textbf{(13.141)}$$

In the nineteenth-century days of gas lamps, water gas was frequently used for domestic purposes. This was fraught with danger, however, because of the extreme toxicity of carbon monoxide (see p. 864). Today, interest in synthesis gas has been renewed with increasing interest in coal gasification and increased economies related to the use of various petroleum fractions and various organic chemical "scraps," etc.

Typical of the current interest in sythesis gas is a process being developed by Texaco to increase the chain length ("homologation") of carboxylic acids.[130] For example, acetic acid can be converted to propanoic acid as shown in Eq. 13.142:

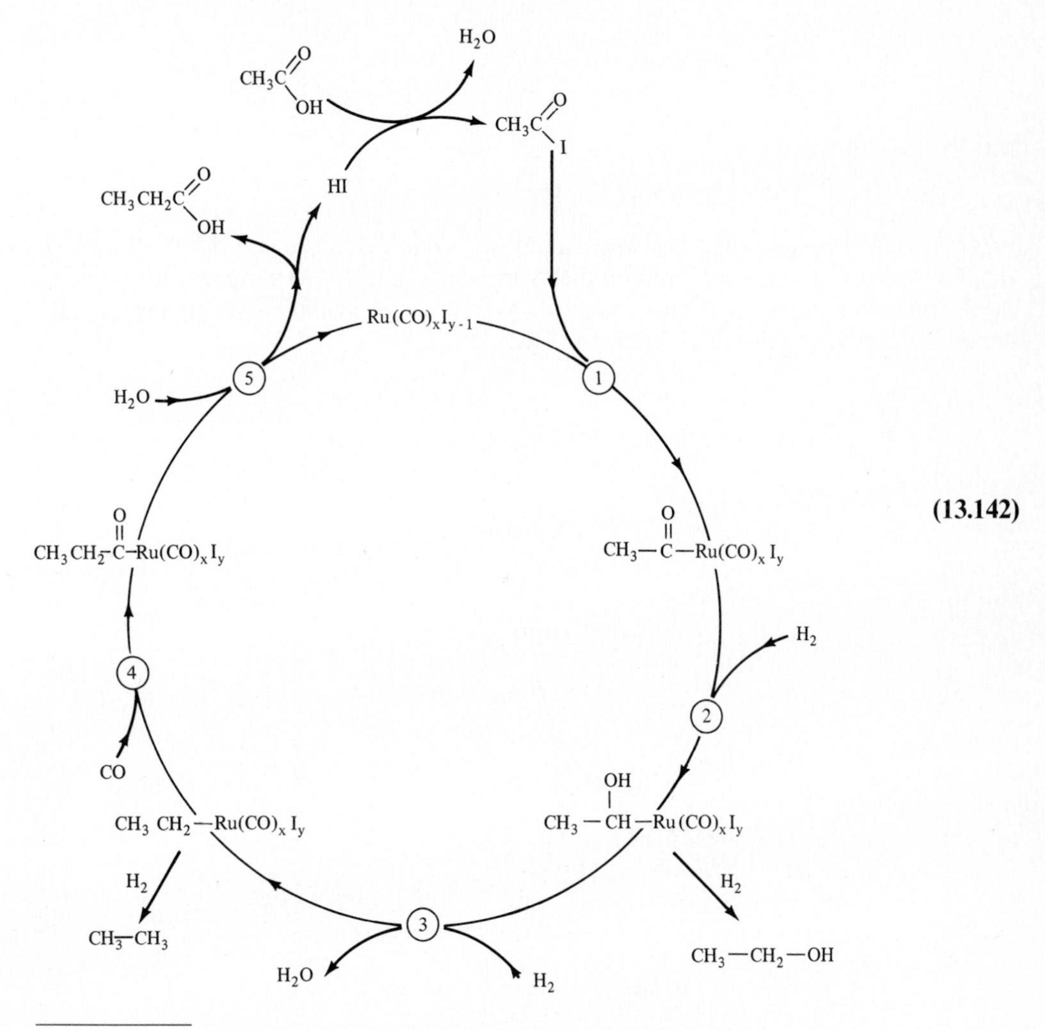

(13.142)

[130] J. F. Knifton, *Chem. Commun.*, **1981**, 41; *J. Mol. Cat.*, **1981**, *11*, 91.

Considerable interest has been shown in the *water gas shift reaction*, the process of forcing Eq. 13.143 to the right:

$$CO + H_2O \xrightarrow[\text{oxides}]{\text{metal}} CO_2 + H_2 \qquad \textbf{(13.143)}$$

There are several reasons for wishing to do this. First, hydrogen is a more versatile industrial chemical than is water gas. Second, small organic molecules tend to have roughly three to four times as many hydrogen atoms as carbon atoms, so if the H_2/CO mole ratio can be changed to about two, a good feedstock is obtained. Commercially, this reaction is usually carried out over Fe_3O_4.[131] However, current interest centers on homogeneous catalysts. The various possible mechanisms have not been worked out completely, but the reaction may be viewed as a nucleophilic attack on carbon monoxide:

$$M \xrightarrow{CO} M{-}C{\equiv}O \ (\leftarrow HO^-) \longrightarrow M{-}\underset{\displaystyle OH}{\underset{|}{C}}{=}O \xrightarrow{-CO_2} MH \qquad \textbf{(13.144)}$$

The hydridic hydrogen in MH can now attack water:

$$MH + H_2O \longrightarrow M + OH^- + H_2 \qquad \textbf{(13.145)}$$

Alternatively (and equivalently) a water molecule can attack in Eq. 13.144 (freeing an H^+ ion) followed by attack of a proton on the hydridic ion in 13.145. Although most work has been done in alkaline solution, some catalysts work in acidic solution. Two typical, simplified catalytic cycles can be presented as:

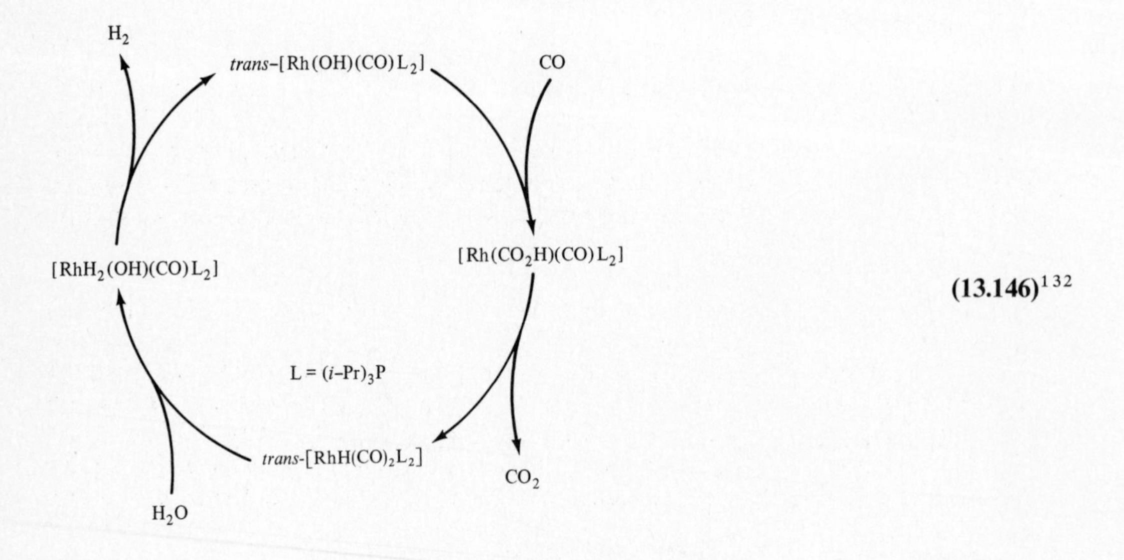

(13.146)[132]

[131] P. C. Ford, *Acc. Chem. Res.*, **1981**, *14*, 31.

[132] T. Yoshida et al., *J. Am. Chem. Soc.*, **1981**, *103*, 3411.

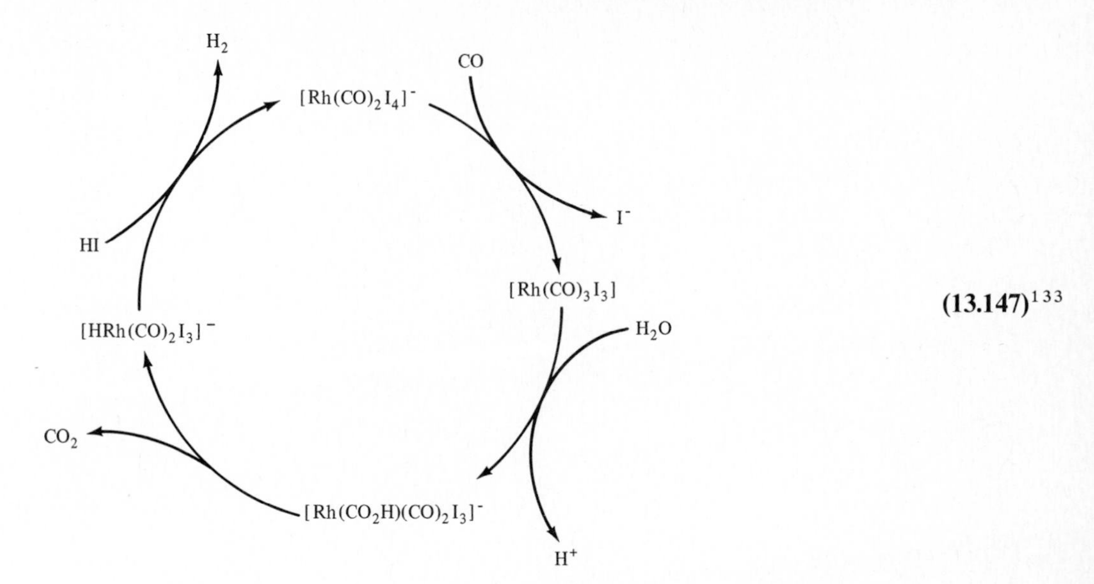

(13.147)[133]

Synthetic gasoline

With an appropriate catalyst, a 2:1 synthesis gas mixture can be converted directly to methanol:[134]

$$CO + 2H_2 \xrightarrow{\text{CuO}} CH_3OH \qquad \textbf{(13.148)}$$

The latter holds great potential as Mobil has recently developed a method for converting methanol to gasoline.[135] It utilizes a zeolite known as ZSM-5 as catalyst (Fig. 13.38). The zeolite acts as an acid catalyst and effectively eliminates a molecule of water from the methanol. The carbene may be stabilized by the zeolite before it is inserted into another molecule of methanol. The details of the mechanism are not clear, but the overall reaction may be written:

$$xCH_3OH \xrightarrow{\text{ZSM-5}} (CH_2)_x + xH_2O \qquad (x = 5\text{–}10) \qquad \textbf{(13.149)}$$

The important aspect of ZSM-5 catalysts that makes them essentially unique is the pore size, intermediate between the smallest and the largest known in zeolites, and the fact that there are interconnecting channels. The pore size is just right for a "gasoline" molecule. Thus the insertion of methylene groups continues until the molecule is too large to grow further, and the growth of the molecules terminates abruptly at C_{10}. The process is also

[133] F. C. Baker et al., *J. Am. Chem. Soc.*, **1980**, *102*, 1020.

[134] F. Marschner et al., "Chemical Feedstocks from Coal," J. Falbe, ed., Wiley, New York, **1982**, Chapter 9.

[135] S. L. Meisel et al., *Chemtech.*, **1976**, Feb., p. 86; I. Berry, *Chem. Eng.*, April 21, **1980**, p. 86; B. M. Harney and G. A. Mills, *Hydrocarbon Proc.*, **1980**, *60*, 67; V. S. U. Rao and R. J. Gormley, ibid., **1980**, *60*, 139.

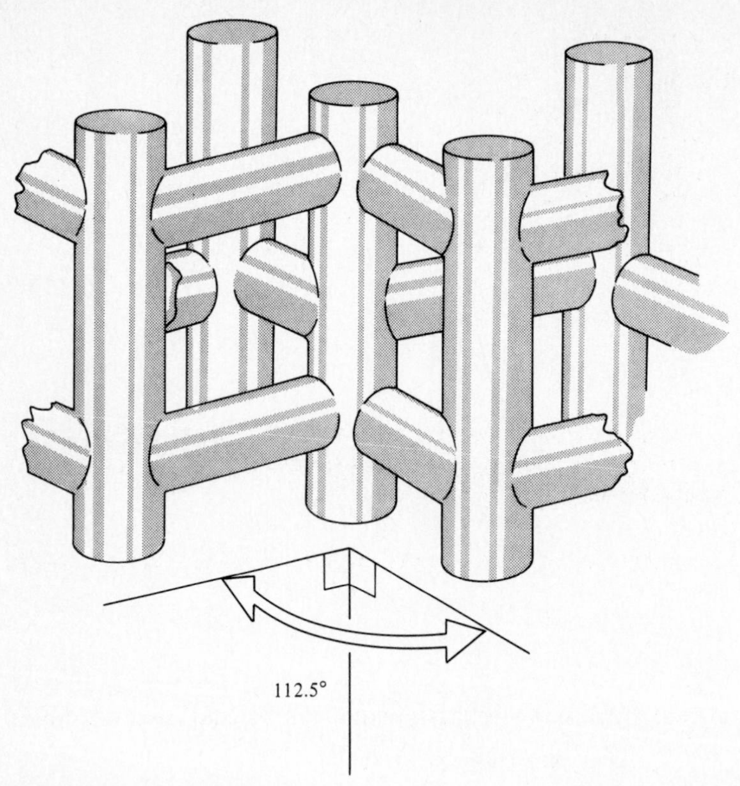

Fig. 13.38 Schematic sketch of the channel system of the structure of ZSM-5 catalyst. The vertical channels are straight and elliptical in cross section with diameters of 510 to 580 pm. The horizontal channels are zig-zag and nearly circular with diameters of 540 to 560 pm. [Courtesy of J. E. Nagy, H. Resing, and G. R. Miller.]

efficient. Over 99% of the methanol is converted in a single pass, 75% of which is of the appropriate molecular weight for gasoline, with over 90% of the latter consisting of highly branched alkanes, highly branched alkenes, and aromatics. All of the latter improve the quality of gasoline so the product has an octane number of 90–100.

The Fischer-Tropsch reaction holds great potential for the organic chemicals industry. Using synthesis gas as a feedstock over a metal catalyst, the overall reaction is rather complicated both as to mechanism and end products.[136] The reaction may be viewed as the reductive polymerization of carbon monoxide, with the molecular hydrogen of the synthesis gas as the reducing agent. The ultimate end-product is the formation of alkanes. Some insight into this *heterogeneous catalysis* (see below) can be obtained by mimicking a catalytic metal surface with a metal cluster compound (see Chapter 14). One catalytic analogue consists of triosmium dodecacarbonyl (see p. 596) and the reaction proceeds

[136] C. K. Rofer-DePoorter, *Chem. Rev.*, **1981**, *81*, 447; C. Masters, *Adv. Organometal. Chem.*, **1979**, *17*, 61; see also ref. 134, Chapter 8.

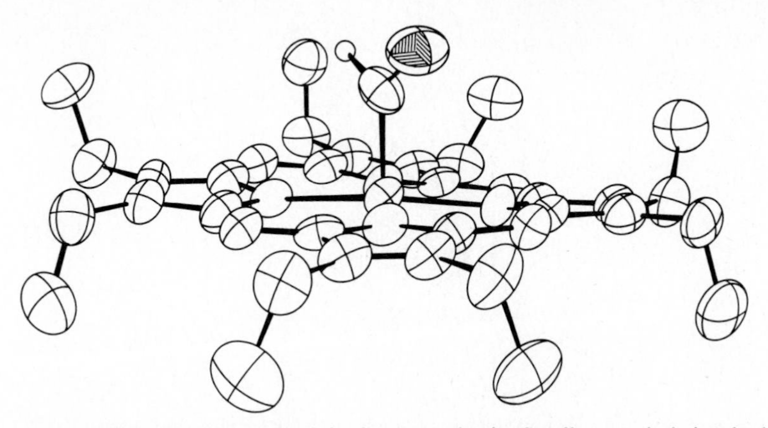

Fig. 13.39 Formyl complex of rhodium porphyrin. Small open circle is a hydrogen atom, four larger, open circles are nitrogen atoms, and sectioned ellipsoid is the oxygen atom. [From B. B. Wayland et al., *J. Am. Chem. Soc.*, **1982**, *104*, 302. Reproduced with permission.]

from carbon monoxide through formyl and hydroxymethyl intermediates to form methane (Eq. 13.150).[137]

$$\underset{\mathbf{1}}{(O{=}C)M_3} \xrightarrow{[H]} \underset{\mathbf{2}}{(HC{=}O)M_3} \xrightarrow{[H]} \underset{\mathbf{3}}{(H(OH)C)M_3} \xrightarrow[-H_2O]{[H]} \underset{\mathbf{4}}{(H_2C)M_3} \xrightarrow{[H]} CH_4 \qquad \mathbf{(13.150)}$$

Hydroxycarbonyl (formyl) derivatives have often been formulated as intermediates in reactions of this sort but much of the evidence has been indirect. However, it has been observed that rhodium octaethyl porphyrin hydride (see Chapter 18 for more on porphyrins) reacts with carbon monoxide to form a metalloformyl complex that has been characterized by X-ray diffraction (see Fig. 13.39).[138] The C—H and C=O bond distances agree well with organic formyl compounds (see Appendix E).

Alternatively, evidence may be obtained from a catalytic carbonyl carbide that provides support for a "carbide" mechanism. This assumes that a metallic carbide forms as an intermediate in the Fischer-Tropsch process. A model compound for this type of

[137] G. R. Steinmetz and G. L. Geoffroy, *J. Am. Chem. Soc.*, **1981**, *103*, 1278.

[138] B. B. Wayland and B. A. Woods, *Chem. Commun.*, **1981**, 701; B. B. Wayland et al., *J. Am. Chem. Soc.*, **1982**, *104*, 302.

reaction is a butterfly cluster formed from an unusual 6-coordinate carbide:[139]

$Fe_6C(CO)_{16}^{-2}$ $\xrightarrow{[O]}$ (butterfly $Fe_4C^{\delta+}$) **(13.151)**

The exposed carbide carbon will react successively with carbon monoxide and methyl alcohol:

$Fe_4C^{\delta+} \xrightarrow{CO} Fe_4C—CO \xrightarrow{MeOH} Fe_4C—C(=O)OMe$ **(13.152)**

followed by hydrogenation to yield methyl acetate. The net synthesis provides an organic oxygenate from synthesis gas, which may also be the product of Fischer-Tropsch reaction.

The Fischer-Tropsch reaction can be modified by varying the feedstock and conditions of the reaction to give the more highly hydrogenated compounds (as in the methane synthesis above). In view of the increasing cost of crude petroleum, one attractive possibility is the conversion of coal to synthesis gas and thence via the Fischer-Tropsch reaction to hydrocarbon fuels. Germany moved portions of its armies in World War II with such synthetic fuels, and recent technological developments have improved the attractiveness of the process. South Africa is currently operating a modern plant, SASOL I, to produce Fischer-Tropsch fuels from coal. A second plant, SASOL II, is under construction. The product is relatively expensive compared to petroleum sources, but its attractiveness is certain to increase with changes in technology, world economic conditions, and petroleum supplies.[140]

---

[139] J. S. Bradley et al., *J. Am. Chem. Soc.*, **1981**, *103*, 4968. Each of the iron atoms has several attached carbonyl groups, both bridging and terminal.

[140] For further references to processes discussed here or mentioned only in passing, the interested reader is referred to "Chemical Feedstocks from Coal," J. Falbe, ed., Wiley, New York, **1982**, Chapter 3, *Acetylene from Calcium Carbide*; Chapter 4, *Hydrogenation of Coal*; Chapter 5, *Gasification of Coal*; Chapter 6, *Methane—Substitute Natural Gas*; Chapter 10, *Higher Alcohols from Synthesis Gas.*

Heterogeneous catalysis

All of the reactions we have seen thus far have been examples of homogeneous catalysis. The reactants and catalyst are soluble in the solvent used. This offers some definite advantages, and most catalysis by organometallic compounds is of this type. Alternatively, catalysis can take place at a surface as in the well-known hydrogenation of alkenes over platinum. In general, this is less closely related to organometallic chemistry, but two important topics will be included.

The Ziegler-Natta catalyst[141] is made by treating titanium tetrachloride with triethylaluminum to form a fibrous material that is partially alkylated. The titanium does not have a filled coordination sphere and acts like a Lewis acid, accepting ethylene as another ligand. The reaction proceeds somewhat after the manner of Wilkinson's catalyst discussed above except that the ethyl group (instead of a hydrogen atom) migrates to the ethylene:

$$\text{—Ti—CH}_2\text{—CH}_3 \xrightarrow{C_2H_4} \text{—Ti(}\leftarrow \text{H}_2\text{C=CH}_2\text{)—CH}_2\text{—CH}_3 \longrightarrow \text{—Ti—CH}_2\text{—CH}_2\text{—CH}_2\text{—CH}_3 \xrightarrow{C_2H_4} \text{—Ti(CH}_2\text{CH}_2\text{CH}_2\text{CH}_3\text{)} \leftarrow \text{CH}_2\text{=CH}_2 \quad \textbf{(13.153)}$$

Recently several workers have tried to combine the catalytic advantages of homogeneous catalysis with the "mechanical advantages" of heterogeneous catalysts, i.e., greater recoverability. When dealing with precious metals such as rhodium, this can be a very important factor in the production process. One way to effect this combination of advantages is to attach the "homogeneous" catalysts to the surface of a polymer such as polystyrene. Wilkinson's catalyst, for example, can be treated as follows:[142]

$$\text{+CH}_2\text{—CH(C}_6\text{H}_5\text{)+}_n \xrightarrow[(2)\ Li]{(1)\ Br_2,\ BF_3} \text{+CH}_2\text{—CH(C}_6\text{H}_4\text{—Li)+}_n \xrightarrow{\phi_2PCl} \text{+CH}_2\text{—CH}_2\text{(C}_6\text{H}_4\text{—P}\phi_2\text{)+}_n \xrightarrow{(\phi_3P)_3\ RhCl} \text{+CH}_2\text{—CH(C}_6\text{H}_4\text{—P}\phi_2\text{—Rh(P}\phi_3\text{)}_2\text{Cl)+}_n \quad \textbf{(13.154)}$$

[141] H. Sinn and W. Kaminsky, *Adv. Organometal. Chem.*, **1980**, *18*, 99.

[142] R. H. Grubbs et al., *J. Macromol. Sci. Chem.*, **1973**, *A7*, 1047; K. Kochloefl et al., *Chem. Commun.*, **1977**, 510.

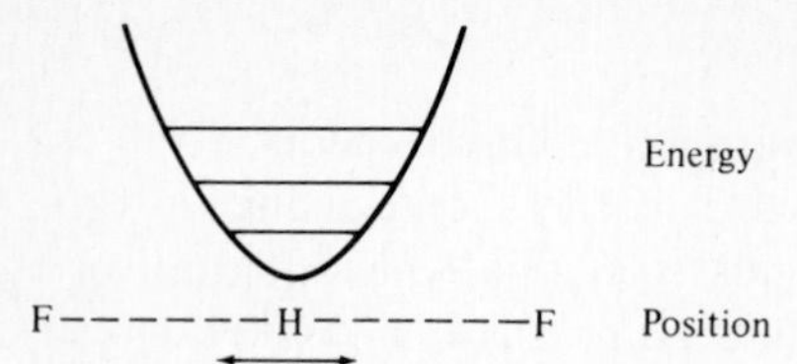

Fig. 13.40 Energy and position of the hydrogen atom in the symmetrical HF_2^- system. The average position of the hydrogen atom is midway between the fluoride ions.

A heterogeneous analogue of the rhodium acetic acid catalyst has been prepared,[143] and in principle it appears that most homogeneous catalysts will function in a heterogeneous system when appropriately handled.[144]

FLUXIONAL MOLECULES[145]

Chemists are prone to think in terms of static molecular structures. This viewpoint is initiated by stick-and-ball models and reinforced by constant inspection of molecular structures determined by "instantaneous" methods such as X-ray diffraction.[146] Although the importance of molecular vibrations has been realized for some time, it is only recently that the importance of *fluxional*[147] molecules has been appreciated. Fluxional molecules differ from others in possessing more than a single configuration representing an energy minimum. Several such minima may be present and accessible with ordinary thermal energies. As a very simple example, consider the symmetric and unsymmetric hydrogen bonds (pp. 269–271). If in the symmetric HF_2^- ion the fluoride ions are considered relatively immobile with respect to the much lighter hydrogen, the motion of the latter can be considered, to a first approximation, as a vibration in a potential well (Fig. 13.40) with an average position midway between the fluorides. In contrast, the unsymmetrical hydrogen bonds possess two potential wells in which the hydrogen can vibrate (Fig. 13.41), occasionally being sufficiently excited thermally to jump to the other well.[148] In such a system

[143] K. K. Robinson et al., *J. Catalysis*, **1972**, *27*, 389.

[144] For reviews, see J. C. Bailar, Jr., *Cat. Rev. Sci. Eng.*, **1974**, *10*, 17; Z. M. Michalska and D. E. Webster, *Plat. Metals Rev.*, **1974**, *18*, 65.

[145] See F. A. Cotton, *Acc. Chem. Res.*, **1968**, *1*, 257; K. Vrieze and P. W. N. M. van Leeuwen, *Progr. Inorg. Chem.*, **1971**, *14*, 1.

[146] Obviously a structure is not determined "instantly" with X-ray diffraction techniques. "Instantaneous" refers to the fact that the time period over which the diffracted wave interacts with the electrons of the molecule is infinitesimally short with respect to the frequency of atomic motions. See discussion on p. 669 for further explanation of the importance of the time scale.

[147] This definition includes all stereochemically nonrigid molecules having identical minima and configurations at those minima including molecules such as PF_5 discussed in Chapter 5. Some chemists would not extend the term to such molecules and restrict it to the molecules discussed here where bonds are broken and re-formed, not just pushed around.

[148] For simplicity's sake, both potential wells are shown to be the same depth. In general for a hydrogen bond (such as —O—H · · · N) this is not true. All of the configurations of a truly fluxional system are energetically equivalent, however, and have equivalent potential wells.

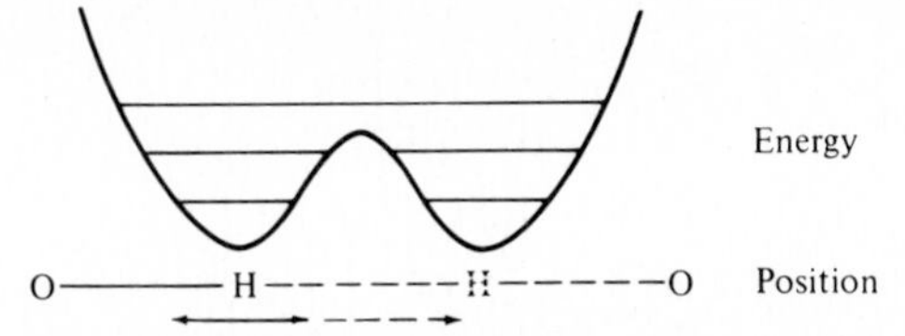

Fig. 13.41 Energy and position of the hydrogen atom in an unsymmetrical hydrogen-bonding system. The positional sketch represents the hydrogen in the left potential well; the dotted H represents the average position of the hydrogen atom the other half of the time. The height of the energy barrier is qualitative and not meant to represent any particular system.

Table 13.6 Time scale for structural techniques

Techniques	Approx. time scale (s)
Electron diffraction	10^{-20}
Neutron diffraction	10^{-18}
X-ray diffraction	10^{-18}
Ultraviolet	10^{-15}
Visible	10^{-14}
Infrared-Raman	10^{-13}
Electron spin resonance[a]	10^{-4} to 10^{-8}
Nuclear magnetic resonance[a]	10^{-1} to 10^{-9}
Quadrupole resonance[a]	10^{-1} to 10^{-8}
Mössbauer (iron)	10^{-7}
Molecular beam	10^{-6}
Stop-flow kinetics	10^{-3} to 10^{2}
Experimental separation of isomers	$>10^{2}$

[a] Time scale sensitively defined by chemical systems under investigation.

SOURCE: E. L. Muetterties, *Inorg. Chem.*, **1965**, *4*, 769. Used with permission.

the hydrogen would be "found" in one potential well or another by "instantaneous methods." If the barrier between the two configurations is thermally accessible, the system is fluxional.

Muetterties[149] has discussed the relationship between the lifetime of a particular molecular structure and the various physical methods for studying the structures of molecules (Table 13.6). The diffraction methods (10^{-18} to 10^{-20} s) are "instantaneous" in the sense that the interaction between the molecule and the diffracted wave takes place over a period of time that is very short with respect to molecular motions. Obviously this time period relates to a single interaction, and the actual experiment must take a considerably longer period of time for the collection of the data. The resulting structure is thus a weighted average of all the molecular configurations present, and this is commonly encountered as *thermal ellipsoids* of motion. In a molecule such as that described by Fig. 13.41, if the higher vibrational levels were occupied, the hydrogen atom would be

[149] E. L. Muetterties, *Inorg. Chem.*, **1965**, *4*, 769; *Acc. Chem. Res.*, **1970**, *3*, 266.

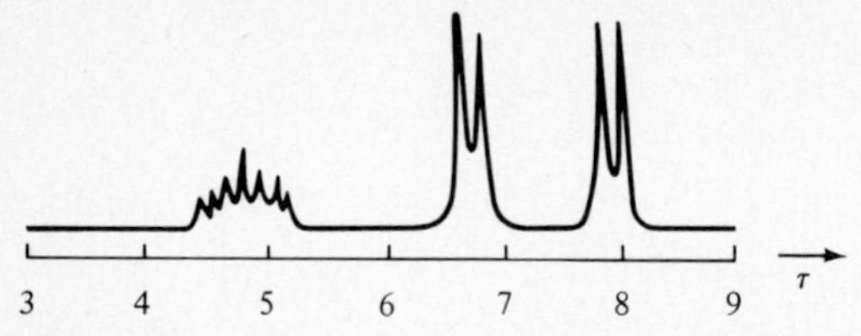

Fig. 13.42 Typical spectrum of *trihapto*-allyl system. [From G. E. Coates et al., "Principles of Organometallic Chemistry," Methuen, London, **1968**. Reproduced with permission.]

"smeared" over the entire vibrational amplitude. If only the lowest vibrational level is occupied, the hydrogen will show up as a "half atom" at one minimum and a "half atom" at the other.

In the same way, spectroscopic methods using ultraviolet, visible, or infrared light are generally much faster than molecular vibrations or interconversions, and the spectrum reflects a weighted average of the species present (cf. the broad absorption bands in the visible spectrum of transition metal complexes: see p. 372).

The remaining spectroscopic methods are slower, and the time period of the interaction may be comparable to that of the lifetime of the molecules present. The nature of the changes taking place can now be studied by such techniques. Nuclear magnetic resonance has probably been used more than the other methods, but all are potentially useful in studying fluxional molecules.

Nuclear magnetic resonance techniques have proved invaluable in the study of fluxional molecules. There are two limiting cases depending upon the rate of interconversion.[150] Consider a molecule in which there are two environments for the protons and no interconversion at all. There will be two resonance lines[151] separated by a difference (Δ) that reflects the difference in the environment of the two sites. At the other extreme, if interconversion takes place with a frequency which is large with respect to the separation Δ, the spin of the proton hardly loses phase from the time it is in a particular site until it is at that site again. As far as the response of the proton is concerned, it behaves as though it were in an average environment and a single line is found at the average frequency.

To return to the limiting case of fixed sites, if it were possible to "freeze" a molecule in a particular configuration for a short while, we could measure the distinct spectrum with individual lines for each site. If we then allowed the molecule to undergo fluxional behavior, we would observe that the formerly split lines would collapse to the single, average line. This type of thought experiment cannot be realized in the laboratory, but we can approach it by varying the temperature and hence the rapidity of movement of the molecule under study. For an example of the application of this technique, consider a *trihapto*-allyl complex. At low temperatures it would typically yield a proton magnetic resonance spectrum as in Fig. 13.42. The two large doublets represent the two hydrogen atoms (H_s) in *syn* or "*cis*" positions relative to the fifth hydrogen and the two hydrogen atoms (H_a) in *anti* or

[150] For a more thorough discussion, see A. Carrington and A. D. McLachlan, "Introduction to Magnetic Resonance, " Harper & Row, New York, **1967**, Chapter XII.

[151] These lines may be split by spin–spin interactions.

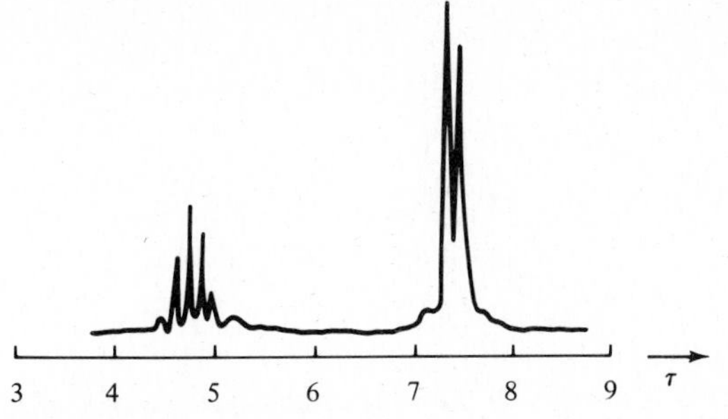

Fig. 13.43 Spectrum of fluxional allyl system showing time averaging of the four terminal hydrogen atoms. [From G. E. Coates et al., "Principles of Organometallic Chemistry, Methuen, London, **1968**. Reproduced with permission.]

"*trans*" positions relative to the fifth hydrogen. The molecule is rigid, so each of the three types of hydrogen produces a distinctive chemical shift depending upon its environment. The terminal hydrogen atoms are split by the fifth, nonterminal hydrogen into doublets. The multiplet represents the fifth hydrogen split by the other four.

Upon warming, the spectrum shown in Fig. 13.42 often changes with collapse of the two doublets into one (Fig. 13.43). This spectrum indicates that on a time average there are only two types of hydrogen: four terminal hydrogen atoms and one nonterminal hydrogen atom; the four terminal hydrogen atoms are equivalent. This could result from the molecule rapidly interconverting:

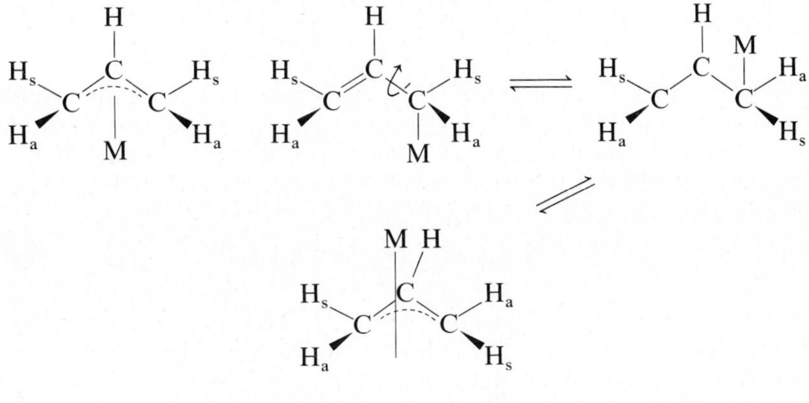

(13.155)

In such a rapidly interconverting system the distinction between H_s and H_a is lost.

A somewhat similar system has been found for the complex between tetramethylallene and iron carbonyl (Fig. 13.44a). Below -60° the proton magnetic resonance spectrum shows three peaks in the ratio 1:1:2, representing the three *cis* hydrogen atoms, three *trans* hydrogen atoms, and six hydrogen atoms in a plane perpendicular to the carbon–iron bond. As the temperature is raised, the spectrum collapses to a single resonance for the

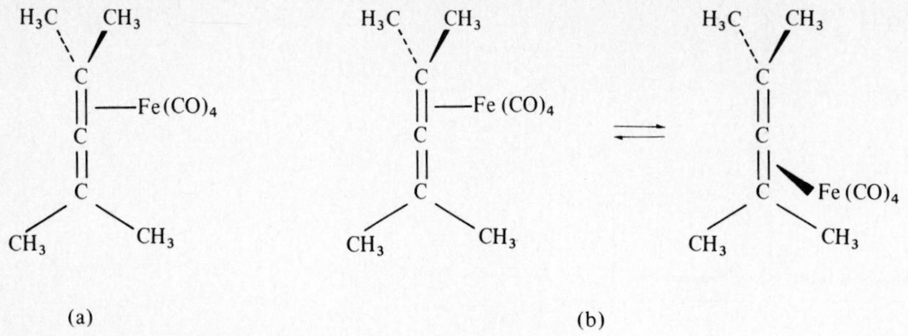

(a) (b)

Fig. 13.44 Tetramethylalleneiron tetracarbonyl.

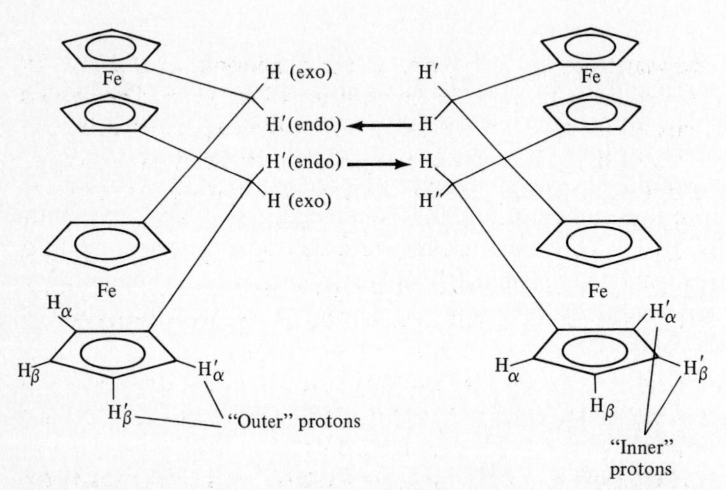

Fig. 13.45 The fluxional, "inside-out" behavior of the ferrocenophane molecule. Note the interconversion of the methylenic hydrogen atoms: exo ⇌ endo. Note also that the "outer" protons become "inner" protons. As a result of this averaging, there are only three magnetically distinct kinds of protons: α, β, and bridge. [From R. Dagani, *Chem. Eng. News*, March 1, **1982**, *23*. Reproduced with permission.]

average environment of the 12 hydrogens as the iron presumably migrates around the allene π system (Fig. 13.44b).

Cyclic systems can also undergo fluxional behavior. We have seen that the rings in ferrocene are free to rotate. Ordinarily this is not observable as all of the protons (for example) are equivalent whether the molecule is static or fluxional. Mueller-Westerhoff and co-workers[152] have described a ferrocenophane (two ferrocene molecules linked by methylene bridges) that undergoes unusual fluxional behavior (Fig. 13.45). The methylene bridges "flip" over to the opposite conformation faster than can be followed on the NMR time scale. Thus the NMR spectrum is unexpectedly simple: a sharp singlet for the methylenic protons and two narrow multiplets for the α- and β-protons of the rings.

---

[152] U. T. Mueller-Westerhoff et al., *J. Am. Chem. Soc.*, **1981**, *103*, 7678; R. Dagani, *Chem. Eng. News*, March 1, **1982**, *23*.

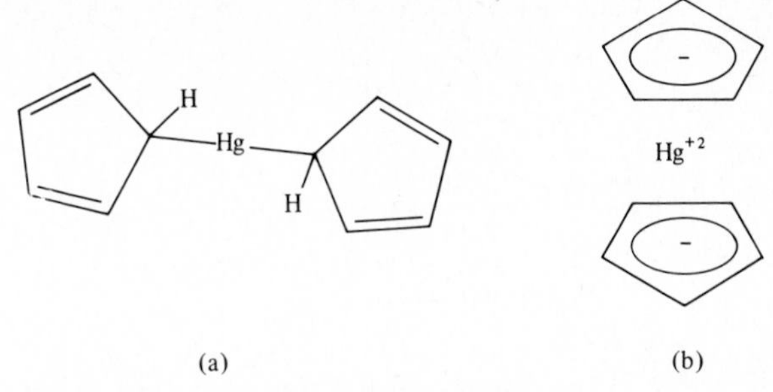

Fig. 13.46 Possible structures for bis(cyclopentadienyl)mercury: (a) *monohapto* complex; (b) *pentahapto* complex.

Another example of fluxional behavior in cyclopentadienyl compounds is shown by Cp_2Hg, which could have either of two structures as shown in Fig. 13.46. One would correspond to divalent mercury σ-bonded to two C_5H_5 groups (*monohapto* complex), the second to an ionic[153] structure analogous to the magnesium compound (*pentahapto* complex). Structure 13.46a might be expected to exhibit three types of hydrogen resonances in the ratio 1:2:2, whereas all of the hydrogen atoms are equivalent in 13.46b. The NMR spectrum of the compound down to -70 °C consists of a single peak, indicating equivalence of the protons. On the other hand, the infrared spectrum is much more complex than is consistent for the highly symmetrical structure 13.46b. It was thus proposed[154] that the mercury-ring bonds are undergoing a constant precession, and this point is supported by more recent work.[155] A similar process is believed to occur in *monohapto*-cyclopentadienyl*pentahapto*cyclopentadienyldicarbonyliron,[154] and it is thought that the migration of the iron atom takes place via a series of 1,2 shifts.[145]

One of the systems studied extensively is the cyclooctatetraene metal tricarbonyl complex, such as that synthesized in Eq. 13.108. The first compound of this type that was synthesized was the iron complex. The earliest interpretations of structure, based on the equivalence of protons, suggested planar cyclooctatetraene rings. However, the X-ray structure was shown to be that of a 1,3-diene adduct (Fig. 13.47a). There followed a series of suggestions on possible fluxional mechanisms for averaging the protonic environments in the molecule. For convenience the various structures and mechanisms (Fig. 13.47) may be named: (1) the 1,2 shift, circular ring migration, based on a structure similar to that in the solid; (2) a structure bonded through transannular (1,5) double bonds interconverting through a flipping mechanism; and (3) a transannular jump (1,5 shift) of the iron atom in a structure similar to that in the solid. Nuclear magnetic resonance studies of the iron complex were inconclusive because the spectrum could not be resolved even at -145 °C.

[153] Mercury cannot form a *covalent* metallocene structure without invoking high-energy orbitals (EAN would equal 90, greater than 86 for Rn).

[154] G. Wilkinson and T. S. Piper, *J. Inorg. Nucl. Chem.*, **1956**, *2*, 32; T. S. Piper and G. Wilkinson, *J. Inorg. Nucl. Chem.*, **1956**, *3*, 104.

[155] E. Maslowsky, Jr., and K. Nakamoto, *Inorg. Chem.*, **1969**, *8*, 1108; F. A. Cotton and T. J. Marks, *J. Am. Chem. Soc.*, **1969**, *91*, 7281.

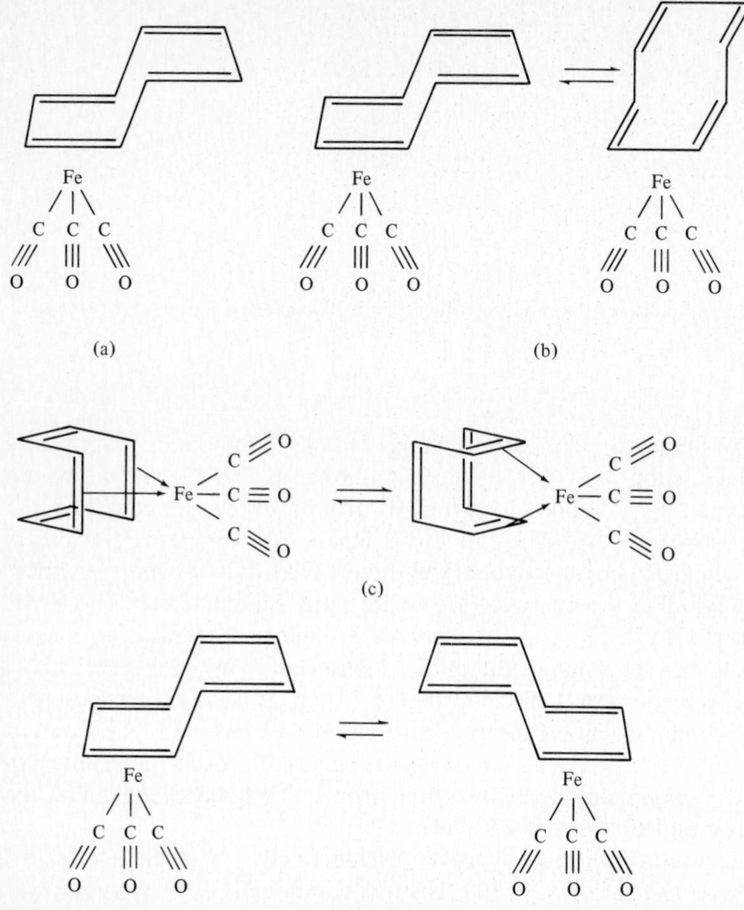

Fig. 13.47 Structures and mechanisms suggested for [(COT)Fe$(CO)_3$]: (a) molecular structure in the solid; (b) 1,2 shift; (c) flip mechanism; (d) transannular jump.

Cotton and co-workers[145,156] prepared the analogous ruthenium compound, which rearranges at a slower rate than the iron compound. On cooling, the NMR spectrum was resolved, confirming the 1,2 shift mechanism for the ruthenium complex. Presumably the same mechanism is operative in the iron complex as well.

CONCLUSION

Organometallic chemistry, as its name implies, has links to both organic chemistry and transition metal chemistry. The ties to organic chemistry, at first in compounds that were more or less laboratory curiosities, have burgeoned into indispensable components of the

[156] F. A. Cotton et al., *J. Am. Chem. Soc.*, **1969**, *91*, 6598, and references therein.

petroleum chemicals industry. The transition metal chemistry is largely an extension of the coordination chemistry of strongly bonded ligands, but the cluster chemistry anticipates the polyhedral chemistry of boranes with extreme delocalization in molecular orbitals (Chapter 14). The concept of isolobal fragments also ties the two disciplines together, with fragments as seemingly disparate as $CH_3C\leftarrow$ and $(OC)_3Co\leftarrow$ being isolobal. The entire field is growing rapidly, with two new journals, *Organometallics* and *Polyhedron*, having been established in 1982, and it will continue to grow as energy and organic feedstocks become ever more expensive.

PROBLEMS

13.1. Formulate neutral complexes of chromium containing
a. cyclopentadienyl and nitrosyl ligands
b. cyclopentadienyl, carbonyl, and nitrosyl ligands
[P. L. Pauson, "Organometallic Chemistry," St. Martin's Press, New York, **1967**.]

13.2. Suggest a structure for $(Cp)_2Fe(CO)_2$. [See p. 634.]

13.3. Suggest reasonable formulas for molecules containing one or more
a. chromium atom(s)
b. cyclopentadienyl group(s)
c. carbonyl group(s)
d. nitrosyl group(s)

13.4. Suggest the most reasonable formula for a molecule containing one or more
a. iron atom(s)
b. cyclopentadienyl group(s)
c. carbonyl group(s)

13.5. Suggest reasonable syntheses for
a. $Mo(C_6H_6)(CO)_3$
b. $Mo(C_6H_6)_2$
c. a compound containing $[Co(C_5H_5)_2]^+$
d. $SiH_3Co(CO)_4$

13.6. Calculate the effective atomic number for $[HRu(P\phi_3)_3]^+$ assuming four σ-bonded ligands. On the basis of the 18-electron Rule, suggest an alternative structure. [J. C. McConway et al., *Chem. Commun.*, **1974**, 327.]

13.7. Suggest a reasonable structure for $[Co(C_6H_6)(CO)_2]^+$.

13.8. Using Fig. 13.19, predict the number of unpaired electrons in
a. Cp_2Ti^+
b. Cp_2Cr^+
c. Cp_2Cr

13.9. Read through the first portion of this chapter again and
a. Determine what fraction of the cyclopentadienyl compounds mentioned obey the 18-electron Rule.
b. Name all of the cyclopentadienyl compounds by the *hapto* nomenclature.

13.10. R. E. David and R. Pettit have shown that a number of aromatic hydrocarbon–iron carbonyl complexes have short and long C—C bond lengths. Examples are:

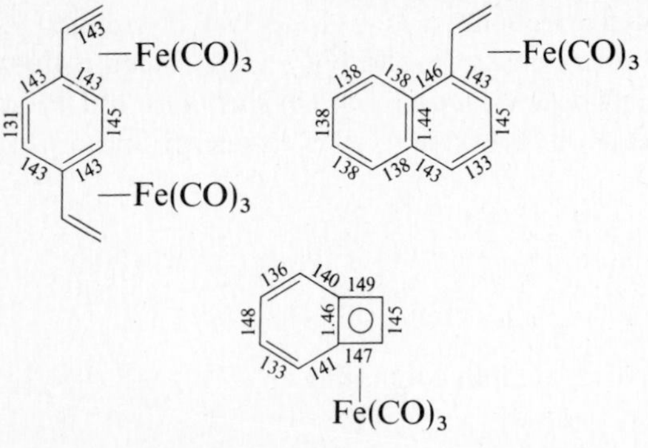

Suggest an explanation for this effect. [*J. Am. Chem. Soc.*, **1970**, *92*, 716.] Distances in pm.

13.11. Treatment of Cp_2Mn with nitric oxide yields $Cp_3Mn_2(NO)_3$. The latter shows infrared absorptions at 1497 cm^{-1} and 1720 cm^{-1} and two broad bands in the proton magnetic resonance spectrum with relative intensities of 2:1. Suggest a structure. [T. S. Piper and G. Wilkinson, *J. Inorg. Nucl. Chem.*, **1956**, *2*, 38, 136; R. B. King and M. B. Bisnette, *Inorg. Chem.*, **1964**, *3*, 791.]

13.12. Organometallic chemistry seems especially prone to the development of descriptive words and phrases. Although much of this language is often considered jargon and not allowed to enter the formal literature, it is common in oral usage. See if you can identify the following terms, each applicable to material discussed in this chapter:

a. open-faced sandwich
b. supersandwich
c. club sandwich
d. ringwhizzer
e. molecular broad jump
f. piano stool molecules
g. fly-over bridge
h. triple-decker sandwich

13.13. The structures of some arene compounds of chromium are:

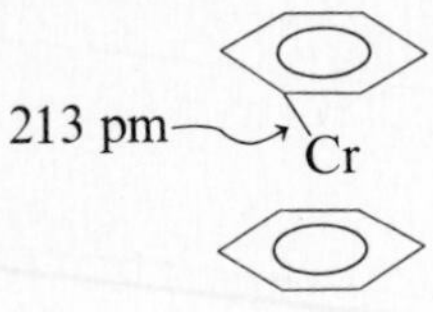

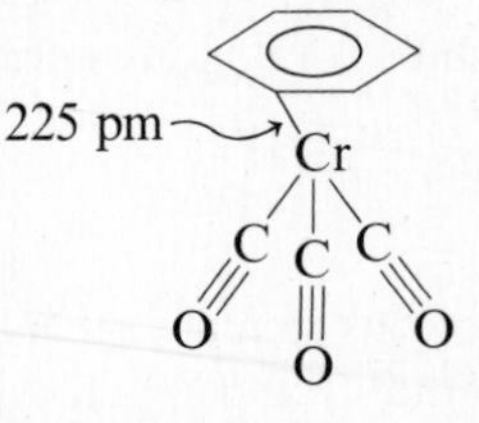

Discuss the bond lengths in terms of a π-bonding model. [G. E. Coates et al., "Organometallic Compounds," Methuen, London, **1968**, Vol. II, pp. 171–173.]

13.14. Predict the products of the reaction:

$$(C_6H_6)Cr(C_6H_6) + CH_3\overset{\overset{O}{\|}}{C}Cl \xrightarrow{AlCl_3}$$

[For the surprising answer, see P. L. Pauson, "Organometallic Chemistry," St. Martin's Press, New York, **1967**, p. 140.]

13.15. Discuss the difference between formulating $Cr(C_6H_6)_2$ as dibenzenechromium (a) or as bis(hexa*hapto*cyclohexatriene)chromium (b):

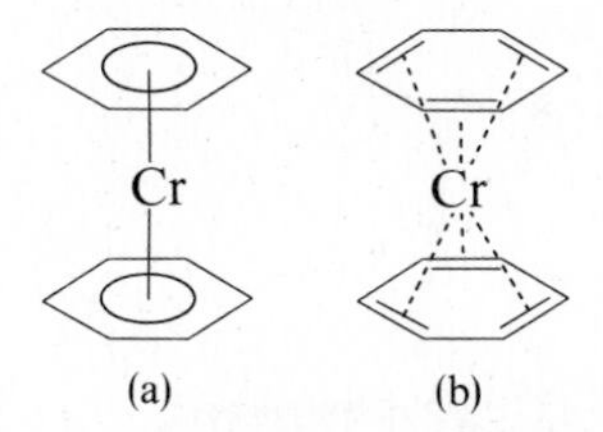

Do these represent different chemical species or a semantic problem? Suggest experimental methods that might distinguish between (a) and (b).

13.16. Treatment of $Co_2(CO)_8$ with lithium in diethyl ether yields red crystals of empirical composition $LiCo_3(CO)_{10}$ and evolution of carbon monoxide. The lithium compound exhibits three distinct types of carbonyl bands in the regions 2080–2000, 1850, and 1600 cm^{-1}. Suggest a possible structure. [S. A. Fieldhouse et al., *Chem. Commun.*, **1970**, 181.]

13.17. The following compounds are known: $(C_7H_7)Cr(C_5H_5)$, $(C_6H_6)Mn(C_5H_5)$, $(C_5H_5)_2Fe$, $(C_4H_4)Co(C_5H_5)$. Suppose you wished to make a compound which contained both a cyclopentadienyl ring and a three-membered ring. Suggest how you might attempt to synthesize such a compound. [M. D. Rausch et al., *J. Am. Chem. Soc.*, **1970**, *92*, 4981; R. M. Tuggle and D. L. Weaver, *Inorg. Chem.*, **1971**, *10*, 1504.]

13.18. Do the compounds $Rh_6(CO)_{16}$ and $Co_6(CO)_{16}$ obey the 18-electron Rule or not? What assumptions must be made? [E. R. Corey and L. F. Dahl, *J. Am. Chem. Soc.*, **1963**, *85*, 1202.]

13.19. The first edition of this book had the following sketch of tris(cyclopentadienyl)-titanium:

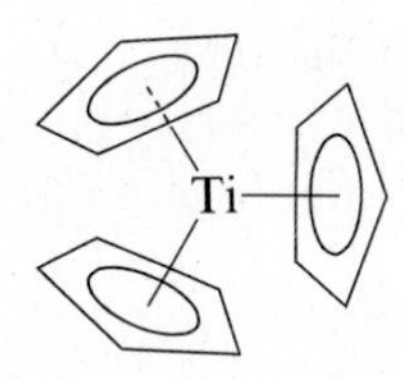

Taken literally, why is this perhaps a naive picture of this molecule? Suggest an appropriate modification. [C. R. Lucas, *Chem. Commun.*, **1973**, 97.]

13.20. Give an example of

a. an oxidative addition reaction

b. a reductive elimination

c. an insertion reaction

d. a fluxional molecule

e. homogeneous catalysis

13.21. Figure 13.48 illustrates the molecular structure of the $(C_5H_5)Rh(C_2F_4)(C_2H_4)$ molecule. Discuss the stoichiometry and geometry of this compound. [L. J. Guggenberger and R. Cramer, *J. Am. Chem. Soc.*, **1972**, *94*, 3779.]

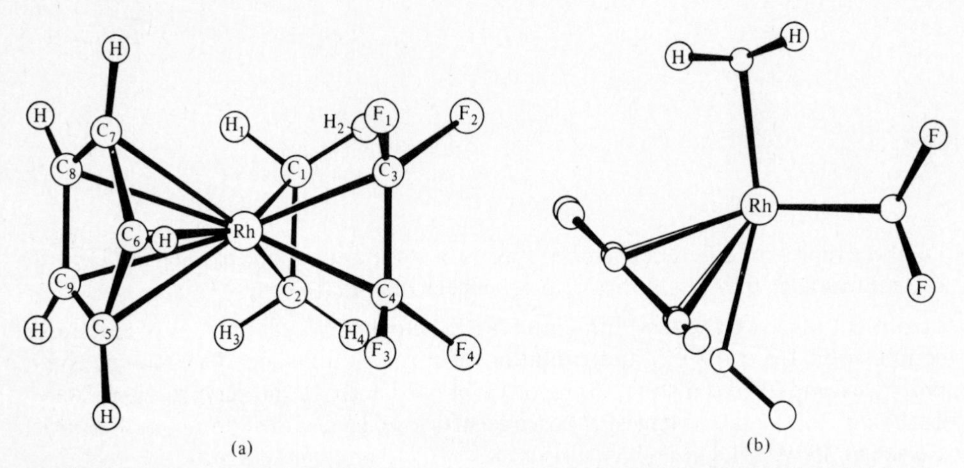

Fig. 13.48 The molecular structure of $(C_5H_5)Rh(C_2F_4)(C_2H_4)$. (a) Perspective view showing cyclopentadienyl and two olefin ligands. (b) View along an axis parallel to the C=C bonds. The atoms of the lower half of the molecule are thus eclipsed by the upper ones. Note that the fluorine atoms of C_2F_4 are bent further away from the metal than are the hydrogen atoms of C_2H_4. [From L. J. Guggenberger and R. Cramer, *J. Am. Chem. Soc.*, **1972**, *94*, 3779. Reproduced with permission.]

13.22. When one mole of $(\phi_3P)_2Pt(C_2H_4)$ is treated with two moles of BF_3, the ethylene is quantitively released and the BF_3 completely absorbed. The product consists of a single compound, which is monomeric in dichloromethane. Formulate the product and account for the bonding in it. [M. Fishwick et al., *Inorg. Chem.*, **1976**, *15*, 490.]

13.23. Upon encountering $Ni_3Cp_3(CO)_2$ for the first time, you might suppose that a typographical error had occurred. Why?

13.24. Metals in low oxidation states are usually strong reducing agents. Give an example of a metal in a zero oxidation state acting as an *oxidizing agent.*

13.25. It appears that $Ru_2(CO)_9$ and $Os_2(CO)_9$ are not isostructural with $Fe_2(CO)_9$. Suggest possibilities.

13.26. Convince yourself that the statement made on p. 601 concerning the interrelationships among M—M bond length, M—C bond length, and M—C—M bond angles is correct. A ruler and a protractor will help.

13.27. Draw the structures of all of the molecules shown in Eq. 13.49.

13.28. Formulate the cyclopentadienylnickel carbonyl shown in Fig. 13.23 as an alkyldiene-like structure (cf. Eq. 13.122). Discuss this structure in terms of isolobal fragments.

13.29. For each of the Tolman catalytic loops given for homogeneous catalysis:

a. Determine the effective atomic number of each metallic species.

b. Clearly identify: reactants, products, oxidative addition, reductive elimination, and insertion.

13.30. Consider the insertion reaction in Eq. 13.138. Would you consider the alkyl group better characterized as a carbocation or as a carbanion? Give reasons for your choice. [G. W. Parshall, "Homogeneous Catalysis," Wiley-Interscience, New York, **1980**, p. 16.]

13.31. Suggest the details (mechanism if possible) of what is happening at each numbered step in Eq. 13.142.

13.32. Using data given in this book, calculate the enthalpies of reaction for:

$$C_{(s)} + H_2O_{(g)} \longrightarrow CO_{(g)} + H_{2(g)} \qquad \textbf{(13.156)}$$

$$H_2O_{(g)} + CO_{(g)} \longrightarrow CO_{2(g)} + H_{2(g)} \qquad \textbf{(13.157)}$$

13.33. Where did the hydrogen atom come from in Eq. 13.48?

13.34. When the ferrocenophane discussed above is treated with butyllithium, a H^+ ion is abstracted from one of the methylenic bridges (Fig. 13.49). In contrast to the parent compound, *the ferrocenophanyl anion is not fluxional*. Discuss.

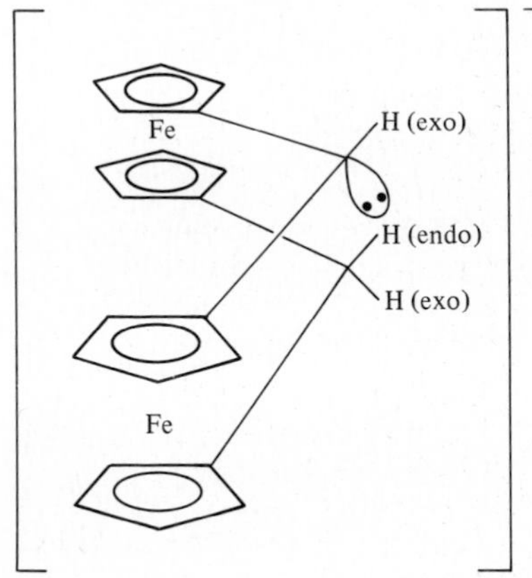

Fig. 13.49 Ferrocenophanyl carbanion. The anionic carbon is thought to be sp^3 hybridized with the lone pair directed as shown. [From R. Dagani, *Chem. Eng. News*, March 1, **1982**, *23*. Reproduced with permission.]

13.35. Figure 13.50 displays the infrared spectra of the mixtures obtained from the co-condensation of vanadium atoms and carbon monoxide molecules in a carbon monoxide or CO/rare gas matrix. What conclusions and speculations can you make from them? Carefully distinguish between your conclusions and your speculations.

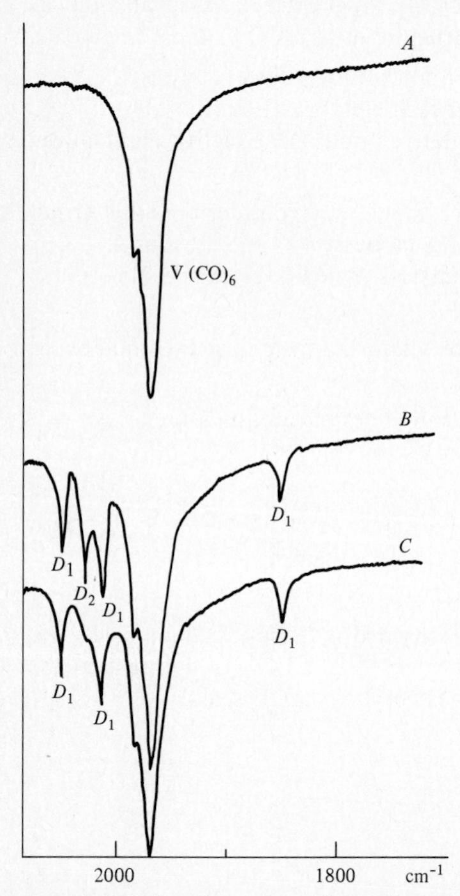

Fig. 13.50 Infrared spectrum of vanadium carbonyl matrices at 6–10 K: (a) at low vanadium concentration; the product here is $V(CO)_6$, which can be verified by producing it independently in other ways; (b) at high vanadium concentrations, conducive to V–V bond formation; (c) sample (b) warmed to 35 K. [From T. A. Ford et al., *Inorg. Chem.*, **1976**, *15*, 1666. Reproduced with permission.]

13.36. In this chapter examples of polynuclear metal carbonyls have been examined as well as simple metal carbonyl hydrides. Consider the complex polynuclear carbonyl hydride $H_2Os_3(CO)_{12}$. Rationalize the formulation of this species. From your application of the 18-electron Rule, what can you say about the structure of this molecule? How is it similar or different from the similar $Os_3(CO)_{12}$ shown in Fig. 13.4? (M. R. Churchill and H. J. Wasserman, *Inorg. Chem.*, **1980**, *19*, 2391.)

13.37. On p. 605 the somewhat casual statement was made: "The large difference in ionization constants for the two protons provided the first evidence that the hydrogen atoms in $H_2Fe(CO)_4$ were both bonded to the same atom. . . ." Present theoretical support for this line of thinking.

13.38. Why is such a large fraction of the products from ZSM-5 composed of alkenes and aromatics?

14

Inorganic chains, rings, cages, and clusters

From topics discussed previously in Chapters 4 to 13 it should be obvious that there is no sharp distinction between inorganic and organic chemistry. Nowhere is the borderline less distinct than in the compounds of the nonmetals. Some, such as the halides and oxides, are typical inorganic compounds, but others, such as compounds of nonmetals with organic substituents, are usually called the "organic compounds."[1] The situation is further complicated by the tendency of some nonmetals to resemble carbon in certain properties. This chapter discusses the chemistry of nonmetals in terms of one such property: their propensity to form chains, rings, and cages. Most metals show much less tendency to form compounds of this type, and the length of the chains and size of the rings thus formed are severely restricted.

CHAINS

Catenation

If there is a single feature of carbon which makes possible a unique branch of chemistry devoted to a single element,[2] it is its propensity to form extensive chains. This phenomenon is rare in the remainder of the periodic table, although the congeners of carbon and related nonmetals exhibit it to a reduced extent. Despite the fact that there appears to be no thermodynamic barrier to the formation of long-chain silanes, Si_nH_{2n+2}, their synthesis and

[1] In this area the nomenclature of the chemist is imprecise to say the least. Thus it is common to distinguish between "organophosphorus compounds," such as ϕ_3P, $Me_2P(S)P(S)Me_2$, and "inorganic phosphorus compounds," such as PCl_5 and $Na_3P_3O_9$. While the latter undoubtedly belong to the province of inorganic chemistry, a chemist interested in the former is as apt to consider himself an inorganic chemist as an organic chemist.

[2] This is a poor definition of organic chemistry, of course. Even the simplest organic compounds involve carbon and hydrogen, and inorganic chemists cannot resist pointing out that the most interesting parts of organic chemistry usually involve "inorganic" functional groups containing oxygen, nitrogen, sulfur, etc.

characterization is a formidable task. They are difficult to handle since they differ from paraffins in reactivity. Thus reactions 14.1 and 14.2 appear similar:

$$C_nH_{2n+2} + \frac{3n+1}{2}O_2 \longrightarrow nCO_2 + (n+1)H_2O \quad \textbf{(14.1)}$$

$$Si_nH_{2n+2} + \frac{3n+1}{2}O_2 \longrightarrow nSiO_2 + (n+1)H_2O \quad \textbf{(14.2)}$$

And in fact both are thermodynamically favored to proceed to the right. The important difference, not apparent in the stoichiometric equation, is the energy of activation which causes the paraffins to be kinetically inert in contrast to the reactive silanes.[3] Further complications with silanes arise from lack of convenient syntheses and difficulties in separation. Nevertheless, compounds from $n = 1$ to $n \simeq 8$ have been characterized, including both straight-chain and branched-chain compounds. Fluorinated and chlorinated long-chain compounds are known up to and including $Si_{10}Cl_{22}$ (which can be distilled under reduced pressure at 215–220° without decomposition). A further illustration that several factors other than inherent Si—Si bond strength are involved is the *greater* stability of silane derivatives when bulky substituents are present. For example, Si_2Br_6 may be distilled without decomposition at 265°C, but C_2Br_6 decomposes to C_2Br_4 and Br_2 at 200°C.[5] Further evidence of the inherent stability of silicon–silicon bonds is the isolation of a larger number of permethylcyclosilanes (see p. 722).

The chemistry of germanes is similar to that of silanes with moderate chain lengths reported. Heavier congeners of carbon, however, have severely restricted catenation properties. Distannane, Sn_2H_6, is known, although it is unstable. Plumbane, PbH_4, is of marginal stability itself, and hence a large number of heavier analogues is not expected, although the interesting compound $Pb(Pb\phi_3)_4$ has been synthesized.

Some other nonmetals such as nitrogen, phosphorus, and sulfur form chains of a few atoms in length, although in general their chemistry is relatively unimportant. The series of sulfanes, HS_nH, is most extensive, but chains with more than about a half-dozen atoms have not been well characterized.[6] Diphosphine, P_2H_4, is well known, but only recently has firm evidence of triphosphine, PH_2PHPH_2, and tetraphosphine, $PH_2PHPHPH_2$, been obtained.[7] These oligophosphines readily polymerize to yellow solids, which presumably contain polyphosphine chains:

$$P_2H_4 \longrightarrow \frac{1}{x}(PH)_x + PH_3 \quad \textbf{(14.3)}$$

---

[3] The difference in activation energy can be qualitatively described as the application of a match to Eq. 14.1 compared with merely allowing oxygen to contact silanes in Eq. 14.2. One author[4] has stated that one of the prerequisites for the synthesis of higher silanes is a large amount of courage.

[4] B. J. Aylett, *Adv. Inorg. Chem. Radiochem.*, **1968**, *2*, 249. This article may be consulted for further information on current work in silicon hydride chemistry.

[5] For further discussion of the stability of catenated silicon compounds, see A. G. MacDiarmid, in "New Pathways in Inorganic Chemistry," E. A. V. Ebsworth et al., eds., Cambridge University Press, London, 1968, Chap. 8.

[6] O. Foss, *Adv. Inorg. Chem. Radiochem.*, **1960**, *2*, 237.

[7] M. Baudler et al., *Z. Anorg. Allg. Chem.*, **1972**, *388*, 125.

but these have not been well characterized.[8] Oxygen forms no chains longer than three, and only a few compounds of this type are known; all of them are bis(perfluoroalkyl)trioxides such as CF_3OOOCF_3.[9]

Allotropes of both sulfur and selenium are known in which helical chains are present. Red phosphorus is thought to involve long chains, but its exact structure is not known.

The halogens are known to form reasonably stable chains in polyhalide anions, the best known example being the triiodide ion, I_3^-. They are considered separately in Chapter 15. A great number of nonmetals form simple X—X bonded molecules: B_2Cl_4, N_2H_4, P_2I_4, As_2R_4, HOOH, ClSSCl, X_2 (halogens), etc. In general these are relatively stable, although all are susceptible to attack by reagents (even catalysts) that can cleave the X—X bond:

$$R_2AsAsR_2 + X_2 \longrightarrow 2R_2AsX \quad (14.4)$$

$$Cl_2BBCl_2 + CH_2{=}CH_2 \longrightarrow Cl_2BCH_2CH_2BCl_2 \quad (14.5)$$

$$Cl_2 + OH^- \longrightarrow ClOH + Cl^- \quad (14.6)$$

$$HOOH + [H] \text{ (reducing agents)} \longrightarrow 2H_2O \quad (14.7)$$

$$H_2PPH_2 \xrightarrow[NH_3]{\text{liq.}} PH_3 + (PH)_x \text{ (approximate)} \quad (14.8)$$

$$H_2NNH_2 + O_2 \xrightarrow{[Cu^{+2}]} N_2 + 2H_2O \quad (14.9)$$[10]

Finally, there is an extensive chemistry of metal–metal bonds, developed only recently, that is sufficiently interesting to warrant separate treatment (p. 740).

Heterocatenation

Although there is a paucity of inorganic compounds exhibiting true catenation, the phenomenon of *heterocatenation*, or chains built up of alternating atoms of different elements, is quite widespread. The simplest are those formed by the dehydration of acids or their salts:

$$2\left[O{=}\underset{\substack{|\\O^-}}{\overset{\substack{O^-\\|}}{P}}{-}OH\right] \xrightarrow{\text{heat}} \left[O{=}\underset{\substack{|\\O^-}}{\overset{\substack{O^-\\|}}{P}}{-}O{-}\underset{\substack{|\\O^-}}{\overset{\substack{O^-\\|}}{P}}{=}O\right] + H_2O \quad (14.10)$$

Such dehydrations were first effected by the action of heat on the simple acid phosphate salts and hence the resulting product was termed pyrophosphate (Gr. πυρ, fire). With an increased number of *polyphosphates* ($P_nO_{3n+1}^{-n-2}$) known, however, the preferred nomenclature has become *diphosphate* ($n = 2$), *triphosphate* ($n = 3$), etc. For many heterocatenated

[8] See R. B. Callen and T. P. Fehlner, *J. Am. Chem. Soc.*, **1969**, *91*, 4122, and references cited therein.

[9] F. A. Hohorst et al., *J. Am. Chem. Soc.*, **1973**, *95*, 3866.

[10] The formation of N_2 rather than cleavage as in the other examples can be attributed to the extremely strong triple bond in molecular nitrogen.

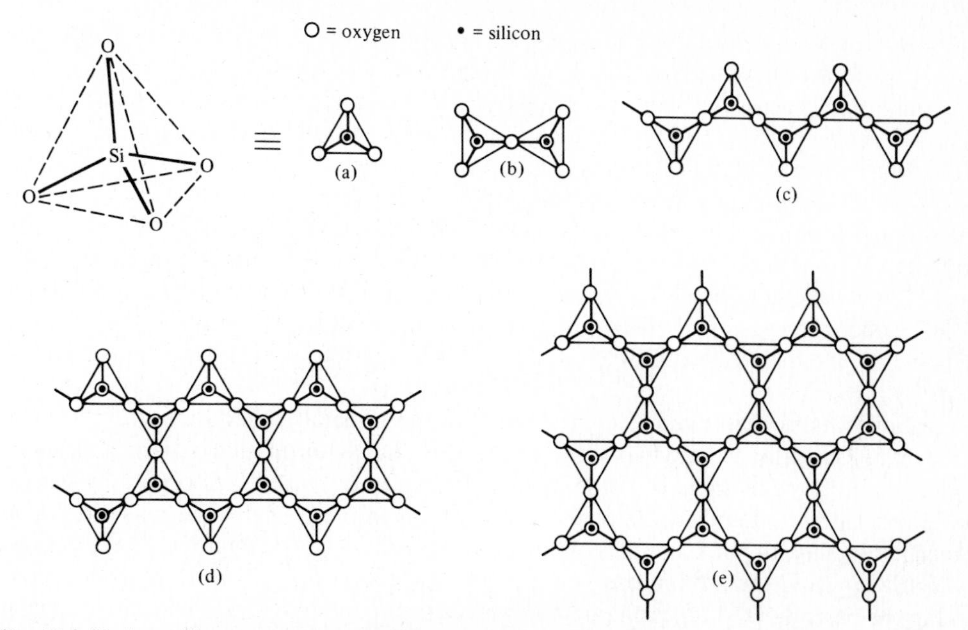

Fig. 14.1 Various silicate structures: (a) SiO_4 tetrahedron. When carrying a -4 charge, this is the orthosilicate ion. (b) The disilicate anion. (c) Portion of an infinite single chain, $(SiO_3)_n^{-2n}$. (d) Portion of an infinite double chain or band, $(Si_4O_{11})_n^{-6n}$. (e) Portion of a sheet or layer structure, $(Si_2O_5)_n^{-2n}$.

compounds there are simpler synthetic routes than the thermal elimination of water:

$$H_2SO_4 + SO_3 \longrightarrow HO-\overset{\overset{\displaystyle O}{\|}}{\underset{\underset{\displaystyle O}{\|}}{S}}-O-\overset{\overset{\displaystyle O}{\|}}{\underset{\underset{\displaystyle O}{\|}}{S}}-OH \qquad \textbf{(14.11)}$$

$$5\,Na_3PO_4 + P_4O_{10} \xrightarrow[\text{cool.}]{\text{melt,}} 3\,Na_5P_3O_{10} \qquad \textbf{(14.12)}$$

Condensed polyphosphates such as sodium triphosphate are of great industrial importance since they are currently used in large tonnages as "builders" in the manufacture of detergents. As such they function to adjust the pH and to complex water-hardening ions such as Ca^{+2} and Mg^{+2}.[11]

Silicon forms a very large number of compounds containing heterocatenated anions. These are of great importance in the makeup of various minerals since about three-fourths of the earth's crust is silicon and oxygen. Simple silicate anions, SiO_4^{-4} ("orthosilicates"; Fig. 14.1a), are not common in minerals, although they are present in olivine, $(Mg, Fe)_2SiO_4$,

[11] Sodium triphosphate is currently the largest tonnage chemical used in the manufacture of detergents. A high-molecular-weight phosphate known as Graham's salt (often erroneously called "hexametaphosphate") is used as the water softener Calgon. For further discussion of polyanions of this type, see E. Thilo, *Adv. Inorg. Chem. Radiochem.*, **1962**, *4*, 1, and J. R. Van Wazer, "Phosphorus and Its Compounds," Wiley (Interscience), New York, **1958**, Vols, I., II. The second volume of the latter is devoted to the industrial chemistry of phosphorus.

an important constituent of basalt, which in turn is the most voluminous of the extrusive rocks formed from outpouring of magma. Other minerals containing discrete[12] orthosilicate ions are phenacite (Be_2SiO_4), willemite (Zn_2SiO_4), and zircon ($ZrSiO_4$). The large class of *garnets* is composed of minerals of the general formula $M_3^{II}M_2^{III}(SiO_4)_3$, where M^{II} can be Ca^{+2}, Mg^{+2}, or Fe^{+2}, and M^{III} is Al^{+3}, Cr^{+3}, or Fe^{+3}.

Minerals containing the pyrosilicate or disilicate anion, $Si_2O_7^{-6}$ (Fig. 14.1b), are not common, although some are known: thortveitite, $Sc_2Si_2O_7$, hemimorphite, $Zn_4(OH)_2Si_2O_7$, as well as vesuvianite and epidote which contain both SiO_4^{-4} and $Si_2O_7^{-6}$ ions. The next higher order of complexity consists of the so-called metasilicate anions, which are cyclic structures[13] of general formula $(SiO_3)_n^{-2n}$, occurring in benitoite, $BaTiSi_3O_9$; catapleite, $Na_2ZrSi_3O_9 \cdot 2H_2O$; dioptase, $Cu_6Si_6O_{18} \cdot 6H_2O$; and beryl, $Be_3Al_2Si_6O_{18}$.

Infinite chains of formula $(SiO_3)_n^{-2n}$ are found in minerals called *pyroxenes*. In these chains the silicon atoms share two of the four tetrahedrally coordinated oxygen atoms with adjacent silicon atoms (Fig. 14.1c). If further sharing of oxygen atoms occurs by half of the silicon atoms, a double chain or band structure is formed. This is the structure found in *amphiboles* (Fig. 14.1d). Examples of the former include enstatite, $MgSiO_3$, and diopside, $CaMg(SiO_3)_2$, and the lithium ore, spodumene. Amphiboles are more complicated, containing the basic $Si_4O_{11}^{-6}$ repeating unit as well as metal and hydroxide ions, for example, tremolite, $Ca_2Mg_5(OH)_2(Si_4O_{11})_2$.

Further linkage by the complete sharing of three oxygen atoms per silicon (analogous to edge bonding between many amphibole bands) results in layer or sheet structures (Fig. 14.1e). This yields an empirical formula of $(Si_2O_5)_n^{-2n}$. Many important minerals have silicate sheet structures, such as *kaolin*, $Al_2(OH)_4Si_2O_5$; *talc*, $Mg_3(OH)_2Si_4O_{10}$; and the micas: *biotite*, $K(Mg, Fe)_3(OH)_2(AlSi_3O_{10})$, and *muscovite*, $KAl_2(OH)_2(AlSi_3O_{10})$. The latter exemplify a common phenomenon in silicate minerals: the replacement of one or more silicon atoms by aluminum. As long as electroneutrality is maintained by the concomitant adjustment of the number and/or charges of the cations, the structures remain perfectly stable.

The ultimate in cross-linking and sharing of oxygen atoms by silicon is the complete sharing of all four oxygen atoms per SiO_4 tetrahedron in a framework structure. This results in a formula of $(SiO_2)_n$ or silicon dioxide. Silicon dioxide can exist in several forms such as quartz (thermodynamically stable at room temperature), tridymite, and cristobalite, as well as dense varieties formed under high pressure. All of these contain silicate tetrahedra with complete sharing but with different linking arrangements of the tetrahedra. Finally, silicon dioxide also occurs as a glass with disordering of the tetrahedra so that no long-range order exists.

---

[12] These minerals contain "discrete" silicate anions in the sense that there are no Si—O—Si—O—Si—O chains. There is considerable covalent character in the metal–silicate bond, however, and therefore these compounds cannot be considered $2M^{+2}[SiO_4]^{-4}$ analogues of, for example, $M^+[ClO_4]^-$. In some ways they resemble chain (or higher-structured) silicates in which some of the silicon atoms have been replaced by other metal atoms.

[13] The cyclic silicates are considered later in the section on inorganic ring systems.

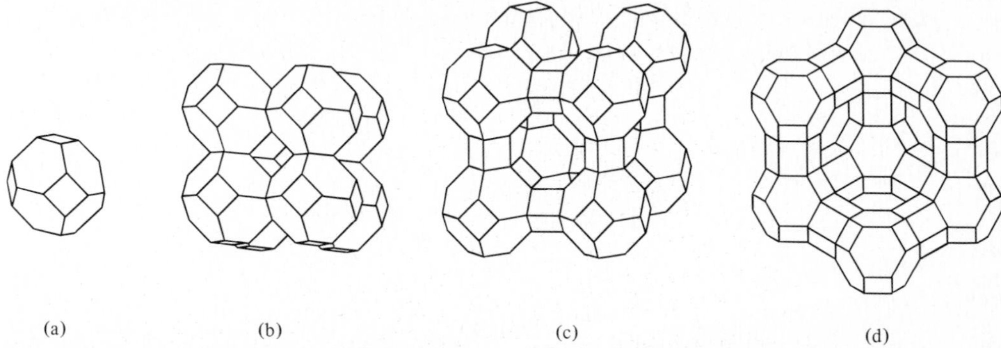

Fig. 14.2 Some typical zeolite structures: (a) a sodalite unit; (b) sodalite (260 pm); (c) zeolite A (420 pm); and (d) faujasite (1140 pm). Lines represent oxygen bridges; intersections of lines show the positions of silicon or aluminum atoms. Note increasing size of pore apertures (diameters given in parentheses). [From D. A. Whan, *Chem. Brit.* **1981**, *17*, 532. Reproduced with permission.]

If aluminum atoms are substituted for some of the silicon atoms, the framework is no longer electrically neutral but is anionic, and the charge distribution must be compensated for by corresponding cations. Three classes of aluminosilicate framework minerals are of importance: the feldspars, zeolites, and ultramarines. The feldspars, of general formula $M(Al, Si)_4O_8$, are the most important rock-forming minerals, comprising some two-thirds of igneous rocks, such as granite, which is a mixture of quartz, feldspars, and micas. Those in which M is a larger ion, such as K^+ or Ba^{+2}, crystallize in the monoclinic system: *orthoclase*, $K(AlSi_3O_8)$, and *celsian*, $Ba(Al_2Si_2O_8)$. The triclinic or plagioclase feldspars contain smaller ions such as Na^+ and/or Ca^{+2}. Examples are albite, $Na(AlSi_3O_8)$, and anorthite, $Ca(Al_2Si_2O_8)$. These are practically isomorphous, and intermediate compositions are known in which Ca^{+2} and Al^{III} replace Na^+ and Si^{IV}.

The zeolites are aluminosilicate framework minerals of general formula $M_{x/n}^{+n}[Al_xSi_yO_{2x+2y}]^{-x} \cdot zH_2O$. They are characterized by their open structures that permit exchange of cations and water molecules (Fig. 14.2). Both natural and synthetic zeolites find wide application as cation exchangers since the ions can migrate rather freely through the open structure. Some cations will fit more snugly in the cavities than others. In addition, certain zeolites may behave as molecular sieves if the water adsorbed in the cavities is completely removed. Various uncharged molecules such as CO_2, NH_3, and organic compounds can be selectively adsorbed in the cavities depending upon their size. The zeolite framework thus behaves in some respects similarly to a clathrate cage except that the adsorbed molecules must be capable of squeezing through the preformed opening rather than having the cage formed about them. Molecules small enough to enter, yet large enough to fit with reasonably large dipolar and London forces will be selectively adsorbed.

Zeolites may also behave as acidic catalysts. The acidity may be of the Brønsted type if hydrogen ions are exchanged for mobile cations (such as Na^+) by washing with acid. If the zeolite is heated, water may then be eliminated from the Brønsted sites leaving aluminum atoms coordinated to only three oxygen atoms:

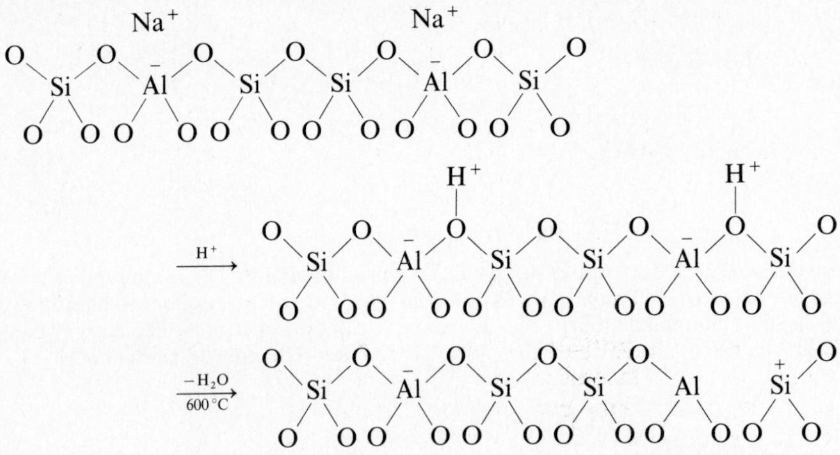

```
          Na+                       Na+
 O     O  -    O     O     O  -  O     O
  \   / \     / \   / \   / \   / \   /
   Si    Al      Si    Si    Al    Si
  / \   / \     / \   / \   / \   / \
 O   O O   O   O   O O   O O   O O   O

                          H+                          H+
                          |                           |
                O      O  -   O      O      O   -     O      O
          H+     \    / \    / \    / \    / \       / \    /
        ----->    Si     Al     Si     Si     Al       Si
                 / \    / \    / \    / \    / \      / \
                O   O  O   O  O   O  O   O  O   O    O   O

                O      O  -   O      O      O            +  O
         -H2O    \    / \    / \    / \    / \              /
        ----->    Si     Al     Si     Si     Al         Si
        600 °C   / \    / \    / \    / \    / \        / \
                O   O  O   O  O   O  O   O  O   O      O   O   (14.13)[14]
```

The latter will act as Lewis acids. These catalytic sites occur at high density and are uniform in their activity (as opposed to amorphous solids) because of the microcrystalline nature of the zeolites. If cations of high charge such as Ce^{+3} are exchanged for Na^{+}, high electric field gradients will be present that can activate molecules. It has even been suggested that zeolites may be considered as solid ionizing solvents[15] (see p. 191).

We have seen previously the shape-selective catalysis of ZSM-5 in the conversion of methanol to gasoline (p. 663). The high performance of ZSM-5 catalyst has been attributed to its high acidity and to the peculiar shape, arrangement, and dimensions of the channels. It may be that the reactants enter by one type of channel and the products exit by another.[16] This would facilitate the flow of reactants and products through the channels with little "back pressure."

Another class of framework aluminosilicates consists of the ultramarines. They are characterized by an open framework and intense colors. They differ from the previous examples by having "free" anions and no water in the cavities. Ultramarine blue (lapis lazuli) probably contains the radical anion, S_3^-, and ultramarine green both S_3^- and S_2^-. Although these two anions occur in ultramarine red, the characteristic color is due to a third species, perhaps S_4.[17] Structurally related, but colorless, minerals such as sodalite (containing chloride anions) and noselite (containing sulfate anions) are sometimes included in the broad category of ultramarines.

---

[14] D. A. Whan, *Chem. Brit.*, **1981**, *17*, 532.

[15] J. A. Rabo et al., *Acta Physica et Chemica.*, **1978**, *24*, 39; J. A. Rabo, in "Catalysis by Zeolites," B. Imelik et al., eds., Elsevier, Amsterdam, **1980**, pp. 341–342.

[16] A. Auroux et al., in "Proceedings of the Fifth International Conference on Zeolites," L. V. C. Rees, ed., Heyden, London, **1980**, pp. 433–439; C. Naccache and Y. Ben Taarit, *ibid.*, pp. 592–606; E. G. Derouane, in "Catalysis by Zeolites," B. Imelik et al., eds., Elsevier, Amsterdam, **1980**, pp. 5–18.

[17] R. J. H. Clark and D. G. Cobbold, *Inorg. Chem.*, **1978**, *17*, 3169.

The study of silicaceous minerals is an important one not only with respect to better understanding of the conditions of formation and their relation to the geochemistry of these minerals, but also with respect to the structural principles and the synthesis of new structures not found in nature (synthetic zeolites and ultramarines).[18]

In all of the silicates discussed above, the sharing of oxygen atoms between tetrahedra is by an apex only:

$$O_3Si{-}O{-}SiO_3$$

No cases are known in which edges or faces are shared:

$$O_2Si(\mu\text{-}O)_2SiO_2 \qquad O{-}Si(\mu\text{-}O)_3Si{-}O$$

Pauling[19] has listed a set of rules based on an ionic model. Although no one now accepts a purely electrostatic model for silicates and similar compounds, Pauling's rules are still reasonably accurate as long as the partial charges on the atoms are sufficiently large to make electrostatic repulsions significant. Such repulsions militate against the sharing of edges or faces by tetrahedra since this places the positive centers too near each other.

Intercalation chemistry[20]

Intercalation compounds consist of layers ("sandwiches") of different chemical species. The name comes from that describing the insertion of extra days (such as February 29th) into the calendar to make it match the solar year. Most work on intercalation compounds has been on synthetic systems in which atoms, ions, or molecules have been inserted between layers of the host material. However, some aluminosilicates that we have encountered above provide useful examples. Thus talc and micas form layered structures with ions between the silicate sheets (Fig. 14.3). Some minerals, including *all* clays, have water molecules intercalated between the framework sheets. In some, such as vermiculite, the water may rapidly and dramatically be evacuated by heating. The water molecules leave faster than they can diffuse along the layers—exfoliation occurs. The result is the familiar expanded vermiculite used as a packing material and as a potting soil conditioner.

---

[18] For a more thorough discussion of silicate minerals, see A. F. Wells, "Structural Inorganic Chemistry," 4th ed., Oxford University Press, London, **1975**, Chap. 23. Wells also includes stereoscopic photographs of models illustrating the packing of the silicate tetrahedra.

[19] For a complete discussion see L. Pauling, "The Nature of the Chemical Bond," 3rd ed., Cornell University Press, Ithaca, N. Y., **1960**, pp. 544–562.

[20] M. S. Whittingham and M. B. Dines, *Surv. Progr. Chem.*, **1980**, *9*, 55.

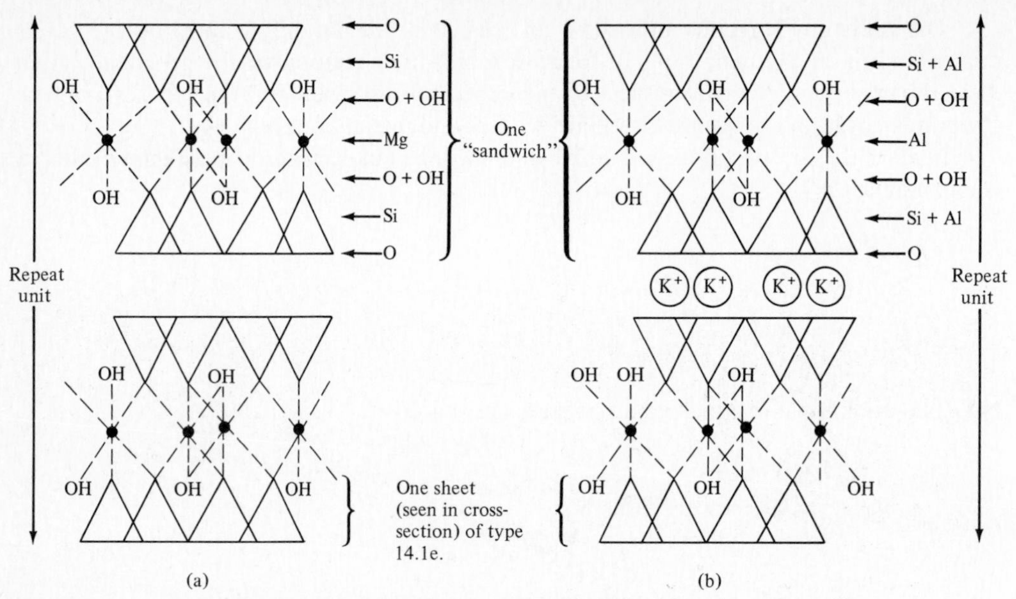

Fig. 14.3 Layered silicate structures: (a) Talc, $Mg_3(OH)_2(Si_4O_{10})$; (b) a mica, muscovite, $KAl_2(OH)_2(Si_3AlO_{10})$.[*Note*: (1) Electroneutrality is maintained by balance of K(I), Mg(II), Al(III), and Si(IV). (2) The repeating layers in muscovite are bound together by the K^+ cations.] [From D. M. Adams, "Inorganic Solids," Wiley, New York, **1974**.]

Another example of this type of intercalation compound is sodium beta alumina where the sodium ions are free to move between the spinel layers (see p. 194). The sodium ions can be replaced by almost any +1 cation such as: Li^+, K^+, Rb^+, Cs^+, NH_4^+, H_3O^+, Tl^+, Ga^+, NO^+, etc., though the conductivity varies with the size of the ions moving between the fixed-distance (Al—O—Al) layers.

Graphite is perhaps the simplest layered structure. The intralayer C—C distance (142 pm) is twice the covalent radius of aromatic carbon (cf. 139 pm in benzene) and the interlayer C—C distance is 335 pm, twice the van der Waals radius of carbon. The sheets are held together by weak van der Waals forces. Many substances can be intercalated between the layers of graphite, but one of the longest known and best studied is potassium. Potassium can be intercalated into graphite until a limiting formula of C_8K is reached. This is known as the *first stage compound*. The earlier, lower stages have the general formula of $C_{12n}K$. The stages form stepwise as new layers of potassium are added, giving well characterized compounds with $n = 4, 3, 2$ (Fig. 14.4a–c). The final step (yielding Stage 1) includes filling in all of the remaining available sites (Fig. 14.5) in addition to forming the maximum number of layers (Fig. 14.4d). Presumably further intercalation cannot take place because of electrostatic repulsion.

Upon intercalation, the layers move apart somewhat (205 pm), though less than expected as estimated from the diameter of the potassium ion (304 pm or greater). This indicates that the K^+ ion "nests" within the hexagonal carbon net, and one can even speculate about weak complexing from the π cloud.

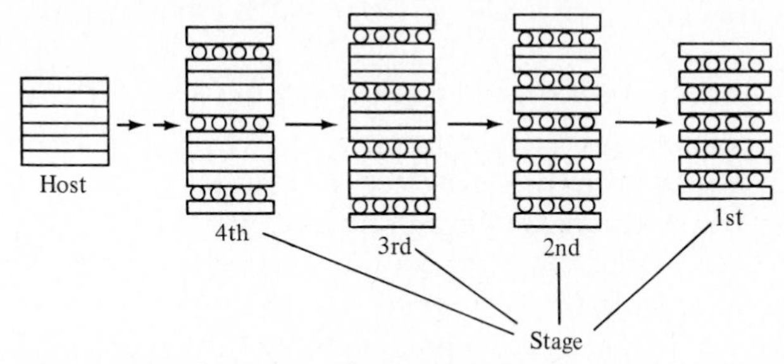

Fig. 14.4 Staging in graphite intercalation compounds, $C_{12n}K$. Addition of potassium proceeds through $n = 4, 3, 2 \ldots$ to the limit in stage 1: $C_{12}K \rightarrow C_8K$. [From M. S. Whittingham and M. B. Dines, *Surv. Progr. Chem.*, **1980**, *9*, 55.]

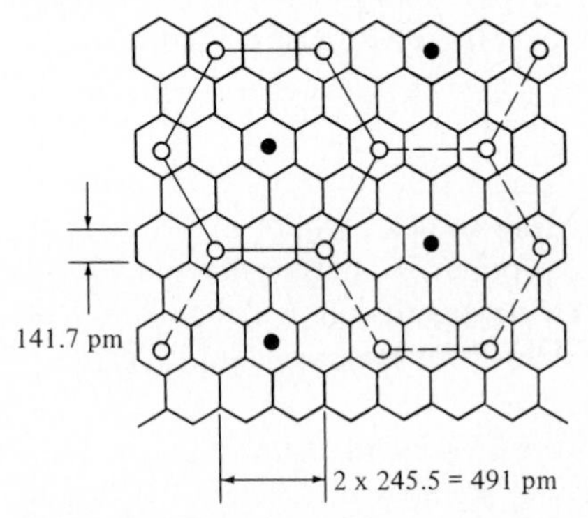

Fig. 14.5 Filling of available hexagonal sites in each layer of graphite: (1) For the limit of C_8K, K = ● + ○; (2) for $C_{12n}K$, K = ○ only. [From M. S. Whittingham and M. B. Dines, *Surv. Progr. Chem.*, **1980**, *9*, 55.]

The mention of the K^+ ion presupposes knowledge of the nature of the potassium species present. Because of the similarity of energy of the valence and conduction bands (see p. 199), graphite can be either an electron donor or acceptor. Intercalation of potassium atoms into graphite results in the formation of K^+ ions and free electrons in the conduction band. Graphite will react with an electron acceptor such as bromine to form C_8Br in which electrons have transferred from the valence band of the graphite to the bromine. Apparently, simple bromide ions are not formed but polybromide chains instead (see Chapter 15 for similar polyhalide chains). The expected Br—Br distance in such a polyhalide[21] (254 pm) compares well with the repeating hexagons of the graphite (256 pm). In contrast, the expected Cl—Cl (224 pm) and I—I (292 pm) distances do not fit well with the graphite structure, and, in fact these two halogens do *not* intercalate into graphite! In contrast, iodine monochloride (expected to average 255 pm, I—Cl) does.

Another example where graphite donates an electron is provided by nitryl salts such as the tetrafluoroborate, hexafluorophosphate, and hexafluoroantimonate. If these compounds are dissolved in an inert solvent such as nitromethane and added to graphite, a

[21] A. F. Wells, footnote 18, p. 336.

red-brown gas, NO_2 is soon evolved:

$$NO_2^+X^- + C_x \longrightarrow NO_2\uparrow + C_x^+X^- \qquad (X^- = BF_4^-, PF_6^-, \text{and } SbF_6^-) \qquad \textbf{(14.14)}$$

An electron from the valence band of graphite has reduced the nitryl ion. The anions from the original salt intercalate in the graphite layers to preserve electrical neutrality.[22]

All three of these intercalation compounds are good conductors of electricity. In the potassium intercalant, the electrons in the conduction band can carry the current directly, as in a metal. In the compounds of graphite with polybromide and fluoro anions, the holes in the valence band can conduct by the mechanism discussed previously (see "semiconductors," Chapter 4).

There is an unusual hetero chain, $(SN)_x$, discovered in 1910 but only recently receiving detailed attention. Interest in it centers on the fact that although it is composed of atoms of two nonmetals, polymeric sulfur nitride (also called polythiazyl) has some physical properties of a *metal*. The preparation is from tetrasulfur tetranitride (see p. 716):

$$S_4N_4 \xrightarrow{Ag} S_2N_2 \longrightarrow (SN)_x \qquad \textbf{(14.15)}$$

The S_4N_4 is pumped in a vacuum line over silver wool at 220 °C, and S_2N_2 is collected on a cold finger at −195 °C. It is then sublimed to a trap at 0 °C, where it polymerizes slowly to a lustrous golden material.[23] The resulting product is analytically pure, as is necessary for it to show metallic properties to a high degree: it has a conductivity of 25 $(\Omega\ m)^{-1}$ at room temperature compared with 104 $(\Omega\ m)^{-1}$ for mercury, and it becomes a superconductor at low temperatures.[24]

X-ray diffraction studies show that the SN chains have the structure shown in Fig. 14.6. This chain can be generated from adjacent square planar S_2N_2 molecules. The S—N bonds in the latter have bond order $1\frac{1}{2}$ and a bond length of 165.4 pm intermediate between single (174 pm) and double (154 pm) sulfur–nitrogen bonds. A free radical mechanism has been suggested[23] leading to the linear chains of the polymer (Fig. 14.7). Since polymerization can take place with almost no movement of the atoms, the starting material and product are pseudomorphs and the crystallinity of the former is maintained.[25]

If one attempts to draw a unique Lewis structure for the $(SN)_x$ chain, one is immediately frustrated by the odd number of electrons available. Many resonance structures can be drawn and they contribute to the hybrid structure, but the single structure:

```
 ·      ·
:S=N: :S=N:
 |   |  |   |  |
    :S=N: :S=N:
     ·      ·
```

illustrates many features: a conjugated single-bond–double–bond resonance system with *nine* electrons on each sulfur atom rather than a Lewis octet; every S—N unit will thus

[22] W. C. Forsman and H. E. Mertwoy, *Synth. Metals*, **1980**, *2*, 171; D. Billaud et al., *ibid.*, **1980**, *2*, 177.

[23] C. M. Mikulski et al., *J. Am. Chem. Soc.*, **1957**, *97*, 6358.

[24] R. L. Greene et al., *Phys. Rev. Lett.*, **1975**, *34*, 577.

[25] As opposed to the more commonly observed result of solid state reactions: beautiful crystals turning into an amorphous powder.

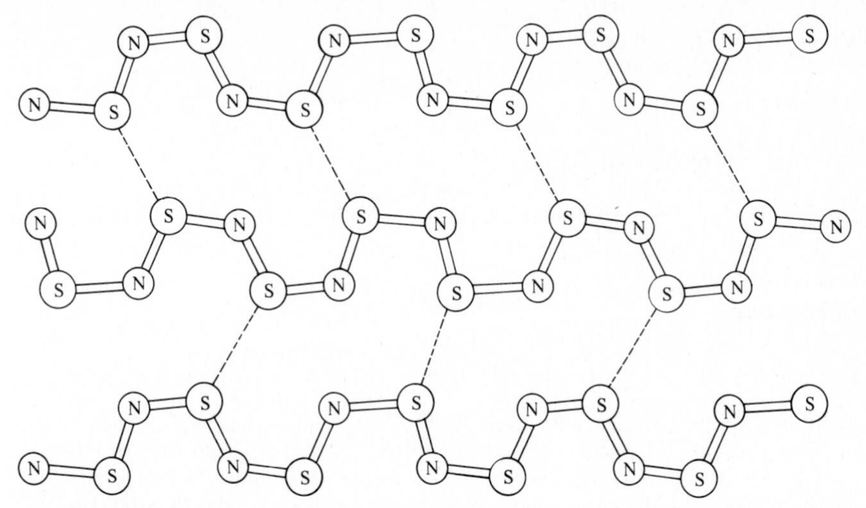

Fig. 14.6 $(SN)_x$ chains in one layer of polymeric sulfur nitride. [From A. G. MacDiarmid et al., in "Inorganic Compounds with Unusual Properties," R. B. King, ed., Advances in Chemistry Series No. 150, **1976.** Reproduced with permission.]

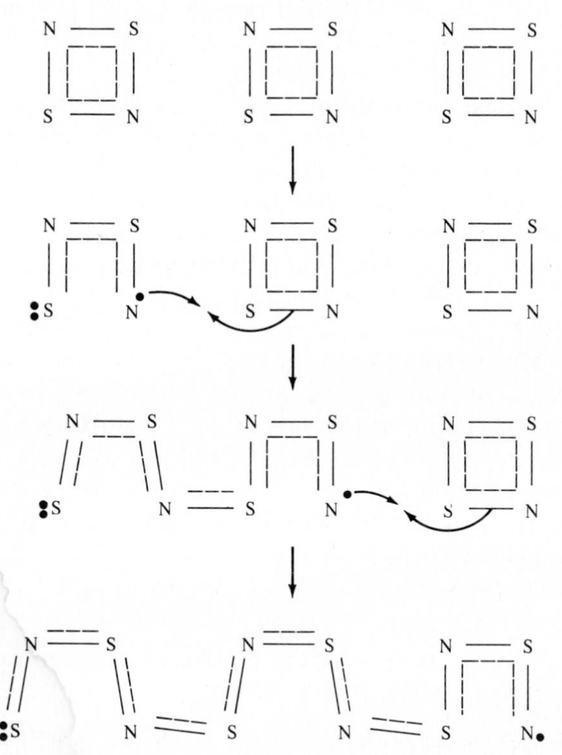

Fig. 14.7 Polymerization of S_2N_2 to form $(SN)_x$ chains with minimal movement of atoms.

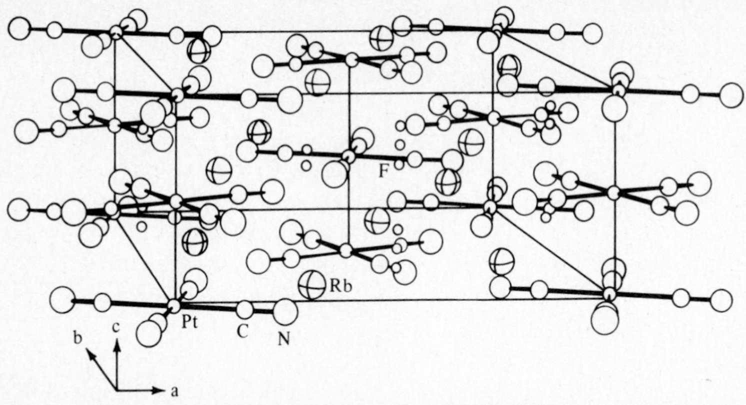

Fig. 14.8 Perspective view of the unit cell of $Rb_2[Pt(CN)_4]_{0.40}$. The one-dimensional chains of staggered $Pt(CN)_4^{-2}$ ions occupy the corners and the center of the unit cell. The triad of small circles represents the partially occupied positions of the FHF^- ions. Note the very short Pt-Pt distance (279.8 pm). [From A. J. Schultz et al., *Inorg. Chem.*, **1977**, *16*, 2129. Reproduced with permission.]

have one antibonding, π^*, electron. The half-filled, overlapping π^* orbitals will combine to form a half-filled conduction band in much the same way as we have seen half-filled 2*s* orbitals on a mole of lithium atoms form a conduction band (see pp. 195–197). Note however, that this conduction band *will lie only along the direction of the* $(SN)_x$ *fibers*; the polymer is a "one-dimensional metal."[26]

In many ways similar in one-dimensional conductivity properties are the "stacked" columnar complexes typified by $[Pt(CN)_4]^{-2}$. In such an arrangement of closely stacked square planar ions, there will be considerable interaction among the d_{z^2} orbitals of the platinum atoms. These orbitals are normally filled with electrons, so in order to get a conduction band some oxidation (removal of electrons) must take place. This may readily be accomplished by adding a little elemental chlorine or bromine to the pure tetrachloroplatinate salt to get stoichiometries such as $K_2Pt(CN)_4Br_{0.3}$, in which the platinum has an average oxidation state of +2.3. The oxidation may also be accomplished electrolytically, as in the preparation of $Rb_2Pt(CN)_4(FHF)_{0.4}$. The latter compound (Fig. 14.8) has the shortest known Pt–Pt separation for compounds of this type. The Pt–Pt distance is only 280 pm, almost as short as that found in platinum metal itself (277 pm).[27] Gold-bronze materials of this type were discovered as early as 1842, though they have been little understood until recently. The complexes behave not only as one-dimensional conductors, but their electrical properties are considerably more complex. Detailed discussion of the many interesting aspects of these materials is beyond the scope of this book, but fortunately there are recent and thorough reviews of the subject.[28]

---

[26] A. G. MacDiarmid et al., in "Inorganic Compounds with Unusual Properties," R. B. King, ed., Advances in Chemistry Series, No. 150, American Chemical Society, Washington, D. C., **1976**, pp. 63–72.

[27] A. J. Schultz et al., *Inorg. Chem.*, **1977**, *16*, 2129.

[28] B. M. Hoffman et al., in "Extended Linear Chain Compounds," J. S. Miller, ed., Plenum, New York, **1982**; J. M. Williams et al., ibid., Chapter 3; J. A. Ibers et al., *Struct. Bond.*, **1982**, *50*, 3.

Isopoly anions

Transition metals in their higher oxidation states are formally similar to nonmetals with corresponding group numbers: V^V and P^V in VO_4^{-3} and PO_4^{-3}, Cr^{VI} and S^{VI} in CrO_4^{-2} and SO_4^{-2}, Mn^{VII} and Cl^{VII} in MnO_4^- and ClO_4^-. The analogy may be extended to polyanions, such as dichromate, $Cr_2O_7^{-2}$; however, the differences in behavior between the metal anions and the nonmetal anions are often more important than their similarities. Whereas polyphosphoric acids and polysulfuric acids form only under rather stringent dehydrating conditions, polymerization of some metal anions occurs spontaneously upon acidification. For example, the chromate ion is stable only at high pHs. As the pH is lowered, protonation and dimerization occur:

$$CrO_4^{-2} + H^+ \longrightarrow (HO)CrO_3^- \tag{14.16}$$

$$(HO)CrO_3^- + H^+ \longrightarrow (HO)_2CrO_2 \tag{14.17}$$

$$2(HO)_2CrO_2 \longrightarrow Cr_2O_7^{-2} + H_2O + 2H^+ \tag{14.18}$$

Treatment with concentrated sulfuric acid completes the dehydration process and red chromium(VI) oxide ("chromic acid") precipitates:

$$\frac{n}{2}Cr_2O_7^{-2} + nH^+ \longrightarrow (CrO_3)_n + \frac{n}{2}H_2O \tag{14.19}$$

The structure of CrO_3 consists of infinite linear chains of CrO_4 tetrahedra.

Other metals such as vanadium have more complicated chemistry. The vandate ion, VO_4^{-3}, exists in extremely basic solution. As the pH is lowered protonation and dehydration occur spontaneously to form divanadate and trivanadate ions.[29] Further polymerization occurs until hydrous V_2O_5 precipitates as the pH passes the neutral point. The precipitation of vanadium(V) oxide from aqueous solution as well as the similar behavior of other metal oxides, such as MoO_3 and WO_3, stands in sharp contrast to the extremely hygroscopic behavior of the analogous nonmetal compounds P_2O_5 and SO_3.

The polymerization of vanadate, molybdate, and tungstate ions forming isopolyanions has received a great deal of attention. Early in the condensation process the coordination number of the metals changes from 4 to 6, and the basic building unit in the polymerization process is an octahedron of six oxygen atoms surrounding each metal atom.[29] Unlike the case of linked tetrahedra, the resulting octahedra may link by sharing either an apex or edge (rarely a face) as a result of the relaxation of electrostatic repulsions in the larger octahedra. As a result, the structures tend to be small clusters of octahedra in the discrete polyanions, culminating in infinite structures in the oxides. When the sharing of edges takes place, the structure may be stabilized (relative to electrostatic repulsions) if some distortion occurs such that the metal atoms move away from each other. As the polymerization increases, it becomes more and more difficult to have all of the metal atoms capable of moving to accomplish this reduction in electrostatic repulsion. Ultimately the sharing of

---

[29] The nature of these species is still open to question. They are often formulated as though the VO_4 tetrahedra were retained: $[V_2O_6(OH)]^{-3}$ (or $V_2O_7^{-4}$) and $[V_3O_9]^{-3}$. It is probable that the coordination number is higher: $[V_2O_4(OH)_5]^{-3}$ and $[(V(OH)_2O_2)_3]^{-3}$, for example.

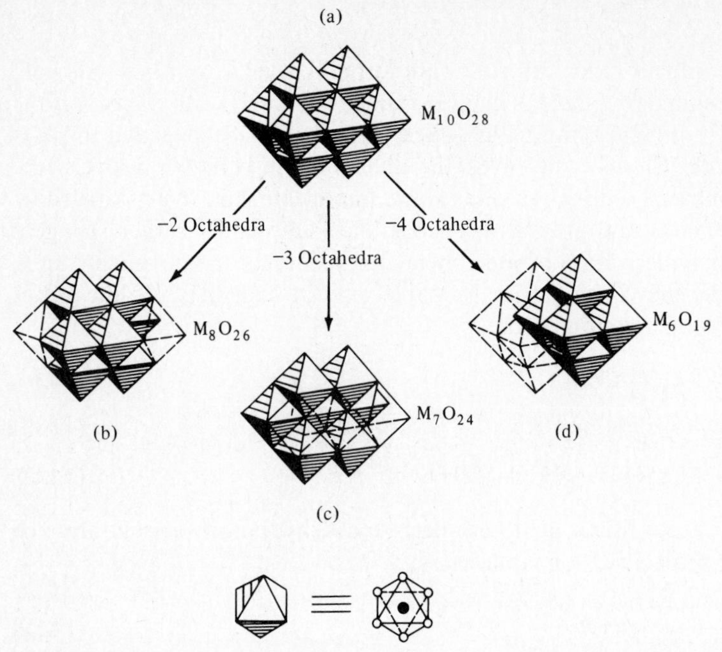

Fig. 14.9 The structures of edge-shared isopolyanions showing their relation to the $M_{10}O_{28}$ structure: (a) M = V; (b) M = Mo; (c) M = Mo; (d) M = Nb, Ta. [From D. L. Kepert, *Inorg. Chem.* **1969**, *8*, 1556. Reproduced with permission.]

edges ceases since the requisite distortion is impossible in the larger ions. The smaller the metal, the less the repulsion and the larger the number of octahedra per unit. For example, the metal radii are V^{+5} (54 pm) < W^{+6} (58 pm) ~ Mo^{+6} (60 pm) < Nb^{+5} (64 pm) = Ta^{+5} (64 pm) and the corresponding edge-shared polyanions are $V_{10}O_{28}^{-6}$, $Mo_8O_{26}^{-4}$, $Mo_7O_{24}^{-6}$, $HW_6O_{21}^{-5}$, $Nb_6O_{19}^{-8}$, and $Ta_6O_{19}^{-8}$. With the possible exception of Mo^{VI} and W^{VI} (the relative sizes of which are somewhat uncertain) this series parallels that of the metal radii. This order also parallels the degree of distortion in the polymeric oxides. To form larger polyanions such as $W_{12}O_{42}^{-12}$ or $H_2W_{12}O_{40}^{-6}$, edge sharing must give way to apex sharing.[30]

The isopolyanions may be considered to be portions of a closest-packed array of oxide ions with the metal ions occupying the octahedral holes. The largest edge-sharing array is found in $V_{10}O_{28}^{-6}$, which consists of ten octahedra stacked as shown in Fig. 14.9a. This seems to be the largest cluster compatible with metal–metal repulsions, and the remaining edge-shared structures represent portions of this unit.[32] Niobium and tantalum also form isopolyanions but these contain fewer octahedra: $Nb_6O_{19}^{-8}$ and $Ta_6O_{19}^{-8}$ (Fig. 14.9d).[33]

[30] There is evidence[31] that the polymerization proceeds *via* diprotonation and diaquation to form monomeric molybdic acid, $Mo(OH)_6$, which reacts with $HMoO_4^-$ anions to form polymers. It is possible that these peripheral molybdenum atoms do not expand their coordination sphere until the heptamer is formed.

[31] J. J. Cruywagen and E. F. C. H. Rohwer, *Inorg. Chem.*, **1975**, *14*, 3136.

[32] D. L. Kepert, *Inorg. Chem.*, **1969**, *8*, 1556.

[33] For a review of isopolyanions of Group VB metals, see M. T. Pope and B. W. Dale, *Quart. Rev. Chem. Soc.*, **1968**, *22*, 527.

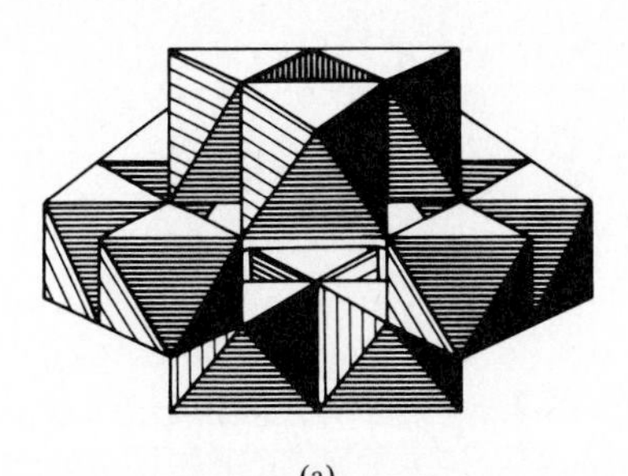

(a)

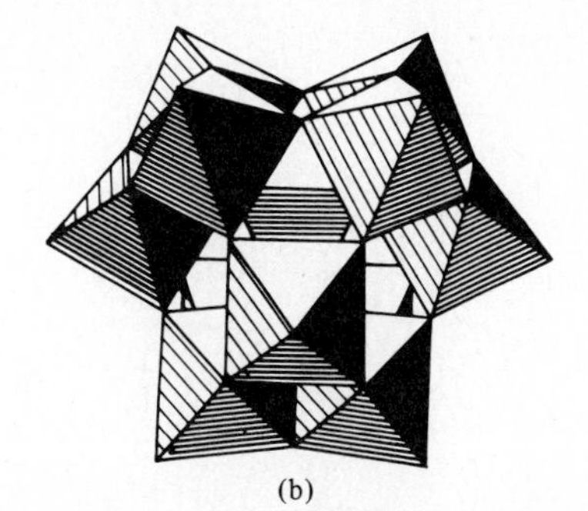

(b)

Fig. 14.10 The structures of two apex-shared dodecatungstate isopolyanions: (a) the "paratungstate Z" ion, $W_{12}O_{42}^{-12}$; (b) the "metatungstate" ion, $H_2W_{12}O_{40}^{-6}$. [From W. N. Lipscomb, *Inorg. Chem.* **1965**, *4*, 132. Reproduced with permission.]

Although elucidation of the various molybdate species continues, three appear to be most important: (1) the simple molybdate, MoO_4^{-2}, stable at high pHs; (2) the heptamolybdate (also known as paramolybdate), $Mo_7O_{24}^{-6}$ (Fig. 14.9c), formed in equilibrium with molybdate down to pH 4–5; and (3) octamolybdate, $Mo_8O_{26}^{-4}$ (Fig. 14.9b), formed in more acidic solutions. There is a possibility that higher polymers (but short of MoO_3) are formed, but their study is hindered by the difficulty in attaining equilibria because of low reaction rates. The amphoteric behavior of Mo^{VI} is shown by the re-solution of MoO_3 in very strongly acidic solution to form simple oxymolybdenum(VI) species.

The formation of isopolytungstates is similar although the chemistry is even more difficult. The simple tungstate, WO_4^{-2} exists in strongly basic solution. Acidification results in the formation of polymers built up from WO_6 octahedra. The nature of the species present depends not only on the present conditions (e.g., pH) but also on the history of the sample since some of the conversions are slow. The most important species appears to be a dodecatungstate ion ("paratungstate Z")[34] $W_{12}O_{42}^{-12}$ (Fig. 14.10a).[35] From more acidic solutions it is possible to crystallize a second dodecatungstate ion (the "metatungstate"), $[H_2(W_{12}O_{40})]^{-6}$.[36] The structure of this ion, although built of the same WO_6 octahedra, is more symmetrical, resulting in a cavity in the center of the ion (Fig. 14.10b).[37]

There are other paratungstate ions such as $HW_6O_{21}^{-5}$ (paratungstate A, structure uncertain; similar(?) to Fig. 14.9d) that are important in solution, but little is known of their exact nature.[38]

---

[34] Prior to any definite knowledge of the structure or even of the empirical formula of each of the various "paratungstate ions," they were arbitrarily assigned letter labels such as A, B, and Z. Much of the early confusion in this field occurred because workers referred to "paratungstate" without specifying which of the many possible species was being studied.

[35] This species is often referred to as $W_{12}O_{41}^{-10}$ or $W_{12}O_{36}(OH)_{10}^{-10}$. The X-ray analysis yields a crystallographic formula of $W_{12}O_{42}$ (W. N. Lipscomb, *Inorg. Chem.*, **1965**, *4*, 132; G. Weiss, *Z. Anorg. Allg. Chem.*, **1969**, *368*, 279).

[36] Although in the past other formulas such as $W_{12}O_{40}^{-8}$ and $H_4O(W_{12}O_{40})^{-6}$ have been applied, it has been shown (M. T. Pope and G. M. Varga, Jr., *Chem. Commun.* **1966**, 653) that there are two unique hydrogen atoms per ion that do not exchange on the NMR time scale, consistent with their occupying positions in the cavity.

[37] Although it might appear that there is a similar, although smaller, cavity in the paratungstate Z ion, the van der Waals radii of the oxygen atoms on the inner apices of the octahedra forming the structure effectively fill the cavity.

[38] For a review of the chemistry of the isopolytungstates, see D. L. Kepert, *Progr. Inorg. Chem.*, **1962**, *4*, 199.

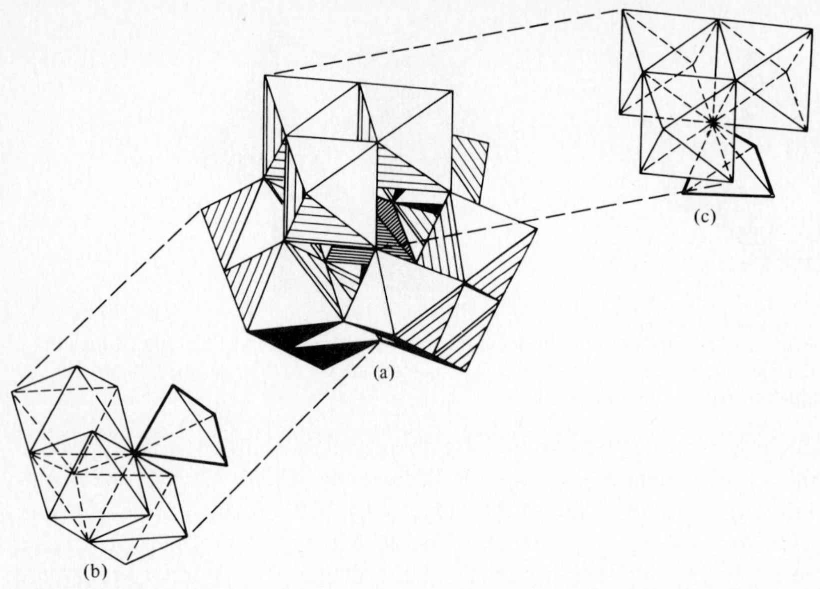

Fig. 14.11 The structures of two heteropoly anions: (a) 12-molybdophosphate or 12-tungstophosphate; (b) and (c) details of coordination of three MO_6 octahedra with one corner of the heteroatom tetrahedron.

Heteropoly anions

It has been noted that there is a cavity in the center of the metatungstate ion. This cavity is surrounded by a tetrahedron of four oxygen atoms (Fig. 14.11a) that is sufficiently large to accommodate a relatively small atom, such as P^{V}, As^{V}, Si^{IV}, Ge^{IV}, Ti^{IV}, and Zr^{IV}. The resulting 12-tungstoheteropoly anions[39] are of general formula $[X^{+n}W_{12}O_{40}]^{-(8-n)}$.[40] Similar 12-molybdoheteropoly anions are known and are of some importance in the qualitative and quantitative analytical chemistry of phosphorus and arsenic.

Between 35 and 40 heteroatoms are known to form heteropoly acids. In addition to the series "A" species discussed above, there is a series "B" of 1:12 heteropoly acids in which the heteroatom is larger—for example, Ce^{IV} and Th^{IV}. The structure of $(NH_4)_2H_6CeMo_{12}O_{42}$ is shown in Fig. 14.12. It is almost unique inasmuch as pairs of MoO_6 octahedra share *faces* to form Mo_2O_9 groups that coordinate to the cerium atom.[41]

Of the many other heteropoly acids, the 6-molybdo species are of some interest. These form with heteroatoms Te^{VI} and I^{VII} and tripositive metal ions such as Rh^{III}. All of these heteroatoms prefer an octahedral coordination sphere, which can be provided by a ring of six MoO_6 octahedra (Fig. 14.13). Note that formally the 6-heteropoly formulation can

---

[39] The prefix 12- may be used to replace the more cumbersome "dodeca-" to indicate the number of metal-atom octahedra coordinated to the heteroatom.

[40] One or more protons may be affixed to the anion with corresponding reduction of anionic charge.

[41] D. D. Dexter and J. V. Silverton, *J. Am. Chem. Soc.*, **1968**, *90*, 3589.

Fig. 14.12 Idealized sketch of the $[CeMo_{12}O_{42}]^{8-}$ ion showing the linkage of the MoO_6 octahedra.

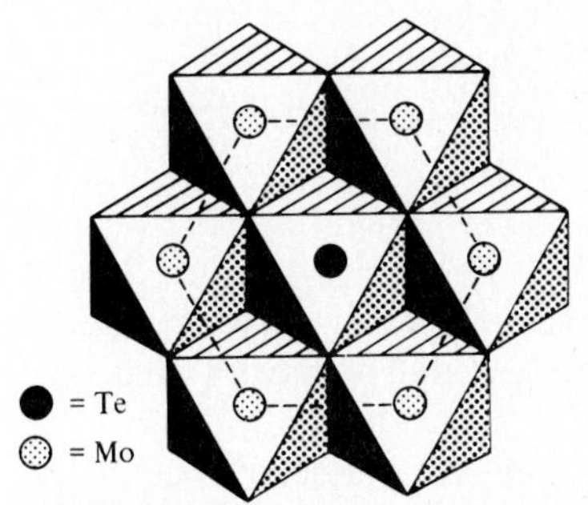

Fig. 14.13 The structure of the 6-molybdotellurate anion, $TeMo_6O_{24}^{-6}$. [Adapted, in part, from D. L. Kepert, *Progr. Inorg. Chem.* **1965**, *4*, 199. Reproduced with permission.]

be applied to the heptamolybdate species discussed earlier if the seventh molybdenum atom is considered to be a pseudo-heteroatom. At one time it was felt that the 6-heteropoly acids should be isomorphous with the heptamolybdate. This is not the case as may be seen by comparison of Figs. 14.9c and 14.13. The structures do not differ as much as might be supposed, however, differing principally in whether the heteroatom is surrounded by a planar ring of molybdenum atoms (6-heteropoly species, Fig. 14.14) or a puckered ring

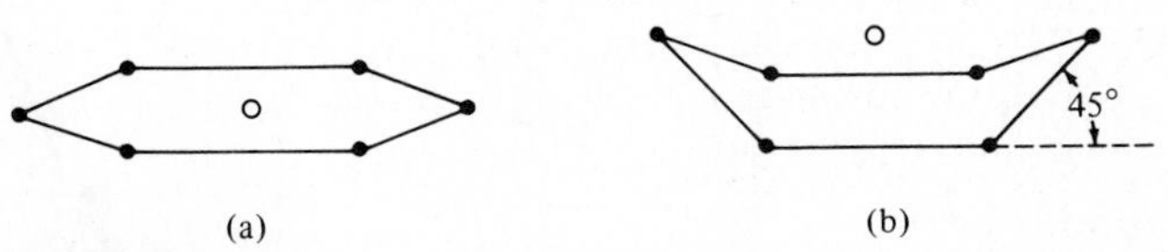

Fig. 14.14 (a) Planar ring of Mo atoms surrounding heteroatom in 6-heteropoly acids. (b) Puckered ring of Mo atoms surrounding seventh Mo atom in $Mo_7O_{24}^{-6}$.

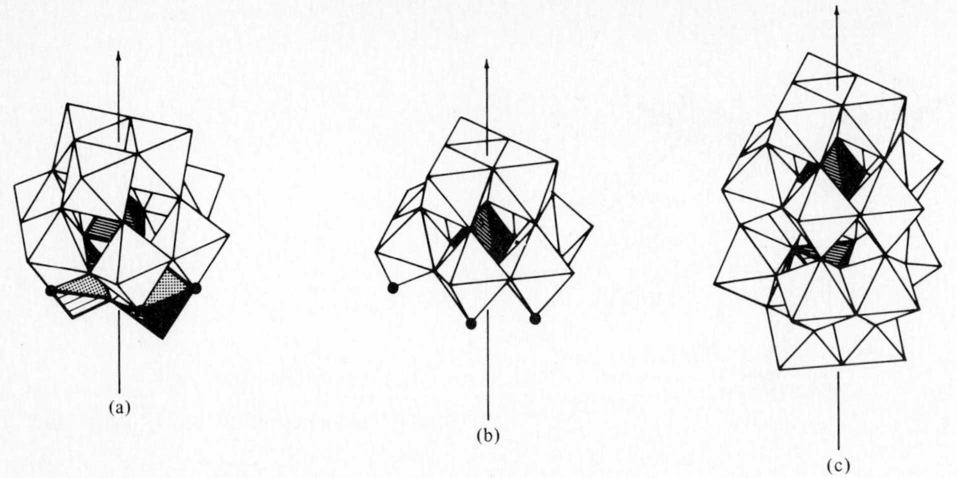

Fig. 14.15 Relation between the structures of 12-heteropoly and dimeric 9-heteropoly acids: (a) the $PW_{12}O_{40}^{-3}$ anion (see Fig. 12.4a); (b) the $PW_9O_{31}^{-3}$ half-unit formed by removal of shaded octahedra from (a); (c) the dimeric $P_2W_{18}O_{62}^{-6}$ ion formed from two half-units, (b). [From A. F. Wells, "Structural Inorganic Chemistry," 4th ed., Oxford University Press, London, **1975**. Reproduced with permission.]

(heptamolybdate, Fig. 14.10c). There are more complicated heteropoly acids, including diheteropoly acids such as $[P_2W_{18}O_{62}]^{-6}$, which has been found to have a structure related to the 12-heteropoly acids (Fig. 14.15).

In some heteropoly anions the heteroatom is not completely surrounded by octahedra of MoO_6 or WO_6 units. For example, Fig. 14.16a,b illustrates such an anion: $P_2Mo_5O_{23}^{-6}$. The fourth oxygen of the PO_4^{-3} tetrahedron is uncoordinated. If in the preparation of the heteropoly complex an alkylphosphonic acid, $RP(O)(OH)_2$, is used in place of phosphoric acid, a very similar complex is obtained (Fig. 14.16c) having the alkyl group at the fourth position of the phosphorus atoms.[42] This appears to be a general method for putting an organic "handle" on a large oxyanion. For example:

$$(CH_3)_2As(O)OH + 4MoO_4^{-2} + 6H^+ \xrightarrow[pH = 4-5]{} (CH_3)_2AsMo_4O_{14}(OH)]^{-2} \qquad \textbf{(14.20)}$$

This anion[43] (Fig. 14.16d,e) is the second known example of a heteropoly anion containing octahedra with shared faces.

As a class, the isopoly and heteropoly anions offer several interesting facets for study. They may be considered small "chunks" of metal oxide lattices. As anions they show very low surface charge density and low basicities. For example, we generally think of the perchlorate ion, ClO_4^-, as having a very low basicity. One study[44] has shown that the hexamolybdate(-2) ion and the 12-tungstophosphate(-3) ion have *lower* basicities than perchlorate, and the 12-molybdophosphate(-3) ion is only slightly more basic. Furthermore,

[42] W. Kwak, et al., *J. Am. Chem. Soc.*, **1975**, *97*, 5735.
[43] K. M. Barkigia et al., *J. Am. Chem. Soc.*, **1975**, *97*, 4146.
[44] L. Barcza and M. T. Pope, *J. Phys. Chem.*, **1975**, *79*, 92.

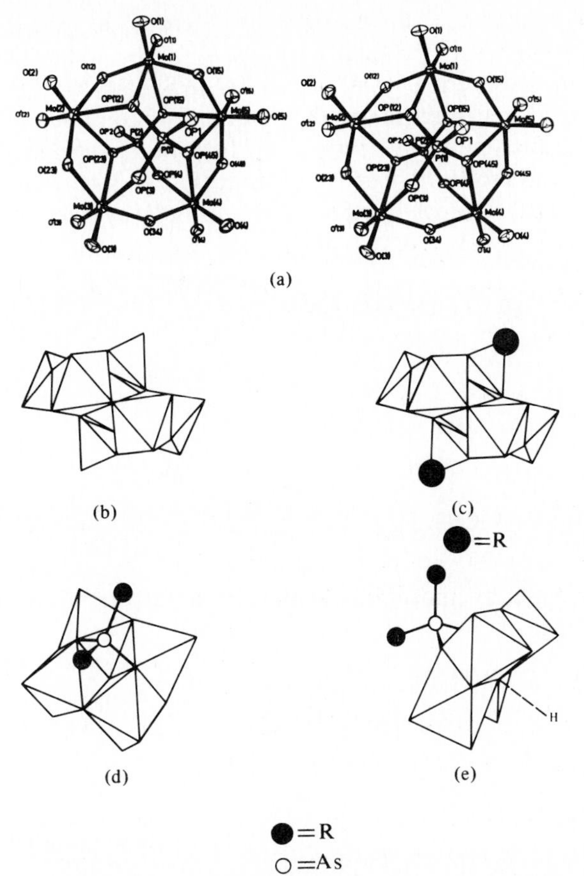

Fig. 14.16 Heteropolyphosphates with organic groups. (a) Stereoview of parent $P_2Mo_5O_{23}^{-6}$ ion. (b) Conventional view of parent $P_2Mo_5O_{23}^{-6}$ ion. (c) Structure of the $(RP)_2Mo_5O_{21}^{-4}$ anion. (d) Structure of the $R_2AsMo_4O_{14}OH^{-2}$ anion. (e) Alternative view of the $R_2AsMo_4O_{14}OH^{-2}$ anion showing probable position of the hydrogen atom. [Courtesy of R. Strandberg and M. T. Pope et al., *J. Am. Chem. Soc.*, **1975**, *97*, 4146, 5735. Reproduced with permission.]

these systems give rise to highly colored mixed-oxidation-state species (which can only be briefly mentioned here): the tungsten bronzes[45] and the heteropoly blues.[46]

RINGS

Borazines

The most important ring system of organic chemistry is the benzene ring, either as a separate entity or in polynuclear hydrocarbons such as naphthalene, anthracene, and

[45] See P. Hagenmuller, in "Comprehensive Inorganic Chemistry," Vol. 4, pp. 541–567, A. F. Trotman-Dickenson, ed., Pergamon, Oxford, 1973; M. S. Whittingham and M. B. Dines, *Surv. Progr. Chem.*, **1980**, *9*, 66–69.

[46] J. J. Altenau et al., *Inorg. Chem.*, **1975**, *14*, 417, and references therein.

phenanthrene. Inorganic chemistry has two (at least) analogues of benzene: borazine, $B_3N_3H_6$, and trimeric phosphonitrilic compounds, $P_3N_3X_6$.

Borazine has been known since the pioneering work of Alfred Stock early in this century. Stock's work was important in two regards: He was the first to study compounds such as the boranes, silanes, and other similar nonmetal compounds; he perfected vacuum line techniques for the handling of air- and moisture-sensitive compounds, an invaluable technique of the modern inorganic chemist.[47] Stock synthesized borazine[48] by heating the adduct[49] of diborane and ammonia:

$$3B_2H_6 + 6NH_3 \longrightarrow 3(B_2H_6)(NH_3)_2 \xrightarrow{\text{heat}} 2B_3N_3H_6 + 12H_2 \tag{14.21}$$

More efficient syntheses are:[50]

$$3NH_4Cl + 3BCl_3 \longrightarrow 9HCl + Cl_3B_3N_3H_3 \xrightarrow{NaBH_4} B_3N_3H_6 + \tfrac{3}{2}B_2H_6 + 3NaCl \tag{14.22}$$

$$3NH_4Cl + 3LiBH_4 \longrightarrow B_3N_3H_6 + 3LiCl + 9H_2 \tag{14.23}$$

N- or B-substituted borazines may be made by appropriate substitution on the starting materials prior to the synthesis of the ring:

$$3RNH_3Cl + 3BCl_3 \longrightarrow 9HCl + Cl_3B_3N_3R_3 \xrightarrow{NaBH_4} H_3B_3N_3R_3 + \tfrac{3}{2}B_2H_6 + 3NaCl \tag{14.24}$$

or substitution after the ring has formed:

$$Cl_3B_3N_3R_3 + 3LiR' \longrightarrow R'_3B_3N_3R_3 + 3LiCl \tag{14.25}$$

Borazine is isoelectronic with benzene (Fig. 14.17). In physical properties, borazine is indeed a close analogue of benzene. The similarity of the physical properties of the alkyl-substituted derivatives is more remarkable. The ratio of the absolute boiling points of the substituted borazines to similarly substituted benzene derivatives is 0.93 ± 0.01.[51] This similarity of physical properties has unfortunately been overemphasized and led to a description of borazine as "inorganic benzene." The *chemical* properties of borazine and benzene are quite different. Both compounds have aromatic π clouds of electron density delocalized over all of the ring atoms. Because of the difference in electronegativity between boron and nitrogen, the π cloud in borazine is "lumpy" with more electron density localized on the nitrogen.[52] This partial localization weakens the π bonding in the ring. In addition,

[47] See D. F. Shriver, "The Manipulation of Air-Sensitive Compounds," McGraw-Hill, New York, **1969**.
[48] A. Stock and E. Pohland, *Chem. Ber.*, **1926**, *59*, 2215.
[49] For the structure of the adduct, see p. 732.
[50] R. A. Geanangel and S. G. Shore, *Prep. Inorg. React.*, **1966**, *3*, 123.
[51] See T. Moeller, "Inorganic Chemistry," Wiley, New York, **1952**, Table 17.16, p. 802.
[52] This is confirmed both by the nature of the addition reactions to be described and by molecular orbital calculations (R. J. Boyd et al., *Chem. Phys. Letters*, **1968**, *2*, 227). It is interesting to note that the universal assump-

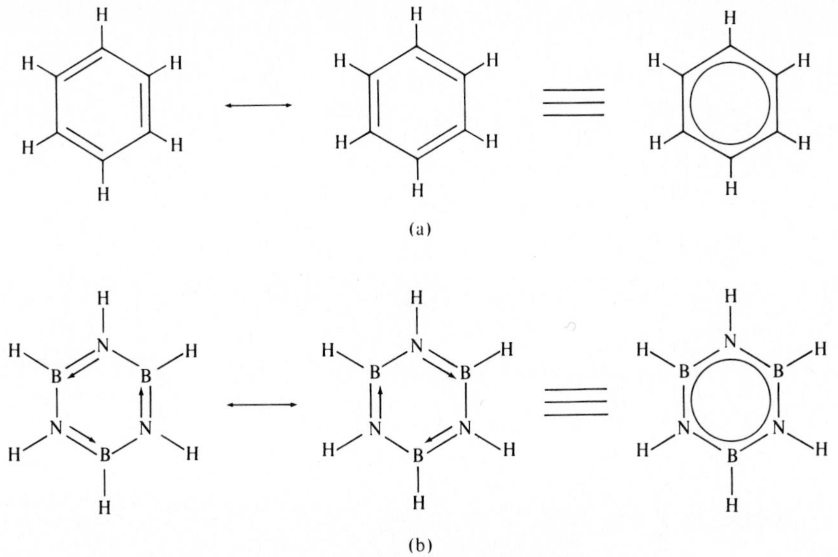

(a)

(b)

Fig. 14.17 Electronic structures of: (a) benzene; (b) borazine.

nitrogen retains some of its basicity and the boron some of its acidity. Polar species such as HCl can therefore attack the "double bond" between nitrogen and boron. The different electronegativities of boron and nitrogen tend to stabilize bonding to boron by *electronegative substituents* and to nitrogen by *electropositive substituents.* Thus, in contrast to benzene, borazine readily undergoes addition reactions:

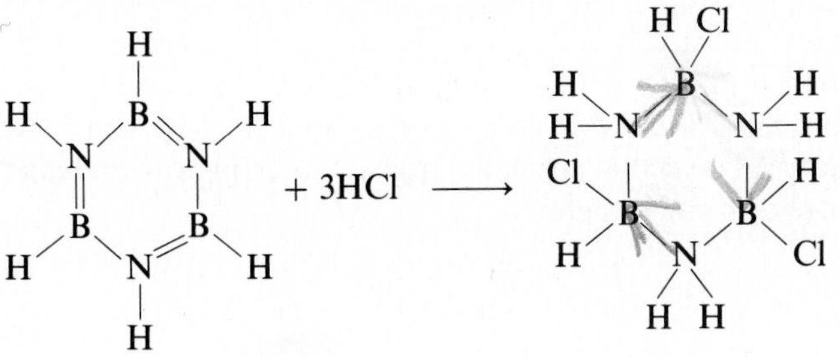

$$B_3N_3H_6 + 3HCl \longrightarrow (H_2N\text{—}BHCl)_3 \qquad (14.26)$$

$$C_6H_6 + HCl \longrightarrow \text{No reaction} \qquad (14.27)$$

The tendency to undergo addition rather than aromatic substitution is well contrasted by

tion that borazine is planar has been extremely difficult to prove experimentally. For a complete discussion of the difficulties involved in the structure determination as well as a discussion of the previous literature, see W. Harshbarger et al., *Inorg. Chem.*, **1969**, *8*, 1683.

the reactions of these two compounds toward bromine:

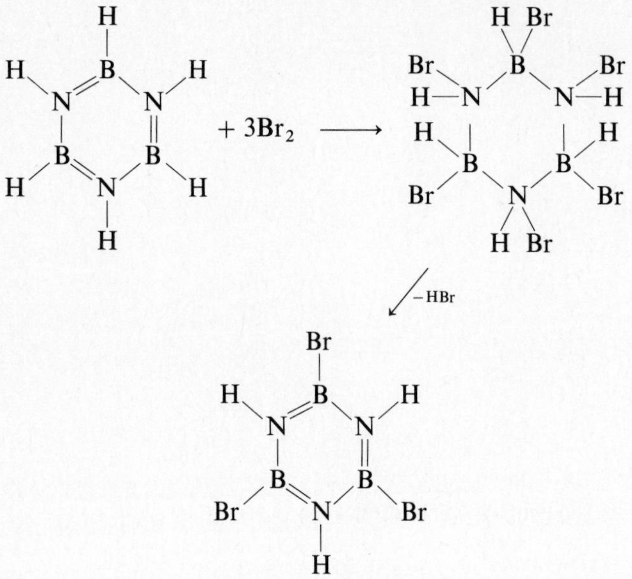

$$\text{B}_3\text{N}_3\text{H}_6 + 3\text{Br}_2 \longrightarrow \text{B}_3\text{N}_3\text{H}_6\text{Br}_6 \xrightarrow{-\text{HBr}} \text{B}_3\text{N}_3\text{H}_3\text{Br}_3 \qquad (14.28)$$

$$\text{C}_6\text{H}_6 + \text{Br}_2 \xrightarrow{[\text{FeBr}_3]} \text{C}_6\text{H}_5\text{Br} + \text{HBr} \qquad (14.29)$$

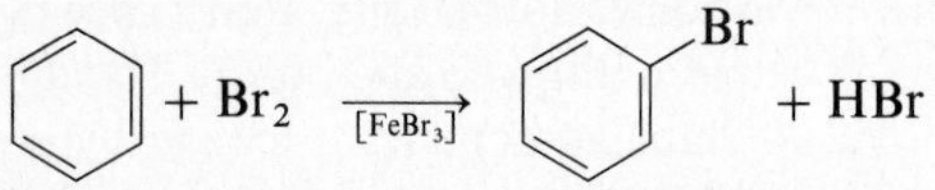

This electronic difference between benzene and borazine is further supported by the properties of the type $(R_6B_3N_3)Cr(CO)_3$. Although these are formally analogous to $(C_6H_6)Cr(CO)_3$, the bonding is not nearly so strong. The dissociation energy of the ring from the metal appears to be about one-half as strong in the borazine complex as in the arene complex.[53] In addition, there is considerable evidence that in these complexes the borazine molecule is puckered.[54] The actual structure appears to be intermediate between a true π complex and the extreme σ-only model:

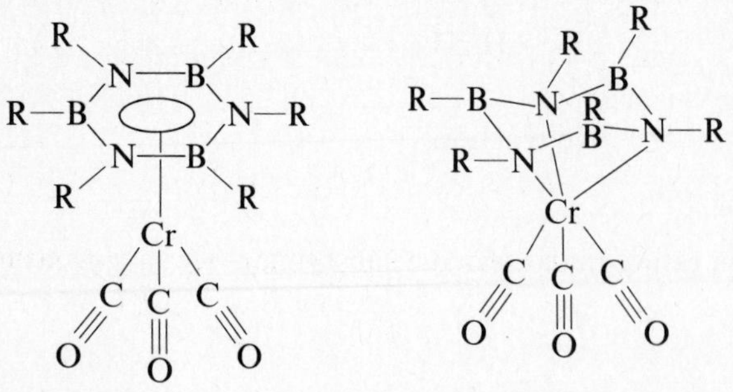

[53] M. Scotti et al., *Inorg. Chem.*, **1977**, *25*, 261.
[54] J. J. Lagowski, *Coordin. Chem. Rev.*, **1977**, *22*, 185.

Borazine analogues of naphthalene and related hydrocarbons have been made by the pyrolysis of borazine[55] or its passage through a silent electric discharge:[56]

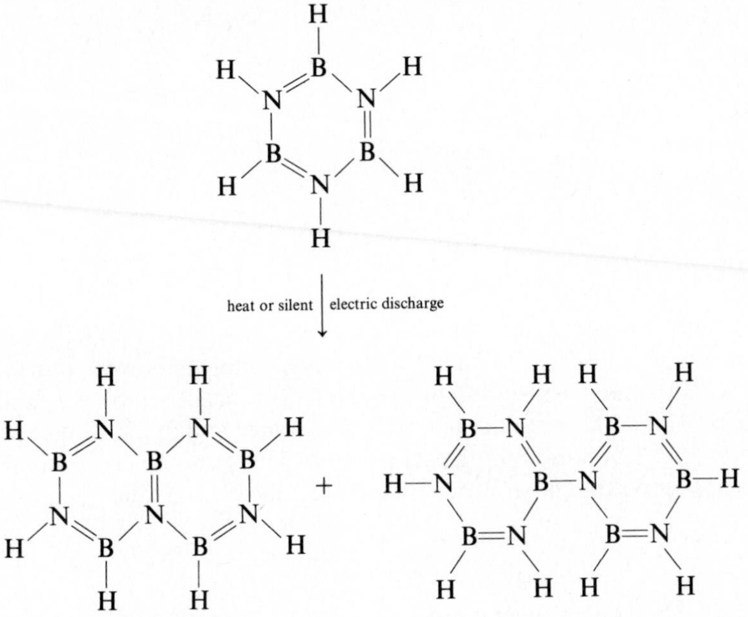

heat or silent electric discharge **(14.30)**

Related four-membered rings, $R_2B_2N_2R'_2$, and eight-membered rings, $R_4B_4N_4R'_4$, are also known, but considerably less work has been done on them than on borazine.

Benzene may be hydrogenated to produce the saturated compound cyclohexane. Hydrogenation of borazine results in polymeric materials of indefinite compositions. Substituted derivatives of the saturated cycloborazane, $B_3N_3H_{12}$, form readily by addition to borazine (Eqs. 14.26 and 14.28), but special techniques are necessary to prepare the parent compound. It was first prepared by the reduction of the chloro derivative:

$$2B_3N_3H_6 + 6HCl \longrightarrow 2H_3Cl_3B_3N_3H_6 \xrightarrow{6NaBH_4} 2H_6B_3N_3H_6 + 3B_2H_6 + 6NaCl \quad \textbf{(14.31)}$$

A complete series of cycloborazanes may be prepared by the treatment of the adduct of ammonia and diborane with a very strong base:

$$2NH_3 + B_2H_6 + NaNH_2 \longrightarrow \frac{1}{n}(BH_2NH_2)_n + 2NH_3 + NaBH_4 \quad \textbf{(14.32)}$$

The principal product of the reaction is the cyclopentaborazane ($n = 5$), an unexpectedly stable white microcrystalline solid. It is hydrolyzed only slowly by boiling water and

[55] A. W. Laubengayer et al., *J. Am. Chem. Soc.*, **1961**, *83*, 1337.

[56] A. W. Laubengayer and O. T. Beachley, Jr., in "Boron–Nitrogen Chemistry," R. F. Gould, ed., Advances in Chemistry Series, No. 42, American Chemical Society, Washington, D.C., 1964, p. 281.

complete hydrolysis is effected only with acid solutions at 160°C over an extended period of time.

Cyclodiborazane, $H_4B_2N_2H_4$, may be isolated in small quantities from reaction 14.30 or by the pyrolysis of cyclopentaborazane. It spontaneously isomerizes to cyclotriborazane ("borazane") on standing at room temperature:

$$3\ \begin{array}{c} H_2B{-}NH_2 \\ | \quad\ \ | \\ H_2N{-}BH_2 \end{array} \longrightarrow 2\ \begin{array}{ccc} & H_2B & \\ H_2N & & NH_2 \\ | & & | \\ H_2B & & BH_2 \\ & NH_2 & \end{array} \qquad (14.33)$$

Evidence has been obtained for the existence of the corresponding tetramer, $(H_2BNH_2)_n$, but it was not possible to isolate it. No higher polymers were observed.[57]

Isoelectronic with the borazine is boroxine, $H_3B_3O_3$. It is produced by the explosive oxidation of B_2H_6 or B_5H_9. The structure has been determined and found to be perfectly planar.[58] Boroxine is even less stable (and presumably has even less π delocalization) than borazine, decomposing at room temperature to diborane and boron oxide.

Phosphazenes

Early workers noted the extreme reactivity of phosphorus pentachloride toward basic reagents such as water or ammonia. With the former the reaction is reasonably straightforward, at least for certain stoichiometries:

$$PCl_5 + 2H_2O \longrightarrow OPCl_3 + 2HCl \qquad (14.34)$$

$$PCl_5 + 4H_2O \longrightarrow H_3PO_4 + 5HCl \qquad (14.35)$$

In the case of ammonia early workers proposed compounds such as $HN{=}PCl_3$ and $HN{=}P(NH_2)_3$ by analogy, but this work was obscured by difficulties of incomplete reactions, separations, sensitivity to moisture, and similar problems. Furthermore, gradual polymerization occurred with loss of ammonia to yield "phospham," a poorly characterized solid of approximate formula $(PN_2H)_x$, as the ultimate deammonation product.

If instead of free ammonia the less reactive conjugate acid is used, the reaction proceeds at a moderate rate and the results are more meaningful:

$$NH_4Cl + PCl_5 \xrightarrow[\text{refluxing } HCl_2CCHCl_2]{146^\circ} \text{"}PNCl_2\text{"} \qquad (14.36)$$

[57] K. W. Böddeker et al., *J. Am. Chem. Soc.*, **1966**, *88*, 4396. For reviews of boron–nitrogen ring systems, see R. A. Geanangel and S. G. Shore, *Prep. Inorg. React.*, **1966**, *3*, 123; K. Niedenzu and J. W. Dawson, "Boron–Nitrogen Compounds," Academic, New York, **1965**; and E. K. Mellon, Jr., and J. J. Lagowski, *Advan. Inorg. Chem. Radiochem.*, **1963**, *5*, 259.

[58] For the determination of the structure of boroxine as well as references to earlier synthetic work, see C. H. Chang et al., *Inorg. Chem.*, **1969**, *8*, 1689.

If the latter were a monomer, a structure could be drawn for it, $Cl_2P{\equiv}N$, analogous to organic nitriles, $R{-}C{\equiv}N$. For this reason the older names used for these compounds were "phosphonitriles," "phosphonitrilic chloride," etc. As a matter of fact the products are either cyclic or linear polymers of general formula $[NPCl_2]_n$. Thus the name *phosphazene*[59] has been suggested by analogy with benzene, borazine, etc. The major product[60] and easiest to separate is the trimer, $n = 3$. Smaller amounts of the tetramer and other oligomers up to $n = 8$ have been characterized and higher polymers exist (see below). Analogous bromo compounds may be prepared in the same manner, except that bromine should be added to suppress the decomposition of phosphorus pentabromide:

$$PBr_5 \longrightarrow PBr_3 + Br_2 \qquad (14.37)$$

$$PBr_5 + NH_4Br \xrightarrow{\text{excess } Br_2} [NPBr_2]_n \qquad (14.38)$$

The fluoride must be prepared indirectly by fluorination of the chloride:[61]

$$[NPCl_2]_n + 2nNaF \xrightarrow{\text{nitrobenzene}} [NPF_2]_n + 2nNaCl \qquad (14.39)$$

The corresponding iodides are unknown, but a phosphazene with a single P—I bond has recently been reported:[62]

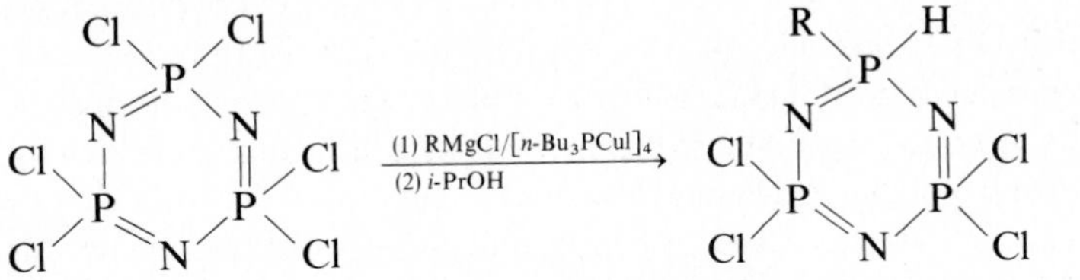

$$N_3P_3Cl_6 \xrightarrow[(2)\ i\text{-PrOH}]{(1)\ RMgCl/[n\text{-}Bu_3PCuI]_4} N_3P_3Cl_4(R)(H) \qquad (14.40)$$

$$N_3P_3Cl_4(R)(H) \xrightarrow{I_2/CCl_4} N_3P_3Cl_4(R)(I) \qquad (14.41)$$

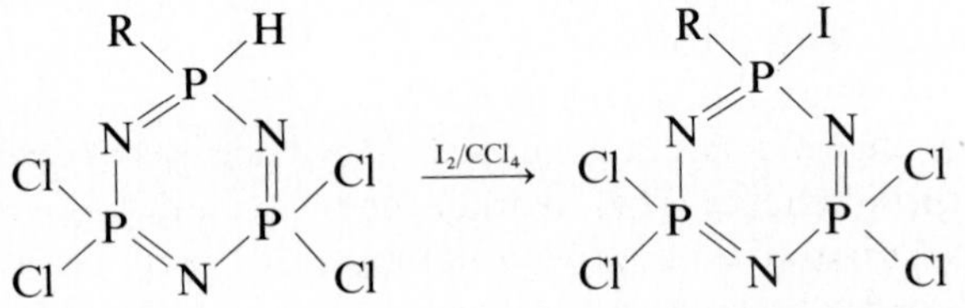

---

[59] Shaw and co-workers (S. K. Ray and R. A. Shaw, *J. Chem. Soc.* **1961**, 872; R. A. Shaw et al., *Chem. Rev.*, **1962**, *62*, 247) have suggested the "phosphazene" system of nomenclature based upon the names phosphazatriene and phosphazatetraene for the cyclic trimers and tetramers, respectively. While the system has considerable merit, it has not yet been universally adopted. Some inorganic chemists use older terms like triphosphonitrile and tetraphosphonitrile (see C. D. Schmulbach, *Progr. Inorg. Chem.*, **1962**, *4*, 275).

[60] The relative amounts of the various products can be altered by changing the mole ratio of the reactants and the temperature (by changing the solvent). See R. A. Shaw et al., *Prep. Inorg. React.*, **1965**, *2*, 1. For specific laboratory procedures, see W. L. Jolly, "Synthetic Inorganic Chemistry," Prentice-Hall, Englewood Cliffs, N. J., **1960**, p. 171.

[61] T. Moeller et al., *Chem. Ind. (London)*, **1961**, 347.

[62] H. R. Allcock and P. J. Harris, *Inorg. Chem.*, **1981**, *20*, 2844.

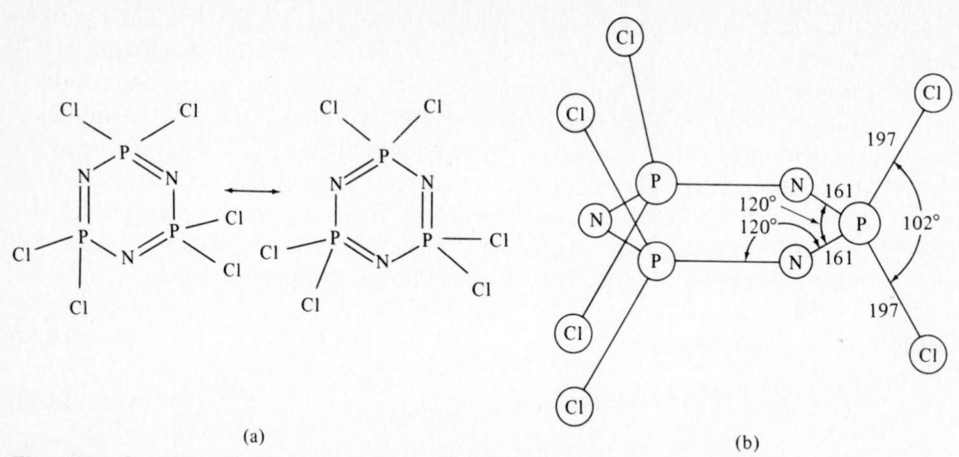

Fig. 14.18 Structure of trimeric phosphonitrilic chloride, $P_3N_3Cl_6$: (a) contributing resonance structures; (b) molecular structure as determined by X-ray diffraction. [From C. D. Schmulbach, *Progr. Inorg. Chem.*, **1962**, *4*, 275. Reproduced with permission.] Distances in pm.

The trimer consists of a planar[63] six-membered ring (Fig. 14.18). The bond angles are consistent with sp^2 hybridization of the nitrogen and approximately sp^3 hybridization of the phosphorus. As shown in Fig. 14.18, resonance structures can be drawn analogous to those for benzene indicating aromaticity in the ring. The situation is more complex than these simple resonance structures indicate, however. Unlike benzene, π bonding in phosphonitriles involves d–p π bonds. There have been two descriptions offered for such d_π–p_π bonding in phosphonitrilic systems. Craig and Paddock[64] suggested the following model: The d_{xz} orbital on each phosphorus atom overlaps with the p_z orbitals on the nitrogen atoms adjacent to it (Fig. 14.19a,b). As a result of the *gerade* symmetry of the d orbital, an inevitable mismatch of the sign of the wave function occurs in the trimer (see Fig. 3.27). This node reduces the stability of the delocalized molecular orbital.

Dewar and co-workers[65] have offered an alternative view. They believe that both the d_{xz} and d_{yz} orbitals participate in the π bonding. They hybridize these two orbitals to form the hybrids d^a and d^b directed toward the adjacent nitrogen atoms (Fig. 14.19c). This allows delocalized, three-center bonds[66] to form about each nitrogen atom (Fig. 14.19d). This scheme results in delocalization over selected three-atom segments of the ring, but nodes

[63] The small deviations from planarity found for the bromide (H. Zoer et al., *Acta Crystallogr.*, *A*, **1969**, *25*, S107), the chloride [A. Wilson and D. F. Carroll, *J. Chem. Soc.*, **1960**, 2548), and the fluoride [M. W. Dougill, *J. Chem. Soc.*, **1963**, 3211) are probably the result of packing effects rather than an inherent tendency toward nonplanarity on the part of the phosphonitrile ring. Unsymmetrical substitution, however, appears capable of forcing the ring out of a planar conformation (N. V. Mani et al., *Acta Crystallogr.*, **1965**, *15*, 539; C. W. Allen et al., *J. Am. Chem. Soc.*, **1967**, *89*, 6361) and so the resistance of the ring to deformation (especially when appropriately substituted) may be considerably lower than in benzene.

[64] D. P. Craig and N. L. Paddock, *Nature*, **1958**, *181*, 1052; D. P. Craig, *Chem. Ind. (London)*, **1958**, 3, *J. Chem. Soc.*, **1959**, 997: D. P. Craig et al., *Chem. Soc.*, **1961**, 1376: D. P. Craig and K. A. R. Mitchell, *J. Chem. Soc.*, **1965**, 4682.

[65] M. J. S. Dewar et al., *J. Chem. Soc.*, **1960**, 2423.

[66] See p. 767 for a discussion of three-center bonding.

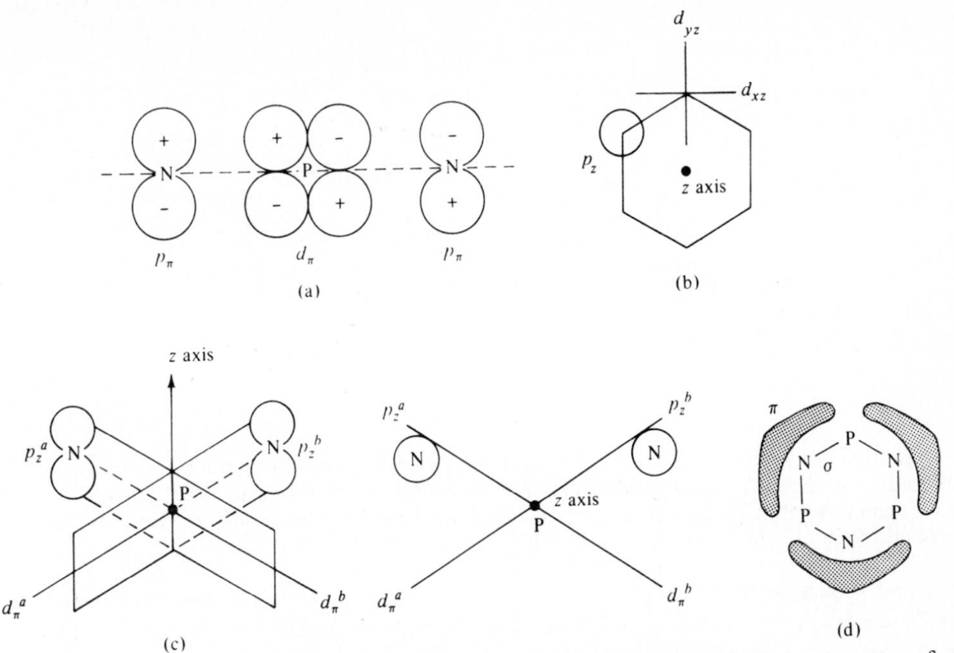

Fig. 14.19 Π bonding in phosphonitrilic system. Theory of Craig and Paddock: (a) symmetry of p(N) and d(P) orbitals forming π bonds, (b) convention used for the orientation of the d orbitals on phosphorus. The d orbitals lie in the planes denoted by the straight lines. Theory of Dewar; formation of delocalized three-center bond over a P—N—P segment of a phosphonitrilic ring: (c) the relation of the orthogonal d^a and d^b orbitals to the nitrogen p_z orbitals as seen perpendicular to the z axis and parallel to the z axis, respectively; (d) the three-center π bond model for $P_3N_3Cl_3$. [From C. D. Schmulbach, *Prog. Inorg. Chem.*, **1965**, *4*, 275. Reproduced with permission.]

are present at each phosphorus atom since the d^a and d^b orbitals of the phosphorus atom are orthogonal to each other. Both groups of workers have offered experimental evidence in support of their models, but neither theory has been confirmed to the exclusion of the other. Both agree that the phosphazene ring will be stabilized by delocalization (= "aromaticity" or "resonance") and that this delocalization will be imperfect.

The structures of tetrameric phosphazenes are more flexible than those of the trimers. Only one studied thus far has been planar, $(NPF_2)_4$.[67] Instead, two other conformers appear to be favored: (1) the "tub" structure found for the tetrameric chlorophosphazene (Fig. 14.20a); (2) the "crown" form found in octamethyltetraphosphazatetraene and octakis(dimethylamino)tetraphosphazatetraene (Fig. 14.20b). The interesting feature of these compounds is that the nonplanar structures do not militate against extensive delocalization in the rings. The corresponding organic compound, cyclooctatetraene, C_8H_8, is nonaromatic for two reasons: (1) Its nonplanar, "tub" structure[68] precludes efficient p_π–p_π overlap; and (2) it does not obey the Hückel rule of $(4n + 2)$ π electrons. The Hückel rule was formulated on the basis of p_π–p_π bonding and holds for cyclic organic compounds

[67] H. McD. McGeachin and F. R. Tromans, *J. Chem. Soc.*, **1961**, 4777.

[68] O. Bastiansen et al., *J. Chem. Phys.*, **1957**, *27*, 1311.

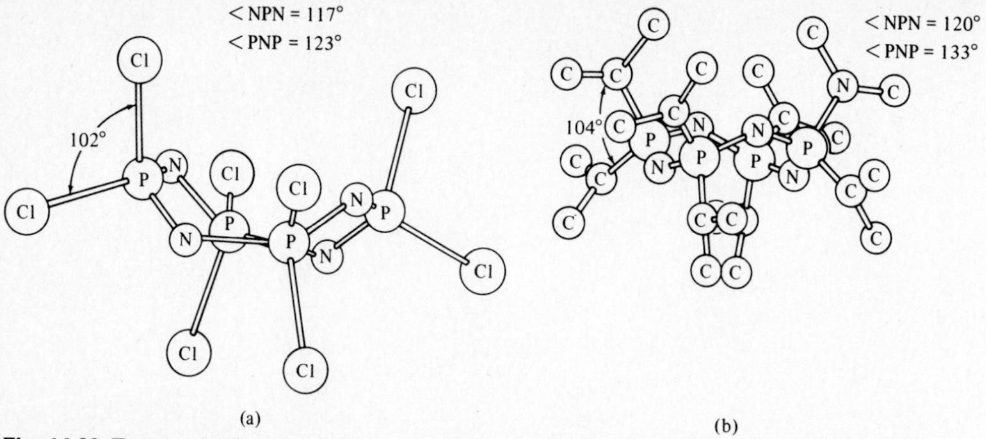

$<$NPN = 117°
$<$PNP = 123°
102°

(a)

$<$NPN = 120°
$<$PNP = 133°
104°

(b)

Fig. 14.20 Tetrameric phosphonitriles: (a) molecular structure of octachlorotetraphosphonitrile ("tub" conformation); (b) molecular structure of octakis(dimethylamino)tetraphosphonitrile ("crown" conformation). [Structure (a) from R. Hazekanap et al., *Acta Crystallogr.*, **1962**, *15*, 539; structure (b) from G. J. Bullen, *J. Chem. Soc.*, **1962**, 3193. Reproduced with permission.]

from $n = 1$ (benzene) to $n = 4$ ([18]annulene).[69] The use of d orbitals removes the restrictions of the Hückel rule and also allows greater flexibility of the ring since the diffuse d orbitals are more amenable to bonding in nonplanar systems. Either the Craig or Dewar model predicts that the tetramer is stabilized by delocalization (unlike cyclooctatetraene) either as much as the trimer (Dewar) or more so (Craig).

When hexachlorobis(ethylamino)cyclotetraphosphazatetraene is treated with an excess of dimethylamine, one obtains, in addition to the expected product having dimethylamide groups in place of the chlorine atoms, a product that arises from internal transannular attack by one of the ethylamino groups:

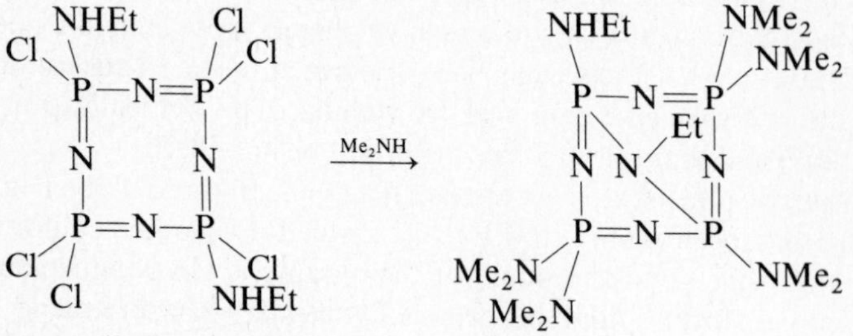

$$\text{(EtNH)Cl}_2\text{P}_4\text{N}_4\text{Cl}_4\text{(NHEt)} \xrightarrow{\text{Me}_2\text{NH}} \text{(EtNH)P}_4\text{N}_4(\text{NEt})(\text{NMe}_2)_5 \qquad (14.42)$$

Recently the benzylamino analogue has been reported.[70]

Higher cyclic polymers have been characterized, but none has been studied so extensively as the trimer and tetramer. Linear polymers are also known. If excess phosphorus

[69] G. M. Badger, "Aromatic Character and Aromaticity," Cambridge University Press, London, **1969**, and M. J. S. Dewar and G. J. Gleicher, *J. Am. Chem. Soc.*, **1965**, *87*, 685.

[70] T. S. Cameron et al., *Chem. Commun.*, **1975**, 975. S. S. Krishnamurthy et al., *Inorg. Chem.*, **1982**, *21*, 407.

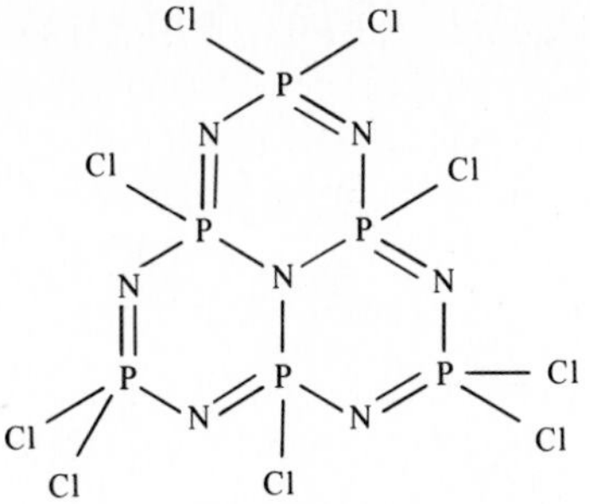

Fig. 14.21 Presumed structure of $Cl_9P_6N_7$.

pentachloride is used in the preparation, it is possible to isolate linear polymers of the type:

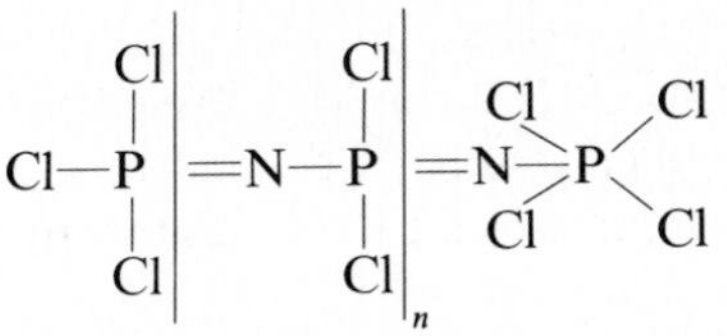

$$Cl{-}PCl_2\left[{=}N{-}PCl_2\right]_n{=}N{-}PCl_3$$

A great deal of interest has been expressed in these rubbery polymers and in cross-linked derivatives of the ring systems in the hopes of obtaining thermally stable elastomers for use at high temperatures. Although the thermal stability is encouraging, all of these systems are subject to hydrolytic degradation and hence of limited usefulness (see below).

The "phosphams" studied by early workers are examples of such high polymers. Audrieth and Sowerby[71] showed that the careful ammonolysis of phosphonitrilic chloride results in an isolable phosphonitrilic amide:

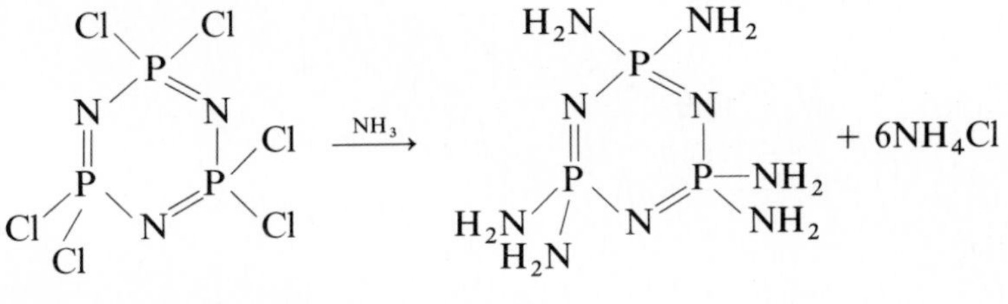

$$(NPCl_2)_3 \xrightarrow{NH_3} [NP(NH_2)_2]_3 + 6NH_4Cl \qquad \mathbf{(14.43)}$$

which, however, undergoes elimination of ammonia with increasing temperature to produce cross-linking between rings (or possible rupture of rings to form linear polymers), finally approaching the stoichiometry $(NPNH)_n$. Little is known of the structure of such polymers, although early workers isolated a relatively low-molecular-weight polymer believed to be a three-ringed species (Fig. 14.21), and it is possible that the higher polymers contain more extensive ring systems.

There has been considerable interest in developing inorganic polymers that would have advantages over the usual polyolefin or polyester polymers. One system that has

[71] L. F. Audrieth and D. B. Sowerby, *Chem. Ind. (London)*, **1959**, 748. See also M. C. Miller and R. A. Shaw, *J. Chem. Soc.*, **1963**, 3233, for corresponding work on the tetramer.

been extensively developed is that of the siloxanes, compounds related to the silicate structures given in the first part of this chapter but with organic side chains. Polyphosphazenes have long held promise in this area. The earliest studies were hampered by the moisture susceptibility of the phosphorus–chlorine bond. However, it has been found possible to polymerize phosphonitrilic chloride thermally:

$$[PNCl_2]_3 \xrightarrow[\text{heat}]{} [PNCl_2]_x \tag{14.44}$$

If this is done carefully, extensive cross-linking does not take place and the polymer ($x \simeq$ 15 000) remains soluble in organic solvents. The reactive chlorine atoms are still susceptible to nucleophilic attack and displacement:

$$[PNCl_2]_x + 2x\text{NaOR} \longrightarrow [PN(OR)_2]_x + 2x\text{NaCl} \tag{14.45}$$

By varying the nature of the side chain, R, various elastomers and other useful polymers have been obtained.[72] Some of the fluoroalkoxy substituted polymers (R = CH_2CF_3) are so water repellent that they do not interact with living tissues and promise to be useful in fabrication of artificial blood vessels and prosthetic devices. In some ways these polymers resemble the hydrophobic portions of protein chains and may well prove useful as "pseudoproteins" in biomedicine. Some promising attempts to reproduce the hydrophobic environment in hemoglobin (see Chapter 18) have already been made.

Although the hydrolytic stability of some phosphazene polymers makes them attractive as structural materials, it is possible to create hydrolytically sensitive phosphazenes that may be useful medically as slow-release drugs. Thus steroids and the anti-cancer moiety, *cis*-dichloroplatinum (see Chapter 18 for the action of the latter) have been linked to a phosphazene skeleton. Slow hydrolysis would provide these drugs in a therapeutic steady-state system.[73]

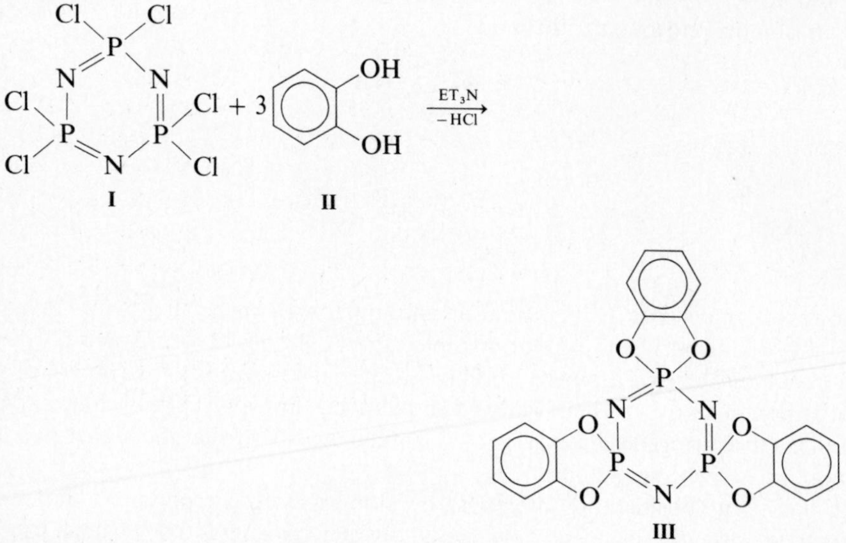

I + 3 II $\xrightarrow[-HCl]{ET_3N}$ III (14.46)

[72] H. R. Allcock, *Science*, **1976**, *193*, 1214.

[73] H. R. Allcock, in "Contemporary Topics in Polymer Science," Vol. 3, M. Sten, ed., Plenum, New York, **1979**, pp. 55–78.

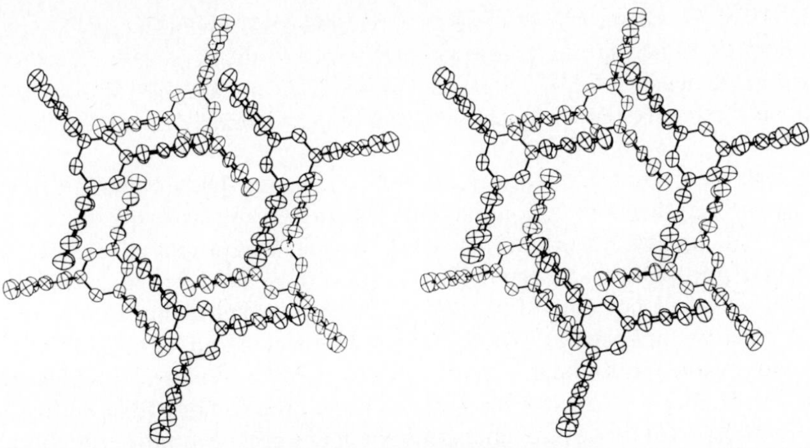

Fig. 14.22 Stereoview of the tunnels present in clathrate structures formed by tris(*o*-phenylenedioxy)cyclotriphosphazene with benzene or *o*-xylene (guest molecules not shown). [From H. R. Allcock et al., *J. Am. Chem. Soc.*, **1976**, *98*, 5120. Reproduced with permission.]

Another interesting property of some substituted phosphonitrilic trimers relates directly to their molecular structures. The compounds are large polycyclic molecules that stack in certain ways. First, let's look at the data that Allcock and co-workers[74] had to work with for one of these compounds. (1) Purification of the spirocyclophosphazene (**III**) proved unexpectedly difficult. Recrystallization of **III** from benzene yielded white, needle-shaped crystals which showed the expected molecular weight by mass spectrometry and the anticipated IR, UV, and NMR spectra, but consistently gave aberrant C—H microanalytical data. Despite its apparent purity from spectroscopy, a broad melting point range (222–245 °C) confirmed the indication from microanalytical data that an impure compound was in hand. (2) Repeated sublimation of the product, however, improved both the mp (sharpened to 245 °C) and the microanalytical data. (3) X-ray powder patterns of the unsublimed and sublimed materials showed that the former had a hexagonal crystal structure while the latter had a monoclinic or triclinic structure. The explanation? The molecules of **III** naturally tend to form paddle-like structures. These can stack up to form tunnels in the solid that will entrap small molecules (but not larger ones) to form channel clathrates (see Fig. 14.22). Solvent molecules (such as benzene) were trapped during each successive recrystallization of these compounds and the purity could not be improved. This clathrate naturally has aberrant properties when studied as a solid (i.e., microanalysis, melting point, or X-ray structure when first crystallized, etc.) but the expected properties of the pure host are observed when studied in solution or upon sublimation as the clathrated guest molecules are released.

Other heterocyclic inorganic ring systems

Polymeric chain, band, and sheet silicate structures have been discussed previously, and it is not surprising to learn that cyclic silicate anions are known, such as $[Si_3O_9]^{-6}$ and

[74] H. R. Allcock et al., *J. Am. Chem. Soc.*, **1976**, *98*, 5120.

$[Si_6O_{18}]^{-12}$ (Fig. 14.23). These anions are sometimes referred to as "metasilicates" in line with the older system of nomenclature, which assigned *ortho* to the most fully hydrated species (as in "orthosilicic acid, $Si(OH)_4$") and *meta* to the acid (and anion) from which 1 mole of water has been removed (either in fact or formally; for example, "metasilicic acid, $OSi(OH)_2$").

Isoelectronic with the cyclic silicates are cyclic metaphosphates. The simplest member of the series is the trimetaphosphate anion, $[P_3O_9]^{-3}$. The tetrametaphosphate anion, $[P_4O_{12}]^{-4}$, is also well known. By careful chromatographic separations of the glassy mixture of polymeric phosphates and metaphosphates known as Graham's salt it is possible to show the existence not only of tri- and tetrametaphosphates, but also penta-, hexa-,[75] hepta-, and octametaphosphate. The separation is effected and some qualitative knowledge of structure gained from the fact that two factors play a role in the mobility of phosphate anions: (1) Higher-molecular-weight anions move more slowly than do lower members of the series; and (2) the ring or metaphosphate anions move more rapidly in basic solution than do the straight-chain anions of comparable complexity (Fig. 14.24).

In progressing from silicon to phosphorus the increase of one in atomic number resulted in a corresponding decrease of one per central atom in the anionic charge of the rings. Further progression from trimetaphosphate to sulfur trioxide results in a neutral molecule, trimeric sulfur trioxide. This form is known as γ-SO_3 and is isoelectronic and isostructural with the analogous trimetasilicate and trimetaphosphate anions. It is thermodynamically unstable with respect to two other forms: β-SO_3, which consists of infinite chains, and α-SO_3, which probably consists of infinite sheets (cf. silicate structures).[76]

In many cases it is possible to prepare ammono analogues in which $-NH_2$ and $=NH$ groups replace $-OH$ and $=O$ (see p. 331). For example, imidophosphates may be prepared by hydrolysis of phosphonitrilic chlorides accompanied by a tautomeric shift of hydrogen atoms to form a ring isostructural with those discussed above:

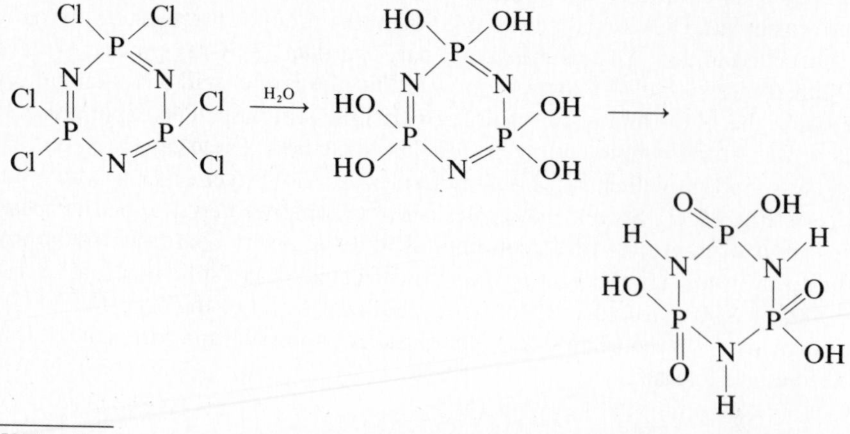

$$(\text{Cl}_2\text{PN})_3 \xrightarrow{H_2O} (\text{(HO)}_2\text{PN})_3 \longrightarrow (\text{HN}-\text{PO(OH)})_3 \qquad (14.47)$$

[75] There is confusion in the literature over the name "hexametaphosphate." On the basis of erroneous reasoning concerning the nature of "double salts," the term "hexametaphosphate" was assigned to Graham's salt of empirical composition $NaPO_3$. It has also been applied to the related commercial product (Calgon) in which the Na/P ratio is 1:1. The true hexametaphosphate contains a twelve-membered phosphorus–oxygen ring and is but a very minor component of the mixture known as "Graham's salt."

[76] See F. A. Cotton and G. Wilkinson, "Advanced Inorganic Chemistry," 4th ed., Wiley (Interscience), New York, **1980**, p. 530.

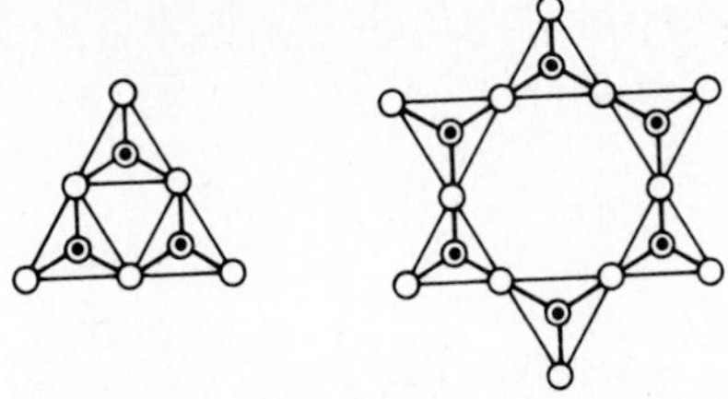

Fig. 14.23 Diagrammatic structures of the "metasilicate" anions: (a) trisilicate, $[Si_3O_9]^{-6}$; (b) hexasilicate, $(Si_6O_{18}]^{-2}$.

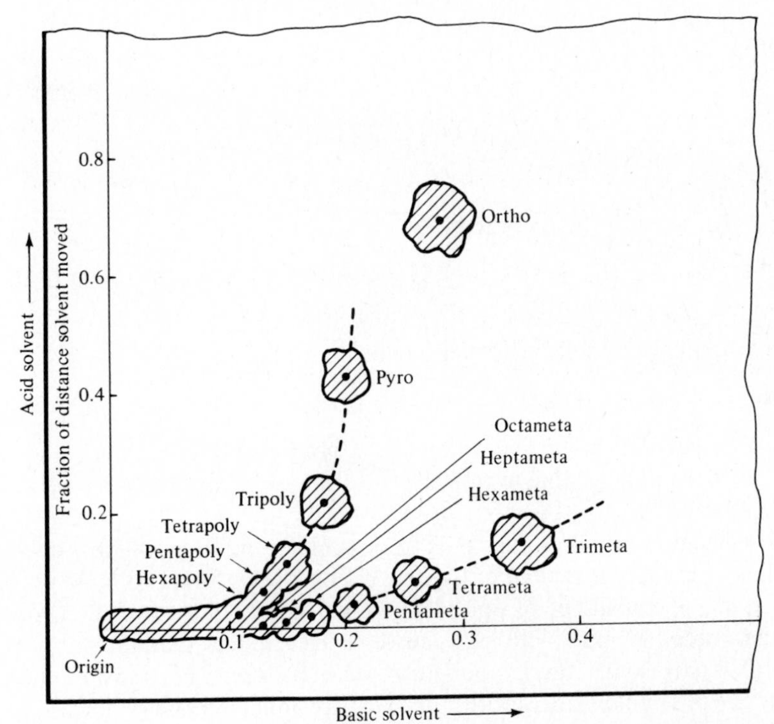

Fig. 14.24 Separation of polyphosphate anions by paper chromatography. The ions are first allowed to migrate in a basic solvent and subsequently in an acidic solvent. The straight-chain polyphosphates lie on the ascending branch of the "Y," the metaphosphates on the lower branch. [From J. R. Van Wazer, "Phosphorus and Its Compounds," Wiley, New York, **1958**, Vol. I, p. 702. Reproduced with permission.]

Sulfur–nitrogen rings may be prepared by ammonolysis of sulfuryl chloride:

$$SO_2Cl_2 + 4NH_3 \longrightarrow SO_2(NH_2)_2 + 2NH_4Cl \tag{14.48}$$

followed by deammonation of the resulting sulfamide:

$$3SO_2(NH_2)_2 \xrightarrow[\text{heat}]{} 3NH_4^+ + [(SO_2N)_3]^{-3} \tag{14.49}$$

Several heavy metal slats have been prepared of which the silver salt is useful in preparing the free acid (known in solution only) and organic derivatives:

$$3Ag^+ + [(SO_2N)_3]^{-3} \xrightarrow{HCl} 3AgCl\downarrow + \text{(cyclic } S_3N_3 \text{ ring: each S bearing two O, each N bearing H)} \tag{14.50}$$

$$3Ag^+ + [(SO_2N)_3]^{-3} \xrightarrow{CH_3I} 3AgI\downarrow + \text{(cyclic } S_3N_3 \text{ ring: each S bearing two O, each N bearing } CH_3\text{)} \tag{14.51}$$

The corresponding tetramer can also be prepared:

$$2SO_2(NH_2)_2 + 2SO_2Cl_2 \xrightarrow[CH_3CN]{} \xrightarrow[H_2O]{NH_3} 4NH_4^+ + [(SO_2N)_4]^{-4} \tag{14.52}$$

The ammonolysis of sulfur monochloride, S_2Cl_2, either in solution in an inert solvent or heated over solid ammonium chloride, yields tetrasulfur tetranitride:

$$6S_2Cl_2 \xrightarrow[CCl_4]{NH_3} S_4N_4 + S_8 + 12NH_4Cl \tag{14.53}$$

The product is a bright orange solid insoluble in water but soluble in some organic solvents. Although the crystals are reasonably stable to attack by air, they are explosively sensitive to shock or friction.

A few moments' reflection will show that it is impossible to write a simple Lewis structure for S_4N_4. Furthermore, the structure (Fig. 14.25a) has been found to have two "nonbonding" sulfur atoms at a distance of only 258 pm, considerably shorter than the sum of the van der Waals radii (360 pm). Although this distance is longer than the normal S—S bond length (~206 pm), some interaction must occur between the transannular sulfur atoms. All of the S—N bond distances within the ring are equal (163 pm), indicating extensive delocalization rather than alternating discrete single and double bonds (see below). The situation is similar to but more complicated than that of the phosphonitrilic system, and molecular orbital interpretations have been proposed.[77]

Fluorination of S_4N_4 produces tetrathiazyl tetrafluoride:

$$S_4N_4 + 4AgF_2 \longrightarrow N_4S_4F_4 + 4AgF \tag{14.54}$$

The presence of the fluorine substituents appears to destroy the delocalization of the electrons in the ring (at least in part) since the bond lengths in the ring alternate in length (Fig. 14.25b).

Reduction of tetrasulfur tetranitride with tin(II) chloride produces tetrasulfur tetraimide, $S_4(NH)_4$, isoelectronic with rhombic sulfur (see p. 719):

$$S_4N_4 \xrightarrow{SnCl_2} S_4N_4H_4 \tag{14.55}$$

[77] P. S. Braterman, *J. Chem. Soc.*, **1965**, 2297. See also footnote 64.

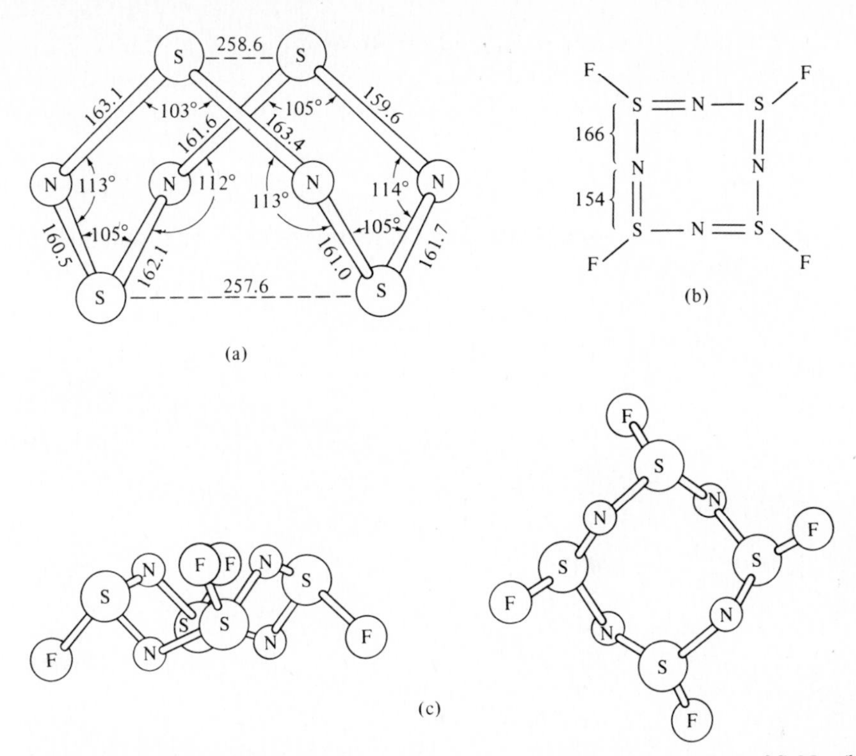

Fig. 14.25 Eight-membered sulfur–nitrogen rings: (a) molecular structure of S_4N_4; (b) diagrammatic structure of $N_4S_4F_4$ illustrating alternating bond lengths; (c) perspective drawings of the $N_4S_4F_4$ molecule. [From B. D. Sharma and J. Donohue, *Acta Crystallogr.*, **1963**, *16*, 891; G. A. Wiegers and A. Vos, *ibid.*, **1962**, *14*, 562; **1963**, *16*, 152. Reproduced with permission.] Distances in pm.

Like the latter, it exists in a crown configuration. Related, isoelectronic S—NH rings have been studied extensively during the past fifteen years, including isomers of pentasulfurtriimide, hexasulfurdiimide, and heptasulfurimide:[78]

```
H       H     H       H     H       H     H
 \     /       \     /       \     /       \
  N—S—N         N—S—N         N—S—N         N—S—S
  |   |         |   |         |   |         |   |
  S   S         S   S         S   S         S   S
  |   |         |   |         |   |         |   |
  N—S—N         S—S—N         S—N—S         S—S—N
 /     \             \          |                \
H       H             H         H                 H

   H             H       H             H
    \             \     /             /
     N—S—S         N—S—N         S—S—N
     |   |         |   |         |   |
     S   N—H       S   S         S   S
     |   |         |   |         |   |
     S—S—S         S—S—S         S—S—S
```

[78] See B. A. Olsen and F. P. Olsen, *Inorg. Chem.*, **1969**, *8*, 1736, and references therein.

Note that all possible $S_x(NH)_{8-x}$ isomers are known with the exception of N—N bonded structures. The latter may be unstable as a result of the weak N—N bond (see Appendix E).

Six-membered S—N rings are also known. Oxidation of S_4N_4 with chlorine produces trithiazyl trichloride:

$$3S_4N_4 + 6Cl_2 \longrightarrow 4N_3S_3Cl_3 \qquad (14.56)$$

This compound may be converted into the corresponding fluoride or oxidized to sulfanuryl chloride:

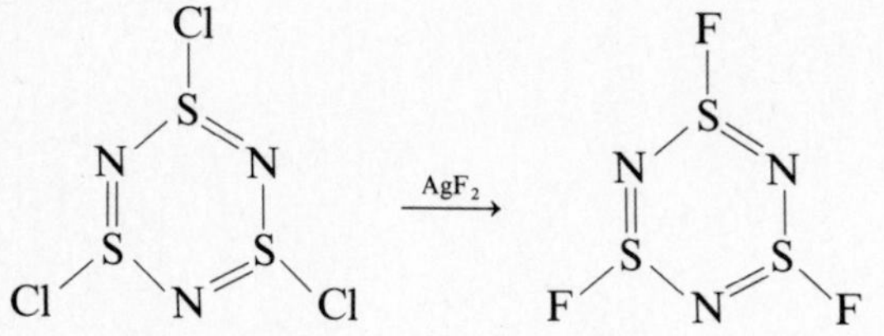

$$\xrightarrow{AgF_2} \qquad (14.57)$$

$$\xrightarrow{SO_3} \qquad (14.58)$$

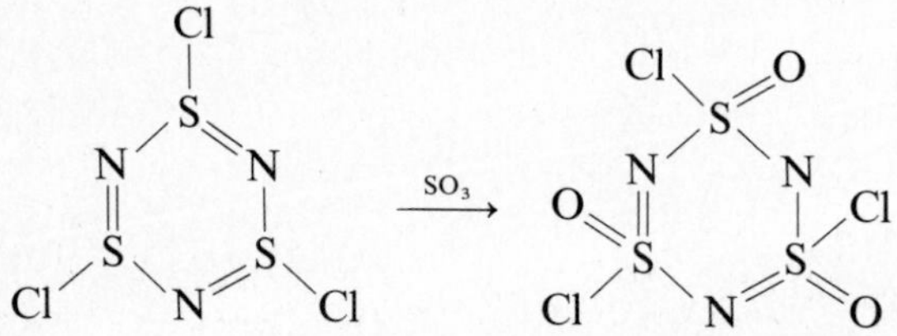

The latter compound is more readily made from sulfamic acid:[79]

$$H_2NSO_2OH + 2PCl_5 \xrightarrow{\Delta} Cl_3N{=}PSO_2Cl + OPCl_3 + 3HCl \qquad (14.59)$$

$$3Cl_3N{=}PSO_2Cl \xrightarrow{\Delta} (NSOCl)_3 + 3OPCl_3 \qquad (14.60)$$

The relative stability of six- and eight-membered rings is governed by factors that are not fully understood. When the relatively unstable thiazyl fluoride, NSF, polymerizes, only the trimer, $N_3S_3F_3$, is formed. Likewise, the formation of the trimeric chloride in reaction 14.56 (from a *tetrameric* reactant) indicates a preferential stability of the six-membered ring. On the other hand, reduction of the trimeric chloride with tetrasulfurtetraimide results only in products containing eight-membered rings:[80,81]

$$4N_3S_3Cl_3 + 3S_4N_4H_4 \longrightarrow 6S_4N_4 + 12HCl \qquad (14.61)$$

When S_4N_4 is sublimed over silver wool, the planar four-membered ring, S_2N_2, is formed. It is the precursor of the $(SN)_x$ polymer (see p. 692).

---

[79] T. Moeller and R. L. Dieck, *Prep. Inorg. Reactions*, **1971**, *6*, 63.

[80] Note that simple splitting off of HCl *might* result in $4S_3N_3 + 3S_4N_4$. Apparently the trimer, if formed at all, is unstable and rearranges to S_4N_4.

[81] For further reading on S—N ring compounds, see O. Glemser and M. Fild in *Halogen Chem.*, **1967**, *2*, 1, and M. Becke-Goehring, *Adv. Inorg. Chem. Radiochem.*, **1960**, *2*, 159.

Homocyclic inorganic systems

Several elements form homocyclic rings. Rhombic sulfur, the thermodynamically stable form at room temperature, consists of S_8 rings in the crown conformation. Unstable modifications containing six-, seven-, nine-, ten-, twelve-, and eighteen-membered sulfur rings are also known.[82] Selenium also forms eight-membered rings, but they are unstable with respect to the chain form.

Oxidation of several nonmetals in strongly acidic systems produces polyatomic cationic species of the general type Y_n^{+m}. The best characterized of these are the S_4^{+2}, Se_4^{+2}, and Te_4^{+2} ions.[83]

The structures of the Se_4^{+2} and Te_4^{+2} ions have been shown to be square planar[84]

```
Se—Se+2        Te—Te+2
| O |          | O |
Se—Se          Te—Te
```

and it is probable that all three species have the same structure, stabilized to a certain extent by a Hückel sextet of π electrons. Another homoatomic cation, Se_8^{+2}, has been found[85] to be bicyclic (Fig. 14.26). The transannular bond is longer (284 pm) than those around the ring (232 ± 2 pm) but considerably less than the sum of the van der Waals radii (380 pm). Other ions of this type, presumably cyclic though less thoroughly studied, S_4^{+}, S_8^{+2}, S_{16}^{+2}, Sb_4^{+2}, Sb_8^{+2}, Sb_n^{+n}, and Te_6^{+2}, have been suggested as products of mild oxidation of some nonmetals.[86] The synthesis, characterization, and explanation of rings of this type will undoubtedly result in considerable experimental and theoretical work in the future.

Several cyclopolyphosphines are known.[87] The reduction of trifluoromethylphosphorus halides (X = Br, I) results in approximately equal amounts of four- and five-membered compounds:

$$CF_3PX_2 + Hg \longrightarrow HgX_2 + \frac{1}{n}(CF_3P)_n \qquad n = 4, 5 \tag{14.62}$$

---

[82] M. Schmidt and E. Eilhelm, *Chem. Commun.*, **1970**, 1111, and references therein. Recently, equilibrium concentrations (~1%) of S_6 and S_7 have been reported in solutions of S_8 in polar solvents, F. N. Tebbe et al., *J. Am. Chem. Soc.*, **1982**, *104*, 4971.

[83] J. Barr et al., *Can. J. Chem.*, **1968**, *46*, 3607 [Se_4^{+2}]; D. J. Prince et al., *Inorg. Chem.*, **1970**, *9*, 2731 [Se_4^{+2}, Se_8^{+2}, Te_4^{+2}]; N. J. Bjerrum, *Inorg. Chem.*, **1970**, *9*, 1965 [Te_4^{+2}]; J. Barr et al., *Inorg. Chem.*, **1971**, *10*, 362 [Te_4^{+2}, Te_6^{+2}]; and references therein.

[84] I. D. Brown et al., *Inorg. Chem.*, **1971**, *10*, 2319; T. W. Couch et al., ibid., **1972**, *11*, 357.

[85] R. K. McMullan et al., *Inorg. Chem.*, **1971**, *10*, 1749.

[86] W. F. Giggenbach, *Chem. Commun.*, **1970**, 852 [S_4^{+2}, S_8^{+2}]; P. A. W. Dean and R. J. Gillespie, *Chem. Commun.* **1970**, 853 [Sb_n^{+n}]; K. C. Malhotra and J. K. Puri, *Inorg. Nucl. Chem. Letters*, **1971**, *7*, 209 [Sb_4^{+2}, Sb_8^{+2}]; R. J. Gillespie et al., *Inorg. Chem.*, **1971**, *10*, 1327 [S_4^{+2}, S_8^{+2}, S_{16}^{+2}]. Two recent reviews cover some of these species: R. J. Gillespie and J. Passmore, *Adv. Inorg. Chem. Radiochem.*, **1975**, *17*, 49; J. D. Corbett, *Progr. Inorg. Chem.*, **1976**, *21*, 129. See also footnote 83.

[87] For a selection papers on this topic together with editorial comment on them, see A. H. Cowley, ed., "Compounds Containing Phosphorus–Phosphorus Bonds," Dowden, Hutchinson & Ross, Inc., Stroudsburg, Pa., **1972**.

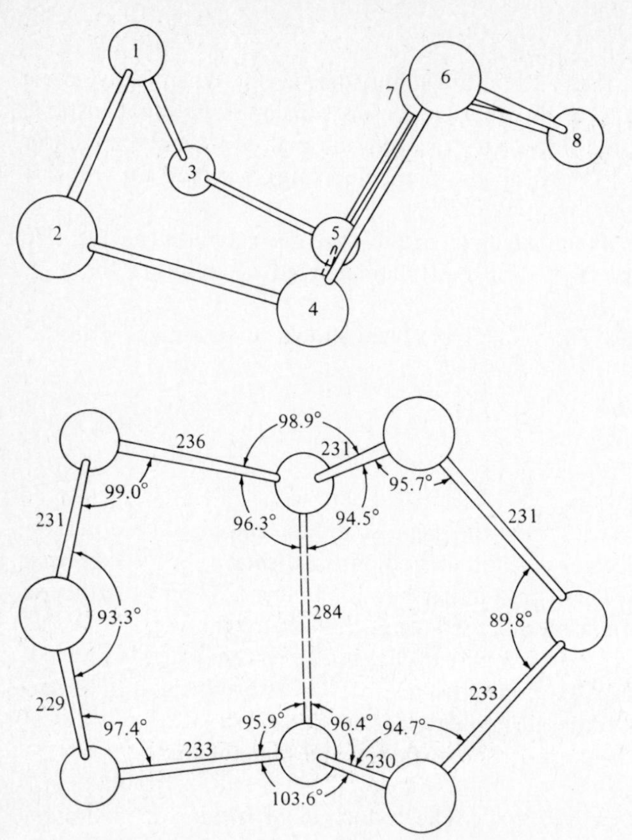

Fig. 14.26 Molecular structure of the bicyclic Se_8^{+2} cation. [From R. K. McMullan et al., *Inorg. Chem.*, **1971**, *10*, 1749. Reproduced with permission.] Distances in pm.

Similar compounds may be obtained by elimination of hydrogen chloride:

$$\frac{n}{2}\,\phi PH_2 + \frac{n}{2}\,\phi PCl_2 \longrightarrow (\phi P)_n + n HCl \qquad \textbf{(14.63)}$$

Although pentaphenylcyclopentaphosphine and hexaphenylcyclohexaphosphine are isolated from this reaction, there is evidence that in solution both forms exist as the tetramer. This facile interconversion has hindered the elucidation of the nature of these compounds. Earlier workers[88] assigned a tetrameric structure based on freezing-point depression in solution, but X-ray studies have confirmed the presence of five-[89] and six-membered[90] rings in the solid. Other alkylcyclopolyphosphines, $(RP)_n$, are known with $R = CH_3$, C_2H_5, C_3H_7, C_4H_9. Arsenic is also known to form several homocyclic rings such as $(CF_3As)_4$, $(CH_3As)_5$, and $(\phi As)_5$, but relatively little work has been done on these systems.[91]

[88] W. Kuchen and H. Buchwald, *Chem. Ber.*, **1958**, *91*, 2296. See also W. A. Henderson et al., *J. Am. Chem. Soc.*, **1963**, *85*, 2462, concerning the conversion to the tetramer in solution.

[89] J. J. Daly and L. Maier, *Nature*, **1964**, *203*, 1167; J. J. Daly, *J. Chem. Soc.*, **1964**, 6147.

[90] J. J. Daly and L. Maier, *Nature*, **1965**, *208*, 383; J. J. Daly, *J. Chem. Soc.*, **1965**, 4789.

[91] See A. H. Cowley et al., *J. Am. Chem. Soc.*, **1966**, *88*, 3178, and references therein.

The oxidation of red phosphorus with hypohalites in alkaline solution produces the anion of an interesting phosphorus acid:

$$\frac{6}{x}P_x + 9XO^- + 6OH^- \longrightarrow (PO_2)_6^{-6} + 9X^- + 3H_2O \qquad (14.64)$$

The acid has been shown to be cyclic with a structure:[92]

```
         O   OH
          \\ /
 HO        P     O
   \      / \   //
  O=P         P—OH
 HO  |        |  O
   \ |        | //
     P        P
   //  \     /   \
  O      P        OH
        / \\
      HO    O
```

There is a series of analogous cyclic thiophosphoric acids with the formula $(HS_2P)_n$ that may be prepared by the oxidation of red or white phosphorus with polysulfides under a variety of conditions. For example, the reaction of white phosphorus with a mixture of sulfur and hydrogen sulfide dissolved in triethylamine (which acts as a base) and chloroform opens the phosphorus cage (see p. 723) to form the tetrameric cyclic anion:

$$P_4 + 4Et_3N + 2H_2S + 6S \xrightarrow[CHCl_3]{} (Et_3NH)_4P_4S_8 \qquad (14.65)$$

The square structure of the anion has been confirmed by X-ray crystallography.[93]

A series of cyclopolyarsines is known. They may be prepared by a generally useful reaction that is reminiscent of the Wurtz reaction of organometallic chemistry:

$$RAsCl_2 + Na \longrightarrow (RAs)_{4,5} \qquad (14.66)$$

Although the four- and five-membered rings can be made in this way, the three-membered ring requires a special, though related, reaction:

$$(t\text{—Bu})\text{—}\underset{\displaystyle K}{\underset{|}{As}}\text{—}\underset{\displaystyle K}{\underset{|}{As}}\text{—}(t\text{—Bu}) + t\text{—Bu—AsCl}_2 \xrightarrow{-78\,^\circ C} \text{cyclo-}(t\text{-BuAs})_3 \qquad (14.67)$$

```
t-Bu        t-Bu
    \       /
     As——As
      \   /
       As
        |
      t-Bu
```

This compound is stable only at $-30\ ^\circ C$ in the dark and in the absence of air. It spontaneously ignites on exposure to air.[94]

An unusual "triple-decker sandwich" (see Fig. 13.22) compound, stable at room temperature, can be made from pentamethylcyclopentaarsine:

$$[CpMo(CO)_3]_2 + (CH_3As)_5 \xrightarrow{190^\circ} CpMo(As)_5MoCp \qquad (14.68)$$

[92] B. Blaser and K.-H. Worms, *Z. Anorg. Allg. Chem.*, **1959**, *300*, 237.

[93] H. Falius et al., *Angew. Chem. Int. Ed. Eng.*, **1981**, *20*, 103.

[94] M. Baudler and P. Bachmann, *Angew. Chem. Int. Ed. Eng.*, **1981**, *20*, 123.

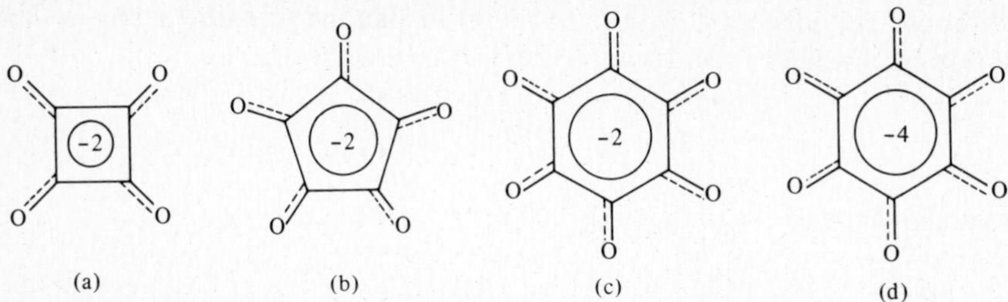

Fig. 14.27 Cyclic oxocarbon anions: (a) squarate; (b) croconate; (c) rhodizonate; (d) tetraanion of tetrahydroquinone.

Although the reactant pentaarsine is fully methylated, the resulting product contains an unsubstituted, five-membered ring of arsenic atoms. The molybdenum carbonyl appears to catalyze the demethylation reaction.[95]

There is an interesting series of oxocarbon anions of general formula $[(CO)_n]^{-2,-4}$ (Fig. 14.27). The croconate ion, $C_5O_5^{-2}$, was the first member of the series to be synthesized. From a historical point of view it is especially interesting: (1) It was isolated in 1825 by Gmelin and thus shares with benzene (isolated from coal tar by Faraday the same year) the honor of being the first aromatic compound discovered. (2) It was the first "*inorganic*" substance discovered that is aromatic, although its importance was unrealized until recently. (3) It is a bacterial metabolic product and was possibly the first "organic" compound synthesized, predating Wöhler's synthesis of urea by three years, although here too the significance was unappreciated at the time.

All of these oxocarbon anions are aromatic according to simple molecular orbital calculations. The aromatic stabilization of the anion is apparently responsible for the fact that squaric acid ($H_2C_4O_4$) is about as strong as sulfuric acid.[96] There is a considerable and growing body of knowledge of the chemistry of these systems, but most of it is probably more appropriate to a discussion of organic chemistry.[97]

In a formal sense, silicon might be expected to parallel the extensive alicyclic and aromatic chemistry of carbon, but we have already seen the reduced stability of silanes. Substitution of hydrogen atoms by methyl groups seems to stabilize these systems. A large series of permethylcyclosilanes can be synthesized by treatment of chlorosilanes with an active metal over a prolonged period of time:

$$Me_2SiCl_2 + Na/K \xrightarrow[C_{10}H_8]{THF} (Me_2Si)_x + (Me_2Si)_n \qquad \textbf{(14.69)}$$

The product consists of various amounts of high polymer (x is very large) and discrete cyclosilanes with $n = 5$–35. This is the largest homologous series of cyclic compounds now

[95] A. L. Rheingold et al., *J. Am. Chem. Soc.*, **1982**, *104*, 4727.

[96] Note that oxalic acid containing carbon in a comparable oxidation state but not aromatic has a K_1 approximately equal to K_2 for squaric and sulfuric acids, and K_2 for oxalic acid is three orders of magnitude smaller. For a dissident, partially semantic difference of opinion concerning the aromaticity of the oxocarbon anions, see J. Aihara, *J. Am. Chem. Soc.*, **1981**, *103*, 1633.

[97] A. H. Schmidt and W. Ried, *Synthesis*, **1978**, 869; R. West, *Isr. J. Chem.*, **1980**, *20*, 390; R. West et al., *J. Am. Chem. Soc.*, **1981**, *103*, 5073; R. West, ed., "Chemistry of Oxocarbons," Academic Press, New York, **1980**.

known except for the cycloalkanes. Although these compounds are formally "saturated," they behave in some ways as aromatic hydrocarbons. They can be reduced to anion radicals, and ESR spectra indicate that the unpaired electron is delocalized over the entire ring.[98]

CAGES

Cage structures range from clathrate compounds on the one hand to metal–metal clusters and boranes on the other. These classes are discussed elsewhere,[99] and this section will be restricted to certain nonmetal compounds having cage structures.

The simplest cage-type molecule is found in white phosphorus, P_4. Although P_2 molecules, isoelectronic with N_2, are found in phosphorus vapor at higher temperatures, P_4 is more stable at room temperature.[100] This molecule is a tetrahedron of phosphorus atoms:

P 221 pm
P P
P

Such a structure requires bond angles of 60°. Inasmuch as the lowest interorbital angle available using only *s* and *p* orbitals is 90° (pure *p* orbitals), the smaller bond angle in P_4 must be accomplished either through the introduction of considerable *d* character or through the use of "bent bonds." The former requires considerable promotion energy and is therefore unlikely. The latter results in a loss in bonding energy of some 96 kJ mol^{-1} due to strain but is still thought to be favored energetically. In any event the molecule is destabilized and quite reactive. It reacts readily with oxygen to form a mixture of oxides and can be converted readily to more stable allotropes:

$$P_4 + O_2 \longrightarrow P_4O_6, P_4O_{10} \tag{14.70}$$

$$\frac{x}{4} P_4 \xrightarrow[\text{heat}]{h\nu \text{ or}} P_x \quad \text{(red phosphorus)} \tag{14.71}$$

$$\frac{x}{4} P_4 \xrightarrow[\text{Hg catalyst}]{\text{pressure or}} P_x \quad \text{(black phosphorus)} \tag{14.72}$$

Crystalline black phosphorus has a corrugated layer structure.[101] "Red phosphorus" does not appear to be a well-defined substance but differs according to the method of preparation.

---

[98] R. West and E. Carberry, *Science*, **1975**, *189*, 179; L. F. Brough, *J. Organometal. Chem.*, **1980**, *194*, 139; L. F. Brough, *J. Am. Chem. Soc.*, **1981**, *103*, 3049.

[99] Clathrate compounds are discussed both in Chapter 6 and earlier in this chapter, while metal clusters and boranes are found later in this chapter.

[100] See Chapter 17 for a discussion of the difference between the chemistry of nitrogen and phosphorus.

[101] For this and several other interesting elemental structures, see J. Donahue, "The Structures of the Elements," Wiley (Interscience), New York, **1974**.

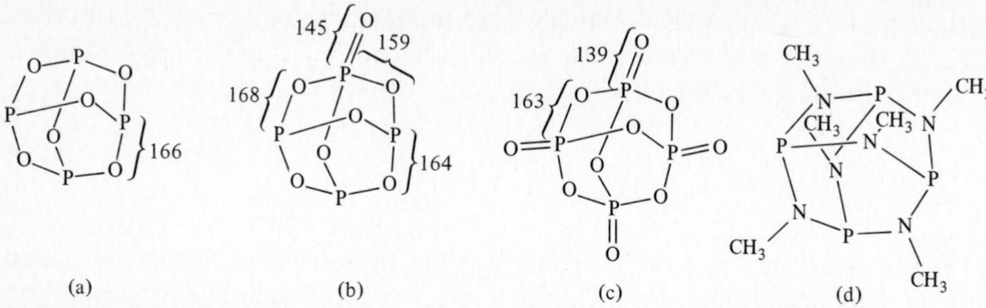

Fig. 14.28 Phosphorus cage molecules: (a) P_4O_6; (b) P_4O_7; (c) P_4O_{10}, (d) $P_4(NCH_3)_6$. Distances in pm.

It probably consists of random chains. The rate of formation is increased by certain substances such as iodine which appear to be incorporated into the product.

There are three known phosphorus oxides with cage structures. All have tetrahedral or pseudotetrahedral (C_{3v} in the case of P_4O_7)[102] symmetry. Although the molecular formulas of two of them are P_4O_6 and P_4O_{10}, they are often referred to as the "trioxide" and "pentoxide." All are anhydrides that react readily with water to form the corresponding acids:

$$P_4O_6 \xrightarrow[\text{vigorous agitation}]{\text{cold water}} 4HPO(OH)_2 \qquad \textbf{(14.73)}$$

$$P_4O_{10} + 6H_2O \longrightarrow 4OP(OH)_3 \qquad \textbf{(14.74)}$$

(The P_4O_7 molecule reacts to form both phosphoric acid and phosphorous acid.) In addition to the discrete cage molecule pictured above, phosphorus pentoxide also exists in several polymeric forms.

Reaction of phosphorus trichloride with methylamine[103] produces a cage phosphorus imide (Fig. 14.28d) isoelectronic and isostructural with P_4O_6:

$$4PCl_3 + 18CH_3NH_2 \longrightarrow P_4(NCH_3)_6 + 12CH_3NH_3Cl \qquad \textbf{(14.75)}$$

The chemistry of phosphorus and sulfur is considerably more complicated than phosphorus–oxygen chemistry. Only one phosphorus sulfide, P_4S_{10}, is isoelectronic and isostructural with a phosphorus oxide. It may be prepared by allowing stoichiometric amounts of phosphorus and sulfur to react:

$$4P_4 + 5S_8 \longrightarrow 4P_4S_{10} \qquad \textbf{(14.76)}$$

By mixing phosphorus and sulfur in appropriate stoichiometric quantities, two other sulfides may be obtained:

$$8P_4 + 3S_8 \longrightarrow 8P_4S_3 \qquad \textbf{(14.77)}$$

$$8P_4 + 7S_8 \longrightarrow 8P_4S_7 \qquad \textbf{(14.78)}$$

[102] M. Jansen and M. Voss, *Angew. Chem. Int. Ed. Eng.*, **1981**, *20*, 100.
[103] R. R. Holmes, *J. Am. Chem. Soc.*, **1961**, *83*, 1334.

A fourth phosphorus sulfide may be most conveniently prepared by the slow oxidation of P_4S_3 with sulfur:

$$4P_4S_3 + S_8 \xrightarrow[\text{daylight, } CS_2]{\text{diffuse}} 4P_4S_5 \quad \textbf{(14.79)}$$

Two cage phosphorus sulfides may be synthesized by the formation of sulfide bridges through the action of bis(trimethyltin) sulfide:[104]

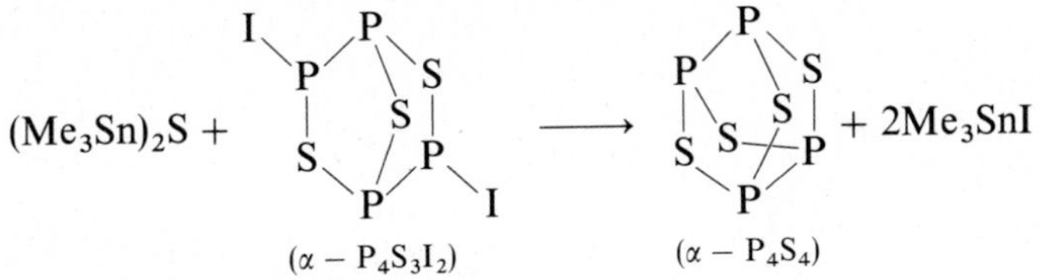

$$(Me_3Sn)_2S + (\alpha - P_4S_3I_2) \longrightarrow (\alpha - P_4S_4) + 2Me_3SnI \quad \textbf{(14.80)}$$

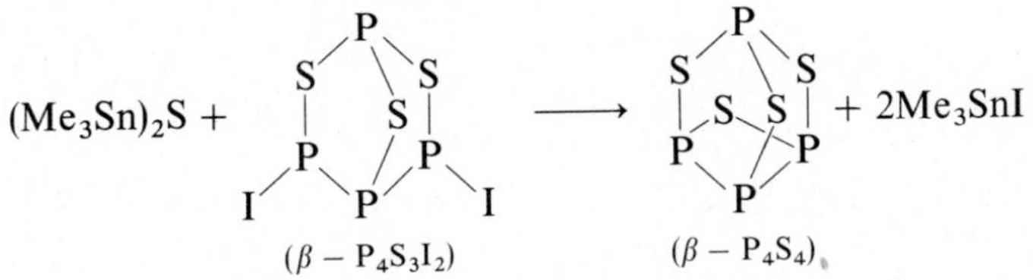

$$(Me_3Sn)_2S + (\beta - P_4S_3I_2) \longrightarrow (\beta - P_4S_4) + 2Me_3SnI \quad \textbf{(14.81)}$$

The structures of all of these sulfides are known (Fig. 14.29). They are all derived from a tetrahedron of phosphorus atoms with sulfur atoms bridging along various edges. All except P_4S_{10} retain one or more P—P bonds.[105]

The heavier congeners of phosphorus resemble it in a tendency to form cages. Both arsenic and antimony form unstable tetrameric molecules which readily revert to polymeric structures. Cage molecules are also known for As_4O_6 and Sb_4O_6 as well as polymeric forms. In addition several sulfides form, some of which are known to exist as cages (Fig. 14.30).

By extension of the reactions involved in the formation of cyclopolysilanes, West[98] has synthesized six new bicyclic and cage permethylpolysilanes such as:

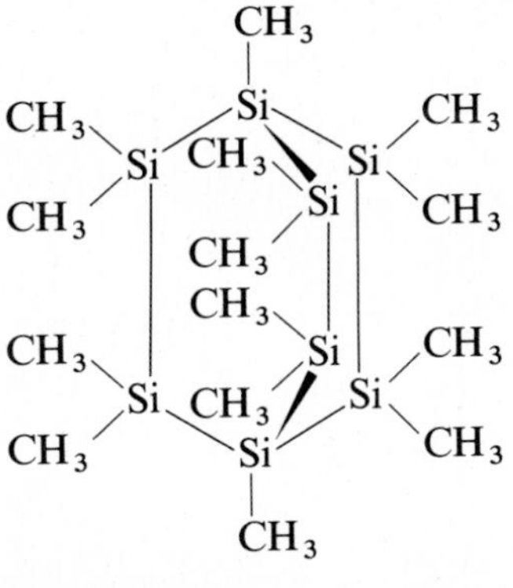

[104] A. M. Griffin et al., *Chem. Commun.*, **1976**, 809.

[105] For a more extensive discussion of the sulfides of phosphorus as well as references to the structural work, see J. R. Van Wazer, "Phosphorus and Its Compounds," Wiley (Interscience), New York, **1958**, Vol. I, pp. 289–305.

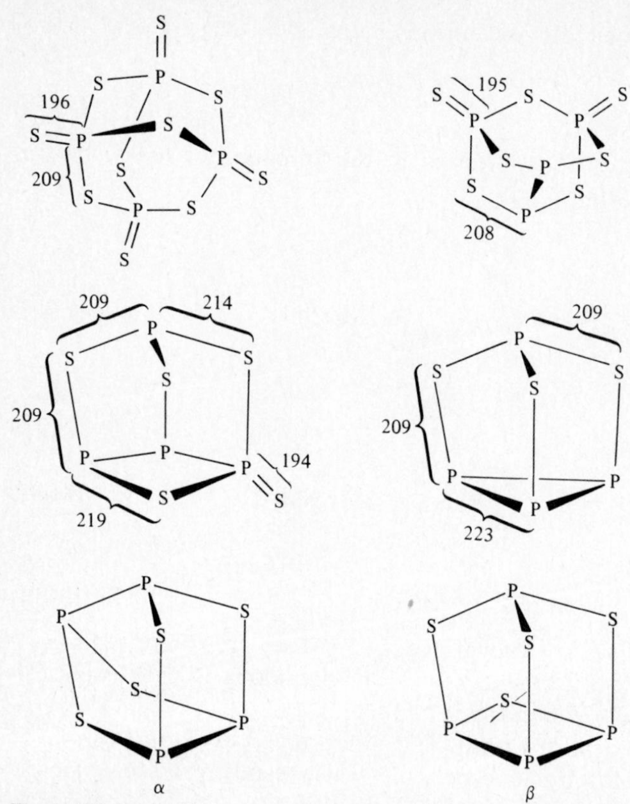

Fig. 14.29 Molecular structures of some phosphorus sulfides. Distances in pm.

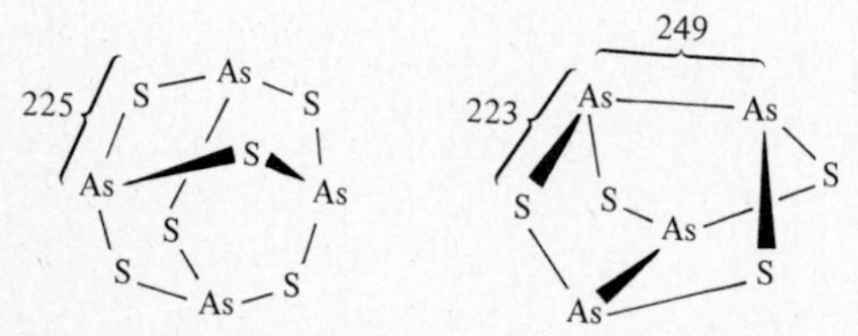

Fig. 14.30 Molecular structures of two arsenic sulfides. Distances in pm.

In order to get branching to form cages (bridgehead silicon atoms) some methyltrichlorosilane is added to the dimethyldichlorosilane in the reaction.

BORON CAGE COMPOUNDS

Boranes

Reduction of boron halides might be expected to produce borane, BH_3. However, it is found to be impossible to isolate the monomer, all syntheses resulting in diborane,

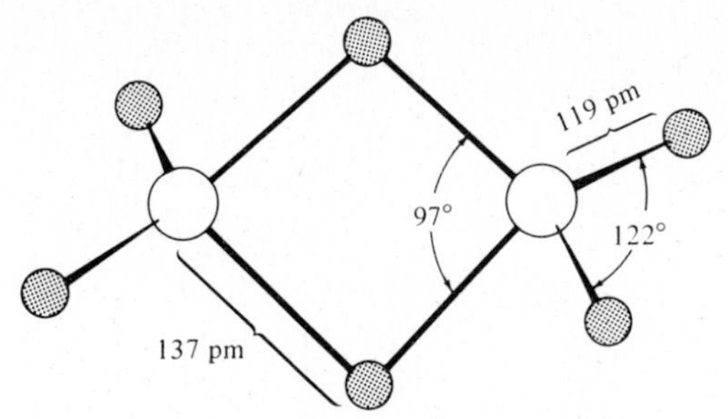

Fig. 14.31 Molecular structure of diborane, B_2H_6.

B_2H_6:[106]

$$8BF_3 + 6NaH \xrightarrow{\text{polyethers}} 6NaBF_4 + B_2H_6 \quad (14.82)$$

$$4BCl_3 + 3LiAlH_4 \xrightarrow{(C_2H_5)_2O} 3LiCl + 3AlCl_3 + 2B_2H_6 \quad (14.83)$$

$$4BF_3 + 3NaBH_4 \xrightarrow{(C_2H_5)_2O} 3NaBF_4 + 2B_2H_6 \quad (14.84)$$

$$2NaBH_4 + H_2SO_4 \longrightarrow Na_2SO_4 + 2H_2 + B_2H_6 \quad (14.85)$$

Although BH_3 exists in the form of Lewis acid–base adducts and as a presumable intermediate in reactions of diborane, it exists in extremely small quantities as a free molecule. The equilibrium constant for dimerization is approximately 10^5 and the enthalpy of dissociation of the dimer to the monomer is about $+150$ kJ mol^{-1} or slightly more. Diborane is the simplest of what have come to be known as *electron deficient compounds.* They are electron deficient only in a formal sense since BH_3 satisfies all the requirements of simple valence theory[107] and the formation of tetracoordinate boron would seem to require additional electrons. Since these compounds show absolutely no tendency to accept electrons when offered by reducing agents, it is obvious that there is less deficiency of electrons than a deficiency of theory. Thus there have been several attempts to rationalize the bonding in these compounds. The most successful and extensive work in this area, as we shall see, has been that of William N. Lipscomb. His work has been of such value that he received the 1976 Nobel Prize in Chemistry for it.[108] It has aptly been said, "Boron hydrides are the children of this century yet the discovery of the polyhedral boranes, carboranes, and metalloboranes and the subsequent elaboration of chemistry, structure, and theory, and the incredibly rapid one considering the small number of investigators, are among the major developments in inorganic chemistry."[109]

Before proceeding with an examination of the bonding in diborane it will be helpful to examine its structure (Fig. 14.31). Each boron atom is surrounded by an approximate

---

[106] For a discussion of the synthetic chemistry of B—H compounds, see R. W. Parry and M. K. Walter, *Prep. Inorg. React.*, **1968**, *5*, 45.

[107] To be sure, BH_3 lacks a Lewis octet of electrons, but no more so than $B(CH_3)_3$, which shows no tendency to dimerize.

[108] See p. 359.

[109] E. L. Muetterties, ed., "Boron Hydride Chemistry," Academic Press, New York, **1975**.

tetrahedron of hydrogen atoms. The *bridging* hydrogen atoms are somewhat further from the boron atom and form a smaller bond angle than the other four nonbridging or terminal hydrogen atoms.

The earliest attempt at rationalizing the dimerization of borane invoked resonance in a valence bond (VB) context:

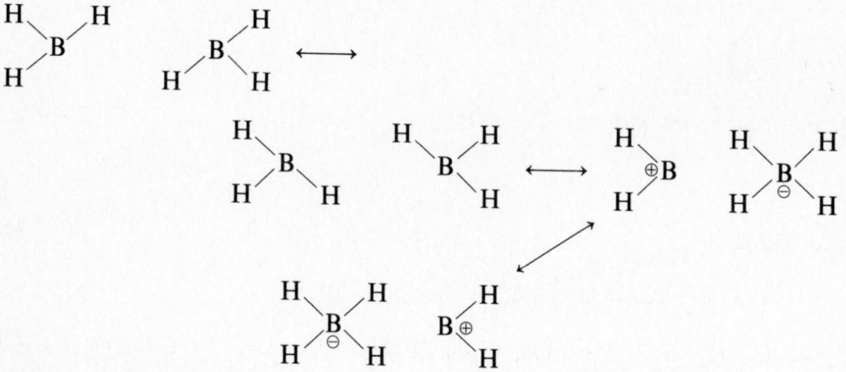

H H H
B B ⟷
H H H

H H H H H H H
B B ⟷ ⊕B B (14.86)
H H H H H ⊖ H

H H H
B B⊕
H ⊖ H H

Although adequate from a formal point of view, it suffers from the usual unwieldiness of VB terminology when extensive delocalization exists. A second attempt considered the $B_2H_4^{-2}$ anion as isoelectronic and isostructural with ethylene, C_2H_4. Such an ion would have a cloud of electron density above and below the B—H plane.[110] The neutral B_2H_6 molecule could then be formally produced by embedding a proton in the electronic cloud above and below the plane of the $B_2H_4^{-2}$ ion. Although this may appear to be somewhat far-fetched, it is but a simplistic way of describing the bonding situation which is currently accepted as best—the three-center, two-electron bond.

Consider each boron atom to be sp^3 hybridized.[111] The two terminal B—H bonds on each boron atom presumably are simple σ bonds involving a pair of electrons each. This accounts for eight of the total of twelve electrons available for bonding. Each of the bridging B—H—B linkages then involves a delocalized or three-center bond as follows. The appropriate combinations of the three orbital wave functions, ϕ_{B_1}, ϕ_{B_2} (approximately sp^3 hybrids), and ϕ_H (an *s* orbital) result in three molecular orbitals:

$$\Psi_b = \tfrac{1}{2}\phi_{B_1} + \tfrac{1}{2}\phi_{B_2} + \frac{1}{\sqrt{2}}\phi_H \tag{14.87}$$

$$\Psi_n = \frac{1}{\sqrt{2}}\phi_{B_1} - \frac{1}{\sqrt{2}}\phi_{B_2} \tag{14.88}$$

$$\Psi_a = \tfrac{1}{2}\phi_{B_1} + \tfrac{1}{2}\phi_{B_2} - \frac{1}{\sqrt{2}}\phi_H \tag{14.89}$$

[110] Note that this is true whether the σ–π model or the bent-bond model is employed for the double bond.

[111] This is only approximately correct (see Problem 14.2). The argument here does not rest upon the exact nature of the hybridization.

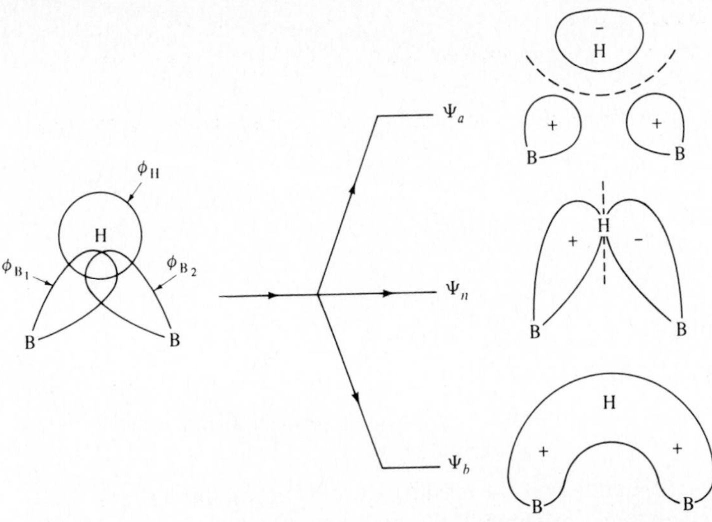

Fig. 14.32 Qualitative description of atomic orbitals (left), resulting three-center molecular orbitals (right), and the approximate energy level diagram (center) for one B—H—B bridge in diborane.

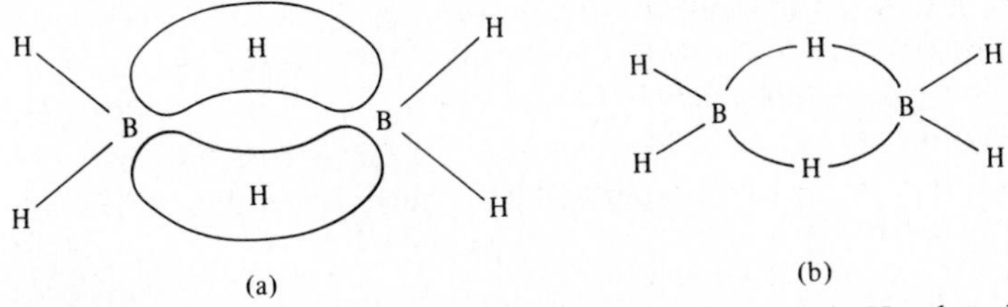

Fig. 14.33 (a) Qualitative picture of bonding in diborane. (b) Usual method of depicting B—H—B bridges.

where Ψ_b is a bonding MO, Ψ_a is an antibonding MO, and Ψ_n is, to a first approximation, a nonbonding MO.[112] The diagrammatic possibilities of overlap together with sketches of the resulting MOs and their relative energies are given in Fig. 14.32.

Each bridging bond thus consists of a bonding MO containing two electrons. Although the nonbonding orbital could conceivably accept an additional pair of electrons, this would not serve to stabilize the molecule beyond that achieved by the configuration Ψ_b^2. The second B—H—B bridge may likewise be considered as above and to have a configuration $\Psi_{b'}^2$. This accounts for the total of twelve bonding electrons and provides the rationale for the existence of the dimer (Fig. 14.33).

Diborane provides examples of two types of bonds found in higher boranes: the two-center, two-electron B—H terminal bond and the three-center, two-electron bridging B—H—B bond. Two other bonds are of importance in the higher analogues: (1) the

[112] The exact energy of this orbital varies depending upon the nature of the bonding properties of the atoms involved. For our purposes it does no harm (and simplifies the model) if it is assumed to be nonbonding.

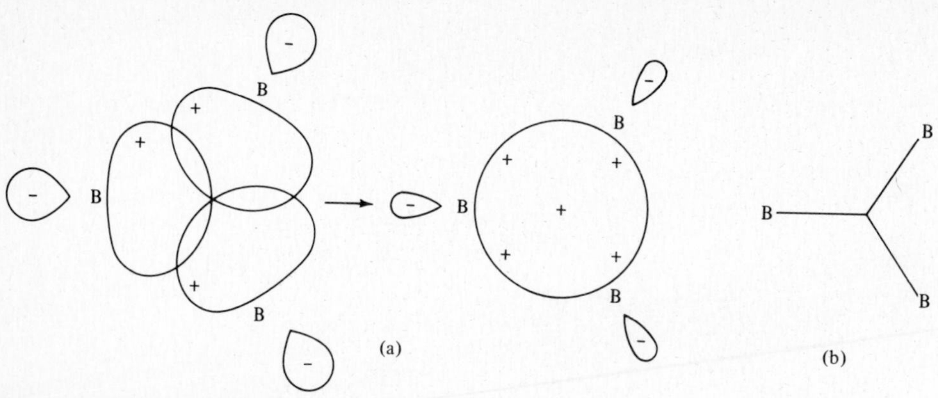

Fig. 14.34 The closed three-center, two-electron B—B—B bond: (a) formation from three boron atomic orbitals; (b) simplified representation.

two-center, two-electron B—B bond, best exemplified by the boron subhalides, X_2B—BX_2; and (2) the three-center, two-electron B—B—B bond, which may be formed by overlap of three orbitals from three corners of an equilateral triangle of boron atoms (Fig. 14.34).[113] Like the three-center B—H—B bond, three molecular orbitals will result, of which only the lowest energy or bonding one will be occupied by a pair of electrons.

With this repertoire of bonding possibilities at our disposal we can construct the molecular structure of various boron–hydrogen compounds, both neutral species and anions. The simplest is the tetrahydroborate[114] or borohydride ion, BH_4^-. Although borane is unstable with respect to dimerization, the addition of a Lewis base, H^-, satisfies the fourth valency of boron and provides a stable entity. Other Lewis bases can coordinate as well:

$$B_2H_6 + 2NaH \xrightarrow{\text{diglyme}} 2NaBH_4 \tag{14.90}$$

$$B_2H_6 + 2CO \longrightarrow 2H_3BCO \tag{14.91}$$

$$B_2H_6 + 2R_3N \longrightarrow 2H_3BNR_3 \tag{14.92}$$

Although the ammonia adduct of BH_3 is stable, it must be prepared by a method that does not involve B_2H_6 such as:

$$(CH_3)_2OBH_3 + NH_3 \longrightarrow H_3BNH_3 + (CH_3)_2O \tag{14.93}$$

---

[113] N. Lipscomb (*Adv. Inorg. Chem. Radiochem.*, **1959**, *1*, 117, and "Boron Hydrides," Benjamin, New York, **1963**) has provided an extensive discussion of the structural properties of boron–hydrogen compounds. He has suggested an additional type of bond, the "open" or "linear" B—B—B bond which more nearly resembles the open B—H—B bond discussed above. More recently (I. R. Epstein and W. N. Lipscomb, *Inorg. Chem.*, **1971**, *10*, 1921) it has been suggested that open B—B—B bonds may be dismissed with little or no loss in effectiveness but with a considerable gain in simplicity. For recent discussions of theory, see E. L. Muetterties (Chapter 1) and W. N. Lipscomb (Chapter 2) in "Boron Hydride Chemistry," E. L. Muetterties, ed., Academic Press, New York, **1975**. For Professor Lipscomb's Nobel Laureate address, including further discussion of boranes, see *Science*, **1977**, *196*, 1047.

[114] Tetrahydroborate is the preferred name.

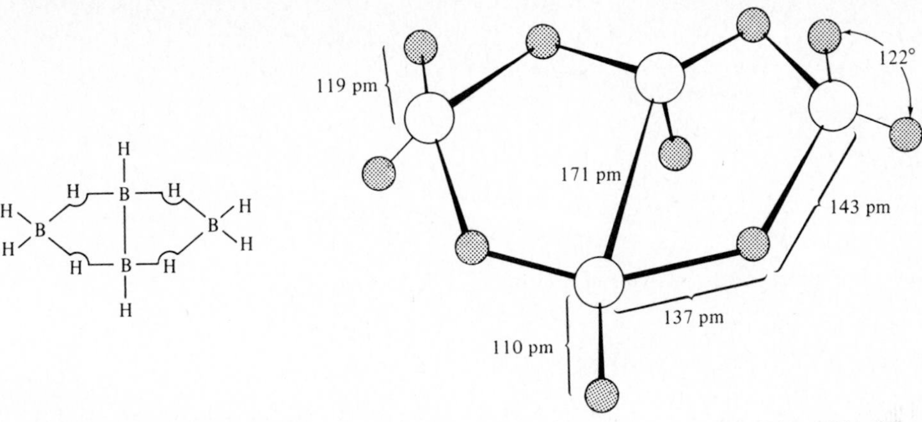

Fig. 14.35 Bonding and structure of tetraborane, B_4H_{10}. [From E. L. Muetterties, "The Chemistry of Boron and Its Compounds," Wiley, New York, **1967**. Reproduced with permission.]

Direct reaction of ammonia and diborane results in the "ammoniate" of diborane, which has been shown to be ionic:

$$B_2H_6 + 2NH_3 \longrightarrow [BH_2(NH_3)_2]^+ + [BH_4]^- \qquad \textbf{(14.94)}$$

This "unsymmetrical cleavage"

is typical of small, hard Lewis bases and further examples will be discussed below. Larger bases, such as phosphines, promote "symmetrical cleavage":

All of the compounds discussed thus far (except diborane) contain only simple two-center, two-electron bonds. A simple boron hydride containing three types of bonds is tetraborane, B_4H_{10} (Fig. 14.35). It is formed by the slow decomposition of diborane:

$$2B_2H_6 \longrightarrow B_4H_{10} + H_2 \qquad \textbf{(14.95)}$$

In addition to terminal and bridging B—H bonds this compound contains a direct B—B bond.

Tetraborane undergoes both symmetrical and unsymmetrical cleavage (Fig. 14.36). Larger Lewis bases tend to split off BH_3 moieties, which are either complexed or allowed to dimerize to form diborane:

$$B_4H_{10} + 2(CH_3)_3N \longrightarrow (CH_3)_3NB_3H_7 + (CH_3)_3NBH_3 \qquad \textbf{(14.96)}$$

$$2B_4H_{10} + 2(C_2H_5)_2O \longrightarrow 2(C_2H_5)_2OB_3H_7 + B_2H_6 \qquad \textbf{(14.97)}$$

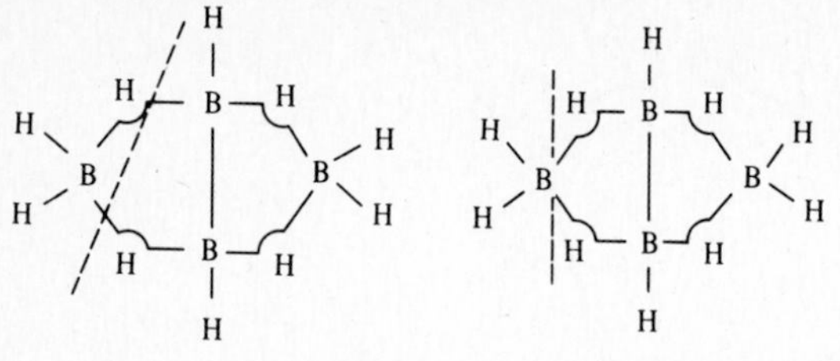

Fig. 14.36 Symmetrical (a) and unsymmetrical (b) cleavage of tetraborane.

$$2B_4H_{10} + 2(CH_3)_2S \longrightarrow 2(CH_3)_2SB_3H_7 + B_2H_6 \tag{14.98}$$

Small, hard Lewis bases such as ammonia and the hydroxide ion result in unsymmetrical cleavage, i.e., the splitting off of the BH_2^+ moiety:

$$B_4H_{10} + 2NH_3 \longrightarrow [H_2B(NH_3)_2]^+ + [B_3H_8]^- \tag{14.99}$$

$$2B_4H_{10} + 4OH^- \longrightarrow [B(OH)_4]^- + [BH_4]^- + 2[B_3H_8]^- \tag{14.100}$$

The latter reaction can be considered as an abstraction of BH_2^+ if it is assumed that $[H_2B(OH)_2]^-$ is unstable and disproportionates:

$$2[H_2B(OH)_2]^- \longrightarrow [B(OH)_4]^- + [BH_4]^- \tag{14.101}$$

Rather than continue to progress from less complex to more complex boron–hydrogen compounds it will be more convenient to jump to a complex but highly symmetrical borohydride ion, $B_{12}H_{12}^{-2}$. It may be synthesized from the pyrolysis of the $B_3H_8^-$ ion:

$$5B_3H_8^- \xrightarrow[\text{diglyme}]{\text{refluxing}} B_{12}H_{12}^{-2} + 3BH_4^- + 8H_2 \tag{14.102}$$

It is not necessary to use $B_3H_8^-$ directly; it may be formed *in situ* from diborane and borohydride:

$$2BH_4^- + 5B_2H_6 \xrightarrow[\text{100–108°C}]{\text{pyridine}} B_{12}H_{12}^{-2} + 13H_2 \tag{14.103}$$

The structure of the $B_{12}H_{12}^{-2}$ is a regular icosahedron with 20 equilateral triangles forming the faces (Fig. 14.37a). All of the hydrogen atoms are external to the boron icosahedron and are attached by terminal B—H bonds. The icosahedron itself involves a resonance hybrid of several canonical forms of the type shown in Fig. 14.37b,c. Both B—B and three-center B—B—B bonding are involved.

An icosahedral framework of boron atoms if of considerable importance in boron chemistry. Three forms of elemental boron as well as several nonmetal borides contain discrete B_{12} icosahedra. For example, α-rhombohedral boron consists of layers of icosahedra linked within the layer by three-center B—B—B bonds and between layers by B—B bonds (Fig. 14.38). β-Rhombohedral boron consists of twelve B_{12} icosahedra arranged icosahedrally about a central B_{12} unit, i.e., $B_{12}(B_{12})_{12}$. Tetragonal boron

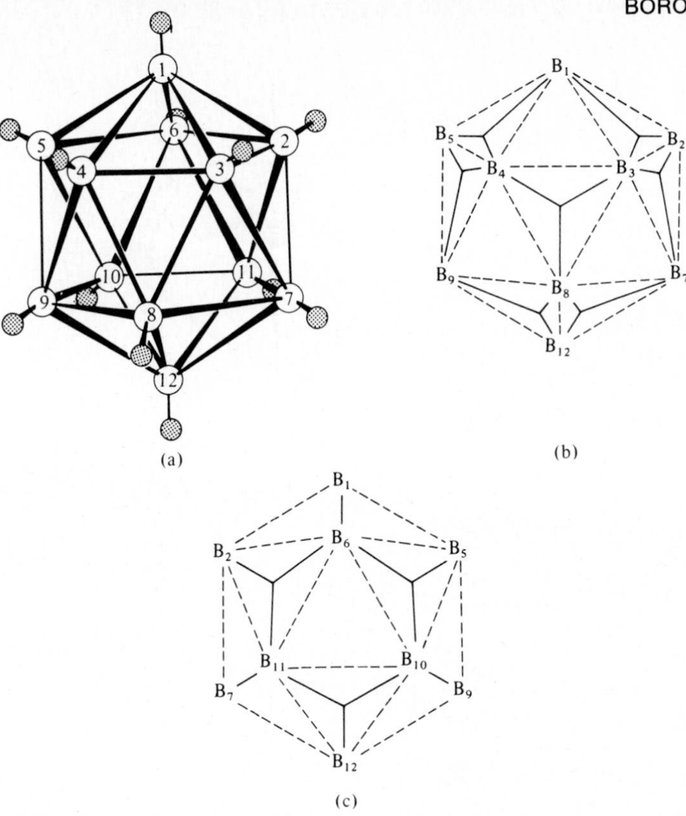

Fig. 14.37 (a) Molecular structure of the $B_{12}H_{12}^{-2}$ anion; (b), (c) front and back sides of the $B_{12}H_{12}^{-2}$ framework showing one of many canonical forms contributing to the resonance hybrid. [(a) From E. L. Muetterties, "The Chemistry of Boron and Its Compounds," Wiley, New York, **1967**; (b) from W. L. Jolly, "The Chemistry of the Non-Metals," Prentice-Hall, Englewood Cliffs, N.J., **1966**. Reproduced with permission.]

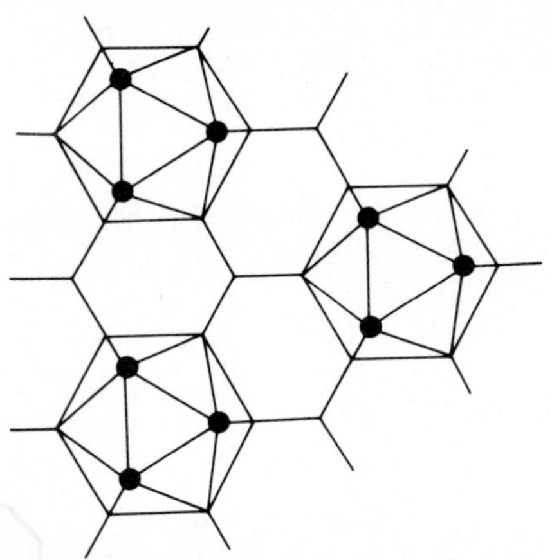

Fig. 14.38 Structure of α-rhombohedral boron. The icosahedra are linked within the layer via three-center bonds. This layer is linked to the layer above by B—B bonds arising from the boron atoms marked ● and to the layer below by three additional boron atoms (not seen) on the opposite face.

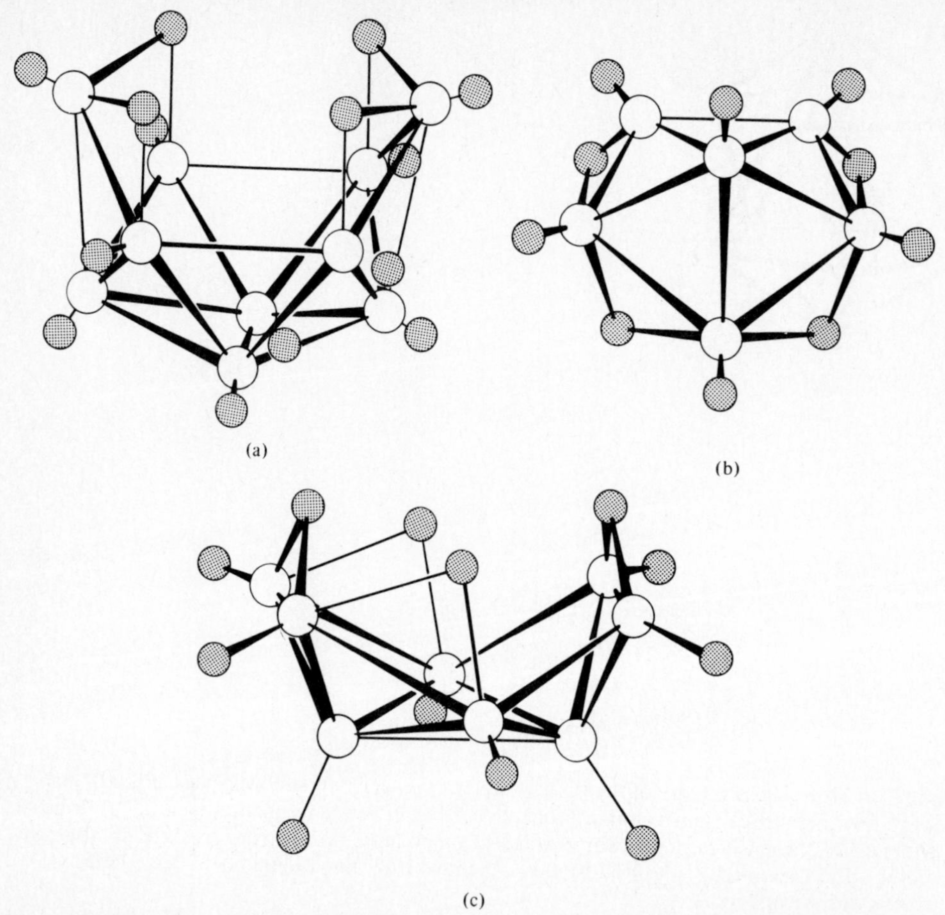

Fig. 14.39 Molecular structures of boranes related to $B_{12}H_{12}^{-2}$; (a) decaborane(14) formed by removal of atoms B_1 and B_6; (b) hexaborane(10). Note that the pentagonal pyramid is an apical fragment of an icosahedron. (c) Octaborane(12) related to $B_{12}H_{12}^{-2}$ by removal of B_1, B_2, B_5, B_6. [From E. L. Muetterties, "The Chemistry of Boron and Its Compounds," Wiley, New York, **1967**. Reproduced with permission.]

consists of icosahedra linked not only by B—B bonds between the icosahedra themselves but also by tetrahedral coordination to single boron atoms.[115]

Several boranes may be considered as fragments of a B_{12} icosahedron (or of the $B_{12}H_{12}^{-2}$ ion) in which extra hydrogen atoms are used to "sew up" the unused valences around the edge of the fragment. For example, decaborane(14) (Fig. 14.39a) may be considered to be a $B_{12}H_{12}$ framework from which B_1 and B_6 (Fig. 14.37) have been removed and the "dangling" three-center bonds completed with hydrogen atoms to form B—H—B bridges.

[115] See A. E. Newkirk, in "Boron, Metallo-Boron Compounds, and Boranes," R. M. Adams, ed., Wiley (Interscience), New York, **1964**, Chapter 4, and J. L. Hoard and R. E. Hughes, in "The Chemistry of Boron and Its Compounds," E. L. Muetterties, ed., Wiley, New York, **1967**, Chapter 2.

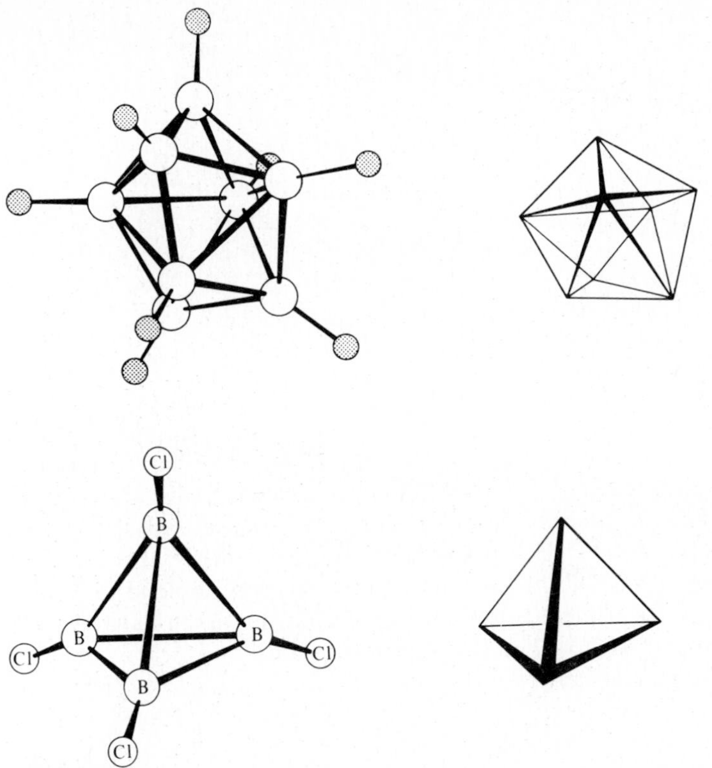

Fig. 14.40 (a) The structure of the $B_8H_8^{-2}$ anion compared to an idealized dodecahedron. [From J. Guggenberger, *Inorg. Chem.*, **1969**, *8*, 2771. Reproduced with permission.] (b) Molecular structure of B_4Cl_4. [From E. L. Muetterties, "The Chemistry of Boron and Its Compounds," Wiley, New York, **1967**. Reproduced with permission.]

Other examples of boranes that are icosahedral fragments are hexaborane(10), which is a pentagonal prism (Fig. 14.39b), pentaborane(11), similar to the former with a basal boron atom missing, octaborane(12) (Fig. 14.39c), and nonaborane(15).

Although relating borane structures to icosahedra was the first successful means of systemizing the structural chemistry of these cages, further experimental work revealed that the icosahedron of $B_{12}H_{12}^{-2}$ was merely the upper limit of a series of regular deltahedra,[116] $B_nH_n^{-2}$, complete from $n = 6$ to $n = 12$, and $n = 4$ in B_4Cl_4 (Fig. 14.40). If all of the vertices of the deltahedron are occupied, as in the $B_nH_n^{-2}$ series, the structure is called a *closo* (Gr., "closed") structure. In the past few years it has been found possible to correlate the structure of boranes and derivatives with the number of electrons involved in the bonding in the framework of the deltahedron.[117] For the *closo* series the number

[116] A deltahedron is a polyhedron with all faces that are equilateral triangles. The deltahedra from $n = 4$ to $n = 12$ are tetrahedron (4), trigonal bipyramid (5), octahedron (6), pentagonal bipyramid (7), bisdisphenoid (dodecahedron) (8), tricapped trigonal prism (9), bicapped square antiprism (10), octadecahedron (11), and icosahedron (12). Most of these are illustrated in Chapters 5 and 10. See also Fig. 14.42.

[117] K. Wade, *Adv. Inorg. Chem. Radiochem.*, **1976**, *18*, 1; R. E. Williams, ibid., **1976**, *18*, 67; R. W. Rudolph, *Acc. Chem. Res.*, **1976**, *9*, 446.

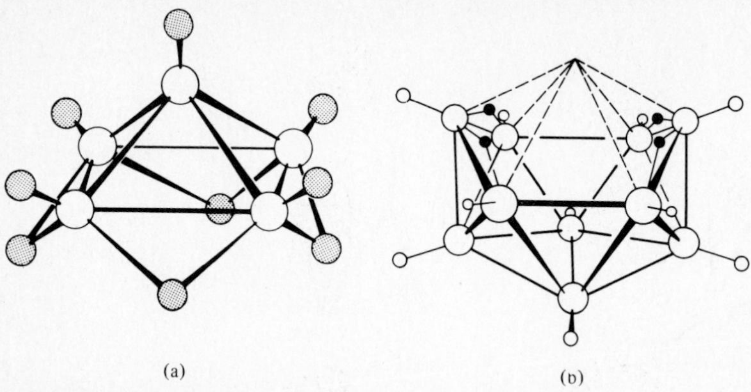

Fig. 14.41 (a) Structure of *nido*-pentaborane(9); (b) structure of *nido*-decaborane(14). Cf. *closo* structures in Fig. 14.42.

of framework electrons equals $2n + 2$. To count framework electrons in, for example, $B_{12}H_{12}^{-2}$, one notes that each boron atom has one of its three valence electrons tied up with the *exo* B—H bond (an *exo* B—H bond is one extending radially outward from the center of the cluster; see Figs. 14.37, 14.40); it thus has two to contribute to the framework: $2n$, or in this case 24. No neutral B_nH_n species are known, but we have seen the array of dianions corresponding to $2n + 2$, in the present case, 26.

If one (in a thought experiment) removes a boron atom from a vertex of a *closo* structure, a cup-like or nest-like structure remains (Fig. 14.41). Such structures are termed *nido* (Gr., "nest") structures. We have seen that structures such as this contain extra hydrogen atoms to "sew up" the loose valencies around the opening. The *nido* structures obey the electron formula $2n + 4$. Consider B_5H_9, for example. Five *exo* B—H groups will contribute two electrons each and the four "extra" hydrogen atoms will contribute four electrons for a total of 14 ($2n + 4$, $n = 5$). The structure is thus a square pyramid *nido* form derived from the *closo* octahedron. The "extra" four hydrogen atoms form bridges across the "open" edges of the "nest" (Fig. 14.41).

If two vertex boron atoms are removed, the resulting structure is an *arachno* (Gr., "spider's web") structure. With two vertices missing, the structure is even more open than in the *nido* structures and the resemblance to the "parent" *closo* structure less apparent. *Arachno* structures obey the electronic formula $2n + 6$. Pentaborane(11), B_5H_{11} must therefore have an *arachno* structure. In the *arachno* series the "extra" hydrogen atoms form *endo* B—H bonds (lying in close to the framework) as well as bridges.

There is some evidence that there may be one or more additional series of this type. Only the *hypho* (Gr., "tissue," as in "tissue paper") series, $2n + 8$, has been suggested to augment the *closo*, *nido*, and *arachno* series.

The complete structural relationships among the *closo*, *nido*, and *arachno* species are shown in Fig. 14.42. The diagonal lines connecting the *closo*/*nido*/*arachno* species represent the hypothetical transformations discussed above, removal of boron vertices moving lower left to upper right.[118]

[118] Only the *exo* hydrogen atoms are shown in Fig. 14.42. The number of bridging and *endo* hydrogen atoms will vary depending upon whether the species is neutral, anionic, etc., has heteroatoms (see below), has a Lewis base coordinated to it, etc.

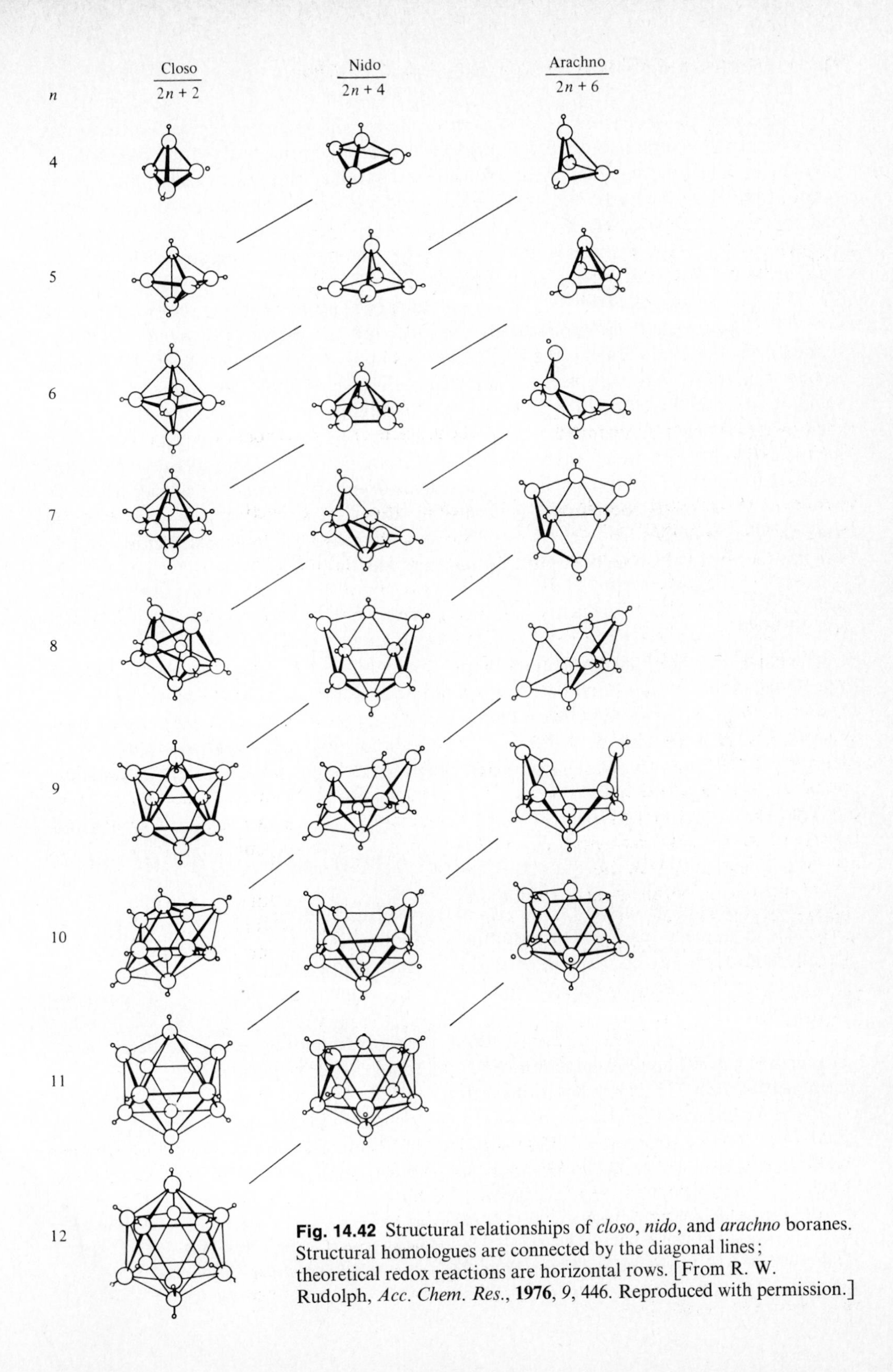

Fig. 14.42 Structural relationships of *closo*, *nido*, and *arachno* boranes. Structural homologues are connected by the diagonal lines; theoretical redox reactions are horizontal rows. [From R. W. Rudolph, *Acc. Chem. Res.*, **1976**, *9*, 446. Reproduced with permission.]

The horizontal series represent structures having the same number of boron atoms but differing in the total number of framework electrons to correspond to the *closo* $(2n + 2)$, *nido* $(2n + 4)$, or *arachno* $(2n + 6)$ electronic specifications. In a few cases (see below) the change from one structure to another on the same line can be effected by a simple redox reaction, but in most cases this is not so. However, we can probably anticipate more examples of this type of transformation now that the general principles of framework electron count and structure are better understood.

What is the source of the $2n + 2$, $2n + 4$, and $2n + 6$ rules and their correspondence with the *closo*, *nido*, and *arachno* structures? Space does not allow derivation of the molecular orbitals for deltahedra, but the results may be simply stated: *For a regular deltahedron having n vertices, there will be n + 1 bonding molecular orbitals.* The capacity of these bonding MOs is therefore $2n + 2$. This gives the highly symmetrical *closo* structure. If there are two more electrons $(2n + 4)$, one bonding MO and one vertex must be used for the extra electrons rather than a framework atom, and a *nido* with a "missing" vertex results. Although the electrons in boranes are delocalized and cannot be assigned a specific region of localization, the parallel between the "extra" electrons in *nido* structures and lone pairs in molecules like NH_3 is real. The $2n + 6$ formula of *arachno* structures simply extends these ideas by one more electron pair and one more vertex.

Carboranes

Carbon has one more electron than boron, so the C—H moiety is isoelectronic with the $B—H^-$ moiety. In a formal sense it should be possible to replace a boron atom in a borane with a carbon atom (with an increase of one in positive charge) and retain an isoelectronic system. This is indeed the case. The best-studied system, $B_{10}H_{12}$, is isoelectronic with $B_{12}H_{12}^{-2}$ and may be synthesized readily from decaborane in diethyl sulfide as solvent:

$$B_{10}H_{14} + 2(C_2H_5)_2S \longrightarrow B_{10}H_{12}\cdot 2S(C_2H_5)_2 + H_2 \quad \textbf{(14.104)}$$

$$B_{10}H_{12}\cdot 2S(C_2H_5)_2 + RC{\equiv}CR \longrightarrow R_2C_2B_{10}H_{10} + 2(C_2H_5)_2S + H_2 \quad \textbf{(14.105)}$$

The acetylene may be unsubstituted (R = H) or substituted, in which case the reaction proceeds even more readily. The resulting compound is known as 1,2-dicarbaclosododecaborane, or the "ortho" carborane, and is isoelectronic and isostructural with $B_{12}H_{12}^{-2}$. It isomerizes on heating to the 1,7-("meta" or "neo" isomer) and the 1,12-("para" isomer) (Fig. 14.43).

Other dicarboranes are derived from the corresponding pentaborane(5), hexaborane(6), hexaborane(8), heptaborane(7), octaborane(8), nonaborane(9), and decaborane(10) dianions $(B_nH_n^{-2})$. The monocarboranes CB_5H_7, and CB_5H_9 are also known.[119]

The carboranes conform to the electronic rules given above and are known in *closo*, *nido*, and *arachno* structures. When applying the formulas, each C—H group donates *three* electrons to the framework count. Some carboranes provide interesting examples

---

[119] For a more complete discussion of the preparation and reactions of the carboranes, see M. F. Hawthorne in "The Chemistry of Boron and Its Compounds," E. L. Muetterties, ed., Wiley, New York, **1967**, Chapter 5; G. B. Dunks and M. F. Hawthorne, *Inorg, Chem.*, **1969**, *8*, 2667; *Acc. Chem. Res.*, **1973**, *6*, 124.

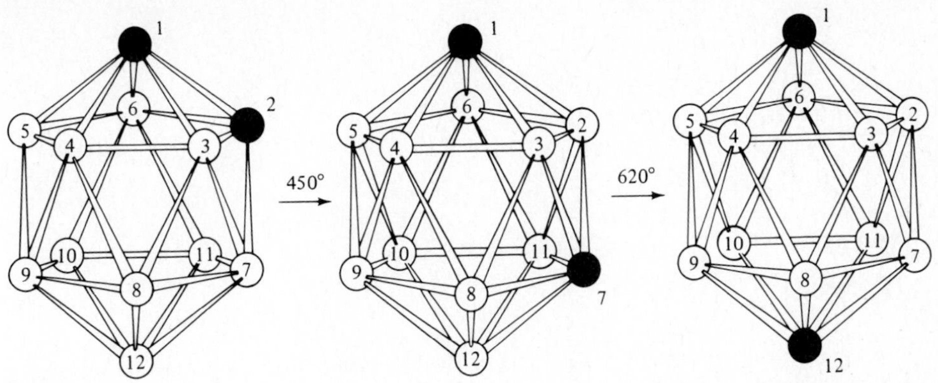

Fig. 14.43 Structure and isomerizations of the three isomers of dicarbaclosododecaborane. ○ = B, ● = C.

of the possible horizontal transformations of Fig. 14.42 mentioned above. For example:[120]

$$closo\text{-}C_2B_9H_{11} + 2e^- \longrightarrow [nido\text{-}C_2B_9H_{11})^{-2} \quad \textbf{(14.106)}$$

$$[nido\text{-}C_2B_9H_{11}]^{-2} \longrightarrow closo\text{-}C_2B_9H_{11} + 2e^- \quad \textbf{(14.107)}$$

Metallocene carboranes

Strong bases attack 1,2-dicarbaclosododecaborane(12) with the splitting out of a boron atom:

$$B_{10}C_2H_{12} + CH_3O^- + 2CH_3OH \longrightarrow B_9C_2H_{12}^- + H_2 + B(OCH_3)_3 \quad \textbf{(14.108)}$$

The resulting anion is the conjugate base of a strong acid which may be obtained by acidification:

$$B_9C_2H_{12}^- + HCl \longrightarrow B_9C_2H_{13} + Cl^- \quad \textbf{(14.109)}$$

Conversely, treatment with the very strong base sodium hydride abstracts a second proton:

$$B_9C_2H_{12}^- + NaH \xrightarrow{\text{tetrahydrofuran}} B_9C_2H_{11}^{-2} + H_2 + Na^+ \quad \textbf{(14.110)}$$

The probable structure of the $B_9C_2H_{11}^{-2}$ anion is shown in Fig. 14.44. Each of the three boron atoms and the two carbon atoms on the open face of the cage directs an orbital (taken as sp^3 for convenience) toward the apical position occupied formerly by the twelfth boron atom. Furthermore, these orbitals contain a total of six electrons. They thus bear a striking resemblance to the *p* orbitals in the π system of cyclopentadienide anion.[121] Noting this resemblance, Hawthorne suggested that the $B_9C_2H_{11}^{-2}$ could be considered

[120] V. Chowdry et al., *J. Am. Chem. Soc.*, **1973**, *95*, 4560.

[121] The *p* orbitals in cyclopentadienide anion have both positive and negative lobes, of course, but this has no effect in the present comparison since only one side of the C_5H_5 ring is used in bonding. The orbitals in the carborane anion converge rather than being perfectly parallel, but this should increase the strength of the bonding.

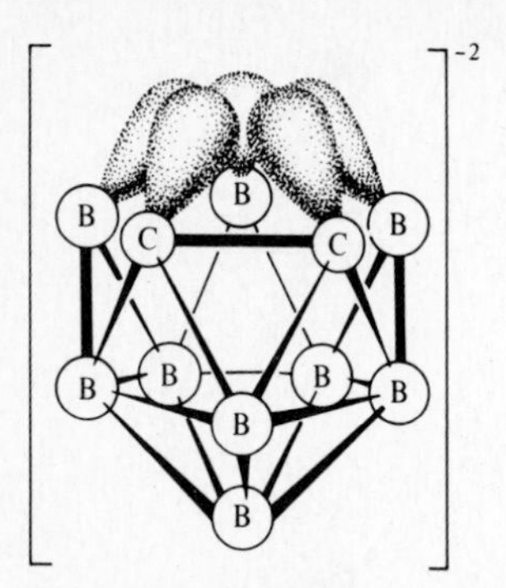

Fig. 14.44 Probable structure of the $B_9C_2H_{11}^{-2}$ anion. The five orbitals directed toward the missing apical boron are included. [From R. G. Adler and M. F. Hawthorne, *J. Amer. Chem. Soc.*, **1970**, *92*, 6174. Reproduced with permission.]

isoelectronic with the cyclopentadienide anion and should therefore be capable of acting as a π ligand in metallocene compounds. He then succeeded in synthesizing several metallocene carboranes:[122]

$$2B_9C_2H_{11}^{-2} + FeCl_2 \longrightarrow [(B_9C_2H_{11})_2Fe]^{-2} + 2Cl^- \qquad \textbf{(14.111)}$$

$$B_9C_2H_{11}^{-2} + C_5H_5^- + FeCl_2 \longrightarrow [B_9C_2H_{11}FeC_5H_5]^- + 2Cl^- \qquad \textbf{(14.112)}$$

$$B_9C_2H_{11}^{-2} + BrMn(CO)_5 \longrightarrow [B_9C_2H_{11}Mn(CO)_3]^- + Br^- + 2CO\uparrow \qquad \textbf{(14.113)}$$

The ferrocene analogues, like ferrocene, are oxidizable with the loss of one electron. In those cases in which structures have been determined they have been found to correspond to that expected on the basis of metallocene chemistry (Fig. 14.45).

METAL CLUSTERS

Occurrence of metal–metal bonds

Compounds containing metal–metal bonds are as old as chemistry (calomel was known to the chemists of India as early as the twelfth century). The dimeric nature of the mercurous ion was not confirmed until the turn of the century and the next half century passed in discussions on the possibility that zinc and cadmium might possess similar species. It was only some 20–25 years ago that the study of other metal–metal bonds began in earnest; yet this branch of inorganic chemistry has grown at a phenomenal rate.

Metal cluster compounds can be conveniently grouped into two classes: (1) polynuclear carbonyls, nitrosyls, and related compounds; and (2) lower halides and oxides. The former

[122] See M. F. Hawthorne in "The Chemistry of Boron and Its Compounds," E. L. Muetterties, ed., Wiley, New York, **1967**, pp. 308–315, and E. L. Muetterties and W. L. Knoth, "Polyhedral Boranes," Dekker, New York, **1968**, pp. 27–31. For a list of recent references, see L. F. Warren, Jr., and M. F. Hawthorne, *J. Am. Chem. Soc.*, **1970**, *92*, 1157.

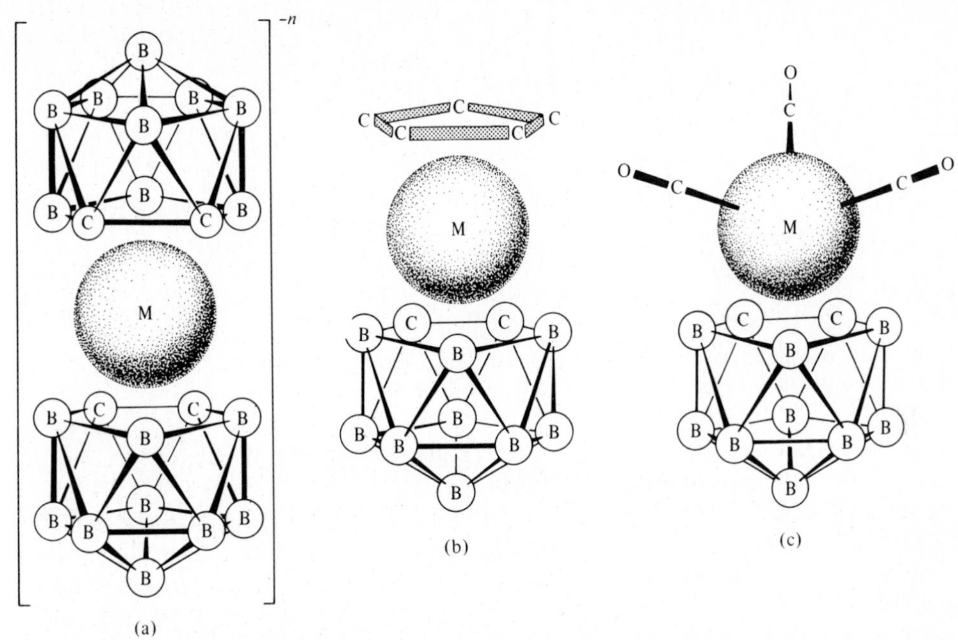

Fig. 14.45 Structures of some carbollyl metallocene compounds. (a) Dicarbollyl species: M = Fe, $n = 2$ and M = Co, $n = 1$ are isoelectronic with ferrocene and cobalticinium ion; M = Fe, $n = 1$ is isoelectronic with ferricinium ion; (b) mixed carbollyl-cyclopentadienyl analogue of ferrocene; (c) mixed carbollyl–carbonyl compound, M = Mn, Re. [(a) and (b) from M. F. Hawthorne and T. D. Andrews, *J. Am. Chem. Soc.*, **1965**, *87*, 2496; (c) from R. G. Adler and M. F. Hawthorne, *J. Am. Chem. Soc.*, **1970**, *92*, 6174. Reproduced with permission.]

group was included in Chapter 13. The second class will be discussed briefly in this section.[123]

One of the prerequisites for the formation of metal–metal bonds is a low formal oxidation state. In this regard metal clusters may be regarded as small bits of the elemental metals themselves. The latter can only be discussed adequately by theories which account for the extensive delocalization in the metal lattice ("band theory"). Although Pauling has applied valence bond theory to metals, the relative abundance of orbitals to valence electrons requires extensive numbers of resonance forms. We might therefore expect to find that treatment of metal clusters by valence bond theory will prove to be cumbersome. A molecular orbital treatment with its concomitant delocalization of electrons will be preferable.

The tendency to retain clusters of metal atoms will predominate in those metals with very large energies of atomization (and hence very high melting and boiling points).

---

[123] Among the many reviews of various aspects of this topic, some of the best are: F. A. Cotton, *Chem. Soc. Rev.*, **1975**, *4*, 27; *Acc. Chem. Res.*, **1978**, *11*, 225; Chapter 1 in "Reactivity of Metal–Metal Bonds," M. H. Chisholm, ed., *ACS Symp. Ser.*, No. 155, Washington, D.C., **1981**; M. H. Chisholm and I. P. Rothwell, *Progr. Inorg. Chem.*, **1982**, *29*, 1; F. A. Cotton and M. H. Chisholm, *Chem. Eng. News*, June 28, **1982**, p. 40; F. A. Cotton and R. A. Walton, "Multiple Bonds Between Metal Atoms," Wiley, New York, **1982**.

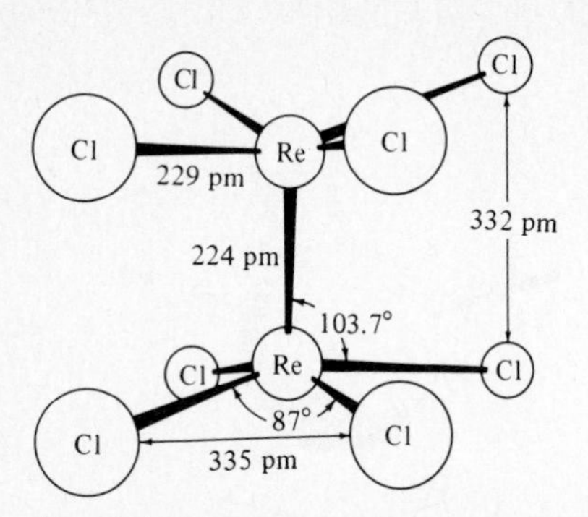

Fig. 14.46 The structure of the octachlorodirhenate(III) ion, $Re_2Cl_8^{-2}$. [From F. A. Cotton and C. B. Harris, *Inorg. Chem.*, **1965**, *4*, 330. Reproduced with permission.]

Thus the most refractory metals (Zr, Nb, Mo, Tc, Ru, Rh, Hf, Ta, W, Re, Os, Ir, and Pt) have the greatest tendency to form metal clusters.

A second factor necessitating low oxidation states is the nature of the *d* orbitals. The size of the *d* orbitals is inversely related to the effective nuclear charge. Since effective overlap of *d* orbitals appears necessary to stabilize metal clusters, excessive contraction of them will destabilize the cluster. Hence large charges resulting from high oxidation states are unfavorable. For the first transition series, the *d* orbitals are relatively small, and even in moderately low oxidation states (+2 and +3) they apparently do not extend sufficiently for good overlap.

Binuclear compounds

The best studied species are $Re_2X_8^{-2}$ ions. They may be prepared by reduction of perrhenate in the presence of X^-:

$$2ReO_4^- \xrightarrow[HX]{H_3PO_2} Re_2X_8^{-2} \qquad X = Cl, Br \tag{14.114}$$

The most interesting aspect of these compounds is their structure (Fig. 14.46), which possesses two unusual features. The first is the extremely short Re—Re distance of 224 pm compared with an average Re—Re distance of 275 pm in rhenium metal and 248 pm in Re_3Cl_9. The second unexpected feature is the eclipsed configuration of the chlorine atoms. One might have supposed that since the short Re—Re bond requires that the chlorine atoms lie at distances (~330 pm) which are less than the sum of their van der Waals radii (~340–360 pm), the staggered configuration would be preferred (the chlorine atoms would form a square antiprism rather than a cube). Cotton has explained both phenomena by invoking a quadruple bond which is unexpectedly short and unexpectedly strong.

Cotton's explanation is as follows.[124] The *z* axis is taken as the line joining the rhenium atoms. Each rhenium atom is bonded to four chlorine atoms almost in a square planar

[124] The argument presented here is a little more restricted and a little less general than that by Cotton, merely to simplify the situation somewhat. Cotton's discussions of quadruple bonds can be found in *Inorg. Chem.*, **1965**, *4*, 334, and *Rev. Pure Appl. Chem.*, **1967**, *17*, 25.

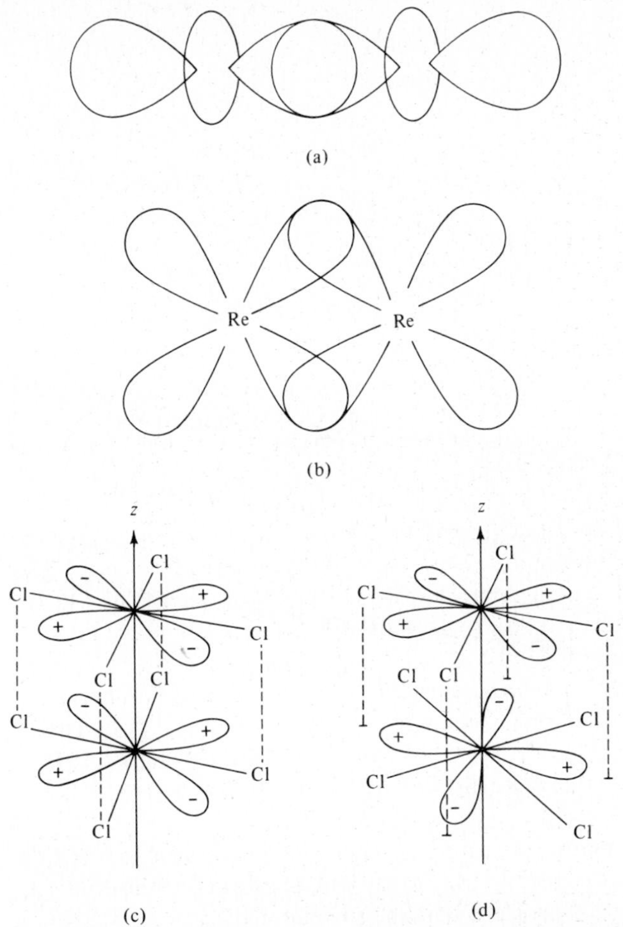

Fig. 14.47 Multiple bonding between rhenium atoms: (a) Formation of a σ bond from overlap of d_{z^2} orbital of each rhenium atom. (b) Formation of a π bond from overlap of the d_{xz} orbital of each rhenium atom. A second π bond forms in the yz plane. (c) Positive overlap from d_{xz} orbitals to form a δ bond in the eclipsed conformation. (d) Zero overlap occurring in the staggered conformation. [In part from F. A. Cotton, *Acc. Chem.*, **1969**, *2*, 240. Reproduced with permission.]

array (the Re is 50 pm out of the plane of the $4Cl^-$). We may take the Re—Cl bonds to be approximate dsp^2 hybrids utilizing the $d_{x^2-y^2}$ orbital. The d_{z^2} and p_z orbitals lie along the bond axis and may be hybridized to form one orbital directed to the other rhenium atom and one orbital directed in the opposite direction. The former can overlap with the similar orbital on the second rhenium atom to form a σ bond (Fig. 14.47a) while the second hybrid orbital forms an approximately nonbonding orbital.[125]

The d_{xz} and d_{yz} orbitals of each rhenium are directed obliquely toward the other rhenium atom, overlapping to form d–d π bonds (Fig. 14.47b). This results in two π bonds,

[125] Essentially the same result could have been obtained by forming a d^2sp^3 hybrid on each rhenium atom initially.

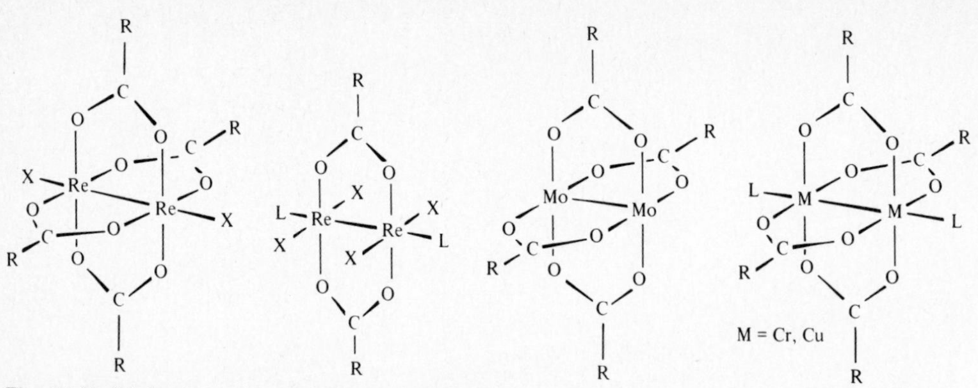

Fig. 14.48 Molecular structures of some carboxylate complexes containing metal–metal bonding; (a) Re–Re = 220 pm, X = Cl, Br, I; (b) Re–Re = 220 pm, X = Br, Cl, L = H_2O; (c) Mo–Mo = 210 pm; (d) Cr–Cr = 236 pm, Cu–Cu = 264 pm, L = H_2O. Bond lengths are averages; individual values are known to greater accuracy.

one in the xz plane and one in the yz plane. A fourth bond can now form by overlap of the remaining d orbital on each atom, the d_{xy}. This "sideways" overlap of two d orbitals results in a δ bond. Overlap can only occur if the chlorine atoms are eclipsed (Fig. 14.47c). If the chlorine atoms are staggered, the two d_{xy} orbitals will likewise be staggered with resulting zero overlap (Fig. 14.47d).

Re(III) is a d^4 species. The Re—Cl bonds may be considered dative bonds from Cl^- ions to Re^{+3} ions. The formation of one σ bond, two π bonds, and one δ bond causes the pairing of the four electrons in the quadruple bond; hence the complex is diamagnetic. The strength of the bond and the short Re—Re distance are accounted for as is the eclipsed configuration.

Subsequently there have been many compounds discovered which seem to resemble the $Re_2X_8^{-2}$ ions in possessing extremely short M—M distances, eclipsed conformations, and, presumably, quadruple bonds. The isoelectronic molybdenum(II) species, $Mo_2Cl_8^{-4}$, is known and both Re(III) and Mo(II) form a large series of carboxylate complexes of formulas $Re_2(RCO_2)_4X_2$, $Re_2(RCO_2)_2X_4L_2$, and $Mo_2(RCO_2)_4$:

$$Re_2Cl_8^{-2} + 2CH_3COOH \xrightarrow[25^\circ C]{} Re_2(CH_3CO_2)_2Cl_4 + 4Cl^- + 2H^+ \quad (14.115)$$

$$Re_2Cl_8^{-2} + 4CH_3COOH \xrightarrow[\text{reflux}]{} Re_2(CH_3CO_2)_4Cl_2 + 6Cl^- + 4H^+ \quad (14.116)$$

$$Re_2(CH_3CO_2)_4Cl_2 + 4\phi COOH \xrightarrow[250^\circ C]{} Re_2(\phi CO_2)_4Cl_2 + 4CH_3COOH\uparrow \quad (14.117)$$

$$2Mo(CO)_6 + 4CH_3COOH \longrightarrow Mo_2(CH_3CO_2)_4 + 12CO + 2H_2 \quad (14.118)$$

Structurally these complexes (Fig. 14.48) are clearly related to $Re_2Cl_8^{-2}$, the only difference being the addition of a ligand to the nonbonding orbitals present in $Re_2Cl_8^{-2}$. The absence of a ligand in this position in the octahalodirhenates is most likely attributable to steric hindrance of the large halide ions compared with the smaller oxygen atoms in the carboxylate groups.

Although Cotton's molecular orbital scheme was largely qualitative, based on an approach involving a combination of atomic orbitals, a self-consistent-field calculation

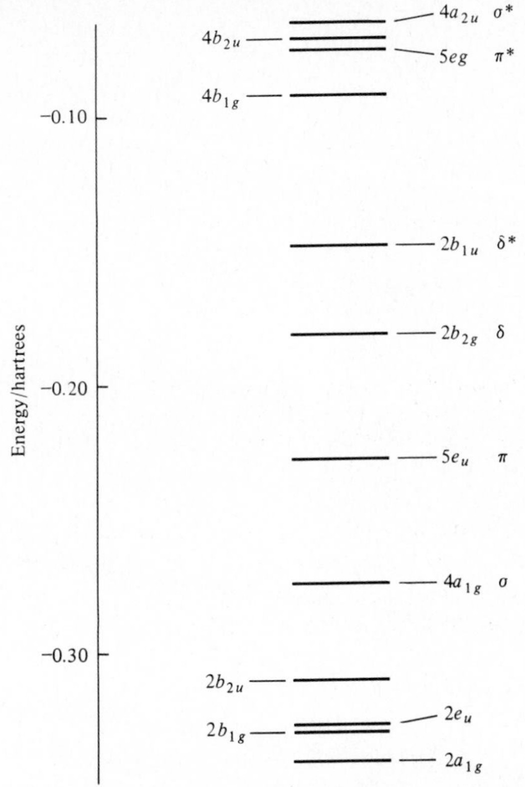

Fig. 14.49 Valence energy levels of $Mo_2Cl_8^{-4}$ with appreciable molybdenum character. Levels are occupied through the $2b_{2g}$ level. Note the $5e_u$ is doubly degenerate. [From J. G. Norman, Jr., and H. J. Kolari, *J. Am. Chem. Soc.*, **1975**, *97*, 33. Reproduced with permission.]

based on first principles[126] confirms the essential correctness of the σ-π-δ bonding (Fig. 14.49). Furthermore, solution of the one-electron Schrödinger equation obtained by this method gives contour maps for electron density that are remarkably similar to what we would predict based upon combination of the atomic orbitals on the two molybdenum atoms involved (Fig. 14.50). Furthermore, electron densities as determined experimentally[127] are consistent with the quadruple bond picture.

Rather interestingly, and not surprisingly, Pauling[128] has provided an alternative, valence bond treatment of the quadruple bond involving *spd* hybrid orbitals and four equivalent bent bonds (cf. pp. 233–237 for similar treatments of double and triple bonds). His method explains the same experimental facts discussed above and provides a good estimate of the bond length.

The strength of the quadruple bond in dirhenium and dimolybdenum compounds has been a matter of considerable differences of opinion. Estimates of the bond energy

---

[126] J. G. Norman, Jr., and H. J. Kolari, *Chem. Commun.*, **1974**, 303; *J. Am. Chem. Soc.*, **1975**, *97*, 33; J. G. Norman et al., *Inorg. Chem.*, **1977**, *16*, 987.

[127] K. Hino et al., *Acta Cryst.*, **1981**, *B37*, 2164.

[128] L. Pauling, *Proc. Nat. Acad. Sci., U.S.*, **1975**, *72*, 3799.

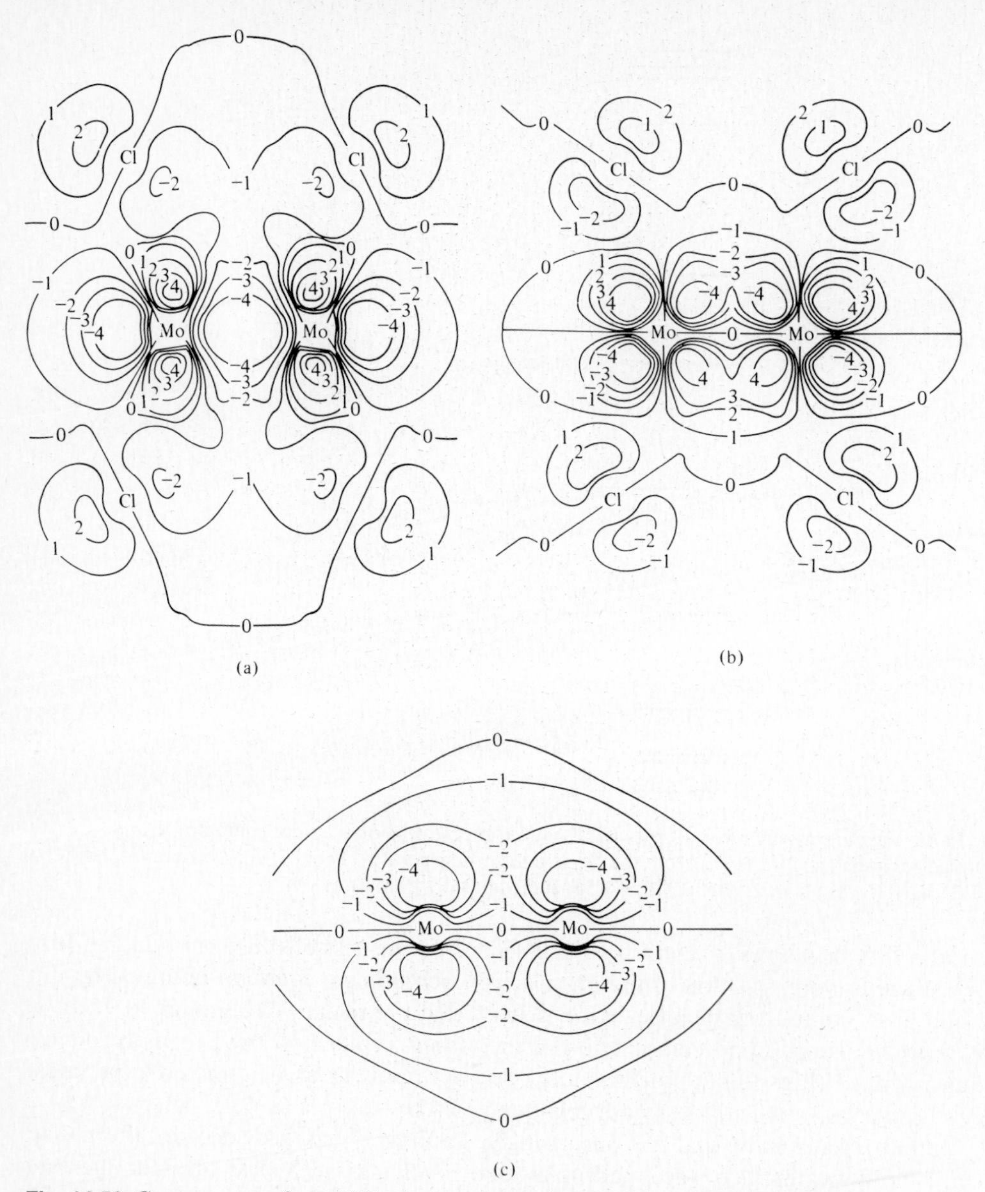

Fig. 14.50 Contour maps from SCF solution of one-electron Schrödinger equation for $Mo_2Cl_8^{-4}$. (a) The $4a_{1g}$ level (σ bond). Cf. Fig. 14.47a. (b) The $5e_u$ levels (one of the two π bonds). Cf. Fig. 14.47b. (c) The $2b_{2g}$ level (δ bond). Cf. Fig. 14.47c. The plane is a dihedral one halfway between *xz* and *yz*. [From J. G. Norman, Jr., and H. J. Kolari, *J. Am. Chem. Soc.*, **1975**, *97*, 33. Reproduced with permission.]

have ranged from as low as 300 kJ mol^{-1} (less than a C—C single bond) to as high as 1500 kJ mol^{-1} (stronger than any other bond). There now seems to be some general agreement that the upper limit is probably no higher than 500 kJ mol^{-1}. Recent estimates include 408 kJ mol^{-1} ($Re_2Br_8^{-2}$), 489 kJ mol^{-1} ($Mo_2(O_2CCH_3)_4$), and 305 kJ mol^{-1} ($Mo_2Cl_8^{-2}$).[129] The relative weakness of these quadruple bonds may seem paradoxical, but it should be recalled that we tend to compare them with multiple bonds in small elements where bonds are inherently stronger because of the better overlap in small atoms (see p. 321).

There are two metals, Cu^{II} and Cr^{II}, in the first transition series which form acetate complexes similar in structure to the rhenium and molybdenum carboxylate complexes (Fig. 14.48d). Like the latter they are diamagnetic, indicating a pairing of spins. They differ from the complexes of the heavier metals, however. The Cu—Cu distance in the Cu^{II}, d^9 complex is 264 pm, which is actually somewhat longer than the Cu—Cu distance (256 pm) in metallic copper. It would appear that the Cu—Cu bond in copper(II) acetate is only a weak single bond resulting from the pairing of the odd electron on each copper atom.

The chromium(II) acetate molecule was long thought to have the same bond length as the copper compound and thus have a similar weak bond. Recently, however, the structure was redetermined,[130] and the Cr—Cr distance was found to be 236.2 pm, which is considerably shorter than that found in metallic chromium (249.8 pm). In fact, the Cr—Cr bond has been estimated to be about 45 kJ mol^{-1}, more stable than the Cu—Cu bond.[131] The chief interest in dichromium compounds has been in the bond lengths that vary over quite a range (185–254 pm). Some of these are the shortest metal–metal bonds known and have been dubbed "super-short" bonds. The wide variation in bond length, dependent upon the nature of the substituent ligands, is in sharp contrast to the relative uniformity in length of quadruple bonds in the heavier congeners ($Mo \equiv Mo$ = 204–218 pm; $W \equiv W$ = 216–230 pm).[123]

Descriptive aside on chromium(II) acetate (dichromium(II) tetraacetate)

If the reader would like a sample of a compound containing a quadruple bond to see and touch, chromium(II) acetate is highly recommended. It is easy to prepare and requires no expensive chemicals or equipment: simply the reduction of Cr(III) to Cr(II) followed by precipitation by acetate anions.[132] As explained in Chapter 1, for many, many years before it was realized that the compound contained a quadruple bond it was assigned

[129] W. C. Trogler et al., *J. Am. Chem. Soc.*, **1977**, *99*, 2993; J. A. Connor and H. A. Skinner, Chapter 10, "Reactivity of Metal–Metal Bonds," M. H. Chisholm, ed., *ACS Symp. Ser.*, No. 155, Washington, D.C., **1981**; J. G. Norman, Jr., and P. B. Ryan, *J. Comput. Chem.*, **1980**, *1*, 59.

[130] F. A. Cotton et al., *J. Am. Chem. Soc.*, **1970**, *92*, 2926.

[131] R. D. Cannon, *Inorg. Chem.*, **1981**, *20*, 2341.

[132] G. Pass and H. Sutcliffe, "Practical Inorganic Chemistry," Chapman and Hall, London, **1968**, p. 205; W. L. Jolly, "The Synthesis and Characterization of Inorganic Compounds," Prentice-Hall, Englewood Cliffs, N.J., **1970**, p. 442.

frequently in inorganic prep courses as a test of skill and patience—unless it is absolutely dry it is readily oxidized in air (see p. 4).[133]

Other examples of metal–metal quadruple bonds have been slow in coming. After a rather long delay, technetium analogues of the rhenium compounds have been prepared and characterized. Likewise, for an unexpectedly long time, tungsten refused to form molybdenum analogues with quadruple bonds, though they have recently been reported.[134] There has been some work on heteronuclear quadruple bonds between chromium and molybdenum.

There are other compounds known containing triple bonds and double bonds between atoms of elements such as vanadium, iron, and rhodium (Fig. 14.51).[123,135] The recent characterization of $W_2Cl_4(OR)_4(HR)_2$ ($R = CH_3, C_2H_5$) with a W=W double bond completes a series of ditungsten compounds with bond orders of 4, 3, 2, and 1.[136]

Two other interesting compounds containing metal–metal bonds are mercury(I) chloride ("calomel," mentioned above):

$$HgCl_2 + Hg \longrightarrow Cl{-}Hg{-}Hg{-}Cl \qquad \textbf{(14.119)}$$

and an organogold compound synthesized recently containing a gold-gold bond:[137]

$$Me_2P\begin{matrix} \diagup CH_2{-}Au{-}CH_2 \diagdown \\ \diagdown CH_2{-}Au{-}CH_2 \diagup \end{matrix}PMe_2 + Cl_2 \longrightarrow Me_2P\begin{matrix} & Cl & \\ & | & \\ \diagup CH_2{-}&Au&{-}CH_2 \diagdown \\ & | & \\ \diagdown CH_2{-}&Au&{-}CH_2 \diagup \\ & | & \\ & Cl & \end{matrix}PMe_2 \qquad \textbf{(14.120)}$$

In the latter compound, the structural geometry of the organic ligands and the gold-gold bond are ideal for the square planar coordination expected.

Trinuclear clusters

The best known examples of clusters containing three metal atoms are the derivatives of rhenium trihalides. The basic structural unit is shown in Fig. 14.52a. Each rhenium atom is bonded to the other two rhenium atoms directly by metal–metal bonds and indirectly by a bridging halide ion. In addition, each rhenium atom in the triangular array is coordinated by two more halide ions above and below the plane defined by the three rhenium atoms.

---

[133] Even skilled chemists have not perfected the technique inasmuch as all chromium(II) acetate exhibits some residual paramagnetism, which is probably attributable to Cr^{+3} impurities.

[134] D. DeMarco et al., *Inorg. Chem.*, **1980**, *19*, 575.

[135] F. A. Cotton and M. Millar, *J. Am. Chem. Soc.*, **1977**, *99*, 7886.

[136] L. B. Anderson et al., *J. Am. Chem. Soc.*, **1981**, *103*, 5078.

[137] H. Schmidbauer et al., *Chem. Ber.*, **1977**, *110*, 2236, 2751, 2758; *Z. Naturforsch.*, **1978**, 32B, 1325.

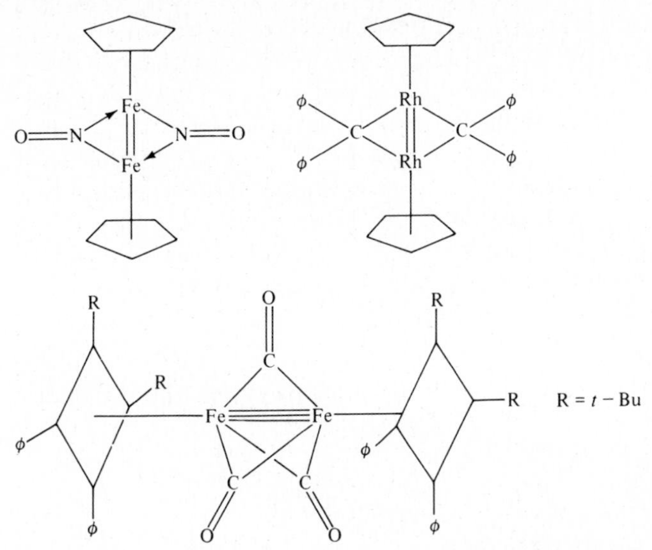

Fig. 14.51 Iron and rhodium organometallic compounds containing double and triple bonds.

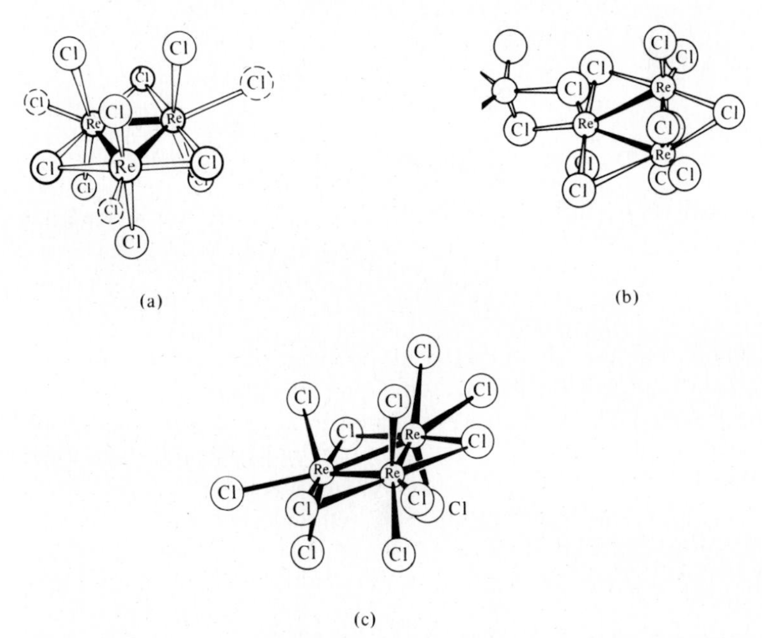

Fig. 14.52 Rhenium(III) clusters: (a) The structural unit present in Re(III) trinuclear cluster. The positions marked ○ are empty in the trihalides in the gas phase but have coordinating groups in other situations. [From B. R. Penfold, *Perspectives in Structural Chemistry*, **1968**, *2*, 71, J. D. Dunitz and J. A. Ibers, eds., Wiley, New York. Reproduced with permission.] (b) The structure of solid $(ReCl_3)_x$. [From F. A. Cotton and J. T. Mague, *Inorg. Chem.*, **1964**, *3*, 1402. Reproduced with permission.] (c) The $Re_3Cl_{12}^{-3}$ anion. [From J. A. Bertrand et al., *Inorg. Chem.*, **1963**, *2*, 1166. Reproduced with permission.]

In the solid state the halides retain this basic unit but further bridging between rhenium atoms by chloride ions results in a polymeric structure (Fig. 14.52b). Likewise, dissolving the halides in solutions of the hydrohalic acids results in the formation of dodecahalotrirhenate(III) ions, $Re_3X_{12}^{-3}$ (Fig. 14.52c), in which additional halide ions have coordinated to the empty positions present in the Re_3X_9 units. Other ligands can also coordinate to these positions. The Re_3 cluster is persistent in many chemical transformations. The bond length is 240–250 pm, which is indicative of strong bonding although less so than in $Re_2X_8^{-2}$. A bond order of two exists between each of the pairs of rhenium atoms in the cluster corresponding to the pairing of the four electrons on each atom.

Tetranuclear clusters

Quadruply bonded binuclear compounds can dimerize through a cycloaddition reaction to give a tetrameric molecule:

$$2K_4Mo_2Cl_8 \xrightarrow{R_3P} \quad [\text{structure}] \quad + KCl \tag{14.121}$$

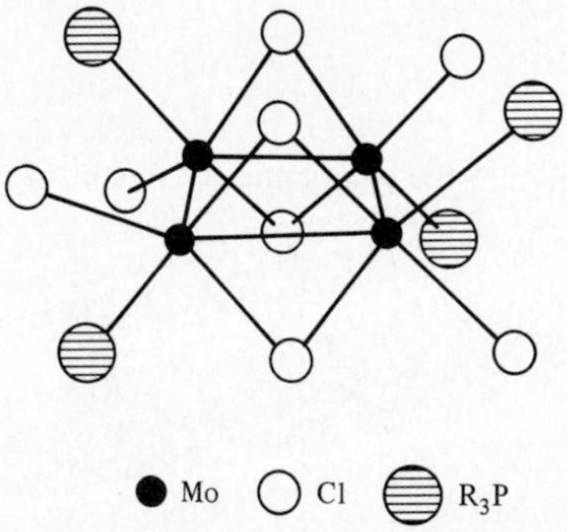

● Mo ○ Cl ⊜ R_3P

The resulting four-membered ring is not square. It appears that there are alternating single and triple Mo–Mo bonds as evidenced by the bond lengths.[138]

Octahedral clusters

Clusters of six molybdenum, niobium, or tantalum atoms have been known for over 30 years and thus predate the later work with rhenium. There are two types: In the first, an octahedron of six molybdenum(II) atoms is coordinated by eight chloride ions, one on each face of the octahedron (Fig. 14.53a). This is found in "molybdenum dichloride, Mo_6Cl_{12}," better formulated as $[Mo_6Cl_8]Cl_4$. The microenvironment about each molybdenum atom is approximately a square antiprism with four chloride ions above and four molybdenum atoms below. The Mo(II) atom can use its four electrons to form four bonds with the adjacent molybdenum atoms and receive dative bonds from the four chloride ions.[139]

Cotton has pointed out that a metal in a low oxidation state can adopt one of two strategies in forming clusters. It can form multiple bonds to another metal, as in $Re_2X_8^{-2}$, or it can form several single bonds to several other metal atoms, as in the octahedral

---

[138] R. E. McCarley et al., "Reactivity of Metal–Metal Bonds," M. H. Chisholm, ed., *ACS Symp. Ser.*, No. 155, Washington, D.C., **1981**, p. 41.

[139] Square antiprismatic coordination is often found in the heavier metals. (See p. 509.)

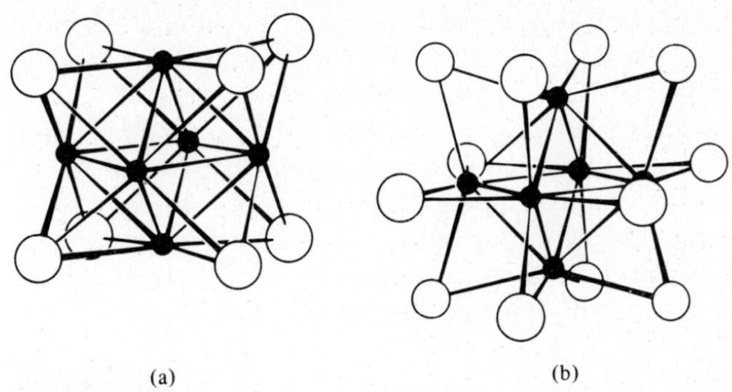

Fig. 14.53 (a) The structure of the $Mo_6Cl_8^{+4}$ ion; ● = Mo, ○ = Cl. (b) The structure of the $M_6X_{12}^{+2}$ ions; ● = Nb, Ta, ○ = Cl, Br. [From F. A. Cotton, *Acc. Chem. Res.*, **1969**, *2*, 240. Reproduced with permission.]

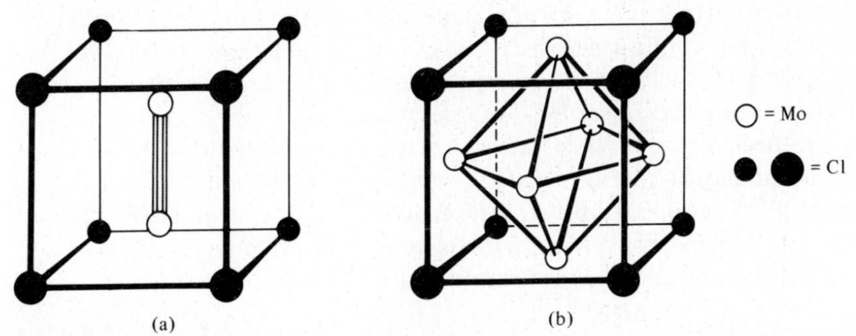

Fig. 14.54 A comparison of two chloro complexes of Mo(II): (a) quadruply bonded $Mo_2Cl_8^{-4}$; (b) singly bonded $Mo_6Cl_8^{+4}$. [From F. A. Cotton, *Acc. Chem. Res.*, **1969**, *2*, 240. Reproduced with permission.]

clusters. It is interesting that Mo(II) adopts both methods (Fig. 14.54) and that both structures have a cubic arrangement of chloride ions.

The second class also contains an octahedron of metal atoms but coordinated by twelve halide ions along the edges (Fig. 14.53b). Niobium and tantalum form these clusters. Here the bonding situation is somewhat more complicated: The metal atoms are surrounded by a very distorted square prism of four metal and four halogen atoms. Furthermore, these compounds are electron deficient in the sense of the boranes—there are fewer pairs of electrons than orbitals to receive them and so fractional bond orders of $\frac{2}{3}$ are obtained.

Bonding in metal clusters

Brief statements have been included above concerning the bonding in each of the clusters discussed. No attempt will be made to explore the possibilities of application of molecular orbital theory to these compounds, but a few interesting aspects of the bonds in these compounds should be mentioned. The bond lengths correlate quite well with the assigned multiplicity of the bonds: $Ta_6Cl_{12}^{+2}$ (order = $\frac{2}{3}$), 290 pm; Re_3X_9 (order = 2), 250 pm; $Re_2Cl_5(DTH)_2$ (order = 3),[140] 230 pm; $Re_2X_8^{-2}$ (order = 4), 220 pm.

It should be noted that the bonding rules developed for the polyhedral boranes (see p. 735) are also applicable to large metal clusters such as those encountered in Chapter 13. It should be noted that just as when the number of electrons outnumber skeletal atoms the number of vertices decreases (because MOs are filled with nonbonding electrons), when metal atoms outnumber electron pairs there is a corresponding *increase* in the number of vertices, namely by the process of "capping" the faces of the polyhedron with the extra metal atoms (see Fig. 13.2).[141]

"Metal-only" clusters

There is a small group of ions that do not fall neatly into any classification. They are exemplified by Bi_9^{+5}, Ge_9^{-2}, Ge_9^{-4}, Sn_9^{-4}, Pb_5^{-2}, Pb_9^{-4}, and Sb_7^{-3}. They exist as regular polyhedra (Fig. 14.55). In general, there is a good correlation between electronic structure and geometry as predicted by the rules discussed above, though some exceptions are known. Thus the Bi_9^{+5} and Sn_9^{-4} are isoelectronic, yet have quite different structures. However, for the other ions listed above, Wade's rules work.

The strategy for the isolation of these ions is of some interest. They had been known for some time in liquid ammonia solutions of alkali–metal alloys. Isolation proved difficult, however, because the alloy ⇄ salt equilibrium is reversible. Isolation can often be achieved by stabilizing the alkali metal ion as cryptate complexes,[142] $[Na(crypt)^+]_2Pb_5$ and $[Na(crypt)^+]_4Sn_9$, which reduces the tendency to revert back to the metal alloy.[143]

Metal–metal bonding in extended phases

Ternary molybdenum chalcides, $M_xMo_6X_8$, are of special interest. These compounds, often called *Chevrel phases*, have both unusual structures and interesting electrical and magnetic properties. An example is $PbMo_6S_8$, which is a superconductor at temperatures below 13.3 K. The idealized structure may be thought of as an octahedral cluster of molybdenum atoms (as in Fig. 14.54b) surrounded by a cubic cluster of sulfur atoms, which in turn is surrounded by a cubic lattice of lead atoms. However, in the actual structure the inner Mo_6S_8 cube is rotated with respect to the Pb lattice (Fig. 14.56).[144] It appears that this rotation is the result of very strong repulsions between the negatively charged sulfur atoms (or sulfide ions) in one S_8 cube with those in an adjacent cube. Thus if lead is replaced by a more electropositive metal (e.g., Eu^{+2}) the calculated charges on

---

[140] DTH = dithiahexane. Cotton has suggested that this complex contains a triple Re—Re bond [M. J. Bennett et al., *J. Am. Chem. Soc.*, **1966**, *88*, 3886; F. A. Cotton, *Acc. Chem. Res.*, **1969**, *2*, 240.

[141] K. Wade, *Adv. Inorg. Chem. Radiochem.*, **1976**, *18*, 1.

[142] "Crypt" is the "football" ligand discussed on p. 532.

[143] J. D. Corbett and P. A. Edwards, *J. Am. Chem. Soc.*, **1977**, *99*; 3313; D. G. Adolphson et al., ibid. **1976**, *98*, 7234; J. D. Corbett, pers. comm.

[144] F. S. Delk, II, and M. J. Sienko, *Inorg. Chem.* **1980**, *19*, 1352; M. Potel et al., *Acta Crystallogr., Sect. B*, **1980**, *36B*, 1319.

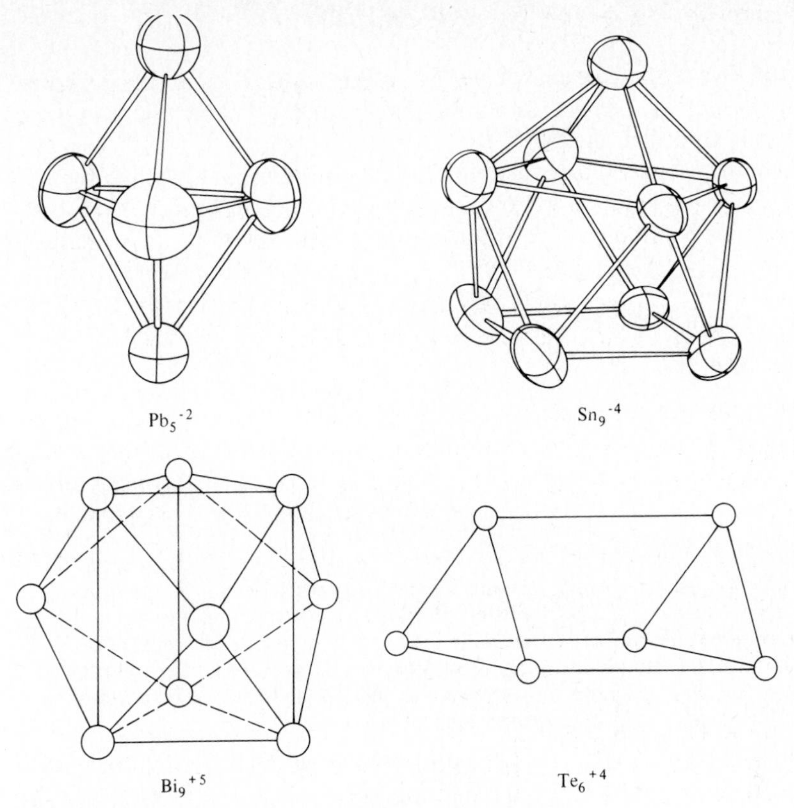

Fig. 14.55 Some representative "metal-only" cluster ions.

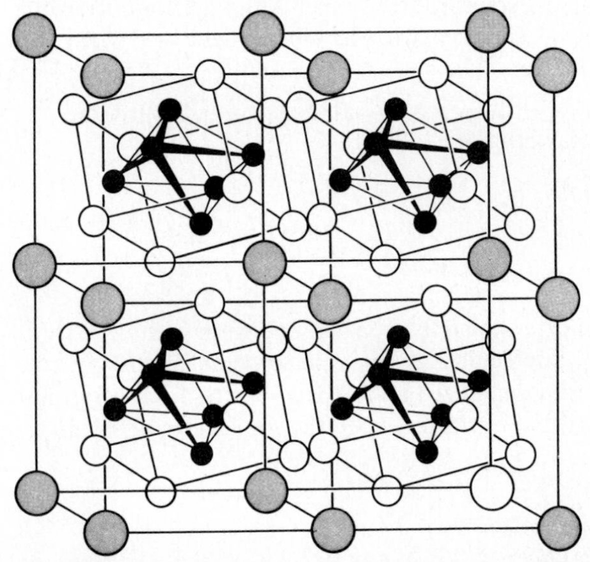

Fig. 14.56 Structure of one of the Chevrel compounds $PbMo_6S_8$. This drawing was supplied by M. J. Sienko of Cornell University. (●) Mo; (○) S; (◎) Pb. [From F. A. Cotton, in "Reactivity of Metal–Metal Bonds," Chapter 1, M. H. Chisholm, ed., *ACS Symp. Ser.*, No. 155, Washington, D.C., **1981**. Reproduced with permission.]

sulfur increase and the turn angle increases. Since the superconductivity is thought to be dependent upon the overlap of the *d* orbitals on molybdenum, this property may be "tunable" by appropriate choice of metals.[145]

The study of metal halides with extended metal–metal bonding was essentially nonexistent until about 10 years ago; yet recently many interesting structures have been described.[146] Some of the possibilities are listed, together with reference to structures, in Table 14.1.

Synthesis of metal clusters

In the preceding description of metal clusters, synthetic reactions were given for some, but not for many others. The paucity of reactions reflects in part the fact that the synthetic chemistry is not as yet systematized—often an attempt to make a quadruply bonded compound, for example, may lead to a doubly bonded one instead. A few years ago Cotton described the situation as follows:

> There does not yet appear to be any instance in which a synthetic reaction was deliberately designed to produce a particular cluster compound from mononuclear starting materials. On the contrary, all known clusters were discovered by chance or prepared unwittingly. Thus the student of cluster chemistry is in somewhat the position of the collector of lepidoptera or meteorites, skipping observantly over the countryside and exclaiming with delight when fortunate enough to encounter a new specimen.[147]

If the author may be allowed to change his hat for a moment from that of inorganic chemist to field biologist, it might be suggested that the discovery of new species, whether of Lepidoptera or metal clusters, is enhanced *slightly* by a knowledge of the terrain, a benefit presumably *not* enjoyed by the collector of meteorites! Although the rationality and predictability of cluster compounds is still somewhat less than might be desired, it is obvious that a prediction made a decade ago:[148]

> if as much work is done on them in the next ten years as in the last decade they will be one of the best understood areas of inorganic chemistry . . ."

is indeed coming true.

Conclusion

An extremely wide variety of compounds, ranging from metal–only to nonmetal–only to those molecules that are mixed metal/nonmetal, has been encountered in this chapter. All have had their descriptive chemistry systematized on the basis of structural principles. Perhaps surprisingly (and perhaps *not*!) metal clusters obey the same rules as nonmetal

---

[145] J. K. Burdett and J. H. Lin, *Inorg. Chem.*, **1982**, *21*, 5.

[146] J. D. Corbett, Chapter 18 in "Solid State Chemistry: A Contemporary Overview," J. B. Milstein and M. Robbins, eds., *ACS Adv. Chem. Ser.*, No. 186, Washington, D.C., **1980**; *Acc. Chem. Res.*, **1981**, *14*, 239; E. L. Muetterties, *Chem. Eng. News*, Aug. 30, **1982**.

[147] F. A. Cotton, *Quart. Rev. Chem. Soc.*, **1966**, *20*, 397.

[148] This text, First Edition. A large portion of this advance has come out of Cotton's own labs at MIT and Texas A & M. See Fig. 1.2.3 in F. A. Cotton and F. A. Walton, "Multiple Bonds Between Metal Atoms," Wiley, New York, **1982**.

Table 14.1 Some representative extended metal–metal structures[a]

Formula[b]	Cl/M ratio	Structural description
$Sc(Sc_6Cl_{12})$	1.714	*Discrete* Sc_6Cl_{12} *clusters* (cf. Fig. 14.54b) plus Sc(III) in octahedral holes.
${}^{1}_{\infty}(ScCl_2)^+ + (Sc_4Cl_6)^-$	1.600	*Single chains of metal octahedra* (edge-bridged by Cl) plus parallel chains of $Sc(III)Cl_6$ octahedra, both sharing *trans* edges.
${}^{1}_{\infty}(ScCl_2) + (Sc_6Cl_8)$	1.429	*Double chains of metal octahedra* sharing edges (exposed faces capped by Cl) with a parallel chain of $Sc(III)Cl_6$ octahedra sharing *trans* edges.
${}^{2}_{\infty}ScCl$	1.000	*Double metal layers* alternating with double chlorine layers.

[a] Adapted from J. D. Corbett, *Acc. Chem. Res.*, **1981**, *14*, 239.

[b] The left superscript indicates one-dimensional chains or two-dimensional layers; the subscript indicates infinite chains or layers.

borane clusters; metal–metal multiple bonds follow the same symmetry as those in organic chemistry, then go the latter one better; catenation, long thought to be an almost exclusive province of organic chemistry has proved to be an extremely important aspect of inorganic chemistry. For those interested in exploring these topics beyond the present short chapter, an entire book has been devoted to the subject.[149]

PROBLEMS

14.1. **a.** Asbestos is the name used for various commercially useful inorganic fibers: chrysotile and amphiboles. From your knowledge of the amphibole structure offer an explanation for the fibrous nature of asbestos.

b. Suggest an explanation for the cleavage properties of the micas.

14.2. Assuming that the external H—B—H angle in B_2H_6 accurately reflects the interorbital angle:

a. Calculate the *s* and *p* character in these bonds.

b. Calculate the *s* and *p* character remaining for the bridging orbitals.

c. Compare the value from (b) with the experimental internal angles.

14.3. **a.** Determine the number of isomers of the compound $N_3P_3Cl_4\phi_2$.

b. Determine the number of isomers of $N_3P_3\phi_3(NHR)_3$ [see T. Moeller and P. Nannelli, *Inorg. Chem.*, **1962**, *1*, 721].

c. Determine the number of isomers of $N_4P_4Cl_6R_2$ [see L. F. Audrieth, *Rec. Chem. Progr.*, **1959**, *20*, 57].

14.4. Suggest the most likely product(s) from the reaction of phosphorus trichloride and excess 1,1-dimethylhydrazine, $H_2NN(CH_3)_2$. By what methods might you attempt to characterize the product? [D. B. Whigan et al., *Inorg. Chem.* **1970**, *9*, 1279].

14.5. **a.** Construct paper models of paratungstate Z and metatungstate (Fig. 14.10).

b. Construct paper models of a 12-heteropoly anion (Fig. 14.11a) and show relation to metatungstate and dimeric 9-heteropoly anions (Fig. 14.16).

c. Convince yourself that the tetrahedral hole in the metatungstate ion has an edge identical in length to that of the octahedra composing that structure. (See Appendix H for help.)

14.6. Thermal decomposition of the hydrazine adduct of *t*-butylborane might be expected to proceed as follows:

$$t\text{-}C_4H_9BH_2 \leftarrow NH_2NH_2 \xrightarrow{-H_2} t\text{-}C_4H_9B\begin{matrix} HN{-}NH \\ / \quad\quad \backslash \\ \quad\quad\quad \\ \backslash \quad\quad / \\ HN{-}NH \end{matrix}B{-}t\text{-}C_4H_9$$

The product isolated had the expected empirical composition but twice the expected

[149] A. L. Rheingold, ed., "Homoatomic Rings, Chains and Macromolecules of Main-Group Elements," Elsevier, Amsterdam, **1977**.

molecular weight. Suggest a structure. [J. J. Miller and F. A. Johnson, *Inorg. Chem.*, **1970**, *9*, 69].

14.7. Diphosphines discussed in this chapter have the structure shown in (a), where R can be an alkyl group, a halogen, a hydroxy group, etc. Rationalize the rather odd-appearing structures of diphosphines (b) and (c):

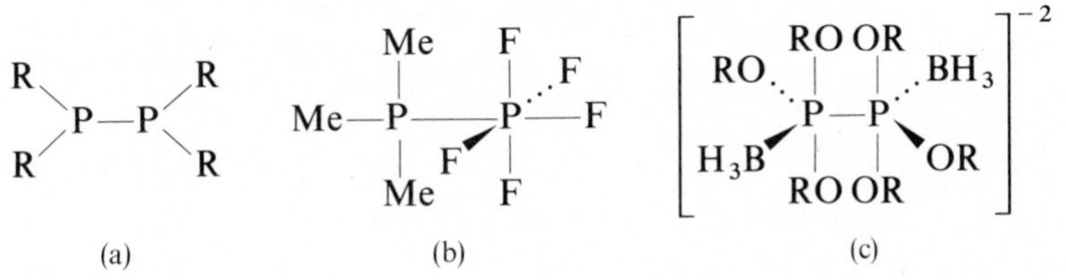

(a) (b) (c)

[*Hint:* Think in terms of acid–base chemistry. See C. W. Schultz and R. W. Rudolph, *J. Am. Chem. Soc.*, **1971**, *93*, 1898; L. A. Peacock and R. A. Geanangel, *Inorg. Chem.*, **1976**, *15*, 244].

14.8. From an inspection of the figures, or even better, from molecular models, determine the geometry of the coordination sphere (cavity) in each of the heteropoly anions discussed in the text.

14.9. As discussed in the text, trimeric phosphonitriles are usually planar but can be forced out of the planar arrangement. In contrast, benzene derivatives are strictly planar. Discuss the reasons for the greater flexibility of the phosphonitriles.

14.10. The classical argument concerning the equivalence of the positions on the benzene ring is based on the existence of three (*ortho, meta, para*) isomers of xylene (dimethylbenzene). How many isomers are there of dimethylborazine?

14.11. It has been suggested [J. D. Corbett, *Progr. Inorg. Chem.*, **1976**, *21*, 129] that Se_8^{+2} exists in the *endo* form (Fig. 14.26) rather than an *exo* "crown" form with Se(1) flipped down (see Fig. 14.20) because of reduced lone-pair repulsions between Se(2), Se(3), Se(6), and Se(7). Sketch Se_8^{+2} as shown in Fig. 14.26 and also in a flipped Se(1) *exo* conformation. Sketch in the lone pairs and indicate how the stabilization occurs in the *endo* form.

14.12. Complete the following equations:

a. $[PNCl_2]_3 + \text{excess } (CH_3)_2NH \longrightarrow$

b. $B_2H_6 + 2R_3P \longrightarrow$

c. $B_2H_6 + 2NH_3 \longrightarrow$

d. $P_2Cl_4 + \text{excess } Cl_2 \longrightarrow$

14.13. Predict the molecular structures of some of the X—X species mentioned in the text, such as P_2I_4, HOOH, Cl_2BBCl_2, and $PH_2PHPHPH_2$.

14.14. Draw out the intermediates in the nucleophilic attack shown in Eq. 14.42.

14.15. Predict the products of the reactions of pentaborane(11) with

a. NH_3

b. $F_2PN(CH_3)_2$
[G. Kodama et al., *J. Am. Chem. Soc.*, **1971**, *93*, 3372].

H H H
H B B H
H—B—B—B—H
H H H H

14.16. If a solution of $Hg_4(AsF_6)_2$ in SO_2 is cooled to -30 °C over several days, crystals of $Hg_{2.86}(AsF_6)$ are deposited on the side of the container. Crystallographic analysis indicates that the structure consists of chains of mercury atoms running through a lattice of AsF_6^- ions (Fig. 14.57). What compounds discussed in this chapter should be resembled by $Hg_{2.86}(AsF_6)$? Discuss ways in which the properties might be similar and ways in which they might be different. [B. D. Cutforth et al., in "Inorganic Compounds with Unusual Properties," R. B. King, ed., Advances in Chemistry Series No. 150, American Chemical Society, Washington, D.C., **1976**, pp. 56–62.]

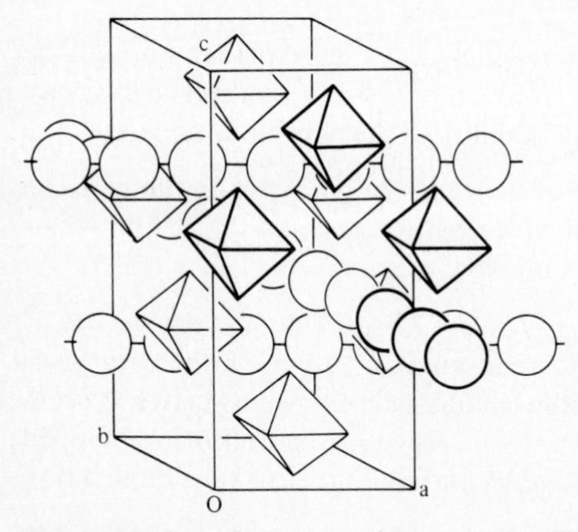

Fig. 14.57 Crystal structure of $Hg_{2.86}AsF_6$. The circles represent mercury atoms (each bearing a formal charge of $+0.35$) and the octahedra represent $[AsF_6]^-$ anions. [From I. D. Brown et al., *Can. J. Chem.*, **1974**, *52*, 791. Reproduced with permission.]

14.17. Consider the Au—Au compound pictured in Eq. 14.120.

a. What is the electron configuration of the gold atoms?

b. What is the oxidation state of the gold?

c. What is the formal charge on each gold atom?

d. Why is the gold expected to favor a square planar geometry in the product but a linear geometry in the reactant?

e. How can we infer the presence of a Au—Au bond since the structure would be essentially the same in its absence?

14.18. Note that the product of Eq. 14.121 contains Mo—Mo pairs that are doubly bridged by chlorine and Mo—Mo pairs that are not. If this molecule contains alternating single and triple bonds, which are which?

14.19. **a.** Why are such seemingly disparate substances as talc, clay, and graphite slippery and useful as lubricants?

b. Although the structures of talc and muscovite are rather similar, the latter is much harder and unsuitable as a lubricant. Why? Should these minerals have any properties in common?

14.20. The reactant in the oxidation of $Rh_2(bridge)_4$ (containing Rh(I), Eq. 11.45) has a Rh—Rh distance of 326 pm. Do you expect the product of Eq. 11.45 (see also Eq. 11.46) to have a shorter or a longer Rh—Rh distance? Explain.

15

The chemistry of the halogens and the noble gases

At first thought it might appear incongruous to discuss the chemistry of the halogens and the noble gases together. The former includes the violently reactive fluorine which will oxidize all save a half dozen elements, even reacting explosively with a compound as stable as water, while the latter family contains the inert gases[1] neon (differing from fluorine by one proton and one electron per atom) and argon (Gr. *argos* = lazy, useless). In one important aspect, however, they are very similar, namely their ionization energies: F = 1681 kJ mol^{-1} (17.42 eV), Ne = 2081 kJ mol^{-1} (21.56 eV), Ar = 1520 kJ mol^{-1} (15.75 eV).[2] The difference between these two families lies in the inordinate disparity in electron affinities. The common tendency to emphasize the differences between the families and dismiss the similarities results from a lack of recognition of the two types of behavior (gain vs. loss of electrons) and the fact that the noble gases are unique only in the discontinuity of their ionization energy–electron affinity (or electronegativity) function. Thus it is often said or implied that the electronegativities of the noble gases are low or nonexistent, equating electronegativity with electron affinity. It is true that the noble gases have no negative oxidation states. But, on the other hand, electronegativity also means a reluctance to release electrons and in this regard the noble gases, as a group, are unsurpassed. In fact, the limiting factor with regard to which noble gas compounds form and which do not seems to be the extent to which a given noble gas atom is willing to share electrons with an atom of another electronegative element.

[1] Several names have been applied to the Group VIIIA elements: noble gases, inert gases, aerogens. The term "inert" is inapplicable to the group as a whole (it is used above as a specific adjective of neon and argon, not as a group appellation) because at least three members of the family are *not* inert. The term *aerogen* has some merit especially since it resembles halogen, chalcogen, etc., but it has been but little used. The name "noble gas" is probably best because it implies a reluctance to react rather than complete abstention, thus paralleling the use of this term in describing the chemistry of certain metals such as gold and platinum.

[2] It would be equally valid to draw a comparison between the noble gases and the metals based on their mutually low electron affinities, but since neither of these groups has any chemistry based on acceptance of electrons such a comparison would be of little use.

The halogens (except fluorine) exhibit both electron-accepting and electron-releasing behaviors. Since they resemble the noble gas elements only in regard to electron-releasing, this aspect of the chemistry of the two families will be discussed first. The more familiar acceptance of electrons is discussed later in the chapter.

NOBLE GAS CHEMISTRY

The discovery of the noble gases

Although the observation of a line in the sun's spectrum as early as 1868 led workers to postulate the existence of an unknown element in the sun's atmosphere, the isolation in 1889 of helium from the mineral cleveite by heating was not recognized as a related phenomenon. The first definitive work was by Lord Rayleigh, who noticed a discrepancy between the density of "chemical nitrogen" and that of "atmospheric nitrogen." The former was obtained by chemically removing nitrogen from various nitrogen oxides, ammonia, or other compounds. The latter was obtained by removal of oxygen, carbon dioxide, and water vapor from air. The difference in density was not great: 1.2572×10^{-3} g cm^{-3} for "atmospheric nitrogen" compared with 1.2506×10^{-3} g cm^{-3} for "chemical nitrogen" under the same conditions. The careful work necessary to establish this difference has often been pointed to, quite rightly, as an example of the importance of precise measurements. Unfortunately, too often the emphasis has been upon the number of significant figures rather than the realization by Rayleigh and Ramsay that the difference was chemically significant. Their arguments concerning the significance of the ratio of specific heats of the noble gases[3] ($C_p/C_v = 1.66$) and their rebuttal of various arguments advanced by their critics show as much chemical insight as the gas density argument, if not more.[4]

Ramsay and Rayleigh succeeded in isolating all of the noble gases except radon and in showing that they were inert to all common reagents. They also discovered the identity of alpha particles and ionized helium.

The early chemistry of the noble gases

It is often assumed that the noble gases had no chemistry prior to 1962. This is true only if one restricts the definition of a chemical compound to (1) something containing "ordinary" covalent or ionic bonds *and* (2) something which may be isolated and placed in a bottle on the reagent shelf. If either of the criteria is dismissed, much important chemistry of the noble gases can be recognized prior to the last decade.

If an aqueous solution of hydroquinone is cooled while under a pressure of several atmospheres of a noble gas [X = Ar, Kr, Xe], a crystalline solid of approximate composition

---

[3] Indicating that the gas must be monatomic since energy is absorbed only by translational modes, not by vibration or rotation (cf. $C_p/C_v = 1.40$, 1.36, and 1.32 for N_2, Cl_2, and Br_2).

[4] For a discussion of the earliest work on the noble gases, see J. H. Wolfenden, *J. Chem. Educ.*, **1969**, *43*, 569, and E. N. Hiebert, in "Noble-Gas Compounds," H. H. Hyman, ed., University of Chicago Press, Chicago, **1963**, p. 3.

Fig. 15.1 The structure of the xenon hydrate clathrate. The xenon atoms occupy the centers of regular pentagonal dodecahedra of water molecules (cf. Fig. 6.7). [From Linus Pauling, "General Chemistry," 3rd ed., W. H. Freeman, San Francisco. Copyright © **1970**. Reproduced with permission.]

$[C_6H_4(OH)_2]_3X$ is obtained. These solids are β-hydroquinone clathrates with noble gas atoms filling most of the cavities.[5] Similar noble gas hydrates are known (Fig. 15.1). These clathrates are of some importance since they provide a stable, solid source of the noble gases, especially the radioactive forms. They have also been used to effect separations of the noble gases since there is a certain selectivity exhibited by the clathrates.

Of particular interest is the effect of noble gases in biological systems. For example, xenon has an anaesthetic effect. This is somewhat surprising in that the conditions present in biological systems are obviously not sufficiently severe to effect chemical combination of the noble gas (in the ordinary sense of that word). It has been proposed that the structure of water might be altered via a clathrate-type interaction.

Although clathrate formation and dipole interactions are perfectly acceptable subjects for chemical discussions, chemists feel more at ease when they can find stable compounds formed from the species being studied. A logical approach would be the investigation of the noble gases for possible Lewis basicity. Since the noble gases are isoelectronic with halide ions which can be strong Lewis bases, it seemed reasonable that noble gas adducts of strong Lewis acids might likewise exist:

$$F^- + BF_3 \longrightarrow BF_4^- \quad \textbf{(15.1)}$$

$$Ne + BF_3 \longrightarrow NeBF_3\ (?) \quad \textbf{(15.2)}$$

$$Xe + BF_3 \longrightarrow XeBF_3\ (?) \quad \textbf{(15.3)}$$

[5] The percentages of available cavities that are filled by noble gas atoms are 67% (Ar), 67–74% (Kr), and 88% (Xe).

A thorough study of solutions of xenon in boron trichloride and boron tribromide was undertaken.[6] A phase study of the melting point of these systems as a function of composition showed no evidence of compound formation.[7] Even stronger evidence comes from the Raman spectra of these mixtures, which are identical to those of pure BX_3. Since the vibrational modes of the boron trihalide molecule will differ depending upon whether it is absolutely planar or not, any deviations from planarity will be observable, even if the product does not reach tetrahedral symmetry.

Strong compound formation:

$$H_3N + BX_3 \longrightarrow H_3N\text{—}BX_3 \tag{15.4}$$

microsymmetry of $BX_3 \simeq T_d$

Weak compound formation, hypothetical:

$$Xe + BX_3 \longrightarrow Xe\text{---}BX_3 \tag{15.5}$$

microsymmetry of $BX_3 = C_{3v}$

No compound formation:

$$Xe + BX_3 \longrightarrow Xe + BX_3 \tag{15.6}$$

microsymmetry of $BX_3 = D_{3h}$

As a result it was concluded that no interaction occurred between xenon and boron trichloride or boron tribromide.[7] Although neon is isoelectronic with fluoride ion, which *does* form adducts, xenon was thought a more likely candidate because it is more polarizable. Two factors favor the formation of halide ion adducts that the noble gases do not enjoy: (1) a negative charge on the halides which results in very low electronegativities and hence good electron-donating abilities (see pp. 151–155); (2) a negative charge which results in some Madelung interaction with the positive center of the Lewis acid helping to stabilize the adduct (see pp. 321–322).

An alternative approach to the formation of true chemical compounds of the noble gases is suggested by two lines of thought. (1) From an acid–base point of view the strongest Lewis acid is the bare nucleus, such as the proton, H^+, so if any of the noble gases is capable

[6] The noble gases are insoluble in boron trifluoride, indicating no interaction. Since BCl_3 and BBr_3 are stronger acids than BF_3 towards most bases, it was decided to investigate them.

[7] N. N. Greenwood and A. J. Osborn, in "Advances in the Chemistry of the Coordination Compounds," S. Kirschner, ed., Macmillan, New York, **1961**, p. 366.

of exhibiting basic behavior it might be expected to do so with H^+ (*not* H_3O^+):

$$He + H^+ \longrightarrow HeH^+ \tag{15.7}$$

$$Ar + H^+ \longrightarrow ArH^+ \tag{15.8}$$

(2) From a simple molecular orbital diagram such as given in Fig. 3.24 it might be supposed that four electrons would result in a nonbonding condition but that any number of electrons less than four would result in some bonding though not necessarily in an integral bond order. (See p. 106 for a discussion of bonding of helium in such species.)

Spectroscopic evidence for species such as HeH^+ and ArH^+ has been obtained from a mixture of hydrogen and noble gases passed through gas discharge tubes. Similar reactions can take place between two noble gas atoms if energy is supplied to remove the necessary electron:

$$He + He - e^- \longrightarrow He_2^+ \tag{15.9}$$

$$Kr + Kr - e^- \longrightarrow Kr_2^+ \tag{15.10}$$

$$Ne + Xe - e^- \longrightarrow NeXe^+ \tag{15.11}$$

The noble gas hydride ions should have a bond order of one and the diatomic noble gas ions should have a bond order of one-half. Neither type can be isolated in the form of salts of the type HeH^+X^- or $He_2^+X^-$ since each electron affinity is greater than that of any appropriate species X^-, and so such salts would spontaneously decompose:

$$2He_2^+F^- \longrightarrow 2He + F_2 \tag{15.12}$$

$$ArH^+Cl \longrightarrow Ar + HCl \tag{15.13}$$

Although the above reactions may seem to be of little interest to the chemist, it has been found[8] that in similar gas-phase reactions xenon behaves as a nucleophile, forming the methylxenonium ion, CH_3Xe^+. The C—Xe bond in this ion has a strength of 180 ± 30 kJ mol^{-1}. (For further reactions of this type, see pp. 301, 336.)

The discovery of stable, isolable noble gas compounds

Although there had been suggestions that some of the noble gases might form compounds, unsuccessful attempts to oxidize krypton and xenon with fluorine in the 1930s essentially put a halt to further speculation, especially in view of the success of valency theory in relating stability to filled octets. This worship of the octet is all the more surprising in view of the fact that compounds with expanded valence shells were already known for over two-thirds of the remaining nonmetals![9]

[8] D. Holtz and J. L. Beauchamp, *Science*, **1971**, *173*, 1237.

[9] This point will be discussed further in this chapter and in Chapter 17, but for now it may be noted that all the nonmetals that form compounds are known with valence shells (by conventional electron-pairing formalisms) containing from 10 to 14 electrons except B, C, N, O, and F.

In the early 1960s Neil Bartlett was studying the properties of platinum hexafluoride, an extremely powerful oxidizing agent. In fact, by merely mixing oxygen with platinum hexafluoride it is possible to remove one electron from the oxygen molecule and isolate the product:

$$O_2 + PtF_6 \longrightarrow O_2^+[PtF_6]^- \quad \textbf{(15.14)}$$

Bartlett realized that the first ionization energy of molecular oxygen, 1180 kJ mol^{-1} (12.2 eV), is almost identical to that of xenon, 1170 kJ mol^{-1} (12.1 eV). Furthermore, the dioxygenyl cation should be roughly the same size as a Xe^+ ion, and hence the lattice energies of the corresponding compounds should be similar. He mixed xenon and platinum hexafluoride with the result that an immediate reaction took place with formation of a yellow solid.[10] Based on the volumes of gases that reacted, Bartlett suggested the formulation $Xe^+[PtF_6]^-$. However, the reaction is not so simple as was once thought, and the product has been formulated variously as $Xe[PtF_6] + Xe[PtF_6]_2$, $Xe[PtF_6]_x$ $(1 < x < 2)$, or $FXe^+[PtF_6]^- + FXe^+[Pt_2F_{11}]^-$.[11] If a large excess of inert SF_6 is added as a diluent, pure $Xe^+PtF_6^-$ may be isolated.[11] Despite the experimental difficulties in characterizing the products of the reaction there was no doubt whatever that a reaction had taken place and the myth of inertness was shattered forever.

It is of some interest to note that several groups were thinking about the possibility of compound formation by the noble gases at the time of Bartlett's discovery. R. Hoppe's research group in Germany had analyzed bond energies of related nonmetal compounds (see below) and concluded that XeF_2 and XeF_4 should exist. However, they did not attempt any preparations until after the publication of Bartlett's work. Clifford and Zeilenga in the United States were reinvestigating the earlier attempts to synthesize xenon fluorides. Working with Pyrex apparatus presented difficulties due to contamination from silicon, but they later isolated a product believed to be Xe_2SiF_6. Meanwhile, workers at Oak Ridge Laboratories were investigating the curious lack of xenon in the gas above molten fluoride reactor fuels. Fission of UF_4 in this fuel should have produced both Kr (which was observed) and Xe (which was not). With 20–20 hindsight[12] it is easy for us to ascribe this discrepancy to the formation of xenon fluorides.[13]

Once the initial breakthrough had been made, a flood of experiments followed, and within a year eight compounds were known and many physical measurements had been made.[14]

---

[10] N. Bartlett, *Proc. Chem. Soc.* (*London*), **1962**, 218. For a review of the early chemistry of xenon compounds, see N. Bartlett, *Endeavour*, **1963**, *88*, 3.

[11] For a review of recent developments in noble gas chemistry, see K. Seppelt and D. Lentz, *Progr. Inorg. Chem.*, **1982**, *29*, 167.

[12] Also known as "retrospective wisdom."

[13] For a longer discussion of this topic, see J. H. Holloway, "Noble-Gas Chemistry," Methuen, London, **1968**, Chap. I, Part 3.

[14] Much of this work was gathered into a single volume: "Noble-Gas Compounds," H. H. Hyman, ed., University of Chicago Press, Chicago, **1963**. Although some of the material therein is speculative and some has been refuted as further work was done, this book is still of considerable value as an insight into the spirit of the workers in the midst of a chemical revolution.

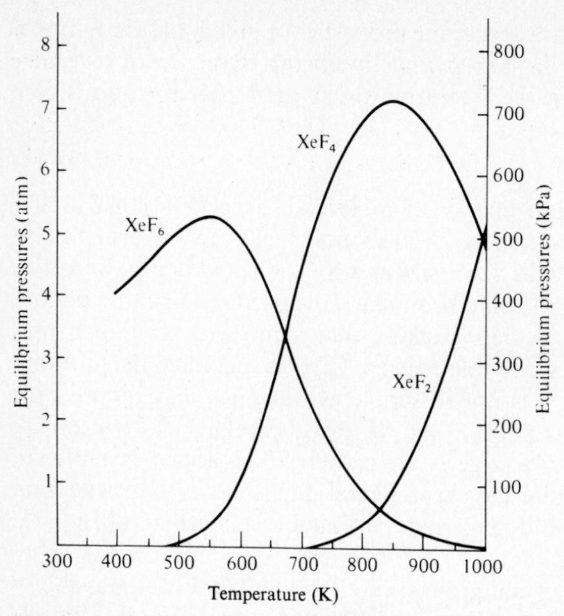

Fig. 15.2 Equilibrium pressures of xenon fluorides as a function of temperature. Initial conditions: 125 mmol Xe, 1225 mmol F_2 per dm^3. At higher Xe/F_2 ratios the XeF_6 diminishes considerably and the remaining two curves shift to the left. [From H. Selig, *Halogen Chem.*, **1967**, *1*, 403, V. Gutmann, ed., Academic Press, New York. Reproduced with permission.]

The fluorides of the noble gases[11]

Mixing xenon and fluorine and activating the mixture by thermal, photochemical, or similar means result in the production of fluorides:

$$Xe + F_2 \longrightarrow XeF_2 \tag{15.15}$$

$$Xe + 2F_2 \longrightarrow XeF_4 \tag{15.16}$$

$$Xe + 3F_2 \longrightarrow XeF_6 \tag{15.17}$$

The chief difficulties in these reactions (once the proper equipment for handling elemental fluorine has been assembled) are not the syntheses but the separations. All three products tend to form (Fig. 15.2). Xenon difluoride can be obtained either by separating it rapidly before it has a chance to react further (by freezing it out on a cold-finger, for example) or by keeping a high Xe/F_2 ratio. The hexafluoride is favored by large excesses of fluorine and low temperatures, but some XeF_4 is present which must be separated. The most difficult compound to prepare pure is xenon tetrafluoride since even optimum conditions for its formation will result in concomitant formation of XeF_2 and XeF_6.[15]

---

[15] For details of the synthetic chemistry of xenon, see E. H. Appelman and J. G. Malm, *Prep. Inorg. Reactions*, **1965**, *2*, 341, and J. H. Holloway, "Noble-Gas Chemistry," Methuen, London, **1968**, Chap. II, Part 3.

The chemistry of krypton is much more limited than that of xenon. The difluoride is known, and the tetrafluoride has been claimed but is doubtful. Radon should react even more readily than xenon, but its chemistry is complicated by the difficulty in working with a compound of such exceedingly high radioactivity. Nevertheless, compound formation with fluorine was shown conclusively shortly after the discovery of xenon compounds, but their exact natures were not elucidated. More recently radon chemistry in solution has been studied (see below).

Bonding in noble gas fluorides

There are currently two approaches to the problem of bonding in noble gas compounds. Neither is completely satisfactory, but between the two they account adequately for the properties of these compounds. The first might be termed a valence bond approach. It would treat the xenon fluorides by means of expanded valence shells through promotion of electrons to the 5*d* orbitals:

$$\text{Ground state: Xe} = [\text{Kr}]5s^2 4d^{10} 5p^6$$

$$\text{Valence state: Xe} = [\text{Kr}]5s^2 4d^{10} 5p^{6-n} 5d^n$$

For XeF_2, $n = 1$ and two bonds form; for XeF_4, $n = 2$ and four bonds form; and for XeF_6, $n = 3$ and six bonds form. Using the arguments of Gillespie (see Chapter 5) the resulting electronic arrangements and structures are as follows:

Compound	Electron pairs	Hybridization	Predicted structure[16]	Experimental structure
XeF_2	5	sp^3d	Linear ("TBP")	Linear
XeF_4	6	sp^3d^2	Square ("octahedral")	Square planar
XeF_6	7	sp^3d^3	Nonoctahedral ("capped octahedron"?)	Unknown but *not* octahedral

The use of Gillespie's VSEPR theory has allowed the rationalization of these as well as several other structures of noble gas compounds (Fig. 15.3). One of the signal successes of this approach was the early successful prediction[17] that XeF_6 was nonoctahedral (see pp. 216–218, 769–770). The most serious objection to it is the required promotion of electrons. This has been estimated to be 1000 kJ mol^{-1} or more for xenon, a serious difficulty. Furthermore, *d* orbitals tend to be diffuse and their importance in nonmetal chemistry is a matter of some controversy (see pp. 824–837).

An alternative approach to bonding in noble gas compounds is the molecular orbital approach involving three-center, four-electron bonds. Consider the linear F—Xe—F molecule. A 5*p* orbital on the xenon can overlap with fluorine bonding orbitals (either pure *p* orbitals or hybrids) to form the usual trio of three-centered orbitals: bonding,

[16] The first structure is that of the atoms; the structure in parentheses refers to the approximate arrangement of all of the valence shell electrons.

[17] R. J. Gillespie, in "Noble-Gas Compounds," H. H. Hyman, ed., University of Chicago Press, Chicago, **1963**, pp. 333–339.

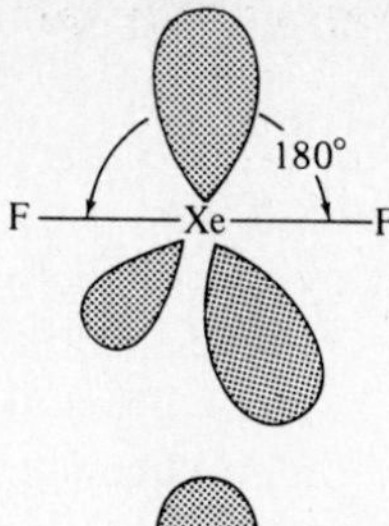

Linear molecule with three nonbonding electron pairs at the points of an equilateral triangle

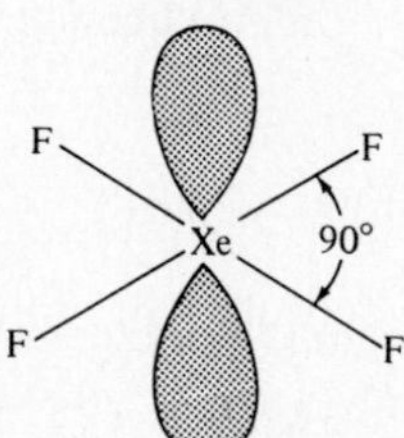

Square planar molecule with two nonbonding electron pairs, one above and one below the plane of the molecule

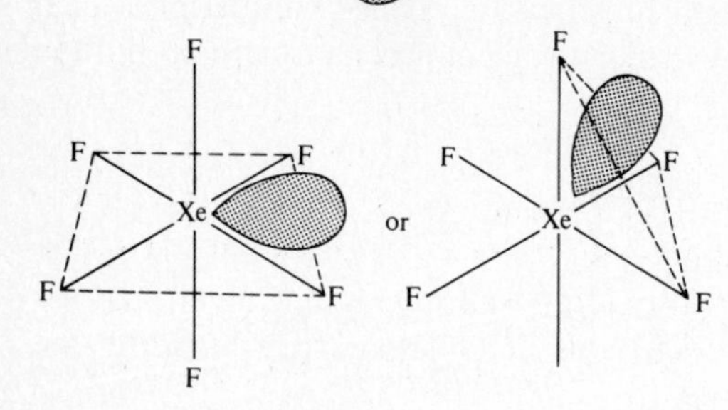

Distorted octahedron with a nonbonding electron pair either at the center of a face or the midpoint of an edge

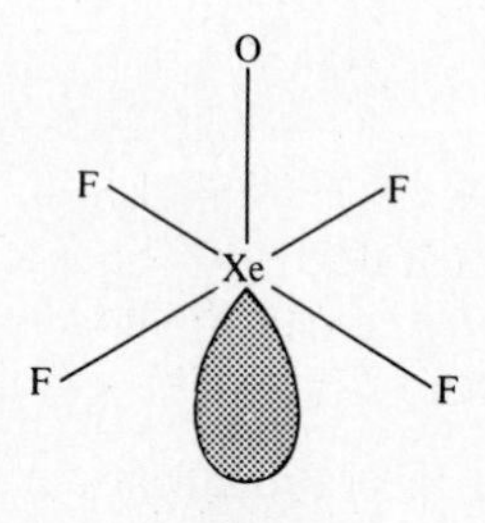

Square pyramidal molecule with a nonbonding electron pair protruding from the base of the pyramid

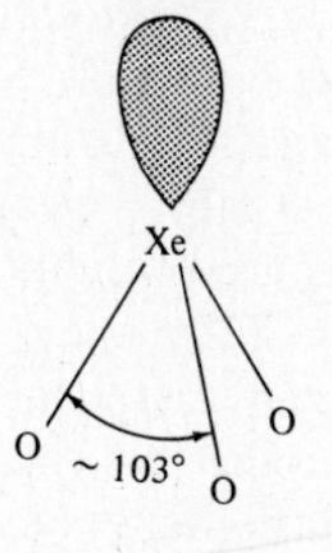

Trigonal pyramidal molecule with a nonbonding electron pair protruding from the apex of the pyramid

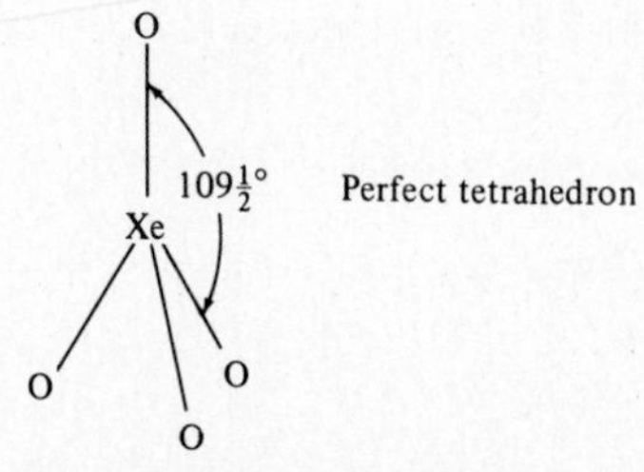

Fig. 15.3 Molecular shapes predicted by simple VSEPR theory. Bond angle values represent experimental results where known. [Modified from J. H. Holloway, *Progr. Inorg. Chem.*, **1964**, *6*, 241, F. A. Cotton, ed., Wiley (Interscience), New York. Reproduced with permission.]

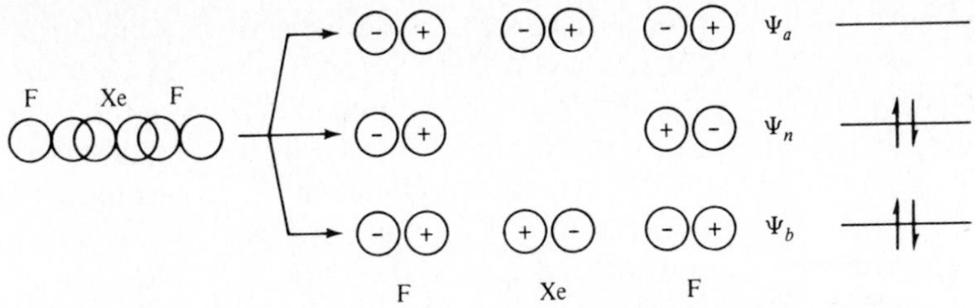

Fig. 15.4 Molecular orbital diagram for F—Xe—F three-center bonds.

nonbonding, and antibonding (Fig. 15.4). Filling in the four valence electrons (Xe_{5p^2} + F_{2p^1} + F_{2p^1}) results in a filled bonding orbital and a filled nonbonding (to a first approximation) orbital. A single bond (or bonding MO) is thus spread over the F—Xe—F system. A second *p* orbital at right angles to the first can form a second three-center F—Xe—F bond (XeF_4), and the third orthogonal *p* orbital can form a third three-center bond (XeF_6). The nature of the *p* orbitals involved in the bonding allows one to predict that XeF_2 will be linear, XeF_4 square planar, and XeF_6 octahedral. The first two predictions are correct, but the last is not. On the other hand, difficulties with promotion energies are avoided.

A comparison of the VSEPR model and the MO model as applied to MX_6 species

Before discussing the difficulties inherent in these two models and their possible resolution, a brief survey of the experimental facts is in order. The number of species isoelectronic with XeF_6 is quite limited. The anions $SbBr_6^{-3}$, $TeCl_6^{-2}$, and $TeBr_6^{-2}$ are octahedral. Both IF_6^- and XeF_6 are nonoctahedral, though their exact molecular structures are unknown. Iodine heptafluoride and rhenium heptafluoride may be considered isoelectronic with these species if they are all considered to have 14 valence shell electrons of approximately equal steric requirements. Both have a pentagonal bipyramidal structure (see Fig. 5.12).

The perfectly octahedral species conform to the expectations based on the simple MO derivation given above. The nonoctahedral fluoride species do not, but this difficulty is a result of the oversimplifications in the method. There is no inherent necessity for delocalized MOs to be restricted to octahedral symmetry. Bartell and co-workers[18] have attacked this problem and have shown that results comparable to those obtained from the VSEPR model can be obtained. Furthermore, it is possible to transform delocalized molecular orbitals into localized molecular orbitals[19] or to predict the symmetry of nonlocalized molecular orbitals from simple localized molecular orbitals.[20] Gillespie has suggested that the electron pairs in his VSEPR model may be in localized MOs. Although the VSEPR theory is normally couched in VB terms, it depends basically on the repulsion of electrons of like spins, and if these are in localized orbitals the results should be comparable.

[18] L. S. Bartell, *Inorg. Chem.*, **1966**, *5*, 1635; *J. Chem. Educ.*, **1968**, *45*, 754; H. B. Thompson and L. S. Bartell, *Inorg. Chem.*, **1968**, *7*, 488; R. M. Gavin, Jr., *J. Chem. Educ.*, **1969**, *46*, 413.

[19] S. F. A. Kettle, *Theoret. Chim. Acta*, **1965**, *3*, 282.

[20] H. B. Thompson, *Inorg. Chem.*, **1968**, *7*, 604.

The molecular structure of XeF_6 continues to be a vexing problem. In the solid, XeF_5^+ and F^- ions exist. The former have five bonding and one nonbonding pair and is therefore expected to be a square pyramid with the lone pair occupying the sixth position of the idealized octahedron. Experimentally, this is found to be the case, in accord with VSEPR theory.[21] In the gas phase, however, the structure is much more perplexing, from both an experimental and theoretical view. Electron diffraction studies[22] indicated that the molecule was a slightly distorted octahedron that was probably "soft" with respect to deformation. It was found that dipole moments were not measurable.[23] This would rule out large, static distortions that might have been expected from VSEPR theory, but would not rule out a stereochemically nonrigid molecule that rapidly passed from one nonoctahedral configuration to another, and this interpretation is currently accepted.[24]

Gillespie[25] has discussed the problem presented to the VSEPR theory by the perfectly octahedral species such as $SbBr_6^{-3}$, $TeCl_6^{-2}$, and $TeBr_6^{-2}$. He pointed out that steric interactions between the large halide ligands will be of considerable importance (the Br—Br distance is approximately equal to the sum of the van der Waals radii and a "7-coordinate" structure with a large lone pair occupying one position would be unfavorable). He therefore suggested that as a result the seventh pair of electrons resides in an unhybridized *s* orbital *inside* the valency shell. As such it would be sterically inactive except for shielding the valence electrons and loosening them from the nucleus. The somewhat lengthened bond in $[TeBr_6]^{-2}$, 275 pm, compared with that expected from addition of covalent radii 250 pm, is consonant with this interpretation (as it is also with a bond order of less than one from a three-center bond). In the fluorides the reduction of steric factors allows the lone pair to emerge to the surface of the molecule, although perhaps less than it would in a 4- or 5-coordinate molecule; hence these molecules appear less distorted than might have been expected (see also p. 216).

Bond strengths in noble gas compounds

As might be expected, xenon does not form strong bonds with any elements, but, contrary to the beliefs prior to 1962, it forms exothermic compounds with fluorine. Some typical bond strengths are listed in Table 15.1. Bartlett[26] has shown that such values might have been expected by extrapolation of known bond energy in related nonmetal compounds, again indicating that chemists should have been farsighted with respect to these compounds.

---

[21] R. D. Burbank and G. R. Jones, *J. Am. Chem. Soc.*, **1974**, *96*, 43.

[22] L. S. Bartell and R. M. Gavin, *J. Chem. Phys.*, **1968**, *48*, 2460, 2466.

[23] R. F. Code et al., *J. Chem. Phys.*, **1971**, *47*, 4955; W. E. Falconer et al., ibid., **1968**, *48*, 312.

[24] For an interpretation of all preceding experimental work in terms of a nonoctahedral molecule undergoing pseudorotation, see L. S. Bernstein and K. S. Pitzer, *J. Chem. Phys.*, **1975**, *62*, 6230; K. S. Pitzer and L. S. Bernstein, ibid., **1975**, *63*, 3849.

[25] R. J. Gillespie, *J. Chem. Educ.*, **1970**, *47*, 18.

[26] N. Bartlett, *Endeavour*, **1963**, *88*, 3. It is interesting to note that Bartlett did *not* use isoelectronic series in his extrapolation. Furthermore, although his extrapolations provide quite reasonable values for XeF_6 and XeO_3, they led to much too high a value for KrF_2.

Table 15.1 Bond strengths in noble gas compounds (kJ mol^{-1})

Compounds	Bond	Bond energy
XeF_2, XeF_4, XeF_6	Xe—F	130 ± 4
XeO_3	Xe—O	84
KrF_2	Kr—F	50

Other compounds of xenon

Attempts to isolate a stable xenon chloride have not been very successful. Two chlorides have been identified, both apparently unstable species observable as a result of trapping in a matrix. The radioactive decay of ^{129}I in $KICl_4$:

$$^{129}ICl_4^- \longrightarrow {}^{129}XeCl_4 + \beta^- \tag{15.18}$$

has been used to produce xenon tetrachloride, characterized by means of the Mössbauer effect (see p. 242) of the gamma emission from the resulting excited state. Mixtures of xenon and chlorine passed through a microwave discharge and immediately frozen on a CsI window at 20 K gave infrared evidence for the existence of $XeCl_2$.

Careful hydrolysis of xenon hexafluoride produces xenon oxide tetrafluoride:

$$XeF_6 + H_2O \longrightarrow XeOF_4 + 2HF \tag{15.19}$$

Complete hydrolysis of xenon hexafluoride or hydrolysis and disproportionation of xenon tetrafluoride produces the trioxide:

$$XeF_6 + 3H_2O \longrightarrow XeO_3 + 6HF \tag{15.20}$$

$$6XeF_4 + 12H_2O \longrightarrow 2XeO_3 + 4Xe + 3O_2 + 24HF \tag{15.21}$$

By comparison, hydrolysis of xenon difluoride results only in decomposition:

$$2XeF_2 + 2H_2O \longrightarrow 2Xe + O_2 + 4HF \tag{15.22}$$

Xenon dioxide difluoride cannot be isolated as an intermediate between the partial hydrolysis to form $XeOF_4$ and complete hydrolysis to XeO_3, but may be prepared by combining these two species:

$$XeOF_4 + XeO_3 \longrightarrow 2XeO_2F_2 \tag{15.23}$$

Huston[27] has systematized much of this type of chemistry in terms of the Lux-Flood definition of acids and bases (see p. 287). The latter is one of those highly specialized definitions that may be very useful in a restricted area but not elsewhere. In the present instance the xenon fluorides are the oxide acceptors while fluoridating the oxide donor. It is possible to construct a scale of relative acidity, $XeF_6 > XeO_2F_4 > XeO_3F_2 > XeO_4 > XeOF_4 >$

[27] J. L. Huston, *Inorg. Chem.*, **1982**, *21*, 685.

$XeF_4 > XeO_2F_2 > XeO_3 > XeF_2$, wherein any acid can react with any base below it to produce an intermediate acid. Thus, in general, XeF_6 is the most useful and XeF_2 the least useful fluoridators. When XeF_4 is used in place of XeF_6, the reactions are slower.[27]

Xenon fluorides are also excellent fluorinators, though not so reactive as KrF_2 (see below). They are often "clean," the only by-products being xenon gas:

$$2(SO_3)_3 + 3XeF_2 \longrightarrow 3S_2O_6F_2 + 3Xe \tag{15.24}$$

$$(C_6H_5)_2S + XeF_2 \longrightarrow (C_6H_5)_2SF_2 + Xe \tag{15.25}$$

Sometimes the fluorination occurs with displacement of a by-product:

$$2R_3SiCl + XeF_2 \longrightarrow 2R_3SiF + Cl_2 + Xe \tag{15.26}$$

Even more interesting is the production of the Xe_2^+ cation in antimony pentafluoride as solvent. It was first prepared by reduction of Xe(II).[28] Many reducing agents are suitable, including metals such as lead and mercury, or phosphorus trifluoride, lead monoxide, arsenic trioxide, sulfur dioxide, carbon monoxide, silicon dioxide, and water. Surprisingly, even *gaseous xenon* may be used as the reducing agent:[29]

$$3Xe + XeF^+ \xrightarrow{SbF_5} 2Xe_2^+ + F^- \tag{15.27}$$

Alternatively, one can view this as an acid–base (instead of a redox) reaction of a basic xenon *atom* undergoing a nucleophilic attack on an acidic xenon *cation* to form the diatomic cation (cf. the reaction of $Xe + CH_3^+$, p. 764).

Xenon trioxide is an endothermic compound which explodes violently at the slightest provocation. Aqueous solutions are stable but powerfully oxidizing. These solutions are weakly acidic ("xenic acid") and contain molecular XeO_3. When these solutions are made basic, $HXeO_4^-$ ions are formed and alkali xenates, $MHXeO_4$, may be isolated from them.[30]

Xenate ions disproportionate in alkaline solution to yield perxenates:

$$2HXeO_4^- + 2OH^- \longrightarrow XeO_6^{-4} + Xe + O_2 + 2H_2O \tag{15.28}$$

Xenate solutions may also be oxidized directly to perxenate with ozone. Solid perxenates are rather insoluble and are unusually stable for xenon–oxygen compounds—most do not decompose until heated above 200°C. X-ray determinations have been made for several perxenates, and they have been found to contain the octahedral XeO_6^{-4} ion, which persists in aqueous solution (possibly with protonation to $HXeO_6^{-3}$).[30]

Treatment of a perxenate salt with concentrated sulfuric acid results in the most unusual xenon tetroxide:

$$Ba_2XeO_6 + 2H_2SO_4 \xrightarrow[-5°C]{} 2BaSO_4 + 2H_2O + XeO_4 \tag{15.29}$$

The tetroxide is the most volatile xenon compound known with a vapor pressure of 3.3 kPa at 0 °C. The structure of the molecule is tetrahedral like the isoelectronic IO_4^- ion.

---

[28] L. Stein et al., *Chem. Comm.*, **1978**, 502.

[29] L. Stein and W. W. Henderson, *J. Am. Chem. Soc.*, **1980**, *102*, 2856.

[30] J. L. Peterson et al., *Inorg. Chem.*, **1970**, *9*, 619, and references cited therein.

The xenon–oxygen compounds are extremely powerful oxidizing agents in acid solution as shown by the following *emf* values:

$$H_4XeO_6 \xrightarrow{2.3-3.0} XeO_3 \xrightarrow{1.8-2.1} Xe$$

Like the fluorides, they are relatively "clean" reagents. Unfortunately, the explosive properties of XeO_3 have resulted in less work being done with them.

Xenon forms stable compounds only with the most electronegative elements: fluorine ($X = 4.0$) and oxygen ($X = 3.5$), or with groups such as $OSeF_5$ and $OTeF_5$ that contain these elements. Reasonably stable, though uncommon, bonds are known between xenon and both chlorine ($X = 3.0$) and nitrogen ($X = 3.0$). Bis(trifluoromethyl)xenon, $Xe(CF_3)_2$ ($X_{CF_3} = 3.3$), is known but decomposes in a matter of minutes. The least electronegative element suggested to form Xe—X bonds is boron ($X = 2.0$), thought to be present in $F—Xe—BF_2$, the product of the reaction between xenon and dioxygenyl tetrafluoroborate; however, there is some doubt concerning the composition of the product—oxygen may be present.[31] Even if the $Xe—BF_2$ bond is confirmed, it will not violate the rule stated above since the presence of the two highly electronegative fluorine atoms raises the electronegativity of the boron atom to about 3.0.[32]

Xenon hexafluoride can act as a Lewis acid. It reacts with the heavier alkali fluorides to form 7-coordinate anions, which in turn can disproportionate to form 8-coordinate species:

$$MF + XeF_6 \longrightarrow M^+[XeF_7]^- \quad (M = Na, K, Rb, Cs) \tag{15.30}$$

$$M^+[XeF_7]^- \xrightarrow[\Delta]{} 2M^+[XeF_8]^{-2} + XeF_6 \quad (M = Rb, Cs) \tag{15.31}$$

The octafluoroxenates are the most stable xenon compounds known; they can be heated to 400 °C without decomposition.[33] The anions have square antiprismatic geometry.[34] They, too, present a problem to VSEPR theory analogous to that of XeF_6 since they should also have a stereochemically active lone pair of electrons that should lower the symmetry of the anion. If Gillespie's theory of steric crowding is correct, however, the presence of eight ligand atoms could force the lone pair into an *s* orbital.

Xenon fluorides can also act as fluoride ion donors. Strong Lewis acids react with xenon fluorides to yield the expected compounds, but since both the cationic and anionic species can form fluoride bridges, the stoichiometries may appear strange at times:

$$XeF_6 + PtF_5 \longrightarrow XeF_5^+PtF_6^- \tag{15.32}$$

$$2XeF_2 + AsF_5 \longrightarrow Xe_2F_3^+AsF_6^- \tag{15.33}$$

$$XeF_4 + 2SbF_5 \longrightarrow XeF_3^+Sb_2F_{11}^- \tag{15.34}$$

Even compounds with deceptively simple stoichiometries may be more complex, as in

[31] For a discussion of the entire question of Xe—X bond stability, see footnote 11.

[32] J. E. Huheey, *J. Phys. Chem.*, **1965**, *69*, 3284.

[33] R. D. Peacock et al., *Proc. Chem. Soc.*, **1964**, *295*; *J. Inorg. Nucl. Chem.*, **1966**, *28*, 2561.

[34] F. W. Peterson et al., *Science*, **1971**, *173*, 1238.

the case of $XeF_2 \cdot AsF_5$, which has both bridged cations and anions:

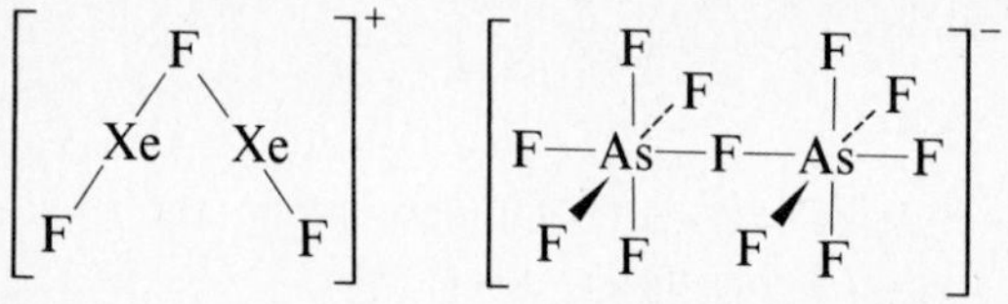

```
[      F      ]+    [   F        F     ]-
[     / \     ]     [   | ,F     | ,F  ]
[   Xe   Xe   ]     [ F—As—F—As—F      ]
[   /     \   ]     [   ◢ |      ◢ |   ]
[ F         F ]     [ F  F     F  F    ]
```

Bartlett's compound (p. 765), though still incompletely understood, is thought to be of this type. Krypton difluoride forms analogous compounds such as $KrF^+SbF_6^-$, $KrF^+Sb_2F_{11}^-$, and $Kr_2F_3^+SbF_6^-$,[35] and these together with the parent KrF_2 appear to be the only compounds of krypton. Although it has not been possible to form krypton–oxygen compounds, krypton fluoride has proven extremely useful as a fluorinating agent—it is 50 kJ mol^{-1} more exothermic than F_2! It may be used to raise metals to unusual oxidation states:[36]

$$7KrF_2 + 2Au \longrightarrow 2KrF^+AuF_6^- + 5Kr + F_2 \quad \textbf{(15.35)}$$

$$KrF^+AuF_6^- \xrightarrow{60\text{–}65\,^\circ C} AuF_5 + Kr + F_2 \quad \textbf{(15.36)}$$

The chemistry of radon

Since radon is the heaviest member of the noble gas family it has the lowest ionization energy, 1030 kJ mol^{-1} (10.7 eV), and might be expected to be the most reactive. The radioactivity of this element presents problems not only with respect to the chemist (who can be shielded) but also with respect to the possible compounds (which cannot be shielded). On the other hand, this radioactivity provides a built-in tracer since the position of the radon in a vacuum line can be ascertained by the γ radiation of ^{214}Bi, one of the decay products.[37] It was found that when a mixture of radon and fluorine was heated, a non-volatile product was formed, possibly an ionic radon fluoride. Similar experiments with chlorine mixtures left the radon in volatile form, presumably unreacted. More recently it has been found that radon reacts with various halogen fluoride solvents (BrF_3, BrF_5, and ClF_3) to form a species in solution which remains behind when the solvent is volatilized. It is quite possible that the Rn^{+2} species is present. Although the charge is not known with certainty, the radon is present as a cation in solution since it migrates to the negative electrode.[38]

HALOGENS IN POSITIVE OXIDATION STATES

Interhalogen compounds

In addition to the diatomic species X_2 known for all of the halogens in the elemental state, all possible combinations XY are also known containing two different halogen

[35] R. J. Gillespie and G. J. Schrobilgen, *Inorg. Chem.*, **1976**, *15*, 22.

[36] J. H. Holloway and G. J. Schrobilgen, *Chem. Comm.*, **1975**, *623*.

[37] The α and β radiation of ^{222}Rn and the first and second daughters will not penetrate the vacuum line.

[38] L. Stein, *J. Am. Chem. Soc.*, **1969**, *91*, 5396; *Science*, **1970**, *168*, 362. Note that iodine also forms cations in solution (see p. 784).

Table 15.2 Interhalogen compounds

	Increasing oxidation state →			
$\Delta\chi^a$	XY	XY_3	XY_5	XY_7
1.38	IF(66.4)[b]	IF_3(~65)	IF_5(64.0)	IF_7(55.2)
1.28	BrF(59.6)	BrF_3(48.1)	BrF_5(44.7)	
0.95	ClF(59.5)	ClF_3(41.2)	ClF_5(~34)	
0.43	ICl(49.7)	$(ICl_3)_2$		
0.33	BrCl(51.6)			
0.10	IBr(41.9)			
0	F_2(37.0)			
0	Cl_2(57.3)			
0	Br_2(45.4)			
0	I_2(35.6)			

(Decreasing $\Delta\chi$ ↓)

[a] Based on p orbital electronegativities, Table 3.12.
[b] Values in parentheses are bond energies from Appendix E (kJ mol^{-1}).

atoms.[39] In addition there are many compounds in which a less electronegative halogen atom is bound to three, five, or seven more electronegative halogen atoms to form stable molecules. The known interhalogens are listed in Table 15.2.

Several trends are noticeable from the data in Table 15.2. The bond strengths of the interhalogens are clearly related to the difference in electronegativity between the component halogen atoms, as expected on the basis of Pauling's ideas on ionic character (p. 144). Furthermore, the tendency to form the higher fluorides and chlorides depends upon the initial electronegativity of the central atom.[40] Only iodine forms a heptafluoride or trichloride. Not shown in Table 15.2 (except by computation from the values) is the instability of certain lower oxidation states to disproportionation:

$$5IF \longrightarrow 2I_2 + IF_5 \qquad \textbf{(15.37)}$$

$$\text{B.E.}: 1390 \longrightarrow 300 + 1340 \qquad \text{Gain in bond energy} = 250 \text{ kJ mol}^{-1} \qquad \textbf{(15.38)}$$

This tendency toward disproportionation is common among the lower fluorides of iodine and bromine. The behavior of the four iodine fluorides presents a good picture of the factors important in the relative stabilities. Both IF and IF_3 tend to disproportionate (the former to the extent that it cannot be isolated), not because of weakness in bonding, but because of the greater number of bonds in the pentafluoride to which they disproportionate. At the other extreme, the heptafluoride, while stable, is a reactive species (it is a stronger fluorinating agent than IF_5) because of the weaker bond energy (resulting from both steric factors and resistance to the extremely high oxidation state on the part of the iodine). Bromine fluoride likewise disproportionates, but BrF_3 and BrF_5 are stable.

[39] Since the most stable isotope of astatine has a half-life of only 8.3 hours the chemistry of this halogen has not been studied extensively. In the following discussion generalities made about the halogens may or may not include astatine. In the present instance AtBr and AtCl have been prepared.

[40] Steric factors may also be important since stability also parallels the size of the central atom compared to that of the surrounding atoms.

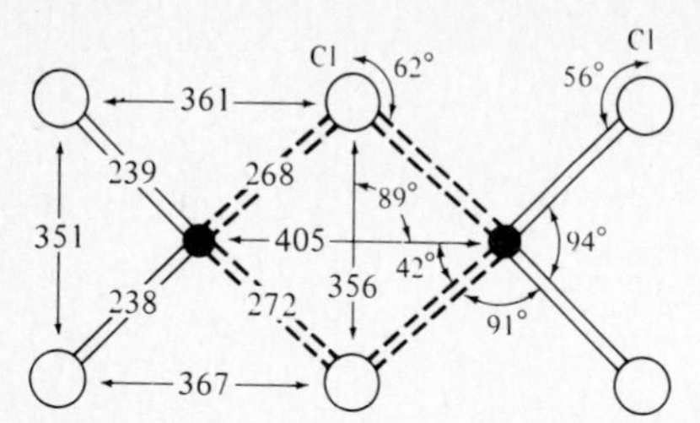

Fig. 15.5 Molecular structure of I_2Cl_6. Distances in pm. [From K. H. Boswijk and E. H. Wiebenga, *Acta Crystallogr.*, **1956**, *7*, 417. Reproduced with permission.]

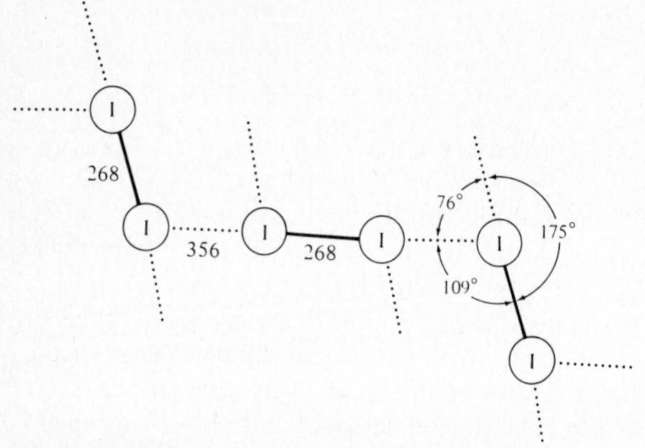

Fig. 15.6 Crystal structure of iodine. The molecules lie in parallel layers, one of which is pictured here. Distances in pm.

In the case of chlorine both the monofluoride and trifluoride are stable. Less is known about ClF_5, but it is probably somewhat less stable and more reactive in view of the fact that it was not prepared prior to 1963.[41]

The interhalogen compounds obey the expectations based on the VSEPR theory, and typical structures are given in Chapter 5. One compound not included there is the dimeric iodine trichloride, in which the monomeric species appears to act as a Lewis acid and accept an additional pair of electrons from a chlorine atom (Fig. 15.5).

The molecules Br_2, I_2, and ICl show an interesting effect in the solid. Although discrete diatomic molecules are still distinguishable there appears to be some intermolecular bonding. For example, the molecules pack in layers with the intermolecular distance *within* a layer 20–80 pm smaller than the distance between layers. Within layers "molecules" approach each other much closer than would be indicated by addition of van der Waals radii but less than that of normal covalent radii (Fig. 15.6). At the same time there is a slight lengthening of the bond between the two atoms forming the nominally diatomic molecule (50 pm in I_2 and ICl). It would appear that in the solid some delocalization of

[41] For a more comprehensive review of the preparation and properties of the halogen fluorides, see L. Stein, *Halogen Chem.*, **1967**, *1*, 133.

Table 15.3 Polyhalogen anions in crystalline salts

X_3^-	X_4^{-2}	X_5^-	X_7^-	X_8^{-2}	X_9^-	X_{16}^{-4}
I_3^-	I_4^{-2}	I_5^-	I_7^-	I_8^{-2}	I_9^-	I_{16}^{-4}
I_2Br^-	Br_4^{-2}	I_4Cl^-	I_6Br^-			
I_2Cl^-		I_4Br^-	IF_6^-			
IBr_2^-		$I_2Br_3^-$	Br_6Cl^-			
ICl_2^-		$I_2Br_2Cl^-$				
$IBrCl^-$		$I_2BrCl_2^-$				
$IBrF^-$		$I_2Br_3^-$				
Br_3^-		$IBrCl_3^-$				
Br_2Cl^-		ICl_4^-				
$BrCl_2^-$		ICl_3F^-				
Cl_3^-		IF_4^-				
		BrF_4^-				
		ClF_4^-				

SOURCE: A. I. Popov, *Halogen Chem.*, **1967**, *1*, 225; K. F. Tebbe in "Homoatomic Rings, Chains and Macromolecules of Main-Group Elements," A. L. Rheingold, ed., Elsevier, New York, **1977**. Used with permission.

electrons takes place, making a simple single-bonded molecular structure no longer completely appropriate.

Polyhalide ions

It has long been known that although iodine has a rather low solubility in water (0.3 g kg^{-1} at 20 °C) it is readily soluble in aqueous solutions of potassium iodide. The molecular iodine behaves as a Lewis acid toward the iodide ion (as it does to other Lewis bases; see pp. 302, 304):

$$I_2 + I^- \longrightarrow I_3^- \tag{15.39}$$

Similar reactions occur with other halogens and every possible combination of bromine, chlorine, and iodine exists under appropriate conditions in aqueous solution:

$$Br_2 + I^- \longrightarrow Br_2I^- \tag{15.40}$$

The evidence for participation of fluorine is uncertain, and although all of the other trihalide ions have been isolated, only one fluoride-containing trihalide ion is known as a crystalline salt (Table 15.3).

The triodide ion presents exactly the same problem to classical bonding theory as xenon difluoride, and although the triodide ion was discovered in 1819, only eight years after the discovery of iodine itself, chemists managed to live with this problem for almost a century and a half without coming to grips with it. The explanation offered most often was that the interaction was electrostatic—an ion-induced dipole interaction. The existence of symmetrical triiodide ions as well as unsymmetrical triiodide ions makes this interpretation suspect, and the existence of ions such as BrF_4^- and IF_6^- makes it untenable.

Currently two points of view are applicable to these species, as they also are to the isoelectronic noble gas fluorides: (1) a valence bond approach with promotion of electrons to *d* orbitals; and (2) three-center, four-electron bonds. The same arguments, pro and con, apply as given previously. Independent of the alternative approaches via VB or MO theory, all are agreed that Madelung energy ("ionic character") is very important in stabilizing both the polyhalide ions and the polyhalogens.[42]

The polyhalide ions may conveniently be classified into two groups (X_3^--type ions belong to both groups): (I) those which are isoelectronic with polyhalogen and noble gas compounds and have general formulas XY_n^-, where n = an even number; (II) polyiodide ions, I_x^-, where x can have various values, usually odd.

Group I polyhalide ions generally consist of a central atom surrounded by two, four, or six more electronegative atoms, with linear, square planar, and distorted octahedral (?) structures, respectively. Their structures obey the VSEPR rules and are closely related to polyhalogen structures, differing by a lone pair in place of a bonding pair.

The polyiodide ions (Group II) present some unusual bonding situations, and so it is simplest to start with I_2 and the two limiting structures of the I_3^- ion. The bond length in the iodine molecule is 267 pm. We can imagine that upon the approach of an I^- anion the charge cloud of the I_2 will distort, with a resulting induced dipole. The greater the distortion of the iodine charge cloud, the weaker the original I—I covalent bond is apt to be,[43] and we might expect that bond to lengthen somewhat. This is the situation in ammonium triiodide—the bond lengths are 282 pm and 310 pm. The "molecular" bond has lengthened by 15 pm, while the long "intermolecular" bond is 40 pm longer than a "normal" covalent bond and only about 80–100 pm shorter than a simple van der Waals contact. Further approach by the iodide ion eventually will result in a symmetrical triiodide system. Such is found in tetraphenylarsonium triiodide, $\phi_4As^+I_3^-$, in which the two bond lengths are equal at 290 pm (Fig. 15.7).[44]

The question as to why both symmetrical and unsymmetrical triiodide ions are found in crystal structures has not been completely resolved. In many ways it parallels the problem of symmetrical and unsymmetrical hydrogen-bonding systems, $[F—H—F]^-$ versus $[F—H\cdots F^-]$ (Chapter 6). Slater[45] has pointed out (in a more rigorous manner than that given above) that at large distances the asymmetric form is favored and the symmetric form is stable only at short distances. We might equate the symmetrical arrangement with optimum, very strong bonding. Indeed, in the hydrogen-bonding system only the very strongest bonding systems F—H—F and O—H—O (and not all of them) exhibit the symmetrical structure. Slater[45] attributed the asymmetrical structure to the lack of pressure from certain ions, but it is more likely that polarization effects from cations can produce an imbalance in the triiodide system and make the asymmetrical form more stable.

---

[42] For calculations, see E. H. Wiebenga and D. Knacht, *Inorg. Chem.*, **1969**, *8*, 738.

[43] Since the unperturbed I—I molecule represents the minimum in the energy curve, any distortion of the molecule will result in a *decrease* in bond energy. In other words, if distortion *could* strengthen the bond, the molecule *would* distort *without* the influence of the iodide ion.

[44] Consult A. I. Popov, *Halogen Chem.*, **1967**, *1*, 225, and E. H. Wiebenga et al., *Adv. Inorg. Chem. Radiochem.*, **1961**, *3*, 133, for the original references and further discussion.

[45] J. C. Slater, *Acta Crystallogr.*, **1959**, *12*, 197.

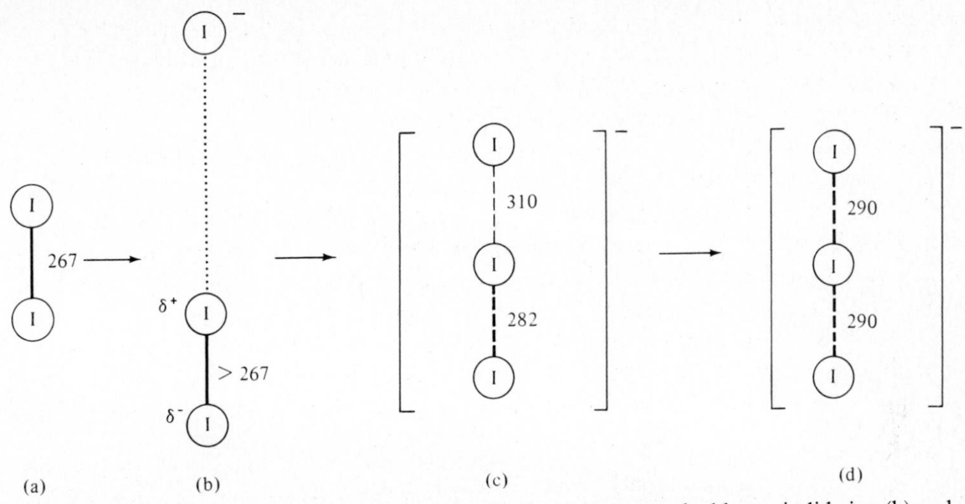

Fig. 15.7 Structural changes as an iodine molecule, I_2, (a) is approached by an iodide ion (b) and changes to an unsymmetrical (c) or symmetrical (d) triiodide ion. Distances in pm.

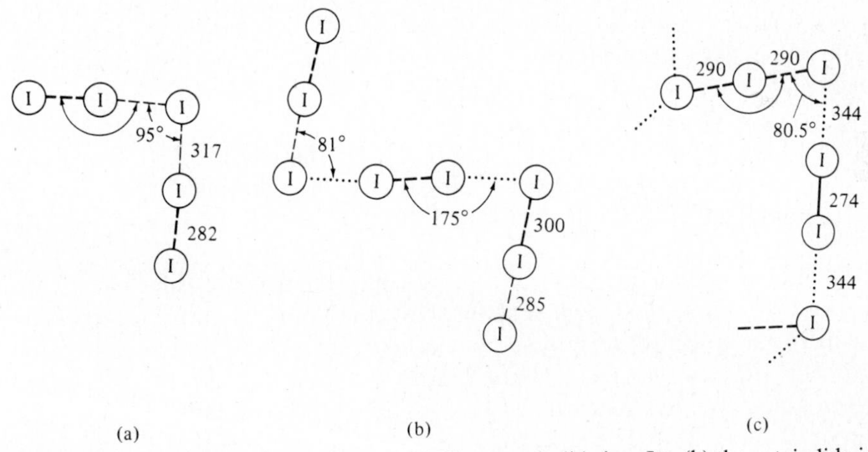

Fig. 15.8 Structures of polyiodide ions: (a) The pentaiodide ion, I_5^-; (b) the octaiodide ion, I_8^{-2}; (c) portion of the infinite polymeric structure of the "heptaiodide" ion as found in Et_4NI_7. Solid lines represent essentially "normal" covalent bonds, dashed lines represent weakened or partial bonds, and dotted lines represent very long and weak interaction. Distances in pm.

The higher polyiodides provide a more complicated picture than the triiodide. The pentaiodide ion, I_5^-, is an L-shaped molecule which may be considered to be two iodine molecules coordinated to a single iodide ion. Alternatively, it can be considered to be two asymmetrical triiodide ions sharing a common iodine atom. There are two bond lengths, 1,2 and 4,5 (282 pm) and 2,3 and 3,4 (317 pm), which correspond very well to the bond lengths in the asymmetrical triiodide (Fig. 15.8a).

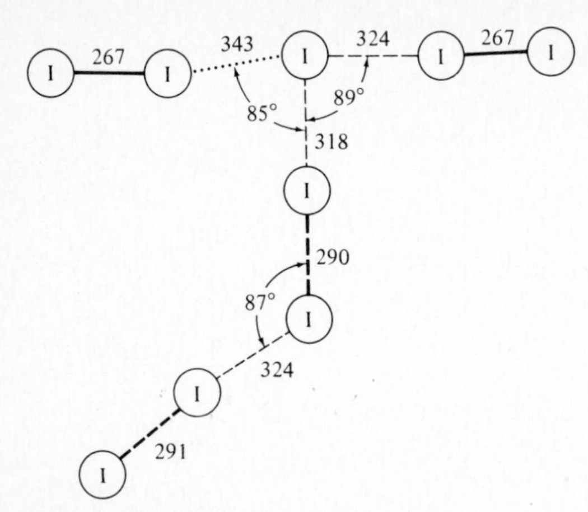

Fig. 15.9 Portion of the structure of the "enneaiodide" ion found in Me_4NI_9. The shortest distance between this moiety and the next is only 349 pm. Distances in pm.

The "tetraiodide ion" as found in CsI_4 has been found to be dimeric, I_8^{-2}. It consists of another leg added to the I_5^- ion to form a Z arrangement. It can be considered to be a central I_2 molecule with an asymmetrical triiodide coordinated to each end. The "short" bonds of the triiodide groups (bonds 1,2 and 7,8) are 285 pm and the "long" bonds (2,3 and 6,7) are 300 pm, in fair agreement with the preceding values. The bonds joining these triiodide moieties to the central iodine "molecule" (3,4 and 5,6) are longer than any encountered thus far—342 pm (Fig. 15.8b).

The so-called heptaiodide ion is found in $(NH_4)I_7$. No discrete I_7^- ions are present. The structure is an infinite three-dimensional framework of "I_2 molecules" (274 pm) and symmetrical "triiodide ions" (290 pm) coordinated through "very long" bonds (344 pm; Fig. 15.8c).

Finally, the enneaiodide ion, I_9^-, is yet a more complex structure, which has been characterized as $I_7^- + I_2$ or $I_5^- + 2I_2$, but probably best considered as a three-dimensional structure similar to the so-called heptaiodide but of more irregular structure with bond lengths of 267, 290, 318, 324, 343, and 349 pm. Unless the latter is arbitrarily considered too long to be a true bond, the system must be considered to be an infinite polymer. A portion of the structure is shown in Fig. 15.9.

Fluorine–oxygen chemistry

There is no evidence that fluorine ever exists in a positive oxidation state. This is reasonable in view of the fact that it is the most electronegative element.[46] Certainly the oxygen compounds of fluorine come the *closest* to achieving a positive charge on fluorine, and since their chemistry is in some ways comparable to that of the other oxyhalogen compounds it is convenient to include them here.

---

[46] Of course since electronegativity is a function of hybridization and charge, it is not impossible that fluorine might find itself in a compound with an element in a particular valence state that was more electronegative than fluorine.

A few oxygen fluorides are known. The most stable of these is oxygen difluoride. It may be prepared by the electrolysis of aqueous hydrofluoric acid:

$$2HF + H_2O \xrightarrow[\text{electrolysis}]{} 2H_2 + OF_2 \qquad \textbf{(15.41)}$$

Thermodynamically it is a stronger oxidizing agent than mixtures of oxygen and fluorine. However, it is relatively unreactive unless activated by an electric discharge or similar high-energy source. In contrast to most molecules containing fluorine, it has a very small dipole moment and thus one of the lowest boiling points of inorganic compounds.[47] Other oxygen fluorides have been suggested of which only O_2F_2 has been well characterized. Little is known of O_4F_2 which supposedly forms as a red-brown solid at 77 K but decomposes on warming. It is possible that the very unstable "O_3F_2" is a mixture.[48]

Several workers have investigated the possibility of synthesizing fluorine–oxygen acids such as HOF, HOFO, $HOFO_2$, or $HOFO_3$. Despite the various claims to have formed species such as these it is the present opinion that only one, HOF, has been synthesized. It may be prepared[49] by passing fluorine over wet surfaces:

$$F_2 + H_2O \xrightarrow{0\,^\circ\text{C}} HOF + HF \qquad \textbf{(15.42)}$$

It is difficult to prepare and isolate because of its reactivity even toward water:

$$HOF + H_2O \longrightarrow HF + H_2O_2 \qquad \textbf{(15.43)}$$

Other compounds containing the —OF group are known: CF_3OF, O_2NOF, F_5SOF, and O_3ClOF. These are all rather unstable compounds with strong oxidizing properties.[50]

Oxyacids of the heavier halogens

The series of acids HOCl, HOClO, $HOClO_2$, and $HOClO_3$ (or HClO, $HClO_2$, $HClO_3$, $HClO_4$) is well known, arising from the disproportionation of chlorine and related reactions:

$$Cl_2 + H_2O \xrightarrow[\text{cold}]{} HOCl + H^+ + Cl^- \qquad \textbf{(15.44)}$$

$$Cl_2 + HCO_3^- \xrightarrow[\text{cold}]{} HOCl + CO_2 + Cl^- \qquad \textbf{(15.45)}$$

$$3Cl_2 + 6OH^- \xrightarrow[\text{hot}]{} ClO_3^- + 5Cl^- + 3H_2O \qquad \textbf{(15.46)}$$

$$4KClO_3 \xrightarrow[\text{careful heating, } 400\text{–}500\,^\circ\text{C}]{} 3KClO_4 + KCl \qquad \textbf{(15.47)}$$

Chlorous acid and chlorite salts cannot be formed in this way but must be formed indirectly from chlorine dioxide which in turn is formed from chlorates:

$$KClO_3 \xrightarrow[\text{oxalic acid}]{\text{moist}} K^+ + H_2O + ClO_2 \qquad \textbf{(15.48)}$$

---

[47] This results from the convergence of the hybridized oxygen and unhybridized fluorine and the effect of lone pair moments (see pp. 160–162). The dipole moment of OF_2 is 0.991×10^{-30} C m. R. D. Nelson et al., Jr., NSRDS-NBS 10, Washington, D.C., **1967**.

[48] See M. Schmeisser and K. Brändle, *Adv. Inorg. Chem. Radiochem.*, **1963**, *5*, 41.

[49] M. H. Studier and E. A. Appelman, *J. Am. Chem. Soc.*, **1971**, *93*, 2349.

[50] C. J. Hoffman, *Chem. Rev.*, **1964**, *64*, 91.

The latter disproportionates with formation of chlorite and chlorate which may in turn be used to prepare the free acid:

$$2ClO_2 + 2OH^- \longrightarrow ClO_2^- + ClO_3^- + H_2O \quad \textbf{(15.49)}$$

$$Ba(ClO_2)_2 + H_2SO_4 \longrightarrow 2HOClO + BaSO_4\downarrow \quad \textbf{(15.50)}$$

The heavier halogens form similar series of compounds although less complete. In all probability neither HOBrO or HOIO exists. The periodate ion has a higher coordination number (resulting from the increase in radius of iodine over chlorine) of 6, IO_6^{-5},[51] as well as 4, IO_4^-. For many years it proved to be impossible to synthesize the perbromate ion or perbromic acid. The apparent nonexistence of perbromate coincided with decreased stability of other elements of the first long period in their maximum oxidation states. This reluctance to exhibit maximum valence has been correlated with promotion energies and with stabilization through π bonding (see Chapter 17).

The first synthesis of perbromate resulted from the *in situ* production of bromine by β decay (cf. $XeCl_4$, p. 771). It was soon found that it could be readily synthesized by chemical means:

$$NaBrO_3 + XeF_2 + H_2O \longrightarrow NaBrO_4 + 2HF + Xe\uparrow \quad \textbf{(15.51)}$$

providing a good example of the use of noble gas compounds as oxidizing agents. They are extraordinary "clean," providing a convenient source of fluorine with only the "inert" xenon given off as a gas.

A more practical synthesis[52] of perbromate is to use fluorine directly as the oxidizing agent:

$$NaBrO_3 + F_2 + 2NaOH \longrightarrow NaBrO_4 + 2NaF + H_2O \quad \textbf{(15.52)}$$

Once formed, perbromate is reasonably stable. The enthalpy of formation and the *emf* of the BrO_3^-/BrO_4^- electrode have been determined.[53] Although perbromate is a stronger oxidizing agent ($\mathscr{E}^0 = -1.8$ V for $H_2O + BrO_3^- \rightarrow BrO_4^- + 2H^+ + 2e^-$) than either perchlorate ($\mathscr{E}^0 = 1.23$ V) or periodate ($\mathscr{E}^0 = -1.64$ V), the difference is not great, and from a thermodynamic point of view it is difficult to explain the lack of oxidation to perbromate by oxidants such as peroxydisulfate ($\mathscr{E}^0 = -2.01$ V) or ozone ($\mathscr{E}^0 = -2.07$ V). The reduction of perbromate proceeds only slowly, and it may be that perbromate is "isolated" from formation and destruction by a high activation energy barrier.

The crystal structure of potassium perbromate has been determined,[54] and it was found that the perbromate ion is tetrahedral as expected from the isoelectronic ClO_4^-, IO_4^-, and XeO_4 species.

---

[51] The chemistry of the orthoperiodate ion is more complicated than implied by the formula IO_6^{-5}. Periodic acid often behaves as a dibasic acid forming salts of $H_3O_6^{-2}$. Furthermore, pyro-type salts are known with the IO_6 octahedra sharing edges and faces. See A. F. Wells, "Structural Inorganic Chemistry," 4th ed., Oxford University Press, London, **1975**, pp. 343–345.

[52] E. H. Appelman, *J. Am. Chem. Soc.*, **1968**, *90*, 1900; *Inorg. Chem.*, **1969**, *8*, 223.

[53] J. R. Brand and S. A. Bunck, *J. Am. Chem. Soc.*, **1969**, *91*, 6500; G. K. Johnson et al., *Inorg. Chem.*, **1970**, *9*, 119.

[54] S. Siegel et al., *Inorg. Chem.*, **1969**, *8*, 1190.

Table 15.4 Halogen oxides and oxyfluorides

Cl_2O	Br_2O	
ClO_2	Br_2O_4	I_2O_4?
Cl_2O_4	Br_3O_8?	I_4O_9?
		I_2O_5
ClO_2F	BrO_2F	IO_2F
$ClOF_3$		IOF_3
Cl_2O_6		
Cl_2O_7		
ClO_3F	BrO_3F	IO_3F
ClO_2F_3		IO_2F_3
		IOF_5

Halogen oxides and oxyfluorides

The heavier halogens form a large number of oxides and oxyfluorides (Table 15.4). Most are rather strong oxidizing agents and some are extremely unstable. These will not be discussed here except to call attention to the use of chlorine dioxide above (Eqs. 15.49 and 15.50) and to its use commercially as a bleaching agent.[55] One exception to the general reactivity of this class of compounds is perchloryl fluoride, ClO_3F. Although it is inherently a strong oxidizing agent it behaves as such only at elevated temperatures. It has a dipole moment of 0.077×10^{-30} C m, lower than any other polar substance. Perbromyl and periodyl fluorides are also known and share the lessened reactivity and low dipole moment but are somewhat less stable.

Halogen cations

In addition to the polyhalide ions discussed previously, which were all anionic, there are comparable cationic species known,[56] although they have been studied considerably less. Many pure interhalogen compounds are thought to undergo autoionization (see Chapter 8) with the formation of appropriate cationic species:

$$2ICl_3 \rightleftharpoons ICl_2^+ + ICl_4^- \tag{15.53}$$

$$2IF_5 \rightleftharpoons IF_4^+ + IF_6^- \tag{15.54}$$

In many cases these cationic species have been postulated on the basis of chemical intuition coupled with knowledge that the pure interhalogen compounds are slightly conducting. In one case X-ray evidence has revealed the existence of an interhalogen cation, ICl_2^+, which is known to exist in $[ICl_2^+][SbCl_6^-]$ and $[ICl_2^+][AlCl_4^-]$.

---

[55] For a review of the chemistry of these compounds see M. Schmeisser and K. Brändle, *Adv. Inorg. Chem. Radiochem.*, **1963**, *5*, 41.

[56] See R. J. Gillespie and J. Passmore, *Adv. Inorg. Chem. Radiochem.*, **1975**, *17*, 49, and J. Shamir, *Structure and Bonding*, **1979**, *37*, 141.

It has been shown that chlorine fluoride reacts with Lewis acids with the formation of an interhalogen cation:[57]

$$2ClF + AsF_5 \longrightarrow [Cl_2F^+][AsF_6^-] \quad \mathbf{(15.55)}$$

When the latter is treated with an excess of chlorine a yellow solid is formed:

$$Cl_2 + [Cl_2F^+] \longrightarrow [Cl_3^+] + ClF \quad \mathbf{(15.56)}$$

suggesting the formation of the Cl_3^+ ion analogous to the previously known Br_3^+ and I_3^+ ions.[58] The iodine system is probably best understood. The I_3^+ forms readily in 100% sulfuric acid:

$$HIO_3 + 7I_2 + 8H_2SO_4 \longrightarrow 5I_3^+ + 3H_2O + 8HSO_4^- \quad \mathbf{(15.57)}$$

These solutions[59] will dissolve one additional mole of iodine with characteristic changes in the absorption spectrum but with no changes in freezing point or conductivity. These data are consistent with

$$I_3^+ + I_2 \longrightarrow I_5^+ \quad \mathbf{(15.58)}$$

The existence of X_3^+ ions which can be considered to be $X_2 \cdot X^+$ raises the question of the possible existence of free X^+ ions. Although the halogens have a much greater tendency to accept electrons than to lose them, the existence of species such as Sb^{+3} and Te_4^{+2} for adjacent elements increases the likelihood that iodine might yield similar species. When iodine is dissolved in 60% oleum (SO_3/H_2SO_4), it is oxidized (by the SO_3 present) to yield a deep blue species, which has been formulated as I^+ or I_2^+. Conductometric, spectrophotometric, cryoscopic, and magnetic susceptibility measurements[60] are all compatible with I_2^+.

Although the presence of the monatomic cation, I^+, has not been demonstrated in systems discussed above, there are conditions under which it is stabilized through coordination and is well characterized. The dichloroiodate(I) anion, ICl_2^-, which has been encountered previously is a simple example. Other complexes of I^+ may be prepared, as through the disproportionation of iodine in the presence of pyridine:

$$Ag^+ + I_2 + 2py \longrightarrow [I(py)_2]^+ + AgI\downarrow \quad \mathbf{(15.59)}$$

When chlorine dioxide fluoride reacts with Lewis acids, the chloryl cation is formed:

$$ClO_2F + AsF_5 \longrightarrow [ClO_2]^+ + [AsF_6]^- \quad \mathbf{(15.60)}$$

The ClO_2^+ cation is bent as expected since it is isoelectronic with SO_2 and NO_2^-.[61]

In an attempt to make salts of the OF^+ cation, OF_2 was allowed to react with a fluoride ion acceptor. The desired reaction would be:

$$OF_2 + AsF_5 \longrightarrow [OF]^+[AsF_6]^- \quad \mathbf{(15.61)}$$

---

[57] K. O. Christe and W. Sawodny, *Inorg. Chem.*, **1969**, *8*, 212.

[58] R. J. Gillespie and M. J. Morton, *Inorg. Chem.*, **1970**, *9*, 811, and references therein.

[59] R. A. Garrett et al., *Inorg. Chem.*, **1965**, *4*, 563.

[60] R. J. Gillespie and J. B. Milne, *Inorg. Chem.*, **1966**, *5*, 1577; R. J. Gillespie and K. C. Malhotra, *Inorg. Chem.*, **1969**, *8*, 1751.

[61] K. O. Christe et al., *Inorg. Chem.*, **1969**, *8*, 2489.

Table 15.5 Polyhalogen cations

X_2^+	X_3^+	X_4^+	X_4^{+2}	X_5^+	X_7^+	XY_2^+	X_2Y^+	XYZ^+	XY_4^+	XY_6^+
	Cl_3^+	Cl_4^+				ClF_2^+	Cl_2F^+		ClF_4^+	ClF_6^+
Br_2^+	Br_3^+					BrF_2^+			BrF_4^+	BrF_6^+
I_2^+	I_3^+		I_4^{+2}	I_5^+	I_7^+	IF_2^+			IF_4^+	IF_6^+
						ICl_2^+	I_2Cl^+			
						IBr_2^+	I_2Br^+	$BrICl^+$		

SOURCE: J. Shamir, *Struct. Bond.*, **1979**, *37*, 141.

Table 15.6 Radii of halogen atoms and halide ions (pm)

Element	r_{cov}	r_{X^-}	r_{VDW}
F	71	136	150
Cl	99	181	190
Br	114	195	200
I	133	216	210

The anticipated reaction did not take place, but dioxygenyl salts (see p. 765) were isolated instead:[62]

$$4OF_2 + 2AsF_5 \longrightarrow 2[O_2]^+ + 2[AsF_6]^- + 3F_2 \qquad \textbf{(15.62)}$$

This brief overview has not included all of the polyhalogen cations known but merely discussed a few typical examples. See Table 15.5 for a listing.

HALIDES

Although many of the compounds of the halogens discussed thus far have exhibited a halogen in a positive oxidation state, most of the chemistry of this family involves either halide ion or covalent molecules in which the halogen is the most electronegative atom.

Physical inorganic chemistry of the halogens[63]

The pertinent trends in the Group VIIA elements are size (Table 15.6) and tendency to attract electrons (Table 15.7). It is only when both these factors are considered that the chemistry of these elements can be rationalized.

The most obvious trend in the family is the attraction for electrons. The ionization energy decreases from fluorine to iodine as expected (see p. 40). There is an apparent anomaly in the case of the electron affinity of fluorine, which is lower than that of chlorine.

[62] J. B. Beal, Jr., et al., *Inorg. Chem.*, **1969**, *8*, 828.

[63] For an excellent review of this topic, see A. G. Sharpe, *Halogen Chem.*, **1967**, *1*, 1.

Table 15.7 Ionization energy, electron affinity, and electronegativity of the halogens

Element	IE[a]	EA[a]	Electronegativity[b]
F	17.4	3.4	3.90
Cl	13.0	3.6	2.95
Br	11.8	3.4	2.62
I	10.4	3.1	2.52

[a] Electronvolts.
[b] Mulliken-Jaffé values in Pauling units.

The small size of the fluorine atom causes it to be saturated quickly with electron density, and the addition of a unit charge causes some destabilization (see pp. 46–49, 155).

The electronegativity of the three heavier members of the family does not change rapidly compared with the difference between Cl and F. This large difference in electronegativity results from the greater contribution of ionization energy (in which fluorine differs considerably) than of electron affinity (which changes little among the halogens).[64] The great electronegativity of fluorine combined with its small size (which enhances Madelung energy from charge separation, see p. 321) results in a much greater reactivity of fluorine than the remaining halogens. In covalent molecules it is exhibited by compounds of fluorine without other halogen analogues, for example, AsF_5, XeF_2, and IF_5. In aqueous solution it is exhibited by the high *emf* of the fluorine electrode resulting from large hydration energy of the small fluoride ion.[65] This much greater reactivity of fluorine has led to its characterization as a "superhalogen."

The anomaly of fluorine

Quite often the first member of a periodic group differs from the remaining members of the group (see Chapter 17). In the case of fluorine the anomaly is quite pronounced. Politzer[66] has illuminated this odd behavior by documenting the weakening of bonding by fluorine to other elements compared with that expected on the basis of extrapolations from the heavier halogens. For example, we have seen that the electron affinity of fluorine is less than might have been expected from the trend of the other halogens. If this trend is extrapolated to fluorine, a value of 440 kJ mol^{-1} is obtained, 110 kJ mol^{-1} greater than the experimental values. As a result of the lower electron affinity, ionic compounds of fluorine have bond energies which are about 108 kJ mol^{-1} lower than values extrapolated from the other halides: LiF (104 kJ mol^{-1} lower), NaF (108 kJ mol^{-1} lower), KF (117 kJ mol^{-1} lower), RbF (104 kJ mol^{-1} lower), and CsF (130 kJ mol^{-1} lower). This

[64] Note that this is true only for small partial charges. As complete ionicity is reached, fluorine is less electronegative than chlorine. Such a situation is unimportant in covalent bonding or in most ionic compounds.

[65] Alternatively the electrode potential can be ascribed to ease with which the F—F bond is broken. As shown by the following discussion the interrelation between bond energy, size, electronegativity energy, etc., is complex and attributing everything to one factor is unwise.

[66] P. Politzer, *J. Am. Chem. Soc.*, **1969**, *91*, 6235; *Inorg. Chem.*, **1977**, *16*, 3350; "Homoatomic Rings, Chains and Macromolecules of Main Group Elements," Chap. 4, A. L. Rheingold, ed., Elsevier, New York, **1977**, pp. 95–115.

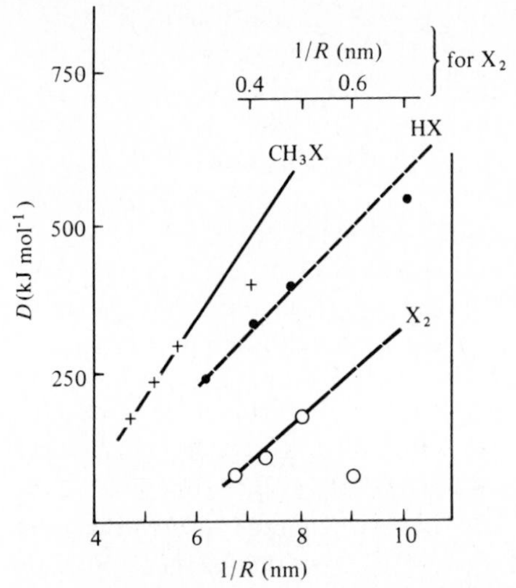

Fig. 15.10 Bond dissociation energies and bond lengths of the hydrogen halides, methyl halides, and halogen molecules. [Modified from P. Politzer, *J. Am. Chem. Soc.*, **1969**, *91*, 6235. Reproduced with permission.]

destabilization can be attributed to forcing a full electronic charge (or nearly one) onto the small fluorine atom.

The surprising fact pointed out by Politzer is that covalent compounds of fluorine seem to show the same destabilization. In Fig. 15.10 the dissociation energies of the hydrogen halides and of the C—X bonds in the methyl halides are plotted against the reciprocals of their bond lengths.[67] The compounds of the three heavier halogens fall on a straight line which, when extrapolated, predicts values for the fluorine compound that are 113 kJ mol^{-1} (HF) and 96 kJ mol^{-1} (CH_3F) too high. This indicates that even when *sharing* an electron from another atom fluorine is destabilized by its small size. Finally, the fluorine molecule itself has a notoriously weak bond (155 kJ mol^{-1}) compared with chlorine (243 kJ mol^{-1}), and it is some 226 kJ mol^{-1} (= 2 × 113) weaker than the extrapolated value. The weak bond in F_2 has traditionally been interpreted in terms of lone-pair repulsion (see p. 320) between the adjacent fluorine atoms. Inspection of the bond energies above as well as certain quantum-mechanical calculations[68] indicates that the weakening of the F—F bond is merely an extension of the tendency toward a weakening of X—F bonds in general, whether X = F, H, C, or a metal.

How do we resolve this apparent paradox: Is fluorine a "superhalogen" or a "subhalogen"? Does it bond better than the other halogens or worse than expected? There really is no conflict here—it depends upon what one is using for a reference. In comparison with the heavier halogens, fluorine is by far the most active, the most electronegative, and provides the most strongly exothermic reactions. In this regard the weakening present in any X—F bond is offset by the weak F—F bond, so the overall enthalpy of the reaction

[67] Note that this figure which is taken directly from Politzer's work portrays in a different way relationships that are closely related to Fig. 7.6.

[68] G. L. Caldow and C. A. Coulson, *Trans. Faraday Soc.*, **1962**, *58*, 633.

is not affected and the effect of a high electronegativity on Madelung energy and electronegativity energy terms is dominant. Fluorine has three factors favoring it over the larger halogens, all resulting from its small size: (1) the largest electronegativity, at least for small partial charges; (2) large Madelung energies in polar molecules; and (3) good covalent bonding resulting from the ability to "get in close" to the adjacent atom.[69] Except for F_2, in which factors 1 and 2 cannot come into play, fluorine typically forms stronger bonds than chlorine (all kJ mol^{-1}): HF = 564, HCl = 430; LiF = 573, LiCl = 506; C—F = 443, C—Cl = 326 (in CX_4), etc. Thus fluorine displays all the properties expected of the smallest halogen. The deficit of 109 kJ mol^{-1} is simply an example of the law of diminishing returns—at some point decreasing size no longer provides increasing bonding benefits *in proportion to further decreasing in size.* Were it not for the saturation effect coming into play, fluorine would probably be a "super-superhalogen"!

Whatever one's interpretation of fluorine chemistry, i.e., sub-, super-, or super-superhalogen, it is obvious that thermochemistry is perhaps more important in understanding the chemistry of fluorine than that of any other element, though, of course, thermodynamics is of supreme importance in the understanding of the chemistry of *any* element. Because of fluorine's unique energetic propensities, it is indeed fortunate that a recent review[70] provides good, thorough thermodynamic data on fluorine compounds.

ASTATINE

It was noted above that discussion of astatine together with the other halogens is inconvenient. Although it is, as expected, the most "metallic" of the halogens, there are few values or experimental data to cite in support of this. (Note for example that such fundamental quantities as ionization energies are unavailable.) Astatine is produced only in tracer amounts, and its chemistry is essentially the descriptive chemistry obtained by tracer methods; macroscopic amounts are not available for other methods.

The best known oxidation state of astatine is −1. The astatide ion behaves as expected for a heavy halide, and "its similarity to the iodide ion is very striking."[71] Astatine may be readily reduced to astatide:

$$2At + SO_2 + 2H_2O \longrightarrow 2At^- + 3H^+ + HSO_4^- \quad \textbf{(15.63)}$$

It forms an insoluble silver astatide which precipitates quantitatively with silver iodide as a carrier.

Study of elemental astatine is complicated by the fact that the small amounts of astatine present are readily attacked by impurities that normally would not be considered important. Most studies of At(0) involve an excess of iodine which ties the astatine up in

[69] Note that with the exception of hydrogen, fluorine is the smallest bonding atom known, and so almost every bond it makes will be with a larger atom.

[70] A. A. Woolf, *Adv. Inorg. Chem. Radiochem.*, **1981**, *24*, 1.

[71] A. H. W. Aten, Jr., *Adv. Inorg. Chem. Radiochem.*, **1964**, *6*, 207. This review should be consulted for additional details concerning the chemistry of astatine.

AtI molecules.[72] It behaves much as might be expected from the known behavior of I_2: It is readily extractable into CCl_4 or $CHCl_3$ and may be oxidized to positive oxidation states by reasonably mild oxidizing agents.

The best characterized positive oxidation state is At(V). Astatate ions may be formed by oxidation of At(0) by persulfate, ceric ion, or periodate:

$$At^- + 6Ce^{+4} + 3H_2O \longrightarrow AtO_3^- + 6Ce^{+3} + 6H^+ \tag{15.64}$$

As such it may be quantitatively precipitated with insoluble iodates such as $Pb(IO_3)_2$ and $Ba(IO_3)_2$.

It appears that perastatate, AsO_4^-, has not been prepared. When astatate was treated with very strong oxidizing agents a negligible amount of the activity precipitated with KIO_4, most precipitating instead with $Ba(IO_3)_2$. The apparent absence of At(VII) is surprising in view of the lower electronegativity and larger size of astatine. If perastatic acid *does* exist, it is probably with coordination number 6: H_5AtO_6.

At least one more oxidation state, presumably At(I) or At(III), is known in aqueous solution, but it has not been characterized. It can be produced by reduction of astatate by chloride ion or oxidation of At(0) by Fe^{+3}. Little is known about it except that it is different from the other forms. It does not precipitate with silver (At^-) or barium (AtO_3^-), nor extract into CCl_4(AtI).

PSEUDOHALOGENS

There are certain inorganic radicals which have the properties of existing either as monomeric anions or as neutral dimers. In many ways these groups display properties analogous to single halogen atoms, and hence the terms *pseudohalogen* or *halogenoid* have been applied to them. A list of some of the properties of the halogens which are mimicked by the pseudohalogens is as follows:[73]

1. They form dimeric molecules, X—X, which are usually quite volatile.
2. They react with many metals to give salts; the Ag(I), Hg(I), and Pb(II) salts are insoluble in water.
3. They form acids with hydrogen, HX, which are highly dissociated in aqueous solution.
4. They form interhalogen compounds and polyhalide ions, XY, XY_n, XY_m^-, etc.
5. They form complexes with various metal ions, MX_4^-.

[72] Strictly speaking one might argue that the astatine in AtI should be considered At(I) instead of At(0) since At should be less electronegative than iodine. The difference in electronegativities is certain to be very small, however, and AtI can probably best be considered as an almost nonpolar molecule with essentially zero charge on both atoms.

[73] P. Walden and L. F. Audrieth, *Chem. Rev.*, **1928**, *5*, 339; T. Moeller, "Inorganic Chemistry," Wiley, New York, **1952**, p. 464.

6. The halide anion, X^-, may be oxidized to the free halogen, X_2.

To the above we may add one more:

7. They may be divided into two types: hard bases (F^-) and soft bases (Cl^-, Br^-, and I^-).

The extent to which the various pseudohalogens obey the above rules is in general quite high, although there are several exceptions. Thus thiocyanogen, $(SCN)_2$, is stable only at low temperatures and at room temperature polymerizes to $(SCN)_x$, violating rule 1. Hydrocyanic acid, HCN, is a very weak acid (pK > 9), and others such as hydrazoic and cyanic are considerably weaker than the hydrohalic acids. With respect to division into hard and soft species, most pseudohalogens are composed of several nonmetal atoms, often with multiple bonding, and so are quite polarizable. As such they tend to resemble iodine considerably more than fluorine. Some are ambidentate, however, and can behave as reasonably hard bases by coordination via a hard nitrogen or oxygen atom (see pp. 312–315).

Some typical reactions of the pseudohalogens which show their resemblance to the halogens are as follows:

1. Formation by the oxidation of X^- ions:

$$2SCN^- + 4H^+ + MnO_2 \longrightarrow (SCN)_2 + 2H_2O + Mn^{+2} \quad \textbf{(15.65)}$$

$$2SCSN_3^- + I_2 \longrightarrow (SCSN_3)_2 + 2I^- \quad \textbf{(15.66)}$$

2. Disproportionation of the free pseudohalogen by a base:

$$(CN)_2 + 2OH^- \longrightarrow CN^- + OCN^- + H_2O \quad \textbf{(15.67)}$$

$$(SCSN_3)_2 + 2OH^- \longrightarrow SCSN_3^- + OSCSN_3^- + H_2O \quad \textbf{(15.68)}$$

3. Precipitation by certain metal ions:

$$Ag^+ + N_3^- \longrightarrow AgN_3\downarrow \quad \textbf{(15.69)}$$

$$Pb^{+2} + 2CN^- \longrightarrow Pb(CN)_2\downarrow \quad \textbf{(15.70)}$$

4. Formation of complex ions:

$$Zn^{+2} + 4SCN^- \longrightarrow [Zn(NCS)_4]^{-2} \quad \textbf{(15.71)}$$

$$Hg^{+2} + 4SCN^- \longrightarrow [Hg(SCN)_4]^{-2} \quad \textbf{(15.72)}$$

Finally, pseudohalogens can be compared with the halogens on the basis of their relative oxidizing power:[74]

$$SCSN_3^- \longrightarrow \tfrac{1}{2}(SCSN_3)_2 + e^- \qquad \mathscr{E}^0 = -0.275 \text{ V} \quad \textbf{(15.73)}$$

$$I^- \longrightarrow \tfrac{1}{2}I_2 + e^2 \qquad \mathscr{E}^0 = -0.535 \text{ V} \quad \textbf{(15.74)}$$

[74] For further discussion of the pseudohalogens, see T. Moeller, "Inorganic Chemistry," Wiley, New York, 1952.

$$SCN^- \longrightarrow \tfrac{1}{2}(SCN)_2 + e^- \qquad \mathcal{E}^0 = -0.77\ V \tag{15.75}$$

$$Br^- \longrightarrow \tfrac{1}{2}Br_2 + e^- \qquad \mathcal{E}^0 = -1.07\ V \tag{15.76}$$

PROBLEMS

15.1. Consider the formation of $O_2^+PtF_6^-$. Why is the ionization potential of O_2 less than that of O? Is it likely that a compound $N_2^+PtF_6^-$ will form? [See pp. 106–107.]

15.2. Show how the FHF^- ion (Chapter 6) can be treated as a three-center, four-electron bond.

15.3. Suggest autoionization possibilities for BrF_3, ICl, and BrF_5, and probable structures for the ions formed.

15.4. Christe and Sawodny (footnote 57) suggested that the Cl_2F^+ ion had the symmetric $Cl—F—Cl^+$ structure since they presumed it was formed from Cl^+ coordinating to the negative (basic) end of the $Cl^{\delta+}—F^{\delta-}$ dipole. It has been argued that while F is negative in ClF it does not follow that it is a better electron donor and the alternative, asymmetric structure $F—Cl—Cl^+$ has been suggested. What are the arguments involved here? [R. J. Gillespie and M. J. Morton, *Inorg. Chem.* **1970**, *9*, 811.]

15.5. Pure iodine is purple in color as are its solutions in CCl_4, $CHCl_3$, and benzene. Aqueous solutions of $K^+I_3^-$ are brown. Solutions of iodine in acetone, dimethyl sulfoxide, and diethyl ether are brown. Suggest an explanation.

15.6. Suggest syntheses for
a. XeO_4
b. $HClO_3$
c. $KBrO_4$

15.7. Suggest probable structures for I_4Cl^- and ICl_4^-, and give reasons why the two are probably not isostructural.

15.8. The production of pseudohalogens requires mild oxidizing conditions (Eqs. 15.65 and 15.66). Recall a metal ion discussed recently that, though not normally noted for its oxidizing action, readily oxidized I^- to I_2. Might it be suitable?

15.9. Why are the halogen cations Cl_3^+, Br_3^+, and I_3^+ best isolated as salts of AsF_6^-, SbF_6^-, and similar anions?

15.10. Predict the molecular structure of the following species encountered in this chapter: ICl_2^+, IF_4^+, IF_6^-, NCSSCN, I_3^+, I_5^+.

15.11. The melting points of the fluorides, MF, MF_2, and MF_3 are generally somewhat lower than the corresponding oxides, M_2O, MO, M_2O_3, because of the greater lattice energy resulting from the dinegative oxide ion, O^{-2}. Yet all of the following reactions are exothermic. Explain.

$$Li_2O + F_2 \longrightarrow 2LiF + \tfrac{1}{2}O_2 \qquad \Delta G^0_{500} = -602\ kJ\ mol^{-1}$$

$$MgO + F_2 \longrightarrow MgF_2 + \tfrac{1}{2}O_2 \qquad \Delta G^0_{500} = -740\ kJ\ mol^{-1}$$

$$Fe_2O_3 + 3F_2 \longrightarrow 2FeF_3 + \tfrac{3}{2}O_2 \qquad \Delta G^0_{500} = -1162 \text{ kJ mol}^{-1}$$

$$ZrO_2 + 2F_2 \longrightarrow ZrF_4 + O_2 \qquad \Delta G^0_{500} = -740 \text{ kJ mol}^{-1}$$

[J. Portier, *Angew. Chem. Int. Ed. Engl.*, **1976**, *15*, 475.]

15.12. If you ask an organic chemist which element can form the largest number of compounds (not that they have necessarily all been synthesized yet), you will usually get one of two answers: carbon from some, hydrogen from others. If you ask an inorganic chemist, you may get a third answer. What is *this* answer? Who, of the three, is "right"? Discuss.

15.13. On p. 773 it is stated that xenon forms bonds with only the most electronegative elements such as the very active fluorine and oxygen. How can you reconcile this with the formation of the Xe—Xe bond in Xe_2^+? [*Hint:* Rethink Problem 3.44.]

15.14. Xe_2^+ obviously will have a fairly high electron affinity (see the ionization energy of atomic xenon), and if it gains an electron, it will dissociate (see p. 106). Combine these facts with the choice of SbF_5 as solvent and acid–base theory to provide a self-consistent interpretation.

15.15. The $OSeF_5$ and $OTeF_5$ groups are very electronegative as shown by the stability of the xenon compounds. Lentz and Seppelt[75] believe, on the basis of several kinds of data such as those shown in Fig. 15.11, that these groups may be more electro-

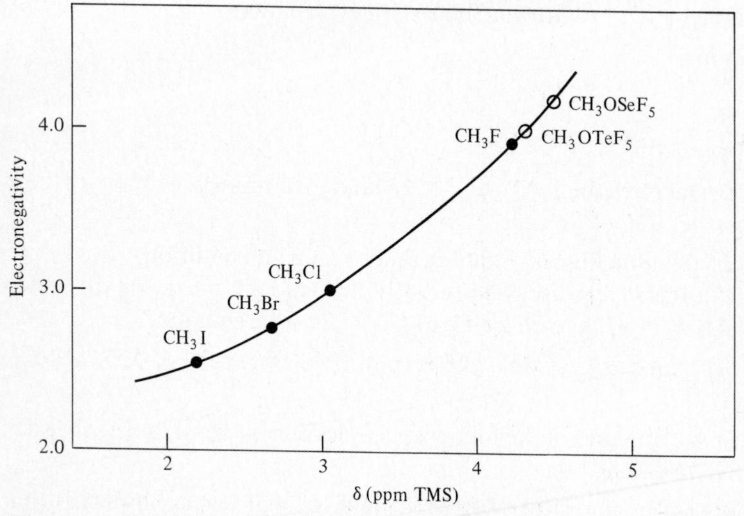

Fig. 15.11 Correlation of ^{1}H chemical shifts (with respect to tetramethylsilane) of methyl compounds CH_3—X with the Allred-Rochow electronegativity of X. Note that both —$OSeF_5$ and —$OTeF_5$ appear to have higher electronegativities than fluorine. [From P. Huppmann et al., *Z. anorg. allg. Chem.*, **1981**, *472*, 26. Reproduced with permission.]

[75] D. Lentz and K. Seppelt, *Z. anorg. allg. Chem.*, **1980**, *460*, 5; P. Huppmann et al., *Z. anorg. allg. Chem.*, **1981**, *472*, 26.

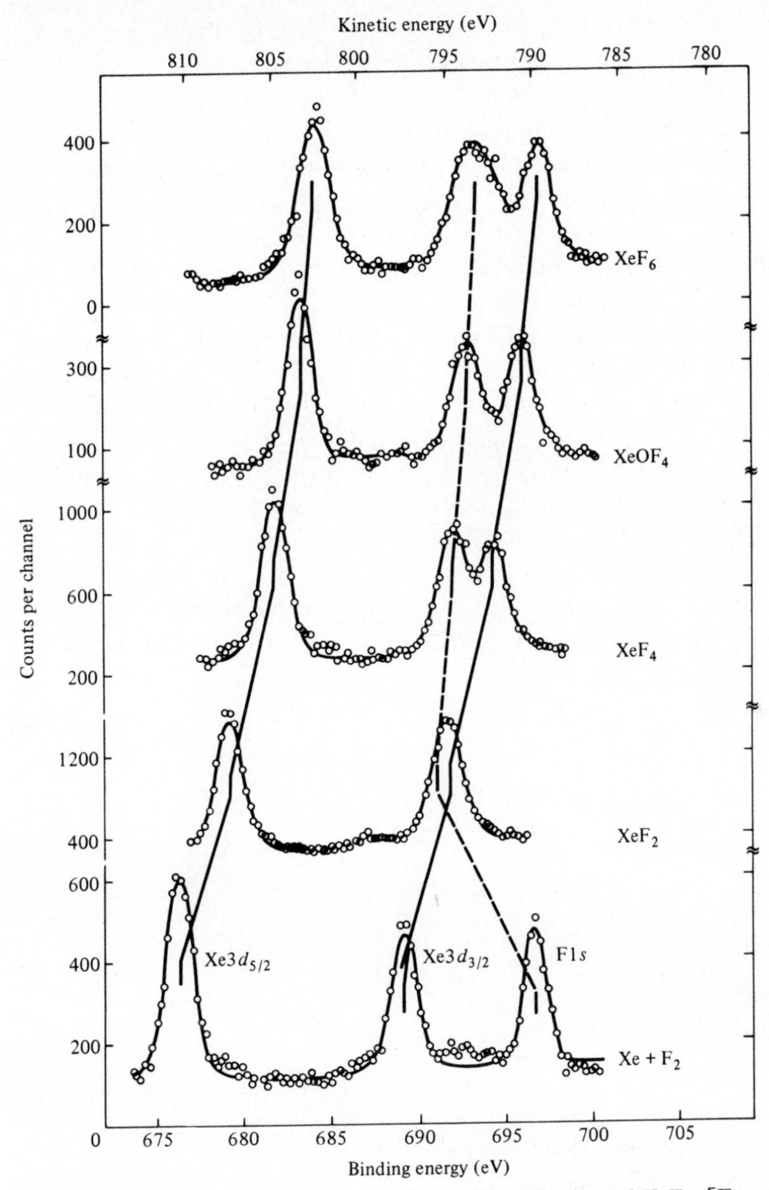

Fig. 15.12 ESCA spectra of Xe, F_2, XeF_2, XeF_4, $OXeF_4$, and XeF_6. [From T. X. Carroll et al., *J. Am. Chem. Soc.*, **1974**, *96*, 1989. Reproduced with permission.]

negative than fluorine itself. Admitting that the interpretation of data of this sort is fraught with difficulties, it is still obvious that the two groups are very electronegative. Can they actually be more electronegative than fluorine? Discuss.

15.16. The ESCA spectra of Xe, F_2, XeF_2, XeF_4, $OXeF_4$, and XeF_6 are shown in Fig. 15.12.

What information can you obtain from these spectra? [The appearance of two peaks for Xe is attributable to the ejection of 3*d* electrons of different *j* values and is irrelevant to the question being asked.]

15.17. Consider the series of xenon oxyfluorides and their relative acidity. Discuss the reasons for the ordering of these compounds. Can you semi-quantify your answer?

15.18. **a.** Predict the most likely by-products of the reduction of XeF_2 to Xe_2^+ as discussed on p. 772.

b. Write balanced equations for all of these reactions.

15.19. The xenon fluorides are described in this chapter as both flurori*d*ators and fluori*n*ators. Is this a typographical error? Can these terms be differentiated? Discuss.

16

The lanthanide, actinide, and transactinide elements

This chapter includes the chemistry of the elements Sc, Y, La to Lu, and Ac to Lr.[1] In addition, some speculations are made concerning heavier elements that may be synthesized in the future.

The lanthanides are characterized by the gradual filling of the $4f$ subshell and the actinides by the filling of the $5f$ subshell. The relative energies of the nd and $(n-1)f$ orbitals are very similar and sensitive to the occupancy of these orbitals (Fig. 16.1).[2] The electron configurations of the neutral atoms (see Table 2.2) thus show some irregularities. Notable is the stable f^7 configuration found in Eu, Gd, Am, and Cm. For the $+3$ cations of both the lanthanides and actinides, however, there is strict regularity; all have $4f^n5d^06s^0$ or $5f^n6d^07s^0$ configurations.

In many ways the chemical properties of the lanthanides are repeated by the actinides. Much use of this similarity was made during the early work on the chemistry of the synthetic actinides, which were often handled in very small quantities and are also radioactive. Prediction of their properties by analogy to the lanthanide series proved very helpful. On the other hand, it should not be thought that the actinide series is merely a replay of the lanthanides. There are several significant differences between the two series related principally to the differences between the $4f$ and $5f$ orbitals.

Stable oxidation states

The characteristic oxidation state of the lanthanide elements is $+3$. The universal preference for this oxidation state together with the notable similarity in size led to great difficulties[3] in the separation of these elements prior to the development of chromat-

[1] The elements Sc, Y, La, and Ac are properly members of Group IIIB and not lanthanides or actinides. They are included here for convenience since they resemble the lanthanides and actinides.

[2] Compare the discussion of the similar problem (*d* versus *s* orbitals) in transition metals, pp. 41–46.

[3] For a discussion of the difficulties and confusion surrounding the early chemistry of these elements, see T. Moeller, "The Chemistry of the Lanthanides," Van Nostrand-Reinhold, New York, **1963**. This is also a short, readable account of other aspects of lanthanide and actinide chemistry.

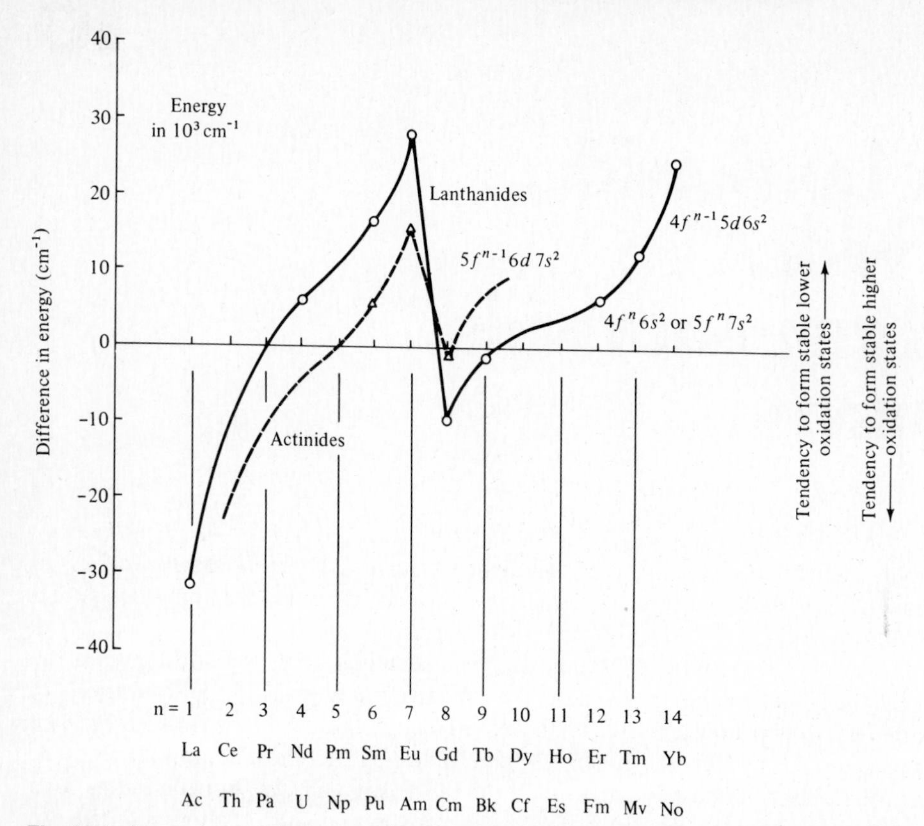

Fig. 16.1 Approximate relative energies of the $f^{n-1}ds^2$ and f^ns^2 electron configurations. [From M. Fred, "Lanthanide/Actinide Chemistry," Advances in Chemistry Series, No. 71, American Chemical Society, Washington, D.C., **1967**. Reproduced with permission.]

ographic methods. Despite the propensity to form stable +3 cations, the lanthanides do not closely resemble transition metals such as chromium or cobalt. The free lanthanide metals are more reactive and in this respect are more similar to the alkali or alkaline earth metals than to most of the transition metals. They all react with water with evolution of hydrogen. One difference lies in the sum of the first three ionization energies—from 3500 to 4200 kJ mol^{-1} (36 to 44 eV) for the lanthanides,[4] compared with 5230 kJ mol^{-1} (54.2 eV) for Cr^{+3} and 5640 kJ mol^{-1} (58.4 eV) for Co^{+3}. A second factor is the heat of atomization necessary to break up the metal lattice: Transition metals with *d* electrons available for bonding are much harder and have higher heats of atomization than the alkali, alkaline earth, and lanthanide metals. Two lanthanides, europium and ytterbium, are particularly similar to the alkaline earth elements. They have the lowest enthalpies of vaporization and the largest atomic radii of the lanthanides (Fig. 16.2), more similar to barium than to typical lanthanides. Presumably these elements donate only two electrons to the bonding orbitals ("bands") in the metal and may be said to be in the "divalent"

[4] M. M. Faktor and R. Hanks, *J. Inorg. Nucl. Chem.*, **1969**, *31*, 1649.

state in the metal as opposed to their congeners. These same two elements resemble the alkaline earth metals in another respect—they dissolve in liquid ammonia to yield conducting blue solutions (see Chapter 8).

Although +3 is the characteristic oxidation state of the lanthanides, the +2 oxidation state is also important. As might be anticipated from the above discussion, Eu^{+2} and Yb^{+2} are the most stable dipositive species. These ions are stabilized by $4f^7$ and $4f^{14}$ configurations enjoying the special stability (from exchange energy, see p. 379) of half-filled and filled subshells. Other lanthanides form Ln(II) compounds which are stable as solids, and there is evidence that *all* the lanthanides can be made to form stable divalent cations in a host lattice.[5]

Not all "divalent" lanthanide compounds are truly such; i.e., some do not contain Ln(II) ions. For example, LaI_2, Ce_2, etc. have been formulated as $Ln^{+3}(I^-)_2e^-$. Although this formulation appears strange because of the free electron, it is no more so than Na^+e^- encountered in Chapter 8. However, in contrast to the electrolytic behavior of the latter in ammonia solution, because of the delocalized electrons these lanthanide diiodides are actually metallic phases.[6]

Higher oxidation states are unusual for the lanthanides. Cerium forms a stable +4 species which is a very strong oxidizing agent in aqueous solution ($\mathscr{E}^0_{red} = +1.74$V), although the reaction with water is sufficiently slow to allow its existence.[7]

Although all of the actinides with the exception of protactinium and possibly thorium exhibit a +3 oxidation state, it is not the most stable oxidation state for several of them. In contrast to the lanthanides, the actinides utilize their *f* electrons more readily and thus exhibit positive oxidation states equal to the sum of the 7*s*, 6*d*, and 5*f* electrons: Ac^{III}, Th^{IV}, Pa^{V}, U^{VI}, and Np^{VII}. As in the first transition series this trend reaches a maximum at +7, and thereafter there is a tendency toward reduction in maximum oxidation state (see Table 16.1). The reduced tendency to use 5*f* electrons as one progresses along the actinide series is clear: U(III) may be oxidized with water, Np(III) requires air, and Pu(III) requires a strong oxidizing agent such as chlorine. The +4 state is the highest known for curium and berkelium, and beyond these elements only +2 and +3 oxidation states are known. Nobelium is actually more stable in solution as No^{+2} than as No^{+3} (cf. Yb^{+2}).

The lanthanide and actinide contractions

As a consequence of the poor shielding of the 4*f* and 5*f* electrons there is a steady increase in effective nuclear charge and concomitant reduction in size. Although this trend is

[5] P. N. Yocom, in "Lanthanide/Actinide Chemistry," R. F. Gould, ed., *Advances in Chemistry Series*, No. 71, American Chemical Society, Washington, D.C., **1967**, p. 51.

[6] J. D. Corbett et al., in "Lanthanide/Actinide Chemistry," Chapter 5, P. R. Fields and T. Moeller, ed., *ACS Adv. Chem. Ser.*, No. 71, Washington, D.C., **1967**; J. D. Corbett, in "Solid State Chemistry: A Contemporary Overview," Chapter 18, S. L. Holt et al., eds., *ACS Adv. Chem. Ser.*, No. 186, Washington, D.C., **1980**.

[7] For a review of the less common oxidation states of the lanthanides, see D. A. Johnson, *Adv. Inorg. Chem. Radiochem.*, **1977**, *20*, 1. L. R. Morss, (*Chem. Rev.*, **1976**, *76*, 827), has provided a review of the thermochemical properties of the lanthanides that is useful in interpreting their chemistry.

Table 16.1 Oxidation states of lanthanides and actinides[a]

Lanthanides				Actinides						
Symbol	+2	+3	+4	Symbol	+2	+3	+4	+5	+6	+7
La		+		Ac		+				
Ce	(+)	+	+	Th	(?)	(?)	+			
Pr		+	(+)	Pa			+	+		
Nd	(+)	+	(+)	U		+	+	+	+	
Pm		+		Np		+	+	+	+	+
Sm	(+)	+		Pu		+	+	+	+	(+)
Eu	+	+		Am	(+)	+	+	+	+	
Gd		+		Cm		+	+			
Tb		+	(+)	Bk		+	+			
Dy		+	(+)	Cf	(?)	+				
Ho		+		Es	(?)	+				
Er		+		Fm	(?)	+				
Tm	(+)	+		Md	+	+				
Yb	+	+		No	+	+				
Lu		+		Lr		+				

[a] Abbreviations: +, in solution; (+), in solid only; (?), in doubt.
SOURCE: T. Moeller, *J. Chem. Educ.*, **1970**, *47*, 417. Used with permission.

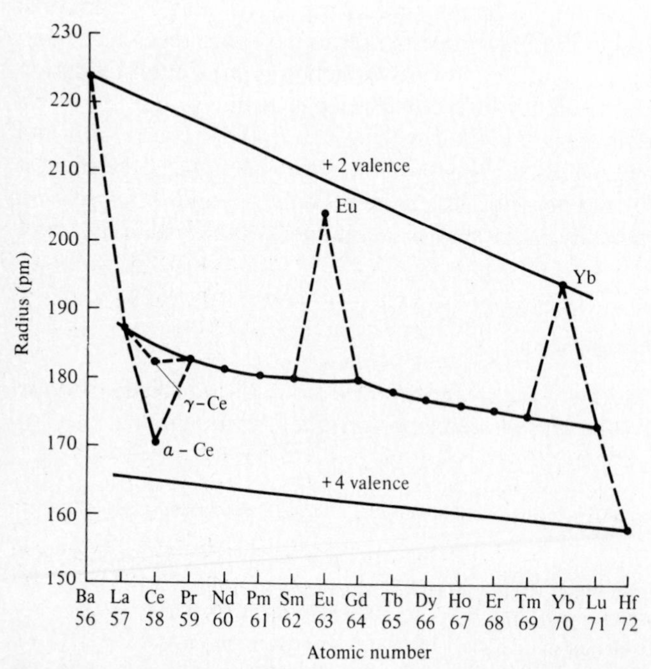

Fig. 16.2 Atomic radii of barium, lanthanum, the lanthanides, and hafnium. [From F. H. Spedding and A. H. Daane, "The Rare Earths," Wiley, New York, **1963**. Reproduced with permission.]

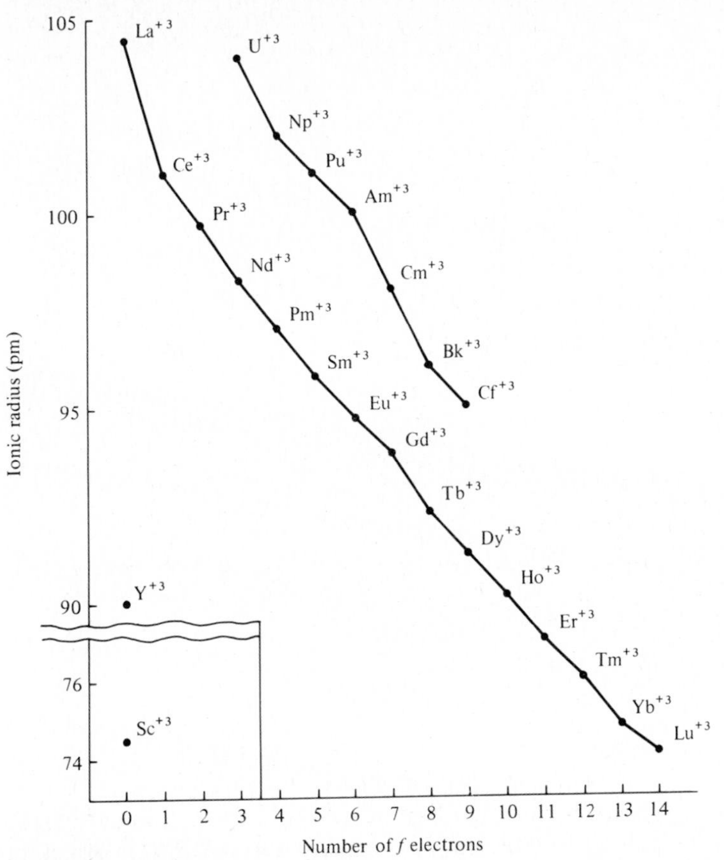

Fig. 16.3 Ionic radii (C.N. = 6) of Sc^{+3}, Y^{+3}, La^{+3}, Ln^{+3}, and An^{+3} ions.

apparent from the atomic radii (Fig. 16.2), it is best shown by the radii of the +3 cations (Fig. 16.3). There are two noticeable differences between the two series of ions: (1) although the actinide contraction initially parallels that of the lanthanides, the elements from curium on are smaller than might have been expected, probably resulting from poorer shielding by 5*f* electrons in these elements; (2) the lanthanide curve consists of two very shallow arcs with a discontinuity at the spherically symmetrical Gd^{+3} ($4f^7$) ion. A similar discontinuity is not clearly seen at Cm^{+3}.

A consequence of this contraction[8] is that when holmium is reached, the increase in size from $n = 5 \rightarrow n = 6$ has been lost and Ho^{+3} is the same size as the much lighter Y^{+3} (~90 pm, M^{+3}) with correspondingly similar properties (see Fig. 16.4). The contraction does not proceed sufficiently far to include scandium ($r_{Sc+3} = 73$ pm), but the properties of the latter may be extrapolated from the lanthanide series and in some ways provide a bridge between the strictly lanthanide metals and the transition metals.

[8] For other consequences, see Chaps. 12 and 17.

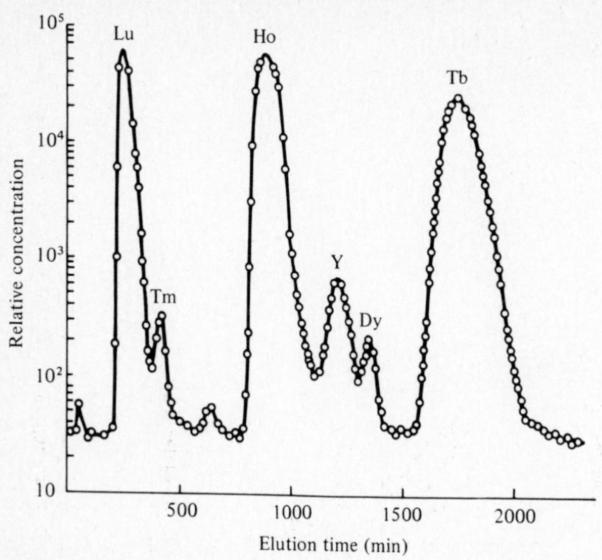

Fig. 16.4 Chromatographic separation of the lanthanides using 5% citrate at pH 3.20, showing relative speed of elution. Modified from B. H. Ketelle and G. E. Boyd, *J. Amer. Chem. Soc.*, **1947**, *69*, 2800. Reproduced with permission.]

The *f* orbitals

The *f* orbitals have not been considered previously except to note that they are *ungerade* (p. 24) and that they are split by an octahedral field into three levels, t_1, t_2, and a_2 (p. 442). The complete set of seven 4*f* orbitals is shown in Fig. 16.5. As with the *d* orbitals, there is no unique way of representing these orbitals, nor is there even a way which is optimum for all problems.[9] Thus Fig. 16.5 presents two sets, a "general set" and a "cubic set." The latter is advantageous in considering the properties of the orbitals in cubic (i.e., octahedral and tetrahedral) fields.

Differences between 4*f* and 5*f* orbitals

As with other types of orbitals, the 4*f* and 5*f* orbitals do not differ in the angular part of the wave function but only in the radial part. The 5*f* orbitals have a radial node which the 4*f* orbitals lack, but this is hardly likely to be of chemical significance. The chief difference between the two seems to depend upon the relative energies and spatial distribution of the orbitals. The 4*f* orbitals populated in the lanthanides are sufficiently low in energy that the electrons are seldom ionized or shared (hence the rarity of Ln^{IV} species). Furthermore, the 4*f* electrons seem to be buried so deeply within the atom that they are unaffected by the environment to any great degree.[10] This point will be discussed further below. In

[9] H. G. Friedman, Jr., et al., *J. Chem. Educ.*, **1964**, *41*, 354; C. Becker, *J. Chem. Educ.*, **1964**, *41*, 358; E. A. Ogryzlo, *J. Chem. Educ.*, **1965**, *42*, 150; H. G. Friedman et al., *J. Chem. Educ.*, **1965**, *42*, 151.

[10] For discussion of the possibilities of weak covalent bonding in lanthanide complexes, see C. K. Jørgensen et al., *J. Chem. Phys.*, **1963**, *39*, 1422, and D. E. Henrie and G. R. Choppin, *J. Chem. Phys.*, **1968**, *49*, 477.

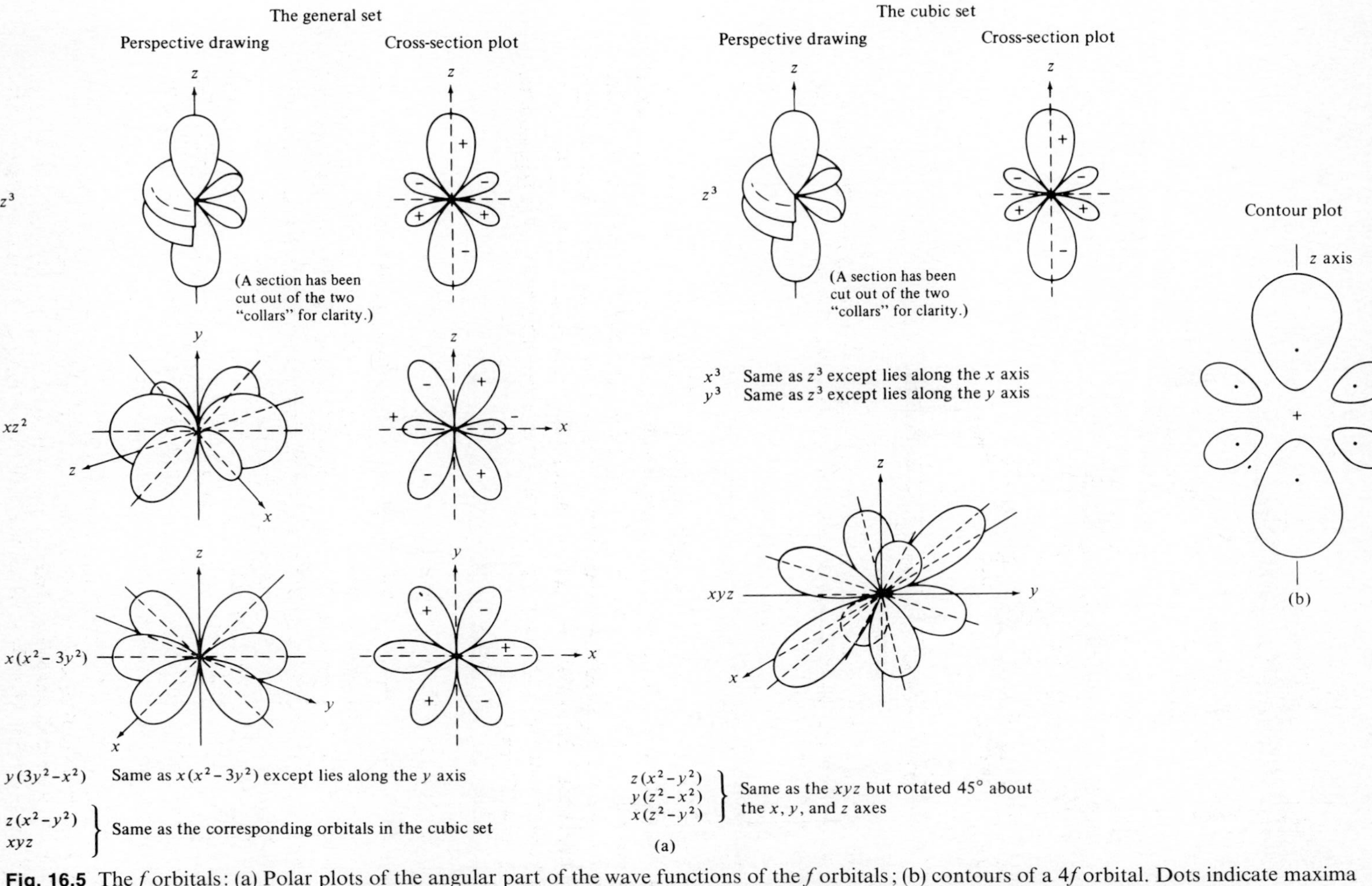

Fig. 16.5 The f orbitals: (a) Polar plots of the angular part of the wave functions of the f orbitals; (b) contours of a $4f$ orbital. Dots indicate maxima in electron density. The lines are drawn for densities which are 10% of maximum. [(a) From H. G. Friedman et al., *J. Chem. Educ.*, **1964**, *41*, 354; (b) from E. A. Ogryzlo, ibid., **1965**, *42*, 150. Reproduced with permission.]

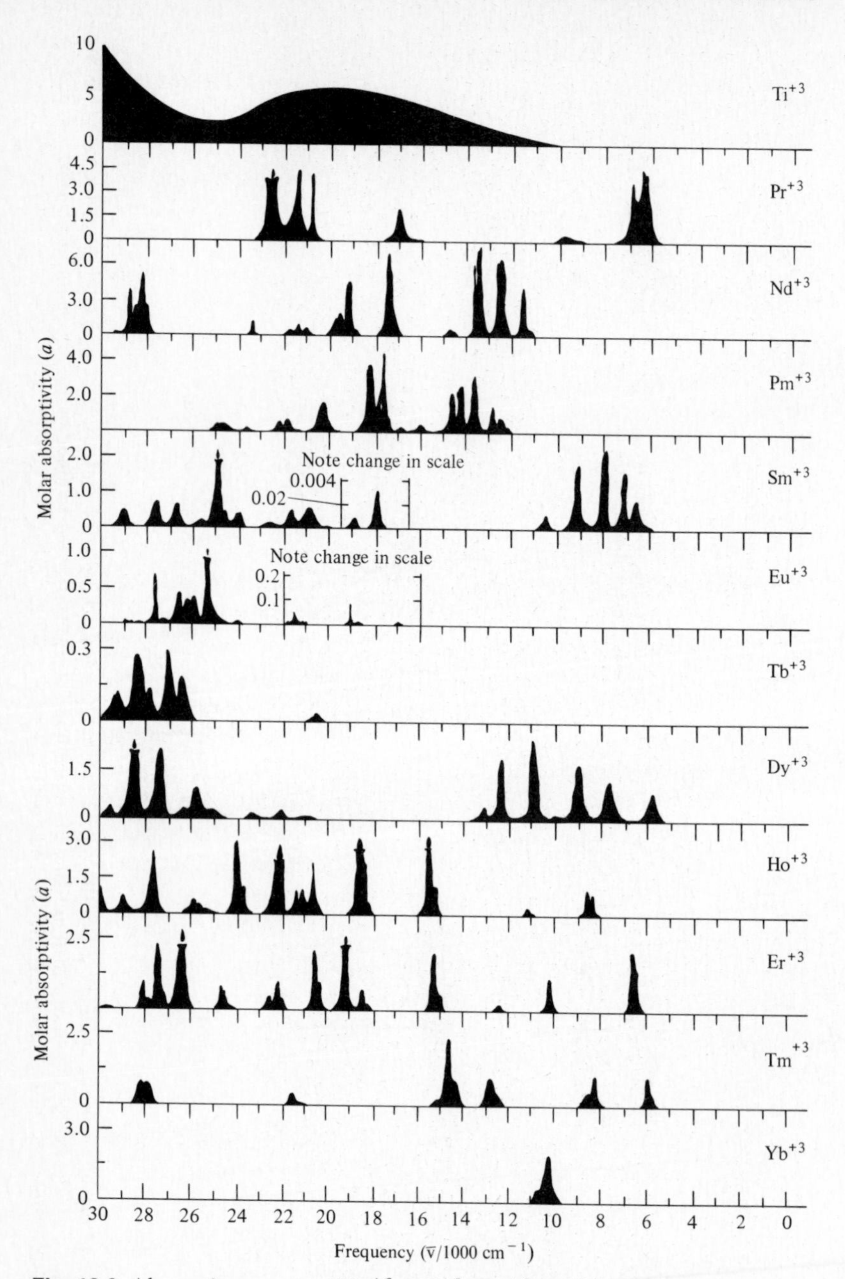

Fig. 16.6 Absorption spectra of Pr^{+3}, Nd^{+3}, Pm^{+3}, Sm^{+3}, Eu^{+3}, Tb^{+3}, Dy^{+3}, Ho^{+3}, Er^{+3} Tm^{+3}, Yb^{+3} in dilute acid solution compared with spectrum of Ti^{+3} (cf. Fig. 8.10). [Modified from W. T. Carnall and P. R. Fields, "Lanthanide/Actinide Chemistry," Advances in Chemistry Series, No. 71, American Chemical Society, Washington, D.C., **1967**. Reproduced with permission.]

contrast, the 5*f* electrons, at least in the earlier elements of the series, Th to Bk, are available for bonding allowing oxidation states up to +7. In this respect these electrons resemble *d* electrons of the transition metals. Because of the higher oxidation states in the early actinides it was once popular to assign these elements to transition metal families: thorium to IVB, protactinium to VB, and uranium to VIB. In 1944 Seaborg suggested that this arrangement was incorrect and that the elements following actinium formed a new "inner transition" series analogous to the lanthanides.[11] This suggestion, known as the "actinide hypothesis," was useful in elucidating the properties of the heavier actinides and was fully substantiated by their behavior (notably their lower oxidation states) and electron configurations. Nevertheless, we should not lose sight of the fact that in the earlier actinides the 5*f* electrons *are* available for use and that these elements do show certain resemblances to the transition metals.

Absorption spectra of the lanthanides and actinides

The absorption spectra of several lanthanide +3 cations are shown in Fig. 16.6. These spectra result from *f–f* transitions in a manner analogous to the *d–d* transitions of the transition metals. In contrast to the latter, however, the 4*f* orbitals in the lanthanides are buried deep within the atom and the broadening effect of ligand vibrations is minimized. Absorption spectra of the lanthanide cations are thus typically sharp, line-like spectra as opposed to the broad absorptions of the transition metals.

The absorption spectra of some typical actinide elements are shown in Fig. 16.7. They may be conveniently divided into two groups: (1) Am^{+3} and heavier actinides which have spectra that resemble those of the lanthanides; and (2) Pu^{+3} and lighter actinides which have spectra which are similar in some ways but have a tendency toward broadening of the absorption peaks, somewhat like the broadening seen in transition metal ions. Apparently the greater "exposure" of the 5*f* orbitals in the lighter actinide elements results in greater ligand–metal orbital interaction and some broadening from vibrational effects. As the nuclear charge increases, the 5*f* orbitals behave more like the 4*f* orbitals in the lanthanides and the spectra of the heavier actinides become more lanthanide-like.[12]

COORDINATION CHEMISTRY[13]

Comparison of inner transition and transition metals

The lanthanides behave as typical hard acids, bonding preferentially to fluoride and oxygen donor ligands. In the presence of water, complexes with nitrogen, sulfur, and halogen (except F^-) donors are not stable. The absence of extensive interaction with the 4*f* orbitals

---

[11] For a discussion of this topic, see G. T. Seaborg, "Man-Made Transuranium Elements," Prentice-Hall, Englewood Cliffs, N.J., **1963**, Chapter 3, or *Actinides Rev.*, **1967**, *1*, 3.

[12] For a thorough discussion of the absorption spectra of lanthanide and actinide ions, see Chapter 7 by W. T. Carnal and P. R. Fields, in "Lanthanide/Actinide Chemistry," P. R. Fields and T. Moeller, eds., *Advances in Chemistry Series*, No. 71, American Chemical Society, Washington, D.C., **1967**.

[13] The following discussion barely scratches the surface of the extensive work done on complexes of the lanthanides and actinides. For complete reviews see T. Moeller et al., *Chem. Rev.*, **1965**, *65*, 1, for the lanthanides, and A. D. Jones and G. R. Choppin, *Actinides Rev.*, **1969**, *1*, 311, for the actinides.

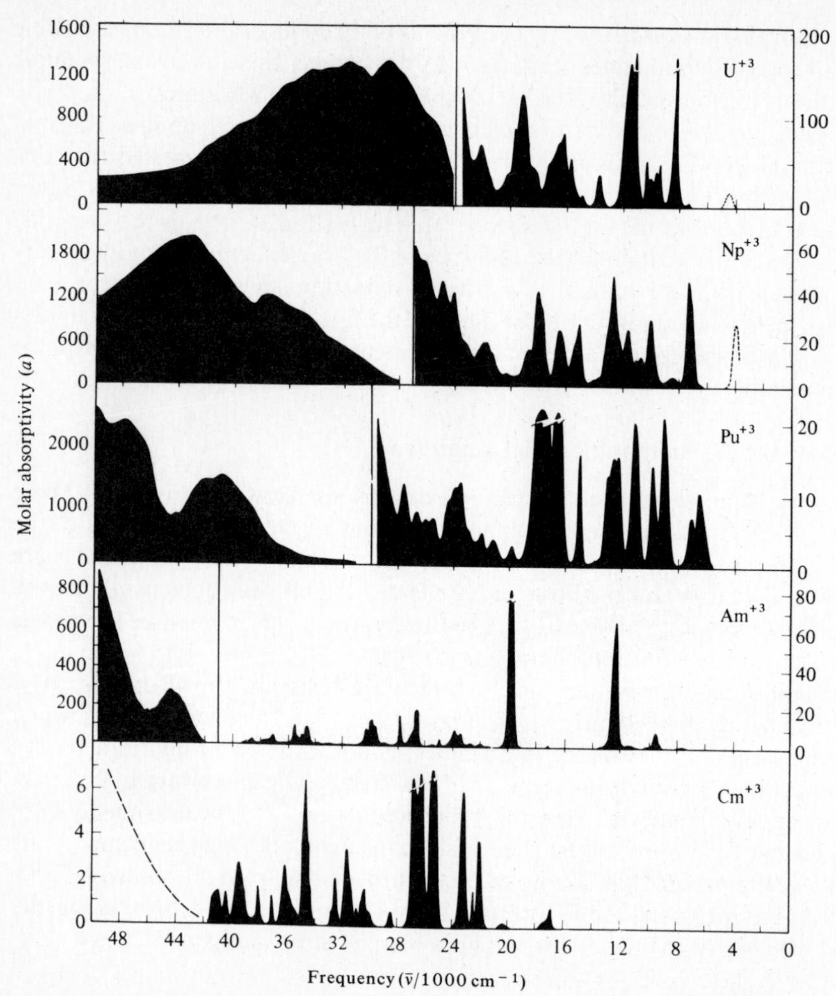

Fig. 16.7 Absorption spectra of U^{+3}, Np^{+3}, Pu^{+3}, Am^{+3}, and Cm^{+3} in dilute acid solution. [Modified from W. T. Carnall and P. R. Fields, "Lanthanide/Actinide Chemistry," Advances in Chemistry Series, No. 71, American Chemical Society, Washington, D.C., **1967**. Reproduced with permission.]

minimizes ligand field stabilization effects (LFSE). The lack of LFSE reduces overall stability but on the other hand provides a greater flexibility in geometry and coordination numbers since LFSE is not lost, for example, when an octahedral complex is transformed to trigonal prismatic or square antiprismatic geometry. Furthermore, the complexes tend to be labile in solution. Karraker[14] has summarized these differences from the properties of typical transition metal complexes (Table 16.2).

One noticeable difference is the tendency toward increased coordination numbers in the lanthanide and actinide complexes. This is shown most readily by the early (and hence

[14] D. G. Karraker, *J. Chem. Educ.*, **1970**, *47*, 424.

Table 16.2 Comparison of transition metal ions and lanthanide ions

	Lanthanide ions	First series Transition metal ions
Metal orbitals	$4f$	$3d$
Ionic radii	106–85 pm	75–60 pm
Common coordination numbers	6, 7, 8, 9	4, 6
Typical coordination polyhedra	Trigonal prism Square antiprism Dodecahedron	Square planar Tetrahedron Octahedron
Bonding	Little metal–ligand orbital interaction	Strong metal–ligand orbital interaction
Bond direction	Little preference in bond direction	Strong preference in bond direction
Bond strengths	Ligands bond in order of electronegativity: F^-, OH^-, H_2O, NO_3^-, Cl^-	Bond strengths determined by orbital interaction normally in following order: CN^-, NH_3, H_2O, OH, F^-
Solution complexes	Ionic, rapid ligand exchange	Often covalent; covalent complexes may exchange slowly

SOURCE: Modified from D. G. Karraker, *J. Chem. Educ.*, **1970**, *47*, 424. Used with permission.

largest) members of the series coordinated to small ligands. The structures of the crystalline lanthanide halides, LnX_3, exhibit this effect. For lanthanum, coordination number 9 is obtained for all of the halides except LaI_3. For lutetium, only the fluoride exhibits a coordination number greater than 6.[15] In solution the degree of hydration also appears to decrease in progressing along the series. Evidence comes from several kinds of data such as the partial molal volumes of the hydrated Ln^{+3} ions.[16] As the central ion decreases in size the partial molal volume decreases as expected until crowding of the ligands becomes too intense. At this point (Sm) a water molecule is expelled from the coordination sphere and the molal volume increases temporarily before resuming (Tb) a steady decrease (Fig. 16.8).

Separation of the lanthanides and actinides

The early separation of the lanthanides was beset by difficulties as a result of the similarity in size and charge of the lanthanide ions. The separations were generally based on slight differences in solubility, which were exploited through schemes of fractional crystallization. The differences in behavior resulting from decrease in ionic radius are commonly referred to as a decrease in basicity along the series, reflected by a decrease in solubility

[15] For a discussion of the coordination numbers of the lanthanide halides as well as some other aspects of the coordination chemistry of the lanthanides, see footnote 14.

[16] F. H. Spedding et al., *J. Phys. Chem.*, **1966**, *70*, 2440.

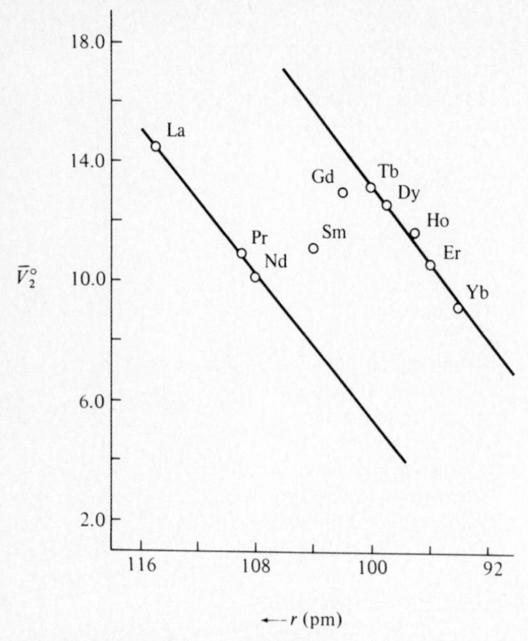

Fig. 16.8 Partial molal volume of hydrated Ln^{+3}. Lines represent 9- and 8-coordination. Sm^{+3} and Gd^{+3} exist as equilibria between the two species. [From F. H. Spedding et al., *J. Phys. Chem.*, **1966**, *70*, 2440. Reproduced with permission.]

of the hydroxides, oxides, carbonates, and oxalates. The fractional crystallization and fractional precipitation methods are extremely tedious and have been largely replaced by more modern methods.

The decrease in basicity (or more realistically, the *increase* in acidity) of the lanthanides provides an opportunity for employing coordinating ligands to effect separations. Other things being equal,[17] the more acidic a cationic species the more readily it will form a complex. In practice the lanthanides are placed upon an ion-exchange resin and eluted with a complexing agent, such as the citrate ion. Ideally, the ions should come off the column with a minimum of overlapping of the various bands (see Fig. 16.4). Such processes have increased the amounts of lanthanides available and opened up possibilities for commercial use (e.g., "rare earth phosphors" for color television). The initial separations of many of the actinide elements (as they were synthesized) were effected by similar methods.

Lanthanide chelates

The low stability of lanthanide complexes can be increased by means of the chelate effect, and much work has been directed to the elucidation of the stability of the lanthanide chelates.[18] The results are only partially interpretable in terms of simple models. Figure 16.9 portrays the relative stability of various lanthanide chelates. Two[19] types of behavior may

[17] As we shall see below, these "other things" may be extremely complicated.

[18] See T. Moeller et al., *Chem. Rev.*, **1965**, *1*, 1.

[19] This classification is adequate for the present discussion, but some authors further subdivide the second category into two classes; see footnote 18.

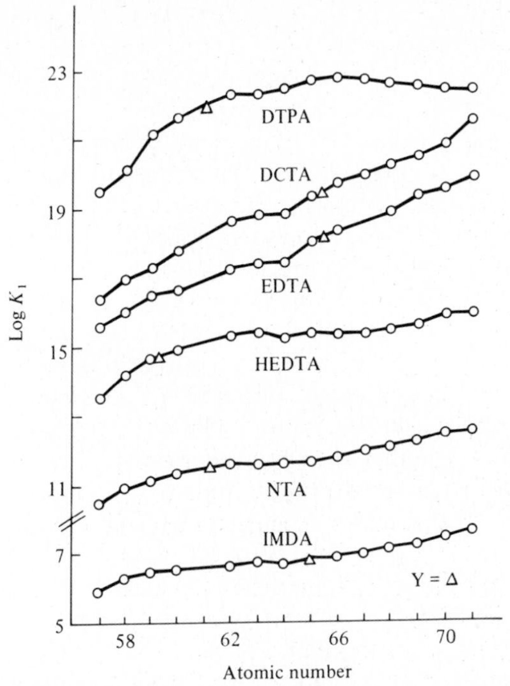

Fig. 16.9 Formation constants at 25 °C for 1:1 chelates of Ln^{+3} ions with various aminepolycarboxylate ions (IMDA, iminodiacetate; NTA, nitrilotriacetate; HEDTA, *N*-hydroxyethylethylenediaminetri-acetate; EDTA, ethylenediaminetetraacetate; DCTA, *trans*-1,2-diaminocyclohexanetetraacetate; DTPA, diethylenetriaminepentaacetate). [From T. Moeller, *J. Chem. Educ.*, **1970**, *47*, 418. Reproduced with permission.]

be noted: (1) "ideal"[20] behavior exemplified by chelates of ethylenediaminetetraacetic acid (EDTA) and the closely related *trans*-1,2-diaminocyclohexanetetraacetic acid (DCTA); and (2) "nonideal" behavior as exemplified by diethylenetriaminepentaacetic acid (DTPA) complexes. The former conforms to our expectations based on simple electrostatic or acid–base concepts of size and charge (a more or less uniform increase in stability accompanying the decrease in ionic radii). The discontinuity at gadolinium (the "gadolinium break") could be attributable to the discontinuity in crystal radii at this ion or, more plausibly, both may reflect small LFSEs associated with splitting of the partially filled *f* orbitals. The position of yttrium on these stability curves is that expected on the basis of its size—it falls very close to dysprosium.

Unfortunately, about half of the ligands that have been studied in complexes with all of the lanthanides show discrepancies from the simple picture presented above and must be considered type 2 ligands. In general, these may be characterized as giving stability/atomic number curves similar to type 1 for the lighter lanthanides, usually with a break at gadolinium. The behavior of the heavier lanthanides is variable, however, often showing essentially no change in stability, sometimes even showing *decreased* stability with increasing atomic number. Furthermore, the placement of yttrium on these curves is

[20] This behavior is "ideal" in the sense that it follows our preconceived notions of what *should* occur.

variable, often falling with the pre-gadolinium elements rather than immediately after gadolinium as expected on the basis of size and charge alone.

Several factors have been advanced to account for the unusual behavior of the type 2 complexes. First, ligand field effects might be expected to affect the position of yttrium, since it has a noble gas configuration with no *d* or *f* electrons to provide LFSE as opposed to all of the lanthanide ions except Gd^{+3} and Lu^{+3}. Obviously, however, this is insufficient to account for the variable results for the Tb^{+3}–Lu^{+3} complexes. A second factor is the possibility of coordination numbers greater than 6, which may also vary along the series. Thus it is entirely possible that an effect similar to that seen previously for the degree of hydration is taking place. At some point along the series the decrease in metal ion size might cause the expulsion of one of the donor groups from a multidentate ligand and decreased stability. This point could be reached at different places along the series depending upon the geometry and steric requirements of the multidentate ligand. It should be remembered that the thermodynamic stability of complexes in aqueous solution reflects the ability of the ligand to compete with water as a ligand and that the observed trends will be a summation of the effects of the lanthanide contraction, etc., on the stability of both the complex in question and the aquo complex. For this reason it is not surprising that the situation is rather complex.[21]

Chelates of the sterically hindered β-diketones 2,2,6,6-tetramethylheptane-3,5-dione (thd) and 1,1,1,2,2,3,3-heptafluoro-7,7-dimethyloctane-4,6-dione (fod)[22] are of considerable interest because of their volatility. Complexes of the type $Ln(thd)_3$,[23] $Ln(fod)_3$,[23] $An(thd)_4$,[24] and $An(fod)_4$[24] have been synthesized, and in spite of molecular weights of about 950 to 1050 (LnY_3) and 1300 to 1400 (AnY_4) they have measurable vapor pressures at temperatures below the boiling point of water. The volatility of the lanthanide chelates has been exploited in separating the lanthanides by means of gas chromatography. They are potential antiknock additives and homogeneous catalysts,[25] but their greatest usefulness to date has been that of europium, praseodymium, and ytterbium complexes as NMR shift reagents.[25,26]

Organometallic chemistry

Organometallic compounds of the lanthanides and actinides are sparse compared with those of the transition elements. Early attempts to make alkyl derivatives in the Manhattan project in the hope that they would be volatile and useful in isotope separation were unsuccessful. In contrast to the typical behavior of the transition metals, both the lanthanides and actinides yield ionic cyclopentadienides, $LnCp_3$ and $AnCp_2$, instead of π complexes.[27] The lanthanide complexes are strong and soft Lewis acids forming adducts

[21] See T. Moeller et al., in "Progress in the Science and Technology of the Rare Earths," L. Eyring ed., Pergamon, Elmsford, N.Y., **1968**, Vol. 3, pp. 61–128.

[22] Although these ligands are structurally complex in order to increase the steric effects of the organic "shrubbery," they behave in the same way as the simpler acetylacetone, p. 528.

[23] K. J. Eisentraut and R. E. Sievers, *J. Am. Chem. Soc.*, **1965**, *87*, 5254; C. S. Springer, Jr., et al., *Inorg. Chem.*, **1967**, *6*, 1105.

[24] H. A. Swain, Jr., and D. G. Karraker, *Inorg. Chem.*, **1970**, *9*, 1766.

[25] R. E. Sievers et al., in "Inorganic Compounds with Unusual Properties," R. B. King, ed., *Advances in Chemistry Series*, No. 150, American Chemical Society, Washington, D.C., **1976**, pp. 222–231.

[26] C. C. Hinckley, *J. Am. Chem. Soc.*, **1969**, *91*, 5160.

[27] See P. G. Laubereau and J. H. Burns, *Inorg. Chem.*, **1970**, *9*, 1091, and references therein.

with phosphines and sulfides[28] and with carbonyl and nitrosyl groups in organometallic compounds.[29]

Although the lanthanides and actinides do not form the usual type of metallocene compounds, a novel type of complex has been synthesized. Cyclooctatetraene (COT) accepts two electrons to form the aromatic, ten-electron π-system $C_8H_8^{-2}$:

$$C_8H_8(=COT) \xrightarrow[THF]{K} 2K^+ + [C_8H_8]^{-2}(=COT^{-2}) \qquad \textbf{(16.1)}$$

The cyclooctatetraene dianion is similar to the cyclopentadienide anion except that it has ten electrons to donate. When it is allowed to react with certain tetrapositive actinide ions such as U^{+4}, Np^{+4}, and Pu^{+4}, a neutral complex precipitates:

$$U^{+4} + 2COT^{-2} \longrightarrow U(COT)_2 \qquad \textbf{(16.2)}$$

The uranium compound was the first to be synthesized,[30] and a sandwich structure was proposed[30] and later verified (Fig. 16.10).[31] By analogy to ferrocene, the name *uranocene* was proposed. The neptunium(IV) and plutonium(IV) compounds are also known.[32] It has been proposed that 5*f* orbitals are involved in bonding these compounds. A simple molecular orbital scheme analogous to that provided previously (see pp. 619–622) for ferrocene is shown in Fig. 16.11. The a_{1g}, a_{2u}, e_{1u}, e_{1g}, and e_{2g} orbitals are qualitatively similar to the orbitals in ferrocene. In addition there are the e_{2u} and e_{3u} orbitals formed from appropriate combinations with the 5*f* orbitals. The first six of these represent a total of ten COT^{-2} orbitals donating (formally) 20 electrons to the metal. Any 5*f* electrons which the metal may possess will thus occupy the doubly degenerate e_{3u} orbitals. For the three known species the electron configurations are $U^{IV} = 5f^2$, $Np^{IV} = 5f^3$, and

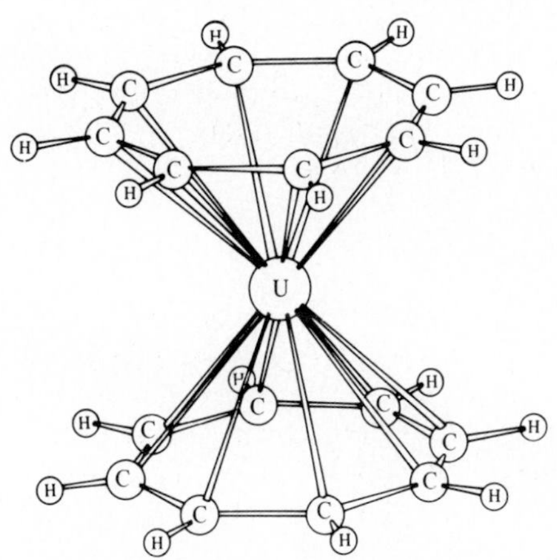

Fig. 16.10 Structure of bis(cyclooctatetraenyl) uranium ("uranocene"). [From A. Zalkin and K. N. Raymond, *J. Amer. Chem. Soc.*, **1969**, *91*, 5667. Reproduced with permission.]

[28] T. J. Marks et al., in "Nuclear Magnetic Shift Reagents," R. E. Sievers, ed., Academic Press, New York, **1973**, p. 247.

[29] A. E. Crease and P. Legzdins, *J. Chem. Soc., Dalton Trans.*, **1973**, 1501.

[30] A. Streitwieser, Jr., and U. Müller-Westerhoff, *J. Am. Chem. Soc.*, **1968**, *90*, 7364; A. Streitwieser, Jr., et al., *J. Am. Chem. Soc.*, **1973**, *95*, 8644; A. Streitwieser, Jr., and C. A. Harmon, *Inorg. Chem.*, **1973**, *12*, 458.

[31] A. Zalkin and K. N. Raymond, *J. Am. Chem. Soc.*, **1969**, *91*, 5667.

[32] D. G. Karraker et al., *J. Am. Chem. Soc.*, **1970**, *92*, 4841.

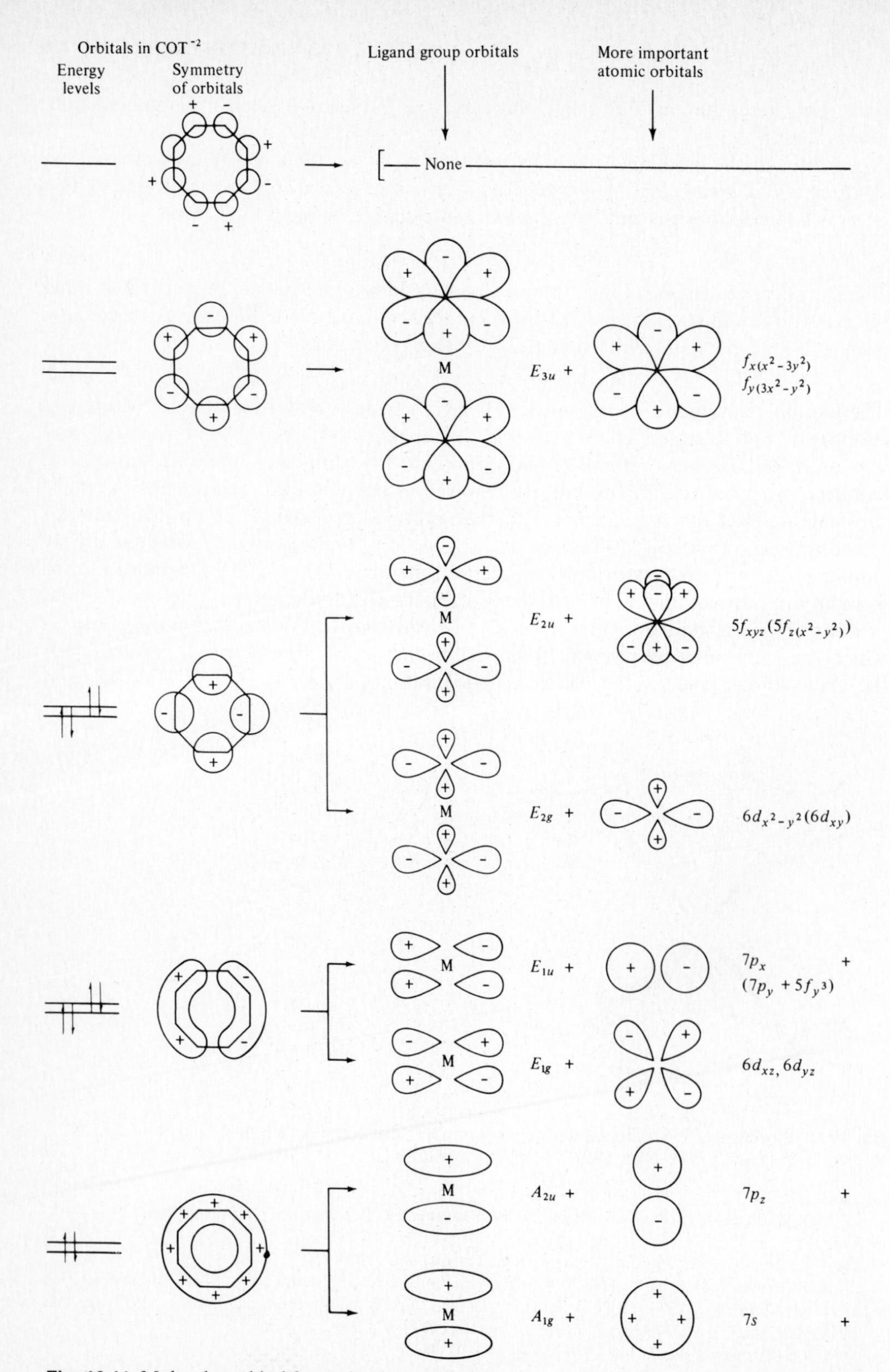

Fig. 16.11 Molecular orbital formation in uranocene.

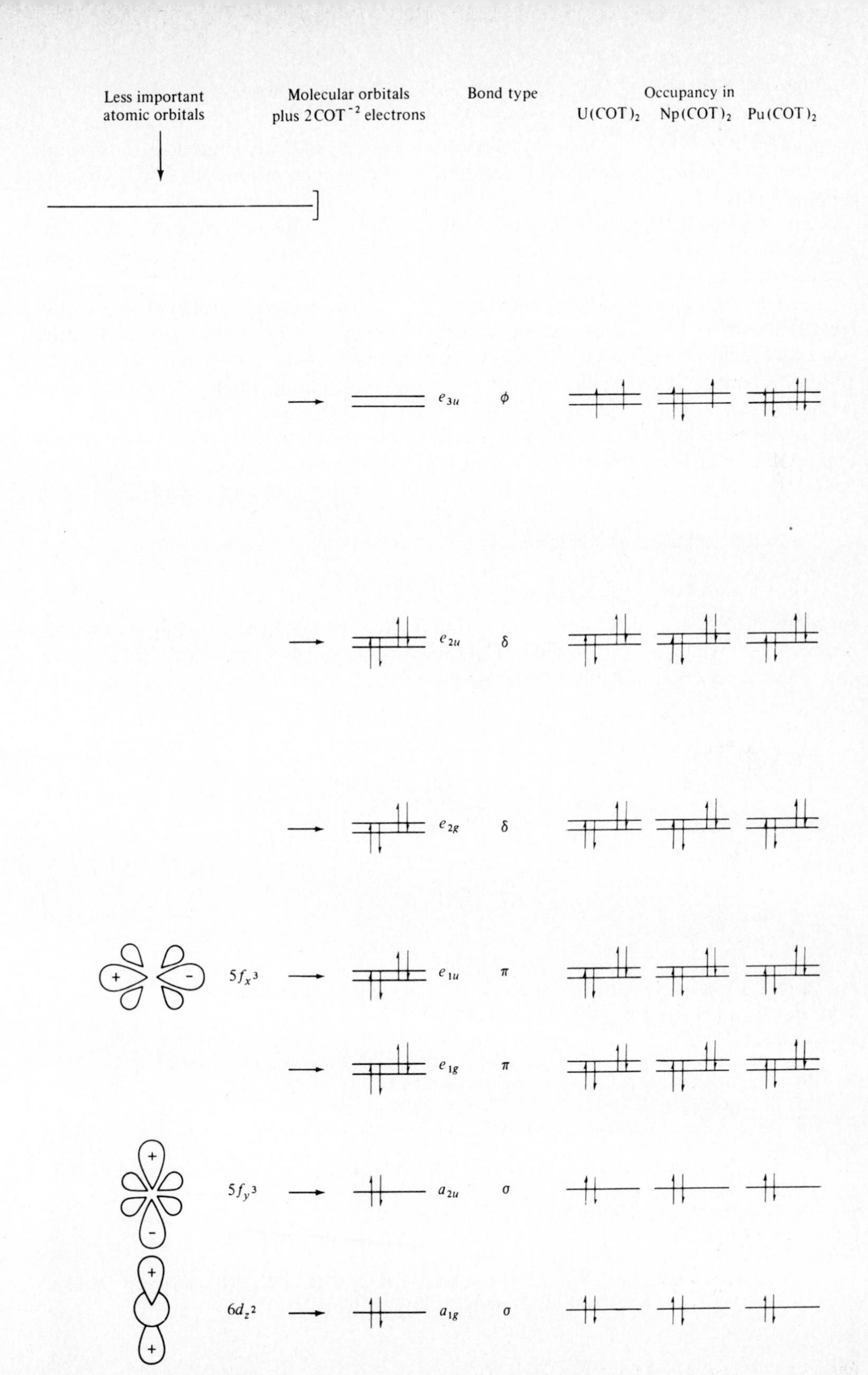

Less important atomic orbitals
Molecular orbitals plus 2COT^{-2} electrons
Bond type
Occupancy in
$U(COT)_2$
$Np(COT)_2$
$Pu(COT)_2$
e_{3u}
ϕ
e_{2u}
δ
e_{2g}
δ
+
-
$5f_{x^3}$
e_{1u}
π
e_{1g}
π
+
-
$5f_{y^3}$
a_{2u}
σ
+
+
$6d_{z^2}$
a_{1g}
σ

$Pu^{IV} = 5f^4$. We would thus expect uranocene, neptunocene, and plutonocene to have two, one, and no unpaired electrons, respectively. These expectations have been verified experimentally.

The lanthanides might be expected to form similar complexes, using $4f$ orbitals in place of the $5f$ orbitals of the actinides. Some $[Ln(COT)_2]^-$ complexes are known,[33] but monocyclooctatetraenide complexes are also known.[34,35]

As we have seen several times, the fact that something is *possible* does not necessarily mean that it *will occur*, and two workers can look at the same data and come to diametrically opposite conclusions. Thus in contrast to the molecular orbital approach of Streitwieser (above), Raymond[36] has emphasized the *ionic* nature of the bonding in lanthanide and actinide organometallic compounds. While arguments can and have been marshaled pro and con, stimulating further experimentation and interpretation, one must also remember that resolution of differences of opinion of this kind may be elusive because they include a further component: Just as one person will say a bucket is half empty, another will say that it is half full!

Recently, unexpectedly stable actinide alkyls were prepared by the reaction

$$Cp_3UCl \xrightarrow[\text{or RMgX}]{\text{RLi}} Cp_3UR \qquad \textbf{(16.3)}$$

In addition to R = methyl, ethyl, propyl, butyl, etc., the added group may be phenyl, benzyl, or phenylethynyl.[37] The stability of these compounds has been attributed[38] to the coordinate congestion of these complexes (Fig. 16.12). Indeed, the crowding of the three

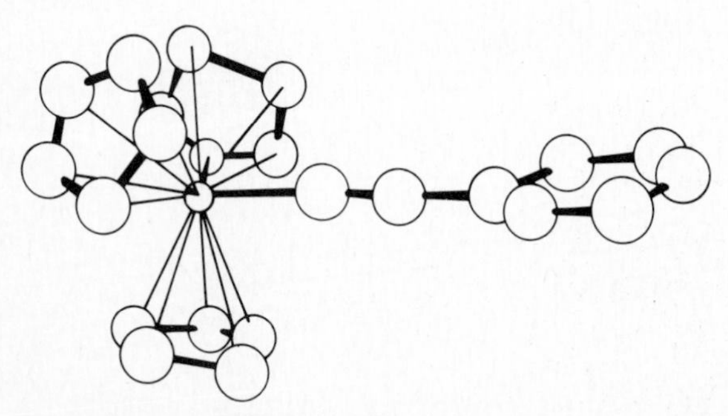

Fig. 16.12 The molecular structure of phenylethynyltris(cyclopentadienyl)uranium(IV). [From J. L. Atwood et al., *Chem. Commun.*, **1973**, 542. Reproduced with permission.]

---

[33] F. Mares et al., *J. Organometal. Chem.*, **1970**, *24*, C68; K. O. Hodgson et al., *J. Am. Chem. Soc.*, **1973**, *95*, 8650.

[34] R. G. Haynes and J. L. Thomas, *J. Am. Chem. Soc.*, **1969**, *71*, 6876.

[35] F. Mares et al., *J. Organometal. Chem.*, **1971**, *25*, C24; K. O. Hodgson et al., *J. Am. Chem. Soc.*, **1973**, *95*, 8650.

[36] K. N. Raymond and C. W. Eigenbrot, Jr., *Acc. Chem. Res.*, **1980**, *13*, 276. As with many questions of this type, both views are correct: These molecules are highly ionic with some important contributions from overlap of f orbitals. See J. C. Green, *Struct. Bond.*, **1981**, *43*, 37, and T. J. Marks, *Science*, **1982**, *217*, 989.

[37] A. E. Gebala and M. Tsutsui, *J. Am. Chem. Soc.*, **1973**, *95*, 91; T. J. Marks et al., ibid, **1973**, *95*, 5529; G. Brandi et al., *Inorg. Chim. Acta*, **1973**, *7*, 319.

[38] T. J. Marks, in "Inorganic Compounds with Unusual Properties," R. B. King, ed., *Advances in Chemistry Series*, No. 150, American Chemical Society, Washington, D.C., **1976**, pp. 232–255.

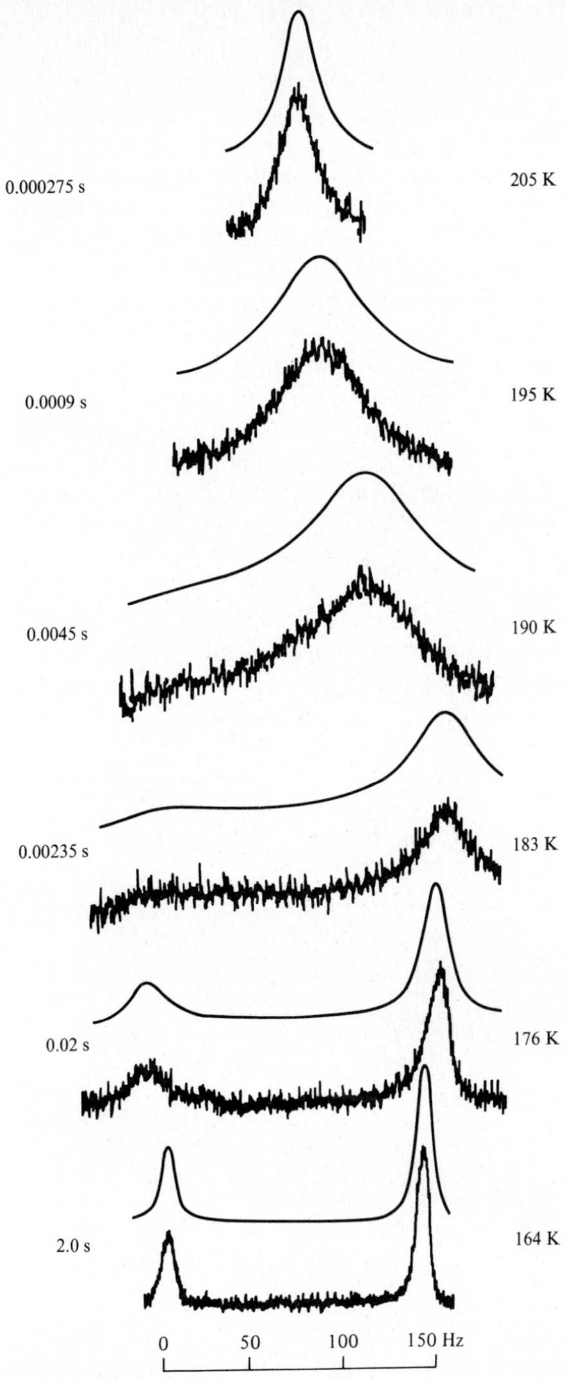

Fig. 16.13 PMR spectra (at 90 MHz) in the C_5H_5 region of Cp_3U(i-Pr) as a function of temperature (dimethyl ether/toluene solvent). Computer generated spectra are for mean pre-exchange lifetimes. [From T. J. Marks et al., *J. Am. Chem. Soc.*, **1973**, *95*, 5529. Reproduced with permission.]

cyclopentadienyl groups is such that in the *i*-propyl compound, there is restricted rotation about the U—C σ bond (Fig. 16.13). Furthermore, the structure of the tetrakis(cyclopentadienyl)uranium molecule is known,[39] and it appears that forcing in a fourth cyclopentadienyl group causes the original three to back off about 10 pm from the uranium atom.

The analogous thorium compounds have also been prepared,[40] and they are thermally even more stable than the uranium compounds. If they are carefully heated at 170 °C, a colorless compound is formed with the elimination of RH. The molecular structure was found to have two cyclopentadienyl rings that are monohapto (σ) bonded to one thorium atom and pentahapto (π) bonded to the other (Fig. 16.14).[41]

We have seen that transition metals can form metallocarborane complexes that resemble metallocenes in many ways (pp. 739–740). Although the lanthanides and actinides do not, in general, form particularly stable cyclopentadienyl complexes, one might expect the actinocarborane complexes to be more stable. The dicarbollide anion ($C_2B_9H_{11}^{-2}$) has an open face that is larger than the cyclopentadienide ion and carries a higher formal negative charge. Indeed, reaction of uranium tetrachloride with the dicarbollide anion produces the expected complex:[42]

$$UCl_4 + 2C_2B_9H_{11}^{-2} \longrightarrow [U(C_2B_9H_{11})_2Cl_2]^{-2} + 2Cl^- \quad \textbf{(16.4)}$$

The structure (Fig. 16.15) is reminiscent of the "bent" or "tilted" metallocenes found for the early transition metals (cf. Fig. 13.15, $Ticp_2Cl_2$).

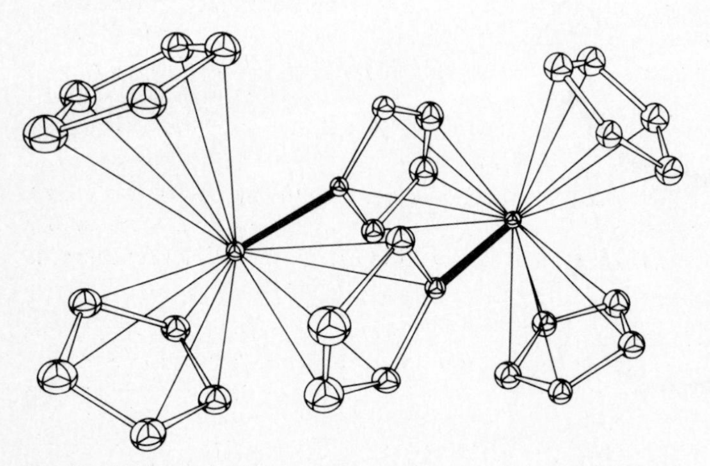

Fig. 16.14 The molecular structure of monohapto/pentahapto cyclopentadienyl thorium complex (σ bonds drawn more heavily for emphasis). [From E. C. Baker et al., *J. Am. Chem. Soc.*, **1974**, *96*, 7586. Reproduced with permission.]

[39] J. H. Burns, Jr., *J. Am. Chem. Soc.*, **1973**, *95*, 3815.
[40] T. J. Marks and W. A. Wachter, *J. Am. Chem. Soc.*, **1976**, *98*, 703.
[41] E. C. Baker et al., *J. Am. Chem. Soc.*, **1974**, *96*, 7586.
[42] F. R. Fronczek et al., *J. Am. Chem. Soc.*, **1977**, *99*, 1769. For reviews of organoactinide chemistry, see E. C. Baker et al., *Struct. Bond.*, **1976**, *25*, 23; T. J. Marks, *Acc. Chem. Res.*, **1976**, *9*, 223.

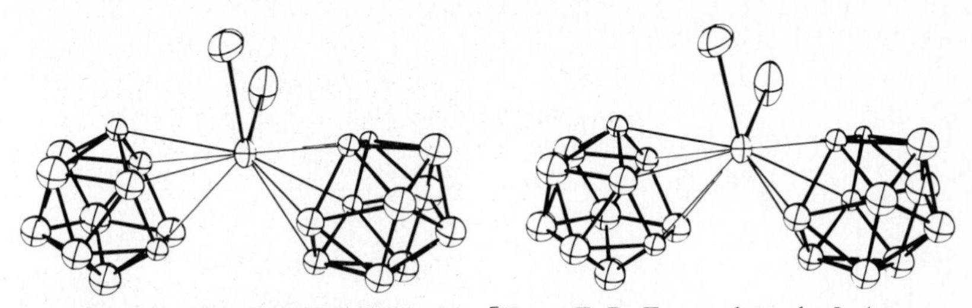

Fig. 16.15 Stereoview of dicarbollyluranium(IV) dichloride. [From F. R. Fronczek et al., *J. Am. Chem. Soc.*, **1977**, *99*, 1769. Reproduced with permission.]

THE TRANSACTINIDE ELEMENTS

At one time it was considered extremely unlikely that there would be any significant chemistry for elements with atomic numbers greater than about 100. The nuclear stability of the transuranium elements decreases rapidly with atomic number so that the half-lives for these elements become too short for fruitful chemical studies, (i.e., $t_{1/2} \approx$ seconds). However, advanced chemical techniques have helped elucidate information from these difficult systems. More promising is the outlook for further synthesis of transactinide elements. With the synthesis of the elements $_{101}$Md, $_{102}$No, and $_{103}$Lr the actinide series was completed and three transactinide elements, $_{104}$Rf, $_{105}$Ha, and $_{106}$[43] congeners of hafnium, tantalum, and tungsten, have been synthesized.[45] There has been an increasing amount of speculation concerning the possibility of stable species of relatively high atomic number. Theoretical calculations[46] on the stability of the nuclei predict unusual stability for atomic numbers 50, 82, 114, and 164. Such stability is found for $_{50}$Sn, which has more stable isotopes than any other element, and for $_{82}$Pb and $_{83}$Bi, which are the heaviest elements with nonradioactive isotopes. The stability of nuclei in the regions of the "magic numbers" has been described allegorically by Seaborg as mountains in a sea of instability as shown in Fig. 16.16. The expected stability is proportional to the elevation of the islands

---

[43] There is intense national rivalry and a certain amount of chauvinism in this area. The above names are those suggested by the American workers at Berkeley who have claimed their discovery. Russian workers had earlier claimed the synthesis of these elements as well and proposed the following names: joliotium (= nobelium), kurchatovium (= rutherfordium), and nielsbohrium (= hahnium). They have also tended to use "element 103" in place of the American name. The Russian claims are almost certainly in error. The IUPAC, traditionally responsible for resolving questions of this kind, has effectively admitted its inability to act as arbitrator in this matter by recently recommending a new system of naming elements.[44] Under this system the name is no longer selected by the discoverer but is to be composed of a three-bit "translation" of the atomic number into a Latinized form: If element 114, mentioned above, were discovered, its name would be uniuniquadium. An interesting, if startling, feature of this system is that the corresponding symbols contain *three* letters, one for each bit of the systematic name: Uuq. At this point, with the one-to-one translation, one wonders what has been gained by the system and why it is not sufficient just to stick with "element 114"!

[44] J. Chatt, *Pure Appl. Chem.*, **1979**, *51*, 381.

[45] A. Ghiorso et al., *Phys. Rev. Lett.*, **1974**, *33*, 1490.

[46] Such theory is, of course, beyond the scope of this book. For a very readable introduction to the study of nuclear stability, see G. T. Seaborg, *J. Chem. Educ.*, **1969**, *46*, 626. See also, C. E. Bemis and J. R. Nix, *Comm. Nucl. Part. Phys.*, **1977**, *7A*, 65.

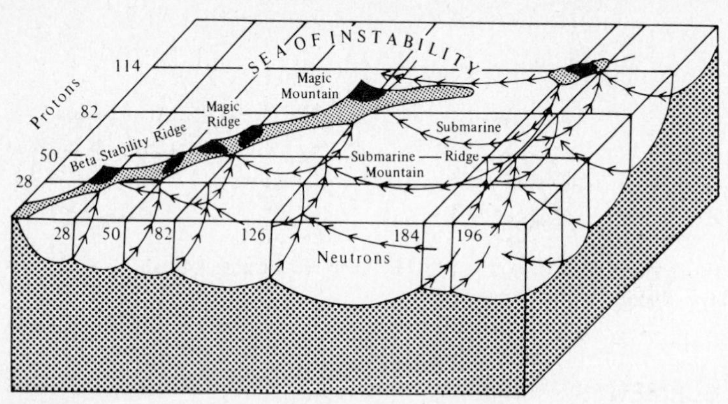

Fig. 16.16 Known and predicted regions of nuclear stability, surrounded by a sea of instability. [From G. Seaborg, *J. Chem. Educ.*, **1969**, *46*, 626. Reproduced with permission.]

above "sea level." The peninsula running "northeast" from lead represents the decreasing stability of the actinide elements. The predicted island of stability at atomic number 114 and with 184 neutrons may be accessible with new methods to "jump" the unstable region and form these nuclei directly. Thus, one might expect to find a group of relatively stable nuclei in the region of elements 105–115. The possibility of jumping to the next island (now shown) at atomic number 164 provides even more exciting (and more improbable) possibilities to extend our knowledge of the chemistry of heavy elements.

Periodicity of the translawrencium elements

Lawrencium completes the actinide series and fills the $5f$ set of orbitals. Rutherfordium and hahnium would be expected to be congeners of hafnium and tantalum and to be receiving electrons into the $6d$ orbitals. This process should be complete at element 112 (eka-mercury), and then the $7p$ orbitals should fill from element 113 to 118, which should be another noble gas element. Elements 119, 120, and 121 should belong to Groups IA, IIA, and IIIB. The former two will undoubtedly have $8s^1$ and $8s^2$ configurations. If the following elements parallel their lighter congeners, we might expect 121, eka-actinium, to accept one $7d$ electron and the following elements ("eka-actinides") to proceed with the filling of $6f$ orbitals. Unfortunately, we know very little about the relative energy levels for these hypothetical atoms except that they will be extremely close. Thus, although Fig. 2.13 would predict an order of filling $8s$, $5g$, $6f$, $7d$, $8p$, etc., it is not known whether this would be followed or not. Calculations[47] indicate that the levels are so close together that "mixed" configurations (analogous to the $5d^14f^n$ configurations found in the lanthanides) such as $8s^26f^15g^n$ or $8s^27d^16f^35g^n$ may occur. For this reason it does not seem profitable to speculate on the separate existence of the $5g^{18}$ and $6f^{14}$ series. Seaborg[46] has suggested that the two series be combined into a larger series of 32 elements and called the *super-actinides*. His revised form of the periodic chart is shown in Fig. 16.17. The $6f$ and $5g$

[47] J. T. Waber et al., *J. Chem. Phys.*, **1969**, *51*, 664; W. Greiner and B. Fricke, *Phys. Letters*, **1969**, *30B*, 317; J. B. Mann and J. T. Waber, *J. Chem. Phys.*, **1970**, *53*, 2397; B. Fricke et al., *Theoret. Chim. Acta* (Berl.), **1971**, *21*, 235; B. Fricke and J. T. Waber, *J. Chem. Phys.*, **1972**, *57*, 371.

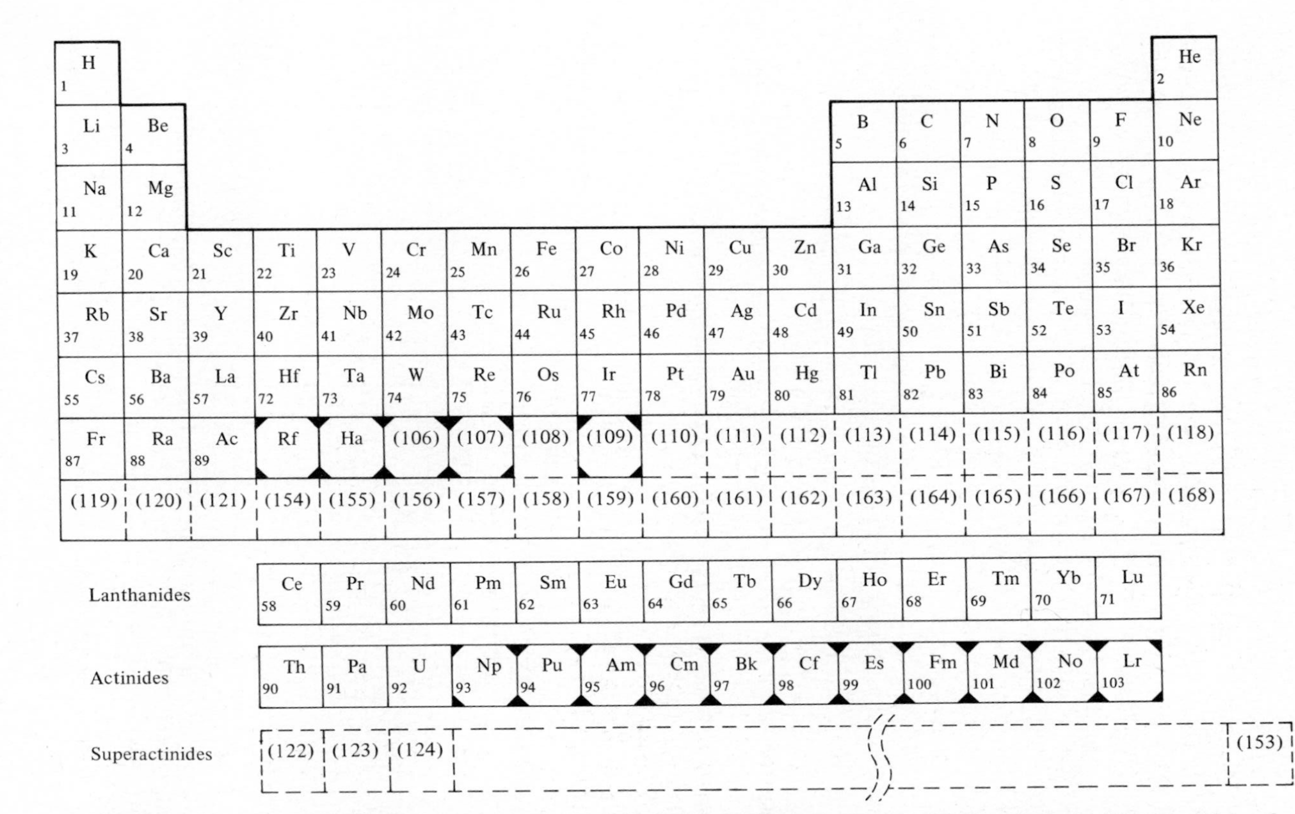

Fig. 16.17 Long form of the periodic chart extended to include hypothetical translawrencium elements. Blocks with shaded corners represent synthetic elements and those with numbers represent undiscovered elements. [Modified from G. Seaborg, *J. Chem. Educ.* **1969**, *46*, 626. Reproduced with permission.]

elements form an extra long "inner transition" series followed by (presumably) a series of ten transition elements (154–162) with filling of 7*d* orbitals, etc. Magic number nucleus 164 would thus be a congener of lead ("dvi-lead").[48]

While much of the preceding is speculative, it is no more speculative chemically than Mendeleev's predictions of gallium (eka-aluminum) and germanium (eka-silicon). The speculation centers on the possible or probable stability of nuclei with up to twice as many protons as the heaviest stable nucleus. The latter falls outside the realm of inorganic chemistry but the synthesis and characterization of some of these elements would be most welcome.

Several workers[49] have predicted the properties of certain translawrencium elements. For example, the "inert pair effect" should be accentuated, and thus the most stable oxidation states of eka-thallium, eka-lead, and eka-bismuth should be +1, +2, and +3, respectively. The possibility of forming elements with atomic numbers in the region of 164 is of considerable interest. As discussed in Chapter 17, elements following the completion of each new type of subshell (e.g., $2p^6$, $3d^{10}$, $4f^{14}$) show "anomalous" properties, and thus the chemical properties of dvi-lead should be equally interesting.[50]

PROBLEMS

16.1. **a.** Discuss the ways in which the actinide elements resemble their lanthanide congeners.

b. Discuss the ways in which the actinides more closely resemble "normal" transition elements.

16.2. Figure 16.18 shows the distribution of the lanthanides, barium, and strontium in certain lunar samples. Without getting involved in the intricacies of geochemistry,

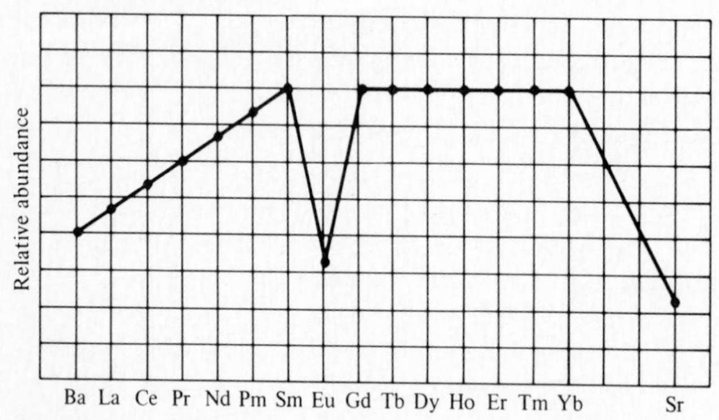

Fig. 16.18 Typical relative abundance curve for certain lunar samples. [Modified from H. Watika and R. A. Schmitt, *Science*, **1970**, *170*, 969. Reproduced with permission. Copyright 1970 by the American Association for the Advancement of Science.]

[48] Since element 164 would have an atomic number exactly twice that of lead (82), it has been suggested that it be dubbed "zwei blei"!

[49] O. L. Keller, Jr., and G. T. Seaborg, *Ann. Rev. Nucl. Sci.*, **1977**, *27*, 139; A. V. Grosse, *J. Inorg. Nucl. Chem.*, **1965**, *27*, 509; D. Bonchev and V. Kamenska, *J. Phys. Chem.*, **1981**, *85*, 1177.

[50] *Note added in press*: The approach to "magic number" $Z = 114$ has been augmented during the past year by the synthesis of elements 107 and 109 by West German physicists at Darmstadt. See R. Rawls, *Chem. Eng. News*, October 11, **1982**, p. 27.

suggest reasons why the behavior of europium may not be as anomalous as it first appears. [K. Rankama and T. G. Sahama, "Geochemistry," University of Chicago Press, Chicago, **1950**, p. 529.]

16.3. Suggest a possible reason for the negative deviation of α-cerium in Fig. 16.2.

16.4. Didymium (a mixture of Pr and Nd) goggles are preferred for glassworking to absorb the glare from sodium. Explain why these goggles are more suitable than a pair made with Ti^{+3}, for example.

16.5. **a.** Seaborg has suggested that element 168 is better described as a "noble liquid" rather than a noble gas. Explain.

b. From an inspection of the diagonal relationship show how 168 might almost be a "noble metal."

16.6. Biochemists sometimes think of Ln^{+3} ions as analogues of Ca^{+2}. Discuss reasons why there might be a resemblance between Ln^{+3} and Ca^{+2}. Any notable differences? In what ways might biochemists exploit a resemblance?

16.7. Historically, and continuing today, there has been much discussion of whether the 4*f* orbitals in the lanthanides were involved in or responsible for a given observed phenomenon. Discuss how the use of Y^{+3} can provide evidence on questions of this sort.

16.8. A mass spectral study of $LnCp_3$ compounds [G. G. Devyatykh et al., *Dokl. Akad. Nauk. S.S.S.R.*, **1973**, 1094] concluded that metal–ligand bond strength *decreases* with increasing atomic number of the lanthanide. Would you have expected this? Discuss and offer possible explanations.

16.9. Figure 16.19 illustrates a portion of the spectrum of $K[Am(COT)_2]$ in THF. Compare this spectrum with that shown in Fig. 16.7 for $Am(H_2O)_n^{+3}$ ions. Ignore the shoulder on the left, which is caused by charge-transfer absorptions. Carefully compare frequency and shape of the other major band. What does this imply concerning the bonding? [D. G. Karraker, *J. Inorg. Nucl. Chem.*, **1977**, *39*, 87.]

16.10. Studies[51] of radioisotopes, both natural and from fallout, in Mono Lake, California, showed that ^{90}Sr, ^{226}Ra, and ^{210}Pb occur at lower levels than might have been expected. Some actinides such as ^{234}Th, ^{231}Pa, ^{238}U, and ^{240}Pu occurred at higher levels than expected, but others such as ^{227}Ac and ^{241}Am did not. The most notable characteristic of Mono Lake is its high alkalinity (pH = 10) caused by large amounts of carbonate ion (~0.3 mol dm^{-3}). Suggest factors that may be responsible for these relative abundances. Depending upon the complexity and completeness of your discussion, in addition to material from this chapter you may wish to consult Chapters 7–9 and Appendix F.

---

[51] H. J. Simpson et al., *Science*, **1982**, *216*, 513; R. F. Anderson et al., *Science*, **1982**, *216*, 514.

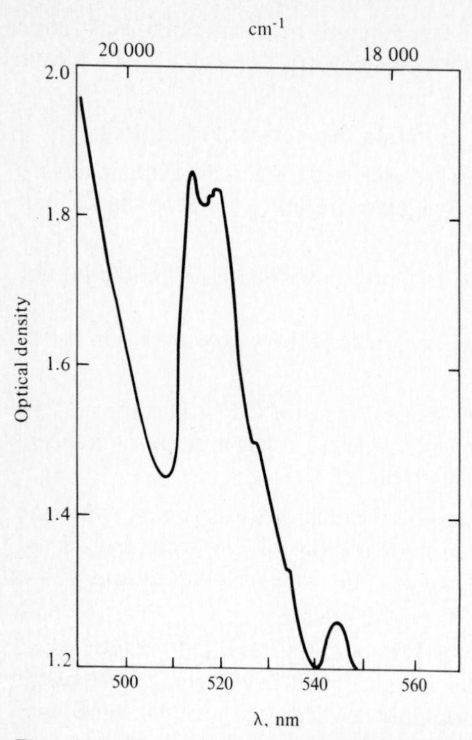

Fig. 16.19 Absorption spectrum of $K[Am(COT)_2]$ between 500 and 560 nm. Shoulder at left is a charge transfer absorption. [Modified from D. G. Karraker, *J. Inorg. Nucl. Chem.*, **1977**, *39*, 87. Reproduced with permission.]

17

Periodicity

The most fascinating aspect of inorganic chemistry as well as its most difficult problem is the diversity of reactions and structures encountered in the chemistry of somewhat over one hundred elements. The challenge is to be able to treat adequately the chemistry of boranes and noble gas fluorides and ferrocene and lanthanum compounds without developing a separate set of rules and theory for each element. The tool that has kept the inorganic chemist from throwing up his hands in despair is the periodic table, now slightly over one hundred years old.[1] It is of such overwhelming importance in the correlation of the properties of the elements that entire books have been written on this theme.[2] It is considered so essential that no general chemistry textbook would be complete without a discussion of the trends summed up in Chapter 2 (pp. 38–50). Unfortunately the impression is often given that all of the periodic properties vary smoothly.

Fundamental trends

The fundamental trends of the periodic chart have been discussed in Chapter 2. They may be summarized as follows. Within a family there are increases in size and decreases in ionization energy, electron affinity, electronegativity, etc. Increasing the atomic number across a given period results in concomitant increases in ionization energy, electron affinity, and electronegativity but a decrease in size. The change in effective atomic number within a period is reasonably smooth, but the various periods differ in length (8, 18, and 32 elements). The properties of an element will depend upon whether it follows an 8, 18, or 32 sequence. One of the best known examples is the similarity in properties of

[1] Periodic classifications of the elements by Dmitri Mendeleev and by Lothar Meyer appeared in 1869. For a centennial-celebrating discussion of the periodic table, see J. W. van Spronsen, "The Periodic System of Chemical Elements," Elsevier, Amsterdam, **1969**.

[2] R. Rich, "Periodic Correlations," Benjamin, New York, **1965**; R. T. Sanderson, "Chemical Periodicity," Van Nostrand-Reinhold, New York, **1960**. The latter book was rewritten but kept the periodicity theme and appeared under the title "Inorganic Chemistry" in **1967**.

hafnium, tantalum, tungsten, and rhenium to those of zirconium, niobium, molybdenum, and technetium, respectively, as a result of the lanthanide contraction (see Chapter 12). Similar effects follow the filling of the *d* levels (very rarely referred to as the "scandide" contraction). A second factor tending to affect the regularity of properties is the absence of *d* orbitals in the elements lighter than sodium. This results in a discontinuity of properties from the second-row elements, Li–F, to the heavier congeners.

FIRST- AND SECOND-ROW ANOMALIES

In many ways the first ten elements differ considerably from the remaining 90%. Hydrogen is a classic example—it belongs neither with the alkali metals nor with the halogens although it has some properties in common with both. Thus it has a +1 oxidation state in common with the alkali metals, but the bare H^+ has no chemical existence[3] and hydrogen tends to form covalent bonds that have properties more closely resembling those of carbon than those of the alkali metals. With the halogens it shares the tendency to form a −1 oxidation state, even to the extent of forming the hydride ion, H^-; however, the latter is a curious chemical species. In contrast to the proton which was anomalous because of its vanishingly small size, the hydride ion is unusually large. It is larger than any of the halide ions except iodide![4] The source of this apparent paradox lies in the lack of control of a single nuclear proton over two mutually repelling electrons. Since the hydride ion is large and very polarizable it certainly does not extend the trend of I^- through F^- of decreasing size and increasing basicity and hardness.

The elements of the second row also differ from their heavier congeners. Lithium is anomalous among the alkali metals and resembles magnesium more than its congeners. In turn, in Group IIA beryllium is more closely akin to aluminum than to the other alkaline earths. The source of this effect is discussed below. We have already seen that fluorine has been termed a superhalogen on the basis of its differences from the remainder of Group VIIA.

One simple difference that the elements Li to F have with respect to their heavier congeners is in electron-attracting power. Thus fluorine is much more reactive than chlorine, bromine, or iodine; lithium is less reactive than its congeners.[5] The most electronegative and smallest element of each family will be those of the second row.

The great polarizing power of the Li^+ cation was commented upon in Chapter 3. As a result of its small size and higher electronegativity this ion destabilizes salts that

[3] Those who disapprove of writing H_3O^+ often point out that the hydration number of the H^+ is uncertain and "all cations are hydrated in solution." To treat H^+ (rather than H_3O^+) as a cation similar to Na^+, for example, is to equate nuclear particles with atoms, a discrepancy by a factor of about 10^5.

[4] Pauling, "The Nature of the Chemical Bond," 3rd ed., Cornell University Press, **1960**, p. 514, has provided an estimate of 208 pm for the hydride ion compared to 216 pm for I^-. To be sure, the existence of an unpolarized hydride ion is even less likely than an unpolarized anion of some other kind, but insofar as ionic radii have meaning this would be the best estimate of the size of a free hydride ion.

[5] The inherent unreactivity of lithium is offset in aqueous solution by the exothermic hydration of the very small Li^+ ion. Nevertheless, in general, lithium is a less reactive metal than Na, K, Rb, or Cs.

are stable for the remaining alkali metals:

$$2LiOH \xrightarrow[\text{red heat}]{} Li_2O + H_2O \tag{17.1}$$

$$2NaOH \xrightarrow[\text{red heat}]{} \text{No reaction} \tag{17.2}$$

$$2LiSH \longrightarrow Li_2S + H_2S \tag{17.3}$$

$$Li_2CO_3 \longrightarrow Li_2O + CO_2 \tag{17.4}$$

In contrast, for the large polarizable hydride ion which can bond more strongly by a covalent bond the lithium compound is the *most* stable:

$$LiH \xrightarrow[\text{heat}]{} \text{No reaction} \tag{17.5}$$

$$2NaH \xrightarrow[\text{heat}]{} Na_2 + H_2 \tag{17.6}$$

The diagonal relationship

It was mentioned previously that a strong resemblance obtained between Li and Mg, Be and Al, C and P, and other "diagonal elements," and it was pointed out that this could be related to a size-charge phenomenon (see p. 130). Some examples of these resemblances are as follows:

Lithium–magnesium. There is a large series of lithium alkyls and lithium aryls which are useful in organic chemistry in much the same way as the magnesium Grignard reagents. Unlike Na, K, Rb, or Cs, but like Mg, lithium reacts directly with nitrogen to form a nitride:

$$3Li_2 + N_2 \longrightarrow 2Li_3N \tag{17.7}$$

$$6Mg + 2N_2 \longrightarrow 2Mg_3N_2 \tag{17.8}$$

Finally, the solubility of several lithium compounds more nearly resembles those of the corresponding magnesium salts than of other alkali metal salts.

Beryllium–aluminum. These two elements resemble each other in several ways. The oxidation *emf*s of the elements are similar ($\mathscr{E}^0_{Be} = 1.85$; $\mathscr{E}^0_{Al} = 1.66$), and although reaction with acid is thermodynamically favored, it is rather slow, especially if the surface is protected by the oxide. The similarity of the ionic potential (see p. 130) for the ions is remarkable ($Be^{+2} = 6.45$, $Al^{+3} = 6.00$) and results in similar polarizing power and of the cations. For example, the carbonates are unstable, the hydroxides dissolve readily in excess base, and the Lewis acidities of the halides are comparable.

Boron–silicon. Boron differs from aluminum in showing almost no metallic properties and its resemblance to silicon is greater. Both boron and silicon form volatile, very reactive hydrides; the hydride of aluminum is a polymeric solid. The halides (except BF_3) hydrolyze to form boric acid and silicic acid. The oxygen chemistry of the borates and silicates also has certain resemblances.

Table 17.1 Maximum coordination numbers of the nonmetals as shown by the fluorides

CF_4	NF_3[a]	OF_2[a]	$FF(F_3^-)$
SiF_6^{-2}	PF_6^-	SF_6	ClF_5[a]
			⋮
			$IF_7(IF_8^-)$

[a] N, O, and other elements can achieve higher coordination in onium salts, e.g., NH_4^+.

Carbon–phosphorus, nitrogen–sulfur, and oxygen–chlorine. All metallic properties have been lost in these elements, and so charge-to-size ratios have little meaning. However, the same effects appear in the electronegativities of these elements, which show a strong diagonal effect:[6]

$$\begin{array}{cccc} C = 2.55 & N = 3.04 & O = 3.44 & F = 3.98 \\ Si = 1.90 & P = 2.19 & S = 2.58 & Cl = 3.16 \end{array}$$

The similarities in electronegativities are not so close as those of the ionic potential for Be^{+2} and Al^{+3}. The heavier element in the diagonal pair always has a lower electronegativity, but the effect is still noticeable. Thus, when considering elements that resemble carbon, phosphorus is as good a choice as silicon, and the resemblance is sufficient to establish a base from which notable *differences* can be formulated.

THE USE OF *d* ORBITALS BY NONMETALS

It is an obvious fact that the elements Li to F are restricted to the set of *s* and *p* orbitals, but their heavier congeners such as Na to Cl can use 3*s*, 3*p*, and 3*d* orbitals. This provides extra opportunities for bonding in the heavier elements that their light congeners do not enjoy. The extent to which these elements use valence shell *d* orbitals is a matter of some controversy. Some of the experimental evidence will be discussed first, followed by theoretical interpretations.

Experimental evidence for σ bonds involving *d* orbitals

It is a fact that the second-row elements, Li to F, show a maximum covalence of 4,[7] corresponding to a maximum hybridization of sp^3. In contrast, third-row and heavier elements show 5, 6, and 7 coordination (Table 17.1) consistent with use of *d* orbitals.

[6] These values are Pauling thermochemical electronegativities rather than those based on ionization energy–electron affinity. This choice of empirical values was made to obviate the necessity to choose (arbitrarily) the proper hybridization.

[7] In the metallic state lithium and beryllium have coordination numbers greater than 4, as does elemental boron in icosahedral structures. In each of these examples multicenter bonding occurs and the concept of fixed bonds is not applicable.

A second factor which may affect the coordination number and which does not require the assumption that *d* orbitals participate is size. One would expect that the coordination number would increase upon progressing down the chart, and indeed it does. Thus the number of σ bonds a nonmetal forms may be determined as much by the number of substituents that can fit as by the number of orbitals available.

Experimental evidence for π bonding; the phosphorus–oxygen bond in phosphoryl compounds

In the case of π bonding we again find the old problem of detecting the existence of a bond. We can infer the presence of a σ bond when we find two elements at distances considerably shorter than the sum of their van der Waals radii. The detection of a π bond depends on more subtle criteria: shortening or strengthening of a bond, stabilization of a charge distribution, etc., experimental data which may be equivocal.

One example of the apparent existence of π bonding is in phosphine oxides. Most tertiary phosphines are unstable relative to oxidation to the phosphine oxide:

$$2R_3P + O_2 \longrightarrow 2R_3PO \tag{17.9}$$

This reaction takes place so readily that aliphatic phosphines must be protected from atmospheric oxygen. The triarylphosphines are more stable in this regard but still can be oxidized readily:

$$\phi_3P \xrightarrow[KMnO_4]{HNO_3\text{ or}} \phi_3PO \tag{17.10}$$

In contrast, aliphatic amines do not have to be protected from the atmosphere although they can be oxidized:

$$R_3N + HOOH \longrightarrow [R_3NOH]^+OH^- \xrightarrow{-H_2O} R_3NO \tag{17.11}$$

However, the amine oxides decompose upon heating:

$$Et_3NO \xrightarrow[heat]{} Et_2NOH + CH_2{=}CH_2 \tag{17.12}$$

a reaction completely unknown for the phosphine oxides, which are thermally stable. In fact, it has been said that "tertiary phosphine oxides have the reputation of being the most stable chemical structures in the family of organophosphorus compounds."[8] They are not reduced even by heating with metallic sodium. The tendency of phosphorus to form such $P \rightarrow O$ or $P{=}O$ linkages is one of the driving forces of phosphorus chemistry and may be used to rationalize and predict reactions and structures. For example, the lower phosphorus acids exist in the 4-coordinate structures even though they are prepared by the hydrolysis of 3-coordinate halides:

$$PX_3 \xrightarrow{H_2O} [P(OH)_3] \longrightarrow HPO(OH)_2 \tag{17.13}$$

[8] J. R. Van Wazer, "Phosphorus and Its Compounds," Wiley (Interscience), New York, **1958**, Vol. I, p. 287.

$$R-\underset{X}{\overset{P}{\wedge}}-X \xrightarrow{H_2O} \left[R-\underset{OH}{P}-OH \right] \longrightarrow R-\underset{H}{\overset{\overset{O}{\uparrow}}{P}}-OH \tag{17.14}$$

$$X_2P-PX_2 \xrightarrow{H_2O} \left[(HO)_2P-P(OH)_2 \right] \longrightarrow H-\underset{HO}{\overset{\overset{O}{\uparrow}}{P}}-\underset{OH}{\overset{\overset{O}{\uparrow}}{P}}-H \tag{17.15}$$

The tendency to form P=O bonds is responsible for the Arbusov reaction, described as "one of the most useful reactions in organophosphorus chemistry."[9] The typical reaction is the rearrangement of a trialkyl phosphite to a phosphonate:

$$(RO)_3P \xrightarrow{RX} (RO)_2\overset{\overset{O}{\uparrow}}{P}R \tag{17.16}$$

If the catalytic amounts of RX in Eq. 17.16 are replaced by equimolar amounts of R′X, the role of the alkyl halide in the formation of a quasiphosphonium salt is revealed:

$$(RO)_3P + R'X \longrightarrow \left[RO-\underset{\underset{R}{O}}{\overset{\overset{R}{O}}{P}}-R' \right]^+ + X^- \longrightarrow RO-\underset{\underset{R}{O}}{\overset{\overset{O}{\uparrow}}{P}}-R' + RX \tag{17.17}$$

Oxidation of trialkyl phosphites by halogens illustrates the same principle:

$$(RO)_3P + Cl_2 \longrightarrow [(RO)_3PCl]^+Cl^- \longrightarrow (RO)_2P(O)Cl + RCl \tag{17.18}$$

A final difference between amine oxides and phosphine oxides lies in the polarity of the molecules. The dipole moment of trimethylamine oxide is 16.7×10^{-30} C m compared with 14.6×10^{-30} C m for triethylphosphine oxide. A consequence of this polarity is the tendency of the amine oxides to form hydrates, $R_3NO \cdot H_2O$, and their greater basicity relative to the phosphine oxides.

The difference between the behavior of the amine oxides and phosphine oxides can be rationalized in terms of the possibility of back bonding in the latter. Whereas amine oxides are restricted to a single structure containing a dative N—O bond, $R_3N \rightarrow O$, the phosphine oxides can have contributions from d_π–p_π bonding between the phosphorus and oxygen atoms:

$$\underset{(I)}{R_3P^+ \rightarrow O^-} \longleftrightarrow \underset{(II)}{R_3P{=}O} \tag{17.19}$$

[9] R. F. Hudson, "Structure and Mechanism in Organo-Phosphorus Chemistry," Academic Press, New York, **1965**, pp. 135–141, 216. See also R. G. Harvey and E. R. De Sombre, *Topics in Phosphorus Chemistry*, **1964**, *1*, 57.

Table 17.2 Infrared stretching frequencies of some phosphoryl compounds

Compound	$\bar{\nu}_{PO}$ (cm^{-1})	$\Sigma_{\chi_{MJ}}$
F_3PO	1404	11.70
F_2ClPO	1358	10.75
Cl_3PO	1295	8.85
Cl_2BrPO	1285	8.52
$ClBr_2PO$	1275	8.19
Br_3PO	1261	7.86
ϕ_3PO	1190	(7.2)
Me_3PO	1176	(6.0)

The double bond character introduced by the latter strengthens the bond and accounts for the extraordinary stability of the phosphorus oxygen linkage. Note that this extra stability cannot be attributed to ionic resonance energy (*a priori* a reasonable suggestion since the difference in electronegativity is greater in P—O than N—O) because the dipole moment of the *nitrogen* compound is greater than that of the phosphorus compound, a result completely unexpected on the basis of electronegativities, unless consideration is taken of canonical form 17.19 (II), which would be expected to lead to a reduced moment.

A comparison of the bond energies also supports the above intepretation. The dissociation energies of P=O bonds in a variety of compounds lie in the range of 500–600 kJ mol^{-1} compared with values for N → O of about 200–300 kJ mol^{-1}.[10] The value for the latter is typical of what we might expect for a single bond, but 600 kJ mol^{-1} is stronger than any known single bond (see Chapter 6, p. 263). A closer examination of the strengths of various P=O bonds in terms of infrared stretching frequencies shows some interesting trends. For a series of similar molecules, such as the phosphine oxides, the stretching frequency provides an indication of the strength of the bond (Table 17.2).[11] The highest stretching frequency among the phosphoryl compounds is that of F_3PO, and the lowest of the halides is that of Br_3PO (the iodo compound is unknown). When the stretching frequencies (in the form of wavelength, $\lambda = 1/\bar{\nu}$) are plotted as a function of the sum of the electronegativities of the substituents, a straight line is obtained (Fig. 17.1). This relationship has been used to obtain group electronegativities from the stretching frequencies of phosphine oxides, for example, which are internally self-consistent and which agree with those obtained by other means (see Table 3.13) except in the case of groups such as $—NH_2$ which can compete in the π bonding.[12]

[10] R. F. Hudson, "Structure and Mechanism in Organo-Phosphorus Chemistry," Academic Press, New York, **1965**, pp. 67–70.

[11] Note that the dissociation energy, $R_3PO \rightarrow R_3P + O$, is not a sensitive measure of the P=O *bond energy* since the remaining three bonds may be strengthened or weakened in the dissociation process. The IR stretching frequency is a function of the force constant, k, and the reduced mass, μ, of the molecule. If the molecule is assumed to be a light oxygen atom vibrating on a "fixed" larger mass of the R_3P group, the reduced mass is constant, and so changes in frequency will reflect corresponding changes in the force constant. For similar molecules the force constant will be related to the total bond energy.

[12] J. V. Bell et al., *J. Am. Chem. Soc.*, **1954**, *76*, 5185. See also M. A. Davis, *J. Org. Chem.*, **1967**, *32*, 1161, and papers cited therein.

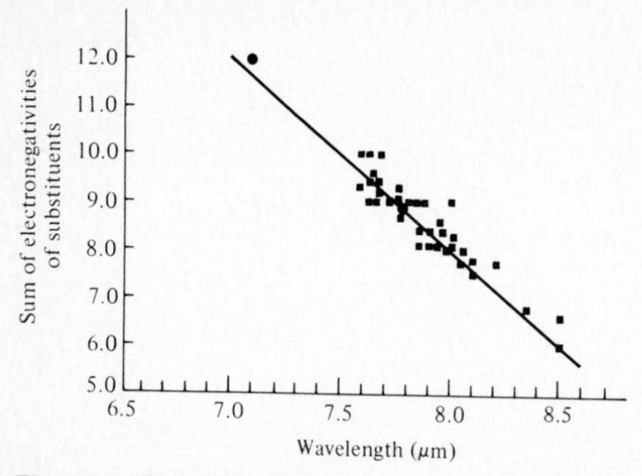

Fig. 17.1 The relation between the stretching frequency (expressed as wavelength) and the electronegativity of the substituents. Circles represent halides, squares other substituted P = O compounds. [From J. V. Bell et al., *J. Am. Chem. Soc.*, **1954**, *76*, 5185. Reproduced with permission.]

The correlation between the electronegativity of substituent groups and the strength of the P=O bond provides support for a π-bonding model but not for the alternative dative σ-only model. The latter might be expected to be destabilized as electron density is removed from the phosphorus, requiring it to withdraw electrons from the $P \rightarrow O$ bond, weakening it. In contrast, if the oxygen can back bond to the phosphorus through a d–p π bond, the induced charge on the phosphorus can be diminished and the P=O bond strengthened.

The bond lengths in phosphoryl compounds are in accord with the concept of double bond character.[13] In the simplest case, that of P_4O_{10}, there are two P—O bond lengths. There are 12 relatively long ones (163 pm) within the cage framework proper and 4 shorter ones (139 pm) between the phosphorus atoms and the oxygen atoms external to the cage. It is interesting to note that the ratio of these two bond lengths ($\sim$0.85) is about the same as C=C to C—C or C=O to C—O.

Isoelectronic with the phosphine oxides are the phosphonium ylids, R_3PCH_2.[14] Like the oxides, two resonance forms

$$\underset{\text{(I)}}{R_3\overset{+}{P}-\overset{-}{C}H_2} \longleftrightarrow \underset{\text{(II)}}{R_3P=CH_2} \tag{17.20}$$

contribute to the stability of the phosphonium ylids but not the corresponding ammonium ylids, $R_3N^+-C^-H_2$. This difference is reflected in the reactivity. The ammonium ylids are generally quite basic and quite reactive; the phosphonium ylids are much less so,

[13] The complete discussion, including partial ionic character, is too long to present here. See R. F. Hudson, *Adv. Inorg. Chem. Radiochem.*, **1963**, *5*, 347, and K. A. R. Mitchell, *Chem. Rev.*, **1969**, *69*, 157.

[14] The nomenclature is unfortunate since the "-onium" name used with the ylids focuses attention on the polar nature (resonance structure (I)) whereas the "-ine oxide" name seems to focus more on the covalent nature (resonance structure (II)). Electronically, both types of molecules are resonance hybrids of (I) and (II).

$H_3\overset{-}{Si}{=}\overset{+}{N}(SiH_3)_2 \quad H_3Si{-}\overset{+}{N}(SiH_3){=}\overset{-}{Si}H_3 \quad H_3Si{-}\overset{+}{N}(SiH_3){=}\overset{-}{Si}H_3$

(a)

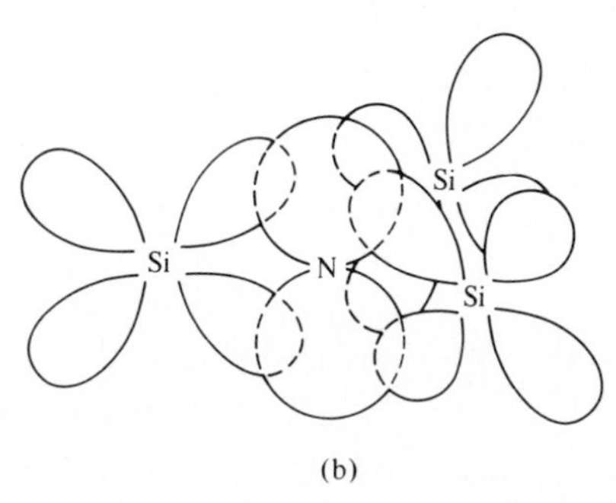

(b)

Fig. 17.2 Delocalization of the lone pair in trisilylamine. (a) Resonance structures. (b) Overlap of d_{Si} and p_N orbitals.

many not being sufficiently basic to abstract a proton from water and, in fact, not dissolving in water unless strong acids are present.[15]

Evidence from bond angles

The trimethylamine molecule has a pyramidal structure much like that of ammonia with a $CH_3{-}N{-}CH_3$ bond angle of $107.8° \pm 1°$. In contrast, the trisilylamine molecule is planar. Although steric effects of the larger silyl groups might be expected to open up the bond angles, it seems hardly possible that they could force the lone pair out of a fourth "tetrahedral" orbital and make the molecule perfectly planar (even ϕ_3N has bond angles of 116°). It seems more likely that the lone pair adopts a pure *p* orbital on the nitrogen atom because orbitals on the three silicon atoms can overlap with it and delocalize the lone pair over the entire system (Fig. 17.2).

Rather similar results are obtained by comparing the bond angles in the silyl and methyl ethers (Fig. 17.3) and isothiocyanates (Fig. 17.4). In the former the oxygen is hybridized approximately sp^3 with two lone pairs on the oxygen atom in dimethyl ether as compared to an approximate sp^2 hybrid in disiloxane with π bonding. In the same way the methyl isothiocyanate molecule, $CH_3N{=}C{=}S$, has a lone pair localized on the nitrogen atom, hence is bent ($N \sim sp^2$), but the delocalization of this lone pair into a back-bonding π orbital to the silicon atom of $H_3SiN{=}C{=}S$ leads to a linear structure for this molecule.

The hypothesized delocalization of lone pair electrons in the above silicon compounds is supported by the lowered basicity of the silyl compounds as compared to the

[15] A. W. Johnson, "Ylid Chemistry," Academic Press, New York, **1966**.

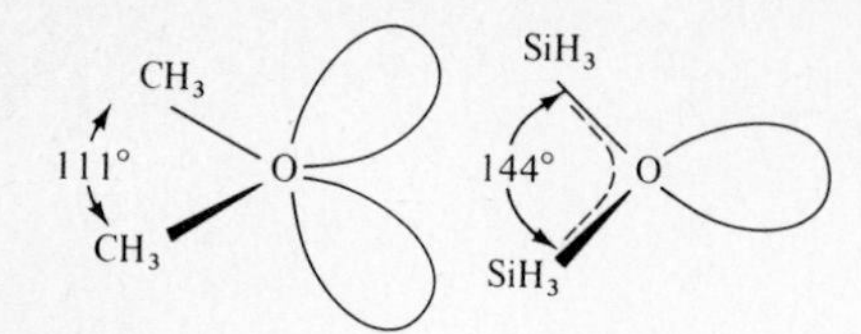

Fig. 17.3 Comparison of the molecular structure of dimethyl ether and disiloxane.

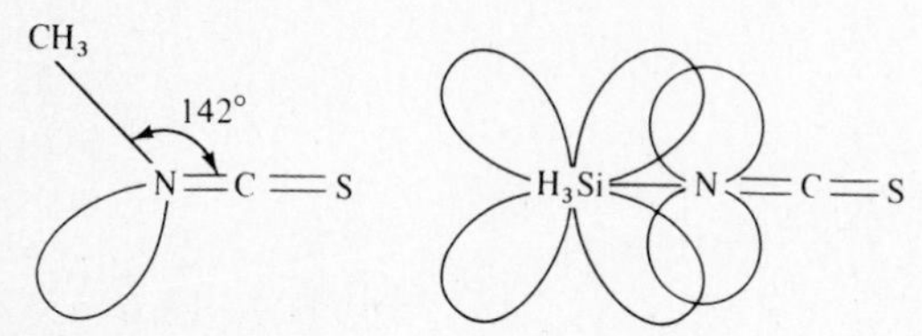

Fig. 17.4 Comparison of the molecular structure of methyl isothiocyanate and silyl isothiocyanate.

corresponding carbon compounds. This reduced basicity is contrary to that expected on the basis of electronegativity effects operating through the σ system since silicon is less electronegative than carbon. It is consistent with an "internal Lewis acid–base" interaction between the nitrogen and oxygen lone pairs and empty acceptor d orbitals on the silicon. Experimentally this reduced basicity is shown by the absence of disiloxane adducts with BF_3 and BCl_3:

$$(CH_3)_2O + BF_3 \longrightarrow (CH_3)_2O \rightarrow BF_3 \tag{17.21}$$

$$(SiH_3)_2O + BF_3 \longrightarrow \text{No adduct} \tag{17.22}$$

and by the absence of trisilylammonium salts. Instead of onium salt formation trisilylamine is cleaved by hydrogen chloride:

$$(SiH_3)_3N + 4HCl \longrightarrow NH_4Cl + 3SiH_3Cl \tag{17.23}$$

Theoretical arguments against *d* orbital participation in nonmetals

Several workers have objected to the inclusion of d orbitals in bonding in nonmetals. The principal objection is to the large promotion energy required to effect:

$$s^2p^nd^0 \longrightarrow s^1p^{n-m}d^{m+1} \tag{17.24}$$

where $m = 0$ (P), 1 (S), or 2 (Cl), to achieve a maximum multiplicity and availability of electrons for bonding. A second factor which does not favor the utilization of d orbitals is the poor overlap that they make with the orbitals of neighboring atoms. The $3d$ orbitals of the free sulfur atoms, for example,[16] are shielded completely[17] by the lower-lying

[16] These same general arguments apply to the other nonmetals as well.

[17] In the simplified Slater scheme (p. 36) the d orbitals are assumed to be shielded to the extent of 1.0 electronic unit for each electron lying "below" them.

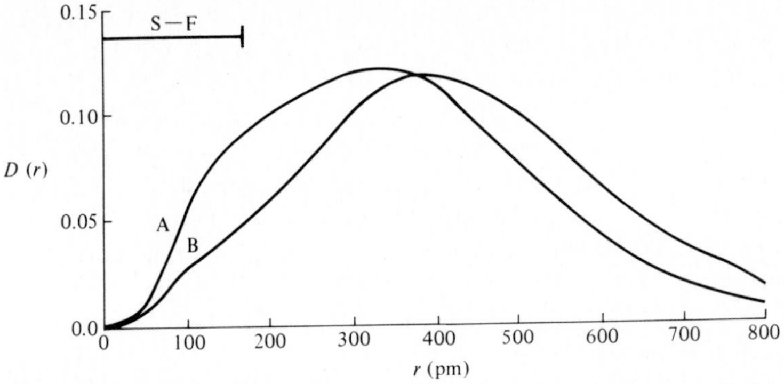

Fig. 17.5 The 3*d* orbital distribution functions in d^1 configurations: (A) in the ^6D term of $P(sp^3d)$; (B) in the ^5D term of $S(s^2p^3d)$. Line represents a typical S—F bond length. [Modified from K. A. R. Mitchell, *Chem. Rev.*, **1969**, *69*, 157. Reproduced with permission.]

electrons and hence do not feel the nuclear charge as much as the 3*s* and 3*p* electrons. As a result they are extremely diffuse, having radial distribution maxima at a distance which is approximately *twice* a typical bond distance (Fig. 17.5). This results in extremely poor overlap and weak bonding.[18]

Two alternatives have been provided to account for the higher oxidation states of the nonmetals without invoking the use of the high energy *d* orbitals. Pauling has suggested that resonance of the following type could take place.[19]

PCl_5 (I): Cl—P(Cl)(Cl)(Cl)(Cl) ⟷ (II): Cl^- Cl—P^+(Cl)(Cl)(Cl) ⟷ Four more forms like (II)

Only structure I involves *d* orbitals, and so the *d* character of the total hybrid is small. Each P—Cl bond has 20% ionic character and 80% covalent character from resonance structures such as II. Pauling has termed the "extra" bonds formed (over and above the

[18] A point that may easily be overlooked in this regard is that the poor overlap results not only from the poor spatial arrangement of a diffuse orbital but also from the concomitant low value of the wave function, Ψ_s, at any particular point in space, and so a low buildup of electron density from $\Psi_s \cdot \Psi_x$.

[19] Pauling listed a third type of canonical structure,

Cl—P(Cl)(Cl) with the remaining two Cl atoms bonded to each other (Cl—Cl)

but pointed out that the Cl—Cl interaction is a very "long bond" and does little to stabilize the molecule. [L. Pauling, "The Nature of the Chemical Bond," 3rd ed., Cornell University Press, Ithaca, N.Y., **1960**, pp. 177–179.]

four in a noble gas octet or "argononic" structure) as "transargononic" bonds and pointed out that they tend to be weaker than "normal" or "argononic" bonds and form only with the most electronegative ligands. Thus the average bond energy in PCl_3 is 317 kJ mol^{-1} but in PCl_5 it is only 165 kJ mol^{-1}.[20] The same effect is found in PF_3 and PF_5, but in this case the difference in bond energy is only 61.1 kJ mol^{-1}, corresponding to the stabilization of the structure by increased importance of the ionic structures in the fluorides. The stabilization of these structures by differences in electronegativity is exemplified by the tendency to form the higher halogen fluorides.[21] The enthalpies of fluorination of the halogen monofluorides are:

$$ClF_{(g)} + 2F_{2(g)} \longrightarrow ClF_{5(g)} \qquad \Delta H = -152.7 \text{ kJ mol}^{-1} \qquad \textbf{(17.25)}$$

$$BrF_{(g)} + 2F_{2(g)} \longrightarrow BrF_{5(g)} \qquad \Delta H = -376.1 \text{ kJ mol}^{-1} \qquad \textbf{(17.26)}$$

$$IF_{(g)} + 2F_{2(g)} \longrightarrow IF_{5(g)} \qquad \Delta H = -751.4 \text{ kJ mol}^{-1} \qquad \textbf{(17.27)}$$

The second alternative is the three-center, four-electron bond developed by simple molecular orbital theory for the noble gas fluorides (see pp. 767–769). Since this predicts that each bonding pair of electrons (each "bond") is spread over three nuclei, the bond between two of the nuclei is less than that of a normal two-center, two-electron bond. Furthermore, since the nonbonding pair of electrons is localized on the fluorine atoms, there is a separation of charge ("ionic character"). In both respects, then, this interpretation agrees with Pauling's approach and with the experimental facts.

Theoretical arguments in favor of *d* orbital participation

In contrast to the arguments presented against participation by *d* orbitals in the bonding of nonmetals, several workers have pointed out that the large promotion energies and diffuse character described above are *properties of an isolated sulfur atom*. What we need to know are the *properties of a sulfur atom in a molecule*, such as SF_6. This is an exceedingly difficult problem and cannot be dealt with in detail here.[22] However, we have seen how it is possible to calculate such properties as electronegativity on isolated atoms as charge is added or withdrawn (see pp. 153–155) and how this might *approximate* such properties in a molecular environment.

It is apparent from the preceding discussions that participation of *d* orbitals, if it occurs at all, is found only in the nonmetals when in high oxidation states with electronegative substituents. The partial charge induced on the central P or S atom will be large merely from the electronegativity of the fluorine (as in PF_5, SF_6) or oxygen (as in OPX_3, O_2SX_2) irrespective of any bonding model (such as Pauling's or the three-center bond) invoked.

---

[20] For a more complete discussion of this point, see L. Pauling, "General Chemistry," 3rd ed., Freeman, San Francisco, **1970**, pp. 244–247.

[21] See Table 15.2 and attendant discussion. The values in Eqs. 17.25–17.27 differ slightly from those given by Pauling (footnote 20) in order to be self-consistent with the values in Table 15.2.

[22] For a detailed discussion of this complex problem, see K. A. R. Mitchell, *Chem. Rev.*, **1969**, 69, 157, and references therein.

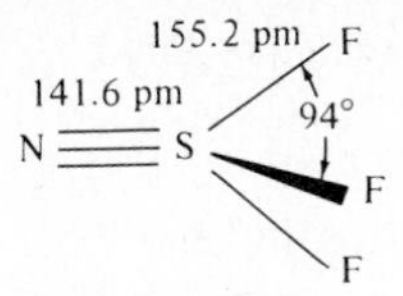

Fig. 17.6 Molecular structure of thiazyl trifluoride, NSF_3.

We have seen in Chapter 2 that the $3d$ orbital lies above the $4s$ orbital at atomic number 20 but falls below as the atomic number increases. Furthermore, in the M^{+2} ions the $3d$ orbital lies below the $4s$ orbital at least as soon as atomic number 22 (Ti^{+2} is d^2, not s^2). This is a general phenomenon: Increasing effective nuclear charge makes the energy levels of an atom approach more closely the degenerate levels of the hydrogen atom. We might expect, in general, that increasing the effective nuclear charge on the central atom as a result of inductive effects would result in the lowering of the *d* orbitals more than the corresponding *s* and *p* orbitals since the former are initially shielded more and hence will be more sensitive to changes in electron density. The promotion energy would thus be lowered. A second effect of large partial charges on the central atom will be a shrinking of the large, diffuse *d* orbitals into smaller, more compact orbitals that will be more effective in overlapping neighboring atomic orbitals. For example, sample calculations indicate that in SF_6 the *d* orbitals have been contracted to an extent that the radius of maximum probability is only 130 pm compared with the large values in the free sulfur atom.

Experimental evidence for *d* orbital contraction and participation

One of the most remarkable molecules is thiazyl trifluoride, NSF_3 (Fig. 17.6). This compound is very stable. It does not react with ammonia at room temperature, with hydrogen chloride even when heated, or with metallic sodium at temperatures below 400°C. The S—N bond, 141.6 pm, is the shortest known between these two elements. The FSF bond angles of 94° are compatible with approximate sp^3 bonding[23] and the presence of an sp^3 hybrid σ bond and two p–d π bonds between the sulfur and the nitrogen. The contraction of the *d* orbitals by the inductive effect of the fluorine atoms presumably permits effective overlap and π-bond formation. The alternative explanation would require a double dative bond from the sulfur atom, extremely unlikely in view of the positive character of the sulfur atom.

The bond length is consistent with a triple bond. Bond lengths of 174 pm for single S—N bonds (in NH_2SO_3H) and 154 pm for double S═N bonds (in $N_4S_4F_4$) are consistent with a bond order of 2.7 in thiazyl trifluoride. This value is also in agreement with an estimate based upon the force constant.[24] The relative bond lengths of S—N, S═N,

---

[23] Some *d* orbital hybridization may enter into the σ system, reducing the bond angles.

[24] Of course, this involves certain assumptions about the relation between the force constant and bond order. For a more complete discussion of bond length and bond order in these compounds, see O. Glemser and M. Fild, *Halogen Chem.*, **1967**, *2*, 1. See also O. Glemser and R. Mews, *Adv. Inorg. Radiochem.*, **1970**, *14*, 33.

and S═N bonds are thus 1.00:0.88:0.81 compared with similar shortenings of 1.00:0.87:0.78 for corresponding C—N, C═N, and C≡N bonds.

Two other molecules indicating the influence of fluorine substitution on *d* orbital participation are $S_4N_4H_4$ and $N_4S_4F_4$ (see pp. 716–718). Tetrasulfur tetraimide is isoelectronic with the S_8 molecule and so the structure:

```
         H
         |
     S—N—S
     |       |
H—N       N—H
     |       |
     S—N—S
         |
         H
```

and corresponding crown conformation appear quite reasonable. The fluoride, however, has an isomeric structure with substitution on the sulfur atoms:

```
F—S═N—S—F
    |       ‖
    N       N
    ‖       |
F—S—N═S—F
```

Double bonding in this molecule is clearly shown by the alternation in S—N bond lengths in the ring (see Fig. 14.21b). Now both the above electronic structure for $S_4N_4F_4$ and that for $S_4N_4H_4$ are reasonable but raise the question: Why doesn't tetrasulfur tetraimide isomerize from the N-substituted form to the S-substituted form isoelectronic with the fluoride:

```
H—S═N—S—H
    |       ‖
    N       N
    ‖       |
H—S—N═S—H
```

retaining the same number of σ bonds and gaining four π bonds? Apparently the reason the isomerism does not take place is that although π bonding is feasible in the presence of the electronegative fluorine atoms, it is so weak with electropositive hydrogen substituents that it cannot compensate for the weakening of the σ bonding as the hydrogen atom shifts from the more electronegative nitrogen atom to the less electronegative sulfur atom.[25]

Presumably substitution by halogens in the phosphonitrilic series results in contracted *d* orbitals and more efficient π bonding in the ring. It was noted previously (see

[25] This argument is based on the stabilization of the H—X σ bonds by ionic resonance energy and is supported by the bond energies in NH_3 (391 kJ mol^{-1}) and H_2S (347 kJ mol^{-1}).

p. 708) that unsymmetrical substitution may allow the normally planar trimeric ring to bend. A good example of this is found in 1,1-diphenylphosphonitrilic fluoride trimer:[26]

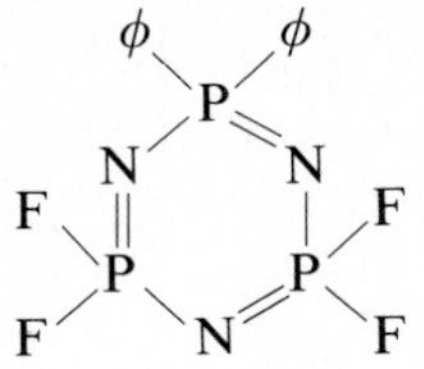

The three nitrogen atoms and the fluoro-substituted phosphorus atoms are coplanar (within 2.5 pm), but the phenyl-substituted phosphorus atom lies 20.5 pm above this plane. The explanation offered is that the more electropositive phenyl groups cause an expansion of the phosphorus *d* orbitals, less efficient overlap with the *p* orbitals of the nitrogen atom, and a weakening of the π system at that point. This allows[27] the ring to deform and the ϕ_2P moiety to bend out of the plane.

A further example of the jeopardy involved in casually dismissing *d* orbital participation is the findings of Haddon and co-workers[28] that *d*-orbital participation is especially important in S_4F_4, which is nonplanar, and also that it accounts for about one-half of the delocalization energy in the one-dimensional conductor $(SN)_x$. In the latter case, the low electronegativity of the *d* orbitals (see p. 181) increases the ionicity of the S—N bond and stabilizes the structure.

Finally, it will be recalled that the existence of strong P=O bonds in OPF_3 (see p. 828) is consistent with enhanced back donation of electron density from the oxygen atom to the phosphorus atom bearing a positive partial charge from the four σ bonds to electronegative atoms. In light of the above discussion of the contraction of phosphorus and sulfur *d* orbitals when bearing a positive charge, better overlap may be added to the previous discussion as a second factor stabilizing this molecule.

The question of *d* orbital participation in nonmetals is still an open controversy. In the case of σ-bonded species such as SF_6 the question is not of too much importance since all of the models predict an octahedral molecule with very polar bonds. Participation in π bonding is of considerably more interest, however. Inorganic chemists of a more theoretical bent tend to be somewhat skeptical, feeling that the arguments regarding promotion energies and poor overlap have not been adequately solved. On the other hand, chemists interested in synthesis and characterization tend to favor the use of *d* orbitals in describing these compounds, pointing to the great heuristic value that has been provided by such descriptions in the past and feeling that until rigorous *ab initio* calculations[29] on these molecules show the absence of significant *d* orbital participation it is too soon to abandon a useful model.

---

[26] C. W. Allen et al., *J. Am. Chem. Soc.*, **1967**, *89*, 6361.

[27] Note that this explanation provides a rationale which allows the bending but does not provide a driving force to make the ring bend; however, if the ring is sufficiently weakened, small perturbing forces such as those of crystal packing might be sufficient to deform the ring even though the same forces would have little effect in an unweakened ring system.

[28] R. C. Haddon et al., *J. Am. Chem. Soc.*, **1980**, *102*, 6687.

[29] Not likely in the near future.

Π bonding in the heavier congeners

In view of the uncertainty with which π bonding is known in the very well studied phosphorus and sulfur systems, it is not surprising that little can be said concerning the possibility of similar effects in arsenic, antimony, selenium, tellurium, etc. In general it is thought that the problems faced in phosphorus and sulfur chemistry concerning promotion energies and diffuse character may be even larger in the heavier congeners. In the latter regard it is interesting to note the apparent effectiveness of π bonding in metal complexes. To the extent that softness in a ligand can be equated with the ability to accept electrons from soft metal ions in d_π–d_π "back bonds," information can be obtained from the tendency to complex with "b" metal ions (see p. 313): P > As > Sb. This order would indicate that the smaller phosphorus atom can more effectively π bond with the metal atom. With the halide ions the order is reversed. $I^- > Br^- > Cl^-$. Whether this is a result of the importance of polarizability in this series or more effective use of d orbitals by iodine is unknown.

REACTIVITY AND *d* ORBITAL PARTICIPATION

It has been pointed out above that the elements of the second row (Li to F) not only resemble their heavier congeners to a certain extent (as far as formal oxidation state, at least) but also the lower right diagonal element (as far as charge, size, and electronegativity are concerned). For example, both silicon and phosphorus form hydrides that have some properties in common with alkanes, although they are much less stable. As a result of the electronegativity relationships the P—H bond more closely approaches the polarity of the C—H bond than does the Si—H bond. The resemblance of phosphorus to carbon has even been extended to the suggestion that a discipline be built around it in the same manner as organic chemistry is built on carbon.[30]

There is one important aspect of the chemistry of both silicon and phosphorus which differs markedly from that of carbon. Consider the following reactions:

$$CCl_4 + H_2O \longrightarrow \text{No reaction} \tag{17.28}$$

$$SiCl_4 + 4H_2O \longrightarrow Si(OH)_4 + 4HCl \quad \text{rapid} \tag{17.29}$$

$$PCl_5 \xrightarrow[\text{rapid}]{H_2O} 2HCl + OPCl_3 \xrightarrow[\text{slower}]{3H_2O} OP(OH)_3 \tag{17.30}$$

In contrast to the inertness of carbon halides, the halides of silicon and phosphorus are extremely reactive with water, to the extent that they must be protected from atmospheric moisture. A clue to the reactivity of these halides is provided by the somewhat similar reactivity of acid halides in which the carbon–halogen bond is readily attacked:

$$R\overset{O}{\overset{\|}{C}}Cl + H_2O \longrightarrow R-\underset{\underset{H\diagup\overset{|}{O^+}\diagdown H}{|}}{\overset{O^-}{\overset{|}{C}}}-Cl \longrightarrow R-\overset{O}{\overset{\|}{C}}-OH + HCl \tag{17.31}$$

---

[30] J. R. Van Wazer, "Phosphorus and Its Compounds," Wiley (Interscience), New York, **1958**, Vol. I, pp. vii–ix.

The unsaturation of the carbonyl group provides the possibility of the carbon expanding its coordination shell from 3 to 4, thereby lowering the activation energy. Carbon tetrahalide cannot follow a similar path, but the halides of silicon and phosphorus can employ 3*d* orbitals to expand their octets:

$$PCl_5 + H_2O \longrightarrow PCl_5(OH_2) \longrightarrow OPCl_3 + 2HCl, \text{ etc.} \qquad \textbf{(17.32)}$$

This enhanced reactivity of compounds of silicon and phosphorus is typical of all of the heavier nonmetals in contrast to the elements of the second row.

THE USE OF *p* ORBITALS IN π BONDING

Carbon-silicon similarities and contrasts

In view of the extensive chemistry of alkenes it was only natural for organic and inorganic chemists to search for analogous Si═Si doubly bonded structures. For a long time such attempts proved to be fruitless. Recently the first stable C═Si[31] and Si═Si[32] compounds were synthesized. One synthesis involves the rearrangement of cyclotrisilane:

$$\text{Cl—Si(Cl)(Mes)}_2 \longrightarrow (Mes_2Si)_3 \xrightarrow{h\nu} Mes_2Si{=}SiMes_2 \qquad \textbf{(17.33)}$$

It is possible to add reagents across the Si═Si double bond in some ways analogous to

[31] A. G. Brook et al., *Chem. Commun.*, **1981**, 191.
[32] R. West et al., *Science*, **1981**, *214*, 1343; S. Matsumune et al., *J. Am. Chem. Soc.*, **1982**, *104*, 1153.

the C=C bond in alkenes:

$$\text{Si}{=}\text{Si} \xrightarrow{\text{HCl}} -\underset{\text{H}}{\text{Si}}-\underset{\text{Cl}}{\text{Si}}- \tag{17.34}$$

$$\text{Si}{=}\text{Si} \xrightarrow[50\,^\circ\text{C}]{\text{EtOH}} -\underset{\text{H}}{\text{Si}}-\underset{\text{OEt}}{\text{Si}}- \tag{17.35}$$

$$\text{Si}{=}\text{Si} \xrightarrow{\text{O}_2} -\underset{\text{O}}{\text{Si}}-\underset{\text{O}}{\text{Si}}- \;(\text{O}-\text{O}) \tag{17.36}$$

The successful isolation of these compounds is more a tribute to the persistence with which they were pursued than to any inherent stability of the Si=Si double bond. To invert George Leigh Mallory's remark about Mt. Everest, the extraordinary efforts expended on this class of compounds stemmed from the fact that they *weren't* there.

Compounds that are formally analogous to carbon compounds are found to have quite different structures. Thus carbon dioxide is a gaseous monomer but silicon dioxide is an infinite single-bonded polymer. In a similar manner, *gem*-diols are unstable relative to ketone

$$(CH_3)_2C(OH)_2 \longrightarrow CH_3C(O)CH_3 + H_2O \tag{17.37}$$

and the analogous silicon compounds are also unstable, but the "dimethylsilicone"[33] that forms is a linear polymer:

$$(CH_3)_2SiCl_2 \xrightarrow{2H_2O} [(CH_3)_2Si(OH)_2] \xrightarrow{-H_2O} -O-\underset{CH_3}{\overset{CH_3}{Si}}-O-\underset{CH_3}{\overset{CH_3}{Si}}-O-\underset{CH_3}{\overset{CH_3}{Si}}- \tag{17.38}$$

[33] The term "silicone" was coined by analogy to ket*one* under the mistaken belief that monomeric $R_2Si{=}O$ compounds could be isolated.

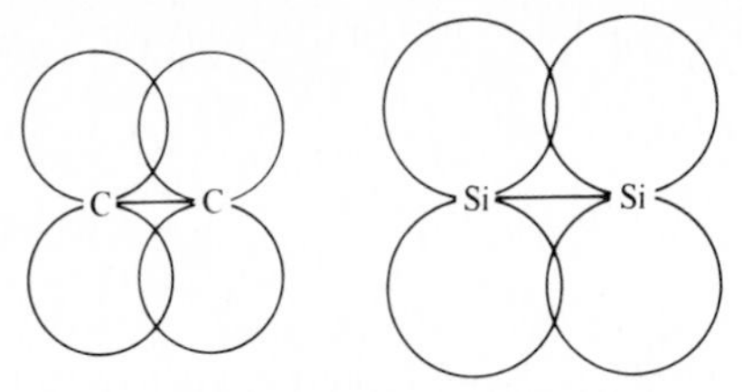

Fig. 17.7 Diagrammatic representation of the possibly poorer overlap of the *p* orbitals in C—C and Si—Si to form π bonds.

There are two possible sources of instability of doubly bonded silicon structures involving p_π–p_π bonds. It may be that the longer σ bond in the silicon system reduces the sideways overlap of the *p* orbitals (Fig. 17.7). Mulliken has objected that the overlap is *not* reduced and has advanced the idea that *d* orbital participation may favor two "single" bonds over the one double bond. This viewpoint would emphasize that all Si—X bonds (where X = N, O, halogens) will have some double bond character (even though much less than that of a full bond) from back bonding using the *d* orbitals. Hence formal "single bonds" actually would be stronger than a simple σ bond and more than half as strong as the hypothetical p_π–p_π double bonds.[34]

Nitrogen-phosphorus analogies and contrasts

The stable form of nitrogen at room temperature is N_2, which has an extraordinarily strong (946 kJ mol^{-1}) triple bond. In contrast, white phosphorus consists of P_4 molecules (see p. 723), and the thermodynamically stable form is black phosphorus, a polymer. At temperatures above 800°C dissociation to P_2 molecules does take place, but these are considerably less stable than N_2 with a bond energy of 488 kJ mol^{-1}. In this case, too, in the heavier element several single bonds are more effective than one multiple bond.

In 1961 the phosphorus analogue of hydrogen cyanide was prepared:[35]

$$CH_4 + PH_3 \xrightarrow[\text{arc}]{\text{electric}} HC{\equiv}P + 3H_2 \qquad \textbf{(17.39)}$$

In contrast to the stability of hydrogen cyanide, HCP is a highly pyrophoric gas which polymerizes above −130 °C.

During the past five years the number of molecules containing C≡P bonds has increased to about a dozen. One method of obtaining them is dehydrohalogenation:[36]

$$CF_3PH_2 \xrightarrow{-HF} CF_2{=}PH \xrightarrow{-HF} FC{\equiv}P \qquad \textbf{(17.40)}$$

[34] For further discussion, see W. E. Dasent, "Nonexistent Compounds," Dekker, New York, **1965**, Chap. 4; B. J. Aylett, *Adv. Inorg. Chem. Radiochem.*, **1968**, *11*, 249. For a discussion of bond energies in silicon compounds, see R. Walsh, *Acc. Chem. Res.*, **1981**, *14*, 246.

[35] T. E. Gier, *J. Am. Chem. Soc.*, **1961**, *83*, 1769.

[36] H. W. Kroto et al., *J. Am. Chem. Soc.*, **1978**, *100*, 446.

A stable, liquid phosphaalkyne has recently been synthesized:[37]

$$(R_3Si)_3P \longrightarrow R_3Si{-}P{=}C(OSiR_3)(t\text{-}Bu) \longrightarrow t\text{-}BuC{\equiv}P \qquad (17.41)$$

For a considerable period of time it seemed impossible to prepare doubly bonded phosphorus analogues of nitrogen. Finally, phosphorus analogues of pyridine of the type:

were synthesized,[38] and now all of the group VA analogues of pyridine have been prepared:[39]

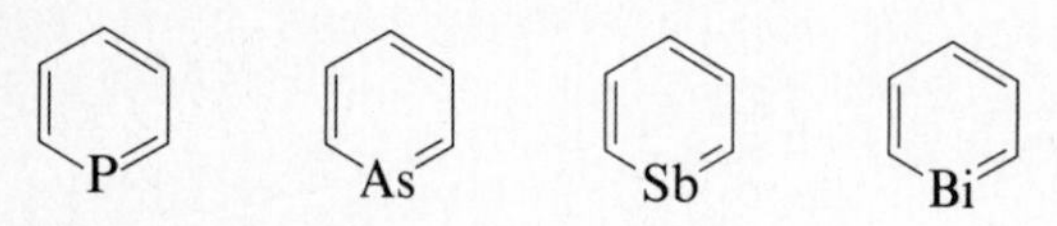

However, these compounds must be considered the exception rather than the rule as far as the heavier elements are concerned. In fact, when the barrier to inversion of the pyramidal phosphole:

was found to be only half as large as the usual value for pyramidal phosphines, R_3P, it was immediately reported as evidence that aromaticity (analogous to that of pyrrole) was responsible for lowering the activation energy. Note, however, that the aromatic stabilization is insufficient to keep the phosphole molecule planar—the ground state pyramidal with the lone pair localized on the phosphorus atom rather than delocalized in the π_p system.[40]

---

[37] See H. W. Kroto and J. F. Nixon, "Phosphorus Chemistry," L. D. Quin and J. G. Verkade, eds., *ACS Symp. Ser. 171*, **1981**, Paper 79.

[38] G. Märkl, *Angew. Chem. Int. Ed. Engl.*, **1966**, *5*, 846; A. J. Ashe, III, *J. Am. Chem. Soc.*, **1971**, *93*, 3293.

[39] A. J. Ashe, III, *J. Am. Chem. Soc.*, **1971**, *93*, 6690; A. J. Ashe, III, and M. D. Gordon, *J. Am. Chem. Soc.*, **1972**, *94*, 7596; A. J. Ashe, III et al., *J. Am. Chem. Soc.*, **1976**, *98*, 5451.

[40] W. Egan et al., *J. Am. Chem. Soc.*, **1970**, *92*, 1442. The evidence from the barrier to inversion is somewhat ambiguous—other factors may also be important (see R. D. Baechler and K. Mislow, *J. Am. Chem. Soc.*, **1971**, *93*, 773).

The isolation of compounds containing simple C=P double bonds parallels the triple-bond work. The first stable phosphaalkene was synthesized but five years ago. Again, base-induced dehydrohalogenation and stabilization by bulky groups is important:[41]

$$\mathrm{H{-}C(R')_2{-}P(R'')Cl} \xrightarrow[-\mathrm{HCl}]{\text{base}} \mathrm{(R')_2C{=}P{-}R''} \qquad \textbf{(17.42)}$$

The steric hindrance is critical: If R = phenyl or 2-methylphenyl, the bulkiness is insufficient to stabilize the molecules, but the 2,6-dimethylphenyl and 2,4,6-trimethyl derivatives are stable.

PERIODIC ANOMALIES OF THE NONMETALS AND POSTTRANSITION METALS

It is generally assumed that the properties of the various families of the periodic chart change smoothly from less metallic (or more electronegative) at the top of the family to more metallic (or less electronegative) at the bottom of the family. Certainly for the extremes of the chart—the alkali metals on the left and the halogens and noble gases on the right—this is true; the ionization potentials, for example, vary in a rather monotonous way. This is not true for certain central parts of the chart, however.

Reluctance of fourth-row nonmetals to exhibit maximum valence

There is a definite tendency for the nonmetals of the fourth row—As, Se, and Br—to be unstable in their maximum oxidation state. For example, the synthesis of arsenic pentachloride eluded chemists until recently[42] ($AsBr_5$ and AsI_5 are still unknown) although both PCl_5 and $SbCl_5$ are stable. The only stable arsenic pentahalide is AsF_5.

In Group VIA the same phenomenon is encountered. Selenium trioxide is thermodynamically unstable relative to sulfur trioxide and tellurium trioxide. The enthalpies of formation of SF_6, SeF_6, and TeF_6 are -1210, -1030, and -315 kJ mol^{-1}, respectively. This indicates comparable bond energies for S—F and Te—F bonds (326 kJ mol^{-1}), which are 42 kJ mol^{-1} more stable than Se—F bonds.

The best known exceptions to the general reluctance of bromine to accept a +7 oxidation state are perbromic acid and the perbromate ion, which were unknown prior to 1968 (see p. 782). Their subsequent synthesis has made their "nonexistence" somewhat less crucial as a topic of immediate concern to inorganic chemists, but bromine certainly continues the trend started by arsenic and selenium. Thus the perbromate ion is a stronger oxidizing agent than either perchlorate or periodate.

---

[41] See T. A. van der Knaap et al., "Phosphorus Chemistry," L. D. Quin and J. G. Verkade, eds., *ACS Symp. Ser. 171*, **1981**, Paper 82.

[42] K. Seppelt, *Z. anorg. Chem.*, **1977**, *434*, 5.

Anomalies of Groups IIIA and IVA

Before seeking an explanation of the reluctance of As, Se, and Br to exhibit maximum oxidation states, a related phenomenon will be explored. This involves a tendency for germanium to resemble carbon more than silicon. Some examples are:[43]

1. *Reduction of halides (X) with zinc and hydrochloric acid.* Germanium resembles carbon and tin resembles silicon:

$$>\text{C—X} \xrightarrow[\text{HCl}]{\text{Zn}} >\text{C—H} \tag{17.43}$$

$$>\text{Si—X} \xrightarrow[\text{HCl}]{\text{Zn}} \text{No} >\text{Si—H} \tag{17.44}$$

$$>\text{Ge—X} \xrightarrow[\text{HCl}]{\text{Zn}} >\text{Ge—H} \tag{17.45}$$

$$>\text{Sn—X} \xrightarrow[\text{HCl}]{\text{Zn}} \text{No} >\text{Sn—H} \tag{17.46}$$

2. *Hydrolysis of the tetrahydrides.* Silane hydrolyzes in the presence of catalytic amounts of hydroxide. In contrast, methane, germane, and stannane do not hydrolyze even in the presence of large amounts of hydroxide ion.

3. *Reaction of organolithium compounds with* $(C_6H_5)_3MH$. Triphenylmethane and triphenylgermane differ in their reaction with organolithium compounds from triphenylsilane and triphenylstannane:

$$\phi_3CH + LiR \longrightarrow LiC\phi_3 + RH \tag{17.47}$$

$$\phi_3SiH + LiR \longrightarrow \phi_3SiR + LiH \tag{17.48}$$

$$\phi_3GeH + LiR \longrightarrow LiGe\phi_3 + RH \xrightarrow{\phi_3GeH} \phi_3GeGe\phi_3 + LiH \tag{17.49}$$

$$\phi_3SnH + LiR \longrightarrow \phi_3SnR + LiH \tag{17.50}$$

4. *Alteration in enthalpies of formation.* There is a tendency for the enthalpies of formation of compounds of the Group IVA elements to alternate from C—Si—Ge—Sn—Pb. Although closely related to the previous phenomena, this variation is also related to the "inert pair effect" and will be discussed further below.

The elements of Group IIIA show similar properties, although, in general, the differences are not so striking as for Group IVA.[44] It may be noted that the covalent radius of gallium appears to be slightly smaller than that of aluminum in contrast to what might have been expected. The first ionization energies of the two elements are surprisingly close (578 and 579 kJ mol^{-1}), and if the sum of the first three ionization energies is taken, there is an alternation in the series: B = 6887, Al = 5044, Ga = 5521, In = 5084, Tl = 5439 kJ mol^{-1}.

---

[43] A. L. Allred and E. G. Rochow, *J. Inorg. Nucl. Chem.*, **1958**, *5*, 269.

[44] R. T. Sanderson, *J. Am. Chem. Soc.*, **1952**, *74*, 4792. This was the first paper to call attention to the anomalies discussed in this chapter and may be consulted for further details on several of the topics.

Table 17.3 Ionization energies of *s* electrons in kJ mol^{-1} (*eV*)

Element	$IE_2 + IE_3$	Element	$IE_3 + IE_4$
B	**6 090** (*63.1*)	C	**10 820** (*112.1*)
Al	**4 550** (*47.2*)	Si	**7 580** (*78.6*)
Ga	**4 940** (*51.2*)	Ge	**7 710** (*79.9*)
In	**4 520** (*46.9*)	Sn	**6 870** (*71.2*)
Tl	**4 840** (*50.2*)	Pb	**7 160** (*74.2*)

The "inert *s*-pair effect"

Among the heavy posttransition metals there is a definite reluctance to exhibit the highest possible oxidation state or the greatest covalence. Thus, although carbon is universally tetravalent except as transient carbene or methylene intermediates, it is possible to prepare divalent germanium, tin, and lead compounds. For example, if bulky substituents [R = $CH(SiMe_2)_2$] are present, the compounds R_2Ge, R_2Sn, and R_2Pb exist as diamagnetic monomers in solution, although there is a tendency to dimerize in the solid.[45] The molecular structure of the tin dimer has been determined and found to be in the *trans* conformation:

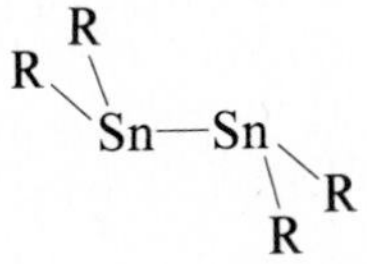

```
   R
R \
   Sn—Sn
        \ R
        R
```

Further examples of the stability of divalent compounds of the heavier congeners may be found in the oxidation state in principally ionic compounds. Thus in Group IVA, tin has a stable +2 oxidation state in addition to +4 and for lead the +2 oxidation state is far more important. Other examples are stable Tl^+ (Group IIIA) and Bi^{+3} (Group VA). These oxidation states, two less than the group number, have led to the suggestion that the pair of *s* electrons is "inert" and only the *p* electrons are employed in the bonding. It has even been suggested that the unreactivity of metallic mercury is a result of the fact that the only bonding electrons it has are the "inert 6*s* electrons." Although the term "inert pair" provides a convenient label for the phenomenon, it does little to promote understanding. Indeed, it has been pointed out that the pair of *s* electrons is certainly not stereochemically inert in the tin(II) halides—they are bent as expected for a divalent molecule with a lone pair. On the other hand, the *s* pair *does* seem to be stereochemically inert in $SbBr_6^{-3}$, $TeCl_6^{-2}$, and $TeBr_6^{-2}$ (see p. 217).

It can readily be shown that there is no exceptional stability of the *s* electrons in the heavier elements. Table 17.3 lists the ionization energies of the *s* electrons of the valence

[45] M. F. Lappert, in "Inorganic Compounds with Unusual Properties," R. B. King, ed., *Advances in Chemistry Series*, No. 150, American Chemical Society, Washington, D.C., **1976**, pp. 256–265. Also see P. J. Davidson et al., *J. Chem. Soc., Dalton*, **1976**, 2268.

Table 17.4 Enthalpies of dissociation, $MX_n \rightarrow MX_{n-2} + X_2$ in kJ mol^{-1}

Element	Fluorides	Chlorides	Bromides	Iodides
B		+301		
Al		+335		
Ga		+343		
In		+305		
Tl		+209		
Ge	+694	+381	+259	+167
Sn	+544	+276	+243	+142
Pb	+385	+121	+88	+17

shell of the elements of Groups IIIA and IVA. Although the 6*s* electrons are "stabilized" to the extent of ~300 kJ mol^{-1} (3 eV) over the 5*s* electrons, this cannot be the source of the "inert pair effect" since the 4*s* electrons of Ga and Ge have even greater ionization energies and these elements do not show the effect—the lower valence Ga^{I} and Ge^{II} compounds are obtained only with difficulty.

The pragmatic criterion of the presence or absence of an "inert pair effect" can be given as the tendency (or lack thereof) for the following reaction to proceed to the right:[46]

$$MX_{n(\text{solid})} \longrightarrow MX_{n-2(\text{solid})} + X_2 \qquad \textbf{(17.51)}$$

We might then inquire into the possibility of systematic variations in the lattice energies of either the higher or lower halides. Although such lattice effects could be responsible for differential stability of the solids, in those cases for which accurate data for the gas phase are available the same trends exist. In Table 17.4 are listed some enthalpies for the following reaction:

$$MX_{n(\text{gas})} \longrightarrow MX_{n-2(\text{gas})} + X_2 \qquad \textbf{(17.52)}$$

There is a tendency for the reaction to proceed more readily (although endothermic for all the compounds listed) with the heavier elements, especially thallium and lead.

The relative instability of the higher oxidation state for the heavier elements stems from a general weakening of the bonds (1) in the higher oxidation state and (2) with higher atomic number (Table 17.5). As a result of these two trends the difference between n weaker bonds and $n - 2$ slightly stronger bonds becomes less and less. Tending to promote dissociation is the driving force of the bond energy of the halogen molecule X_2.

Drago has provided an analysis of the factors important in the bonding which are responsible for observations (1) and (2) above.[47] He assumed the Pauling values for the electronegativity of the heavy elements: Ga = 1.6, In = 1.7, Tl = 1.8; Ge = 1.8, Sn = 1.8, Pb = 1.8. If the electronegativities of these elements are essentially constant, the ionic character of the bonds should be nearly constant, and hence any changes in bond energy are a result of changes in the covalent portion of the bonding.

[46] Although the tendency for lower oxidation states to be stable is found in many series of compounds in addition to the halides, the latter will be used in the following discussion since we have better thermodynamic data for them.

[47] R. S. Drago, *J. Phys. Chem.*, **1958**, *62*, 353.

Table 17.5 Bond energies of some group IVA halides in kJ mol^{-1}

Element	MF_2	MF_4	MCl_2	MCl_4	MBr_2	MBr_4	MI_2	MI_4
Si	—	565	—	380	—	310	—	230
Ge	481	452	385	354	326	275	264	218
Sn	481	414	386	323	329	273	262	205
Pb	394	330	304	240	260	200	205	140

The relatively small decrease in average bond energy upon going from the lower oxidation state to the higher oxidation state can be attributed to the promotion energy ($s^2p^n \rightarrow s^1p^{n+1}$), which is almost compensated by the improved bonding of hybrid orbitals. Data on promotion energies for the heavier elements are rather uncertain, but it does not appear that they are an important factor in the problem at hand other than causing all elements to have weaker bonds in the higher oxidation state compound. More important appears to be the inherent weakness of the bonding of the heavier elements. Drago ascribed this to two factors: (1) poorer overlap of the orbitals of the larger atoms; and (2) repulsion of inner electrons, most pronounced in Ga and Ge (the first posttransition metals) and Tl and Pb (the first postlanthanide posttransition metals).

The electronegativities of the elements of Group IVA

Before discussing the bonding in Groups IIIA and IVA further, it is necessary to digress momentarily to examine work on the electronegativity of the Group IVA elements that turns out to be intimately involved in the "inert pair effect." From various physical measurements[48] Allred and Rochow[49] began to suspect that the "conventional" order of electronegativities for the Group IVA elements was not correct. They suggested that a constant value[50] of 1.8 for the elements Si, Ge, Sn, and Pb did not adequately account for the properties of these elements. Consider the "anomalous" reactions of the Group IIIA and IVA elements discussed previously. The general tendency for germanium to parallel carbon indicates that the electronegativity of germanium is higher than that of silicon, making it more nearly approach that of carbon. Consider the hydrolysis of silane. Since only catalytic amounts of hydroxide ion are required, it is suggested that it forms a complex ion to effect the reaction and is then regenerated:

$$SiH_4 + 2OH^- \longrightarrow [SiH_4(OH)_2]^{-2} \tag{17.53}$$

Hydrogen is more electronegative than silicon and so will carry a negative charge in silane (which will be increased in the negative dihydroxo anion) and behave as a hydridic hydrogen

[48] One important source of data was the measurement of NMR chemical shifts which give an indication of the shielding of the nucleus of an atom by electrons. Electronegativity effects obviously are important in determining electron density about a nucleus, but unfortunately other factors are also involved in chemical shifts and it is impossible to assign unequivocally the role of electronegativity.

[49] A. L. Allred and E. G. Rochow, *J. Inorg. Nucl. Chem.*, **1958**, *5*, 269.

[50] L. Pauling, "The Nature of the Chemical Bond," 3rd ed., Cornell University Press, Ithaca, N.Y., **1960**.

reacting with a positive hydrogen from water:

$$\begin{array}{l} \;\;\delta^- \quad \delta^+ \quad \delta^- \\ \;\;\mathrm{H}\cdots\mathrm{H{-}OH} \\ \;\;\;| \\ \mathrm{H_3Si(OH)_2^{-2}} \\ \;\;\;\delta^+ \end{array} \longrightarrow \mathrm{H_2 + H_3Si(OH)_3^{-2}} \longrightarrow \text{Further hydrolysis} \qquad \textbf{(17.54)}$$

The lack of reactivity of germane and stannane could result from the absence of a $M^{\delta+}$—$H^{\delta-}$ dipole of sufficient magnitude to induce reaction, i.e., if the electronegativities of Ge and Sn are closer to that of hydrogen than that of Si. The behavior of the reactions of the triphenyl derivatives with lithium alkyls supports a similar interpretation: When bonded to a more electronegative carbon or germanium atom,[51] a hydrogen atom will be more acidic than when bonded to a less electronegative silicon or tin atom.[51]

Finally, Allred and Rochow examined the bond energies of the Group IVA halides (see Table 17.5). They noted that although the homopolar bond energies of Si—Si and Ge—Ge are almost identical (188 and 186 kJ mol^{-1}), Ge—X bonds are from 21 to 113 kJ mol^{-1} weaker than Si—X bonds. This difference in energy can be assigned to differences in electronegativity *only if germanium has a higher electronegativity than silicon and when bonded to electronegative halogens yields smaller ionic resonance energies.* By a series of Pauling-type calculations[52] they obtained the following electronegativities for the Group IVA elements (all sp^3 hybrids): C = 2.60 (assumed), Si = 1.91, Ge = 2.00, Sn = 1.94, Pb = 2.23.

Two of the Allred-Rochow experimental values[53] (Ge and Pb) for the electronegativities deserve special comment. That the electronegativity of germanium is greater than that of either silicon or tin is supported by three electronegativity scales: the Allred-Rochow electrostatic scale, the Sanderson relative compactness scale, and the Mulliken-Jaffé ionization energy scale.[54] The former two are based on size and reveal the results of the contraction across the first transition series. The latter, based on ionization and electron affinity energies, reflects the increased ionization energy resulting from poor shielding by d electrons. They are thus in essential agreement over the nature of the increase in electronegativity for germanium. This coupled with the experimental evidence given above makes most inorganic chemists accept the higher value for germanium.

The electronegativity of lead is still a matter of considerable disagreement. On all counts the data available for calculations for lead are poor. Exact values for its radius, effective atomic number, ionization energy, electron affinity, and bond energies are much more difficult to obtain than for the lighter elements. On the basis of other work, Allred and Rochow suggested an even higher value than that derived from bond energies, 2.45!

---

[51] The relative electronegativities of Ge and Sn vary somewhat but are obviously close to hydrogen and greater than silicon. Substituent effects could be responsible for these small variations.

[52] The interested reader is referred to the original paper for details of the calculations.

[53] These are the values obtained from bond energies and other experimental work described above and should not be confused with the Allred-Rochow electrostatic values. See p. 149.

[54] Although this scale has the soundest theoretical basis, in the present instance it is of limited usefulness since all of these elements except carbon utilize some (unknown) amount of d character and the d orbitals have low (unknown) electronegativities. The values for the sp^3 hybrids are all somewhat high therefore.

Many workers, including Drago,[55] could not accept such a high value (yet note that even the "conservative" Pauling scale gave Au a value of 2.4, 0.5 unit higher than Ag, and both Hg and Tl were given values higher than their lighter congeners). In support of a higher value for lead (not necessarily as high as 2.45) it may be pointed out that both the ionization energy and electron affinity (see Table 2.5) of lead are relatively high as a result of the addition of 14 poorly shielded protons across the lanthanide series.[56]

The controversy has not as yet been completely resolved. It is interesting to note (and to oversimplify somewhat) that Drago assumed the constancy of electronegativity (Si—Pb), the concomitant constancy of ionic bonding, and hence was forced to the conclusion that something was peculiar with the overlap and/or repulsions in the bonds of Ge and Pb. On the other hand, Allred and Rochow assumed that all of the bonds were more or less normal in behavior as far as the covalent contribution was concerned—hence differences in bonding reflected differences in ionic character and higher electronegativities for Ge and Pb.[57] By extension, this would ascribe the "inert pair effect" to weak bonds in the heavy metals because they are not stabilized by ionic character. It is interesting to note that both points of view agree on one point, an empirical observation: Those elements such as Ga, Ge, and As that follow the first transition series and those elements such as Tl, Pb, and Bi that follow the first lanthanide or inner transition series are affected by the increased effective nuclear charge and exhibit unusual properties. When exhibited by the fourth-row elements it may be referred to as a "reluctance to exhibit the maximum oxidation state (for As, Se, and Br)," and when shown by the sixth-row elements it may be termed the "inert pair effect." In both cases it probably rests ultimately on the effects of unshielded nuclear charge.[58]

This behavior has been generalized as follows. Those elements that follow the first filling of a given type of sublevel (*p*, *d*, *f*, etc.) will exhibit a lowered tendency to form stable compounds in their highest oxidation states.[59] This correlates the behavior of the post-lanthanide (first filling of *f* levels) elements (Hf to Rn) and the postscandide (first filling of *d* levels) elements (Ga to Kr). In addition, more subtle effects can also be found: Both sodium and magnesium form less stable compounds than would be expected on the basis of the behavior of their lighter (Li, Be) and heavier (K, Ca) congeners.[60] These elements are those that follow immediately after the first filling of a set of *p* orbitals (Ne), and the same effects of incomplete shielding (though less pronounced to be sure) presumably are operative

---

[55] R. S. Drago, *J. Inorg. Nucl. Chem.*, **1960**, *15*, 237. This was followed by a rebuttal from Allred and Rochow, and the debate has continued. The interested reader is urged to study this controversy as an excellent example of a situation in which intelligent and knowledgeable workers can take the same data and interpret them in diametrically opposite ways! See A. L. Allred and E. G. Rochow, *J. Inorg. Nucl. Chem.*, **1960**, *20*, 167; R. S. Drago and N. A. Matwiyoff, *J. Organometal. Chem.*, **1965**, *3*, 62.

[56] Unfortunately the spectroscopic data necessary to calculate promotion energies and valence state ionization energies and electron affinities are not available, so it is impossible to provide a Mulliken-Jaffé electronegativity.

[57] Both sets of authors have done a considerably better job of arguing their points of view than can possibly be represented in the space available here.

[58] For a discussion of these problems, see W. E. Dasent, "Nonexistent Compounds," Dekker, New York, **1965**, Chapters 5 and 6. For a discussion of the use of *d* orbitals in these elements, see D. S. Urch, *J. Inorg. Nucl. Chem.*, **1963**, *25*, 771.

[59] J. E. Huheey and C. L. Huheey, *J. Chem. Educ.*, **1972**, *49*, 227.

[60] R. S. Evans and J. E. Huheey, *J. Inorg. Nucl. Chem.*, **1970**, *32*, 777.

here as well as in the postlanthanide and postscandide elements. This principle has been used to predict some of the chemical properties of the superheavy transactinide elements.[59]

"Anomalous" ionization energies and electron affinities

Many introductory chemistry books give simple rules for remembering the periodic changes of ionization energies and electron affinities. The rules usually follow some modification of "Ionization energies and electron affinities increase as one moves to the right in the periodic chart; they decrease as one moves from top to bottom." These generalizations, as well as the shielding rules that account for the atomic behavior, were discussed in Chapter 2, along with some of the exceptions. Unfortunately for simplicity, the exceptions are somewhat more numerous than is generally realized. Many of the problems discussed in the preceding sections result from these "exceptions."

The horizontal behavior of atoms follows the general rule with good regularity as might be expected from adding a single proton at a time with expected monotonic changes in properties. We have already seen the exception of the inversion of the ionization potentials of the VA and VIA groups related to the stability associated with half-filled subshells. A similar inversion of electron affinities takes place, for the same reason, between groups IVA and VA.

The vertical exceptions to the generalizations are much more widespread: If we count every time that a heavier element has a higher ionization potential or higher electron affinity than its next lighter congener, we find that about one-third of the elements show "electron affinity anomalies"[61] and a somewhat higher fraction of the elements show "ionization energy anomalies." With such a high fraction of exceptions, one wonders why the rules were formulated as they were originally. The answer seems to lie in the lack of data available until recently; most of the good data were for familiar elements, such as the alkali metals and the halogens. For these main-group elements, with the exception of the lower electron affinity of fluorine resulting from electron–electron repulsion (and paralleled by oxygen and nitrogen), the rules work fairly well; however, the poorer shielding *d* and *f* electrons upset the simple picture. For the transition metals, higher ionization energies with increasing atomic number in a group are the *rule*, not the exception. As we have seen in the preceding discussion, this carries over somewhat into the posttransition elements, causing some of the problems associated with families IIIA and IVA.

The increased ionization energies of the heavier transition metals should not be unexpected by anyone who has had a modicum of laboratory experience with any of these elements. Although none of the coinage metals is very reactive, gold has a well-deserved reputation for being less reactive than copper or silver;[62] iron, cobalt, and nickel rust and corrode, but osmium, iridium, and platinum are noble and unreactive and therefore are used in jewelry; platinum wires are the material of choice for flame tests without contamination; and one generates hydrogen with zinc and simple acids, not with mercury.

---

[61] E. C. M. Chen and W. E. Wentworth (*J. Chem. Educ.*, **1975**, *52*, 486) have reviewed the experimental values of atomic electron affinities and plotted them periodically.

[62] The legend of *aqua regia* seems to persist even in the absence of student contact with this powerful elixir.

Conclusion

The periodic chart is the inorganic chemist's single most powerful weapon when faced with the problem of relating the physical and chemical properties of over 100 elements. In addition to knowing the general trends painted in broad brush strokes by the simple rules, the adept chemists should know something of the "fine structure" that is at the heart of making inorganic chemistry diverse and fascinating.

PROBLEMS

17.1. Compare Figs. 14.28 and 14.29. The difference in P—O bond lengths in P_4O_{10} is $162 - 139 = 23$ pm but only 13 pm in P_4S_{10}. Explain.

17.2. The separation of zirconium and hafnium has been of considerable interest because of the low neutron cross section of zirconium and the high neutron cross section of hafnium. Unfortunately, the separation of these two elements is perhaps the most difficult of any pair of elements. Explain why.

17.3. Carbon tetrachloride is inert toward water but boron trichloride hydrolyzes in moist air. Suggest a reason.

17.4. Discuss the possibility that there is an alternation of electronegativities in Group VA. [A. L. Allred and A. L. Hensley, Jr., *J. Inorg. Nucl. Chem.*, **1961**, *17*, 43.]

17.5. "Gallium dichloride," $GaCl_2$, is a diamagnetic compound that conducts electricity when fused. Suggest a structure.

17.6. If a major breakthrough in nuclear synthesis is achieved, two elements that are hoped for are those with atomic numbers 114 and 164, both congeners of lead. Look at the periodic chart on p. 817 and suggest properties (such as stable oxidation states) for these two elements. How do you suppose their electronegativities will compare with the other Group IVA elements? [G. T. Seaborg, *J. Chem. Educ.*, **1969**, *46*, 626; also see footnote 59.]

17.7. The small F—S—F bond angles in $F_3S{\equiv}N$ can be rationalized by
a. Bent's rule
b. Gillespie-type rules
c. Bent bonds
Discuss each and explain their usefulness (or lack) in the present case.

17.8. One of the challenges of teaching (or text writing) is the problem of presenting material (especially to a person new to the subject) in a simplifying and nondistracting way without oversimplifying (or even falsifying) the material for the sake of "helping" the student. Read the chapter or section on periodicity in your freshman text and criticize it in light of what you now know. Could you rewrite it to improve it?

17.9. It is sometimes said that nitrogen cannot form a pentafluoride because it has no *d* orbitals in its valence shell. C. T. Goetschel et al., [*Inorg. Chem.*, **1972**, *11*, 1696] believe they have synthesized NF_5. Discuss the bonding and structural possibilities.

17.10. Either look up the article by Chen and Wentworth[61] or plot the electron affinities from Table 2.5 onto a periodic chart. Discuss the reasons for the "exceptions" you observe.

17.11. Plot the first ionization energies from Table 2.4A onto a periodic chart. Discuss the reasons for the "exceptions" you observe.

17.12. On p. 843 the statement is rather casually made that "R_2Ge, R_2Sn, and R_2Pb exist as diamagnetic monomers in solution." What experiments must an inorganic chemist do to substantiate these statements? [See footnote 45.]

17.13. The compound R_4Sn_2 shown on p. 843 is diamagnetic. Draw out the most reasonable electronic structure for it, and compare it with the geometric structure. Discuss.

17.14. Lithium carbonate is often administered orally in the treatment of mania or depression or both. From what you have learned of the diagonal relationship of the periodic chart, predict one possible unpleasant side effect of lithium therapy.

17.15. Zinc is a much more reactive metal than cadmium, as expected from the discussion on pp. 847–848. Yet *both* are used to protect iron from rusting. How is this possible?

17.16. Sodium hypophosphite, NaH_2PO_2, has been suggested as a replacement of sodium nitrite, $NaNO_2$, as a meat preservative to prevent botulism. Draw the structure of each anion.

17.17. The simplest relationship between electronegativity and dipole moments is a linear one: The greater the difference in electronegativity, the greater the dipole. How can you reconcile this with the N—O and P—O dipoles cited in this chapter (p. 826)?

18

Inorganic chemistry in biological systems

The chemistry of life can ultimately be referred to two chemical processes: (1) the use of radiant solar energy to drive chemical reactions that produce oxygen and reduced organic compounds from carbon dioxide and water, and (2) the oxidation of the products of (1) with the production of carbon dioxide, water, and energy. Alternatively, living organisms have been defined as systems capable of reducing their own entropy at the expense of their surroundings (which must gain in entropy).[1] An important feature of living systems is thus their unique dependence upon kinetic stability for their existence. All are thermodynamically unstable—they would burn up immediately to carbon dioxide and water if the system came to thermodynamic equilibrium. Life processes depend upon the ability to restrict these thermodynamic tendencies by controlled kinetics to produce energy as needed.[2] Two important aspects of life will be of interest to us: (1) the ability to capture solar energy; (2) the ability to employ catalysts for the controlled release of that energy. Examples of such catalysts are the enzymes which control the synthesis and degradation of biologically important molecules. Many enzymes depend upon a metal ion for their activity. Metal-containing compounds are also important in the process of chemical and energy transfer, reactions which involve the transport of oxygen to the site of oxidation and various redox reactions resulting from its use.

ENERGY SOURCES FOR LIFE

It is somewhat surprising that most of the reactions for obtaining energy for living systems are basically inorganic. Of course, the reactions are mediated and made possible by complex

[1] These reductionist definitions of life are not meant to imply that life processes or living organisms are simplistic or any the less interesting. A similar definition of physics and chemistry might be "the study of the interactions of matter and energy." None of these definitions hints at the fascination of some of the problems presented by these branches of science.

[2] For a discussion of thermodynamics and kinetics in biological systems including the significance of oxidation–reduction energetics, see C. S. G. Phillips and R. J. P. Williams, "Inorganic Chemistry," Oxford University Press, London, **1965**, Vol. I, pp. 639–646.

biochemical systems. As we shall see later, isolated inorganic systems are only now reaching the point of challenging biological systems in terms of energy capture.

Nonphotosynthetic processes

Even though almost all living organisms depend directly (green plants) or indirectly (saprophytes and animals) upon photosynthesis to capture the energy of the sun, there are a few reactions, relatively unimportant in terms of scale but extremely interesting in terms of chemistry, utilizing *inorganic* sources of energy. Even these may be indirectly dependent upon photosynthesis, since it is believed that all free oxygen on earth has been formed by photosynthesis.

Chemolithotrophic[3] bacteria obtain energy from various sources. For example, *iron bacteria* produce energy by the oxidation of iron(II) compounds:

$$2Fe^{+2} \xrightarrow{[O]} Fe_2O_3 + \text{energy} \qquad \textbf{(18.1)}$$

Nitrifying bacteria are of two types, utilizing ammonia and nitrite ion as nutriments:

$$2NH_3 \longrightarrow 2NO_2^- + 3H_2O + \text{energy} \qquad \textbf{(18.2)}$$

$$NO_2^- \longrightarrow NO_3^- + \text{energy} \qquad \textbf{(18.3)}$$

Though they are photolithotrophs (Gr., *photos*, "light") and thus more closely related to the chemistry of normal photosynthesis (see p. 855), the *green sulfur bacteria* and the *purple sulfur bacteria* are included here to demonstrate the diverse bacterial chemistry based on sulfur paralleling the more common biochemistry involving water and oxygen. Light energy is used to split hydrogen sulfide into sulfur, which is stored in the cells, and hydrogen which forms carbohydrates, etc., from carbon dioxide.

To return to the chemolithotrophs, there are species of sulfur bacteria that obtain energy from the oxidation of various states of sulfur:

$$8H_2S \longrightarrow 8H_2O + S_8 + \text{energy} \qquad \textbf{(18.4)}$$

$$S_8 + 8H_2O \longrightarrow 8SO_4^{-2} + 16H^+ + \text{energy} \qquad \textbf{(18.5)}$$

These latter reactions are the source of energy for a unique fauna, one completely isolated from the sun on the floor of the oceans. Recently it has been discovered that at certain rifts in the earth's crust on the ocean's floor, large amounts of sulfide minerals are spewed forth from hydrothermal vents.[4] The source of energy for the community based on the hydrothermal vents is the hydrogen sulfide, which is oxidized by bacteria as shown above. It is of considerable interest that the enzymes, mechanisms, and products of this chemically driven synthesis are essentially identical to that of photosynthesis (see p. 855), except that

[3] That is, feeding (Gr. *trophos*) on inorganic (Gr. *lithos* = stone) *chem*icals.

[4] F. N. Spiess et al., *Science*, **1980**, *207*, 1421; R. Heikinian et al., ibid., **1980**, *207*, 1433; J. M. Edmond et al., *Nature*, **1982**, *297*, 187.

the source of electrons for the reduction of water to carbohydrates is sulfur(−II) rather than photoactivated chlorophyll. Whether this parallelism results from an adaptation of the cycle in photosynthetic bacteria, or whether these chemolithotrophic bacteria are possibly ancestral to photosynthetic organisms presents the age-old phylogenetic problem—which came first, the chicken or the egg? These bacteria often live in close symbiosis with larger organisms, forming the basis of a food chain consisting of several species, including giant tube worms, clams, and crabs, none of which are dependent upon photosynthesis for survival except for use of by-product oxygen.[5] There is even evidence[6] that some animals such as pogonophoran worms can use hydrogen sulfide directly as their energy source.

Photosynthesis

The photosynthetic process in green plants consists of splitting the elements of water, followed by reduction of carbon dioxide:

$$2H_2O \longrightarrow [2H_2] + O_2 \qquad (18.6)$$

$$CO_2 + [2H_2] \longrightarrow \frac{1}{x}(CH_2O)_x + H_2O \qquad (18.7)$$

where $[2H_2]$ does not imply free hydrogen but the reducing capacity formed by the oxidation–reduction of water. The details of the reactions involved in photosynthesis are not known, although the broad outlines are fairly clear.[7] In green plants there are two photosynthetic systems. The two differ in the type of chlorophyll present and in the accessory chemicals for processing the trapped energy of the photon. Chlorophyll is the first example we have encountered of a very important group of bioinorganic compounds containing metals, the metalloporphyrins.

METALLOPORPHYRINS, PHOTOSYNTHESIS, AND RESPIRATION

The porphyrin ring system

Perhaps the most important class of metal-containing compounds in biological systems is that comprising complexes between metal ions and porphyrin ligands. The latter are macrocyclic tetrapyrrole systems with conjugated double bonds (Fig. 18.1) and various groups attached to the perimeter. We shall not be concerned with the nature and variety of these substituents except to note that by their electron-donating or electron-withdrawing

[5] J. B. Corliss and R. B. Ballard, *Nat. Geogr.*, **1977**, *152*, 441; R. D. Ballard and J. F. Grassle, **1979**, *154*, 689.

[6] M. L. Jones, *Science*, **1981**, *213*, 333; H. Felbeck, ibid., **1981**, *213*, 336; G. H. Rau, ibid., **1981**, *213*, 338; C. M. Cavanaugh et al., ibid. **1981**, *213*, 340.

[7] H. R. Mahler and E. H. Cordes, "Biological Chemistry," 2nd ed., Harper & Row, New York, **1971**, Chapter 12; D. G. Bishop and M. L. Reed, *Photochem. Photobiol. Rev.*, **1976**, *1*, 1.

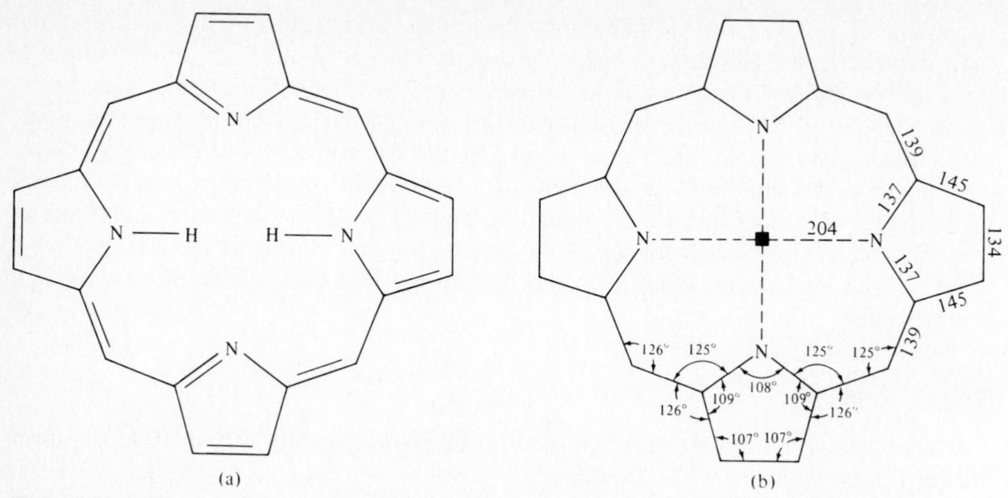

Fig. 18.1 (a) The porphin molecule. Porphyrins have substituents at the eight pyrrole positions. (b) A "best" set of parameters for an "average" porphyrin skeleton. Distance in pm. [From E. B. Fleischer, *Acc. Chem. Res.*, **1970**, *3*, 105. Reproduced with permission.]

ability they can "tune" the delocalized molecular orbitals of the complex and thus vary the properties of the complex. The porphyrins can accept two hydrogen ions to form the +2 diacid or donate two protons and become the −2 anion. It is in the latter form that the porphyrins complex with metal ions, usually dipositive, to form metalloporphyrin complexes.

From covalent bond radii (Table 6.1) we can estimate that a bond between a nitrogen atom and an atom of the first transition series should be about 200 pm long. The size of the "hole" in the center of the porphyrin ring is ideal for accommodating metals of the first transition series (Fig. 18.1b). The porphyrin system is fairly rigid and the metal–nitrogen bond distance does not vary greatly from 193–196 pm in nickel porphyrins to 210 pm in high-spin[8] iron(II) porphyrins.[9] The rigidity of the ring derives from the delocalization of the π electrons in the pyrrole rings. Nevertheless, if the metal atom is too small, as in nickel porphyrinates, the ring becomes ruffled to allow closer approach of the nitrogen atoms to the metal.[10] At the other extreme, if the metal atom is too large, it cannot fit into the "hole" and sits above the ring (see p. 875).

A group of related compounds, the phthalocyanines (Fig. 18.2), are isoelectronic with the porphyrins. They are of interest not only as model compounds for the biologically important porphyrins but also because the intensely colored metal complexes are of commercial importance as dyes and pigments.[11]

---

[8] High-spin metal atoms and ions are larger than those of low spin (see Table 3.4) because of repulsion of the ligands by electrons occupying antibonding orbitals directed at the ligands.

[9] E. B. Fleischer, *Acc. Chem. Res.*, **1970**, *3*, 105.

[10] J. L. Hoard, *Ann. N. Y. Acad. Sci.*, **1973**, *206*, 18.

[11] A. B. P. Lever, *Adv. Inorg. Chem. Radiochem.*, **1965**, *7*, 27.

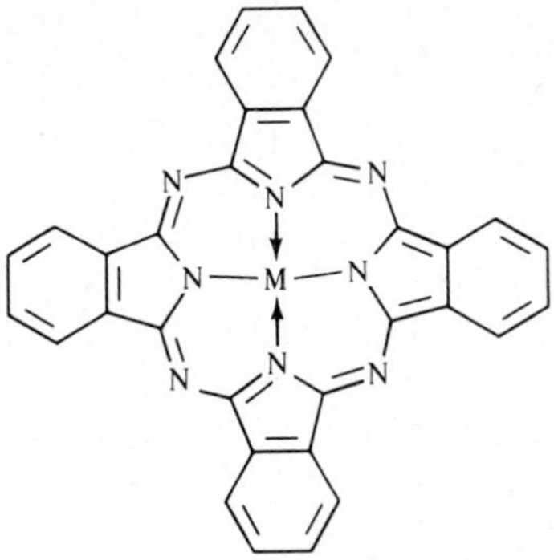

Fig. 18.2 Molecular structure of a metal phthalocyanine complex.

Recently "molecular metals" (see p. 694) have been prepared from stacked metal phthalocyanines.[12] The "hole" in the phthalocyanine ring is about 10 pm smaller than in the porphyrins as a result of the smaller size of the nitrogen atom compared with carbon.

The order of stability of complexes of porphyrins with +2 metal ions is that expected on the basis of the Irving-Williams series (see p. 316) except that the square planar ligand favors the d^8 configuration of Ni^{+2}. The order is $Ni^{+2} > Cu^{+2} > Co^{+2} > Fe^{+2} > Zn^{+2}$. The kinetics of formation of these metalloporphyrins have also been measured and found to be in the order $Cu^{+2} > Co^{+2} > Fe^{+2} > Ni^{+2}$.[13] If this order holds in biological systems, it poses interesting problems related to the much greater abundance of iron porphyrins (see below) than cobalt porphyrins. What might have been the implications for the origin and evolution of biological systems if the natural abundance of iron were not over a thousandfold greater than that of cobalt?

Two metalloporphyrins and a related metallocorrin will be discussed to illustrate the scope of the structures and functions of these molecules. One of the most widespread, both in point of view of occurrence in a variety of organisms and a variety of functions, is *heme*, an iron(II) porphyrin. Chlorophyll, present in one form or another in green plants, is a magnesium complex containing a modified ring system. Finally, vitamin B_{12} is a related cobalt complex of somewhat different structure. The reason for the importance of porphyrin complexes in a variety of biological systems is probably twofold: (1) Metalloporphyrins are biologically accessible compounds whose functions can be varied by changing the metal, its oxidation state, or the nature of the organic substituents on the porphyrin structure; (2) it is a general principle that evolution tends to proceed by modifying structures and functions that are already present in the organism rather than producing new ones *de novo*.

Chlorophyll

The chlorophyll ring system is a porphyrin in which a double bond in one of the pyrrole rings has been reduced. A fused cyclopentanone ring is also present (Fig. 18.3). Chlorophyll absorbs low-energy light in the far red region (~700 nm). The exact frequency depends on

[12] C. J. Schramm et al., *J. Am. Chem. Soc.*, **1980**, *102*, 6702.

[13] D. J. Kingham and D. A. Brisbin, *Inorg. Chem.*, **1970**, *9*, 2034.

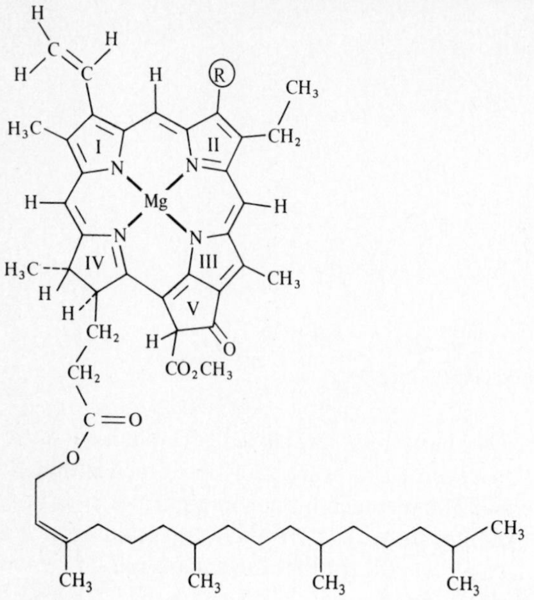

Chlorophyll *a* (R = CH_3)
Chlorophyll *b* (R = CHO)

Fig. 18.3 Structure of chlorophyll. The long alkyl chain at the bottom is the phytyl group mentioned in the text.

the nature of the substituents on the chlorophyll. In addition, other pigments such as carotenoids are present which absorb higher energy light. Such absorption serves a twofold function: (1) The energy may be passed along to the chlorophyll system and used in photosynthesis; and (2) it protects the biological system from photochemical damage.[14]

A photon of light hitting a molecule of chlorophyll in either of the photosynthetic systems mentioned above provides the energy for a series of redox reactions (see Fig. 18.4). System I produces a moderately strong reducing species (RED_I) and a moderately strong oxidizing species (OX_I).[15] System II provides a stronger oxidizing agent (OX_{II}) but a weaker reducing agent (RED_{II}).

OX_{II} is responsible for the production of molecular oxygen. A manganese complex, probably containing two atoms of manganese per molecule, reduces OX_{II} which is recycled for use by another excited chlorophyll. In the redox reactions the manganese cycles between Mn(II) and Mn(III), and Mn(IV) may also take part.[16] A suggested scheme for this redox chemistry is shown in Fig. 18.5.

[14] See C. S. Foote et al., *J. Am. Chem. Soc.*, **1970**, *92*, 5216, 5218, and references therein.

[15] The terms "strong," "stronger," and "weaker" are only relative. In system I, both *emfs* have an absolute value of about $\frac{1}{2}$ V.

[16] D. T. Sawyer and M. E. Bodini, *J. Am. Chem. Soc.*, **1975**, *97*, 6588; D. T. Sawyer et al., "Bioinorganic Chemistry—II," K. N. Raymond, ed., *ACS Adv. Chem. Ser. No. 162*, **1977**, Chapter 19; K. Sauer, *Acc. Chem. Res.*, **1980**, *13*, 249; J. Livorness and T. D. Smith, *Struct. Bond.*, **1982**, *48*, 1.

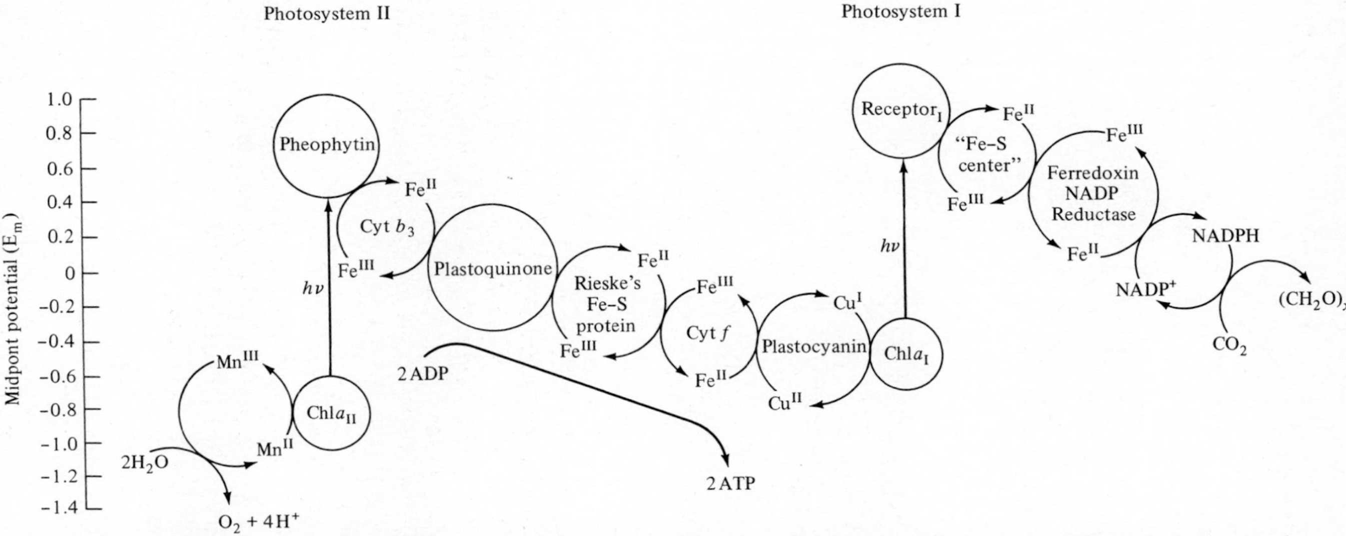

Fig. 18.4 Electron flow in photosystems I and II ("Z-scheme"). Vertical axis gives mid-point redox potential with reducing species (top) and oxidizing species (bottom). [Modified from J. Livorness and T. D. Smith, *Struct. Bond.*, **1982**, *48*, 1.]

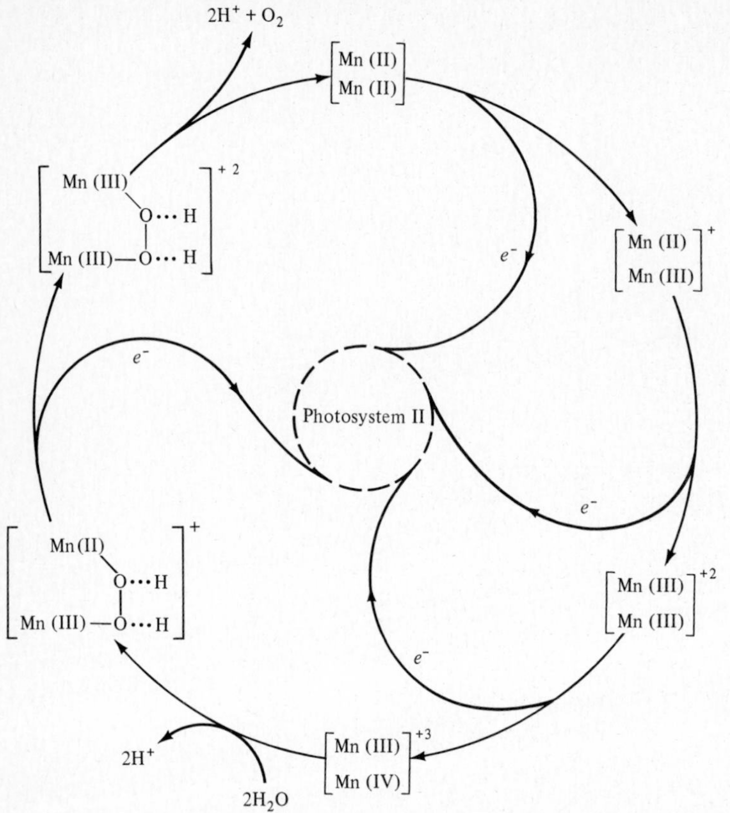

Fig. 18.5 Suggested scheme for the manganese-mediated oxidation of water to free molecular oxygen in photosynthesis (Photosystem II).

RED_I transfers its electron through carriers such as ferredoxin (Fe^{+2}/Fe^{+3}), eventually to form NADPH.[17] The latter serves as a stable source of reducing capacity to convert carbon dioxide to carbohydrate.

The acceptor in photosystem II is pheophytin. The initial product of photon excitation, the Chl a^+ $Pheo^-$ radical-ion pair, has been observed, although its lifetime is only 10 ns.[18] Actually this has been described as a dramatically *long* time and is the single most unusual aspect of the PSI or PSII reaction centers. OX_I and RED_{II} react with each other to complete the regeneration of the original species present prior to irradiation. The electron flows through plastoquinone (a purely organic compound), a series of cytochromes (see p. 864), iron-sulfur compounds, and plastocyanin (Cu^+/Cu^{+2}). Two moles of energy-rich ATP[19] are synthesized from the energy released in the process.

[17] NADPH is reduced nicotinamide adenine dinucleotide phosphate, a key reducing species in biochemical reactions. In photosynthesis it yields carbohydrate.

[18] J. R. Norris et al., *Proc. Nat. Acad. Sci. U.S.A.*, **1982**, in press.

[19] Adenine triphosphate, an important energy-rich species in metabolism. See p. 916.

Based on our knowledge of the structure of chlorophyll as well as the results of studies on the photo behavior of chlorophyll *in vitro*, it is possible to summarize some of the features of the chlorophyll system which enhance its usefulness as a pigment in photosynthesis.[20] First, there is extensive conjugation of the porphyrin ring. This lowers the energy of the electronic transitions and shifts the absorption maximum into the region of visible light. Conjugation also helps make the ring rigid, and thus less energy is wasted in internal thermal degradation (via molecular vibrations).

The maximum intensity of light *reaching the earth's surface* is in the visible region; ultraviolet light is absorbed in the earth's atmosphere by species such as oxygen and ozone, infrared light is absorbed by carbon dioxide, water, etc. The absorption spectra of photosynthetic systems fall nicely within that portion of the sun's spectrum that reaches the earth. Some of the light not absorbed by the chlorophyll itself is made available for use by accessory pigments. An unrelated system that also shows considerable adaptation to the "window" provided by the earth's atmosphere is implied in the term "visible region." This is the octave of the electromagnetic spectrum from about 380 to 760 nm. Outside these limits the intensity drops to less than one-half of the maximum intensity. The region of relatively high intensity (75% of maximum) lies between about 420 and 750 nm, which corresponds closely to the sensitivity of the human eye.

A second important factor is the phosphorescent behavior of chlorophyll. In order for phosphorescence to occur there must be an excited state with a finite lifetime. If such an excited state is available, then there is time for a chemical reaction to take place to take advantage of the energy prior to phosphorescence. If fluorescence occurs exclusively, the energy will be lost via an immediate transition and will not be available for chemical use. The presence of a metal atom is necessary in order that phosphorescence take place. The free porphyrins show only fluorescent emission. Spin-orbit coupling by the metal ion allows mixing of the excited singlet and triplet states and promotes the formation of the relatively stable triplet state, which is the source of the phosphorescence (and of the energy for photosynthesis).[21]

It is of interest to note that a completely synthetic model that simulates the photosynthetic process of chlorophyll has been devised.[22] The photon-capturing pigment is zinc tetraphenylporphyrin deposited on a clean aluminum surface (Fig. 18.6). The electron carriers in solution are potassium ferricyanide and potassium ferrocyanide. The zinc tetraphenylporphyrin was activated by orange light, and the captured energy was used to reduce NADP and oxidize water to oxygen gas.

Studies of a "synthetic leaf," similar to that shown in Fig. 18.6, but utilizing natural chlorophyll have provided further insight into the photosynthetic process. A chlorophyll–water adduct is either impregnated in a polymer membrane or deposited on one side of a metal foil. The membrane or foil is then placed in the cell with an electron-accepting solution on one side and an electron-donating solution on the other side. Irradiation with light produces a potential.[23] It is postulated that *in vivo* photosynthesis involves hundreds of "antenna" molecules that capture photons and pass the energy along to the "reactive

[20] G. M. Maggiora and L. L. Ingraham, *Struct. Bonding*, **1967**, *2*, 126.

[21] See footnote 20; also J. C. Hindman et al., *Proc. Nat. Acad. Sci. U.S.*, **1977**, *74*, 5.

[22] J. H. Wang, *Proc. Nat. Acad. Sci. U.S.*, **1969**, *62*, 653.

[23] *Chem. Eng. News*, Feb. 16, **1976**, pp. 32–34.

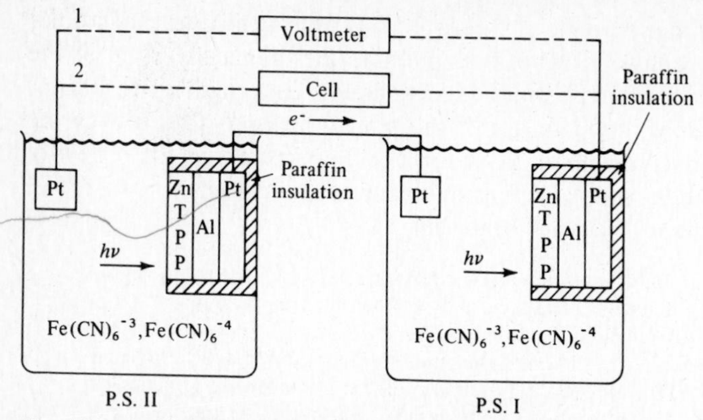

Fig. 18.6 Artificial *in vitro* photosynthetic model system using zinc tetraphenylporphyrin (ZnTPP) as the photosynthetic pigment. (1) If the circuit shown is closed to the voltmeter, a potential of 2.2–2.6 V can be obtained. (2) If the circuit is closed to an electrolytic cell containing NADP and NADP-reductase, the NADP is reduced and oxygen gas is evolved. In practice, a microcell is used (footnote 22).

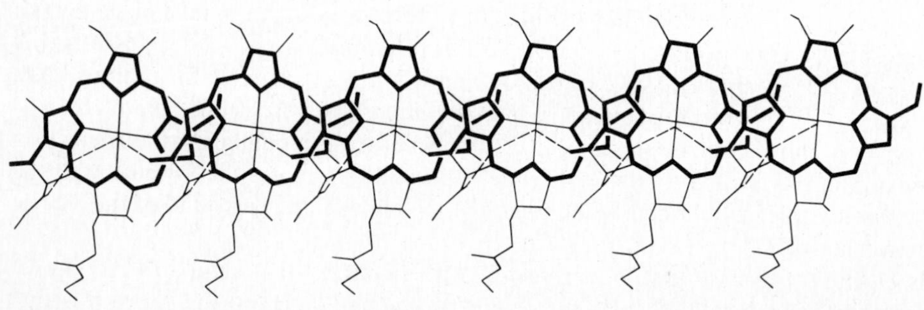

Fig. 18.7 One-dimensional chain found in crystalline ethyl chlorophyllide $a \cdot 2H_2O$. The *heavy lines* indicate the conjugated portion of each molecule. The *dashed lines* represent hydrogen bonds. The molecules are twisted almost 90° counterclockwise with respect to Fig. 18.3. [From H-C. Chow et al., *J. Am. Chem. Soc.*, **1975**, *97*, 9230. Reproduced with permission.]

center." There is no universal agreement on the structures of the antenna molecules or the reactive center. Some workers[24] have postulated a hydrophobic environment for the chlorophyll and have therefore studied chlorophyll in hydrocarbon solvents. They postulated coordination of the carbonyl group in ring V (see Fig. 18.3) to the magnesium atom in an adjacent molecule. An alternative structure has been formulated on the basis of the molecular structure of ethyl chlorophyllide *a* dihydrate, a chlorophyll molecule in which the phytyl side chain has been replaced by an ethyl group. One water molecule is co-

[24] K. Ballschmiter and J. J. Katz, *J. Am. Chem. Soc.*, **1969**, *91*, 2661; J. J. Katz and J. R. Norris, *Curr. Topics BioEnerg.*, **1973**, *5*, 41; M. R. Wasielewski et al., *Proc. Nat. Acad. Sci. U.S.A.*, **1976**, *74*, 4282; J. J. Katz et al., *Proc. Brookhaven Symp. No. 28*, **1976**, pp. 16–55.

ordinated to the magnesium atom and forms a hydrogen bond with the ring V keto group of an adjacent molecule, which, in turn, has a magnesium-coordinated water molecule hydrogen-bonded to a third molecule, and so on. This forms a one-dimensional array of partially overlapping chlorophyll molecules that is not unlike a row of dominoes that has been toppled (Fig. 18.7). It is this one-dimensional array that has been suggested as a model for the aggregation in the antenna system.[25] In the crystalline ethyl chlorophyllide *a* dihydrate, the second water molecule forms hydrogen bonds with the magnesium-coordinated water molecule, the methyl ester of the same chlorophyll molecule, and the ethyl ester in another strand—although the latter interaction presumably would not take place in antenna systems. The nature of the reactive site is also a matter of some disagreement, though it is generally agreed that two chlorophyll molecules and two water molecules are involved. Two proposed structures are shown in Fig. 18.8.[26,27,28]

Cytochromes

One of the important classes of electron carriers involved in photosynthesis is comprised of the various cytochromes. The active center of the cytochromes is a *heme* group (Fig. 18.9). It consists of a porphyrin ring chelated to an iron atom. The oxidation state [29] of the iron may be either +2 or +3, and the importance of the cytochromes lies in their ability to act as redox intermediates in electron transfer. They are present not only in the chloroplasts for photosynthesis but also in mitochondria to take part in the reverse process of respiration.

The heme group in cytochrome *c* has a polypeptide chain attached and wrapped around it (Fig. 18.10). This chain contains a variable number of amino acids, ranging from 103 in some fish and 104 in other fish and terrestrial vertebrates to 112 in some green plants. A nitrogen atom from a histidine segment and a sulfur atom from a methionine segment of this chain are coordinated to the fifth and sixth coordination sites of the iron (Fig. 18.11).[30] Thus, unlike the iron in hemoglobin and myoglobin (see below), there is no position for further coordination. Cytochrome *c* therefore cannot react by simple coordination but must react indirectly by an electron-transfer mechanism. It can reduce the oxygen and transmit its oxidizing power toward the burning of food and release of energy in respiration (the reverse process to complement photosynthesis).

The importance of cytochrome *c* in photosynthesis and respiration indicates that it is probably one of the oldest (in terms of evolutionary history) of the chemicals involved in biological processes. An interesting "family tree" of the evolution of living organisms can be constructed from the differences in amino acid sequences in the peptide chains between the various types of cytochrome *c* found, for example, in yeasts, higher plants, insects, and

---

[25] H-C. Chow et al., *J. Am. Chem. Soc.*, **1975**, *97*, 7230.

[26] F. K. Fong et al., *J. Am. Chem. Soc.*, **1976**, *98*, 6406.

[27] S. G. Boxer and G. L. Closs, *J. Am. Chem. Soc.*, **1976**, *98*, 5406; L. L. Shipman et al., *Proc. Nat. Acad. Sci., U.S.*, **1976**, *73*, 1791; M. R. Wasielewski et al., *J. Am. Chem. Soc.*, **1977**, *99*, 4172.

[28] For a review of chlorophyll and photosynthesis, see C. E. Strouse, *Progr. Inorg. Chem.*, **1976**, *21*, 159.

[29] The term *heme* usually refers to the neutral group containing Fe(II). When oxidized to Fe(III), there will be a net positive charge and an associated anion, often the chloride ion. When oxidized, the term *hemin* is applied, as in hemin chloride. *Hematin*, long thought to be "hemin hydroxide," is actually a μ-oxo dimer (see p. 870).

[30] For the complete structures of ferrocytochrome *c* (Fe^{+2}) and ferricytochrome *c* (Fe^{+3}), see T. Takano et al., *J. Biol. Chem.*, **1977**, *252*, 776.

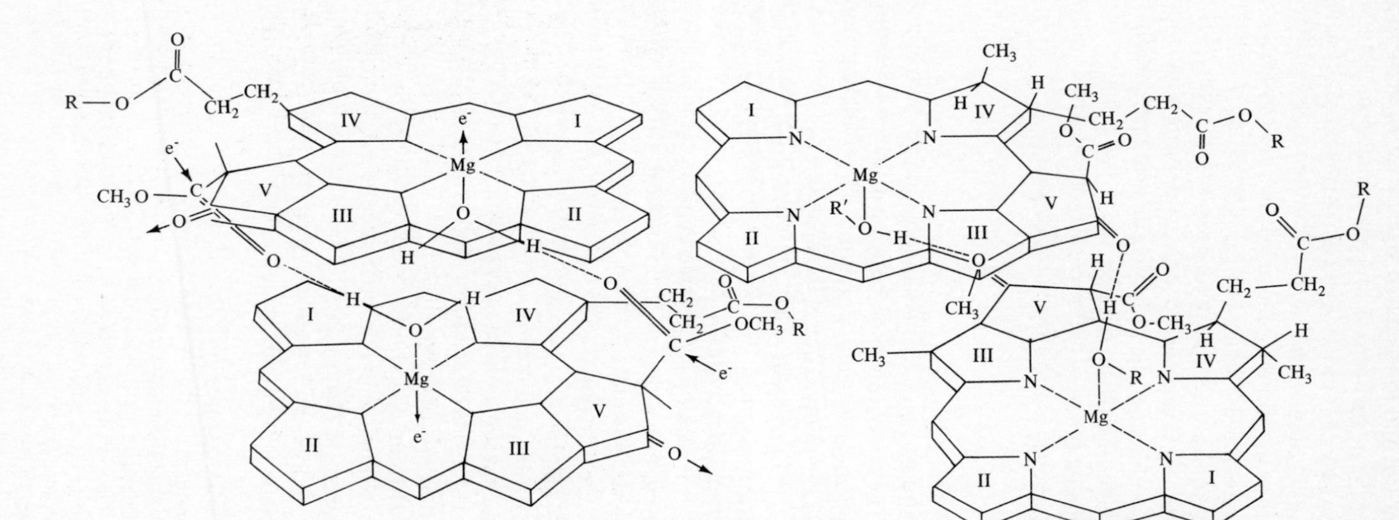

Fig. 18.8 Proposed structures of the reactive center (chlorophyll dimer). Left: Structure proposed by Fong and co-workers. [Modified from F. K. Fong et al., *J. Am. Chem. Soc.*, **1976**, *98*, 6406. Reproduced with permission.] Right: Structure proposed by Katz and co-workers. [From L. L. Shipman et al., *Proc. Nat. Acad. Sci., U.S.*, **1976**, *73*, 1791. Reproduced with permission.]

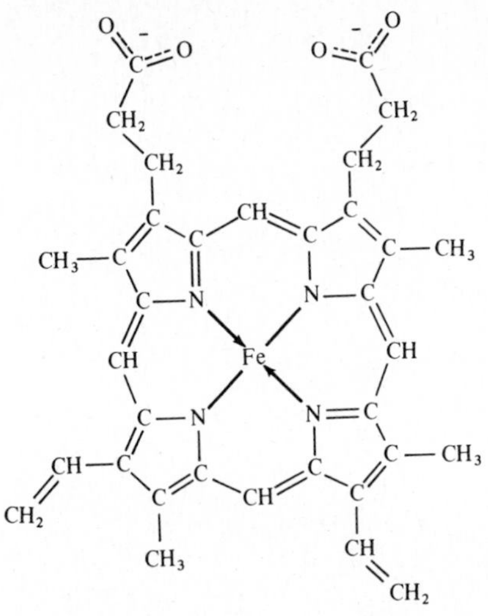

Fig. 18.9 The structure of the heme group.

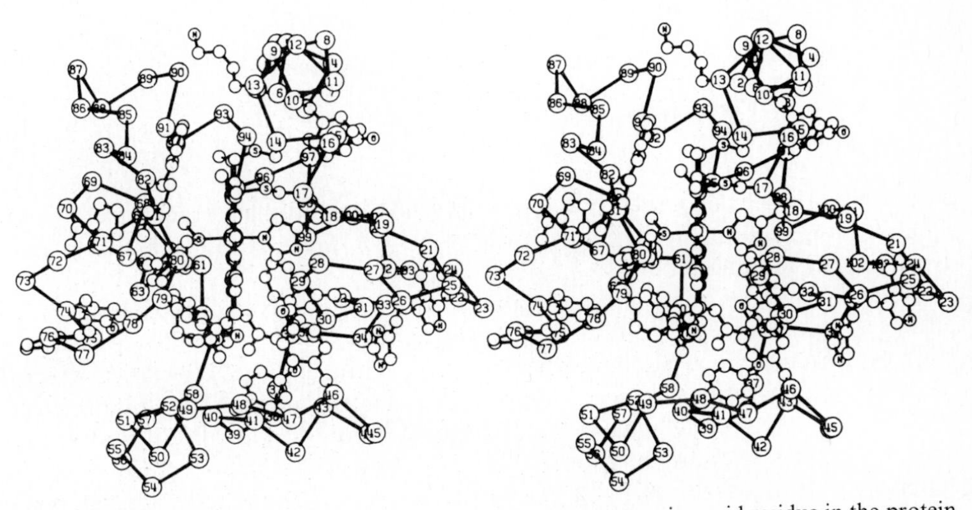

Fig. 18.10 The cytochrome *c* molecule. Each number represents an amino acid residue in the protein chain. The heme group, seen edge on, is located vertically between the sulfur atom above number 61 and the nitrogen atom to the left of number 99. Note the complete coordination sphere of the iron ion (see also Fig. 18.11) as well as the protection afforded by the encircling protein chains. [From T. Takano et al., *J. Biol. Chem.*, **1977**, *252*, 776. Reproduced with permission.]

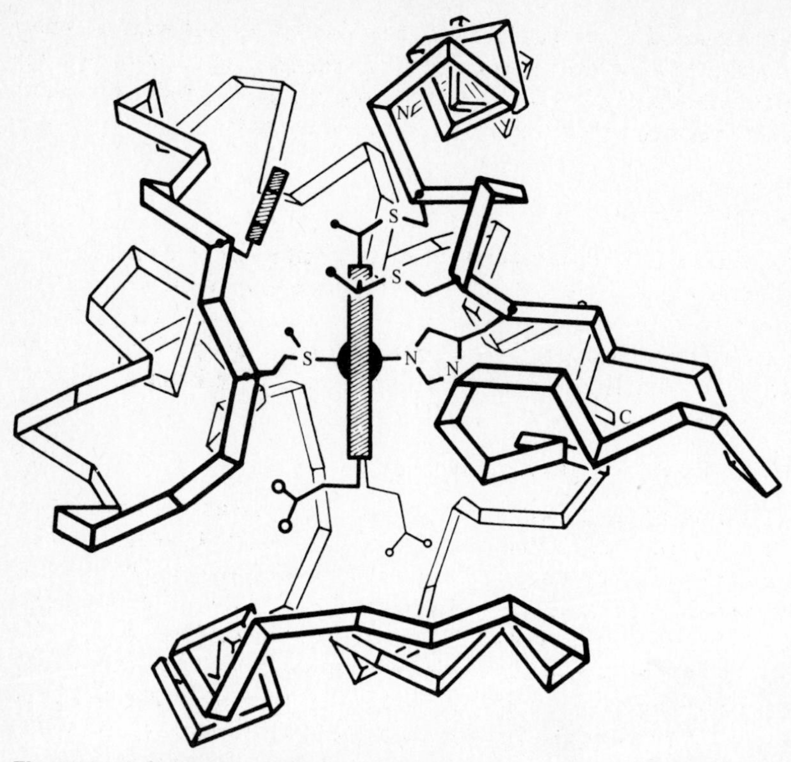

Fig. 18.11 Schematic view of cytochrome *c*. The heme group is viewed edge on, with the iron atom (large black atom) coordinated to a sulfur atom from a methionine residue and a nitrogen atom from a histidine residue. [Courtesy of R. E. Dickerson.]

man. Despite these differences, however, it should be noted that cytochrome *c* is evolutionarily conservative—cytochrome *c* from plants will react in animal systems and vice versa, though at reduced rates.[31]

There is quite a variety of cytochromes, most of which have not been as well characterized as cytochrome *c*. Depending upon the ligands present, the redox potential of a given cytochrome can be tailored to meet the specific need in the electron transfer scheme, whether in photosynthesis or in respiration. The potentials are such that the electron flow is $b \rightarrow c \rightarrow a \rightarrow O_2$. At least some of the *a* type are thought to be capable of binding oxygen (and reducing it). They are thus normally 5-coordinate in contrast to cytochrome *c*. They account for the unusual toxicity of the cyanide ion, CN^-. The latter binds tightly to the sixth position and stabilizes the Fe(III) to such an extent that it can no longer be readily reduced and take part in the electron shuttle. It is of some interest that cyanide, like the isoelectronic carbon monoxide molecule, binds to hemoglobin, but the inhibition of cytochrome *a* by the cyanide ion is much more serious than the interference with oxygen

[31] See R. E. Dickerson and I. Geis, "The Structure and Action of Proteins," Harper & Row, New York, **1969**, pp. 62–66; R. E. Dickerson, *Sci. Amer.*, April, **1972**, pp. 58–72.

transport. In fact, the standard treatment for cyanide poisoning is inhalation of amyl nitrite or injection of sodium nitrite to oxidize some of the hemoglobin to methemoglobin (see p. 870). The latter, although useless for oxygen transport, binds cyanide even more tightly than hemoglobin or cytochrome *a* and removes it from the system.[32]

Ferredoxins and rubredoxins[33]

There are several nonheme iron–sulfur proteins that are involved in electron transfer. We have seen that one, ferredoxin, is the first electron acceptor in photosynthesis. The iron–sulfur proteins have received considerable attention in the last few years.[34] They contain distinct iron–sulfur clusters composed of iron atoms, sulfhydryl groups from cysteine residues, and "inorganic" or "labile" sulfur atoms or sulfide ions. The latter are readily removed by washing with acid:

$$(RS)_4Fe_4S_4 + 8H^+ \longrightarrow (RS)_4Fe_4^{-4} + 4H_2S \qquad \textbf{(18.8)}$$

The cysteine moieties are incorporated within the protein chain and are thus not labile. The clusters are of several types. The simplest is bacterial rubredoxin, $(Cys{-}S)_4Fe$ (often abbreviated Fe_1S_0, where S stands for inorganic sulfur) and contains only nonlabile sulfur. It is a bacterial protein of uncertain function with a molecular weight of about 6000. The single iron atom is at the center of a tetrahedron of four cysteine ligands (Fig. 18.12a). The cluster in the ferredoxin molecule associated with photosynthesis in higher plants is thought to have the bridged structure Fe_2S_2 shown in Fig. 18.12b. The most interesting cluster is found in certain bacterial ferredoxins involved in anaerobic metabolism. It consists of a cubane-like cluster of four iron atoms, four labile sulfur atoms, thus Fe_4S_4, and four cysteine ligands (Fig. 18.12c).

Because of the inherent chemical interest in clusters of this sort as well as their practical significance to biochemistry, there has been considerable effort expended in making model compounds for study as well as making theoretical calculations[35] of their properties. Three examples are shown in Figs. 18.13–18.15. These model compounds allow direct experimentation on the cluster in the absence of the protein chain.[36]

Bio-redox agents and mechanisms

Perhaps the three most important redox systems in bioinorganic chemistry are: (1) high-spin, tetrahedral Fe(II)/Fe(III) in rubredoxin, ferredoxin, etc. (see above); (2) low-spin, octahedral Fe(II)/Fe(III) in the cytochromes (see p. 861); and (3) pseudotetrahedral

[32] See R. P. Hanzlik, "Inorganic Aspects of Biological and Organic Chemistry," Academic Press, New York, **1976**, p. 152; E-I. Ochiai, "Bioinorganic Chemistry," Allyn & Bacon, Boston, **1977**, p. 483.

[33] A. Bezkorovainy, "Biochemistry of Nonheme Iron," Plenum, New York, **1980**, Chapter 8.

[34] W. H. Orme-Johnson, *Ann. Rev. Biochem.*, **1973**, *42*, 159; L. H. Jensen, ibid., **1974**, *43*, 461; W. Lovenberg, "Iron-Sulfur Proteins," Vols. 1 & 2, Academic Press, New York, **1973**; R. H. Holm, *Endeavour*, **1975**, *34*, 38.

[35] C. Y. Yang et al., *J. Am. Chem. Soc.*, **1975**, *97*, 6596; R. A. Bair and W. A. Goddard, ibid., **1977**, *99*, 3505.

[36] I. Bernal et al., *J. Coord. Chem.*, **1972**, *2*, 61; L. Que et al., *J. Am. Chem. Soc.*, **1974**, *96*, 4168; J. J. Mayerle et al., ibid., **1975**, *97*, 1032; W. O. Gillum et al., *Inorg. Chem.*, **1976**, *15*, 1095; R. W. Lane et al., *Proc. Nat. Acad. Sci., U.S.*, **1975**, *72*, 2868.

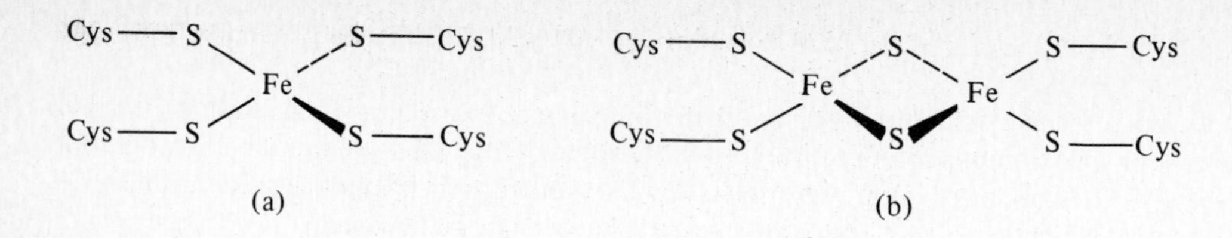

Fig. 18.12 Iron clusters in ferredoxins: (a) Fe_1S_0 in bacterial rubredoxin; (b) Fe_2S_2 in photosynthetic ferredoxin; (c) Fe_4S_4 in cubane-like ferredoxin. The symbol S here represents an "inorganic" or labile sulfur atom.

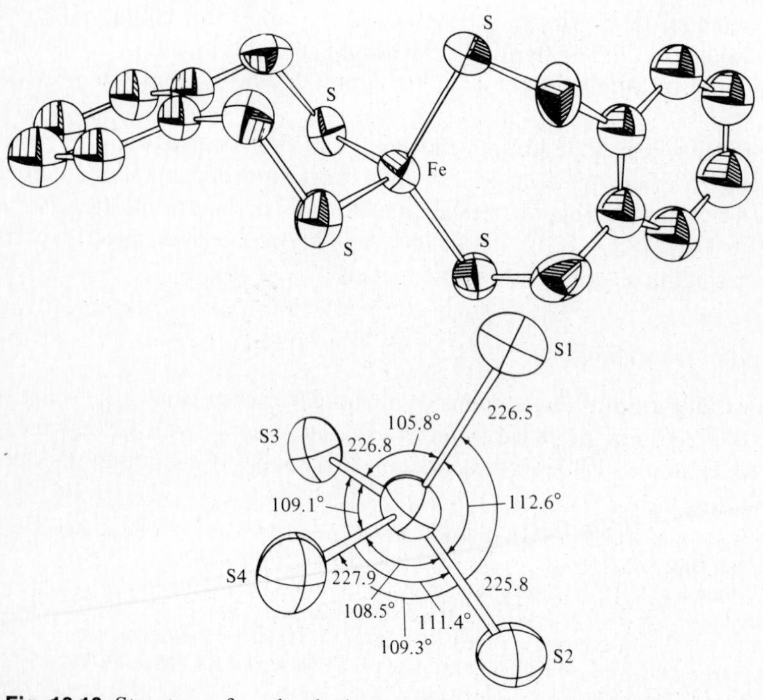

Fig. 18.13 Structure of a rubredoxin analogue, $Fe(S_2\text{-}o\text{-xyl})_2$. Hydrogen atoms have been omitted for clarity. Distances in pm. [From R. W. Lane et al., *Proc. Nat. Acad. Sci. U.S.A.*, **1975**, *72*, 2868. Reproduced with permission.]

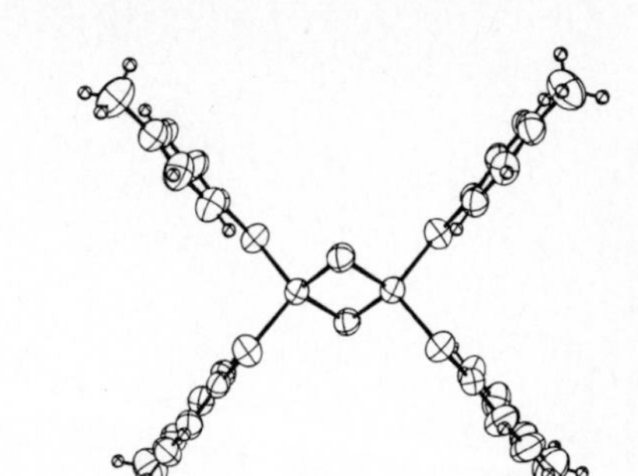

Fig. 18.14 Stereoview of a 2Fe-Ferredoxin analogue, $Fe_2S_2(S\text{-}p\text{-}tol)_4^{-2}$. [From J. J. Mayerle et al., *J. Am. Chem. Soc.*, **1975**, *97*, 1032. Reproduced with permission.]

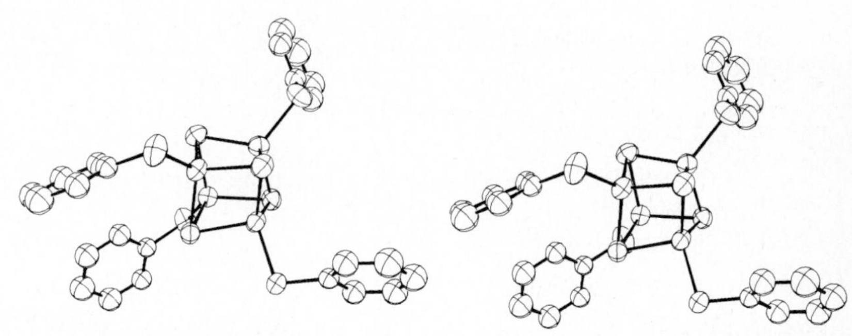

Fig. 18.15 Stereoview of cubane-like ferredoxin analogue. Hydrogen atoms have been omitted for clarity. [From L. Que, Jr., et al., *J. Am. Chem. Soc.*, **1974**, *97*, 4168.]

Cu(I)/Cu(II) in the *blue copper proteins*, such as stellacyanin, plastocyanin, and azurin (see below). Gray[37] has pointed out that these redox centers are ideally adapted for electron exchange in that no change in spin state occurs. Thus there is little or no movement of the ligands—the Franck-Condon activation barriers will be small. The rate of electron exchange is related to how deeply the active site is buried in the protein mass, and this has undoubtedly been adapted for optimum efficiency.

The structure of plastocyanin[38] is especially instructive in this regard. Copper(I) is d^{10} and thus provides no ligand field stabilization energy in any geometry. Since it is relatively small (74 pm), it is usually found in a tetrahedral environment. In contrast, Cu(II) is d^9 and is usually octahedrally coordinated with Jahn-Teller distortion, often to the point of square planar coordination. In the case of plastocyanin (Fig. 18.16), the copper is situated in a flattened tetrahedron, essentially "half way" between the two idealized geometries. This facilitates electron transfer over that of a system that is at either essentially

[37] A. G. Mauk et al., *J. Am. Chem. Soc.*, **1980**, *102*, 4360.

[38] P. M. Colman et al., *Nature*, **1978**, *272*, 319.

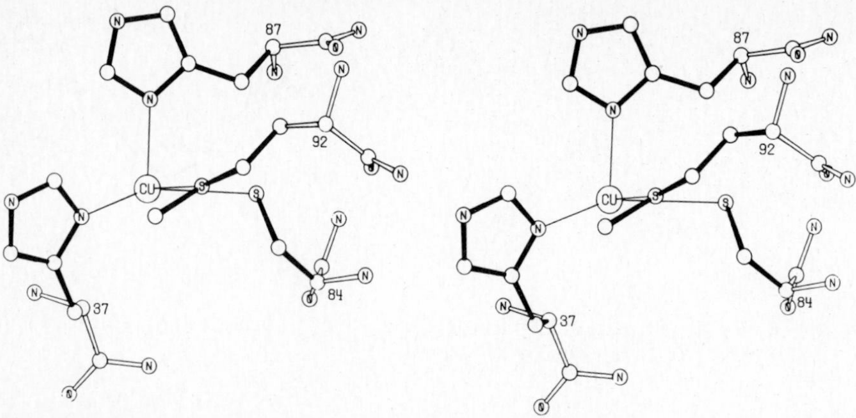

Fig. 18.16 Stereoview of the copper-binding site of plastocyanin. The four ligand residues are His 37, Cys 84, His 87, and Met 92. Note that the geometry about the copper is neither tetrahedral nor square planar, but intermediate. [From P. M. Colman et al., *Nature*, **1978**, *272*, 319. Reproduced with permission.]

the tetrahedral or square planar extremes, requiring considerable reorganization upon electron transfer.

Hemoglobin and myoglobin

The function of oxygen transport and storage in higher animals[39] is provided by hemoglobin and myoglobin. The former transports the oxygen from its source (lungs, gills, or skin) to the site of use inside the muscle cells. There the oxygen is transferred to myoglobin for use in respiration.

Several roles have been suggested for myoglobin. They are not mutually exclusive, and more than one may be important. The myoglobin may serve as a simple storage reservoir. This appears to be the reason for the increased concentrations of myoglobin in diving mammals such as whales. Other suggested functions of myoglobin include facilitation of oxygen flow within the cell and a "buffering" of the partial pressure of oxygen within the cell.[40]

Before discussing the structures of these molecules the physiology of blood should be briefly reviewed. (1) Myoglobin must have a greater affinity for oxygen than hemoglobin in order to effect the transfer of oxygen at the cell. (2) The equilibrium constant for myoglobin–oxygen complexation is given by the simple equilibrium expression:

$$K_M = \frac{[MbO_2]}{[Mb][O_2]} \tag{18.9}$$

[39] Small organisms require no oxygen transport system beyond simple diffusion. Some worms and mollusks have hemoproteins related to hemoglobin. Others employ nonheme iron proteins (see *hemerythrin*). Lobsters, crabs, and some snails, etc., use a copper-containing protein for oxygen transport (see *hemocyanin*).

[40] See R. P. Cole, *Science*, **1982**, *216*, 523.

If the total amount of myoglobin ([Mb] + [MbO_2]) is held constant while the concentration of oxygen is varied (in terms of partial pressure) the curve shown in Fig. 18.17 is obtained. Myoglobin is largely converted to oxymyoglobin even at low oxygen concentrations such as occur in the cells. (3) The equilibrium constant for the formation of oxyhemoglobin is somewhat more complicated. The expression for the curve in the range of physiological importance in the tissues is:

$$K_H = \frac{[HbO_2]}{[Hb][O_2]^{2.8}} \tag{18.10}$$

The 2.8 exponent for oxygen results from the fact that a single hemoglobin molecule can accept four oxygen molecules and *the binding of the four is not independent*. It is not the presence of four heme groups to bind four oxygen molecules *per se* that is important. If they acted independently, they would give a curve identical to that of myoglobin. It is the *cooperativity* of the four heme groups that produces the curves shown in Fig. 18.17. The presence of several bound oxygen molecules favors the addition of more oxygen molecules; conversely, if only one oxygen molecule is present, it dissociates more readily than from a more highly oxygenated species. The net result is that at low oxygen concentrations hemoglobin is less oxygenated and at high oxygen concentrations hemoglobin is more

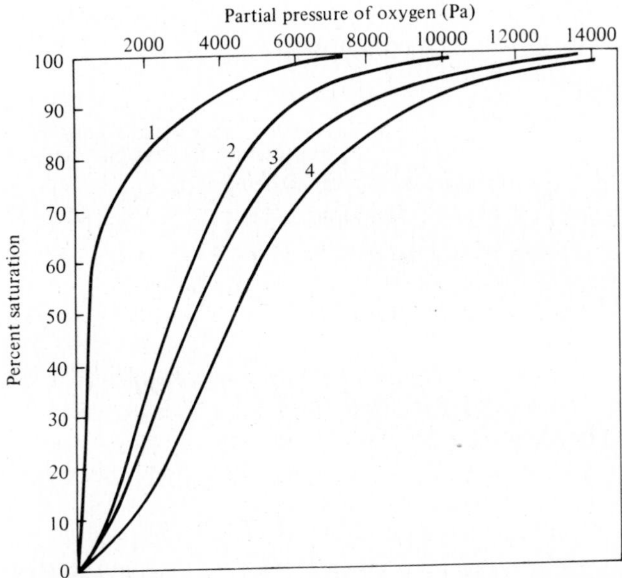

Fig. 18.17 Oxygen binding curves for (1) myoglobin and for hemoglobin at various partial pressures of carbon dioxide: (2) 2700 Pa; (3) 5300 Pa; (4) 10 700 Pa. Note that myoglobin has a stronger affinity for oxygen than hemoglobin and that this effect is more pronounced in the presence of large amounts of carbon dioxide. [Modified from A. V. Bock et al., *J. Biol. Chem.*, **1924**, *59*, 353. Reproduced with permission.]

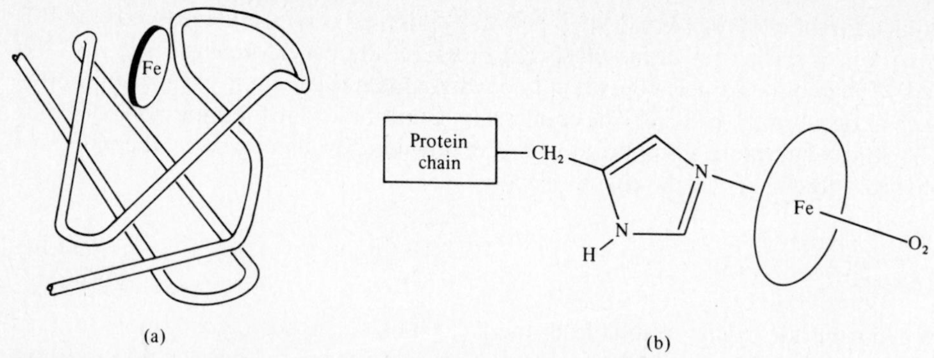

Fig. 18.18 The myoglobin molecule: (a) the folding of the polypeptide chain about the heme group (represented by the disk); (b) close-up view of the heme environment. [Modified from J. C. Kendrew et al., *Nature*, **1960**, *185*, 422. Reproduced with permission.]

oxygenated than if the exponent were 1. This results in a sigmoid curve for oxygenation of hemoglobin (Fig. 18.17). This effect favors oxygen transport since it helps the hemoglobin get saturated in the lungs and deoxygenated in the capillaries. (4) There is a pH dependence shown by hemoglobin, known as the Bohr effect, the net result of which is to enhance the release of oxygen at low pH values (high carbon dioxide concentrations, in the tissues).

The iron in myoglobin and hemoglobin is in the +2 oxidation state. The oxidized forms containing iron(III), called metmyoglobin and methemoglobin, *will not bind oxygen.* It is of special interest to note that *free heme* is immediately oxidized in the presence of oxygen and water and thus rendered useless for oxygen transport:

$$\text{Heme(Fe}^{\text{II}}) \xrightarrow[\text{water}]{O_2} \text{Hematin(Fe}^{\text{III}}) \tag{18.11}$$

In a biological system this would be fatal. The stability of heme(Fe^{II}) in myoglobin and hemoglobin is a result of the globin or protein portion of the molecule. Myoglobin has a molecular weight of 17 000, of which most is a protein chain folded about the heme reducing access to the iron and simultaneously producing a hydrophobic environment (Fig. 18.18).[41] This steric and chemical control allows access to and coordination by an oxygen molecule but does not allow the simultaneous presence of oxygen and one or more water molecules, which seems to be necessary for electron transfer.

The stabilization of heme by the presence of hydrophobic surfaces has been nicely illustrated by embedding heme in a matrix of polystyrene containing 1-(2-phenylethyl)-imidazole (Fig. 18.19). The imidazole molecule approximated the function of a histidine group in myoglobin and hemoglobin (see below). This "synthetic hemoglobin" was found to combine reversibly with oxygen even in the presence of water.[42]

More recent work has indicated that the situation is not one merely of providing a hydrophobic environment but also perhaps one of keeping heme groups apart so that

---

[41] See R. E. Dickerson and I. Geis, "The Structure and Action of Proteins," Harper & Row, New York, **1969**, pp. 17–23, for a discussion of the use of amino acids with hydrophobic side chains to create "nonaqueous" environments in polypeptides which must operate in aqueous systems.

[42] J. H. Wang, *J. Am. Chem. Soc.*, **1958**, *80*, 3168; *Intern. Union Biochem. Symp. Series*, **1961**, *19*, 98.

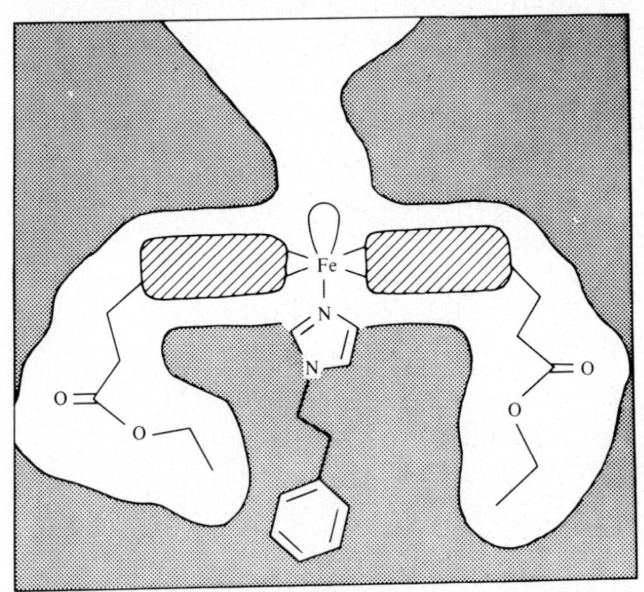

Fig. 18.19 A synthetic model for hemoglobin or myoglobin. [From J. H. Wang, *Acc. Chem. Res.*, **1970**, *3*, 90. Reproduced with permission.]

peroxo-bridged dimers that degenerate to $Fe^{III}—O—Fe^{III}$ can be avoided.[43] Further insight into the importance of steric hindrance about the heme group comes from the study of "picket fence" compounds in which bulky rings[44] or *t*-butyl groups[45] protect the binding site allowing reversible oxygen addition without the presence of the globin chain (Fig. 18.20a,b).

The microenvironment of the iron in myoglobin and hemoglobin is similar to that found in cytochrome *c*. The heme is coordinated to the nitrogen atom of a histidine group in the protein chain. Unlike cytochrome *c*, there are no links from the outer, organic portions of the heme to the chain; also, unlike cytochrome *c*, there is no sixth ligand from the chain to complete the coordination sphere of the iron atom. This provides a site for the coordination of an oxygen molecule.

Note how the differences in structure between the oxygen carriers and the electron carrier specify their respective functions. In cytochrome *c*, coordination by oxygen is prevented and thus redox behavior is favored. The ligands are held rigidly in place preventing their dissociation and replacement. In myoglobin and hemoglobin the redox

---

[43] I. A. Cohen and W. S. Caughey, *Biochemistry*, **1968**, *7*, 636; T. G. Traylor et al., *J. Am. Chem. Soc.*, **1979**, *101*, 6716; D-H. Chin et al., *J. Am. Chem. Soc.*, **1980**, *102*, 4344.

[44] R. G. Little et al., *J. Am. Chem. Soc.*, **1975**, *97*, 7049; T. G. Traylor, "Biomimetic Chemistry," D. Dolphin et al., eds., *ACS Adv. Chem. Ser. No. 191*, **1980**.

[45] J. P. Collman, *Acc. Chem. Res.*, **1977**, *10*, 265; J. P. Collman and K. S. Suslick, *Pure Appl. Chem.*, **1978**, *50*, 951.

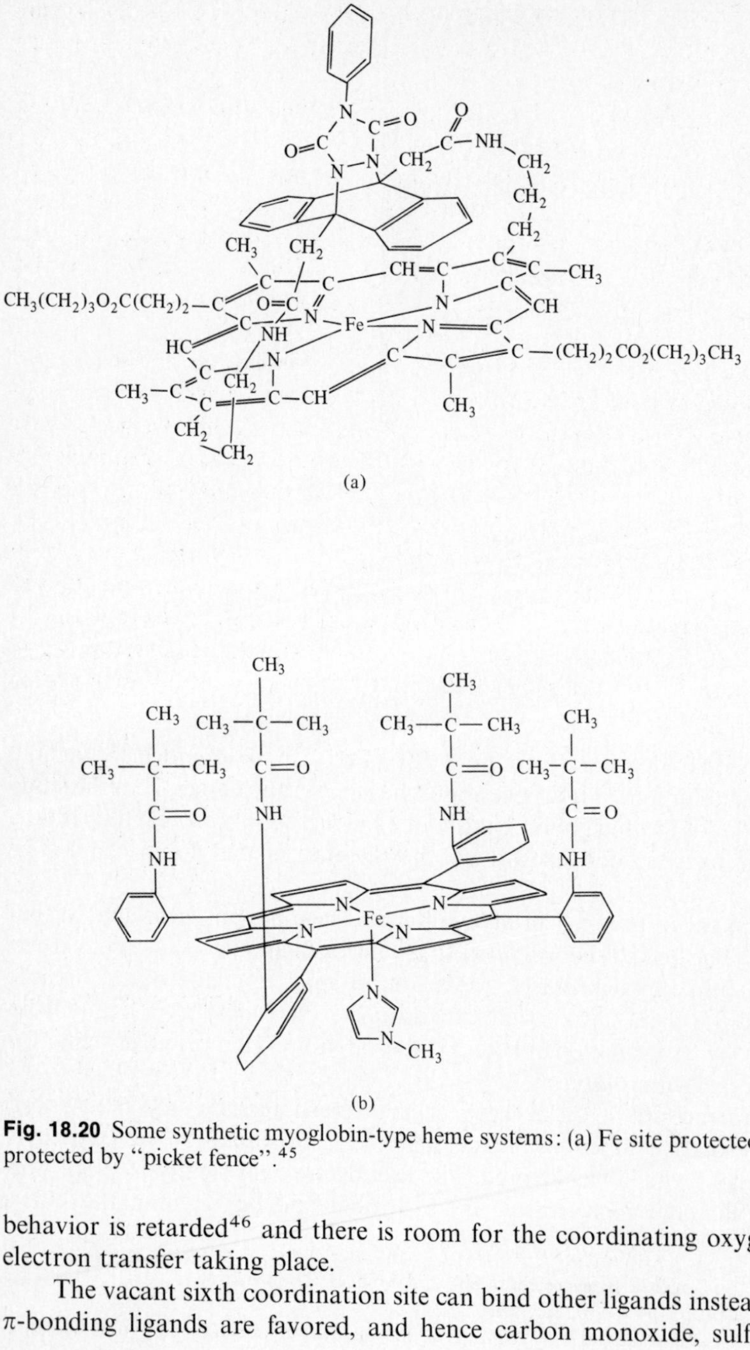

(a)

(b)

Fig. 18.20 Some synthetic myoglobin-type heme systems: (a) Fe site protected in cleft;[44] (b) Fe site protected by "picket fence".[45]

behavior is retarded[46] and there is room for the coordinating oxygen molecule without electron transfer taking place.

The vacant sixth coordination site can bind other ligands instead of oxygen. Strongly π-bonding ligands are favored, and hence carbon monoxide, sulfur- and phosphorus-

[46] Note that the oxygen molecule is a poor one-electron acceptor (see Problem 18.18). See also W. J. Wallace and W. S. Caughey, *Biochem. Biophys. Res. Commun.*, **1975**, *62*, 561.

containing ligands, and other soft ligands bind more tightly than the weakly binding oxygen and in the presence of large amounts of any of these the hemoglobin is tied up and unavailable for oxygen transport.

Hemoglobin and myoglobin contain high-spin iron(II) in the absence of a sixth ligand. Coordination by oxygen results in diamagnetic oxymyoglobin and oxyhemoglobin. This change in spin state is somewhat puzzling since molecular oxygen is not expected to be a particularly strong-field ligand. It is probable that the heme molecule is properly "tuned" by its substituents to make this spin pairing more likely. It has been suggested that the formation of a low-spin complex makes a redox reaction less likely.[47] The spin pairing of the normally paramagnetic oxygen molecule is also a problem of interest, although often overlooked (see Problem 18.9).

The exact molecular arrangement of oxygen in oxymyoglobin and oxyhemoglobin has been of great interest in understanding the chemistry of oxygen transport and storage. Unfortunately, it has been difficult to achieve because of the high molecular weight of these molecules and the low resolution of the X-ray determined structure. The oxygen molecule is coordinated via an angular, $\sim sp^2$ hybrid lone pair. Note that molecular oxygen is isoelectronic with the NO^- ion, and a bent Fe—O=O arrangement would correspond to the bent nitrosyl group discussed previously (see p. 612) and thought to be present in nitrosylmyoglobin and nitrosylhemoglobin. However, it is also possible that the oxygen molecule coordinates through a μ bond in a manner similar to ethylene. Fortunately, it has been possible to crystallize and perform an X-ray crystallographic analysis of one of the "picket fence" compounds, and it has been found that the oxygen molecule coordinates in the bent arrangement (Fig. 18.21).[45]

The structure of oxymyoglobin has been refined to a resolution of 160 pm. The oxygen is bonded angularly with a bond angle of ~115° and an Fe—O bond length of ~180 pm.[48] Oxyhemoglobin has not been resolved as accurately (210 pm), but the Fe—O bond length has been estimated as ~175 pm.[49] Both are compatible with the more accurate value of 175 (±2) pm in the "picket fence" adduct.[45] However the Fe—O—O bond angles vary considerably:

$$O_2 \cdot Mb = 112^\circ \qquad O_2 \cdot Hb = 156^\circ \qquad \text{“Picket fence”} = 131^\circ$$

The source of the differences is not clear, but calculations[50] have indicated that the bond energy changes but little with bond angle, and so other factors such as steric effects or hydrogen bonding with a neighboring group could become important (Fig. 18.22). Even more surprising, however, than the disparity between oxymyoglobin and oxyhemoglobin is the finding that the Fe—O—O bond angle in another pigment, oxyerythrocruorin, is almost 180°.[51]

---

[47] H. R. Mahler and E. H. Cordes, "Biological Chemistry," 2nd ed., Harper & Row, New York, **1971**, p. 669.

[48] S. E. V. Phillips, *Nature*, **1978**, *273*, 247; *J. Mol. Biol*., **1980**, *142*, 531.

[49] B. Shaanan, *Nature*, **1982**, *296*, 683.

[50] R. Hoffmann et al., *Inorg. Chem*., **1977**, *16*, 503; R. F. Kirchner and G. H. Loew, *J. Am. Chem. Soc*., **1977**, *99*, 4639.

[51] W. Stiegemann and F. Weber, *J. Mol. Biol*., **1979**, *127*, 309. Erythrocruorin is a form of myoglobin found in chironomid midges (flies).

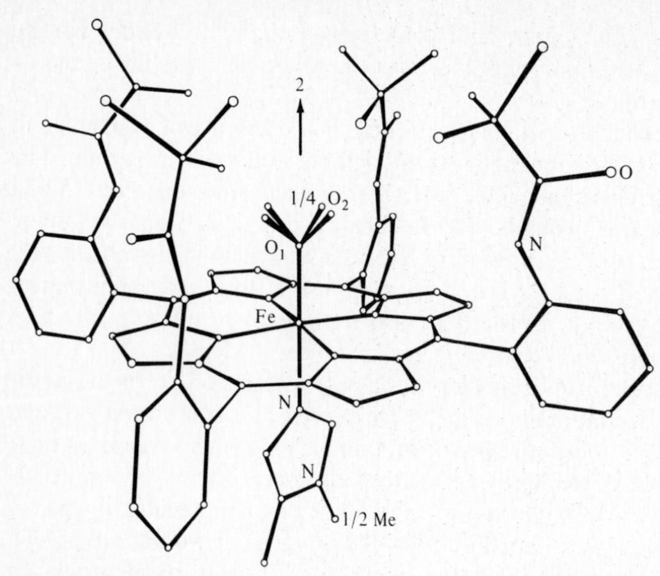

Fig. 18.21 Perspective view of picket-fence–dioxygen adduct. The apparent presence of four different O2 atoms results from a four-way statistical disorder of the oxygen atoms on different molecules responding to the X-ray diffraction. [From J. P. Collman et al., *Proc. Nat. Acad. Sci. U.S.*, **1974**, *71*, 1326. Reproduced with permission.]

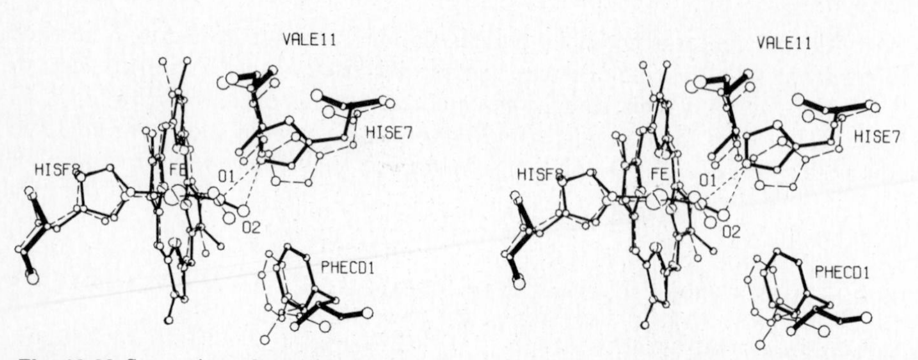

Fig. 18.22 Stereoview of superimposed heme environments in oxyhemoglobin and oxymyoglobin. Solid lines denote $Hb \cdot O_2$ and dashed lines $Mb \cdot O_2$. Note the difference in the Fe ∴ O ∴ O bond angles and the presumed hydrogen bond (dotted line) to the histidine residue (His E7). [From B. Shaanan, *Nature*, **1982**, *296*, 683. Reproduced with permission.]

Fig. 18.23 The tetrameric hemoglobin molecule. The four heme groups are represented by disks. [Modified from M. F. Perutz et al., *Nature*, **1960**, *185*, 416. Reproduced with permission.]

Structure and mechanism of hemoglobin

Hemoglobin may be considered an approximate tetramer of myoglobin. It has a molecular weight of 64 500 and contains four heme groups bound to four protein chains (Fig. 18.23). The differences between hemoglobin and myoglobin in their behavior towards oxygen (particularly 3 and 4 on p. 869) are related to the structure and movements of the four chains. If the tetrameric hemoglobin is broken down into dimers or monomers, these effects are lost. Upon oxygenation of hemoglobin, two of the heme groups move about 100 pm toward each other while two others separate by about 700 pm.[52]

These movements seem responsible for the cooperative effects observed (see below). In addition, hemoglobin has a channel, 2000 pm wide, lined with polar groups. Protonation or deprotonation of these polar groups with changes in pH can contribute to the Bohr effect.

Perutz[53] has suggested a mechanism to account for the cooperativity of the four heme groups in hemoglobin. Basically it is founded on the idea that an interaction between an oxygen molecule and a heme group can effect the position of the protein chain attached to it, which in turn affects the other protein chains through hydrogen bonds, etc. It has been dubbed the Rube Goldberg effect[54] after the marvelous mechanisms consisting of ropes, pulleys, and levers in Goldberg's cartoons.

The key or "trigger" in the Perutz mechanism is the high-spin Fe(II) atom in an unbound (i.e., oxygenless) heme. The radius of Fe^{+2} is 78 pm. Perutz estimates that the

---

[52] R. E. Dickerson and I. Geis, "The Structure and Action of Proteins," Harper & Row, New York, **1969**, p. 59.

[53] M. F. Perutz, *Nature*, **1970**, *228*, 726; *Brit. Med. Bull.*, **1976**, *32*, 195.

[54] J. E. Huheey, in "REACTS 1973, Proceedings of the Regional Annual Chemistry Teaching Symposium," K. Egolf et al., eds., University of Maryland, College Park, **1973**, pp. 52–78.

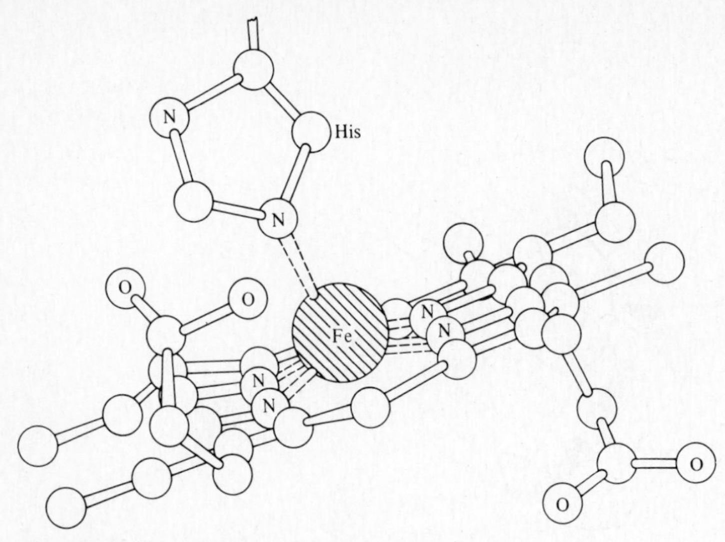

Fig. 18.24 Close-up of the heme group in myoglobin and hemoglobin. Note that the iron atom does not lie in the plane of the heme group.

Fe—N bond distance in heme must therefore be 218 pm. Since there is only room for a bond length of about 200–205 pm, the iron atom must sit about 80 pm above the plane of the heme group (Fig. 18.24). We have seen that coordination of the iron by oxygen results in spin-pairing of the electrons. Low-spin Fe(II) is 17 pm smaller than high-spin Fe(II) (Table 3.4). The Fe—N bond distance should therefore be almost exactly 200 pm and the low-spin iron atom should fit snugly in the porphyrin "hole." Coordination with oxygen will therefore cause the iron atom to "drop" about 60 pm into the plane of the heme group (Fig. 18.25). The imidazole group of the histidine residue attached to the iron atom must follow, and the tertiary structure of the protein chain of which it is a part will be rather

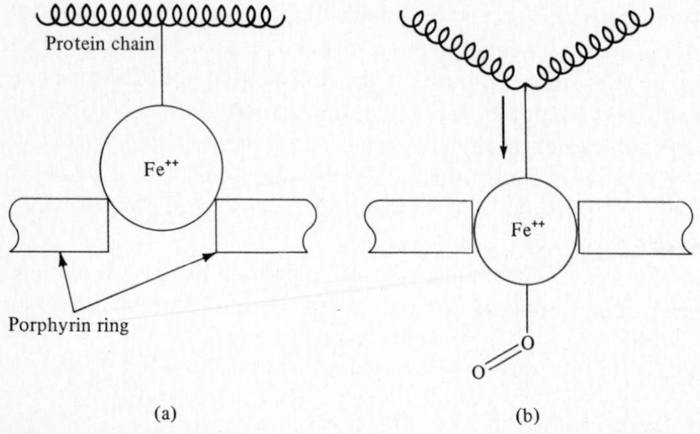

Fig. 18.25 Mechanical depiction of shrinkage of Fe^{+2}, movement into the porphyrin ring, and subsequent tension on the protein system.

drastically altered. The latter is more the realm of the biochemist than the inorganic chemist, but to make a long story short, the movement of the protein chains predisposes another heme group to react. When the fully saturated (four O_2 molecules) hemoglobin molecule reaches the tissues, the reverse takes place. As each oxygen molecule comes off, the molecule rearranges with increasing tendency to push off the remaining oxygen molecules. This is the vital process without which man would asphyxiate in pure oxygen.[53]

The high-spin–low-spin "trigger" was first suggested by Hoard[55] and is an interesting example of how a simple "inorganic" change, i.e., high-spin state to low-spin state, can be responsible for a highly important biological function. It is also interesting to note how the heme group acts as an "amplifier." The change in spin state results in a modest change of 17 pm in the radius of the iron atom, but because this 17 pm is the critical difference between the iron "fitting" or not fitting, the total movement of the iron atom is 50–60 pm. This movement of the iron atom and the accompanying histidine residue is sufficient to disrupt the configuration of the globin chains and cause the movements of 100 and 700 pm on the part of the protein chains, and more important, to effect the changes in bonding tendency on the part of the other heme groups. The change in configuration of the protein chains from the T ("tense" or deoxy) to the R ("relaxed" or oxy) form, triggered by the change of spin state of the iron atom, results in an estimated difference of 12–14 kJ mol^{-1} in oxygen affinity by the two forms. This fundamental difference in the energetics of oxygen binding is responsible for the cooperativity shown in Fig. 18.17. The word "tense" should not be taken too literally (although it was first proposed as such), but merely to mean that there are one or more factors inhibiting the histidine from following the iron, thus inhibiting the iron from dropping into the porphyrin ring and accepting an oxygen molecule. In the oxy (R) form these inhibitions are relaxed. Lest we get overly involved in the details of the mechanism, recall that the *reduced affinity* of the T form is nature's device that makes it possible for hemoglobin *to push the oxygen molecule off* ($R \rightarrow T$) *in the tissues and transfer it to myoglobin.*[56]

No discussion of the Perutz mechanism would be complete without brief mention of two other aspects. One is that the changes in the tertiary structure of the hemoglobin molecule upon deoxygenation are responsible for the polymerization of hemoglobin S and for the resulting crises of sickle cell anemia.[53,57] Also it should be noted that the Perutz mechanism is not without difficulties and has been criticized strongly, and various steric and electronic factors, all probably involved to some extent, have been suggested.[58] Nevertheless, the cooperativity of the heme groups is very real, and the "trigger" for the $T \rightleftharpoons R$

[55] J. L. Hoard, in "Hemes and Hemoproteins," B. Chance et al., eds., Academic Press, New York, **1966**; J. L. Hoard and W. R. Scheidt, *Proc. Nat. Acad. Sci., U.S.*, **1973**, *70*, 3919.

[56] W. R. Scheidt and C. A. Reed, *Chem. Rev.*, **1981**, *81*, 543.

[57] Return to Fig. 1.6, and you will see that the red blood cell in the lower right hand corner of the illustration is badly sickled. Erythrocytes that are sickled cannot flow as readily through the capillaries as normal red blood cells and they are more susceptible to damage. These factors are responsible for the symptoms of sickle cell anemia.

[58] F. Basolo et al., *Acc. Chem. Res.*, **1975**, *8*, 384; D. M. Scholler et al., *J. Am. Chem. Soc.*, **1976**, *98*, 7866, 7868. For papers in support of the Perutz mechanism (perhaps with modifications) see K. Imai et al., *J. Mol. Biol.*, **1977**, *109*, 83; J. M. Baldwin, *Brit. Med. Bull.*, **1976**, *32*, 213. For a thorough discussion of many aspects of the hemoglobin problem, see J. P. Collman, *Acc. Chem. Res.*, **1977**, *10*, 265.

interconversion *must* be something connected with the coordination of the oxygen to the iron. Therefore the Perutz mechanism involving Hoard's high-spin ⇌ low-spin equilibrium is the only one proposed thus far that can account for it in a realistic way.[56,59]

ENZYMES

Enzymes[60] are the catalysts of biological systems. They not only control the rate of reactions but by favoring certain geometries in the transition state can lower the activation energy for the formation of one product rather than another. The basic structure of enzymes is built of proteins. Those of interest to the inorganic chemist are composed of a protein structure (called an apoenzyme) and a small *prosthetic group* which may be either a simple metal ion or a complexed metal ion. For example, heme is the prosthetic group in hemoglobin. A reversibly bound group that combines with an enzyme for a particular reaction and then is released to combine with another is termed a *coenzyme*. Both prosthetic groups and coenzymes are sometimes called *cofactors*.[61]

Vitamin B_{12} and the B_{12} coenzymes[62]

In 1948 an "anti-pernicious anemia factor" was isolated, crystallized, and named vitamin B_{12} or cyanocobalamin. The molecule is built around a corrin ring containing a cobalt(III) atom. The corrin ring is a modified porphyrin ring in which one of the =CH— bridges between two of the pyrrole-type rings is missing, contracting the ring. The fifth and sixth coordination sites on the cobalt are filled by a nitrogen atom from an imidazole ring and a cyanide ion. The latter is an artifact of the isolation procedure and is not present in the biological system, where the sixth position appears to hold a loosely bound water molecule.

Vitamin B_{12} may be reduced by one electron ("vitamin B_{12r}") or two electrons ("vitamin B_{12s}") to form the Co^{II} and Co^{I} complexes, respectively. The two-electron reduction may be accomplished by NADH and flavin adenine dinucleotide (FAD). The cobalt(I) complex is strongly nucleophilic and may be important in the biological functioning of B_{12}. It readily undergoes alkylation via oxidative addition (pp. 654–658):

$$[B_{12}(Co^{I})] + CH_3I \longrightarrow [B_{12}(Co^{III})-CH_3]^+ + I^- \qquad \textbf{(18.12)}$$

Two important recognized functions of vitamin B_{12} involve the reduced Co^{I} form. The first is reduction of organic species. The second involves the acceptance of a methyl

---

[59] M. F. Perutz, "Biomimetic Chemistry," D. Dolphin et al., eds., *ACS Adv. Chem. Ser. No. 191*, **1980**, Chapter 11.

[60] See M. Dixon and E. C. Webb, "Enzymes," 2nd ed., Academic, New York, **1964**; J. Westley, "Enzymic Catalysis," Harper & Row, New York, **1969**; R. E. Dickerson and I. Geis, "The Structure and Action of Proteins," Harper & Row, New York, **1969**.

[61] These terms are not always used in exactly the same way. See footnote 60 for discussions of usage.

[62] M. N. Hughes, "The Inorganic Chemistry of Biological Processes," 2nd ed., Wiley, New York, **1981**, pp. 177–181, 307–309; E-I. Ochiai, "Bioinorganic Chemistry," Allyn & Bacon, Boston, **1977**, Chapter 12.

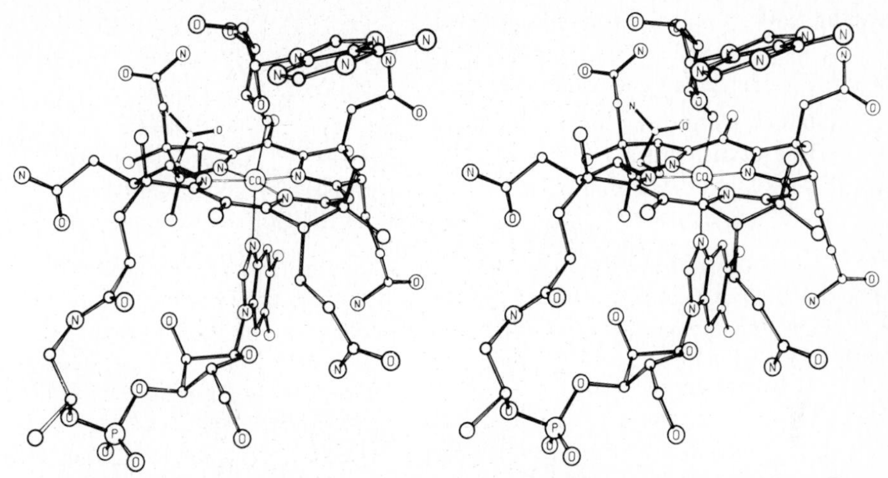

Fig. 18.26 A stereoview of the molecular structure of vitamin B_{12}. [From P. G. Lenhert, *Proc. Roy. Soc.*, **1968**, *A303*, 45. Reproduced with permission.]

group from N^5-methyltetrahydrofolate (CH_3–THF). The cobalt(III) methylcorrinoid can then partake in biomethylation reactions:

$$\begin{array}{ccccc} CH_3\text{–}THF & \searrow & B_{12s}(Co^{I}) & \leftarrow\nearrow & \text{methionine} \\ THF & \swarrow & CH_3\text{–}B_{12}^{+} & \nwarrow & \text{homocysteine} \end{array} \qquad (18.13)$$

Vitamin B_{12} is unusual in several ways. The ability to form a metal–carbon bond is unique—these are nature's only organometallic compounds. It is the only vitamin known to contain a metal. It is essential for all higher animals, but it is not found in higher plants, nor can it be synthesized by animals. It appears to be synthesized exclusively by bacteria. The formation of methane and the methylation of various heavy metals such as Hg, As, Tl, Cr, Se, S, Pt, and Pd is effected by bacteria in anaerobic sludges. The latter may pose serious environmental problems (see p. 914).

When vitamin B_{12s} reacts with adenosine triphosphate (ATP), alkylation takes place as in Eq. 18.13 with the formation of a direct carbon–cobalt bond between adenosine and cobalt (Fig. 18.26). This species is called B_{12} coenzyme and is involved in acting in concert with several other enzymes to effect 1,2 shifts of the general type:

$$-\overset{\displaystyle H}{\underset{|}{\overset{|}{C_1}}}-\overset{|}{\underset{\displaystyle X}{\underset{|}{C_2}}}- \rightleftharpoons -\overset{|}{\underset{\displaystyle X}{\underset{|}{C_1}}}-\overset{\displaystyle H}{\underset{|}{\overset{|}{C_2}}}- \qquad (18.14)$$

Examples of reactions of this sort are:

Enzyme	Reaction
Glutamate mutase	Glutamic acid $\longrightarrow$ β-methylaspartic acid
Glycerol dehydrase	Glycerol $\longrightarrow$ 3-hydroxypropanal
Ethanolamine deaminase	Ethanolamine $\longrightarrow$ acetaldehyde
Methylmalonyl CoA mutase	Methylmalonyl CoA $\longrightarrow$ succinyl CoA

One of the indicators of pernicious anemia, a disease caused by inability to absorb B_{12} through the gut wall, is an increase in excretion of methylmalonic acid.

Until the proposed mechanism is examined, some of these rearrangements appear unusual:

$$\mathrm{HO(O)C{-}CH(CH_3){-}C(O){-}S{-}R} \longrightarrow \mathrm{HO(O)C{-}CH_2{-}CH_2{-}C(O){-}S{-}R} \tag{18.15}$$

It is believed[63] that the reaction starts with homolytic cleavage of the cobalt–carbon bond (at a cost of perhaps 100 kJ mol^{-1})[64] to yield a Co(II) atom and a 5′-deoxyadenosyl radical. This radical then abstracts a hydrogen atom (in Eq. 18.15 from the methyl group). Migration of the –C(O)SR group takes place, followed by return of the hydrogen atom from 5′-deoxyadenosine to the substrate. This regenerates the 5′-deoxyadenosyl radical, which can recombine with the Co(II) atom to form the coenzyme.

The "fitness" of the B_{12} system in performing the above duties has been attributed[65] to the fact that (1) the geometric state of the Co^{III} can be such that one of the six ligands is held weakly or is absent; (2) the electronic state of the cobalt is such that oxidative-addition and reductive-elimination reactions take place without large activation energies; and (3) the corrin ring is flexible and can take up a number of conformations to fit the environment. It is of interest to note that although cobalt porphyrin analogues of B_{12} have been synthesized, they cannot be reduced to cobalt(I) in aqueous (and hence biological) media. Hence corrins were selected in place of porphyrins in the evolution of the B_{12} cobalt complexes.[66]

Structure and function

To illustrate the structure of an enzyme and its relation to function, consider carboxypeptidase A. This pancreatic enzyme cleaves the carboxyl terminal amino acid from a

[63] B. M. Babior, *Acc. Chem. Res.*, **1975**, *8*, 376; A. Eschenmoser and C. E. Wintner, *Science*, **1977**, *196*, 1410.

[64] J. Halpern, in "Biomimetic Chemistry," D. Dolphin et al., eds., *ACS Adv. Chem. Ser. No. 191*, Washington, D.C., **1980**, Chapter 9.

[65] H. A. O. Hill et al., *Chem. Brit.*, **1968**, *4*, 397. See also H. R. Mahler and E. H. Cordes, "Biological Chemistry," 2nd ed., Harper & Row, New York, **1971**, pp. 806–817.

[66] See G. N. Schrauzer, *Angew. Chem. Int. Ed. Engl.*, **1976**, *15*, 417, and references therein.

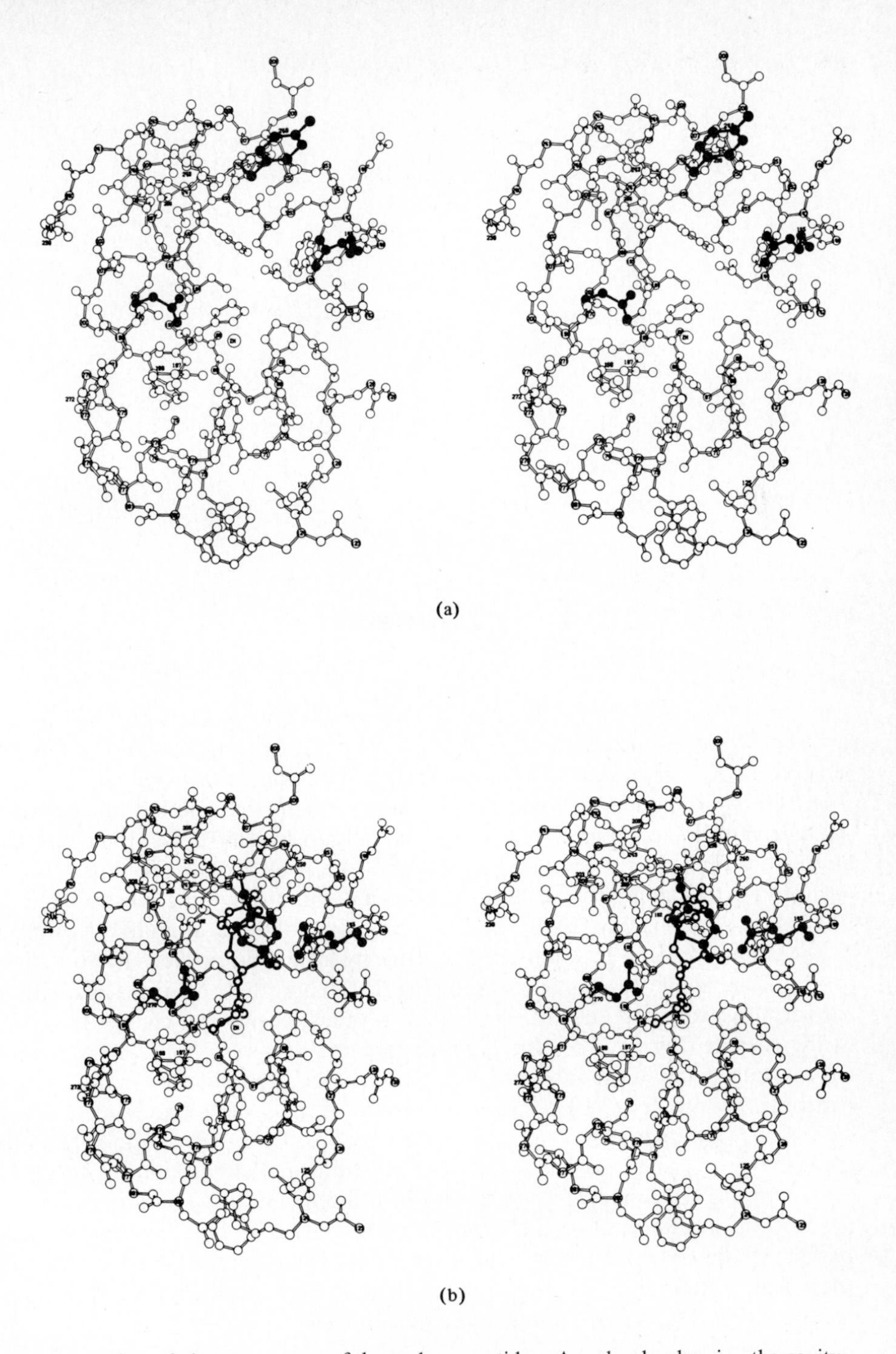

Fig. 18.27 (a) Stereoview of about a quarter of the carboxypeptidase A molecule, showing the cavity, the zinc atom, and the functional groups (shown with black atoms) Arg-145 (right), Tyr-248 (above), and Glu-270 (left). (b) Stereoview of the same region, after the addition of glycyl-L-tyrosine (heavy open circles), showing the new positions of Arg-145, Tyr-248, and Glu-270. The quanidinium movement of Arg-145 is 200 pm, the hydroxyl of Tyr-248 moves 1200 pm, and the carboxylate of Glu-270 moves 200 pm when Gly-Tyr binds to the enzyme. [From W. N. Lipscomb, *Chem. Soc. Rev.*, **1972**, *1*, 319. Reproduced with permission.]

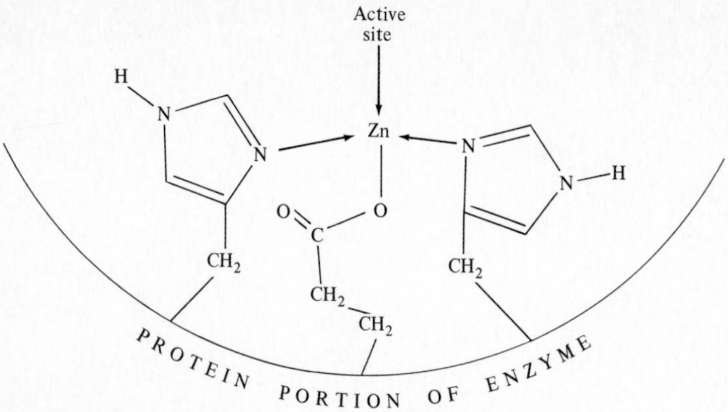

Fig. 18.28 The active site in carboxypeptidase A.

peptide chain by hydrolyzing the amide linkage:

$$\ldots\text{Pro-Leu-Glu-Phe} \xrightarrow[\text{carboxypeptidase A}]{H_2O} \ldots\text{Pro-Leu-Glu} + \text{phenylalanine} \qquad \textbf{(18.16)}$$

The enzyme consists of a protein chain of 307 amino acid residues plus one Zn^{++} ion for a molecular weight of about 34 600. The molecule is roughly egg-shaped, with a maximum dimension of approximately 5000 pm and a minimum dimension of about 3800 pm (Fig. 18.27a). There is a cleft on one side that contains the zinc ion, the active site. The metal is coordinated approximately tetrahedrally to two nitrogen atoms and an oxygen atom from three amino acids in the protein chain (Fig. 18.28). The fourth coordination site is free[67] to accept a pair of electrons from a donor atom in the substrate to be cleaved. The enzyme is thought to act through coordination of the zinc atom to the carbonyl group of the amide linkage. In addition, a nearby hydrophobic pocket envelops the organic group of the amino acid to be cleaved (Fig. 18.29) and those amino acids with aromatic side groups react most readily. Accompanying these events is a change in conformation of the enzyme (Fig. 18.27b): The arginine side chain moves about 200 pm closer to the carboxylate group of the substrate, and the phenolic group of the tyrosine comes within hydrogen bonding distance of the imido group of the C-terminal amino acid, a shift of 1200 pm.[68] The hydrogen bonding to the free carboxyl group (by arginine) and the amide linkage (by tyrosine) not only holds the substrate to the enzyme but helps break the N—C bond. Nucleophilic displacement of the amide group by an attacking carboxylate group from a glutamate group could form an anhydride link to the remainder of the peptide chain. Hydrolysis of this anhydride could then complete the cycle and regenerate the original enzyme. More likely, the glutamate acts indirectly by polarizing a water molecule (Fig. 18.29b) that attacks the amide linkage.

[67] There is probably a loosely bound water molecule at this position when the enzyme is not engaged in active catalysis.

[68] W. N. Lipscomb, *Chem. Soc. Rev.*, **1972**, *1*, 319; *Tetrahedron*, **1974**, *30*, 1725; *Acc. Chem. Res.*, **1982**, *15*, 232.

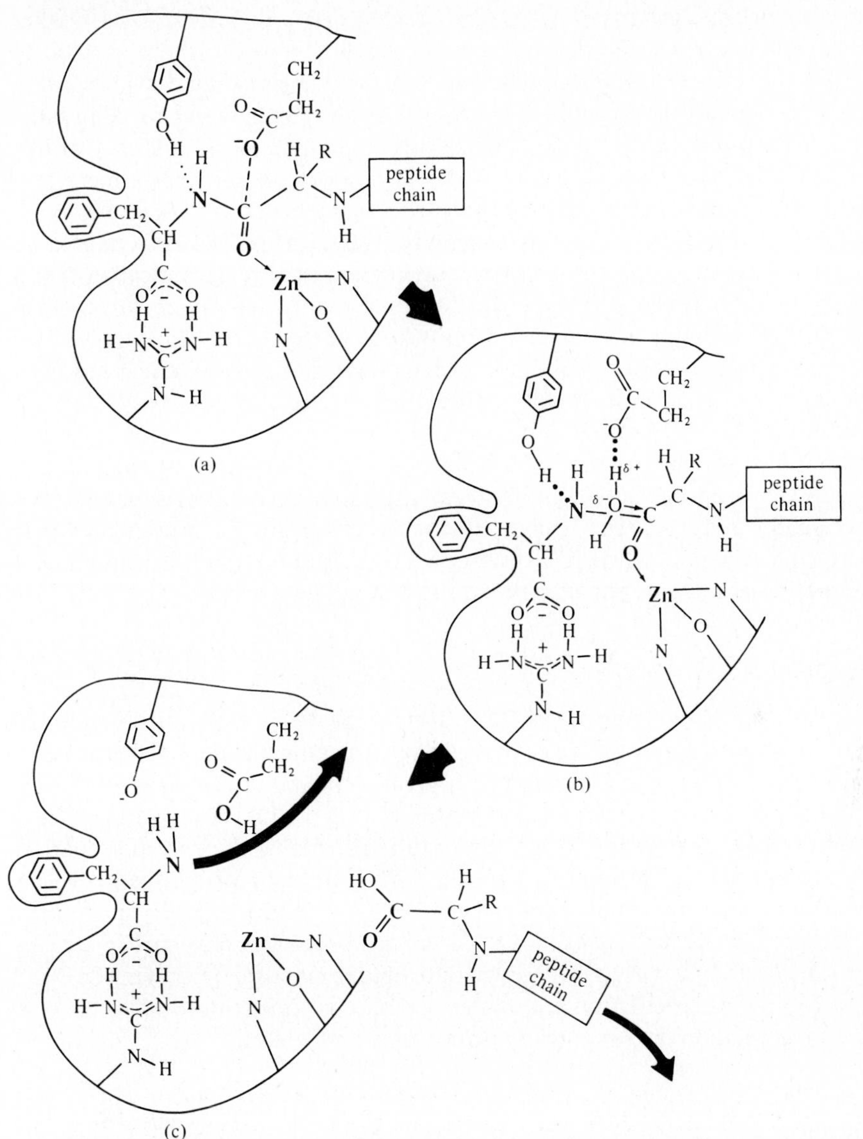

Fig. 18.29 Suggested mode of action of carboxypeptidase A in the hydrolysis of an amide linkage in a polypeptide. (a) Positioning of the substrate on the enzyme. Interactions are (1) coordinate covalent bond, carbonyl to zinc; (2) hydrogen bonds, arginine to carboxylate and tyrosine to amide; (3) van der Waals attraction, hydrophobic pocket to aromatic ring; (4) dipole attraction and *possible* incipient bond formation, carboxylate (glutamate) oxygen to carbonyl group (amide linkage). Drawing is diagrammatic portrayal in two dimensions of the three-dimensional structure. (b) *Probable* intervention of polarized water molecule in the incipient breaking of the amide (N ∴ C bond) linkage. (c) Completed reaction, removal of the products (amino acid and shortened peptide chain). Original configuration of enzyme results upon proton shift from glutamic acid to tyrosine. For more detailed discussion of the mechanism of catalysis by carboxypeptidase A, see R. Breslow and D. L. Wernick, *Proc. Nat. Acad. Sci. U.S.*, **1977**, *74*, 1303.

This example illustrates the basic key-and-lock theory first proposed by Emil Fischer in which the enzyme and substrate fit each other sterically. However, there is more to enzymatic catalysis than merely bringing reactants together. There is good evidence that the enzyme also encourages the reaction by placing a strain on the bond to be broken. Evidence comes from spectroscopic studies of enzymes containing metal ions that, unlike Zn^{+2}, show *d–d* transitions. The spectrum of the enzyme containing such a metal ion provides information on the microsymmetry of the site of the metal. For example, Co^{+2} can replace the Zn^{+2} and the enzyme retains its activity. The spectrum of carboxypeptidase A(Co^{II}) is "irregular" and has a high absorptivity (extinction coefficient), indicating that a regular tetrahedron is not present.[69] The distortion presumably aids the metal to effect the reaction. It has been suggested that the metal in the enzyme is peculiarly poised for action and that this lowers the energy of the transition state. The term *entatic*[70] has been coined[69] to describe this state of the metal in an enzyme.

The substitution of a different metal into an enzyme provides a very useful method for studying the immediate environment of the metal site. In addition to the use of Co^{+2} for spectral studies, appropriate substitution allows the use of physical methods such as electron spin resonance (Co^{+2}, Cu^{+2}), the Mössbauer effect (Fe^{+2}), proton magnetic resonance relaxation techniques (Mn^{+2}), or X-ray crystallography (with a heavy metal atom to aid in the structure solution).[71]

A synthetic model of enzyme action

Wang[72] has devised a synthetic system which illustrates the importance of strain in catalysis. The decomposition of hydrogen peroxide is catalyzed by transition metal ions such as Fe^{+3} that can undergo redox reactions. The catalysis is greatly enhanced, however, if the iron is incorporated into a complex such as $[Fe(trien)(OH)_2]^+$ (Fig. 18.30a). The hydroperoxide ion can replace the two hydroxide ions (Fig. 18.30c), but in spanning the two coordination sites on the iron atom the O—O bond is strained. The bond distance in free hydrogen peroxide (149 pm) is only about half of that expected (~290 pm)[73] for an unstrained complex. This strain weakens and breaks the O—O bond. The positive oxygen (Fig. 18.30d) abstracts a hydride ion from a second hydroperoxide ion, regenerating $[Fe(trien)(OH)_2]^+$ and releasing oxygen. The triethylenetetraamine complex is much more efficient as a catalyst for this reaction than the square planar iron atom in hematin.[74]

---

[69] B. L. Vallee and R. J. P. Williams, *Proc. Nat. Acad. Sci. U.S.A.*, **1968**, *59*, 498; *Chem. Brit.*, **1968**, *4*, 397; D. D. Ulmer and B. L. Vallee, "Bioinorganic Chemistry," R. F. Gould, ed., *ACS Adv. Chem. Ser. No. 100*, Washington, D.C., **1971**, Chapter 10.

[70] Gr. εντεινω, to stretch, strain, or bend.

[71] R. J. P. Williams, *Endeavour*, **1967**, *26*, 96.

[72] J. H. Wang, *J. Am. Chem. Soc.*, **1955**, *77*, 822, 4715; *Acc. Chem. Res.*, **1970**, *3*, 90.

[73] Estimated from the Fe—O distance in aqua complexes.

[74] Hematin is heme in which the iron has been oxidized to Fe^{III}. Note, however, that the most efficient catalyst of all, that found in biological systems, is *catalase* which contains hematin. Presumably a different type of coordination is present in this enzyme. The rate constants are phenomenally high: $10^6\ dm^3\ mol^{-1}\ s^{-1}$.

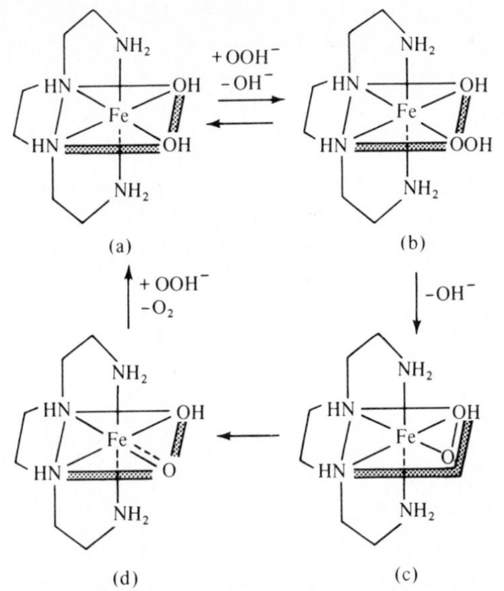

Fig. 18.30 The catalytic decomposition of hydrogen peroxide by $[Fe(trien)(OH)_2]^+$. [From J. H. Wang, *Acc. Chem. Res.*, **1970**, *3*, 90. Reproduced with permission.]

Inhibition and poisoning

The study of the factors which enable an apoenzyme to select the appropriate metal ion is of importance to the proper understanding of enzyme action.[75] The same factors that favor the formation of certain complexes in the laboratory probably are also important in biological systems. For example, the Irving-Williams series and the HSAB principle should be helpful guides. Thus we expect to find the really hard metal ions (Groups IA and IIA) preferring ligands with oxygen donor atoms. The somewhat softer metal atoms of the first transition series (Co to Zn) may prefer coordination to nitrogen atoms (cf. Fig. 7.5). The important thiol group, —SH, should have a particularly strong affinity for soft metal ions.

The usual structural principles of coordination chemistry such as the chelate effect, the preference for five- and six-membered rings, and of certain conformations should hold in biological systems. In addition, however, enzymes present structural effects not observed in other complexes. An interesting example is carbonic anhydrase which catalyzes the interconversion of carbon dioxide and carbonates. Like carboxypeptidase, carbonic anhydrase contains zinc atoms, in this case coordinated to three histidine residues (His 94, His 96, His 119) and a water molecule or hydroxide ion. The active site (Fig. 18.31) contains

[75] H. Sigel and D. B. McCormick, *Acc. Chem. Res.*, **1970**, *3*, 201.

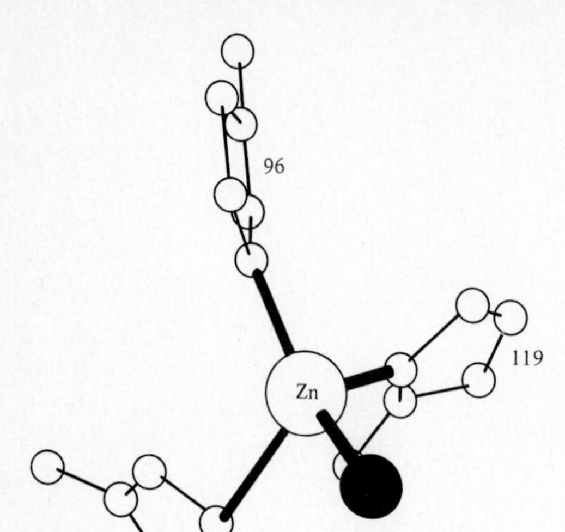

Fig. 18.31 Active site of carbonic anhydrase. In the resting enzyme a water molecule (O = ●) coordinates to the zinc atom. All hydrogen atoms have been omitted for clarity.

other amino acids that may function through hydrogen bonding, proton transfer, etc. The relative binding power of the zinc ion toward halide ions is reversed in the enzyme ($I^- > Br^- > Cl^- > F^-$) compared with the free Zn^{+2} ions ($F^- > Cl^- > Br^- > I^-$). This reversal could be interpreted as some sort of "softening" effect on the zinc by the apoenzyme were it not that the soft ligand CN^- is bound equally well by the free ion as by the complex.[76] Furthermore, NO_3^-, CNO^-, and N_3^-, none of which is known for exceptional softness, are bound with exceptional strength. They are, however, isoelectronic and isostructural with the reactants and products of the enzyme reaction, CO_2, CO_3^{-2}, and HCO_3^-, respectively. The explanation appears to be a tailoring of the structure of the enzyme molecule to form a pocket about 450 pm long next to the zinc ion, perhaps containing an additional positive center, to stabilize ions of appropriate size.[77]

The mechanism of the reversible hydration of carbon dioxide to carbonic acid (actually the hydrogen carbonate ion at physiological pHs) is thought to follow the pathway shown in Eq. 18.17. Like all truly catalytic processes, it is a closed loop:[78]

---

[76] It should be noted that this discussion is based on a *comparison* of the equilibrium constants of enzymic zinc–cyanide complexation versus aqueous zinc–cyanide complexation. Cyanide has a high affinity for zinc under *both* conditions (stability constant of $[Zn(CN)_4]^{-2} = 7.7 \times 10^{16}$); hence it should not be concluded that there is any lack of affinity for cyanide in the enzyme.

[77] B. L. Vallee and R. J. P. Williams, *Chem. Brit.*, **1968**, *4*, 397, and references therein.

[78] I. Bertini et al., *Struct. Bond.*, **1982**, *48*, 45; S. Lindskog, *Adv. Inorg. Biochem.*, **1982**, *4*, 115.

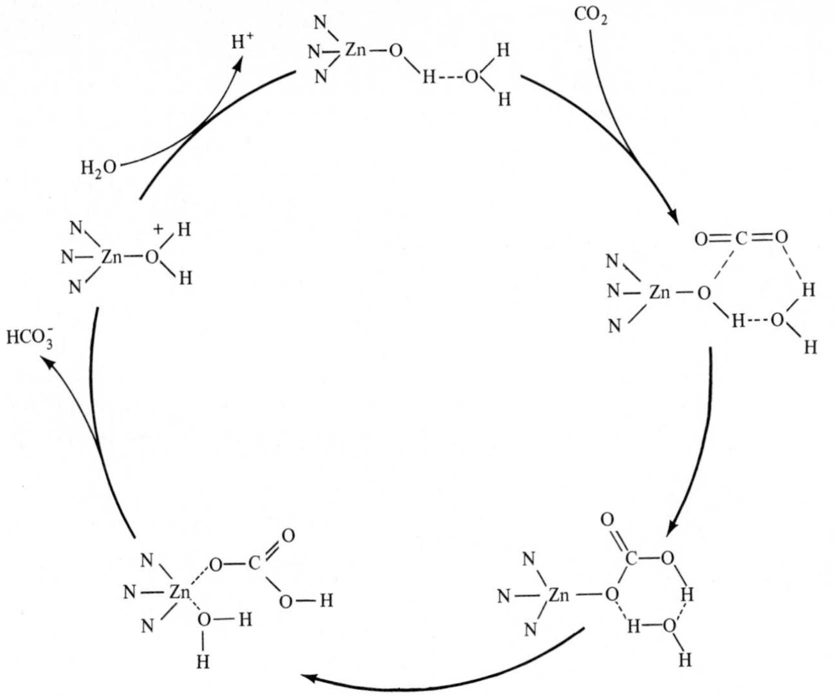

(18.17)

It may operate either clockwise (as drawn) or counterclockwise depending upon the concentrations of the reactants.

Ligands which can coordinate to an active center in an enzyme and prevent coordination by the substrate will tend to inhibit the action of that enzyme.[79] Azide ion thus inhibits the activity of carbonic anhydrase to a moderate extent and as little as 4×10^{-6} mol dm^{-3} cyanide or hydrogen sulfide inhibits the enzymatic activity 85%.[80]

Inhibition may also be effected by metal ions. Most prosthetic groups involve metals of the first transition series (molybdenum seems to be the sole exception). Coordination of the apoenzyme to a heavier metal ion may destroy the enzymatic activity. Particularly poisonous in this regard are metal ions such as Hg^{+2}. The latter has a special affinity for sulfur (see p. 314) and thus tends to form extremely stable complexes with amino acids containing sulfur such as cysteine, cystine, and methionine. The inhibition of an enzyme by Hg^{+2} has been taken as an indication of the presence of thiol groups[81] but is not

[79] Compare the action of carbon monoxide on hemoglobin, p. 872.

[80] For a general discussion of inhibition effects, see E. J. Hewitt and D. J. D. Nicholas, in "Metabolic Inhibitors," R. M. Hochster and J. H. Quastel, eds., Academic Press, New York, **1963**, Vol. II, Chapter 29.

[81] For a discussion of the subject, see N. B. Madsen, in "Metabolic Inhibitors," R. M. Hochster and J. H. Quastel, eds., Academic Press, New York, **1963**, Vol. II, Chapter 21.

infallible. (For example, Hg^{+2} completely abolishes the activity of carboxypeptidase A in hydrolyzing amide linkages.) Nevertheless, the affinity of sulfur and mercury is responsible for many of the poisonous effects of mercury in biological systems. Often these effects may be reversed by addition of sulfur-containing compounds such as cysteine or glutathione. Another sulfur donor, 2,3-dimercaptopropanol, has a strong affinity for soft metal ions. Developed during World War I as an antidote for the organoarsenic war gas, lewisite, it was dubbed British antilewisite (BAL). It has proved to be extremely useful as an antidote for arsenic, cadmium, and mercury poisoning.

The inhibition of enzyme systems does not necessarily cause unwanted effects. Consider the enzyme xanthine oxidase. It contains two atoms of molybdenum, four Fe_2S_2, and two FAD (flavin adenine dinucleotide) moieties, and it has a molecular weight of 275 000–300 000. There is no evidence that the two units ($Mo/2Fe_2S_2/FAD$) are near each other or interact in any way. It is believed that the immediate environment of each molybdenum atom consists of one oxygen and three sulfur atoms (additional ligands may be present):[82]

168 pm S—H 238 pm
O=Mo—S—R
S—R
Reduced

$\underset{+e^-}{\overset{-e^-}{\rightleftharpoons}}$

170 pm S 215 pm
O=Mo—S—R
S—R 247 pm
Oxidized

Knowledge of the nature and number of the ligands of this sort can be gathered by a relatively new method called *extended X-ray absorption fine structure* (EXAFS). It elaborates upon the long known fact that X-ray absorption spectra show element-specific "edges" that correspond to quantum jumps of core electrons to unoccupied orbitals or to the continuum. By choosing X-ray frequencies near the X-ray edge of a particular element, atoms of that element can be excited to emit photoelectrons. The wave of each electron will be backscattered by the nearest neighbors in proportion to the number and kind of the ligands and inversely proportional to the interatomic distance. If the backscattered wave is in phase with the original wave, reinforcement will occur, yielding a maximum in the X-ray absorption spectrum. Out-of-phase waves will cancel and give minima. The EXAFS spectrum consists, then, of the X-ray absorption plotted against the energy of the incident X-ray photon. The amplitudes and frequencies of the oscillations in the absorption are related to the number, type, and spacing of the ligands. Thus if one bombards heme with an X-ray frequency characteristic of an iron "edge," it should, in principle, be possible to learn that there are four atoms of atomic number 7 equidistantly surrounding the iron atom.[83]

---

[82] S. P. Cramer et al., *J. Am. Chem. Soc.*, **1981**, *103*, 7721.

[83] The present discussion has been greatly simplified to give the general technique as well as the information obtained without going into the details of the analysis. For the latter as well as the experimental technique, see S. P. Cramer and K. O. Hodgson, *Progr. Inorg. Chem.*, **1979**, *25*, 1.

In the same way, for xanthine oxidase the determination of a single oxygen atom at 168–170 pm (and therefore doubly bonded) and three sulfur atoms at 238 pm (single-bonded HS— and RS—) in the reduced form, or two at 247 (RS—) and one at 215 pm (S=) in the oxidized form, can be made from the EXAFS spectrum.

This enzyme catalyzes the oxidation of xanthine to uric acid:

(18.18)

The electron flow may be represented as:

$$\text{Xanthine} \longrightarrow \text{Mo} \longrightarrow 2\text{Fe}_2\text{S}_2 \longrightarrow \text{FAD} \longrightarrow \text{O}_2 \qquad \textbf{(18.19)}$$

Uric acid is the chief end product of purine metabolism in primates, birds, lizards, and snakes. An inborn metabolic error in humans results in increased levels of uric acid and its deposition as painful crystals in the joints. This condition (gout) may be treated by the drug allopurinol which is also oxidized by xanthine oxidase to alloxanthine (dashed line in Eq. 18.18). However, alloxanthine binds so tightly to the molybdenum that the enzyme is inactivated, the catalytic cycle broken, and uric acid formation is inhibited.

The extra stability of the alloxanthine complex may be a result of strong N—H—N hydrogen bonding by the nitrogen in the 8-position:

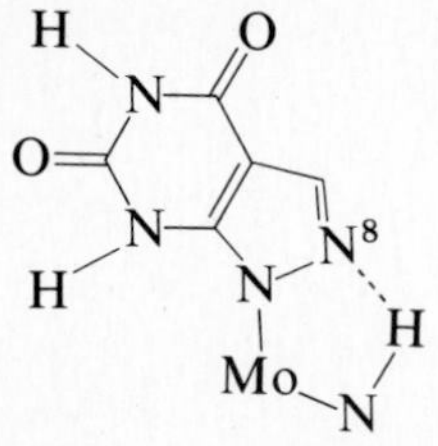

This structure resembles the hydrogen-bonded transition state for the nucleophilic attack of hydroxide ion (Eq. 18.18) where the hydrogen bond promotes the attack on the carbon. With a nitrogen atom at the 8-position there is no way for the alloxanthine to leave.[84]

A closely related enzyme is aldehyde oxidase. It also contains two ($Mo/2Fe_2S_2/FAD$) units with a molecular weight of about 300 000. It converts acetaldehyde to acetic acid via electron flow:

$$\text{Acetaldehyde} \longrightarrow \text{Mo(VI)} \longrightarrow 2\text{Fe}_2\text{S}_2 \longrightarrow \text{FAD} \longrightarrow \text{O}_2 \qquad \textbf{(18.20)}$$

When ethanol is consumed, the initial metabolic product is the extremely poisonous acetaldehyde, which is kept in low concentration by the oxidase-catalyzed conversion to harmless acetic acid. The drug Antabuse, used for treating alcoholism, is a sulfur-containing ligand, disulfiram:

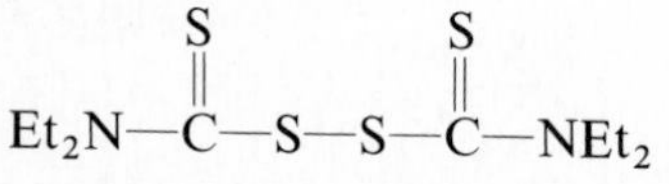

$$Et_2N{-}\overset{\overset{\displaystyle S}{\|}}{C}{-}S{-}S{-}\overset{\overset{\displaystyle S}{\|}}{C}{-}NEt_2$$

In the body, Antabuse inhibits acetaldehyde oxidase, presumably via the soft–soft molybdenum–sulfur interaction.[85] Any alcohol ingested will be converted to acetaldehyde which, in the absence of a pathway to destroy it, will build up with severely unpleasant effects, discouraging further consumption.

NITROGEN FIXATION

An enzyme system of particular importance is that which promotes the fixation of atmospheric nitrogen. This is of considerable interest for a variety of reasons. It is a very

[84] E. I. Stiefel, *Progr. Inorg. Chem.*, **1977**, *22*, 1.

[85] The above surmise is based on the known chemistry between molybdenum and sulfur-containing ligands. The possibility of multiple pathways for aldehyde oxidation and other actions of disulfiram make the situation unclear. Disulfiram is also used to prevent renal toxicity from platinum when *cis*-diamminedichloroplatinum(II) (see p. 921) is used to treat neoplasms and trypanosomiasis (see M. A. Wysor et al., *Science*, **1982**, *217*, 454). The complexing agent is thought to be diethylcarbamate, a metabolite of disulfiram.

important step in the nitrogen cycle, providing available nitrogen for plant nutrition. It is an intriguing process since it occurs readily in various bacteria, blue-green algae, yeasts, and in symbiotic bacteria–legume associations under mild conditions. However, nitrogen stubbornly resists ordinary chemical attack, even under stringent conditions.

Molecular nitrogen, N_2, is so unresponsive to ordinary chemical reactions that it has been characterized as "almost as inert as a noble gas."[86] The very large triple bond energy (945 kJ mol^{-1}) tends to make the activation energy prohibitively large. Thus, in spite of the fact that the overall enthalpy of formation of ammonia is exothermic by about 50 kJ mol^{-1}, the common Haber process requires about 20 MPa pressure and 500 °C temperature to proceed, even in the presence of the best Haber catalyst. In addition to the purely pragmatic task of furnishing the huge supply of nitrogen compounds necessary for industrial and agricultural uses as cheaply as possible, the chemist is intrigued by the possibility of discovering processes that will work under less drastic conditions. We *know* they exist: We can *watch* a clover plant growing at 100 kPa and 25°C!

In vitro nitrogen fixation

The discovery that molecular nitrogen was capable of forming stable complexes with transition metals (p. 615) led to extensive investigation of the possibility of fixation of nitrogen via such complexes. Of the various systems investigated, that employing titanium(II) was the first to be successful. Titanium(II) alkoxides form dinitrogen complexes which may then be reduced with subsequent release of ammonia or hydrazine:

$$\mathrm{Ti(OR)_4} + 2e^- \longrightarrow \mathrm{Ti(OR)_2} + 2\mathrm{RO}^- \quad \textbf{(18.21)}$$

$$\mathrm{Ti(OR)_2} + \mathrm{N_2} \longrightarrow [\mathrm{Ti(OR)_2N_2}] \quad \textbf{(18.22)}$$

$$[\mathrm{Ti(OR)_2N_2}] + 4e^- \longrightarrow [\mathrm{Ti(OR)_2N_2}]^{-4} \quad \textbf{(18.23)}$$

$$[\mathrm{Ti(OR)_2N_2}]^{-4} \begin{cases} \xrightarrow{4\mathrm{H}^+} \mathrm{N_2H_4} + \mathrm{Ti(OR)_2} \\ \xrightarrow{2e^-} [\mathrm{Ti(OR)_2N_2}]^{-6} \end{cases} \quad \textbf{(18.24)}$$

$$[\mathrm{Ti(OR)_2N_2}]^{-6} + 6\mathrm{H}^+ \longrightarrow 2\mathrm{NH_3} + \mathrm{Ti(OR)_2} \quad \textbf{(18.25)}$$

The exact nature of the dinitrogen complexes is unknown. Under certain conditions the starting materials may be regenerated (in part), making the reaction catalytic, although thus far only a few cycles are possible before the "catalytic materials" are depleted. Such a process is not apt to be commercially competitive with the Haber process for the synthesis of ammonia but promises to be useful in the synthesis of other nitrogen compounds such as hydrazine (Eq. 18.24) and organic nitrogen compounds.[87]

---

[86] W. L. Jolly, "The Chemistry of the Non-Metals," Prentice-Hall, Englewood Cliffs, N.J., **1966**, p. 72.

[87] See E. E. Van Tamelen, *Acc. Chem. Res.*, **1970**, *3*, 361; E. E. van Tamelen and H. Rudler, *J. Am. Chem. Soc.*, **1970**, *92*, 5253.

Until recently, all methods for converting dinitrogen complexes into ammonia required very powerful reducing agents—the dinitrogen in the complex was almost as unreactive as atmospheric N_2. An important development was the discovery[88] that certain phosphine complexes of molybdenum and tungsten containing dinitrogen readily yield ammonia in acidic media:

$$[MoCl_3(thf)_3] + 3e^- + 2N_2 + \text{excess dpe} \longrightarrow [Mo(N_2)_2(dpe)_2] + 3Cl^- \quad \textbf{(18.26)}$$

$$[Mo(N_2)_2(dpe)_2] + 6H^+ \longrightarrow 2NH_3 + N_2 + Mo^{VI}\text{ products} \quad \textbf{(18.27)}$$

where thf = tetrahydrofuran and dpe = 1,2-bis(diphenylphosphino) ethane, ϕ_2PCH_2-$CH_2P\phi_2$. Both reactions take place at room temperature and atmospheric pressure. The reducing agent is a Grignard reagent. This reaction sequence is important for two reasons: (1) It models the *in vivo* nitrogenase systems that appear to employ molybdenum (see below); (2) it may provide insight into the development of useful catalysts for the industrial fixation of nitrogen.

With respect to the latter process, it is a certainty that the chemist will not be able to keep pace with the lively imagination of the journalist. As an interesting aside on the inherent inability of the scientist to match ever increasing expectations, the reader is directed to the following selection of titles and headlines. The first is the title of the initial research report by Chatt's group in England and the remainder are headlines of various reports of it in the popular press:[88]

The reduction of mono-co-ordinated molecular nitrogen to ammonia in a protic environment	*Nature* (Jan. 3, 1975)
Fuel-saving way to make fertiliser	*The Times* (Jan. 3, 1975)
Fuel break-through	*The Guardian* (Jan. 3, 1975)
More progress in nitrogen fixation	*New Scientist* (Jan. 9, 1975)
Cheaper nitrogen by 1990	*Farmer's Weekly* (Jan. 10, 1975)
Basic life process created in UK lab	*The Province* (British Columbia, Jan. 15, 1975)

With each re-telling the story grew, until by the time it reached British Columbia, it appeared that the press was *almost* able in 12 days to duplicate what is recorded as a 6-day event in Genesis! The resultant disappointment when the scientist is not able to meet expectations benefits neither him nor the public.

In vivo nitrogen fixation

There are several bacteria and blue-green algae that can fix molecular nitrogen *in vivo*. Both free-living species and symbiotic species are involved. There are the strictly anaerobic

[88] J. Chatt et al., *Nature*, **1975**, *253*, 39; J. Chatt, *J. Organometal. Chem.*, **1975**, *100*, 17. For a very readable account of "The Nitrogen Problem," including the headlines given above, see J. Chatt, *Proc. Roy. Instn. Gt. Br.*, **1976**, *49*, 281.

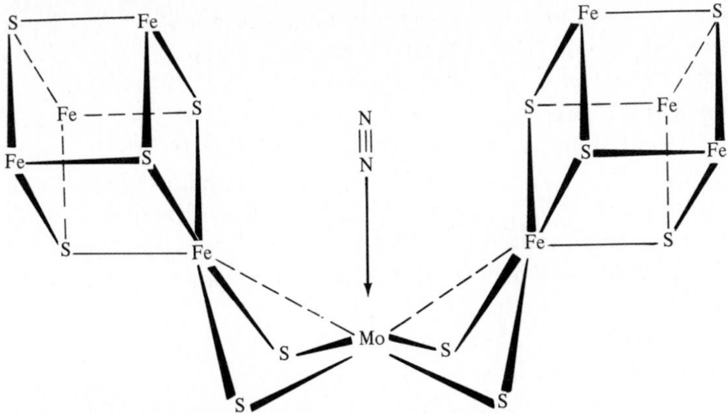

Fig. 18.32 Proposed structure of active site in nitrogenase. The dinitrogen molecule can coordinate to the molybdenum atom and be supplied with electrons from the ferredoxin-like Fe–S clusters.

Clostridium pasteurianum,[89] facultative aerobes like *Klebsiella pneumoniae*, and strict aerobes like *Azotobacter vinelandii*. Even in the aerobic forms it appears that the nitrogen fixation takes place under essentially anaerobic conditions (see below). The most important nitrogen-fixing species are the mutualistic species of *Rhizobium* living in root nodules of various species of legumes (clover, alfalfa, beans, peas, etc.).

The active enzyme in nitrogen fixation is nitrogenase. It is not a unique enzyme but appears to differ somewhat from species to species. Nevertheless the various enzymes are very similar. Two proteins are involved. The smaller has a molecular weight of 57 000–73 000. It contains an Fe_4S_4 cluster. The larger protein is an $\alpha_2\beta_2$ tetramer with a molecular weight of 220 000–240 000 containing two molybdenum atoms, about 30 iron atoms, and about 30 labile sulfide ions.[90] The iron–sulfur clusters probably act as redox centers. A suggested arrangement of these components that accounts for observed properties is shown in Fig. 18.32. Neither protein by itself shows any nitrogen-fixing ability, but recombination of them gives *immediate* activity, indicating that no slow winding of protein chains of the two components is necessary. Interestingly, bacteria grown in the presence of WO_4^{-2} instead of MoO_4^{-2} incorporate the tungsten into the nitrogenase, but the latter is inactive.[91]

It seems most likely that the active site involves one or both molybdenum atoms. It has been established by EXAFS[92] that the coordination sphere of the molybdenum

---

[89] *Clostridium* includes, in addition to the useful nitrogen-fixing *C. pasteurianum*, the other dangerous anaerobic species: *C. tetani* (causes tetanus, "lockjaw"), *C. botulinum* (causes botulism), and *C. welchi* (causes "gas gangrene").

[90] E. I. Stiefel et al., "Bioinorganic Chemistry—II," *Advances in Chemistry Series, No. 162*, K. N. Raymond, ed., American Chemical Society, Washington, D. C., **1977**, pp. 353–388; *Progr. Inorg. Chem.*, **1977**, *22*, 1; M. J. Nelson et al., *Adv. Inorg. Biochem.*, **1982**, *4*, 1.

[91] W. J. Brill et al., *J. Bact.*, **1974**, *118*, 986; H. H. Nagatani et al., ibid., **1974**, *120*, 697.

[92] S. P. Cramer et al., *J. Am. Chem. Soc.*, **1978**, *100*, 3398, 3814.

consists of several sulfur atoms at distances of about 235 pm. A Mo=O double bond, so common in complexes of Mo(IV) and Mo(VI), is *not* present. There are other heavy atoms, perhaps iron, nearby (~270 pm). The ultimate source of reductive capacity is pyruvate, and the electrons are transferred via ferredoxin (see p. 865) to nitrogenase.[93] There is some evidence, not strong, that Mo(III) is involved. Two Mo(III) atoms cycling through Mo(VI) would provide the six electrons necessary for reduction of dinitrogen. Alternatively, since the enzyme is rich in ferredoxin-type clusters, there should be a ready flow of electrons, and the molybdenum may stay in the one or two oxidation states that most readily bind dinitrogen and its intermediate reductants. The overall catalytic cycle may resemble that shown in Eq. 18.28:[94]

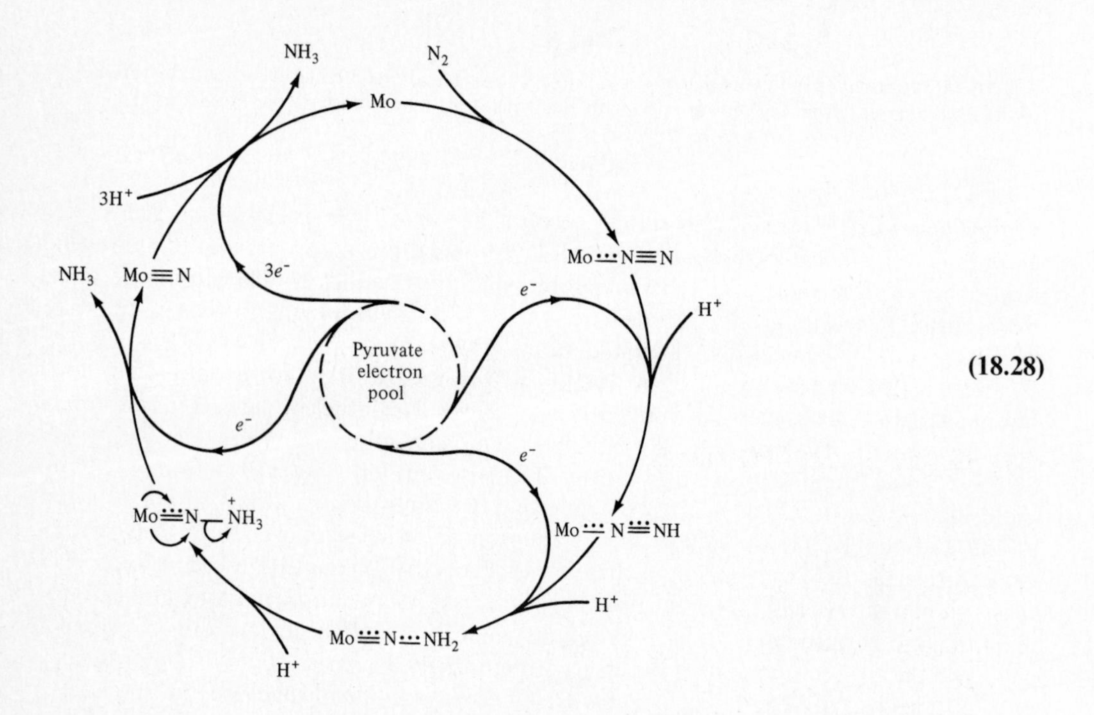

(18.28)

A schematic diagram for the production of fixed nitrogen compounds, including the sources of materials and energy, and the overall reactions, is given in Fig. 18.33. Note the presence of leghemoglobin. This is a monomeric, oxygen-binding molecule rather closely resembling myoglobin. It is felt that the leghemoglobin binds any oxygen that is present very tightly and thus protects the nitrogenase, which cannot operate in the presence of oxygen. On the other hand, it allows a reservoir of oxygen for respiration to supply

[93] H. R. Mahler and E. H. Cordes, "Biological Chemistry," 2nd ed., Harper & Row, New York, **1971**, pp. 757–758; A. L. Lehninger, "Biochemistry," 2nd ed., Worth, New York, **1970**, p. 722.

[94] J. Chatt, in "Biomimetic Chemistry," D. Dolphin et al., eds., *ACS Adv. Chem. Ser. No. 191*, Washington, D.C., **1980**, Chapter 21.

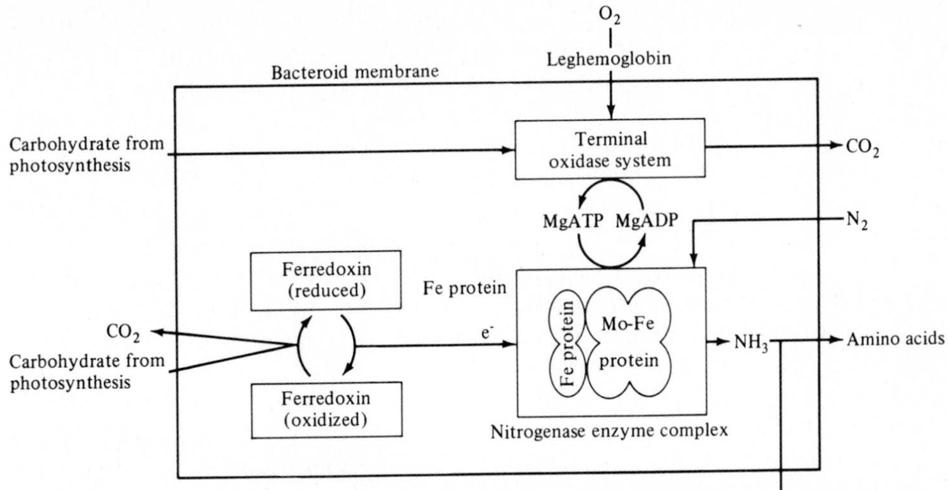

Fig. 18.33 Schematic diagram of nitrogenase activity in a bacterial cell. Carbohydrate provides reducing capacity (ferredoxin), energy (MgATP), and organic precursors for the manufacture of amino acids. [From K. J. Skinner, *Chem. Eng. News*, Oct. 4, **1976**, p. 23. Reproduced with permission.]

energy to keep the fixation process going.[95] Research on the nitrogenase system is just reaching the point where we are beginning to understand the processes involved.[96]

THE BIOCHEMISTRY OF IRON

It is impossible to cover adequately the chemistry of various elements in biological systems in a single chapter. Before discussing the salient points of other essential and trace elements, the biochemistry of iron will be discussed briefly. Iron is the most abundant transition element and serves more biological roles than any other metal. It can therefore serve to illustrate the possibilities available for the absorption, storage, handling, and use of an essential metal. Iron has received much study, and similar results can be expected for other metals as studies of the chemistry of trace elements in biological systems advance.

Availability of iron

Although iron is the fourth most abundant element in the earth's crust, it is not always readily available for use. Both $Fe(OH)_2$ and $Fe(OH)_3$ have very low solubilities, the

[95] J. B. Wittenberg et al., *J. Biol. Chem.*, **1974**, *249*, 4056; *Ann. N. Y. Acad. Sci.*, **1975**, *244*, 28; C. A. Appleby, in "Biological Nitrogen Fixation," A. Quispel, ed., North-Holland, Amsterdam, **1974**.

[96] See R. Murray and D. C. Smith, *Coordin. Chem. Rev.*, **1968**, *3*, 429; J. T. Spence, *Coordin. Chem. Rev.*, **1969**, *4*, 475; W. E. Newton et al., *J. Am. Chem. Soc.*, **1971**, *93*, 268; R. H. Burris, *Proc. Roy. Soc.*, **1969**, *B172*, 339; S. L. Streicher and R. C. Valentine, *Ann. Rev. Biochem.*, **1973**, *42*, 279; R. A. D. Wentworth, *Coordin. Chem. Rev.*, **1976**, *18*, 1; E. I. Stiefel, *Progr. Inorg. Chem.*, **1977**, *22*, 1; M. N. Hughes, "The Inorganic Chemistry of Biological Processes," 2nd ed., Wiley, New York, **1981**, Chapter 6.

latter especially so ($K_{sp}^{II} = 8 \times 10^{-16}$; $K_{sp}^{III} = 2 \times 10^{-39}$).[97] An extreme example is iron deficiency in pineapples grown on rust-red soil on Oahu Island containing over 20% Fe, but none of it available because it is kept in the +3 oxidation state by the presence of manganese dioxide and the absence of organic reducing agents.[98] Similarly under alkaline conditions in the soil (e.g., in geographic regions where the principal rocks are limestone and dolomite) even iron(II) is not readily available to plants. The stress is especially severe on those species such as rhododendron and azalea that naturally live in soils of low pH. Under these circumstances gardeners and farmers often resort to the use of "iron chelate," and EDTA complex. The latter is soluble and makes iron available to the plant for the manufacture of cytochromes, ferredoxins, etc. The clever application of coordination chemistry by the chemical agronomist was predated by some hundreds of millions of years by certain higher plants. Some have evolved the ability to exude various chelating organic acids through the root tips to solubilize the iron so that it may be absorbed.

The presence of organic chelates of iron in surface waters has been related to the "red tide." The latter is an explosive "bloom" of algae (*Gymnodium breve*) that results in mass mortality of fish. It is possible to correlate the occurrence of these outbreaks with the volume of stream flow and the concentrations of iron and humic acid.[99] At least one of the dinoflagellates in the red tide possesses an iron-binding siderophore (see below).[99a]

Within the organism a variety of complexing agents are used to transport the iron. In higher animals it is carried in the bloodstream by the *transferrins*. These iron-binding proteins are responsible for the transport of iron to the site of synthesis of other iron-containing compounds (such as hemoglobin and the cytochromes) and its insertion via enzymes into the porphyrin ring.[100] The iron is present in the +3 oxidation state (Fe^{+2} does not bind) and is coordinated to two or three tyrosyl residues, a couple of histidyl residues, and perhaps a tryptophanyl residue in a protein chain of molecular weight about 80 000.[101] There are two iron-binding sites per molecule.

Most aerobic microorganisms have analogous compounds, called siderophores, which solubilize and transport iron(III). They have relatively low molecular weights (500–1000) and, depending upon their molecular structure and means of chelating iron, are classified into several groups such as the ferrichromes, ferrioxamines, and enterobactins. Some examples are shown in Fig. 18.34. It is obvious that these molecules are polydentate ligands with many potential ligating atoms to form chelates. They readily form extremely stable octahedral complexes with high-spin Fe(III). Although the complexes are very stable, which is extremely important to their biological function (see below), they are

---

[97] K. N. Raymond, in "Bioinorganic Chemistry—II," K. N. Raymond, ed., *ACS Adv. Chem. Ser. No. 162*, Washington, D.C., **1977**, Chapter 2.

[98] R. C. Brasted, *J. Chem. Educ.*, **1947**, *25*, 495; **1970**, *47*, 634.

[99] R. M. Ingle and D. F. Martin, *Environ. Letters*, **1971**, *1*, 69; D. F. Martin et al., *Univ. South Florida Prof. Pap. No. 12*, **1971**, 1; M. T. Doig III and D. F. Martin, *Water Res.*, **1974**, *8*, 601; Y. S. Kim and D. F. Martin, ibid., **1974**, *8*, 607; D. F. Martin and B. B. Martin, *J. Chem. Educ.*, **1976**, *53*, 614.

[99a] C. G. Trick et al., *Science*, **1983**, *219*, 306.

[100] A. Bezkorovainy, "Biochemistry of Nonheme Iron," Plenum Press, New York, **1980**, Chapter 4; R. J. P. Williams, *Endeavour*, **1967**, *26*, 96; E. D. Weinberg, *Science*, **1974**, *184*, 952; I. Kochan, in "Bioinorganic Chemistry—II," K. N. Raymond, ed., *ACS Adv. Chem. Ser. No. 162*, Washington, D.C., **1977**, pp. 55–77.

[101] M. Llinás, *Struct. Bond.*, **1973**, *17*, 135.

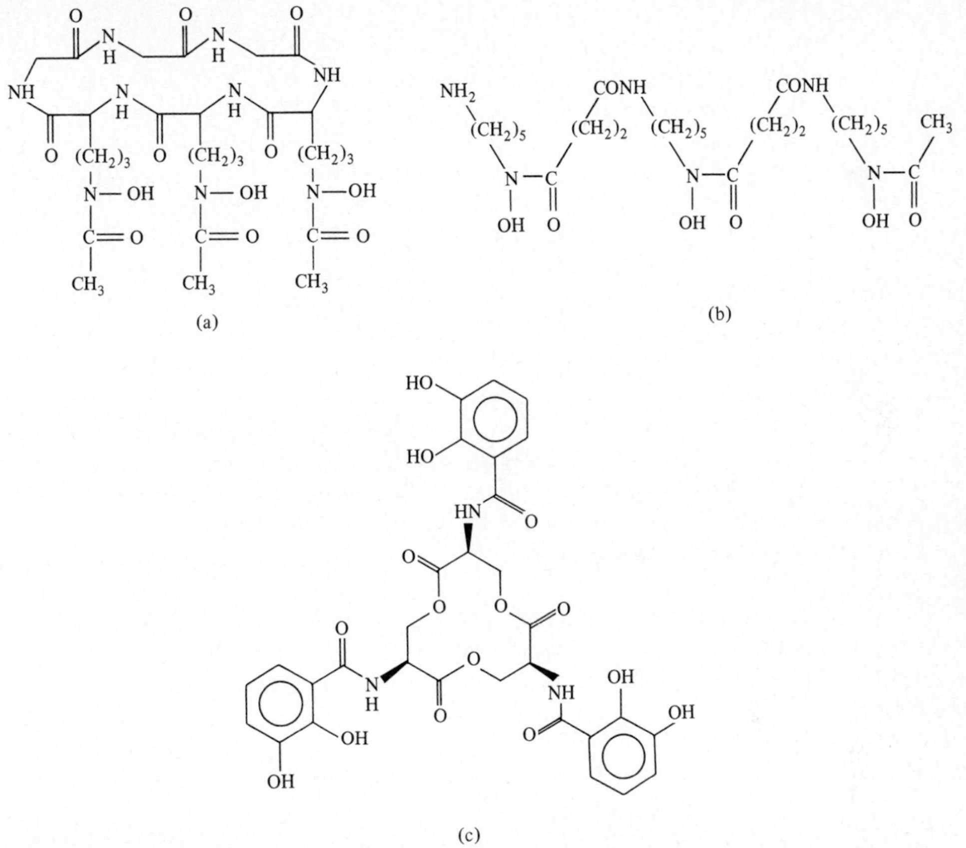

(a)

(b)

(c)

Fig. 18.34 Three types of bacterial siderophores: (a) desferrichrome; (b) desferrioxamine B; (c) enterobactin.

labile, which allows the iron to be transported and transferred within the bacteria.[102] The ferrichromes and ferrioxamines are trihydroxamic acids which form neutral trischelates from three bidentate hydroxamate monoanions. Enterobactin contains a different chelating functional group, *o*-dihydroxybenzene ("catechol"). Each catechol group in enterobactin behaves as a dianion for a total charge of −6 for the ligand. A characteristic of all of these is that, in addition to the natural tendency of trischelates to form globular complexes, the remainder of the siderophore molecule consists of a symmetric, hydrophilic portion that presumably aids in transport across the cell membrane[102] (Fig. 18.35).

Competition for iron

In addition to the transport of iron, the transferrins of higher animals and the siderophores of bacteria show another interesting parallel. It is most readily shown by conalbumin (= ovatransferrin) of egg white, though we shall see several other examples. There is a large amount, up to 16%, in the protein in egg white, although it has not been possible

[102] J. B. Neilands, in "Bioinorganic Chemistry—II," K. N. Raymond, ed., *ACS Adv. Chem. Ser. No. 162*, Washington, D.C., **1977**, Chapter 1.

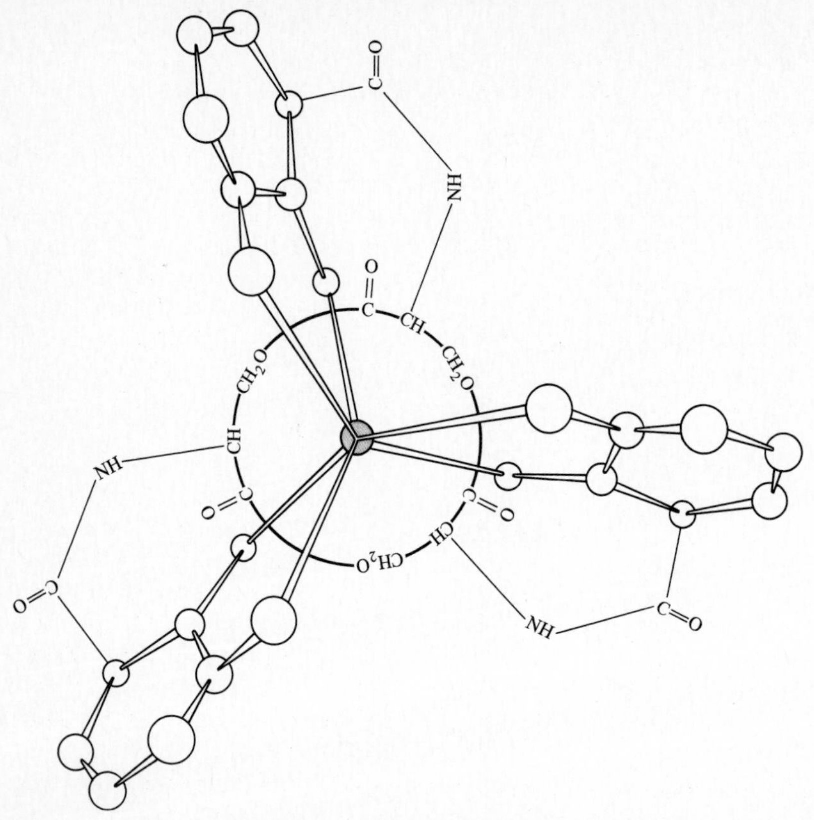

Fig. 18.35 The △-*cis* isomer of iron(III) enterobactin. The metal lies at the center of a distorted octahedron of the six oxygen atoms of the three catechol ligands. [From S. S. Isied et al., *J. Am. Chem. Soc.*, **1976**, *98*, 1763. Reproduced with permission.]

to find an iron-transporting function for it here. In general, the transferrins have larger stability constants toward iron(III) than do the various siderophores. It is thus quite likely that conalbumin acts as an antibacterial agent. In the presence of excess conalbumin, bacteria would be iron deficient since the siderophore cannot compete successfully for the iron.[100]

Lactoferrin, found in mother's milk, appears to be the most potent antibacterial transferrin and seems to play a role in the protection of breast-fed infants from certain infectious diseases. It has been claimed that milk proteins remain intact in the infant's stomach for up to 90 minutes and then pass into the small intestine unchanged, thus retaining their iron-binding capacity. In guinea pigs, addition of hematin to the diet abolishes the protective effects of the mother's milk.[103]

---

[103] A. Bezkorovainy, "Biochemistry of Nonheme Iron," Plenum Press, New York, **1980**, pp. 336–337.

The question of iron chelation as an antibacterial defense is receiving increasing attention. It appears to be far more general than had previously been supposed.[104] An interesting sidelight is that the fever that often accompanies infection enhances the bacteriostatic action of the body's transferrins.[105]

Another interesting example of this sort is the competition between bacteria and the roots of higher plants. Both use chelators to win iron from the soil. However, higher plants have one more mechanism with which to compete: The Fe(III) is reduced and absorbed by the roots in the uncomplexed Fe(II) form. When EDTA and other chelating agents are used to correct chlorosis in plants due to iron deficiency, the action is merely one of solubilizing the Fe(III) and making it physically accessible to the roots—the chelates are not absorbed intact. Indeed, chelates that strongly bind Fe(II) may actually inhibit iron uptake from the root medium.[106]

Iron toxicity and nutrition

Exactly the opposite problem may occur for plants whose roots are growing in anaerobic media. In flooded soils the roots may be exposed to high levels of Fe(II), posing potential problems of iron toxicity. Rice plants and water lilies with roots in anaerobic soils transport gaseous oxygen (from the air or photosynthesis, or both) to the periphery of the roots where it oxidizes the Fe(II) to Fe(III). In this case the insolubility of iron(III) hydroxide is utilized *to protect* the plant from iron poisoning.[106,107] A similar problem from too *much* iron occurs in parts of Africa where cooking and brewing is done in iron pots. Fermenting mixtures of low pH solubilize the iron and an excessive amount enters the diet, resulting in siderosis.[108]

It should be noted in this connection that the body has no mechanism for the excretion of iron, and except for women in the child-bearing years, the dietary requirement of iron is extremely low.

The absorption of iron in the gut, preferentially in the +2 oxidation state, was once thought to be a result of special physiological mechanisms, but now is generally agreed to be merely another aspect of the differential solubility of $Fe(OH)_2$ and $Fe(OH)_3$.[109] However, there *is* a significant differential in the absorption of heme versus nonheme iron: Heme iron is absorbed 5–10 times more readily than nonheme iron.[110] Since meat contains large quantities of hemoglobin, myoglobin, and cytochromes, this difference could be nutritionally significant.

It is conceivable that iron could be stored in the form of a complex such as transferrin or even hemoglobin, and in lower organisms ferrichrome apparently serves this purpose.

[104] E. D. Weinberg, *Microbiol. Rev.*, **1978**, *42*, 45; K. N. Raymond, in "Bioinorganic Chemistry—II," K. N. Raymond, ed., *ACS. Adv. Chem. Ser. No. 162*, Washington, D.C., **1977**, Chapter 8.

[105] M. J. Kluger and B. A. Rothenburg, *Science*, **1979**, *203*, 374.

[106] R. A. Olsen et al., *Am. Sci.*, **1981**, *69*, 378.

[107] J. W. H. Dacey, *Science*, **1980**, *210*, 1017.

[108] C. L. Rollinson and M. G. Enig, "Kirk-Othmer: Encyclopedia of Chemical Technology," 3rd ed., **1981**, Vol. 15, pp. 570–603.

[109] T. G. Spiro and P. Saltman, *Struct. Bond.*, **1969**, *6*, 116.

[110] D. Narins, in A. B. Beskorovainy, "Biochemistry of Nonheme Iron," Plenum Press, New York, **1980**, Chapter 3.

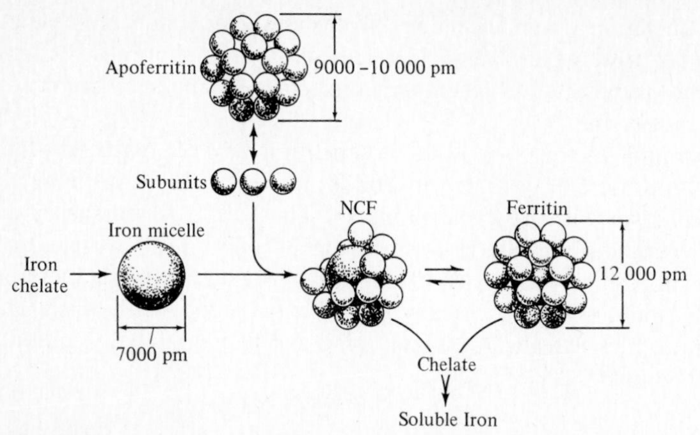

Fig. 18.36 A model for the synthesis of ferritin from protein subunits and the role of the iron micelle in the ultimate structural formation. Mobilization of the iron from the micelle by chelates is also indicated. [From T. G. Spiro and P. Saltman, *Struct. Bonding*, **1969**, *6*, 116. Reproduced with permission.]

Such storage is wasteful, however, and higher animals have evolved a simpler method of storing the iron. The iron is stored in the form of *ferritin*, which in a sense is simply a small particle of iron(III) hydroxide about 7000 pm in diameter. The structure of the polymeric core involves phosphate groups as well as hydroxide and appears to be such as to bind a proteinaceous covering (called *apoferritin*) to the iron core. The complete ferritin particle provides an available source from which chelates in the biological system can pick up iron and transfer it to usable sites (Fig. 18.36). It thus provides high-density storage of inorganic iron combined with ready availability.[109,111]

The importance of iron-containing proteins such as hemoglobin, myoglobin, the cytochromes, leghemoglobin, and ferredoxin in respiration has been mentioned previously. The structure and function of the heme-containing compounds have been discussed on pp. 861–878. The two nonheme compounds will be discussed briefly.

Hemerythrin is a nonheme, oxygen-binding pigment utilized by several phyla of marine invertebrates. Its chief interest to the chemist lies in certain similarities to and differences from hemoglobin and myoglobin. Like both of the latter, hemerythrin contains iron(II) which binds oxygen reversibly, but when oxidized to methemerythrin (Fe^{+3}) it does not bind oxygen.[112] There is a monomeric form, analogous to myoglobin, in the tissues (Fig. 18.37). An octameric form with a molecular weight of about 100 000 transports oxygen. And just as hemoglobin consists of four chains (2α and 2β), each of which is very similar to the single chain of myoglobin, octameric hemerythrin consists of eight subunits very similar to myohemerythrin, differing in only five amino acids. However, these similarities should not be pushed too far. The quaternary structure of hemoglobin is quite

---

[111] R. R. Crichton, ibid., **1973**, *17*, 67; footnote 100, Chapter 5.

[112] I. M. Klotz et al., *Science*, **1976**, *192*, 335; W. A. Hendrickson et al., *Proc. Nat. Acad. Sci. U.S.A.*, **1975**, *72*, 2160; K. B. Ward et al., *Nature*, **1975**, *257*, 818; W. A. Hendrickson and K. B. Ward, *Biochem. Biophys. Res. Commun.*, **1975**, *66*, 1349; *J. Biol. Chem.*, **1977**, *252*, 3012.

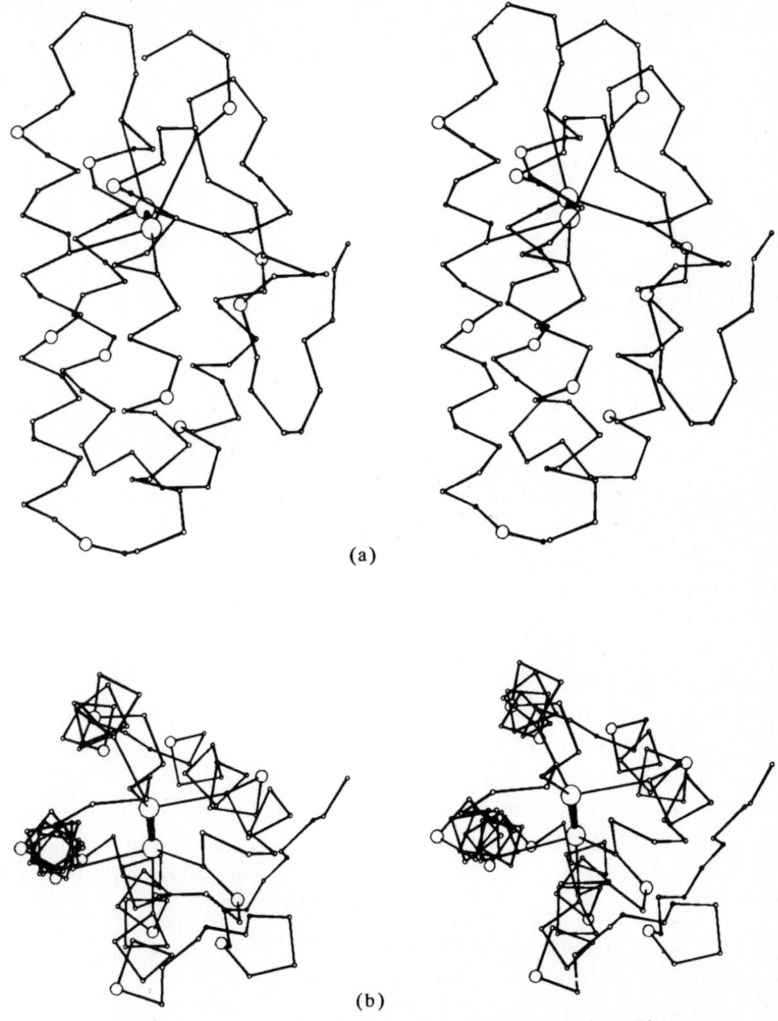

Fig. 18.37 Stereoviews of the myohemerythrin molecule. (a) Front view. (b) Top view. The two large circles represent iron atoms, smaller circles represent α-carbon atoms in the protein chain, every tenth one enlarged. [From W. A. Hendrickson and K. B. Ward, *Biochem. Biophys. Res. Commun.*, **1975**, *66*, 1349. Reproduced with permission.]

stable, but the octameric form of hemerythrin appears to be in dissociative equilibrium with the monomeric subunits in solution. More interesting from the inorganic chemist's point of view is the binding of oxygen. Each oxygen-binding site (whether monomer or octamer) contains *two* iron(II) atoms, and the reaction takes place via a redox reaction to form iron(III) and peroxide (O_2^{-2}). Oxyhemerythrin is diamagnetic, indicating spin coupling of the odd electrons on the two iron(III) atoms. Mössbauer data indicate that the two iron(III) atoms are in different environments in oxyhemerythrin. This could result from the peroxide ion coordinating one iron atom and not the other, or from each of

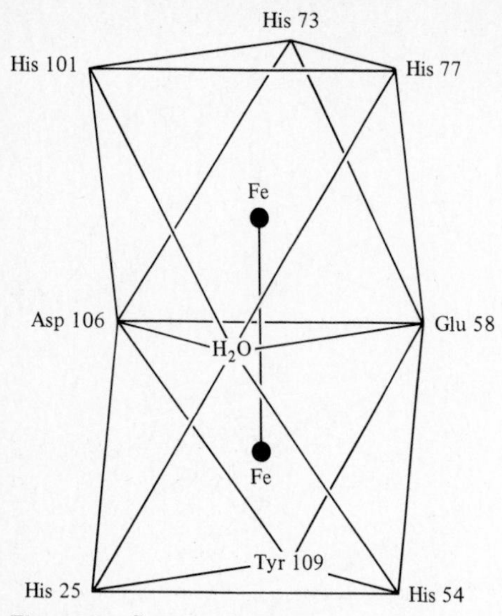

Fig. 18.38 Suggested coordination sphere of the diiron center of hemerythrin. Both iron atoms have approximately octahedral coordination.

the iron atoms having different ligands in its coordination sphere. Indeed, a recent X-ray study has found that in methemerythrin the two iron atoms exist with approximately octahedral coordination with the octahedra sharing a triangular face of water, aspartic acid, and glutamic acid as the three ligands (Fig. 18.38).[113] The remaining ligands are three histidine residues on one iron atom and two histidines and one tyrosine on the other. This is a rather small difference, but can be reconciled with the other data, and so until recently the consensus opinion has tended to favor a simple peroxo bridge between the two iron atoms:

$$\underline{Fe^{II} \quad\quad Fe^{II}} + O_2 \longrightarrow \underline{Fe^{III}\text{---}O\text{—}O^{-2}\text{--}Fe^{III}}$$

I

where the continuous line connecting the two iron atoms is a simplified representation of the coordination spheres and the protein chain holding the iron atoms in place. Militating against the above structure is the fact that the Mössbauer spectrum is not able to distinguish the iron atoms in *deoxyhemerythrin*. If the non-O_2, i.e., amino acid, ligands can differentiate the iron atoms in oxyhemerythrin, why not in deoxyhemerythrin? The strongest data at present come from the Raman spectrum of oxy($^{16}O^{18}O$)hemerythrin, which shows the two oxygen atoms to be in *nonequivalent* positions.[114] Of the various

[113] R. E. Stenkamp et al., *Proc. Nat. Acad. Sci., U.S.A.*, **1976**, *73*, 349.
[114] D. M. Kurtz, Jr., et al., *J. Am. Chem. Soc.*, **1976**, *98*, 5033.

alternative structures that have been proposed, the Raman data are compatible with only two:

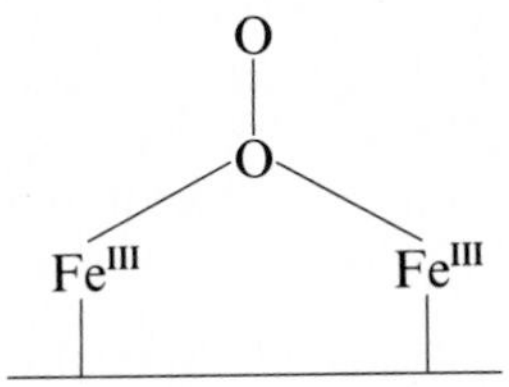

O
O
Fe^{III} Fe^{III}

II

O
O
Fe^{III}--------Fe^{III}

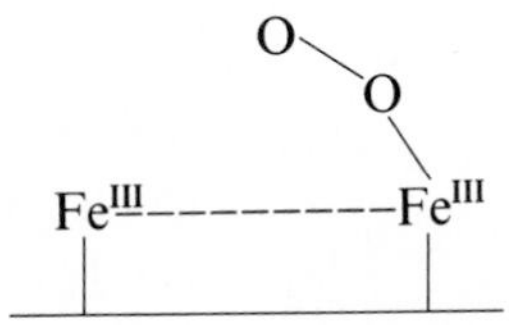

III

Structure **II** suffers from the same problem as **I** above: How can a symmetrically bonded peroxide anion cause differentiation between the iron atoms that is not observable in the deoxy form? Structure **III** is strongly reminiscent of the "bent" coordination of oxygen to hemoglobin seen previously. The dashed line between the two iron atoms does not indicate a full-strength metal–metal bond, but merely that there is spin coupling of the odd electrons, perhaps through another ligand. The summation of the Mössbauer data for oxy- and deoxyhemerythrin, the X-ray structure of methemerythrin, and the Raman $^{16}O^{18}O$ data, as well as recent spectroscopic evidence,[115] is most compatible with structure **III**, but the question is still open to further experimentation.[116] Also, one should not forget that what the chemist glibly calls "hemerythrin" may not be the same from one species to the next: The phyla in which hemerythrin is found comprise thousands of species.

The parallels and differences between hemoglobin and hemerythrin[112] illustrate the way evolution has often solved what is basically the same problem in different ways under different circumstances. A further example of parallel functions by different molecules is provided by the confusingly named *hemocyanin*, which simply means "blue blood"; it contains neither the heme group nor the cyanide ion. It is a copper protein with a molecular weight of 50 000 to 75 000 and, analogously to hemerythrin, requires two copper atoms to bind a single oxygen molecule. It is found in many species in the phyla Mollusca and Arthropoda.[117]

The same problems arise with hemocyanin as with hemerythrin with regard to structure and function; however, less work has been done, and less is known about hemocyanin. It is thought that the coordination of oxygen takes place with a structure similar to **I** above. The coordination sphere about the copper atoms appears to have three or four nitrogen-containing ligands.[118]

---

[115] R. R. Gay and E. I. Solomon, *J. Am. Chem. Soc.*, **1978**, *100*, 1972.

[116] D. M. Kurtz, Jr., et al., *Coord. Chem. Rev.*, **1977**, *24*, 145; R. E. Stenkamp and L. H. Jensen, *Adv. Inorg. Biochem.*, **1979**, *1*, 219; J. S. Loehr and T. M. Loehr, ibid., **1979**, *1*, 235.

[117] See N. M. Senozan, *J. Chem. Educ.*, **1976**, *53*, 684; W. A. Hendrickson, *Trends Biochem. Sci.*, **1977**, *2*, 108.

[118] A. Nakahara et al., in "Biomimetic Chemistry," D. Dolphin et al., eds., *ACS Adv. Chem. Ser. No. 191*, **1980**, Chapter 19; J. Brown et al., *J. Am. Chem. Soc.*, **1980**, *102*, 4210; J. A. Larrabee and T. G. Spiro, ibid., **1980**, *102*, 4217.

ESSENTIAL AND TRACE ELEMENTS IN BIOLOGICAL SYSTEMS

The discussion of metalloporphyrins and metalloenzyme systems has indicated the importance of certain metals in chemical reactions within living organisms. Certain elements are *essential* in that they are absolutely necessary (perhaps in large, perhaps in small quantities) for life processes. Other elements are nonessential since if they are absent other elements may serve the same function. Obviously, determining the "essentiality" of an element is difficult. The term "trace element" although widely used is not precisely defined. For example, molybdenum averages about 1–2 ppm in rocks, soils, plants, and marine animals and even lower in land animals. Yet it is an essential trace metal. At the other extreme, iron, which averages about 5% in rocks and soils and 0.02–0.04% in plants and animals, might or might not be considered a "trace" metal.

Although the role of iron in various heme derivatives and zinc in carboxypeptidase and carbonic anhydrase is clear, there are many instances in which little is known of the function of the trace metal. For example, it was known that ascidians ("sea squirts") concentrated vanadium from sea water by a factor of a millionfold long before it was suggested that vanadium is associated with respiration. However, even *more* recently it has definitely been demonstrated that the vanadium does *not* act as an oxygen carrier.[119] The true function remains obscure. There are many elements that at the present time are known to be useful although no specific function has been proved. The list of known functions is expanding rapidly at the present time, however.

The problem of toxicity is difficult to quantify. There are so many synergistic effects between various components of biological systems that it is almost impossible to define the limits of beneficial and detrimental concentrations. There is also endless variation among organisms. Truly, "one man's meat is another man's poison." The phenomenon of an essential element becoming toxic at higher than normal concentrations is not rare. Selenium is an essential element in mammals yet one of the most vexing problems is the poisoning of livestock from eating plants that concentrate this element (see p. 914).

Space does not permit an extensive discussion of other elements of importance to biological systems. Fortunately, there is an excellent summary available of (1) the availability of the elements in the earth's crust; (2) the cycling of elements in biosystems; (3) uptake and excretion by organisms; (4) essentiality, deficiency, and toxicity of the elements; and (5) function of the elements in biological systems.[120]

The importance of trace elements is manifold and, unfortunately, previously hampered by relatively insensitive analytical methods. Good methods for determining concentrations of 1 ppm or less have been available for relatively few elements; yet these may be the optimum concentrations for a particular trace element. When the responses of living organisms are more sensitive than the laboratory "black boxes," the chemist naturally develops an inferiority complex. Fortunately, the recent development of analytical techniques capable of determining *parts per billion* has opened new vistas for the study of these problems. Some of these techniques are atomic absorption, atomic fluorescence, activation analysis, and X-ray fluorescence.

---

[119] I. G. Macara et al., *Biochem. J.*, **1979**, *181*, 457; *Comments Inorg. Chem.*, **1982**, *2*, 1.

[120] H. J. M. Bowen, "Environmental Chemistry of the Elements," Academic Press, New York, **1979**.

Although the previous discussion has been limited (with the exception of chlorophyll) entirely to transition metals, some of the alkali and alkaline earth metals are important in biological systems as well. These tend to be somewhat more loosely bound than transition metals, and so they are free to move more readily. Sodium and potassium ions are "pumped" across cell membranes against an electrochemical potential. This concentration gradient is intimately involved in the "firing" of nerve impulses. Magnesium and calcium are involved in the process of conversion of chemical energy into work in muscle contraction.[121] Finally, some nonmetals are extremely important as well (see p. 915).

Periodic survey of essential and trace elements

The number of elements that are known to be biologically important comprises a relatively small fraction of the 108 known elements. Natural abundance limits the availability of the elements for such use. Molybdenum ($Z = 42$) is the heaviest metal, and iodine ($Z = 53$) is the heaviest nonmetal of known biological importance. The metals of importance in enzymes are principally those of the first transition series, and the other elements of importance are relatively light: sodium, potassium, magnesium, calcium, carbon, nitrogen, phosphorus, oxygen, chlorine, and, of course, hydrogen.

Some of the elements that have been found to be essential or to be poisonous are listed with their actions in Table 18.1. It is certain that this list will be expanded as the present techniques and theory are improved.[122]

Biological importance and relative abundance

It is perfectly proper to suggest that iron functions well in cytochromes and ferredoxins because the Fe^{+2}/Fe^{+3} redox *emf* has a value in the appropriate range for life processes and, conversely, that mercury is poisonous because it irreversibly binds with enzymes, destroying their activity. One could go through the list, element by element, function by function, and attempt to perceive the correlations. Such attempts are indeed instructive, but a different perspective emerges if one correlates the biological activity with the crustal abundance of a given element.[123] If we look at some typical essential transition elements, we find in addition to Fe, Co, Zn, and Cu mentioned previously, V, Cr, Mn, Ni, and Mo. Representative essential metals are Na, K, Mg, and Ca, and nonmetals (see p. 915) found to be essential are C, N, O, P, S, and Cl. All of these elements except Mo are relatively abundant in the earth's crust (see Table 18.2).[124] When we look for abundant elements

[121] R. J. P. Williams, in "Bioinorganic Chemistry," R. F. Gould, ed., *ACS Adv. Chem. Ser. No. 100*, Washington, D.C., **1971**, Chapter 8.

[122] For a complete list for all elements including also information on abundances in rocks, soil, sea and fresh water, animals, and plants, see footnote 120.

[123] J. E. Huheey, "REACTS 1973, Proceedings of the Regional Annual Chemistry Teaching Symposium," K. Egolf et al., eds., University of Maryland, College Park, **1973**, pp. 52–78.

[124] Because almost all of the earth's crust is silicon dioxide either in pure form (quartz) or in silicates, only about two dozen elements occur with a frequency of one atom per 10 000 atoms of silicon; these are considered "abundant" or "relatively abundant."

Table 18.1 Function and toxicity of the elements in biological systems

Atomic number	Element	Biological functions	Toxicity[a]	Comments
1	Hydrogen	Molecular hydrogen metabolized by some bacteria.		Constituent of water and all organic molecules. D_2O is toxic to mammals.
2	Helium	None known.		Used to replace nitrogen as an O_2 diluent in breathing mixtures to prevent the "bends" in high-pressure work.
3	Lithium	None known.	Slightly toxic.	Used pharmacologically to treat manic-depressive patients.
4	Beryllium	None known.	Very toxic.	Pollution occurs from industrial smokes. There are some fears concerning poisoning from camping lantern mantles.[b]
5	Boron	Unknown, but essential for green algae and higher plants.	Moderately toxic to plants; slightly toxic to mammals.	
6	Carbon	Synthesis of all organic molecules and of biogenetic carbonates.	Carbon monoxide is slightly toxic to plants and very toxic to mammals; CN^- is very toxic to all organisms.	Carbon dioxide and CO are global pollutants from burning fossil fuels; CN^- is a local pollutant of rivers near mines.
7	Nitrogen	Synthesis of proteins, nucleic acids, etc. Steps in the nitrogen cycle (organic $N \rightarrow NH_3 \rightarrow NO_2^- \rightarrow NO_3^- \rightarrow N_2 \rightarrow$ organic N) are important activities of certain microorganisms.	Ammonia is toxic at high concentrations.	Leaching of nitrogenous fertilizers from agricultural land and nitrogenous materials in sewage cause serious water pollution.
8	Oxygen	Structural atom of water and most organic molecules in biological systems; required for respiration by most organisms.	Induces convulsions at high P_{O_2}; very toxic as ozone, superoxide, peroxide, and hydroxyl radicals.[d]	
9	Fluorine	Essential element, 2.5 ppm in diet for optimal growth.[c] Strengthens teeth in mammals and used as CaF_2 in some molluscs.	Moderately toxic; may cause mottled teeth.	Pollution by fluoride present in superphosphate fertilizers.
10	Neon	None known.		

11	Sodium	Important in nerve functioning in animals. Major component of vetebrate blood plasma. Activates some enzymes.	Relatively harmless except in excessive amounts. Associated with some forms of hypertension.	Tolerance of and/or dependence upon sodium chloride can be an important consideration in the survival of plants and aquatic animals. This depends upon osmotic regulation rather than sodium specificity.
12	Magnesium	Essential to all organisms. Present in all chlorophylls. Has other electrochemical and enzyme-activating functions.		May cause deficiencies of other elements (e.g., Fe) by the effect of the alkalinity of dolomite.
13	Aluminum	May activate succinic dehydrogenase and δ-aminolevulinate hydrolase.[e] The latter is involved in porphyrin synthesis.[f]	Moderately toxic to most plants; slightly toxic to mammals. Suggested as involved in the etiology of Alzheimer's disease.[g]	Relatively inaccessible except in acidic media as a result of insolubility of $Al(OH)_3$.
14	Silicon	Essential element for growth and skeletal development in chicks and rats;[h] probably essential in higher plants. Used in the form of silicon dioxide for structural purposes in diatoms, some protozoa, some sponges, limpets, and one family of plants.	Not chemically toxic, but large amounts of finely divided silicates or silica are injurious to the mammalian lung.	Finely divided asbestos from construction work may pose a health problem. Some evidence for a negative correlation between silicon content of drinking water and heart disease.[h]
15	Phosphorus	Important constituent of DNA, RNA, bones, teeth, some shells, membrane phospholipids, ADP and ATP, and metabolic intermediates.	Inorganic phosphates are relatively harmless; P_4 and PH_3 are very toxic in mammals and fish. Phosphate esters are used as insecticides (nerve poisons).	Leached from fertilizers applied to agricultural land; present in detergents and other sewage sources.
16	Sulfur	Essential element in most proteins; important in tertiary structure (through S—S links) of proteins; involved in vitamins, fat metabolism, and detoxification processes.[i] H_2SO_4 in digestive fluid in ascidians ("sea squirts"); H_2S replaces H_2O in photosynthesis of some bacteria; H_2S and S_8 are oxidized by other bacteria.	Elemental sulfur is highly toxic to most bacteria and fungi, relatively harmless to higher organisms. H_2S is highly toxic to mammals; SO_2 is highly toxic.	Sulfur dioxide is a serious atmospheric pollutant, especially serious when it settles in undisturbed pockets; oxidized to H_2SO_4. Also causes acid drainage from mines.

Table 18.1 *(Continued)*

Atomic number	Element	Biological functions	Toxicity[a]	Comments
17	Chlorine	Essential for higher plants and mammals. NaCl electrolyte; HCl in digestive juices; impaired growth in infants has been linked to chloride deficiency.	Relatively harmless as Cl^-. Highly toxic in oxidizing forms: Cl_2, ClO^-, ClO_3^-.	
18	Argon	None known.		
19	Potassium	Essential to all organisms with the possible exception of blue-green algae; major cation of cytoplasm; important in nerve action and cardiac function.	Moderately toxic to mammals when injected intravenously; otherwise harmless.	Possible pollution problem from leaching of fertilizers from agricultural land.
20	Calcium	Essential for all organisms; used in cell walls, bones, and some shells as structural component; important electrochemically and involved in blood clotting.	Relatively harmless.	May cause deficiencies of other elements (e.g., Fe) by effect of alkalinity of limestone.
21	Scandium	None known.	Scarcely toxic.	
22	Titanium	None known, but it tends to be accumulated in siliceous tissues.[e]	Relatively harmless.	Relatively unavailable because of insolubility of TiO_2.
23	Vanadium	Essential to ascidians ("sea squirts"), which concentrate it a millionfold from sea water. Essential to chicks and rats. Deficiencies cause reduced growth, impaired reproduction and survival of young, impaired tooth and bone metabolism.[c,f] Thought to inhibit cholesterol biosynthesis in mammals; has a beneficial effect against tooth decay.	Highly toxic to mammals if injected intravenously.	Possible pollutant from industrial smokes—may cause lung disease.

24	Chromium	Essential, functioning as a glucose tolerance factor. It is related to insulin in its biological role and thus to sugar metabolism and diabetes.[c]	Highly toxic as Cr^{VI}, moderately so as Cr^{III}; carcinogenic.	Potential pollutant since amount used industrially is large compared with normal biological levels; normally relatively unavailable because of low solubility.
25	Manganese	Essential to all organisms; activates numerous enzymes; deficiencies in soils lead to infertility in mammals, bone malformation in growing chicks.	Moderately toxic.	
26	Iron	Essential to all organisms. See text.	Normally only slight toxicity, but excessive intake can cause siderosis and damage to organs through excessive iron storage (hemochromatosis).[i]	A very abundant element (5% of earth's crust); may not be available at high pHs.
27	Cobalt	Essential for many organisms including mammals; activates a number of enzymes; vitamin B_{12}.	Very toxic to plants and moderately so injected intravenously in mammals.	Extensive areas are known where low soil cobalt affects the health of grazing animals.[e]
28	Nickel	Essential trace element. Chicks and rats raised on deficient diet show impaired liver function and morphology;[c,k] stabilizes coiled ribosomes.[e]	Very toxic to most plants, moderately so to mammals; carcinogenic.	Local industrial pollutant of air and water.
29	Copper	Essential to all organisms; constituent of redox enzymes and hemocyanin.	Very toxic to most plants; highly toxic to invertebrates, moderately so to mammals.	Pollution from industrial smoke and possibly from agricultural use. Wilson's disease, genetic recessive, results in toxic increase in copper storage.
30	Zinc	Essential to all organisms; used in enzymes; stabilizes coiled ribosomes. Plays a role in sexual maturation and reproduction.	Moderately to slightly toxic; orally causes vomiting and diarrhea.[l]	Pollution from industrial smoke may cause lung disease; use of zinc promotes cadmium pollution. Certain areas (e.g., Iran and Egypt) are zinc deficient.[h]
Most of the heavier elements are comparatively unimportant biologically. Some of the exceptions are:				
33	Arsenic	Essential element in red algae and mammals; function unknown—improves hair growth??[e]	Moderately toxic to plants, highly toxic to mammals.	Serious pollution problems in some areas; sources include mining, burning coal, impure sulfuric acid, insecticides, and herbicides.

Table 18.1 *(Continued)*

Atomic number	Element	Biological functions	Toxicity[a]	Comments
34	Selenium	Essential to mammals and some higher plants. Component of glutathione peroxidase, protects against free-radical oxidant stressors; protects against heavy ("soft") metal ions.[e,h,i]	Moderately toxic to plants, highly toxic to mammals.	Livestock grown on soils high in selenium are poisoned by eating *Astragalus* ("locoweed"), which concentrates it; sheep grown on land deficient in selenium develop "white muscle disease." Deficiency of selenium involved in Keyshan's disease in China.
35	Bromine	May be essential in red algae and mammals.	Nontoxic except in oxidizing forms, e.g., Br_2.	Function unknown, but found in the molluscan pigment, royal purple.
42	Molybdenum	Essential to all organisms with the possible exception of green algae; used in enzymes connected with nitrogen fixation and nitrate reduction.	Moderately toxic and antagonistic to copper—molybdenum excesses in pasturage can cause copper deficiency.[i] Excessive exposure in parts of U.S.S.R. associated with a gout-like syndrome.[h]	Pollution from industrial smoke may be linked with lung disease.
48	Cadmium	Possibly essential in rats.[e]	Moderately toxic to all organisms; a cumulative poison in mammals, causing renal failure; possibly linked with hypertension in man.	Has caused serious disease ("itai itai") in Japan from pollution. May also pose pollution problem associated with industrial use of zinc, e.g., galvanization.
50	Tin	Essential element necessary for growth in rats at levels of 1–2 ppm in diet. Biological function unknown.[c]	Organotin compounds used as bacteriostats and fungistats; hardly toxic in higher plants and animals.	
53	Iodine	Essential in many organisms; thyroxin important in metabolism and growth regulation, amphibian metamorphosis.	Scarcely toxic as the iodide; low iodide availability in certain areas increases the incidence of goiter; largely eliminated by the use of iodized salt. Elemental iodine is toxic like Cl_2 and Br_2.	Concentrated up to 2.5 ppt by some marine algae.
78	Platinum	None known.	Moderately toxic to mammals by intravenous injection.	*Cis*-diamminedichloroplatinum(II) used as an anticancer drug.
79	Gold		Scarcely toxic.	Some use in the treatment of arthritis.

80	Mercury	None known.	Very toxic to fungi and green plants, and to mammals if in soluble form; a cumulative poison in mammals.	Serious pollution problems from use of organomercurials as fungicides and from industrial uses of mercury.
82	Lead	None known.	Very toxic to most plants; cumulative poison in mammals. Inhibits δ-aminolevulinate dehydrolyase and thus hemoglobin synthesis in mammals (see Al).[e] One of the symptoms of lead poisoning is anemia. Toxic to central nervous system.	Worldwide pollutant of the atmosphere, concentrated in urban areas from the combustion of tetraethyl lead in gasoline; local pollutant from mines; some poisoning from lead-based paint pigments.
88–92	Radium and Actinides	None known.	May be concentrated in organisms and toxic as a result of radioactivity.	Potential pollutants from use of nuclear fuel as energy source.

[a] Toxic effects often are exhibited only at concentrations above those naturally occurring in the environment. See p. 904.

[b] K. Griggs, *Science*, **1973**, *181*, 842.

[c] K. Schwartz, *Fed. Proc. Am. Soc. Exp. Biol.*, **1974**, *33*, 1748.

[d] I. Fridovich, *Am. Sci.*, **1975**, *63*, 54; W. H. Bannister and J. V. Bannister, eds., "Biological and Clinical Aspects of Superoxide and Superoxide Dismutase," Elsevier/North Holland, New York, **1980**.

[e] H. J. M. Bowen, "Environmental Chemistry of the Elements," Academic Press, New York, **1979**.

[f] H. A. Harper, "Review of Physiological Chemistry," Lange Medical Publications, Los Altos, Calif., **1971**.

[g] D. R. Crapper et al., *Brain*, **1976**, *99*, 67; G. L. Eichorn et al., in "Inorganic Chemistry in Biology and Medicine," A. E. Martell, ed., *ACS Symp. Ser. No. 140*, **1980**, Chapter 4.

[h] W. Mertz, *Science*, **1981**, *213*, 1332.

[i] C. L. Rollinson and M. G. Enig, "Kirk-Othmer: Encyclopedia of Chemical Technology," 3rd ed., **1981**, Vol. 15, pp. 570–603.

[j] L. L. Hopkins, Jr., and H. E. Mohr, *Fed. Proc. Am. Soc. Exp. Biol.*, **1974**, *33*, 1773.

[k] F. H. Nielsen and D. A. Ollerich, ibid., **1974**, *33*, 1767.

[l] A. M. Fiabane and D. R. Williams, "The Principles of Bio-Inorganic Chemistry," *Chem. Soc. Monogr. Teach. No. 31*, London, **1977**.

[m] C. F. Shaw III, in "Inorganic Chemistry in Biology and Medicine," A. E. Martell, ed., *ACS Symp. Ser. No. 140*, Washington, D.C., **1980**, Chapter 20.

Table 18.2 Abundances of the elements in the earth's crust, river, and sea water

Element	Earth's crust $g\ kg^{-1}$	Earth's crust atoms/10^4 atoms Si	River water $mg\ dm^{-3}$	Ocean water $mg\ dm^{-3}$
Hydrogen			1.119×10^5	1.078×10^5
Helium				7.2×10^{-6}
Lithium	0.02	3	0.003	0.18
Beryllium	0.028	3	$< 1 \times 10^{-4}$	6×10^{-7}
Boron	0.01	1	0.01	4.5
Carbon[a]			1.2	28
Nitrogen			0.25[b]	0.5[b,c]
Oxygen	474[d]	9600	8.8×10^5	8.56×10^5
Fluorine	0.625	32.7	0.1	1.4
Neon				0.00012
Sodium	24	1040	9	1.105×10^4
Magnesium	20	820	4.1	1.326×10^3
Aluminum	82	3020	0.4	0.005[c]
Silicon	282	10000	4	1[c]
Phosphorus	1	32	0.02	0.07[c]
Sulphur	0.26	8.1	3.7	928
Chlorine	0.13	3.6	8	1.987×10^4
Argon				0.45
Potassium	24	610	2.3	416
Calcium	42	1040	1.5	4.22
Scandium	0.022	0.48	4×10^{-6}	1.5×10^{-6}
Titanium	5.7	120	0.003	0.001
Vanadium	0.135	2.64	0.001	0.0015
Chromium	0.1	2	0.001	0.0006[c]
Manganese	0.95	17	~0.005	0.002[c]
Iron	56	1000	0.67	0.003[c]
Cobalt	0.025	0.42	0.0002	8×10^{-5}[c]
Nickel	0.075	1.3	0.0003	0.002
Copper	0.055	0.86	0.005	0.003[c]
Zinc	0.070	1.1	0.01	0.005
Gallium	0.015	0.21	1×10^{-4}	3×10^{-5}
Germanium	0.0015	0.021		6×10^{-5}
Arsenic	0.0018	0.024	~0.001	0.0023
Selenium	5×10^{-5}	6×10^{-4}	0.0002	0.00045
Bromine	0.0025	0.031	~0.02	68
Krypton				0.00021
Rubidium	0.09	1	0.001	0.12
Strontium	0.375	4.26	0.050	8.5
Yttrium	0.033	0.37	0.04	1.3×10^{-5}
Zirconium	0.165	1.80	0.003	2.6×10^{-6}
Niobium	0.02	0.2		1×10^{-6}
Molybdenum	0.0015	0.016	0.001	0.01
Technetium				
Ruthenium	1×10^{-6}[d]	1×10^{-5}[d]		7×10^{-7}
Rhodium	2×10^{-7}[d]	2×10^{-6}[d]		
Palladium	8×10^{-7}[d]	8×10^{-6}[d]		
Silver	7×10^{-5}	6×10^{-4}	0.0003	0.0001
Cadmium	0.0002	0.0018		5×10^{-5}
Indium	0.0001	9×10^{-4}		1×10^{-7}

Table 18.2 *(Continued)*

Element	Earth's crust g kg^{-1}	Earth's crust atoms/10^4 atoms Si	River water mg l^{-1}	Ocean water mg l^{-1}
Tin	0.002	0.02	4×10^{-5}	1×10^{-5}
Antimony	0.0002	0.002	0.001	0.0002
Tellurium	4×10^{-6}[d]	4×10^{-5}[d]		
Iodine	0.0005	0.004	~0.005	0.06[c]
Xenon				5×10^{-6}
Cesium	0.003	0.02	5×10^{-5}	0.0005
Barium	0.425	3.08	0.01	0.03
Lanthanum	0.03	0.2	0.0002	3.4×10^{-6}
Cerium	0.06	0.4		1.2×10^{-6}
Praseodymium	0.0082	0.058		6×10^{-7}
Neodymium	0.028	0.19		2.8×10^{-6}
Promethium				
Samarium	0.006	0.04		4.5×10^{-7}
Europium	0.0012	0.08		1.3×10^{-7}
Gadolinium	0.0054	0.034		7×10^{-7}
Terbium	0.0009	0.006		1.4×10^{-7}
Dysprosium	0.003	0.02		9.1×10^{-7}
Holmium	0.0012	0.007		2×10^{-7}
Erbium	0.0028	0.017		9×10^{-7}
Thulium	0.0005	0.003		2×10^{-7}
Ytterbium	0.003	0.02		8×10^{-7}
Hafnium	0.003	0.02		
Tantalum	0.002	0.01		2×10^{-5}
Tungsten	0.0015	0.008	3×10^{-5}	0.00012
Rhenium	5×10^{-6}	3×10^{-5}		1×10^{-6}
Osmium	1×10^{-8}[d]	5×10^{-8}[d]		
Iridium	1×10^{-8}[d]	5×10^{-8}[d]		
Gold	4×10^{-6}	2×10^{-5}	2×10^{-6}	5×10^{-5}[c]
Mercury	8×10^{-5}	4×10^{-4}	7×10^{-5}	5×10^{-5}[c]
Thallium	0.00045	0.0022		1×10^{-6}
Lead	0.0125	0.06	0.003	3×10^{-5}[c]
Bismuth	0.00014	7×10^{-4}		2×10^{-5}
Polonium				2×10^{-14}
Astatine				
Radon			2×10^{-16}	6×10^{-16}[c]
Francium				
Radium			4×10^{-10}	1×10^{-10}[c]
Actinium				
Thorium	0.0096	0.041	0.0001	4×10^{-8}[c]
Protactinium				2×10^{-19}[c]
Uranium	0.0027	0.011	4×10^{-5}	0.0033

[a] Inorganic carbon.
[b] Combined nitrogen; about 15 mg L^{-1} dissolved N_2.
[c] Considerable variation occurs.
[d] H. J. M. Bowen, "Environmental Chemistry of the Elements," Academic Press, New York, **1979**, Chapter 13.

SOURCE: J. P. Riley and R. Chester, "Introduction to Marine Chemistry," Academic Press, New York, **1971**, except as noted.

that are not essential elements, we find only four—Si,[125] Al, Ti, and Zr—all of which form extremely insoluble oxides at biologically reasonable pH values. No common element is toxic at levels normally encountered, though of course almost anything can be harmful at too high levels. When we consider the elements that are currently causing problems in the environment, we find that they are all extremely rare in their crustal abundances: As (0.024), Pb (0.08), Cd (0.0018), and Hg (4×10^{-5}).[126] The conclusion is inescapable: *Life evolved utilizing those elements that were abundant and available to it and became dependent upon them.* Those elements that are rare were not used by living systems because they were not available; neither did these systems evolve mechanisms to cope with them.

A closely related corollary of this thesis is that many elements that are *essential* when occurring at ambient concentrations are *toxic* at higher concentrations (and, of course, cause deficiency symptoms at lower concentrations). Interesting examples are copper, selenium, and even sodium—all oceanic organisms are adapted to live in 0.6 mol dm^{-3} NaCl and our blood has been described as a sample of the primeval seas. Yet too high concentrations of NaCl are toxic through simple hypertonicity, i.e., osmotic dehydration. Selenium is a problem when it is either too rare or too abundant in the environment: Livestock grown on selenium-deficient pasture suffer from "white muscle disease"; when grazing plants (*Astragalus* sp., "locoweed") that concentrate selenium from the soil, they suffer central nervous system toxinosis. Copper is essential to many of the redox enzymes necessary to both plants and animals; yet too much copper is severely toxic to most green plants. Life used and adapted to those elements and those concentrations available to it. When man started mining, using, and releasing these elements into the environment, the ecosystem was faced with hazards it had never before encountered, and to which it had, therefore, never adapted.

A slightly different view of this idea has been presented by Egami,[127] who has pointed out that the three enzyme systems in the most primitive bacterium, *Clostridium*, are involved in electron transfer (e.g., ferredoxin), reduction of small molecules (e.g., nitrogenase), and hydrolysis (e.g., carboxypeptidase and carbonic anhydrase), and employ, respectively, iron, molybdenum, and zinc, the three most common transition elements in sea water. It is postulated that these enzyme systems arose from protoenzymes that utilized these most common metals in primitive seas. One puzzle is copper, which is fairly abundant in sea water, and although it has been thought to be essential for all organisms, apparently no requirement for it has been found in strict anaerobes. Egami postulates that copper, with a positive reduction *emf*, was incorporated into living systems only when the atmosphere shifted from reducing (CH_4, H_2, NH_3) to oxidizing (O_2).[128] This indicates the importance of considering changes that have occurred with time (including the advent of terrestrialism) and perhaps considering the microabundance of the various elements in different habitats.

---

[125] Silicon is used in the form of SiO_2 for structural purposes in some specialized plants and animals and recently has been found to be essential in some higher animals (see p. 916).

[126] All figures in atoms per 10 000 atoms silicon.

[127] F. Egami, *J. Mol. Evol.*, **1974**, *4*, 113; *J. Biochem.*, **1975**, *77*, 1165.

[128] See E. Broda, *J. Mol. Evol.*, **1975**, *7*, 87, for a discussion of this and related problems in the primitive biosphere.

ADAPTATIONS TO NATURAL ABUNDANCES

When the abundance of an element is unusually high or unusually low, organisms develop mechanisms to handle the stress. The first and perhaps best documented example of this phenomenon is the adaptation of various plants to exceptionally high concentrations of various heavy metals in mine dumps and tailings. Not only have some species adapted to extremely high concentrations of normally toxic metals, but they have also evolved a high level of self-fertilization to prevent pollination from nearby populations that are not metal-tolerant.

The hydrothermal vents discussed previously (p. 852) provide a parallel, *natural* environment with unusually large amounts of various metals—iron, copper, zinc—dissolved from the crustal rocks by the superheated water. It will be of interest to learn how the animals in the hydrothermal ecosystem have developed mechanisms to avoid toxicity from these metals.

At the other extreme are adaptations to very low concentrations of a particular element. We have already seen mechanisms directed toward the sequestration of iron when it is present in small amounts. The ability *to detect* extremely small amounts of an element can be a useful adaptation for an animal if that element is important to it. For example, hermit crabs recognize shells suitable for occupation not only by tactile stimuli but apparently also by the minute amount of calcium carbonate that is dissolved in the water around a shell. They can readily distinguish natural shells ($CaCO_3$), calcium-bearing replicas ($CaSO_4$), and naturally containing calcium minerals (calcite, aragonite, and gypsum) from non-calcium minerals (celestite, $SrSO_4$; rhodochrosite, $MnCO_3$; siderite, $FeCO_3$; and quartz, SiO_2).[129] Inasmuch as the solubility product of calcium carbonate is only 10^{-18}, the concentration of calcium detected by the hermit crab is of the order of 5 ppm or less. Almost nothing is known about the chemical mechanisms used by organisms in detecting various elements.

BIOCHEMISTRY OF THE NONMETALS

Many of the nonmetals such as hydrogen, carbon, nitrogen, oxygen, phosphorus, sulfur, chlorine, and iodine are essential elements, and most are used in quantities far beyond the trace levels. Nevertheless, most of the chemistry of these elements in biological systems is more closely associated with organic chemistry than with inorganic chemistry.

Structural uses[129a]

There are three important minerals used by organisms to form hard tissues such as bones and shells. The most widespread of these is calcium carbonate, an important structural component in animals ranging from Protozoa to Mollusca and Echinodermata. It is also a minor component of vertebrate bones. Its widespread use is probably related to the

---

[129] K. A. Mesce, *Science*, **1982**, *215*, 993.

[129a] J. F. V. Vincent, "Structural Biomaterials," Wiley, New York, **1982**.

generally uniform distribution of dissolved calcium bicarbonate. Animals employing calcium carbonate are most abundant in fresh waters containing large amounts of calcium and magnesium ("hard water") and in warm, shallow seas where the partial pressure of carbon dioxide is low (e.g., the formation of coral reefs by coelenterates). The successful precipitation of calcium carbonate depends upon the equilibrium:

$$Ca^{+2} + 2HCO_3^- \rightleftharpoons CaCO_3 + CO_2 + H_2O \qquad \textbf{(18.29)}$$

and is favored by high $[Ca^{+2}]$ and low $[CO_2]$. Nevertheless, organisms exhibit a remarkable ability to deposit calcium carbonate from hostile environments. A few freshwater clams and snails are able to build reasonably large and thick shells in lakes with a pH of 5.7–6.0 and as little as 1.1 ppm dissolved calcium carbonate.[130]

It is of interest that two thermodynamically unstable forms of calcium carbonate, aragonite and vaterite, are found in living organisms as well as the more stable calcite. There appears to be no simple explanation for the distribution of the different forms in the various species.

Tissues containing silica are found in the primitive algal phyla Pyrrhophyta (dinoflagellates) and Chrysophyta (diatoms and silicoflagellates). One family of higher plants, the Equisetaceae, or horsetails, contains gritty deposits of silica—hence their name "scouring rushes." Some Protozoa (radiolarians), Gastropoda (limpets), and Porifera (glass sponges) employ silica as a structural component. Silicon is an essential trace element in chicks and rats[131] and is probably necessary for proper bone growth in all higher animals.

The third type of compound used extensively as a structural component is apatite, $Ca_5(PO_4)_3X$. Hydroxyapatite (X = OH) is the major component of bone tissue in the vertebrate skeleton. It is also the principal strengthening material in teeth. Partial formation of fluorapatite (X = F) strengthens the structure and causes it to be less soluble in the acid formed from fermenting organic material. Application of stannous fluoride to the teeth, especially to carious enamel, probably results in the conversion of the hydroxyapatite to a new material. In the laboratory, at least two compounds have been characterized: $Sn_2(OH)PO_4$ formed at low concentrations of SnF_2, and $Sn_3F_3PO_4$ formed at higher concentrations.[132] Presumably these conversions of hydroxyapatite are involved in the reduction of caries by fluoride ions and stannous fluoride. Fluorapatite is also used structurally in certain Brachiopod shells.

ADP and ATP

Adenosine triphosphate (ATP) is formed to a certain extent directly in photosynthesis and is the end product of respiration. It may then be used as an energy source to drive the many chemical reactions of the cell. The energy of ATP comes from the hydrolysis of a

[130] For a discussion of this point as well as other examples of organisms living on limiting concentrations of nutrients, see A. C. Allee et al., "Principles of Animal Ecology," Saunders, Philadelphia, **1949**, pp. 164–167, 189–206; R. W. Pennak, "Freshwater Invertebrates of the United States," Ronald Press, New York, **1953**, pp. 681, 705f.

[131] E. M. Carlisle, *Science*, **1972**, *178*, 619; *Fed. Proc. Am. Soc. Exp. Biol.*, **1974**, *33*, 1758.

[132] T. H. Jordan et al., *Inorg. Chem.*, **1976**, *15*, 1810; ibid., **1980**, *19*, 2551; *J. Dent. Res.*, **1978**, *57*, 933.

polyphosphate linkage (see p. 684):

$$\mathrm{NH_2}\text{-adenine(N, N, N, N)-ribose(O; OH, OH)-}\mathrm{CH_2O{-}\overset{O}{\overset{\|}{P}}(O^-){-}O{-}\overset{O}{\overset{\|}{P}}(O^-){-}O{-}\overset{O}{\overset{\|}{P}}(O^-){-}O^-} + \mathrm{H_2O} \longrightarrow$$

$$\mathrm{NH_2}\text{-adenine(N, N, N, N)-ribose(O; OH, OH)-}\mathrm{CH_2O{-}\overset{O}{\overset{\|}{P}}(O^-){-}O{-}\overset{O}{\overset{\|}{P}}(O^-){-}OH} + \mathrm{HPO_4^{-2}} \qquad \textbf{(18.30)}$$

Although this reaction is usually described as given in Eq. 18.30, magnesium is necessary for the conversion to be effected, and therefore Mg^{+2} complexes are involved. The standard free energy change for the hydrolysis of ATP to ADP is estimated to be $-31.0\ \mathrm{kJ\ mol^{-1}}$. Because of concentrations obtaining in the cell, the available free energy for biosynthesis is thought to be about -40 to $-50\ \mathrm{kJ\ mol^{-1}}$.[133]

MEDICINAL CHEMISTRY

Antibiotics and related compounds

The suggested antibiotic action of transferrin is typical of the possible action of several antibiotics in tying up essential metal ions. Streptomycin, aspergillic acid, usnic acid, the tetracyclines, and other antibiotics are known to have chelating properties. Presumably some antibiotics are delicately balanced so as to be able to compete successfully with the metal-binding agents of the bacteria while not disturbing the metal processing by the host. There is evidence that at least some bacteria have developed resistance to antibiotics

[133] See H. R. Mahler and E. H. Cordes, "Biological Chemistry," 2nd ed., Harper & Row, New York, **1971**, pp. 22–23, 377–384.

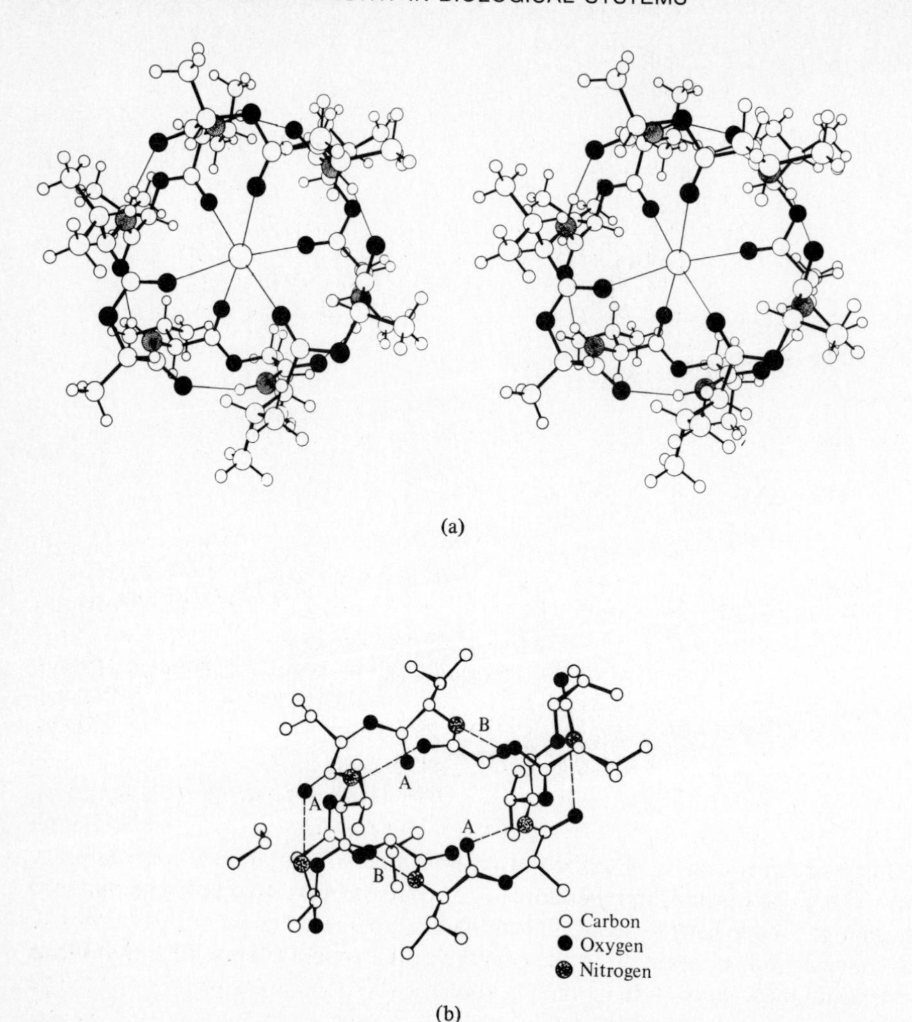

Fig. 18.39 (a) Molecular structure of valinomycin coordinated to the K^+ ion. (b) Molecular structure of free valinomycin molecule. The carbonyl groups marked (A) are free to coordinate to K^+. Hydrogen bonding is shown by dotted lines, with those marked (B) thought to be most susceptible to breaking. [From K. Neupert-Laves and M. Dobler, *Helv. Chem. Acta*, **1975**, *58*, 432, and G. D. Smith et al., *J. Am. Chem. Soc.*, **1975**, *97*, 7242. Reproduced with permission.]

through the development of altered enzyme systems that can compete successfully with the antibiotic.[134] The action of the antibiotic need not be a simple competitive one. The chelating properties of the antibiotic may be used in metal transport across membranes or to attach the antibiotic to a specific site from which it can interfere with the growth of bacteria.

[134] H. B. Woodruff and I. M. Miller, in "Metabolic Inhibitors," R. M. Hochster and J. H. Quastel, eds., Academic, New York, **1963**, Vol. II, Chapter 17.

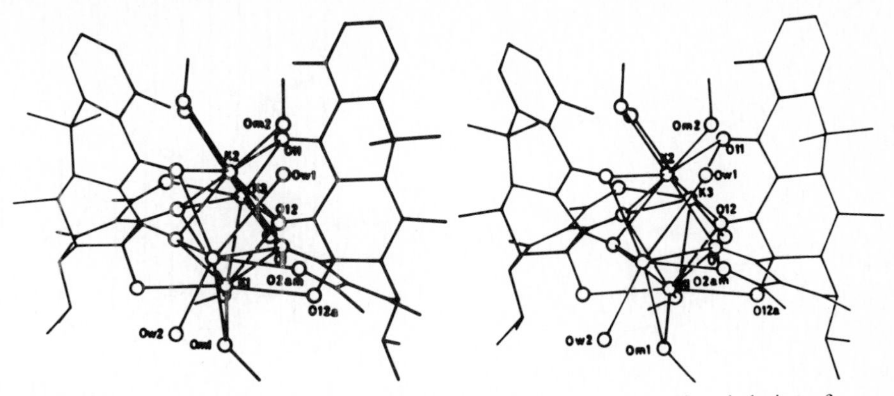

Fig. 18.40 Stereoview of the dipotassium salt of oxytetracycline. Note the extensive chelation of the potassium ions by oxygen atoms on the tetracycline molecule. [From K. H. Jogun and J. J. Stezowski, *J. Am. Chem. Soc.*, **1976**, *98*, 6018. Reproduced with permission.]

The behavior of valinomycin is typical of a group known as "ionophore antibiotics."[135] These compounds resemble the crown ethers and cryptates (pp. 530–534) by having several oxygen or nitrogen atoms spaced along a chain or ring that can wrap around a metal ion (Fig. 18.39a). These antibiotics are useless in man because they are toxic to mammalian cells, but some of them find use in treating coccidiosis in chickens. The toxicity arises from the ion-transporting ability. Cells become "leaky" with respect to potassium, which is transported across the cell membrane by valinomycin. In the absence of a metal ion, valinomycin has a quite different conformation (Fig. 18.39b), one stabilized by hydrogen bonds between amide and carbonyl groups. It has been postulated[136] that the potassium ion can initially coordinate to the four free carbonyl groups (A) and that this can provide sufficient stabilization to break two of the weaker hydrogen bonds (B). This provides two additional carbonyl groups to coordinate and complete the change in conformation to that shown in Fig. 18.39a. Such a stepwise mechanism would indicate that the whole system is a balanced one and that the reverse process can be readily triggered by a change in environment such as at a membrane surface or if there is a change in hydrogen-bonding competition.

The tetracyclines form an important group of antibiotics. Their activity appears to result from their ability to chelate metals since the extent of antibacterial activity parallels the ability to form stable chelates. The metal in question appears to be magnesium since addition of large amounts of magnesium can inhibit the antibiotic effects of the tetracycline. It is not certain how the tetracyclines coordinate magnesium, but the molecular structure of the potassium salt of oxytetracycline is known and shows extensive chelation (Fig. 18.40).[137]

---

[135] See B. C. Pressman in "Inorganic Biochemistry," G. L. Eichorn, ed., Elsevier, Amsterdam, **1973**, Vol. I, p. 203, and R. P. Hanzlik, "Inorganic Aspects of Biological and Organic Chemistry," Academic Press, New York, **1976**, pp. 33–34.

[136] G. D. Smith et al., *J. Am. Chem. Soc.*, **1975**, *97*, 7242.

[137] K. H. Jogun and J. J. Stezowski, *J. Am. Chem. Soc.*, **1976**, *97*, 6018.

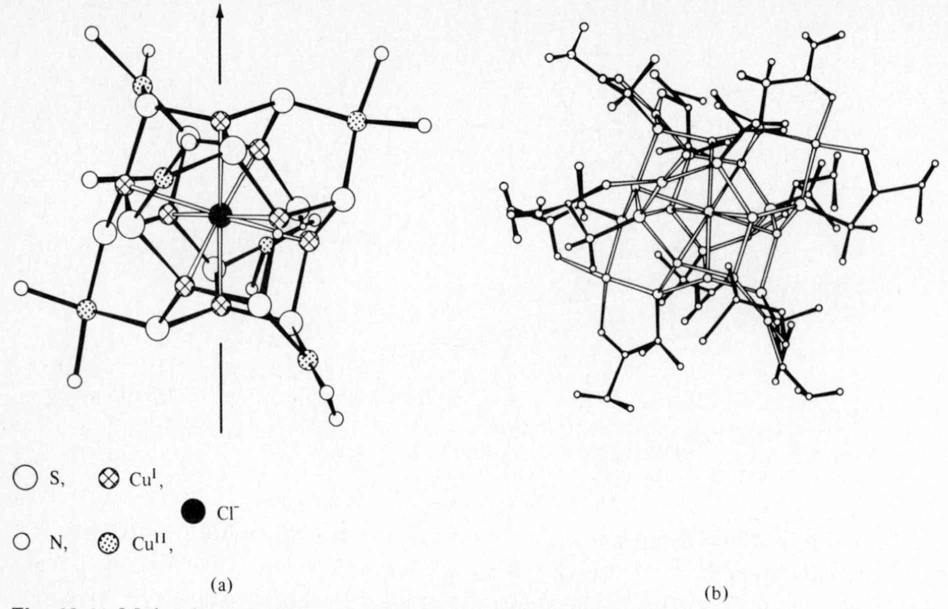

Fig. 18.41 Molecular structure of copper complex of D-penicillamine. The $[Cu^{I}_{8}\,Cu^{II}_{6}$ $(penicillaminate)_{12}Cl]$: (a) The central cluster of Cu and ligating atoms only; (b) the entire ion with the central cluster oriented as in (a). [From P. J. M. W. L. Birker and H. C. Freeman, *Chem. Commun.*, **1976**, 312. Reproduced with permission.]

Chelate therapy

We have seen previously that chelating agents can be used therapeutically to treat problems caused by the presence of toxic elements. We have also seen that an essential element can be toxic if present in too great a quantity. This is the case in Wilson's disease (hepatolenticular degeneration), a genetic disease involving the buildup of excessive quantities of copper in the body. Many chelating agents have been used to remove the excess copper, but one of the best is D-penicillamine, $HSC(CH_3)_2CH(NH_2)COOH$. This chelating agent forms a complex with copper ions that is colored an intense purple and, surprisingly, has a molecular weight of 2600. Another surprising finding is that the complex will not form unless chloride or bromide ions are present and the isolated complex always contains a small amount of halide. These puzzling facts were explained when the X-ray crystal structure was done.[138] The structure (Fig. 18.41) consists of a central halide ion surrounded by eight copper(I) atoms bridged by sulfur ligands. These are in turn coordinated to six copper(II) atoms. Finally, the chelating amino groups of the penicillamine complete the coordination sphere of the copper(II) atoms.

As we have seen, the body has essentially no means of eliminating iron, so an excessive intake of iron causes various problems known as siderosis. Chelating agents are used to treat the excessive build-up of iron. In many cases these chelates resemble the analogous

[138] P. J. M. W. L. Birker and H. C. Freeman, *Chem. Comm.*, **1976**, 213.

compounds used by bacteria to chelate iron.[139] The ideal chelating agent will be specific for the metal to be detoxified since a more general chelating agent is apt to cause problems by altering the balance of other essential metals. The concepts of hard and soft metal ions and ligands can be used to aid in this process of designing therapeutic chelators.[140]

A slightly different mode of therapy involves the use of *cis*-diamminedichloroplatinum(II), $Pt(NH_3)_2Cl_2$, in the treatment of cancer. The exact action of the drug is not known, but only the *cis* isomer is active, not the *trans* isomer. It is thought that the platinum binds to the DNA, with a base such as guanosine replacing the chloride ions. The difference between the action of the two isomers could be either "structural" or "chemical" (or both). Structurally, a *cis*-diammineplatinum moiety can bind to groups about 330 pm apart, and the two groups may be on the same purine, giving chelate-effect stability. The *trans* isomer bonds to groups about 500 pm apart that approach the platinum from opposite directions. The chemical effect is that the *cis* isomer will be more reactive because of the *trans* effect: Replacement of the chloride ion by a more *trans*-activating heterocyclic nitrogen atom in the purine might labilize the ammine groups, giving complete replacement of the original ligands (compare Eqs. 11.9 and 11.10; let tu represent a purine nitrogen atom). In this case, cross-linking of the DNA chains might be possible.[141,142] It has also been suggested that hydrogen bonding by the hydrogen atoms of the ammonia molecules in the complex might hold the latter in a stereoselective position.[143] This would seriously interfere with the ability of the guanine and adjacent bases to undergo Watson-Crick base pairing. This view is supported by the higher activity of ammonia and primary amines over more highly substituted amine complexes.

For *cis*-diamminedichloroplatinum(II) to work according to the proposed mechanism, it must hydrolyze (or, perhaps better, aminolyze) *in the right place*; if it hydrolyzes in the blood before it gets to the chromosomes within the cell, it will lose its high reactivity. Fortunately, for the stability of the complex, the blood is approximately 0.1 mol dm^{-3} in chloride ion, forcing the hydrolysis equilibrium back to the chloro complex. Once the drug crosses the cell membrane into the cytoplasm, it finds a chloride ion concentration of only 4 mmol dm^{-3}: Reactions with the biological targets can take place.[142]

PROBLEMS IN BIOLOGICAL SYSTEMS

It is an unfortunate fact that most of the problems in biological systems, especially the ecosystem, that are potentially solvable by the application of inorganic chemistry are those dealing with the deleterious effects of inorganic chemicals. Thus we find that many of the

[139] W. F. Andersen, "Inorganic Chemistry in Biology and Medicine," A. E. Martell, ed., *ACS Symp. Ser. No. 140*, Washington, D.C., **1980**, Chapter 15.

[140] C. G. Pitt and A. E. Martell, "Inorganic Chemistry in Biology and Medicine," A. E. Martell, ed., *ACS Symp. Ser. No. 140*, Washington, D.C., **1980**, Chapter 17.

[141] M. N. Hughes, "The Inorganic Chemistry of Biological Processes," 2nd ed., Wiley, New York, **1981**; B. Rosenberg, "Inorganic Chemistry in Biology and Medicine," A. E. Martell, ed., *ACS Symp. Ser. No. 140*, Washington, D.C., **1980**, Chapter 8.

[142] S. J. Lippard, *Acc. Chem. Res.*, **1978**, *11*, 211; *Science*, **1982**, *218*, 1075; "Inorganic Chemistry in Biology and Medicine." A. E. Martell, ed., *ACS Symp. Ser. No. 140*, Washington, D.C., **1980**, Chapter 9.

[143] M. Gullotti et al., *Inorg. Chem.*, **1982**, *21*, 2006.

major sources of concern are inorganic species such as mercury in the water, sulfur dioxide and nitrogen oxides in the air, and lead practically everywhere. To a large extent many of the key issues fall outside the scope of pure chemistry, even outside the scope of pure science, into the areas of medicine, law, sociology, and politics. Nevertheless, in this section a brief survey of the inorganic chemistry relevant to these problems will be given.

The historical perspective

Until man developed cultural traits such as language, the use of tools and fire, and agriculture, he was little different from the other predators[144] that roamed the plains and forests. The development of agriculture resulted in modification of the ecosystem, but in a manner not unlike recurring natural phenomena. An abandoned field returns to the forest as rapidly as an area burnt over by a forest fire. The advent of civilization resulted in increased consumption, and irreversible changes began to occur. One example is the deforestation of the lands of the eastern Mediterranean to provide timber for construction, a major component of commerce (e.g., the famous cedars of Lebanon). Deforestation was followed by erosion and in some areas intensive grazing. Fundamental changes in the flora, fauna, soil, and even climate can take place under these circumstances.[145]

Despite the fact that such stresses on the ecosystem are as old as recorded history, it was not until the Industrial Revolution that man had sufficient control of his environment to destroy it. This was relatively recent as man's history goes; yet the problem continued to develop for about 200 years before solutions were sought. Even today there is a lot of whistling-in-the-dark "theorizing" by some people, such as: "If flies can become resistant to DDT why can't man?"[146]

Since there appears to be little prospect that man is willing to reduce his population level to that which existed when he was a hunter and food-gatherer,[147] and even less that he will voluntarily return to that earlier level of creature comforts, the problem facing us is how to maximize the availability of the products of civilization while minimizing the cost in terms of pollution, depletion of natural resources, and related problems.[148]

Agriculture

In view of the overwhelming population problem there has been a tendency to seek simplistic solutions by increasing the food supply via the application of fertilizers and insecti-

---

[144] Although man is accustomed to thinking of his ancestors in terms of hunting, it is probable that early in his evolution man was principally a vegetarian and scavenger.

[145] As examples, compare the present stark and inhospitable landscape of Greece, Crete, and certain areas of the Levant with their earlier heavily forested condition.

[146] *Answer:* (1) Flies became resistant to DDT and bacteria immune to certain antibiotics by the death of 99 per 100, 999 per 1000, or more, in each generation, a mortality that the human race finds ethically unacceptable. (2) Even ignoring (1), most of the damage occurring to man at the present time does not shorten life to the extent that it prevents reproduction. A factor which causes, for example, death at 35 but allows the reproduction of two or more children may never be eliminated!

[147] As late as 25 000 years ago the human population was probably about 3 million. Today it is over 4 billion with no signs of stabilizing in the foreseeable future.

[148] The ultimate problem, that a finite earth cannot support an infinite population, does not seem to be amenable to treatment by inorganic chemistry.

cides. The former have proved valuable in increasing crop yields but have been a source of environmental degradation via run-off and leaching. In this way large quantities of nitrogen, phosphate, and potassium enter the ecosystem resulting in (1) explosive growth of algae in rivers and lakes, and (2) stagnant sloughs as the excess algae die and rot without normal consumption and utilization in the food chain.

There is no panacea for the problem of pollution from agricultural fertilizers. It is apparent that more selectivity will have to be used in the future—dumping huge amounts of lime or a nitrate–phosphate–potassium fertilizer per square kilometre may not solve a nutritional problem resulting from a trace element deficiency. It may even worsen the problem since several metal ions are rendered unavailable by precipitation by phosphate or hydroxide:

$$3Mn^{+2} + 2PO_4^{-3} \longrightarrow Mn_3(PO_4)_2\downarrow \qquad (18.31)$$

$$Fe^{+n} + nOH^- \longrightarrow Fe(OH)_n\downarrow \qquad n = 2,3 \qquad (18.32)$$

There is a large and growing literature on the importance of trace elements in plant and animal nutrition, and the future should see increasing applications of the principle of selective nutritional adjustments. The use of "slow-release" fertilizers can also reduce the portion wasted in run-off.

The use of herbicides and insecticides has also posed a problem. Their exclusive use in control is seldom practical. The target develops resistance and even larger doses are applied. Sooner or later, nontarget organisms become involved. Often these chemicals owe their usefulness to the inhibition of enzyme action and when concentrated by the food chain in nontarget organisms, they can cause unforeseen effects.

In contrast to problems arising from the lack of essential elements, problems may also arise through the overabundance of certain elements. Obviously land may be poisoned and made unfit for agriculture through the natural presence or application of elements which are toxic in high concentrations. More subtle is the production of nonarable land simply through agricultural use, not via depletion but by accumulation. Continuous irrigation in arid lands leads to a buildup in the soil of ionic solutes as the water evaporates without sufficient rainfall to leach out these solutes. This type of poisoning is merely a matter of ionic strength and osmotic competition with the plants rather than the specific action of any particular element.

Gaseous air pollution

It is somewhat disconcerting to the inorganic chemist to learn that with the exception of incompletely burned hydrocarbons, the common pollutants of the atmosphere are inorganic molecules. Although there are other sources, they commonly result from combustion processes. Roasting processes in the preparation of metal sulfide ores can be a serious source of pollution if no effort is made to trap the sulfur dioxide by-product:

$$Cu_2S + 2O_2 \longrightarrow 2CuO + SO_2 \qquad (18.33)$$

The classic example is Copperhill, Tennessee, where the products of Eq. 18.33 have disturbed the forest cover over an area of approximately 200 km^2. The forest has been completely eradicated and only grassland remains in a belt comprising 70 km^2, and in the center of the basin there was an area of 28 km^2 completely denuded of vegetation.[149] Now, over two-thirds of a century later, the area is still a wasteland. Less well known is the damage done (Fig. 18.42) by a zinc smelter at Lehigh Gap, Pennsylvania, which in addition to sulfur dioxide has been emitting zinc, cadmium, copper, and lead.[150] Soil samples collected approximately two kilometres from a smelter contained up to 8% zinc!

One of the most serious problems in urban areas is the formation of smog and related forms of pollution from nitrogen oxides. The latter form in combustion processes, chiefly those occurring in internal combustion engines. A simplified scheme for the inorganic photochemistry of smog is

$$N_2 + O_2 \longrightarrow 2NO \text{ (engine cylinder)} \tag{18.34}$$

$$2NO + O_2 \longrightarrow 2NO_2 \tag{18.35}$$

$$NO_2 + h\nu \longrightarrow NO + O \tag{18.36}$$

$$O + O_2 \longrightarrow O_3 \tag{18.37}$$

The ozone produced in this reaction is a strong oxidant and irritant. The above set of reactions can be further complicated by the formation of free radicals from incompletely burned hydrocarbons.

A product of civilization that is not normally considered to be a pollutant is carbon dioxide. The per capita consumption of energy is a good index of the extent to which civilization has advanced technologically, and most methods of generating power, such as steam-driven electrical generating plants and internal combustion engines, produce carbon dioxide from burning fossil fuels. Since carbon dioxide is a natural component of the atmosphere, direct deleterious effects to life are not expected (plants grow better in increased CO_2 concentrations). Instead, natural processes should take place to "buffer" the concentration of carbon dioxide in the atmosphere. For example, for a given amount of carbon dioxide added to the atmosphere, about five-sixths should dissolve in the ocean. There it should slowly react with limestone sediments:

$$CO_2 + H_2O + CaCO_3 \longrightarrow Ca^{+2} + 2HCO_3^- \tag{18.38}$$

to "fix" the carbon dioxide. Further consumption of carbon dioxide results from weathering of exposed rocks on land. All of these processes are slow, however, compared to the production of carbon dioxide from combustion. For the United States the annual emission of carbon dioxide appears to double every ten to fifteen years. It has been estimated that it would take about five years for the oceans to come to equilibrium with each new injection of carbon dioxide in the atmosphere. The chemical equilibria involved in the ocean and in surface weathering are even slower, measured in terms of thousands of years. It is possible, therefore, to build up temporarily large concentrations of carbon dioxide in the atmosphere.[151]

---

[149] C. R. Hursh, "Local Climate of Copper Basin of Tennessee as Modified by Removal of Vegetation," USDA Circular No. 774, **1935**.

[150] M. J. Jordan, *Ecology*, **1975**, *56*, 78.

[151] M. Stiver, *Science*, **1978**, *199*, 253.

Fig. 18.42 Lehigh Gap before and after construction of zinc smelters. The upper photograph was taken in the 1880s; the lower in the 1930s. The light gray area in the upper right corner of the upper photograph is an artifact of film development, not a natural feature. [From M. Jordan, *Ecology*, **1975**, *56*, 78. Reproduced with permission.]

Increasing concentrations of carbon dioxide in the atmosphere are of some concern because of the "greenhouse effect."[152] Carbon dioxide is transparent in the visible and near ultraviolet regions of the spectrum in which the radiation reaching the earth from the sun lies. On the other hand, the wavelength of radiation emitted by the earth (acting as a blackbody) is considerably longer, and some of this infrared radiation is absorbed by the carbon dioxide in the atmosphere, holding the energy in the atmosphere rather than allowing it to be dissipated into space. Since the temperature of the earth is a balanced equilibrium between the energy received from the sun and that radiated back into space, such absorption by carbon dioxide could affect the earth's climates, resulting in the melting of the polar icecaps and concomitant problems.[153] Although there is considerable debate about the *effect* and the *rate* of carbon dioxide accumulation, there is general agreement that the phenomenon is real.[154]

Acid rain

Much less spectacular than the denuded Copperhill Basin or smog-bound Los Angeles, but more important in the long run because of the larger areas involved, is the steady emission of sulfur oxides from combustion of high-sulfur fuels. In addition, some nitrogen in the air combines with oxygen upon passing through a flame, the more so the hotter the flame. These oxides eventually form sulfuric and nitric acids. Most industrialized nations (and those downwind of them) now experience rainfall of low pH. The usual criterion for "acid rain" is having a pH below 5.6. The average pH of rainfall in the northeastern United States is now 4.3, with a pH as low as 1.5[155] recorded at Wheeling, W. Va. The Los Angeles basin routinely has fogs with water droplets of pH 2.2–4.0.[156] As a consequence of the acid precipitation, marble statuary and buildings, often of great artistic and historic value, are being corroded and etched at an alarming rate.

Acid rain may have harmful effects on the ecosystem as well. If it falls in a limestone area, the buffering action of the calcium carbonate tends to neutralize its adverse effects. Most other rocks and soils lack this buffering action, and the pH of the water in lakes may drop below 5. Not only is the low pH harmful *per se*, but the acidified water dissolves metal oxides in the soil and releases zinc, iron, manganese, and aluminum, all toxic if present in too great a quantity.[157] Unlike naturally occurring sphagnum bogs/lakes, which have unusual but thriving flora and fauna, the artificially acidified lakes lack the high concentration of organic matter that provides humic acid to chelate these metals (see p. 896).[158]

---

[152] There has recently been considerable controversy over the aptness of this term. Regardless of the pros and cons involved in the argument, it is likely that it will stick.

[153] V. Gornitz et al., *Science*, **1982**, *215*, 1611.

[154] S. B. Idso, *Science*, **1980**, *207*, 1462; R. A. Madden and V. Ramanathan, ibid., **1980**, *209*, 763; J. Hansen et al., ibid., **1981**, *213*, 957; G. Kukla and J. Gavin, ibid., **1981**, *214*, 497; R. Etkins and E. S. Epstein, ibid., **1982**, *215*, 287; J. Hansen, *Chem. Eng. News*, Jan. 11, **1982**, p. 21; R. Revelle, *Sci. Amer.*, August, **1982**, pp. 35–43.

[155] Note that the hydrogen ion concentration in these solutions is about one-third that of standard laboratory 0.1 mol dm^{-3} titrating acid!

[156] J. M. Waldman et al., *Science*, **1982**, *218*, 677.

[157] It has also been suggested that acid precipitation can solubilize metals in dust accumulated on leaves, making them available for foliar absorption [S. E. Lindberg et al., *Science*, **1982**, *215*, 1609].

[158] L. R. Ember, *Chem. Eng. News*, Sept. 14, **1981**, p. 20; R. Patrick et al., *Science*, **1981**, *211*, 446; F. D'Itri, "Acid Precipitation: Effects on Ecological Systems," Ann Arbor Science, Woburn, Mass., **1982**.

Nitrogen oxides, chlorofluorocarbons, and the upper atmosphere

One might suppose that pollution of the upper atmosphere would be much less serious than that of the lower atmosphere simply because the latter is in contact with the ecosystem and the former is not. We have seen in Chapter 5 some of the reactions that tend to destroy ozone in the upper atmosphere:

$$O + O_3 \xrightarrow{NO_x} 2O_2 \tag{18.39}$$

$$O + O_3 \xrightarrow{ClO_x} 2O_2 \tag{18.40}$$

The chief artificial source of the nitrogen oxides is from SST-type aircraft of the Concorde type,[159] and the chlorine oxides result from the photolysis of chlorofluorocarbons.[160]

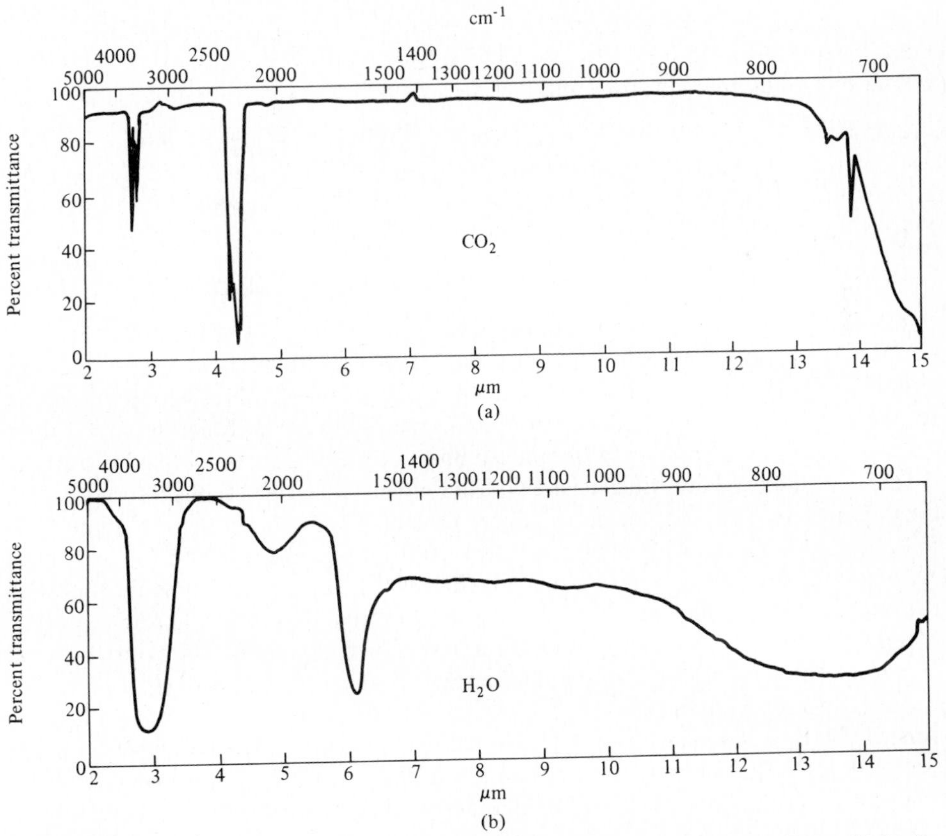

Fig. 18.43 Infrared spectra of principal absorbing species in the atmosphere: (a) CO_2; (b) H_2O. [© Sadtler Research Laboratories, Inc. (1962, 1976). Reproduced with permission.]

---

[159] H. S. Johnston, *Acc. Chem. Res.*, **1975**, *8*, 289; F. N. Alyea et al., *Science*, **1975**, *188*, 117.
[160] J. G. Anderson et al., *Science*, **1977**, *198*, 501.

The absorption spectrum of ozone in the ultraviolet region is such that it is chiefly responsible for the sharp cutoff of solar radiation at about 300 nm. It is in the region from 280 to 320 nm that biological damage from solar radiation dramatically increases through sunburn (erythema) and damage to DNA, for example. This could create serious problems for humans and for other forms of life.[159,161]

The chlorofluorocarbons also pose a problem in terms of the greenhouse effect. They absorb infrared radiation in the 8–12-μm region. Even though quantities of the species in the atmosphere could never come close to those of other IR-absorbing species such as carbon dioxide and water, the latter have always been present in the atmosphere in relatively large amounts. The chlorofluorocarbons absorb in a portion of the spectrum that was formerly transparent (Fig. 18.43). It has been estimated[162] that if chlorofluorocarbon usage were to continue at the present rates plus a 10% increment per year (now highly unlikely because of concern over the ozone layer), the greenhouse effect from them would surpass that of carbon dioxide by the year 2000.

Particulate air pollution

Lead alkyls are used extensively to increase the octane rating of gasoline used in automobile engines. Dibromoethane and dichloroethane ("ethylene halides") are used to promote the formation of the volatile (at combustion temperatures) lead halides rather than nonvolatile compounds that would foul the engine:

$$(C_2H_5)_4Pb + C_2H_4X_2 + 16O_2 \longrightarrow PbX_2 + 10CO_2 + 12H_2O \qquad \textbf{(18.41)}$$

The lead halides form aerosols in the atmosphere to the extent of 10–50 μg m^{-3} of air in urban areas. The concentrations outside of urban areas are considerably lower, but contamination is worldwide. Although it is now not possible to estimate the concentration of lead that existed in the atmosphere before man contributed from other, less important industrial processes, it is possible to obtain estimates of preautomotive levels from snows which were deposited and held (in glaciers, for example) prior to 1900. It has been estimated that the present level of lead in the atmosphere of the Northern Hemisphere is about 1000 times the natural level. At the present time there is little evidence on the biological effect of these relatively large concentrations of lead, although it seems improbable that the lead is not entering the food chain and perhaps being concentrated in some organisms.

More typically, we think of air pollution in terms of sooty smoke emerging from tall smoke stacks. The soot, *per se*, is more of a nuisance than a direct hazard (though it may contain carcinogenic hydrocarbons), but often industrial smokes contain relatively large amounts of metal oxides. These are potential sources of problems since dispersed metal oxides can pose health hazards, as for example in lung irritation, and any widespread dispersal of a heavy metal is likely to be unfavorable. The precipitation and collection of such smoke presents an interesting and potentially profitable problem. For example, the

---

[161] "Fluorocarbons and the Environment," Council on Environmental Quality, Washington, D.C., **1975**.

[162] V. Ramanathan, *Science*, **1975**, *190*, 50; *J. Atmos. Sci.*, **1976**, *33*, 1330; "Protection Against Depletion of Stratospheric Ozone by Chorofluorocarbons," National Academy of Science, Washington, D.C., **1979**.

first commercial source of rhenium metal was flue dust from the processing of molybdenite ores.

Other industrial pollution

Although the picture of clouds of smoke drawn in the preceding paragraph is a highly visible source of pollution, there are many others that are somewhat less visible. Dumping of waste acid baths, used electrolytic baths (containing cyanide), and various organic pollutants all contribute to poor water quality. Even more insidious is the release of relatively small amounts of toxic materials which may be concentrated in living organisms. The recent discovery of mercury in fish intended for human consumption is an example. It had been known that mercury was being lost from electrolytic cells in the manufacture of sodium hydroxide ("caustic soda") and chlorine, but the amount lost is small (of the order of 0.1 ppm in the waste water, which is further diluted upon discharge into a river). It has now been found that microorganisms are able to transform it into methylmercury cation.[163] The latter is strongly held in the biological system (as might have been supposed from the binding of mercury by sulfur in proteins; see pp. 313–314, 885–890) and concentrated in the food chain. In view of the known effects of mercury on enzyme systems, this provides a particularly serious contamination of the ecosystem. It is merely another example of the principle that dilution does not solve pollution problems since living organisms can concentrate pollutants again to dangerously high levels.

Mining problems

The process of winning a metal from its ore and using it in a technological application always is a potential source of pollution even aside from the various purification, smelting, and manufacturing processes. Life has evolved and adapted to the availability of elements in the earth's crust. Basically, every time man exploits relatively rare deposits of concentrated metal compounds a potential problem exists. Mining metals such as Co, Cu, Zn, Pb, and Hg, and making no effort to prevent these elements from entering the ecosystem provide the possibility of a drastic overload of the biological systems affected. In some cases, such as iron, no problem is posed (save for the aesthetic one of rusting automobile bodies) from extravagant use of the metal. The iron can do no more than rust and reenter the earth's crust, already containing relatively large amounts of iron. In other cases the amounts that could potentially enter the ecosystem are far larger than the "natural amounts" that are available from weathering of rock, recycling of organic matter, etc. Although estimates of both the "natural" and man-made amounts of metals involved are difficult to make,[164] the best estimates indicate, for example, that for the elements Ag, Au, Cd, Cr, Cu, Hg, Pb, Sb, Sn, Tl, and Zn the amounts entering the ocean yearly are from 4

---

[163] W. P. Ridley et al., *Science*, **1977**, *197*, 329.

[164] See H. J. M. Bowen, "Environmental Chemistry of the Elements," Academic Press, New York, **1979**, Chapter 12.

to 600 times the amount lost from the ocean, and thus these elements present particularly likely pollution problems.[165]

A second problem associated with the exploitation of mineral resources is that even after the mining activities cease, problems may still persist. For example, coal mines, both active and abandoned, release the equivalent of 8 million metric tons of sulfuric acid annually which pollutes over 15 000 km of streams in Appalachia alone. The sulfuric acid results from the oxidation of iron sulfide both chemically and biochemically by chemolithotrophic bacteria (see p. 852).

$$\text{FeS (or FeS}_2) \xrightarrow[\text{H}_2\text{O}]{[\text{O}]} \tfrac{1}{2}\text{Fe}_2\text{O}_3 + (2)\text{H}_2\text{SO}_4 \qquad \textbf{(18.42)}$$

The resulting acid can leach out metals from the surrounding rocks before eventually draining into surface waters. Current methods of treating the problem (in addition to ignoring it) involve neutralization with lime or limestone. There have been suggestions to cap the spent mines and prevent the entry of oxygen or to fill the shaft with limestone to neutralize the acid formed, but such methods are expensive. Even more expensive are some suggestions for working active mines in an oxygen-free atmosphere.[166]

A related problem shows how easily unexpected complications can arise. Thus the building of a new road with extensive road cuts in the Great Smoky Mountains National Park was not expected to create any difficulties beyond the aesthetic scars to the landscape. However, some of the road cuts were blasted out of rock belonging to the Anakeesta formation, which as a result of its composition leaches sulfuric acid into surface waters similar to mine drainage. It was found that all brook trout and salamanders were extirpated from Beech Flats Creek for a distance of 8 km downstream of the cut.[167]

As with many environmental phenomena, the chemolithotrophic action of bacteria is a two-sided coin. The same bacteria then cause acid drainage from mines and other sulfide-bearing rocks can be used to solubilize metals such as copper from low-grade ores. The resulting solution is ideally suited for the recovery of the copper by reduction with scrap iron (see "Hydrometallurgy," Chapter 8).[168] Interestingly, it may be possible to use these and related bacteria to precipitate metal ions from waste waters, thereby improving their quality.

The future

The problem of the effect, either positive or negative, of various chemical species in the ecosystem is all the more difficult because of the synergistic and antagonistic effects of various trace elements on one another. Thus the recurring "startling revelations" that fruit flies (white rats, chimpanzees, or daisies, *choose one*) when raised on the effluent of

---

[165] The problems will probably not arise in the ocean as a whole but in smaller, local bodies of water with particularly high concentrations of these elements.

[166] *Chem. Eng. News*, May 18, **1970**, pp. 33–35.

[167] J. W. Huckabee et al., *Trans. Amer. Fish. Soc.*, **1975**, *104*, 677.

[168] C. L. Brierly, *Sci. Amer.*, August, **1982**, pp. 44–53.

the world's dirtiest industrial process show $X\%$ less infertility (loss of weight, paranoid tendencies, or leaf drop, *choose one*) are meaningless until placed in the perspective of *all* of the effects on *all* of the organisms involved.

An example is a recent defense of DDT despite the abundance of evidence of deleterious effects in many organisms. In one case it was shown that DDT presumably stimulates the production of hepatic microsomal enzymes that tend to prevent the toxic effects of aflatoxin. Probably similar enzyme stimulation is responsible for low hormone levels in female birds and their inability to deposit sufficient calcium in the egg shell. Consider the relative merits of being resistant to aflatoxin (which you get from eating moldy peanuts) to being able to lay a hatchable egg, or the feminization of any *male* embryos that do hatch to the extent that they develop ovarian tissue and oviducts, even showing aberrant reproductive behavior.[169]

It is highly unlikely that man (or any other organism) will, after adapting for millions of years to the concentrations of elements available in the earth's crust, suddenly benefit from random addition of various elements to the ecosystem. While judicious application of *small* amounts of certain elements *can* be of benefit (e.g., fluoridation), it is impossible that the haphazard dispersal of chemicals which happen to be the by-products of various processes employed to provide the "necessities" and "luxuries" of "life as we know it" will provide the touchstone for the golden millenium of health and prosperity for all mankind.

SUMMARY

It is true that most of the facts described in this chapter were gathered by biologists, biochemists, X-ray crystallographers, and environmental chemists, not inorganic chemists. But presumably such factors as (1) alteration of *emf* by complexation; (2) stabilization of complexes by ligand field effects; (3) hardness and softness of acids and bases; (4) the thermodynamics and kinetics of both "natural" and "unnatural" (i.e., pollutant) species; (5) catalysis by metal ions; (6) preferred geometry of metal complexes; and (7) energetics of (a) complex formation, (b) redox reactions, and (c) polyanion formation come within the ken of the inorganic chemist, and he should be able to contribute fully to the future study of these systems. The effect is already being felt. One need only compare a recent biochemistry text with one of a decade ago to note the emphasis on high-spin vs. low-spin metal ions, coordination geometry and configuration, and redox reactions and thermodynamics.

The present convergence of physical and analytical techniques combined with inorganic theory makes this one of the most exciting times to be involved in this area of chemistry. One can combine the hard facts and principles of our discipline with the ever-elusive yet fascinating mystery of life.

[169] D. M. Fry and C. K. Toone, *Science*, **1981**, *213*, 922.

POSTSCRIPT

"I say that it touches a man that his blood is sea water and his tears are salt, that the seed of his loins is scarcely different from the same cells in a seaweed, and that of stuff like his bone is coral made. I say that physical and biological law lies down with him, and wakes when a child stirs in the womb, and that the sap in a tree, uprushing in the spring, and the smell of loam, where the bacteria bestir themselves in darkness, and the path of the sun in the heaven, these are facts of first importance to his mental conclusions, and that a man who goes in no consciousness of them is a drifter and a dreamer, without a home or any contact with reality."

Donald Culross Peattie

PROBLEMS

18.1. Why was the covalent radius of the metal used on p. 854 instead of that of the +2 ion?

18.2. Why are transition metals such as Mn, Fe, Co, and Cu needed in photosynthesis and respiration rather than metals such as Zn, Ga, or Ca?

18.3. Calculate the energy available from one photon of light at wavelength 700 nm. If the potential difference generated is 1 V, what is the conversion efficiency?

18.4. Discuss how the use of simple model systems can aid our understanding of biochemical systems. Is there any way they might detract? [J. H. Wang, *Acc. Chem. Res.*, **1970**, *3*, 90.]

18.5. There are two ways in which photosynthesis increases the energy available: (1) by using two light capturing mechanisms, PS I and PS II; (2) by stacking the chlorophyll in the grana which are in turn stacked in the chloroplast (see A. L. Lehninger, *Sci. Am.*, September, **1961**, for the structures). Which corresponds to hooking batteries in parallel and which to hooking them in series?

18.6. Common ions in enzyme systems are those that have low site preference energies (from LFSE) such as Co^{+2}, Zn^{+2}, and Mn^{+2} rather than Fe^{+2}, Ni^{+2}, or Cu^{+2}. Discuss this phenomenon in terms of the entatic hypothesis. [R. J. P. Williams, *Endeavour*, **1967**, *26*, 96.]

18.7. Discuss the probable difference in the pockets present in carboxypeptidase and carbonic anhydrase.

18.8. The toxicity of metals has been variously correlated with their (1) electronegativity; (2) insolubility of the sulfides; (3) stability of chelates. Discuss.

18.9. Show how coordination of an O_2 molecule to a heme group can result in pairing of the electron *on the oxygen molecule* whether the bonding is

a. Through a μ bond

b. Through a lone pair on *one* oxygen atom

18.10. Directions for the use of the antibiotic tetracycline advise aganist drinking milk or taking antacids with the medication. In addition, warnings are given concerning its use (teeth may be mottled in certain cases). Suggest the chemical property of tetracycline that may be involved in these effects.

18.11. High mercury levels in terminal food chain predators like tuna fish have caused considerable worry. It has been found that tuna also contain larger than average amounts of selenium (H. E. Ganther et al., *Science*, **1972**, *175*, 1122). Discuss the possible role of selenium with respect to the presence of mercury. [See also, R. W. Chen et al., *Bioinorg. Chem.*, **1975**, *4*, 125; G. N. Schrauzer, *Bioinorg. Chem.*, **1976**, *5*, 275; M. L. Scott, in "Organic Selenium Compounds: Their Chemistry and Biology," D. L. Klayman and W. H. H. Günther, eds., Wiley (Interscience), New York, **1973**, pp. 646–647.]

18.12. Although the hypothesis of Egami (p. 914) may be an oversimplification, it is certainly true that Fe^{+2}/Fe^{+3} is widely used in redox systems, Zn^{+2} in hydrolysis, esterification, and similar reactions, and molybdenum in nitrogenase, xanthine oxidase, nitrate reductase, etc. Putting abundance aside, discuss the specific chemical properties of these metals that make them well suited for their tasks.

18.13. Carboxypeptidase A (Co^{+2}) not only retains the activity of carboxypeptidase A (Zn^{+2}) (p. 884), *it is actually a more active enzyme.* This being the case, why do you suppose that Co^{+2} is *not* used in the natural system?

18.14. If you did not answer Problem 16.6 when you read Chapter 16, do so now.

18.15. When patients are treated with D-penicillamine for schleroderma, cystinuria, rheumatoid arthritis, and idiopathic pulmonary fibrosis, 32% show decreased taste acuity (hypogeusia). In contrast, only 4% of the patients being treated with D-penicillamine for Wilson's disease exhibit hypogeusia. Discuss a possible mechanism. How might the hypogeusia be treated? [R. I. Henkin and D. F. Bradley, *Proc. Nat. Acad. Sci., U.S.*, **1969**, *62*, 30.]

18.16. Predict which way the following equilibrium will lie:

$$Hb + Hb(O_2)_4 = 2Hb(O_2)_2$$

Explain.

18.17. Although sickle cell anemia causes problems in many organ systems, the chief cause of death of children with SCA is bacterial infection. Discuss.

18.18. Using the reduction *emf*'s given in Appendix F, construct a Latimer diagram, complete with skip-step *emf*'s for one-, two-, and four-electron reduction of oxygen to superoxide, peroxide, and hydroxide. Discuss the biological significance of these *emf*'s. Recall that a living cell is basically a reduced system threatened by oxidizing agents. [I. Fridovich, *Am. Sci.*, **1975**, *63*, 54.]

18.19. Biochemists tend to speak of "dismutation reactions" such as

$$2H^+ + O_2^- + O_2^- \longrightarrow H_2O_2 + O_2$$

What term do inorganic chemists use for this phenomenon?

18.20. Gray and co-workers [*J. Am. Chem. Soc.*, **1975**, *97*, 2092] have prepared copper(II) carboxypeptidase A and compared its spectrum, that of the enzyme with inhibitor present, and those of several other copper(II) complexes with nitrogen and oxygen ligating atoms. Some of these data together with the geometry about the copper ion are:

Set of ligating atoms[a]	Structure	ν_{max} (cm^{-1})
N, N, N, N	Planar	19 200–19 600
N, N, O, O	Planar	14 100–17 500
N, N, O, O	Pseudotetrahedral	12 900–14 700
O, O, O, O	Planar	13 500–15 000
Cu^{II}CPA:		
N, N, O, O?	?	12 580
CuC^{II}PA · βPP[b]:		
N, N, O, O?	?	11 400

[a] These are the four atoms in the coordination sphere of the copper(II) ion.
[b] βPP is β-phenylpropionate, an inhibitor of carboxypeptidase A.

a. Account for the trends in the values of ν_{max} in the first four rows of the table listing literature values and known geometries.
b. Predict the geometry of the ligating atoms about the Cu^{II} ion in Cu^{II}CPA and in Cu^{II}CPA · βPP.
c. Comparing the values of ν_{max} of Cu^{II}CPA with and without the inhibitor present, suggest what effect the inhibitor may be having on the geometry of the copper ion.

18.21. For simplicity the iron–oxygen interaction in myoglobin and hemoglobin (but not hemerythrin) was discussed in terms of neutral oxygen molecules binding to Fe^{II}. However, much of the current literature discusses these phenomena in terms of superoxo and peroxo complexes and one sees $Fe^{III} \cdot O_2^-$. Discuss what these formulations and terms mean, and describe the related consequences in terms of charges, electron spins, etc. [L. Vaska, *Acc. Chem. Res.*, **1976**, *9*, 175; C. A. Reed and S. K. Cheung, *Proc. Nat. Acad. Sci., U.S.*, **1977**, *74*, 1780.]

18.22. Using your knowledge of periodic relationships, predict which element might come closest to reproducing the behavior of molybdenum in nitrogenase. Recall that nitrogen fixation involves both complexation and redox reactions. [E. I. Stiefel, *Proc. Nat. Acad. Sci., U.S.*, **1973**, *70*, 988.]

18.23. In order to study the function of oxygen-binding by myoglobin and its effect on muscle function, Cole (footnote 40) perfused an isolated muscle with hydrogen peroxide. Why did he do this?

18.24. a. One problem encountered in the manufacture and preservation of H_2O_2 is its spontaneous decomposition:

$$2H_2O_2 \longrightarrow 2H_2O + O_2 \quad \textbf{(18.43)}$$

which is exothermic ($\Delta H_r = -196$ kJ mol^{-1}). To reduce this decomposition, copper ions are carefully excluded or chelated.[170]

b. Nature's protection against the destructive oxidative powers of H_2O_2 are the enzymes catalase and peroxidase, both of which contain iron.

[170] C. A. Crampton et al., in "The Modern Inorganic Chemicals Industry," R. Thompson, ed., Chemical Society Spec. Publ. No. 31, London, **1977**, p. 244.

c. Superoxide, $\cdot O_2H^-$, is perhaps even more dangerous than hydrogen peroxide as an oxidant *and* free radical, though otherwise somewhat similar.

d. Superoxide dismutase contains both zinc and copper. Zinc may be replaced by Co(II) and Hg(II) and the enzyme retains its activity; no substitution of copper with retention of activity has yet been found.[171] Discuss. [*Hint:* You can discuss this on several levels, from the most offhand conjectures to a quantitative demonstration with numbers. Choose an appropriate level (or ask your professor) and attack the problem accordingly.]

18.25. One of the compounds in this chapter has been dubbed "The Iron Pagoda." Which compound and why?

18.26. $Fe(OH)_2$ and $Fe(OH)_3$ have serendipitous K_{sp} values: If one can remember the exponents, one can "instantly" estimate the pH necessary to make a 1 mol dm^{-3} solution of Fe^{+2} and Fe^{+3}.

a. Explain.

b. Calculate the exact pH of the solution approximated above.

18.27. Bezkorovainy cites several studies indicating reduced iron absorption during febrile illnesses. Frame a hypothesis for the adaptiveness of such an effect. [A. Bezkorovainy, "Biochemistry of Nonheme Iron," Plenum, New York, **1980**, p. 90.]

18.28. Acid rain has been defined as any precipitation with a pH lower than 5.6. Why 5.6 instead of 7.0? Can you perform a calculation to reproduce this value?

18.29. Explain what effect acid rain would have on the condition of each of the following, and why:

a. The Taj Mahal, at Agra, India

b. A limestone barn near Antietam Battlefield, Maryland, dating from the Civil War

c. A granite statue of S. P. L. Sørensen in Stockholm, Sweden

d. The Karyatides, the Acropolis, Athens, Greece

e. The asbestos-shingled roof on my house in College Park, Maryland

f. The integrity of the copper eaves-troughs and downspouting on that house

g. The integrity of the brick siding of that house

h. The growth of the azaleas planted along the foundation of that house

i. The integrity of the aluminum siding on a neighbor's house

j. The slate roof on another neighbor's house

k. The longevity of galvanized steel fencing in the neighborhood

l. The ability of an aquatic snail to form its shell in a lake in the Adirondack Mountains

18.30. For each of the above for which you predicted an adverse effect, speculate as to the likelihood that there actually will *be* an effect, i.e., whether there will be acid rain at that particular geographic site or not.

18.31. Page 896 refers to "the rust-red soils of Oahu." What is the chemical origin of the "rust-red" color? What is the physical source of the color?

18.32. Niebohr and Richardson have written an extremely interesting article entitled "The replacement of the nondescript term 'heavy metals' by a biologically and chemically

---

[171] I. Fridovich, *Adv. Inorg. Biochem.*, **1979**, *1*, 67.

significant classification of metal ions." Their abstract states, in part:

> It is proposed that the term "heavy metals" be abandoned in favor of a classification which separates metals . . . according to their binding preferences . . . related to atomic properties . . . A review of the roles of metal ions in biological systems demonstrates the potential of the proposed classification for interpreting the biochemical basis for metal-ion toxicity . . .

Discuss in terms of the suggestions provided by the abstract. Propose a theme for the Niebohr/Richardson article (as though it were your own) and give some illustrative examples. [E. Niebohr and D. H. Richardson, *Environ. Pollut*., **1980**, *1*, 3.]

18.33. Review your knowledge of coordination chemistry with respect to nomenclature: Why is the molecule shown in Fig. 18.35 the "Δ" isomer?

18.34. The parallelism of sunlight-driven photosynthesis/respiration and the chemolithotophic oxidation of sulfide and sulfur by bacteria (p. 852), as well as the possibility of metal toxicity near hydrothermal vents (p. 915), has been noted. Suggest other problems and possible solutions to be expected from hydrothermal vent organisms. [A. J. Arp and J. J. Childress, *Science*, **1983**, *219*, 295; M. A. Powell and G. N. Somero, ibid., **1983**, *219*, 297; R. C. Terwilliger et al., ibid., **1983**, *219*, 981.]

A

Questions to be considered after finishing the course of study

These questions are not to be considered especially hard, but they are questions that generally involve more than a single chapter and thus could not be placed at the end of any one of them. It is also hoped that they will provide an overview of inorganic chemistry.

1. On the basis of your study of inorganic chemistry, define it in your own terms. Find the definition you wrote at the beginning of the course (see Problem 1.2) and compare the two.

2. Reread Chapter 1 and compare your understanding of the phenomena mentioned there with your initial impressions recorded at the beginning of the course (see Problem 1.1).

3. Consider all of the inorganic molecules you have read about in this book. List the half dozen that you think will have the lowest dipole moments. Molecules such as O=C=O and SF_6 that have zero dipole moments because of their symmetry are trivial and should not be included in your list.

4. Heating rhenium metal at 600 °C in chlorine gas results in a dark red-brown vapor that condenses to a dark red solid, mp = 261 °C. Thermal decomposition of **A** produces a nonvolatile, mauve solid (**B**). The compound **B** dissolves in hydrochloric acid to give a red solution. If cesium chloride is present in high concentration, crystals of **C** form. If **B** is dissolved in a melt of $Et_2NH_2^+Cl^-$, a blue color attributable to the anion **D** is produced. A compound, **E**, also containing the anion **D** may be crystallized from the solution produced by the reduction of $KReO_4$ with hypophosphorous acid if cesium chloride is added. Analytical data are:

Compound	%Re	%Cl	MW	FW/ν[1]
A	51	48	727	—
B	63	35	—	—
C	40	31	—	346
E	39	30	—	307

[1] The factor ν is the number of particles per "molecule" upon dissolution in a solvent, in this case water. See p. 334.

Suggest the appropriate compositions and structures of these species. Describe what you believe would be definitive experimental methods to confirm your suggestions.

5. Name ten contemporary inorganic chemists and discuss their research specialities in general or a particular definitive research accomplishment of each.

6. Re*write* Chapter 1, emphasizing *your* prejudices (i.e., likes and dislikes concerning inorganic chemistry).

7. In the bromine pentafluoride molecule the bromine atom is *below* the basal plane of the square pyramid, but in the chlorobis[*o*-phenylenebis(dimethylarsine)]palladium(II) cation the palladium atom is *above* the basal plane. Explain.

8. Discuss in what way the seemingly disparate species HF_2^-, XeF_2, and I_3^- resemble each other from a bonding point of view.

9. Discuss each of the following techniques as they apply to inorganic chemistry:
a. Mössbauer spectroscopy
b. nuclear magnetic resonance spectroscopy
c. electron spin resonance
d. nuclear quadrupole resonance
e. dipole moment measurements
f. infrared and Raman spectroscopy
g. magnetic susceptibility measurements
h. visible and ultraviolet spectroscopy
i. X-ray photoelectron spectroscopy
j. ultraviolet photoelectron spectroscopy
k. X-ray, electron, and neutron diffraction

10. Write equations (and note any special conditions necessary) for the synthesis of the following compounds:
a. $Mo(CO)_3(P\phi_3)_3$
b. $P_3N_3(NMe_2)_6$
c. $(OC)_5MnHgMn(CO)_5$
d. $(Me_2N)_3PO$
e. XeO_4
f. $K_2SnCl_4I_2$
g. $[Co(NH_3)_6]Cl_3$
h. $Fe(C_5H_4COOH)_2$
i. *trans*-$[Pt(P\phi_3)_2Cl_2]$
j. *cis*-$[Pt(P\phi_3)_2Cl_2]$

11. The terms clathrate, clathro-chelate, cryptate, and zeolite have appeared at scattered places in the text. Write a one-page summary illustrating the similarities and differences between these species.

12. Complete the following equations:
a. $Y + F_2 \longrightarrow$
b. $R_2P(O)Cl + (CH_3)_2NH \longrightarrow$
c. $NO + O_3 \longrightarrow$
d. $PCl_3 + (CH_3)_3N \longrightarrow$
e. $PCl_3 + (CH_3)_2NH \longrightarrow$
f. $NH_4^+ + S^{-2} \longrightarrow$
g. $NOCl + xsBrF_3 \longrightarrow$
h. $CoCl_2 + NH_3 + NH_4Cl \xrightarrow{O_2}$
i. $PtCl_4^{-2} + 2\phi_2PCH_3 \longrightarrow$

j. $PtCl_4^{-2} + \phi_2PCH_2CH_2P\phi_2 \longrightarrow$
k. $Fe(CO)_5 + CH_2{=}CH{-}CH{=}CH_2 \longrightarrow$
l. $Ir(CO)(\phi_3P)_2Cl + H_2 \longrightarrow$
m. $Cl_2 + OH^- \xrightarrow{\text{cold}}$
n. $Cl_2 + OH^- \xrightarrow{\text{hot}}$
o. $Ar + xsF_2 \xrightarrow{200^\circ}$
p. $La + F_2 \longrightarrow$
q. $Ce + F_2 \longrightarrow$
r. $U + F_2 \longrightarrow$
s. $Hb(O_2)_4 + CO \longrightarrow$

13. From the information given, tell as much as possible about the identification of the products described in the following research. Specifically, deduce as much as possible concerning the (1) formula; (2) structure; (3) related properties of the compounds synthesized. Remember that these are actual experimental data rather than "theoretical values."

a. Phosphorus trichloride was allowed to react with 2-hydroxymethyl-2-methyl-1,3-propanediol in dry tetrahydrofuran in the presence of pyridine:

$$PCl_3 + CH_3{-}C(CH_2OH)_3 + 3Py \longrightarrow 3PyH^+Cl^-\downarrow + \text{soluble product} \qquad \textbf{(A.1)}$$

The pyridinium chloride was separated, the tetrahydrofuran removed by distillation, and the products heated slowly *in vacuo*. A sublimate, **A**, formed on the cold finger and a residue, **B**, remained. The elemental analysis of **A** is 20.6% P, 40.55% C, and 6.07% H. **A** has a mp of 97–98°C and is soluble in heptane, ether, ethyl alcohol, and nitrobenzene. Cryoscopic determination of the molecular weight in nitrobenzene solution yielded a value of 157. **B** is a glassy, nonmelting, insoluble solid containing 20.6% P, 40.7% C, and 6.10% H.

Hydrogen peroxide was added to an ethanolic solution of **A** yielding a product **C**, mp = 249–250°C. Elemental analysis of **C** yielded 18.9% P, 36.90% C, 5.48% H. Cryoscopic determination of the molecular weight in nitrobenzene solution gave a value of 171. **C** has a characteristic infrared absorption at 1325 cm^{-1}.

When **A** was heated with sulfur in a sealed tube, a white powder, **D**, mp 224–225°C, was obtained. Elemental analysis of **D**: 16.9% P, 33.56% C, 5.18% H, and 17.8% S. The cryoscopically found molecular weight was 174. [J. G. Verkade and L. T. Reynolds. *J. Org. Chem.*, **1960**, *25*, 663.]

A may be synthesized more readily by reaction of 2-hydroxymethyl-2-methyl-1, 3-propanediol with trimethyl phosphite:

$$P(OCH_3)_3 + CH_3{-}C(CH_2OH)_3 \xrightarrow{\text{reflux}} A + 3CH_3OH \qquad \textbf{(A.2)}$$

[C. W. Heitsch and J. G. Verkade, *Inorg. Chem.*, **1962**, *1*, 392; J. G. Verkade et al., *Inorg. Chem.*, **1965**, *4*, 83.]

b. Reaction of **A** with Lewis acids produces the following compounds:

$$\mathbf{A} + (CH_3)_3B \longrightarrow \mathbf{E} \quad \textbf{(A.3)}$$

$$2\mathbf{A} + B_2H_6 \longrightarrow \mathbf{F} \quad \textbf{(A.4)}$$

$$2\mathbf{A} + B_4H_{10} \longrightarrow \mathbf{F} + \mathbf{G} \quad \textbf{(A.5)}$$

Elemental analyses and molecular weight determinations yielded:

Compound	% B	% C	% H	MW
F	6.59	37.7	7.49	152 ± 15
G	16.8	32.2	8.16	—

Solutions of **F** in methylene chloride were found to be nonconducting. [C. W. Heitsch and J. G. Verkade, *Inorg. Chem.*, **1962**, *1*, 392.]

c. Reaction of **A** with certain transition metal chlorides and nitrates resulted in the formation of the following products:

$$Cu(NO_3)_2 \cdot 3H_2O \xrightarrow[\text{abs. EtOH}]{\mathbf{A}} \mathbf{H} + \mathbf{C} \quad \textbf{(A.6)}$$

$$PdCl_2 \cdot 4H_2O \xrightarrow[\text{abs. EtOH}]{\mathbf{A}} \mathbf{I} \quad \textbf{(A.7)}$$

$$PtCl_4^{-2} \xrightarrow[\text{EtOH/H}_2\text{O}]{\mathbf{A}} \mathbf{J} \quad \textbf{(A.8)}$$

Properties of these complexes are:

Compound	% Metal	% C	% H	Other	Exp. MW	Color	Molar conductivity
H	—	33.60	5.02	1.76% N	—	White	147
I	23.0	25.48	3.21	15.2% Cl	—	White	2.6
J	34.7	21.53	3.56	12.4% Cl	—	White	5.9

SOURCE: J. G. Verkade and T. S. Piper, *Inorg. Chem.*, **1962**, *1*, 453.

d. In a second series of reactions with transition metals, **A** was allowed to react with with $Co(ClO_4)_2$ in absolute ethanol. Two compounds, **K** and **L**, can be obtained from this reaction. They always form in equimolar amounts. Some data on these complexes are:

Compound	Color	P/Co Ratio	Magnetic moment	Molar conductivity
K	White	6	0.0 BM	410
L	Yellow	5	0.0 BM	98

SOURCE: J. G. Verkade and T. S. Piper, *Inorg. Chem.*, **1963**, *2*, 944.

e. A third series of transition metal complexes was prepared by allowing **A** to react with metal carbonyls. A mixture of 1.0 mol $Ni(CO)_4$ with 1.5 mol **A** yielded compound **M**. Reaction of **M** with an equimolar amount of **A** produced **N**. A mixture of 1.0 mol **N** refluxed ethylbenzene (bp = 136°) for six hours with 4.0 mol of **A** yielded **O** as a product. Refluxing a mixture of 1.0 mol of **O** with 2.0 mol of **A** in chlorobenzene (bp = 132°) for 18 hours yielded **P**. Analytical data for these complexes are

Compound	% Ni	% P	% C	% H	MW
M	20.3	10.3	33.1	3.2	293
N	14.4	14.9	34.9	4.6	409
O	11.0	17.2	36.1	5.4	—
P	9.1	18.9	37.0	5.6	—

SOURCE: J. G. Verkade et al., *Inorg. Chem.*, **1965**, *4*, 228.

f. Displacement of the ligands in hexakis(dimethylsulfoxide)nickel(II) by **A** resulted in the formation of **R**. A suspension of **R** in water when heated with two moles of sodium bicarbonate evolved CO_2 and produced **Q** and an oxidized hydrolysate of **A** (including phosphate). Data for these complexes are:

Compound	% Ni	P/Ni ratio	Color	Molar conductivity
Q	9.1	4	Colorless	Insoluble: related compounds yield values of 2–3
R	5.13	6 ± 1[a]	Yellow	238

[a] There is some uncertainty in the analytical data for this complex.
SOURCE: T. J. Hutteman et al., *Inorg. Chem.*, **1965**, *4*, 950.

g. Reduction of NiI_2 in the presence of excess **A** led to the formation of **P**. When **P** is dissolved in chloroform or dichloromethane and some fluorosulfonic acid or trifluoroacetic acid is added, a pale yellow solution is immediately formed. NMR spectroscopy indicates that in addition to a phosphorus singlet plus the expected proton spectrum deriving from **A**, new peaks formed on the addition of acid: a new phosphorus doublet and a new hydrogen quintet. Treatment of a solution of **P** with HBF_4 caused immediate formation of a precipitate **S** of composition 32.3% C, 5.0% H, 15.6% P; a similar reaction using HPF_6 yielded **T**(29.9% C, 4.7% H, 19.5% P).

h. Discussion questions based on facts elucidated in preceding sections:

1. Do reactions (A–4) and (A–5) go as you would expect? [pp. 730–732.]
2. Predict the products of reaction of **A** with (a) CH_3I; (b) Br_2. [p. 826; T. A. Beineke, *Acta Crystallogr.*, **1964**, *B25*, 413.]
3. Do you consider the fact that **M** is white at all unusual for a Co(III) complex? What does this say about $10Dq$ and the place of **A** in the spectrochemical series? [p. 384.]

4. Compare the infrared stretching frequency of P=O in **C** (1325 cm^{-1}) with those in trimethyl phosphate (1274 cm^{-1}) and tributyl phosphate (1260 cm^{-1}). Suggest how the bonding may differ in these compounds [pp. 150–151, 826–828].
5. What evidence is given above that indicates that **A** is a strong π-bonding ligand? Using compounds similar to those described in part e, describe how further evidence on π bonding might be obtained.
6. Trialkyl phosphites are, in general, very poor ligands, especially when compared with the behavior of **A**. Suggest factors that may favor **A**. [T. L. Brown et al., *J. Phys. Chem.*, **1961**, *65*, 2051; J. G. Verkade and T. S. Piper, *Inorg. Chem.*, **1962**, *1*, 453.]
7. In contrast to the behavior of the phosphites [h(6)], **C** is a poorer extractant of lanthanide ions than tributyl phosphate. Suggest an explanation. [S. C. Goodman and J. G. Verkade, *Inorg. Chem.*, **1966**, *5*, 498.]
8. Compound **R** is diamagnetic. How does this compare with the magnetic behavior of other known 6-coordinate Ni(II) complexes? [See p. 576.] In view of the uncertainty in stoichiometry, suggest the correct formulation and structure of this compound. [K. J. Coskran et al., *ACS Adv. Chem. Ser. No. 62*, R. F. Gould, ed., Washington, D.C., **1966**, p. 590.]

14. The photolysis product (**A**) of *cis*-$Pt(N_3)_2(P\phi_3)_2$, trapped in a matrix at 77 K, evolves N_2 upon warming. Photolysis of the *trans* isomer under similar conditions yields **B**. ESR signals from **B** indicate the presence of free radicals ($\cdot N_3$?). There is no ESR signal from **A**.

Consider the catalysis of acetylenes (alkynes) to benzenes by organometallic catalysts (Chapter 13). With this in mind as a lead, suggest a possible structure for **A**.

Discuss the instability of **A** compared to its organic analogue.

[A. Vogler et al., *Angew. Chem. Int. Eng. Ed.*, **1980**, *19*, 717; H. Huber, *Angew. Chem. Int. Eng. Ed.*, **1982**, *21*, 64.]

15. The compound $U(OTeF_5)_6$ is a curiosity. In spite of a molecular weight of 1669, possibly the highest for a nonpolymeric inorganic compound, it melts at only 160 °C, and can be sublimed at 60 °C and a pressure of 1 Pa. Explain the basis of this low mp and high volatility. [Courtesy of F. A. Schroeder and D. R. Stein, eds., *Gmelin Newsletter*, **1982**, No. 2.]

16. Reaction of $(CH_3)_3SiN_3$ with S_3N_3Cl or S_3N_2Cl at -15 °C in acetonitrile produces the evolution of nitrogen gas, the formation of $(CH_3)_3SiCl$, and golden crystals that contain 70% sulfur. Suggest the nature of this third product. [F. A. Kenneth et al., *J. Chem. Soc., Dalton Trans.*, **1982**, 851.]

17. Several times in this text the words "hindsight," "retrospective wisdom," etc., have appeared. Discuss these various references and draw parallels from your own experience or reading. Why is the phenomenon so common, i.e., why does the seemingly impossible appear to be almost trivial after it is done? (Cf. the comment on Sherlock Holmes, ref. 40, Chapter 13.) What does that tell you about your future research?

B

Symmetry and term symbols

Symmetry is becoming increasingly important in inorganic chemistry. The discussion presented here is intended as an introduction to the subject sufficient to understanding the use of symmetry in this book. For a thorough understanding of symmetry the reader is referred to several longer accounts.[1]

SYMMETRY ELEMENTS AND SYMMETRY OPERATIONS

There are two types of symmetry of interest to the inorganic chemist. One, *translational symmetry*, is exemplified by a picket fence or a stationary row of ducks in a shooting gallery. If we turn on the mechanism so that the ducks start to move and then blink our eyes just right, the ducks appear motionless – the ducks move the distance between them while we blink and all ducks are identical. Under these conditions we could not tell if the ducks were moving or not because they would appear identical *after* the change to the way they appeared *before* the change. Translational symmetry is of importance to crystallographers (a stacked array of unit cells is like the row of ducks), but we shall be more interested in the symmetry of isolated molecules.

If we take as our reference point for symmetry operations the center of gravity of the molecule, the resultant symmetry is referred to as *point symmetry*. If we now perform some action, such as rotation about an axis passing through that point, and if at the end of that action, the molecule is indistinguishable from before, we have performed a *symmetry operation*. Consider *trans*-dinitrogen difluoride (**I**). If we construct an axis normal to the

[1] Three very good introductory articles have been prepared by M. Orchin and H. H. Jaffé, *J. Chem. Educ.*, **1970**, *47*, 246, 372, 510. There are several books: F. A. Cotton, "Chemical Applications of Group Theory," 2nd ed., Wiley (Interscience), New York, **1971**; P. B. Dorain, "Symmetry in Inorganic Chemistry," Addison-Wesley, Reading, Mass, **1965**; J. P. Fackler, Jr., "Symmetry in Coordination Chemistry," Academic, New York, **1971**; W. E. Hatfield and W. E. Parker, "Symmetry in Chemical Bonding and Structure," Merrill, Columbus, Ohio, **1974**; M. Orchin and H. H. Jaffé, "Symmetry, Orbitals, and Spectra," Wiley, New York, **1971**; D. S. Urch, "Orbitals and Symmetry," Penguin, Baltimore, **1970**.

I $\qquad$ C_2

plane of the paper and midway between the nitrogen atoms, we can rotate the molecule by 180° and obtain an identical configuration. (See the diagram above.) Rotation by 180° is thus a symmetry operation. The axis about which the rotation takes place is a *symmetry element*. In this case, **I** is said to have a *twofold rotational axis* because if the 180° operation is performed *twice*, all atoms are back at their initial positions.[2]

Consider *cis*-dinitrogen difluoride (**II**). It possesses no axis perpendicular to the plane of the paper that allows rotation (other than the trivial 360° one) that qualifies as a symmetry element. However, it is possible to draw an axis that lies in the plane of the paper and equidistant between the two nitrogen atoms and also equidistant between the two fluorine atoms. This is also a twofold axis. (See the diagram below.)

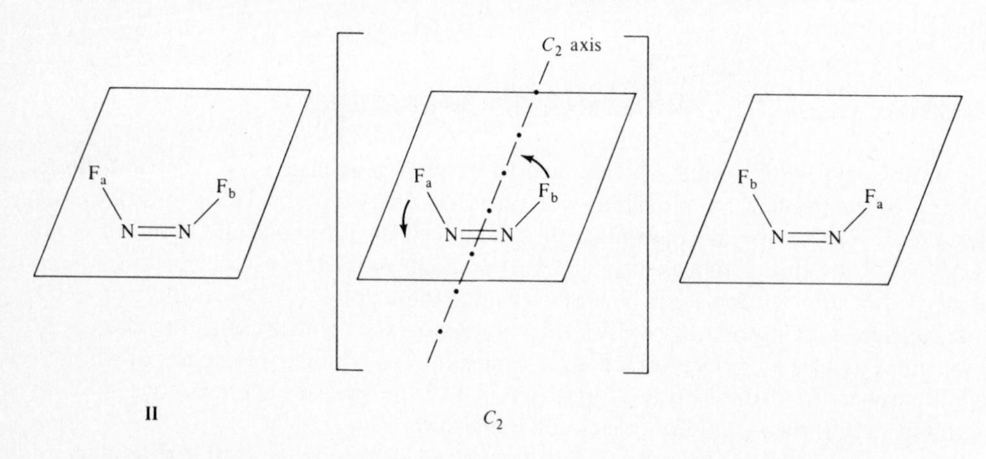

Rotational axes are denoted by the symbol C_n representing an n-fold axis. Thus both *cis*- and *trans*-dinitrogen difluoride have a C_2 axis.

Center of symmetry, *i*

We have already considered the symmetry operation known as *inversion about a center of symmetry* in Chapter 2. We found that it was possible to invert *gerade* orbitals without

[2] Some confusion can arise here. We can only perform symmetry operations if the atoms are indistinguishable from each other; yet in order to perform symmetry operations, we must attach little invisible labels to the atoms in order to recognize when the atoms are back to their initial positions.

changing them (the symmetry operation) and that they thus had a center of symmetry (the symmetry element)—but not so for *ungerade* orbitals. Of the three most common geometries encountered in inorganic chemistry (**III**–**V**), one has a center of symmetry (**V**) and two do not (**III**–**IV**). The center of symmetry is designated i for inversion.

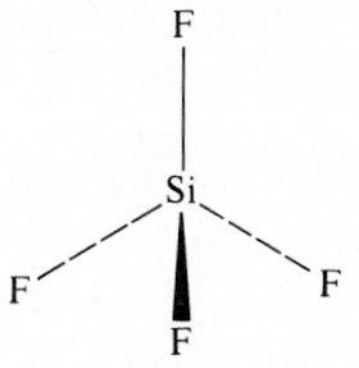

III 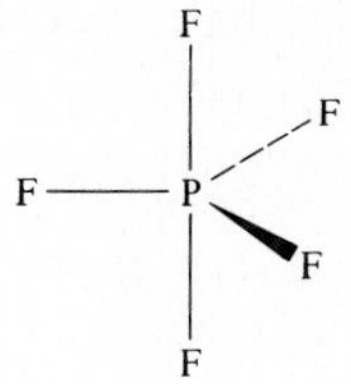 IV 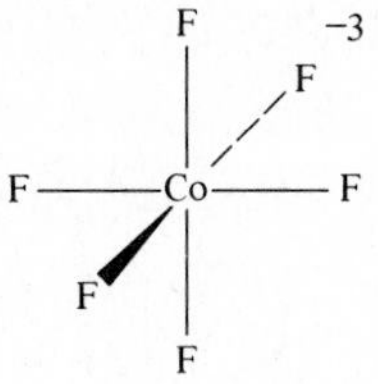 V

Rotational axis, C_n

We have seen rotational symmetry above, and in this section we shall simply look over a few molecules having such axes. Molecule **III** has a threefold rotational axis, C_3. In fact, it has four of them, each lying along an Si—F bond. The threefold axis is somewhat less obvious in ammonia, **VI**. It lies on an imaginary line running through the center of the lone pair and equidistant from the three hydrogen atoms. If all the bond angles change, as from 107° in NH_3 to 102° in NF_3, does the symmetry change? Note that iron pentacarbonyl (**VII**) also has a C_3 axis. (Are there any other molecules, **I**–**VII**, that have one or more C_3 axes?[3]) Note that it also has three C_2 axes, one through each of the equatorial carbonyl groups.

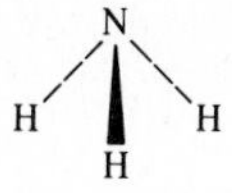

VI 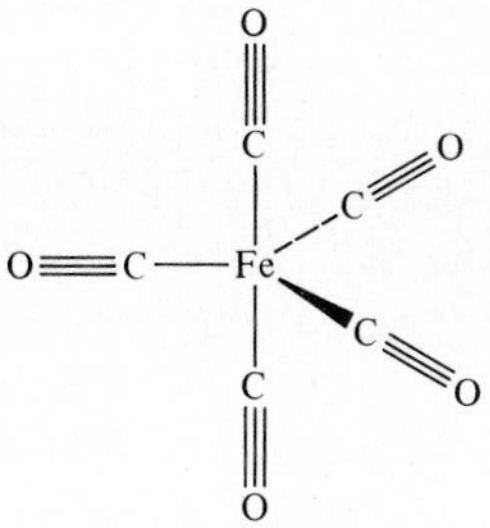 VII 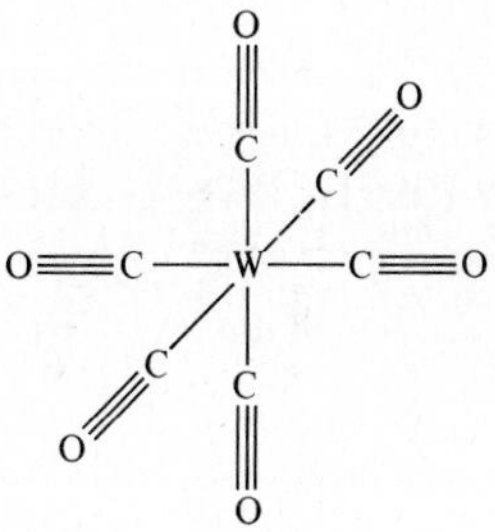 VIII

In contrast, tungsten hexacarbonyl (**VIII**) has a fourfold axis. In fact, it has three C_4 axes: (1) one running top to bottom; (2) one running from left to right; (3) one running front to back. (In addition, it has other rotational axes. Can you find them?[4])

---

[3] *Answer*: Both NH_3 and NF_3 have a single C_3 axis. SiF_4, PF_5, $[CoF_6]^{-3}$, NH_3, and $Fe(CO)_5$ all have one or more C_3 axes.

[4] *Answer*: $W(CO)_6$ has three C_4 axes, four C_3 axes (through the octahedral faces), and six C_2 axes (through the octahedral edges).

There may be higher order rotational axes. Consider the molecule osmocene (**IX**), which has a C_5 axis through the osmium atom and perpendicular to the cyclopentadienyl rings. Consider the related though staggered molecule ferrocene (**X**). Does it have a fivefold rotational axis?[5] Next consider borazine (**XI**). Does it have a C_6 axis? No, because a rotation of only 60° places a boron atom where there formerly was a nitrogen atom and vice versa. However, benzene has a true C_6 axis, as do the six-carbon anions illustrated in Fig. 14.27.

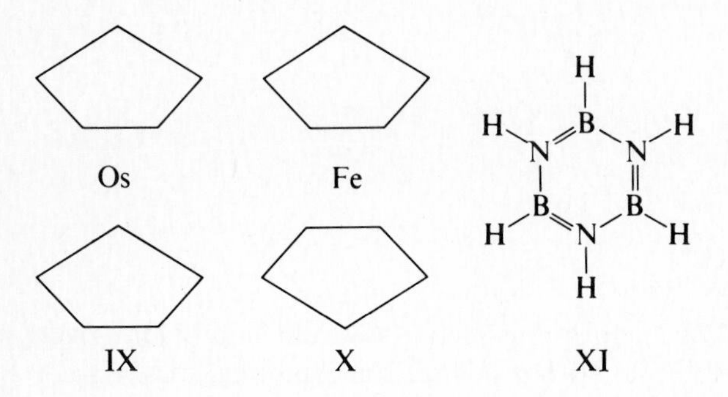

An axis designated C_n is often called a "proper" rotational axis and the rotation about it a "proper" rotation, in contrast to a different symmetry element and an operation operation about it (see below), which are termed "improper."

Identity

We have seen above that a C_1 operation results in the same molecule we started with. It is therefore an identity operation. Identity operations are denoted by I (or sometimes E). It would appear that such operations would be unimportant inasmuch as they accomplish nothing. Nevertheless for completeness they are included and some useful relationships can be constructed. For example, we have seen that two consecutive C_2 operations about the same axis result in identity. We may therefore write $C_2 \times C_2 = I$, and likewise $C_3 \times C_3 \times C_3 = I$. This may also be expressed as $C_2^2 = I$ and $C_3^3 = I$.

The mirror plane, σ

Students who have had experience in organic chemistry will be familiar with the mirror plane, since one criterion for optical activity is the absence of a mirror plane. More rigorously, a molecule will be optically active if it possesses neither a mirror plane nor a center of symmetry. Most inorganic molecules are *not* optically active and contain one or more mirror planes. For example, all planar molecules (**I**, **II**, **XI**) will contain a mirror plane, σ, through all of the atoms (an atom lying in the mirror plane is considered to be reflected) in the molecule. Mirror planes can be classified as horizontal, σ_h, that is, perpendicular to the highest order n-fold axis, and vertical, σ_v, in which the highest order

[5] *Answer*: Ferrocene has a fivefold rotational axis.

n-fold axis lies in the mirror plane. Horizontal mirror planes exist in **I**, **IV**, **V**, **VII**, **VIII**, **IX**, and **XI**. Vertical mirror planes exist in **II**, **III**, **IV**, **V**, **VI**, **VII**, **VIII**, **IX**, **X**, and **XI**.

An improper rotation may be visualized as occurring in two steps: rotation by $360°/n$ *followed* by reflection across a plane perpendicular to the rotational axis. Neither the rotation axis nor the mirror plane need to be true symmetry elements that can stand alone. For example, we have seen that SiF_4 (**III**) has no C_4 though it has four C_3 axes. Nevertheless it has an S_4 axis:

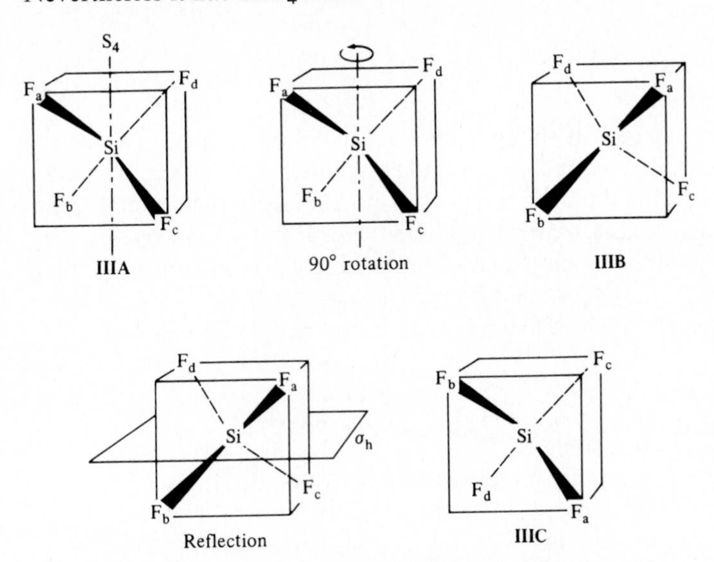

IIIA 90° rotation IIIB

Reflection IIIC

Consider the *trans* configuration of hydrazine, **XII**. If we perform a C_2 operation followed by a σ_h operation, we will have a successful S_2 operation. Note, however, that the same result could have been obtained by an inversion operation:

XIIA 180° rotation XIIB

i

Reflection in plane of paper XIIC

Thus S_2 is equivalent to *i*. Confirm this to your satisfaction with molecule **II** (which contains a center of symmetry) and molecule **III** (which, although it possesses true C_2 axes, does *not* have a center of symmetry and thus cannot have an S_2).

Furthermore, S_1 is equivalent to σ, because we have seen that $C_1 = I$, and therefore the second step, reflection, is σ. The chief reason for pointing out these relationships is for systemization: All symmetry operations can be included in C_n and S_n. Taken in the order in which they were introduced, we have: $i = S_2$; $I = C_1$; $\sigma = S_1$.

Dipole moments

Dipole moments of molecules are vectors that must lie on each of the symmetry elements. This is because the dipole moment cannot change during a symmetry operation. If a molecule has a center of symmetry, *i*, the dipole moment must be zero, since the center of symmetry is a point and a point is a vector with magnitude zero. If more than one C_n axes ($n > 1$) are present, a dipole moment cannot exist since the dipole vector cannot lie along more than one axis at a time. A mirror plane does not prevent the possibility of a dipole moment as long as any C_n axis lies in the plane. The dipole moment lies along the C_n axis.

Point groups

If we analyze the symmetry elements of a molecule such as water (**XIII**), we find that it has one C_2 axis, two σ_v,[6] and of course, *I*. (See the diagram below.) This set of four symmetry elements is said to form a *symmetry group*, or *point group*. Many point groups are important only for organic molecules, so the following list will be restricted to some point groups that the inorganic chemist commonly encounters.[7]

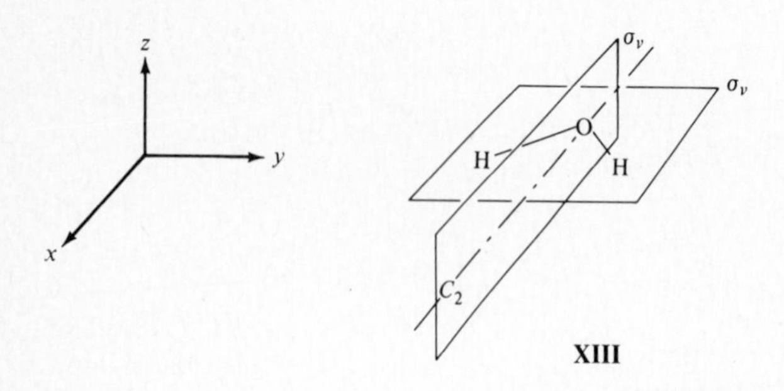

XIII

[6] Alternatively, one can label these in terms of the coordinate system: C_2^x, σ_{xz}, and σ_{yz}.

[7] Procedures for working out the correct point group using flowchart techniques are given by M. Orchin and H. H. Jaffé, *J. Chem. Educ.*, **1970**, *47*, 372, and W. E. Hatfield and W. E. Parker, "Symmetry in Chemical Bonding and Structure," Merrill, Columbus, Ohio, **1974**.

Examples of some common point groups

Point group	Examples	Explanation
No symmetry. C_1	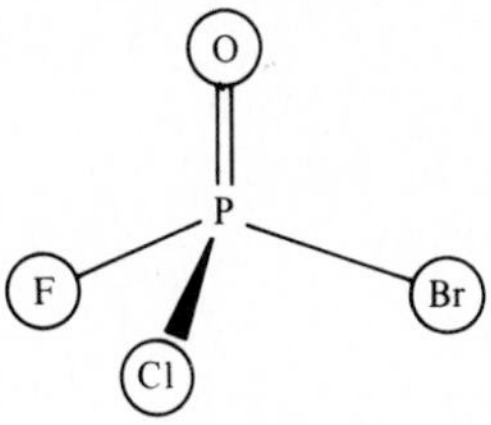	Completely asymmetric molecules.
Rotational symmetry. C_{2v}	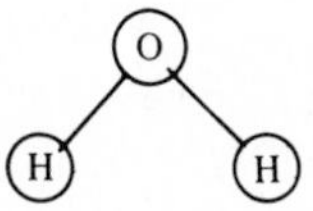 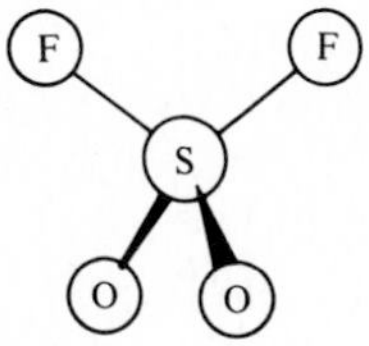	Twofold rotational axis plus two vertical mirror planes containing the rotational axis.
Dihedral symmetry. D_3	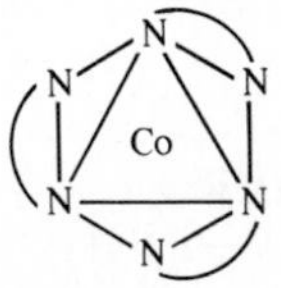	Threefold rotational axis plus three twofold axes perpendicular to C_3.
D_{5d}	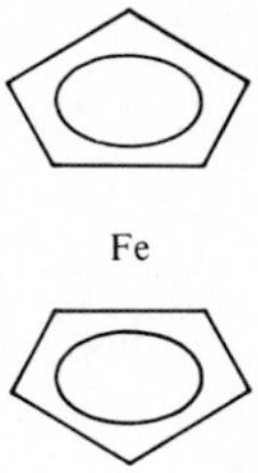	Fivefold rotational axis, five vertical mirror planes containing C_5, and five twofold axes perpendicular to C_5.
D_{3h}	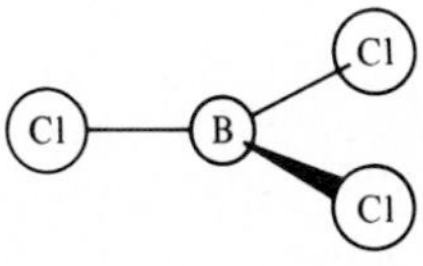 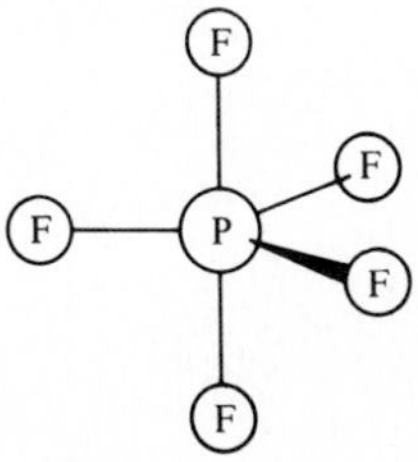	As for D_3 (threefold rotational axis; three twofold axes perpendicular to C_3) plus horizontal (*i.e.*, perpendicular to the principal axis) mirror plane.
D_{4h}	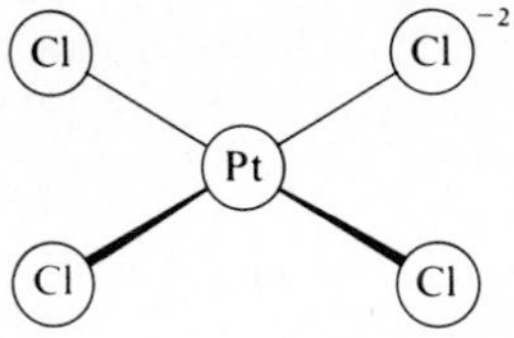 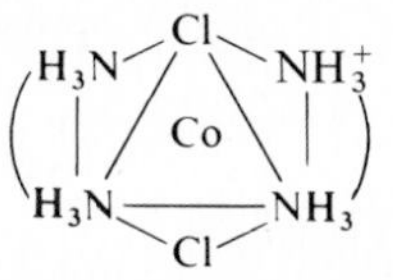	Fourfold rotational axis, four twofold axes perpendicular to C_4, and a horizontal mirror plane.
$D_{\infty h}$		Totally symmetric about the principal axis plus a horizontal mirror plane.

Examples of some common point groups (*Continued*)

Point group	Examples	Explanation
Higher symmetries.		
T_d	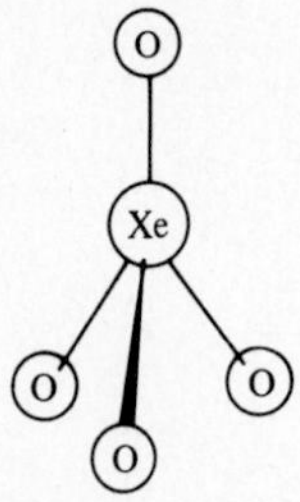 XeO_4	Tetrahedral symmetry.
O_h	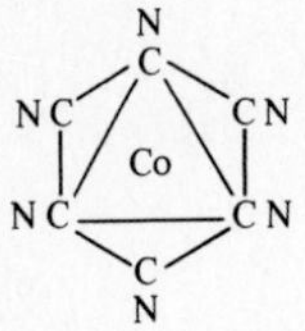 $[Co(CN)_6]$	Octahedral symmetry.
I_h	$[B_{12}H_{12}]^{-2}$(See Fig. 14.37.)	Icosahedral symmetry.

Symbols of irreducible representations and examples of atomic orbitals[a]

Symbol	Example	Explanation
a_{1g}	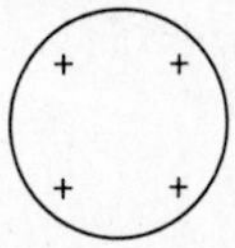	Singly degenerate and symmetric about the principal axis (*a*), symmetric with respect to a second axis or to reflection in a vertical plane[b] (1) and *gerade* (*g*).
a_{1u}	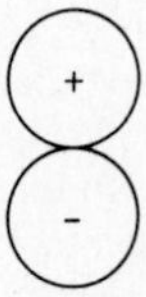	As above but *ungerade* (*u*).
a_{2u}	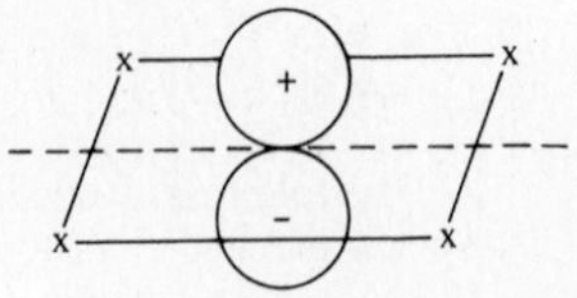	Singly degenerate and symmetric about the principal axis (*a*), but antisymmetric with respect to a second (C_2) axis (dashed line) (2) and *ungerade* (*u*).
b_{1u}	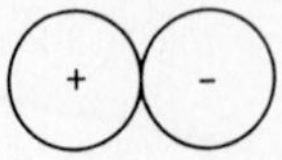	Singly degenerate but antisymmetric about the principal axis (*b*), symmetric with respect to reflection in a vertical plane[b] (1), and *ungerade* (*u*).
b_{2u}	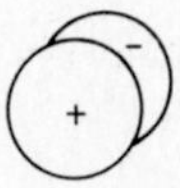	As above except antisymmetric with respect to reflection in a vertical plane (2).

Symbols of irreducible representations and examples of atomic orbitals[a] (*Continued*)

Symbol	Example	Explanation
e_{1g}	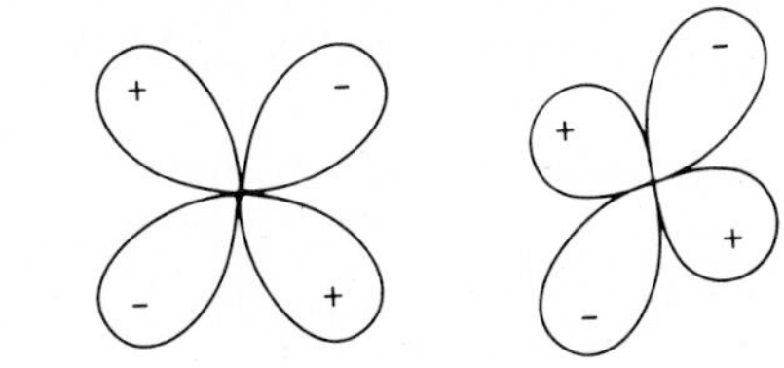	Doubly degenerate (e), antisymmetric with respect to 180° rotation about the z axis (1), and *gerade* (g).
e_{2g}	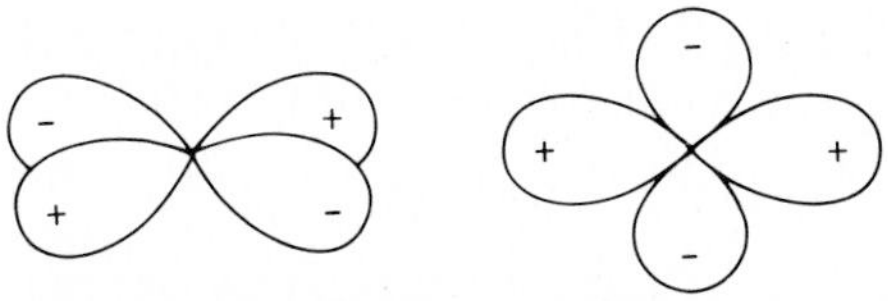	Doubly degenerate (e), antisymmetric with respect to 90° rotation about the z axis (2), and *gerade* (g).

[a] The x axis (vertical) is taken as the principal axis. Note that the assignment of subscripts 1 and 2 is different for the a and b representations from that of the e and t.

[b] This sheet of paper may be taken as the vertical plane.

ATOMIC STATES AND TERM SYMBOLS

We have seen (p.11) that the energy of a spectral transition is given by the Rydberg formula, which consists of two terms. It is common for spectroscopists to apply the word *term* to the *energies* associated with the states of an atom involved in a transition. Term symbols are an abbreviated description of the energy, angular momentum, and spin multiplicity of an atom in a particular state. Although the inorganic chemist generates the term symbols he uses from his knowledge of atomic orbitals, the historical process was the reverse: S, P, D, and F states were observed spectroscopically and named after *sharp* ($S \rightarrow P$), *principal* ($P \rightarrow S$), *diffuse* ($D \rightarrow P$), and *fundamental* ($F \rightarrow D$) characteristics of the spectra. Later the symbols s, p, d, and f were applied to orbitals.

Atoms in $S, P, D, F, \ldots$ states have the same orbital angular momentum as a hydrogen atom with its single electron in an $s, p, d, f, \ldots$ orbital. Thus we can define a quantum number L, which has the same relationship to the atomic state as l has to an atomic orbital (e.g., $L = 2$ describes a D state). L is given by:

$$L = l_1 + l_2, l_1 + l_2 - 1, l_1 + l_2 - 2, \ldots, |l_1 - l_2| \quad \textbf{(B.1)}$$

We can also define the component of the total angular momentum along a given axis:

$$\underline{\overline{M_L}} = L, L - 1, L - 2, \ldots, 0, \ldots, -L \quad \textbf{(B.2)}$$

The number of possible values of M_L is given by $2L + 1$. M_L is also given by:

$$M_L = m_{l_1} + m_{l_2} + \cdots + m_{l_n} \quad \textbf{(B.3)}$$

Likewise we can define an atomic spin quantum number representing the total spin:

$$S = \sum_i s, \quad \textbf{(B.4)}$$

For a given value of S, there will be $2S + 1$ spin states characterized by M_S:

$$M_S = S, S - 1, S - 2, \ldots, -S \quad \textbf{(B.5)}$$

or

$$M_S = m_{s_1} + m_{s_2} + \cdots + m_{s_n} \quad \textbf{(B.6)}$$

Now the total angular momentum of an electron is the resultant of the orbital angular momentum vector and the electron-spin angular momentum vector. Both of these are quantized, and we can define a new quantum number, j:

$$j = l + s \quad \textbf{(B.7)}$$

Since $s = \pm\frac{1}{2}$, it is obvious that every value of l will have two values of j equal to $l + \frac{1}{2}$ and $l - \frac{1}{2}$. The only exception is $l = 0$, for which $j = \pm\frac{1}{2}$; these values are identical since it is the absolute magnitude of j that determines the angular momentum.

We can now couple the resultant orbital angular momentum (L) with the spin angular momentum (S). The new quantum number J is obtained:

$$J = L + S, L + S - 1, L + S - 2, \ldots, |L - S| \quad \textbf{(B.8)}$$

The origin of the J values can be seen from a pictorial representation of the vectors involved.

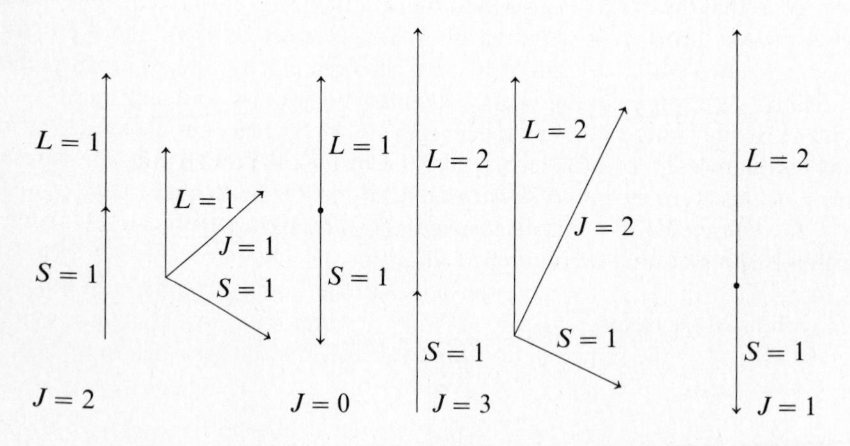

The number of J values available is termed the *multiplicity* of a state. In both of the examples pictured above the multiplicity is three. In general,[8] the multiplicity is given by $2S + 1$. The multiplicity is appended to the upper left of the symbol of the state and the J to the lower right. The above examples are thus ^3P and ^3D states (pronounced "triplet

[8] The exception is the ns^1 configuration, where $L = 0$ and $S = \frac{1}{2}$. The only value J can have is $+\frac{1}{2}$. [Note absolute magnitude symbol in last term of Eq. B.8.]

P" and "triplet D"). The individual terms are $^3P_2$, $^3P_1$, and $^3P_0$ (left) and $^3D_3$, $^3D_2$, and $^3D_1$ (right).

To turn again momentarily from the abstractions of orbitals and quantum numbers back to the spectra that generated them, consider the transition from a 1*s* orbital in hydrogen to a 2*p* orbital. The terms and transitions are:[9]

$$^2S \longrightarrow {}^2P_{3/2}$$

$$^2S \longrightarrow {}^2P_{1/2}$$

The $^2P_{3/2}$ term lies slightly higher in energy than the $^2P_{1/2}$, and therefore the spectral line is split into a "doublet"; hence the origin of the usage. It may be noted that in respect to these transitions the following selection rules operate: Δn, arbitrary; $\Delta l = \pm 1$; $\Delta j = \pm 1, 0$.

Assigning term symbols

We have seen that the term symbol for the ground state of the hydrogen atom is $^2S_{1/2}$. For a helium atom $L = 0$, $S = 0$, $J = 0$, and the term symbol for the ground state is $^1S_0$. For an atom such as boron, we can make use of the fact that all closed shells and subshells (such as the He example just given) contribute nothing to the term symbol. Hence both the $1s^2$ and $2s^2$ electrons give $L = S = J = 0$. The $2p^1$ electron has $L = 1$, $S = \frac{1}{2}$, and $J = 1 \pm \frac{1}{2}$, yielding $^2P_{1/2}$ and $^2P_{3/2}$. For carbon there are two p electrons. The spins may be paired or unpaired, so $L = 2, 1, 0$; $S = 1, 0$; and $J = 3, 2, 1, 0$. To work out the appropriate states for this atom requires a systematic approach. Note, however, that when neon is reached we have again a $^1S_0$: sodium repeats the $^2S_{1/2}$, magnesium $^1S_0$, etc.

A systematic approach to term symbols

In Chapter 2 it was shown how m_l and m_s values could be summed to give M_L and M_S values to yield terms for the spectroscopic states of an atom. If there are two or more electrons, it is usually necessary to proceed in a systematic fashion in generating these terms. The following is one method of doing so. The p^2 configuration of carbon is used.

1. *Determine the possible values of M_L and M_S.* For the p^2 configuration, L can have a maximum value of 2 and M_L can have values of $-2, -1, 0, +1, +2$. The electrons can be paired ($M_S = 0$) or parallel ($M_S = +1, -1$).
2. *Determine the electron configurations that are allowed by the Pauli principle.* The easiest way to do this is to draw up a number of sets of p orbitals as in Fig. B.1 (each

[9] Despite the fact that there is but a single value of J for the ground state of hydrogen, spectroscopists write its term symbol as $^2S_{1/2}$. The reason for this is that transitions between states of different spin multiplicities are *spin-forbidden*; thus transitions from a spin-paired singlet (such as ^1S) to a spin-unpaired triplet (^3S) are not allowed. However, with hydrogen (and the alkali metals as well) the ground state has an unpaired electron, and transition to doublet states containing a single unpaired electron are allowed. For ease in noting which transitions are allowed and which are spin-forbidden, spectroscopists write ^2S, though admittedly it can lead to confusion.

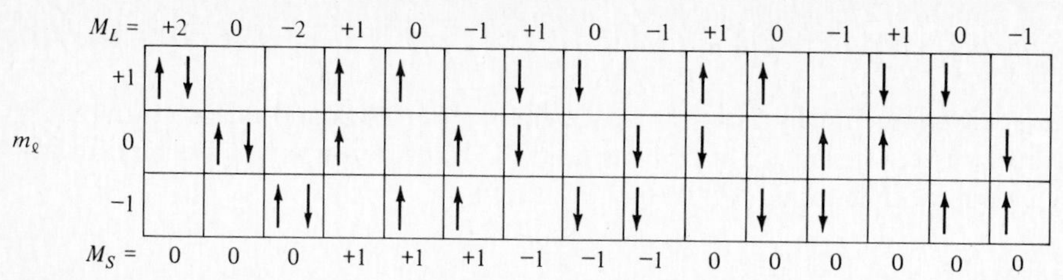

Fig. B.1 Term splitting in the ground-state ($1s^2 2s^2 2p^2$) configuration of carbon. All energies are in cm^{-1}. The ^3P, ^1D, and ^1S terms are split as a result of electron–electron repulsion. The ^3P term is further split with $J = 0, 1, 2$ as a result of spin-orbit coupling. The scale of the latter is exaggerated in this figure. [From R. L. DeKock and H. B. Gray, "Chemical Structure and Bonding," Benjamin/Cummings, Menlo Park, Calif., **1980**. Reproduced with permission.]

vertical column represents a set of three p orbitals) and fill in electrons until all possible arrangements have been found. The M_L value for each arrangement can be found by summing m_1 and M_S from the sum of m_s (spin-up electrons have arbitrarily been assigned $m_s = +\frac{1}{2}$). Each *microstate* consists of one combination of M_L and M_S.

3. *Set up a chart of microstates.* For example, the microstate corresponding to the first vertical column in Fig. B.1 has $M_L = +2$ and $M_S = 0$. It is then entered into the table below under those values. Sometimes the m_l and m_s values are entered directly into the table,[10] but if the electron configurations have been carefully worked out, there is no need of this. The fifteen microstates of p^2 yield:

		M_s +1	M_s 0	M_s −1
M_L	−2			
	−1	×	×	×
	0	×	×	×
	+1	×	×	×
	+2			

4. *Resolve the chart of microstates into appropriate atomic states.* An atomic state forms an array of microstates consisting of $2S + 1$ columns and $2L + 1$ rows. For example, a ^3P state requires a 3×3 array of microstates. A ^1D state requires a single column of 5 and a ^5D requires a 5×5 array, etc. Looking at the arrays of microstates, it is easy to spot the unique third microstate at $M_L = 0$ and $M_S = 0$; this must be a ^1S. A central column of $M_S = 0$ provides a ^1D. Removing these two states from the table, one is left with an obvious 3×3 array of a ^3P state. The states of carbon

[10] To ensure that all microstates have been written, the total number, N, of microstates associated with an electronic configuration, l^x, having x electrons in an orbital set with an azimuthal quantum number, l, is

$$N = \frac{N_l!}{x!(N_l - x)!}$$

where $N_l = 2(2_l + 1)$, the number of m_l, m_s combinations for a *single* electron in the orbital set. [From E. U. Condon and G. H. Shortley, "The Theory of Atomic Spectra," 1st ed., Cambridge University Press, **1963**.]

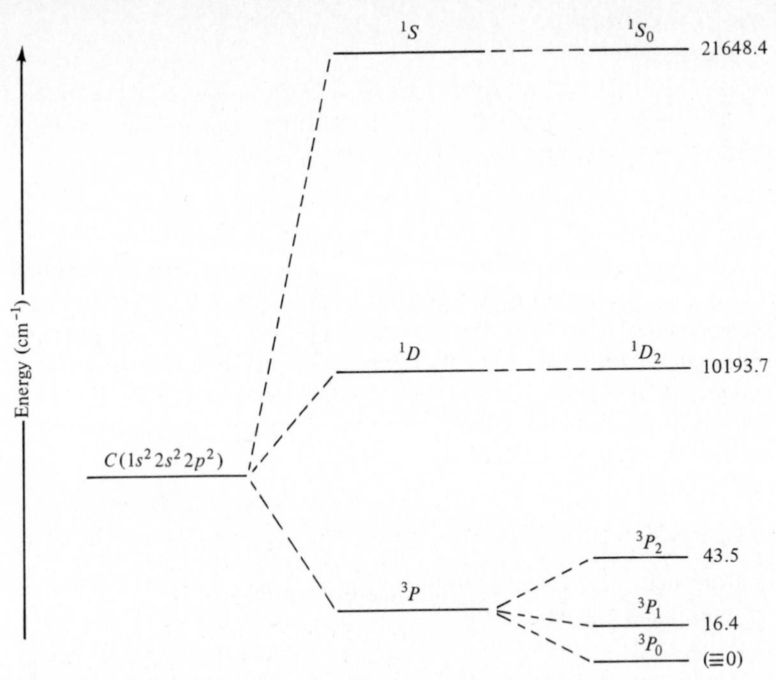

Fig. B.2 The fifteen microstates and resultant values of M_L and M_S for the $1s^2 2s^2 2p^2$ electron configuration of carbon.

Table B.1 Multiple terms of various electron configurations

	Equivalent electrons
s^2, p^6, and d^{10}	^1S
p and p^5	^2P
p^2 and p^4	^3P, ^1D, ^1S
p^3	^4S, ^2D, ^2P
d and d^9	^2D
d^2 and d^8	^3F, ^3P, ^1G, ^1D, ^1S
d^3 and d^7	^4F, ^4P, ^2H, ^2G, ^2F, ^2D, ^2D, ^2P
d^4 and d^6	^5D, ^3H, ^3G, ^3F, ^3F, ^3D, ^3P, ^3P, ^1I, ^1G, ^1G, ^1F, ^1D, ^1D, ^1S, ^1S
d^5	^6S, ^4G, ^4F, ^4D, ^4P, ^2I, ^2H, ^2G, ^2G, ^2F, ^2F, ^2D, ^2D, ^2D, ^2P, ^2S
	Nonequivalent electrons
$s\ s$	^1S, ^3S
$s\ p$	^1P, ^3P
$s\ d$	^1D, ^3D
$p\ p$	^3D, ^1D, ^3P, ^1P, ^3S, ^1S
$p\ d$	^3F, ^1F, ^3D, ^1D, ^3P, ^1P
$d\ d$	^3G, ^1G, ^3F, ^1F, ^3D, ^1D, ^3P, ^1P, ^3S, ^1S
$s\ s\ s$	^4S, ^2S, ^2S
$s\ s\ p$	^4P, ^2P, ^2P
$s\ p\ p$	^4D, ^2D, ^2D, ^4P, ^2P, ^2P, ^2P, ^4S, ^2S, ^2S
$s\ p\ d$	^4F, ^2F, ^2F, ^4D, ^2D, ^2D, ^4P, ^2P, ^2P

SOURCE: Jeff C. Davis, Jr., "Advanced Physical Chemistry: Molecules, Structure, and Spectra." Copyright © 1965, The Ronald Press Company, New York. Used with permission.

are therefore ^1S, ^1D, and ^3P.[11] The ^3P is further split by differing J values to the terms $^3P_0$, $^3P_1$, and $^3P_2$. The relative magnitude of these splittings can be seen in Fig. B.2. States for various electron configurations are shown in Table B.1.

Although the complexity of determining the appropriate terms increases with the number of electrons and with higher L values, the method outlined above (known as Russell-Saunders coupling) may be applied to atoms with more electrons than the carbon atom in the foregoing example. Russell-Saunders coupling (also called LS coupling because it assumes that the individual values of l and s couple to form L and S, respectively) is normally adequate, especially for lighter atoms. For heavier atoms with higher nuclear charges, coupling occurs between the spin and orbit for each electron ($j = l + s$). The resultant coupling is known as jj coupling. In general, LS coupling is usually assumed and deviations are discussed in terms of the effects of spin–orbit interactions.

Hund's rules

The ground state of an atom may be chosen by application of *Hund's rules*. Hund's first rule is that of *maximum multiplicity*. It states that the ground state will be that having the largest value of S, in the case of carbon the ^3P. Such a system having a maximum number of parallel spins will be stabilized by the *exchange energy* resulting from their more favorable spatial distribution compared with that of paired electrons (see Pauli principle, p. 29).

The second rule states that if two states have the same multiplicity, the one with the higher value of L will lie lower in energy. Thus the ^1D lies lower in energy than the ^1S.[12] The greater stability of states in which the electrons are coupled to produce maximum angular momentum is also related to the spatial distribution and movement of the electrons.

The third rule states that for subshells that are less than half full, states with *lower* J are lower in energy; for subshells that are more than half full, states with *higher* J values are more stable. Applied to carbon, this rule predicts the ground state to be $^3P_0$.

---

[11] For alternative approaches and discussions, see K. E. Hyde, *J. Chem. Educ.*, **1975**, *52*, 87; J. H. Wise, ibid., **1976**, *53*, 496; and the inorganic textbooks listed on pp. 997–998.

[12] Hund's rules are inviolate in predicting the correct *ground state* of an atom. There are occasional exceptions when the rules are used to predict the ordering of *excited states*.

C

Units and conversion factors

THE INTERNATIONAL SYSTEM OF UNITS (SI)

SI base units

Physical quantity	Unit	Symbol
Length	meter	m
Mass	kilogram	kg
Time	second	s
Electric current	ampere	A
Thermodynamic temperature	kelvin	K
Amount of substance	mole	mol
Luminous intensity	candela	cd

Common derived units

Physical quantity	Unit	Symbol	Definition
Frequency	hertz	Hz	s^{-1}
Energy	joule	J	$kg\ m^2\ s^{-2}$
Force	newton	N	$J\ m^{-1}$
Pressure	pascal	Pa	$N\ m^{-2}$
Power	watt	W	$J\ s^{-1}$
Electric charge	coulomb	C	A s
Electric potential difference	volt	V	$J\ A^{-1}\ s^{-1}$
Electric resistance	ohm	Ω	$V\ A^{-1}$
Electric capacitance	farad	F	$A\ s\ V^{-1}$
Magnetic flux	weber	Wb	V s
Inductance	henry	H	$V\ s\ A^{-1}$
Magnetic flux density	tesla	T	$V\ s\ m^{-2}$

Prefixes

Prefix	Symbol	Multiply by
atto	a	10^{-18}
fempto	f	10^{-15}
pico	p	10^{-12}
nano	n	10^{-9}
micro	μ	10^{-6}
milli	m	10^{-3}
centi	c	10^{-2}
deci	d	10^{-1}
deka	da	10
hecto	h	10^{2}
kilo	k	10^{3}
mega	M	10^{6}
giga	G	10^{9}
tera	T	10^{12}
peta	P	10^{15}
exa	E	10^{18}

Physical and chemical constants

Electronic charge	$e = 1.60210 \times 10^{-19}$ C
Planck constant	$h = 6.6262 \times 10^{-34}$ J s
Speed of light	$c = 2.997925 \times 10^{8}$ m s^{-1}
Rydberg constant	$R = 1.09737312 \times 10^{7}$ m^{-1}
Boltzmann constant	$k = 1.38062 \times 10^{-23}$ J K^{-1}
Gas constant	$R = 8.3143$ J K^{-1} mol^{-1}
	8.2053×10^{-2} dm^{3} atm K^{-1} mol^{-1}
Avogadro number	$N_A = 6.022169 \times 10^{23}$ mol^{-1}
Faraday constant	$\mathscr{F} = 9.648670 \times 10^{4}$ C mol^{-1}
Electronic rest mass	$m_e = 9.109558 \times 10^{-31}$ kg
Proton mass	$m_p = 1.672614 \times 10^{-27}$ kg
Bohr radius	$a_0 = 5.2917715 \times 10^{-11}$ m
Bohr magneton	$\mu_B = 9.274096 \times 10^{-24}$ A m^{2}
Permittivity of vacuum	$\varepsilon_0 = 8.854185 \times 10^{-12}$ C^{2} m^{-1} J^{-1}
Pi	$\pi = 3.1415926536$
Base, natural logarithms	$e = 2.71828$

Conversion factors

Multiply	by	to obtain
Length		
cm	10^{8}	Å
cm	10^{7}	nm
cm	10^{10}	pm
Å	100	pm
Energy		
kcal mol^{-1}	4.184	kJ mol^{-1}
eV	96.49	kJ mol^{-1}

D

Tanabe-Sugano diagrams

[Originally from Y. Tanabe and S. Sugano, *J. Phys. Soc.* (*Japan*), **1954**, 753, 766, these figures are from B. N. Figgis, "Introduction to Ligand Fields," Wiley, New York, **1966**. Reproduced by permission of John Wiley and Sons, Inc. An extensive set of diagrams of this sort may be found in E. König and S. Kremer, "Ligand Field Diagrams," Plenum, New York, **1977**.]

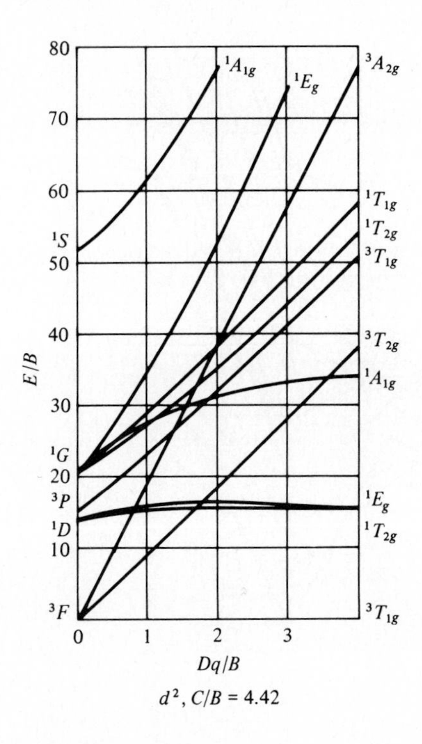

d^2, $C/B = 4.42$

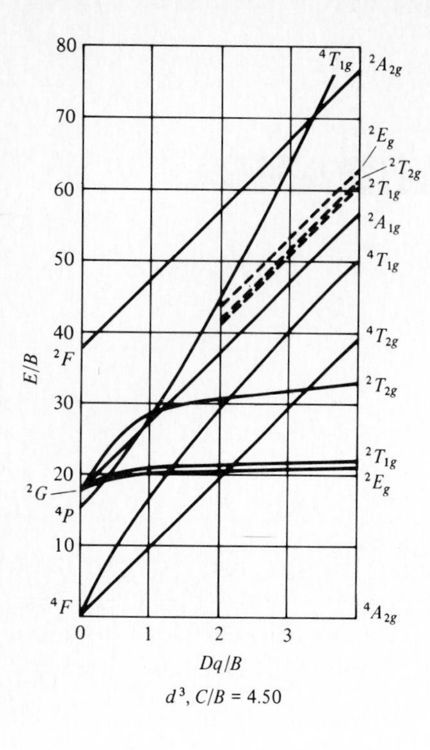

d^3, $C/B = 4.50$

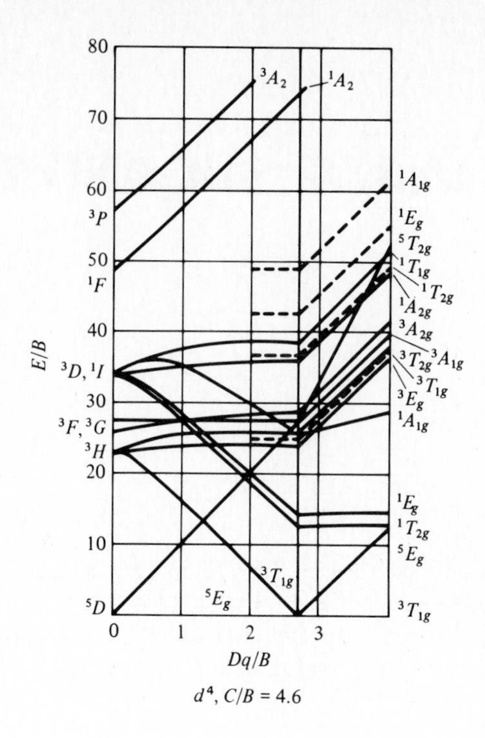

d^4, $C/B = 4.6$

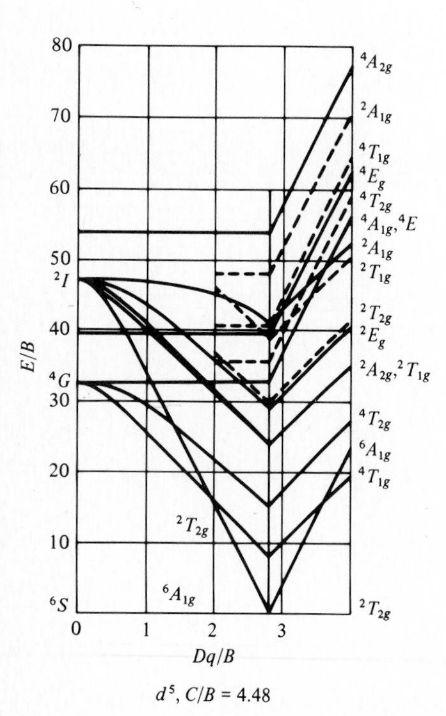

d^5, $C/B = 4.48$

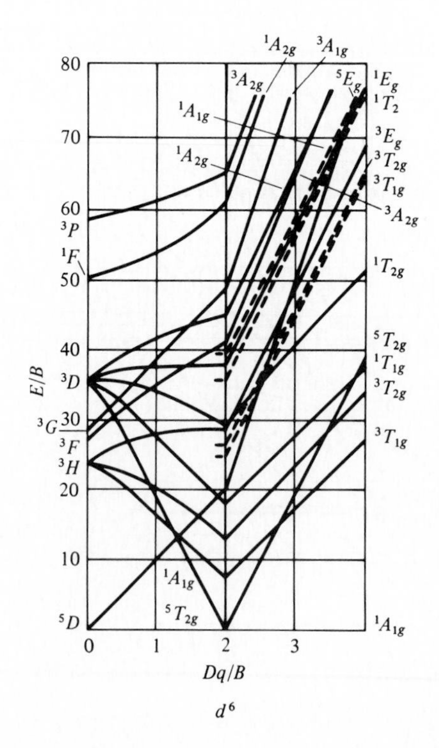

d^6

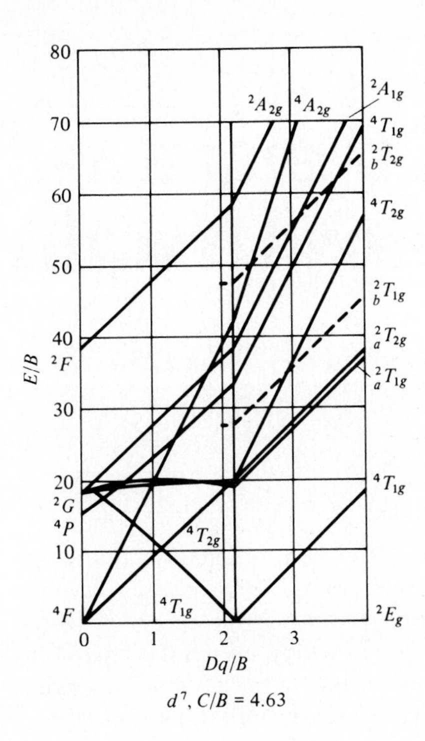

d^7, $C/B = 4.63$

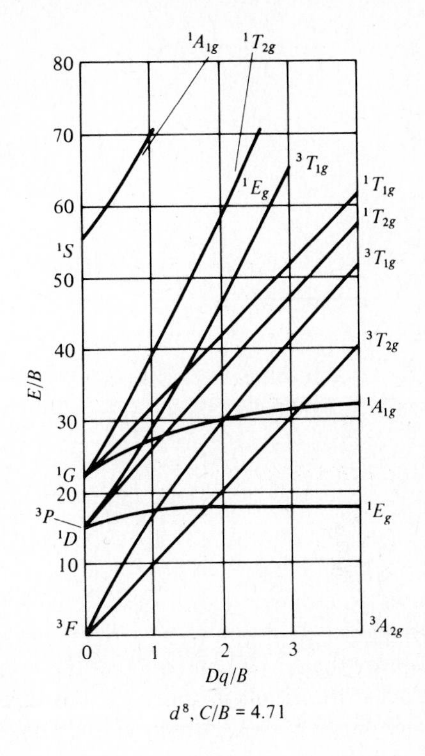

d^8, $C/B = 4.71$

E

Bond energies and bond lengths

Although the concept of bond energy seems intuitively simple, it is actually rather complicated when inspected closely. Consider a simple molecule A—B dissociating. It might be thought that it would be a relatively simple matter to measure the energy necessary to rupture the A—B bond and get the bond energy. Unfortunately, even if the experiment is feasible the result is generally not directly interpretable in terms of "bond energies" without further work. Among the factors to be considered are the vibrational, rotational, and translational energies of the reactants and products, the zero-point energy, and pressure–volume work if enthalpies are involved. The interested reader is referred to books on thermodynamics for a complete discussion (especially W. E. Dasent, "Inorganic Energetics," Penguin, Harmondsworth, England, **1970**). The following is meant as a brief outline of the problem.

Consider the energy of a diatomic molecule as shown in the figure. The concept of bond energy may be equated with the difference between the bottom of the energy curve and the energy of the completely separated atoms (ΔU_{el}). However, as a result of the zero-point vibrational energy of the AB molecule, even at 0 K, the energy necessary to separate the atoms ΔU is somewhat less (by a quantity of $\frac{1}{2}h\nu$). The zero-point energy is greatest in molecules containing light atoms such as hydrogen (25.9 kJ mol^{-1} in H_2) and somewhat less in molecules containing heavier atoms.

There is a corresponding difference between two estimates of the bond distance in a molecule A—B. One, r_e, corresponds to the minimum in the energy distance curve (see figure). The second, r_0, corresponds to the average distance in a molecule vibrating with zero-point energy. Since the curve is not perfectly parabolic the two values are not identical.

If the dissociation is to take place at some temperature, T, other than 0 K, the energy necessary to accomplish the dissociation must include a quantity sufficient to provide the separated atoms with the translational energy at that temperature ($\frac{3}{2}RT$). Compensating in part for this will be the translational, rotational, and vibrational energy of the molecule AB at temperature T ($\sim$6.3 kJ mol^{-1} for H_2 at 298 K). The difference between the dissociation energy at 298 K (ΔU_{298}) and that at 0 K (ΔU_0) is very small ($\sim$1 kJ mol^{-1} for H_2).

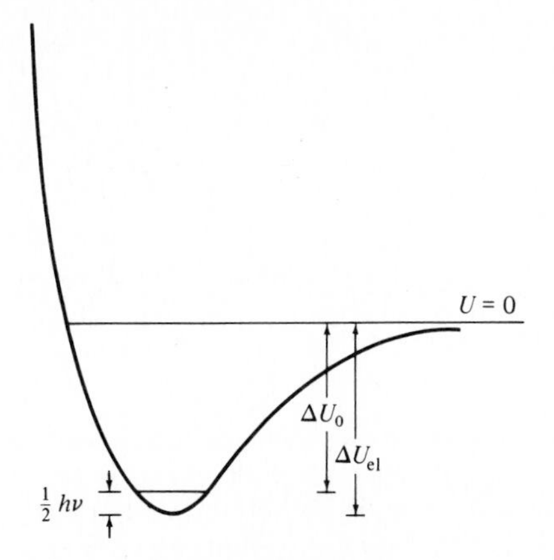

The quantity which is generally more accessible experimentally is the enthalpy. The enthalpy of dissociation at a given temperature differs from the energy of dissociation by $P\Delta V$ work: ~2.5 kJ mol^{-1} at room temperature. Some examples of the various quantities (in kJ mol^{-1}) for H_2 are:

		ΔU_{el}	ΔU_0	ΔU_{298}	ΔH_{298}
$H_2 \longrightarrow$	2H	458.1	432.00	433.21	435.93

Since the last three values—those most often quoted for "bond energies"—differ by very little, the difference may be ignored except in precise work.

The electronic energy, ΔU_{el}, is of interest mainly in connection with bonding theory since it is not an experimentally accessible quantity. A second quantity of this type is the "intrinsic bond energy," the difference in energy between the atoms in the molecule and the separated atoms *in the valence state*, i.e., with all of the atoms in the same condition (with respect to spin and hybridization) as in the molecule. It is a measure of the strength of the bond after all other factors except the bringing together of valence state atoms have been eliminated [cf. the discussion of methane, p. 113; C. A. Coulson, "Valence," 2nd ed., Oxford University Press, London, **1961**, pp. 205–208].

The situation becomes further complicated in polyatomic molecules. The energy of interest to chemists, generally, is that associated with breaking the bond without any change in the remaining parts of the molecule. For example, if we are interested in the bond energies in CH_4 or CF_4, we wish to know the energy of the reaction:

$$CX_4 \longrightarrow \cdot CX_3 + X\cdot$$

In general, the quasitetrahedral species CX_3 is not observed. In $\cdot CF_3$ a pyramidal molecule approaching this configuration is found, but in $\cdot CH_3$ the resulting species is planar with

sp^2 hybridization instead of sp^3. The energy associated with various dissociative steps for methane are:

$$CH_4 \longrightarrow CH_3 + H \qquad \Delta U = 421.1 \text{ kJ mol}^{-1}$$
$$CH_3 \longrightarrow CH_2 + H \qquad \Delta U = 469.9 \text{ kJ mol}^{-1}$$
$$CH_2 \longrightarrow CH + H \qquad \Delta U = 415 \text{ kJ mol}^{-1}$$
$$CH \longrightarrow C + H \qquad \Delta U = 334.7 \text{ kJ mol}^{-1}$$

We can associate the greater energy of the second dissociation step with a presumed greater bonding strength of trigonal: sp^2 hybrids over sp^3 hybrids. Whether we understand (or at least believe we do) the reasons for each of the quantities listed above or not, it is obvious that *none* represents the bond energy in methane. However, since the summation of these four experimentally observable processes must be identical to the energy for the nonobservable (but desirable) reaction:

$$CH_4 \longrightarrow C + 4H$$

the average of these four quantities (411 kJ mol^{-1}) can be taken as the *mean bond energy* for the C—H bond in methane.

The mean bond energy is a useful quantity, but it should be remembered that it is derived from a particular molecule and may not be exactly correct in application to another molecule. Thus if the total bond energy in dichloromethane does not equal two times the average bond energy in methane plus two times the average bond energy in carbon tetrachloride, we should not be surprised. The presence of bonds of one type may have an effect in strengthening or weakening bonds of another type. As a matter of fact, there is no unequivocal way of assigning bond energies for molecules such as dichloromethane by means of thermodynamics. The summation of all of the bond energies may be determined as above for methane, but the assignment to individual bonds must be made by secondary assumptions, e.g., the bond C—H energies are comparable to those in methane. Alternatively, the bond energies in molecules containing more than one type of bond may be assigned on the basis of some other type of information such as infrared stretching frequencies.

One of the most serious problems hindering the assignment of bond energies arises for bonds such as N—N and O—O. The nitrogen triple bond and oxygen double bond may be evaluated directly from the dissociation of the gaseous N_2 and O_2 molecules. Single bonds for these elements present special problems because additional elements are always present. For example, consider the following dissociation energies accompanying splitting of the N—N bond:

$$N_2H_4 \longrightarrow 2NH_2 \qquad D_0 = 247 \text{ kJ mol}^{-1}$$
$$N_2F_4 \longrightarrow 2NF_2 \qquad D_0 = 88 \text{ kJ mol}^{-1}$$
$$N_2O_4 \longrightarrow 2NO_2 \qquad D_0 = 57.3 \text{ kJ mol}^{-1}$$

None of these represents the breaking of a hypothetical, isolated N—N bond:

$$\cdot\ddot{\underset{\cdot}{N}}-\ddot{\underset{\cdot}{N}}\cdot \longrightarrow 2\cdot\ddot{\underset{\cdot}{N}}\cdot$$

By using N—H and N—F bond energies from NH_3 and NF_3 it is possible to estimate the inherent strength of the N—N bond:

Total energy of atomization, N_2H_4	1703 kJ mol^{-1}
4N—H bonds (assumed from NH_3)	1544 kJ mol^{-1}
Difference (equated with N—N)	159 kJ mol^{-1}
Total energy of atomization, N_2F_4	1305 kJ mol^{-1}
4N—F bonds (assumed from NF_3)	1134 kJ mol^{-1}
Difference (equated with N—N)	172 kJ mol^{-1}

The results of this calculation are gratifyingly congruent, and we feel reasonably confident in a value of about 167 kJ mol^{-1} for the N—N bond. Similar results can be obtained for hydrogen peroxide and dioxygen difluoride to obtain an estimate of about 142 kJ mol^{-1} for the O—O bond.

Although the calculations for N—N and O—O bonds are self-consistent, there is always the possibility that a wider series of compounds would show greater variability. This is especially probable in molecules in which the electronegativity of the substituents is believed to affect the bonding by particular orbitals, e.g., overlap by diffuse *d* orbitals. Thus on the basis of observed stabilities the presence of electronegative substituents such as —F and —CF_3 seems to stabilize the P—P bond relative to H_2P—PH_2, although there are not enough data to investigate this possibility quantitatively.

There is at present no convenient, self-consistent source of all bond energies. The standard work is T. L. Cottrell, "The Strengths of Chemical Bonds," 2nd ed., Butterworths, London, **1958**, but it suffers from a lack of recent data. Darwent (National Bureau of Standards publication NSRDS-NBS 31, **1970**) has summarized recent data on dissociation energies but did not include some earlier work or values known only for total energies of atomization rather than for stepwise dissociation. Three useful references of the latter type are: L. Brewer and E. Brackett, *Chem. Rev.*, **1961**, *61*, 425; L. Brewer et al., *Chem. Rev.*, **1963**, *63*, 111; R. C. Feber, Los Alamos Report LA-3164, **1965**. The book by Darwent mentioned above also lists bond energy values for some common bonds.

Table E.1 has been compiled from the above sources. The ordering is as follows: hydrogen, Group IA, Group IIA, Group IIIB, transition elements, Group IIIA, Group IVA, Group VA, Group VIA, Group VIIA, Group VIIIA. For a given group (such as IA) the compounds are listed in the following order: halides, chalcogenides, etc. Unless specified otherwise, the bond energies are for compounds representing group number oxidation states, such as BCl_3, SiF_4, and SF_6. Other compounds are listed in parentheses, such as (TlCl) and (PCl_3). Values are for the dissociation energies of molecules A–B and mean dissociation values for AB_n. For molecules such as N_2H_4 two values are given: H_2N—NH_2 represents the dissociation energy to two amino radicals, and N—N (N_2H_4) represents the estimated N—N bond energy in hydrazine obtained by means of assumed N—H bond energies as shown above.

Table E.1 Bond energies and bond lengths

Bond	D_0		r	
	kJ mol^{-1}	kcal^{-1} mol^{-1}	pm	Å
Hydrogen Compounds				
H—H	432.00 ± 0.04	103.25 ± 0.01	74.2	0.742
H—F	565 ± 4	135 ± 1	91.8	0.918
H—Cl	428.02 ± 0.42	102.3 ± 0.1	127.4	1.274
H—Br	362.3 ± 0.4	86.6 ± 0.1	140.8	1.408
H—I	294.6 ± 0.4	70.4 ± 0.1	160.8	1.608
H—O	458.8 ± 1.4	109.6 ± 0.4	96	0.96
H—S	363 ± 5	87 ± 1	134	1.34
H—Se	276?	66?	146	1.46
H—Te	238?	57?	170	1.7
H—N	386 ± 8	92 ± 2	101	1.01
H—P	~322	~77	144	1.44
H—As	~247	~59	152	1.52
H—C	411 ± 7	98.3 ± 0.8	109	1.09
H—CN	531 ± 21	127 ± 5	106.6	1.066
H—Si	318	76	148	1.48
H—Ge			153	1.53
H—Sn			170	1.70
H—Pb				
H—B	389?	93?	119	1.19
H—Cu	276 ± 8	66 ± 2		
H—Ag	230 ± 13	55 ± 3		
H—Au	285 ± 13	68 ± 3		
H—Be				
H—Mg				
H—Ca				
H—Sr				
H—Ba				
H—Li	~243	~58	159.5	1.595
H—Na	197	47	188.7	1.887
H—K	180	43	224.4	2.244
H—Rb	163	39	236.7	2.367
H—Cs	176	42	249.4	2.494
Group IA				
Li—Li	105	25	267.2	2.672
Li—F	573 ± 21	137 ± 5	154.7	1.547
Li—Cl	464 ± 13	111 ± 3	202	2.02
Li—Br	418 ± 21	100 ± 5	217.0	2.170
Li—I	347 ± 13	83 ± 3	239.2	2.392
Na—Na	72.4	17.3	307.8	3.078
NaF	477.0	114.0	184	1.84
NaCl	407.9	97.5	236.1	2.361
NaBr	362.8	86.7	250.2	2.502
NaI	304.2	72.7	271.2	2.712
K_2	49.4	11.8	392.3	3.923
KF	490 ± 21	117 ± 5	213	2.13
KCl	423 ± 8	101 ± 2	266.7	2.667
KBr	378.7 ± 8	90.5 ± 2	282.1	2.821
KI	326 ± 13	78 ± 3	304.8	3.048
Rb—Rb	45.2	10.8		

Table E.1 *(Continued)*

Bond	D_0		r	
	kJ mol^{-1}	kcal^{-1} mol^{-1}	pm	Å
Rb—F	**490 ± 21**	**117 ± 5**	*226.6*	*2.266*
Rb—Cl	**444 ± 21**	**106 ± 5**	*278.7*	*2.787*
Rb—Br	**385 ± 25**	**92 ± 6**	*294.5*	*2.945*
Rb—I	**331 ± 13**	**79 ± 3**	*317.7*	*3.177*
Cs—Cs	43.5	10.4		
Cs—F	**502 ± 42**	**120 ± 10**	*234.5*	*2.345*
Cs—Cl	**435 ± 21**	**104 ± 5**	*290.6*	*2.906*
Cs—Br	**416.3 ± 13**	**99.5 ± 3**	*307.2*	*3.072*
Cs—I	**335 ± 21**	**80 ± 5**	*331.5*	*3.315*
Group IIA				
Be—Be	(208)[a]	(51)[a]		
Be—F	**632 ± 53**	**151 ± 13**	*140*	*1.40*
Be—Cl	**461 ± 63**	**110 ± 15**	*175*	*1.75*
Be—Br	724	89	*191*	*1.91*
Be—I	289	69	*210*	*2.10*
Be=O	**444 ± 21**	**106 ± 5**	*133.1*	*1.331*
Mg—Mg	(129)[a]	(31)[a]		
Mg—F	**513 ± 42**	**123 ± 10**	*177*	*1.77*
Mg—Cl	***406***	***97***	*218*	*2.18*
Mg—Br	***~339***	***~81***	*234*	*2.34*
Mg—I	***264***	***63***	*254*	*2.54*
Mg=O	**377 ± 42**	**90 ± 10**	**174.9**	**1.749**
Ca—Ca	(105)[a]	(25)[a]		
Ca—F	**550 ± 42**	**132 ± 10**	*210*	*2.10*
Ca—Cl	**429 ± 42**	**103 ± 10**	*251*	*2.51*
Ca—Br	***402***	***96***	*267*	*2.67*
Ca—I	***~326***	***~78***	*288*	*2.88*
Ca=O	**460 ± 84**	**110 ± 20**	**182.2**	**1.822**
Ca=S	**310 ± 21**	**74 ± 5**		
Sr—Sr	(84)[a]	(20)[a]		
Sr—F	**553 ± 42**	**132 ± 10**	*220*	*2.20*
Sr—Cl	***~469***	***~112***	*267*	*2.67*
Sr—Br	***405***	***97***	*282*	*2.82*
Sr—I	***~335***	***~80***	*303*	*3.03*
Sr=O	347?	83?	**192.1**	**1.921**
Sr=S	222?	53?		
Ba—F	**578 ± 42**	**138 ± 10**	*232*	*2.32*
Ba—Cl	**475 ± 21**	**114 ± 10**	*282*	*2.82*
Ba—Br	***427***	***102***	*299*	*2.99*
Ba—I	***~360***	***~86***	*320*	*3.20*
Ba=O	**561 ± 42**	**134 ± 10**	**194.0**	**1.940**
Ba—OH	**467 ± 63**	**112 ± 15**		
Ba=S	**396.2 ± 18.8**	**94.7 ± 4.5**		
Group IIIB				
Sc—Sc	**108.4 ± 21**	**25.9 ± 5**		
Sc—F	*~594*	*~142*		
Sc—Cl	*460.6*	*110.1*		
Sc—Br	*391.2*	*93.5*		
Sc—I	*~322*	*~77*		
Y—Y	**156.1 ± 21**	**37.3 ± 5**		

Table E.1 *(Continued)*

Bond	D_0 / kJ mol^{-1}	r / pm
Y—F	*~628*	
Y—Cl	*~494*	
Y—Br	*~431*	
Y—I	**362.8**	
La—La	**241.0 ± 21**	
La—F	*~657*	
La—Cl	*513.0*	
La—Br	*~439*	
La—I	*363.6*	

Lanthanides and actinides (kJ mol^{-1})

Metal	MF_3	MCl_3	MBr_3	MI_3	MF_4	MF_5	MF_6
Ce	*~644*	*499.6*	*~431*	*~356*			
Pr	*~623*	*482.4*	*~418*	*338.5*			
Nd	*~611*	*470.7*	*~406*	*323.8*			
Pm	*~590*	*~452*	*~385*	*~305*			
Sm	*~565*	*~427*	*~360*	*~285*			
Eu	*~552*	*~414*	*~347*	*~272*			
Gd	*~619*	*~485*	*~418*	*~343*			
Tb	*~615*	*~477*	*~410*	*~339*			
Dy	*~577*	*~444*	*~377*	*~305*			
Ho	*~577*	*~444*	*~381*	*~305*			
Er	*~582*	*~448*	*~381*	*~310*			
Tm	*~548*	*~418*	*~351*	*~251*			
Yb	*~519*	*~385*	*~322*	*~251*			
Lu	*~607*	*~477*	*~410*	*~343*			
Th					*641.0*		
Pa							
U	*~619*	*495.4*	*424.3*	*~343*	*598.0*	*564.8*	*522.2*
Np	*~586*	*~460*	*~397*	*~326*	*~561*	*~519*	*~477*
Pu	*562.7*	*442.2*	*378.7*	*~310*	*~536*	*~485*	*~427*
Am	*~573*	*~452*	*~389*	*~318*	*~519*		

Table E.1 (*Continued*)
Lanthanides and actinides (kcal mol^{-1})

Bond	D_0 / kJ mol^{-1}	r / pm
Transition metals		
Ti—F(TiF_4)		
Ti—Cl($TiCl_4$)	*584.5*	
Ti—Cl($TiCl_3$)	*429.3*	**218**
Ti—Cl($TiCl_2$)	*460.2*	
Ti—Br($TiBr_4$)	*504.6*	
Ti—I(TiI_4)	*366.9*	**231**
Zr—F	*296.2*	
Zr—Cl	*646.8*	
Zr—Br	*489.5*	**232**
Zr—I	*423.8*	
Hf—F	*345.6*	
Hf—Cl	~*649.4*	
Hf—Br	*494.9*	
Hf—I	~*431*	
Mn—F(MnF_2)	~*360*	
Mn—Cl($MnCl_2$)	*457.7*	
Mn—Br($MnBr_2$)	*392.5*	
Mn—I(MnI_2)	*332.2*	
Fe—Fe	*267.8*	
Fe—F(FeF_3)	**156 ± 25**	
Fe—F(FeF_2)	~*456*	
Fe—Cl($FeCl_3$)	*481*	
Fe—Cl($FeCl_2$)	*341.4*	
Fe—Br($FeBr_3$)	*400.0*	
Fe—Br($FeBr_2$)	*291.2*	
Fe—I(FeI_3)	*339.7*	
Fe—I(FeI_2)	*233.5*	
Ni—Ni	*279.1*	
Ni—F(NiF_2)	**228.0 ± 2.1**	
Ni—Cl($NiCl_2$)	*462.3*	
Ni—Br($NiBr_2$)	*370.7*	
Ni—I(NiI_2)	*312.5*	
Cu—Cu	*254.8*	
Cu—F(CuF_2)	**190.4 ± 13**	
Cu—F(CuF)	**365 ± 38**	
Cu—Cl($CuCl_2$)	~*418*	
	293.7	

Table E.1 (*Continued*)

	D_0		r	
Bond	kJ mol^{-1}	kcal mol^{-1}	pm	Å
Cu—Cl(CuCl)	*360.7*	*86.2*		
Cu—Br($CuBr_2$)	*~259*	*~62*		
Cu—Br(CuBr)	*330.1*	*78.9*		
Cu—I(CuI_2)	*~192*	*~46*		
Cu—I(CuI)	~142	~34		
Ag—F(AgF_2)	~268	~64		
Ag—F(AgF)	348.9	83.4		
Ag—Cl	**314 ± 21**	**75 ± 5**	**225**	**2.25**
Ag—Br	**289 ± 42**	**69 ± 10**		
Ag—I	*256.9*	*61.4*	**254.4**	**2.544**
Au—F	*~305*	*~73*		
Au—Cl	**289 ± 63**	**69 ± 15**		
Au—Br	*~251*	*~60*		
Au—I	*~230*	*~55*		
Zn—F	*400.0*	*95.6*	**181**	**1.81**
Zn—Cl	*319.7*	*76.4*		
Zn—Br	*268.6*	*64.2*		
Zn—I	*207.5*	*49.6*		
Cd—F	*326*	*78.0*		
Cd—Cl	*280.7*	*67.1*		
Cd—Br	*242.7*	*58.0*		
Cd—I	*190.8*	*45.6*		
Hg—F	*~268*	*~64*		
Hg—Cl	*224.7*	*53.7*	**229**	**2.29**
Hg—Br	*184.9*	*44.2*	**241**	**2.41**
Hg—I	*145.6*	*34.8*	**259**	**2.59**
Group IIIA				
B—B	**293 ± 21**	**70 ± 5**		
B—F	**613.1 ± 53**	**146.5 ± 13**		
B—Cl	456	109	**175**	**1.75**
B—Br	377	90		
B—I				
B—OR	536	128		
B…N($B_3N_3H_3Cl_3$)	445.6	106.5		
B—C	372	89		
Al—F	**583.0 ± 31**	**139.3 ± 8**		
Al—Cl	**420.7 ± 10**	**100.5 ± 2**		
Al—C	255	61		
Ga—Ga	**113 ± 17**	**27 ± 4**		
Ga—F	*~469*	*~112*	**188**	**1.88**
Ga—Cl	*354.0*	*84.6*		
Ga—Br	*301.7*	*72.1*		
Ga—I	*237.2*	*56.7*	**244**	**2.44**
In—In	**100 ± 13**	**24 ± 3**		
In—F(InF_3)	*~444*	*~106*		
In—F(InF)	**523 ± 8**	**125 ± 2**		
In—Cl($InCl_3$)	*328.0*	*78.4*		
In—Cl(InCl)	**435 ± 8**	**104 ± 2**	**240.1**	**2.401**
In—Br($InBr_3$)	*279.1*	*66.7*		

Table E.1 *(Continued)*

Bond	D_0		r	
	kJ mol^{-1}	kcal mol^{-1}	pm	Å
In—Br(InBr)	**406 ± 21**	**97 ± 5**	**254.3**	**2.543**
In—I(InI_3)	*225.1*	*53.8*		
In—I(InI)			**275.4**	**2.754**
Tl—F(TlF)	**439 ± 21**	**105 ± 5**		
Tl—Cl(TlCl)	**364 ± 8**	**87 ± 2**	**248.5**	**2.485**
Tl—Br(TlBr)	**326 ± 21**	**78 ± 5**		
Tl—I(TlI)	**280 ± 21**	**67 ± 5**	**281.4**	**2.814**
Group IVA				
C—C	345.6	82.6	**154**	**1.54**
C=C(C_2)	**602 ± 21**	**144 ± 5**	**134**	**1.34**
C≡C	835.1	199.6	**120**	**1.20**
C—F	485	116	**135**	**1.35**
C—Cl	327.2	78.2	**177**	**1.77**
C—Br	285	68	**194**	**1.94**
C—I	213	51	**214**	**2.14**
C—O	357.7	85.5	**143**	**1.43**
C=O	**798.9 ± 0.4**	**190.9 ± 0.1**	**120**	**1.20**
C≡O	**1071.9 ± 0.4**	**256.2 ± 0.1**	**112.8**	**1.128**
C—S	272	65	**182**	**1.82**
C=S	**573 ± 21**	**137 ± 5**	**160**	**1.60**
C—N	304.6	72.8	**147**	**1.47**
C=N	615	147		
C≡N	887	212	**116**	**1.16**
C—P	264	63	**184**	**1.84**
C—Si(SiC)	318	76	**185**	**1.85**
C—Ge($GeEt_4$)	213?	51?	**194**	**1.94**
C—Sn($SnEt_4$)	226	54	**216**	**2.16**
C—Pb($PbEt_4$)	130	31	**230**	**2.30**
Si—Si	222	53	**235.2**	**2.352**
Si—F	565	135	**157**	**1.57**
Si—Cl	381	91	**202**	**2.02**
Si—Br	310	74	**216**	**2.16**
Si—I	234	56	**244**	**2.44**
Si—O	452	108	**166**	**1.66**
Si—S	293?	70?	**~200**	**~2.0**
Ge—Ge	188	45	**241**	**2.41**
Ge=Ge	**272 ± 21**	**65 ± 5**		
Ge—F(GeF_4)	*~452*	*~108*	**168**	**1.68**
Ge—F(GeF_2)	*481*	*115*		
Ge—Cl($GeCl_4$)	*348.9*	*83.4*	**210**	**2.10**
Ge—Cl($GeCl_2$)	*~385*	*~92*		
Ge—Br($GeBr_4$)	*276.1*	*66.0*	**230**	**2.30**
Ge—Br($GeBr_2$)	*325.5*	*77.8*		
Ge—I(GeI_4)	*211.7*	*50.6*		
Ge—I(GeI_2)	*264.0*	*63.1*		
Sn—Sn	146.4	35.0		
Sn—F(SnF_4)	*~414*	*~99*		
Sn—F(SnF_2)	*~481*	*~115*		
Sn—Cl($SnCl_4$)	*323.0*	*77.2*	**233**	**2.33**
Sn—Cl($SnCl_2$)	*385.8*	*92.2*	**242**	**2.42**

Table E.1 *(Continued)*

Bond	D_0		r	
	kJ mol^{-1}	kcal mol^{-1}	pm	Å
Sn—Br($SnBr_4$)	*272.8*	*65.2*	**246**	**2.46**
Sn—Br($SnBr_2$)	*329.3*	*78.7*	**255**	**2.55**
Sn—I(SnI_4)	*~205*	*~49*	**269**	**2.69**
Sn—I(SnI_2)	*261.5*	*62.5*	**273**	**2.73**
Pb—F(PbF_4)	*~331*	*~79*		
Pb—F(PbF_2)	*394.1*	*94.2*		
Pb—Cl($PbCl_4$)	*~243*	*~58*		
Pb—Cl($PbCl_2$)	*303.8*	*72.6*	**242**	**2.42**
Pb—Br($PbBr_4$)	*~201*	*~48*		
Pb—Br($PbBr_2$)	*260.2*	*62.2*		
Pb—I(PbI_4)	*~142*	*~34*		
Pb—I(PbI_2)	*205.0*	*49.0*	**279**	**2.79**
Group VA				
N—N(N_2H_4)	**~167**	**~40**		
H_2N—NH_2	**247 ± 13**	**59 ± 3**	**145**	**1.45**
N=N	418	100	**125**	**1.25**
N≡N	**941.69 ± 0.04**	**225.07 ± .01**	**109.8**	**1.098**
N—F	**283 ± 24**	**68 ± 6**	**136**	**1.36**
N—Cl	**313**	**72**	**175**	**1.75**
N—O	201	48	**140**	**1.40**
N=O	607	145	**121**	**1.21**
$N \overset{\cdots}{=} N$	678	162	**115**	**1.15**
P—P(P_4)	201	48	**221**	**2.21**
Cl_2P—PCl_2	**239**	**57**		
P≡P	**481 ± 8**	**115 ± 2**	**189.3**	**1.893**
P—F(PF_3)	490	117	**154**	**1.54**
P—Cl(PCl_3)	326	78	**203**	**2.03**
P—Br(PBr_3)	264	63		
P—I(PI_3)	184	44		
P—O	335?	80?	**163**	**1.63**
P=O	~544	~130	**~150**	**~1.5**
P=S	~335	~80	**186**	**1.86**
As—As(As_4)	146	35	**243**	**2.43**
As≡As	**380 ± 21**	**91 ± 5**		
As—F(AsF_5)	*~406*	*~97*		
As—F(AsF_3)	*484.1*	*115.7*	**171.2**	**1.712**
As—Cl($AsCl_3$)	*321.7*	*76.9*	**216.1**	**2.161**
As—Br($AsBr_3$)	*458.2*	*61.7*	**233**	**2.33**
As—I(AsI_3)	*200.0*	*47.8*	**254**	**2.54**
As—O	301	72	**178**	**1.78**
As=O	~389	~93		
Sb—Sb(Sb_4)	121?	29?		
Sb≡Sb	**295.4 ± 6.3**	**70.6 ± 1.5**		
Sb—F(SbF_5)	*~402*	*~96*		
Sb—F(SbF_3)	*~440*	*~105*		
Sb—Cl($SbCl_5$)	*248.5*	*59.4*		
Sb—Cl($SbCl_3$)	*314.6*	*75.2*	**232**	**2.32**
Sb—Br($SbBr_5$)	*184*	*44*		
Sb—Br($SbBr_3$)	*259.8*	*62.1*	**251**	**2.51**
Sb—I(SbI_3)	*195.0*	*46.6*		

Table E.1 *(Continued)*

Bond	D_0		r	
	kJ mol^{-1}	kcal mol^{-1}	pm	Å
Bi≡Bi	**192 ± 4**	**46 ± 1**		
Bi—F(BiF_5)	*~297*	*~71*		
Bi—F(BiF_3)	*~393*	*~94*		
Bi—Cl($BiCl_3$)	*274.5*	*65.6*	**248**	**2.48**
Bi—Br($BiBr_3$)	*232.2*	*55.5*	**263**	**2.63**
Bi—I(BiI_3)	*168.2*	*40.2*		
Group VIA				
O—O(H_2O_2)	**~142**	**~34**		
HO—OH	**207.1 ± 2.1**	**49.5 ± 0.5**	**148**	**1.48**
O=O	**493.59 ± 0.4**	**117.97 ± 0.1**	**120.7**	**1.207**
O—F	189.5	45.3	**142**	**1.42**
S—S(S_8)	226	54	**205**	**2.05**
S—S(H_2S_2)	**268 ± 21**	**64 ± 5**	**205**	**2.05**
S=S	**424.7 ± 6.3**	**101.5 ± 1.5**	**188.7**	**1.887**
S—F	284	68	**156**	**1.56**
S—Cl(S_2Cl_2)	255	61	**207**	**2.07**
S—Br(S_2Br_2)	217?	52?	**227**	**2.27**
S=S(SO)	**517.1 ± 8**	**123.6 ± 2**	**149.3**	**1.493**
S$\cdots$O(SO_2)	**532.2 ± 8**	**127.2 ± 2**	**143.2**	**1.432**
S$\cdots$O(SO_3)	**468.8 ± 8**	**112.1 ± 2**	**143**	**1.43**
Se—Se(Se_6)	172	41		
Se=Se	272	65	**215.2**	**2.152**
Se—F(SeF_6)	*284.9*	*68.1*		
Se—F(SeF_4)	*~310*	*~74*		
Se—F(SeF_2)	*~351*	*~84*		
Se—Cl($SeCl_4$)	*~192*	*~46*		
Se—Cl($SeCl_2$)	*~243*	*~58*		
Se—Br($SeBr_4$)	*~151*	*~36*		
Se—Br($SeBr_2$)	*~201*	*~48*		
Se—I(SeI_2)	*~151*	*~36*		
Te—Te	(126)[a]	(30)[a]		
Te=Te	**218 ± 8**	**52 ± 2**		
Te—F(TeF_6)	*329.7*	*78.8*		
Te—F(TeF_4)	*~335*	*~80*		
Te—F(TeF_2)	*~393*	*~94*		
Te—Cl($TeCl_4$)	*310.9*	*74.3*	**233**	**2.33**
Te—Cl($TeCl_2$)	*~284*	*~68*		
Te—Br($TeBr_4$)	*~176*	*~42*	**268**	**2.68**
Te—Br($TeBr_2$)	*~243*	*~58*	**251**	**2.51**
Te—I(TeI_4)	*~121*	*~29*		
Te—I(TeI_2)	*~192*	*~46*		
Group VIIA				
I—I	**148.95 ± 0.04**	**35.60 ± 0.01**	**266.6**	**2.666**
I—F(IF_7)	231.0[b]	55.2[b]	~183	~1.83
I—F(IF_5)	267.8[b]	64.0[b]	175,186[c]	1.75,1.86[c]
I—F(IF_3)	~272[b]	~65[b]		
I—F(IF)	**277.8 ± 4**	**66.4 ± 1**	191[c]	1.91[c]
I—Cl(ICl)	**207.9 ± 0.4**	**49.7 ± 0.1**	**232.1**	**2.321**
I—Br(IBr)	**175.3 ± 0.4**	**41.9 ± 0.1**		
I—O(I—OH)	201	48		

Table E.1 *(Continued)*

	D_0	r
Bond	kJ mol^{-1}	pm
Br—Br	**190.16 ± 0.04**	**228.4**
Br—F(BrF_5)	187.0[b]	
Br—F(BrF_3)	201.2[b]	172,184[d]
Br—F(BrF)	249.4[b]	**175.6**
Br—Cl(BrCl)	**215.9 ± 0.4**	**213.8**
Br—O(Br—OH)	201	
Cl—Cl	**239.7 ± 0.4**	**198.8**
Cl—F(ClF_5)	~142[b]	
Cl—F(ClF_3)	172.4[b]	**169.8**
Cl—F(ClF)	**248.9 ± 2.1**	**162.8**
Cl—O(Cl—OH)	218	
F—F	**154.8 ± 4**	**141.8**
At—At	**115.9**	
Xe—F(XeF_6)	126.2[e]	190[e]
Xe—F(XeF_4)	130.4[e]	195[e]
Xe—F(XeF_2)	130.8[e]	200[e]
Xe—O(XeO_3)	84[e]	175[e]
Kr—F(KrF_2)	50[e]	190[e]

[a] J. E. Huheey and R. S. Evans, *J. Inorg. Nucl. Chem.*, **1970**, *32*, 383.
[b] L. Stein, *Halogen Chemistry*, **1967**, *1*, 133.
[c] W. Harshberger et al., *J. Am. Chem. Soc.*, **1967**, *89*, 6466.
[d] R. D. Burbank and F. N. Beaney, Jr., *J. Chem. Phys.*, **1957**, *27*, 982.
[e] J. H. Holloway, "Noble-Gas Chemistry," Methuen, London, **1968**.

Values in boldface type are from Darwent and represent his estimates of the "best value" and uncertainties for the energies required to break the bonds at 0 K. Where values are not available from Darwent, they are taken from Brewer et al., for metal halides and dihalides (boldface italics) or from Feber for transition metal, lanthanide, and actinide halides (italics). These values represent enthalpies of atomization at 298 K. The remaining values are from Cottrell (Arabic numerals) and other sources (Arabic numerals with superscripts keyed to references at end of table).

The table is intended for quick reference for reasonably accurate values for rough calculations. No effort has been made to convert the values from 298 K to 0 K, and in many cases the errors in the estimates are greater than the correction term anyway. The accuracy of the values can be graded in a descending scale: (1) those giving ± uncertainties; (2) those giving "exact" values to the nearest 0.1 or 1; (3) those expressed as "about" a certain value (~); and (4) those which are almost pure guesses, followed by a question mark. All values are experimental except for a few for which A—A bond energies are not known but would be helpful (as for electronegativity calculations). Estimates for these hypothetical bonds (such as Be—Be) are listed in parentheses.

The bond lengths are mainly from two sources: *Tables of Interatomic Distances and Configuration in Molecules and Ions* (The Chemical Society, Special Publication No. 11, 1958) and *Supplement* (Special Publication No. 18, 1965) with values given in boldface, and Brewer et al. (italics). As in the case of the energies the purpose of the table is for quick reference and the bond length is given as a typical value that should be accurate for most purposes ± 2 pm.

For special purposes and precise computations, the original sources should be consulted for the value, nature, and source of the bond energies, and for the accuracy, experimental method, and variability of the bond lengths. Space requirements prohibit extensive tabulation of information of this type here.

Energies of acid-base interactions

Values of the Drago-Wayland *E-C* parameters[1] are given in Table E.2 and those of the Kroeger-Drago *e-c-t* parameters[2] are given in Table E.3. These parameters allow the estimation of the enthalpy of adduct formation of acids (A) and bases (B) by the equations:

$$-\Delta H = E_A E_B + C_A C_B \quad \text{in kcal mol}^{-1} \text{ (multiply by 4.184 kJ kcal}^{-1} \text{ for SI values)}$$

and

$$-\Delta H = e_A e_B + c_A c_B + t_A t_B \quad \text{in kcal mol}^{-1} \text{ (multiply by 4.184 kJ kcal}^{-1} \text{ for SI values)}$$

The original papers should be consulted for sources of experimental data, methods, uncertainties, and precautions when using these data.

Table E.2 Acid parameters

Acid	C_A	E_A	c_A	e_A	t_A
Cations					
H^+ (proton)	—	—	8.554	8.654	15.040
Li^+	—	—	0.968	23.066	1.715
Na^+	—	—	0.546	21.798	1.030
K^+	—	—	0.148	21.378	0.300
Rb^+	—	—	0.073	21.951	**0.116**[a]
Cs^+	—	—	0.010	21.534	**0.012**[a]
Methyl cation	—	—	3.592	20.326	6.627
Ethyl cation	—	—	1.546	26.500	3.106
iso-Propyl cation	—	—	0.744	28.080	1.697
t-Butyl cation	—	—	**0.010**[a]	31.931	0.442
Sr^{+2}	—	—	1.023	22.494	1.742
Cu^+	—	—	1.653	23.621	3.096
Nitrosyl cation	—	—	**0.103**[a]	22.683	2.867
Bi^+	—	—	0.700	25.846	1.513
Pb^{+2}	—	—	0.836	24.133	1.659
$(C_5H_5)Ni^+$	—	—	2.285	1.977	3.890
Neutral acceptors					
Group IA acceptors:					
Hexafluoroisopropyl alcohol	0.623	5.93	0.426	5.214	0.475
Perfluoro-*t*-butyl alcohol	0.816	5.82	—	—	—

[1] R. S. Drago and B. B. Wayland, *J. Am. Chem. Soc.*, **1965**, *87*, 3571.
[2] M. K. Kroeger and R. S. Drago, *J. Am. Chem. Soc.*, **1981**, *103*, 3250.

Table E.2 *(Continued)*

Acid	C_A	E_A	c_A	e_A	t_A
Trifluoroethanol	0.451	3.88	—	—	—
t-Butyl alcohol	0.300	2.04	0.408	3.747	**0.010**[a]
Phenol	0.442	4.33	0.274	4.561	0.315
m-Trifluoromethylphenol	0.300	2.04	0.408	3.747	0.010
Fluoromethylphenol	0.506	4.42	0.291	4.426	0.347
p-Chlorophenol	0.478	4.34	—	—	—
p-Fluorophenol	0.446	4.17	—	—	—
p-Methylphenol	0.404	4.18	—	—	—
p-*t*-Butylphenol	0.387	4.06	—	—	—
Thiophenol	0.198	0.987	—	—	—
Water	—	—	0.372	1.649	0.196
Chloroform	0.159	3.02	**0.010**[a]	2.247	0.267
1-Hydroperfluoroheptane	0.226	2.45	—	—	—
Isothiocyanic acid	0.227	5.30	—	—	—
Isocyanic acid	0.258	3.22	—	—	—
Pyrrole	0.295	2.54	—	—	—
Transition metal acceptors					
Methylcobaloxime	1.53	9.14	—	—	—
Bis(hexafluoroacetylacetonato)copper(II)	1.32	3.46	1.672	3.186	0.235
Zinc tetraphenylporphyrin	0.624	5.15	0.442	5.543	0.371
Bis(hexamethyldisilylamino)zinc	1.07	5.16	—	—	—
Bis(hexamethyldisilylamino)cadmium	1.07	5.16	—	—	—
Bis(1,5-cyclooctadiene)chlororhodium(I)	1.25	4.93	1.394	2.321	0.457
Group IIIA acceptors					
Boron trifluoride	1.62	9.88	1.668	**2.000**[a]	0.914
Boron trimethyl	1.70	6.14	1.601	8.590	0.292
Aluminum trimethyl	1.43	16.9	0.685	16.28	1.255
Aluminum triethyl	2.04	12.5	—	—	—
Gallium trimethyl	0.881	13.3	—	—	—
Gallium triethyl	0.593	12.6	—	—	—
Indium trimethyl	0.654	15.3	—	—	—
Group IVA acceptor					
Trimethyltin chloride	0.0296	5.76	0.300	6.668	0.307
Group VA acceptor					
Antimony pentachloride	5.13	7.38	—	—	—
Group VIA acceptor					
Sulfur dioxide	0.808	0.920	0.721	3.777	**0.010**[a]
Group VIIA acceptors					
Iodine	**1.000**[a]	**1.000**[a]	**1.200**[a]	0.231	0.122
Iodine monochloride	0.830	5.10	—	—	—
Iodine monobromide	1.56	2.41	—	—	—

[a] These are fixed values that are assumed as reference points for the scales.

SOURCE: R. S. Drago et al., *J. Am. Chem. Soc.*, **1971**, *93*, 6014; ibid., **1977**, *99*, 3203; R. S. Drago, *Struct. Bond.*, **1973**, *15*, 73; A. J. Pribula and R. S. Drago, *J. Am. Chem. Soc.*, **1976**, *98*, 2784; M. P. Li and R. S. Drago, ibid., **1976**, *98*, 5129; M. K. Kroeger and R. S. Drago, ibid., **1981**, *103*, 3250.

Table E.3 Base parameters

Base	C_B	F_B	c_B	e_B	t_B
Anions					
H^-	—	—	**−267.005**[a]	5.334	175.307
F^-	—	—	34.768	6.154	1.108
Cl^-	—	—	2.730	5.111	17.508
Br^-	—	—	4.863	4.897	4.863
I^-	—	—	−2.502	4.517	19.608
OH^-	—	—	34.817	6.599	2.394
CH_3^-	—	—	−8.763	7.564	28.476
CH_3O^-	—	—	39.827	6.756	−1.298
CN^-	—	—	2.585	5.420	18.949
NH_2^-	—	—	−59.777	6.715	56.684
NO^-	—	—	−65.781	5.197	57.159
NO_2^-	—	—	3.396	5.896	18.200
Neutral donors					
Group VA donors					
Ammonia	3.46	1.36	2.713	0.694	11.587
Methylamine	5.88	1.30	4.655	0.786	11.281
Dimethylamine	8.73	1.09	6.934	0.755	10.574
Trimethylamine	11.54	0.808	9.272	**0.745**[a]	9.003
Ethylamine	6.02	1.37	4.866	0.824	11.017
Diethylamine	8.83	0.866	6.898	0.579	10.551
Triethylamine	11.04	0.991	8.830	0.781	9.786
Piperidine	—	—	7.223	0.770	10.277
Quinoline	—	—	10.546	0.735	8.786
N-Methylimidazole	—	—	6.854	0.672	10.704
Pyridine	6.40	1.17	5.112	0.629	11.486
Hydrogen cyanide	—	—	0.173	0.727	11.653
Acetonitrile	1.34	0.886	0.352	0.153	12.230
Chloroacetonitrile	0.530	0.940	0.013	0.085	11.969
Trimethylphosphine	—	—	3.398	0.870	12.490
4-Ethyl-1-Phospha-2,6,7-trioxabicyclooctane	—	—	7.085	**0.026**[a]	9.78
Group VIA donors					
Dimethylcyanamide	1.81	1.10	—	—	—
Dimethylformamide	2.48	1.23	1.665	0.405	12.698
Dimethylacetamide	2.58	1.32	2.011	0.485	12.803
Methyl acetate	1.61	0.903	0.787	0.155	12.384
Ethyl acetate	1.74	0.975	0.892	0.133	12.612
Formaldehyde	—	—	1.505	0.228	10.285
Acetone	2.33	0.987	1.463	0.252	12.029
Water (A)[b]	—	—	10.582	0.259	4.417
Water (B)[b]	—	—	19.203	0.649	**0.100**[a]
Methanol	—	—	2.197	0.061	10.684
Dimethyl ether	—	—	2.170	0.330	10.978
Diethyl ether	3.25	0.963	2.289	0.400	11.587
Di-*iso*-propyl ether	3.19	1.11	2.289	0.400	11.966
Di-*n*-butyl ether	3.30	1.06	—	—	—
p-Dioxane	2.38	1.09	1.636	0.317	11.645
Tetrahydrofuran	4.27	0.978	3.216	0.359	11.160

Table E.3 *(Continued)*

Base	C_B	F_B	c_B	e_B	t_B
Tetrahydropyran	3.91	0.949	2.820	0.389	11.132
Triethyl phosphate	—	—	1.198	0.475	13.479
Dimethylsulfoxide	2.85	1.34	2.329	0.584	11.969
Tetramethylene sulfoxide	3.16	1.38	—	—	—
Dimethylsulfide	7.46	0.343	5.448	**0.024**[a]	9.986
Diethylsulfide	7.40	0.339	5.448	**0.010**[a]	10.210
Trimethylene sulfide	6.84	0.343	—	—	—
Tetramethylene sulfide	7.90	0.341	—	—	—
Pentamethylene sulfide	7.40	0.375	—	—	—

[a] These are fixed values that are assumed as reference points for the scales.

[b] The (A) values are for water bonding covalently through a lone pair as with H^+, CH_3^+, Cu^+, etc. The (B) values are for interactions that are principally electrostatic in nature, such as with Li^+, Na^+, Sr^{+2}, etc.

SOURCE: R. S. Drago et al., *J. Am. Chem. Soc.*, **1971**, *93*, 6014; ibid., **1977**, *99*, 3203; R. S. Drago, *Struct. Bond.*, **1973**, *15*, 73; A. J. Pribula and R. S. Drago, *J. Am. Chem. Soc.*, **1976**, *98*, 2784; M. P. Li and R. S. Drago, ibid., **1976**, *98*, 5129; M. K. Kroeger and R. S. Drago, ibid., **1981**, *103*, 3250.

F

Electrode potentials and electromotive forces

This appendix contains values of *standard reduction electromotive force* (*emf*) for water and ammonia. For a thermodynamic calculation they may be written as *oxidation emfs* with change in sign. Since the listed half-reactions are written as *reduction* processes, they are therefore also the *standard electrode potentials.* Refer to the discussion on pp. 350–352 concerning the conventions relating to these terms, particularly to footnote 35, where it is pointed out that although the European convention may be a useful mnemonic for someone wiring a circuit, it is logically faulty in the thermodynamic sense and has probably engendered more confusion than it has alleviated, even among chemists who should know better. As has been clearly and succinctly pointed out recently,[1] there would never have been any possibility of confusion if authors had merely stated their usage. All values in this book are treated as *emf* values (thermodynamic convention).

Except as noted, all of the values in Table F.1 are from de Bethune and Loud[2] for aqueous solution and those in Table F.2 from Jolly[3] for ammonia. The interested reader should also consult Milazzo and Caroli,[4] and Antelman.[5] Latimer's excellent book,[6] while somewhat dated in terms of some of the experimental data, provides unsurpassed insight into the entire subject of the thermodynamic treatment of electrochemistry.

---

[1] T. Moeller, "Inorganic Chemistry, A Modern Introduction," Wiley, New York, **1982**, pp. 555–556.

[2] A. J. de Bethune and N. A. S. Loud, "Standard Aqueous Electrode Potentials and Temperature Coefficients at 25 °C," C. A. Hampel, Skokie, Ill., **1964**. Used with permission.

[3] W. L. Jolly, *J. Chem. Educ.*, **1956**, *33*, 512. Used with permission.

[4] G. Milazzo and S. Caroli, "Tales of Standard Electrode Potentials," Wiley, New York, **1978**.

[5] M. S. Antelman, "The Encyclopedia of Chemical Electrode Potentials," Plenum, New York, **1982**.

[6] W. M. Latimer, "The Oxidation States of the Elements and Their Potentials in Aqueous Solution," 2nd ed., Prentice-Hall, Englewood Cliffs, N.J., **1952**.

Table F.1 Table of standard aqueous reduction *emfs* at 25 °C

Electrode	$\mathscr{E}^0$ (V)
Acid solutions	
$\frac{3}{2}N_2 + H^+ + e^- = HN_3(g)$	−3.40
$\frac{3}{2}N_2 + H^+ + e^- = HN_3(aq)$	−3.09
$Li^+ + e^- = Li$	−3.045
$K^+ + e^- = K$	−2.925
$Rb^+ + e^- = Rb$	−2.925
$Cs^+ + e^- = Cs$	−2.923
$Ra^{+2} + 2e^- = Ra$	−2.916
$Ba^{+2} + 2e^- = Ba$	−2.906
$Sr^{+2} + 2e^- = Sr$	−2.888
$Ca^{+2} + 2e^- = Ca$	−2.866
$Na^+ + e^- = Na$	−2.714
$Ac^{+3} + 3e^- = Ac$	−2.6
$La^{+3} + 3e^- = La$	−2.522
$Ce^{+3} + 3e^- = Ce$	−2.483
$Pr^{+3} + 3e^- = Pr$	−2.462
$Nd^{+3} + 3e^- = Nd$	−2.431
$Pm^{+3} + 3e^- = Pm$	−2.423
$Sm^{+3} + 3e^- = Sm$	−2.414
$Eu^{+3} + 3e^- = Eu$	−2.407
$Gd^{+3} + 3e^- = Gd$	−2.397
$Tb^{+3} + 3e^- = Tb$	−2.391
$Y^{+3} + 3e^- = Y$	−2.372
$Mg^{+2} + 2e^- = Mg$	−2.363
$Dy^{+3} + 3e^- = Dy$	−2.353
$Am^{+3} + 3e^- = Am$	−2.320
$Ho^{+3} + 3e^- = Ho$	−2.319
$Er^{+3} + 3e^- = Er$	−2.296
$Tm^{+3} + 3e^- = Tm$	−2.278
$Yb^{+3} + 3e^- = Yb$	−2.267
$Lu^{+3} + 3e^- = Lu$	−2.255
$\frac{1}{2}H_2 + e^- = H^-$	−2.25
$H^+ + e^- = H(g)$	−2.1065
$Sc^{+3} + 3e^- = Sc$	−2.077
$AlF_6^{-3} + 3e^- = Al + 6F^-$	−2.069
$Pu^{+3} + 3e^- = Pu$	−2.031
$Th^{+4} + 4e^- = Th$	−1.899
$Np^{+3} + 3e^- = Np$	−1.856
$Be^{+2} + 2e^- = Be$	−1.847
$U^{+3} + 3e^- = U$	−1.789
$Hf^{+4} + 4e^- = Hf$	−1.70
$Al^{+3} + 3e^- = Al$	−1.662
$Ti^{+2} + 2e^- = Ti$	−1.628

Table F.1 *(Continued)*

Electrode	$\mathscr{E}^0$ (V)
Acid solutions	
$Zr^{+4} + 4e^- = Zr$	-1.529
$SiF_6^{-2} + 4e^- = Si + 6F^-$	-1.24
$Yb^{+3} + e^- = Yb^{+2}$	-1.21
$TiF_6^{-2} + 4e^- = Ti + 6F^-$	-1.191
$V^{+2} + 2e^- = V$	-1.186
$Mn^{+2} + 2e^- = Mn$	-1.180
$Sm^{+3} + e^- = Sm^{+2}$	-1.15
$Nb^{+3} + 3e^- = Nb$	-1.099
$PaO_2^+ + 4H^+ + 5e^- = Pa + 2H_2O$	-1.0
$Po + 2H^+ + 2e^- = H_2Po$	> -1.00
$TiO^{+2} + 2H^+ + 4e^- = Ti + H_2O$	-0.882
$H_3BO_3(aq) + 3H^+ + 4e^- = B + 3H_2O$	-0.8698
$H_3BO_3(c) + 3H^+ + 3e^- = B + 3H_2O$	-0.869
$SiO_2(quartz) + 4H^+ + 4e^- = Si + 2H_2O$	-0.857
$Ta_2O_5 + 10H^+ + 10e^- = 2Ta + 5H_2O$	-0.812
$Zn^{+2} + 2e^- = Zn$	-0.7628
$Zn^{+2} + Hg + 2e^- = Zn(Hg)$	-0.7627
$TlI + 3e^- = Tl + I$	-0.752
$Cr^{+3} + 3e^- = Cr$	-0.744
$Te + 2H^+ + 2e^- = H_2Te(aq)$	-0.739
$Te + 2H^+ + 2e^- = H_2Te(g)$	-0.718
$TlBr + e^- = Tl + Br^-$	-0.658
$Nb_2O_5 + 10H^+ + 10e^- = 2Nb + 5H_2O$	-0.644
$U^{+4} + e^- = U^{+3}$	-0.607
$As + 3H^+ + 3e^- = AsH_3(g)$	-0.607
$TlCl + e^- = Tl + Cl^-$	-0.5568
$Ga^{+3} + 3e^- = Ga$	-0.529
$Sb + 3H^+ + 3e^- = SbH_3(g)$	-0.510
$H_3PO_2 + H^+ + e^- = P(white) + 2H_2O$	-0.508
$H_3PO_3(aq) + 2H^+ + 2e^- = H_3PO_2(aq) + H_2O$	-0.499
$Fe^{+2} + 2e^- = Fe$	-0.4402
$Eu^{+3} + e^- = Eu^{+2}$	-0.429
$Cr^{+3} + e^- = Cr^{+2}$	-0.408
$Cd^{+2} + 2e^- = Cd$	-0.4029
$Se + 2H^+ + 2e^- = H_2Se(aq)$	-0.399
$Ti^{+3} + e^- = Ti^{+2}$	-0.369
$PbI_2 + 2e^- = Pb + 2I^-$	-0.365
$PbSO_4 + 2e^- = Pb + SO_4^{-2}$	-0.3588
$Cd^{+2} + Hg + 2e^- = Cd(Hg)$	-0.3516
$PbSO_4 + Hg + 2e^- = Pb(Hg) + SO_4$	-0.3505
$In^{+3} + 3e^- = In$	-0.343
$Tl^+ + e^- = Tl$	-0.3363
$HCNO + H^+ + e^- = \frac{1}{2}C_2N_2 + H_2O$	-0.330

Table F.1 *(Continued)*

Electrode	$\mathscr{E}^0$ (V)
Acid solutions	
$PtS + 2H^+ + 2e^- = Pt + H_2S(aq)$	−0.327
$PtS + 2H^+ + 2e^- = Pt + H_2S(g)$	−0.297
$PbBr_2 + 2e^- = Pb + 2Br^-$	−0.284
$Co^{+2} + 2e^- = Co$	−0.277
$H_3PO_4(aq) + 2H^+ + 2e^- = H_3PO_3(aq) + H_2O$	−0.276
$PbCl_2 + 2e^- = Pb + 2Cl^-$	−0.268
$V^{+3} + e^- = V^{+2}$	−0.256
$V(OH)_4^+ + 4H^+ + 5e^- = V + 4H_2O$	−0.254
$SnF_6^{-2} + 4e^- = Sn + 6F^-$	−0.25
$Ni^{+2} + 2e^- = Ni$	−0.250
$N_2 + 5H^+ + 4e^- = N_2H_5^+$	−0.23
$2SO_4^{-2} + 4H^+ + 2e^- = S_2O_6^{-2} + 2H_2O$	−0.22
$Mo^{+3} + 3e^- = Mo$	−0.20
$CO_2(g) + 2H^+ + 2e^- = HCOOH(aq)$	−0.199
$CuI + e^- = Cu + I^-$	−0.1852
$AgI + e^- = Ag + I^-$	−0.1518
$GeO_2 + 4H^+ + 4e^- = Ge + 2H_2O$	−0.15
$Sn^{+2} + 2e^- = Sn(white)$	−0.136
$O_2 + H^+ + e^- = HO_2$	−0.13
$Pb^{+2} + 2e^- = Pb$	−0.126
$WO_3(c) + 6H^+ + 6e^- = W + 3H_2O$	−0.090
$2H_2SO_3 + H^+ + 2e^- = HS_2O_4^- + 2H_2O$	−0.082
$P(white) + 3H^+ + 3e^- = PH_3(g)$	−0.063
$Hg_2I_2 + 2e^- = 2Hg + 2I^-$	−0.0405
$HgI_4^{-2} + 2e^- = Hg + 4I^-$	−0.038
$2D^+ + 2e^- = D_2$	−0.0034
$2H^+ + 2e^-(SHE) = H_2$	±0.0000
$2H^+ + 2e^- = H_2$(sat'd, t.p. 1 atm)	+0.0004
$Ag(S_2O_3)_2^{-3} + e^- = Ag + 2S_2O_3^{-2}$	+0.017
$CuBr + e^- = Cu + Br^-$	+0.033
$UO_2^{+2} + e^- = UO_2^+$	+0.05
$HCOOH(aq) + 2H^+ + 2e^- = HCHO(aq) + H_2O$	+0.056
$AgBr + e^- = Ag + Br^-$	+0.0713
$TiO^{+2} + 2H^+ + e^- = Ti^{+3} + H_2O$	+0.099
$Si + 4H^+ + 4e^- = SiH_4(g)$	+0.102
$C(graphite) + 4H^+ + 4e^- = CH_4(g)$	+0.1316
$CuCl + e^- = Cu + Cl^-$	+0.137
$Hg_2Br_2 + 2e^- = 2Hg + 2Br^-$	+0.1397
$S(rhombic) + 2H^+ + 2e^- = H_2S(aq)$	+0.142
$Np^{+4} + e^- = Np^{+3}$	+0.147
$Sn^{+4} + 2e^- = Sn^{+2}$	+0.15
$Sb_2O_3 + 6H^+ + 6e^- = 2Sb + 3H_2O$	+0.152
$Cu^{+2} + e^- = Cu^+$	+0.153

Table F.1 *(Continued)*

Electrode	$\mathscr{E}^0$ (V)
Acid solutions	
$BiOCl + 2H^+ + 3e^- = Bi + H_2O + Cl^-$	+0.160
$SO_4^{-2} + 4H^+ + 2e^- = H_2SO_3 + H_2O$	+0.172
$At_2 + 2e^- = 2At^-$	+0.2
$AgCl + e^- = Ag + Cl^-$	+0.2222
$HgBr_4^{-2} + 2e^- = Hg + 4Br^-$	+0.223
$(CH_3)_2SO_2 + 2H^+ + 2e^- = (CH_3)_2SO + H_2O$	+0.23
$HAsO_2(aq) + 3H^+ + 3e^- = As + 2H_2O$	+0.2476
$ReO_2 + 4H^+ + 4e^- = Re + 2H_2O$	+0.2513
$Hg_2Cl_2 + 2e^- = 2Hg + 2Cl^-$	+0.2676
$BiO^+ + 2H^+ + 3e^- = Bi + H_2O$	+0.320
$UO_2^{+2} + 4H^+ + 2e^- = U^{+4} + 2H_2O$	+0.330
$Cu^{+2} + 2e^- = Cu$	+0.337
$AgIO_3 + e^- = Ag + IO_3^-$	+0.354
$SO_4^{-2} + 8H^+ + 6e^- = S + 4H_2O$	+0.3572
$VO^{+2} + 2H^+ + e^- = V^{+3} + H_2O$	+0.359
$Fe(CN)_6^{-3} + e^- = Fe(CN)_6^{-4}$	+0.36
$ReO_4^- + 8H^+ + 7e^- = Re + 4H_2O$	+0.362
$\frac{1}{2}C_2N_2(g) + H^+ + e^- = HCN(aq)$	+0.373
$H_2N_2O_2 + 6H^+ + 4e^- = 2NH_3OH^+$	+0.387
$Tc^{+2} + 2e^- = Tc$	+0.4
$2H_2SO_3 + 2H^+ + 4e^- = S_2O_3^{-2} + 3H_2O$	+0.400
$RhCl_6^{-3} + 3e^- = Rh + 6Cl^-$	+0.431
$H_2SO_3 + 4H^+ + 4e^- = S + 3H_2O$	+0.450
$Ag_2CrO_4 + 2e^- = 2Ag + CrO_4^{-2}$	+0.464
$Sb_2O_5(c) + 2H^+ + 2e^- = Sb_2O_4(c) + H_2O$	+0.479
$Ag_2MoO_4 + 2e^- = 2Ag + MoO_4^{-2}$	+0.486
$ReO_4^- + 4H^+ + 3e^- = ReO_2 + 2H_2O$	+0.510
$4H_2SO_3 + 4H^+ + 6e^- = S_4O_6^{-2} + 6H_2O$	+0.51
$C_2H_4(g) + 2H^+ + 2e^- = C_2H_6(g)$	+0.52
$Cu^+ + e^- = Cu$	+0.521
$TeO_2(c) + 4H^+ + 4e^- = Te + 2H_2O$	+0.529
$I_2(c) + 2e^- = 2I^-$	+0.5355
$I_3^- + 2e^- = 3I^-$	+0.536
$Cu^{+2} + Cl^- + e^- = CuCl$	+0.538
$AgBrO_3 + e^- = Ag + BrO_3^-$	+0.546
$TeOOH^+ + 3H^+ + 4e^- = Te + 2H_2O$	+0.559
$H_3AsO_4(aq) + 2H^+ + 2e^- = HAsO_2 + 2H_2O$	+0.560
$AgNO_2 + e^- = Ag + NO_2^-$	+0.564
$MnO_4^- + e^- = MnO_4^{-2}$	+0.564
$S_2O_6^{-2} + 4H^+ + 2e^- = 2H_2SO_3$	+0.57
$PtBr_4^{-2} + 2e^- = Pt + 4Br^-$	+0.581
$Sb_2O_5(c) + 6H^+ + 4e^- = 2SbO^+ + 3H_2O$	+0.581

Table F.1 *(Continued)*

Electrode	$\mathscr{E}^0$ (V)
Acid solutions	
$CH_3OH(aq) + 2H^+ + 2e^- = CH_4(g) + H_2O$	+0.588
$TcO_2 + 4H^+ + 2e^- = Tc^{+2} + 2H_2O$	+0.6
$PdBr_4^{-2} + 2e^- = Pd + 4Br^-$	+0.60
$RuCl_5^{-2} + 3e^- = Ru + 5Cl^-$	+0.601
$Hg_2SO_4 + 2e^- = 2Hg + SO_4^{-2}$	+0.6151
$UO_2^+ + 4H^+ + e^- = U^{+4} + 2H_2O$	+0.62
$PdCl_4^{-2} + 2e^- = Pd + 4Cl^-$	+0.62
$Cu^{+2} + Br^- + e^- = CuBr$	+0.640
$AgC_2H_3O_2 + e^- = Ag + C_2H_3O_2^-$	+0.643
$Po^{+2} + 2e^- = Po$	+0.65
$Ag_2SO_4 + 2e^- = 2Ag + SO_4^{-2}$	+0.654
$Au(CNS)_4^- + 3e^- = Au + 4CNS^-$	+0.655
$PtCl_6^{-2} + 2e^- = PtCl_4^{-2} + 2Cl^-$	+0.68
$O_2(g) + 2H^+ + 2e^- = H_2O_2(aq)$	+0.6824
$HN_3(aq) + 11H^+ + 8e^- = 3NH_4^+$	+0.695
$C_6H_4O_2 + 2H^+ + 2e^- = C_6H_4(OH)_2$	+0.6994
$2HAtO + 2H^+ + 2e^- = At_2 + 2H_2O$	+0.7
$TcO_4^- + 4H^+ + 3e^- = TcO_2 + 2H_2O$	+0.7
$H_2O_2(aq) + H^+ + e^- = OH_g + H_2O$	+0.71
$2NO + 2H^+ + 2e^- = H_2N_2O_2$	+0.712
$PtCl_4^{-2} + 2e^- = Pt + 4Cl^-$	+0.73
$C_2H_2(g) + 2H^+ + 2e^- = C_2H_4(g)$	+0.731
$H_2SeO_3(aq) + 4H^+ + 4e^- = Se(gray) + 3H_2O$	+0.740
$NpO_2^+ + 4H^+ + e^- = Np^{+4} + 2H_2O$	+0.75
$(CNS)_2 + 2e^- = 2CNS^-$	+0.77
$IrCl_6^{-3} + 3e^- = Ir + 6Cl^-$	+0.77
$Fe^{+3} + e^- = Fe^{+2}$	+0.771
$Hg_2^{+2} + 2e^- = 2Hg$	+0.788
$Ag^+ + e^- = Ag$	+0.7991
$PoO_2 + 4H^+ + 2e^- = Po^{+2} + 2H_2O$	+0.80
$Rh^{+3} + 3e^- = Rh$	+0.80
$2NO_3^- + 4H^+ + 2e^- = N_2O_4(g) + 2H_2O$	+0.803
$OsO_4(c, yellow) + 8H^+ + 8e^- = Os + 4H_2O$	+0.85
$2HNO_2 + 4H^+ + 4e^- = H_2N_2O_2 + 2H_2O$	+0.86
$Cu^{+2} + I^- + e^- = CuI$	+0.86
$Rh_2O_3 + 6H^+ + 2e^- = 2Rh + 3H_2O$	+0.87
$AuBr_4^- + 3e^- = Au + 4Br^-$	+0.87(60 °C)
$2Hg^{+2} + 2e^- = Hg_2^{+2}$	+0.920
$PuO_2^{+2} + e^- = PuO_2^+$	+0.93
$NO_3^- + 3H^+ + 2e^- = HNO_2 + H_2O$	+0.94
$AuBr_2^- + e^- = Au + 2Br^-$	+0.956
$NO_3^- + 4H^+ + 3e^- = NO + 2H_2O$	+0.96

Table F.1 *(Continued)*

Electrode	$\mathscr{E}^0$ (V)
Acid solutions	
$Pu^{+4} + e^- = Pu^{+3}$	+0.97
$Pt(OH)_2 + 2H^+ + 2e^- = Pt + 2H_2O$	+0.98
$Pd^{+2} + 2e^- = Pd$	+0.987
$IrBr_6^{-3} + e^- = IrBr_6^{-4}$	+0.99
$HNO_2 + H^+ + e^- = NO + H_2O$	+1.00
$AuCl_4^- + 3e^- = Au + 4Cl^-$	+1.00
$V(OH)_4^+ + 2H^+ + e^- = VO^{+2} + 3H_2O$	+1.00
$IrCl_6^{-2} + e^- = IrCl_6^{-3}$	+1.017
$H_6TeO_6(c) + 2H^+ + 2e^- = TeO_2 + 4H_2O$	+1.02
$N_2O_4 + 4H^+ + 4e^- = 2NO + 2H_2O$	+1.03
$PuO_2^{+2} + 4H^+ + 2e^- = Pu^{+4} + 2H_2O$	+1.04
$ICl_2^- + e^- = 2Cl^- + \frac{1}{2}I_2$	+1.056
$Br_2(l) + 2e^- = 2Br^-$	+1.0652
$N_2O_4 + 2H^+ + 2e^- = 2HNO_2$	+1.07
$Br_2(aq) + 2e^- = 2Br^-$	+1.087
$PtO_2 + 2H^+ + 2e^- = Pt(OH)_2$	ca. +1.1
$PuO_2^+ + 4H^+ + e^- = Pu^{+4} + 2H_2O$	+1.15
$SeO_4^{-2} + 4H^+ + 2e^- = H_2SeO_3 + H_2O$	+1.15
$NpO_2^{+2} + e^- = NpO_2^+$	+1.15
$CCl_4 + 4H^+ + 4e^- = 4Cl^- + C + 4H^+$	+1.18
$O_2 + 4H^+ + 4e^- = 2H_2O(g)$	+1.185
$IO_3^- + 6H^+ + 5e^- = \frac{1}{2}I_2 + 3H_2O$	+1.196
$Pt^{+2} + 2e^- = Pt$	ca. +1.2
$ClO_3^- + 3H^+ + 2e^- = HClO_2 + H_2O$	+1.21
$O_2 + 4H^+ + 4e^- = 2H_2O(l)$	+1.229
$S_2Cl_2 + 2e^- = 2S + 2Cl^-$	+1.23
$MnO_2 + 4H^+ + 2e^- = Mn^{+2} + 2H_2O$	+1.23
$ClO_4^- + 2H^+ + 2e^- = ClO_3^- + H_2O$	+1.230
$Tl^{+3} + 2e^- = Tl^+$	+1.25
$AmO_2^+ + 4H^+ + e^- = Am^{+4} + 2H_2O$	+1.261
$N_2H_5^+ + 3H^+ + 2e^- = 2NH_4^+$	+1.275
$ClO_2 + H^+ + e^- = HClO_2$	+1.275
$PdCl_6^{-2} + 2e^- = PdCl_4^{-2} + 2Cl^-$	+1.288
$2HNO_2(aq) + 4H^+ + 4e^- = N_2O(g) + 3H_2O$	+1.29
$Cr_2O_7^{-2} + 14H^+ + 6e^- = 2Cr^{+3} + 7H_2O$	+1.33
$NH_3OH^+ + 2H^+ + 2e^- = NH_4^+ + H_2O$	+1.35
$Cl_2 + 2e^- = 2Cl^-$	+1.3595
$HAtO_3 + 4H^+ + 4e^- = HAtO + 2H_2O$	+1.4
$2NH_3OH^+ + H^+ + 2e^- = N_2H_5^+ + 2H_2O$	+1.42
$Au(OH)_3(c) + 3H^+ + 3e^- = Au + 3H_2O$	+1.45
$HIO + H^+ + e^- = \frac{1}{2}I_2 + H_2O$	+1.45
$PbO_2 + 4H^+ + 2e^- = Pb^{+2} + 2H_2O$	+1.455
$HO_2(aq) + H^+ + e^- = H_2O_2(aq)$	+1.495

Table F.1 *(Continued)*

Electrode	$\mathscr{E}^0$ (V)
Acid solutions	
$Au^{+3} + 3e^- = Au$	+1.498
$Mn^{+3} + e^- = Mn^{+2}$	+1.51
$MnO_4^- + 8H^+ + 5e^- = Mn^{+2} + 4H_2O$	+1.51
$BrO_3^- + 6H^+ + 5e^- = \frac{1}{2}Br_2(l) + 3H_2O$	+1.52
$PoO_3 + 2H^+ + 2e^- = PoO_2 + H_2O$	+1.52?
$Bi_2O_4 + 4H^+ + 2e^- = 2BiO^+ + 2H_2O$	+1.593
$HBrO + H^+ + e^- = \frac{1}{2}Br_2(l) + H_2O$	+1.595
$Bk^{+4} + e^- = Bk^{+3}$	ca. +1.6
$Ce^{+4} + e^- = Ce^{+3}$	+1.61
$HClO + H^+ + e^- = \frac{1}{2}Cl_2 + H_2O$	+1.63
$AmO_2^{+2} + e^- = AmO_2^+$	+1.639
$H_5IO_6 + H^+ + 2e^- = IO_3^- + 3H_2O$	+1.644
$HClO_2 + 2H^+ + 2e^- = HClO + H_2O$	+1.645
$NiO_2 + 4H^+ + 2e^- = Ni^{+2} + 2H_2O$	+1.678
$PbO_2 + SO_4^{-2} + 4H^+ + 2e^- = PbSO_4 + 2H_2O$	+1.682
$Au^+ + e^- = Au$	+1.691
$AmO_2^{+2} + 4H^+ + 3e^- = Am^{+3} + 2H_2O$	+1.694
$MnO_4^- + 4H^+ + 3e^- = MnO_2 + 2H_2O$	+1.695
$AmO_2^+ + 4H^+ + 2e^- = Am^{+3} + 2H_2O$	+1.721
$BrO_4^- + 2H^+ + 2e^- = BrO_3^- + H_2O$	+1.763[a]
$H_2O_2 + 2H^+ + 2e^- = 2H_2O$	+1.776
$XeO_3 + 6H^+ + 6e^- = Xe + 3H_2O$	+1.8
$Co^{+3} + e^- = Co^{+2}$	+1.808
$HN_3 + 3H^+ + 2e^- = NH_4^+ + N_2$	+1.96
$Ag^{+2} + e^- = Ag^+$	+1.980
$S_2O_8^{-2} + 2e^- = 2SO_4^{-2}$	+2.01
$O_3 + 2H^+ + 2e^- = O_2 + H_2O$	+2.07
$F_2O + 2H^+ + 4e^- = 2F^- + H_2O$	+2.15
$Am^{+4} + e^- = Am^{+3}$	+2.18
$FeO_4^{-2} + 8H^+ + 3e^- = Fe^{+3} + 4H_2O$	+2.20
$H_4XeO_6 + 2H^+ + 2e^- = XeO_3 + 3H_2O$	+2.3[b]
$O(g) + 2H^+ + 2e^- = H_2O$	+2.422
$H_2N_2O_2 + 2H^+ + 2e^- = N_2 + 2H_2O$	+2.65
$OH + H^+ + e^- = H_2O$	+2.85
$Pr^{+4} + e^- = Pr^{+3}$	+2.86
$F_2(g) + 2e^- = 2F^-$	+2.87
$H_4XeO_6 + 2H^+ + 2e^- = XeO_3 + 3H_2O$	+3.0
$F_2(g) + 2H^+ + 2e^- = 2HF(aq)$	+3.06

[a] G. K. Johnson et al., *Inorg. Chem.*, **1970**, *9*, 119.
[b] J. H. Holloway, "Noble-Gas Chemistry," Methuen, London, **1968**, p. 143.

Table F.1 *(Continued)*

Electrode	$\mathscr{E}^0$ (V)
Base solutions	
$Ca(OH)_2 + 2e^- = Ca + 2OH^-$	−3.02
$Ba(OH)_2 \cdot 8H_2O + 2e^- = Ba + 2OH^- + 8H_2O$	−2.99
$H_2O + e^- = H(g) + OH^-$	−2.9345
$La(OH)_3 + 3e^- = La + 3OH^-$	−2.90
$Sr(OH)_2 + 2e^- = Sr + 2OH^-$	−2.88
$Ce(OH)_3 + 3e^- = Ce + 3OH^-$	−2.87
$Pr(OH)_3 + 3e^- = Pr + 3OH^-$	−2.85
$Nd(OH)_3 + 3e^- = Nd + 3OH^-$	−2.84
$Pm(OH)_3 + 3e^- = Pm + 3OH^-$	−2.84
$Sm(OH)_3 + 3e^- = Sm + 3OH^-$	−2.83
$Eu(OH)_3 + 3e^- = Eu + 3OH^-$	−2.83
$Gd(OH)_3 + 3e^- = Gd + 3OH^-$	−2.82
$Ba(OH)_2 + 2e^- = Ba + 2OH^-$	−2.81
$Y(OH)_3 + 3e^- = Y + 3OH^-$	−2.81
$Tb(OH)_3 + 3e^- = Tb + 3OH^-$	−2.79
$Dy(OH)_3 + 3e^- = Dy + 3OH^-$	−2.78
$Ho(OH)_3 + 3e^- = Ho + 3OH^-$	−2.77
$Er(OH)_3 + 3e^- = Er + 3OH^-$	−2.75
$Tm(OH)_3 + 3e^- = Tm + 3OH^-$	−2.74
$Yb(OH)_3 + 3e^- = Yb + 3OH^-$	−2.73
$Lu(OH)_3 + 3e^- = Lu + 3OH^-$	−2.72
$Mg(OH)_2 + 2e^- = Mg + 2OH^-$	−2.690
$Be_2O_3^{-2} + 3H_2O + 4e^- = 2Be + 6OH^-$	−2.63
$BeO + H_2O + 2e^- = Be + 2OH^-$	−2.613
$Sc(OH)_3 + 3e^- = Sc + 3OH^-$	−2.61
$HfO(OH)_2 + H_2O + 4e^- = Hf + 4OH^-$	−2.50
$Th(OH)_4 + 4e^- = Th + 4OH^-$	−2.48
$Pu(OH)_3 + 3e^- = Pu + 3OH^-$	−2.42
$UO_2 + 2H_2O + 4e^- = U + 4OH^-$	−2.39
$H_2ZrO_3 + H_2O + 4e^- = Zr + 4OH^-$	−2.36
$H_2AlO_3^- + H_2O + 3e^- = Al + 4OH^-$	−2.33
$Al(OH)_3 + 3e^- = Al + 3OH^-$	−2.30
$U(OH)_4 + e^- = U(OH)_3 + OH^-$	−2.20
$U(OH)_3 + 3e^- = U + 3OH^-$	−2.17
$H_2PO_2^- + e^- = P + 2OH^-$	−2.05
$H_2BO_3^- + H_2O + 3e^- = B + 4OH^-$	−1.79
$SiO_3^{-2} + 3H_2O + 4e^- = Si + 6OH^-$	−1.697
$Na_2UO_4 + 4H_2O + 2e^- = U(OH)_4 + 2Na^+ + 4OH^-$	−1.618
$HPO_3^{-2} + 2H_2O + 2e^- = H_2PO_2^- + 3OH^-$	−1.565
$Mn(OH)_2 + 2e^- = Mn + 2OH^-$	−1.55
$MnCO_3(c) + 2e^- = Mn + CO_3^{-2}$	−1.50
$MnCO_3(ppt) + 2e^- = Mn + CO_3^{-2}$	−1.48

Table F.1 *(Continued)*

Electrode	$\mathscr{E}^0$ (V)
Base solutions	
$Cr(OH)_3(c) + 3e^- = Cr + 3OH^-$	−1.48
$ZnS(wurtzite) + 2e^- = Zn + S^{-2}$	−1.405
$Cr(OH)_3(hydr) + 3e^- = Cr + 3OH^-$	−1.34
$CrO_2^- + 2H_2O + 3e^- = Cr + 4OH^-$	−1.27
$Zn(CN)_4^{-2} + 2e^- = Zn + 4CN^-$	−1.26
$Zn(OH)_2 + 2e^- = Zn + 2OH^-$	−1.245
$H_2GaO_3^- + H_2O + 3e^- = Ga + 4OH^-$	−1.219
$ZnO_2^{-2} + 2H_2O + 2e^- = Zn + 4OH^-$	−1.215
$CdS + 2e^- = Cd + S^{-2}$	−1.175
$HV_6O_{17}^{-3} + 16H_2O + 30e^- = 6V + 33OH^-$	−1.154
$Te + 2e^- = Te^{-2}$	−1.143
$PO_4^{-3} + 2H_2O + 2e^- = HPO_3^{-2} + 3OH^-$	−1.12
$2SO_3^{-2} + 2H_2O + 2e^- = S_2O_4^{-2} + 4OH^-$	−1.12
$ZnCO_3 + 2e^- = Zn + CO_3^{-2}$	−1.06
$WO_4^{-2} + 4H_2O + 6e^- = W + 8OH^-$	−1.05
$MoO_4^{-2} + 4H_2O + 6e^- = Mo + 8OH^-$	−1.05
$Zn(NH_3)_4^{+2} + 2e^- = Zn + 4NH_3(aq)$	−1.04
$NiS(\gamma) + 2e^- = Ni + S^{-2}$	−1.04
$HGeO_3^- + 2H_2O + 4e^- = Ge + 5OH^-$	−1.03
$Cd(CN)_4^{-2} + 2e^- = Cd + 4CN^-$	−1.028
$In(OH)_3 + 3e^- = In + 3OH^-$	−1.00
$CNO^- + H_2O + 2e^- = CN^- + 2OH^-$	−0.970
$Pu(OH)_4 + e^- = Pu(OH)_3 + OH^-$	−0.963
$FeS(\alpha) + 2e^- = Fe + S^{-2}$	−0.95
$PbS + 2e^- = Pb + S^{-2}$	−0.93
$Sn(OH)_6^- + 2e^- = HSnO_2^- + H_2O + 3OH^-$	−0.93
$SO_4^{-2} + H_2O + 2e^- = SO_3^{-2} + 2OH^-$	−0.93
$Se + 2e^- = Se^{-2}$	−0.92
$HSnO_2^- + H_2O + 2e^- = Sn + 3OH^-$	−0.909
$Tl_2S + 2e^- = 2Tl + S^{-2}$	−0.90
$Cu_2S + 2e^- = 2Cu + S^{-2}$	−0.89
$P(white) + 3H_2O + 3e^- = PH_3 + 3OH^-$	−0.89
$Fe(OH)_2 + 2e^- = Fe + 2OH^-$	−0.877
$SnS + 2e^- = Sn + S^{-2}$	−0.87
$NiS(\alpha) + 2e^- = Ni + S^{-2}$	−0.830
$2H_2O + 2e^- = H_2 + 2OH^-$	−0.82806
$Cd(OH)_2 + 2e^- = Cd + 2OH^-$	−0.809
$FeCO_3 + 2e^- = Fe + CO_3^{-2}$	−0.756
$CdCO_3 + 2e^- = Cd + CO_3^{-2}$	−0.74
$Co(OH)_2 + 2e^- = Co + 2OH^-$	−0.73
$Ni(OH)_2 + 2e^- = Ni + 2OH^-$	−0.72
$Fe_2S_3 + 2e^- = 2FeS(\alpha) + S^{-2}$	−0.715
$HgS(black) + 2e^- = Hg + S^{-2}$	−0.69

Table F.1 *(Continued)*

Electrode	$\mathscr{E}^0$ (V)
Base solutions	
$AsO_4^{-3} + 2H_2O + 2e^- = AsO_2^- + 4OH^-$	−0.68
$AsO_2^- + 2H_2O + 3e^- = As + 4OH^-$	−0.675
$Ag_2S(\alpha) + 2e^- = 2Ag + S^{-2}$	−0.66
$SbO_2^- + 2H_2O + 3e^- = Sb + 4OH^-$	−0.66
$CoCO_3 + 2e^- = Co + CO_3^{-2}$	−0.64
$Cd(NH_3)_4^{+2} + 2e^- = Cd + 4NH_3(aq)$	−0.613
$ReO_4^- + 2H_2O + 3e^- = ReO_2 + 4OH^-$	−0.594
$ReO_4^- + 4H_2O + 7e^- = Re + 8OH^-$	−0.584
$PbO(r) + H_2O + 2e^- = Pb + 2OH^-$	−0.580
$ReO_2 + 2H_2O + 4e^- = Re + 4OH^-$	−0.577
$2SO_3^{-2} + 3H_2O + 4e^- = S_2O_3^{-2} + 6OH^-$	−0.571
$TeO_3^{-2} + 3H_2O + 4e^- = Te + 6OH^-$	−0.57
$Fe(OH)_3 + e^- = Fe(OH)_2 + OH^-$	−0.56
$O_2 + e^- = O_2^-$	−0.563
$HPbO_2^- + H_2O + 2e^- = Pb + 3OH^-$	−0.540
$PbCO_3 + 2e^- = Pb + CO_3^{-2}$	−0.509
$PoO_3^{-2} + 3H_2O + 4e^- = Po + 6OH^-$	−0.49
$Ni(NH_3)_6^{+2} + 2e^- = Ni + 6NH_3(aq)$	−0.476
$Bi_2O_3 + 3H_2O + 6e^- = 2Bi + 6OH^-$	−0.46
$NiCO_3 + 2e^- = Ni + CO_3^{-2}$	−0.45
$S + 2e^- = S^{-2}$	−0.447
$Cu(CN)_2^- + e^- = Cu + 2CN^-$	−0.429
$Hg(CN)_4^{-2} + 2e^- = Hg + 4CN^-$	−0.37
$SeO_3^{-2} + 3H_2O + 4e^- = Se + 6OH^-$	−0.366
$Cu_2O + H_2O + 2e^- = 2Cu + 2OH^-$	−0.358
$Tl(OH)(c) + e^- = Tl + OH^-$	−0.343
$Ag(CN)_2^- + e^- = Ag + 2CN^-$	−0.31
$Cu(CNS) + e^- = Cu + CNS^-$	−0.27
$HO_2^- + H_2O + e^- = OH(g) + 2OH^-$	−0.262
$HO_2^- + H_2O + e^- = OH(aq) + 2OH^-$	−0.245
$CrO_4^{-2} + 4H_2O + 3e^- = Cr(OH)_3(hydr) + 5OH^-$	−0.13
$Cu(NH_3)_2^+ + e^- = Cu + 2NH_3$	−0.12
$2Cu(OH)_2 + 2e^- = Cu_2O + 2OH^- + H_2O$	−0.080
$O_2 + H_2O + 2e^- = HO_2^- + OH^-$	−0.076
$Tl(OH)_3 + 2e^- = TlOH + 2OH^-$	−0.05
$MnO_2 + 2H_2O + 2e^- = Mn(OH)_2 + 2OH^-$	−0.05
$AgCN + e^- = Ag + CN^-$	−0.017
$2AtO^- + 2H_2O + 2e^- = At_2 + 4OH^-$	0.0
$NO_3^- + H_2O + 2e^- = NO_2^- + 2OH^-$	+0.01
$HOsO_5^- + 4H_2O + 8e^- = Os + 9OH^-$	+0.015
$Rh_2O_3 + 3H_2O + 6e^- = 2Rh + 6OH^-$	+0.04
$SeO_4^{-2} + H_2O + 2e^- = SeO_3^{-2} + 2OH^-$	+0.05
$Pd(OH)_2 + 2e^- = Pd + 2OH^-$	+0.07

Table F.1 *(Continued)*

Electrode	$\mathscr{E}^0$ (V)
Base solutions	
$S_4O_6^{-2} + 2e^- = 2S_2O_3^{-2}$	+0.08
$HgO(r) + H_2O + 2e^- = Hg + 2OH^-$	+0.098
$Ir_2O_3 + 3H_2O + 6e^- = 2Ir + 6OH^-$	+0.098
$Co(NH_3)_6^{+3} + e^- = Co(NH_3)_6^{+2}$	+0.108
$Pt(OH)_6^{-2} + 2e^- = Pt(OH)_2 + 4OH^-$	−0.1 to −0.4
$N_2H_4 + 4H_2O + 2e^- = 2NH_4OH + 2OH^-$	+0.11
$Mn(OH)_3 + e^- = Mn(OH)_2 + OH^-$	+0.15
$Pt(OH)_2 + 2e^- = Pt + 2OH^-$	+0.15
$Co(OH)_3 + e^- = Co(OH)_2 + OH^-$	+0.17
$PuO_2(OH)_2 + e^- = PuO_2OH + OH^-$	+0.234
$PbO_2 + H_2O + 2e^- = PbO(r) + 2OH^-$	+0.247
$IO_3^- + 3H_2O + 6e^- = I^- + 6OH^-$	+0.26
$Ag(SO_3)_2^{-3} + e^- = Ag + 2SO_3^{-2}$	+0.295
$ClO_3^- + H_2O + 2e^- = ClO_2^- + 2OH^-$	+0.33
$Ag_2O + H_2O + 2e^- = 2Ag + 2OH^-$	+0.345
$ClO_4^- + H_2O + 2e^- = ClO_3^- + 2OH^-$	+0.36
$Ag(NH_3)_2^+ + e^- = Ag + 2NH_3$	+0.373
$TeO_4^{-2} + H_2O + 2e^- = TeO_3^{-2} + 2OH^-$	ca. +0.4
$O_2 + 2H_2O + 4e^- = 4OH^-$	+0.401
$O_2^- + H_2O + e^- = OH^- + HO_2^-$	+0.413
$Ag_2CO_3 + 2e^- = 2Ag + CO_3^{-2}$	+0.47
$IO^- + H_2O + 2e^- = I^- + 2OH^-$	+0.485
$NiO_2 + 2H_2O + 2e^- = Ni(OH)_2 + 2OH^-$	+0.490
$AtO_3^- + 2H_2O + 4e^- = AtO^- + 4OH^-$	+0.5
$MnO_4^- + 2H_2O + 3e^- = MnO_2(pyrolusite) + 4OH^-$	+0.588
$MnO_4^{-2} + 2H_2O + 2e^- = MnO_2 + 4OH^-$	+0.60
$RuO_4^- + e^- = RuO_4^{-2}$	+0.6
$2AgO + H_2O + 2e^- = Ag_2O + 2OH^-$	+0.607
$BrO_3^- + 3H_2O + 6e^- = Br^- + 6OH^-$	+0.61
$ClO_2^- + H_2O + 2e^- = ClO^- + 2OH^-$	+0.66
$H_3IO_6^{-2} + 2e^- = IO_3^- + 3OH^-$	+0.7
$FeO_4^{-2} + 4H_2O + 3e^- = Fe(OH)_3 + 5OH^-$	+0.72
$2NH_2OH + 2e^- = N_2H_4 + 2OH^-$	+0.73
$Ag_2O_3 + H_2O + 2e^- = 2AgO + 2OH^-$	+0.739
$BrO^- + H_2O + 2e^- = Br^- + 2OH^-$	+0.761
$HO_2^- + H_2O + 2e^- = 3OH^-$	+0.878
$ClO^- + H_2O + 2e^- = Cl^- + 2OH^-$	+0.89
$HXeO_4^- + 3H_2O + 6e^- = Xe + 7OH^-$	+0.9
$HXeO_6^{-3} + 2H_2O + 2e^- = HXeO_4^- + 4OH^-$	+0.9
$Cu^{+2} + 2CN^- + e^- = Cu(CN)_2^-$	+1.103
$ClO_2(g) + e^- = ClO_2^-$	+1.16
$O_3(g) + H_2O + 2e^- = O_2 + 2OH^-$	+1.24
$OH(g) + e^- = OH^-$	+2.02

Alphabetical index for Table F.1

Under each element are listed, in order of increasing *emf* (i.e., from the most active to the most noble), the *emfs* associated with the oxidation–reduction reactions of each element. Those marked B will be found in the basic solutions section of Table F.1.

Actinium
- −2.6

Aluminum
- −2.33B
- −2.30B
- −2.069
- −1.662

Americium
- −2.320
- +1.261
- +1.639
- +1.694
- +1.721
- +2.18

Antimony
- −0.66B
- −0.510
- +0.152
- +0.479
- +0.581

Arsenic
- −0.68B
- −0.675B
- −0.607
- +0.2476
- +0.560

Astatine
- 0.0B
- +0.2
- +0.5B
- +0.7
- +1.4

Barium
- −2.99B
- −2.906
- −2.81B

Berkelium
- +1.6

Beryllium
- −2.63B
- −2.613B
- −1.847

Bismuth
- −0.46B
- +0.160
- +0.320
- +1.593

Boron
- −1.79B
- −0.8698
- −0.869

Bromine
- +0.61B
- +0.761B
- +1.0652
- +1.087
- +1.52
- +1.595
- +1.763

Cadmium
- −1.175B
- −1.028B
- −0.809B
- −0.74B
- −0.613B
- −0.4029
- −0.3516

Calcium
- 3.02B
- −2.866

Carbon
- −0.970B
- −0.330
- −0.199
- +0.056
- +0.1316
- +0.23
- +0.373
- +0.52
- +0.588
- +0.6994
- +0.731
- +0.77
- +1.18

Cerium
- −2.87B
- −2.483
- +1.61

Cesium
- −2.923

Chlorine
- +0.33B
- +0.36B
- +0.66B
- +0.89B
- +1.16B
- +1.21
- +1.230
- +1.275
- +1.3595
- +1.63
- +1.645

Chromium
- −1.48B
- −1.34B
- −1.27B
- −0.744
- −0.408
- −0.13B
- +1.33

Cobalt
- −0.73B
- −0.64B
- −0.277
- +0.108B
- +0.17B
- +1.808

Copper
- −0.89B
- −0.429B
- −0.358B
- −0.27B
- −0.1852
- −0.12B
- −0.080B
- +0.033
- +0.137
- +0.153
- +0.337
- +0.521
- +0.538
- +0.640
- +0.86
- +1.103B

Dysprosium
- −2.78B
- −2.353

Erbium
- −2.75B
- −2.296

Europium
- −2.83B
- −2.407
- −0.429

Fluorine
- +2.15
- +2.87
- +3.06

Gadolinium
- −2.82B
- −2.397

Gallium
- −1.219B
- −0.529

Germanium
- −1.03B
- −0.15

Gold
- +0.655
- +0.87
- +0.956
- +1.00
- +1.45
- +1.498
- +1.691

Hafnium
- −2.50B
- −1.70

Holmium
- −2.77B
- −2.319

Hydrogen
- −2.9345B
- −2.25
- −2.1065
- −0.82806B
- −0.0034
- 0.0000
- +0.0004

Indium
- −1.00B
- −0.343

Iodine
- +0.26B
- +0.485B
- +0.5355
- +0.536
- +0.7B
- +1.056
- +1.195
- +1.45
- +1.644

Iridium
- +0.098B
- +0.77
- +0.99
- +1.017

Iron
- −0.95B
- −0.877B
- −0.756B
- −0.715B
- −0.56B
- −0.4402

+0.36
+0.72B
+0.771
+2.20

Lanthanum
−2.90B
−2.522

Lead
−0.93B
−0.580B
−0.540B
−0.509B
−0.365
−0.3588
−0.3505
−0.284
−0.268
−0.126
+0.247B
+1.455
+1.682

Lithium
−3.045

Lutetium
−2.72B
−2.255

Magnesium
−2.690B
−2.363

Manganese
−1.55B
−1.50B
−1.48B
−1.180
−0.05B
+0.15B
+0.564
+0.588B
+0.60B
+1.23
+1.51
+1.51
+1.695

Mercury
−0.69B
−0.37B
−0.0405
−0.038
+0.098B
+0.1397
+0.223
+0.2676
+0.6151
+0.788
+0.920

Molybdenum
−1.05B
−0.20

Neodymium
−2.84B
−2.431

Neptunium
−1.856
+0.147
+0.75
+1.15

Nickel
−1.04B
−0.830B
−0.72B
−0.476B
−0.45B
−0.250
+0.490B
+1.678

Niobium
−1.099
−0.644

Nitrogen
−3.40
−3.09
−0.970B
−0.330
−0.23
+0.01B
+0.11B
+0.373
+0.387
+0.695
+0.712
+0.73B
+0.77
+0.803
+0.86
+0.94
+0.96
+1.00
+1.03
+1.07
+1.275
+1.29
+1.35
+1.42
+1.96
+2.65

Osmium
+0.015B
+0.85

Oxygen
−0.563B
−0.262B
−0.245B
−0.13
−0.076B
+0.401B
+0.413B
+0.6824
+0.71
+0.878B
+1.185
+1.229
+1.24B
+1.495
+1.776
+2.02B
+2.07
+2.422
+2.85

Palladium
+0.07B
+0.60
+0.62
+0.987
+1.288

Phosphorus
−2.05B
−1.565B
−1.12B
−0.89B
−0.508
−0.499
−0.276
−0.063

Platinum
−0.327
−0.297
+0.1B
+0.15B
+0.581
+0.68
+0.73
+0.98
+1.1
+1.2

Plutonium
−2.42B
−2.031
−0.963B
+0.234B
+0.93
+0.97
+1.04
+1.15

Polonium
−1.00
−0.49B
+0.65
+0.80
+1.52

Potassium
−2.925

Praseodymium
−2.85B
−2.462
+2.86

Promethium
−2.48B
−2.423

Protactinium
−1.0

Radium
−2.916

Rhenium
−0.594B
−0.584B
−0.577B
+0.2513
+0.362
+0.510

Rhodium
+0.04B
+0.431
+0.80
+0.87

Rubidium
−2.925

Ruthenium
+0.6B
+0.601

Samarium
−2.83B
−2.414
−1.15

Scandium
−2.61B
−2.077

Selenium
−0.92B
−0.399
−0.366B
+0.05B
+0.740
+1.15

Silicon
−1.697B
−1.24
−0.857
+0.102

Silver
−0.66B

Alphabetical index for Table F.1 (*Continued*)

−0.31B
−0.1518
−0.017B
+0.017
+0.0713
+0.2222
+0.295B
+0.345B
+0.354
+0.373B
+0.464
+0.47B
+0.486
+0.546
+0.564
+0.607B
+0.643
+0.654
+0.739B
+0.7991
+1.980

Sodium
−2.714

Strontium
−2.888
−2.88B

Sulfur
−1.12B
−0.93B
−0.571B
−0.447B
−0.22
−0.082
+0.08B
+0.142
+0.172
+0.23
+0.3572
+0.400
+0.450
+0.51
+0.57
+0.77
+1.23
+2.01

Tantalum
−0.812

Technetium
+0.4
+0.6
+0.7

Tellurium
−1.143B
−0.739
−0.718
−0.57B
+0.4B
+0.529
+0.559
+1.02

Terbium
−2.79B
−2.391

Thallium
−0.90B
−0.752
−0.658
−0.5568
−0.343B
−0.3363
−0.05B
+1.25

Thorium
−2.48B
−1.899

Thulium
−2.74B
−2.278

Tin
−0.93B
−0.909B
−0.87B
−0.25
−0.136
+0.15

Titanium
−1.628
−1.191
−0.882
−0.369
+0.099

Tungsten
−1.05B
−0.090

Uranium
−2.39B
−2.20B
−2.17B
−1.789
−1.618B
−0.607
+0.05
+0.330
+0.62

Vanadium
−1.186
−1.154B
−0.256
−0.254
+0.359
+1.00

Xenon
+0.9B
+0.9B
+1.8
+2.3
+3.0

Ytterbium
−2.73B
−2.267
−1.21

Yttrium
−2.81B
−2.372

Zinc
−1.405B
−1.26B
−1.245B
−1.215B
−1.06B
−1.04B
−0.7628

Zirconium
−2.36B
−1.529

Table F.2 Reduction *emfs* in liquid ammonia at 25°

Half-reaction		Acid solutions	$\mathscr{E}^0$
$NH_4^+ + e^-$	⟶	$\frac{1}{2}H_2 + NH_3$	0.00
$Li^+ + e^-$	⟶	Li	−2.34
$Na^+ + e^-$	⟶	Na	−1.89
$K^+ + e^-$	⟶	K	−2.04
$Rb^+ + e^-$	⟶	Rb	−2.06
$Cs^+ + e^-$	⟶	Cs	−2.08
$Mg^{+2} + 2e^-$	⟶	Mg	−1.74
$Ca^{+2} + 2e^-$	⟶	Ca	−2.17
$Sr^{+2} + 2e^-$	⟶	Sr	−2.3
$Ba^{+2} + 2e^-$	⟶	Ba	−2.2
$Mn^{+2} + 2e^-$	⟶	Mn	−0.56
$Fe^{+2} + 2e^-$	⟶	Fe	~0.0

Table F.2 *(Continued)*

Half-reaction		Acid solutions	$\mathscr{E}^0$
$NH_4^+ + e^-$	$\longrightarrow$	$\frac{1}{2}H_2 + NH_3$	0.00
$Co^{+2} + 2e^-$	$\longrightarrow$	Co	~ 0.0
$Co^{+3} + e^-$	$\longrightarrow$	Co^{+2}	$\sim +0.6$
$Ni^{+2} + 2e^-$	$\longrightarrow$	Ni	~ -0.1
$Cu^+ + e^-$	$\longrightarrow$	Cu	$+0.36$
$Cu^{+2} + e^-$	$\longrightarrow$	Cu^+	$+0.44$
$Ag^+ + e^-$	$\longrightarrow$	Ag	$+0.76$
$Zn^{+2} + 2e^-$	$\longrightarrow$	Zn	-0.54
$Cd^{+2} + 2e^-$	$\longrightarrow$	Cd	~ -0.2
$Hg_2^{+2} + 2e^-$	$\longrightarrow$	$2Hg$	$\sim +1.5$
$2Hg^{+2} + 2e^-$	$\longrightarrow$	Hg_2^{+2}	~ -0.2
$Hg^{+2} + 2e^-$	$\longrightarrow$	Hg	$+0.67$
$Hg_2I_2 + 2e^-$	$\longrightarrow$	$2Hg + 2I^-$	$+0.68$
$2Hg^{+2} + 2I^- + 2e^-$	$\longrightarrow$	Hg_2I_2	$+0.66$
$Tl^+ + e^-$	$\longrightarrow$	Tl	$+0.25$
$Pb^{+2} + 2e^-$	$\longrightarrow$	Pb	$+0.28$
$N_2H_4 + 2H^+ + 2e^-$	$\longrightarrow$	$2NH_3$	$\sim +0.9$
$2N_3^- + 12H^+ + 10e^-$	$\longrightarrow$	$3N_2H_4$	$\sim +0.1$
$3N_2 + 2e^-$	$\longrightarrow$	$2N_3^-$	~ -2.7
$N_2 + 6H^+ + 6e^-$	$\longrightarrow$	$2NH_3$	$+0.04$
$N_3^- + 9H^+ + 8e^-$	$\longrightarrow$	$3NH_3$	$\sim +0.4$
$N_2 + 4H^+ + 4e^-$	$\longrightarrow$	N_2H_4	~ -0.4
$N_2O + H^+ + 2e^-$	$\longrightarrow$	$N_2 + OH^-$	$+1.82$
$3N_2O + 6H^+ + 8e^-$	$\longrightarrow$	$2N_3^- + 3H_2O$	$\sim +0.7$
$2NO + H^+ + 2e^-$	$\longrightarrow$	$N_2O + OH^-$	$+1.64$
$NO_2^- + H^+ + e^-$	$\longrightarrow$	$NO + OH^-$	$\sim +0.6$
$NO_2^- + 5H^+ + 6e^-$	$\longrightarrow$	$NH_3 + 2OH^-$	$\sim +0.7$
$2NO_2^{-2} + 2H^+ + 2e^-$	$\longrightarrow$	$N_2O_2^{-2} + 2OH^-$	$\sim +1.1$
$2NO_2^{-2} + 4H^+ + 4e^-$	$\longrightarrow$	$N_2 + 4OH^-$	$\sim +1.8$
$NO_2^- + e^-$	$\longrightarrow$	NO_2^{-2}	$\sim +0.5$
$N_2O_4 + 2e^-$	$\longrightarrow$	$2NO_2^-$	$\sim +1.6$
$2NO_3^- + 2H^+ + 2e^-$	$\longrightarrow$	$N_2O_4 + 2OH^-$	$+0.2$
$NO_3^- + H^+ + 2e^-$	$\longrightarrow$	$NO_2^- + OH^-$	$\sim +0.9$
$NO_3^- + H^+ + 3e^-$	$\longrightarrow$	$NO_2^{-2} + OH^-$	$\sim +0.8$
$2NO_3^- + 6H^+ + 10e^-$	$\longrightarrow$	$N_2 + 6OH^-$	$+1.17$
$F_2 + 2e^-$	$\longrightarrow$	$2F^-$	$+3.50$
$Cl_2 + 2e^-$	$\longrightarrow$	$2Cl^-$	$+1.91$
$2ClNH_2 + 2H^+ + 2e^-$	$\longrightarrow$	$Cl_2 + 2NH_3$	$\sim +0.9$
$ClNH_2 + H^+ + 2e^-$	$\longrightarrow$	$Cl^- + NH_3$	$\sim +1.4$
$ClO_3^- + NH_3 + 2H^+ + 4e^-$	$\longrightarrow$	$NH_2Cl + 3OH^-$	$\sim +1.5$
$\frac{1}{2}O_2 + 2H^+ + 2e^-$	$\longrightarrow$	H_2O	$+1.28$
$ClO_3^- + 3H^+ + 6e^-$	$\longrightarrow$	$Cl^- + 3OH^-$	$+1.47$
$Br_2 + 2e^-$	$\longrightarrow$	$2Br^-$	$+1.73$
$I_2 + 2e^-$	$\longrightarrow$	$2I^-$	$+1.26$
e^-	$\longrightarrow$	e^- (am)	-1.95

Half reaction		Base solutions	$\mathscr{E}^0$
$NH_3 + e^-$	$\longrightarrow$	$\frac{1}{2}H_2 + NH_2^-$	-1.59
$LiNH_2 + e^-$	$\longrightarrow$	$Li + NH_2^-$	-2.70
$NaNH_2 + e^-$	$\longrightarrow$	$Na + NH_2^-$	-2.02
$K^+ + e^-$	$\longrightarrow$	K	-2.04

Table F.2 *(Continued)*

Half reaction		Base solutions	$\mathscr{E}^0$
$NH_3 + e^-$	$\longrightarrow$	$\frac{1}{2}H_2 + NH_2^-$	-1.59
$Rb^+ + e^-$	$\longrightarrow$	Rb	-2.06
$Cs^+ + e^-$	$\longrightarrow$	Cs	-2.08
$Ca(NH_2)_2 + 2e^-$	$\longrightarrow$	$Ca + 2NH_2^-$	-2.83
$Mn(NH_2)_4^{-2} + 2e^-$	$\longrightarrow$	$Mn + 4NH_2^-$	~ -1.7
$Ni(NH_2)_4^{-2} + 2e^-$	$\longrightarrow$	$Ni + 4NH_2^-$	~ -1.3
$Cu(NH_2)_3^{-2} + e^-$	$\longrightarrow$	$Cu + 3NH_2^-$	~ -1.4
$Cu(NH_2)_4^{-2} + e^-$	$\longrightarrow$	$Cu(NH_2)_3^{-2} + NH_2^-$	~ -0.0
$Ag(NH_2)_2^- + e^-$	$\longrightarrow$	$Ag + 2NH_2^-$	~ -1.0
$Zn(NH_2)_4^{-2} + 2e^-$	$\longrightarrow$	$Zn + 4NH_2^-$	~ -1.8
$Cd(NH_2)_4^{-2} + 2e^-$	$\longrightarrow$	$Cd + 4NH_2^-$	~ -1.4
$Hg_3N_2 + 4NH_3 + 6e^-$	$\longrightarrow$	$3Hg + 6NH_2^-$	~ -1.1
$Tl(NH_2)_2^- + e^-$	$\longrightarrow$	$Tl + 2NH_2^-$	~ -1.3
$Pb(NH_2)_3^- + 2e^-$	$\longrightarrow$	$Pb + 3NH_2^-$	~ -1.4
$N_2H_4 + 2e^-$	$\longrightarrow$	$2NH_2^-$	~ -0.7
$2N_3^- + 12NH_3 + 10e^-$	$\longrightarrow$	$3N_2H_4 + 12NH_2^-$	~ -1.8
$N_3^- + 6NH_3 + 8e^-$	$\longrightarrow$	$9NH_2^-$	~ -1.4
$3N_2 + 2e^-$	$\longrightarrow$	$2N_3^-$	~ -2.7
$N_2 + 4NH_3 + 6e^-$	$\longrightarrow$	$6NH_2^-$	-1.55
$N_2 + 4NH_3 + 4e^-$	$\longrightarrow$	$N_2H_4 + 4NH_2^-$	~ -2.0
$N_2O + NH_3 + 2e^-$	$\longrightarrow$	$N_2 + OH^- + NH_2^-$	$+0.95$
$3N_2O + 3NH_3 + 8e^-$	$\longrightarrow$	$2N_3^- + 3OH^- + 3NH_2^-$	~ 0.0
$2NO + NH_3 + 2e^-$	$\longrightarrow$	$N_2O + OH^- + NH_2^-$	$+0.55$
$NO_2^- + NH_3 + e^-$	$\longrightarrow$	$NO + OH^- + NH_2^-$	~ -1.6
$2NO_2^{-2} + 2NH_3 + 2e^-$	$\longrightarrow$	$N_2O_2^{-2} + 2OH^- + 2NH_2^-$	~ -0.8
$2NO_2^{-2} + 4NH_3 + 4e^-$	$\longrightarrow$	$N_2 + 4OH^- + 4NH_2^-$	~ -0.8
$NO_2^- + e^-$	$\longrightarrow$	NO_2^{-2}	$\sim +0.5$
$N_2O_4 + 2e^-$	$\longrightarrow$	$2NO_2^-$	$\sim +1.6$
$2NO_3^- + 2NH_3 + 2e^-$	$\longrightarrow$	$N_2O_4 + 2OH^- + 2NH_2^-$	-2.0
$NO_3^- + NH_3 + 2e^-$	$\longrightarrow$	$NO_2^- + OH^- + NH_2^-$	~ -0.2
$NO_3^- + NH_3 + 3e^-$	$\longrightarrow$	$NO_2^{-2} + OH^- + NH_2^-$	~ 0.0
$2NO_3^- + 6NH_3 + 10e^-$	$\longrightarrow$	$N_2 + 6OH^- + 6NH_2^-$	-0.14
$\frac{1}{2}O_2 + NH_3 + 2e^-$	$\longrightarrow$	$OH^- + NH_2^-$	-0.06
$F_2 + 2e^-$	$\longrightarrow$	$2F^-$	$+3.50$
$Cl_2 + 2e^-$	$\longrightarrow$	$2Cl^-$	$+1.91$
$2NHCl^- + 2NH_3 + 2e^-$	$\longrightarrow$	$Cl_2 + 4NH_2^-$	~ -2.1
$NHCl^- + NH_3 + 2e^-$	$\longrightarrow$	$Cl^- + 2NH_2^-$	~ -0.1
$ClO_3^- + 2NH_3 + 4e^-$	$\longrightarrow$	$NHCl^- + 3OH^- + NH_2^-$	$\sim +0.2$
$ClO_3^- + 3NH_3 + 6e^-$	$\longrightarrow$	$Cl^- + 3OH^- + 3NH_2^-$	$+0.13$
$Br_2 + 2e^-$	$\longrightarrow$	$2Br^-$	$+1.73$
$I_2 + 2e^-$	$\longrightarrow$	$2I^-$	$+1.26$
e^-	$\longrightarrow$	e^- (am)	-1.95

G

The literature of inorganic chemistry

The following is not meant to be an exhaustive list of all of the books and journals of interest to an inorganic chemist, but it is a reasonably complete list of useful titles. To keep it to reasonable length, most titles fifteen or more years old have been deleted—they may be found in the 1st or 2nd editions of this book. The list has been subdivided as follows for convenience:

- Texts and general reference books
- Classical and comprehensive reference works
- Topical monographs
  - I. Acids, bases, and solvent systems
  - II. Coordination chemistry
  - III. Kinetics and mechanisms
  - IV. Organometallic chemistry
  - V. Structure and bonding
  - VI. Bioinorganic chemistry
  - VII. Other topics
- Monographs on single elements or groups
- Works dealing with the laboratory: synthesis and physical methods
- Journals
- Books from other areas of use to inorganic chemists
- Series of annual reviews

The library user should also be familiar with an article entitled "Retrieval and Use of the Literature of Inorganic Chemistry," by H. M. Woodburn, *J. Chem. Educ.*, **1972**, *49*, 689.

TEXTS AND GENERAL REFERENCE BOOKS

Aylett, B. J., and B. C. Smith, "Problems in Inorganic Chemistry," American Elsevier, New York, 1966.

Bell, C. F., and A. K. Lott, "Modern Approach to Inorganic Chemistry," 3rd ed., Butterworths, London, 1972.
Cotton, F. A., and G. Wilkinson, "Advanced Inorganic Chemistry," 4th ed., Wiley, New York, 1980.
Cotton, F. A., and G. Wilkinson, "Basic Inorganic Chemistry," Wiley, New York, 1976.
Day, M. C., and J. Selbin, "Theoretical Inorganic Chemistry," 2nd ed., Van Nostrand-Reinhold, New York, 1969.
Demitras, G. C., C. R. Russ, J. F. Salmon, J. H. Weber, and G. R. Weiss, "Inorganic Chemistry," Prentice-Hall, Englewood Cliffs, N.J., 1972.
Douglas, B. E., D. H. McDaniel, and J. Alexander, "Concepts and Models of Inorganic Chemistry," 2nd ed., Wiley, New York, 1983.
Duffy, J. A., "General Inorganic Chemistry," 2nd ed., Longman, London, 1973.
Durrant, P. J., and B. Durrant, "Introduction to Advanced Inorganic Chemistry," 2nd ed., Wiley (Interscience), New York, 1970.
Emeléus, H. J., and A. G. Sharpe, "Modern Aspects of Inorganic Chemistry," 4th ed., Van Nostrand-Reinhold, New York, 1973.
Heslop, R. B., "Numerical Aspects of Inorganic Chemistry," American Elsevier, New York, 1970.
Heslop, R. B., and K. Jones, "Inorganic Chemistry: A Guide to Advanced Study," American Elsevier, New York, 1976.
Holderness, A., and M. Berry, "Advanced Level Inorganic Chemistry," 3rd ed., Heinneman Educ. Books, Exeter, N. H., 1980.
Jolly, W. L., "The Principles of Inorganic Chemistry," McGraw-Hill, New York, 1976.
Lagowski, J. J., "Modern Inorganic Chemistry," Dekker, New York, 1973.
Latimer, W. M., and J. H. Hildebrand, "Reference Book of Inorganic Chemistry," 3rd ed., Macmillan, New York, 1951.
Lingren, W. E., "Inorganic Nomenclature: A Programmed Approach," Prentice-Hall, Englewood Cliffs, N.J., 1980.
Moeller, T., "Inorganic Chemistry: A Modern Introduction," Wiley, New York, 1982.
Phillips, C. S. G., and R. J. P. Williams, "Inorganic Chemistry," Vol. I, *Principles and Non-Metals*, 1965; Vol. II, *Metals*, 1966; Oxford University Press, London.
Purcell, K. F., and J. C. Kotz, "An Introduction to Inorganic Chemistry," Saunders, Philadelphia, 1980.
Purcell, K. F., and J. C. Kotz, "Inorganic Chemistry," Saunders, Philadelphia, 1977.
Sanderson, R. T., "Inorganic Chemistry," Van Nostrand-Reinhold, New York, 1967.
Williams, A. F., "A Theoretical Approach to Inorganic Chemistry," Springer, New York, 1979.

CLASSICAL AND COMPREHENSIVE REFERENCE WORKS

Abegg, R., "Handbuch der anorganischen Chemie," Hirzel, Leipzig, 1905–1936.
Bailar, J. C., Jr., et al., eds., "Comprehensive Inorganic Chemistry," Pergamon Press, Oxford, 1973.

Emeléus, H. J., ed., "MTP International Review of Science: Inorganic Chemistry, Series I," Butterworths, London, 1972.

Gmelin, L., "Handbuch der anorganischen Chemie," Verlag Chemie, Weinheim, 1924–1971.

Mellor, J. W., "A Comprehensive Treatise of Inorganic and Theoretical Chemistry," Longmans, Green, London, 1922–1937; Supplements, 1958–1967.

Pascal, P., ed., "Nouveau Traite de Chimie Minerale," Masson, Paris, 1956.

Sneed, M. C., J. L. Maynard, and R. C. Brasted, "Comprehensive Inorganic Chemistry," Van Nostrand-Reinhold, New York, 1953–1961.

Wilkinson, G., ed., "Comprehensive Organometallic Chemistry," Pergamon, Oxford, 1982.

TOPICAL MONOGRAPHS

I. Acids, bases, and solvent systems

Addison, C. C., "Chemistry in Nonaqueous Ionizing Solvents," Vol. III, Part 1, *Chemistry in Dinitrogen Tetroxide*, Pergamon, Elmsford, N.Y., 1967.

Albert, A., and E. P. Serjeant, "The Determination of Ionization Constants: A Laboratory Manual," 2nd ed., Chapman and Hall, London, 1971.

Amis, E. S., and J. F. Hinton, "Solvent Effects on Chemical Phenomena," New York, Academic, 1973.

Audrieth, L. F., and J. Kleinberg, "Non-Aqueous Solvents," Wiley, New York, 1953.

Bell, R. P., "Acids and Bases: Their Quantitative Behavior," 2nd ed., Methuen, London, 1969.

Bloom, H., "The Chemistry of Molten Salts: An Introduction to the Physical and Inorganic Chemistry of Molten Salts and Salt Vapors," Benjamin, New York, 1967.

Braunstein, J., G. Mamantov, and G. P. Smith, eds., "Advances in Molten Salt Chemistry," Vol. 1, Plenum, New York, 1971.

Charlot, G., and B. Tremillon, "Chemical Reactions in Solvents and Melts," Pergamon, Elmsford, N.Y., 1969.

Coetzee, J. F., and C. D. Ritchie, eds., "Solute-Solvent Interactions," Dekker, New York, 1969.

Covington, A. K., and P. Jones, "Hydrogen-Bonded Solvent Systems," Methuen, London, 1968.

Drago, R. S., and N. A. Matwiyoff, "Acids and Bases," Heath, Lexington, Mass., 1968.

Fowles, G. W. A., and D. Nicholls, "Reactions in Liquid Ammonia," American Elsevier, New York, 1971.

Gutmann, V., "Coordination Chemistry in Nonaqueous Solvents," Springer, New York, 1968.

Gutmann, Viktor, "The Donor-Acceptor Approach to Molecular Interactions," Plenum, New York, 1978.

Hart, E. J., and M. Anbar, "The Hydrated Electron," Wiley, New York, 1970.

Hills, G. J., and D. Kerridge, "Fused Salts," American Elsevier, New York, 1971.

Ho, Tse-Lok, "Hard and Soft Acids and Bases Principle in Organic Chemistry," Academic, New York, 1977.

Holliday, A. K., and A. G. Massey, "Inorganic Chemistry in Non-Aqueous Solvents," Pergamon, Elmsford, N.Y., 1965.

Inman, D., and D. G. Lovering, "Ionic Liquids," Plenum, New York, 1981.

Jander, J., and C. Lafrenz, "Ionizing Solvents," Verlag Chemie, Weinheim, 1970.

Janz, G. J., "Molten Salts Handbook," Academic, New York, 1967.

Jensen, W. B., "The Lewis Acid–Base Concepts: An Overview," New York, Wiley, 1980.

Jolly, William L., ed., "Metal-Ammonia Solutions," Dowden, Hutchinson & Ross, Stroudsburg, Pa., 1972.

Jortner, J., and N. R. Hestner, eds., "Electrons in Fluids: The Nature of Metal–Ammonia Solutions," Springer, New York, 1973.

Karcher, W., and H. Hecht, "Chemistry in Nonaqueous Ionizing Solvents," Vol. III, Part 2, *Chemie in flüssigem Schwefeldioxid*, Pergamon, Elmsford, N.Y., 1968.

Kevan, L., and B. C. Webster, eds., "Electron-Solvent and Anion-Solvent Interactions," Elsevier, New York, 1976.

Lagowski, J. J., ed., "The Chemistry of Nonaqueous Solvents," Vols. 1–3, Academic, New York, 1966–70.

Lister, M. W., "Oxyacids," American Elsevier, New York, 1965.

Mamantov, G., ed., "Molten Salts: Characterization and Analysis," Dekker, New York, 1969.

Mamantov, G., ed., "Characterization of Solutes in Nonaqueous Solvents," Plenum, New York, 1978

Mulliken, R. S., and W. B. Person, "Molecular Complexes," Wiley, New York, 1969.

Nicholls, D., "Inorganic Chemistry in Liquid Ammonia," Topics in Inorganic and General Chemistry, Monograph 17, Elsevier, Amsterdam, 1979.

Pearson, R. G., 'Hard and Soft Acids and Bases," Dowden, Hutchinson & Ross, Stroudsburg, Pa., 1973.

Popovych, O., and R. P. T. Tomkins, "Nonaqueous Solution Chemistry," Wiley, New York, 1981.

Sisler, H. H., "Chemistry of Nonaqueous Solvents," Van Nostrand-Reinhold, New York, 1971.

Tanabe, K., "Solid Acids and Bases: Their Catalytic Properties," New York, Academic, 1970.

Thompson, J. C., "Electrons in Liquid Ammonia," Oxford University, London, 1976.

Waddington, T. C., ed., "Non-Aqueous Solvent Systems," Academic, New York, 1965.

Zingaro, R. A., "Nonaqueous Solvents," Heath, Lexington, Mass., 1968.

II. Coordination chemistry

Adamson, A. W., and P. D. Fleischauer, "Concepts of Inorganic Photochemistry," Wiley, New York, 1975.

Advances in Chemistry, No. 62, "Werner Centennial," American Chemical Society, Washington, D.C., 1967.

Ashcroft, S. F., and C. T. Mortimer, "Thermochemistry of Transition Metal Complexes," Academic, New York, 1970.

Ballhausen, C. J., "Molecular Electronic Structures of Transition Metal Complexes," McGraw-Hill, New York, 1979.

Balzani, V., and V. Carassiti, "Photochemistry of Coordination Compounds," Academic, New York, 1970.

Basolo, F., J. L. Bunnett, and J. Halpern, "Transition Metal Chemistry," American Chemical Society, 1973.

Bell, C. F., "Principles and Applications of Metal Chelation," Oxford University, London, 1978.

Cais, M., "Progress in Coordination Chemistry," American Elsevier, New York, 1968.

Candlin, J. P., K. A. Taylor, and D. T. Thompson, "Reactions of Transition-Metal Complexes," American Elsevier, New York, 1969.

Carlin, R. L., and A. J. Van Duyneveldt, "Magnetic Properties of Transition Metal Compounds," Springer, Berlin, 1977.

The Chemical Society, London, "Stability Constants of Metal-Ion Complexes." Sect. I: *Inorganic Ligands*, comp. by L. G. Sillén, Sect. II: *Organic Ligands*, comp. by A. E. Martell. 2nd ed., 1964.

Douglas, B. E. and Y. Saito, eds., "Stereochemistry of Optically Active Transition Metal Compounds," ACS Symposium Series, Vol. 119, American Chemical Society, 1980.

Figgis, B. N., "Introduction to Ligand Fields," Wiley, New York, 1966.

Graddon, D. P., "An Introduction to Coordination Chemistry," 2nd ed., Pergamon, Elmsford, N.Y., 1968.

Hargiattai, M., and I. Hargiattai, "The Molecular Geometries of Coordination Compounds in the Vapour Phase," Elsevier, New York, 1977.

Hawkins, C. J., "Absolute Configuration of Metal Complexes," Wiley, New York, 1971.

Henrici-Olivé, G. and S. Olivé, "Coordination and Catalysis," Monographs in Modern Chemistry, Vol. 9, Verlag Chemie, New York, 1977.

Jones, M. M., "Ligand Reactivity and Catalysis," Academic, New York, 1968.

Jørgensen, C. K., "Absorption Spectra and Chemical Bonding in Complexes," Pergamon, Elmsford, N.Y., 1962.

Jørgensen, C. K., "Inorganic Complexes," Academic, New York, 1964.

Kauffman, G. B., ed., "Classics in Coordination Chemistry," Part I: *The Selected Papers of Alfred Werner*, Dover, New York, 1968.

Kauffman, G. B., "Classics in Coordination Chemistry" Part 2, Selected Papers, Dover, New York, 1976.

Kauffman, G. B., "Classics in Coordination Chemistry, Part 3," Dover, New York, 1978.

Kettle, S. F. A., "Coordination Compounds," Nelson, London, 1969.

Kirschner, S., ed., "Coordination Chemistry," [John C. Bailar, Jr., Symposium on Coordination Chemistry], Plenum, New York, 1969.

Malatesta, L., and F. Bonati, "Isocyanide Complexes of Metals," Wiley, New York, 1969.

Marcus, Y., "Ion Exchange and Solvent Extraction of Metal Complexes," Wiley, New York, 1969.

Martell, A. E., ed., "Coordination Chemistry," ACS Monograph 168, Van Nostrand-Reinhold, New York, 1971.

Martell, A. E., ed., "Coordination Chemistry," Vol. 2, ACS Monograph No. 174, American Chemical Society, Washington, 1978.

McAuliffe, C. A., and W. Levason, "Phosphine, Arsine and Stibine Complexes of the Transition Elements," Elsevier, New York, 1979.

Melson, G. A., ed., "Coordination Chemistry of Macrocyclic Compounds, Plenum, New York, 1979.

Orgel, L., "An Introduction to Transition Metal Chemistry," 2nd ed., Wiley, New York, 1966.

Quagliano, J. V., and L. M. Vallarino, "Coordination Chemistry," Heath, Lexington, Mass., 1969.

Saito, Y., "Inorganic Molecular Dissymmetry," Springer, New York, 1979.

Schläfer, H. L., and G. Gliemann, "Basic Principles of Ligand Field Theory," Wiley, New York, 1969.

Sharpe, A. G., "The Chemistry of Cyano Complexes of the Transition Metals," Academic, New York, 1976.

Taube, H., "Complex Ions," Academic, New York, 1969.

Wendlandt, W. W., and J. P. Smith, "The Thermal Properties of Transition Metal-Ammine Complexes," American Elsevier, New York, 1967.

Yatsimirskii, K. B., and V. P. Vasil'ev, "Instability Constants of Complex Compounds," Plenum, New York, 1960.

III. Kinetics and mechanisms

Basolo, F., and R. G. Pearson, "Mechanisms of Inorganic Reactions," 2nd ed., Wiley, New York, 1967.

Benson, D. "Mechanisms of Inorganic Reactions in Solution: An Introduction," McGraw-Hill, New York, 1968.

Edwards, J. O., ed., "Inorganic Reaction Mechanisms," Wiley, New York, 1970.

Langford, C. H., and H. B. Gray, "Ligand Substitution Processes," Benjamin, New York, 1966.

Lockart, J. C., "Introduction to Inorganic Reaction Mechanisms," Van Nostrand-Reinhold, New York, 1966.

Sykes, A. G., "Kinetics of Inorganic Reactions," Pergamon, Elmsford, N.Y., 1966.

Taube, H., "Electron Transfer Reactions of Complex Ions in Solution," Academic, New York, 1970.

Tobe, Martin L., "Inorganic Reaction Mechanisms," Thames and Nelson, London, 1972.

Wilkins, R. G., "The Study of Kinetics and Mechanism of Reactions of Transition Metal Complexes," Allyn and Bacon, Boston, 1974.

IV. Organometallic chemistry

Alper, H., ed., "Transition Metal Organometallics in Organic Synthesis," Academic, New York, 1976.

Aylett, B. J., "Organometallic Compounds," 4th ed., Vol. 1, Part 2, Wiley, New York, 1979.

Becker, E. I., and M. Tsutsui, "Organometallic Reactions," Vols. 1, 2, Wiley, New York, 1970.

Braterman, P. S., "Metal Carbonyl Spectra," Academic, New York, 1975.

Brester, J. H., ed., "Aspects of Mechanism and Organometallic Chemistry," Plenum, New York, 1978.

Chien, James C. W., ed., "Coordination Polymerization, A Memorial to Karl Ziegler," Academic, New York, 1975.

Coates, G. E., and K. Wade, "Organometallic Compounds," Vol. I, *The Main Group Elements*, 3rd ed., Methuen, London, 1967.

Coates, G. E., M. L. H. Green, and K. Wade, "Organometallic Compounds," Vol. II, *The Transition Elements*, 3rd ed., Methuen, London, 1968.

Coates, G. E., M. L. H. Green, P. Powell, and K. Wade, "Principles of Organometallic Chemistry," Methuen, London, 1968.

Collman, J. P., and L. S. Hegedus, "Principles and Applications of Organotransition Metal Chemistry," University Science Books, Mill Valley, Calif., 1980.

Daganello, G., "Transition Metal Complexes of Cyclic Polyolefins," Academic, New York, 1979.

Doak, G. O., and L. D. Freedman, "Organometallic Compounds of Arsenic, Antimony and Bismuth," Wiley, New York, 1970.

Dub, M., "Organometallic Compounds," Springer, New York, 1972.

Eisch, J. J., "The Chemistry of Organometallic Compounds," Macmillan, New York, 1967.

Fischer, E. O., and H. Werner, "Metal π-Complexes with Di- and Oligo-olefinic Ligands," American Elsevier, New York, 1966.

Geoffroy, G. L., and M. S. Wrighton, "Organometallic Photochemistry," Academic, New York, 1979.

George, W. O., ed., "Spectroscopic Methods in Organometallic Chemistry," C.R.C. Press, Cleveland, 1970.

Hagihara, N., M. Kumada, and R. Okawara, eds., "Handbook of Organometallic Compounds," Benjamin, New York, 1968.

Heck, Richard F., "Organotransition Metal Chemistry," Academic, New York, 1974.

Herberhold, Max, "Metal π-Complexes," American Elsevier, New York, 1972.

Houghton, R. P., "Metal Complexes in Organic Chemistry," Cambridge University, Cambridge, 1979.

Ishii, Yoshio, and Minoru Tsutsui, eds., "Organotransition-Metal Chemistry," Plenum, New York, 1975.

Khan, M. M. T., and A. E. Martell, "Homogeneous Catalysis by Metal Complexes," Vols. 1 and 2, Academic, New York, 1974.

King, R. B., "Transition-Metal Organometallic Chemistry," Academic, New York, 1969.

Kochi, J. K., "Organometallic Mechanisms and Catalysis," Academic, New York, 1978.

MacDiarmid, A. G., ed., "The Bond to Carbon," Vol. I, *Organometallic Compounds of the Group IV Elements*, Dekker, New York, 1968.

Malatesta, L., and S. Genini, "Zerovalent Compounds of Metals," Academic, New York, 1974.

Masters, C., "Homogeneous Transition-metal Catalysis," Chapman and Hall, London, 1981.

Matteson, D. S., "Organometallic Reaction Mechanisms of the Nontransition Elements," Academic, New York, 1974.

Parshall, G. W., "Homogeneous Catalysis, The Applications and Chemistry of Catalysis by Soluble Transition Metal Complexes," Wiley, New York, 1980.

Pauson, P. L., "Organometallic Chemistry," St. Martin's, New York, 1967.

Ramsey, B. "Organometalloids," Academic, New York, 1969.

Rochow, E. G., "The Metalloids," Heath, Lexington, Mass., 1966.

Seyferth, D., ed., "Organometallic Chemistry Reviews," Elsevier, New York, 1980.

Seyferth, D., and R. B. King, "Annual Survey of Organometallic Chemistry," Vol. 1 (1965)–3 (1967), Elsevier, New York.

Tsutsui, M., "Characterization of Organometallic Compounds," Part I (1969) and Part II (1970), Wiley, New York.

Tsutsui, M., et al., "Introduction to Metal π-Complex Chemistry," Plenum, New York, 1970.

Tsutsui, M., and A. Nakamura, "Principles and Applications of Homogeneous Catalysis," Wiley, New York, 1980.

Zeiss, H., P. T., Wheatley and H. J. S. Winkler, "Benzenoid Metal Complexes—Structural Determinations and Chemistry," Ronald, New York, 1966.

V. Structure and bonding

Adams, D. M., "Inorganic Solids," Wiley, New York, 1974.

Anderson, J. R., "Structure of Metallic Catalysts," Academic, New York, 1975.

Atkins, P. W., and M. C. R. Symons, "The Structure of Inorganic Radicals," American Elsevier, New York, 1967.

Ballhausen, C. J., and H. B. Gray, "Molecular Electronic Structures, an Introduction," Benjamin/Cummings, Reading, Mass., 1980.

Burdett, J. K., "Molecular Shapes: Theoretical Models of Inorganic Stereochemistry," Wiley, New York, 1980.

Cartmell, E., and G. W. A. Fowles, "Valency and Molecular Structure," 4th ed., Butterworths, Boston, 1977.

The Chemical Society, London, *Tables of Interatomic Distances and Configurations in Molecules and Ions*, comp. by H. J. M. Bowen et al., 1958. Supplement edited by L. E. Sutton et al., 1965.

Companion, A. L., "Chemical Bonding," 2nd ed., McGraw-Hill, New York, 1979.

DeKock, R., and H. Gray, "Chemical Structure and Bonding," Benjamin/Cummings, Reading, Mass., 1980.

Donohue, J., "The Structures of the Elements," Wiley, New York, 1974.

Fergusson, J. E., "Stereochemistry and Bonding in Inorganic Chemistry," Prentice-Hall, Englewood Cliffs, N.J., 1974.

Galasso, F. S., "Structure and Properties of Inorganic Compounds," Pergamon, Oxford, 1970.

Geoffroy, G. L., ed., "Topics in Inorganic and Organometallic Stereochemistry," "Topics in Stereochemistry," Vol. 12, Wiley, New York, 1981.

Gimarc, B. M., "Molecular Structure and Bonding," Academic, New York, 1979.

Hamilton, W. C., and J. A. Ibers, "Hydrogen Bonding in Solids," Benjamin, New York, 1968.

Hannay, N. B., ed., "Treatise on Solid State Chemistry," Vols. 1–6, Plenum, New York, 1973–1976.

Joesten, M. D., and L. J. Schaad, "Hydrogen Bonding," Dekker, New York, 1974.

Jørgensen, C. K., "Orbitals in Atoms and Molecules," Academic, New York, 1962.

Jørgensen, C. K., "Oxidation Numbers and Oxidation States," Springer, New York, 1969.

Kwart, H., and K. King, "d-Orbitals in the Chemistry of Silicon, Phosphorus and Sulfur," Springer, New York, 1977.

Ladd, M. F. C., "Structure and Bonding in Solid State Chemistry," Halsted, New York, 1979.

Minkin, V. I., O. A. Osipov, and Y. A. Zhdanov, "Dipole Moments in Organic Chemistry," Plenum, New York, 1970.

Moody, G. J., and J. D. R. Thomas, "Dipole Moments in Inorganic Chemistry," Edward Arnold, London, 1971.

Newnham, R. E., "Structure-Property Relations," New York, Springer, 1975.

O'Keefe, M., and A. Navrotsky, eds., "Structure and Bonding in Crystals," Vols. I and II, Academic, New York, 1981.

Orville-Thomas, W. J., "The Structure of Small Molecules," Elsevier, New York, 1966.

Rheingold, A., ed., "Homoatomic Rings, Chains, and Macromolecules of Main-Group Elements," Elsevier, New York, 1977.

Rich, A., and N. Davidson, "Structure Chemistry and Molecular Biology," Freeman, San Francisco, 1968.

Schuster, P., G. Zundel, and C. Sandorfy, eds., "The Hydrogen Bond," Vols. I, II, III, North-Holland, Amsterdam, 1976.

Stranks, D. R., M. L. Heffernam, K. C. Lee Dow, P. T. McTigue, and G. R. A. Withers, "Chemistry, A Structural View," 2nd ed., Cambridge University Press, New York, 1974.

Taylor, M. J., "Metal-to-Metal Bonded States of the Main Group Elements," Academic, New York, 1975.

Waugh, John L. T., "The Constitution of Inorganic Compounds," Wiley, New York, 1972.

Wells, A. F., "The Third Dimension in Chemistry," Clarendon, Oxford, 1956.

Wells, A. F., "Structural Inorganic Chemistry," 4th ed., Clarendon, Oxford, 1975.

VI. Bioinorganic chemistry

Addison, A. W., W. R. Cullen, D. Dolphin, and B. R. James, eds., "Biological Aspects of Inorganic Chemistry, Wiley, New York, 1977.

Babior, Bernard M., "Cobalamin: Biochemistry and Pathophysiology," Wiley, New York, 1975.

Bezkorovainy, A., "Biochemistry of Non-Heme Iron," Plenum, New York, 1980.

Brill, A. S., "Transition Metals in Biochemistry," Springer, New York, 1977.

Caughey, W. S., "Biochemical and Clinical Aspects of Oxygen," Academic, New York, 1979.

Chatt, J., L. M. da Camara Pina, and R. L. Richards, "New Trends in the Chemistry of Nitrogen Fixation," Academic, New York, 1980.

Coughlan, M. P., ed., "Molybdenum and Molybdenum-Containing Enzymes," Pergamon, Elmsford, N.Y., 1980.

Dessy, R., J. Dillard, and L. Taylor, eds., "Bioinorganic Chemistry," Advances in Chemistry, No. 100, American Chemical Society, Washington, D.C., 1971.

Dhar, S. K., ed., "Metal Ions in Biological Systems," Plenum, New York, 1973.

Dolphin, D., C. McKenna, Y. Murakami, and I. Tabushi, eds., "Biomimetic Chemistry," Advances in Chemistry, No. 191, American Chemical Society, Washington, D.C., 1980.

Eichhorn, Gunther L., ed., "Inorganic Biochemistry," Elsevier, Amsterdam, 1973.

Fiabane, A. M., and D. R. Williams, "Principles of Bio-inorganic Chemistry," The Chemical Society, London, 1977.

Hamilton, E. I., "The Chemical Elements and Man," Charles C. Thomas, Springfield, Ill., 1979.

Hanzlik, Robert P., "Inorganic Aspects of Biological and Organic Chemistry," Academic, New York, 1976.

Harrison, P. M., and R. J. Hoare, "Metals in Biochemistry," Chapman and Hall, New York, 1980.

Hill, M. A. O., ed., "Inorganic Biochemistry," Vol. 1: *A Review of the Recent Literature Published to Late 1977*, The Chemical Society, London, 1979.

Hoekstra, W. G., et al., eds., "Trace Element Metabolism in Animals-2," University Park, Baltimore, 1974.

Hughes, M. N., "The Inorganic Chemistry of Biological Processes," Wiley, New York, 1973.

Irving, J. T., "Calcium and Phosphorus Metabolism," Academic, New York, 1973.

Lovenberg, W., ed., "Iron-Sulfur Proteins," Vols. 1 and 2, Academic, New York, 1973.

Luckey, T. D., and B. Venugopal, "Metal Toxicity in Mammals," Vols. 1 (1977), 2 (1978), Plenum, New York.

Martell, A. E., ed., "Inorganic Chemistry in Biology and Medicine," ACS Symposium Series, No. 140, American Chemical Society, Washington, D.C., 1980.

Mathesa, J., and E. T. Degens, "Structural Molecular Biology of Phosphates," Gustav Fischer, Stuttgart, 1971.

McAuliffe, C. A., "Techniques and Topics in Bioinorganic Chemistry," Halsted, New York, 1975.

Mendels, Joseph, and Secunda, St. K., "Lithium in Medicine," Gordon and Breach, New York, 1972.

Newton, W. E., and S. Otsuka, eds., "Molybdenum Chemistry of Biological Significance," Plenum, New York, 1980.

Nicholas, D. J. D., and Adrian R. Egan, "Trace Elements in Soil-Plant-Animal Systems," Academic, New York, 1975.

Ochiai, E. I., "Bioinorganic Chemistry: An Introduction," Allyn and Bacon, Boston, 1977.

Phipps, D. A., "Metals and Metabolism," Oxford University, London, 1978.

Pratt, J. M., "Inorganic Chemistry of Vitamin B_{12}," Academic, New York, 1972.

Raymond, K. N., "Bioinorganic Chemistry—II," Advances in Chemistry No. 162, American Chemical Society, Washington, D.C., 1977.

Roy, A. B., and P. A. Trudinger, "The Biochemistry of Inorganic Compounds of Sulphur," Cambridge University, Cambridge, 1970.

Sanadi, D. Rao, and Lester Packer, eds., "Current Topics in Bioenergetics," Academic, New York, 1973.

Schroeder, A., "Elements in Living Systems," Plenum, New York, 1975.

Smith, K. M., ed., "Porphyrins and Metalloporphyrins," Elsevier, Amsterdam, 1975.

Spiro, T. G., ed., "Metal Ion Activation of Dioxygen," Wiley, New York, 1980.

Spiro, T., ed., "Copper Proteins," Wiley, New York, 1981.

Wacker, W. E. C., "Magnesium and Man," Harvard, Cambridge, Mass., 1980.

Williams, R. J. P., and J. R. R. F. da Silva, "New Trends in Bio-Inorganic Chemistry," Academic, New York, 1978.

VII. Other topics

Ball, M. C., and A. H. Norbury, "Physical Data for Inorganic Chemists," Longman, London, 1974.

Barin, I., and O. Knacke, "Thermochemical Properties of Inorganic Substances," Springer, New York, 1973.

Bhatnagar, V. M., "Clathrate Compounds," Chemical Publishing Co., New York, 1970.

Brand, J. C. D., and J. C. Speakman, "Molecular Structure: The Physical Approach," 2nd ed., Wiley, New York, 1975.

Cotton, F. A., "Chemical Applications of Group Theory," 2nd ed., Wiley, New York, 1971.

Dasent, W. E., "Nonexistent Compounds: Compounds of Low Stability," Dekker, New York, 1965.

Ebsworth, E. A. V., A. G. Maddock, and A. G. Sharpe, eds., "New Pathways in Inorganic Chemistry," Cambridge University, Cambridge, 1968.

Eichhorn, G. L., and L. G. Marzilli, eds., "Metal Ions in Genetic Information Transfer," Advances in Inorganic Biochemistry, Vol. 3, Elsevier North-Holland, New York, 1981.

Hagan, M. M., "Clathrate Inclusion Compounds," Van Nostrand-Reinhold, New York, 1962.

Haiduc, I., "The Chemistry of Inorganic Ring Systems," Parts 1 and 2, Wiley, New York, 1970.

Hedvall, J. A., "Solid State Chemistry," American Elsevier, New York, 1966.

WORKS DEALING WITH THE LABORATORY: SYNTHESIS AND PHYSICAL METHODS

Abragam, A., and B. Bleaney, "Electron Paramagnetic Resonance of Transition Ions," Oxford University, New York, 1970.

Adams, D. M., and J. B. Raynor, "Advanced Practical Inorganic Chemistry," Wiley, New York, 1965.

Adamson, Arthur W., and Paul D. Fleischauer, eds., "Concepts of Inorganic Photochemistry," Wiley, New York, 1975.

Advances in Chemistry, No. 68, "The Mössbauer Effect and Its Application to Chemistry," American Chemical Society, Washington, D.C., 1967.

Angelici, R. J., "Synthesis and Technique in Inorganic Chemistry," Saunders, Philadelphia, 1969.

Brauer, G., "Handbook of Preparative Inorganic Chemistry," Academic, New York, 1963.

The Chemical Society, London, *Inorganic Reaction Mechanisms*, see "Series of Annual Reviews."

Crow, D. R., "Polarography of Metal Complexes," Academic, New York, 1969.

Drago, R. S., "Physical Methods of Chemistry," Saunders, Philadelphia, 1977.

Goldanskii, V. I., and R. H. Herber, "Chemical Applications of Mössbauer Spectroscopy," Academic, New York, 1968.

Headridge, J., "Electrochemical Techniques for Inorganic Chemists," Academic, New York, 1969.

Herber, R., "Inorganic Isotope Syntheses," Benjamin, New York, 1962.

Hill, H. O., and P. Day, "Physical Methods in Advanced Inorganic Chemistry," Wiley, New York, 1968.

Jolly, W. L., "Synthetic Inorganic Chemistry," Prentice-Hall, Englewood Cliffs, N.J., 1960.

Jolly, W. L., "The Synthesis and Characterization of Inorganic Compounds," Prentice-Hall, Englewood Cliffs, N.J., 1970.

Jolly, W. L., *Preparative Inorganic Chemistry*, see "Series of Annual Reviews."

Jonassen, H. B., and A. Weissberger, *Techniques of Inorganic Chemistry*, see "Series of Annual Reviews."

King, R. B., "Organometallic Syntheses," Academic, New York, 1969.

Lever, A. B. P., "Inorganic Electronic Spectroscopy," American Elsevier, New York, 1968.

Litzow, M. R., and T. R. Spalding, "Mass Spectrometry of Inorganic and Organometallic Compounds," American Elsevier, New York, 1973.

Margrave, J. L., Chairman, "Mass Spectrometry in Inorganic Chemistry," Advances in Chemistry, No. 72, American Chemical Society, Washington, D.C., 1968.

Nakamoto, K., "Infrared Spectra of Inorganic and Coordination Compounds," Wiley, New York, 1963.

Palmer, W. G., "Experimental Inorganic Chemistry," Cambridge University, Cambridge, 1954.

Pass, G., and H. Sutcliffe, "Practical Inorganic Chemistry: Preparations, Reactions, and Instrumental Methods," Chapman and Hall, London, 1968.

Rao, C. N. R., and J. R. Ferraro, eds., "Spectroscopy in Inorganic Chemistry," Vols. I (1970) and II (1971), Academic, New York.

Sanderson, R. T., "Vacuum Manipulation of Volatile Compounds," Wiley, New York, 1948.

Schlessinger, G. G., "Inorganic Laboratory Preparations," Chemical Publishing Co., New York, 1962.

Shriver, D. F., "The Manipulation of Air-Sensitive Compounds," McGraw-Hill, New York, 1969.

Walton, H. F., "Inorganic Preparations," Prentice-Hall, Englewood Cliffs, N.J., 1948.

R. B. King, ed., "Inorganic Compounds with Unusual Properties," Advances in Chemistry, No. 150, American Chemical Society, Washington, D.C., 1976.

R. B. King, ed., "Inorganic Compounds with Unusual Properties—II," Advances in Chemistry, No. 173, American Chemical Society, Washington, D.C., 1979.

Kleinberg, J., "Unfamiliar Oxidation States and Their Stabilization," University of Kansas, Lawrence, Kansas, 1950.

Krebs, H., "Fundamentals of Inorganic Crystal Chemistry," McGraw-Hill, New York, 1970.

Latimer, W. M., "The Oxidation States of the Elements and Their Potentials in Aqueous Solutions," 2nd ed., Prentice-Hall, Englewood Cliffs, N.J., 1952.

O'Dwyer, M. F., J. E. Kent, and R. D. Brown, "Valency," Springer, New York, 1978.

Ray, N. H., "Inorganic Polymers," Academic, New York, 1978.

Rich, R. L., "Periodic Correlations," Benjamin, New York, 1965.

Sanderson, R. T., "Chemical Bonds and Bond Energy," 2nd ed., Academic, New York, 1976.

Weeks, M. E., "The Discovery of the Elements," 7th ed., Chemical Education Publishing Co., Easton, Pa., 1968.

MONOGRAPHS ON SINGLE ELEMENTS OR GROUPS

Group I: Hydrogen

Libowitz, G. C., "The Solid-State Chemistry of Binary Hydrides," Benjamin, New York, 1965.

Mackay, K. M., "Hydrogen Compounds of the Metallic Elements," Barnes and Noble, New York, 1966.

Mueller, W., J. Blackledge, and G. C. Libowitz, "Metal Hydrides," Academic, New York, 1968.

Shaw, B. L., "Inorganic Hydrides," Pergamon, Elmsford, N.Y., 1967.

Wiberg, E., and E. Amberger, "Hydrides," American Elsevier, New York, 1971.

Metals

Parish, R. V., "The Metallic Elements," Longman, London, 1977.

Group IA: Li, Na, K, Rb, Cs

The Chemical Society, "The Alkali Metals," London, 1967.

Wakefield, B. J., "The Chemistry of Organolithium Compounds," Pergamon, Elmsford, N.Y., 1974.

Group IIA: Be, Mg, Ca, Sr, Ba

Everest, D. A., "The Chemistry of Beryllium," American Elsevier, New York, 1964.
Floyd, D. R., and J. N. Lowe, eds., "Beryllium Science and Technology," Vol. 2, Plenum, New York, 1979.
Webster, D., and G. J. London, "Beryllium Science and Technology," Vol. 1, Plenum, New York, 1979.

Group IB: Cu, Ag, Au

Puddephatt, R. M., "The Chemistry of Gold," Topics in Inorganic and General Chemistry, 16, Elsevier/North-Holland, New York, 1978.

Group IIB: Zn, Cd, Hg

Farnsworth, M., "Cadmium Chemicals," International Lead Zinc Organization, New York, 1980.
McAuliffe, C. A., "The Chemistry of Mercury," Macmillan, Toronto, 1977.

Group IIIB: Sc, Y, La-Lu, Ac-Lw

Bagnall, K. W., "The Actinide Elements," American Elsevier, New York, 1972.
Brown D., "Halides of the Lanthanides and Actinides," Wiley, New York, 1969.
Chernyaev, I. I., ed., "Complex Compounds of Uranium," Daniel Davey, New York, 1966.
Cleveland, J. M., "The Chemistry of Plutonium," Gordon and Breach, New York, 1970.
Cordfunke, E. H. P., "The Chemistry of Uranium," American Elsevier, New York, 1969.
Edelstein, N. M., ed., "Lanthanide and Actinide Chemistry and Spectroscopy," ACS Symposium Series, No. 131, American Chemical Society, Washington, D.C., 1980.
Eyring, L. R., *Rare Earths*, see "Series of Annual Reviews."
Fields, P. R., and T. Moeller, eds., "Lanthanide/Actinide Chemistry," Advances in Chemistry, No. 71, American Chemical Society, Washington, D.C., 1967.
Freeman, A. J., and J. B. Darby, Jr., eds., "The Actinides: Electronic Structure and Related Properties," Vols. 1, 2, Academic, New York, 1974.
Gol'danskii, V. I., and S. M. Polidanov, "The Transuranium Elements," Plenum, New York, 1973.
Horovitz, C. T., et al., eds., "Scandium: Its Occurrence, Chemistry, Physics, Metallurgy, Biology and Technology," Academic, New York, 1975.
Makarov, E. S., "Crystal Chemistry of Simple Compounds (of U, Th, Pu, Np)," Consultants Bureau, New York, 1959.
Samsonov, G. V., "High-Temperature Compounds of Rare Earth Metals With Nonmetals," Consultants Bureau, New York, 1965.
Schulz, W. W., "The Chemistry of Americium," Technical Information Center, Energy Research and Development Administration, 1976.

Sinha, S. P., "Complexes of the Rare Earths," Pergamon, Elmsford, N.Y., 1966.
Sinha, S. P., "Europium," Springer, New York, 1967.
Spitsyn, V. I., and J. J. Katz, eds., "The Chemistry of Transuranium Elements," Pergamon, Elmsford, N.Y., 1976.
Topp, N. E., "The Chemistry of the Rare-Earth Elements," American Elsevier, New York, 1965.

Transition metals, general (see also "Coordination Chemistry")

Colton, R., and J. Canterford, "Halides of the Second and Third Row Transition Metals," American Elsevier, New York, 1968.
Cotton, F. A., and L. A. Hart, "The Heavy Transition Elements," Wiley, New York, 1975.
Kepert, D. L., "The Early Transition Metals," Academic, New York, 1972.
Larsen, E. M., "Transitional Elements," Benjamin, New York, 1965.

Group IVB: Ti, Zr, Hf

Barksdale, J., "Titanium—Its Occurrence, Chemistry, and Technology," Ronald, New York, 1966.
Blumenthal, W. B., "The Chemical Behavior of Zirconium," Van Nostrand-Reinhold, New York, 1958.
Clark, R. J. H., "The Chemistry of Titanium and Vanadium," American Elsevier, New York, 1968.
Skinner, G., H. L. Johnston, and C. Beckett, "Titanium and Its Compounds," Herrick L. Johnston, Columbus, Ohio, 1964.
Wailes, P. C., R. S. P. Coutts, and H. Weigold, "Organometallic Chemistry of Titanium, Zirconium, and Hafnium," Academic, New York, 1974.

Group VB: V, Nb, Ta

Clark, R. J. H., "The Chemistry of Titanium and Vanadium," American Elsevier, New York, 1968.
Fairbrother, F., "The Chemistry of Niobium and Tantalum," American Elsevier, New York, 1967.

Group VIB: Cr, Mo, W

Quarrell, A. G., "Niobium, Tantalum, Molybdenum, and Tungsten," American Elsevier, New York, 1961.
Rieck, G. D., "Tungsten and Its Compounds," Pergamon, Elmsford, N.Y., 1967.
Sneeden, R. P. A., "Organochromium Compounds," Academic, New York, 1975.
Udy, M. J., "Chromium," Vol. I, *Chemistry of Chromium and Its Compounds*, American Chemical Society Monograph No. 132, Van Nostrand-Reinhold, New York, 1956.

Group VIIB: Mn, Tc, Re

Colton, R., "The Chemistry of Rhenium and Technetium," Wiley, New York, 1965.

Peacock, R. D., "Chemistry of Technetium and Rhenium," American Elsevier, New York, 1966.

Group VIIIB: Fe, Co, Ni, Ru, Rh, Pd, Os, Ir, Pt

Advances in Chemistry, No. 98, "Platinum Group Metals and Compounds," American Chemical Society, Washington, D.C., 1971.

Belluco, V., "Organometallic and Coordination Chemistry of Platinum," Academic, New York, 1974.

Griffiths, W. P., "The Chemistry of the Rare Platinum Metals," Wiley, New York, 1967.

Hartley, F. R., "The Chemistry of Platinum and Palladium," Halsted, New York, 1973.

Jolly, P. W., and G. Wilke, "Organic Chemistry of Nickel," Vols. 1 and 2, Academic, New York, 1974, 1975.

Koerner von Gustorf, E. A., F. W. Grevels, and I. Fischler, eds., "The Organic Chemistry of Iron," Vol. 1, Academic, New York, 1978.

Maitlis, P. M., "The Organic Chemistry of Palladium," Vol. I, *Metal Complexes*, and Vol. II, *Catalytic Reactions*, Academic, New York, 1971.

Group IIIA: B, Al, Ga, In, Tl

Aronsson, B., T. Lundstrom, and S. Rundqvist, "Borides, Silicides, and Phosphides," Wiley, New York, 1965.

Brotherton, R. J., and H. Steinberg, *Boron*, see "Series of Annual Reviews."

Brotherton, R. J., and H. Steinberg, "Progress in Boron Chemistry," Vols. 2 and 3, Pergamon, New York, 1970.

Eaton, G. R., and W. N. Lipscomb, "NMR Studies of Boron Hydrides and Related Compounds," Benjamin, New York, 1969.

Gaule, G., "Boron," Vol. I (1960), Vol. II (1966), Plenum, New York.

Grimes, R. N., "Carboranes," Academic, New York, 1970.

Hughes, R. L., I. C. Smith, and E. W. Lawless, in "Production of the Boranes and Related Research," R. T. Holzman, ed., Academic, New York, 1967.

Lee, A. G., "The Chemistry of Thallium," American Elsevier, New York, 1971.

Muetterties, E. L., "The Chemistry of Boron and Its Compounds," Wiley, New York, 1967.

Muetterties, E. L., "Boron Hydride Chemistry," Academic, New York, 1975.

Muetterties, E. L., and W. H. Knoth, "Polyhedral Boranes," Dekker, New York, 1968.

Niedenzu, K., "Boron Nitrogen Compounds," Academic, New York, 1965.

Onak, T., "Organoborane Chemistry," Academic, New York, 1975.

Sheka, I. A., I. A. Chaus, and T. T. Mityureva, "The Chemistry of Gallium," American Elsevier, New York, 1966.

Steinberg, H., R. J. Brotherton, and W. G. Woods, "Organoboron Chemistry," Vol. I, *Boron–Oxygen and Boron–Sulfur Compounds*, by H. Steinberg (1964), Vol. II, *Boron–Nitrogen and Boron–Phosphorus Compounds*, by H. Steinberg, and R. J. Brotherton (1966), Wiley, New York.

Group IVA: C, Si, Ge, Sn, Pb

Aronsson, B., T. Lundstrom, and S. Rundqvist, "Borides, Silicides, and Phosphides," Wiley, New York, 1965.

Borisov, S. N., M. G. Voronkov, and E. Ya. Lukevits, "Organosilicon Derivatives of Phosphorus and Sulfur," Plenum, New York, 1971.

Borisov, S. N., M. G. Voronkov, and E. Ya. Lukevits, "Organosilicon Heteropolymers and Heterocompounds," Plenum, New York, 1970.

Davydov, V. I., "Germanium," Gordon and Breach, New York, 1967.

Eitel, W., "Silicate Science," Vol. I, *Silicate Structures*, Vol. II, *Properties and Constitution of Silicate Glasses*, Vol. III, *Dry Silicate Systems*, Vol. IV, *Hydrothermal Silicate Systems*, Vol. V, *Ceramics and Hydraulic Binders*, Vol. VI, *Silicate Structures*, Vol. VII, *Constitution and Technology of Glasses*, Academic, New York, 1964–1975.

Glockling, F., "The Chemistry of Germanium," Academic, New York, 1969.

Lesbre, M., P. Mazerolles, and J. Satge, "The Organic Compounds of Germanium," Wiley (Interscience), New York, 1971.

Neuman, W. P., "The Organic Chemistry of Tin," Wiley, New York, 1970.

Poller, R. C., "The Organic Chemistry of Tin," Academic, New York, 1970.

Rijkens, F., and G. J. M. Van der Kerk, "Investigations in the Field of Organogermanium Chemistry," Schotonus & Jens, Utrecht, 1964.

Shapiro, H., and F. W. Frey, "The Organic Compounds of Lead," Wiley, New York, 1968.

Sawyer, A., "Organotin Compounds," Vols. 1 (1971) and 3 (1972), Dekker, New York.

Walker, P. L., *Carbon*, see "Series of Annual Reviews."

Weiss, R. W., ed., "Organometallic Compounds," Vol. II, *Compounds of Germanium, Tin, and Lead*, Springer, New York, 1967.

Zuckermann, J. J., "Organotin Compounds: New Chemistry and Applications," Advances in Chemistry, No. 157, American Chemical Society, Washington, D.C., 1976.

Group VA: N, P, As, Sb, Bi

Aronsson, B., T. Lundstrom, and S. Rundqvist, "Borides, Silicides, and Phosphides," Wiley, New York, 1965.

Colburn, C. B., "Developments in Inorganic Nitrogen Chemistry," American Elsevier, New York, Vol. 1, 1966, Vol. 2, 1973.

Corbridge, "The Structural Chemistry of Phosphorus," Elsevier, New York, 1974.

Corbridge, D. E. C., "Phosphorus, an Outline of its Chemistry, Biochemistry, and Technology," Elsevier, New York, 1978.

Cowley, A. H., ed., "Compounds Containing Phosphorus–Phosphorus Bonds," Dowden, Hutchinson & Ross, Stroudsburg, Pa., 1972.

Doak, G. O., and L. D. Freedman, "Organometallic Compounds of Arsenic, Antimony, and Bismuth," Wiley, New York, 1970.

Dub, M., ed., "Organometallic Chemistry," Vol. III, *Compounds of Arsenic, Antimony, and Bismuth*, Springer, New York, 1968.

Emsley, J., and D. Hall, "The Chemistry of Phosphorus," Wiley, New York, 1976.

Goldwhite, H., "Introduction to Phosphorus Chemistry," Cambridge University, New York, 1981.

Grayson, M., and E. Griffith, *Phosphorus*, see "Series of Annual Reviews."

Griffith, E. J., and M. Grayson, "Phosphorus Chemistry," Wiley, New York, 1977.

Heal, H. G., "The Inorganic Heterocyclic Chemistry of Sulfur, Nitrogen and Phosphorus," Academic, New York, 1980.

Holmes, R. R., "Spectroscopy and Structure of Pentacoordinated Phosphorus Compounds," American Chemical Society, Washington, D.C., April 1979.

Jolly, W. L., "The Inorganic Chemistry of Nitrogen," Benjamin, New York, 1964.

Kosolapoff, G. M., and L. Maier, eds., "Organic Phosphorus Compounds," Vols. 1–7, Wiley, New York, 1972–1975.

Krannich, L. K., "Compounds Containing As–N Bonds," Halsted, New York, 1976.

Niedenzu, K., "Boron Nitrogen Compounds," Academic, New York, 1965.

Thomas, L. C., "The Identification of Functional Groups in Organophosphorus Compounds," Academic, New York, 1974.

Thomas, L. C., "Interpretation of the Infrared Spectra of Organophosphorus Compounds," Hayden, New York, 1974.

Van Wazer, J. R., "Phosphorus," Vol. I, *Chemistry*, Wiley, New York, 1958.

Walker, B. J., "Organophosphorus Chemistry," Penguin, Baltimore, 1972.

Wright, A. N., and C. A. Winkler, "Active Nitrogen," Academic, New York, 1968.

Group VIA: O, S, Se, Te, Po

Ardon, M., "Oxygen: Elementary Forms and Hydrogen Peroxide," Benjamin, New York, 1965.

Bagnall, K. W., "The Chemistry of Selenium, Tellurium, and Polonium," American Elsevier, New York, 1966.

Chizhikov, D. M., and V. P. Shchastlivyi, "Selenium and Selenides," Collet's, London, 1968.

Clive, D. L. J., "Modern Organoselenium Chemistry," Pergamon, Elmsford, N.Y., 1979.

Gattow, G., and W. Pehrendt, "Carbon Sulfides and Their Inorganic and Complex Chemistry," George Thieme, Stuttgart, 1977.

Hogg, D. R., "Organic Compounds of Sulphur, Selenium, and Tellurium," Vol. 5, The Chemical Society, London, 1979.

Irgolic, K. J., "The Organic Chemistry of Tellurium," Gordon and Breach, New York, 1975.

Janssen, M. J., ed., "Organosulfur Chemistry," Wiley, New York, 1968.

Klayman, D. L., and W. H. H. Gunter, eds., "Organic Selenium Compounds: Their Chemistry and Biology," Wiley, New York, 1973.

Nickless, G., "Inorganic Sulphur Chemistry," American Elsevier, New York, 1969.

Oae, S., ed., "Organic Chemistry of Sulfur," Plenum, New York, 1977.

Senning, A., ed., "Sulfur in Organic and Inorganic Chemistry," Vol. 1, Dekker, New York, 1971.

Stirling, C. J. M., ed., "Organic Sulphur Chemistry," Butterworths, Boston, 1975.

Tobolsky, A. V., ed., "The Chemistry of the Sulfides," Wiley, New York, 1968.

Vol'nov, I., "Peroxides, Superoxides, and Ozonides of Alkali and Alkaline Earth Metals," Plenum, New York, 1966.

Zingaro, R. A., and W. C. Cooper, eds., "Selenium," Van Nostrand-Reinhold, New York, 1974.

Group VIIA: F, Cl, Br, I, As

Emeléus, H. J., "The Chemistry of Fluorine and Its Compounds," Academic, New York, 1969.

Gutmann, V., "Halogen Chemistry," Vols. I–III, Academic, New York, 1967.

Jolles, Z. E., ed., "Bromine and Its Compounds," Academic, New York, 1966.

Masschelein, W. J., "Chlorine Dioxide (Chemistry and Environmental Impact of Oxychlorine Compounds)," Ann Arbor Science, Ann Arbor, Mich., 1979.

Neumark, H. R., "Fluorine and Fuorine Compounds," Wiley, New York, 1967.

Schilt, A. A., "Perchloric Acid and Perchlorates," G. Frederick Smith, Columbus, Ohio, 1979.

Solymosi, F., "Structure and Stability of Salts of Halogen Oxyacids in the Solid Phase," Wiley, New York, 1977.

Stacey, M., et al., *Fluorine*, see "Series of Annual Reviews."

Group VIIIA: He, Ne, Ar, Kr, Xe, Rn

Bartlett, N., "The Chemistry of the Noble Gases," American Elsevier, New York, 1971.

Claassen, H. H., "The Noble Gases," Heath, Lexington, Mass., 1966.

Cook, G. A., ed., "Argon, Helium and the Rare Gases," Vol. 1, *History, Occurrence, and Properties*, Wiley, New York, 1961.

Hawkins, D. T., Falconer, W. E., and N. Bartlett, "Noble Gas Compounds," (A Bibliography: 1962–1976), Plenum, New York, 1978.

Holloway, J. H., "Noble-Gas Chemistry," Methuen, London, 1968.

Hyman, H. H., ed., "Noble-Gas Compounds," University of Chicago, Chicago, 1963.

Moody, G. J., and J. D. R. Thomas, "Noble Gases and Their Compounds," Pergamon, Elmsford, N.Y., 1964.

Smith, B. L., "The Inert Gases-Model Systems for Science," Springer, New York, 1971.

Nonmetals, general discussions

Emsley, J., "The Inorganic Chemistry of the Non-Metals," Methuen, London, 1971.

Johnson, R. C., "Introductory Descriptive Chemistry; Selected Nonmetals, Their Properties and Behavior," Benjamin, New York, 1966.

Jolly, W. L., "The Chemistry of the Nonmetals," Prentice-Hall, Englewood Cliffs, N.J., 1964.

Lister, M. W., "Oxyacids," American Elsevier, New York, 1965.

Powell, P., and P. L. Timms, "The Chemistry of Non-Metals," Methuen, New York, 1974.

Sherwin, E., and G. J. Weston, "Chemistry of the Non-metallic Elements," Pergamon, Elmsford, N.Y., 1966.

Steudel, R., "Chemistry of the Non-metals," Walter De Gruyter, New York, 1977.

Rare elements, general discussions

Filyand, M. A., and E. I. Semenova, "Handbook of Rare Elements," Chemical Publishing Co., New York, 1968.
Hampel, C. A., "Rare Metals Handbook," Van Nostrand-Reinhold, New York, 1964.
Prakash, S., "Advanced Chemistry of Rare Elements," Chemical Publishing Co., New York, 1967.

JOURNALS

The following list consists of the principal inorganic chemistry journals, although items of interest to inorganic chemists are often published in journals in other areas such as the *Journal of Chemical Physics*, the *Journal of Organic Chemistry*, and *Acta Crystallographica*. In addition to the review journals listed below, check the series of annual reviews for appropriate review articles.

Primary journals

Angew. Chem. Int. Ed. Engl. (1962)[1]
Bioinorg. Chem. (1971)
Chem. Ber. (1868)
Chem. Commun. (1965)
Inorg. Chem. (1962)
Inorg. Chim. Acta (1967)
Inorg. Macromolecules (1971)
Inorg. Nucl. Chem. Letters (1965–1981)
J. Amer. Chem. Soc. (1879)
J. Chem. Soc., London (1849)
J. Coord. Chem. (1971)
J. Cryst. Mol. Struct. (1971)
J. Fluorine Chem. (1971)
J. Inorg. Nucl. Chem. (1955–1981)
J. Organometal. Chem. (1963)
Organometal. Chem. Synth. (1971)
Organometallics (1982)
Phosphorus (1971)
Polyhedron (1982)
Synth. Inorg. Metalorg. Chem. (1971)
Z. Anorg. Chem. (1892)
Zh. Neorg. Khim. (1956)

Review journals

Actinides Rev. (1967)
Chem. Rev. (1924)
Chem. Soc. Rev. (1972)
Coord. Chem. Rev. (1966)
Inorg. Macromolecules Rev. (1971)
Organometallic Rev. (1966)
Quart. Rev. Chem. Soc. (1947)

[1] The data refers to the first year of publication of the journal.

BOOKS FROM OTHER AREAS OF USE TO INORGANIC CHEMISTS

The following normally would not be considered "inorganic books" but they prove useful as sources of information on bonding, structure, and thermodynamics.

Advances in Chemistry, No. 18, "Thermodynamic Properties of the Elements," American Chemical Society, Washington, D.C., 1956.

Bellamy, L. J., "The Infrared Spectra of Complex Molecules," Chapman and Hall, Vols. 1, 2, 3rd ed., London, 1975.

Burger, K., "Coordination Chemistry: Experimental Methods," Butterworths, London, 1973.

Cotton, F. A., "Chemical Application of Group Theory," Wiley, New York, 1963.

Cottrell, "The Strengths of Chemical Bonds," 2nd ed., Butterworths, London, 1958.

Gray, H. B., "Chemical Bonds," Benjamin, New York, 1973.

Guettich, P., R. Link, and A. Frautwein, "Moessbauer Spectroscopy and Transition Metal Chemistry," Springer, New York, 1978.

Johnson, D. A., "Some Thermodynamic Aspects of Inorganic Chemistry," Cambridge University, 2nd ed., Cambridge, 1982.

McClellan, A. L., "Tables of Experimental Dipole Moments," Freeman, San Francisco, 1963.

McWeeny, R., "Coulson's Valence," Oxford University, Oxford, 1979.

Nelson, R. D., Jr., D. R. Lide, Jr., and A. A. Maryott, "Selected Values of Electric Dipole Moments for Molecules in the Gas Phase," National Bureau of Standards, Washington, D.C., 1967.

Orchin, M., and H. H. Jaffé, "The Importance of Antibonding Orbitals," Houghton Mifflin, Boston, 1957.

Orchin, M., and H. H. Jaffé, "Symmetry, Orbitals, and Spectra," Wiley, New York, 1971.

Pauling, L., "The Nature of the Chemical Bond," 3rd ed., Cornell University, Ithaca, N.Y., 1960.

Pilar, F. L., "Elementary Quantum Chemistry," McGraw-Hill, New York, 1968.

Pimentel, G. C., and R. D. Spratley, "Chemical Bonding Clarified Through Quantum Mechanics," Holden-Day, San Francisco, 1969.

Rossini, F. D., "Selected Values of Chemical Thermodynamic Properties," National Bureau of Standards Circular No. 500, Washington, D.C., 1952.

Tetrahedral, octahedral, and icosahedral models and the use of stereopsis

It is convenient for many purposes to have models available for inspection in order to realize fully the three-dimensional aspect of molecular and lattice structures. "Ball-and-stick" models of various stages of sophistication are useful when it is necessary to be able to see through the structure under consideration. Space-filling models of atoms with both covalent and van der Waals radii are particularly helpful when steric effects are important. The space-filling models and the more sophisticated stick models tend to be rather expensive, but there are several inexpensive modifications of the "ball-and-stick" type available. It is extremely useful to have such a set at hand when considering molecular structures.

Simple tetrahedral or octahedral models are useful in connection with basic structural questions (as, for example, the first time you try to convince yourself that the two isomers of CHFClBr or of $[Co(en)_3]^{+3}$ are *really* nonsuperimposable). If stick models are not available, such simple models can be constructed in a few minutes from paper. In addition, models having bond angles not normally found in ball-and-stick kits—for example, the icosahedral boranes and carboranes—can also be readily constructed from paper. Paper models are especially useful when large numbers of models are necessary as, for example, in constructing models of the iso- and heteropolyanions.

On the following page a generalized outline is given which may be used to construct tetrahedra, octahedra, and icosahedra. This outline (Fig. H.1) may be reproduced as many times as desired by means of Xerox or similar copying machines. Instructions for cutting are given below. It should be noted that the positions of the tabs for gluing may be changed to suit the convenience of the constructor.

Tetrahedral models

Cut out the four triangles enclosed by the T_d brackets (Fig. H.2) and marked with vertical lines in the drawing. Glue or tape tabs on to adjacent faces to form the tetrahedron. (If many tetrahedra are desired and few octahedra, other sections of four triangles may be cut from the octahedral fragments.)

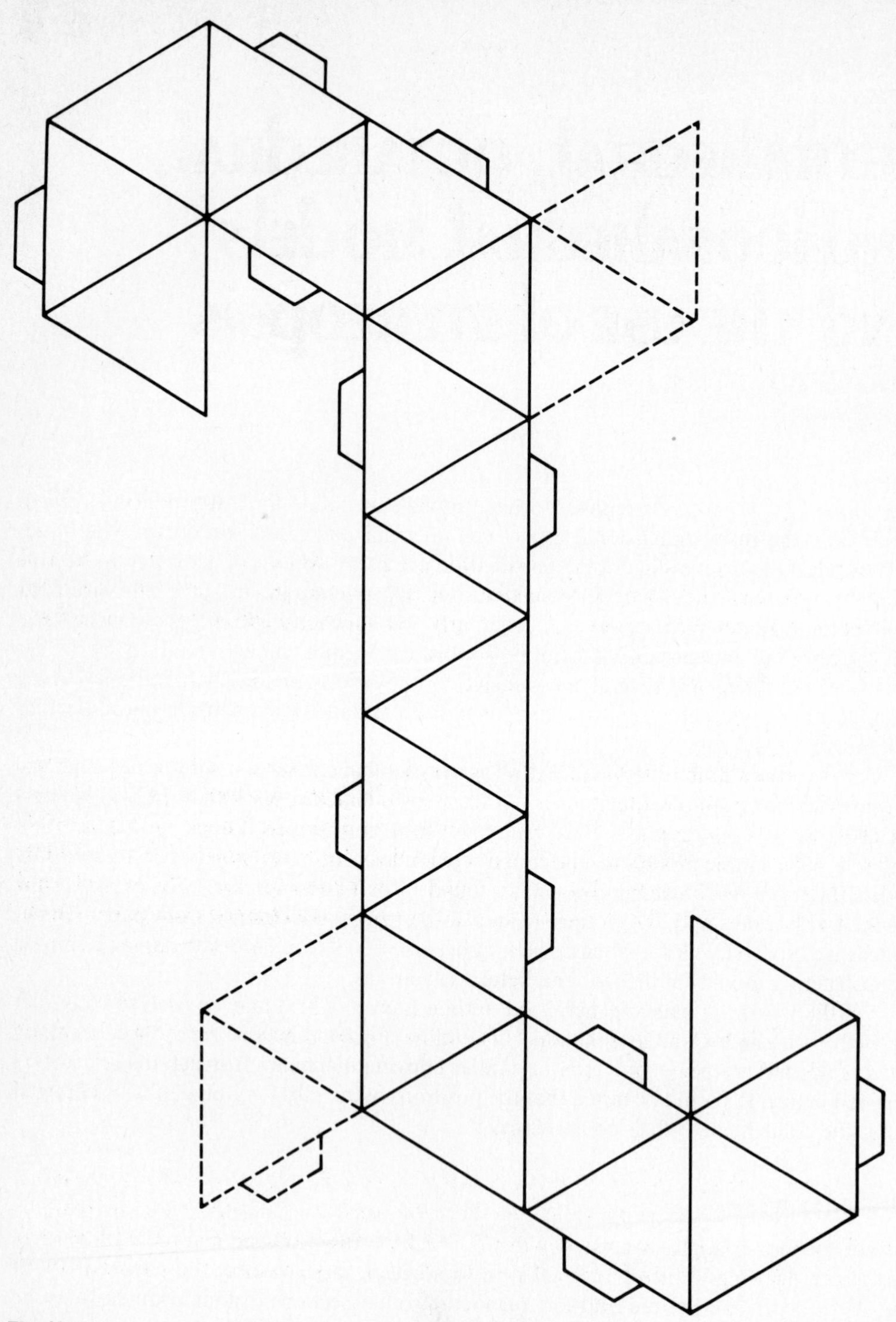

Fig. H.1

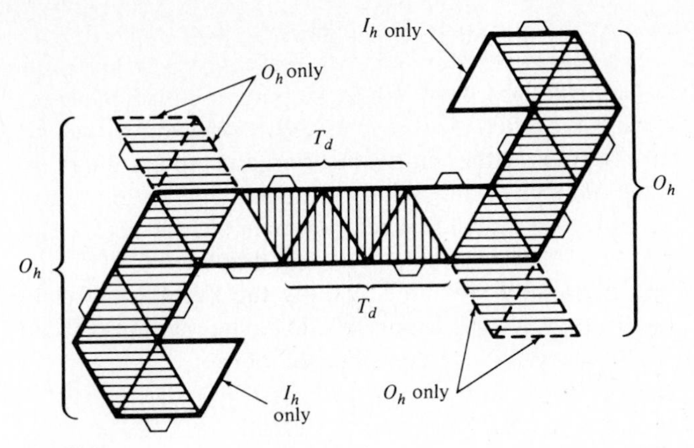

Fig. H.2

Octahedral models

Cut out the two sets of eight triangles enclosed by the O_h brackets (Fig. H.2) and marked with horizontal lines in the drawing. Include the dotted triangles marked "O_h only," but omit the triangles marked "I_h only." Glue or tape the tabs on to adjacent surfaces to form the octahedron.

Icosahedral models

Cut out the entire figure drawn with solid lines in Fig. H.2. Omit the faces marked "O_h only." Construction is facilitated by bending and gluing the two end sections into the "capping" and "foundation" pentagonal pyramids first. Then the remaining "equatorial" band of ten faces can be wound around and fastened to these "end groups." The complete icosahedron represents the $B_{12}H_{12}^{-2}$ ion or dicarbaclosododecaboranes. Other polyhedral boranes can be formed by removal of the appropriate faces from the complete icosahedron.

Other paper models

Several other polyhedral models constructed from simple materials have been described in a series of articles.[1]

Stereoviews

Closely related to the use of models is that of stereoviews to illustrate molecular perspective. An increased number of stereoviews has been included in this edition. If difficulty is encountered in achieving stereopsis, do not feel frustrated—either half of a stereoview conveys all of the information of the usual line drawing. In fact many of the latter used in

[1] S. Yamana and M. Kawaguchi, *J. Chem. Educ.*, **1968**, *45*, 245; **1980**, *57*, 434; **1982**, *59*, 196, 197, 578.

this book were drawn from one-half of a stereo pair. However, *most people can achieve stereopsis with a little practice.* It is *certainly* worth the effort. The easiest way for most people is to use a viewer. Inexpensive cardboard viewers may be purchased from sources listed below, or more expensive viewers are available if much stereo work will be done. If the reader is interested in structural chemistry, either inorganic or organic, such a purchase might be considered, inasmuch as the number of stereoviews is increasing constantly, and they may already account for 40% of the structures in current journals.[2] Many people can achieve stereopsis with a small card placed between the individual drawings. Others find that focusing the eyes on an object at infinity (and thus making the two lines of sight parallel) and then dropping the lines of sight to the stereoview allows stereovision without any "tools" whatsoever. For further discussion of stereovision, see footnote 2.

Sources of viewers
Hubbard Scientific Co., Northbrook, IL.
Taylor-Mechant Corp., 25 West 45th St., New York, NY 10036.
C. F. Casella and Co., Ltd., Regent House, Britannia Walk, London, N1 7ND.

It should also be noted that one popular freshman text, J. E. Brady and G. E. Humiston, "General Chemistry: Principles and Structure," 2nd ed., Wiley, New York, **1980**, comes with a stereoviewer and might be available in used condition.

[2] J. C. Speakman, *New Scientist*, **1978**, *78*, 827; A. H. Johnstone et al., *Educ. Chem.*, **1980**, *17*, 172. See also W. B. Jensen, *J. Chem. Educ.*, **1982**, *59*, 385.

I

Traditional ionic radii

In Chapters 3 and 4 the method of determining ionic crystal radii by X-ray diffraction was discussed and a set of Shannon-Prewitt crystal radii presented (Table 3.4). These are universally agreed to be the best currently available, but much of the older literature lists Pauling,[1] Goldschmidt,[2] Ahrens,[3] or other values. The reader should therefore be aware of their existence and the relationship they bear to the Shannon-Prewitt crystal radii. Furthermore, the assumptions and methods implicit in them are still worth considering even though the derived values have been superseded.

The first attempt to obtain the radius of a single ion was by Landé in 1920.[4] Lande assumed that Li^+ would be the smallest ion found in crystals and that in combination with the larger halide ions, it would be too small to keep the halide ions from touching. The larger the halide ion, of course, the more likely that this is true. Figure I.1 shows how in crystals of LiI, the I^- ions might possibly touch if they are sufficiently larger than the Li^+

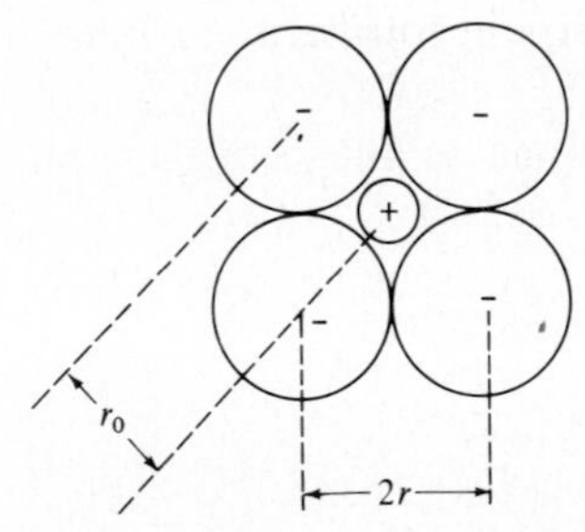

Fig. I.1 Relative sizes of cations and anions as required by Landé hypothesis.

---

[1] L. Pauling, "The Nature of the Chemical Bond," 3rd ed., Cornell University Press, Ithaca, N.Y., **1960**.

[2] V. M. Goldschmidt, *Skrifter Norske Videnskaps-Adad. Oslo*, **1926**, *2*.

[3] L. H. Ahrens, *Geochim. Acta*, **1952**, *2*, 155.

[4] A. Landé, *Z. Physik*, **1920**, *1*, 191.

ions. If the I^- ions touch, simple geometry allows us to determine the I—I distance and take one half of it to obtain an estimate of 213 pm for the radius of the I^- ion. The other halide ions then become: $Br^- = 188$, $Cl^- = 172$, $F^- = 132$. Using these values for X^- and subtracting from r_0 in Table I.1 allows us to assign values of $Na^+ = 99$, $K^+ = 142$, and $Rb^+ = 155$. Since CsF has the same crystal structure as RbF and has an r_0 value which is 18 pm larger, we can tentatively assume $Cs^+ = 173$ pm. Although all of these values are dependent upon the truth of Landé's original assumption (if the iodide ions do not touch in LiI, the values are meaningless!) and the uniformity of ionic radii from one crystal to the next, the results differ by only a few picometers from the commonly accepted values of today.

Other workers such as Bragg[5] and Goldschmidt[2] used similar methods to estimate the radius of O^{-2} (135 pm) and extend the coverage to over 80 ions.

Pauling[1] has described a theoretical method for apportioning the r_0 value between the cation and the anion. He has assumed that the size of an ion should be inversely proportional to the effective nuclear charge, other factors being equal. In order to make the comparison as accurate as possible, Pauling chose the isoelectronic ion pairs NaF, KCl, RbBr, and CsI. Pauling calculated his screening constants from theory,[6] molar refraction,[6] and X-ray term values.[7] The details need not concern us here and we may for convenience employ the method of Slater, pp. 37–38, instead.

Consider the isoelectronic ion pair, Na^+F^-. The r_0 value for sodium fluoride is 231 pm. Using Slater's rules the effective nuclear charge for the ions may be calculated:

$$Z^*_{Na^+} = 11 - (2 \times 0.85) - (8 \times 0.35) = 6.50 \quad \textbf{(I.1)}$$

$$Z^*_{F^-} = 9 - (2 \times 0.85) - (8 \times 0.35) = 4.50 \quad \textbf{(I.2)}$$

The reader may be concerned as to why all eight electrons in the valence shell were counted in the shielding instead of only seven electrons shielding the valence ("the last") electron. The answer must be a pragmatic one: If *all* the electrons are counted, the resulting effective nuclear charges, 6.50 and 4.50, resemble those of Pauling (6.48 and 4.48) very closely. Actually, if the shielding is calculated in the alternate way to give $Z^* = 6.85, 4.85$, the resulting values of r_+ and r_- change to an insignificant extent. Furthermore, one is neither adding nor removing electrons.

Since the size of each ion is assumed to be inversely proportional to its effective nuclear charge, we can write:

$$6.5r_+ = 4.5r_-$$

and **(I.3)**

$$r_0 = r_+ + r_- = 231$$

These two equations in two unknowns may readily be solved by substitution to yield

[5] W. L. Bragg and J. West, *Proc. Roy. Soc. London*, **1927**, *A114*, 450.
[6] L. Pauling, *Proc. Roy. Soc. London*, **1927**, *A114*, 181.
[7] L. Pauling and J. Sherman, *Z. Kristallogr.*, **1932**, *81*, 1.

$r_+ = 95$ and $r_- = 136$. Similar calculations yield values of $K^+ = 133$, $Cl^- = 181$; $Rb^+ = 148$, $Br^- = 195$; ($Cs^+ = 169$, $I^- = 216$).[8]

A consequence of the inverse relationship assumed between $r_\pm$ and Z^* is that their product should be a constant. For ions isoelectronic with Ne (O^{-2}, F^-, Na^+, Mg^{+2}, Al^{+3}) the value of the constant is 614. Hence for all of these ions, radii are obtainable by the formula:

$$r_\pm = \frac{614}{Z^*} \tag{I.4}$$

Thus for Mg^{+2} and O^{-2}, we can estimate radii as follows:

$$r_{Mg^{+2}} = \frac{614}{7.50} = 82 \text{ pm} \tag{I.5}$$

$$r_{O^{-2}} = \frac{614}{3.50} = 176 \text{ pm} \tag{I.6}$$

If we add these values for Mg^{+2} and O^{-2} we obtain an estimate for the Mg—O distance of 258 pm. The observed distance in MgO is only 210 pm. Since MgO cystallizes in the sodium chloride lattice, crystal structure cannot be responsible for our 23% error in predicting the MgO distance. Rather, the above values are in error because they correct for only *one* of the two ways in which Mg^{+2} and O^{-2} differ from Na^+ and F^-. The values obtained from Eqs. I.5 and I.6 were termed *univalent* radii by Pauling. They are too large because they are uncorrected for the additional compression which ions undergo in a 2:2 lattice (roughly four times as great as 1:1 lattice). If univalent radii have any physical meaning at all, it is that they represent the size of divalent ions when compressed by the lattice energy of a 1:1 lattice, as, for example, if a Mg^{+2} ion were "doped" into a NaF lattice.

In order to compensate for the compression effect, Pauling suggested that the correction factor should be a function of the Born exponent, n, since that parameter measures compressibility. Pauling's derivation was similar to the following. We can rearrange Eq. 3.10 to obtain the equilibrium radius, r_0:

$$r_0 = \left(\frac{-4\pi\varepsilon_0 nB}{AZ^+Z^-e^2}\right)^{1/n-1} \tag{I.7}$$

$$r_0 = \left(\frac{4\pi\varepsilon_0 nB}{Ae^2}\right)^{1/n-1} \cdot \left(\frac{-1}{Z^+Z^-}\right)^{1/n-1} \tag{I.8}$$

The ratio of the equilibrium radius in a crystal with $Z^+ = +i$ and $Z^- = -j$ to that in a uni-univalent crystal will be obtained by substituting the proper values for Z^+ and Z^-:

$$r_{ij}/r_{11} = \left(\frac{-1}{(+i)(-j)}\right)^{1/n-1} \tag{I.9}$$

[8] To compensate for the different lattice structure of CsI, Pauling subtracted about 3% from the observed interionic distance.

Table I.1 Values (in pm) of Pauling crystal radii and (in parentheses) univalent radii of ions[a]

														H^{-} 208 (208)
He (93)	Li^{+} 60 (60)	Be^{+2} 31 (44)								B^{+3} 20 (35)	C^{+4} 15 (29)	N^{+5} 11 (25)	O^{+6} 9 (22)	F^{+7} 7 (19)
											C^{-4} 260 (414)	N^{-3} 171 (247)	O^{-2} 140 (176)	F^{-} 136 (136)
Ne (112)	Na^{+} 95 (95)	Mg^{+2} 65 (82)								Al^{+3} 50 (72)	Si^{+4} 41 (65)	P^{+5} 34 (59)	S^{+6} 29 (53)	Cl^{+7} 26 (49)
											Si^{-4} 271 (384)	P^{-3} 212 (279)	S^{-2} 184 (219)	Cl^{-} 181 (181)
Ar (154)	K^{+} 133 (133)	Ca^{+2} 99 (118)	Sc^{+3} 81 (106)	Ti^{+4} 68 (96)	V^{+5} 59 (88)	Cr^{+6} 52 (81)	Mn^{+7} 46 (75)	Cu^{+} 96 (96)	Zn^{+2} 74 (88)	Ga^{+3} 62 (81)	Ge^{+4} 53 (76)	As^{+5} 47 (71)	Se^{+6} 42 (66)	Br^{+7} 39 (62)
											Ge^{-4} 272 (371)	As^{-3} 222 (285)	Se^{-2} 198 (232)	Br^{-} 195 (195)
Kr (169)	Rb^{+} 148 (148)	Sr^{+2} 113 (132)	Y^{+3} 93 (120)	Zr^{+4} 80 (109)	Nb^{+5} 70 (100)	Mo^{+6} 62 (93)		Ag^{+} 126 (126)	Cd^{+2} 97 (114)	In^{+3} 81 (104)	Sn^{+4} 71 (96)	Sb^{+5} 62 (89)	Te^{+6} 56 (82)	I^{+7} 50 (77)
											Sn^{-4} 294 (370)	Sb^{-3} 245 (295)	Te^{-2} 221 (250)	I^{-} 216 (216)
Xe (190)	Cs^{+} 169 (169)	Ba^{+2} 135 (153)	La^{+3} 115 (139)					Au^{+} 137 (137)	Hg^{+2} 110 (125)	Tl^{+3} 95 (115)	Pb^{+4} 84 (106)	Bi^{+5} 74 (98)		
				Ce^{+4} 101 (127)										

[a] Adapted from L. Pauling, "The Nature of the Chemical Bond," 3rd ed., Cornell University Press, Ithaca N.Y., **1960**, p. 514.

For Mg^{+2} and O^{-2}, we now have the following expression:

$$r_{\text{crystal}} = r_{\text{univalent}}\left(\tfrac{1}{4}\right)^{1/6} \quad \textbf{(I.10)}$$

The resulting values for the crystal radii are $Mg^{+2} = 65$ and $O^{-2} = 140$ pm. The sum of these two radii comes much closer to the experimental value of 210 pm, the error being of the order of 2%.

By these methods Pauling was able (1) to obtain values for the radii of the alkali metal cations and halide anions, (2) to extend this type of calculation to obtain *univalent radii* for elements in other periodic groups, and (3) to use Eq. I.9 to obtain *crystal radii* for all of the main group elements. From the anion crystal radii, crystal radii of transition metal ions could then be obtained. The results are shown in Table I.1. These values are for crystals in which the ions have the same coordination number (6) as in the reference crystals NaF, KCl, etc. The radius of an ion changes somewhat in different environments. For example, the radius of Al^{+3} is about 50 pm when octahedrally coordinated. In a tetrahedral environment (C.N. = 4) it is somewhat smaller, about 40 pm. Pauling[9] has given factors for the conversion from one coordination number to another, but the mechanics need not concern us here. Ordinarily one may use values already computed and tabulated.

[9] L. Pauling, "The Nature of the Chemical Bond," 3rd ed., Cornell University Press, Ithaca, N.Y., **1960**, pp. 537–540.

J

The rules of inorganic nomenclature

The standards of nomenclature in chemistry are the Rules published by the International Union of Pure and Applied Chemistry (IUPAC). Since these often represent a compromise of conflicting views they are sometimes not completely acceptable to all chemists and American chemists occasionally employ usages not officially sanctioned by the IUPAC Rules (the latest publication of an ACS commentary on American usage is *J. Am. Chem. Soc.*, **1960**, *82*, 5523). The Rules should not be viewed as a rigid code but as an evolving attempt to clarify the process of naming. It should be realized that one purpose of any set of rules is to prevent the increase in the number of nonsystematic names that have been invented *ad hoc* for particular compounds. Nevertheless many trivial names will occur in common usage. For example, it is not anticipated that *ethanoate* will soon displace *acetate* for the anion CH_3COO^-.

The usage in this book has been as close to IUPAC nomenclature as is consistent with good pedagogy. I have followed the IUPAC Rules except in those cases in which they conflict directly with current American usage. No purpose is served by confusing the student by using one type of nomenclature in a text and then encouraging him to read the original literature in which he finds a radically different nomenclature. The prime purpose of this book is to illustrate inorganic chemistry rather than the details of nomenclatural technique. The nomenclature has therefore been that which would serve best in teaching inorganic chemistry and help the student in reading the original literature. Those interested in the official IUPAC Rules can find a synopsis here.

This Appendix consists of short excerpts from the Preamble and Rules of the 1971 Report of the Commission on the Nomenclature of Inorganic Chemistry, International Union of Pure and Applied Chemistry.[1] These represent but a small fraction of the entire rules, and the interested reader is referred to the original for any specific problems.[2] The

[1] "Nomenclature of Inorganic Chemistry—Definitive Rules 1970," 2nd ed., Butterworths, London, **1971**; also, *Pure Appl. Chem.*, **1971**, *27*, 67.

[2] A continuing series of articles, "Notes on Nomenclature," is very useful in interpreting the IUPAC Rules as well as other aspects of nomenclature: W. C. Fernelius et al., *J. Chem. Educ.*, **1974**, *51*, 468, 603; **1975**, *52*, 583, 793; **1976**, *53*, 354, 495; **1977**, *54*, 299.

following is meant as a brief outline to acquaint the reader with the general principles of inorganic nomenclature. The author's comments and other additions have been placed in square brackets, [ ]. Otherwise the following are verbatim extracts of the Rules with the exception of minor editorial changes to conform to American usage in the spellings of stoichiometry (Engl.: stoicheiometry), center (Engl.: centre), etc.

PREAMBLE

Often the general principles of nomenclature do not stand out clearly in the details of specific rules. The purpose of this preamble is to point out some general practices and to provide illustrative examples of the ways in which they are applied.

OXIDATION NUMBER

The concept of oxidation number is interwoven in the fabric of inorganic chemistry in many ways, including nomenclature. The oxidation number of an element in any chemical entity is the charge which would be present on an atom of the element if the electrons in each bond were assigned to the more electronegative atom, thus:

Oxidation numbers		
MnO_4^- = One Mn^{7+} and four O^{2-} ions	Mn = VII	O = −II
ClO^- = One Cl^+ and one O^{2-} ion	Cl = I	O = −II

COORDINATION NUMBER

The coordination number of the central atom in a compound is the number of atoms which are directly linked to the central atom. The attached atoms may be charged or uncharged or part of an ion or molecule. In some types of coordination compounds, the two atoms of a multiple bond in an attached group are assigned to a single coordination position. Crystallographers define the coordination number of an atom or ion in a lattice as the number of near neighbors to that atom or ion.

USE OF MULTIPLYING AFFIXES, ENCLOSING MARKS, NUMBERS, AND LETTERS

Chemical nomenclature uses multiplying affixes, numbers (both Arabic and Roman), and letters to indicate both stoichiometry and structure.

Multiplying affixes

The simple multiplying affixes, mono, di, tri, tetra, penta, hexa, octa, nona (ennea), deca, undeca (hendeca), dodeca, etc., indicate

a. Stoichiometric proportions:

CO	carbon monoxide
CO_2	carbon dioxide

b. Extent of substitution:

$SiCl_2H_2$	dichlorosilane
$PO_2S_2^{3-}$	dithiophosphate ion

c. Number of identical coordinated groups:

$[Co(NH_3)_4Cl_2]^+$ tetraamminedichlorocobalt(III) ion

The affixes have somewhat different uses in designating (1) the number of identical central atoms in condensed acids and their characteristic anions:

H_3PO_4	(mono)phosphoric acid	[The parentheses indicate that the "mono" may be omitted; they are not part of the name.]
$H_4P_2O_7$	diphosphoric acid	
$H_2S_3O_{10}$	trisulfuric acid	

or (2) the number of atoms of the same element forming the skeleton of some molecules or ions:

Si_2H_6	disilane
$S_4O_6^{2-}$	tetrathionate ion

The multiplicative affixes bis, tris, tetrakis, pentakis, etc., were originally introduced into organic nomenclature to indicate a set of identical radicals each substituted in the same way:

$(ClCH_2CH_2)_2NH$ bis(2-chloroethyl)amine

or to avoid ambiguity:

$P(C_{10}H_{21})_3$ tris(decyl)phosphine

Tris(decyl) avoids any ambiguity with the organic radical tridecyl, $C_{13}H_{27}$. The use of these affixes has been extended:

$[P(CH_2OH)_4]^+Cl^-$	tetrakis(hydroxymethyl)phosphonium chloride
$[Fe(CN)_2(CH_3NC)_4]$	dicyanotetrakis(methyl isocyanide)iron(II)

In the second case one wishes to avoid any doubt that the ligand is CH_3NC.

Enclosing marks

Enclosing marks are used in formulae to enclose sets of identical groups of atoms: $Ca_3(PO_4)_2$, $B[N(CH_3)_2]_3$. In names, enclosing marks generally are used following bis, tris, etc., around all complex expressions, and elsewhere to avoid any possibility of ambiguity. In the formulae of coordination compounds, square brackets are used to enclose a complex ion or a neutral coordination entity:

$K_3[Co(C_2O_4)_3]$
$[Co(NH_3)_3(NO_2)_3]$

Numbers

In names of inorganic compounds, Arabic numerals are used to designate the atoms at which there is a substituent or addition in a chain:

$H_3Si—ClSiH—SiH_2SiH_3$ 2-chlorotetrasilane

[This system is similar to that used in organic nomenclature and will not be elaborated further.]

Arabic numerals followed by + and − and enclosed in parentheses also are used to indicate the charge on a free or coordinated ion [Ewens-Bassett system]. Zero is not used with the name of an uncharged coordination compound. Finally, Arabic numerals are also sometimes used in place of multiplying affixes.

$AlK(SO_4)_2 \cdot 12H_2O$	Aluminum potassium sulfate 12-water [in place of aluminum potassium sulfate dodecahydrate]
B_6H_{10}	Hexaborane(10) [in place of hexaboron decahydride]

Roman numerals are used in parentheses to indicate the oxidation number of an element [Stock system]. The cipher 0 is used to indicate an oxidation state of zero. A negative oxidation state is indicated by the use of the negative sign with a Roman numeral.

Italic letters

The symbols of the elements, printed in italics, are used to designate

a. The element in a heteroatomic chain or ring at which there is substitution:

CH_3ONH_2 *O*-methylhydroxylamine

b. The element in a ligand which is coordinated to a central atom:

```
  S———CH2
 /      |
M       |
 ↖      |
  NH2—C—COOH      cysteinato—S,N . . .
        |
        H
```

c. The presence of bonds between two metal atoms:

$(OC)_3Fe(C_2H_5S)_2Fe(CO)_3$ bis(μ-ethylthio)bis(tricarbonyliron)(Fe—Fe)

d. The point of attachment in some addition compounds:

$CH_3ONH_2 \cdot BH_3$ *O*-methylhydroxylamine(*N*—*B*)borane

1. ELEMENTS

[A complete list of accepted names of elements, their atomic numbers, and their atomic weights may be found in Appendix K.]

The mass number, atomic number, number of atoms, and ionic charge of an element may be indicated by means of four indices placed around the symbol. The positions are to be occupied thus:

Left upper index	mass number
Left lower index	atomic number
Right lower index	number of atoms
Right upper index	ionic charge

2. FORMULAE AND NAMES OF COMPOUNDS IN GENERAL

In formulae the *electropositive constituent* (cation) should always be placed first, e.g., KCl, $CaSO_4$. In the case of binary compounds between nonmetals, in accordance with established practice, that constituent should be placed first which appears earlier in the sequence: Rn, Xe, Kr, B, Si, C, Sb, As, P, N, H, Te, Se, S, At, I, Br, Cl, O, F.

Examples: XeF_2, NH_3, H_2S, S_2Cl_2, Cl_2O, OF_2

The name of the *electropositive constituent* will not be modified [in forming systematic names]. If the electronegative constituent is monatomic or homopolyatomic its name is modified to end in -ide. For binary compounds the name of the element standing later in the sequence [given above] is modified to end in -ide: sodium plumbide, sodium chloride, calcium sulfide, lithium nitride, arsenic selenide, nickel arsenide, boron hydrides, hydrogen chloride, hydrogen sulfide, silicon carbide, carbon disulfide, sulfur hexafluoride, chlorine dioxide, oxygen difluoride.

If the electronegative constituent is heteropolyatomic it should be designated by the termination -ate. In certain exceptional cases the termination -ide and -ite are used [see below].

In inorganic compounds it is generally possible in a polyatomic group to indicate a *characteristic atom* (as in ClO^-) or a *central atom* (as in ICl_4^-). Such a polyatomic group is designated as a *complex*, and the atoms, radicals, or molecules bound to the characteristic or central atom are termed *ligands*.

In this case the name of a negatively charged complex should be formed from the name of the characteristic or central element modified to end in -ate. [For examples, see Section 7.]

Binary hydrogen compounds may be named by the principles [given above]. Volatile hydrides, except those of Group VII and of oxygen and nitrogen, may also be named by citing the root name of the element followed by the suffix -ane. If the molecule contains more than one atom of that element, the number is indicated by the appropriate Greek prefix.

Recognized exceptions are water, ammonia, [and] hydrazine, owing to long usage. Phosphine, arsine, stibine, and bismuthine are also allowed. However, for all molecular hydrides containing more than one atom of the element, "-ane" names should be used.

B_2H_6	diborane	PbH_4	plumbane
Si_3H_8	trisilane	H_2S_n	polysulfane

3. NAMES FOR IONS AND RADICALS

Monatomic cations should be named as the corresponding element, without change or suffix:

Cu^+	the copper(I) ion
Cu^{2+}	the copper(II) ion
I^+	the iodine(I) cation

The preceding principle should also apply to polyatomic cations corresponding to radicals for which special names are given [below], i.e., these names should be used without change or suffix: the nitrosyl cation (NO^+); the nitryl cation (NO_2^+).

Names for polyatomic cations derived by addition of more protons than required to give a neutral unit to monatomic anions are formed by adding the ending -onium to the root of the name of the anion element: phosphonium, arsonium, sulfonium, iodonium. The name ammonium for the ion NH_4^+ does not conform to [the above rule] but is retained. Substituted ammonium ions derived from nitrogen bases with names ending in -amine will receive names formed by changing -amine to -ammonium. For example, $HONH_3^+$, the hydroxylammonium ion. When the nitrogen base is known by a name ending otherwise than in -amine, the cation name is to be formed by adding the ending -ium to the name of the case (if necessary omitting a final e or other vowel): hydrazinium, anilinium, glycinium, pyridinium.

The names for monatomic anions shall consist of the name (sometimes abbreviated) of the elements with the termination -ide:

H^-	hydride ion
F^-	fluoride ion
N^{3-}	nitride ion
O^{2-}	oxide ion

Certain polyatomic anions have names ending in -ide. [Some are:]

OH^-	hydroxide ion	N_3^-	azide ion
O_2^{2-}	peroxide ion	NH^{2-}	imide ion
S_2^{2-}	disulfide ion	NH_2^-	amide ion
I_3^-	triiodide ion	CN^-	cyanide ion
HF_2^-	hydrogen difluoride ion	C_2^{2-}	acetylide ion

It is quite practical to treat oxygen also in the same manner as other ligands but it has long been customary to ignore the name of this element altogether in anions and to indicate its presence and proportion by means of a series of prefixes and sometimes also by the suffix -ite in place of -ate. The termination -ite has been used to denote a lower state of oxidation and may be retained in trivial names in the following cases:

NO_2^-	nitrite	SO_3^{2-}	sulfite	ClO_2^-	chlorite
$N_2O_2^{2-}$	hyponitrite	$S_2O_6^{2-}$	disulfite	ClO^-	hypochlorite
NOO_2^-	peroxonitrite	$S_2O_4^{2-}$	dithionite	BrO^-	hypobromite
AsO_3^{3-}	arsenite	SeO_3^{2-}	selenite	IO^-	hypoiodite

[Without explanation the Commission on the Nomenclature of Inorganic Chemistry dismissed the name phosphite for PO_3^{3-} (while retaining arsenite!) and replaced it with phosphonate, which is already used for $RP(O)(OH)_2$. It is not likely that most American inorganic chemists would agree that this was an improvement.]

A radical is here regarded as a group of atoms which occurs repeatedly in a number of different compounds. Certain neutral and cationic radicals containing oxygen or other chalcogens have, irrespective of charge, special names ending in -yl, and the Commission approves the provisional retention of the following:

HO	hydroxyl	SO	sulfinyl (thionyl)
CO	carbonyl	SO_2	sulfonyl (sulfuryl)
NO	nitrosyl	S_2O_5	disulfuryl
NO_2	nitryl	SeO	seleninyl
PO	phosphoryl	SeO_2	selenonyl
ClO	chlorosyl	CrO_2	chromyl
ClO_2	chloryl	UO_2	uranyl
ClO_3	perchloryl	NpO_2	neptunyl
(similarly for other halogens)		(similarly for other actinides)	

Radicals analogous to the above containing other chalcogens in place of oxygen are named by adding the prefixes thio-, seleno-, etc.

These polyatomic radicals are always treated as forming the positive part of the compound:

$COCl_2$	carbonyl chloride
$PSCl_3$	thiophosphoryl chloride
SO_2NH	sulfonyl (sulfuryl) imide

[In the past, the names sulfinyl and sulfonyl chloride have been more popular with organic chemists, the names thionyl and sulfuryl with inorganic chemists.]

It should be noted that the same radical may have different names in inorganic and organic chemistry. Names of purely organic compounds [as ligands, for example] should be in accordance with the nomenclature of organic chemistry.

Organic chemical nomenclature is to a large extent built upon the scheme of substitution, i.e., replacement of hydrogen atoms by other atoms or groups. Such "substitutive names" are extremely rare in inorganic chemistry; they are used, e.g., in the following cases: NH_2Cl is called chloramine, and $NHCl_2$ dichloramine. Other substitutive names are fluoro- and chloro-sulfonic acid, etc. These names should be replaced by:

HSO_3F	fluorosulfuric acid	NH_2SO_3H	amidosulfuric acid
HSO_3Cl	chlorosulfuric acid		[often called "sulfamic acid"]

4. ISO- AND HETEROPOLYANIONS

Anions of polyacids derived by condensation of molecules of the same monoacid, containing the characteristic element in the oxidation state corresponding to its Group number, are named by indicating with Greek numeral prefixes the number of atoms of that element:

$Na_2B_4O_7$	disodium tetraborate
$Mo_7O_{24}^{6-}$	heptamolybdate

Cyclic and strain-chain structures may be distinguished by means of the prefixes *cyclo* and *catena*, although the latter may usually be omitted:

```
    O     O     O
    |     |     |
1. OP—O—P—O—PO⁵⁻        triphosphate
    |     |     |
    O     O     O
```

```
      O₂      ]3−
      P
    /   \
   O     O
2. |     |               cyclo-triphosphate
  O₂P    PO₂
    \   /
      O
```

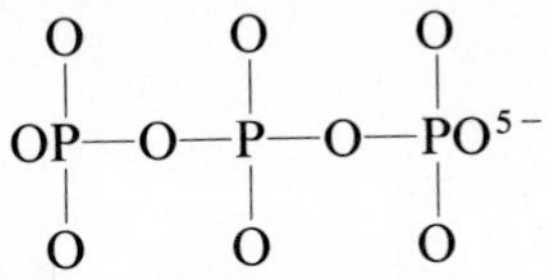

2. 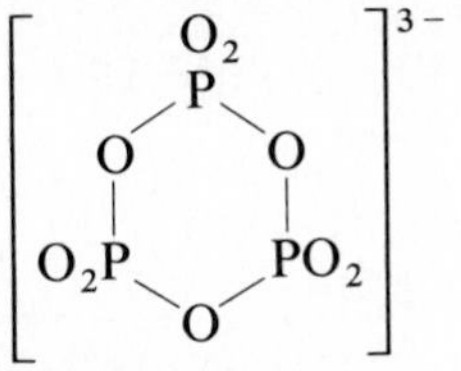 *cyclo*-triphosphate

3. $[O(PO_3)_n]^{(n+2)-}$ *catena*-polyphosphate

4. $\left[\begin{array}{ccccc} O & & O & & O \\ | & & | & & | \\ OSi & -O- & Si & -O- & SiO \\ | & & | & & | \\ O & & O & & O \end{array}\right]^{8-}$ trisilicate

Heteropolyanions with a chain or ring structure

Dinuclear anions are named by treating the anion which comes first in alphabetical order as the ligand on the characteristic atoms of the second:

$[O_3P{-}O{-}SO_3]^{3-}$ phosphatosulfate(3−)

Cyclic heteropolyanions are named in a manner similar to those with a chain structure, the starting point and direction of citation of the units being chosen according to alphabetical priority:

$\left[\begin{array}{ccc} & O_2 & \\ & As{-}O & \\ O & & PO_2 \\ | & & | \\ O_2Cr & & O \\ & O{-}S & \\ & O_2 & \end{array}\right]^{2-}$ *cyclo*-arsenatochromatosulfatophosphate (2−)

In polyanions with an obvious central atom the peripheral anions are named as ligands on the central atom and cited in alphabetical order:

$\left[\begin{array}{ccc} & O & \\ & \| & \\ O_3CrO- & P & -OAsO_3 \\ & | & \\ & O & \\ & SO_3 & \end{array}\right]^{4-}$ (arsenato)(chromato)(sulfato)phosphate(4−)

Condensed heteropoly anions

The three-dimensional frameworks of linked WO_6, MoO_6, etc., octahedra surrounding a central atom are designated by the prefixes wolframo, molybdo, etc., e.g., wolframophosphate (tungstophosphate), *not* phosphowolframate (phosphotungstate). The numbers of atoms of the characteristic element are indicated by Greek prefixes or numerals.

$[PW_{12}O_{40}]^{3-}$ dodecawolframophosphate(3−)
12-wolframophosphate(3−)

$[P_2Mo_{18}O_{62}]^{6-}$ 18-molybdodiphosphate(V)(6−)

[American chemists almost never use the combining form *wolframo-*, preferring the form *tungsto-*.]

5. ACIDS

Acids giving rise to the -ide anions (Section 3) will be named as binary and pseudobinary compounds of hydrogen, e.g., hydrogen chloride, hydrogen sulfide, hydrogen cyanide.

For the oxoacids the ous-ic notation to distinguish between different oxidation states is applied in many cases. The -ous names are restricted to acids corresponding to the -ite anions listed [above].

The prefix hypo- is used to denote a lower oxidation state, and may be retained in the following cases:

$H_2N_2O_2$	hyponitrous acid	HOCl	hypochlorous acid
$H_4P_2O_6$	hypophosphoric acid	HOBr	hypobromous acid
		HOI	hypoiodous acid

The prefix per- has been used to designate a higher oxidation state and is retained for $HClO_4$, perchloric acid, and corresponding acids of the other elements in Group VII. This use of the prefix per- should not be extended to elements of other groups, and such names as perxenonate and perruthenate are not approved. [The name "perxenate" is fairly well established in the American literature.]

The prefixes ortho- and meta- have been used to distinguish acids differing in the "content of water." The following names are approved:

H_3BO_3	orthoboric acid	$(HBO_3)_n$	metaboric acid
H_4SiO_4	orthosilicic acid	$(H_2SiO_3)_n$	metasilicic acid
H_3PO_4	orthophosphoric acid	$(HPO_3)_n$	metaphosphoric acid
H_5IO_6	orthoperiodic acid		
H_6TeO_6	orthotelluric acid		

[Note that the "removal of water" has resulted in polymerization to form chains or rings.]

Names for [Some] Oxo Acids

H_3BO_3	orthoboric acid or boric acid
H_2CO_3	carbonic acid
HOCN	cyanic acid
HNCO	isocyanic acid
HONC	fulminic acid
H_4SiO_4	orthosilicic acid
$(H_2SiO_3)_n$	metasilicic acids
HNO_3	nitric acid
HNO_2	nitrous acid
H_3PO_4	orthophosphoric acid or phosphoric acid
$H_4P_2O_7$	diphosphoric or pyrophosphoric acid
$H_5P_3O_{10}$	triphosphoric acid
$(HPO_3)_n$	metaphosphoric acids
$(HO)_2OP—PO(OH)_2$	hypophosphoric acid or diphosphoric(IV) acid
H_2SO_4	sulfuric acid
$H_2S_2O_7$	disulfuric acid

Names for [Some] Oxo Acids	
H_2SO_5	peroxomonosulfuric acid
$H_2S_2O_8$	peroxodisulfuric acid
$H_2S_2O_3$	thiosulfuric acid
$H_2S_2O_6$	dithionic acid
H_2SO_3	sulfurous acid
$H_2S_2O_4$	dithionous acid
$HClO_4$	perchloric acid
$HClO_3$	chloric acid
$HClO_2$	chlorous acid
$HClO$	hypochlorous acid

6. SALTS AND SALT-LIKE COMPOUNDS

Simple salts

Simple salts fall under the broad definition of binary compounds given in Section 2, and their names are formed from those of the constituent ions (given in Section 3) in the manner set out in Section 2.

Salts containing acid hydrogen

Names are formed by adding the word "hydrogen," with numerical prefix where necessary, to denote the replaceable hydrogen in the salt. The word hydrogen shall be placed immediately in front of the name of the anion:

$NaHCO_3$	sodium hydrogencarbonate
LiH_2PO_4	lithium dihydrogenphosphate
KHS	potassium hydrogensulfide

[American usage is strongly in favor of separating the hydrogen: sodium hydrogen carbonate, etc.]

7. COORDINATION COMPOUNDS

In *formulae* the usual practice is to place the symbol for the central atom(s) *first* (except in formulae which are primarily structural), with the ionic and neutral ligands following and the formula for the whole complex enclosed in square brackets.

In *names* the central atom(s) should be placed after the ligands.

Indication of oxidation number and proportion of constituents

The names of coordination entities always have been intended to indicate the charge of the central atom (ion) from which the entity is derived. Since the charge on the coordination entity is the algebraic sum of the charges of the constituents, the necessary information may be supplied by giving either the STOCK number (formal charge on the central ion, i.e., oxidation number) or the EWENS-BASSETT number:

$K_3[Fe(CN)_6]$	potassium hexacyanoferrate(III) potassium hexacyanoferrate(3−) tripotassium hexacyanoferrate
$K_4[Fe(CN)_6]$	potassium hexacyanoferrate(II) potassium hexacyanoferrate(4−) tetrapotassium hexacyanoferrate

Structural information may be given in formulae and names by prefixes such as *cis*, *trans*, *fac*, *mer*, etc. [see below].

Anions are given the termination -ate. Cations and neutral molecules are given no distinguishing termination.

The ligands are listed in alphabetical order regardless of the number of each. The name of a ligand is treated as a unit. Thus, "diammine" is listed under "a" and "dimethylamine" under "d." [Until the present Rules were published in 1971, this rule required the listing of negative ligands first (e.g., dichlorodiammineplatinum(II) rather than the proposed diamminedichloroplatinum(II). The older usage is found in much of the literature cited.]

The names of anionic ligands, whether inorganic or organic, end in -o

In general, if the anion name ends in -ide, -ite, or -ate, the final -e is replaced by -o, giving -ido, -ito, and -ato, respectively. Enclosing marks are required for inorganic anionic ligands containing numerical prefixes, as (triphosphato), and for thio, seleno, and telluro analogues of oxo anions containing more than one atom, as (thiosulfato). Examples of organic anionic ligands which are named in this fashion [are]:

CH_3COO^-	acetato
$(CH_3)_2N^-$	dimethylamido

The anions listed below do not follow exactly the above rule and modified forms have become established:

	Ion	Ligand
F^-	fluoride	fluoro
Cl^-	chloride	chloro
Br^-	bromide	bromo
I^-	iodide	iodo
O^{2-}	oxide	oxo
H^-	hydride	hydrido (hydro)[3]
OH^-	hydroxide	hydroxo
O_2^{2-}	peroxide	peroxo
CN^-	cyanide	cyano

The letter in each of the ligand names which is used to determine the alphabetical listing is given in boldface type in the following examples to illustrate the alphabetical

[3] Both hydrido and hydro are used for coordinated hydrogen, but the latter term usually is restricted to boron compounds.

arrangement. For many compounds, the oxidation number of the central atom and/or the charge on the ion are so well known that there is no need to use either a Stock number or a Ewens-Bassett number. However, it is not wrong to use such numbers and they are included here.

Formula	Name
$Na[B(NO_3)_4]$	sodium tetranitratoborate(1 −) sodium tetranitratoborate(III)
$K_2[OsCl_5N]$	potassium penta**ch**loro**ni**tridoosmate(2 −) potassium penta**ch**loro**ni**tridoosmate(VI)
$[Co(NH_2)_2(NH_3)_4]OC_2H_5$	di**ami**dotetra**amm**inecobalt(1 +) ethoxide di**ami**dotetra**amm**inecobalt(III) ethoxide
$[CoN_3(NH_3)_5]SO_4$	penta**amm**ine**azi**docobalt(2 +) sulfate penta**amm**ine**azi**docobalt(III) sulfate
$Na_3[Ag(S_2O_3)_2]$	sodium bis**(th**iosulfato)argentate(3 −) sodium bis**(th**iosulfato)argenate(I)
$NH_4[Cr(NCS)_4(NH_3)_2]$	ammonium diamminetetrakis**(i**sothiocyanato) chromate(1 −) ammonium diamminetetrakis**(i**sothiocyanato) chromate(III)
$Ba[BrF_4]_2$	barium tetrafluorobromate(1 −) barium tetrafluorobromate(III)

Although the common hydrocarbon radicals generally behave as anions when they are attached to metals and in fact are sometimes encountered as anions, their presence in coordination entities is indicated by the customary radical names even though they are considered as anions in computing the oxidation number.

Formula	Name
$K[B(C_6H_5)_4]$	potassium tetraphenylborate(1 −) potassium tetraphenylborate(III)
$K[SbCl_5C_6H_5]$	potassium pentachloro(phenyl)antimonate(1 −)[4] potassium pentachloro(phenyl)antimonate(V)

The name of the coordinated molecule or cation is to be used without change. All neutral ligands are set off with enclosing marks. (Four exceptions are listed below.)

Formula	Name
cis-$[PtCl_2(Et_3P)_2]$	*cis*-dichlorobis(triethylphosphine)platinum *cis*-dichlorobis(triethylphosphine)platinum(II)
$[CuCl_2(CH_3NH_2)_2]$	dichlorobis(methylamine)copper dichlorobis(methylamine)copper(II)
$[Pt(py)_4][PtCl_4]$	tetrakis(pyridine)platinum(2 +)tetrachloroplatinate(2 −) tetrakis(pyridine)platinum(II) tetrachloroplatinate(II)
$[Co(en)_3]_2(SO_4)_3$	tris(ethylenediamine)cobalt(3 +) sulfate tris(ethylenediamine)cobalt(III) sulfate
$K[PtCl_3(C_2H_4)]$	potassium trichloro(ethylene)platinate(1 −) potassium trichloro(ethylene)platinate(II) or potassium trichloromonoethyleneplatinate(II)
$[Ru(NH_3)_5(N_2)]Cl_2$	pentaammine(dinitrogen)ruthenium(2 +)chloride pentaammine(dinitrogen)ruthenium(II) chloride

[4] Normallyphenyl would not be placed within enclosing marks. They are used here to avoid confusion with a chlorophenyl radical.

Water and ammonia as neutral ligands in coordination complexes are called "aqua" (formerly "aquo") and "ammine," respectively. The groups NO and CO, when linked directly to a metal atom, are called "nitrosyl" and "carbonyl," respectively. In computing the oxidation number these ligands are treated as neutral.

$[Cr(H_2O)_6]Cl_3$	hexaaquachromium(3+) chloride hexaaquachromium trichloride
$Na_2[Fe(CN)_5NO]$	sodium pentacyanonitrosylferrate(2−) sodium pentacyanonitrosylferrate(III)
$K_3[Fe(CN)_5CO]$	potassium carbonylpentacyanoferrate(3−) potassium carbonylpentacyanoferrate(II)

Alternative modes of linkage of some ligands

The different points of attachment of a ligand may be denoted by adding the italicized symbol(s) for the atom or atoms through which attachment occurs at the end of the name of the ligand. Thus the dithiooxalato anion

^-S S^-
C—C
O O

conceivably may be attached through S or O, and these are distinguished as dithiooxalato-*S,S'* and dithiooxalato-*O,O'*, respectively.

In some cases different names are already in use for alternative modes of attachment as, for example, thiocyanato (—SCN) and isothiocyanato (—NCS), nitro (—NO_2), and nitrito (—ONO).

$Na_3[Co(NO_2)_6]$	sodium hexanitrocobaltate(3−) sodium hexanitrocobaltate(III)
$[Co(ONO)(NH_3)_5]SO_4$	pentaamminenitritocobalt(2+) sulfate pentaamminenitritocobalt(III) sulfate
$[Co(NCS)(NH_3)_5]Cl_2$	pentaammineisothiocyanatocobalt(2+) chloride pentaammineisothiocyanatocobalt(III) chloride

Use of abbreviations

In the literature of coordination compounds, abbreviations for ligand names are used extensively, especially in formulae. A list of common abbreviations is given below. [The IUPAC list is rather short. Additional abbreviations used in this book or in the current literature are included in the following list (italicized). It should be noted that often these are used as capital letters (e.g., EDTA) but should be lower case (edta) according to the Rules.]

Name of ligand	Abbreviation
acetylacetonate ion	acac
alanine anion	*alan*
benzene	*bz*
benzenedithiolate anion	*bdt*
benzoylacetate ion	*benzac*
biguanide	Hbq
1,2-bis(diphenylphosphino)ethane	diphos
cyclooctatetraene	cot
1,2,trans-cyclohexanediamine	*chxn*
cyclopentadienyl radical	*cp*
1,2,trans-cyclopentanediamine	*cptn*
diallylamine	*dlm*
trans-1,2-diaminocyclohexanetetraacetate anion	*dcta*
diethylenetriamine	dien
diethylenetriaminepentaacetate	*dtpa*
diethylselenophosphate anion	*dsep*
diethylthiocarbamate anion	*dtc*
diethylthiophosphate anion	*dtp*
dimethylacetamide	*dma*
dimethylformamide	*dmf*
dimethylglyoxime anion	*dmg*
dimethylsulfoxide	*dmso*
bipyridine	*bpy*
ethylenediamine	en
ethylenediaminetetraacetic acid	edta
ethylenethiourea	*etu*
glycine anion	*gly*
halide	*X*
1,1,1,2,2,3,3-heptafluoro-7,7-dimethyloctane-4,6-dione	*fod*
N-hydroxyethylethylenediaminetriacetate anion	*hedta*
any unspecified ligand	*L*
maleonitriledithiolate anion	*mnt*
methylbis(3-dimethy arsinopropyl)arsine	*tas*
nitrilotriacetate anion	*nta*
oxalate dibasic anion	ox
1,10-phenanthroline	phen
o-phenylenebis(dimethylarsine)	diars
1-phosphabicyclo[2.2.2]octane	*phosph*
phthalocyanine (dinegative group)	*pc*
propylenediamine (1,2-diaminopropane)	pn
pyridine	py
stilbenediamine (1,2-diphenylethylenediamine)	*stien*
2,2,6,6-tetramethylheptane-3,5-dione	*thd*
2,2′,2″,2‴-tetrapyridyl	*tetrpy*
thenoyltrifluoroacetone	*tta*
thiourea	*tu*
toluene-3,4-dithiolate anion	*tdt*
1,2,3-triaminopropane	*tn*
2,2′,2″-triaminotriethylamine	tren
trimethylenetetramine	trien
trimethylenediamine	*trim*

Name of ligand	Abbreviation
2,2′,2″-tripyridyl	*tripy*
tris(3-dimethylarsinopropyl)phosphine	*tap*
urea	*ur*

Other abbreviations commonly used; these are almost always capitalized.

R = *alkyl groups*
R_f = *perfluoroalkyl groups*
Me = *methyl*
Et = *ethyl*
Pr = *propyl*
Bu = *butyl*
ϕ = *phenyl*
Ac = *acetyl*
AcO = *acetate*
M = *metal*

Complexes with unsaturated molecules or groups

The name of the ligand group is given with the prefix η. η may be read as *eta* or *hapto* (from the Greek *haptein*, $\eta\alpha\pi\tau\varepsilon\iota\nu$, to fasten) [F. A. Cotton, *J. Am. Chem. Soc.*, **1968**, *90*, 6230]. [When] some, but not all, ligand atoms in a chain or ring, or some, but not all ligand atoms involved in double bonds are bound to the central atom, locant designators are inserted preceding η. [This is somewhat at variance with Cotton's original proposal to designate the number of "bonding atoms." Both systems have advantages. Examples of Cotton's method are given on p. 645.

$[PtCl_2(NH_3)(C_2H_4)]$	amminedichloro(η-ethylene)platinum amminedichloro(η-ethylene)platinum(II)
$K[PtCl_3(C_2H_4)]$	potassium trichloro(η-ethylene)platinate(1 −) potassium trichloro(η-ethylene)platinate(II)
$[ReH(C_5H_5)_2]$	bis(η-cyclopentadienyl)hydridorhenium bis(η-cyclopentadienyl)hydridorhenium(I)

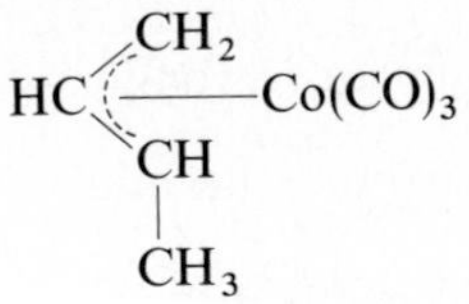

CH_2 / HC — $Co(CO)_3$ / CH / CH_3

tricarbonyl(1-3-η-2-butenyl)cobalt
tricarbonyl(1-3-η-2-butenyl)cobalt(I)

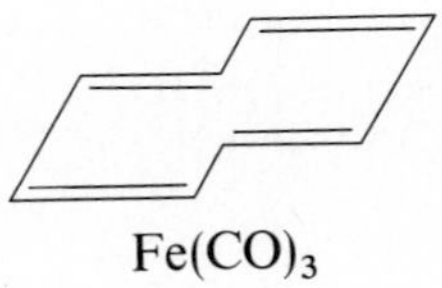

$Fe(CO)_3$

tricarbonyl(1-4-η-cyclooctatetraene)iron

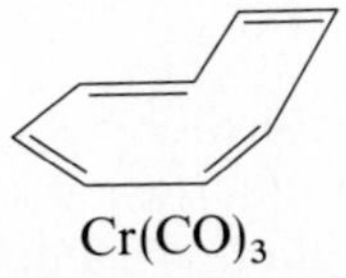

$Cr(CO)_3$

tricarbonyl(1-6-η-cyclooctatetraene)chromium

Designation of isomers

Among coordination compounds, isomerism may arise in a number of ways:

a. Different atoms of the ligand through which coordination to a central atom occurs.

b. Coordination of isomeric ligands.

c. Interchange of ions between coordination sphere and ionic sphere.

d. Geometrical arrangement of two or more kinds of ligands in the coordination sphere: *cis*-, *trans*-; *fac*-, *mer*-.

e. Chiral (asymmetrical) arrangement of ligands in the coordination sphere.

f. Asymmetry of an atom in a ligand that originates in the coordination process.

[Examples of the above are included in the text.]

Compounds with bridging atoms or groups

a. A bridging group is indicated by adding the Greek letter μ immediately before its name and separating the name from the rest of the complex by hyphens.

b. Two or more bridging groups of the same kind are indicated by di-μ- (or bis-μ-), etc.

c. The bridging groups are listed with the other groups in alphabetical order *unless the symmetry of the molecule permits simpler names by the use of multiplicative prefixes.*

d. Where the same ligand is present as a bridging ligand and as a non-bridging ligand, it is cited first as a bridging ligand.

Bridging groups between two centers of coordination are of two types: (1) the two centers are attached to the same atom of the bridging group and (2) the two centers are attached to different atoms of the bridging group. For bridging groups of the first type it is often desirable to indicate the bridging atom. This is done by adding the italicized symbol for the atom at the end of the name of the ligand. For bridging groups of the second type, the symbols of all coordinated atoms are added.

$[(NH_3)_5Cr{-}OH{-}Cr(NH_3)_5]Cl_5$	μ-hydroxobis(pentaamminechromium)(5+) chloride
	μ-hydroxobis[pentaamminechromium(III)] chloride
$[(CO)_3Fe(CO)_3Fe(CO)_3]$	tri-μ-carbonylbis(tricarbonyliron)
$[Br_2Pt(SMe_2)_2PtBr_2]$	bis(μ-dimethylsulfide)bis-[dibromoplatinum(II)]

Di- and polynuclear compounds without bridging groups; direct linking between centers of coordination

There are a number of compounds containing metal–metal bonds. Such compounds, when symmetrical, are named by the use of multiplicative prefixes; when unsymmetrical, one central atom and its attached ligands shall be treated as a ligand on the other center.

$[Br_4Re—ReBr_4]^{2-}$	bis(tetrabromorhenate)(2−) bis[tetrabromorhenate(II)]
$[(CO)_5Mn—Mn(CO)_5]$	bis(pentacarbonylmanganese)
$[(CO)_4Co—Re(CO)_5]$	pentacarbonyl(tetracarbonylcobaltio)- rhenium
$[\eta\text{-}C_5H_5(CO)_3Mo—Mo(CO)_3\text{-}\eta\text{-}C_5H_5]$	bis(tricarbonyl-η-cyclopentadienyl- molybdenum)
$[(Cl_3Sn)_2RhCl_2Rh(SnCl_3)_2]^{4-}$	di-μ-chloro-bis[bis(trichlorostannyl)- rhodate](4−) ion di-μ-chloro-bis[bis(trichlorostannyl)- rhodate(I)] ion

Homoatomic aggregates

There are several instances of a finite group of metal atoms with bonds directly between the metal atoms but also with some nonmetal atoms or groups (ligands) intimately associated with the *cluster*. The geometrical shape of the cluster is designated by *triangulo, quadro, tetrahedro, octahedro*, etc., and the nature of the bonds to the ligands by the conventions for bridging bonds and simple bonds. Numbers are used as locant designators as they are for homoatomic chains and boron clusters (cf. Rules for Boron Compounds).

$Os_3(CO)_{12}$	dodecacarbonyl-*triangulo*-triosmium
$Cs_3[Re_3Cl_{12}]$	cesium dodecachloro-*triangulo*-trirhenate(3−) tricesium dodecachloro-*triangulo*-trirhenate
B_4Cl_4	tetrachloro-*tetrahedro*-tetraboron
$[Nb_6Cl_{12}]^{2+}$	dodeca-μ-chloro-*octahedro*-hexaniobium(2+) ion

8. ADDITION COMPOUNDS

This rule covers some donor–acceptor complexes and a variety of lattice compounds. It is particularly relevant to compounds of uncertain structure; new structural information often makes possible naming according to [the rules for coordination compounds].

The ending -ate is now the accepted ending for anions and should generally not be used for addition compounds. Alcoholates are the *salts* of alcohols and this name should not be used to indicate alcohol of crystallization. Analogously addition compounds containing ether, ammonia, etc., should *not* be termed etherates, ammoniates, etc.

However, one exception has to be recognized. According to the commonly accepted meaning of the ending -ate, "hydrate" would be, and was formerly regarded as, the name for a *salt* of water, i.e., what is now known as a hydroxide; the name hydrate has now a very firm position as the name of a compound containing water of crystallization and is allowed also in these Rules to designate water bound in an unspecified way. It is considered to be preferable even in this case to avoid the ending -ate by using the name "water" (or its equivalent in other languages) when possible.

The names of addition compounds may be formed by connecting the names of individual compounds by hyphens and indicating the number of molecules after the name by Arabic numerals separated by the solidus. Boron compounds and water are always cited last in that order. Other molecules are cited in order of increasing number;

any which occur in equal numbers are cited in alphabetical order.

sodium carbonate 10–water
sodium carbonate–water (1/10)
sodium carbonate decahydrate
aluminum sulfate–potassium sulfate–water (1/1/24)
calcium chloride–ammonia (1/8)
aluminum chloride–ethanol (1/4)
methanol–boron trifluoride (2/1)
ammonia–boron trifluoride (1/1)

9. CRYSTALLINE PHASES OF VARIABLE COMPOSITION

10. POLYMORPHISM

[These sections have been omitted since there is little in them dealing directly with the contents of this text.]

11. BORON HYDRIDES[5]

The name of BH_3 is borane and it and higher boron hydrides are called boranes. The number of boron atoms in the molecule is indicated by a Greek numerical prefix.

B_2H_6	diborane(6) [Fig. 14.31]
B_4H_{10}	tetraborane(10) [Fig. 14.35]
B_5H_9	pentaborane(9) [Fig. 14.41]
B_5H_{11}	pentaborane(11)
B_6H_{10}	hexaborane(10) [Fig. 14.39]
B_9H_{15}	nonaborane(15)
$B_{10}H_{14}$	decaborane(14) [Fig. 14.39]
$B_{20}H_{16}$	icosaborane(16)

The polyboranes and their derivatives may be considered to consist of two general classes: (1) closed structures (that is, structures with boron skeletons that are polyhedra having all triangular faces) and (2) nonclosed structures. The members of the first class are designated by the prefix *closo*. Some members of the second class have structures very close to a closed structure. These may be denoted by the prefix *nido* (from Latin *nidus*, nest). [Usage has extended these definitions—see pp. 735–738.]

PREFIXES OR AFFIXES USED IN INORGANIC NOMENCLATURE

Multiplying affixes (a) mono, di, tri, tetra, penta, hexa, hepta, octa, nona(ennea), deca, undeca (hendeca), dodeca, etc., used by direct joining without hyphens (b) bis, tris, tetrakis,

[5] For extended boron rules, see *Pure Appl. Chem.*, **1972**, *30*, 683.

	pentakis, etc., used by direct joining without hyphens but usually with enclosing marks around each whole expression to which the prefix applies
Structural affixes	italicized and separated from the rest of the name by hyphens
antiprismo	eight atoms bound into a rectangular antiprism
asym	asymmetrical
catena	a chain structure; often used to designate linear polymeric substances
cis	two groups occupying adjacent positions; sometimes used in the sense of *fac*
closo	a cage or closed structure, especially a boron skeleton that is a polyhedron having all triangular faces
cyclo	a ring structure[6]
dodecahedro	eight atoms bound into a dodecahedron with triangular faces
fac	three groups occupying the corners of the same face of an octahedron
hexahedro	eight atoms bound into a hexahedron (e.g., cube)
hexaprismo	twelve atoms bound into a hexagonal prism
icosahedro	twelve atoms bound into a triangular icosahedron
mer	meridional; three groups on an octahedron in such a relationship that one is *cis* to the two others which are themselves *trans*
nido	a nest-like structure, especially a boron skeleton that is very close to a closed or closo structure
octahedro	six atoms bound into an octahedron
pentaprismo	ten atoms bound into a pentagonal prism
quadro	four atoms bound into a quadrangle (e.g., square)
sym	symmetrical
tetrahedro	four atoms bound into a tetrahedron
trans	two groups directly across a central atom from each other; i.e., in the polar position on a sphere
triangulo	three atoms bound into a triangle
triprismo	six atoms bound into a triangular prism
η	signifies that two or more contiguous atoms of the group are attached to a metal
μ	signifies that the group so designated bridges two centers of coordination
σ	signifies that one atom of the group is attached to a metal

[6] Cyclo here is used as a modifier indicating structure and hence is italicized. In organic nomenclature, cyclo is considered to be part of the parent name and therefore is not italicized.

Index

85 86 9 8 7 6 5 4 3

THE NAMES, SYMBOLS, ATOMIC NUMBERS, AND ATOMIC WEIGHTS OF THE ELEMENTS[a]

Name[b]	Symbol	Atomic number	Atomic weight[c]
Actinium	Ac	89	227.0278
Aluminum	Al	13	26.98154
Americium	Am	95	(243)
Antimony	Sb	51	121.75
Argon	Ar	18	39.948
Arsenic	As	33	74.9216
Astatine	At	85	(210)
Barium	Ba	56	137.33
Berkelium	Bk	97	(247)
Beryllium	Be	4	9.01218
Bismuth	Bi	83	208.9804
Boron	B	5	10.81
Bromine	Br	35	79.904
Cadmium	Cd	48	112.41
Calcium	Ca	20	40.08
Californium	Cf	98	(251)
Carbon	C	6	12.011
Cerium	Ce	58	140.12
Cesium	Cs	55	132.9054
Chlorine	Cl	17	35.453
Chromium	Cr	24	51.996
Cobalt	Co	27	58.9932
Copper (*cuprum*)	Cu	29	63.546
Curium	Cm	96	(247)
Dysprosium	Dy	66	162.50
Einsteinium	Es	99	(254)
Erbium	Er	68	167.26
Europium	Eu	63	151.96
Fermium	Fm	100	(257)
Fluorine	F	9	18.998403
Francium	Fr	87	(223)

Name	Symbol	Atomic number	Atomic weight
Gadolinium	Gd	64	157.25
Gallium	Ga	31	69.72
Germanium	Ge	32	72.59
Gold (*aurum*)	Au	79	196.9665
Hafnium	Hf	72	178.49
Hahnium[d]	Ha	105	(260)
Helium	He	2	4.00260
Holmium	Ho	67	164.9304
Hydrogen	H	1	1.0079
Indium	In	49	114.82
Iodine	I	53	126.9045
Iridium	Ir	77	192.22
Iron (*ferrum*)	Fe	26	55.847
Krypton	Kr	36	83.80
Lanthanum	La	57	138.9055
Lawrencium	Lr	103	(260)
Lead (*plumbum*)	Pb	82	207.2
Lithium	Li	3	6.941
Lutetium	Lu	71	174.97
Magnesium	Mg	12	24.305
Manganese	Mn	25	54.9380
Mendelevium	Md	101	(258)
Mercury	Hg	80	200.59
Molybdenum	Mo	42	95.94
Neodymium	Nd	60	144.24
Neon	Ne	10	20.179
Neptunium	Np	93	237.0482
Nickel ("niccolum")	Ni	28	58.70
Niobium	Nb	41	92.9064
Nitrogen	N	7	14.0067
Nobelium[e]	No	102	(259)

Name	Symbol	Atomic number	Atomic weight	Name	Symbol	Atomic number	Atomic weight
Osmium	Os	76	190.2	Silver (*argentum*)	Ag	47	107.868
Oxygen	O	8	15.9994	Sodium	Na	11	22.98977
Palladium	Pd	46	106.4	Strontium	Sr	38	87.62
Phosphorus	P	15	30.97376	Sulfur	S	16	32.06
Platinum	Pt	78	195.09	Tantalum	Ta	73	180.9479
Plutonium	Pu	94	(244)	Technetium	Tc	43	(97)
Polonium	Po	84	(209)	Tellurium	Te	52	127.60
Potassium	K	19	39.0983	Terbium	Tb	65	158.9254
Praseodymium	Pr	59	140.9077	Thallium	Tl	81	204.37
Promethium	Pm	61	(145)	Thorium	Th	90	232.0381
Protactinium	Pa	91	231.0359	Thulium	Tm	69	168.9342
Radium	Ra	88	226.0254	Tin (*stannum*)	Sn	50	118.69
Radon	Rn	86	(222)	Titanium	Ti	22	47.90
Rhenium	Re	75	186.207	Tungsten (*wolfram*)[g]	W	74	183.85
Rhodium	Rh	45	102.9055	Uranium	U	92	238.029
Rubidium	Rb	37	85.4678	Vanadium	V	23	50.9414
Ruthenium	Ru	44	101.07	Xenon	Xe	54	131.30
Rutherfordium[f]	Rf	104	(261)	Ytterbium	Yb	70	173.04
Samarium	Sm	62	150.4	Yttrium	Y	39	88.9059
Scandium	Sc	21	44.9559	Zinc	Zn	30	65.38
Selenium	Se	34	78.96	Zirconium	Zr	40	91.22
Silicon	Si	14	28.0855				

[a] From the Table of Atomic Weights 1975 by the Commission on Atomic Weights, IUPAC. Weights are scaled to the relative atomic mass $A_r(^{12}C) = 12$.

[b] The names in parentheses are the Latin forms used in complex formation (except for wolfram): gold (*aurium*); $[AuCl_4]^{-1}$ is tetrachloroaurate(III).

[c] Atomic weights in parentheses are those of the most stable radioisotope.

[d] This name has been suggested by American workers and has not yet been approved by IUPAC. Russian workers have suggested the name *nielsbohrium*.

[e] Although the name nobelium has official IUPAC sanction, some Russian workers have used the name *joliotium*.

[f] This name has been suggested by American workers and has not yet been approved by IUPAC. Russian workers have suggested the name *kurchatovium*.

[g] *Wolfram* is the German word for tungsten.